DEVELOPMENT THROUGH THE LIFESPAN

ANNOTATED INSTRUCTOR'S EDITION

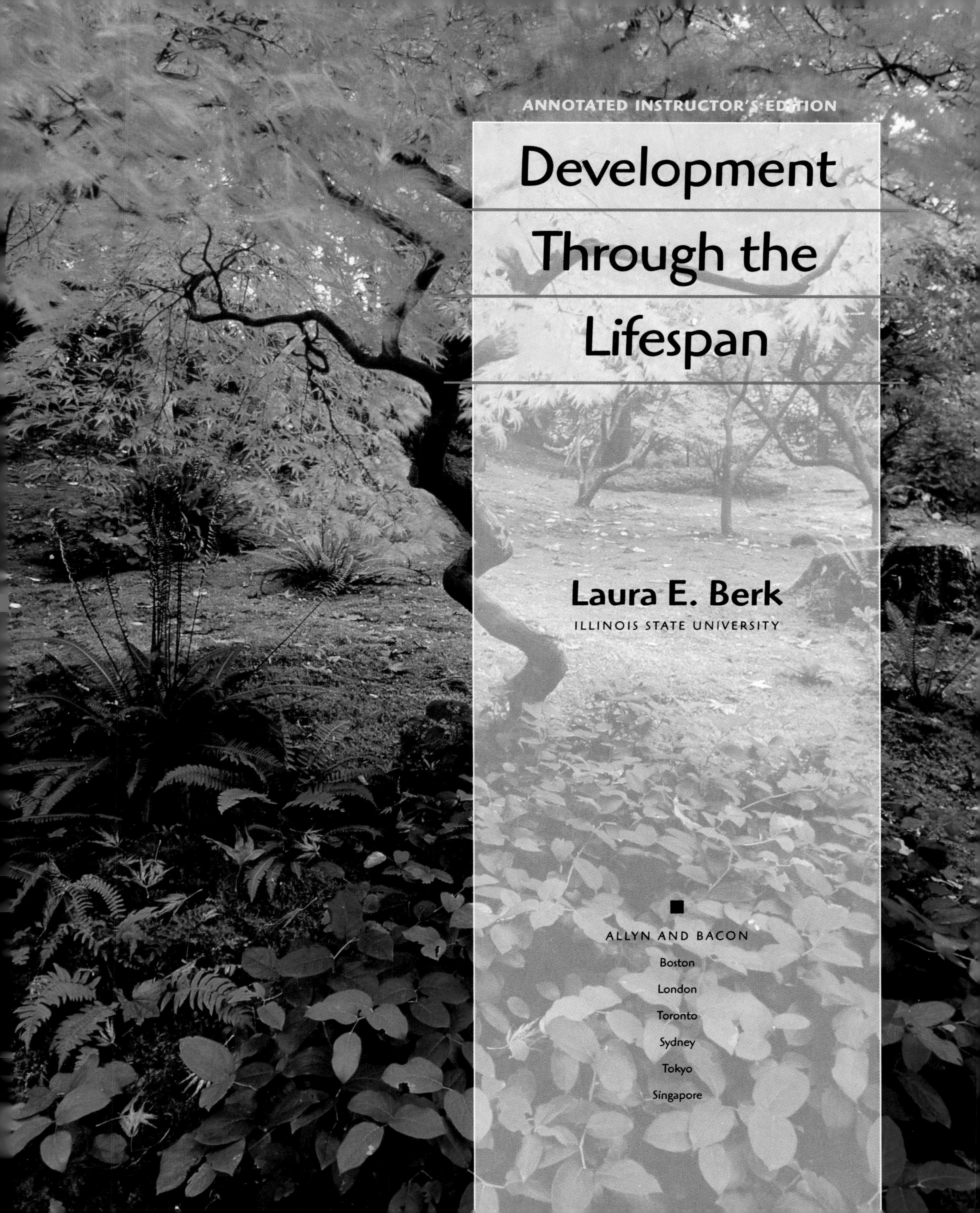

Development Through the Lifespan

Laura E. Berk
ILLINOIS STATE UNIVERSITY

ALLYN AND BACON
Boston
London
Toronto
Sydney
Tokyo
Singapore

Vice-President and Editor-in-Chief, Social Sciences: Sean W. Wakely
Developmental Editors: Anne A. Reid, Sue Gleason
Vice-President/Director of Field Marketing: Joyce Nilsen
Editorial Assistant: Jessica Barnard
Editorial-Production Service: Thomas E. Dorsaneo
Text Design and Composition: Seventeenth Street Studios
Cover Designer: Linda Knowles
Composition and Prepress Buyer: Linda Cox
Manufacturing Buyer: Megan Cochran

Copyright © 1998 by Allyn & Bacon
A Viacom Company
160 Gould Street
Needham Heights, MA 02194-2310

Internet: www.abacon.com
America Online: Keyword: College online

ISBN 0-205-19291-2

Printed in the United States of America
10 9 8 7 6 5 4 3 2 1 VHP 01 00 99 98 97

BRIEF CONTENTS

Now Berk takes your students further...

Development Through the Lifespan

by Laura E. Berk, Illinois State University

*a journey of
human growth and development
through the lifespan.*

For the past decade, Laura Berk's child development textbooks have helped thousands of students grasp even the most complex topics of development *and* successfully apply what they have learned to their own lives. Now this highly respected teacher, parent, and researcher invites your students on an extended journey through the *lifespan* — to help them better appreciate where they have been and where they will be going in the years ahead.

This long-awaited exploration of human development offers the research, relevance, and readability you've come to expect from Dr. Laura Berk. Her approach blends *scholarship, a distinctive, accessible writing style, and an abundance of clear, lively examples* that capture students' interest from the very first page. In this chronologically organized text, Berk combines a highly effective pedagogical program with an unparalleled appreciation for the interrelatedness of theory, research, and applications, to help students understand the many factors that *have influenced* and *will influence* the paths their lives take.

Take a look at the other exciting features that this new text offers...

Several strong emphases guide your students toward a better understanding of lifespan development:

■ the multidimensional nature of development (physical, cognitive, emotional, and social) with special attention to diversity in individual life paths

■ the multiple interacting contextual influences on development (biological, psychological, social, community, societal, cultural, and historical)

■ gender issues in development

A dynamic lifespan development Observation Video program brings the developmental journeys of children, adolescents, and adults to life in your classroom.

An Annotated Instructor's Edition offers useful notes and helpful ways to maximize the **many teaching and learning resources** available with this text.

Berk takes your students further...

with the most important trends and issues that shape our individual life courses.

It's no wonder that Laura Berk's development textbooks are acclaimed for their student, professional, and practitioner appeal. This teacher/scholar/parent with over a quarter century of experience in instruction and published research has brought to bear her considerable expertise to produce this complete, current, easy-to-use textbook.

A Solid Foundation in History, Theory, and Research

Chronologically spanning development from conception to death, *Development Through the Lifespan* begins with a clear presentation of important themes, research strategies, and an overview of the history of the field in **PART I**, with strong emphasis on the interrelatedness of theory, research, and applications. Berk introduces your students to a variety of important classic and contemporary perspectives that serve as a background for analysis of many controversial issues throughout the text.

A Discussion of Development and the Processes That Underlie It

Two chapters in **PART II** introduce students to the foundations of development, with an important discussion of processes of change. Biological and environmental influences are covered in Chapter 2, while Chapter 3 offers a thorough discussion of prenatal development, birth, and the newborn. Important topics, such as *pros and cons of reproductive technologies* and *cross-national comparisons of prenatal health care and infant mortality,* are covered.

An Integrated, Chronological Approach

The remaining chapters of *Development Through the Lifespan* follow the seven major age periods of change with a special emphasis on providing continuity across chapters. Topics that consistently recur (physical, cognitive, emotional, social) are presented in a similar organization, making it easier for students to draw connections across the age periods and construct a continuous view of developmental change.

Coverage at the Forefront of the Field

"*This* is an outstanding and timely text in the field...Berk very eloquently invites the reader to accompany [her] on a personal journey into the lifespan. The case vignettes and other stimulating pedagogical tools contribute to this fascinating...exploration of lifespan material."
— *Joan Cannon, University of Massachusetts Lowell*

Cutting-edge topics throughout the text underscore the book's major themes —

- the lifespan perspective
- interdisciplinary contributions to the study of development (psychology, sociology, anthropology, education, and other fields)
- multiple contexts for development (including family, community, culture, social policy, and historical time period)
- integration of physical, cognitive, emotional, and social domains
- applications of theory and research
- gender differences, with special attention given to the distinctive roles and life paths of males and females.

Berk takes your students further...

by portraying the complexities of development in a way that captures students' interest and helps them learn.

From prenatal development and the miracle of birth, to death and the rituals of grieving, no other text conveys the beauty and wonder of the physical and emotional journey of human development the way Berk does — in a way that helps students learn and understand. In this remarkable new text, you'll find a highly accessible writing style — a hallmark of Laura Berk's texts — that is clear, lively, and full of interesting examples to clarify difficult topics, yet avoids being too simplistic.

The Text That Teaches as It Tells a Story

To truly engage your students, Berk has created a lifespan development text that conveys content through an engaging story-telling style. *Development Through the Lifespan* is brought to life with stories and vignettes about children, adolescents, adults, and families, many of whom Laura Berk has known personally. For each chronological age division, your students will be able to construct a clearer image of human life by following the stories of real people from chapter to chapter, observing the dramatic changes in their lives. In addition to this set of main characters, numerous other real-life examples throughout represent a wide diversity of age periods and contexts to allow students to easily — and accurately — link theory and research to applications.

> "Pedagogically, the organization and outline structure of Berk's text is exceptional. Brief reviews and chapter summaries reinforce the student's reading. The stories and vignettes make developmental principles concrete for students."
> — *Ellen Pastorino, Gainesville College*

A Close Link Between Theory, Research, and Applications

Development Through the Lifespan offers a comprehensive, research-based approach to the subject matter and integrates applications throughout the entire text to provide a real-world context for abstract ideas. For instance, "Caregiving Concerns" tables provide easily accessible practical advice on the importance of caring for oneself and others throughout the lifespan.

Special attention is paid throughout to a current focus in the field — harnessing knowledge of human development to shape social policies that support human needs throughout the lifespan. The text also addresses the current condition of children, adolescents, and adults in the United States and around the world and shows how theory and research have combined with public interest to spark successful interventions.

The back seat and trunk piled high with belongings, 22-year-old Sharese hugged her mother and brother good-bye, jumped in the car, and headed toward the interstate with a sense of newfound independence mixed with apprehension. Three months earlier, the family had watched proudly as Sharese received her bachelor's degree in chemistry from a small university 40 miles from her home. Her college years had been ones of gradual release from economic and psychological dependence on her family. She returned home on weekends as often as she desired and lived there during the summer months. Her mother supplemented Sharese's loans with a monthly allowance. But this day marked a turning point. She was moving to her own apartment in a new city 800 miles away, with plans to begin working on a master's degree the following week. In charge of all her educational and living expenses, Sharese felt more self-sufficient than at any previous time in her life.

The college years had been ones in which Sharese made important lifestyle changes and settled on a vocational direction. Overweight throughout high school, she lost 20 pounds during her freshman year, revised her diet, and began a regimen of exercise by joining the university's Ultimate Frisbee team. The sport helped her acquire healthier habits and leadership skills as team captain. A summer spent as a counselor at a camp for chronically ill children, combined with personal events we will take up later, convinced Sharese to apply her background in science to a career in public health.

Still, she wondered whether her choice was right. Two weeks before she was scheduled to leave, Sharese confided to her mother that she had doubts and might not go. Her mother advised, "Sharese, we never know ahead of time whether the things we choose are going to suit us just right, and most times they aren't perfect. It's what we make of them—how we view and mold them—that turns a choice into a success." And so Sharese embarked on her journey and found herself face to face with a multitude of exciting challenges and opportunities.

In this chapter, we take up the physical and cognitive sides of early adulthood—the period of the twenties and thirties. In Chapter 1, we emphasized that the adult years are difficult to divide into discrete periods, since the timing of important milestones varies greatly among individuals—much more so than

societal contexts, most young adults make the best of wrong turns and solve problems successfully. These are energetic decades that, more than any phase, offer the potential for living to the fullest.

PHYSICAL DEVELOPMENT IN EARLY ADULTHOOD

In earlier chapters, we saw that the body grows larger and stronger, coordination improves, and sensory systems gather information more effectively during childhood and adolescence. Once body structures reach maximum capacity and efficiency, *biological aging* begins—genetically influenced declines in the functioning of organs and systems that are universal in all members of our species (Williams, 1992). However, like physical growth, biological aging is *asynchronous* (see Chapter 7, pages 207–208). Change varies widely across parts of the body, with some structures not affected at all. In addition, individual differences are great—variation that the lifespan perspective helps us understand. Biological aging is influenced by a host of contextual factors, each of which can accelerate or slow age-related declines. These include the person's unique genetic makeup, lifestyle, living environment, and historical period (Arking, 1991). As a result, the physical changes of the adult years are, indeed, multidimensional and multidirectional.

In the following sections, we examine the process of biological aging, first at the larger level of human life expectancy and limits of the human lifespan. Then we turn to theories of aging and specific physical and motor changes,

Berk takes your students further...

with coverage of important real-life factors that affect development across the country ... and around the world.

Throughout *Development Through the Lifespan*, in vignettes, discussions, and special boxed sections, Berk pays strong attention to social, educational, and multicultural issues that are critical to the overall condition of children, adolescents, and adults in today's world. In each chapter, Berk takes your students to distant parts of the world, reviewing a growing body of cross-cultural evidence and contemporary social and policy concerns. The text presents an excellent perspective on the diversity of children, adolescents, and adults and their experiences, as well as extensive coverage of multiculturalism in the United States and its effects on development. You'll also find expanded information on such important issues as poverty, ethics in research, public policy, and developmentally appropriate practice.

"Social Issues" features

discuss the condition of children, adolescents, and adults in the United States and around the world, and emphasize the need for sensitive social policies to ensure their well-being. Topics include lead poisoning in childhood, children's eyewitness testimony, and elder suicide.

"Try This!" activities

In each *Social Issues* feature, these suggested activities actively involve students in opportunities for observation, discussion, and self-reflection.

"Cultural Influences" features

highlight the impact of context and culture on all aspects of development. Topics include the African-American extended family, cultural variations in the experience of aging, and menopause as a biocultural event.

"*B*erk is superior in organization, readability, and coverage. It can compete effectively with any market and any level, 2-year or 4-year college."
— *Ed Brady, Belleville Area College*

Berk takes your students further...

with a variety of learning aids that ensure maximum comprehension.

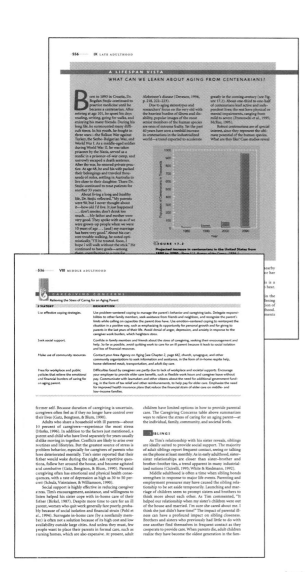

"Lifespan Vista" features

This special series of boxes is devoted to topics that have long-term implications for development or involve intergenerational issues.

"Ask Yourself ..." critical thinking questions

The focus of these critical thinking questions is divided between theory and applications. Many describe problematic situations faced by parents, teachers, and children and ask students to resolve them in light of what they have learned.

"Milestones" tables

These tables appear at the end of each chronological age division to summarize, in one location, the major physical, cognitive, language, emotional, and social developments of each age span. They provide a convenient device for reviewing the chronology of development across the lifespan.

"Caregiving Concerns" tables

These tables provide easily accessible practical advice on the importance of caring for oneself and others throughout the lifespan.

Also included:

Chapter introductions • End-of-book Glossary •
Brief Reviews • End-of-chapter Summaries •
"Important Terms and Concepts" lists •
"FYI... For Further Information and Help" sections

"*B*erk has done an admirable job of making this material succinct, current in terms of research, and very readable... Also, the brief summaries are superb."
— Joe Tinnin, Richland College

Berk takes your students further...

by providing everything you need to make learning more accessible — and teaching easier.

Development Through the Lifespan offers teaching tools that help you organize the material, plan classes and examinations, and present the subject matter for maximum student comprehension.

Annotated Instructor's Edition (AIE)

The Annotated Instructor's Edition is your guide to all of the supplements for the course and includes useful notes in the margins, for your eyes only.

The annotations, printed in blue for easy identification, include...

⊕ icons denoting cross-cultural studies or data in the text

👪 icons denoting a lifespan perspective on a developmental issue in the text

⚥ icons denoting gender-related issues in the text

References to Test Bank items

An **Instructor's Section** is bound into the front of the AIE. For each chapter, you'll find a Chapter-at-a-Glance table with a Chapter Outline and a listing of all the chapter's Learning Objectives, test items, Lecture Extensions, Learning Activities, transparencies, and media materials.

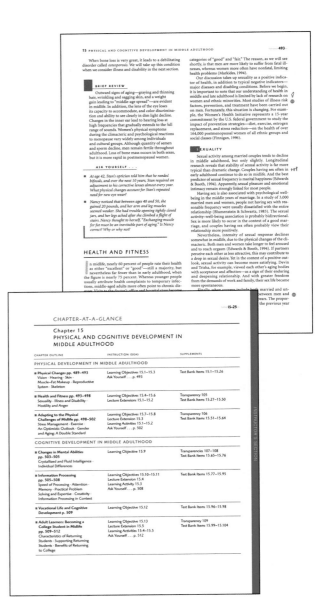

All information is accurate as of date of printing and is subject to change without notice. Some restrictions may apply to some supplements — check with your publisher's representative for details.

Instructor's Resource Manual (IRM)

Prepared by *Laura E. Berk, Illinois State University; Belinda M. Wholeben, Rockford College;* and *Heather A. Bouchey, University of Denver*, the Instructor's Resource Manual contains additional material to make the text an even more effective and enjoyable teaching tool. Organized by chapter, the IRM includes:

Chapter-at-a-Glance; Brief Chapter Summary; Learning Objectives; Lecture Outline; Lecture Extensions; Learning Activities; answers to the text's Ask Yourself questions; Suggested Student Readings; a list of Transparencies; and a list of Media Materials. All features are page-referenced to relevant sections of the text.

Test Bank

Prepared by Celia C. Reaves, Monroe Community College, the test bank contains over 2,000 multiple-choice and essay questions, each of which is answered, cross-referenced to learning objectives, page-referenced to chapter content, and classified according to type (factual, applied, or conceptual). Questions that appear in the Practice Tests booklet and on the website are keyed with "PT".

Computerized Test Bank

This computerized version of the test bank is available in IBM (Windows or DOS) and Macintosh versions using ESATEST III.

Transparencies

More than 120 full-color transparencies taken from the text and various other sources.

Study Guide

Prepared by *Jenny L. Churchill* and *Laura E. Berk, Illinois State University*, this helpful guide offers chapter summaries, learning objectives, study questions organized according to major headings in the text, "Ask Yourself" questions that also appear in the textbook, suggested readings, crossword puzzles for mastering important terms, and multiple-choice self-tests.

Practice Tests

Twenty multiple-choice items per chapter plus an answer key with justifications are drawn from the test bank for extra practice for your students.

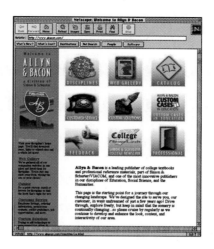

Visit us in Cyberspace!

http://www.abacon.com/berk

Designed for students and faculty of Lifespan or Human Development and Child Development classes, this comprehensive website not only encourages online and interactive learning but also offers the most current links and information about development.

Your students will love...

the **Online Study Guide** with chapter summaries, and multiple-choice and short answer questions drawn from the Test Bank.

the **"How to Use This Text" area** with guidelines for the most effective use of the pedagogy.

the **Student Supplements listing** that describes each study aid with links to ordering information.

the **Hot Links** to other relevant websites for research purposes.

the **Games, Biographical Sketches** of characters from the text, and much more.

You will appreciate...

the **Teaching Aids section** with Online Handbook and Lecture Notes.

the **Instructor's Supplements listing** that describes each teaching aid with links to order information.

Berk takes your students further...

with an *Observation Video* program that brings textbook concepts to life in your classroom!

Development Through the Lifespan in Action
Observation Video and Guide

Observation Videotape

Laura Berk created this real-life look at human development in conjunction with the *Illinois State University Television Production Studio.*  Hundreds of observation segments are carefully filmed and edited into this insightful videotape to illustrate the many theories, concepts, and milestones of development through the lifespan. It's the perfect solution to time-consuming and hard-to-organize observation assignments outside the classroom!

Observation Guide

This Observation Guide helps your students use the Observation Videotape in conjunction with *Development Through the Lifespan,* deepening their understanding of the material and applying  what they have learned to everyday life. In addition to chapter-by-chapter summaries of video content, questions, and activities, the guide features perforated pages that students can hand in as assignments.

Packaging Options

The videotape and guide are free to instructors who adopt the text and are available in a discounted package, shrinkwrapped with the textbook, for students.

Can newborn babies imitate facial expressions of adults? Indeed, they can. Here 2-week-old Anna Marie imitates Professor Berk as she opens her mouth. Imitation grants newborns a powerful means for relating to and learning from their social world.

Early in life, individual differences in temperament are evident. Some children, like 21-month-old Ben who is determined to stack these cups, are highly persistent and goal directed. Others are easily distracted.

George and Dorothy have been married for 56 years. Dorothy is in her mid-seventies, and George is in his early eighties. They can't move as quickly as they could in earlier years, but good health, a close-knit family, and a multitude of leisure interests make their late life rich and rewarding.

When discussing the story of quality in Laura Berk's *Development Through the Lifespan,* we are referring both to the text and to the supplements. This Annotated Instructor's Edition (AIE) is designed to provide a complete description of the total package.

THE WALKTHROUGH

Step by step, the Walkthrough that precedes this introduction provides the instructor with examples of the various features, pedagogy, and supplements that form the complete Allyn and Bacon teaching package for your lifespan development course. Sample pages demonstrate the layout of the text and how to use the complete package most effectively. A description of the supplements can be found at the end of the Walkthrough.

THE INSTRUCTOR'S SECTION

For each chapter, the Instructor's Section of the AIE provides a Chapter-at-a-Glance table with a chapter outline linked to ideas for instruction in the Instructor's Resource Manual and to other supplements available with this text. Each Chapter-at-a-Glance table is followed by the chapter's Learning Objectives.

THE ANNOTATIONS

The Annotated Instructor's Edition for *Development Through the Lifespan* has been specially prepared with margin annotations, printed in blue. The annotations do not appear in the student edition. They fall into four categories.

Test Bank Items. These annotations, at the bottom of pages, are designed to help you coordinate the text and the Test Bank. They refer to a corresponding range of items in the printed and computerized test bank versions. The computerized test bank is available in IBM (Windows or DOS) and Macintosh formats.

Cultural coverage. Small globe icons appear in the margins next to cross-cultural and multicultural information.

Lifespan perspective. Icons depicting three joined figures appear in the margins next to information emphasizing a broad, lifespan view of development, including research focusing on constancy and change across several age periods.

Gender-related topics. Male/female icons appear in the margins adjacent to material addressing gender issues.

Great care has gone into the preparation of these teaching materials. We hope we have met our goal of creating the most effective teaching and learning package available for your lifespan development course.

CHAPTER-AT-A-GLANCE

Chapter 1
HISTORY, THEORY, AND RESEARCH STRATEGIES

CHAPTER OUTLINE	INSTRUCTION IDEAS	SUPPLEMENTS
■ Human Development as an Interdisciplinary, Scientific, and Applied Field p. 5	Learning Objective 1.1 Lecture Extensions 1.1–1.2	Test Bank Items 1.1–1.2
■ Basic Themes and Issues pp. 5–10 View of the Developing Person · View of the Course of Development · View of the Determinants of Development · The Lifespan Perspective: A Balanced Point of View	Learning Objectives 1.2–1.3 Lecture Extensions 1.3–1.5	Transparency 6 Test Bank Items 1.3–1.17
■ A Overview of the Lifespan pp. 10–11	Learning Objective 1.4 Learning Activity 1.1	Test Bank Items 1.18–1.21
■ Domains of Development as Interwoven p. 11	Ask Yourself . . . p. 11	
■ Historical Foundations pp. 11–15 Philosophies of Childhood · Philosophies of Adulthood and Aging · Early Scientific Beginnings	Learning Objective 1.5 Lecture Extension 1.6 Ask Yourself . . . p. 15	Test Bank Items 1.22–1.30
■ Mid-Twentieth-Century Theories pp. 15–21 The Psychoanalytic Perspective · Behaviorism and Social Learning Theory · Piaget's Cognitive Developmental Theory	Learning Objective 1.6 Lecture Extension 1.7 Ask Yourself . . . p. 21	Transparencies 1–3, 5 Test Bank Items 1.31–1.48
■ Recent Perspectives pp. 21–26 Information Processing · Ethology · Ecological Systems Theory · Cross-Cultural Research and Vygotsky's Sociocultural Theory	Learning Objective 1.7 Learning Activity 1.2 Ask Yourself . . . p. 26	Transparency 7 Test Bank Items 1.49–1.64
■ Comparing and Evaluating Theories p. 26	Learning Objective 1.8	Test Bank Items 1.65–1.69

CHAPTER OUTLINE	INSTRUCTION IDEAS	SUPPLEMENTS
■ **Studying Development** pp. 26–36 Common Research Methods · General Research Designs · Designs for Studying Development	Learning Objectives 1.9–1.11 Lecture Extensions 1.8–1.9 Learning Activities 1.3–1.4 Ask Yourself . . . p. 36	Transparencies 8–12 Test Bank Items 1.70–1.96
■ **Ethics in Lifespan Research** pp. 36–38	Learning Objective 1.12 Lecture Extension 1.10 Learning Activity 1.5	Test Bank Items 1.97–1.99
■ **A Special Note to Readers** p. 38	Ask Yourself . . . p. 38	

LEARNING OBJECTIVES

After reading this chapter, students should be able to:

■ 1.1 Explain the importance of the terms *interdisciplinary* and *applied* as they help to define the field of human development. (p. 5)

■ 1.2 Explain the role of theories in understanding human development, and describe three basic issues on which major theories take a stand. (p. 6)

■ 1.3 Describe factors that sparked the emergence of the lifespan perspective, and explain the four assumptions that make up this point of view. (p. 7)

■ 1.4 Cite your text's division of the lifespan into age periods, and describe the domains of development to be considered in each. (p. 10)

■ 1.5 Trace historical influences on modern theories of human development, from medieval times through the early twentieth century. (p. 11)

■ 1.6 Describe theoretical perspectives that influenced human development research in the mid-twentieth century, and cite the contributions and limitations of each. (p. 15)

■ 1.7 Describe four recent theoretical perspectives on human development, noting the contributions of major theorists. (p. 21)

■ 1.8 Identify the stand that each modern theory takes on the three basic issues presented earlier in this chapter. (p. 26)

■ 1.9 Describe commonly used methods in research on human development. (p. 28)

■ 1.10 Contrast correlational and experimental research designs, and cite the strengths and limitations of each. (p. 30)

■ 1.11 Describe three research designs for studying development, and cite the strengths and limitations of each. (p. 32)

■ 1.12 Discuss special ethical concerns in lifespan research. (p. 37)

CHAPTER-AT-A-GLANCE

Chapter 2
BIOLOGICAL AND ENVIRONMENTAL FOUNDATIONS

CHAPTER OUTLINE	INSTRUCTION IDEAS	SUPPLEMENTS
■ **Genetic Foundations pp. 44–52** The Genetic Code · The Sex Cells · Conception · Male or Female? · Multiple Births · Patterns of Genetic Inheritance	Learning Objectives 2.1–2.6 Lecture Extensions 2.1–2.3 Learning Activity 2.1	Transparencies 13–18 Test Bank Items 2.1–2.38
■ **Chromosomal Abnormalities** **pp. 52–53** Down Syndrome · Abnormalities of the Sex Chromosomes	Learning Objective 2.7 Ask Yourself . . . p. 53	Test Bank Items 2.39–2.44
■ **Reproductive Choices** **pp. 54–58** Genetic Counseling · Prenatal Diagnosis and Fetal Medicine · Adoption	Learning Objective 2.8 Lecture Extensions 2.4–2.6 Try This . . . p. 57 Ask Yourself . . . p. 58	Test Bank Items 2.45–2.50 Transparency 19 FHS: Genetic Counseling and Prenatal Testing
■ **Environmental Contexts for** **Development pp. 58–70** The Family · Social Class and Family Functioning · The Impact of Poverty · Beyond the Family: Neighborhoods, Towns, and Cities · The Cultural Context	Learning Objectives 2.9–2.13 Lecture Extensions 2.7–2.10 Learning Activities 2.2–2.6 Try This . . . p. 69 Ask Yourself . . . p. 70	Test Bank Items 2.51–2.76
■ **Understanding the Relationship** **between Heredity and** **Environment pp. 70–73** The Question of "How Much?" · The Question of "How?"	Learning Objectives 2.14–2.15 Lecture Extension 2.11 Ask Yourself . . . p. 73	Test Bank Items 2.77–2.94

Note: The abbreviation FHS refers to Allyn & Bacon videos from the Films for the Humanities and Sciences.

LEARNING OBJECTIVES

After reading this chapter, students should be able to:

- 2.1 Describe the relationship between phenotype and genotype. (p. 44)

- 2.2 Describe the structure of the DNA molecule, and explain the process of mitosis. (p. 45)

- 2.3 Explain the process of meiosis. (p. 45)

- 2.4 Describe the process of human reproduction, and explain how the sex of the new individual is determined. (p. 46)

- 2.5 Identify two types of twins, and explain how each is created. (p. 46)

- 2.6 Describe basic patterns of genetic inheritance, and indicate how harmful genes are created. (p. 47)

- 2.7 Describe Down syndrome and common abnormalities of the sex chromosomes. (p. 52)

- 2.8 Discuss reproductive options available to prospective parents and the controversies related to them. (p. 54)

- 2.9 Describe the social systems perspective on family functioning, including direct and indirect influences and the family as dynamic and ever changing. (p. 59)

- 2.10 Discuss the impact of social class and poverty on family functioning. (p. 60)

- 2.11 Summarize the role of neighborhoods, schools, towns, and cities in lives of children and adults. (p. 62)

- 2.12 Cite ways in which cultural values and practices affect human development. (p. 63)

- 2.13 Discuss the impact of the political and economic conditions of a nation on lifespan development. (p. 64)

- 2.14 Describe ways in which researchers determine "how much" heredity and environment influence complex human characteristics. (p. 70)

- 2.15 Describe concepts that indicate "how" heredity and environment work together to influence complex human characteristics. (p. 71)

CHAPTER-AT-A-GLANCE

Chapter 3
PRENATAL DEVELOPMENT, BIRTH, AND THE NEWBORN BABY

CHAPTER OUTLINE	INSTRUCTION IDEAS	SUPPLEMENTS
■ **Prenatal Development pp. 78–82** Period of the Zygote · Period of the Embryo · Period of the Fetus ·	Learning Objective 3.1 Ask Yourself. . . p. 82	Test Bank Items 3.1–3.19 Transparencies 21–22
■ **Prenatal Environmental Influences pp. 82–93** Teratogens · Other Maternal Factors · The Importance of Prenatal Health Care	Learning Objectives 3.2–3.4 Lecture Extensions 3.1–3.4 Learning Activities 3.1–3.2 Try This . . . p. 88 Try This . . . p. 90 Ask Yourself . . . p. 93	Test Bank Items 3.20–3.61 FHS:Fetal Alcohol Syndrome
■ **Childbirth pp. 94–96** The Stages of Childbirth ·The Baby's Adaptation to Labor and Delivery · The Newborn Baby's Appearance · Assessing the Newborn's Physical Condition: The Apgar Scale	Learning Objectives 3.5–3.7 Lecture Extension 3.5 Learning Activity 3.3	Test Bank Items 3.62–3.75 Transparency 23
■ **Approaches to Childbirth pp. 96–98** Natural, or Prepared, Childbirth · Home Delivery	Learning Objective 3.8 Lecture Extensions 3.6–3.8	Test Bank Items 3.76–3.81 FHS: Prepared Childbirth

CHAPTER OUTLINE	INSTRUCTION IDEAS	SUPPLEMENTS
■ **Medical Interventions pp. 98–99** Fetal Monitoring · Labor and Delivery · Medication · Cesarean Delivery	Learning Objective 3.9 Lecture Extensions 3.9–3.10 Ask Yourself . . . p. 99	Test Bank Items 3.82–3.93
■ **Preterm and Low-Birth-Weight Infants pp. 99–101** Preterm versus Small-for-Date · Consequences for Caregiving · Intervening with Preterm Infants	Learning Objective 3.10 Lecture Extension 3.11 Learning Activity 3.40 Ask Yourself . . . p. 101	Test Bank Items 3.94–3.99
■ **Precious Moments After Birth pp. 102–103**	Learning Objective 3.11 Lecture Extension 3.12	Test Bank Items 3.100-3.104
■ **The Newborn Baby's Capacities pp. 103–109** Newborn Reflexes · Sensory Capacities · Newborn States · Neonatal Behavioral Assessment	Learning Objectives 3.12–3.15 Lecture Extension 3.13 Learning Activities 3.5–3.6 Ask Yourself . . . p. 109	Test Bank Items 3.105–3.128 Transparencies 24–25 FHS: Reflexes of Newborns
■ **Adjusting to the New Family Unit p. 109**	Learning Objective 3.16 Learning Activity 3.7 Ask Yourself . . . p. 109	Test Bank Items 3.129–3.130

LEARNING OBJECTIVES

After reading this chapter, students should be able to:

■ 3.1 List the three phases of prenatal development, and describe the major milestones of each (p. 79)

■ 3.2 Define the term *teratogen*, and summarize factors that affect the impact of teratogens. (p. 82)

■ 3.3 List agents known to be or suspected of being teratogens, and discuss evidence supporting the harmful impact of each. (p. 82)

■ 3.4 Discuss maternal factors other than teratogens that can affect the developing embryo or fetus. (p. 87)

■ 3.5 Describe the three stages of labor. (p. 94)

■ 3.6 Discuss the baby's adaptation to labor and delivery, and describe the appearance of the newborn. (p. 95)

■ 3.7 Explain the purpose and main features of the Apgar Scale. (p. 95)

■ 3.8 Describe and evaluate various approaches to childbirth, including delivery in freestanding birth centers, natural childbirth, and home delivery. (p. 96)

■ 3.9 Describe circumstances that justify use of fetal monitoring, labor and delivery medication, and cesarean delivery, and explain any risks associated with each. (p. 98)

■ 3.10 Discuss risks associated with low birth weight, and cite factors that can help infants who survive a traumatic birth develop. (p. 99)

■ 3.11 Discuss parents' emotional involvement with their newborn babies, including findings on bonding. (p. 102)

■ 3.12 Name and describe major newborn reflexes, noting their functions and the importance of assessing them. (p. 103)

■ 3.13 Describe the newborn baby's responsiveness to touch, taste, smell, sound, and visual stimulation. (p. 104)

■ 3.14 Describe newborn states of arousal, including characteristics of sleep and crying and ways to soothe a crying newborn. (p. 106)

■ 3.15 Describe Brazelton's Neonatal Behavioral Assessment Scale (NBAS), and explain its usefulness. (p. 108)

■ 3.16 Describe typical changes in the family after the birth of a new baby. (p. 109)

CHAPTER-AT-A-GLANCE

Chapter 4
PHYSICAL DEVELOPMENT IN INFANCY AND TODDLERHOOD

CHAPTER OUTLINE	INSTRUCTION IDEAS	SUPPLEMENTS
■ **Body Growth in the First Two Years pp. 116–118** Individual and Group Differences · Patterns of Body Growth	Learning Objective 4.1 Learning Activity 4.1	Test Bank Items 4.1–4.10 Transparency 15 SL: Infancy and Early Childhood
■ **Brain Development pp. 118–123** Development of Neurons · Development of the Cerebral Cortex · Changing States of Arousal	Learning Objectives 4.2–4.3 Lecture Extension 4.1 Ask Yourself . . . p. 123	Test Bank Items 4.11–4.23 Transparencies 24–25, 48, 50–54, 58
■ **Influences on Early Physical Growth pp. 123–126** Heredity · Nutrition · Malnutrition · Emotional Well-Being	Learning Objectives 4.4–4.6 Lecture Extensions 4.2–4.3 Learning Activities 4.2–4.4 Ask Yourself. . . p. 126	Test Bank Items 4.24–4.40
■ **Motor Development During the First Two Years pp. 127–131** The Sequence of Motor Development · Motor Skills as Dynamic Systems of Action · Cultural Variations in Motor Development · Fine Motor Development: The Special Case of Voluntary Reaching	Learning Objectives 4.7–4.9 Lecture Extensions 4.4–4.6 Learning Activity 4.5 Ask Yourself . . . p. 131	Test Bank Items 4.41–4.55 Transparencies 32, 40–41
■ **Learning Capacities pp. 131–135** Classical Conditioning · Operant Conditioning · Habituation and Dishabituation · Imitation	Learning Objective 4.10 Lecture Extensions 4.7–4.8 Learning Activities 4.6–4.7 Try This . . . p. 134 Ask Yourself . . . p. 135	Test Bank Items 4.56–4.72 Transparencies 26–27, 31
■ **Perceptual Development in Infancy pp. 135–140** Hearing · Vision · Intermodal Perception	Learning Objectives 4.11–4.12 Lecture Extensions 4.9–4.12 Learning Activity 4.8 Ask Yourself . . . p. 140	Test Bank Items 4.73–4.91 Transparencies 34–36
■ **Understanding Perceptual Development pp. 140–141**	Learning Objective 4.13	Test Bank Items 4.92–4.95

Note: The abbreviation SL refers to the *Seasons of Life* video series from the Annenberg/CPB Collection.

Each videotape applies to an entire chapter or chapters.

LEARNING OBJECTIVES

After reading this chapter, students should be able to:

- **4.1** Describe changes in body size and muscle–fat makeup during the first 2 years, along with individual and group differences and patterns of body growth. (p. 116)

- **4.2** Describe brain development during infancy and toddlerhood, at the level of individual brain cells and at the level of the cerebral cortex. (p. 118)

- **4.3** Describe changes in the organization of sleep and wakefulness during the first 2 years, noting the contributions of brain maturation and the social environment. (p. 121)

- **4.4** Describe evidence indicating that heredity contributes to body size and rate of physical growth, and discuss the nutritional needs of infants and toddlers. (p. 123)

- **4.5** Describe two dietary diseases caused by malnutrition during infancy and toddlerhood, along with their consequences for physical growth and brain development. (p. 125)

- **4.6** Discuss the origins and symptoms of nonorganic failure to thrive. (p. 126)

- **4.7** Describe the overall sequence of motor development during the first 2 years, and explain how motor development involves acquiring increasingly complex, dynamic systems of action. (p. 127)

- **4.8** Discuss the influence of movement opportunities, a stimulating environment, and infant-rearing practices on motor development, as indicated by cross-cultural research. (p. 128)

- **4.9** Describe the development of voluntary reaching, citing evidence on the impact of early experience. (p. 129)

- **4.10** Describe four basic infant learning mechanisms, the conditions under which they occur, and the unique value of each. (p. 131)

- **4.11** Summarize the development of hearing and vision during infancy, giving special attention to depth and pattern perception. (p. 136)

- **4.12** Describe evidence supporting the conclusion that, from the start, babies are capable of intermodal perception. (p. 139)

- **4.13** Explain differentiation theory of perceptual development. (p. 140)

CHAPTER-AT-A-GLANCE

Chapter 5
COGNITIVE DEVELOPMENT IN INFANCY AND TODDLERHOOD

CHAPTER OUTLINE	INSTRUCTION IDEAS	SUPPLEMENTS
■ **Piaget's Cognitive-Developmental Theory** pp. 146–153 Key Piagetian Concepts · The Sensorimotor Stage · Recent Research on Sensorimotor Development · Evaluation of the Sensorimotor Stage	Learning Objectives 5.1–5.3 Lecture Extensions 5.1–5.3 Learning Activities 5.1–5.3 Ask Yourself. . . p. 153	Test Bank Items 5.1–5.26 SL: Infancy and Early Childhood
■ **Information Processing in the First Two Years** pp. 153–157 A Model of Human Information Processing · Attention and Memory · Categorization · Evaluation of Information-Processing Findings	Learning Objective 5.4 Lecture Extension 5.4 Learning Activity 5.4	Test Bank Items 5.27–5.41 FHS: Learning in Infants
■ **The Social Context of Early Cognitive Development** pp. 157–158	Learning Objective 5.5 Lecture Extension 5.5 Ask Yourself. . . p. 158	Test Bank Items 5.42–5.47

CHAPTER OUTLINE	INSTRUCTION IDEAS	SUPPLEMENTS
■ **Individual Differences in Early Mental Development** pp. 158–164 Infant Intelligence Tests · Early Environment and Mental Development · Early Intervention for At-Risk Infants and Toddlers	Learning Objectives 5.6–5.7 Lecture Extensions 5.6–5.8 Learning Activity 5.5 Try This . . . p. 163 Ask Yourself. . . p. 164	Test Bank Items 5.48–5.59 Transparencies 73, 76
■ **Language Development During the First Two Years** pp. 164–170 The Behaviorist Perspective · The Nativist Perspective · The Interactionist Perspective · Getting Ready to Talk · First Words · The Two-Word Utterance Phase · Individual and Group Differences · Supporting Early Language Development	Learning Objectives 5.8–5.12 Lecture Extensions 5.9–5.11 Learning Activities 5.6–5.8 Ask Yourself . . . p. 170	Test Bank Items 5.60–5.96 Transparencies 77, 79

LEARNING OBJECTIVES

After reading this chapter, students should be able to:

■ 5.1 Explain Piaget's view of what changes with development and how cognitive change takes place. (p. 146)

■ 5.2 Name Piaget's six sensorimotor substages, and describe the major cognitive achievements in each. (p. 147)

■ 5.3 Discuss recent research on sensorimotor development and its implications for the accuracy of Piaget's sensorimotor stage. (p. 150)

■ 5.4 Describe Atkinson and Shiffrin's information-processing model, the development of cognitive strategies over the first 2 years, and the contributions and limitations of the information-processing approach to our understanding of early cognitive development. (p. 153)

■ 5.5 Explain how Vygotsky's concept of the zone of proximal development expands our understanding of early cognitive development. (p. 157)

■ 5.6 Describe the mental testing approach, the meaning of intelligence test scores, and the extent to which infant tests predict later performance. (p. 158)

■ 5.7 Discuss environmental influences on early mental development, including home, day care, and early intervention for at-risk infants and toddlers. (p. 160)

■ 5.8 Describe three major theories of language development, indicating the emphasis each places on innate abilities and environmental influences. (p. 164)

■ 5.9 Describe the ways in which babies prepare for language, and explain how adults support their emerging capacities. (p. 165)

■ 5.10 Describe toddlers' first words and two-word combinations, and explain why language comprehension develops ahead of production. (p. 166)

■ 5.11 Describe individual and group differences in early language development and factors that influence these differences. (p. 168)

■ 5.12 Explain how motherese and conversational give-and-take support early language development. (p. 168)

CHAPTER-AT-A-GLANCE

Chapter 6
EMOTIONAL AND SOCIAL DEVELOPMENT IN INFANCY AND TODDLERHOOD

CHAPTER OUTLINE	INSTRUCTION IDEAS	SUPPLEMENTS
■ **Theories of Infant and Toddler Personality pp. 176–179** Erik Erikson: Trust and Autonomy · Margaret Mahler: Separation-Individuation · Similarities Between Erikson's and Mahler's Theories ·	Learning Objective 6.1 Learning Activity 6.1 Ask Yourself. . . p. 178	Test Bank Items 6.1–6.12 Transparencies 1–2 SL: Infancy and Early Childhood
■ **Emotional Development During the First Two Years pp. 179–183** Development of Some Basic Emotions · Understanding and Responding to the Emotions of Others · Emergence of Self-Conscious Emotions · Beginnings of Emotional Self-Regulation	Learning Objectives 6.2–6.5 Lecture Extensions 6.1–6.3 Ask Yourself . . . p. 182	Test Bank Items 6.13–6.37
■ **Temperament and Development pp. 183–186** Measuring Temperament · Stability of Temperament · Genetic Influences · Environmental Influences · Temperament and Child Rearing: The Goodness-of-Fit Model	Learning Objectives 6.6–6.7 Lecture Extensions 6.4–6.6 Learning Activity 6.2 Ask Yourself . . . p. 186	Test Bank Items 6.38–6.55 FHS: Temperament
■ **Development of Attachment pp. 186–196** Ethological Theory of Attachment · Measuring the Security of Attachment · Cultural Variations · Factors That Affect Attachment Security · Multiple Attachments · Attachment and Later Development	Learning Objectives 6.8–6.12 Lecture Extensions 6.7–6.9 Learning Activity 6.3 Ask Yourself . . . p. 196	Test Bank Items 6.56–6.95 Transparency 81
■ **Self-Development During the First Two Years pp. 196–199** Self-Recognition · Categorizing the Self · Emergence of Self-Control	Learning Objective 6.13 Lecture Extension 6.10 Learning Activities 6.4–6.5 Ask Yourself . . . p. 199	Test Bank Items 6.96–6.103 Transparency 88

LEARNING OBJECTIVES

After reading this chapter, students should be able to:

- 6.1 Explain Erikson's and Mahler's theories of infant and toddler personality, noting similarities between them. (p. 176)

- 6.2 Describe the development of happiness, anger, and fear over the first year, noting the adaptive functions of each. (p. 179)

- 6.3 Summarize changes in infants' ability to understand and respond to the emotions of others. (p. 181)

- 6.4 Explain why self-conscious emotions emerge during the second year, and indicate their role in development. (p. 181)

- 6.5 Trace the development of emotional self-regulation during the first 2 years. (p. 182)

- 6.6 Describe the meaning of temperament, the ways in which it is measured, and three temperamental styles identified in the New York Longitudinal Study. (p. 183)

- 6.7 Discuss the role of heredity and environment in the stability of temperament, including the goodness-of-fit model. (p. 184)

- 6.8 Describe the unique features of ethological theory of attachment and the development of attachment during the first 2 years. (p. 187)

- 6.9 Describe the Strange Situation, the four attachment patterns assessed by it, and cultural variations in infants' reaction to this procedure. (p. 189)

- 6.10 Discuss factors that affect attachment security. (p. 190)

- 6.11 Compare fathers' and mothers' attachment relationships with infants, and note factors that affect early sibling relationships. (p. 192)

- 6.12 Describe and interpret the relationship between secure attachment in infancy and cognitive and social competence in childhood. (p. 195)

- 6.13 Discuss the development of self-awareness in the second year, and explain its role in the development of empathy, categorizing the self, and self-control. (p. 196)

CHAPTER-AT-A-GLANCE

Chapter 7
PHYSICAL AND COGNITIVE DEVELOPMENT IN EARLY CHILDHOOD

CHAPTER OUTLINE	INSTRUCTION IDEAS	SUPPLEMENTS
PHYSICAL DEVELOPMENT IN EARLY CHILDHOOD		
■ **Body Growth pp. 206–208** Skeletal Growth · Asynchronies in Physical Growth	Learning Objective 7.1 Learning Activity 7.1	Test Bank Items 7.1–7.7 Transparencies 43, 46 SL: Infancy and Early Childhood
■ **Brain Development pp. 208–210** Lateralization and Handedness Other Advances in Brain Development	Learning Objective 7.2 Lecture Extension 7.1 Ask Yourself . . . p. 210	Test Bank Items 7.8–7.22 Transparency 49
■ **Influences on Physical Growth and Health pp. 210–216** Heredity and Hormones · Emotional Well-Being · Nutrition · Infectious Disease · Childhood Injuries	Learning Objectives 7.3–7.4 Lecture Extensions 7.2–7.5 Learning Activity 7.2 Try This . . . p. 211 Ask Yourself . . . p. 216	Test Bank Items 7.23–7.47 Transparencies 44, 59

CHAPTER OUTLINE	INSTRUCTION IDEAS	SUPPLEMENTS
■ **Motor Development pp. 216–220** Gross Motor Development · Fine Motor Development · Factors That Affect Early Childhood Motor Skills	Learning Objective 7.5 Lecture Extensions 7.6 Learning Activities 7.3–7.5 Ask Yourself . . . p. 220	Test Bank Items 7.48–7.59 Transparency 64

COGNITIVE DEVELOPMENT IN EARLY CHILDHOOD

■ **Piaget's Theory:** **The Preoperational Stage** **pp. 220–226** Advances in Mental Representation · Make-Believe Play ·Limitations of Preoperational Thought · Recent Research on Preoperational Thought · Evaluation of the Preoperational Stage · Piaget and Education	Learning Objectives 7.6–7.9 Lecture Extensions 7.7–7.8 Learning Activities 7.6–7.7 Ask Yourself . . . p. 226	Test Bank Items 7.60–7.91 Transparencies 65–67
■ **Vygotsky's Sociocultural Theory** **pp. 226–228** Children's Private Speech · Social Origins of Early Childhood Education · Vygotsky and Education	Learning Objective 7.10 Learning Activity 7.8 Ask Yourself . . . p. 228	Test Bank Items 7.92–7.101
■ **Information Processing** **pp. 229–233** Attention · Memory · The Young Child's Theory of Mind · Early Childhood Literacy · Young Children's Mathematical Reasoning · A Note on Academics in Early Childhood	Learning Objectives 7.11–7.12 Lecture Extensions 7.9–7.11 Learning Activities 7.9–7.11 Ask Yourself . . . p. 233	Test Bank Items 7.102–7.115
■ **Individual Differences in Mental** **Development pp. 233–236** Home Environment and Mental Development · Preschool and Day Care · Educational Television	Learning Objectives 7.13–7.14 Learning Activity 7.12 Ask Yourself . . . p. 236	Test Bank Items 7.116–7.135 Transparencies 62–63
■ **Language Development** **pp. 237–240** Vocabulary · Grammar · Conversation · Supporting Language Learning in Early Childhood	Learning Objective 7.15 Lecture Extension 7.12 Learning Activity 7.13 Ask Yourself . . . p. 240	Test Bank Items 7.136–7.144

LEARNING OBJECTIVES

After reading this chapter, students should be able to:

- 7.1 Describe changes in body size, proportions, and skeletal maturity during early childhood, and discuss asynchronies in physical growth. (p. 206)

- 7.2 Discuss brain development in early childhood, including lateralization and handedness and myelinization of the cerebellum, reticular formation, and corpus callosum. (p. 208)

- 7.3 Describe the impact of heredity, emotional well-being, nutrition, and infectious disease on early childhood growth, and compare child health services in the United States with those of European nations. (p. 210)

- 7.4 Summarize factors related to childhood injuries, and cite preventive measures. (p. 213)

- 7.5 Cite major milestones and factors that affect gross and fine motor development in early childhood. (p. 217)

- 7.6 Describe advances in mental representation during the preschool years, including changes in make-believe play. (p. 220)

- 7.7 Describe the limitations of preoperational thought from Piaget's point of view. (p. 221)

- 7.8 Discuss recent research on preoperational thought and its implications for the accuracy of Piaget's preoperational stage. (p. 223)

- 7.9 Describe three educational principles derived from Piaget's theory. (p. 226)

- 7.10 Contrast Piaget's and Vygotsky's views on the development and significance of children's private speech, and discuss applications of Vygotsky's theory to education. (p. 226)

- 7.11 Describe the development of attention and memory during early childhood. (p. 229)

- 7.12 Discuss preschoolers' awareness of an inner mental life, and explain the limitations of their theory of mind. (p. 230)

- 7.13 Trace the development of preschoolers' literacy and mathematical knowledge, and discuss ways to enhance their academic development. (p. 231)

- 7.14 Describe the impact of home, preschool and day care, and educational television on mental development in early childhood. (p. 233)

- 7.15 Trace the development of vocabulary, grammar, and conversational skills, and cite factors that support language learning in early childhood. (p. 237)

CHAPTER-AT-A-GLANCE

Chapter 8
EMOTIONAL AND SOCIAL DEVELOPMENT IN EARLY CHILDHOOD

CHAPTER OUTLINE	INSTRUCTION IDEAS	SUPPLEMENTS
■ **Erikson's Theory: Initiative versus Guilt pp. 246–247**	Learning Objective 8.1	Test Bank Items 8.1–8.7 SL: Infancy and Early Childhood
■ **Self-Development in Early Childhood pp. 247–248** Foundations of Self-Concept · Emergence of Self-Esteem	Learning Objective 8.2 Learning Activity 8.1	Test Bank Items 8.8–8.14
■ **Emotional Development in Early Childhood pp. 248–251** Understanding Emotion · Improvements in Emotional Self-Regulation · Changes in Self-Conscious Emotions · Development of Empathy	Learning Objective 8.3 Lecture Extension 8.1	Test Bank Items 8.15–8.25
■ **Peer Relations in Early Childhood pp. 251–253** Advances in Peer Sociability · First Friendships	Learning Objectives 8.4–8.5 Lecture Extensions 8.2–8.3 Learning Activities 8.2–8.3 Ask Yourself . . . p. 253	Test Bank Items 8.26–8.38

CHAPTER OUTLINE	INSTRUCTION IDEAS	SUPPLEMENTS
■ Foundations of Morality in Early Childhood pp. 253–259 The Psychoanalytic Perspective · Behaviorism and Social Learning Theory · The Cognitive-Developmental Perspective · The Other Side of Morality: Development of Aggression	Learning Objectives 8.6–8.7 Lecture Extensions 8.4–8.7 Learning Activities 8.4–8.8 Ask Yourself . . . p. 259 Try This . . . p. 260	Test Bank Items 8.39–8.69 Transparency 89
■ Gender Typing in Early Childhood pp. 259–266 Preschoolers' Gender-Stereotyped Beliefs and Behavior · Genetic Influences on Gender Typing · Environmental Influences on Gender Typing · Gender-Role Identity · Reducing Gender Stereotyping in Young Children	Learning Objectives 8.8–8.9 Lecture Extension 8.8 Learning Activity 8.9 Ask Yourself . . . p. 266	Test Bank Items 8.70–8.89
■ Child-Rearing and Emotional and Social Development in Early Childhood pp. 266–270 Child-Rearing Styles · What Makes Authoritative Child-Rearing So Effective? · Cultural and Situational Influences on Child-Rearing Styles · Child Maltreatment	Learning Objectives 8.10–8.11 Lecture Extension 8.9 Ask Yourself . . . p. 270	Test Bank Items 8.90–8.112 FHS: Physical Abuse of Children

LEARNING OBJECTIVES

After reading this chapter, students should be able to:

■ 8.1 Describe Erikson's stage of initiative versus guilt, and explain its relationship to Freud's phallic stage. (p. 246)

■ 8.2 Describe preschoolers' self-concepts and emerging sense of self-esteem. (p. 247)

■ 8.3 Describe changes in understanding and expression of emotion during early childhood, noting achievements and limitations. (p. 248)

■ 8.4 Trace the development of peer sociability in early childhood, noting the special contribution of sociodramatic play to emotional and social development. (p. 251)

■ 8.5 Describe the quality of preschoolers' friendships. (p. 252)

■ 8.6 Compare the central features of psychoanalytic, behaviorist and social learning, and cognitive-developmental approaches to moral development, and trace milestones of morality during early childhood along with child-rearing practices that support or undermine them. (p. 254)

■ 8.7 Describe the development of aggression in early childhood, discuss family and television as major influences, and cite ways to control aggressive behavior. (p. 257)

■ 8.8 Describe preschoolers' gender-stereotyped beliefs and behaviors, and discuss genetic and environmental influences on gender typing. (p. 260)

■ 8.9 Describe and evaluate the accuracy of major theories of the emergence of gender-role identity, and cite ways that adults can reduce rigid gender typing in early childhood. (p. 262)

■ 8.10 Describe the impact of child-rearing styles on development, and discuss cultural and situational influences on child rearing. (p. 266)

■ 8.11 List five forms of child maltreatment, and discuss its multiple origins and consequences for children's development. (p. 268)

CHAPTER-AT-A-GLANCE

Chapter 9
PHYSICAL AND COGNITIVE DEVELOPMENT IN MIDDLE CHILDHOOD

CHAPTER OUTLINE	INSTRUCTION IDEAS	SUPPLEMENTS
PHYSICAL DEVELOPMENT IN MIDDLE CHILDHOOD		
■ **Body Growth pp. 278–279**	Learning Objective 9.1 Learning Activity 9.1	Test Bank Items 9.1–9.4 Transparency 60 SL: Childhood and Adolescence
■ **Common Health Problems pp. 280–282** Vision and Hearing · Malnutrition · Obesity · Bedwetting · Illnesses · Unintentional Injuries	Learning Objective 9.2 Lecture Extensions 9.1–9.2	Test Bank Items 9.5–9.19
■ **Health Education pp. 282–284**	Learning Objective 9.3 Try This . . . p. 284	Test Bank Items 9.20–9.25
■ **Motor Development and Play pp. 285–289** Gross Motor Development · Fine Motor Development · Organized Games with Rules · Physical Education	Learning Objectives 9.4–9.6 Lecture Extensions 9.3–9.5 Learning Activities 9.2–9.3 Ask Yourself . . . p. 289	Test Bank Items 9.26–9.39
COGNITIVE DEVELOPMENT IN MIDDLE CHILDHOOD		
■ **Piaget's Theory: The Concrete Operational Stage pp. 290–292** Conservation · Classification · Seriation · Spatial Reasoning · Limitations of Concrete Operational Thought · Evaluation of the Concrete Operational Stage	Learning Objectives 9.7–9.8 Lecture Extensions 9.6–9.7 Learning Activity 9.4 Ask Yourself . . . p. 292	Test Bank Items 9.40–9.54
■ **Information Processing pp. 292–298** Attention · Memory Strategies · The Knowledge Base and Memory Performance · Culture and Memory Strategies · The School-Age Child's Theory of Mind · Self-Regulation · Applications of Information Processing to Academic Learning	Learning Objectives 9.9–9.11 Lecture Extension 9.8 Learning Activities 9.5–9.9 Ask Yourself . . . p. 298	Test Bank Items 9.55–9.84 Transparency 68

CHAPTER OUTLINE	INSTRUCTION IDEAS	SUPPLEMENTS
■ Individual Differences in Mental Development pp. 298–304 Defining and Measuring Intelligence · Recent Developments in Defining Intelligence · Explaining Individual and Group Differences in IQ	Learning Objectives 9.12–9.14 Lecture Extensions 9.9–9.10 Learning Activity 9.10 Ask Yourself . . . p. 304	Test Bank Items 9.85–9.107 Transparencies 69–71, 75
■ Language Development pp. 304–306 Vocabulary · Grammar · Pragmatics · Learning Two Languages at a Time	Learning Objective 9.15 Learning Activity 9.11 Ask Yourself . . . p. 306	Test Bank Items 9.108–9.114
■ Children's Learning in School pp. 306–312 The Educational Philosophy · Teacher-Pupil Interaction · Computers in the Classroom · Teaching Children with Special Needs · How Well Educated are America's Children?	Learning Objectives 9.16–9.20 Lecture Extension 9.11 Learning Activity 9.12 Ask Yourself . . . p. 312	Test Bank Items 9.115–9.142 FHS: Learning Disabilities

LEARNING OBJECTIVES

After reading this chapter, students should be able to:

■ 9.1 Describe changes in body size, proportions, and skeletal maturity during middle childhood. (p. 278)

■ 9.2 Identify common health problems in middle childhood, discuss their causes and consequences, and cite ways to alleviate them. (p. 280)

■ 9.3 Summarize findings on school-age children's concepts of health and illness, and indicate what parents and teachers can do to encourage good health practices. (p. 284)

■ 9.4 Cite major milestones of gross and fine motor development in middle childhood, noting sex differences. (p. 285)

■ 9.5 Describe qualities of children's play during middle childhood. (p. 287)

■ 9.6 Discuss the importance of high-quality physical education during the school years. (p. 288)

■ 9.7 Describe the major characteristics of concrete operational thought. (p. 290)

■ 9.8 Discuss recent research on concrete operational thought and its implications for the accuracy of Piaget's concrete operational stage. (p. 291)

■ 9.9 Describe the development of attention and memory strategies in middle childhood, and discuss the role of knowledge and culture in memory performance. (p. 292)

■ 9.10 Describe the school-age child's theory of mind and capacity to engage in self-regulation. (p. 294)

■ 9.11 Discuss current controversies in teaching reading and mathematics to elementary school children. (p. 296)

■ 9.12 Describe commonly used intelligence tests in middle childhood. (p. 298)

■ 9.13 Describe applications of information processing to defining intelligence, including Sternberg's triarchic theory and Gardner's theory of multiple intelligences. (p. 300)

■ 9.14 Describe evidence indicating that both heredity and environment contribute to intelligence, and discuss cultural influences on the IQ scores of ethnic minority children. (p. 301)

■ 9.15 Describe changes in vocabulary, grammar, and pragmatics during middle childhood, and discuss the advantages of bilingualism. (p. 304)

■ 9.16 Describe the impact of educational philosophies on children's motivation and academic achievement. (p. 306)

■ 9.17 Discuss the role of teacher–pupil interaction in academic achievement. (p. 307)

■ 9.18 Describe the characteristics of mildly mentally retarded and learning disabled-children, and discuss the conditions under which mainstreaming is successful. (p. 307)

■ 9.19 Describe the characteristics of gifted children and current efforts to meet their educational needs. (p. 309)

■ 9.20 Compare the American cultural climate for academic achievement with that of Japan and Taiwan. (p. 311)

CHAPTER-AT-A-GLANCE

Chapter 10
EMOTIONAL AND SOCIAL DEVELOPMENT IN MIDDLE CHILDHOOD

CHAPTER OUTLINE	INSTRUCTION IDEAS	SUPPLEMENTS
■ **Erikson's Theory: Industry versus Inferiority pp. 318–319**	Learning Objective 10.1	Test Bank Items 10.1–10.5 SL: Childhood and Adolescence
■ **Self-Development in Middle Childhood pp. 319–322** Changes in Self-Concept · Development of Self-Esteem · Influences on Self-Esteem	Learning Objective 10.2 Lecture Extension 10.1 Learning Activity 10.1 Ask Yourself . . . p. 322	Test Bank Items 10.6–10.21 Transparencies 82, 84
■ **Emotional Development in Middle Childhood pp. 322–323**	Learning Objective 10.3 Lecture Extensions 10.2–10.3	Test Bank Items 10.22–10.25
■ **Understanding Others pp. 323–324** Selman's Stages of Perspective Taking · Perspective Taking and Social Skills	Learning Objective 10.4 Lecture Extension 10.4 Learning Activity 10.2	Test Bank Items 10.26–10.30
■ **Moral Development in Middle Childhood pp. 324–326** Learning About Justice Through Sharing · Changes in Moral and Social-Conventional Understanding	Learning Objective 10.5 Learning Activity 10.3 Ask Yourself . . . p. 326	Test Bank Items 10.31–10.36
■ **Peer Relations in Middle Childhood pp. 326–329** Peer Groups · Friendships · Peer Acceptance	Learning Objectives 10.6–10.7 Lecture Extension 10.5	Test Bank Items 10.37–10.58
■ **Gender Typing in Middle Childhood pp. 328–330** Gender-Stereotyped Beliefs · Gender-Role Identity and Behavior · Cultural Influences on Gender Typing	Learning Objective 10.8 Lecture Extension 10.6 Ask Yourself . . . p. 330	Test Bank Items 10.59–10.65 Transparency 95
■ **Family Influences in Middle Childhood pp. 330–338** Parent-Child Relationships · Siblings · Only Children · Divorce · Remarriage · Maternal Employment	Learning Objectives 10.9–10.12 Lecture Extensions 10.7–10.9 Learning Activity 10.4 Ask Yourself . . . p. 338	Test Bank Items 10.66–10.91
■ **Some Common Problems of Development pp. 339–341** Fears and Anxieties · Child Sexual Abuse	Learning Objectives 10.13–10.14 Lecture Extension 10.10 Learning Activity 10.5	Test Bank Items 10.92–10.107 FHS: Sexual Abuse of Children

INSTRUCTOR'S SECTION

■ **Stress and Coping: The Resilient Child pp. 341–343**

Learning Objective 10.15
Try This . . . p. 342

Test Bank Items 10.108–10.112

LEARNING OBJECTIVES

After reading this chapter, students should be able to:

■ **10.1** Describe Erikson's stage of industry versus inferiority, and explain its relationship to Freud's latency stage. (p. 318)

■ **10.2** Describe school-age children's self-concept, self-esteem, and achievement-related attributions, along with factors that affect self-evaluations in middle childhood. (p. 319)

■ **10.3** Describe changes in understanding and expression of emotion in middle childhood. (p. 322)

■ **10.4** Trace the development of perspective taking, and discuss its relationship to social behavior. (p. 323)

■ **10.5** Describe changes in moral understanding during middle childhood, noting cultural influences. (p. 324)

■ **10.6** Describe school-age children's peer groups and friendships and the contributions of each to social development. (p. 326)

■ **10.7** Describe major categories of peer acceptance, the relationship of each to social behavior, and ways to help rejected children. (p. 327)

■ **10.8** Describe changes in gender-stereotyped beliefs and gender-role identity during middle childhood, noting sex differences and cultural influences. (p. 328)

■ **10.9** Describe new child-rearing issues and changes in parent–child communication during middle childhood. (p. 330)

■ **10.10** Discuss changes in sibling relationships during middle childhood and the development of only children. (p. 331)

■ **10.11** Discuss children's adjustment to divorce and remarriage, noting the influence of parent and child characteristics and social supports within the family and surrounding community. (p. 332)

■ **10.12** Discuss the impact of maternal employment on school-age children's development, noting the influence of parent and child characteristics and social supports within the family and surrounding community. (p. 336)

■ **10.13** Discuss common fears and anxieties in middle childhood. (p. 339)

■ **10.14** Discuss factors related to child sexual abuse, its consequences for children's development, and ways to prevent and treat it. (p. 339)

■ **10.15** Cite factors that help children cope with stress and reduce the chances of maladjustment. (p. 341)

CHAPTER-AT-A-GLANCE

Chapter 11
PHYSICAL AND COGNITIVE DEVELOPMENT IN ADOLESCENCE

CHAPTER OUTLINE	INSTRUCTION IDEAS	SUPPLEMENTS
PHYSICAL DEVELOPMENT IN ADOLESCENCE		
■ **Conceptions of Adolescence** pp. 350–351 The Biological Perspective · The Environmental Perspective · A Balanced Point of View	Learning Objective 11.1	Test Bank Items 11.1–11.5 SL: Childhood and Adolescence
■ **Puberty: The Physical Transition to Adulthood** pp. 351–354 Hormonal Changes · Body Growth · Sexual Maturation · Individual and Group Differences	Learning Objectives 11.2–11.3 Learning Activity 11.1	Test Bank Items 11.6–11.19 Transparencies 37–38, 47, 55–57, 61
■ **The Psychological Impact of Pubertal Events** pp. 354–357 Reactions to Pubertal Changes · Pubertal Change, Emotion, and Social Behavior · Early versus Late Maturation	Learning Objectives 11.4–11.5 Lecture Extension 11.1 Learning Activity 11.2–11.4 Ask Yourself . . . p. 357	Test Bank Items 11.20–11.37
■ **Health Issues** pp. 357–367 Nutritional Needs · Serious Eating Disturbances · Sexual Activity · Teenage Pregnancy and Childbearing · Sexually Transmitted Disease · Substance Use and Abuse · Unintentional Injuries	Learning Objectives 11.6–11.13 Lecture Extensions 11.2–11.4 Learning Activities 11.5–11.6 Try This . . . p. 363	Test Bank Items 11.38–11.73 Transparency 91
■ **Motor Development** pp. 367–368	Learning Objective 11.14 Ask Yourself . . . p. 368	Test Bank Items 11.74–11.76 Transparency 42
COGNITIVE DEVELOPMENT IN ADOLESCENCE		
■ **Piaget's Theory: The Formal Operational Stage** pp. 369–371 Hypothetico-Deductive Reasoning · Propositional Thought · Recent Research on Formal Operational Thought	Learning Objectives 11.15–11.16 Learning Activity 11.7	Test Bank Items 11.77–11.88
■ **An Information-Processing View of Adolescent Cognitive Development** p. 371	Learning Objective 11.17	Test Bank Items 11.89–11.93

CHAPTER OUTLINE	INSTRUCTION IDEAS	SUPPLEMENTS
■ Consequences of Abstract Thought pp. 371–373 Argumentativeness · Self-Consciousness and Self-Focusing · Idealism and Criticism · Planning and Decision Making	Learning Objective 11.18 Lecture Extension 11.5 Learning Activity 11.8 Ask Yourself . . . p. 373	Test Bank Items 11.94–11.103
■ Sex Differences in Mental Abilities p. 374	Learning Objective 11.19 Lecture Extension 11.6	Test Bank Items 11.104–11.108
■ Language Development pp. 374–375 Vocabulary and Grammar · Pragmatics · Second-Language Learning	Learning Objective 11.20 Lecture Extension 11.7 Ask Yourself . . . p. 375	Test Bank Items 11.109–11.116 Transparency 78
■ Learning in School pp. 375–381 School Transitions · Academic Achievement · Dropping Out	Learning Objectives 11.21–11.23 Lecture Extensions 11.8–11.11 Learning Activity 11.9 Ask Yourself . . . p. 381	Test Bank Items 11.117–11.137 Transparency 83

LEARNING OBJECTIVES

After reading this chapter, students should be able to:

■ 11.1 Discuss changing conceptions of adolescence over the twentieth century. (p. 350)

■ 11.2 Describe pubertal changes in body size, proportions, and sexual maturity and the hormonal secretions that underlie them. (p. 351)

■ 11.3 Cite factors that influence the timing of puberty. (p. 354)

■ 11.4 Discuss adolescents' reactions to the physical changes of puberty, noting factors that influence their feelings and behavior. (p. 354)

■ 11.5 Discuss the impact of maturational timing on adolescent adjustment, noting sex differences and immediate and long-term consequences. (p. 355)

■ 11.6 Describe the nutritional needs of adolescents. (p. 358)

■ 11.7 Describe the symptoms of anorexia nervosa and bulimia, and cite factors within the individual, the family, and the larger culture that contribute to these disorders. (p. 358)

■ 11.8 Discuss social and cultural influences on adolescent sexual attitudes and behavior. (p. 359)

■ 11.9 Describe factors related to the development of homosexuality and the special adjustment problems of gay and lesbian adolescents. (p. 361)

■ 11.10 Discuss factors related to teenage pregnancy, the consequences of adolescent childbearing for development, and prevention strategies. (p. 361)

■ 11.11 Discuss the high rate of sexually transmitted disease in adolescence, noting the most common illnesses. (p. 365)

■ 11.12 Distinguish between substance use and abuse, describe personal and social factors related to each, and cite prevention strategies. (p. 365)

■ 11.13 Cite common unintentional injuries in adolescence. (p. 367)

■ 11.14 Describe sex differences in motor development during adolescence, and discuss their implications for physical education. (p. 367)

■ 11.15 Describe the major characteristics of formal operational thought. (p. 369)

■ 11.16 Discuss recent research on formal operational thought and its implications for the accuracy of Piaget's formal operational stage. (p. 370)

■ 11.17 Explain how information-processing researchers account for the development of abstract reasoning during adolescence. (p. 371)

■ 11.18 Describe typical reactions of adolescents that result from new abstract reasoning powers. (p. 371)

■ 11.19 Describe sex differences in mental abilities at adolescence, along with factors that influence them. (p. 374)

LEARNING OBJECTIVES (continued)

■ 11.20 Describe changes in vocabulary, grammar, and pragmatics during adolescence, and compare teenagers' capacity for second language learning to that of children. (p. 374)

■ 11.21 Discuss the impact of school transitions on adolescent adjustment, and cite ways to ease the strain of these changes. (p. 375)

■ 11.22 Discuss family, peer, and school influences on academic achievement during adolescence. (p. 377)

■ 11.23 Describe student, family, and school factors related to dropping out, and cite ways to prevent early school leaving. (p. 380)

CHAPTER-AT-A-GLANCE

Chapter 12
EMOTIONAL AND SOCIAL DEVELOPMENT IN ADOLESCENCE

CHAPTER OUTLINE	INSTRUCTION IDEAS	SUPPLEMENTS
■ **Erikson's Theory: Identity versus Identity Diffusion pp. 388–389**	Learning Objective 12.1	Test Bank Items 12.1–12.7 SL: Childhood and Adolescence
■ **Self-Development in Adolescence pp. 389–393** Changes in Self-Concept · Changes in Self-Esteem · Paths to Identity Status and Personality Characteristics · Factors that Affect Identity Development	Learning Objectives 12.2–12.3 Ask Yourself . . . p. 393	Test Bank Items 12.8–12.27 Transparency 85
■ **Moral Development in Adolescence pp. 393–399** Piaget's Theory of Moral Development · Kohlberg's Extension of Piaget's Theory · Environmental Influences on Moral Reasoning · Are There Gender Differences in Moral Reasoning? · Moral Reasoning and Behavior	Learning Objectives 12.4–12.6 Lecture Extensions 12.1–12.4 Learning Activities 12.1–12.4 Ask Yourself . . . pp. 398–399	Test Bank Items 12.28–12.53 Transparencies 86–87
■ **Gender Typing in Adolescence p. 399**	Learning Objective 12.7 Lecture Extension 12.5	Test Bank Items 12.54–12.59
■ **The Family in Adolescence pp. 399–401** Parent-Child Relationships · Siblings	Learning Objective 12.8	Test Bank Items 12.60–12.64
■ **Peer Relations in Adolescence pp. 401–404** Adolescent Friendships · Cliques and Crowds · Dating · Peer Pressure and Conformity	Learning Objectives 12.9–12.10 Lecture Extension 12.6 Learning Activities 12.5–12.6 Ask Yourself . . . p. 404	Test Bank Items 12.65–12.81

CHAPTER OUTLINE	INSTRUCTION IDEAS	SUPPLEMENTS
■ **Problems of Development** pp. 404–409 Depression · Suicide · Delinquency	Learning Objectives 12.11–12.12 Lecture Extensions 12.7–12.8 Learning Activity 12.7 Ask Yourself . . . p. 409	Test Bank Items 12.82–12.98 Transparencies 90, 92, 94 FHS: Teen Suicide

LEARNING OBJECTIVES

After reading this chapter, students should be able to:

- 12.1 Discuss Erikson's account of identity development. (p. 388)

- 12.2 Describe changes in self-concept and self-esteem during adolescence. (p. 389)

- 12.3 Describe the four identity statuses, their relationship to adolescent personality characteristics, and factors that promote identity development. (p. 390)

- 12.4 Describe Piaget's theory of moral development and Kohlberg's extension of it, and evaluate the accuracy of each. (p. 393)

- 12.5 Describe environmental influences on moral reasoning. (p. 396)

- 12.6 Evaluate claims that Kohlberg's theory does not adequately represent the morality of females, and describe the relationship of moral reasoning to behavior. (p. 398)

- 12.7 Explain why early adolescence is a period of gender intensification. (p. 399)

- 12.8 Discuss changes in parent–child and sibling relationships during adolescence. (p. 400)

- 12.9 Discuss changes in friendships and peer groups during adolescence, and describe the contributions of each to emotional and social development. (p. 401)

- 12.10 Describe adolescent dating relationships, and discuss conformity to peer pressure in adolescence. (p. 402)

- 12.11 Discuss factors related to adolescent depression and suicide along with approaches to prevention and treatment. (p. 404)

- 12.12 Discuss factors related to delinquency and ways to prevent and treat it. (p. 407)

CHAPTER-AT-A-GLANCE

Chapter 13
PHYSICAL AND COGNITIVE DEVELOPMENT IN EARLY ADULTHOOD

CHAPTER OUTLINE	INSTRUCTION IDEAS	SUPPLEMENTS
PHYSICAL DEVELOPMENT IN EARLY ADULTHOOD		
■ **Life Expectancy pp. 417–418**	Learning Objective 13.1	Transparencies 96–97 Test Bank Items 13.1–13.6
■ **Maximum Lifespan and Active Lifespan pp. 418–419**	Learning Objective 13.2 Lecture Extensions 13.1–13.2	Test Bank Items 13.7–13.11
■ **Theories of Biological Aging pp. 419–420** Aging at the Level of DNA and Body Cells · Aging at the Level of Organs and Tissues	Learning Objective 13.3	Test Bank Items 13.12–13.18

CHAPTER OUTLINE	INSTRUCTION IDEAS	SUPPLEMENTS
■ **Physical Changes of Aging** pp. 421–425 Cardiovascular and Respiratory Systems · Motor Performance · Immune System · Reproductive Capacity	Learning Objective 13.4 Ask Yourself . . . p. 425	Transparencies 98–99 Test Bank Items 13.19–13.31
■ **Health and Fitness pp. 425–436** Nutrition · Exercise · Substance Abuse · Sexuality · Psychological Stress	Learning Objectives 13.5–13.8 Lecture Extensions 13.3–13.5 Learning Activity 13.1 Try This . . . p. 431 Ask Yourself . . . p. 436	Transparency 100 Test Bank Items 13.32–13.62

COGNITIVE DEVELOPMENT IN EARLY ADULTHOOD

CHAPTER OUTLINE	INSTRUCTION IDEAS	SUPPLEMENTS
■ **Changes in the Structure of Thought pp. 436–438** Perry's Theory · Schaie's Theory · Labouvie-Vief's Theory	Learning Objective 13.9 Learning Activity 13.2	Test Bank Items 13.63–13.71
■ **Information Processing: Expertise and Creativity pp. 438–439**	Learning Objective 13.10 Learning Activity 13.3	Test Bank Items 13.72–13.75
■ **Changes in Mental Abilities pp. 439–440**	Learning Objective 13.11 Lecture Extensions 13.6–13.7 Ask Yourself . . . p. 440	Transparency 101 Test Bank Items 13.76–13.78
■ **The College Experience pp. 440–441** Psychological Impact of Attending College · Dropping Out	Learning Objective 13.12 Learning Activity 13.4	Test Bank Items 13.79–13.82
■ **Vocational Choice pp. 441–445** Selecting a Vocation · Factors Influencing Vocational Choice · Vocational Preparation of Non-College-Bound Young Adults	Learning Objectives 13.13–13.14 Lecture Extension 13.8 Learning Activities 13.5–13.6 Ask Yourself . . . p. 445	Test Bank Items 13.83–13.100

LEARNING OBJECTIVES

After reading this chapter, students should be able to:

■ 13.1 Discuss factors that have contributed to changes in average life expectancy over the twentieth century. (p. 417)

■ 13.2 Discuss the extent to which maximum lifespan and active lifespan can and should be improved. (p. 418)

■ 13.3 Describe current theories of biological aging, including those at the level of DNA and body cells and those at the level of tissues and organs. (p. 419)

■ 13.4 Describe the physical changes of aging, paying special attention to the cardiovascular and respiratory systems, motor performance, the immune system, and reproductive capacity. (p. 421)

■ 13.5 Explain the impact of nutrition and exercise on health, and discuss the problem of obesity in adulthood. (p. 426)

■ 13.6 Name the two most common substance disorders, and discuss the health risks each entails. (p. 428)

■ 13.7 Describe sexual attitudes and behavior of young adults today, and discuss factors related to sexually transmitted disease, sexual coercion, and premenstrual syndrome (PMS). (p. 429)

■ 13.8 Explain how psychological stress affects a person's health. (p. 434)

■ 13.9 Describe the restructuring of thought in adulthood, drawing on three influential theories. (p. 436)

LEARNING OBJECTIVES (continued)

■ 13.10 Discuss the development of expertise and creativity in adulthood. (p. 438)

■ 13.11 Describe changes in mental abilities assessed on intelligence tests during adulthood. (p. 439)

■ 13.12 Describe the impact of a college education on young people's lives, and discuss the problem of dropping out. (p. 440)

■ 13.13 Trace the development of vocational choice, and cite factors that influence it. (p. 441)

■ 13.14 Discuss career options open to the 25 percent of high school graduates who are not college bound. (p. 443)

CHAPTER-AT-A-GLANCE

Chapter 14
EMOTIONAL AND SOCIAL DEVELOPMENT IN EARLY ADULTHOOD

CHAPTER OUTLINE	INSTRUCTION IDEAS	SUPPLEMENTS
■ **Erikson's Theory: Intimacy versus Isolation pp. 452–453**	Learning Objective 14.1	SL: Early Adulthood (Ages 20–40) Test Bank Items 14.1–14.6
■ **Other Theories of Adult Psychosocial Development pp. 453–457** Levinson's Seasons of Life · Vaillant's Adaptation to Life · Limitations of Levinson's and Vaillant's Theories · The Social Clock	Learning Objectives 14.2–14.3 Learning Activity 14.1 Ask Yourself . . . p. 457	Test Bank Items 14.7–14.29
■ **Close Relationships pp. 457–462** Romantic Love · Friendships · Loneliness	Learning Objectives 14.4–14.6 Lecture Extensions 14.1–14.2 Learning Activities 14.2–14.3 Ask Yourself . . . p. 462	Test Bank Items 14.30–14.60
■ **The Family Life Cycle pp. 462–471** Leaving Home · Joining of Families in Marriage · Parenthood	Learning Objective 14.7 Lecture Extension 14.3 Learning Activities 14.4–14.7 Try This . . . p. 467 Ask Yourself . . . p. 471	Transparencies 102–103 Test Bank Items 14.61–14.95
■ **The Diversity of Adult Lifestyles pp. 471–476** Singlehood · Cohabitation · Childlessness · Divorce and Remarriage · Variant Styles of Parenthood	Learning Objectives 14.8–14.10 Lecture Extensions 14.4–14.6 Ask Yourself . . . p. 476	Transparency 104 Test Bank Items 14.96–14.127

CHAPTER OUTLINE	INSTRUCTION IDEAS	SUPPLEMENTS
■ **Vocational Development** pp. 477–479 Establishing a Vocation · Women and Ethnic Minorities · Combining Work and Family	Learning Objective 14.11 Lecture Extensions 14.7–14.8 Ask Yourself . . . p. 479	Test Bank Items 14.128–14.137

LEARNING OBJECTIVES

After reading this chapter, students should be able to:

■ 14.1 Describe Erikson's stage of intimacy versus isolation and related research findings. (p. 452)

■ 14.2 Describe Levinson's and Vaillant's psychosocial theories of adult personality development, noting how they apply to both men's and women's lives. (p. 453)

■ 14.3 Explain how the social clock can affect personality in adulthood. (p. 456)

■ 14.4 Explain the role of romantic love in the young adult's quest for intimacy, and indicate how the balance among its components changes as a relationship develops. (p. 457)

■ 14.5 Describe adult friendships and sibling relationships. (p. 459)

■ 14.6 Explain the role of loneliness in adult development. (p. 461)

■ 14.7 Trace phases of the family life cycle that are prominent in early adulthood, and cite factors that influence these phases. (p. 462)

■ 14.8 Discuss the diversity of adult life-styles, paying special attention to singlehood, cohabitation, and childlessness. (p. 471)

■ 14.9 Discuss today's high rates of divorce and remarriage, and cite factors that contribute to them. (p. 473)

■ 14.10 Discuss challenges associated with variant styles of parenthood, including remarried parents, never-married parents, and gay and lesbian parents. (p. 474)

■ 14.11 Discuss men's and women's patterns of vocational development, and cite difficulties faced by women and ethnic minorities and couples seeking to combine careers and family. (p. 477)

CHAPTER-AT-A-GLANCE

Chapter 15
PHYSICAL AND COGNITIVE DEVELOPMENT IN MIDDLE ADULTHOOD

CHAPTER OUTLINE	INSTRUCTION IDEAS	SUPPLEMENTS
PHYSICAL DEVELOPMENT IN MIDDLE ADULTHOOD		
■ **Physical Changes pp. 489–493** Vision · Hearing · Skin · Muscle–Fat Makeup · Reproductive System · Skeleton	Learning Objectives 15.1–15.3 Ask Yourself . . . p. 493	Test Bank Items 15.1–15.26
■ **Health and Fitness pp. 493–498** Sexuality · Illness and Disability · Hostility and Anger	Learning Objectives 15.4–15.6 Lecture Extensions 15.1–15.2	Transparency 105 Test Bank Items 15.27–15.50
■ **Adapting to the Physical Challenges of Midlife pp. 498–502** Stress Management · Exercise · An Optimistic Outlook · Gender and Aging: A Double Standard	Learning Objectives 15.7–15.8 Lecture Extension 15.3 Learning Activities 15.1–15.2 Ask Yourself . . . p. 502	Transparency 106 Test Bank Items 15.51–15.64
COGNITIVE DEVELOPMENT IN MIDDLE ADULTHOOD		
■ **Changes in Mental Abilities pp. 503–505** Crystallized and Fluid Intelligence · Individual Differences	Learning Objective 15.9	Transparencies 107–108 Test Bank Items 15.65–15.76
■ **Information Processing pp. 505–508** Speed of Processing · Attention · Memory · Practical Problem Solving and Expertise · Creativity · Information Processing in Context	Learning Objectives 15.10–15.11 Lecture Extension 15.4 Learning Activity 15.3 Ask Yourself . . . p. 508	Test Bank Items 15.77–15.95
■ **Vocational Life and Cognitive Development p. 509**	Learning Objective 15.12	Test Bank Items 15.96–15.98
■ **Adult Learners: Becoming a College Student in Midlife pp. 509–512** Characteristics of Returning Students · Supporting Returning Students · Benefits of Returning to College	Learning Objective 15.13 Lecture Extension 15.5 Learning Activities 15.4–15.5 Ask Yourself . . . p. 512	Transparency 109 Test Bank Items 15.99–15.104

LEARNING OBJECTIVES

After reading this chapter, students should be able to:

■ 15.1 Describe the physical changes of middle adulthood, paying special attention to vision, hearing, the skin, muscle–fat makeup, and the skeleton. (p. 489)

■ 15.2 Describe the reproductive changes that occur in women during middle adulthood, and discuss women's psychological reactions to menopause. (p. 491)

■ 15.3 Describe the reproductive changes that occur in men during middle adulthood. (p. 491)

■ 15.4 Discuss sexuality in middle adulthood and its association with psychological well-being. (p. 493)

■ 15.5 Discuss cancer, cardiovascular disease, and osteoporosis, noting risk factors and interventions. (p. 494)

■ 15.6 Discuss the association of hostility and anger with heart disease and other health problems. (p. 497)

■ 15.7 Explain the benefits of stress management, exercise, and an optimistic outlook in dealing effectively with the changes of midlife. (p. 498)

■ 15.8 Describe the double standard of aging. (p. 501)

■ 15.9 Describe changes in crystallized and fluid intelligence in middle adulthood, and discuss individual differences in intellectual development. (p. 503)

■ 15.10 Describe changes in information processing in midlife, paying special attention to speed of processing, attention, and memory. (p. 505)

■ 15.11 Discuss the development of practical problem solving, expertise, and creativity in middle adulthood. (p. 507)

■ 15.12 Identify the relationship between vocational life and cognitive development. (p. 510)

■ 15.13 Discuss challenges facing adults who return to college, ways of supporting returning students, and benefits of earning a degree in midlife. (p. 510)

CHAPTER-AT-A-GLANCE

Chapter 16
EMOTIONAL AND SOCIAL DEVELOPMENT IN MIDDLE ADULTHOOD

CHAPTER OUTLINE	INSTRUCTION IDEAS	SUPPLEMENTS
■ **Erikson's Theory: Generativity versus Stagnation pp. 518–520**	Learning Objective 16.1	SL: Middle Adulthood (Ages 40-60) Test Bank Items 16.1–16.9
■ **Other Theories of Psychosocial Development in Midlife pp. 520–525** Levinson's Seasons of Life ·Vaillant's Adaptation to Life · Is There a Midlife Crisis? · Stage or Life Events Approach	Learning Objectives 16.2–16.4 Learning Activity 16.1 Ask Yourself . . . p. 525	Transparencies 110–111 Test Bank Items 16.10–16.26
■ **Changes in Self-Concept and Personality Traits pp. 525–528** Possible Selves · Self-Acceptance, Autonomy, and Environmental Mastery · Coping Strategies · Gender-Role Identity	Learning Objectives 16.5–16.6 Learning Activities 16.2–16.3	Test Bank Items 16.27–16.44

CHAPTER OUTLINE	INSTRUCTION IDEAS	SUPPLEMENTS
■ **Individual Differences in Personality Traits pp. 528–529**	Learning Objective 16.7 Lecture Extension 16.1 Ask Yourself . . . p. 529	Test Bank Items 16.45–16.51
■ **Relationships at Midlife pp. 529–539** Marriage and Divorce · Changing Parent–Child Relationships · Grandparenthood · Middle-Aged Children and Their Aging Parents · Siblings · Friendships · Relationships Across Generations	Learning Objectives 16.8–16.9 Lecture Extensions 16.2–16.3 Learning Activities 16.4–16.5 Try This . . . p. 534 Ask Yourself . . . pp. 538–539	Transparency 112 Test Bank Items 16.52–16.87
■ **Vocational Life pp. 539–544** Job Satisfaction · Vocational Development · Career Change at Midlife · Unemployment · Planning for Retirement	Learning Objectives 16.10–16.12 Lecture Extensions 16.4–16.5 Ask Yourself . . . p. 544	Test Bank Items 16.88–16.107

LEARNING OBJECTIVES

After reading this chapter, students should be able to:

■ 16.1 Describe Erikson's stage of generativity versus stagnation and related research findings. (p. 518)

■ 16.2 Describe Levinson's and Vaillant's views of psychosocial development in middle adulthood, and discuss similarities and differences in midlife changes for men and women. (p. 520)

■ 16.3 Discuss the extent to which midlife crisis captures most people's experience of middle adulthood. (p. 522)

■ 16.4 Characterize middle adulthood using a life events approach and a stage approach. (p. 524)

■ 16.5 Describe changes in self-concept and personality traits in middle adulthood. (p. 525)

■ 16.6 Describe changes in gender role identity in midlife. (p. 527)

■ 16.7 Discuss individual differences in personality traits in adulthood. (p. 528)

■ 16.8 Describe the middle adulthood phase of the family life cycle, and discuss midlife relationships with a marriage partner, adult children, grandchildren, and aging parents. (p. 529)

■ 16.9 Describe midlife sibling relationships and friendships, and discuss relationships across generations in the United States. (p. 536)

■ 16.10 Discuss job satisfaction and vocational development in middle adulthood, paying special attention to gender differences and the experience of ethnic minorities. (p. 539)

■ 16.11 Discuss career change and unemployment in middle adulthood. (p. 542)

■ 16.12 Discuss the importance of planning for retirement, noting various issues that middle-aged adults should address. (p. 542)

CHAPTER-AT-A-GLANCE

Chapter 17
PHYSICAL AND COGNITIVE DEVELOPMENT IN LATE ADULTHOOD

CHAPTER OUTLINE	INSTRUCTION IDEAS	SUPPLEMENTS
PHYSICAL DEVELOPMENT IN LATE ADULTHOOD		
■ **Longevity pp. 552–554**	Learning Objective 17.1 Learning Activity 17.1	Transparency 113 Test Bank Items 17.1–17.10
■ **Physical Changes pp. 554–563** Nervous System · Vision · Hearing · Taste and Smell · Touch · Cardiovascular and Respiratory Systems · Immune System · Sleep · Physical Appearance and Mobility · Adapting to Physical Changes of Late Adulthood	Learning Objectives 17.2–17.5 Lecture Extension 17.1 Learning Activity 17.2 Ask Yourself . . . p. 563	Transparencies 114–115 Test Bank Items 17.11–17.42
■ **Health, Fitness, and Disability** **pp. 563–575** Nutrition and Exercise · Sexuality · Physical Disabilities · Mental Disabilities · Health Care	Learning Objectives 17.6–17.9 Lecture Extensions 17.2–17.3 Learning Activity 17.3 Try This . . . p. 573 Ask Yourself . . . p. 575	Transparencies 116–117 Test Bank Items 17.43–17.95
COGNITIVE DEVELOPMENT IN LATE ADULTHOOD		
■ **Memory pp. 576–578** Deliberate versus Automatic Memory · Remote Memory · Prospective Memory	Learning Objective 17.10	Test Bank Items 17.96–17.107
■ **Language Processing p. 578**	Learning Objective 17.10 Lecture Extension 17.4 Try This . . . p. 431	Test Bank Items 17.108–17.110
■ **Problem Solving pp. 578–579**	Learning Objective 17.10	Test Bank Items 17.111–17.112
■ **Wisdom p. 579**	Learning Objective 17.11 Learning Activity 17.4	Test Bank Items 17.113–17.115
■ **Factors Related to Cognitive** **Change pp. 579–580**	Learning Objective 17.12	Test Bank Items 17.116–17.119
■ **Cognitive Interventions p. 580**	Learning Objective 17.13 Lecture Extension 17.5	Test Bank Items 17.120–17.121
■ **Lifelong Learning pp. 580–582** Types of Programs · Benefits of Continuing Education	Learning Objective 17.14 Learning Activity 17.5 Ask Yourself . . . p. 581	Test Bank Items 17.122–17.124

LEARNING OBJECTIVES

After reading this chapter, students should be able to:

- 17.1 Discuss aging and longevity among older adults. (p. 552)

- 17.2 Describe changes in the nervous system and the senses in late adulthood. (p. 555)

- 17.3 Describe cardiovascular, respiratory, and immune system changes in late adulthood. (p. 559)

- 17.4 Discuss sleep difficulties in late adulthood. (p. 560)

- 17.5 Describe changes in physical appearance and mobility in late adulthood. (p. 560)

- 17.6 Discuss health and fitness in late life, paying special attention to nutrition, exercise, and sexuality. (p. 563)

- 17.7 Discuss common physical disabilities in late adulthood. (p. 565)

- 17.8 Discuss common mental disabilities in late adulthood, including Alzheimer's disease, cerebrovascular dementia, and misdiagnosed and reversible dementia. (p. 568)

- 17.9 Discuss health care issues that affect older adults, including health care costs and the need for long-term care. (p. 571)

- 17.10 Describe changes in cognitive functioning in late adulthood, including memory, language processing, and problem solving. (p. 575)

- 17.11 Discuss experiences that foster the development of wisdom. (p. 579)

- 17.12 List factors related to cognitive change in late adulthood. (p. 579)

- 17.13 Discuss the effectiveness of cognitive interventions in late adulthood. (p. 580)

- 17.14 Discuss types of programs and benefits of continuing education in late life. (p. 580)

CHAPTER-AT-A-GLANCE

Chapter 18
EMOTIONAL AND SOCIAL DEVELOPMENT IN LATE ADULTHOOD

CHAPTER OUTLINE	INSTRUCTION IDEAS	SUPPLEMENTS
■ **Erikson's Theory: Ego Integrity versus Despair** pp. 588–589	Learning Objective 18.1 Learning Activity 18.1	SL: Late Adulthood (Ages 60+) Test Bank Items 18.1–18.5
■ **Other Theories of Psychosocial Development in Late Adulthood** pp. 589–591 Peck's Theory · Labouvie-Vief's Theory · Reminiscence and Life Review	Learning Objective 18.2 Learning Activity 18.2	Test Bank Items 18.6–18.14
■ **Changes in Self-Concept and Personality Traits** pp. 591–594 Strengthening of Self-Concept · Agreeableness, Sociability, and Acceptance of Change · Spirituality and Religiosity	Learning Objectives 18.3–18.4	Test Bank Items 18.15–18.22
■ **Individual Differences in Psychological Well-Being** pp. 594–598 Control versus Dependency · Health · Negative Life Changes · Social Support and Social Interaction	Learning Objectives 18.5–18.6 Lecture Extensions 18.1–18.2 Try This . . . p. 597 Ask Yourself . . . p. 598	Transparency 118 Test Bank Items 18.23–18.33

CHAPTER OUTLINE	INSTRUCTION IDEAS	SUPPLEMENTS
■ **A Changing Social World** pp. 599–604 Social Theories of Aging · Social Contexts of Aging: Communities, Neighborhoods, and Housing	Learning Objectives 18.7–18.9 Lecture Extension 18.3 Learning Activity 18.3 Ask Yourself . . . pp. 604	Transparency 119 Test Bank Items 18.34–18.58
■ **Relationships in Late Adulthood** pp. 604–612 Marriage · Divorce and Remarriage · Widowhood · Never-Married, Childless Older Adults · Siblings · Friendships · Relationships with Adult Children · Relationships with Adult Grandchildren and Great-Grandchildren · Elder Maltreatment	Learning Objectives 18.10–18.13 Lecture Extension 18.4 Ask Yourself . . . p. 612	Transparency 120 Test Bank Items 18.59–18.87
■ **Retirement and Leisure** pp. 612–615 The Decision to Retire · Adjustment to Retirement · Leisure Activities	Learning Objective 18.14 Lecture Extension 18.5 Learning Activity 18.4	Transparency 121 Test Bank Items 18.88–18.97
■ **Successful Aging** pp. 615–616	Learning Objective 18.15 Learning Activity 18.5 Ask Yourself . . . p. 616	Test Bank Items 18.98–18.100

LEARNING OBJECTIVES

After reading this chapter, students should be able to:

■ 18.1 Describe Erikson's stage of ego integrity versus despair. (p. 588)

■ 18.2 Describe Peck's and Labouvie-Vief's views of development in late adulthood, and discuss the functions of reminiscence and life review in older adults' lives. (p. 589)

■ 18.3 Describe changes in self-concept and personality traits in late adulthood. (p. 591)

■ 18.4 Discuss spirituality and religiosity in late adulthood, and trace the development of faith. (p. 591)

■ 18.5 Discuss individual differences in psychological well-being as older adults respond to challenges posed by issues of control versus dependency, declining health, and negative life changes. (p. 594)

■ 18.6 Describe the role of social support and social interaction in promoting physical health and psychological well-being in late adulthood. (p. 598)

■ 18.7 Describe social theories of aging, including disengagement theory, activity theory, and selectivity theory. (p. 599)

■ 18.8 Discuss the impact of communities, neighborhoods, and fear of crime on elders' social lives. (p. 600)

■ 18.9 Discuss the effects of different housing arrangements on older adults' adjustment. (p. 601)

■ 18.10 Describe changes in social relationships in late adulthood, including marriage, divorce, remarriage, and widowhood, and discuss social experiences and life satisfaction of never-married, childless older adults. (p. 604)

■ 18.11 Describe late-life sibling relationships and friendships. (p. 608)

■ 18.12 Describe older adults' relationships with adult children, adult grandchildren, and great-grandchildren. (p. 610)

■ 18.13 Discuss elder maltreatment, including risk factors and strategies for prevention. (p. 611)

■ 18.14 Discuss retirement and leisure, paying special attention to the decision to retire, adjustment to retirement, and leisure activities. (p. 612)

■ 18.15 Discuss the meaning of successful aging. (p. 615)

CHAPTER-AT-A-GLANCE

Chapter 19
DEATH, DYING, AND BEREAVEMENT

CHAPTER OUTLINE	INSTRUCTION IDEAS	SUPPLEMENTS
■ **How We Die pp. 624–626** Physical Changes · Defining Death · Death with Dignity	Learning Objective 19.1 Lecture Extension 19.1	Test Bank Items 19.1–19.11
■ **The Right to Die pp. 626–631** Passive Euthanasia · Voluntary Active Euthanasia · Assisted Suicide	Learning Objective 19.2 Learning Activity 19.1 Try This . . . p. 630 Ask Yourself . . . p. 631	Test Bank Items 19.12–19.25 Transparencies 122–123
■ **Understanding of and Attitudes Toward Death pp. 631–636** Childhood · Adolescence · Adulthood · Death Anxiety	Learning Objective 19.3 Lecture Extension 19.2 Learning Activity 19.2 Ask Yourself . . . p. 636	Test Bank Items 19.26–19.37
■ **Thinking and Emotions of Dying People pp. 636–640** Do Stages of Dying Exist? · Individual Adaptations to Impending Death · Near-Death Experiences	Learning Objectives 19.4–19.5	Test Bank Items 19.38–19.50
■ **A Place to Die pp. 640–642** Home · Hospital · The Hospice Approach	Learning Objective 19.6 Learning Activity 19.3 Ask Yourself . . . pp. 642	Test Bank Items 19.51–19.58
■ **Bereavement: Coping with the Death of a Loved One pp. 642–646** Phases of Grieving · Personal and Situational Variations · Bereavement Interventions	Learning Objective 19.7 Lecture Extensions 19.3–19.4 Learning Activities 19.4–19.5	Test Bank Items 19.59–19.76
■ **Death Education pp. 646–647**	Learning Objective 19.8 Lecture Extension 19.5 Ask Yourself . . . p. 647	Test Bank Items 19.77–19.79

LEARNING OBJECTIVES

After reading this chapter, students should be able to:

- 19.1 Describe the physical changes of dying, along with their implications for defining death and the meaning of death with dignity. (p. 624)

- 19.2 Discuss controversies surrounding euthanasia and assisted suicide. (p. 626)

- 19.3 Discuss age-related changes in conceptions of and attitudes toward death, and cite factors that influence death anxiety. (p. 631)

- 19.4 Describe and evaluate Kübler-Ross's stage theory, citing factors that influence the responses of dying patients. (p. 636)

- 19.5 Evaluate insights about the transition from life to death gained from near-death experiences. (p. 639)

- 19.6 Evaluate the extent to which homes, hospitals, and the hospice approach meet the needs of dying people and their families. (p. 640)

- 19.7 Describe the phases of grieving, factors that underlie individual variations, and bereavement interventions. (p. 642)

- 19.8 Explain how death education can help people cope with death more effectively. (p. 646)

Development Through the Lifespan

Laura E. Berk
ILLINOIS STATE UNIVERSITY

ALLYN AND BACON

Boston

London

Toronto

Sydney

Tokyo

Singapore

Vice-President and Editor-in-Chief, Social Sciences: Sean W. Wakely
Developmental Editors: Anne A. Reid, Sue Gleason
Vice-President/Director of Field Marketing: Joyce Nilsen
Editorial Assistant: Jessica Barnard
Editorial-Production Service: Thomas E. Dorsaneo
Text Design and Composition: Seventeenth Street Studios
Cover Designer: Linda Knowles
Composition and Prepress Buyer: Linda Cox
Manufacturing Buyer: Megan Cochran

Copyright © 1998 by Allyn & Bacon
A Viacom Company
160 Gould Street
Needham Heights, MA 02194-2310

Internet: www.abacon.com
America Online: Keyword: College online

Library of Congress Cataloging-in-Publication Data

Berk, Laura E.
    Development through the lifespan / Laura E. Berk.
        p.   cm.
    Includes bibliographical references and index.
    ISBN 0-205-14684-8
      1. Developmental psychology.   I. Title
BF713.B465   1998
155—dc21   97-45631
                                         CIP

Printed in the United States of America
10 9 8 7 6 5 4 3 2 1   02 01 00 99 98 97

TO COLLEAGUES AND FRIENDS
AT THE DELISSA INSTITUTE,
UNIVERSITY OF SOUTH AUSTRALIA,
IN GRATITUDE FOR MANY KINDNESSES
AND FOR ENRICHING MY VISION
OF HUMAN DEVELOPMENT

8 EMOTIONAL AND SOCIAL DEVELOPMENT IN EARLY CHILDHOOD 244

■ PART V
MIDDLE CHILDHOOD:
SIX TO ELEVEN YEARS

9 PHYSICAL AND COGNITIVE DEVELOPMENT IN MIDDLE CHILDHOOD 276

10 EMOTIONAL AND SOCIAL DEVELOPMENT IN MIDDLE CHILDHOOD 316

16 EMOTIONAL AND SOCIAL DEVELOPMENT IN MIDDLE ADULTHOOD 516

■ PART IX LATE ADULTHOOD

17 PHYSICAL AND COGNITIVE DEVELOPMENT IN LATE ADULTHOOD 550

18 EMOTIONAL AND SOCIAL DEVELOPMENT IN LATE ADULTHOOD — 586

■ PART X
THE END OF LIFE

19 DEATH, DYING, AND BEREAVEMENT — 622

My decision to write *Development Through the Lifespan* was inspired by a wealth of professional and personal experiences. First and foremost were the interests and concerns of hundreds of students of human development with whom I have worked in more than a quarter-century of college teaching. Each semester, their insights and questions have revealed how an understanding of any single period of development is enriched by an appreciation of the entire lifespan. Second, as I moved through phases of adult development myself, I began to think more intensely about factors that have shaped and reshaped my own life course—family, friends, mentors, co-workers, community, and larger society. My career well established, my marriage having stood the test of time, and my children launched into their adult lives, I felt that a deeper grasp of these multiple, interacting influences would help me better appreciate where I had been and where I would be going in the years ahead. I was also convinced that it could contribute to my becoming a better teacher, scholar, family member, and citizen. And because teaching has been so central and gratifying a part of my work life, I wanted to bring to others a personally meaningful understanding of lifespan development.

In preparing *Development Through the Lifespan,* I aimed for a text that is intellectually stimulating, that provides depth as well as breadth of coverage, and that portrays the complexities of human development in a way that captures students' interest while helping them learn. To achieve these objectives, I have grounded this book in a carefully selected body of classic and current research brought to life with stories and vignettes of children, adolescents, and adults, many of whom I have known personally. In addition, the text discussion emphasizes how the research process helps solve real-world problems and pays special attention to policy issues that are critical to development in today's world. I have also used a clear, engaging writing style and provided a unique pedagogical program that assists students in mastering information, integrating the various aspects of lifespan development, critically examining controversial issues, and applying what they have learned.

TEXT PHILOSOPHY

The approach of this book consists of six philosophical ingredients that I regard as essential for students to emerge from a course with a thorough understanding of lifespan development. Each theme is woven into every chapter:

1. An understanding of major theories and the strengths and shortcomings of each. The first chapter begins by emphasizing that only knowledge of multiple theories can do justice to the richness of human development. As I take up each age sector and aspect of development, I present a variety of theoretical perspectives, indicate how each highlights previously overlooked aspects of development, and discuss research that has been used to evaluate it. Discussion of contrasting theories also serves as the context for an evenhanded analysis of many controversial issues.

2. A grasp of the lifespan perspective as an integrative approach to development. The lifespan perspective—development as lifelong, multidimensional, multidirectional, plastic, and embedded in multiple contexts—serves as a unifying approach to understanding human change. I introduce it as an organizing framework in the first chapter and continually refer to and illustrate its assumptions throughout the text, in an effort to help students construct an overall vision of development from conception to death.

3. Knowledge of the sequence of human development and the processes underlying it. Students are given a description of the sequence of development along with processes of change. An understanding of process—how complex combinations of biological and environmental events produce development—has been the focus of most recent research. Accordingly, the text reflects this emphasis. But new information about the timetable of change has also emerged. In many ways, the very young and the old have proved to be far more competent than they were believed to be in the past. In addition, the occurrence of many milestones of adult development, such as finishing formal education, entering a career, getting married, having children, and retiring, has become less predictable. Current evidence on the sequence and timing of development, along with its implications for process, is presented for all periods of the lifespan.

4. An appreciation of the impact of context and culture on human development. A wealth of new research indicates more powerfully than ever before that people live in rich physical and social contexts that affect all aspects of development. Throughout the book, students travel to distant parts of the world as I review a growing body of cross-cultural evidence. The text narrative also discusses many findings on socioeconomically and ethnically diverse people in the United States. Furthermore, the impact of historical time period and cohort membership receives continuous attention. In this vein, gender issues—the distinctive but continually evolving experiences, roles, and life paths of males and females—are given substantial emphasis. Besides highlighting the effects of immediate settings, such as family, neighborhood, and school, I make a concerted effort to underscore the influence of larger social structures—societal values, laws, and government programs—on lifelong well-being.

5. A sense of the interdependency of all aspects of development—physical, cognitive, emotional, and social. Every chapter emphasizes an integrated approach to human development. I show how physical, cognitive, emotional, and social development are interwoven. Within the text narrative, students are often referred to other sections of the book to deepen their grasp of relationships between various aspects of change.

6. An appreciation of the interrelatedness of theory, research, and applications. Throughout this book, I emphasize that theories of human development and the research stimulated by them provide the foundation for sound, effective practices with children, adolescents, and adults. The link between theory, research, and applications is reinforced by an organizational format in which theory and research are presented first, followed by practical implications. In addition, a current focus in the field—harnessing knowledge of human development to shape social policies that support human needs throughout the lifespan—is reflected in every chapter. The text addresses the current condition of children, adolescents, and adults in the United States and around the world and shows how theory and research have combined with public interest to spark successful interventions. Many important applied topics are considered, such as family planning, infant mortality, maternal employment and day care, teenage pregnancy and childbearing, youth gangs, domestic violence, exercise and adult health, lifelong learning, grandparent visitation rights after parental divorce, adjustment to retirement, and adapting to widowhood.

▬ TEXT ORGANIZATION

I have chosen a chronological organization for *Development Through the Lifespan*. The book begins with an introductory chapter that describes the history of the field, modern theories, and research strategies. It is followed by

two chapters that cover the foundations of development. Chapter 2 combines an overview of biological and environmental contexts into a single, integrated discussion of these multifaceted determinants of development. Chapter 3 is devoted to prenatal development, birth, and the newborn baby. With this foundation, students are ready to look closely at seven major age periods: infancy and toddlerhood (Chapters 4, 5, and 6), early childhood (Chapters 7 and 8), middle childhood (Chapters 9 and 10), adolescence (Chapters 11 and 12), early adulthood (Chapters 13 and 14), middle adulthood (Chapters 15 and 16), and late adulthood (Chapters 17 and 18). Topical chapters within each chronological division cover physical development, cognitive development, and emotional and social development. The book concludes with a chapter on death, dying, and bereavement (Chapter 19).

The chronological approach has the advantage of enabling students to get to know individuals of a given age period very well. It also eases the task of integrating the various aspects of development, since each is discussed in close proximity. At the same time, a chronologically organized book requires that theories covering several age periods be presented piecemeal. This creates a challenge for students, who must link the various parts together. To assist with this task, I frequently remind students of important earlier achievements before discussing new developments, referring back to related sections with page references to encourage students to review. Also, chapters or sections devoted to the same topic (for example, cognitive development) are similarly organized, making it easier for students to draw connections across age periods and construct a continuous view of developmental change.

▬ TOPICAL HIGHLIGHTS

Human development is a fascinating and ever-changing field of study, with constantly emerging discoveries and refinements in existing knowledge. The text represents this burgeoning contemporary literature, with more than 1,400 of its citations appearing in the last 5 years. Cutting-edge topics throughout the text underscore the book's major themes—the lifespan perspective; multiple contexts for development (including family, community, culture, social policy, and historical time period); integration of physical, cognitive, emotional, and social domains; and applications of theory and research. Here is a sampling:

- Cross-cultural research and Vygotsky's sociocultural theory (Chapters 1, 5, and 7)
- Meaning of heritability, concordance, and genetic–environmental correlation (Chapter 2)
- Pros and cons of reproductive technologies (Chapter 2)
- Importance of vitamin–mineral supplements for women of childbearing age (Chapter 3)

- Cross-national comparisons of prenatal health care and infant mortality (Chapter 3)

- Caregiving practices that prevent sudden infant death syndrome (Chapter 4)

- Early motor development as a complex, dynamic system—multiply determined by genetic, environmental, and motivational factors (Chapter 4)

- Sensitive periods in brain development (Chapter 4)

- Infant recognition and recall memory, and the mystery of infantile amnesia (Chapter 5)

- Long-term stability of temperament (Chapter 6)

- Parents' internal working models and infant attachment security (Chapter 6)

- Impact of emotional well-being on physical growth (Chapters 4 and 7)

- Social origins of early childhood cognition (Chapter 7)

- Two-generation models of early intervention (Chapter 7)

- Family and media influences on development of aggression (Chapter 8)

- Sweden's commitment to gender equality (Chapter 8)

- Impact of culture and schooling on cognitive development (Chapter 9)

- Recent developments in defining intelligence—Sternberg's triarchic theory and Gardner's multiple intelligences (Chapter 9)

- Cross-national comparisons of academic achievement (Chapter 9)

- Children's achievement-related attributions (Chapter 10)

- Subtypes of peer-rejected children—rejected-aggressive and rejected-withdrawn (Chapter 10)

- Self-care children (Chapter 10)

- Impact of pubertal maturation on parent–adolescent relationships (Chapter 11)

- Generational change and adolescent sexual attitudes and behavior (Chapter 11)

- Everyday consequences of adolescent abstract thinking (Chapter 11)

- School transitions and adolescent adjustment (Chapter 11)

- Identity development—cognitive, parenting, school, community, and cultural influences (Chapter 12)

- Benefits of adolescent friendships and peer groups (Chapter 12)

- Theories of biological aging, including the roles of free radicals, cross-linkage of tissue fibers, and changes in the immune system (Chapter 13)

- Social-class, ethnic, and sex differences in adult health (Chapters 13, 15, and 17)

- Impact of exercise on physical and mental health (Chapters 9, 13, 15, and 17)

- Sexual choices and lifestyles in early adulthood (Chapter 13)

- Development of pragmatic thought, expertise, and practical problem solving (Chapters 13 and 15)

- Psychological impact of attending college (Chapter 13)

- Vocational preparation of non-college-bound young adults (Chapter 13)

- Influence of parenthood on marital satisfaction (Chapter 14)

- Cultural variations in attitudes toward cohabitation (Chapter 14)

- Variant styles of parenthood, including remarried parents, never-married parents, and gay and lesbian parents (Chapter 14)

- Divergent career paths of men and women (Chapters 14 and 16)

- Relationships across generations (Chapter 16)

- Impact of age-related decline in processing speed on other aspects of cognition (Chapters 15 and 17)

- Becoming a college student at midlife (Chapter 15)

- Career change at midlife (Chapter 16)

- Myth of the midlife crisis (Chapter 16)

- How older adults optimize and compensate in the face of physical and cognitive declines (Chapter 17)

- Selectivity theory as an explanation of the decline in social interaction in late adulthood (Chapter 18)

- A new view of dependency in old age (Chapter 18)

- Ethical controversies surrounding euthanasia and assisted suicide (Chapter 19)

- Variations in the thinking and emotions of dying people (Chapter 19)

PEDAGOGICAL FEATURES

In writing this book, I made a concerted effort to adopt a writing style that is lucid and engaging without being simplistic. I frequently converse with students and encourage them to relate what they read to their own lives. In doing so, I hope to make the study of human development involving and pleasurable.

 STORIES AND VIGNETTES ABOUT REAL PEOPLE. To help students construct a clear image of development and to enliven the text narrative, each chronological age division is unified by case examples woven throughout that set of chapters. For example, within the middle childhood section, students share the experiences and concerns of 10-year-old Joey, 8-year-old Lizzie, their divorced parents Rena and Drake, and their classmates Mona, Terry, and Jermaine. In the chapters on late adulthood, students get to know Walt and Ruth, a vibrant retired couple, along with Walt's older brother Dick and his wife Goldie and Ruth's sister Ida, who suffers from Alzheimer's disease. Besides a set of main characters who bring unity to each age period, many additional vignettes offer vivid examples of development and diversity among children, adolescents, and adults.

■ LIFESPAN VISTA BOXES. A special set of boxes is devoted to topics that have long-term implications for development or involve intergenerational issues. Examples are The Future of Early Intervention (new two-generation models); Sensitive Periods in Brain Development; Attention-Deficit Hyperactivity Disorder (persistence of the disorder from childhood into adulthood); Children of War (long-term impact of wartime trauma and factors that play a protective role); School Desegregation and Life Chances of African-American Adolescents; Psychosocial and Behavioral Predictors of Longevity; Childhood Attachment Patterns and Adult Romantic Relationships; and What Can We Learn About Aging from Centenarians?

■ CULTURAL INFLUENCES AND SOCIAL ISSUES BOXES. These boxes underscore the impact of context on all aspects of development. *Cultural Influences* boxes highlight cross-cultural and multicultural variations—for example, the African-American Extended Family; Child Health Care in the United States and Western European Nations; Only Children in the People's Republic of China; Identity Development Among Ethnic Minority Adolescents; Menopause as a Biocultural Event; Illiteracy and Limited Education: A Lifelong Cul-de-Sac; and Cultural Variations in the Experience of Aging. *Social Issues* boxes discuss the impact of social conditions on children, adolescents, and adults and emphasize the need for sensitive social policies to ensure their well-being—for example, Lead Poisoning in Childhood; Children's Eyewitness Testimony; Should Grandparents Be Awarded Visitation Rights After Parental Divorce?; Squaring of the Population Pyramid and Intergenerational Inequity; and Elder Suicide.

■ TRY THIS . . . In each Social Issues box, I encourage students to become actively involved with the material by suggesting activities that extend their understanding of human development. Students are invited to make observations; talk to children, adolescents, and adults; find out about the status of different age groups in their community and nation; and reflect on their own experiences. Each activity can serve as a course assignment or stimulus for class discussion.

■ ASK YOURSELF . . . Active engagement with the subject matter is also supported by critical thinking questions, which can be found at the end of major sections. The focus of these questions is divided between theory and applications. Many describe problematic situations and ask students to resolve these in light of what they have learned. In this way, the questions inspire high-level thinking and new insights.

■ CAREGIVING CONCERNS TABLES. The relationship of theory and research to practice is woven throughout the text narrative. To accentuate this linkage, Caregiving Concerns tables provide easily accessible practical advice on the importance of caring for oneself and others throughout the lifespan. They include: Do's and Don't's for a Healthy Pregnancy; Building a Foundation for Good Eating Habits; Signs of Developmentally Appropriate Practice in Early Childhood Programs; Helping Children Adjust to Their Parents' Divorce; Keeping Love Alive in a Romantic Partnership; Ways Middle-Aged Parents Can Promote Positive Ties with Their Adult Children; Communicating Effectively with Alzheimer's Victims; and Resolving Grief After a Loved One Dies.

■ MILESTONES TABLES. A Milestones table appears at the end of each age division of the text. These tables summarize major physical, cognitive, language, and emotional and social developments, providing a convenient device for reviewing the chronology of development across the lifespan.

■ FOR FURTHER INFORMATION AND HELP. Students frequently ask where they can go to find out more about high-interest topics or to seek help in areas related to their own lives. To meet this need, an annotated section at the end of each chapter provides the names, addresses, phone numbers, and web sites of organizations that disseminate information about human development and offer special services.

■ CHAPTER INTRODUCTIONS AND END-OF-CHAPTER SUMMARIES. To provide students with a helpful preview of what they are about to read, an outline and overview of chapter content appears in each chapter introduction. Especially comprehensive end-of-chapter summaries, organized according to the major headings in each chapter and including review questions and boldface terms, remind students of key points in the text discussion.

■ BRIEF REVIEWS. Interim summaries of text content appear at the end of major sections in each chapter. They enhance retention by encouraging students to reflect on information they have just read before moving on to a new section.

▨ ADDITIONAL TABLES, ILLUSTRATIONS, AND PHOTOGRAPHS.
Tables are liberally included to help students grasp essential points in the text narrative and extend information on a topic. The many full-color figures and illustrations depict important theories, methods, and research findings. Photos have been carefully selected to portray human development and to represent the diversity of people in the United States and around the world.

▨ END-OF-CHAPTER TERM LIST AND END-OF-BOOK GLOSSARY. Mastery of terms that make up the central vocabulary of the field is promoted through an end-of-chapter term list and an end-of-book glossary. Important terms and concepts appear in boldface type in the text narrative. Each has also been entered into the subject index.

Acknowledgments

The dedicated contributions of many individuals helped make this book a reality. An impressive cast of reviewers, some of whom critiqued manuscript drafts and others who responded to survey questions, provided many helpful suggestions, constructive criticisms, and enthusiasm for the organization and content of the text. I am grateful to each one of them:

Joyce Bishop, Golden West College
Ed Brady, Belleville Area College
Michele Y. Breault, Truman State University
Joan B. Cannon, University of Massachusetts Lowell
Gary Creasey, Illinois State University
Rhoda Cummings, University of Nevada, Reno
Carol Lynn Davis, University of Maine
Clifford Gray, Pueblo Community College
Traci Haynes, Columbus State Community College
Vernon Haynes, Youngstown State University
Paula Hillman, University of Wisconsin—Whitewater
Janet Kalinowski, Ithaca College
Kevin Keating, Broward Community College
Wendy Kliewer, Virginia Commonwealth University
Randy Mergler, California State University
Karla Miley, Black Hawk College
Karen Nelson, Austin College
Bob Newby, Tarleton State University
Peter Oliver, University of Hartford
Ellen Pastorino, Gainesville College
Leslee K. Polina, Southeast Missouri State University
Leon Rappaport, Kansas State University
Randall Russac, University of North Florida
Marie Saracino, Stephen F. Austin State University
Bonnie Seegmiller, City University of New York, Hunter College
Richard Selby, Southeast Missouri State University
Paul S. Silverman, University of Montana
Glenda Smith, North Harris College
Jeanne Spaulding, Houston Community College
Thomas Spencer, San Francisco State University
Vince Sullivan, Pensacola Junior College
Mojisola Tiamiyu, University of Toledo
Joe Tinnin, Richland College
Ursula M. White, El Paso Community College
Lois J. Willoughby, Miami-Dade Community College
Deborah R. Winters, New Mexico State University

I am also indebted to many colleagues for much encouragement and interest in this project. Richard Payne, Department of Political Science, Illinois State University, is a kind and devoted friend with whom I have shared many profitable discussions about the writing process, the condition of children and the elderly, and other topics that significantly influenced my perspective on human development. Gary Creasey and Steven Landau, colleagues in the Department of Psychology, provided helpful consultation in their areas of expertise. Jeff Payne, Illinois State University Television Production Studio, guided development of the Observation Video with great artistry and sensitivity. Freda Briggs, University of South Australia, offered many thought-provoking insights as the adulthood chapters took shape.

Two graduate students contributed significantly to the text and its supplements. Lisa Otte coordinated literature reviews and securing permissions for use of copyrighted material. Dorothy Welty-Rodriguez participated in countless hours of planning and filming for the Observation Video and collaborated with me on its accompanying Observation Guide. Lisa and Dorothy's organizational skills, expertise in human development, and dedication have been invaluable throughout all phases of this project.

The supplement package benefited from the talents and diligence of several other individuals. Belinda Wholeben, Rockford College, and Heather Bouchey, University of Denver, made outstanding contributions to the Instructor's Resource Manual. Celia Reaves, Monroe Community College, prepared the Test Bank with meticulous concern for clarity and accuracy. I am pleased to have collaborated with Jenny Churchill, Illinois State University, on the Study Guide.

I have been fortunate to work with an exceptionally capable editorial team at Allyn and Bacon. Sean Wakely, Vice-President and Editor-in-Chief, Social Sciences, has inspired and energized my work, bringing to bear a combination of qualities unmatched in my experience—keen awareness of instructors' and students' needs, balanced attention to the text's overall concept and to vital details, thorough manuscript reviewing, and a sense of enthusiasm, respect for scholarship, and vision that has prompted me to strive for greater heights. I have especially appreciated his forthrightness, diplomatic problem solving, and day-to-day communication, through which he forges a true editor–author partnership. The simple yet rich image that graces the cover of the book, a radiant metaphor of lifespan development, is Sean's inspiration. I look forward to working with him on future editions and other projects in the years to come.

I would like to express a heartfelt thank you to Joyce Nilsen, Vice-President/Director of Field Marketing, for the outstanding work she has done over the years in marketing my texts. Joyce has made sure that accurate and clear information about my texts and their ancillaries reached Allyn and Bacon's sales force and that the needs of prospective and current adopters were met. Each time I have watched Joyce teach others about my books, I have been impressed with both her knowledge of their content and the vitality with which she conveys her message. She cares deeply about my texts—and about the teaching of human development in colleges and universities. It is a privilege and pleasure to have her in command of marketing activities for *Development Through the Lifespan.*

Sue Gleason, Senior Developmental Editor, managed the preparation of text supplements, ensuring their high quality. Her astute advice and prompt and patient responses to my concerns and queries are very much appreciated. Annie Reid handled the day-to-day development work on the manuscript, as she has done for my other two texts. It is difficult to find words that do justice to her contributions. Annie worked closely with me as I wrote each chapter, making sure that every thought and concept would be precisely expressed and well developed. Her keen visual sense greatly enhanced the book's illustration and photo program. It has been a pleasure to get to know and work with Annie during these past 5 years.

Tom Dorsaneo coordinated the complex production tasks that resulted in a beautiful first edition. His competence, courtesy, and interest in the subject matter as an involved grandfather of an energetic toddler have made working with him a great delight. I thank Elsa Peterson for obtaining the exceptional photographs that so aptly illustrate the text narrative. Jessica Barnard, Editorial Assistant, graciously arranged for manuscript reviews and attended to a wide variety of pressing, last-minute details.

A final word of gratitude goes to my family, whose love, patience, and understanding have enabled me to be wife, mother, teacher, researcher, and text author at the same time. My sons, David and Peter, have taken a special interest in this project. Their reflections on events and progress in their own lives, conveyed over telephone and e-mail and during holiday visits, helped mold the early adulthood chapters. My husband, Ken, willingly made room for yet another time-consuming endeavor in our life together and communicated his belief in its importance in a great many unspoken, caring ways.

—Laura E. Berk

ABOUT THE AUTHOR

Laura E. Berk is a distinguished professor of psychology at Illinois State University, where she teaches human development at both the undergraduate and graduate levels. She received her bachelor's degree in psychology from the University of California, Berkeley, and her masters and doctoral degrees in educational psychology from the University of Chicago. She has been a visiting scholar at Cornell University, UCLA, Stanford University, and the University of South Australia. Berk has published widely on the effects of school environments on children's development and, more recently, on the development of private speech. Her research has been funded by the U.S. Office of Education and the National Institute of Child Health and Human Development. It has appeared in many prominent journals, including *Child Development, Developmental Psychology, Merrill-Palmer Quarterly, Journal of Abnormal Child Psychology,* and *Development and Psychopathology.* Her empirical studies have attracted the attention of the general public, leading to contributions to *Psychology Today* and *Scientific American.* Berk has served as research editor of *Young Children* and is currently consulting editor of *Early Childhood Research Quarterly.* She is author of the chapter on the extracurriculum of schooling for the American Educational Research Association's *Handbook of Research on Curriculum.* Her books include *Private Speech: From Social Interaction to Self-Regulation, Scaffolding Children's Learning: Vygotsky and Early Childhood Education,* and *Landscapes of Development,* an anthology of readings. In addition to *Development Through the Lifespan,* she is author of the two best-selling texts *Child Development* and *Infants, Children, and Adolescents,* published by Allyn and Bacon.

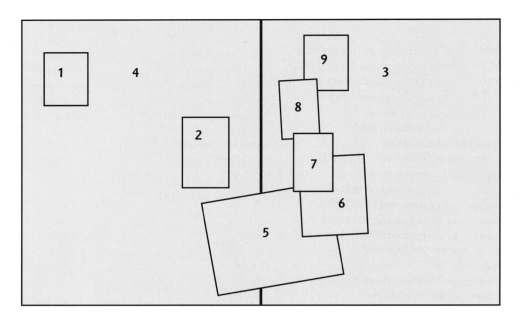

■ SOFIE'S STORY is told in Chapters 1 and 19, from her birth to her death. The photos that appear at the beginning of Chapter 1 follow her through her lifespan. They are:

1. Sofie as a baby, with her mother in 1908.
2. Sofie at age 6, with her brother, age 8, in 1914.
3. Sofie at age 10, before a birthday party in 1919.
4. Sofie at age 18, high school graduation in 1926.
5. Sofie's German passport.
6. Sofie and Phil in their mid-thirties, during World War II, when they became engaged.
7. Sofie, age 60, and daughter Laura on Laura's wedding day in 1968.
8. Sofie and Phil in 1968, less than 2 years before Sofie died.
9. Sofie, age 61, and her first grandchild, Ellen, October 1969, less than 3 months before Sofie died.

DEVELOPMENT THROUGH THE LIFESPAN

History, Theory, and Research Strategies

Sofie Lentschner was born in 1908, the second child of Jewish parents who made their home in Leipzig, Germany, a city of thriving commerce and cultural vitality. Her father was a successful businessman and community leader. Her mother was a socialite well known for her charm, beauty, and hospitality. As a baby, Sofie displayed the determination and persistence that would be sustained throughout her life. She sat for long periods inspecting small objects with her eyes and hands. The single event that consistently broke her gaze was the sound of the piano in the parlor. As soon as Sofie could crawl, she steadfastly pulled herself up to finger its keys and marveled at the tinkling sound.

By the time Sofie entered elementary school, she was an introspective child, often ill at ease at the festive parties that girls of her family's social standing were expected to attend. She immersed herself in her schoolwork, especially in mastering the foreign languages that were a regular part of German elementary and secondary education. Twice a week, she took piano lessons from the finest teacher in Leipzig. By the time Sofie graduated from high school, she spoke English and French fluently and had become an accomplished pianist. Whereas most German girls of her time married by age 20, Sofie postponed serious courtship in favor of entering the university. Her parents began to wonder whether their intense, studious daughter would ever settle into family life.

Sofie wanted marriage as well as education, but her plans were thwarted by the political turbulence of her times. When Hitler rose to power in the early 1930s, Sofie's father feared for the safety of his wife and children and moved the family to Belgium. Conditions for Jews in Europe quickly worsened. The Nazis plundered Sofie's family home and confiscated her father's business. By the end of the 1930s, Sofie had lost contact with all but a handful of her aunts, uncles, cousins, and childhood friends, many of whom (she later learned) were herded into cattle cars and transported to the slave labor and death camps at Auschwitz-Birkenau. In 1939, as anti-Jewish laws and atrocities intensified, Sofie's family fled to the United States.

As Sofie turned 30, her parents concluded she would never marry and would need a career for financial security. They agreed to support her return to school. Over the next 3 years, Sofie earned two master's degrees, one in music and the other in librarianship. Then, on a blind date, she met Philip, a U.S. army officer. Philip's calm, gentle nature complemented Sofie's intensity and worldliness. Within 6 months, they married. Over the next 4 years, two daughters and a son were born. Soon Sofie's father became ill. The strain of uprooting his family and losing his home and business had shattered his health. After months of being bedridden, he died of heart failure.

When World War II ended, Philip left the army and opened a small men's clothing store. Sofie divided her time between caring for her children and helping Philip in the store. Now in her forties, she was a devoted mother, but few women her age were still rearing young children. As Philip struggled with the business, he spent longer hours at work, and Sofie often felt lonely. She rarely touched the piano, which brought back painful memories of youthful life plans shattered by war. Sofie's sense of isolation and lack of fulfillment frequently left her short-tempered. Late at night, she and Philip could be heard arguing.

As Sofie's children grew older and parenting took less of her time, she returned to school once more, this time to earn a teaching credential. At age 50, she finally launched a career. For the next decade, Sofie taught German and French to high school students and English to newly arrived immigrants. Besides easing her family's financial difficulties, she felt a gratifying sense of accomplishment and creativity. These years were among the most energetic and satisfying of Sofie's life. She had an unending enthusiasm for teaching—for transmitting her facility with language, her first-hand knowledge of the consequences of hatred and oppression, and her practical understanding of how to adapt to life in a new land. She watched her children, whose young lives were free of the trauma of war, adopt many of her values and commitments and begin their marital and vocational lives at the expected time.

Sofie approached age 60 with an optimistic outlook. As she and Philip were released from the financial burden of paying for their children's college education, they looked forward to greater leisure. Their affection and respect for one another deepened. Once again, Sofie began to play the piano. But this period of contentment was short-lived.

One morning, Sofie awoke and felt a hard lump under her arm. Several days later, her doctor diagnosed cancer. Sofie's spirited disposition and capacity to adapt to radical life changes helped her meet the illness head on. She defined it as an enemy—to be fought and overcome. As a result, she lived 5 more years. Despite the exhaustion of chemotherapy, Sofie maintained a full schedule of teaching duties and continued to visit and run errands for her elderly mother. But as she weakened physically, she no longer had the stamina to meet her classes. Gradually, she gave in to the ravaging illness. Bedridden for the last few weeks, she slipped quietly into death with Philip at her side.

The funeral chapel overflowed with hundreds of Sofie's students. She had granted each a memorable image of a woman of courage and caring. One of her three children is the author of this book.

Sofie's story raises a wealth of fascinating issues about human life histories. For example:

- What determines the features that Sofie shares with others and those that make her unique—in physical characteristics, mental capacities, interests, and behaviors?

- What led Sofie to retain the same persistent, determined disposition throughout her life but to change in other essential ways?

- How do historical and cultural conditions—for Sofie, the persecution that led to the destruction of her childhood home, the loss of family members and friends, and her flight to the United States—affect well-being throughout life?

- In what way is the timing of events—for example, Sofie's early exposure to foreign languages but her delayed entry into marriage, parenthood, and career—important in development?

- What factors, both intrinsic to the person and present in the environment, led Sofie to die sooner than expected?

These are central questions addressed by **human development,** a field of study devoted to understanding constancy and change throughout the lifespan. Great diversity characterizes the interests and concerns of the thousands of investigators who study human development. But all have a single goal in common: the desire to describe and identify those factors that influence the consistencies and transformations in human beings from conception until death.

HUMAN DEVELOPMENT AS AN INTERDISCIPLINARY, SCIENTIFIC, AND APPLIED FIELD

Look again at the questions just listed, and you will see that they are not just of scientific interest. Each is of *applied,* or practical, importance as well. In fact, scientific curiosity about changes that take place from infancy through old age is just one factor that has led human development to become the exciting field of study it is today. Research about development has also been stimulated by social pressures to better people's lives. For example, the beginning of public education in the early part of this century led to a demand for knowledge about what and how to teach children of different ages. The interest of the medical profession in improving people's health required an understanding of physical development, nutrition, and disease. The social service profession's desire to treat anxieties and behavior problems and to help people adjust to major life events, such as divorce, job loss, or the death of a loved one, required information about person-

ality and social development. And parents have continually asked for expert advice about child-rearing practices and experiences that would foster a happy and successful life for their child.

Our large storehouse of information about human development is *interdisciplinary.* It grew through the combined efforts of people from many fields of study. Because of the need for solutions to everyday problems at all ages, academic scientists from psychology, sociology, anthropology, and biology joined forces in research with professionals from a variety of applied fields, including education, medicine, public health, and social service, to name just a few. Today, the field of human development is a melting pot of contributions. Its body of knowledge is not just scientifically important, but relevant and useful.

BASIC THEMES AND ISSUES

Research on human development is a relatively recent activity. Studies of children did not begin until the early part of the twentieth century. Investigations into adult development and aging emerged only in the 1960s and 1970s (Baltes, 1983). Nevertheless, ideas about how people grow and change have existed for centuries. As these speculations combined with research, they inspired the construction of *theories* of development. Although there are a great many definitions, for our purposes we can think of a **theory** as an orderly, integrated set of statements that describes, explains, and predicts behavior. For example, a good theory of infant–caregiver attachment would *describe* the behaviors and lead up to babies' strong desire to seek the affection and comfort of a familiar adult around 6 to 8 months of age. It would also *explain* why infants have such a strong desire. And it would try to *predict* the consequences of this close emotional bond for relationships throughout life.

Theories are vital tools in human development (and any other scientific endeavor) for two reasons. First, they provide organizing frameworks for our observations of people. In other words, they *guide and give meaning to* what we see. Second, theories that are verified by research provide a sound basis for practical action. Once a theory helps us *understand* development, we are in a much better position to know *what to do* in our efforts to improve the welfare and treatment of children and adults.

As we will see later, theories are influenced by the cultural values and belief systems of their times. But theories differ in one important way from mere opinion and belief: a theory's continued existence depends on scientific verification (Scarr, 1985). This means that the theory must be tested with a fair set of research procedures agreed on by the scientific community.

In the field of human development, there are many theories with very different ideas about what people are like and how they change. The study of development provides no ultimate truth, since investigators do not always agree on the meaning of what they see. In addition, humans are complex beings; they change physically, mentally, emotionally, and socially. As yet, no single theory has been able to explain all these aspects (Baltes, 1987). Finally, the existence of many theories helps advance knowledge, since researchers are continually trying to support, contradict, and integrate these different points of view.

This chapter introduces you to major theories of human development and research strategies that have been used to test them. We will return to each theory in greater detail, as well as introduce a variety of important but less grand theories, in later parts of this book. Although there are many theories, we can easily organize them, since almost all take a stand on three basic issues about human development: (1) How should we describe the developing person? (2) What is the course of human development like? (3) What factors determine, or influence, development? Let's look closely at each of these issues in the following sections.

VIEW OF THE DEVELOPING PERSON

Was Sofie's intense, determined style the result of her own inner tendencies, or was it largely encouraged by others? Did Sofie's talent and passion for music lead her to become a skilled pianist, or was repeated exposure to music responsible? These questions address a puzzling issue about the nature of the human being. They contrast two basic perspectives: the organismic, or *active* position, with the mechanistic, or *passive* point of view.

Organismic theories assume that change is stimulated from *within the organism*—more specifically, that psycho-logical structures exist within the person that underlie and control development. Children and adults are viewed as active, purposeful beings who make sense of their world, select from available experiences, and determine their own learning. For an organismic theorist, the surrounding environment supports development, as Sofie's parents did when they purchased a piano and arranged for her to begin music lessons at an early age. But since people impose their own interpretations and responses on the events of their lives, factors within the individual—interests, mental capacities, personality traits, and talents—are largely responsible for change.

In contrast, **mechanistic theories** focus on relationships between environmental inputs and behavioral outputs. The approach is called *mechanistic* because development is compared to the workings of a machine. Change is stimulated by the environment, which shapes the behavior of the person, who is a passive reactor. According to this view, new capacities result from external forces acting on the person. Development is treated as a straightforward, predictable consequence of events in the surrounding world (Miller, 1993).

VIEW OF THE COURSE OF DEVELOPMENT

How can we best describe the differences in capacities among small infants, young children, adolescents, and adults? As Figure 1.1 illustrates, major theories recognize two possibilities.

On the one hand, infants and preschoolers may respond to the world in much the same way as adults. The difference between the immature and mature being may simply be one of *amount or complexity* of behavior. For example, as a baby, Sofie's inspection of objects, perception

FIGURE 1.1

Is development continuous or discontinuous? (a) Some theorists believe development is a smooth, continuous process. Individuals gradually add more of the same types of skills. (b) Other theorists think development takes place in abrupt, discontinuous stages. People change rapidly as they step up to a new level and then change very little for a while. With each new step, the person interprets and responds to the world in a qualitatively different way.

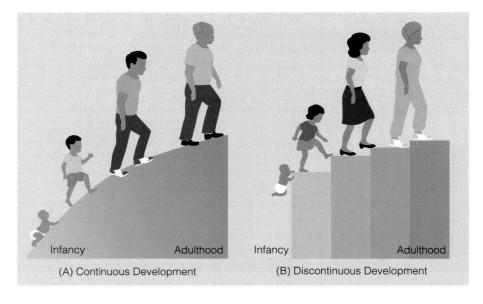

Infancy   Adulthood

(A) Continuous Development

Infancy   Adulthood

(B) Discontinuous Development

of a piano melody, and ability to organize her world may have been much like our own. Perhaps her only limitation was that she could not perform these skills with as many pieces of information as we can. If this is true, then change in her thinking must be **continuous**—a process that consists of gradually adding on more to the same types of skills that were there to begin with.

On the other hand, infants and children may have *unique ways of thinking, feeling, and behaving* that must be understood on their own terms—ones quite different from adults'. If so, then development is a **discontinuous** process in which new understandings emerge at particular time periods. From this perspective, infant Sofie was not yet able to perceive and organize experiences as a mature person could. Instead, she moved through a series of developmental steps, each of which has unique features, until she reached the highest level of functioning.

Theories that accept the discontinuous perspective include a vital developmental concept: the concept of **stage.** Stages are *qualitative changes* in thinking, feeling, and behaving that characterize particular time periods of development. In stage theories, development is much like climbing a staircase, with each step corresponding to a more mature, reorganized way of functioning than the one that came before.

Does development actually take place in a neat, orderly, stepwise sequence that is identical for all human beings? For now, let's note that this is a very ambitious assumption that has not gone unchallenged. We will review some very influential stage theories later in this chapter.

VIEW OF THE DETERMINANTS OF DEVELOPMENT

In addition to describing the course of human development, each theory takes a stand on a major question about its underlying causes: Are genetic or environmental factors most important? This is the age-old **nature–nurture controversy.** By *nature,* we mean inborn biological givens—the hereditary information we receive from our parents at the moment of conception that signals the body to grow and affects all our characteristics and skills. By *nurture,* we mean the complex forces of the physical and social world that people encounter in their homes, neighborhoods, and communities.

Although all theories grant at least some role to both nature and nurture, they vary in the emphasis placed on each. For example, consider the following questions: Is the developing person's ability to think in more complex ways largely the result of an inborn timetable of growth? Or is it primarily influenced by stimulation and encouragement from parents and teachers? Do children acquire language rapidly because they are genetically predisposed to do so, or because parents intensively tutor them from an early age? And what accounts for the vast individual differences

among people—in height, weight, physical coordination, intelligence, personality, and social skills? Is nature or nurture largely responsible?

The stance theories take on nature versus nurture affects their explanations of individual differences. Some emphasize *stability*—that individuals who are high or low in a characteristic (such as verbal ability, anxiety, or sociability) will remain so at later ages. These theorists typically stress the importance of *heredity.* If they regard environment as important, they usually point to *early experience* as establishing a lifelong pattern of behavior. Powerful negative events in the first few years, they argue, cannot be fully overcome by later, more positive ones (Bowlby, 1980; Sroufe, Egeland, & Kreutzer, 1990). Other theorists take a more optimistic view. They believe that *change* is possible and likely if new experiences support it (Chess & Thomas, 1984; Sampson & Laub, 1993; Werner & Smith, 1992).

Throughout this chapter and the remainder of this book, we will see that investigators disagree, often sharply, on the question of **stability versus change.** In addition, answers vary across *domains,* or aspects, of development. Think back to Sofie's story, and you will see that her linguistic ability and persistent approach to challenges were highly stable over the lifespan. In contrast, her psychological well-being and life satisfaction fluctuated considerably.

THE LIFESPAN PERSPECTIVE: A BALANCED POINT OF VIEW

So far, we have discussed basic issues of human development in terms of extremes—solutions on one side or the other. As we trace the unfolding of the field in the rest of this chapter, you will see that the thinking of many theorists has softened. Modern ones, especially, recognize the

Today, older adults are healthier and more active than in previous generations. These elderly marathon participants keep pace with people who are many years younger.

merits of both sides. Some theories take an intermediate stand between an organismic versus mechanistic perspective. They regard both the person and the environment as active and as collaborating to produce development. Similarly, some contemporary researchers believe that both continuous and discontinuous changes characterize development and may alternate with one another. Finally, recent investigators have moved away from asking which is more important—heredity or environment. Instead, they want to know precisely *how nature and nurture work together* to influence the individual's traits and capacities.

These balanced visions owe much to the expansion of research from a nearly exclusive focus on the first two decades to include adult life. In the first half of this century, it was widely assumed that development stopped at adolescence. Infancy and childhood were viewed as periods of rapid transformation, adulthood as a plateau, and aging as a period of decline. The changing character of the American population awakened researchers to the idea that development is lifelong. Due to improvements in nutrition, sanitation, and medical knowledge, the *average life expectancy* (number of years an individual born in a particular year can expect to live) gained more in the twentieth century than in the preceding five thousand years. In 1900, it was just under age 50; today, it is around age 75. As a result, there are more older adults, a trend that has occurred in most of the world but is especially striking in industrialized nations. People age 65 and older accounted for 4 percent of the U.S. population in 1900, 7 percent in 1950, and 13 percent in 1994 (U.S. Bureau of the Census, 1960, 1996). Growth in sheer number of older Americans during the twentieth century has been even more dramatic, as Figure 1.2 reveals.

Older adults are not just more numerous; they are also healthier and more active. They challenge the stereotype of the withering person of earlier years. These observations required a fundamental shift in our conception of human development. Compared to other approaches, the **lifespan**

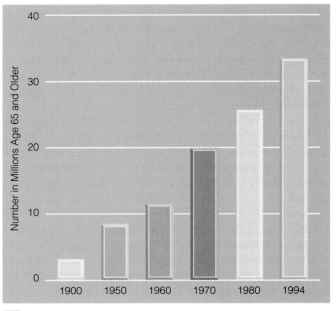

F IGURE 1.2

Increase in number of adults age 65 and older in the United States from 1900 to 1994. In 1900 there were 3 million people over age 65. Today, there are more than 30 million. Since 1950, the elderly population has quadrupled. By the time people born during the "baby boom" (two decades of soaring birth rates following World War II) reach old age, the number of elders is projected to exceed 50 million. *(U.S. Bureau of the Census, 1960, 1996)*

perspective offers a more complex vision of change and the factors that underlie it. Four assumptions, summarized in Table 1.1, make up this broader view (Baltes, 1987; Baltes, Lindenberger, & Staudinger, 1997).

■ DEVELOPMENT AS LIFELONG. According to the life span perspective, no age period is supreme in its impact on the life course. Instead, at all periods, significant changes

TABLE 1.1

Assumptions of the Lifespan Perspective on Human Development

ASSUMPTION ABOUT DEVELOPMENT	DESCRIPTION
Lifelong	No age period is supreme in its impact on the life course. At all periods, significant changes take place.
Multidimensional and multidirectional	Development is affected by an intricate blend of biological, personal, and social forces. At all age periods, development is a joint expression of growth and decline.
Highly plastic	At all ages, individual development can take many forms, depending on how the person's biological and environmental history combines with current life conditions. Development remains modifiable throughout the life course.
Embedded in multiple contexts	Pathways of change are highly diverse because development is affected by multiple contexts. Any particular course of development is the combined result of three types of influences: *age-graded, history-graded,* and *nonnormative.*

Source: Baltes, 1987.

take place that are highly diverse in timing and pattern, as the remaining assumptions make clear.

■ DEVELOPMENT AS MULTIDIMENSIONAL AND MULTI-DIRECTIONAL. Think back to Sofie's life, and notice how she was continually faced with new demands and opportunities. The lifespan perspective regards the challenges and adjustments of development as *multidimensional*—affected by an intricate blend of biological, personal, and social forces.

Lifespan development is also *multidirectional*—in at least two ways. First, development is not limited to improved performance. Instead, at all periods, it is a joint expression of growth and decline. When Sofie directed her energies toward mastering languages and music as a school-age child, she gave up refining other skills to their full potential. When she chose to become a teacher in adulthood, she let go of other career options. Although gains are especially evident early in life and losses during the final years, people of all ages can develop skills that compensate for reduced functioning. One highly accomplished, elderly psychologist who noticed his difficulty remembering people's names devised graceful ways of explaining his memory failure. Sometimes he appealed to his age; at other times, he flattered his listener by remarking that he tended to forget only the names of important people! Under these conditions, he reflected, "forgetting may even be a pleasure" (Skinner, 1983, p. 240).

Second, besides being multidirectional over time, the lifespan perspective emphasizes that change is multidirectional within the same domain of development. Although some qualities of Sophie's intellectual functioning (such as memory) probably declined in her mature years, her knowledge of English and French undoubtedly grew throughout her life. And she also developed new forms of thinking. For example, Sofie's wealth of experience and ability to cope with diverse problems led her to become expert in practical matters—a quality of reasoning called *wisdom*. We will consider the development of wisdom in Chapter 17.

■ DEVELOPMENT AS HIGHLY PLASTIC. *Plasticity* refers to openness to change—the extent to which development can take many forms, depending on how the person's biological and environmental history combines with current life conditions. Plasticity varies across individuals; some people experience more diverse life circumstances, and some adapt to change more easily than others. And over time, development gradually becomes less plastic, as both capacity and opportunity for change are reduced.

Nevertheless, lifespan researchers emphasize that development is highly plastic at all ages. For example, consider Sofie's social reserve in childhood and her decision to study rather than marry as a young adult. As new opportunities arose, Sofie (much to her parents' surprise) moved easily into marriage and childbearing in her thirties. And although parenthood and financial difficulties posed challenges to Philip and Sofie's happiness, their relationship gradually became richer and more fulfilling. In Chapter 17, we will see that intellectual performance also remains flexible with advancing age. Elderly people respond to special training with substantial (but not unlimited) gains in a wide variety of mental abilities (Schaie, 1996b).

Recent evidence on developmental plasticity makes clear that aging is not an eventual "shipwreck," as has often been assumed. Instead, the metaphor of a "butterfly"—of metamorphosis and continued potential—provides a far more accurate picture of lifespan change (Lemme, 1995).

■ DEVELOPMENT AS EMBEDDED IN MULTIPLE CONTEXTS. According to the lifespan perspective, pathways of change are highly diverse because development is *embedded in multiple contexts*. Although these wide-ranging influences can be organized into three categories, they work together, combining in unique ways to fashion each life course (Baltes, 1987).

Age-Graded Influences. Events that are strongly related to age and therefore fairly predictable in when they occur and how long they last are called **age-graded influences.** For example, most individuals walk shortly after their first birthday, acquire their native language during the preschool years, reach puberty around ages 12 to 14, and (for women) experience menopause in their late forties or early fifties. These milestones are largely governed by biology, but social customs can create age-graded influences as well. Starting school around age 6, getting a driver's license at age 16, and entering college around age 18 are events of this kind. Age-graded influences are especially prevalent during childhood and adolescence. During those times, biological changes are rapid, and cultures impose many age-related experiences to ensure that young people acquire the skills needed to become competent, participating members of their society.

History-Graded Influences. Development can also be profoundly affected by forces that are unique to a particular historical era. Epidemics, wars, and periods of economic prosperity or depression; technological advances, such as the introduction of television and computers; and changing cultural values—for example, revised attitudes toward women and ethnic minorities—are examples of **history-graded influences.** These events explain why people born around the same time—called a *cohort*—tend to be alike in ways that set them apart from people born at other times.

Nonnormative Influences. The term *normative* means typical, or average. Age-graded and history-graded influences are normative, since each affects large numbers of people in a similar way. **Nonnormative influences** are events that are irregular, in that they happen to just one or a few people and do not follow a predictable

Development can be profoundly affected by history-graded influences—forces unique to a particular historical era. These members of the "baby-boom" generation, born after the end of World War II, between 1946 and 1964, are alike in ways that set them apart from people born at other times. For example, because of revised attitudes toward women and ethnic minorities, all four hold positions of equal status in their workplace—a circumstance that would have been highly unusual in their parents' generation. (John Coletti)

timetable. Consequently, they enhance the multidirectionality of development. Piano lessons in childhood with an inspiring teacher; a blind date with Philip; delayed marriage, parenthood, and career entry; and a battle with cancer are nonnormative influences that had a major impact on the direction of Sofie's life. Because they occur haphazardly, nonnormative events are among the most difficult for researchers to capture and study. Yet, as each of us can attest from our own experiences, they can affect us in powerful ways.

Lifespan investigators point out that nonnormative influences have become more powerful and age-graded influences less so in modern adult development. Compared to Sofie's era, the ages at which people finish their education, enter careers, get married, have children, and retire are much more diverse (Neugarten & Neugarten, 1987; Schroots & Birren, 1990). Indeed, Sofie's "off-time" accomplishments would have been less unusual had she been born a generation or two later! Age remains a powerful organizer of everyday experiences, and age-related expectations have certainly not disappeared. But age markers have blurred, and they vary considerably across ethnic groups and cultures. The increasing role of nonnormative events in the modern life course adds to the fluid nature of life span development.

AN OVERVIEW OF THE LIFESPAN

As our discussion in the previous section suggests, the life course is not neat and tidy. There are many ways to divide it into separate phases. I have chosen the following eight periods as the structure for this book, since each brings with it new capacities and social expectations that serve as important transitions in major theories:

■ *The prenatal period: from conception to birth.* This 9-month period is the most rapid period of change, during which a one-celled organism is transformed into a human baby with remarkable capacities to adjust to life in the surrounding world.

■ *Infancy and toddlerhood: from birth to 2 years.* This period brings dramatic changes in the body and brain that support the emergence of a wide array of motor, perceptual, and intellectual capacities; the beginnings of language; and first intimate ties to others.

■ *Early childhood: from 2 to 6 years.* During this period, the body becomes longer and leaner, motor skills are refined, and children become more self-controlled and self-sufficient. Make-believe play blossoms and supports every aspect of psychological development. Thought and language expand at an astounding pace, a sense of morality becomes evident, and children start to establish ties with peers.

■ *Middle childhood: from 6 to 11 years.* These are the school years, a phase in which children learn about the wider world and master new responsibilities that increasingly resemble those they will perform as adults. Improved athletic abilities, participation in organized games with rules, more logical thought processes, mastery of basic literacy skills, and advances in understanding of the self, morality, and friendship are hallmarks of this phase.

■ *Adolescence: from 11 to 20 years.* This period is the bridge between childhood and adulthood. Puberty leads to an adult-sized body and sexual maturity. Thought becomes abstract and idealistic, and school achievement becomes more serious as young people prepare for the world of work. Defining personal values and goals and establishing autonomy from the family are major concerns of this phase.

■ *Early adulthood: from 20 to 40 years.* During this period, most people leave home, complete their education, begin full-time work, and attain economic independence. Many young adults are preoccupied with choosing a career, developing a mature intimate relationship, and marrying, rearing children, or establishing other life-styles.

■ *Middle adulthood: from 40 to 60 years.* Middle age is a time of maximum responsibility within the family, at work, and in the community. During this period, many people are at the height of their careers and attain positions of leadership. In addition, they must help their adolescent and young adult children begin independent lives and their aging parents adapt to physical and social changes. Caught between youth and old age, people of this phase become more aware of their own mortality. Many look for new areas of fulfillment in relationships, leisure activities, and second careers.

■ *Late adulthood: from 60 years until death.* During this final phase, people leave the world of work, assume the status of senior citizen, and reflect on the meaning of their lives. The elderly must adjust to retirement, to decreased physical strength and health, and often to the death of a spouse. They must also prepare for the end of their physical existence.

DOMAINS OF DEVELOPMENT AS INTERWOVEN

Within each of the periods just described, we will examine three broad domains of development:

■ *Physical development*—changes in body size, proportions, appearance, and the functioning of various body systems; brain development; perceptual and motor capacities; and physical health.

■ *Cognitive development*—development of a wide variety of intellectual abilities, including attention, memory, academic and everyday knowledge, problem solving, imagination, creativity, and the uniquely human capacity to represent the world through language.

■ *Emotional and social development*—development of emotional communication, self-understanding, knowledge about other people, interpersonal skills, friendships, intimate relationships, and moral reasoning and behavior.

You are already aware from reading the first part of this chapter that these domains are not really distinct: they overlap and interact a great deal. A major advantage of discussing them as a unit within each time period is that we can easily see how they are interwoven. Now let's turn to the historical foundations of the field as a prelude to major theories that address various aspects of change.

ASK YOURSELF . . .

■ *List as many examples of age-graded, history-graded, and nonnormative influences as you can find in Sofie's story at the beginning of this chapter.*

■ *Now list age-graded, history-graded, and nonnormative influences in your own life and one of your parents' lives. How are factors that shaped each life course similar? How are they different?*

HISTORICAL FOUNDATIONS

Modern theories of human development are the result of centuries of change in Western cultural values, philosophical thinking, and scientific progress. To understand the field as it exists today, we must return to its early beginnings—to influences that long preceded scientific study. We will see that many early ideas linger on as important forces in current theory and research.

PHILOSOPHIES OF CHILDHOOD

In medieval times (the sixth through the fifteenth centuries), little importance was placed on childhood as a separate phase of the life course. Instead, once children emerged from infancy, they were regarded as miniature,

In this medieval painting, the young child is depicted as a miniature adult. His dress, expression, and activities resemble those of his elders. Through the fifteenth century, little emphasis was placed on childhood as a unique phase of the life cycle. (Giraudon/Art Resource)

already formed adults, a view called **preformationism** (Ariès, 1962). Some laws recognized that children needed protection from people who might mistreat them, and medical works provided special instructions for their care. However, even though in a practical sense there was some awareness of the vulnerability of children, as yet there were no philosophies of the uniqueness of childhood or separate developmental periods (Borstelmann, 1983; Sommerville, 1982).

By the sixteenth century, a revised image of children sprang from the religious movement that gave birth to Protestantism—in particular, from the Puritan belief in original sin. Harsh, restrictive parenting practices were recommended as the most efficient means for taming the depraved child. Although punitiveness was the prevailing child-rearing philosophy of the times, love and affection for their children prevented most Puritan parents from using extremely repressive measures. Instead, they tried to promote reason in their sons and daughters so they would be able to separate right from wrong and resist temptation (Moran & Vinovskis, 1986; Pollock, 1987).

■ JOHN LOCKE. The seventeenth-century Enlightenment brought philosophies that emphasized ideals of human dignity and respect. The writings of John Locke (1632–1704), a leading British philosopher, served as the forerunner of an important twentieth-century perspective that we will discuss shortly: *behaviorism.* Locke viewed the child as a **tabula rasa.** Translated from Latin, this means "blank slate" or "white piece of paper." According to this idea, children were not basically evil. They were, to begin with, nothing at all, and their characters could be shaped by all kinds of experiences. Locke (1690/1892) described parents as rational tutors who could mold the child in any way they wished, through careful instruction, effective example, and rewards for good behavior. His philosophy prompted a change from harshness toward children to kindness and compassion.

Look carefully at Locke's ideas, and you will see that he took a firm stand on basic issues discussed earlier in this chapter. As blank slates, children were viewed in passive, *mechanistic* terms. Locke also regarded development as *continuous.* Adultlike behaviors are gradually built through the teachings of parents. Finally, Locke was a champion of *nurture*—of the power of the environment to determine whether children become good or bad, bright or dull, kind or selfish.

■ JEAN JACQUES ROUSSEAU. In the eighteenth century, a new theory of childhood was introduced by French philosopher Jean Jacques Rousseau (1712–1778). Children, Rousseau (1762/1955) thought, were not blank slates to be filled by adult instruction. Instead, they were **noble savages,** naturally endowed with a sense of right and wrong and with an innate plan for orderly, healthy growth. Unlike Locke, Rousseau thought children's built-in moral sense and unique ways of thinking and feeling would only be harmed by adult training. His was a permissive philosophy in which the adult should be receptive to the child's needs at each of four stages of development: infancy, childhood, late childhood, and adolescence.

Rousseau's philosophy includes two vitally important concepts that are found in modern theories. The first is the concept of *stage,* which we discussed earlier in this chapter. The second is the concept of **maturation,** which refers to a genetically determined, naturally unfolding course of growth. If you accept the notion that children mature through a sequence of stages, then they are unique and different from adults, and their development is determined by their own inner promptings. Compared to Locke, Rousseau took a very different stand on basic developmental issues. He saw children as *organismic* (active shapers of their own destiny), development as *discontinuous,* and *nature* as having mapped out the path and timetable of change.

PHILOSOPHIES OF ADULTHOOD AND AGING

Shortly after Rousseau devised his conception of childhood, the first lifespan views began to appear. In the eighteenth and early nineteenth centuries, two German philosophers—John Nicolaus Tetens (1736–1807) and Friedrich August Carus (1770–1808)—urged that attention to development be extended through adulthood. Each asked important questions about aging.

Tetens (1777) addressed the origins and extent of individual differences, the degree to which behavior can be changed in adulthood, and the impact of historical eras on the life course. He was ahead of his time in recognizing that intellectual declines in old age can be compensated for and, at times, may reflect hidden gains. For example, Tetens suggested that some memory difficulties among the elderly might be the result of trying to search for a word or name among a lifetime of accumulated information—a possibility acknowledged by modern researchers (Maylor & Valentine, 1992).

Carus (1808) moved beyond Rousseau's stages by identifying four periods that span the life course: childhood, youth, adulthood, and senescence. Like Tetens, Carus viewed aging not only as decline, but also as progression. His writings reflect a remarkable awareness of the assumptions of multidirectionality and plasticity that are at the heart of the contemporary lifespan perspective (Baltes, 1983).

SCIENTIFIC BEGINNINGS

The study of development evolved quickly during the late nineteenth and early twentieth centuries. Early observations of human change were soon followed by improved methods and theories. Each advance contributed to the firm foundation on which the field rests today.

DARWIN: FOREFATHER OF SCIENTIFIC CHILD STUDY.

Charles Darwin (1809–1882), a British naturalist, is often considered to be the forefather of scientific child study. Darwin (1859/1936) observed the infinite variation among plant and animal species. He also saw that within a species, no two individuals are exactly alike. From these observations, he constructed his famous theory of evolution.

The theory emphasized two related principles: *natural selection* and *survival of the fittest*. Darwin explained that certain species are *selected by nature* to survive in particular environments because they have characteristics that *fit with,* or are adapted to, their surroundings. Other species die off because they are not as well suited to their environments. Individuals within a species who best meet the survival requirements of the environment live long enough to reproduce and pass their more favorable characteristics to future generations. We will see that Darwin's emphasis on the adaptive value of physical characteristics and behavior found its way into important twentieth-century theories.

During his explorations, Darwin discovered that the early prenatal growth of many species is strikingly similar. This suggested that all species, including humans, were descended from a few common ancestors. Other scientists concluded from Darwin's observation that the development of the human child followed the same general plan as the evolution of the human species. Although this belief eventually proved to be inaccurate, efforts to chart parallels between child growth and human evolution prompted researchers to make careful observations of all aspects of children's behavior. Out of these first attempts to document an idea about development, scientific child study was born.

THE NORMATIVE PERIOD.

G. Stanley Hall (1846–1924), one of the most influential American psychologists of the early twentieth century, is generally regarded as the founder of the child study movement (Dixon & Lerner, 1992). He also foreshadowed lifespan research by writing one of the few books of his time on aging (Hall, 1922). Inspired by Darwin's work, Hall and his well-known student Arnold Gesell (1880–1961) devised theories of childhood and adolescence based on evolutionary ideas. These early leaders regarded development as a genetically determined process that unfolds automatically, much like a flower (Gesell, 1933; Hall, 1904).

Hall and Gesell are remembered less for their one-sided theories than for their intensive efforts to describe all aspects of development. They launched the **normative approach** to child study. In a normative investigation, measurements of behavior are taken on large numbers of individuals. Then age-related averages are computed to represent typical development. Using this method, Hall constructed elaborate questionnaires asking children of different ages almost everything they could tell about themselves—interests, fears, imaginary playmates, dreams, friendships, everyday knowledge, and more (White, 1992). In the same tradition, Gesell collected detailed normative information on the motor achievements, social behaviors, and personality characteristics of infants and children (Gesell & Ilg, 1943/1949a, 1946/1949b).

Gesell was also among the first to make knowledge about child development meaningful to parents. If, as he believed, the timetable of development is the product of millions of years of evolution, then children are naturally knowledgeable about their needs. His child-rearing advice, in the tradition of Rousseau, recommended sensitivity to children's cues (Thelen & Adolph, 1992). Along with Benjamin Spock's famous *Baby and Child Care*, Gesell's books became a central part of a rapidly expanding literature for parents published over this century (see the Cultural Influences box on page 14).

THE MENTAL TESTING MOVEMENT.

While Hall and Gesell were developing their theories and methods in the United States, French psychologist Alfred Binet (1857–1911) also took a normative approach to child development, but for a different reason. In the early 1900s, Binet and his colleague Theodore Simon were asked to find a way to identify retarded children in the Paris school system who needed to be placed in special classes. The first successful intelligence test, which they constructed for this purpose, grew out of practical educational concerns.

In 1916, at Stanford University, Binet's test was translated into English and adapted for use with American children. It became known as the Stanford-Binet Intelligence Scale. Besides providing a score that could successfully predict school achievement, the Binet test sparked tremendous interest in individual differences in development. Soon many additional tests were devised for both children and adults. Comparisons of the scores of people who vary in gender, ethnicity, birth order, family background, and other characteristics became a major focus of research. Intelligence tests also rose quickly to the forefront of the controversy over nature versus nurture that has continued throughout this century.

BRIEF REVIEW

The modern field of human development has roots dating far back into the past. In medieval times, children were regarded as miniature adults. By the sixteenth century, childhood became a distinct phase of the life course. The Puritan belief in original sin fostered a harsh, authoritarian approach to child rearing. During the Enlightenment,

CULTURAL INFLUENCES

SOCIAL CHANGE AND CHILD-REARING ADVICE TO PARENTS

Almost all parents—new ones especially—feel a need for sound advice on how to rear their children. To meet this need, the field of child development has long been communicating what it knows to the general public through a wide variety of popular books and magazines. One survey looked at the types of advice that experts gave to parents of infants from 1955 to 1984 (Young, 1990). Two widely read sources were carefully examined: *Parents* magazine (to which scholars regularly contribute articles) and *Infant Care* (a pamphlet written by pediatricians and other child development specialists and published at regular intervals by the United States Children's Bureau).

From the 1950s to the 1980s, advice to parents changed in ways that reflected new social realities, cultural beliefs about children, and scientific discoveries. Prior to the 1970s, the publications emphasized the central role of the mother in healthy infant development. Although mothers were encouraged to include fathers in the care of the baby, they were cautioned not to expect fathers to participate equally. The succeeding decade brought considerably fewer references to the primacy of the mother until, in the mid-1980s, an about-face was evident. Experts suggested that fathers might share in the full range of caregiving responsibilities, since new evidence revealed that the father's role is unique and important to all aspects of development.

Around this time, information about maternal employment and day care also appeared in the publications. In contrast to the earlier view of the maternal role as a full-time commitment, experts reassured the modern mother that her baby did not require her continuous presence and offered advice about how to select good day care. Recommendations on this score, however, displayed some ambiva-lence. Mothers were also told that staying with the infant "can be a great human experience," and they were discouraged from entrusting the care of their babies to others. As we will see in Chapter 6, controversy exists in both American culture and in the scientific community about the advisability of placing infants in day care, and it is reflected in contemporary advice to parents (Etaugh, Williams, & Carlson, 1996).

During the three decades studied, some child-rearing themes did not change. Parents were continuously told that they play a large role in guiding their baby's development, that infants are active learners who benefit from a rich variety of physical and social stimulation, and that early experiences have a lasting impact. As you read the rest of this chapter, try to identify major theories of the mid- and late-twentieth century that may have prompted these statements.

Parents often turn to books and magazines for expert advice on how to rear their children. The information they find reflects cultural beliefs and social realities of the times. (Innervisions)

TRY THIS...

■ Visit your library and examine several issues of *Infant Care* published by the federal government earlier in this century and today. How have parenting practices related to feeding, sleeping, and toilet training changed?

■ Check your local bookstore for current parenting titles that extend beyond the period of infancy. Can you find additional examples of how advice to parents is both driven by new discoveries and influenced by the larger social context in which child rearing takes place?

Locke's "blank slate" and Rousseau's "inherently good" child promoted more humane views of children. Tetens and Carus extended conceptions of development to adulthood and anticipated the modern lifespan perspective. Darwin's evolutionary ideas inspired Hall and Gesell's maturational theories and normative investigations. Out of the normative tradition arose Binet's first successful intelligence test and a concern with individual differences in development.

ASK YOURSELF . . .

■ *If you could interview people of medieval times to find out whether they thought development was a continuous or discontinuous process, how do you think they would respond?*

■ *Suppose we could arrange a debate between John Locke and Jean Jacques Rousseau on the nature–nurture controversy. Summarize the argument that each of these historical figures is likely to present.*

MID-TWENTIETH-CENTURY THEORIES

In the mid-twentieth century, human development expanded into a legitimate discipline. As it attracted increasing interest, a variety of mid-twentieth-century theories emerged, each of which continues to have followers today.

THE PSYCHOANALYTIC PERSPECTIVE

Recall that the normative and testing movements had begun to answer the question: What are human beings like at different ages? In the 1930s and 1940s, as more people sought help from professionals in dealing with emotional difficulties, a new question had to be addressed: How and why did people become the way they are? To treat psychological problems, psychiatrists and social workers turned to the **psychoanalytic perspective** on personality development because of its emphasis on understanding the unique life history of each person.

According to the psychoanalytic approach, people move through a series of stages in which they confront conflicts between biological drives and social expectations. The way these conflicts are resolved determines the individual's ability to learn, get along with others, and cope with anxiety. Although many individuals contributed to the psychoanalytic perspective, two have been especially influential: Sigmund Freud, founder of the psychoanalytic movement, and Erik Erikson.

■ FREUD'S THEORY. Freud (1856–1939), a Viennese physician, saw patients in his practice with a variety of nervous symptoms, such as hallucinations, fears, and paralyses, that appeared to have no physical basis. Seeking a cure for these troubled adults, Freud found that their symptoms could be relieved by having them talk freely about painful events of their childhood. On the basis of adult remembrances, he examined the unconscious motivations of his patients and constructed his **psychosexual theory** of development. It emphasized that how parents manage their child's sexual and aggressive drives in the first few years is crucial for healthy personality development.

■ THREE PORTIONS OF THE PERSONALITY. In Freud's theory, three parts of the personality—id, ego, and superego—become integrated during a sequence of five stages of development. The **id,** the largest portion of the mind, is inherited and present at birth. It is the source of basic biological needs and desires. The id seeks to satisfy its impulses head on, without delay. As a result, young babies cry vigorously when they are hungry, wet, or need to be held and cuddled.

The **ego**—the conscious, rational part of personality—emerges in early infancy to ensure that the id's desires are

Sigmund Freud founded the psychoanalytic movement. His psychosexual theory was the first to stress the importance of early experience for later development. Erik Erikson expanded Freud's theory, emphasizing the psychosocial outcomes of each psychosexual stage. By adding three adulthood stages to Freud's model, Erikson emphasized the lifespan nature of development. (Lyrl Ahern)

TABLE 1.2

Freud's Psychosexual Stages

PSYCHOSEXUAL STAGE	PERIOD OF DEVELOPMENT	DESCRIPTION
Oral	Birth–1 year	The new ego directs the baby's sucking activities toward breast or bottle. If oral needs are not met appropriately, the individual may develop such habits as thumb sucking, fingernail biting, and pencil chewing in childhood and overeating and smoking in later life.
Anal	1–3 years	Young toddlers and preschoolers enjoy holding and releasing urine and feces. Toilet training becomes a major issue between parent and child. If parents insist that children be trained before they are ready or make too few demands, conflicts about anal control may appear in the form of extreme orderliness and cleanliness or messiness and disorder.
Phallic	3–6 years	Id impulses transfer to the genitals, and the child finds pleasure in genital stimulation. Freud's *Oedipus conflict* for boys and *Electra conflict* for girls take place. Young children feel a sexual desire for the opposite-sex parent. To avoid punishment, they give up this desire and, instead, adopt the same-sex parent's characteristics and values. As a result, the superego is formed. The relations between id, ego, and superego established at this time determine the individual's basic personality orientation.
Latency	6–11 years	Sexual instincts die down, and the superego develops further. The child acquires new social values from adults outside the family and from play with same-sex peers.
Genital	Adolescence	Puberty causes the sexual impulses of the phallic stage to reappear. If development has been successful during earlier stages, it leads to marriage, mature sexuality, and the birth and rearing of children.

satisfied in accord with reality. Recalling times when parents helped the baby gratify the id, the ego redirects impulses so they are discharged on appropriate objects at acceptable times and places. Aided by the ego, the hungry baby of a few months of age stops crying when he sees his mother unfasten her clothing for breast-feeding or warm a bottle. And the more competent preschooler goes into the kitchen and gets a snack on her own.

Between 3 and 6 years of age, the **superego,** or seat of conscience, appears. It contains the values of society and is often in conflict with the id's desires. The superego develops from interactions with parents, who eventually insist that children control their biological impulses. Once the superego is formed, the ego is faced with the increasingly complex task of reconciling the demands of the id, the external world, and conscience (Freud, 1923/1974). For example, when the ego is tempted to gratify an id impulse by hitting a playmate to get an attractive toy, the superego may warn that such behavior is wrong. The ego must decide which of the two forces (id or superego) will win this inner struggle or work out a reasonable compromise, such as asking for a turn with the toy. According to Freud, the relations established between id, ego, and superego during the preschool years determine the individual's basic personality.

PSYCHOSEXUAL DEVELOPMENT. Freud (1938/1973) believed that over the course of childhood, sexual impulses shift their focus from the oral to the anal to the genital regions of the body. In each stage of development, parents walk a fine line between permitting too much or too little gratification of their child's basic needs. If parents strike an appropriate balance, then children grow into well-adjusted adults with the capacity for mature sexual behavior, investment in family life, and rearing of the next generation. Table 1.2 summarizes Freud's stages.

Freud's psychosexual theory highlighted the importance of family relationships for children's development. It was the first theory to stress the importance of early experience for later development. But Freud's perspective was eventually criticized. First, the theory overemphasized the influence of sexual feelings in development. Second, because it was based on the problems of sexually repressed, well-to-do adults, some aspects of Freud's theory did not apply in cultures differing from nineteenth-century Victorian society. Finally, even close followers disagreed with Freud's emphasis on the first 5 or 6 years (Jung, 1933).

ERIKSON'S THEORY. Several of Freud's followers stretched and rearranged his theory in ways that improved on his vision. The most important of these neo-Freudians for the modern field of human development was Erik Erikson (1902–1994).

Although Erikson (1950) accepted Freud's basic psychosexual framework, he expanded the picture of development at each stage. In his **psychosocial theory,** Erikson emphasized that the ego does not just mediate between id impulses and superego demands. It is also a positive force in development. At each stage, it acquires attitudes and skills that make the individual an active, contributing member of society. A basic psychological conflict, which is resolved along a continuum from positive to negative, determines healthy or maladaptive outcomes at each stage.

Erikson's Psychosocial Stages

PSYCHOSOCIAL STAGE	PERIOD OF DEVELOPMENT	DESCRIPTION	CORRESPONDING PSYCHOSEXUAL STAGE
Basic trust versus mistrust	Birth–1 year	From warm, responsive care, infants gain a sense of trust, or confidence, that the world is good. Mistrust occurs when infants have to wait too long for comfort and are handled harshly.	Oral
Autonomy versus shame and doubt	1–3 years	Using new mental and motor skills, children want to choose and decide for themselves. Autonomy is fostered when parents permit reasonable free choice and do not force or shame the child.	Anal
Initiative versus guilt	3–6 years	Through make-believe play, children experiment with the kind of person they can become. Initiative—a sense of ambition and responsibility—develops when parents support their child's new sense of purpose and direction. The danger is that parents will demand too much self-control, which leads to overcontrol, or too much guilt.	Phallic
Industry versus inferiority	6–11 years	At school, children develop the capacity to work and cooperate with others. Inferiority develops when negative experiences at home, at school, or with peers lead to feelings of incompetence and inferiority.	Latency
Identity versus identity diffusion	Adolescence	The adolescent tries to answer the questions, Who am I, and what is my place in society? Self-chosen values and vocational goals lead to a lasting personal identity. The negative outcome is confusion about future adult roles.	Genital
Intimacy versus isolation	Young adulthood	Young people work on establishing intimate ties to others. Because of earlier disappointments, some individuals cannot form close relationships and remain isolated.	
Generativity versus stagnation	Middle adulthood	Generativity means giving to the next generation through child rearing, caring for other people, or productive work. The person who fails in these ways feels an absence of meaningful accomplishment.	
Ego integrity versus despair	Late adulthood	In this final stage, individuals reflect on the kind of person they have been. Integrity results from feeling that life was worth living as it happened. Older people who are dissatisfied with their lives fear death.	

As Table 1.3 shows, Erikson's first five stages parallel Freud's stages, but Erikson added three adult stages to Freud's model.

Finally, unlike Freud, Erikson pointed out that normal development must be understood in relation to each culture's life situation. For example, among the Yurok Indians (a tribe of fishermen and acorn gatherers on the Northwest coast of the United States), babies are deprived of breast-feeding for the first 10 days after birth and instead are fed a thin soup from a small shell. At age 6 months, infants are abruptly weaned, an event enforced, if necessary, by having the mother leave for a few days. These experiences, from our cultural vantage point, might seem cruel. But Erikson explained that the Yurok live in a world in which salmon fill the river just once a year, a circumstance that requires the development of considerable self-restraint for survival. In this way, he showed that child-rearing experiences can only be understood by making reference to the competencies valued and needed by the individual's society.

■ CONTRIBUTIONS AND LIMITATIONS OF PSYCHOANALYTIC THEORY. A special strength of the psychoanalytic perspective is its emphasis on the individual's unique life history as worthy of study and understanding (Emde, 1992). Consistent with this view, psychoanalytic theorists accept the *clinical method,* which synthesizes information from a variety of sources into a detailed picture of the personality functioning of a single person. (We will discuss the clinical method further at the end of this chapter.) Psychoanalytic theory has also inspired a wealth of research on many aspects of emotional and social development, including infant–caregiver attachment, aggression, sibling relationships, child-rearing practices, morality, gender roles, and adolescent identity.

Despite its extensive contributions, the psychoanalytic perspective is no longer in the mainstream of research on human development. Psychoanalytic theorists may have become isolated from the rest of the field because they were so strongly committed to the clinical approach that they failed to consider other methods. In addition, many psychoanalytic ideas, such as Freud's Oedipus conflict, the psychosexual stages, and certain outcomes of Erikson's stages (for example, generativity and ego integrity) are so vague that they were difficult or impossible to test empirically (Miller, 1993).

Nevertheless, Erikson's broad outline of lifespan change captures the essence of personality development during each major phase of the life course, so we will return to it in later chapters. We will also encounter additional perspectives that clarify the attainments of early, middle, and late adulthood and that are within the tradition of stage models of psychosocial development (Levinson, 1978, 1996; Vaillant, 1977).

BEHAVIORISM AND SOCIAL LEARNING THEORY

At the same time that psychoanalytic theory gained in prominence, human development was also influenced by a very different perspective: **behaviorism,** a tradition consistent with Locke's tabula rasa. American behaviorism began with the work of psychologist John Watson (1878–1958) in the early part of the twentieth century. Watson wanted to create an objective science of psychology. Unlike psychoanalytic theorists, he believed in studying directly observable events—stimuli and responses—rather than the unseen workings of the mind (Horowitz, 1992).

■ TRADITIONAL BEHAVIORISM. Watson was inspired by some studies of animal learning carried out by famous Russian physiologist Ivan Pavlov. Pavlov knew that dogs release saliva as an innate reflex when they are given food. But he noticed that his dogs were salivating before they tasted any food—when they saw the trainer who usually fed them. The dogs, Pavlov reasoned, must have learned to associate a neutral stimulus (the trainer) with another stimulus (food) that produces a reflexive response (salivation). As a result of this association, the neutral stimulus could bring about the response by itself. Anxious to test this idea, Pavlov successfully taught dogs to salivate at the sound of a bell by pairing it with the presentation of food. He had discovered *classical conditioning.*

Watson wanted to find out if classical conditioning could be applied to children's behavior. In a historic experiment, he taught Albert, an 11-month-old infant, to fear a neutral stimulus—a soft white rat—by presenting it several times with a sharp, loud sound, which naturally scared the baby. Little Albert, who at first had reached out eagerly to touch the furry rat, cried and turned his head away when he caught sight of it (Watson & Raynor, 1920). In fact, Albert's fear was so intense that researchers eventually questioned the ethics of studies like this one (an issue we will take up later in this chapter). On the basis of findings like these, Watson concluded that environment was the supreme force in child development. Adults could mold children's behavior in any way they wished, he thought, by carefully controlling stimulus–response associations.

Another form of behaviorism was B. F. Skinner's (1904–1990) *operant conditioning theory.* According to Skinner, behavior can be increased by following it with a wide variety of reinforcers, such as food, praise, or a friendly smile. It can also be decreased through punishment, such as disapproval or withdrawal of privileges. As a result of Skinner's work, operant conditioning became a broadly applied learning principle in child psychology. We will consider these conditioning principles further in Chapter 4.

■ SOCIAL LEARNING THEORY. Psychologists quickly became interested in whether behaviorism might offer a more direct and effective explanation of the development of social behavior than the less precise concepts of psychoanalytic theory. This concern sparked the emergence of **social learning theory.** Social learning theorists built on the principles of conditioning that came before them, offering expanded views of how children and adults acquire new responses.

Several kinds of social learning theory emerged. The most influential was devised by Albert Bandura (1977), who recognized that children acquire many favorable and unfavorable responses simply by watching and listening to others around them. *Modeling,* otherwise known as *imitation* or *observational learning,* became a widely recognized basis for development. The baby who claps her hands after her mother does so, the child who angrily hits a playmate in the same way that he has been punished at home, and the teenager who wears the same clothes and hairstyle as her friends at school are all displaying observational learning.

Bandura's work continues to influence much research on children's social development. However, like the field of human development as a whole, today his theory stresses the importance of *cognition,* or thinking. In fact, the most recent revision of Bandura's (1986, 1989) theory places such strong emphasis on how we think about ourselves and other people that he calls it a *social-cognitive* rather than a social learning approach.

According to this view, children gradually become more selective in what they imitate. From watching others engage in self-praise and self-blame and through feedback about the worth of their own actions, children develop *personal standards* for behavior and a sense of *self-efficacy*—

Applied behavior analysis has been used to relieve a wide range of difficulties in children and adults, such as persistent aggression, language delays, undesirable habits, and extreme fears. A program of modeling and reinforcement helped this woman overcome an intense fear of public speaking. (Will Hart)

beliefs about their own abilities and characteristics—that guide responses in particular situations (Bandura, 1997). For example, imagine a parent who often remarks, "I'm glad I kept working on that challenging task," who explains the value of persistence to her child, and who encourages it by saying, "I know you can do that homework very well!" As a result, the child starts to view himself as hard working and high achieving and, from the many people available in the environment, selects models with these characteristics to copy. In this way, as individuals acquire attitudes, values, and beliefs about themselves, they control their own learning and behavior.

■ CONTRIBUTIONS AND LIMITATIONS OF BEHAVIORISM AND SOCIAL LEARNING THEORY. Like psychoanalytic theory, behaviorism and social learning theory have been helpful in treating emotional and behavior problems. Yet the techniques are decidedly different. **Applied behavior analysis** refers to procedures that combine conditioning and modeling to eliminate undesirable behaviors and in-

crease socially acceptable responses. It has been used to relieve a wide range of difficulties in children and adults, such as persistent aggression, language delays, undesirable habits, and extreme fears (Pierce & Epling, 1995).

Nevertheless, modeling and reinforcement do not provide a complete account of development (Horowitz, 1987). We will see in later sections that many theorists believe that behaviorism and social learning theory offer too narrow a view of important environmental influences. These extend beyond immediate reinforcements and modeled behaviors to the richness of the physical and social worlds. Finally, in emphasizing cognition, Bandura is unique among theorists whose work grew out of the behaviorist tradition. Behaviorism and social learning theory have been criticized for underestimating people's active contributions to their own development.

PIAGET'S COGNITIVE-DEVELOPMENTAL THEORY

If there is one individual who has influenced research on child development more than any other, it is the Swiss cognitive theorist Jean Piaget (1896–1980). Although American investigators had been aware of Piaget's work since 1930, they did not grant it much attention until the 1960s. A major reason is that his ideas were very much at odds with behaviorism, which dominated American psychology during the middle of the twentieth century (Beilin, 1992). Piaget did not believe that knowledge was imposed on a passive, reinforced child. According to his **cognitive-developmental theory,** children actively construct knowledge as they manipulate and explore their world, and their cognitive development takes place in stages.

■ PIAGET'S STAGES. Piaget's view of development was greatly influenced by his early training in biology. Central to his theory is the biological concept of *adaptation* (Piaget, 1971). Just as the structures of the body are adapted to fit with the environment, so the structures of the mind develop during childhood to better fit with, or represent, the external world. In infancy and early childhood, children's understanding is very different from adults'. For example, Piaget believed that young babies do not realize that an object hidden from view—a favorite toy or even the mother—continues to exist. He also concluded that preschoolers' thinking is full of faulty logic and fantasy. For example, children younger than age 7 commonly say that the amount of milk or lemonade changes when it is poured into a differently shaped container. According to Piaget, children eventually revise these incorrect ideas in their ongoing efforts to achieve an *equilibrium*, or balance, between internal structures and information they encounter in their everyday worlds (Beilin, 1992; Kuhn, 1992).

Through careful observations of and clinical interviews with children, Jean Piaget developed his comprehensive theory of cognitive development. His work has inspired more research on children than any other theory. (Yves de Braine/Black Star)

In Piaget's theory, children move through four broad stages of development, each of which is characterized by qualitatively distinct ways of thinking. Table 1.4 provides a brief description of Piaget's stages. In the *sensorimotor stage*, cognitive development begins with the baby's use of the senses and movements to explore the world. These ac-

tion patterns evolve into the symbolic but illogical thinking of the preschooler in the *preoperational stage*. Then cognition is transformed into the more organized reasoning of the school-age child in the *concrete operational stage*. Finally, in the *formal operational stage*, thought becomes the complex, abstract reasoning system of the adolescent and adult.

■ PIAGET'S METHODS OF STUDY. Piaget devised special methods for investigating how children think. In the early part of his career, he carefully observed his three infant children and also presented them with little problems, such as an attractive object that could be grasped, mouthed, kicked, or searched for when hidden from view. From their reactions, Piaget derived his ideas about cognitive changes during the first 2 years of life.

In studying childhood and adolescent thought, Piaget took advantage of children's ability to describe their thinking. He adapted the clinical method of psychoanalysis, conducting open-ended *clinical interviews* in which a child's initial response to a task served as the basis for the next question he would ask. We will look at an example of a Piagetian clinical interview when we discuss research methods later in this chapter.

■ CONTRIBUTIONS AND LIMITATIONS OF PIAGET'S THEORY. Piaget's cognitive-developmental perspective convinced the field that children are active learners whose minds contain rich structures of knowledge. Besides investigating children's understanding of the physical world, Piaget explored their reasoning about the social world. As we will see in later chapters, his stages have sparked a wealth of research on children's conceptions of themselves, other people, and human relationships. Practically speaking, Piaget's theory encouraged the development of educational

TABLE 1.4

Piaget's Stages of Cognitive Development

STAGE	PERIOD OF DEVELOPMENT	DESCRIPTION
Sensorimotor	Birth–2 years	Infants "think" by acting on the world with their eyes, ears, and hands. As a result, they invent ways of solving sensorimotor problems, such as pulling a lever to hear the sound of a music box, finding hidden toys, and putting objects in and taking them out of containers.
Preoperational	2–7 years	Preschool children use symbols to represent their earlier sensorimotor discoveries. Development of language and make-believe play takes place. However, thinking lacks the logical qualities of the two remaining stages.
Concrete operational	7–11 years	Children's reasoning becomes logical. School-age children understand that a certain amount of lemonade or play dough remains the same even after its appearance changes. They also organize objects into hierarchies of classes and subclasses. However, thinking falls short of adult intelligence. It is not yet abstract.
Formal operational	11 years on	The capacity for abstraction permits adolescents to reason with symbols that do not refer to objects in the real world, as in advanced mathematics. They can also think of all possible outcomes in a scientific problem, not just the the most obvious ones.

philosophies and programs that emphasize children's discovery learning and direct contact with the environment.

Despite Piaget's overwhelming contributions, in recent years his theory has been challenged. New evidence indicates that Piaget underestimated the competencies of infants and preschoolers. We will see in later chapters that when young children are given tasks scaled down in difficulty, their understanding appears closer to that of the older child and adult than Piaget believed. Furthermore, many studies show that children's performance on Piagetian problems can be improved with training. This finding raises questions about his assumption that discovery learning rather than adult teaching is the best way to foster development. Finally, some lifespan investigators take issue with Piaget's conclusion that no major cognitive changes occur after adolescence. Several have proposed accounts of *postformal thought* that stress important transformations in adulthood (Arlin, 1984; Labouvie-Vief, 1985).

Today, the field of human development is divided over its loyalty to Piaget's ideas. Those who continue to find merit in the stage approach accept a modified view—one in which changes in thinking are not sudden and abrupt, but take place gradually (Case, 1992; Fischer & Pipp, 1984). Others have given up the idea of cognitive stages in favor of a continuous approach to development—information processing—that we will take up in the next section.

BRIEF REVIEW

Three perspectives dominated research in the middle of the twentieth century. Psychiatrists and social workers turned to Freud's psychoanalytic approach, and Erikson's expansion of it, for help in understanding personality development and the origins of emotional difficulties. Behaviorism and social learning theory use conditioning and modeling to explain the appearance of new responses and to treat behavior problems. Piaget's stage theory revolutionized research on cognitive development with its view of children as active beings who take responsibility for their own learning.

ASK YOURSELF . . .

- *A 4-year-old becomes frightened of the dark and refuses to go to sleep at night. How would a psychoanalyst and a behaviorist differ in their view of how this problem developed?*

- *What biological concept is emphasized in Piaget's cognitive-developmental approach? From which nineteenth-century theory did Piaget borrow this idea?*

RECENT PERSPECTIVES

New ways of understanding the developing person are constantly emerging—questioning, building on, and enhancing the discoveries of earlier theories. Today, a burst of fresh approaches and research emphases is broadening our understanding of the lifespan.

INFORMATION PROCESSING

During the 1970s, researchers became disenchanted with behaviorism and disappointed in their efforts to completely verify Piaget's ideas. They turned to new trends in the field of cognitive psychology for ways to understand the development of thinking. Today, **information processing** is a leading approach to both child and adult cognition. It emerged with the design of complex computers that use mathematically specified steps to solve problems. These systems suggested to psychologists that the human mind might also be viewed as a symbol-manipulating system through which information flows (Klahr, 1992). From presentation to the senses at *input* and behavioral responses at *output,* information is actively coded, transformed, and organized.

Information-processing researchers often use diagrams to map the precise series of steps individuals use to solve problems and complete tasks, much like the plans devised by programmers to get computers to perform a series of "mental operations." Let's look at an example to clarify the usefulness of this approach. Figure 1.3 shows the steps that Andrea, an academically successful 8-year-old, used to complete a two-digit subtraction problem. It also displays the faulty procedure of Jody, who arrived at the wrong answer. The flowchart approach ensures that models of child and adult thinking will be very clear. By comparing the two procedures shown in Figure 1.3, we know what is necessary for effective problem solving and exactly where Jody went wrong in searching for a solution. As a result, we can design an intervention to improve her reasoning.

A wide variety of information-processing models exist. Some (like the flowcharts in Figure 1.3) are fairly narrow in that they track the individual's mastery of one or a few tasks. Others describe the information-processing system as a whole (Atkinson & Shiffrin, 1968; Craik & Lockhart, 1972). These general models are used as guides for asking questions about age changes in thinking. For example, does a child's ability to search the environment for information needed to solve a problem become more organized and planful with age? How much new information can preschoolers hold in memory compared to school-age children and adults? Why is speed of information processing slower among older than younger adults? Are declines in memory during old age evident on only some or all types of tasks?

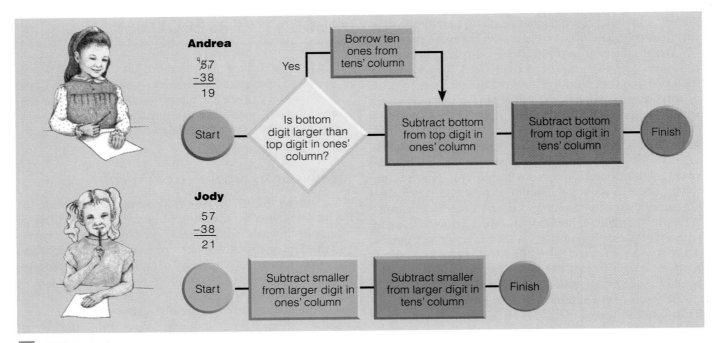

F IGURE 1.3

Information-processing flowcharts showing the steps that two 8-year-olds used to solve a math problem. In this two-digit subtraction problem with a borrowing operation, you can see that Andrea's procedure is correct, whereas Jody's results in a wrong answer.

Like Piaget's theory, information processing regards people as active, sense-making beings who modify their own thinking in response to environmental demands (Klahr, 1992). But unlike Piaget, there are no stages of development. Rather, the thought processes studied—perception, attention, memory, planning strategies, categorization of information, and comprehension of written and spoken prose—are assumed to be similar at all ages but present to a lesser or greater extent. Therefore the view of development is one of continuous rather than stagewise change.

Perhaps you can already tell that information-processing research has important implications for education (Geary, 1994; Hall, 1989; Siegler, 1983). But it has fallen short in some respects. Aspects of cognition that are not linear and logical, such as imagination and creativity, are all but ignored by this approach (Greeno, 1989). In addition, critics complain that information processing isolates thinking from important features of real-life learning situations. Recently, investigators have begun to address this concern by focusing on more realistic activities. Today, they can be found studying conversations, stories, memory for everyday events, and strategies for performing academic tasks.

Fortunately, a major advantage of having many theories is that they can encourage one another to attend to previously neglected dimensions of people's lives. A unique feature of the final three perspectives we will discuss is the emphasis they place on *contexts for development.* The impact of context can be examined at many levels—family, school, community, larger society, and culture. In addition,

human capacities have been shaped by a long evolutionary history in which our brains and bodies adapted to their surroundings. The next theory, ethology, emphasizes this biological side of development.

ETHOLOGY

Ethology is concerned with the adaptive, or survival, value of behavior and its evolutionary history (Hinde, 1989). Its origins can be traced to the work of Darwin. Its modern foundations were laid by two European zoologists, Konrad Lorenz and Niko Tinbergen.

Watching diverse animal species in their natural habitats, Lorenz and Tinbergen observed behavior patterns that promote survival. The most well known of these is *imprinting,* the early following behavior of certain baby birds that ensures that the young will stay close to the mother and be fed and protected from danger. Imprinting takes place during an early, restricted time period of development. If the mother goose is not present during this time, but an object resembling her in important features is, young goslings may imprint on it instead (Lorenz, 1952).

Observations of imprinting led to a major concept in human development: the **sensitive period.** It refers to a time span during which the person is biologically prepared to acquire certain capacities but needs the support of an appropriately stimulating environment. Although it is possible for development to occur later, it is harder to induce it at that time (Bornstein, 1989). Many researchers have con-

Konrad Lorenz was one of the founders of ethology and a keen observer of animal behavior. He developed the concept of imprinting. Here, young geese who were separated from their mother and placed in the company of Lorenz during an early, sensitive period show that they have imprinted on him. They follow him about as he swims through the water, a response that promotes survival. (Nina Leen/Life Magazine © Time Warner)

ducted studies to find out whether complex cognitive and social behaviors must be learned during certain time periods. For example, if children are deprived of adequate food or physical and social stimulation during the early years of life, will their intelligence be impaired? If language is not mastered during the preschool years, is the capacity to acquire it reduced?

Inspired by observations of imprinting, British psychoanalyst John Bowlby (1969) applied ethological theory to the understanding of the human infant–caregiver relationship. He argued that attachment behaviors of babies, such as smiling, babbling, grasping, and crying, are built-in social signals that encourage the parent to approach, care for, and interact with the infant. By keeping the mother near, these behaviors help ensure that the baby will be fed, protected from danger, and provided with stimulation and affection necessary for healthy growth.

The development of attachment in humans is a lengthy process involving changes in psychological structures that lead the baby to form a deep affectional tie with the caregiver. Bowlby (1979) speculated that this bond has lifelong consequences, affecting relationships "from cradle to grave" (p. 129). In later chapters, we will review evidence that supports this conclusion (Bretherton, 1992; Cox et al., 1992; Vormbrock, 1993).

Observations by ethologists have shown that many aspects of social behavior, including emotional expressions, aggression, cooperation, and children's social play, resemble those of our primate ancestors. Although ethology emphasizes the genetic and biological roots of development, learning is also considered important because it lends flexibility and greater adaptiveness to behavior. Consequently, the interests of ethologists are broad. They seek a full understanding of the environment, including physical, social, and cultural aspects (Hinde, 1989; Miller, 1993). The next contextual perspective we will discuss, ecological systems theory, serves as an excellent complement to ethology, since it shows how various aspects of the environment,

from everyday relationships to larger societal forces, work together to affect human development.

ECOLOGICAL SYSTEMS THEORY

Urie Bronfenbrenner, an American psychologist, is responsible for an approach to human development that has risen to the forefront of the field over the last decade. **Ecological systems theory** views the person as developing within a complex system of relationships affected by multiple levels of the surrounding environment. Before Bronfenbrenner's (1979, 1989, 1993) theory, most researchers viewed the environment fairly narrowly—as limited to events and conditions immediately surrounding the individual. As Figure 1.4 shows, Bronfenbrenner expanded this view by envisioning the environment as a series of nested structures that includes but extends beyond the home, school, neighborhood, and workplace settings in which people spend their everyday lives. Each layer of the environment is viewed as having a powerful impact on development.

■ THE MICROSYSTEM. The innermost level of the environment is the **microsystem**, which refers to activities and interaction patterns in the person's immediate surroundings. Bronfenbrenner emphasizes that to understand development at this level, we must keep in mind that all relationships are *bidirectional and reciprocal*. For example, adults affect children's responses, but children's characteristics—their personality styles and ways of thinking—also influence the behavior of adults. A friendly, attentive child is likely to evoke positive and patient reactions from parents, whereas an active, distractible youngster is more likely to be responded to with restriction and punishment (Danforth, Barkley, & Stokes, 1990). But whether either of these children experiences child-rearing practices that enhance or undermine development depends on environmental systems that surround and influence parent–child relationships.

FIGURE **1.4**

Structure of the environment in ecological systems theory. The *microsystem* concerns relations between the developing person and the immediate environment; the *mesosystem,* connections among immediate settings; the *exosystem,* social settings that affect but do not contain the the developing person; and the *macrosystem,* the values, laws, customs, and resources of the culture that affect activities and interactions at all inner layers. The *chronosystem* is not a specific context. Instead, it refers to the dynamic, ever-changing nature of the person's environment.

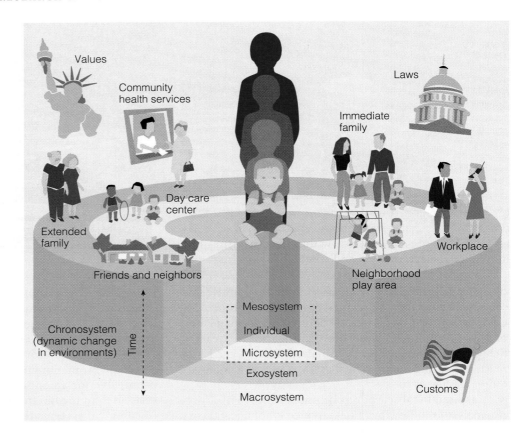

■ THE MESOSYSTEM. The second level of Bronfenbrenner's model is the **mesosystem.** It refers to connections among microsystems that foster development. For example, a child's academic progress depends not just on activities that take place in classrooms. It is also promoted by parent involvement in school life and the extent to which academic learning is carried over into the home (Grolnick & Slowiaczek, 1994). Among adults, how well a person functions as spouse and parent at home is affected by experiences in the workplace, and vice versa (McLoyd, 1989).

Urie Bronfenbrenner is the originator of ecological systems theory. He views the child as developing within a complex system of relationships affected by multiple levels of the surrounding environment, from immediate settings to broad cultural values, laws, and customs. (Courtesy of Urie Bronfenbrenner, Cornell University)

■ THE EXOSYSTEM. The **exosystem** refers to social settings that do not contain the developing person, but that affect experiences in immediate settings. These can be formal organizations, such as the board of directors in the individual's workplace or health and welfare services in the community. For example, flexible work schedules, paid maternity and paternity leave, and sick leave for parents whose children are ill are ways that work settings can help parents in their child-rearing roles and, indirectly, enhance the development of both adult and child. Similarly, communities that provide medical care, nutritious meals, counseling, job training, day care, and other services for the economically disadvantaged reduce the devastating effects of poverty in all phases of the life course.

■ THE MACROSYSTEM. The outermost level of Bronfenbrenner's model is the **macrosystem.** It is not a specific context. Instead, it refers to the values, laws, customs, and resources of a particular culture. The priority that the macrosystem gives to the needs of children and adults affects the support they receive at inner levels of the environment. For example, in countries that require high-quality standards for child care and workplace benefits for employed parents, children are more likely to have favorable experiences in their immediate settings. And when the government provides a generous pension plan for retirees, it supports the well-being of the elderly.

■ A DYNAMIC, EVER-CHANGING SYSTEM. According to Bronfenbrenner (1995, 1997), we must keep in mind that the environment is not a static force that affects people in a uniform way. Instead, it is dynamic and ever-changing. Whenever individuals add or let go of roles or settings in their lives, the breadth of their microsystems changes. These shifts in contexts, or *ecological transitions,* as Bronfenbrenner calls them, take place throughout life and are often important turning points in development. Starting school, entering the work force, marrying, becoming a parent, getting divorced, moving, and retiring are examples.

Bronfenbrenner refers to the temporal dimension of his model as the **chronosystem** (the prefix *chrono* means time). Changes in life events can be imposed externally. Alternatively, they can arise from within the organism since individuals select, modify, and create many of their own settings and experiences. How they do so depends on their age; their physical, intellectual, and personality characteristics; and the environmental opportunities available to them. Therefore, in ecological systems theory, development is neither controlled by environmental circumstances nor driven by inner forces. Instead, people are products and producers of their environments, both of which form a network of interdependent effects. We will see many more examples of these principles in later chapters of this book.

CROSS-CULTURAL RESEARCH AND VYGOTSKY'S SOCIOCULTURAL THEORY

Ecological systems theory, as well as Erikson's psychoanalytic theory, underscores the connection between culture and development. Investigations that make comparisons across cultures, and between ethnic and social-class groups within cultures, provide insight into commonalities and individual differences in development. In doing so, cross-cultural research helps us untangle the contributions of biological and environmental factors to the timing and order of appearance of new behaviors (Greenfield, 1994).

In the past, cross-cultural studies focused on broad cultural differences in development—for example, whether children in one culture are more advanced in motor development or do better on intellectual tasks than another. However, this approach can lead us to conclude incorrectly that one culture is superior in enhancing development, whereas another is deficient. In addition, it does not help us understand the precise experiences that contribute to cultural differences in behavior.

Today, more research is examining the relationship of *culturally specific practices* to development. The contributions of the Russian psychologist Lev Semenovich Vygotsky (1896–1934) have played a major role in this trend. Vygotsky's perspective is called **sociocultural theory.** It focuses on how culture—the values, beliefs, customs, and skills of a social group—is transmitted to the next genera-

According to Lev Semenovich Vygotsky, many cognitive processes and skills are socially transferred from more knowledgeable members of society to children. Vygotsky's sociocultural theory helps us understand the wide variation in cognitive competencies from culture to culture. Vygotsky is pictured here with his daughter. (Courtesy of James V. Wertsch, Washington University)

tion. According to Vygotsky (1934/1987), social interaction—in particular, cooperative dialogues with more knowledgeable members of society—is necessary for children to acquire the ways of thinking and behaving that make up a community's culture (Wertsch & Tulviste, 1992). Vygotsky believed that as adults and more expert peers help children master culturally meaningful activities, the communication between them becomes part of children's thinking. Once children internalize the essential features of these dialogues, they can use the language within them to guide their own actions and acquire skills on their own (Berk, 1994).

Perhaps you can tell from this brief description that Vygotsky's theory has been especially influential in the study of cognition. But Vygotsky's approach to cognitive development is quite different from Piaget's. Recall that Piaget did not regard direct teaching by adults as important. Instead, he emphasized children's active, independent efforts to make sense of their world. Vygotsky agreed with Piaget that children are active, constructive beings. But unlike Piaget, he viewed cognitive development as a *socially mediated process*—as dependent on the support that adults and more mature peers provide as children try new tasks. Finally, Vygotsky did not regard all children as moving through the same sequence of stages. Instead, as soon as children acquire language, their enhanced ability to communicate with others leads to continuous, step-by-step changes in thought and behavior that can vary greatly from culture to culture.

Although most research inspired by Vygotsky focuses on children, his ideas apply to people of any age acquiring new knowledge and skills. A central theme is that cultures select tasks for their members, and social interaction surrounding those tasks leads to competencies essential for success in a particular culture. For example, in industrialized nations, teachers can be seen helping people learn to read, drive a car, or use a computer (Schwebel, Maher, & Fagley, 1990). Among the Zinacanteco Indians of southern Mexico, adult experts guide young girls as they master complicated weaving techniques (Childs & Greenfield, 1982). In Brazil, child candy sellers with little or no schooling develop sophisticated mathematical abilities as the result of buying candy from wholesalers, pricing it in collaboration with adults and experienced peers, and bargaining with customers on city streets (Saxe, 1988).

Findings like these reveal that children and adults develop unique strengths in every culture that are not present in others. A cross-cultural perspective reminds us that research on development is based on only a small minority of humankind. We cannot assume that the developmental sequences we observe are "natural" or that the experiences fostering them are "ideal" without looking around the world.

BRIEF REVIEW

New theories are constantly emerging, questioning and building on earlier discoveries. Using computerlike models of mental activity, information processing has brought exactness and precision to the study of cognitive development. Ethology highlights the adaptive, or survival, value of behavior and its evolutionary history. Ecological systems theory stresses that the developing person is affected by a range of environmental influences, from immediate settings to broad cultural values and programs. Vygotsky's sociocultural theory takes a closer look at relationships that foster development. Through cooperative dialogues with expert partners, children and adults acquire unique, culturally adaptive competencies.

ASK YOURSELF . . .

■ *What shortcoming of the information-processing approach is a strength of ethology, ecological systems theory, and Vygotsky's sociocultural theory?*

■ *Return to the story about Sofie at the beginning of this chapter. Cite examples at each level of Bronfenbrenner's model that influenced her development.*

■ *What features of Vygotsky's sociocultural theory make it very different from Piaget's theory?*

COMPARING AND EVALUATING THEORIES

In the previous sections, we reviewed theoretical perspectives that are major forces in modern human development research. They differ in many respects. First, they focus on different domains of development. Some, such as the psychoanalytic perspective and ethology, emphasize emotional and social development. Others, such as Piaget's cognitive-developmental theory, information processing, and Vygotsky's sociocultural theory, stress changes in cognition. The remaining approaches—behaviorism, social learning theory, and ecological systems theory—discuss factors assumed to affect all aspects of development.

Second, every theory contains a point of view about what the developing person and the process of development is like. As we conclude our review of theoretical perspectives, take a moment to identify the stand that each takes on the three controversial issues presented at the beginning of this chapter. Then check your own analysis against the information given in Table 1.5. If you had difficulty classifying any theory, return to the relevant section of this chapter and reread the description of that theory.

Finally, we have seen that theories have strengths and weaknesses. This may remind you of an important point we made earlier in this chapter—that no theory provides a complete account of development. As you read more about the lifespan in later chapters of this book, you may find it useful to keep a notebook in which you test your own theoretical likes and dislikes against the evidence. Do not be surprised if you revise your ideas many times, just as theorists have done throughout this century. By the end of the course, you will have built your own personal perspective. It might turn out to be a blend of several theories, since each viewpoint we have discussed has contributed to what we know about the life course. And like the field as a whole, you will be left with some unanswered questions. I hope they will motivate you to continue your quest to understand human development in the years to come.

STUDYING DEVELOPMENT

In every science, theories, like those we've just reviewed, guide the collection of information, its interpretation, and its application to real-life situations. In fact, research usually begins with a prediction about behavior drawn from a theory, or what we call a *hypothesis*. But theories and hypotheses are only the beginning of the many activities that result in sound evidence on human development. Conducting research according to scientifically accepted procedures involves many steps and choices. Inves-

Stance of Major Developmental Theories on Three Basic Issues in Human Development

THEORY	VIEW OF THE DEVELOPING PERSON	VIEW OF THE COURSE OF DEVELOPMENT	VIEW OF THE DETERMINANTS OF DEVELOPMENT
Psychoanalytic perspective	*Organismic:* Relations among structures of the mind (id, ego, and superego) determine personality.	*Discontinuous:* Stages of psychosexual and psychosocial development are emphasized.	*Both nature and nurture:* Innate impulses are channeled and controlled through social experiences. *Early experiences* set the course of later development.
Behaviorism and social learning theory	*Mechanistic:* The person consists of connections established between stimulus inputs and behavioral responses.	*Continuous:* Quantitative increase in learned behaviors occurs with age.	*Emphasis on nurture:* Learning principles of conditioning and modeling determine development. *Both early and later experiences* are important.
Piaget's cognitive-developmental theory	*Organismic:* Psychological structures determine the child's understanding of the world. The child actively constructs knowledge.	*Discontinuous:* Stages of cognitive development are emphasized.	*Both nature and nurture:* Children's innate drive to discover reality is emphasized. However, it must be supported by a rich, stimulating environment. *Both early and later experiences* are important.
Information processing	*Both organismic and mechanistic:* Active processing structures combine with a mechanistic, computer-like model of stimulus input.	*Continuous:* A quantitative increase in perception, attention, memory, and problem-solving skills takes place with age.	*Both nature and nurture:* Maturation and learning opportunities affect information-processing skills. *Both early and later experiences* are important.
Ethology	*Organismic:* The individual is biologically prepared with social signals that actively promote survival. Over time, psychological structures develop that underlie attachment and other adaptive behavior patterns.	*Both continuous and discontinuous:* Adaptive behavior patterns increase in quantity over time. But sensitive periods—restricted time periods in which qualitatively distinct capacities and responses emerge fairly suddenly—are also emphasized.	*Both nature and nurture:* Biologically based, evolved behavior patterns are stressed, but an appropriately stimulating environment is necessary to elicit them. Also, learning can improve the adaptiveness of behavior. *Early experiences* set the course of later development.
Ecological systems theory	*Organismic:* Personality characteristics and ways of thinking actively contribute to the person's development.	*Not specified*	*Both nature and nurture:* The person's characteristics and the reactions of others affect each other in a bidirectional fashion. Layers of the environment influence development. *Both early and later experiences* are important.
Vygotsky's sociocultural theory	*Organismic:* Children internalize essential features of social dialogues, forming psychological structures that they use to guide their own behavior.	*Continuous:* Interaction with more expert members of society leads to step-by-step changes in thought and behavior.	*Both nature and nurture:* Maturation and opportunities to interact with more expert members of society affect the development of culturally adaptive knowledge and skills. *Both early and later experiences* are important.

tigators must decide which participants, and how many, to include. Then they must figure out what the participants will be asked to do and when, where, and how many times each will need to be seen. Finally, they must examine relationships and draw conclusions from their data.

In the following sections, we examine research strategies commonly used in the field of human development. We begin with *research methods,* the specific activities in which subjects will be asked to participate, such as taking tests, answering questionnaires, responding to interviews, or being observed. Then we turn to *research designs*—overall plans for research studies that permit the best possible test of the

investigator's hypothesis. Finally, we discuss ethical issues involved in doing research with human participants.

At this point, you may be wondering, Why learn about research strategies? Why not leave these matters to research specialists and concentrate on what is already known about the developing person and how this knowledge can be applied? There are two reasons. First, each of us must be wide and critical consumers of knowledge, not naive sponges who soak up facts about development. A basic appreciation of the strengths and weaknesses of research strategies becomes important in separating dependable information from misleading results. Second, individuals

TABLE 1.6

Strengths and Limitations of Common Research Methods

METHOD	DESCRIPTION	STRENGTHS	LIMITATIONS
Systematic Observation			
Naturalistic observation	Observation of behavior in natural contexts	Observations reflect participant's everyday lives.	Conditions under which participants are observed cannot be controlled.
Structured observation	Observation of behavior in a laboratory	Conditions of observation are the same for all participants.	Observations may not be typical of the way participants behave in everyday life.
Self-Reports			
Clinical interview	Flexible interviewing procedure in which the investigator obtains a complete account of the participant's thoughts	Comes as close as possible to the way participants think in everyday life; great breadth and depth of information can be obtained in a short time.	Participants may not report information accurately; flexible procedure makes comparing individuals' responses difficult.
Structured interview, questionnaires, and tests	Self-report instruments in which each participant is asked the same questions in the same way	Standardized method of asking questions permits comparisons of participants' responses and efficient data collection and scoring.	Does not yield the same depth of information as a clinical interview; responses are still subject to inaccurate reporting.
Clinical method (case study)	A full picture of a single individual's psychological functioning, obtained by combining interviews, observations, and test scores	Provides rich, descriptive insights into processes of development.	May be biased by researcher's theoretical preferences; findings cannot be applied to individuals other than the participant.

who work directly with children or adults are sometimes in a position to carry out research studies, either on their own or with an experienced investigator. At other times, they may have to provide information on how well their goals are being realized to justify continued financial support for their programs and activities. Under these circumstances, an understanding of research becomes essential practical knowledge.

COMMON RESEARCH METHODS

How does a researcher choose a basic approach to gathering information? Common methods include systematic observation, self-reports (such as questionnaires and interviews), and clinical or case studies of a single individual. As you read about these methods, you may find it helpful to refer to Table 1.6, which summarizes the strengths and limitations of each.

■ SYSTEMATIC OBSERVATION. To find out how people actually behave, a researcher may choose systematic observation. Observations can be made in different ways. One approach is to go into the field, or natural environment, and observe the behavior of interest, a method called **naturalistic observation.**

A study of children's social development provides a good example of this technique (Barrett & Yarrow, 1977).

Observing 5- to 8-year-olds at a summer camp, the researchers recorded the number of times each child provided another person with physical or emotional support in the form of comforting, sharing, helping, or expressing sympathy. The great strength of naturalistic observation is that investigators can see directly the everyday behaviors they hope to explain.

Naturalistic observation also has limitations. First, the presence of a watchful, unfamiliar individual may cause children and adults to react in unnatural ways. Second, not all individuals have the same opportunity to display a particular behavior in everyday life. In the study just described, some children happened to be exposed to more cues for positive social responses (such as a tearful playmate), and for this reason they showed more helpful and comforting behaviors. Researchers commonly deal with this difficulty by making **structured observations** in a laboratory. In this approach, the investigator sets up a situation that evokes the behavior of interest so every participant has an equal opportunity to display the response. In one study, children's comforting behavior was observed by playing a tape recording simulating a baby crying in the next room. Using an intercom, children could either talk to the baby or push a button so they did not have to listen (Eisenberg et al., 1993). Notice how structured observation gives investigators more control over the research situation. But its great disadvantage is that people do not necessarily behave in the laboratory as they do in everyday life.

The procedures used to collect systematic observations vary considerably, depending on the nature of the research problem. Some investigators need to describe the entire stream of behavior—everything said and done over a certain time period. In one of my own studies, I wanted to find out how sensitive, responsive, and verbally stimulating caregivers were when they interacted with children in day-care centers (Berk, 1985). In this case, everything each caregiver said and did—even the amount of time she spent away from the children, taking coffee breaks and talking on the phone—was important. In other studies, only one or a few kinds of behavior are needed, and it is not necessary to preserve the entire behavior stream. In these instances, researchers use more efficient observation procedures in which they record only certain events or mark off behaviors on checklists.

Systematic observation provides invaluable information on how children and adults actually behave, but it tells us little about the reasoning that lies behind their responses. For this kind of information, researchers must turn to another type of method: self-reports.

In structured interviews, each participant is asked the same set of questions in the same way, eliminating differences in responses that may be due to the manner of interviewing. (Bob Daemmrich/ The Image Works)

SELF-REPORTS: INTERVIEWS AND QUESTIONNAIRES. Self-reports are instruments that ask participants to answer questions about their perceptions, thoughts, abilities, feelings, attitudes, beliefs, and past experiences. They range from relatively unstructured clinical interviews, the method used by Piaget to study children's thinking, to highly structured interviews, questionnaires, and tests.

Let's look at an example of a **clinical interview** in which Piaget questioned a 5-year-old child about his understanding of dreams:

> Where does the dream come from?—*I think you sleep so well that you dream.*—Does it come from us or from outside?—*From outside.*—What do we dream with?—*I don't know.*—With the hands?. . . With nothing?—*Yes, with nothing.*—When you are in bed and you dream, where is the dream?—*In my bed, under the blanket. I don't really know. If it was in my stomach, the bones would be in the way and I shouldn't see it.*—Is the dream there when you sleep?—*Yes, it is in the bed beside me . . .*—You see the dream when you are in the room, but if I were in the room, too, should I see it?—*No, grownups don't ever dream.*—Can two people ever have the same dream?—*No, never.*—When the dream is in the room, is it near you?—*Yes, there!* (pointing to 30 cm. in front of his eyes). (Piaget, 1926/1930, pp. 97–98)

Notice how Piaget used a flexible, conversational style to encourage the child to expand his ideas. Prompts are given to obtain a fuller picture of the child's reasoning.

The clinical interview has two major strengths. First, it permits people to display their thoughts in terms that are as close as possible to the way they think in everyday life.

Second, the clinical interview can provide a large amount of information in a fairly brief period of time. For example, in an hour-long session, we can obtain a wide range of child-rearing information from a parent—much more than we could capture by observing parent–child interaction for the same amount of time.

A major limitation of the clinical interview has to do with the accuracy with which people report their thoughts, feelings, and experiences. Some participants, desiring to please the interviewer, may make up answers that do not represent their actual thinking. When asked about past events, they may have trouble recalling exactly what happened. And because the clinical interview depends on verbal ability and expressiveness, it may underestimate the capacities of individuals who have difficulty putting their thoughts into words.

The clinical interview has also been criticized because of its flexibility. When questions are phrased differently for each participant, responses may be due to the manner of interviewing rather than real differences in the way people think about a certain topic. **Structured interviews**, in which each participant is asked the same set of questions in the same way, can eliminate this problem. In addition, these techniques are much more efficient. Answers are briefer, and researchers can obtain written responses from an entire group of children or adults at the same time. Also, when structured interviews use multiple-choice, yes-no, and true-false formats, as is done on many tests and questionnaires, responses can be tabulated by machine. However, we must keep in mind that these approaches do not yield the same depth of information as a clinical interview. And they can still be affected by the problem of inaccurate reporting.

■ THE CLINICAL METHOD. Earlier in this chapter, we discussed the **clinical method** (sometimes called the *case study approach*) as an outgrowth of psychoanalytic theory, which stressed the importance of understanding a single life history. The clinical method brings together a wide range of information on one person, including interviews, observations, and sometimes test scores. The aim is to obtain as complete a picture as possible of that individual's psychological functioning and the experiences that led up to it.

Although clinical studies are usually carried out on children and adults who have serious emotional problems, they sometimes focus on well-adjusted individuals. In one recent investigation, researchers wanted to find out what contributes to the accomplishments of people with extraordinary intellectual talents. They selected six child prodigies for intensive study. Among them was Adam, a boy who read, wrote, and composed musical pieces before he was out of diapers. Adam's parents provided a home rich in stimulation and reared him with affection, firmness, and humor. They searched for schools in which he could both develop his abilities and form rewarding social relationships. By age 4, Adam was deeply involved in mastering human symbol systems—BASIC for the computer, French, German, Russian, Sanskrit, Greek, ancient hieroglyphs, music, and mathematics. Would Adam have realized his abilities without the chance combination of his special gift with nurturing, committed parents? Probably not, the investigators concluded (Feldman, 1991). Adam's case illustrates the unique strengths of the clinical method. It yields case narratives that are rich in descriptive detail and that offer valuable insights into development.

The clinical method, like all others, has drawbacks. It is subject to the same problems as the clinical interview. Also, more than other methods, the theoretical preferences of the researcher can bias the interpretations of clinical data. Finally, investigators cannot assume that their conclusions apply to anyone other than the person being studied. The insights drawn from clinical investigations need to be tested further with other research methods.

BRIEF REVIEW

Systematic observation, self-reports, and clinical or case studies are commonly used methods in the field of human development. Naturalistic observation provides information on everyday behaviors. When it is necessary to control the conditions of observation, researchers often make structured observations in a laboratory. The flexible, conversational style of the clinical interview provides a wealth of information on the reasoning behind people's behavior. However, participants may not report their thoughts accurately, and comparing

their responses is difficult. The structured interview is a more efficient method that questions each participant in the same way, but it does not yield the same depth of information as a clinical interview. Clinical studies of single individuals provide rich insights into the processes of development. However, information obtained is often unsystematic and subjective and affected by researchers' theoretical biases.

ASK YOURSELF . . .

■ *Why is it important for students of human development and individuals who work directly with children or adults to understand research strategies?*

■ *A researcher wants to study the thoughts and feelings of recently divorced adults. Which one of the methods described in the preceding sections is best suited for investigating this topic? Explain why.*

GENERAL RESEARCH DESIGNS

In deciding on a research design, investigators choose a way of setting up a study that permits them to test their hypotheses with the greatest certainty possible. Two main types of designs are used in all research on human behavior: correlational and experimental.

■ CORRELATIONAL DESIGN. In a **correlational design,** researchers gather information on already existing groups of individuals without altering their experiences in any way. Suppose we want to answer such questions as, Do parents' styles of interacting with children have any bearing on children's intelligence? How does the arrival of a baby influence a couple's marital satisfaction? Does the death of a spouse in old age affect the surviving partner's physical health and psychological well-being? In these and many other instances, it is either very difficult or ethically impossible to arrange and control the conditions of interest.

The correlational design offers a way of looking at relationships between participants' experiences or characteristics and their behavior or development. But correlational studies have one major limitation: we cannot infer cause and effect. For example, if we find that parental interaction does relate to children's intelligence, we would not know whether parents' behavior actually causes intellectual differences among children. In fact, the opposite is certainly possible. The behaviors of highly intelligent children may be so attractive that they cause parents to interact more favorably. Or a third variable that we did not even think about studying, such as amount of noise and distraction in the home, may be causing both maternal interaction and children's intelligence to change together in the same direction.

Does the death of a spouse in old age affect the surviving partner's physical health and psychological well-being? A correlational design can be used to answer this question, but it does not permit researchers to determine the precise cause of their findings. (Murray/Monkmeyer Press)

In correlational studies, and in other types of research designs, investigators often examine relationships by using a **correlation coefficient.** It is a number that describes how two measures, or variables, are associated with one another. We will encounter the correlation coefficient in discussing research findings throughout this book. So let's look at what it is and how it is interpreted. A correlation coefficient can range in value from +1.00 to −1.00. The *magnitude, or size, of the number* shows the *strength of the relationship.* A zero correlation indicates no relationship, but the closer the value is to +1.00 or −1.00, the stronger the relationship that exists. The *sign of the number* (+ or −) refers to the *direction of the relationship.* A positive sign (+) means that as one variable *increases,* the other also *increases.* A negative sign (−) indicates that as one variable *increases,* the other *decreases.*

Let's take a couple of examples to illustrate how a correlation coefficient works. In one study, a researcher found that a measure of maternal language stimulation at 13 months was positively correlated with the size of children's vocabularies at 20 months, at +.50. This is a moderately high correlation, which indicates that the more mothers talked to their infants, the more advanced their children were in spoken language (Tamis-LeMonda & Bornstein, 1994). In another study, a researcher reported that the extent to which mothers ignored their 10-month-olds' bids for attention was negatively correlated with children's willingness to comply with parental demands one year later—at −.46 for boys and −.36 for girls (Martin, 1981). These moderate correlations reveal that the more mothers ignored their babies, the less cooperative their children were during the second year of life.

Both of these investigations found a relationship between maternal behavior and children's early development. Although the researchers suspected that maternal behavior affected the children's responses, in neither study could they really be sure about cause and effect. However, if we find a relationship in a correlational study, this suggests that it would be worthwhile to track down its cause with a more powerful experimental research strategy, if possible.

■ EXPERIMENTAL DESIGN. Unlike correlational studies, an **experimental design** permits us to make inferences about cause and effect. In an experiment, the events and behaviors of interest are divided into two types: independent and dependent variables. The **independent variable** is the one anticipated by the investigator to cause changes in another variable. The **dependent variable** is the one the investigator expects to be influenced by the independent variable. Inferences about cause-and-effect relationships are possible because the researcher directly *controls* or *manipulates* changes in the independent variable. This is done by exposing participants to two or more treatment conditions and comparing their performance on measures of the dependent variable.

In one *laboratory experiment,* researchers wanted to know if quality of interaction between adults (independent variable) affects young children's emotional reactions while playing with a familiar peer (dependent variable). Pairs of 2-year-olds were brought into a laboratory set up to look much like a family home. One group was exposed to a *warm treatment* in which two adults in the kitchen spoke in a friendly way while the children played in the living room. A second group received an *angry treatment* in which positive communication between the adults was interrupted by an argument in which they shouted, complained, and slammed the door. Children in the angry condition displayed much more distress (such as freezing in place, anxious facial expressions, and crying). They also showed more aggression toward their playmates than did children in the warm treatment (Cummings, Iannotti, & Zahn-Waxler, 1985). The experiment revealed that exposure to even short episodes of intense adult anger can trigger negative emotion and antisocial behavior in very young children.

In experimental studies, investigators must take special precautions to control for unknown characteristics of participants that could reduce the accuracy of their findings. For example, in the study just described, if a greater number of children who had already learned to behave in hostile and aggressive ways happened to end up in the angry treatment, we could not tell whether the independent variable or children's background characteristics produced the results. *Random assignment* of participants to treatment conditions offers protection against this problem. By using an evenhanded procedure, such as drawing numbers out of a hat or flipping a coin, the experimenter increases the chances that participants' characteristics will be equally distributed across treatment groups.

■ MODIFIED EXPERIMENTAL DESIGNS: FIELD AND NATURAL EXPERIMENTS. Most experiments are conducted in laboratories where researchers can achieve the maximum possible control over treatment conditions. But, as we have already indicated, findings obtained in laboratories may not always apply to everyday situations. The ideal solution to this problem is to do experiments in the field as a complement to laboratory investigations. In *field experiments,* investigators capitalize on rare opportunities to randomly assign people to different treatments in natural settings. In the laboratory experiment that we just considered, we can conclude that the emotional climate established by adults affects children's behavior in the laboratory, but does it also do so in daily life?

Another study helps answer this question. This time, the research was carried out in a day-care center. A caregiver deliberately interacted differently with two groups of preschoolers. In one condition (the *nurturant treatment*), she modeled many instances of warmth, helpfulness, and concern for others. In the second condition (the *control,* since it involved no treatment), she behaved as usual, with no special concern for others. Two weeks later, the researchers created several situations that called for helpfulness. For example, a visiting mother asked each child to watch her baby for a few moments, but the infant's toys had fallen out of the playpen. The investigators found that children exposed to the nurturant treatment were much more likely to return toys to the baby than those in the control condition (Yarrow, Scott, & Waxler, 1973).

In testing many hypotheses, researchers cannot randomly assign participants and manipulate conditions in the real world. Sometimes they can compromise by conducting *natural experiments.* Treatments that already exist, such as different day-care centers, school environments, workplaces, or retirement villages, are compared. These studies differ from correlational research only in that groups of participants are carefully chosen to ensure that their characteristics are as much alike as possible. In this way, investigators rule out as best they can alternative explanations for their treatment effects. But despite these efforts, natural experiments are unable to achieve the precision and rigor of true experimental research.

To help you compare correlational and experimental designs, Table 1.7 summarizes their strengths and limitations. It also includes an overview of designs for studying development, to which we now turn.

DESIGNS FOR STUDYING DEVELOPMENT

Scientists interested in human development require information about the way research participants change over time. To answer questions about development, they must extend correlational and experimental approaches to include measurements at different ages. Longitudinal and cross-sectional designs are special *developmental* research strategies. In each, age comparisons form the basis of the research plan.

■ THE LONGITUDINAL DESIGN. In a **longitudinal design,** a group of participants is studied repeatedly at different ages, and changes are noted as they mature. The time spanned may be relatively short (a few months to several years) or it may be very long (a decade or even a lifetime). The longitudinal approach has two major strengths. First, since it tracks the performance of each person over time, researchers can identify common patterns of development as well as individual differences. Second, longitudinal studies permit investigators to examine relationships between early and later events and behaviors. Let's take an example to illustrate these ideas.

A group of researchers wondered whether children who display extreme personality styles—either angry and explosive or shy and withdrawn—retain the same dispositions when they become adults. In addition, they wanted to know what kinds of experiences promote stability or change in personality and what consequences explosiveness and shyness have for long-term adjustment. To answer these questions, the researchers delved into the archives of the Guidance Study, a well-known longitudinal investigation initiated in 1928 at the University of California, Berkeley, and continued over several decades (Caspi, Elder, & Bem, 1987, 1988).

Results revealed that the two personality styles were only moderately stable. Between ages 8 and 30, a good number of individuals remained the same, whereas others changed substantially. When stability did occur, it appeared to be due to a "snowballing effect," in which children evoked responses from adults and peers that acted to maintain their dispositions. In other words, explosive youngsters were likely to be treated with anger and hostility (to which they reacted with even greater unruliness), whereas shy children were apt to be ignored.

Persistence of extreme personality styles affected many areas of adult adjustment. For men, the results of early explosiveness were most apparent in their work lives, in the

TABLE 1.7

Strengths and Limitations of Research Designs

DESIGN	DESCRIPTION	STRENGTHS	LIMITATIONS
General			
Correlational	The investigator obtains information on already existing groups, without altering participants' experiences.	Permits study of relationships between variables.	Does not permit inferences about cause-and-effect relationships.
Experimental	The investigator manipulates an independent variable and looks at its effect on a dependent variable; can be conducted in the laboratory or natural environment.	Permits inferences about cause-and-effect relationships.	When conducted in the laboratory, findings may not apply to the real world; when conducted in the field, control over treatment is usually weaker than in the laboratory.
Developmental			
Longitudinal	The investigator studies the same group of participants repeatedly at different ages.	Permits study of common patterns and individual differences in development and relationships between early and later events and behaviors.	Age-related changes may be distorted because of dropout and test-wiseness of participants and cohort effects.
Cross-sectional	The investigator studies groups of participants differing in age at the same point in time.	More efficient than the longitudinal design.	Does not permit study of individual developmental trends. Age differences may be distorted because of cohort effects.
Longitudinal-sequential	The investigator studies two or more groups of participants born in different years, following each group longitudinally.	Permits both longitudinal and cross-sectional comparisons. Reveals existence of cohort effects.	May have the same problems as longitudinal and cross-sectional strategies, but the design itself helps identify difficulties.

form of conflicts with supervisors, frequent job changes, and unemployment. Since few women in this sample of an earlier generation worked after marriage, their family lives were most affected. Explosive girls grew up to be hot-headed wives and parents who were especially prone to divorce. Sex differences in the long-term consequences of shyness were even greater. Men who had been withdrawn in childhood were delayed in marrying, becoming fathers, and developing stable careers. Because a withdrawn, unassertive style was socially acceptable for females, women who had shy personalities showed no special adjustment problems.

■ PROBLEMS IN CONDUCTING LONGITUDINAL RESEARCH. Despite their many strengths, longitudinal investigations pose a number of problems. For example, participants may move away or drop out of the research for other reasons. This changes the original sample so it no longer represents the population to whom researchers would like to generalize their findings. Also, from repeated study, people may become "test-wise." As a result, the behavior they present to investigators may become unnatural.

But the most widely discussed threat to longitudinal findings is **cohort effects,** another term that researchers use for *history-graded influences* (see page 9). Recall that

cohorts are individuals born in the same time period who are influenced by a particular set of historical and cultural conditions. Longitudinal results based on one cohort may not apply to people developing at other points in time. For example, in the study of personality styles described in the preceding section, we might ask whether the sex differences obtained are still true, in view of recent changes in gender roles in our society. And a longitudinal study of the lifespan would probably result in quite different findings if it were carried out in the 1990s, around the time of World War II, or during the Great Depression of the 1930s. (See the Lifespan Vista box on pages 34–35.)

■ THE CROSS-SECTIONAL DESIGN. The length of time it takes for many behaviors to change, even in limited longitudinal studies, has led researchers to turn toward a more convenient strategy for studying development. In the **cross-sectional design,** groups of people differing in age are studied at the same point in time.

An investigation in which students in grades 3, 6, 9, and 12 filled out a questionnaire about their sibling relationships provides a good illustration. Findings revealed that sibling interaction was characterized by greater equality and less power assertion with age. Also, feelings of sibling companionship declined during adolescence. The

LONG-TERM CONSEQUENCES OF GROWING UP DURING THE GREAT DEPRESSION

Economic disaster, wars, and periods of rapid social change can profoundly affect people's lives. Glen Elder (1974) capitalized on the extent to which families experienced economic hardship during the Great Depression of the 1930s to study its impact on lifespan development. Elder delved into the vast archives of the Oakland Growth Study, a longitudinal investigation begun in the early 1930s that followed 167 adolescents into mature adulthood. He divided the sample into two groups: those whose adolescent years were marked by severe economic deprivation and those whose youth was relatively free of economic strain.

Findings showed that unusual burdens were placed on adolescents from deprived families as their parents' lives changed. Mothers entered the labor force, fathers sought work outside the immediate community, and the stress of economic hardship led to a rising rate of parental divorce and illness. In response, adolescents had to take on family responsibilities. Girls cared for younger siblings and assumed household chores, while

boys tried to find part-time jobs. These changes had major consequences for adolescents' future aspirations. Girls' interests focused on home and family, and they were less likely to think about college and careers. Boys learned that economic resources could not be taken for granted, and they tended to make a very early commitment to an occupational choice.

Relationships also changed in economically deprived homes. As unemployed fathers lost status, mothers were granted greater control over family affairs. This reversal of traditional gender roles often sparked conflict. Fathers became explosive and punitive toward children (Elder, Liker, & Cross, 1984). In response, boys turned toward peers and adults outside the family for emotional support. Because girls were more involved in family affairs, they bore the brunt of their fathers' anger. And some fathers may have been especially resentful of their daughters as family power was transferred to women (Elder, Van Nguyen, & Caspi, 1985).

The impact of the Great Depression continued to be apparent as these

young people entered adulthood. Girls who grew up in economically deprived homes remained committed to domestic life, and many married at an early age. Men had a strong desire for economic security, and they changed jobs less often than those from nondeprived backgrounds. The chance to become parents was especially important to men whose lives had been disrupted by the Depression. Perhaps because they felt a rewarding career could not be guaranteed, these men viewed children as the most enduring benefit of their adult lives.

The fact that the Oakland Growth Study participants were adolescents and, therefore, beyond the early years of intense family dependency may explain why most weathered economic hardship successfully. Elder and his colleagues conducted a similar investigation of another cohort (the Guidance Study sample), who were born later and therefore were much younger when the Great Depression struck. For boys (who, as we will see in later chapters, are especially prone to adjustment problems

---

researchers thought that these age changes were due to several factors. As later-born children become more competent and independent, they no longer need and are probably less willing to accept direction from older siblings. In addition, as adolescents move from psychological dependence on the family to greater involvement with peers, they may have less time and emotional need to invest in siblings (Buhrmester & Furman, 1990). These intriguing ideas about the impact of development on sibling relationships, as we will see in Chapter 12, have been confirmed in subsequent research.

■ PROBLEMS IN CONDUCTING CROSS-SECTIONAL RESEARCH. The cross-sectional design is a very efficient strategy for describing age-related trends. But when researchers choose it, they are short-changed in the kind of information they can obtain about development. Evidence about change at the level at which it actually occurs—the individual—is not available. For example, in the study of sibling relationships that we just discussed, comparisons are limited to age-group averages. We cannot tell if important individual differences exist in the development of sibling relationships, some becoming more supportive and intimate and others becoming increasingly distant with age.

Cross-sectional studies that cover a wide age span have another problem. Like longitudinal research, they can be threatened by cohort effects. For example, comparisons of 10-year-old cohorts, 20-year-old cohorts, and 30-year-old cohorts—groups born and reared in different years—may not really represent age-related changes. Instead, they

in the face of family stress), the impact of economic strain was particularly severe. They showed long-term emotional difficulties and poor attitudes toward school and work (Elder & Caspi, 1988; Elder, Caspi, & Van Nguyen, 1986).

Clearly, historical events do not affect all individuals in the same way. With respect to the Great Depression, the outcomes varied considerably, depending on the young person's sex and the period of development in which social change took place.

The Great Depression of the 1930s left this farm family without a steady income. The adolescent girl (in the back row on the far right) may have been more negatively affected by economic hardship than her brother (second from left). And overall, younger children probably suffered more than older children. (Culver Pictures)

may reflect unique experiences associated with the historical periods in which the age groups were growing up.

■ IMPROVING DEVELOPMENTAL DESIGNS. To overcome some of the limitations of longitudinal and cross-sectional research, investigators sometimes combine the two approaches. One way of doing so is the **longitudinal-sequential design.** It is called a sequential design because it is composed of a sequence of samples (two or more different age groups), each of which is followed longitudinally for a number of years.

The design has two advantages. First, it permits researchers to find out whether cohort effects are operating by comparing people of the same age who were born in different years. Using the example shown in Figure 1.5, we can compare the behaviors of the three samples at ages 20, 30, and 40. If they do not differ, we can rule out cohort effects. Second, it is possible to do both longitudinal and cross-sectional comparisons. If outcomes are similar in each, then we can be especially confident about our findings.

In a study underway that used the design in Figure 1.5, researchers wanted to find out whether adult personality development progresses as Erikson's psychosocial theory predicts (Whitbourne et al., 1992). Questionnaires measuring Erikson's stages were given to three cohorts of 20-year-olds, each born a decade apart. The cohorts have been and will continue to be reassessed every 10 years. Consistent with Erikson's theory, longitudinal and cross-sectional gains in identity and intimacy occurred between ages 20 and 30—a trend unaffected by historical time period. But a

FIGURE 1.5

Example of a longitudinal-sequential design. Three cohorts, born in 1945, 1955, and 1965, are followed longitudinally from 20 to 40 years of age. The design permits the researcher to check for cohort effects by comparing people of the same age who were born in different years. Also, both longitudinal and cross-sectional comparisons can be made.

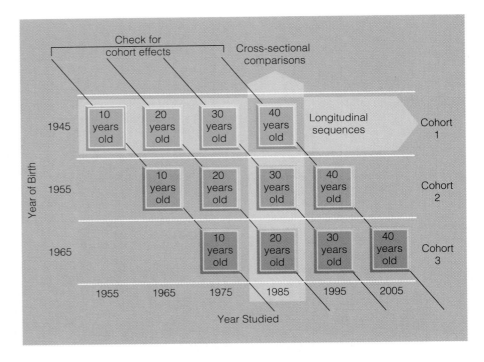

powerful cohort effect emerged for consolidation of the sense of industry: At age 20, Cohort 1 scored substantially below Cohorts 2 and 3. Look at Figure 1.5 again, and notice that members of Cohort 1 reached age 20 in the mid-1960s. As college students, they were part of a political protest movement that reflected disenchantment with the work ethic. Once out of college, they caught up with the other cohorts, perhaps as a result of experiencing the pressures of the work world.

To date, only a handful of longitudinal-sequential studies have been conducted. Yet the design provides researchers with a convenient way to profit from the strengths of both longitudinal and cross-sectional strategies.

BRIEF REVIEW

A variety of research designs are commonly used in the field of human development. In correlational research, information is gathered on existing groups of individuals. Investigators can examine relationships between variables, but they cannot infer cause and effect. Because the experimental design involves random assignment of participants to treatment groups, researchers can find out if an independent variable causes change in a dependent variable. Field and natural experiments permit generalization to everyday life, but they sacrifice rigorous experimental control.

Longitudinal and cross-sectional designs are uniquely suited for studying development. In longitudinal research, participants are studied repeatedly at different ages, an approach that provides information on common patterns as well as individual differences in development and the relationship between early and later events and behaviors. The cross-sectional approach is more efficient because participants differing in age are studied at the same time. However, comparisons are limited to age-group averages. The longitudinal-sequential design permits researchers to reap the benefits of both longitudinal and cross-sectional strategies and to identify cohort effects.

ASK YOURSELF . . .

■ *A researcher compares older adults with chronic heart disease to those who are free of major health problems and finds that the first group scores much lower on mental tests. Should the researcher conclude that heart disease causes declines in intellectual functioning in old age? Why or why not?*

■ *A researcher wants to find out if children who go to day-care centers during the first few years of life do as well in elementary school as those who did not attend day care. Which developmental design, longitudinal or cross-sectional, is appropriate for answering this question? Explain why.*

ETHICS IN LIFESPAN RESEARCH

Research into human behavior creates ethical issues because, unfortunately, the quest for scientific knowledge can sometimes exploit people. For this reason, special guidelines for research have been developed by the federal government, funding agencies, and research-oriented associations, such as the American Psychological Association (1992). Table 1.8 presents a summary of basic research rights drawn from these guidelines. Once you have examined them, read the following research situations, each of which poses a serious ethical dilemma. What precautions do you think should be taken in each instance?

■ In a study of moral development, an investigator wants to assess children's ability to resist temptation by videotaping their behavior without their knowledge. Seven-year-olds are promised an attractive prize for solving some very difficult puzzles. They are also told not to look at a classmate's correct solutions, which are deliberately placed at the back of the room. If the researcher has to tell children ahead of time that cheating is being studied or that their behavior is being closely monitored, he will destroy the purpose of his study.

■ A researcher wants to study the impact of mild daily exercise on the physical and mental health of elderly patients in nursing homes. He consults each resident's doctor to make sure that the exercise routine will not be harmful. But when he seeks the consent of the residents themselves, he finds that many do not comprehend the purpose of the research. And some appear eager to agree simply to relieve feelings of isolation and loneliness.

As these examples indicate, when children or the aged take part in research, the ethical concerns are especially complex. Immaturity makes it difficult or impossible for children to evaluate for themselves what participation in research will mean. And since mental impairment rises with very advanced age, some older adults cannot make voluntary and informed choices. The life circumstances of others make them unusually vulnerable to pressure for participation (Kimmel & Moody, 1990; Society for Research in Child Development, 1993).

Virtually every committee that has worked on developing ethical principles for research has concluded that conflicts arising in research situations often cannot be resolved with simple right or wrong answers (Stanley & Seiber, 1992). The ultimate responsibility for the ethical integrity of research lies with the investigator. However, researchers are advised or, in the case of federally funded research, required to seek advice from others. Special committees exist in colleges, universities, and other institutions for this purpose. These committees weigh the costs of the research to the participant in terms of time, stress, and inconvenience against its value for advancing knowledge

TABLE 1.8

Rights of Research Participants

RESEARCH RIGHT	DESCRIPTION
Protection from harm	Participants have the right to be protected from physical or psychological harm in research. If in doubt about the harmful effects of research, investigators should seek the opinion of others. When harm seems possible, investigators should find other means for obtaining the desired information or abandon the research.
Informed consent	Participants, including children and the elderly, have the right to have explained to them, in language appropriate to their level of understanding, all aspects of the research that may affect their willingness to participate. When children are participants, informed consent of parents as well as others who act on the child's behalf (such as school officials) should be obtained, preferably in writing. Elderly people who are cognitively impaired should be asked to appoint a surrogate decision maker. If they cannot do so, then someone should be named by an ethics committee that includes relatives and professionals who know the person well. All participants have the right to discontinue participation in the research at any time.
Privacy	Participants have the right to concealment of their identity on all information collected in the course of research. They also have this right with respect to written reports and any informal discussions about the research.
Knowledge of results	Participants have the right to be informed of the results of research in language that is appropriate to their level of understanding.
Beneficial treatments	If experimental treatments believed to be beneficial are under investigation, participants in control groups have the right to alternative beneficial treatments if they are available.

Sources: American Psychological Association, 1992; Cassel, 1987, 1988; Society for Research in Child Development, 1993.

and improving conditions of life. If there are any negative implications for the safety and welfare of participants that the worth of the research does not justify, then priority is always given to the research participant.

The ethical principle of *informed consent* requires special interpretation when participants cannot fully appreciate the research goals and activities. Parental consent is meant to protect the safety of children whose ability to decide is not yet mature. For children 7 years and older, their own informed consent should be obtained in addition to parental consent. Around age 7, changes in children's thinking permit them to better understand simple scientific principles and the needs of others. Researchers should respect and enhance these new capacities by providing school-age children with a full explanation of research activities in language that children can understand (Fisher, 1993; Thompson, 1990b).

Most older adults require no more than the usual informed consent procedures. Researchers should not stereotype the elderly as incompetent to decide for themselves about participation (Schaie, 1988). Nevertheless, extra measures must be taken for those who are cognitively impaired or in settings for care of the chronically ill. Sometimes these individuals may agree to participate simply to obtain rewarding social interaction. Yet participation should not be automatically withheld, since it can result in personal as well as scientific benefits (Cassel, 1987). In these instances, potential participants should be asked to appoint a surrogate decision maker. If they cannot do so, then someone should be named by an ethics committee that includes relatives and professionals who know the person well. As an added precaution, if the elderly person is incapable of consenting and the risks of the research are more than minimal, then the study should not be done unless it is likely to result in direct benefits to the participant (Cassel, 1988).

Finally, all ethical guidelines advise that special precautions be taken in the use of deception and concealment, as occurs when researchers observe people from behind one-way mirrors, give them false feedback about their performance, or do not tell them the truth regarding what the research is about. When these kinds of procedures are used, *debriefing,* in which the investigator provides a full account and justification of the activities, occurs after the research session is over. Debriefing should also take place with children, but it does not always work as well. Despite explanations, children may come away from the situation with their belief in the honesty of adults undermined. Ethical standards permit deception if investigators satisfy institutional committees that such practices are necessary. Nevertheless, since deception may have serious emotional consequences for some youngsters, researchers should try to come up with other procedures when children are involved (Cooke, 1982; Ferguson, 1978).

A SPECIAL NOTE TO READERS

With the completion of this overview of theory and research, we are ready to chart the course of development itself. In the following chapters, the story of the life course unfolds in chronological sequence. We begin with a chapter on biological and environmental foundations—the basics of human heredity and how it combines with environmental influences to shape our characteristics and skills. Then we turn to particular time periods of development. As you embark on the study of lifespan development, I wish you a stimulating and rewarding journey.

ASK YOURSELF . . .

- *A researcher wants to recruit preschoolers from low-income families for a study of cognitive development. To ensure enough participants, he decides to offer $50 to each parent who permits his or her child to participate. How might this practice violate research rights?*

- *An investigator decides to conduct an observational study of social interaction in a nursing home. From whom should she seek informed consent?*

SUMMARY

HUMAN DEVELOPMENT AS AN INTERDISCIPLINARY, SCIENTIFIC, AND APPLIED FIELD

What is human development, and what factors stimulated expansion of the field?

■ **Human development** is a field devoted to understanding human constancy and change throughout the lifespan. Research on human development has been stimulated by both scientific curiosity and social pressures to better people's lives.

BASIC THEMES AND ISSUES

Discuss three basic issues on which theories of human development take a stand, and describe the lifespan perspective on development.

■ **Theories** of human development take a stand on three basic issues: (1) How should we describe the developing person? (2) What is the course of human development like? (3) What factors determine, or influence, development?

■ Some theories conceive of the individual as an active, **organismic** being, others as a passive, **mechanistic** being. Sometimes the course of development is viewed as **continuous,** at other times as **discontinuous,** or following a sequence of discrete **stages.** Theories also vary with respect to the **nature–nurture controversy**—whether heredity or environment is most important in determining development. The position they take is related to the issue of **stability versus change** over the life course.

■ The **lifespan perspective** is a balanced view that recognizes great complexity in human change and the factors that underlie it. It assumes that development is multidimensional and multidirectional as well as highly plastic. In addition, it

regards the life course as embedded in multiple contexts. Although these contexts operate in an interconnected fashion, they can be organized into three categories: (1) **age-graded influences** that are predictable in timing and duration; (2) **history-graded influences,** forces unique to a particular era; and (3) **nonnormative influences,** events unique to one or a few individuals.

AN OVERVIEW OF THE LIFESPAN

How can we divide the life course into major periods in order to conveniently describe human development?

■ Although age markers have blurred in adulthood, the life course can be divided into eight periods that serve as important transitions in major theories: (1) the prenatal period; (2) infancy and toddlerhood; (3) early childhood; (4) middle childhood; (5) adolescence; (6) early adulthood; (7) middle adulthood; and (8) late adulthood.

DOMAINS OF DEVELOPMENT AS INTERWOVEN

Cite three domains of development that undergo lifespan change.

■ Within each period, three domains of change—physical development, cognitive development, and emotional and social development—overlap and interact with one another.

HISTORICAL FOUNDATIONS

Describe major historical influences on modern theories of development.

■ Modern theories of human development have roots extending far back into the past. In medieval times, children were regarded as miniature adults, a view called **preformation-**

ism. By the sixteenth century, childhood became a distinct phase of the life course. However, the Puritan belief in original sin led to a harsh philosophy of child rearing. The Enlightenment brought new ideas favoring more humane child treatment. Locke's **tabula rasa** furnished the basis for twentieth-century behaviorism. Rousseau's **noble savage** foreshadowed the concepts of stage and **maturation.**

■ In the eighteenth and early nineteenth centuries, two German philosophers extended conceptions of development through adulthood. Tetens and Carus anticipated many aspects of the contemporary lifespan perspective.

■ Darwin's theory of evolution influenced important twentieth-century theories and inspired scientific child study. In the early twentieth century, Hall and Gesell introduced the **normative approach,** which produced a large body of descriptive facts about children. Binet and Simon constructed the first successful intelligence test, initiating the mental testing movement.

MID-TWENTIETH-CENTURY THEORIES

What theories influenced human development research in the mid-twentieth century?

■ In the 1930s and 1940s, psychiatrists and social workers turned to the **psychoanalytic perspective** for help in treating people's psychological problems. In Freud's **psychosexual theory,** the individual moves through five stages, during which three portions of the personality—**id, ego,** and **superego**—become integrated. Erikson's **psychosocial theory** builds on Freud's theory by emphasizing the development of culturally relevant attitudes and skills and the lifespan nature of development.

- Academic psychology also influenced the study of human development. From **behaviorism** and **social learning theory** came the principles of conditioning and modeling and practical procedures of **applied behavior analysis** to eliminate undesirable behaviors and increase socially acceptable responses.

- In contrast to behaviorism, Piaget's **cognitive-developmental theory** emphasizes an active individual with a mind inhabited by rich structures of knowledge. According to Piaget, children move through five stages, beginning with the baby's sensorimotor action patterns and ending with the elaborate, abstract reasoning system of the adolescent. Piaget's work has stimulated a wealth of research on children's thinking and encouraged educational programs that emphasize discovery learning.

RECENT PERSPECTIVES

Describe four recent theoretical perspectives on human development.

- The field of human development continues to seek new directions. **Information processing** views the mind as a complex, symbol-manipulating system, operating much like a computer. This approach helps investigators achieve a detailed understanding of what individuals of different ages do when faced with tasks and problems.

- Three modern theories place special emphasis on contexts for development. **Ethology** stresses the evolutionary origins and adaptive value of behavior and inspired the **sensitive period** concept. In **ecological systems theory,** nested layers of the environment—**microsystem, mesosystem, exosystem,** and **macrosystem**—are seen as major influences on the developing person. The **chronosystem** represents the

dynamic, ever-changing nature of environmental influences. Vygotsky's **sociocultural theory** has enhanced our understanding of cultural influences, especially on cognitive development. Through cooperative dialogues with expert members of society, children acquire new knowledge and skills.

STUDYING DEVELOPMENT

Describe commonly used research methods in research on human development.

- Common research methods in human development include systematic observation, self-reports, and the clinical or case study approach. **Naturalistic observations** are gathered in children's everyday environments, whereas **structured observations** take place in laboratories, where investigators deliberately set up cues to elicit the behaviors of interest.

- Self-report methods can be flexible and open-ended, like the **clinical interview.** Alternatively, **structured interviews** and questionnaires, which permit efficient administration and scoring, can be given. Investigators use the **clinical method** when they desire an in-depth understanding of a single individual.

Distinguish between correlational and experimental research designs, noting the strengths and limitations of each.

- Two main types of designs are used in all research on human behavior. The **correlational design** examines relationships between variables as they happen to occur, without any intervention. The **correlation coefficient** is often used to measure the association between variables. Correlational studies do not permit statements about cause and effect. However, their use is justified when it is difficult or impossible to control the variables of interest.

- An **experimental design** permits inferences about cause and effect.

Researchers randomly assign participants to treatment conditions and manipulate an **independent variable.** Then they determine what impact this has on a **dependent variable.** To achieve high degrees of control, most experiments are conducted in laboratories, but their findings may not apply to everyday life. Field and natural experiments are strategies used to compare treatments in natural environments.

Describe designs for studying development, noting the strengths and limitations of each.

- Longitudinal and cross-sectional designs are uniquely suited for studying development. The **longitudinal design** permits study of common patterns as well as individual differences in development and the relationship between early and later events and behaviors. The **cross-sectional design** offers an efficient approach to investigating development. However, it is limited to comparisons of age-group averages.

- Findings of longitudinal and cross-sectional research can be distorted by **cohort effects.** To overcome some of the limitations of these designs, investigators sometimes combine the two approaches, as in the **longitudinal-sequential design.**

ETHICS IN LIFESPAN RESEARCH

What special ethical concerns arise in research on human development?

- Research creates ethical issues, since the quest for scientific knowledge can sometimes exploit people. The ethical principle of informed consent requires special safeguards for children and for elderly people who are cognitively impaired or in settings for the care of the chronically ill. The use of deception in research with children is especially risky, since it may undermine their basic faith in the trustworthiness of adults.

IMPORTANT TERMS AND CONCEPTS

human development (p. 5)

theory (p. 5)

organismic theories (p. 6)

mechanistic theories (p. 6)

continuous development (p. 7)

discontinuous development (p. 7)

stage (p. 7)

nature–nurture controversy (p. 7)

stability versus change (p. 7)

lifespan perspective (p. 8)

age-graded influences (p. 9)

history-graded influences (p. 9)

nonnormative influences (p. 9)

preformationism (p. 12)

tabula rasa (p. 12)

noble savage (p. 12)

maturation (p. 12)

normative approach (p. 13)

psychoanalytic perspective (p. 15)

psychosexual theory (p. 15)

id (p. 15)

ego (p. 15)

superego (p. 16)

psychosocial theory (p. 16)

behaviorism (p. 18)

social learning theory (p. 18)

applied behavior analysis (p. 19)

cognitive-developmental theory
  (p. 19)

information processing (p. 21)

ethology (p. 22)

sensitive period (p. 22)

ecological systems theory (p. 23)

microsystem (p. 23)

mesosystem (p. 24)

exosystem (p. 24)

macrosystem (p. 24)

chronosystem (p. 25)

sociocultural theory (p. 25)

naturalistic observation (p. 28)

structured observation (p. 28)

clinical interview (p. 29)

structured interview (p. 29)

clinical method (p. 30)

correlational design (p. 30)

correlation coefficient (p. 31)

experimental design (p. 31)

independent variable (p. 31)

dependent variable (p. 31)

longitudinal design (p. 32)

cohort effects (p. 33)

cross-sectional design (p. 33)

longitudinal-sequential design
  (p. 35)

2

Biological and Environmental Foundations

t's a girl," announces the doctor, who holds up the squalling little creature, while her new parents gaze with amazement at their miraculous creation.

"A girl! We've named her Sarah!" exclaims the proud father to eager relatives waiting by the telephone for word about their new family member.

As we join these parents in thinking about how this wondrous being came into existence and imagining her future, we are struck by many questions. How could this well-formed baby, equipped with everything necessary for life outside the womb, have developed from the union of two tiny cells? What ensures that Sarah will, in due time, roll over, reach for objects, walk, talk, make friends, learn, imagine, and create—just like every other normal member of the human species? Why is she a girl and not a boy, dark-haired rather than blond, calm and cuddly instead of wiry and energetic? What difference will it make that Sarah is given a name and place in one family, community, nation, and culture rather than another?

To answer these questions, this chapter takes a close look at the foundations of development: heredity and environment. Because nature has prepared us for survival, all human beings have many features in common. Yet a brief period of time spent in the company of any person and his or her family reveals that each human being is unique. Take a moment to jot down the most obvious similarities and differences in physical characteristics and behavior for several of your friends and their parents. Did you find that one person shows combined features of both parents, another resembles just one parent, whereas still a third is not like either parent? These directly observable characteristics are called **phenotypes.** They depend in part on the individual's **genotype**—the complex blend of genetic information transmitted from one generation to the next. Yet throughout life, phenotypes are also affected by the person's history of experiences in the environment.

We begin our discussion of development at the moment of conception, an event that establishes the hereditary makeup of the new individual. In the first section of this chapter, we review basic genetic principles that help explain similarities and differences among us in appearance and behavior. Next, we turn to aspects of the environment that play powerful roles throughout the lifespan.

Our discussion will quickly reveal that both nature and nurture are involved in all aspects of development. In fact, some findings and conclusions may surprise you. For example, many people believe that when individuals inherit unfavorable characteristics, not much can be done to help them. Others are convinced that when environments are harmful, the damage done to the developing person can easily be corrected. We will see that neither of these assumptions is true. In the final section of this chapter, we take up the question of how nature and nurture *work together* to shape the course of development.

GENETIC FOUNDATIONS

Each of us is made up of trillions of independent units called cells. Inside every cell is a control center, or nucleus, that contains rodlike structures called **chromosomes,** which store and transmit genetic information. Human chromosomes come in 23 matching pairs (an exception is the XY pair in males, which we will discuss shortly). Each pair member corresponds to the other in size, shape, and genetic functions. One is inherited from the mother and one from the father (see Figure 2.1).

THE GENETIC CODE

Chromosomes are made up of a chemical substance called **deoxyribonucleic acid,** or **DNA.** As Figure 2.2

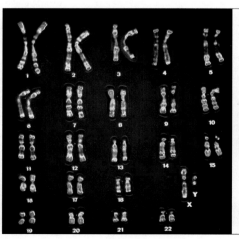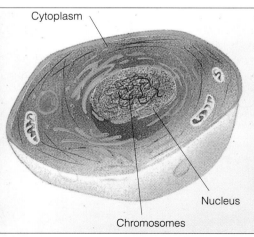

Cytoplasm

Nucleus

Chromosomes

FIGURE 2.1

A photograph, or karyotype, of human chromosomes. The 46 chromosomes shown here were isolated from a body cell, stained, greatly magnified, and arranged in pairs according to decreasing size of the upper arm of each chromosome. Note the twenty-third pair, XY: The cell donor is a male. In females, the twenty-third pair would be XX. *(CNRI/ Science Photo Library/ Photo Researchers)*

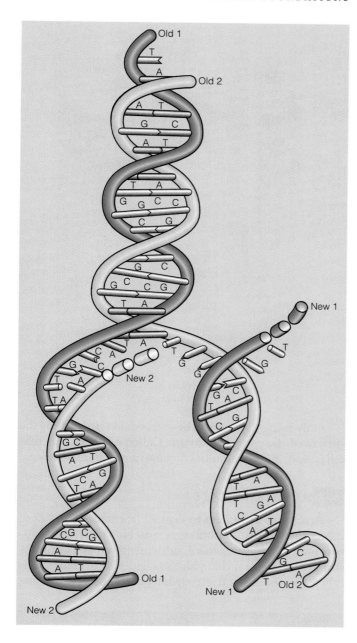

FIGURE 2.2

DNA's ladderlike structure. The pairings of bases across the rungs of the ladder are very specific: adenine (A) always appears with thymine (T), and cytosine (C) always appears with guanine (G). Here, the DNA ladder is shown duplicating by splitting down the middle of its ladder rungs. Each free base picks up a new complementary partner from the area surrounding the cell nucleus.

A unique feature of DNA is that it can duplicate itself through a process called **mitosis.** This special ability permits the one-celled fertilized ovum to develop into a complex human being composed of a great many cells. Refer again to Figure 2.2, and you will see that during mitosis, the chromosomes copy themselves. As a result, each new body cell contains the same number of chromosomes and the identical genetic information.

■ THE SEX CELLS

New individuals are created when two special cells called **gametes,** or sex cells—the sperm and ovum—combine. Gametes contain only 23 chromosomes, half as many as a regular body cell. They are formed through a special process of cell division called **meiosis.** In meiosis, the chromosomes pair up and exchange segments, so that genes from one are replaced by genes from another. Then chance determines which member of each pair will gather with others and end up in the same gamete. These events ensure that no two gametes will ever be alike. Meiosis explains why siblings differ from each other, even though they have features in common, since their genotypes come from the same pool of parental genes.

In the male, four sperm are produced each time meiosis occurs. Also, the cells from which sperm arise are produced continuously throughout life. For this reason, a healthy man can father a child at any age after sexual maturity. In the female, gamete production is much more limited. Each cell division produces just one ovum. Also, the female is born with all her ova already present in her ovaries, and she can only bear children for three to four decades. Still, there are plenty of female sex cells. About 1 to 2 million are present at birth, 40,000 remain at adolescence, and approximately 350 to 450 will mature during a woman's childbearing years (Moore & Persaud, 1993).

■ CONCEPTION

The human sperm and ovum are uniquely suited for the task of reproduction. The ovum is a tiny sphere, measuring 1/175 inch in diameter, about the size of the period at the end of this sentence. But in its microscopic world, it is a giant—the largest cell in the human body. The ovum's size makes it a perfect target for the much smaller sperm, which measure only 1/500 of an inch.

shows, DNA is a long, double-stranded molecule that looks like a twisted ladder. Each rung of the ladder consists of a specific pair of chemical substances called bases, joined together between the two sides. It is this sequence of bases that provides genetic instructions. A **gene** is a segment of DNA along the length of the chromosome. Genes can be of different lengths—perhaps 100 to several thousand ladder-rungs long. Altogether, about 100,000 genes lie along the human chromosomes. Genes accomplish their task by sending instructions for making a rich assortment of proteins to the cytoplasm, the area surrounding the cell nucleus. Proteins, which trigger chemical reactions throughout the body, are the biological foundation from which our characteristics are built.

FIGURE 2.3

Female reproductive organs.
An ovum is released from the ovary
and fertilized high in the fallopian
tube. As the zygote begins to duplicate, it travels toward the uterus and
burrows into the uterine lining.

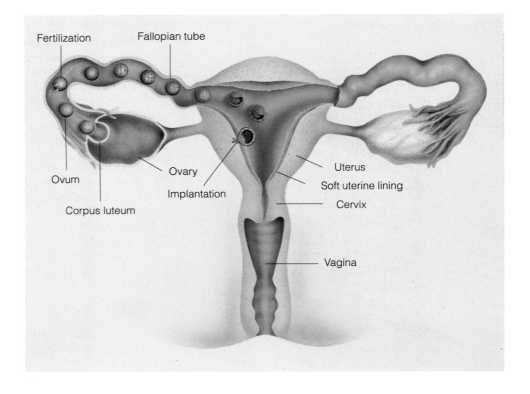

About once every 28 days, in the middle of a woman's menstrual cycle, an ovum bursts from one of her *ovaries* and is drawn into one of two *fallopian tubes*—long, thin structures that lead to the hollow, soft-lined uterus (see Figure 2.3). While the ovum is traveling, the spot on the ovary from which it was released, now called the *corpus luteum,* begins to secrete hormones that prepare the lining of the uterus to receive a fertilized ovum. If pregnancy does not occur, the corpus luteum shrinks, and the lining of the uterus is discarded in 2 weeks with menstruation.

The male produces sperm in vast numbers—an average of 300 million a day—in the *testes,* two glands located in the *scrotum,* sacs that lie just behind the penis (see Figure 2.4). In the final process of maturation, each sperm develops a tail that permits it to swim long distances. During sexual intercourse, about 360 million sperm move through the *vas deferens,* a thin tube in which they are bathed in a protective fluid called *semen.* After semen is ejaculated from the penis into the woman's vagina, the sperm begin to swim upstream in the female reproductive tract, through the *cervix* (opening of the uterus) and into the fallopian tube, where fertilization usually takes place. The journey is difficult, and many sperm die. Only 300 to 500 reach the ovum, if one happens to be present. Sperm live for up to 6 days and can lie in wait for the ovum, which survives for only 1 day after being released into the fallopian tube. However, most conceptions result from intercourse during a 3-day period—on the day of or the 2 days preceding ovulation (Wilcox, Weinberg, & Baird, 1995).

Only a single sperm will be successful in penetrating the surface of the enormous ovum. Called a **zygote,** this first cell is ready to begin multiplying into a new human being.

MALE OR FEMALE?

Return for a moment to Figure 2.1, and note that 22 of the 23 pairs of chromosomes are matching pairs, called **autosomes.** The twenty-third pair consists of **sex chromosomes.** In females, this pair is called XX; in males, it is called XY. The X is a relatively long chromosome, whereas

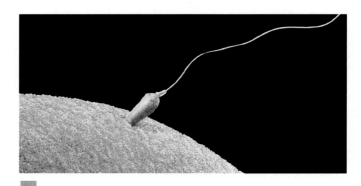

In this photograph of fertilization taken with the aid of a powerful microscope, a tiny sperm completes its journey and starts to penetrate the surface of an enormous-looking ovum, the largest cell in the human body. (Francis Leroy, Biocosmos/Science Photo Library/Photo Researchers)

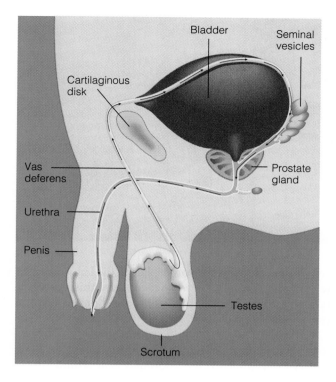

Male reproductive organs. Sperm produced in the testes move through the vas deferens, where they are mixed with semen from the prostate gland and seminal vesicles. Then they are released through the urethra in the penis.

During their early years, children of single births are often healthier and develop more rapidly than do twins (Moilanen, 1989). Jeannie and Jason were born early (as most twins are)—3 weeks before Ruth's due date. Like other premature infants (as we will see in Chapter 3), they required special care after birth. When the twins came home from the hospital, Ruth and Peter had to divide time between them, and neither baby got quite as much attention as the average single infant. As a result, Jeannie and Jason walked and talked several months later than other children their age, although both caught up in development by middle childhood.

PATTERNS OF GENETIC INHERITANCE

Jeannie has her parents' dark, straight hair, whereas Jason is curly-haired and blond. Patterns of genetic inheritance—the way genes from each parent interact—explain these outcomes. Earlier we indicated that except for the XY pair in males, all chromosomes come in matching pairs. Two forms of each gene occur at the same place on the autosomes, one inherited from the mother and one from the father. If the genes from both parents are alike, the child is **homozygous** and will display the inherited trait. If they are different, then the child is **heterozygous,** and relationships between the genes determine the trait that will appear.

the Y is short and carries little genetic material. When gametes are formed in males, the X and Y chromosomes separate into different sperm cells. In females, all gametes carry an X chromosome. Therefore, the sex of the new organism is determined by whether an X-bearing or a Y-bearing sperm fertilizes the ovum.

MULTIPLE BIRTHS

Ruth and Peter, a couple I know well, tried for several years to have a child without success. Ruth's doctor finally prescribed a fertility drug, and twins—Jeannie and Jason—were born. Jeannie and Jason are **fraternal,** or **dizygotic, twins,** the most common type of multiple birth. The drug that Ruth took caused two ova to be released from her ovaries, and both were fertilized. Therefore, Jeannie and Jason are genetically no more alike than ordinary siblings. Fertility drugs are only one cause of fraternal twinning (and occasionally more offspring). As Table 2.1 shows, other genetic and environmental factors are also involved.

There is another way that twins can be created. Sometimes a zygote that has started to duplicate separates into two clusters of cells that develop into two individuals. These are called **identical,** or **monozygotic, twins** because they have the same genetic makeup. The frequency of identical twins is unrelated to the factors listed in Table 2.1. It is about the same around the world—4 out of every 1,000 births. Scientists do not know what causes this type of twinning in humans. In animals, it can be produced by temperature changes, variation in oxygen levels, and late fertilization of the ovum.

TABLE 2.1

Maternal Factors Linked to Fraternal Twinning

FACTOR	DESCRIPTION
Ethnicity	About 8 per 1,000 births among whites, 12 to 16 per 1,000 among blacks, and 4 per 1,000 among Asians.
Age	Rises with maternal age, peaking at 35 years, and then rapidly falls.
Nutrition	Occurs less often among women with poor diets; occurs more often among women who are tall and overweight or of normal weight as opposed to slight body build.
Number of births	Chances increase with each additional birth.
Exposure to fertility drugs	Treatment of infertility with hormones increases the likelihood of multiple fraternal births, from twins to quintuplets.

Source: Cohen, 1984; Little & Thompson, 1988.

These identical, or monozygotic, twins were created when a duplicating zygote separated into two clusters of cells, and two individuals with the same genetic makeup developed. Identical twins look alike, and as we will see later in this chapter, tend to resemble each other in a variety of psychological characteristics. *(Porter/The Image Works)*

DOMINANT–RECESSIVE RELATIONSHIPS. In many heterozygous pairings, only one gene affects the child's characteristics. It is called *dominant;* the second gene, which has no effect, is called *recessive.* Hair color is an example of **dominant–recessive inheritance.** The gene for dark hair is dominant (we can represent it with a capital *D*), whereas the one for blond hair is recessive (symbolized by a lowercase *b*). Children who inherit either a homozygous pair of dominant genes (*DD*) or a heterozygous pair (Db) will be dark-haired, even though their genotype is different. Blond hair (like Jason's) can result only from having two recessive genes (*bb*). Still, heterozygous individuals with just one recessive gene (*Db*) can pass on that trait to their offspring. Therefore, they are called **carriers** of the trait.

In dominant–recessive inheritance, if we know the genetic makeup of the parents, we can predict the percentage of children in a family who are likely to display a trait or be carriers of it. Figure 2.5 shows the pattern of inheritance for hair color. Note that for Jason to be blond, both Peter and Ruth must be carriers of a recessive gene (*b*). The figure also indicates that if Peter and Ruth decide to have more children, most are likely to be dark-haired like Jeannie.

Some human characteristics and disorders that follow the rules of dominant–recessive inheritance are given in Tables 2.2 and 2.3 (see pages 50–51). As you can see, many disabilities and diseases are the product of recessive genes. One of the most frequently occurring recessive disorders is *phenylketonuria,* or *PKU.* PKU is an especially good example, since it shows that inheriting unfavorable genes does not always mean that the condition cannot be treated.

PKU affects the way the body breaks down proteins contained in many foods. Infants born with two recessive genes lack an enzyme that converts one of the basic amino acids that make up proteins (phenylalanine) into a by-product essential for body functioning (tyrosine). Without this enzyme, phenylalanine quickly builds to toxic levels that damage the central nervous system. By 1 year, infants with PKU are permanently retarded. All U.S. states require that each newborn be given a blood test for PKU. If the disease is found, treatment involves placing the baby on a diet low in phenylalanine. Children who receive this treatment attain an average level of intelligence and have a normal lifespan (Mazzocco et al., 1994).

As Table 2.3 suggests, only rarely are serious diseases due to dominant genes. Think about why this is the case. Children who inherited the dominant gene would always develop the disorder. They would seldom live long enough

---

TABLE 2.2

Examples of Dominant and Recessive Characteristics

DOMINANT	RECESSIVE
Dark hair	Blond hair
Normal hair	Pattern baldness
Curly hair	Straight hair
Nonred hair	Red hair
Facial dimples	No dimples
Normal hearing	Some forms of deafness
Normal vision	Nearsightedness
Farsightedness	Normal vision
Normal vision	Congenital eye cataracts
Normal color vision	Red-green color blindness
Normally pigmented skin	Albinism
Double-jointedness	Normal joints
Type A blood	Type O blood
Type B blood	Type O blood
Rh-positive blood	Rh-negative blood

Note. Many normal characteristics that were previously thought to be due to dominant–recessive inheritance, such as eye color, are now regarded as due to multiple genes. For the characteristics listed here, there still seems to be fairly common agreement that the simple dominant–recessive relationship holds.
Source: McKusick, 1995.

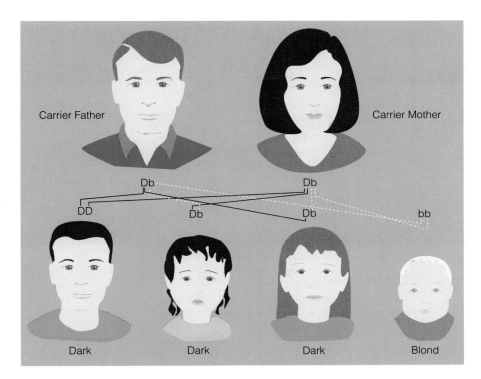

Carrier Father

Carrier Mother

Db   Db

DD   Db   Db   bb

Dark   Dark   Dark   Blond

FIGURE 2.5

Dominant–recessive mode of inheritance as illustrated by hair color. By looking at the possible combinations of the parents' genes, we can predict that 25 percent of their children are likely to inherit two dominant genes for dark hair; 50 percent are likely to receive one dominant and one recessive gene, resulting in dark hair; and 25 percent are likely to receive two recessive genes for blond hair.

to reproduce, and the harmful dominant gene would be eliminated from the family's heredity in a single generation. Some dominant disorders, however, do persist. One of them is *Huntington disease,* a condition in which the central nervous system degenerates. Why has this disorder endured in some families? The reason is that its symptoms usually do not appear until age 35 or later, after the person has passed the dominant gene to his or her children.

■ CODOMINANCE. In some heterozygous circumstances, the dominant–recessive relationship does not hold completely. Instead, we see **codominance,** a pattern of inheritance in which both genes influence the person's characteristics.

The *sickle cell trait,* a heterozygous condition present in many black Africans, provides an example. *Sickle cell anemia* (see Table 2.3) occurs in full form when a child inherits two recessive genes. They cause the usually round red blood cells to become sickle shaped, especially under low oxygen conditions. The sickled cells clog the blood vessels and block the flow of blood. Individuals who have the disorder suffer severe attacks involving intense pain, swelling, and tissue damage. They generally die in the first 20 years of life; few live past age 40. Heterozygous individuals are protected from the disease under most circumstances. However, when they experience oxygen deprivation—for example, at high altitudes or after intense physical exercise—the single recessive gene asserts itself, and a temporary, mild form of the illness occurs (Sullivan, 1987).

The sickle cell gene is common among black Africans for a special reason. Carriers of it are more resistant to

malaria than are individuals with two genes for normal red blood cells. In Africa, where malaria is common, these carriers survived and reproduced more frequently than others, leading the gene to be maintained in the black population.

■ MUTATION AND UNFAVORABLE GENES. At this point, you may be wondering, How are harmful genes created in the first place? The answer is **mutation,** a sudden but permanent change in a segment of DNA. A mutation may affect only one or two genes, or it may involve many genes, as is the case for the chromosomal disorders we will discuss shortly. Some mutations occur spontaneously, simply by chance. Others are caused by a wide variety of hazardous environmental agents that enter our food supply or are present in the air we breathe.

For many years, ionizing radiation has been known to cause mutations. Women who receive repeated doses of radiation before conception are more likely to miscarry and give birth to children with hereditary defects (Zhang, Cai, & Lee, 1992). Genetic abnormalities are also higher when fathers are exposed to radiation in their occupations (Gardner et al., 1990). Does this mean that routine chest and dental X-rays are dangerous to future generations? Research indicates that infrequent and mild exposure to radiation does not cause genetic damage. Instead, high doses over a long period of time appear to be required.

■ X-LINKED INHERITANCE. Males and females have an equal chance of inheriting recessive disorders carried on the autosomes, such as PKU and sickle cell anemia. But when a harmful gene is carried on the X chromosome,

TABLE 2.3

Examples of Dominant and Recessive Diseases

DISEASE	DESCRIPTION	MODE OF INHERITANCE	INCIDENCE	TREATMENT	PRENATAL DIAGNOSIS	CARRIER IDENTIFICATION*
Autosomal Diseases						
Cystic fibrosis	Lungs, liver, and pancreas secrete large amounts of thick mucus, leading to breathing and digestive difficulties.	Recessive	1 in 2,000 to 2,500 Caucasian births, 1 in 16,000 African-American births	Bronchial drainage; prompt treatment of respiratory infections; dietary management. Advances in medical care allow survival with good life quality into adulthood.	Yes	Yes
Phenylketonuria (PKU)	Inability to neutralize the harmful amino acid phenylalanine, contained in many proteins, causes severe central nervous system damage in the first year of life.	Recessive	1 in 8,000 births	Placing the child on a special diet results in average intelligence and normal life span. Subtle difficulties with planning and problem solving are often present.	Yes	Yes
Sickle cell anemia	Abnormal sickling of red blood cells causes oxygen deprivation, pain, swelling, and tissue damage. Anemia and susceptibility to infections, especially pneumonia, occur.	Recessive	1 in 500 African-American births	Blood transfusions, painkillers, prompt treatment of infections. No known cure; 50 percent die by age 20.	Yes	Yes
Tay-Sachs disease	Central nervous system degeneration, with onset at about 6 months, leads to poor muscle tone, blindness, deafness, and convulsions.	Recessive	1 in 3,600 births to Jews of European descent	None. Death by 3 to 4 years of age.	Yes	Yes

X-linked inheritance applies. Males are more likely to be affected because their sex chromosomes do not match. In females, any recessive gene on one X has a good chance of being suppressed by a dominant gene on the other X. But the Y chromosome is only about one third as long and therefore lacks many corresponding genes to overcome those on the X.

Return to Tables 2.2 and 2.3 and review the disorders that are X-linked. A well-known example is red–green color blindness, a condition in which individuals cannot tell the difference between shades of red and green. Another is *hemophilia,* a disease in which the blood fails to clot normally. As Figure 2.6 on page 52 reveals, both of these defects have a much greater likelihood of inheritance by male children whose mothers carry the abnormal gene.

■ GENETIC IMPRINTING. Over 1,000 human characteristics follow the rules of dominant–recessive and codominant inheritance (McKusick, 1995). In these cases, regardless of which parent contributes a gene to the new individual, the gene responds in the same way. Geneticists, however, have identified some exceptions governed by a newly discovered mode of inheritance. In **genetic imprinting,** genes are *imprinted,* or chemically marked, in such a way that one pair member (either the mother's or the father's) is activated, regardless of its makeup. The imprint is often temporary: it may be erased in the next generation, and it may not occur in all individuals (Cassidy, 1995).

Genetic imprinting helps us understand the confusion in genetic inheritance for some disorders. For example, children are more likely to develop diabetes if their father,

TABLE 2.3 CONTINUED

Examples of Dominant and Recessive Diseases

DISEASE	DESCRIPTION	MODE OF INHERITANCE	INCIDENCE	TREATMENT	PRENATAL DIAGNOSIS	CARRIER IDENTIFICATION[a]
Autosomal Diseases (continued)						
Huntington disease	Central nervous system degeneration leads to muscular coordination difficulties, mental deterioration, and personality changes. Symptoms usually do not appear until age 35 or later.	Dominant	1 in 18,000 to 25,000 American births	None. Death occurs 10 to 20 years after symptom onset.	Yes	Not applicable
X-Linked Diseases						
Duchenne muscular dystrophy	Degenerative muscle disease. Abnormal gait, loss of ability to walk between 7 and 13 years of age.	Recessive	1 in 3,000 to 5,000 male births	None. Death from respiratory infection or weakening of the heart muscle usually occurs in adolescence.	Yes	Yes
Hemophilia	Blood fails to clot normally. Can lead to severe internal bleeding and tissue damage.	Recessive	1 in 4,000 to 7,000 male births	Blood transfusions. Safety precautions to prevent injury.	Yes	Yes
Diabetes insipidus	A form of diabetes present at birth caused by insufficient production of the hormone vasopressin. Results in excessive thirst and urination. Dehydration can cause central nervous system damage.	Recessive	1 in 2,500 male births	Hormone replacement.	Yes	No

[a]Carrier status detectable in prospective parents through blood test or genetic analyses.

Sources: Behrman & Vaughan, 1987; Cohen, 1984; Gilfillan et al., 1992; Martin, 1987; McKusick, 1995; Simpson & Harding, 1993.

rather than their mother, suffers from it. And people with asthma or hay fever tend to have mothers, not fathers, with the illness. Imprinting may also explain why Huntington disease, when inherited from the father, tends to emerge at an earlier age and progress more rapidly (Day, 1993; Reik, 1992).

In these examples, genetic imprinting affects traits carried on the autosomes. It can also operate on the sex chromosomes, as *fragile X syndrome* reveals. In this disorder, an abnormal sequence of DNA bases occurs in a special spot on the X chromosome, damaging a particular gene. Fragile X syndrome is a common inherited cause of moderate mental retardation. It has also been linked to 2 to 3 percent of cases of infantile autism, a serious emotional disorder of early childhood involving bizarre, self-stimulating behavior and delayed or absent language and communication. Recent evidence indicates that the defective gene at the fragile site is expressed only when it is passed from mother to child (Rose, 1995; Thapar et al., 1994).

■ POLYGENIC INHERITANCE. So far, we have discussed patterns of inheritance in which people either display a particular trait or do not. These cut-and-dried individual differences are much easier to trace to their genetic origins than characteristics that vary continuously among people, such as height, weight, intelligence, and personality. These

Test Bank Items 2.39 through 2.42

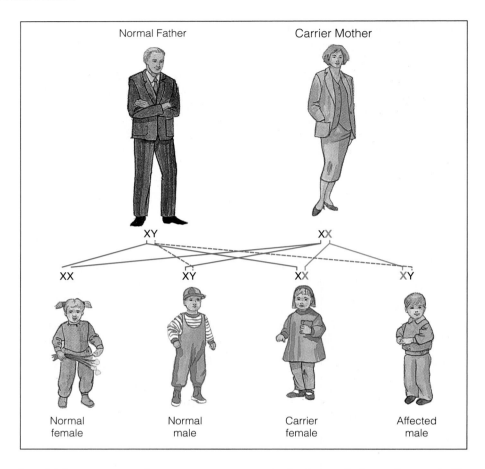

X-linked inheritance. In the example shown here, the gene on the father's X chromosome is normal. The mother has one normal and one abnormal recessive gene on her X chromosomes. By looking at the possible combinations of the parents' genes, we can predict that 50 percent of male children will have the disorder and 50 percent of female children will be carriers of it.

traits are due to **polygenic inheritance,** in which many genes determine the characteristic in question. Polygenic inheritance is complex, and much about it is still unknown. In the final section of this chapter, we will discuss ways that have been used to infer the influence of heredity on human attributes when knowledge of precise patterns of inheritance is unavailable.

CHROMOSOMAL ABNORMALITIES

Besides inheriting harmful recessive genes, abnormalities of the chromosomes are a major cause of serious developmental problems. Most chromosomal defects are the result of mistakes during meiosis when the ovum and sperm are formed. A chromosome pair does not separate properly, or part of a chromosome breaks off. Since these errors involve large amounts of DNA, they usually produce disorders with many physical and mental symptoms.

DOWN SYNDROME

The most common chromosomal disorder, occurring in 1 out of every 800 live births, is *Down syndrome.* In most cases, it results from a failure of the twenty-first pair of chromosomes to separate during meiosis, so the new individual inherits three of these chromosomes rather than the normal two. In other less frequent forms, an extra broken piece of a twenty-first chromosome is present. Or an error occurs during the early stages of mitosis, causing some but not all body cells to have the defective chromosomal makeup (called a *mosaic* pattern). In these instances, since less genetic material is involved, symptoms of the disorder are less extreme (Epstein, 1993; Fishler & Koch, 1991).

The behavioral consequences of Down syndrome include mental retardation, speech problems, limited vocabulary, and slow motor development. Affected individuals also have distinct physical features—a short, stocky build, a flattened face, a protruding tongue, almond-shaped eyes, and an unusual crease running across the palm of the hand. In addition, infants with Down syndrome are often born with eye cataracts and heart and intestinal defects. Because of medical advances, fewer individuals with Down syndrome die in childhood today than in the past. Most live until middle adulthood (Baird & Sadovnick, 1987).

The risk of a Down syndrome baby rises dramatically with maternal age, from 1 in 1,900 births at age 20, to 1 in 300 at age 35, to 1 in 30 at age 45 (Halliday et al., 1995). Why is this so? Geneticists believe that the ova, present in the woman's body since her own prenatal period, weaken over time because of the aging process or increased exposure to harmful environmental agents. As a result, chromosomes do not separate properly during meiosis (Antonarakis, 1992). The mother's gamete, however, is not always the cause of a Down syndrome child. In about 20 percent of cases, the extra genetic material originates with the father. However, Down syndrome and other chromosomal abnormalities are not related to advanced paternal age. In these instances, the mutation occurs for other unknown reasons (Phillips & Elias, 1993).

ABNORMALITIES OF THE SEX CHROMOSOMES

Disorders of the autosomes other than Down syndrome usually disrupt development so severely that miscarriage occurs. When such babies are born, they rarely survive beyond early childhood. In contrast, abnormalities of the sex chromosomes usually lead to fewer problems. In fact, sex chromosome disorders are often not recognized until adolescence when, in some of the deviations, puberty is delayed. The most common problems involve the presence of an extra chromosome (either X or Y) or the absence of one X chromosome in females.

A variety of myths about individuals with sex chromosome disorders exist. For example, males with *XYY syndrome* are not more aggressive and antisocial than XY males, as was once assumed. Also, it is widely believed that children with sex chromosome disorders are retarded. Yet most are not. The intelligence of XYY syndrome boys is similar to that of normal children (Netley, 1986; Stewart, 1982). And the intellectual problems of people with other sex chromosome abnormalities are usually very specific. Verbal difficulties (for example, with reading and vocabulary) are common among girls with triple X syndrome (XXX) and boys with Klinefelter's syndrome (XXY), both of whom inherit an extra X chromosome. In contrast, girls with Turner syndrome (XO), who are missing an X, have trouble with spatial relationships—for example, drawing pictures, telling right from left, and finding their way around the neighborhood (Hall et al., 1982; Netley, 1986; Pennington et al., 1982). These findings tell us that adding to or subtracting from the usual number of X chromosomes results in particular intellectual deficits. At present, geneticists do not know the reason why.

BRIEF REVIEW

Each individual is made up of trillions of cells. Inside each cell nucleus are chromosomes, which contain a chemical molecule called DNA. Genes are segments of DNA that determine our species and unique characteristics. Gametes, or sex cells, are formed through a special process of cell division called meiosis that halves the usual number of

The facial features and short, stocky build of the boy on the left are typical of Down syndrome. Although his intellectual development is impaired, he is doing well because he is growing up in a stimulating home where his special needs are met and he is loved and accepted. (Stephen Frisch/Stock Boston)

chromosomes. A different combination of sex chromosomes establishes whether the organism is male or female. Two types of twins are possible. Fraternal twins are genetically no more alike than other siblings, whereas identical twins have the same genetic makeup. Four patterns of inheritance—dominant–recessive, codominant, X-linked, and genetic imprinting—underlie many traits and disorders. Continuous characteristics, such as intelligence and personality, result from polygenic inheritance, which involves many genes. Chromosomal abnormalities occur when meiosis is disrupted during gamete formation.

ASK YOURSELF . . .

- *Two brothers, Todd and Blake, look strikingly different. Todd is tall and thin; Blake is short and stocky. What events taking place during meiosis contributed to these differences?*

- *Gilbert and Jan are planning to have children. Gilbert's genetic makeup is homozygous for dark hair; Jan's is heterozygous for blond hair. What color is Gilbert's hair? How about Jan's? What proportion of their children are likely to be dark-haired?*

- *Ashley and Harold both carry the defective gene for fragile X syndrome. Explain why Ashley's child inherited the disorder but Harold's did not.*

REPRODUCTIVE CHOICES

Two years after they were married, Ted and Marianne gave birth to their first child. Kendra appeared to be a healthy and lively infant, but by 4 months her growth slowed. Diagnosed as having Tay-Sachs disease (see Table 2.3), Kendra died at 2 years of age. Ted and Marianne were devastated by Kendra's death. Although they did not want to bring another infant into the world who would endure such suffering, they badly wanted to have a child. They began to avoid family get-togethers where little nieces and nephews were constant reminders of the void in their lives.

In the past, many couples with genetic disorders in their families chose not to bear a child at all rather than risk having an abnormal baby. Today, genetic counseling and prenatal diagnosis help people make informed decisions about conceiving, carrying a pregnancy to term, or adopting a child.

GENETIC COUNSELING

Genetic counseling helps couples assess their chances of giving birth to a baby with a hereditary disorder. Individuals likely to seek it are those who have had difficulties bearing children, such as repeated miscarriages, or who know that genetic problems exist in their families (Clarke, 1994). In addition, women who delay childbearing past age 35 are candidates for genetic counseling. After this time,

the overall rate of chromosomal abnormalities rises sharply, from 1 in every 100 to as many as 1 in every 3 pregnancies at age 48 (Hook, 1988).

If a family history of mental retardation, physical defects, or inherited diseases exists, the genetic counselor interviews the couple and prepares a *pedigree,* a picture of the family tree in which affected relatives are identified. The pedigree is used to estimate the likelihood that parents will have an abnormal child, using the same genetic principles we discussed earlier in this chapter. In the case of many disorders, blood tests or genetic analyses can reveal whether the parent is a carrier of the harmful gene. Turn back to pages 50–51, and you will see that carrier detection is possible for most of the diseases listed in Table 2.3. A carrier test has been developed for fragile X syndrome as well (Ryynänen et al., 1995).

When all the relevant information is in, the genetic counselor helps people consider appropriate options. These include "taking a chance" and conceiving, choosing from among a variety of reproductive technologies (see the Social Issues box on pages 56–57), or adopting a child.

PRENATAL DIAGNOSIS AND FETAL MEDICINE

If couples who might bear an abnormal child decide to conceive, several **prenatal diagnostic methods**—medical procedures that permit detection of problems before birth—are available (see Table 2.4). Women of advanced

TABLE 2.4

Prenatal Diagnostic Methods

METHOD	DESCRIPTION
Amniocentesis	The most widely used technique. A hollow needle is inserted through the abdominal wall to obtain a sample of fluid in the uterus. Cells are examined for genetic defects. Can be performed by 11 to 14 weeks after conception; 1 to 2 more weeks are required for test results. Small risk of miscarriage.
Chorionic villus sampling	A procedure that can be used if results are desired or needed very early in pregnancy. A thin tube is inserted into the uterus through the vagina or a hollow needle is inserted through the abdominal wall. A small plug of tissue is removed from the end of one or more chorionic villi, the hairlike projections on the membrane surrounding the developing organism. Cells are examined for genetic defects. Can be performed 6 to 8 weeks after conception, and results are available within 24 hours. Entails a slightly greater risk of miscarriage than does amniocentesis. Also associated with a small risk of limb deformities, which increases the earlier the procedure is performed.
Ultrasound	High-frequency sound waves are beamed at the uterus; their reflection is translated into a picture on a video screen that reveals the size, shape, and placement of the fetus. By itself, permits assessment of fetal age, detection of multiple pregnancies, and identification of gross physical defects. Also used to guide amniocentesis, chorionic villus sampling, and fetoscopy (see below). When used five or more times, may increase the chances of low birth weight.
Fetoscopy	A small tube with a light source at one end is inserted into the uterus to inspect the fetus for defects of the limbs and face. Also allows a sample of fetal blood to be obtained, permitting diagnosis of such disorders as hemophilia and sickle cell anemia as well as neural defects (see below). Usually performed between 15 to 18 weeks after conception, although can be done as early as 5 weeks. Entails some risk of miscarriage.
Maternal blood analysis	By the second month of pregnancy, some of the developing organism's cells enter the maternal bloodstream. An elevated level of alpha-fetoprotein may indicate kidney disease, abnormal closure of the esophagus, or neural defects, such as anencephaly (absence of most of the brain) and spina bifida (bulging of the spinal cord from the spinal column).

Sources: Benacerraf et al., 1988; Burton, 1992; Canick & Saller, 1993; Holmes, 1993; Quintero, Puder, & Cotton, 1993; Shurtleff & Lemire, 1995.

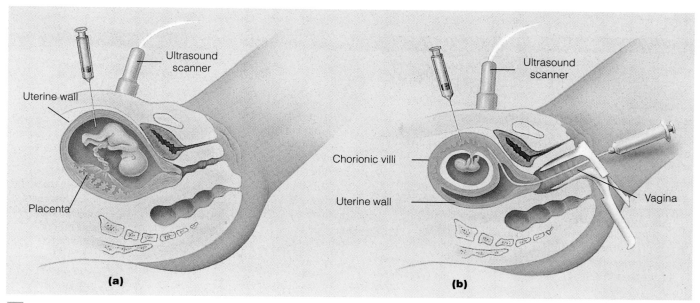

FIGURE 2.7

Amniocentesis and chorionic villus sampling. Today, more than 250 defects and diseases can be detected before birth using these procedures. (a) In amniocentesis, a hollow needle is inserted through the abdominal wall into the uterus. Fluid is withdrawn and fetal cells are cultured, a process that takes about 3 weeks. (b) Chorionic villus sampling can be performed much earlier in pregnancy, at 6 to 8 weeks after conception, and results are available within 24 hours. Two approaches to obtaining a sample of chorionic villi are shown: inserting a thin tube through the vagina into the uterus and inserting a needle through the abdominal wall. In both amniocentesis and chorionic villus sampling, an ultrasound scanner is used for guidance. *(From K. L. Moore & T. V. N. Persaud, 1993,* Before We Are Born, *4th ed., Philadelphia: Saunders, p. 89. Adapted by permission of the publisher and author.)*

maternal age are prime candidates for *amniocentesis* or *chorionic villus sampling* (see Figure 2.7). Except for *ultrasound* and *maternal blood analysis,* prenatal diagnosis should not be used routinely, since other methods have some chance of injuring the developing organism.

Improvements in prenatal diagnosis have led to advances in fetal medicine. Today, some medical problems are being treated before birth. For example, by inserting a needle into the uterus, drugs can be delivered to the fetus. Prenatal surgery has been performed to repair such problems as urinary tract obstructions and neural defects. Nevertheless, these practices remain controversial. Although some babies are saved, the techniques frequently result in complications or miscarriage. Yet parents may be willing to try almost any option, even if there is only a slim chance of success. Currently, the medical profession is struggling with how to help parents make informed decisions about fetal surgery. One suggestion is that the advice of an independent counselor be provided—a doctor or nurse who understands the risks but is not involved in doing research on or performing the procedure (Harrison, 1993).

Advances in *genetic engineering* also offer hope for correcting hereditary defects. Genetic repair of the prenatal organism, once inconceivable, is a goal of today's genetic engineers. Researchers are mapping human chromosomes, finding the precise location of genes for specific traits and cloning (copying) these genes using chemical techniques in the laboratory. Today, DNA markers have been identified for over 16,000 human characteristics and hundreds of inherited diseases, including cystic fibrosis, Huntington disease, Duchenne muscular dystrophy, and some forms of cancer (Cooperative Human Linkage Center, 1996; Cox et al., 1994). Scientists are using this information to identify abnormal conditions with greater accuracy before birth. Eventually, *gene splicing* (replacing a harmful gene with a good one in the early zygote or in cells in the affected part of the body) may permit many defects to be corrected.

ADOPTION

Many parents who cannot have children or who are likely to pass along a genetic disorder decide to adopt. Adoption agencies try to find parents of the same ethnic and religious background as the child. Where possible, they also try to choose parents who are the same age as most natural parents. Because the availability of healthy babies has declined (since fewer young unwed mothers give up their babies than in the past), more people are adopting from foreign countries or taking children who are older or who have developmental problems.

Selection of adoptive parents is important, since sometimes adoptive relationships do not work out. The risk of adoption failure is greatest for children adopted at older ages and children with disabilities, but it is not high. Over

SOCIAL ISSUES

THE PROS AND CONS OF REPRODUCTIVE TECHNOLOGIES

Some couples decide not to risk pregnancy because of a history of genetic disease. And many others—in fact, one-sixth of all couples who try to conceive—discover that they are infertile. Today, increasing numbers of individuals are turning to alternative methods of conception—technologies that, although fulfilling the wish of parenthood, have become the subject of heated debate.

DONOR INSEMINATION AND IN-VITRO FERTILIZATION. For several decades, *donor insemination*—injection of sperm from an anonymous man into a woman—has been used to overcome male reproductive difficulties. In recent years, it has also permitted women without a heterosexual partner to bear children. In the United States alone, 30,000 children are conceived through donor insemination each year (Swanson, 1993).

In vitro fertilization is another reproductive technology that has become increasingly common. Since the first "test tube" baby was born in England in 1978, more than 10,000 infants have been created this way (Ramsay, 1995). With in vitro fertilization, hormones are given to a woman, stimulating ripening of several ova. These are removed surgically and placed in a dish of nutrients, to which sperm are added. Once an ovum is fertilized and begins to duplicate into several cells, it is injected into the mother's uterus, where, hopefully, it will implant and develop.

In vitro fertilization is usually used to treat women whose fallopian tubes are permanently damaged, and it is successful for 20 percent of those who try it. These results have been encouraging enough that the technique has been expanded. By mixing and matching gametes, pregnancies can be brought about when either or both partners have a reproductive problem. In cases where couples might transmit harmful genes, single cells can be plucked from the duplicating zygote and screened for hereditary defects. Fertilized ova can even be frozen and stored in embryo banks for use at some future time, thereby guaranteeing healthy zygotes to older women (Edwards, 1991).

Clearly donor insemination and in vitro fertilization have many benefits. Nevertheless, serious questions have arisen about their use. Many states have no legal guidelines for these procedures. As a result, donors are not always screened for genetic or sexually transmitted diseases. In addition, only a minority of doctors keep records of donor characteristics. Yet the resulting children may someday want information about their genetic background or need it for medical reasons (Nachtigall, 1993).

SURROGATE MOTHERHOOD. A more controversial form of medically assisted conception is *surrogate motherhood*. Typically in this procedure, sperm from a man whose wife is infertile are used to inseminate a woman, who is paid a fee for her childbearing services. In return, the surrogate agrees to turn the baby over to the man (who is the natural father). The child is then adopted by his wife.

Although most of these arrangements proceed smoothly, those that end up in court highlight serious risks for all concerned. In one case, both parties rejected the disabled infant that resulted from the pregnancy. In several others, the surrogate mother changed her mind and wanted to keep the baby. These children came into the world in midst of family conflict that threatened to last for years to come. Most surrogates already have children of their own, who may be deeply affected by the pregnancy. Knowledge that their mother would give away a baby for profit may cause these youngsters to worry about the security of their own family circumstances (McGinty & Zafran, 1988; Ryan, 1989).

NEW REPRODUCTIVE FRONTIERS. Reproductive technologies are evolving faster than societies can weigh the ethics of these procedures. Doctors have used donor ova from younger women in combination with in vitro fertilization to help postmenopausal women become pregnant (see Figure 2.8). Most recipients are in their 40s, but a 62-year-old has given birth in Italy and a 63-year-old in the United States (Beck, 1994; Kalb, 1997). Even though candidates for postmenopausal intervention are selected on the basis of good health, serious questions arise about bringing children into the world whose parents

85 percent do well in their adoptive homes. Of those who do not, 90 percent are successfully placed with a new family (Churchill, 1984; Glidden & Pursley, 1989).

Still, adopted children have more learning and emotional difficulties than do their nonadopted agemates (Verhulst & Versluis-Den Bieman, 1995). There are many reasons for this trend. The biological mother may have been unable to care for the child because of emotional problems believed to be partly genetic, such as alcoholism and schizophrenia.[1] She may have passed this tendency to her offspring. Or perhaps she experienced stress, poor diet,

[1] Schizophrenia is a disorder involving serious difficulty in distinguishing fantasy from reality, frequent delusions and hallucinations, and irrational and inappropriate behaviors.

may not live to see them reach adulthood.

Currently, experts are debating other reproductive options. At donor banks, customers can select ova or sperm on the basis of physical characteristics and even the IQ of potential donors. Some worry that this practice is a dangerous step toward selective breeding of the human species. Researchers have delivered baby mice using the transplanted ovaries of aborted fetuses (Hashimoto, Noguchi, & Nakatsuji, 1992). If the same procedure were eventually applied to humans, it would create babies whose genetic mothers had never been born.

Finally, scientists have successfully cloned (made multiple copies of) fertilized ova in sheep and cattle, and they are working on effective ways of doing so in humans (Kolberg, 1993).

By providing extra ova for injection, cloning might improve the success rate of in vitro fertilization. But it also opens the possibility of mass-producing genetically identical people.

Although new reproductive technologies permit many barren couples to rear healthy newborn babies, laws are needed to regulate them. In the case of surrogate motherhood, the ethical problems are so complex that 18 U.S. states have sharply restricted the practice, and many European governments have banned it (Belkin, 1992; Charo, 1994). Recently, England, France, and Italy took steps to prohibit in vitro fertilization for women past menopause (Beck, 1994). At present, nothing is known about the psychological consequences of being a product of these procedures. Research on how such children grow up, including what they know and how they feel about their origins, is important for weighing the pros and cons of these techniques.

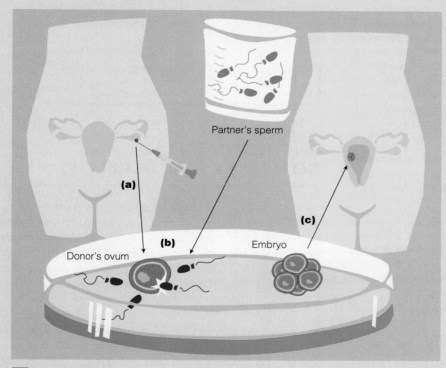

Partner's sperm

(a)

(b)

(c)

Donor's ovum

Embryo

FIGURE 2.8

In vitro fertilization procedure that can help postmenopausal women become pregnant. (a) A young female donor is given hormones to stimulate ovulation. Then a needle is inserted into her ovary, and several ova are extracted. (b) The donor's ova are fertilized in a dish of nutrients using sperm from the recipient's partner or from another male donor. (c) The prospective mother is given hormones to prepare her uterus to receive the fertilized ova, which are inserted.

TRY THIS...

■ Locate newspaper and magazine articles on two highly publicized surrogate motherhood cases: Baby M of New Jersey (1987) and the Calvert case of California (1990). Do you think the problems that arose in each case justify limiting or banning the practice of surrogacy? Why or why not?

or inadequate medical care during pregnancy—factors that (as we will see in Chapter 3) can affect the child. Finally, children adopted after infancy often have a history of conflict-ridden family relationships and lack of parental affection. But despite these risks, most adopted children have happy childhoods and grow up to be well-adjusted, productive citizens.

As we conclude our discussion of reproductive choices, perhaps you are wondering how things turned out for Ted and Marianne. Through genetic counseling, Marianne discovered a history of Tay-Sachs disease on her mother's side of the family. Ted had a distant cousin who died of the disorder. The genetic counselor explained that the chances of giving birth to another affected baby were 1

CAREGIVING CONCERNS

Steps Prospective Parents Can Take Before Conception to Increase the Chances of Having a Healthy Baby

SUGGESTION	RATIONALE
Arrange for a physical exam.	A physical exam before conception permits detection of diseases and other medical problems that might reduce fertility, be difficult to treat after the onset of pregnancy, or affect the developing organism.
Reduce or eliminate toxins under your control.	Since the developing organism is highly sensitive to damaging environmental agents during the early weeks of pregnancy (see Chapter 3), drugs, alcohol, cigarette smoke, radiation, pollution, chemical substances in the home and workplace, and infectious diseases should be avoided while trying to conceive. Furthermore, ionizing radiation and some industrial chemicals are known to cause mutations.
Consider your genetic makeup.	Find out if anyone in your family has had a child with a genetic disease or disability. If so, seek genetic counseling before conception.
Consult your physician after 12 months of unsuccessful efforts at conception.	Long periods of infertility may be due to undiagnosed spontaneous abortions, which can be caused by genetic defects in either partner. If a physical exam reveals a healthy reproductive system, seek genetic counseling.

in 4. Ted and Marianne took the risk. Their son Douglas is now 12 years old. Although Douglas is a carrier of the recessive allele, he is a normal, healthy boy. In a few years, Ted and Marianne will tell Douglas about his genetic history and explain the importance of genetic counseling and testing before he has children of his own. The Caregiving Concerns table above summarizes steps that prospective parents can take before conception to increase their chances of having a healthy baby.

BRIEF REVIEW

Genetic counseling helps couples who have a family history of reproductive problems or hereditary defects make informed decisions about childbearing. For those who decide to conceive, prenatal diagnostic methods permit early detection of fetal

problems. Reproductive technologies, such as donor insemination, in vitro fertilization, and surrogate motherhood, are also available, but they raise serious ethical concerns. Although learning and emotional difficulties are more common among adopted than nonadopted children, careful selection of adoptive parents and family support services make adoption successful in most cases.

ASK YOURSELF . . .

■ *A woman over 35 has just learned that she is pregnant. Although she would like to find out as soon as possible whether her child has a chromosomal disorder, she wants to minimize the risk of injury to the developing organism. Which prenatal diagnostic method is she likely to choose?*

■ *Describe the ethical pros and cons of fetal surgery, surrogate motherhood, and post-menopausal-assisted childbearing.*

ENVIRONMENTAL CONTEXTS FOR DEVELOPMENT

Just as complex as the heredity that sets the stage for development is the person's environment—a many-layered set of influences that combine to help or hinder physical and psychological well-being. Take a moment to think back to your own childhood, and jot down a brief description of events and people that you regard as having a significant impact on your development. Next, do the same for your adult life. When I ask my students to do this, the largest number of items they list involve their families. This emphasis is not surprising, since the family is the first and longest-lasting context for development. But other settings turn out to be important as well. Friends, neighbors, school, workplace, community organizations, and church or synagogue generally make the top ten.

Finally, there is one very important context my students rarely mention. Its influence is so pervasive that we seldom stop to think about it in our daily lives. This is the broad social climate of society—its values and programs that support and protect human development. All people need help with the demands of each phase of the life-span—through well-designed housing, safe neighborhoods, good schools, well-equipped recreational facilities, affordable health services, and high-quality day care and other services that permit them to meet both work and family responsibilities. And some people, because of poverty or special tragedies, need considerably more help than others.

The family is a complex social system in which each person's behavior influences the behavior of others, in both direct and indirect ways. The positive mealtime atmosphere in this family is probably a product of many forces, including parents who respond to children with warmth and patience, aunts and uncles who support parents in their child-rearing roles, and children who have developed cooperative dispositions. *(Michal Heron/Woodfin Camp & Associates)*

In the following sections, we take up these contexts for development. Since they affect every age period and aspect of change, we will return to them in later chapters. For now, our discussion emphasizes that besides heredity, environments can enhance or create risks for development. And when a vulnerable child or adult—an individual with physical or psychological problems—is exposed to unfavorable contexts, then development is seriously threatened.

THE FAMILY

In power and breadth of influence, no context equals the family. The family creates bonds between people that are unique. Attachments to parents and siblings usually last a lifetime and serve as models for relationships in the wider world of neighborhood, school, and community. Within the family, children learn the language, skills, and social and moral values of their culture. And at all ages, people turn to family members for information, assistance, and interesting and pleasurable interaction. Warm, gratifying family ties predict psychological health throughout development. In contrast, isolation or alienation from the family is often associated with developmental problems (Hetherington, 1995).

In the section on *ecological systems theory* in Chapter 1, we saw that modern investigators view the family as a set of interdependent relationships. The **social systems perspective** on family functioning, which has much in common with Bronfenbrenner's (1989, 1995) ecological model, grew out of researchers' efforts to describe and explain the complex patterns of interaction that take place in families. Let's take a close look at its basic features.

■ THE FAMILY AS A SOCIAL SYSTEM. Family systems theorists emphasize that no person is mechanically shaped by the inputs of others. Instead, *bidirectional* influences exist

in which the behaviors of each family member affect those around them. Indeed, the very term *family system* implies that the responses of all family members are interconnected (Kantor & Lehr, 1975; Minuchin, 1988). These system influences operate in both *direct* and *indirect* ways.

Direct Influences. The next time you have a chance to observe family members interacting, watch carefully. You are likely to see that kind, patient communication evokes cooperative, harmonious responses, whereas harshness and impatience engender angry, resistive behavior. Each of these reactions, in turn, prompts a new link in the interactive chain. In the first instance, a positive message tends to follow; in the second, a negative or avoidant one tends to occur.

These observations fit with a wealth of research on the family system. For example, many studies show that when parents' requests are accompanied by warmth and affection, children tend to cooperate. And when children willingly comply, their parents are likely to be warm and gentle in the future (Baumrind, 1983; Lewis, 1981). In contrast, parents who discipline with hostility usually have children who refuse and rebel. And because children's misbehavior is stressful for parents, they may increase their use of punishment, leading to more unruliness by the child (Dodge, Pettit, & Bates, 1994; Patterson, DeBaryshe, & Ramsey, 1989). This principle applies to other two-person family relationships, such as brother and sister, husband and wife, and parent and adult child. In each case, the behavior of one person helps sustain a form of interaction in the other that either promotes or undermines psychological well-being.

Indirect Influences. The impact of family relationships on development becomes even more complicated when we consider that interaction between any two members is affected by others present in the setting. Bronfenbrenner calls these indirect influences the effect of *third parties.*

Third parties can serve as effective supports for development. For example, when their marital relationship is warm and considerate, mothers and fathers praise and stimulate their children more and nag and scold them less. In contrast, when a marriage is tense and hostile, parents are likely to criticize and punish (Hetherington & Clingempeel, 1992; Howes & Markman, 1989; Simons et al., 1992). Similarly, children can affect their parents' relationship in powerful ways. For example, as we will see in Chapter 10, boys especially show lasting emotional problems when parents divorce. But longitudinal research reveals that many sons of divorcing couples were impulsive and undercontrolled long before the marital breakup. These behaviors may have contributed to as well as been caused by their parents' marital problems (Block, Block, & Gjerde, 1988; Cherlin et al., 1991; Hetherington, 1995).

Yet even when family relationships are strained by third parties, other members can help restore effective interaction. Grandparents are a case in point. They can promote children's development in many ways—both directly, by responding warmly to the child, and indirectly, by providing parents with child-rearing advice, models of child-rearing skill, and even financial assistance (Cherlin & Furstenberg, 1986). Of course, like any indirect influence, grandparents can sometimes be harmful. When quarrelsome relations exist between grandparents and parents, parent–child communication may suffer.

A Dynamic, Ever-Changing System. Like the chronosystem in Bronfenbrenner's model (turn back to page 000), the social systems approach views the interplay of forces within the family as dynamic and ever changing. Important events, such as the birth of a baby, a change of jobs, or an elderly parent joining the household due to declining health, create challenges that modify existing relationships. The way such events affect family interaction depends on the support provided by other family members as well as the developmental status of each participant. For example, the arrival of a new baby prompts very different reactions in a toddler than a school-age child. And caring for a very ill elderly parent is more stressful for a middle-aged adult still rearing young children than an adult of the same age without child-rearing responsibilities.

Finally, historical time period contributes to a dynamic family system. In recent decades, a declining birth rate, a high divorce rate, and expansion of women's roles have led to a smaller family size. This, combined with a longer lifespan, means that more generations are alive with fewer members in the youngest ones, leading to a "top-heavy" family structure. Consequently, young people today are more likely to have older relatives than at any time in history—a circumstance that can be enriching as well as a source of tension (Gatz, Bengtson, & Blum, 1990; Jerrome, 1990). In sum, as this complex intergenerational system moves through time, relationships are constantly revised as members adjust to their own and others' development as well as to external pressures.

Despite these variations, some general patterns in family functioning do exist. In the United States and other Western nations, one important source of these consistencies is social class.

SOCIAL CLASS AND FAMILY FUNCTIONING

Social class affects the timing and duration of phases of the family life cycle. People who work in skilled and semi-skilled manual occupations (for example, machinists, truck drivers, and custodians) tend to marry and have children earlier as well as give birth to more children than do people in white-collar and professional occupations. The two groups also differ in values and expectations. For example, when asked about personal qualities they desire for their children, lower-class parents tend to place greater emphasis on external characteristics, such as neatness, cleanliness, and obedience. In contrast, middle-class parents stress psychological traits, such as curiosity, happiness, and self-direction. In addition, middle-class adults expect to occupy a wider range of roles, both within the family and without, than do their lower-class counterparts.

These differences are reflected in family interaction. Lower-class fathers focus on their provider role and tend to devote little time to parenting, whereas middle-class fathers often share in housework and child-rearing responsibilities (although their commitment rarely equals that of mothers). Furthermore, middle-class parents talk to and stimulate their infants more and grant them greater freedom to explore. At older ages, they use more explanations and verbal praise. In contrast, lower-class parents are more likely to be restrictive. Because they think that infants can easily be spoiled, they limit the amount of rocking and cuddling they do (Luster, Rhoades, & Haas, 1989). Later on, commands, such as "You do that because I told you to," as well as criticism and physical punishment occur more often in low-income households (Dodge, Pettit, & Bates, 1994; Hashima & Amato, 1994).

The life conditions of low-income and middle-income families help explain these findings. Low-income adults often feel a sense of powerlessness and lack of influence in their relationships beyond the home. For example, at work, they must obey the rules of others in positions of power and authority. When they get home, their parent–child interaction seems to duplicate these experiences, only with them in the authority roles. Middle-class parents have a greater sense of control over their own lives. At work, they are used to making independent decisions and convincing others of their point of view. At home, they teach these skills to their children (Greenberger, O'Neil, & Nagel, 1994).

Education also contributes to social-class differences in family interaction. Middle-class parents' interest in verbal stimulation and nurturing inner traits is supported by years of schooling, during which they learned to think about abstract, subjective ideas (Richman, Miller, &

LeVine, 1992). Furthermore, the greater economic security of middle-class families frees them from the burden of having to worry about making ends meet on a daily basis. They can devote more time, energy, and material resources to furthering their own and their children's psychological characteristics.

As early as the second year of life, middle-class children tend to be advanced in cognitive and language development over their lower-class agemates. Throughout childhood and adolescence, they do better in school (Brody, 1992; Walker et al., 1994). And by young adulthood, they attain higher levels of education, which greatly enhances opportunities for a prosperous adult life. Researchers believe that social-class differences in family functioning have much to do with these outcomes.

THE IMPACT OF POVERTY

When families become so low income that they slip into poverty, development is seriously threatened. Shirley Brice Heath (1990), an anthropologist who has spent many years studying families of poverty, describes the case of Zinnia Mae, who grew up in Trackton, a close-knit black community located in a small southeastern American city. As unemployment struck Trackton in the 1980s and citizens moved away, 16-year-old Zinnia Mae caught a ride to Atlanta. Two years later, Heath visited her there. By then, Zinnia Mae was the mother of three children—a 16-month-old daughter named Donna and 2-month-old twin boys. She had moved into high-rise public housing.

Each of Zinnia Mae's days was much the same. She watched TV and talked with girlfriends on the phone. The children had only one set meal (breakfast) and otherwise ate whenever they were hungry or bored. Their play space was limited to the living room sofa and a mattress on the floor. Toys consisted of scraps of a blanket, spoons, food cartons, a small rubber ball, a few plastic cars, and a roller skate abandoned in the building. Zinnia Mae's most frequent words were, "I'm so tired." She worried about how to get papers to the welfare office, where to find a baby-sitter so she could go to the laundry or grocery, and what she would do if she located the twins' father, who had stopped sending money. She rarely had enough energy to spend time with her children.

Over the past twenty-five years, economic changes in the United States have caused the poverty rate to climb substantially. Today, nearly 37 million people—14.5 percent of the population—are affected. Those hit hardest are parents under age 25 with young children and elderly people who live alone. Poverty is also magnified among ethnic minorities and women. For example, nearly 22 percent of American children are poor, a rate that climbs to 40 percent for Hispanic children and 47 percent for African-American children. For single mothers with preschool children and black elderly women on their own, the poverty rate is over 60 percent (U.S. Bureau of the Census, 1996). Joblessness, a high divorce rate, a lower remarriage rate among women than men, widowhood, and (as we will see later) inadequate government programs to meet family needs are responsible for these disheartening statistics. The condition of children is particularly worrisome because the earlier poverty begins and the longer it lasts, the more devastating its effects (Chase-Lansdale & Brooks-Gunn, 1994; Children's Defense Fund, 1997). Since the 1970s, the child poverty rate has been higher than that of any age group.

The constant stresses that accompany poverty gradually weaken the family system. Poor families have many daily hassles—bills to pay, the car breaking down, loss of welfare and unemployment payments, something stolen from the house, to name just a few. When daily crises arise, family members become irritable and distracted, and hostile interactions increase (Conger et al., 1992; Garrett, Ng'andu, & Ferron, 1994). These outcomes are especially severe in families that must live in poor housing and dangerous neighborhoods—conditions that make everyday existence even more difficult while reducing social supports that assist in coping with economic hardship (Duncan, Brooks-Gunn, & Klebanov, 1994; McLoyd et al., 1994).

Homelessness in the United States has risen over the past two decades. Families like this one travel from place to place in search of employment and a safe and secure place to live. At night, they sleep in the family car. Because of constant stresses and few social supports, homeless children are usually behind in development, have frequent health problems, and show poor psychological adjustment. (Rick Browne/Stock Boston)

Besides poverty, another problem—one that was quite uncommon two decades ago—has reduced the life chances of many children and adults. By the early 1990s, approximately 3 million people had no place to live. Slightly more than half are adults on their own, some of whom have serious emotional and substance abuse problems (Fischer & Breakey, 1991). But over 40 percent of America's homeless population is made up of families, and 1 in every 4 homeless individuals is believed to be a child. The rise in family homelessness is due to a number of factors, the most important of which is a dramatic decline in the availability of government-supported low-cost housing (Children's Defense Fund, 1997).

Most homeless families consist of women with children under age 5 (Milburn & D'Ercole, 1991). Besides health problems (which affect most homeless people), homeless children suffer from developmental delays and serious emotional stress. An estimated 25 to 30 percent who are old enough do not attend school. Those who do achieve less well than other poverty-stricken children due to poor attendance and health and emotional difficulties (Rafferty & Shinn, 1991). A common factor among the homeless is loss of supportive ties to relatives and friends.

B EYOND THE FAMILY: NEIGHBORHOODS, TOWNS, AND CITIES

Family systems theory emphasizes that ties between family and community are vital for psychological well-being throughout the lifespan. From our discussion of poverty and homelessness, perhaps you can see why. In poverty-stricken urban areas, community life is usually disrupted. Families move often, parks and playgrounds are in disarray, and community centers providing organized leisure time activities do not exist (Wilson, 1991). Family violence and child abuse and neglect are greatest in neighborhoods where residents are dissatisfied with their community, describing it as a socially isolated place to live. In contrast, when family ties to the community are strong—as indicated by regular church attendance and frequent contact with friends and relatives—family stress and adjustment problems are reduced (Cazenave & Straus, 1990; Garbarino & Kostelny, 1993).

■ NEIGHBORHOODS. Let's take a closer look at the functions that communities serve in the lives of children and adults by beginning with the neighborhood. What were your childhood experiences like in the yards, streets, and parks surrounding your home? How did you spend your time, whom did you get to know, and how important were these moments to you? To most children, the neighborhood is not just an outdoor play space—"it is a social universe" (Medrich et al., 1982, p. 33).

The resources offered by neighborhoods play an important part in children's development. One study found that the more varied children's neighborhood experiences—membership in organizations (such as scouting and 4–H), contact with adults of their grandparents' generation, and places to go off by themselves or with friends (a treehouse, a fort, or a neighbor's garage)—the better their emotional and social adjustment (Bryant, 1985). After-school programs that substitute for lack of resources in low-income neighborhoods by providing enrichment activities (scouting, music lessons, and organized sports) are associated with improved school performance, peer relations, and psychological adjustment in middle childhood (Posner & Vandell, 1994). In the United States, the affluence of the surrounding neighborhood affects the quality of its schools. We will devote considerable attention to the powerful impact of schooling on development in later chapters.

Neighborhoods also influence well-being in adulthood. For example, an employed parent who can rely on a neighbor to assist her school-age child in her absence and who lives in an area safe for walking to and from school is granted the peace of mind essential for productive work. During late adulthood, neighborhoods become increasingly important, since the elderly spend more time in their own homes than they did in earlier years. Despite the availability of planned housing for older people, about 88 percent remain in regular housing, usually in the same neighborhood in which they lived during their working lives (Parmelee & Lawton, 1990; U.S. Bureau of the Census, 1996). Proximity to friends and relatives is a significant factor in the decision to move or stay put late in life. In the absence of nearby family members, the elderly mention neighbors as the resource they rely on most for physical and social support (Ward et al., 1981).

■ TOWNS AND CITIES. Neighborhoods are embedded in towns and cities, which also mold children's and adult's daily lives. A well-known study examined the kinds of community settings children entered and the roles they played in a Midwestern town with a population of 700 (Barker, 1955). Many settings existed, and children were granted important responsibilities in them. For example, they helped stock shelves at Kane's Grocery Store, played in the town band, and operated the snow plow when help was short. As children joined in these activities, they did so alongside adults, who taught them the skills they needed to become responsible members of the community. Compared to large urban areas, in small towns connections between settings that influence children's lives are more common. For example, since most citizens know each other and schools serve as centers of community life, contact between teachers and parents occurs often—an important factor in promoting children's academic achievement (Connell, Spencer, & Aber, 1994; Stevenson & Baker, 1987).

Like children, adults in small towns also penetrate more settings and are more likely to occupy positions of leadership, since a greater proportion of residents are

The resources offered by neighborhoods are important for development and well-being at all ages. This city park permits residents to gather for weekend picnics and festivals, where the opportunity to get to know neighbors in a relaxed, safe atmosphere fosters pleasurable interaction and social support. (Jeff Dunne/The Picture Cube)

needed to meet community needs—for example, by serving on the town council, on the school board, or in other civic roles. In old age, people residing in small towns and suburbs find that their neighbors are more willing to provide assistance. As a result, they develop a greater number of warm relationships with nonrelatives (Lawton, 1980; Ward et al., 1981). As one 99-year-old resident of a small midwestern community, living alone and leading an active life, commented, "I don't think I could get along if I didn't have good neighbors." The family next door helps him with grocery shopping, checks each night to make sure his basement light is off (the signal that he is out of the shower and into bed), and looks out in the morning to see that his garage door is raised (the signal that he is up and okay) (Fergus, 1995).

Of course, children and adults in small towns cannot visit museums, go to professional baseball games, or attend orchestra concerts on a regular basis. The variety of settings is somewhat reduced compared to large cities. In small towns, however, active involvement in the community is likely to be greater throughout the lifespan. Also, public places in small towns are safe and secure. Responsible adults are present in almost all settings to keep an eye on children. And the elderly feel a greater sense of personal safety—a strong contributor to how satisfied they are with their place of residence (Parmelee & Lawton, 1990). These conditions are hard to match in today's urban environments.

Think back to the case of Zinnia Mae and her three young children described on page 61. It reveals that community life is especially undermined in high-rise urban housing projects. In these dwellings, social contact is particularly important, since many residents have been uprooted from neighborhoods where they felt a strong sense

of cultural identity and belonging. Typically, high rises are heavily populated with young single mothers, who are separated from family and friends by the cost and inconvenience of cross-town transportation. They report intense feelings of loneliness in the small, cramped apartments. At Heath's (1990) request, Zinnia Mae agreed to tape record her family interactions over a 2-year period. In 500 hours of tape (other than simple directions or questions about what her children were doing), Zinnia Mae started a conversation with Donna and the boys only 18 times. Cut off from community ties, she found it difficult to join in activities with her children. The result was a barren, understimulating environment—one very different from the home and community in which Zinnia Mae herself had grown up.

THE CULTURAL CONTEXT

In Chapter 1, we pointed out that human development can only be fully understood when viewed in its larger cultural context. In the following sections, we expand on this important theme. First, we illustrate ways in which cultural values and practices affect environmental contexts for development. Second, we consider how healthy development throughout the lifespan depends on laws and government programs that shield people from harm and foster their well-being.

■ CULTURAL VALUES AND PRACTICES. Cultures shape family interaction and community settings beyond the home—in short, all aspects of daily life. Many of us remain blind to aspects of our own cultural heritage until we see them in relation to the practices of others.

Each year, I ask my students to think about the following question: "Who should be responsible for rearing young children?" Here are some typical answers: "If parents decide to have a baby, then they should be ready to care for it." "Most people are not happy about others intruding into family life." These statements reflect a widely held opinion in the United States—that the care and rearing of children during the early years is the duty of parents, and only parents (Goffin, 1988). This view has a long history—one in which independence, self-reliance, and the privacy of family life emerged as central American values. It is one reason, among others, that the American public has been slow to accept the idea of publicly supported health insurance and day care for children of employed parents. This strong emphasis on individualism also helps us understand why, among middle-class families (who best represent American cultural values), only a small percentage of grandparents and other relatives participate actively and regularly in the rearing of children (Thompson et al., 1989).

Although American middle-class families value independence and privacy, cooperative family structures can be found in the United States. In large industrialized nations like ours, not all citizens share the same values. **Subcultures** exist—groups of people with beliefs and customs that differ from those of the larger culture. The values and practices of some ethnic minority groups help protect their members from the harmful effects of poverty. A case in point is the African-American family. As the Lifespan Vista box on the following page indicates, the black cultural tradition of **extended family households,** in which three or more generations live together, is a vital feature of black family life that has enabled its members to survive, despite a long history of prejudice and economic deprivation. Within the extended family, grandparents play meaningful roles in guiding younger generations; adults with employment, marital, or child-rearing difficulties receive assistance and emotional support; and caregiving is enhanced for children and the elderly. Active and involved extended families also characterize other American minorities, such as Asian-American, Native-American, and Hispanic subcultures (Harrison et al., 1994).

Consider our discussion so far, and you will see that it reflects a broad dimension on which cultures and subcultures differ: the extent to which *collectivism versus individualism* is emphasized. In collectivist societies, people define themselves as part of a group and stress group over individual goals. In individualistic societies, people think of themselves as separate entities and are largely concerned with their own personal needs (Triandis, 1989; Triandis et al., 1988). Although individualism tends to increase as cultures become more complex, cross-national differences remain. The United States is more individualistic than most other industrialized nations. As we will see in the next sec-

tion, collectivist versus individualistic values have a powerful impact on a nation's approach to protecting human development and well-being.

■ PUBLIC POLICIES AND LIFESPAN DEVELOPMENT. When widespread social problems arise, such as poverty, homelessness, hunger, and disease, nations attempt to solve them by developing **public policies**—laws and government programs designed to improve current conditions. For example, when poverty increases and families become homeless, a country might decide to build more low-cost housing, raise the minimum wage, and increase welfare benefits. When reports indicate that many children are not achieving well in school, federal and state governments might grant more tax money to school districts and make sure that help reaches pupils who need it most. And when senior citizens have difficulty making ends meet because of inflation, a nation might increase its social security benefits.

The United States is among the wealthiest of nations and has the broadest knowledge base for intervening effectively in people's lives. Still, American public policies safeguarding children and youths have lagged behind policies for the elderly. And both sets of policies have been slower to emerge in the United States than in other Western industrialized nations.

Policies for Children, Youths, and Families. We have already seen in previous sections that although many American children fare well, a large number grow up in environments that threaten their development. As Table 2.5 reveals, the United States does not rank among the top countries on any key measure of children's health and well-being.

TABLE 2.5

How Does the United States Compare to Other Nations on Indicators of Child Health and Well-Being?

INDICATOR	U.S. RANK	SOME COUNTRIES THE UNITED STATES TRAILS
Childhood poverty[a]	8th (among 8 industrialized nations studied)	Australia, Canada, Germany, Great Britain, Norway, Sweden, Switzerland
Infant deaths in the first year of life	22nd (worldwide)	Australia, Hong Kong, Ireland, New Zealand, Singapore, Spain
Low-birth-weight newborns	28th (worldwide)	Bulgaria, Egypt, Greece, Iran, Jordan, Kuwait, Paraguay, Romania, Saudi Arabia
Percentage of young children immunized against measles	21st (worldwide)	Chile, Czechoslovakia, Jordan, Poland
Number of school-age children per teacher	12th (worldwide)	Cuba, Lebanon, Libya
Expenditures on education as percentage of gross national product[b]	14th (among 16 industrialized nations studied)	Australia, Canada, France, Great Britain, the Netherlands, Sweden
Teenage pregnancy rate	7th (among 7 industrialized nations considered)	Australia, Canada, England, France, the Netherlands, Sweden

[a] The U.S. child poverty rate of nearly 22 percent is more than twice that of any of these nations. For example, the rate is 9 percent in Australia, 9.3 percent in Canada, 4.6 percent in France, and 1.6 percent in Sweden.

[b] Gross national product is the value of all goods and services produced by a nation during a specified time period. It serves as an overall measure of a nation's wealth.

Sources: Children's Defense Fund, 1997; Danziger & Danziger, 1993; Grant, 1995; Sivard, 1993; Wegman, 1994.

A LIFESPAN VISTA

THE AFRICAN-AMERICAN EXTENDED FAMILY: A LIFESPAN SUPPORT SYSTEM

The African-American extended family can be traced to the African heritage of most black Americans. In many African societies, newly married couples do not start their own households. Instead, they marry into a large extended family that assists its members with all aspects of daily life. This tradition of a broad network of kinship ties traveled to the United States during the period of slavery. Since then, it has served as a protective shield against the destructive impact of poverty and racial prejudice on black family life (McLoyd, 1990). Today, more black than white adults have relatives other than their own children living in the same household. African-American parents also see more kin during the week and perceive them as more important figures in their lives, respecting the advice of relatives and caring deeply about what they think is important (Wilson, 1986).

By providing emotional support and sharing income and essential resources, the African-American extended family helps reduce the stress of poverty and single parenthood. Extended family members often help with the rearing of children (Pearson et al., 1990). Consequently, adolescent mothers who live in extended families are more likely to complete high school and get a job and less likely to be on welfare than mothers living on their own (Furstenberg & Crawford, 1978).

For single mothers who were very young at the time of their child's birth, extended family living is associated with more positive adult–child interaction during infancy and the preschool years. Otherwise, establishing an independent household with the help of nearby relatives is related to improved child rearing. Perhaps this arrangement permits the more mature mother who has developed effective parenting skills to implement them (Chase-Lansdale, Brooks-Gunn, & Zamsky, 1994). In families rearing adolescents, kinship support increases the likelihood of effective parenting, which, in turn, is related to self-reliance, emotional well-being, and reduced delinquency (Taylor & Roberts, 1995).

Finally, African-American elderly report a very high degree of satisfaction from family life. Compared to their white counterparts, black grandmothers are more likely to provide child-rearing assistance and financial help to their children and grandchildren (Taylor, 1985). Relatives hold them in high esteem, describing them as sources of love, strength, and stability. And grandmothers play a central role in transmitting African-American culture by teaching moral and religious values and encouraging cooperation and mutual support (Peterson, 1990; Tolson & Wilson, 1990). These influences strengthen family bonds, protect the development of all family members, and increase the chances that the extended family lifestyle will carry over to the next generation.

Strong bonds with extended-family members have helped to protect the development of many African-American children growing up under conditions of poverty and single parenthood. (Karen Kasmauski/Woodfin Camp & Associates)

Overall, senior citizens in the United States are better off economically than are children. Nevertheless, many older adults—especially women, ethnic minorities, and those living alone—are poverty stricken, and American elders are more likely to be among the "near poor" than other age groups. (Andy Levin/Photo Researchers, Inc.)

The problems of American children and youths extend beyond the indicators in the table. For example, 20 percent have no health insurance, making them the largest segment of the uninsured population (Children's Defense Fund, 1997). Furthermore, the United States has been slow to move toward national standards and funding for day care. According to recent evidence, much day care is substandard in quality (Helburn, 1995; Phillips et al., 1994). In American families affected by divorce, child support payments are poorly enforced. Three fourths of custodial mothers do not receive the full court-ordered allowance from the child's father. By the time American young people finish high school, many do not have the educational preparation they need to contribute fully to society. Noncollege-bound youths generally lack vocational skills required for well-paid jobs. And about 14 percent of adolescents leave high school without a diploma. If they do not return to finish their education, they are at risk for lifelong poverty (Children's Defense Fund, 1997).

Why has the United States not yet created conditions that protect the development of its youngest citizens? A complex set of political and economic forces is involved. The American ideals of self-reliance and privacy have made government hesitant to become involved in family matters. In addition, there is less consensus among American than European citizens on issues of child and family policy, such as public support for day care and health services (Wilensky, 1983). Finally, good social programs are expensive, and they must compete for a fair share of a country's economic resources. Children can easily remain unrecognized in this process, since they cannot vote or speak out to protect their own interests, as adult citizens do. Instead, they must rely on the good will of others to become an important government priority (Garwood et al., 1989).

Policies for the Elderly. Until the mid-twentieth century, the United States had few policies in place to protect its aging population. For example, social security benefits, which address the income needs of retired citizens who contributed to society through prior employment, were not awarded until the 1940s. Yet most western European nations had social security systems in place by the late nineteenth or early twentieth century (DiNitto, 1995). In the 1960s, U.S. federal spending on programs for the elderly rapidly expanded. Medicare, a national health insurance program for older people that pays partial health care costs, was initiated. Today, spending for the aged accounts for 30 percent of all federal expenditures—5 times what was spent 30 years ago (Hooyman & Kiyak, 1993; Koff & Park, 1993).

Social security and Medicare consume 96 percent of the U.S. federal budget for the elderly; only 4 percent is devoted to other programs. Consequently, American programs for the elderly have been criticized for neglecting social services (Hooyman & Kiyak, 1993). To meet this need, a national network for planning, coordinating, and delivering assistance to the aged has been established. Approximately 700 Area Agencies on Aging operate at regional and local levels, assessing community needs and offering communal and home-delivered meals, self-care education, elder abuse prevention, and a wide range of other social services. However, limited funding means that the Area Agencies help far too few people in need.

As we noted earlier, many senior citizens—especially women, ethnic minorities, and those living alone—remain in dire economic straits. Those who had interrupted employment histories, held jobs without benefits, or suffered lifelong poverty are not eligible for social security. Although all people age 65 and older are guaranteed a minimum income, it is less than the poverty line—the amount judged necessary for bare subsistence by the federal government. Furthermore, social security benefits are rarely enough to serve as a sole source of retirement income; they must be supplemented through other pensions and family savings. But a substantial percentage of aging citizens do

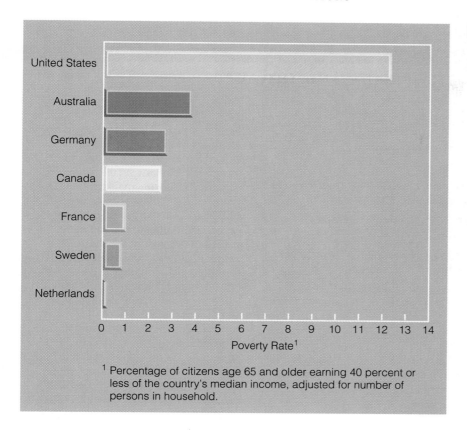

FIGURE 2.9

Percentage of citizens age 65 and older living in poverty in seven industrialized nations, based on the most recent available data. Among countries listed, public expenditures on social security and other income guarantees for senior citizens are highest in the Netherlands, lowest in the United States (United Nations, 1996). Consequently, poverty among the elderly is virtually nonexistent in the Netherlands but greatly exceeds that of other nations in the United States. *(Adapted from National Academy on Aging, 1992.)*

not have access to these resources. Therefore, they are more likely to be among the "near poor" than other age groups (Koff & Park, 1993).

Nevertheless, America's aging population is financially much better off than in the past. Today, the elderly are a large, powerful, well-organized constituency, far more likely than children or low-income and homeless families to attract the support of politicians. As a result, although poverty has risen in the general population, the number of aging poor has declined from 1 out of 3 people in 1960 to 1 out of 7 in the mid-1990s. And modern senior citizens are healthier and more independent than ever before. Still, as Figure 2.9 shows, the elderly in the United States are not as well off as those in Australia, Canada, and Western Europe. Many are inadequately served by current policies.

■ LOOKING TOWARD THE FUTURE. Despite the worrisome state of many children, families, and aging citizens, progress is being made in improving their condition. Throughout this book, we will discuss many successful programs that could be expanded. Another positive sign is that researchers are more involved than ever before in seeking answers to policy-relevant questions about human development and communicating their findings to the country as a whole (Huston, 1994; Koff & Park, 1993).

Finally, growing awareness of the gap between what we know and what we do to better people's lives has led experts in human development to join with concerned citizens as advocates for more effective policies. Over the past few decades, several influential interest groups with the well-being of children or the elderly as their central purpose have emerged. Two of the most vigorous are the Children's Defense Fund and the American Association of Retired Persons. To learn about their activities, refer to the Social Issues box on pages 68–69.

■ BRIEF REVIEW

Just as complex as heredity are the environments in which development takes place. First and foremost is the family—a complex system of mutually influencing relationships that changes over time. Interaction in families is modified by social class, and it is seriously threatened by poverty and homelessness. Development is supported by communities that encourage frequent contact among residents and worthwhile activities in which children and adults participate actively. Cultural values and practices—in particular, the extent to which collectivism versus individualism is emphasized—have a substantial impact on environmental contexts for development. In the complex world in which we live, favorable public policies are essential for protecting development and well-being.

SOCIAL ISSUES

ADVOCACY FOR CHILDREN AND THE AGED

The design and implementation of public policies that enhance human development depend on strong advocacy by highly committed interest groups. In the United States, the Children's Defense Fund and the American Association of Retired Persons are widely known for their efforts to track the status of vulnerable citizens, propose solutions to problems, and disseminate information to lawmakers and the general public. To accomplish its goals, each nonprofit organization engages in research, public education, legal action, drafting of legislation, congressional testimony, and community organizing.

THE CHILDREN'S DEFENSE FUND

- Every 30 seconds a baby is born into poverty.

- Every night 100,000 children go to sleep without homes.

- Every 14 minutes an infant dies in the first year of life.

- Every 59 seconds an infant is born to a teenage mother.

- Every 5 seconds a youth drops out of school.

- Every month at least 225,000 children are abused or neglected.

Founded by Marion Wright Edelman in 1973, the Children's Defense Fund is the most avid interest group for children and adolescents in the United States. To sensitize the public, it presents dramatic images in media ads and nationwide mailings. Besides promoting public awareness, the Children's Defense Fund provides government officials with a steady stream of facts about the status of children and encourages them to support legislation responsive to children's needs. Each year, it publishes *The State of America's Children,* a comprehensive analysis of the current condition of children, government-sponsored programs serving them, and proposals for improving child and family programs.

The Children's Defense Fund also supports an extensive network of state and local organizations that have children's needs as their central purpose. Two of its most significant projects are an adolescent pregnancy prevention program and a prenatal care campaign designed to reduce the high rates of death, illness, and developmental problems among poverty-stricken infants. Dissemination of information on how communities can develop more effective child and family services and preparation of a monthly newsletter on what people across the nation are doing to solve children's problems are among its many activities.

In 1973, Marion Wright Edelman founded the Children's Defense Fund, a private, non-profit organization that provides a strong, effective voice for American children, who cannot vote, lobby, or speak for themselves. Edelman continues to serve as president of the Children's Defense Fund today. (Westenberger/Liaiso, USA)

THE AMERICAN ASSOCIATION OF RETIRED PERSONS

- More than 5.6 million elderly Americans live out their final years in, or close to, poverty.

- Only about 50 percent of the total elderly health care bill is actually paid by Medicare.

- Between 1 and 2 million elderly Americans are victims of abuse each year.

- Suicide among the elderly is the highest of any age group, taking the lives of more than 6,000 people over age 65 annually.

These facts are drawn from the many public policy reports released annually by the American Association of Retired Persons (AARP). Nearly half of Americans over age 50 (both retired and employed)—more than 32 million people—are members of the organization. Founded by Ethel Percy Andrus in 1958, AARP has a large and energetic lobbying staff that works for increased government benefits of all kinds for the aged. Among its programs is an effort to mobilize elderly voters, an initiative that keeps lawmakers highly sensitive to policy proposals favoring older Americans.

A top AARP priority is improved health care. Although Medicare serves over 90 percent of the aging population, it is oriented toward short-term illnesses. Recognizing that millions of elderly cannot afford private health insurance to protect against catastrophic illnesses, AARP has proposed new tax-supported measures to defray the medical expenses of people who require long-term care. In addition, AARP lobbies for consumer services for the elderly, such as regulation of auto insurance and funeral expenses. Also on its extensive agenda are improved social security benefits for women and ethnic minorities, increased funding for elder abuse prevention, and expanded social services for older people who live alone.

Thousands of local AARP groups across the country sponsor community programs on such topics as retirement planning, personal relationships, driver education, and crime prevention. AARP also funds grants to universities for policy-relevant research.

Since 1958, the American Association of Retired Persons has worked to improve quality of life for the elderly by increasing government benefits of all kinds. This older couple enjoys reading *Modern Maturity,* the association's magazine, distributed monthly to over 32 million members. (Hronn Axelsdottir)

TRY THIS...

- Obtain the most recent edition of *The State of America's Children* from the Children's Defense Fund (the address is given at the end of this chapter) or your library. Consult the section entitled "Children in the States." How do children in your state fare on such indicators as poverty, infant mortality, immunization against disease, child abuse and neglect, and teenage childbearing?

- Examine recent issues of AARP's magazine, *Modern Maturity,* at your library. What current policy concerns are represented in the articles? How would you describe the image of aging reflected in this widely distributed publication?

- If you are not a resident of the United States, find out what interest groups advocate for children and the aged in your country. What are their current top priorities and programs?

■ *On one of your trips to the local shopping center, you see a father getting very angry at his young son. Using the family systems perspective, list as many factors as you can that might account for the father's behavior.*

■ *Links between family and community foster development throughout the lifespan. Provide examples and research findings from our discussion that support this idea.*

■ *Check your local newspaper and one or two national news magazines to see how often articles appear on the condition of children, families, and the aged. Why is it important for researchers to communicate with the general public about the well-being of these sectors of the population?*

UNDERSTANDING THE RELATIONSHIP BETWEEN HEREDITY AND ENVIRONMENT

So far in this chapter, we have discussed a wide variety of hereditary and environmental influences, each of which has the power to alter the course of development. Yet people born into the same family (and who therefore share genes and environments) are often quite different in characteristics. We also know that some individuals are affected more than others by their homes, neighborhoods, and communities. Cases exist in which a person provided with all the advantages in life does poorly, whereas another individual exposed to the worst of rearing conditions does well (Hauser, 1995; Werner & Smith, 1992). How do scientists explain the impact of heredity and environment when they seem to work in so many different ways?

All contemporary researchers agree that both heredity and environment are involved in every aspect of development. There is no real controversy on this point because an environment is always needed for genetic information to be expressed (Plomin, 1994). But for polygenic traits (due to many genes) such as intelligence and personality, scientists are a long way from knowing the precise hereditary influences involved. They must study the impact of genes on these characteristics indirectly, and the nature–nurture controversy remains unresolved because researchers do not agree on how heredity and environment influence these complex characteristics.

Some believe that it is useful and possible to answer the question of *how much* each factor contributes to individual differences. These researchers use special methods to find out which factor plays the major role. A second group of investigators regards the question of which factor is more

important as unanswerable. They believe that heredity and environment do not make separate contributions to behavior. Instead, they are always related, and the real question we need to explore is *how* they work together. Let's consider each of these two positions in turn.

THE QUESTION OF "HOW MUCH?"

Two methods—heritability and concordance rates—are used to infer the importance of heredity in complex human characteristics. Let's look closely at the information these procedures yield, along with their limitations.

■ HERITABILITY. **Heritability estimates** measure the extent to which individual differences in continuous traits, such as intelligence and personality, are due to genetic factors. They are obtained from **kinship studies,** which compare the characteristics of family members. The most common type of kinship study compares identical twins, who share all their genes, with fraternal twins, who share only some. If people who are genetically more alike are also more similar in intelligence and personality, then the researcher assumes that heredity plays an important role.

Kinship studies of intelligence provide some of the most controversial findings in the field of human development. Some experts claim a strong role for heredity, whereas others believe that genetic factors are barely involved. When many twin studies are examined, correlations between the scores of identical twins are consistently higher than those of fraternal twins. In a summary of over 30 such investigations, the correlation for intelligence was .86 for identical twins and .60 for fraternal twins (Bouchard & McGue, 1981). Researchers use a complex statistical procedure to compare these correlations, arriving at a heritability estimate ranging from 0 to 1.00. The value for intelligence is about .50 in child and adolescent twin samples. This indicates that half the individual variation in intelligence can be explained by differences in genetic makeup. However, heritability increases in adulthood, with some estimates as high as .80. As we will see in the final section of this chapter, one explanation is that adults exert much greater control over their intellectual experiences than do children (McGue et al., 1993). The fact that the intelligence of adopted children is more strongly related to the scores of their biological parents than adoptive parents offers further support for the role of heredity (Horn, 1983; Scarr & Weinberg, 1983).

Heritability research reveals that genetic factors are important in personality. For traits that have been studied a great deal (such as sociability, emotional expressiveness, and activity level), heritability estimates obtained on child, adolescent, and early adult twins are moderate, at about .40 or .50 (Bouchard, 1994; Braungart et al., 1992). Unlike intelligence, however, there is no evidence for an increase in heritability of personality over the lifespan (Brody, 1994).

■ CONCORDANCE. A second measure that has been used to infer the contribution of heredity to complex characteristics is the **concordance rate.** It refers to the percentage of instances in which both twins show a trait when it is present in one twin. Researchers typically use concordance to study the contribution of heredity to emotional and behavior disorders, which can be judged as either present or absent. A concordance rate ranges from 0 to 100 percent. A score of 0 indicates that if one twin has the trait, the other one never has it. A score of 100 means that if one twin has the trait, the other one always has it. When a concordance rate is much higher for identical twins than for fraternal twins, then heredity is believed to play a major role. Twin studies of schizophrenia and severe depression show this pattern of findings. In the case of schizophrenia, the concordance rate for identical twins is 30 percent, that for fraternal twins only 6 percent. The figures are 69 percent and 13 percent for severe depression (Gershon et al., 1977; Kendler & Robinette, 1983).

These findings suggest that the tendency for schizophrenia and depression to run in families is partly due to genetic factors (Plomin, 1994b). However, we also know that environment is involved, since the concordance rate for identical twins would need to be 100 percent if heredity were the only influence operating. In later chapters, we will see that environmental stresses, such as poverty, family conflict, and a disorganized home life, are often associated with emotional and behavior disorders.

■ LIMITATIONS OF HERITABILITY AND CONCORDANCE. Although heritability estimates and concordance rates provide evidence that genetic factors contribute to complex human characteristics, questions have been raised about their accuracy. Both measures are heavily influenced by the range of environments to which twin pairs are exposed. For example, identical twins reared together have more strongly correlated intelligence test scores than those reared apart. When the former are used to compute heritability estimates, the higher correlation causes the importance of heredity to be overestimated. To overcome this

difficulty, researchers try to find twins who have been reared apart in adoptive families. But few separated twin pairs are available for study, and when they are, social service agencies often place them in advantaged homes that are similar in many ways (Bronfenbrenner & Crouter, 1983; Scarr & Kidd, 1983). Because the environments of most twin pairs do not represent the broad range of environments found in the general population, it is often difficult to generalize heritability and concordance findings to the population as a whole.

Heritability estimates are controversial measures because they can easily be misapplied. For example, high heritabilities have been used to suggest that ethnic differences in intelligence, such as the poorer performance of black children compared to white children, have a genetic basis (Jensen, 1969, 1985b). Yet this line of reasoning is widely regarded as inaccurate. As we will see in Chapter 9, heritabilities computed on mostly white twin samples do not tell us what is responsible for test score differences between ethnic groups.

Perhaps the most serious criticism of heritability estimates and concordance rates has to do with their usefulness. Although they are interesting statistics that tell us heredity is undoubtedly involved in complex traits such as intelligence and personality, they give us no precise information about how these traits develop or how individuals might respond when exposed to environments designed to help them develop as far as possible (Bronfenbrenner & Ceci, 1994). Investigators who conduct heritability research argue that their studies are a first step. As more evidence accumulates to show that heredity underlies important human characteristics, then scientists can begin to ask better questions—about the specific genes involved, the way they affect development, and how their impact is modified by experience.

▇ THE QUESTION OF "HOW?"

According to a second perspective, heredity and environment cannot be divided into separate influences.

Identical twins Jim Lewis and Jim Springer were separated 4 weeks after birth, grew up in different homes, and led separate adult lives until, at age 39, they were reunited. The two Jims discovered that they were alike in many ways. Both drove the same model car, chain smoked, chewed their fingernails, and vacationed at the beach. On personality tests, they scored almost exactly the same. Not all separated twins match up as well as this pair. Nevertheless, the study of identical twins reared apart reveals that heredity contributes to many psychological characteristics. (D. Gordon/ Time Magazine)

FIGURE 2.10

Intellectual ranges of reaction (RR) for three children in environments that vary from unstimulating to highly enriched. *(From I. I. Gottesman, 1963, "Genetic Aspects of Intelligent Behavior," in N. R. Ellis, ed.,* Handbook of Mental Deficiency, *New York: McGraw–Hill, p. 255. Adapted by permission.)*

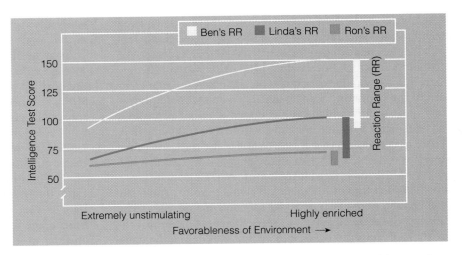

Instead, behavior is the result of a dynamic interplay between these two forces. How do heredity and environment work together to affect development? Several important concepts shed light on this question.

■ REACTION RANGE. The first of these ideas is **range of reaction** (Gottesman, 1963). It emphasizes that each person responds to the environment in a unique way because of his or her genetic makeup. Let's explore this idea by looking at Figure 2.10. Reaction range can apply to any characteristic; here it is illustrated for intelligence. Notice that when environments vary from extremely unstimulating to highly enriched, Ben's intelligence increases dramatically, Linda's only slightly, and Ron's hardly at all.

Reaction range highlights two important points about the relationship between heredity and environment. First, it shows that because each of us has a unique genetic makeup, we respond differently to the same environment. In Figure 2.10, notice how a poor environment results in a lower intelligence test score for Ron than Ben. Also, an advantaged environment raises Ben's score far above what is possible for Ron. Second, sometimes different genetic–environmental combinations can make two people look the same! For example, if Ben is reared in an unstimulating environment, his score will be about 100, or average for people in general. Linda can also obtain this score, but to do so she must grow up in a very advantaged home. The concept of range of reaction tells us that individuals differ in their range of possible responses to the environment. And unique blends of heredity and environment lead to both similarities and differences in behavior.

■ CANALIZATION. The concept of canalization provides another way of understanding how heredity and environment combine. **Canalization** is the tendency of heredity to restrict the development of some characteristics to just one or a few outcomes. A behavior that is strongly canalized follows a genetically set growth plan, and only powerful environmental forces can change it (Waddington, 1957). For example, infant perceptual and motor development

seems to be strongly canalized, since all normal human babies eventually roll over, reach for objects, sit up, crawl, and walk. It takes extreme conditions to modify these behaviors or cause them not to appear. In contrast, intelligence and personality are less strongly canalized, since they respond easily to changes in the environment.

Recently, scientists expanded the notion of canalization to include environmental influences. We now know that environments can also limit development (Gottlieb, 1991). For example, when children are exposed to harmful environments early in life, there may be little that experience can do to change characteristics (such as intelligence) that were quite flexible to begin with (Turkheimer & Gottesman, 1991).

Using the concept of canalization, we learn that genes restrict the development of some characteristics more than others. And over time, even very flexible behaviors can become fixed, depending on the environments to which individuals were exposed.

■ GENETIC–ENVIRONMENTAL CORRELATION. Nature and nurture work together in still another way. Several investigators point out that a major problem in trying to separate heredity and environment is that they are often correlated (Plomin, 1994; Scarr & McCartney, 1983). According to the concept of **genetic–environmental correlation,** our genes influence the environments to which we are exposed. In support of this idea, a recent study showed that the greater the genetic similarity between pairs of adolescents, the more alike they were on many aspects of child rearing, including parental discipline, affection, conflict, and monitoring of the young person's activities (Plomin et al., 1994).

These findings indicate that heredity plays a role in molding experience. The way this happens changes with development.

Passive and Evocative Correlation. At younger ages, two types of genetic–environmental correlation are common. The first is called *passive* correlation because the child has no control over it. Early on, parents provide environments that are influenced by their own heredity. For example, parents who are good athletes are likely to enroll their children in swimming and gymnastics lessons. Besides getting exposed to an "athletic environment," the children may have inherited their parents' athletic ability. As a result, they are likely to become good athletes for both genetic and environmental reasons.

The second type of genetic–environmental correlation is *evocative.* Children evoke responses from others that are influenced by the child's heredity, and these responses strengthen the child's original style of responding. For example, an active, friendly baby is likely to receive more social stimulation from those around her than a passive, quiet infant. And a cooperative, attentive child will probably receive more patient and sensitive interactions from parents than an inattentive, distractible youngster.

Active Correlation. At older ages, *active* genetic–environmental correlation becomes common. As children extend their experiences beyond the immediate family to school, neighborhood, and community and are given the freedom to make more of their own choices, they play an increasingly active role in seeking out environments that fit their genetic tendencies. The well-coordinated, muscular child spends more time at after-school sports, the musically talented youngster joins the school orchestra and practices his violin, and the intellectually curious child is a familiar visitor at her local library.

This tendency to actively choose environments that complement our heredity is called **niche-picking** (Scarr & McCartney, 1983). Infants and young children cannot do much niche-picking, since adults select environments for them. In contrast, older children, adolescents, and adults are much more in charge of their own environments. The niche-picking idea explains why pairs of identical twins reared apart during childhood and later reunited often find, to their great surprise, that they have similar hobbies, food preferences, friendship choices, and vocations (Bouchard et al., 1990; Plomin, 1994). It also helps us understand longi-

tudinal findings indicating that identical twins become somewhat more similar and fraternal twins and adopted siblings less similar in intelligence with age (McGue et al., 1993). The influence of heredity and environment is not constant but changes over time. With age, genetic factors may become more important in determining the environments we experience and choose for ourselves.

A major reason that researchers are interested in the nature–nurture issue is that they want to find ways to improve environments in order to help people develop as far as possible. The concepts of range of reaction, canalization, and genetic–environmental correlation remind us that development is best understood as a series of complex exchanges between nature and nurture. When a characteristic is strongly determined by heredity, it can still be modified. However, people cannot be changed in any way we might desire. The success of any attempt to improve development depends on the characteristics we want to change, the genetic makeup of the individual, and the type and timing of our intervention.

ASK YOURSELF . . .

■ *A researcher wants to know whether alcoholism is influenced by genetic factors. Which method of inferring the importance of heredity in complex human characteristics could help answer this question?*

■ *Bianca's parents are both accomplished musicians. Bianca began taking piano lessons when she was 4 and was accompanying her school choir by age 10. When she reached adolescence, she asked her parents if she could attend a special music high school. After graduation, she entered a music conservatory and became a concert pianist. Explain how genetic and environmental factors worked together to promote Bianca's talent.*

SUMMARY

GENETIC FOUNDATIONS

What are genes, and how are they transmitted from one generation to the next?

■ Development begins at conception, when sperm and ovum unite to form the one-celled **zygote.** Within the cell nucleus are 23 pairs of **chromosomes.** Along their length are **genes,** segments of **DNA** that make us distinctly human and play an important role in determining our development and characteristics.

■ **Gametes,** or sex cells, are produced by the process of cell division know as **meiosis.** Since each zygote receives a unique set of genes from each parent, meiosis ensures that children will be genetically different from one another. Once the zygote forms, it starts to develop into a complex human being through cell duplication, or **mitosis.**

■ If the fertilizing sperm carries an X chromosome, the child will be a girl; if it contains a Y chromosome, a boy

will be born. **Fraternal,** or **dizygotic, twins** result when two ova are released from the mother's ovaries and each is fertilized. In contrast, **identical,** or **monozygotic, twins** develop when a zygote divides in two during the early stages of cell duplication.

Describe various patterns of genetic inheritance.

■ **Dominant–recessive** and **codominant** relationships are patterns of inheritance that apply to many traits controlled by single genes. When recessive disorders are **X-linked** (carried on the X chromosome), males are more likely to be affected. Unfavorable genes arise from **mutations,** which can occur spontaneously or be induced by hazardous environmental agents. **Genetic imprinting** is a newly discovered pattern of inheritance in which one parent's gene is activated, regardless of its makeup.

❚ Human traits that vary continuously, such as intelligence and personality, are **polygenic,** or influenced by many genes. Since the genetic principles involved are unknown, scientists must study the influence of heredity on these characteristics indirectly.

CHROMOSOMAL ABNORMALITIES

Describe major chromosomal abnormalities, and explain how they occur.

❚ Most chromosomal abnormalities are due to errors in meiosis. The most common chromosomal disorder is Down syndrome, which results in physical defects and mental retardation. Disorders of the **sex chromosomes** are milder than defects of the **autosomes.** Contrary to popular belief, males with XYY syndrome are not prone to aggression. Studies of children with triple X, Klinefelter, and Turner syndromes reveal that adding to or subtracting from the usual number of X chromosomes leads to specific intellectual problems.

REPRODUCTIVE CHOICES

What procedures are available to assist prospective parents in having healthy, wanted children?

❚ **Genetic counseling** helps couples at risk for giving birth to children with genetic abnormalities decide whether or not to conceive. **Prenatal diagnostic methods** make early detection of genetic problems possible. Although reproductive technologies, such as donor insemi-

nation, in vitro fertilization, and surrogate motherhood, permit many individuals to become parents who otherwise would not, they raise serious legal and ethical concerns.

❚ Many parents who cannot conceive or who have a high likelihood of transmitting a genetic disorder decide to adopt. Although adopted children have more developmental problems than children in general, most fare quite well.

ENVIRONMENTAL CONTEXTS FOR DEVELOPMENT

Describe the social systems perspective on family functioning, along with aspects of the environment that support family well-being and children's development.

❚ Just as complex as heredity are the environments in which human development takes place. The family is the first and foremost context for development. The **social systems perspective** emphasizes that the behaviors of each family member affect those of others. The family system is also dynamic and ever changing, constantly adjusting to new events and developmental changes in its members.

❚ Despite these variations, one source of consistency in family functioning is social class. Middle-class families tend to be smaller, to place greater emphasis on nurturing psychological traits, and to promote warm, verbally stimulating interaction with children. Lower-class families often stress external characteristics and engage in more restrictive child rearing. Development is seriously undermined by poverty and homelessness.

❚ Supportive ties to the surrounding environment foster well-being throughout the lifespan. Communities that encourage constructive leisure time activities, warm interaction among residents, connections between settings, and children and adults' active participation enhance development.

❚ The values and practices of cultures and **subcultures** influence all aspects of daily life. In contrast to the American middle-class family, in many cultures grandparents, siblings, and other relatives share child-rearing responsibilities. **Extended family households,** in which three or more generations live together, are common among ethnic minorities. They protect development under conditions of high life stress.

❚ In the complex world in which we live, favorable development depends on **public policies.** Effective programs are influenced by many factors, including cultural values, a nation's economic resources, and organizations and individuals that work for an improved quality of life. American policies safeguarding children and their families are not as well developed as those for the elderly. The needs of many people of all ages are inadequately served.

UNDERSTANDING THE RELATIONSHIP BETWEEN HEREDITY AND ENVIRONMENT

Explain the various ways heredity and environment may combine to influence complex traits.

❚ Scientists do not agree on how heredity and environment influence complex characteristics, such as intelligence and personality. Some believe that it is useful and possible to determine "how much" each factor contributes to individual differences. These investigators compute **heritability estimates** and **concordance rates** from **kinship studies.**

❚ According to other researchers, the important question is "how" heredity and environment work together. The concepts of **range of reaction, canalization,** and **genetic–environmental correlation** remind us that development is best understood as a series of complex exchanges between nature and nurture that change over the lifespan.

IMPORTANT TERMS AND CONCEPTS

phenotype (p. 44)
genotype (p. 44)
chromosomes (p. 44)
deoxyribonucleic acid (DNA) (p. 44)
gene (p. 45)
mitosis (p. 45)
gametes (p. 45)
meiosis (p. 45
zygote (p. 46)
autosomes (p. 46)
sex chromosomes (p. 46)
fraternal, or dizygotic, twins (p. 47)

identical, or monozygotic, twins (p. 47)
homozygous (p. 47)
heterozygous (p. 47)
dominant–recessive inheritance (p. 48)
carrier (p. 48)
codominance (p. 49)
mutation (p. 49)
genetic imprinting (p. 50)
X-linked inheritance (p. 50)
polygenic inheritance (p. 52)
genetic counseling (p. 54)
prenatal diagnostic methods (p. 54)

social systems perspective (p. 59)
subculture (p. 64)
extended family household (p. 64)
public policies (p. 64)
heritability estimate (p. 70)
kinship studies (p. 70)
concordance rate (p. 71)
range of reaction (p. 72)
canalization (p. 72)
genetic–environmental correlation (p. 72)
niche-picking (p. 73)

FYI — FOR FURTHER INFORMATION AND HELP

CAUSES OF MUTATION

Environmental Mutagen Society
1730 N. Lynn Street, Suite 502
Arlington, VA 22209
(703) 525-1191
Web site: *www.ornl.gov/ TechResources*
Provides information on environmental agents that cause mutation.

GENETIC DISORDERS

March of Dimes
Birth Defects Foundation
1275 Mamaroneck Avenue
White Plains, NY 10605
(914) 428-7100
Web site: *www.modimes.org*
Works to prevent genetic disorders and other birth defects through public education and community service programs.

PKU Parents
8 Myrtle Lane
San Anselmo, CA 94960
(415) 457-4632
Provides support and education for parents of children with PKU.

Sickle Cell Disease Association
of America
3345 Wilshire Blvd.,
Suite 1106
Los Angeles, CA 90010
(800) 421-8453

Provides information and assists local groups that serve individuals with sickle cell anemia.

National Down
Syndrome Congress
1800 Dempster Street
Park Ridge, IL 60068-1146
(312) 823-7550
Web site: *www.carol.net/~ndsc*
Assists parents in finding solutions to the needs of children with Down syndrome. Local groups exist across the United States.

INFERTILITY

Resolve, Inc.
1310 Broadway
Somerville, MA 02144-1731
(617) 623-0744
Web site: *www.resolve.org*
Offers counseling referral and support to persons with fertility problems.

ADOPTION

National Adoption Information
Clearinghouse
11426 Rockville Pike, Suite 410
Rockville, MD 20852
(800) 332-6347
Web site: *www.naicinfo.com*
Provides information on all aspects of adoption, including children from other

countries, children with special needs, and state and federal adoption laws.

PUBLIC POLICY

Children's Defense Fund
122 C Street, N.W.
Washington, DC 20001
(800) 233-1200
Web site: *www.childrensdefense.org*
An active child advocacy organization. Provides information on the condition of children and government-sponsored programs serving them.

American Association
of Retired Persons
601 E Street, N.W.
Washington, DC 20049
(202) 434-2277
Web site: *www.aarp.org*
An active advocacy organization for older people, with 32 million members age 50 or older, working or retired. Aims to enhance quality of life, independence, and dignity of the elderly.

Gray Panthers
2025 Pennsylvania Avenue, N.W.
Suite 821
Philadelphia, PA 20006
(202) 466-3132
A coalition of young, middle-aged, and older Americans dedicated to overcoming discrimination on the basis of age.

Prenatal Development, Birth, and the Newborn Baby

After months of wondering if the time in their own lives was right, Yolanda and Jay decided to have a baby. I met them one fall in my child development class, when Yolanda was just 2 months pregnant. Both were full of questions: "How does the baby grow before birth? When are different organs formed? Has its heart begun to beat? Can it hear, feel, or sense our presence in other ways?"

Most of all, Yolanda and Jay wanted to do everything possible to make sure their baby would be born healthy. At one time, they believed that the developing organism was completely shielded by the uterus from any dangers in the environment. All babies born with problems, they thought, had unfavorable genes. After browsing through several pregnancy books, Yolanda and Jay realized they were wrong. Yolanda started to wonder about her diet. And she asked me whether an aspirin for a headache, a glass of wine at dinnertime, or a few cups of coffee during study hours might be harmful.

In this chapter, we answer Yolanda and Jay's questions, along with a great many more that scientists have asked about the events before birth. First, we trace prenatal development, paying special attention to environmental factors that can damage the fetus or support healthy growth. Next, we turn to the events of childbirth. We will see that cultures treat labor and delivery in strikingly different ways. Today, women in industrialized nations have many more choices about where and how they give birth than at any time in the past.

Yolanda and Jay's son Joshua reaped the benefits of their careful attention to his needs during pregnancy. Nevertheless, some infants are at serious risk for birth complications, and even when they are not, the birth process does not always go smoothly. We pay special attention to the problems of infants who are born underweight or too early, before the prenatal period is complete. Our discussion will also examine the pros and cons of medical interventions, such as pain-relieving drugs and surgical deliveries, designed to ease a difficult birth and protect the health of mother and baby. We conclude by taking a close look at the remarkable capacities of newborn babies to adapt to the external world and communicate their needs.

PRENATAL DEVELOPMENT

During Yolanda's pregnancy, the one-celled zygote grew into a complex human infant fully ready to be born. The vast changes that take place during these 38 weeks are usually divided into three phases: (1) the period of the zygote, (2) the period of the embryo, and (3) the period of the fetus. As we look at what happens in each,

During the period of the zygote, the fertilized ovum begins to duplicate at an increasingly rapid rate, forming a hollow ball of cells, or blastocyst, by the fourth day of fertilization. Here the blastocyst, magnified thousands of times, burrows into the uterine lining between the seventh and ninth day. (© Lennart Nilsson, *A Child Is Born*/Bonniers)

you may find it useful to refer to Table 3.1, which summarizes milestones of prenatal development.

PERIOD OF THE ZYGOTE

The period of the zygote lasts about 2 weeks, from fertilization until the tiny mass of cells drifts down and out of the fallopian tube and attaches itself to the wall of the uterus. The zygote's first cell duplication is long and drawn out; it is not complete until about 30 hours after conception. Gradually new cells are added at a faster rate. By the fourth day, 60 to 70 cells form a hollow, fluid-filled ball called a **blastocyst.** The cells on the inside, called the **embryonic disk,** will become the new organism; the outer ring will provide protective covering.

IMPLANTATION. Sometime between the seventh and ninth day, **implantation** occurs: the blastocyst burrows deep into the uterine lining. Surrounded by the woman's nourishing blood, now it starts to grow in earnest. The protective outer layer quickly develops into a membrane called the **amnion.** It encloses the developing organism in **amniotic fluid,** which helps regulate temperature and acts as a cushion against jolts caused by the woman's movements. A yolk sac emerges that produces blood cells until the liver, spleen, and bone marrow are mature enough to take over this function (Moore & Persaud, 1993).

Test Bank Items 3.1 through 3.6, 3.8, 3.10

Major Milestones of Prenatal Development

TRIMESTER	PERIOD	WEEKS	LENGTH AND WEIGHT	MAJOR EVENTS
First	Zygote	1		The one–celled zygote multiplies and forms a blastocyst.
		2		The blastocyst burrows into the uterine lining. Structures that feed and protect the developing organism begin to form—amnion, chorion, yolk sac, placenta, and umbilical cord.
	Embryo	3–4	¼ inch	A primitive brain and spinal cord appear. Heart, muscles, backbone, ribs, and digestive tract begin to develop.
		5–8	1 inch	Many external body structures (such as face, arms, legs, toes, fingers) and internal organs form. The sense of touch begins to develop, and the embryo can move.
	Fetus	9–12	3 inches; less than 1 ounce	Rapid increase in size begins. Nervous system, organs, and muscles become organized and connected, and new behavioral capacities (kicking, thumb sucking, mouth opening, and rehearsal of breathing) appear. External genitals are well formed, and the fetus's gender is evident.
Second		13–24	12 inches; 1.8 pounds	The fetus continues to enlarge rapidly. In the middle of this period, fetal movements can be felt by the mother. Vernix and lanugo keep the fetus's skin from chapping in the amniotic fluid. All the neurons that will ever be produced in the brain are present by 24 weeks. Eyes are sensitive to light, and the fetus reacts to sound.
Third		25–38	20 inches; 7.5 pounds	The fetus has a chance of survival if born around this time. Size continues to increase. Lungs gradually mature. Rapid brain development causes sensory and behavioral capacities to expand. In the middle of this period, a layer of fat is added under the skin. Antibodies are transmitted from mother to fetus to protect against disease. Most fetuses rotate into an upside-down position in preparation for birth.

Sources: Moore & Persaud, 1993; Nilsson & Hamberger, 1990.

The events of these first 2 weeks are delicate and uncertain. As many as 30 percent of zygotes do not make it through this phase. In some, the sperm and ovum do not join properly. In others, cell duplication never begins. By preventing implantation in these cases, nature quickly eliminates most prenatal abnormalities (Sadler, 1995).

■ THE PLACENTA AND UMBILICAL CORD. By the end of the second week, another protective membrane, called the **chorion,** surrounds the amnion. From the chorion, tiny fingerlike *villi*, or blood vessels, begin to emerge.[1] As these villi burrow into the uterine wall, a special organ called the

[1] Recall from Chapter 2 that chorionic villus sampling is the prenatal diagnostic method that can be performed earliest, by 6 to 8 weeks after conception. In this procedure, tissue from the ends of the villi are removed and examined for genetic abnormalities.

placenta starts to develop. By bringing the mother's and embryo's blood close together, the placenta permits food and oxygen to reach the developing organism and waste products to be carried away. A special membrane forms that allows these substances to be exchanged but prevents the mother's and embryo's blood from mixing directly.

The placenta is connected to the developing organism by the **umbilical cord,** which first appears as a tiny stalk and eventually grows to a length of 1 to 3 feet. The umbilical cord contains one large vein that delivers blood loaded with nutrients and two arteries that remove waste products. The force of blood flowing through the cord keeps it firm so it seldom tangles while the embryo, like a spacewalking astronaut, floats freely in its fluid-filled chamber (Moore & Persaud, 1993).

By the end of the period of the zygote, the developing organism has found food and shelter. These dramatic

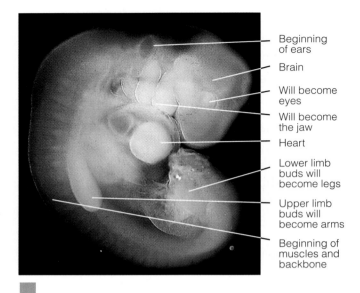

Beginning
of ears

Brain

Will become
eyes

Will become
the jaw

Heart

Lower limb
buds will
become legs

Upper limb
buds will
become arms

Beginning of
muscles and
backbone

This curled embryo is about 4 weeks old. In actual size, it is only
1/4-inch long, but many body structures have begun to form. The
primitive tail will disappear by the end of the embryonic period.
(© Lennart Nilsson, *A Child Is Born*/Bonniers)

beginnings take place before all but the most sensitive
mother knows she is pregnant.

PERIOD OF THE EMBRYO

The period of the **embryo** lasts from the second through
the eighth week of pregnancy. During these brief 6 weeks,
the most rapid prenatal changes take place as the ground-
work is laid for all body structures and internal organs.

LAST HALF OF THE FIRST MONTH. In the first week of
this period, the embryonic disk forms three layers of cells:
(1) the *ectoderm*, which will become the nervous system
and skin; (2) the *mesoderm*, from which will develop the
muscles, skeleton, circulatory system, and other internal
organs; and (3) the *endoderm*, which will become the digestive system, lungs, urinary tract, and glands. These three
layers give rise to all parts of the body.

At first, the nervous system develops fastest. The ectoderm folds over to form a **neural tube**, which will become
the spinal cord and brain. While the nervous system is developing, the heart begins to pump blood around the embryo's circulatory system, and muscles, backbone, ribs, and
digestive tract start to appear. At the end of the first month,
the curled embryo consists of millions of organized groups
of cells with specific functions, although it is only one
fourth of an inch long.

THE SECOND MONTH. In the second month, growth
continues rapidly. The eyes, ears, nose, jaw, and neck form.
Tiny buds become arms, legs, fingers, and toes. Internal or-

gans are more distinct: the intestines grow, the heart develops separate chambers, and the liver and spleen take over
production of blood cells so that the yolk sac is no longer
needed. Changing body proportions cause the embryo's
posture to become more upright. Now 1 inch long and
one-seventh of an ounce in weight, the embryo can already
sense its world. It responds to touch, particularly in the
areas of the mouth and on the soles of the feet. And it can
move, although its tiny flutters are still too light to be felt
by the mother (Nilsson & Hamberger, 1990).

PERIOD OF THE FETUS

Lasting from the ninth week to the end of pregnancy,
the period of the **fetus** is the "growth and finishing" phase.
During this longest prenatal period, the developing organism increases rapidly in size.

THE THIRD MONTH. In the third month, the organs,
muscles, and nervous system start to become organized and
connected. The fetus kicks, bends its arms, forms a fist, curls
its toes, opens its mouth, and even sucks its thumb. The tiny
lungs begin to expand and contract in an early rehearsal of
breathing movements. By the twelfth week, the external
genitals are well formed, and the gender of the fetus can be
determined using ultrasound. Other finishing touches appear, such as fingernails, toenails, tooth buds, and eyelids.
The heartbeat can now be heard through a stethoscope.

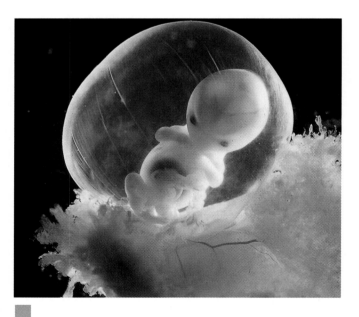

By 7 weeks, the embryo's posture is more upright. Body
structures—eyes, nose, arms, legs, and internal organs—are more
distinct. An embryo of this age responds to touch. It can also
move, although at less than an inch long and an ounce in weight,
it is still too tiny to be felt by the mother. (© Lennart Nilsson,
A Child Is Born/Bonniers)

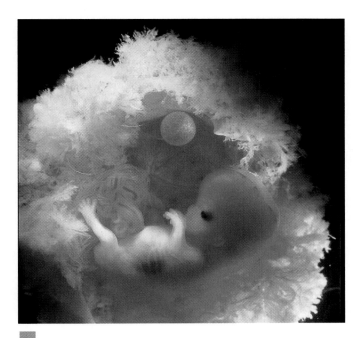

During the period of the fetus, the organism increases rapidly in size, and body structures are completed. At 11 weeks, the brain and muscles are better connected. The fetus can kick, bend its arms, open and close its hands and mouth, and suck its thumb. Notice the yolk sac, which shrinks as pregnancy advances. The internal organs have taken over its function of producing blood cells. (© Lennart Nilsson, *A Child Is Born*/Bonniers)

Prenatal development is sometimes divided into **trimesters,** or three equal time periods. At the end of the third month, the *first trimester* is complete. Two more must pass before the fetus is fully prepared to survive outside the womb.

■ THE SECOND TRIMESTER. By the middle of the *second trimester* (which lasts from 13 to 24 weeks), the new being has grown large enough that its movements can be felt by the mother. A white, cheeselike substance called **vernix** now protects its skin from chapping during the long months spent bathing in the amniotic fluid. White, downy hair called **lanugo** also appears over the entire body, helping the vernix stick to the skin.

By the end of the second trimester, all the brain's *neurons* (nerve cells that store and transmit information) are in place; no more will be produced in the individual's lifetime. However, *glial cells,* which support and feed the neurons, continue to increase at a rapid rate throughout pregnancy, as well as after birth (Nowakowski, 1987).

Brain growth means new behavioral capacities. The 20-week-old fetus can be stimulated as well as irritated by sound. And if a doctor looks inside the uterus with fetoscopy (see Chapter 2, page 54), fetuses try to shield their eyes from the light with the hands, indicating that the sense

of sight has begun to emerge (Nilsson & Hamberger, 1990). Still, a fetus born at this time cannot survive. Its lungs are immature, and the brain cannot yet control breathing movements or regulate body temperature.

■ THE THIRD TRIMESTER. During the *third trimester* (25 to 38 weeks), the fetus has a chance for survival outside the womb. The point at which it can first survive, called the **age of viability,** occurs sometime between 22 and 26 weeks (Moore & Persaud, 1993). If born between the seventh and eighth month, however, the baby usually needs help breathing. Although the brain's respiratory center is mature, tiny air sacs in the lungs are not yet ready to inflate and exchange carbon monoxide for oxygen.

The brain continues to make great strides. The *cerebral cortex,* the seat of human intelligence, enlarges. Now the fetus responds more clearly to sounds, and it even develops a preference for the tone and rhythm of its mother's voice. In one study, mothers read aloud Dr. Seuss's lively poem *The Cat in the Hat* for the last 6 weeks of pregnancy. After birth, their infants learned to turn on recordings of the mother's voice by sucking on nipples. They showed a preference for the familiar poem by sucking harder to hear *The Cat in the Hat* than other rhyming stories (DeCasper & Spence, 1986).

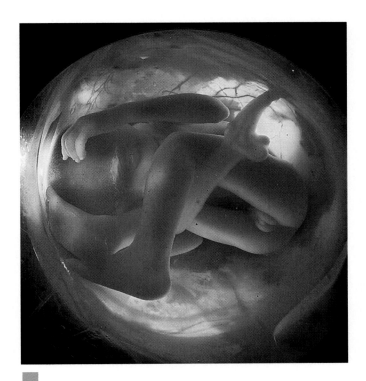

At 22 weeks, this fetus is almost a foot long and slightly over a pound in weight. Its movements can be clearly felt by the mother and by other family members who place a hand on her abdomen. If born at this time, a baby has a slim chance of surviving. (© Lennart Nilsson, *A Child Is Born*/Bonniers)

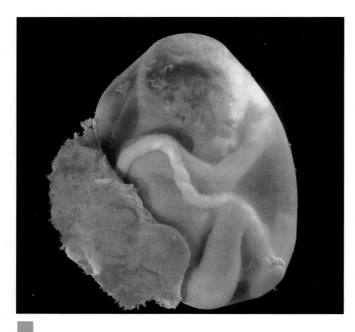

This 36-week-old fetus fills the uterus. To support its need for nourishment, the umbilical cord and placenta have grown very large. Notice the vernix (cheeselike substance), which protects the skin from chapping. The fetus has accumulated a layer of fat to assist with temperature regulation after birth. In another 2 weeks, it will be full term. (© Lennart Nilsson, *A Child Is Born*/Bonniers)

During the final 3 months, the fetus gains more than 5 pounds and grows 7 inches. In the eighth month, a layer of fat is added to assist with temperature regulation. The fetus also receives antibodies from the mother's blood that protect against illnesses, since the newborn's own immune system will not work well until several months after birth. In the last weeks, most fetuses assume an upside-down position, partly because of the shape of the uterus and because the head is heavier than the feet. Growth slows, and birth is about to take place.

BRIEF REVIEW

The vast changes that take place during pregnancy are usually divided into three periods. In the period of the zygote, the tiny one-celled fertilized ovum begins to duplicate and implants itself in the uterine lining. Structures that will feed and protect the developing organism begin to form. During the period of the embryo, the foundations for all body tissues and organs are rapidly laid down. The longest prenatal phase, the period of the fetus, is devoted to growth in size and completion of body systems. Turn back to Table 3.1 on page 79 to review the specific changes during the 9 months before birth.

ASK YOURSELF . . .

■ *Amy, who is 2 months pregnant, wonders how the embryo is being fed and what parts of the body have formed. Amy imagines that very little development has yet taken place. How would you answer Amy's questions? Will she be surprised at your response?*

PRENATAL ENVIRONMENTAL INFLUENCES

Although the prenatal environment is far more constant than the world outside the womb, a great many factors can affect the developing fetus. Yolanda and Jay learned that there was much they could do to create a safe environment for development before birth.

TERATOGENS

The term **teratogen** refers to any environmental agent that causes damage during the prenatal period. It comes from the Greek word *teras*, meaning "malformation" or "monstrosity." This label was selected because medical science first learned about harmful prenatal influences from cases in which babies had been profoundly damaged.

Yet the harm done by teratogens is not always simple and straightforward. It depends on several factors. First, we will see as we discuss particular teratogens that larger doses over a long time period usually have more negative effects than smaller doses over a short period. Second, the genetic makeup of both the mother and the developing organism plays an important role, since (as we saw in Chapter 2) some individuals are better able to withstand harmful environments. Third, the presence of several negative factors at once can worsen the impact of a single harmful agent. Fourth, the effects of teratogens vary with the age of the organism at time of exposure. We can best understand this idea if we think of prenatal development in terms of the *sensitive period* concept introduced in Chapter 1. Recall that a sensitive period is a limited time span during which a part of the body or a behavior is biologically prepared to develop rapidly. During that time, it is especially sensitive to its surroundings. If the environment happens to be harmful, then damage occurs, and recovery is difficult and sometimes impossible.

Figure 3.1 summarizes sensitive periods during prenatal development. In the period of the zygote, before implantation, teratogens rarely have any impact. If they do, the tiny mass of cells is usually so completely damaged that it dies. The embryonic period is the time when serious de-

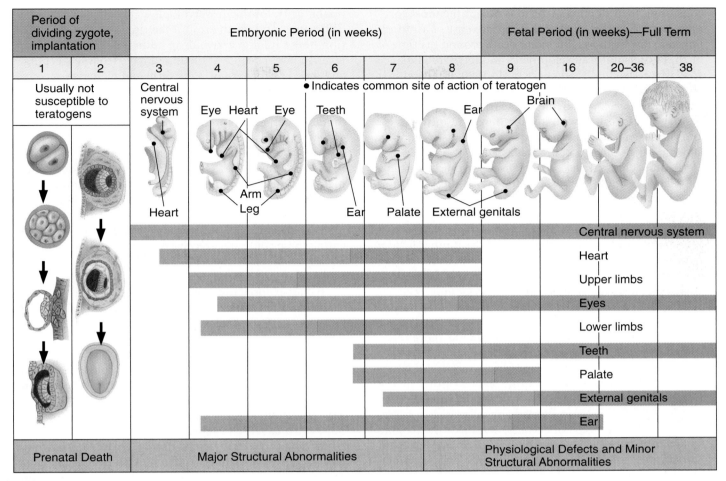

Period of dividing zygote, implantation		Embryonic Period (in weeks)						Fetal Period (in weeks)—Full Term			
1	2	3	4	5	6	7	8	9	16	20–36	38

FIGURE 3.1

Sensitive periods in prenatal development. Each organ or structure has a sensitive period during which its development may be disturbed. Gray horizontal lines indicate highly sensitive periods. Pink horizontal lines indicate periods that are somewhat less sensitive to teratogens, although damage can occur. *(From K. L. Moore & T. V. N. Persaud, 1993, Before We Are Born, 4th ed., Philadelphia: Saunders, p. 130. Reprinted by permission of the publisher and the author.)*

fects are most likely to occur, since the foundations for all body parts are laid down. During the fetal period, damage by teratogens is usually minor. However, some organs, such as the brain, eyes, and genitals, can still be strongly affected. Now let's look at what scientists have discovered about a variety of teratogens.

PRESCRIPTION AND NONPRESCRIPTION MEDICATIONS. In the early 1960s, the world learned a tragic lesson about drugs and prenatal development. At that time, a sedative called thalidomide was widely available in Europe, Canada, and South America. When taken by mothers 4 to 6 weeks after conception, thalidomide produced gross deformities of the arms and legs, affecting about 7,000 infants worldwide (Moore & Persaud, 1993). As children exposed to thalidomide grew older, many scored below average in intelligence. Perhaps the drug damaged the central nervous system directly. Or the child-rearing conditions of these se-

verely deformed youngsters may have impaired their intellectual development (Vorhees & Mollnow, 1987).

Another medication, a synthetic hormone called *diethylstilbestrol (DES)*, was widely prescribed between 1945 and 1970 to prevent miscarriages. Although babies whose mothers had taken DES seemed unharmed at birth, problems with their reproductive systems were sometimes found later on. Young women had unusually high rates of cancer of the vagina and malformations of the uterus. When they tried to have children, their pregnancies more often resulted in prematurity, low birth weight, and miscarriage than those of non-DES-exposed women. Young men showed an increased risk of genital abnormalities and cancer of the testes (Linn et al., 1988; Stillman, 1982).

Just about any drug taken by the mother can enter the embryonic or fetal bloodstream. Yet pregnant women continue to take a wide variety of medications without consulting their doctors. Aspirin is one of the most common. Several studies suggest that repeated use of aspirin is linked to low birth weight, infant death around the time of birth, poorer motor development, and lower intelligence scores in early childhood (Barr et al., 1990; Streissguth et al., 1987). Coffee contains another frequently consumed drug, caffeine. Heavy caffeine intake (over 3 cups of coffee per day) is associated with low birth weight, miscarriage, and newborn withdrawal symptoms, such as irritability and vomiting (Dlugosz & Bracken, 1992; Eskenazi, 1993).

Because children's lives are involved, we must take findings like these quite seriously. At the same time, we cannot be sure that commonly used drugs actually cause the problems just mentioned. Often mothers take more than one kind of drug. If the prenatal organism is injured, it is hard to tell which drug might be responsible or if other factors correlated with drug taking are really at fault. Until we have more information, the safest course of action is one that Yolanda took: cut down or avoid these drugs entirely.

■ ILLEGAL DRUGS. Use of highly addictive mood-altering drugs, such as cocaine and heroin, has become more widespread, especially in poverty-stricken inner cities where drugs provide a temporary escape from a daily life of hopelessness. The number of "cocaine babies" born in the United States has reached crisis levels in recent years. About 400,000 infants are affected annually (Waller, 1993).

Babies born to users of cocaine, heroin, or methadone (a less addictive drug used to wean people away from heroin) are at risk for a wide variety of problems, including

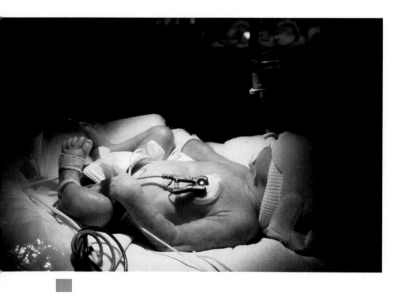

This baby, whose mother took crack during pregnancy, was born many weeks premature. He breathes with the aid of a respirator. His central nervous system may be seriously damaged. Researchers do not yet know if these outcomes are actually caused by crack or by the many other high-risk behaviors of drug users. *(John Giordano/Saba)*

prematurity, low birth weight, physical defects, breathing difficulties, and death around the time of birth. In addition, these infants arrive drug addicted. They are often feverish and irritable at birth and have trouble sleeping, and their cries are abnormally shrill and piercing—a common symptom among stressed newborns (Burkett et al., 1994; Fox, 1994; Miller, Boudreaux, & Regan, 1995). When mothers with many problems of their own must take care of these babies, who are difficult to calm down, cuddle, and feed, behavior problems are likely to persist.

Throughout the first year of life, heroin- and methadone-exposed infants are less attentive to the environment, and their motor development is slow. After infancy, some children get better, whereas others remain jittery and inattentive. Researchers believe the kind of parenting these youngsters receive may explain why there are lasting problems for some but not for others (Vorhees & Mollnow, 1987).

Babies exposed to cocaine seem to be especially at risk. Cocaine is linked to a variety of physical defects, including eye, bone, genital, urinary tract, kidney, and heart deformities as well as brain seizures (Holzman & Paneth, 1994). Babies born to mothers who smoke crack (a cheap form of cocaine that delivers high doses quickly through the lungs) seem worst off in terms of low birth weight and damage to the central nervous system (Kaye et al., 1989). Fathers may also contribute to these effects. Research suggests that cocaine can attach itself to sperm, "hitchhike" its way into the zygote, and cause birth defects (Yazigi, Odem, & Polakoski, 1991). Still, it is difficult to isolate the precise damage caused by cocaine, since users often take several drugs and engage in other high-risk behaviors (Gonzalez & Campbell, 1994).

Marijuana is another illegal drug that is used more widely than heroin and cocaine. Studies examining its relationship to low birth weight and prematurity reveal mixed findings (Fried, 1993). Nevertheless, prenatal marijuana exposure is related to newborn startles, disturbed sleep, an abnormally high-pitched cry, and reduced attention to the environment (Dahl et al., 1995; Fried & Makin, 1987). These outcomes certainly put newborn babies at risk for future problems, even though long-term effects have not been established.

■ CIGARETTE SMOKING. Although smoking has recently declined in the United States, an estimated one-fourth to one-third of adults are regular cigarette users. The rate of smoking is especially high among women under age 25— 30 to 40 percent (Birenbaum-Carmeli, 1995). The most well-known effect of smoking during pregnancy is low birth weight. But the likelihood of other serious consequences, such as prematurity, miscarriage, infant death, and cancer later in childhood is also increased.

Even when a baby of a smoking mother appears to be born in good physical condition, slight behavioral abnormalities may threaten the child's development. Newborns of smoking mothers are less attentive to sounds and display more muscle tension (Fried & Makin, 1987). An unresponsive, restless baby may not evoke the kind of interaction from adults that promotes healthy psychological development. Some studies report that prenatally exposed children have shorter attention spans and poorer mental test scores in early childhood, even after many other factors have been controlled (Fergusson, Horwood, & Lynskey, 1993; Fried & Watkinson, 1990). But other researchers have not been able to confirm these findings, so lasting effects remain uncertain (Barr et al., 1990; Streissguth et al., 1989).

Exactly how can smoking harm the fetus? Nicotine, the addictive substance in tobacco, causes the placenta to grow abnormally. This reduces transfer of nutrients, so the fetus gains weight poorly. Also, smoking raises the concentration of carbon monoxide in the bloodstreams of both mother and fetus. Carbon monoxide displaces oxygen from red blood cells. It damages the central nervous system and reduces birth weight in the fetuses of laboratory animals. Similar effects may occur in humans (Cotton, 1994; Fried, 1993).

The more cigarettes a mother smokes, the greater the likelihood that her baby will be affected. But if a pregnant woman decides to stop smoking at any time, even during the last trimester, she can help her baby. She immediately reduces the chances that the infant will be born underweight and suffer from future problems (Ahlsten, Cnattingius, & Lindmark, 1993; Li, Windsor, & Perkins, 1993).

Finally, Jay gave up cigarettes when Yolanda became pregnant. Newborn infants of fathers who smoke are also at risk for low birth weight and infant death! Jay realized that a smoke-filled home could harm the fetus by turning Yolanda into a "passive smoker" who inhaled nicotine and carbon monoxide from the air around her (Fortier, Marcoux, & Brisson, 1994; Makin, Fried, & Watkinson, 1991).

■ **ALCOHOL.** In *The Broken Cord,* Michael Dorris (1989), a Dartmouth University anthropology professor, described what it was like to raise his adopted son, Adam. The boy's biological mother drank heavily throughout pregnancy and died of alcohol poisoning shortly after his birth. A Sioux Indian, Adam was born with **fetal alcohol syndrome (FAS).** Children with this disorder show mental retardation, poor attention, and overactivity as well as a distinct set of physical symptoms. These include slow physical growth and a particular pattern of facial abnormalities: widely spaced eyes, short eyelid openings, a small

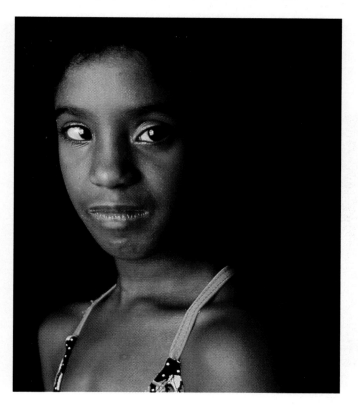

The mother of the severely retarded boy on the left drank heavily during pregnancy. His widely spaced eyes, thin upper lip, and short eyelid openings are typical of fetal alcohol syndrome. The adolescent girl on the right also has these physical symptoms. The brain damage caused by alcohol before she was born is permanent. It has made learning in school and adapting to everyday challenges extremely difficult. (George Steinmetz)

upturned nose, a thin upper lip, and a small head, indicating that the brain has not developed fully. In a related condition, known as **fetal alcohol effects (FAE),** individuals display only some of these abnormalities. Usually, their mothers drank alcohol in smaller quantities (Hoyseth & Jones, 1989).

Even when provided with enriched diets, FAS babies fail to catch up in physical size during infancy or childhood (Hoyseth & Jones, 1989). Mental impairment is also permanent: Now a young adult, Adam's intelligence is still below average, and he has trouble concentrating and keeping a routine job. He also suffers from poor judgment. He might buy something and not wait for change or wander off in the middle of a task. On the reservation where Adam was born, many children show symptoms of prenatal alcohol exposure. Alcohol abuse is higher in poverty-stricken sectors of the population, especially among Native Americans.

How does alcohol produce its devastating effects? First, it interferes with brain development during the early months of pregnancy. Autopsies of FAS babies show a reduced number of brain cells and major structural abnormalities (Moroney & Allen, 1994). Second, the body uses large quantities of oxygen to metabolize alcohol. A pregnant woman's heavy drinking draws away oxygen that the developing organism needs for cell growth (Vorhees & Mollnow, 1987).

How much alcohol is safe during pregnancy? One study linked as little as 2 ounces of alcohol a day, taken very early in pregnancy, to FAS-like facial features (Astley et al., 1992). But recall that other factors—both genetic and environmental—can make some fetuses more vulnerable to teratogenic effects. Therefore, a precise dividing line between safe and dangerous drinking levels cannot be established. Research shows that the more alcohol consumed during pregnancy, the poorer a child's motor coordination, speed of information processing, intelligence, and achievement during the school years (Barr et al., 1990; Jacobson et al., 1993; Streissguth et al., 1994). These dose-related effects indicate that it is best for pregnant women to avoid alcohol entirely.

■ RADIATION. Defects due to radiation were tragically apparent in the children born to pregnant Japanese women who survived the bombing of Hiroshima and Nagasaki during World War II. Miscarriage, slow physical growth, an underdeveloped brain, and malformations of the skeleton and eyes were common (Michel, 1989). Even when an exposed child appears normal at birth, problems may appear later. For example, low-level radiation, as a result of medical X-rays or leakage in the workplace, can increase the risk of childhood cancer (Smith, 1992). Women need to tell their doctors if they are pregnant or trying to become pregnant before X-ray examinations. They should also avoid work environments where X-ray exposure might occur.

■ ENVIRONMENTAL POLLUTION. An astounding number of potentially dangerous chemicals are released into the environment in industrialized nations: over 100,000 are in common use in the United States. Although many chemicals cause serious birth defects in laboratory animals, the impact of only a few on human prenatal development is known.

Mercury is an established teratogen. In the 1950s, an industrial plant released waste containing high levels of mercury into a bay providing food and water for the town of Minimata, Japan. Many children born at the time displayed mental retardation, abnormal speech, difficulty in chewing and swallowing, and uncoordinated movements. Autopsies of those who died revealed widespread brain damage (Vorhees & Mollnow, 1987).

Another teratogen, *lead,* is present in some car exhaust, paint flaking off the walls of old buildings, and materials used in certain industrial occupations. High levels of lead exposure are consistently related to prematurity, low birth weight, brain damage, and a wide variety of physical defects (Dye-White, 1986). Even very low levels of prenatal lead exposure seem to be dangerous. Affected babies show slightly poorer mental development during the first 2 years (Bellinger et al., 1987).

For many years, *polychlorinated biphenyls (PCBs)* were used to insulate electrical equipment until research showed that (like mercury) they found their way into waterways and entered the food supply. In one study, newborn babies of women who frequently ate PCB-contaminated fish had slightly lower than average birth weights, smaller heads (suggesting brain damage), and less interest in their surroundings than did infants whose mothers ate little or no fish (Jacobson et al., 1984). When studied again at 7 months, infants whose mothers ate fish during pregnancy did more poorly on memory tests (Jacobson et al., 1985). A follow-up at 4 years showed persisting memory difficulties and lower verbal intelligence test scores (Jacobson, Jacobson, & Humphrey, 1990; Jacobson et al., 1992).

■ INFECTIOUS DISEASE. On her first prenatal visit, Yolanda's doctor asked if she and Jay had already had measles, mumps, and chicken pox, as well as other illnesses. In addition, Yolanda was checked for the presence of several infections, and for good reason. As you can see in Table 3.2, certain diseases during pregnancy are major causes of miscarriage and birth defects.

Although most viruses, such as the common cold or flu, appear to have little or no impact on the developing fetus, a few can cause extensive damage. In the mid-1960s, a worldwide epidemic of *rubella* (German measles) led to the birth of over 20,000 American babies with serious physical defects. Consistent with the sensitive period concept, rubella causes the greatest damage during the embryonic period. Over 50 percent of infants whose mothers became ill dur-

TABLE 3.2

Effects of Some Infectious Diseases During Pregnancy

DISEASE	MISCARRIAGE	PHYSICAL MALFORMATIONS	MENTAL RETARDATION	LOW BIRTH WEIGHT AND PREMATURITY
Viral				
Acquired immune deficiency syndrome (AIDS)	0	?	+	?
Chicken pox	0	+	+	+
Cytomegalovirus	+	+	+	+
Herpes simplex 2 (genital herpes)	+	+	+	+
Mumps	+	?	0	0
Rubella	+	+	+	+
Bacterial				
Syphilis	+	+	+	?
Tuberculosis	+	?	+	+
Parasitic				
Malaria	+	0	0	+
Toxoplasmosis	+	+	+	+

Note: + = established finding, 0 = no present evidence, ? = possible effect that is not clearly established.

Source: Adapted from F. L. Cohen, *Clinical Genetics in Nursing Practice* (1984), Philadelphia: Lippincott, p. 232. Reprinted by permission.

Additional sources: Chatkupt et al., 1989; Cohen, 1993a; Peckham & Logan, 1993; Qazi et al., 1988; Samson, 1988; Sever, 1983; Vorhees, 1986.

ing that time show heart defects, eye cataracts, deafness, mental retardation, and genital, urinary, and intestinal abnormalities. Infection during the fetal period is less harmful, but low birth weight, hearing loss, and bone defects may still occur (Eberhart-Phillips, Frederick, & Baron, 1993; Samson, 1988). Since 1966, infants and young children have been routinely vaccinated against rubella, so the number of prenatal cases today is much lower than it was a generation ago. Still, 10 to 20 percent of American women of childbearing age lack the rubella antibody, so new outbreaks of the disease are still possible (Lee et al., 1992).

Table 3.2 summarizes the harmful effects of other common viruses. The developing organism is especially sensitive to the family of herpes viruses, for which there is no vaccine or treatment. Among these, cytomegalovirus (the most frequent prenatal infection, transmitted through respiratory or sexual contact) and herpes simplex 2 (which is sexually transmitted) are especially dangerous. The human immunodeficiency virus (HIV), which leads to *acquired immune deficiency syndrome (AIDS)*, is infecting increasing numbers of babies. To find out about its prenatal transmission, refer to the Social Issues box on page 88.

Also included in Table 3.2 are several bacterial and parasitic diseases. Among the most common is *toxoplasmosis,* caused by a parasite found in many animals. Pregnant women may become infected from eating raw or undercooked meat or from contact with the feces of infected cats. About 40 percent of women who have the disease transmit it to the developing organism. If the disease strikes during the first trimester, it often leads to severe eye and brain damage. Later infection is linked to mild visual and cognitive impairments (Peckham & Logan, 1993). Expectant mothers can avoid toxoplasmosis by making sure the meat they eat is well cooked, having pet cats checked for the disease, and avoiding cat litter boxes.

OTHER MATERNAL FACTORS

Besides avoiding teratogens, expectant parents can support the development of the embryo or fetus in other ways. In healthy, physically fit women, regular exercise, in the form of enjoyable activities such as swimming, hiking, and aerobics, is related to increased birth weight (Hatch et al., 1993). (Note, however, that pregnant women with health problems, such as circulatory difficulties or previous miscarriages, should consult their doctors about a physical fitness routine.)

In addition, good nutrition and emotional well-being of the mother are critically important. Blood type differences between mother and fetus sometimes create difficulties. Finally, many expectant parents wonder how a mother's age and previous births affect the course of pregnancy. We examine each of these factors in the following sections.

■ NUTRITION. Children grow more rapidly during the prenatal period than at any other phase of development. During this time, they depend totally on the mother for nutrients to support their growth.

PRENATAL TRANSMISSION OF AIDS

First-born child of Jean and Claire, Ginette was diagnosed with AIDS when she was 6 months old. She died from respiratory infections and a failure to grow normally at 11 months of age. Jean and Claire could not understand the social worker's explanation of why Ginette died. After all, neither parent felt sick. At the time, Claire was pregnant with a second baby. Several weeks after Ginette's death, Jeanine was born. When word spread that AIDS caused Ginette's death, Jean lost his job, and the family was evicted from their apartment. Over the next year, Claire gave birth to a son, Junior, and also became pregnant for a fourth time. During this pregnancy, both Claire and Jeanine began to show symptoms of AIDS. Claire gave birth prematurely, and she and the baby died soon after. Many months later, an uncle brought Jeanine and Junior to a hospital. Junior was tested for HIV infection. Unlike his sisters, he managed to escape it. Jean, heartsick over the death of his wife and two children, left the family and may have died of AIDS. He was never heard from again. (Paraphrased from Siebert et al., 1989, pp. 36–38)

AIDS is a relatively new viral disease that destroys the immune system. Affected individuals like Jean, Claire, and their children eventually die of a wide variety of illnesses that their bodies can no longer fight. Adults at greatest risk include male homosexuals and bisexuals, users of illegal drugs who share needles, and their heterosexual partners. Transfer of body fluids from one person to another is necessary for AIDS to spread.

The percentage of AIDS victims who are women has risen dramatically over the past decade, from 6 to 13 percent in the United States and to over 50 percent in Africa. When they become pregnant, about 20 to 30 percent of the time they pass the deadly virus to the embryo or fetus. The likelihood of transmission is greatest when a woman already has AIDS symptoms, but (as Claire's case reveals) it can occur beforehand. Why only some offspring are affected is not well understood. It may depend on heredity, the timing of maternal infection, the condition of the placenta, and other factors. Infants can also contract HIV during the birth process, when exposure to maternal fluids increases. Nearly 8,000 childhood cases of AIDS have been diagnosed in the United States since 1981. Worldwide, about 1 million children are infected. The large majority (85 percent) are infants who received the virus before or during birth, often from a drug-abusing mother (Cohen, 1993a, 1993b; Grant, 1995).

AIDS symptoms generally take a long time to emerge in older children and adults—up to 5 years after HIV infection. In contrast, the disease proceeds rapidly in infants. By 6 months of age, weight loss, fever, diarrhea, and repeated respiratory illnesses are common. The virus also causes brain damage. Infants with AIDS show a loss in brain weight over time, accompanied by seizures, delayed mental and motor development, and abnormal muscle tone and movements. Like Ginette, most infants survive for only 5 to 8 months after these symptoms appear (Chamberlain, Nichols, & Chase, 1991; Chatkupt et al., 1989).

AIDS babies are generally born to urban, poverty-stricken parents. Lack of money to pay for medical treatment, rejection by relatives and friends who do not understand the disease, and anxiety about the child's future cause tremendous stress in these families. Medical services for young children with AIDS and counseling for their parents are badly needed (Kurth, 1993).

Currently, scientists are searching for ways to interrupt prenatal AIDS transmission. The antiviral drug AZT reduces it by 75 percent but can also cause birth defects (Chadwick & Yogev, 1995). Until a preventive method or cure is found, education of adolescents and adults about the disease, outreach programs that get at-risk women into drug treatment programs, and confidential HIV testing for all pregnant women (with their consent) are the only ways to stop continued spread of the virus to children.

TRY THIS...

■ When is it warranted to test prospective mothers to find out if they are HIV positive? Consult your library to find out the stance of major professional organizations, such as the American Nurses Association and the American Academy of Pediatrics, on this issue. Why should such testing always be done confidentially and with counseling services available?

This government clinic in Kenya prevents early malnutrition by promoting a proper diet for pregnant women and young children, including breast-feeding in infancy (see Chapter 4, page 124). *(Betty Press/ Woodfin Camp & Associates)*

Prenatal malnutrition can damage the central nervous system and other parts of the body. Autopsies of malnourished babies who died at or shortly after birth reveal fewer brain cells, a brain weight as much as 36 percent below average, and abnormal brain organization. The poorer the mother's diet, the greater the loss in brain weight, especially if malnutrition occurred during the last trimester. During that time, the brain is growing rapidly in size, and a maternal diet high in all the basic nutrients is necessary for it to reach its full potential (Morgane et al., 1993). Prenatal malnutrition also distorts the structure of other organs, including the pancreas, liver, and blood vessels, thereby increasing the risk of heart disease and diabetes in adulthood (Barker et al., 1993).

Prenatally malnourished babies enter the world with serious problems. They frequently catch respiratory illnesses, since poor nutrition suppresses development of the immune system (Chandra, 1991). In addition, these infants are irritable and unresponsive to stimulation. Like drug-addicted newborns, they have a high-pitched cry that is particularly distressing to their caregivers. In poverty-stricken families, these effects quickly combine with a stressful home life. With age, low intelligence and serious learning problems become more apparent (Pollitt, 1996).

When the means are available, food supplement programs before birth can prevent miscarriage, premature delivery, and persistent learning problems in childhood and adolescence (Kramer, 1993; Pollitt et al., 1993). Yet the growth demands of the prenatal period require more than just increasing the quantity of a typical diet. As the Social Issues box on the following page reveals, finding ways to optimize maternal nutrition through vitamin–mineral enrichment as early as possible—even before conception—is crucial.

When poor nutrition continues throughout pregnancy, successful intervention in the months after birth must not only provide much-needed nutrients, but also break the cycle of strained and apathetic mother–baby interactions. Some programs intervene by teaching parents how to interact effectively with their infants. Others provide babies with a stimulating and responsive environment through placement in high-quality day care. In either case, the difference in intellectual development between prenatally malnourished babies and their well-nourished counterparts is greatly reduced (Grantham-McGregor et al., 1994; Grantham-McGregor, Schofield, & Powell, 1987).

Although prenatal malnutrition is highest in poverty-stricken regions of the world, it is not limited to developing countries. Each year, 80,000 to 120,000 American infants are born seriously undernourished. Although the federal government provides food packages to low-income pregnant women through its Special Supplemental Food Program for Women, Infants, and Children, funding is limited, and only 72 percent of those eligible are served (Children's Defense Fund, 1997). Furthermore, the diets of some middle-class expectant mothers are also inadequate. In the United States, where thinness is the feminine ideal, women often feel uneasy about adding 25 to 30 pounds—the recommended weight gain for ensuring the health of both mother and baby.

■ EMOTIONAL STRESS. When women experience severe emotional stress during pregnancy, their babies are at risk for a wide variety of difficulties. Intense anxiety is associated with higher rates of miscarriage, prematurity, low birth weight, and newborn respiratory illness. It is also related to certain physical defects, such as cleft palate and pyloric stenosis—tightening of the infant's stomach outlet, which must be treated surgically (Norbeck & Tilden, 1983; Omer & Everly, 1988).

How can maternal stress affect a developing organism? To understand this process, think back to your own body's

IMPORTANCE OF VITAMIN–MINERAL SUPPLEMENTS FOR WOMEN OF CHILDBEARING AGE

The diets of many women of childbearing age do not meet recommended daily allowances of major nutrients. Levels are particularly low for young and poverty-stricken women—those least likely to take regular vitamin–mineral supplements to make up for their poor nutrition (Keen & Zidenberg-Cherr, 1994).

The impact of vitamin-mineral intake on pregnancy outcomes was strikingly revealed when iodine was added to commonly eaten foods, such as table salt. (Iodine is an essential component of thyroxine, a thyroid hormone necessary for central nervous system development and body growth.) As a result, iodine-induced cretinism, involving mental retardation, short stature, and bone deterioration, virtually disappeared (Dunn, 1993).

Other vitamins and minerals also have established benefits. For example, calcium helps prevent maternal high blood pressure and premature births (Repke, 1992). Magnesium and zinc reduce a variety of prenatal and birth complications (Facchinetti et al., 1992; Jameson, 1993).

Recently, the power of folic acid to prevent abnormalities of the neural tube, such as anencephaly and spina bifida (see Table 2.4 on page 54), has captured the attention of medical and public health experts. This member of the vitamin B complex group can be found in green vegetables, fresh fruit, liver, and yeast. In a British study of nearly 2,000 women in seven countries who had previously given birth

to a baby with a neural tube defect, half were randomly selected to receive a folic acid supplement around the time of conception and half a mixture of other vitamins or no supplement. The folic acid group showed a 72 percent reduction in rate of neural tube defects (MCR Vitamin Study Research Group, 1991). Large-scale studies carried out in Australia, Hungary, and the United States confirm these dramatic effects (Bower & Stanley, 1992; Czeizel & Dudas, 1992; Shaw et al., 1995).

Because the average American woman's diet has less than half the recommended amount of folic acid, the U.S. Food and Drug Administration recently proposed that it be added to flour, bread, and other grain products. Yet even under these conditions, we cannot be sure that pregnant women will consume enough. Therefore, an intensive media campaign is underway to get all women of childbearing age to take at least 0.4 mg but not more than 1 mg of folic acid per day through a vitamin–mineral supplement. Special emphasis is being placed on folic acid enrichment around the time of conception and during the early weeks of pregnancy, when the neural tube is forming.

Yet recommending vitamin–mineral supplements for all potential childbearing women is controversial. The greatest concern is that some people might consume dangerously high levels of certain nutrients because they believe, incorrectly, that more is always better. For example, an excess of vitamins A and D (by taking

even two or three multivitamin pills) can result in birth defects (Rosa, 1993; Rothman et al., 1995). Too much folic acid aggravates damage to the central nervous system in individuals with a severe form of anemia (Chanarin, 1994). Without a public education campaign that emphasizes these dangers, a national supplement program runs the risk of serious negative side effects.

Finally, a supplement policy should complement, not replace, economic programs designed to improve diets during pregnancy. For women with low incomes who do not get enough food, or an adequate variety of foods, multivitamin tablets are a necessary, but not sufficient, intervention.

TRY THIS...

■ Contact your local office of the March of Dimes Birth Defects Foundation or the national office at 1275 Mamaroneck Avenue, White Plains, NY 10605. Telephone (914) 428-7100. Ask for copies of brochures aimed at educating prospective mothers about the importance of dietary nutrients for preventing birth defects.
To what extent does the literature describe both the benefits and risks of vitamin-mineral supplements?

response the last time you were under considerable stress. Fear and anxiety cause us to be "poised for action." Extra blood flows to parts of the body involved in the defensive response—the brain, the heart, and muscles in the arms, legs, and trunk. Blood flow to other organs, including the

uterus, is reduced. As a result, the fetus receives less oxygen and nutrients.

Stress affects the fetus in a second way. Stimulant hormones released into the mother's bloodstream cross the placenta, dramatically increasing the fetus's heart rate and activity level. Long-term exposure to these hormones might be responsible for the irritability and digestive disturbances often present in newborn babies of highly stressed mothers (Omer & Everly, 1988). Finally, women who experience long-term anxiety are more likely to smoke, drink, eat poorly, and engage in other behaviors that harm the embryo and fetus.

The risks of emotional stress are greatly reduced when mothers have husbands, other family members, and friends who provide emotional support (McLean et al., 1993; Nuckolls, Cassel, & Kaplan, 1972). Finding ways to provide isolated women with supportive social ties during pregnancy can help prevent prenatal complications.

■ RH BLOOD INCOMPATIBILITY. When inherited blood types of mother and fetus differ, the incompatibility can sometimes cause serious problems. The most common cause of these difficulties is a blood protein called the **Rh factor.** When the mother is Rh negative (lacks the protein) and the father is Rh positive (has the protein), the baby may inherit the father's Rh positive blood type. (Recall from Table 2.2 in Chapter 2 that Rh positive blood is dominant and Rh negative blood is recessive, so the chances are good that a baby will be Rh positive.) If even a little of a fetus's Rh positive blood crosses the placenta into the Rh negative mother's bloodstream, she begins to form antibodies to the foreign Rh protein. If these enter the fetus's system, they destroy red blood cells, reducing the oxygen supply. Mental retardation, miscarriage, heart damage, and infant death can occur.

Since it takes time for the mother to produce Rh antibodies, first-born children are rarely affected. The danger increases with each additional pregnancy. Fortunately, the harmful effects of Rh incompatibility can be prevented in most cases. After the birth of each Rh positive baby, Rh negative mothers are routinely given a vaccine to prevent the buildup of antibodies. In emergency cases, blood transfusions can be performed immediately after delivery or even before birth.

■ MATERNAL AGE AND PREVIOUS BIRTHS. In Chapter 2, we indicated that women who delay childbearing until their thirties or forties face increased risk of infertility, miscarriage, and babies born with chromosomal defects. Are other pregnancy problems more common for older mothers?

For many years, scientists thought that aging and repeated use of the mother's reproductive organs increased the likelihood of a wide variety of pregnancy complications. Recently, these ideas have been questioned. When women without serious health difficulties are considered, even those in their forties do not experience more prenatal problems than do women in their twenties (Ales, Druzin, & Santini, 1990; Spellacy, Miller, & Winegar, 1986). And a large study of over 50,000 pregnancies found no relationship between the number of previous births and prenatal problems (Heinonen, Slone, & Shapiro, 1977). As long as an older woman is in good health, she can carry a baby successfully.

In the case of teenage mothers, does physical immaturity cause prenatal difficulties? Again, research shows that it does not. As we will see in Chapter 11, nature tries to ensure that once a girl can conceive, she is physically ready to carry and give birth to a baby. Infants of teenagers are at risk for quite different reasons. Many pregnant adolescents do not have access to medical care or are afraid to seek it. In addition, most come from low-income backgrounds where stress, poor nutrition, and health problems are common (Ketterlinus, Henderson, & Lamb, 1990).

THE IMPORTANCE OF PRENATAL HEALTH CARE

Yolanda had her first prenatal appointment 3 weeks after missing her menstrual period. After that, she visited the doctor's office once a month until she was 7 months pregnant, then twice during the eighth month. As the due date approached, Yolanda's appointments increased to once a week. The doctor kept track of her general health, weight gain, and the capacity of her uterus and cervix to support the fetus. The fetus's growth was also carefully monitored.

Yolanda's pregnancy, like most, was uneventful. But unexpected difficulties can arise, especially if mothers have health problems to begin with. For example, women with

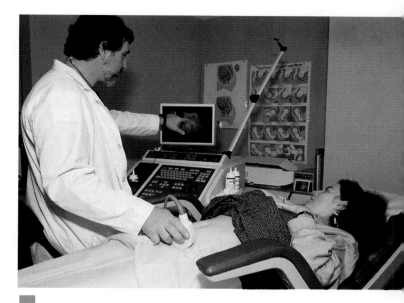

During a routine prenatal visit, this doctor uses ultrasound to show an expectant mother an image of her fetus and to evaluate its development. All pregnant women should receive early and regular prenatal care—to protect their own health and the health of their babies. *(Collins/Monkmeyer Press)*

diabetes need careful monitoring. Extra sugar in the mother's bloodstream causes the fetus to grow larger than average, making pregnancy and birth problems more common. Another complication, *toxemia* (sometimes called *eclampsia)*, is experienced by 5 to 10 percent of pregnant women. In the last half of pregnancy, blood pressure increases sharply and the face, hands, and feet swell. If untreated, toxemia can cause convulsions in the mother and fetal death. Usually, hospitalization, bed rest, and drugs can lower blood pressure to a safe level. If not, the baby needs to be delivered at once.

Unfortunately, 5 percent of pregnant women in the United States wait until the end of pregnancy to seek prenatal care or never get any at all. Most of these mothers are adolescents, unmarried, and members of poverty-stricken ethnic minority groups. Why do they delay going to the doctor? One frequent reason is a lack of health insurance. Although most very poor mothers are eligible for government-sponsored health services, many low-income women do not qualify. Personal problems, including emotional stress, the demands of taking care of other young children, lack of transportation, ambivalence about the pregnancy, and family crises, also keep pregnant women from seeking medical care. Mothers with stressful lives are also likely to engage in high-risk behaviors, such as smoking and drug abuse (Melnikow & Alemagno, 1993). Consequently, women with no medical attention for most of their pregnancies are usually among those who need it most!

Widespread poverty and weak health care programs for mothers and young children are largely responsible for the high American **infant mortality rate**—the number of deaths in the first year of life per 1,000 live births. Although the United States has the most up-to-date health care technology in the world, it ranks only twenty-fourth in infant mortality among nations of the world (see Figure 3.2). Members of poor ethnic minorities are at greatest risk. For example, black infants are two-and-one-half times as likely as white infants to die in the first year of life (Guyer et al., 1995).

A leading cause of infant death is low birth weight, which is largely preventable with good prenatal care. Compared to other developed countries, the number of underweight babies born in the United States is alarmingly high. It is greater than that of 28 other countries (Children's Defense Fund, 1997). Except for the United States, each country listed in Figure 3.2 provides all its citizens with government-sponsored health care benefits. And each takes extra steps to ensure that pregnant mothers and babies have access to good nutrition, high-quality medical care, and social and economic supports that promote effective parenting. For example, all Western European nations guarantee women a certain number of prenatal visits at very low cost. After a baby is born, a health professional routinely visits the home to provide counsel-

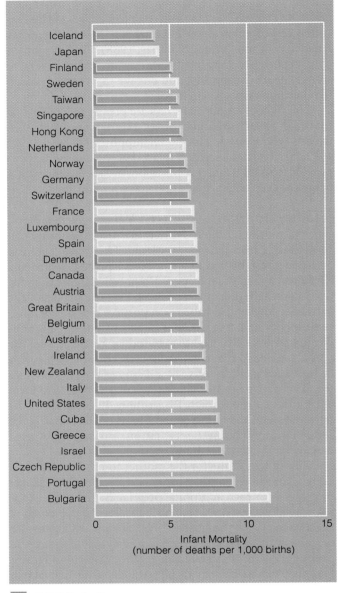

F IGURE 3.2

Infant mortality in 28 nations. Despite its advanced health care technology, the United States ranks poorly. It is twenty-second in the world, with a death rate of 8 infants per 1,000 births. *(Adapted from Central Intelligence Agency, 1996.)*

ing about infant care and to arrange continuing medical services (Kamerman, 1993).

Clearly, public education about the importance of early and sustained high-quality medical services for all pregnant women is badly needed in the United States. The Caregiving Concerns table on the following page summarizes "do's and don't's" for a healthy pregnancy, based on our discussion of the prenatal environment.

CAREGIVING CONCERNS

Do's and Don't's for a Healthy Pregnancy

DO	DON'T
Do make sure you have been vaccinated against infectious diseases dangerous to the embryo and fetus, such as rubella, before you get pregnant. Most vaccinations are not safe during pregnancy.	Don't take any drugs without consulting your doctor.
Do see a doctor as soon as you suspect you are pregnant—within a few weeks after a missed menstrual period. Do continue to get regular medical checkups throughout pregnancy.	Don't smoke cigarettes. If you have already smoked during part of your pregnancy, you can protect your baby by cutting down or (better yet) quitting at any time. If other members of your family are smokers, ask them to smoke outside or in areas of the household that you can easily avoid.
Do obtain literature from your doctor, local library, and bookstore about prenatal development and care. Ask questions about anything you do not understand.	Don't drink alcohol from the time you decide to get pregnant. If you find it difficult to give up alcohol, ask for help from your doctor, local family service agency, or nearest chapter of Alcoholics Anonymous.
Do eat a well-balanced diet. On the average, a woman should increase her intake by 300 calories a day over her usual needs—less at the beginning and more at the end of pregnancy. Gain 25 to 30 pounds gradually.	Don't engage in activities that might expose your baby to environmental hazards, such as radiation or chemical pollutants. If you work in an occupation that involves these agents, ask for a safer assignment or a leave of absence.
Do keep physically fit through mild daily exercise. If possible, join a special exercise class for expectant mothers.	Don't engage in activities that might expose your baby to harmful infectious diseases, such as childhood illnesses and toxoplasmosis.
Do avoid emotional stress. If you are a single parent, find a relative or friend whom you can count on for emotional support.	Don't choose pregnancy as a time to go on a diet.
Do get plenty of rest. An overtired mother is at risk for pregnancy complications.	Don't overeat and gain too much weight during pregnancy. A very large weight gain is associated with complications.
Do enroll in a prenatal and childbirth education class along with the baby's father. When parents know what to expect, the 9 months before birth can be one of the most joyful times of life.	

BRIEF REVIEW

Teratogens—cigarettes, alcohol, certain drugs, radiation, environmental pollutants, and diseases—can seriously harm the embryo and fetus. The effects of teratogens are complex. They depend on amount and length of exposure, the genetic makeup of mother and baby, and the presence of other harmful environmental influences. Teratogens operate according to the sensitive period concept. In general, greatest damage occurs during the embryonic phase, when all parts of the body are forming. Poor maternal nutrition, severe emotional stress, and Rh blood incompatibility can also endanger the developing organism. As long as they are in good health, women in their thirties and forties, women who have given birth to several children, and teenagers have a high likelihood of problem-free pregnancies. Regular medical checkups are important for all expectant mothers, and they are crucial for women with a history of health difficulties.

Widespread poverty and insufficient prenatal health care are largely responsible for the high infant mortality rate in the United States.

ASK YOURSELF . . .

■ *Why is it difficult to determine the effects of some environmental agents, such as over-the-counter drugs and pollution, on the embryo and fetus?*

■ *Nora, who is expecting for the first time at age 40, wonders whether she is likely to have a difficult pregnancy because of her age. How would you respond to Nora's concern?*

■ *Trixie has just learned that she is pregnant. Since she has always been healthy and feels good right now, she cannot understand why the doctor wants her to come in for checkups so often. Why is early and regular prenatal care important for Trixie?*

CHILDBIRTH

Although Yolanda and Jay completed my course 3 months before their baby was born, both agreed to return the following spring to share their experiences with my next class. When the long-awaited day arrived, they brought little Joshua, who was 2 weeks old at the time. Yolanda and Jay's story revealed that the birth of a baby is one of the most dramatic and emotional events in human experience. Jay was present through-out Yolanda's labor and delivery. Yolanda explained,

> By morning, we knew that I was in labor. It was Thursday, so we went in for my usual weekly appointment. The doctor said, yes, the baby was on the way, but it would be a while. He told us to go home and relax or take a leisurely walk and come to the hospital in 3 or 4 hours. We checked in at 3 in the afternoon; Joshua arrived at 2 o'clock the next morning. When, finally, I was ready to deliver, it went quickly; a half hour or so

and some good hard pushes, and there he was! His body had stuff all over it, his face was red and puffy, and his head was misshapen, but I thought, "Oh! he's beautiful. I can't believe he's really here!"

Jay was also elated by Joshua's birth. "I wanted to support Yolanda and to experience as much as I could. It was awesome, indescribable," he said, holding Joshua over his shoulder and patting and kissing him gently. In the following sections, we explore the experience of childbirth, from both the parents' and the baby's point of view.

THE STAGES OF CHILDBIRTH

It is not surprising that childbirth is often referred to as *labor*. It is the hardest physical work a woman may ever do. A complex series of hormonal changes initiates the process. Yolanda's whole system, which for 9 months supported and protected Joshua's growth, now turned toward a new goal: getting him safely out of the uterus.

Stage 1

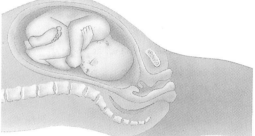

(a) Dilation and Effacement of the Cervix

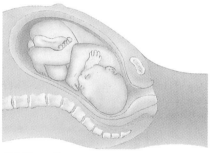

(b) Transition

Stage 2

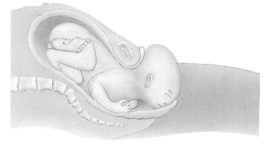

(c) Pushing

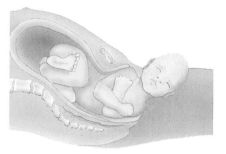

(d) Birth of the Baby

Stage 3

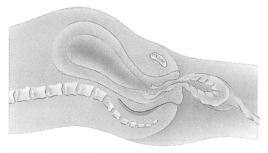

(e) Delivery of the Placenta

FIGURE 3.3

The three stages of labor. Stage 1: (a) Contractions of the uterus cause dilation and effacement of the cervix. (b) Transition is reached when the frequency and strength of the contractions are at their peak and the cervix opens completely. Stage 2: (c) The mother pushes with each contraction, forcing the baby down the birth canal, and the head appears. (d) Near the end of Stage 2, the shoulders emerge and are followed quickly by the rest of the baby's body. Stage 3: (e) With a few final pushes, the placenta is delivered.

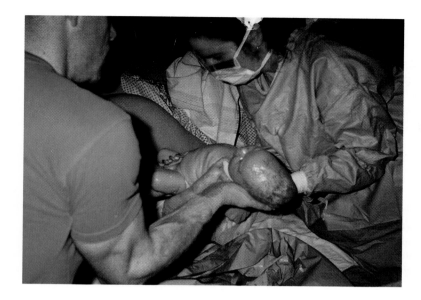

This newborn baby is held by his mother's birthing coach (on the left) and midwife (on the right) just after delivery. The umbilical cord has not yet been cut. Notice how the infant's head is molded from being squeezed through the birth canal for many hours. It is also very large in relation to his body. As the infant takes his first few breaths, his body turns from blue to pink. He is wide awake and ready to get to know his new surroundings. (Courtesy of Dakoda Brandon Dorsaneo)

As Figure 3.3 shows, labor takes place in three stages. Stage 1 is the longest, lasting an average of 12 to 14 hours with a first baby and 4 to 6 hours with later births. **Dilation and effacement of the cervix** occurs—that is, the cervix widens and thins to nothing, creating a clear channel from the uterus to the birth canal, or vagina. Uterine contractions are forceful and regular. Gradually they increase in frequency and strength until they reach a peak called **transition,** when the cervix opens completely.

In Stage 2, which lasts about 50 minutes for a first baby and 20 minutes in later births, the baby is ready to be born. Strong contractions continue, but now the mother feels a natural urge to squeeze and push with her abdominal muscles, forcing the baby down and out.

Stage 3 brings labor to an end with a few final contractions and pushes. These cause the placenta to separate from the wall of the uterus and be delivered, a process that lasts about 5 to 10 minutes.

THE BABY'S ADAPTATION TO LABOR AND DELIVERY

At first glance, labor and delivery seem like a dangerous ordeal for the baby. The strong contractions of Yolanda's uterus exposed Joshua's head to a great deal of pressure, and they squeezed the placenta and the umbilical cord repeatedly. Each time, Joshua's supply of oxygen was temporarily reduced.

Fortunately, healthy babies are well equipped to deal with the trauma of childbirth. The force of the contractions causes the infant to produce high levels of stress hormones. Recall that during pregnancy, the effects of maternal stress can endanger the baby. In contrast, during childbirth the infant's production of stress hormones is adaptive. It helps the baby withstand oxygen deprivation by sending a rich supply of blood to the brain and heart.

In addition, it prepares the baby to breathe effectively by causing the lungs to absorb excess liquid and expanding the bronchial tubes (passages leading to the lungs). Finally, stress hormones arouse the infant into alertness. Joshua was born wide awake, ready to interact with the surrounding world (Emory & Toomey, 1988; Lagercrantz & Slotkin, 1986).

THE NEWBORN BABY'S APPEARANCE

What do babies look like after birth? My students asked Yolanda and Jay this question. "Come to think of it," Jay smiled, "Yolanda and I are probably the only people in the world who thought Joshua was beautiful!"

The average newborn is 20 inches long and 7½ pounds in weight; boys tend to be slightly longer and heavier than girls. The head is very large in comparison to the trunk and legs, which are short and bowed. As we will see in later chapters, the combination of a big head (with its well-developed brain) and a small body means that human infants learn quickly in the first few months of life. But unlike most other mammals, they cannot get around on their own until much later.

Even though newborn babies may look strange, some features do make them attractive. Their round faces, chubby cheeks, large foreheads, and big eyes are just those characteristics that make adults feel like picking them up and cuddling them (Berman, 1980; Lorenz, 1943).

ASSESSING THE NEWBORN'S PHYSICAL CONDITION: THE APGAR SCALE

Infants who have difficulty making the transition to life outside the uterus must be given special help at once. To quickly assess the infant's physical condition, doctors and nurses use the **Apgar Scale.** As Table 3.3 shows, a rating

The Apgar Scale

	SCORE		
SIGN[a]	**0**	**1**	**2**
Heart rate	No heartbeat	Under 100 beats per minute	100 to 140 beats per minute
Respiratory effort	No breathing for 60 seconds	Irregular, shallow breathing	Strong breathing and crying
Reflex irritability (sneezing, coughing, and grimacing)	No response	Weak reflexive response	Strong reflexive response
Muscle tone	Completely limp	Weak movements of arms and legs	Strong movements of arms and legs
Color[b]	Blue body, arms, and legs	Body pink with blue arms and legs	Body, arms, and legs completely pink

[a]To remember these signs, you may find it helpful to use a technique in which the original labels are reordered and renamed as follows: color = **A**ppearance, heart rate = **P**ulse, reflex irritability = **G**rimace, muscle tone = **A**ctivity, and respiratory effort = **R**espiration. Together, the first letters of the new labels spell **Apgar.**

[b]Color is the least reliable of the Apgar signs. The skin tone of nonwhite babies makes it difficult to apply the "pink" criterion. However, newborns of all races can be rated for pinkish glow resulting from the flow of oxygen through body tissues.

Source: Apgar, 1953.

from 0 to 2 on each of five characteristics is made at 1 and 5 minutes after birth. An Apgar score of 7 or better indicates that the infant is in good physical condition. If the score is between 4 and 6, the baby requires assistance in establishing breathing and other vital signs. If the score is 3 or below, the infant is in serious danger and requires emergency medical attention. Two Apgar ratings are given, since some babies have trouble adjusting at first but do quite well after a few minutes (Apgar, 1953).

APPROACHES TO CHILDBIRTH

Childbirth practices, like other aspects of family life, are molded by the society of which mother and baby are a part. In many village and tribal cultures, expectant mothers are well acquainted with the childbirth process. For example, the Jarara of South America and the Pukapukans of the Pacific Islands treat birth as a vital part of daily life. The Jarara mother gives birth in full view of the entire community, including small children. The Pukapukan girl is so familiar with the events of labor and delivery that she can frequently be seen playing at it. Using a coconut to represent the baby, she stuffs it inside her dress, imitates the mother's pushing, and lets the nut fall at the proper moment. In most nonindustrialized cultures, women are assisted during the birth process. Among the Mayans of the Yucatán, the mother leans against the body of a woman called the "head helper," who supports her weight and breathes with her during each contraction (Jordan, 1993; Mead & Newton, 1967).

In large Western nations, childbirth has changed dramatically over the centuries. Before the 1800s, birth usually took place at home and was a family-centered event. Relatives, friends, and children were often present. The nineteenth-century industrial revolution brought greater crowding to cities along with new health problems. Childbirth moved from home to the hospital, where the health of mothers and babies could be protected. Once doctors assumed responsibility for childbirth, women's knowledge of it declined. They were also isolated from the support of friends and relatives (Lindell, 1988).

By the 1950s and 1960s, women started to question the medical procedures that came to be used routinely during labor and delivery. Many felt that frequent use of strong drugs and delivery instruments had robbed them of a precious experience and were often not necessary or safe for the baby. Gradually, a new natural childbirth movement arose in Europe and spread to the United States. Its purpose was to make hospital birth as comfortable and rewarding for mothers as possible. Today, many hospitals carry this theme further by offering birth centers that are family centered in approach and homelike in appearance. *Freestanding birth centers,* which operate independently of hospitals, also exist. They have less rigid rules and encourage early contact between parents and baby but offer less backup medical care. And a small but growing number of American women are rejecting institutional birth entirely by choosing to have their babies at home.

Let's take a closer look at two childbirth approaches that have grown in popularity in recent years: natural childbirth and home delivery.

NATURAL, OR PREPARED, CHILDBIRTH

Yolanda chose **natural,** or **prepared, childbirth** as the way she wanted to have her baby. Although many natural childbirth techniques exist, all try to rid mothers of the idea that birth is a painful ordeal requiring extensive medical intervention. Most programs draw on methods developed by Grantly Dick-Read (1959) in England and

Ferdinand Lamaze (1958) in France. These physicians emphasized that cultural attitudes had taught women to fear the birth experience. An anxious, frightened woman in labor tenses muscles throughout her body, including those of the uterus. This turns the mild pain that sometimes accompanies strong contractions into a great deal of pain.

In a typical natural childbirth program, the expectant mother and a companion (the father, a relative, or a friend) participate in three types of activities:

1. *Classes.* Yolanda and Jay attended a series of classes in which they learned about the anatomy and physiology of labor and delivery. Knowledge about the birth process reduces a mother's fear.

2. *Relaxation and breathing techniques.* After each lecture, Yolanda was taught relaxation and breathing exercises aimed at counteracting the pain of uterine contractions.

3. *Labor coach.* Jay was taught to be a "labor coach." He learned how to help Yolanda during childbirth—by reminding her to relax and breathe, massaging her back, supporting her body, and offering words of encouragement and affection.

Studies comparing mothers who experience natural childbirth with those who do not reveal many benefits. Mothers' attitudes toward labor and delivery are more positive, and they feel less pain. As a result, they require less medication—usually very little or none at all (Hetherington, 1990; Lindell, 1988). Research suggests that social support may be an important part of the success of natural childbirth techniques. In Guatemalan and American hospitals that routinely isolate patients during childbirth, some mothers were randomly assigned a companion who stayed with them throughout labor. These mothers had fewer complications, and their labors were several hours shorter than those of women who did not have companionship. In addition, Guatemalan mothers who received social support were more likely to interact positively with their babies during the first hour after delivery (Kennell et al., 1991; Sosa et al., 1980).

HOME DELIVERY

Home birth has always been popular in certain industrialized nations, such as England, the Netherlands, and Sweden. The number of American women choosing to have their babies at home has grown in recent years, although it is still small, amounting to about 1 percent. Some home births are attended by doctors, but most are handled by certified *nurse-midwives* who have degrees in nursing and additional training in childbirth management (Bastian, 1993; Declercq, 1992).

The joys and perils of home delivery are well illustrated by the story that Don, who painted my house as I worked on this book, told me as we took several coffee breaks together. Two of Don's four children were born at home. "Our first child was delivered in the hospital," he said. "Even though I was present, Kathy and I found the atmosphere to be rigid and insensitive. We wanted a warmer, more personal birth environment." With the coaching and help of a nurse-midwife, Don delivered their second child Cindy at their farmhouse, three miles out of town. Three years later, when Kathy went into labor with Marnie, a heavy snowstorm prevented the midwife from reaching the house on time. Don delivered the baby by himself, but the birth was difficult. Marnie failed to breathe for several minutes; with great effort, Don finally revived her. The frightening memory of Marnie's limp, blue body convinced Don and Kathy to return to the hospital to have their last child. By then, hospital practices had changed, and the event was a rewarding one for both parents.

Don and Kathy's experience raises the question of whether it is just as safe to give birth at home as in a hospital. For healthy women who are assisted by a well-trained

Among the !Kung of Botswana, Africa, a mother gives birth in a sitting position, and she is surrounded by women who encourage and help her. (Shostak/ Anthro-Photo)

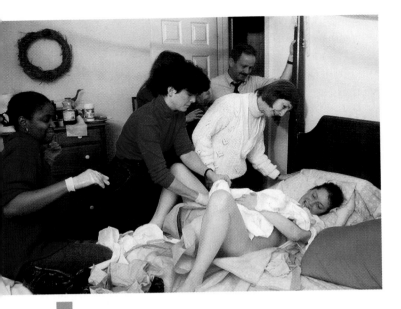

Women who choose home birth want to share the joy of childbirth with family members, avoid unnecessary medical procedures, and exercise greater control over their own care and that of their babies. When assisted by a well-trained doctor or midwife, healthy women can give birth at home safely. (Franck Logue/Stock South)

doctor or midwife, it seems so, since complications rarely occur. However, if attendants are not carefully trained and prepared to handle emergencies, the rate of infant death is high (Schramm, Barnes, & Bakewell, 1987). When mothers are at risk for any kind of complication, the appropriate place for labor and delivery is the hospital, where life-saving treatment is available.

MEDICAL INTERVENTIONS

Two-year-old Melinda walks with a halting, lumbering gait and has difficulty keeping her balance. She has *cerebral palsy,* a general term for a variety of impairments in muscle coordination that result from brain damage before, during, or just after birth.

Like 22 percent of youngsters with cerebral palsy, Melinda's brain damage was caused by **anoxia,** or oxygen deprivation during labor and delivery (Torfs et al., 1990). Her mother got pregnant accidentally, was frightened and alone, and arrived at the hospital at the last minute. Melinda was in **breech position** (turned so that the buttocks or feet would be delivered first), and the cord was wrapped around her neck. As a breech baby, Melinda was at risk for complications. Had her mother come to the hospital earlier, doctors could have monitored the baby's condition and delivered her surgically as soon as squeezing of

the umbilical cord led to distress. Perhaps the damage could have been prevented entirely.

In cases like Melinda's, medical interventions are clearly justified. But in others, they can interfere with delivery and even pose new risks. More than anywhere in the world, childbirth in the United States is a medically monitored and controlled event (Jordan, 1993; Notzon, 1990). In the following sections, we examine the benefits and risks of some commonly used medical techniques.

FETAL MONITORING

Fetal monitors are electronic instruments that track the baby's heart rate during labor. An abnormal heartbeat may indicate that the baby is in distress due to anoxia and needs to be delivered immediately. Fetal monitors are required in almost all American hospitals. The most popular kind is strapped across the mother's abdomen throughout labor. A second, more accurate method involves threading a recording device through the cervix and placing it directly under the baby's scalp.

Fetal monitoring is a safe medical procedure that has saved the lives of many babies in high-risk situations. Nevertheless, the practice is controversial. In healthy pregnancies, it does not reduce the rate of infant brain damage or death. Critics also worry that fetal monitors identify many babies as in danger who, in fact, are not (Prentice & Lind, 1987; Rosen & Dickinson, 1993). Monitoring is linked to an increased rate of cesarean (surgical) deliveries, which we will discuss shortly. In addition, some women complain that the devices are uncomfortable, prevent them from moving easily, and interfere with the normal course of labor.

Still, it is likely that fetal monitors will continue to be used routinely, even though they might not be necessary in most cases. Today, doctors can be sued for malpractice if an infant dies or is born with problems and they cannot show that they did everything possible to protect the baby (McRae, 1993).

LABOR AND DELIVERY MEDICATION

Some form of medication is used in 80 to 95 percent of births in the United States. *Analgesics,* drugs used to relieve pain, may be given in mild doses during labor to help a mother relax. *Anesthetics* are a stronger type of painkiller that blocks sensation. A regional anesthetic may be injected into the spinal column to numb the lower half of the body.

Although pain-relieving drugs enable doctors to perform essential life-saving medical interventions, they can cause problems when used routinely. Anesthesia interferes with the mother's ability to feel contractions during the second stage of labor. As a result, she may not push effectively. In addition, since labor and delivery medication rapidly crosses the placenta, the newborn baby may be sleepy and withdrawn, suck poorly during feedings, and be irritable when awake (Brackbill, McManus, & Woodward, 1985; Brazelton, Nugent, & Lester, 1987).

Does use of medication during childbirth have a lasting impact on physical and mental development? Some researchers claim so (Brackbill, McManus, & Woodward, 1985), but their findings have been challenged (Broman, 1983). Anesthesia may be related to other risk factors that could account for long-term consequences in some studies, and more research is needed to sort out these effects.

CESAREAN DELIVERY

A **cesarean delivery** is a surgical birth; the doctor makes an incision in the mother's abdomen and lifts the baby out of the uterus. Thirty years ago, cesarean delivery was rare in the United States, performed only in emergencies. In 1970, 3 percent of babies were born in this way. Since then, the cesarean rate has climbed steadily. In 1994, the practice accounted for 23.5 percent of American births, the highest rate in the world (U.S. Bureau of the Census, 1996).

Cesareans have always been warranted by medical emergencies, such as Rh incompatibility, premature separation of placenta from the uterus, or serious maternal illness or infection (for example, the herpes simplex 2 virus, which can infect the baby during a vaginal delivery). However, surgical delivery is not always needed in other instances. For example, although the most common reason for a cesarean is a previous cesarean, the technique used today—a small horizontal cut in the lower part of the uterus—makes a vaginal birth possible and safe in later pregnancies (Jakobi et al., 1993). Furthermore, in breech births where the baby risks head injury or anoxia (as in Melinda's case), cesareans are often justified (Cheng & Hannah, 1993). But, the infant's position (which can be felt by the doctor) makes a difference. Certain breech babies fare just as well with a normal delivery as they do with a cesarean. Sometimes the doctor can gently turn the baby into a head-down position during the early part of labor (Collea, Chein, & Quilligan, 1980).

When a cesarean delivery does occur, both mother and baby need extra support. Although the operation is quite safe, it requires more time for recovery. Since anesthetic may have crossed the placenta, newborns are more likely to be sleepy and unresponsive and to have breathing difficulties. These factors can negatively affect the early mother–infant relationship (Cox & Schwartz, 1990).

BRIEF REVIEW

Labor takes place in three stages: dilation and effacement of the cervix, birth of the baby, and delivery of the placenta. Stress hormones help the in-

fant withstand the trauma of childbirth. The Apgar Scale provides a quick rating of the baby's physical condition immediately after birth. Natural, or prepared, childbirth improves mothers' attitudes toward labor and delivery and reduces the need for medication. Home births are safe for healthy women, provided attendants are well trained.

Although often justified, medical interventions during childbirth can cause problems. In some instances, fetal monitoring may mistakenly identify babies as distressed. Pain-relieving drugs can cross the placenta, producing a withdrawn state in the infant. A cesarean delivery requires extra recovery time for the mother, and babies tend to be less alert and more likely to have breathing difficulties.

ASK YOURSELF . . .

■ *What factors help the newborn baby withstand the trauma of labor and delivery?*

■ *Use of any single medical intervention during childbirth increases the likelihood that others will also be used. Provide as many examples as you can to illustrate this idea.*

■ *Sharon, a heavy smoker, has just arrived at the hospital in labor. Which one of the medical interventions discussed in the preceding sections is her doctor justified in using? (Hint: Review our discussion of the prenatal effects of smoking on pages 84–85)*

PRETERM AND LOW-BIRTH-WEIGHT INFANTS

Babies born 3 weeks or more before the end of a full 38-week pregnancy or who weigh less than 5½ pounds (2,500 grams) have, for many years, been referred to as "premature." These infants represent the most common birth complication. A wealth of research indicates that premature babies are at risk for many problems. Birth weight is the best available predictor of infant survival and healthy development. Many newborns who weigh less than 3⅓ pounds (1,500 grams) experience difficulties that are not overcome, an effect that becomes stronger as birth weight decreases (Wilcox & Skjoerven, 1992). Frequent illness, visual impairments, inattention, overactivity, low intelligence test scores, and school learning problems are some of the difficulties that extend into the childhood years (Hack et al., 1994; Liaw & Brooks-Gunn, 1993).

About 1 in 14 infants is born underweight in the United States. Although the problem is not confined to women

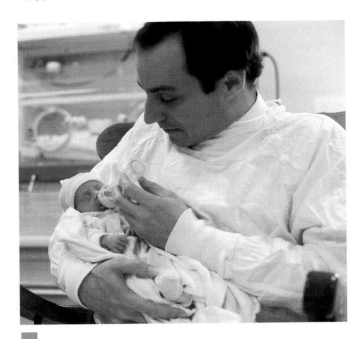

This father feeds his preterm baby in a hospital intensive care nursery. A good parent–infant relationship, the stimulation of touch, and a soft, gentle voice are likely to help this infant recover and catch up in development. *(Joseph Nettis/Photo Researchers)*

with low incomes, it is highest in this group, especially among ethnic minorities (Children's Defense Fund, 1997). These mothers, as we indicated earlier, are more likely to be undernourished and to be exposed to other harmful environmental influences—factors strongly linked to low birth weight. In addition, they often do not receive the prenatal care necessary to protect their vulnerable babies.

Recall from Chapter 2 that prematurity is also common among twins. Because space inside the uterus is restricted, twins gain less weight than single babies after the twentieth week of pregnancy.

PRETERM VERSUS SMALL FOR DATE

Although low-birth-weight infants face many obstacles to healthy development, over half go on to lead normal lives, including some who weighed only a few pounds at birth (Vohr & Garcia-Coll, 1988). To better understand why some of these babies do better than others, researchers have divided them into two groups. The first is called **preterm.** These infants are born several weeks or more before their due date. Although small in size, their weight may still be appropriate for the amount of time they spent in the uterus. The second group is called **small for date.** These babies are below their expected weight when length of the pregnancy is taken into account. Some small-for-

date infants are actually full term. Others are preterm infants who are especially underweight.

Of the two types of babies, small-for-date infants usually have more serious problems. During the first year, they are more likely to die, catch infections, and show evidence of brain damage. By middle childhood, they have lower intelligence test scores, are less attentive, and achieve more poorly in school (Copper et al., 1993; Teberg, Walther, & Pena, 1988). Small-for-date infants probably experienced inadequate nutrition before birth. Perhaps their mothers did not eat properly, the placenta did not function normally, or the babies themselves had defects that prevented them from growing as they should.

CONSEQUENCES FOR CAREGIVING

Imagine a scrawny, thin-skinned infant whose body is only a little larger than the size of your hand. You try to play with the baby by stroking and talking softly, but he is sleepy and unresponsive. When you feed him, he sucks poorly. He is usually irritable during the short, unpredictable periods in which he is awake.

The appearance and behavior of preterm babies can lead parents to be less sensitive and responsive in caring for them. Compared to full-term infants, preterm babies—especially those who are very ill at birth—are less often held close, touched, and talked to gently. At times, mothers of these infants engage in interfering pokes and verbal commands in an effort to obtain a higher level of response from a baby who is a passive, unrewarding social partner (Patteson & Barnard, 1990). Some parents may step up these intrusive acts when ungratifying infant behavior continues. This may explain why preterm babies as a group are at risk for child abuse. When these infants are born to isolated, poverty-stricken mothers who have difficulty managing their own lives, the chances for unfavorable outcomes are increased. In contrast, parents with stable life circumstances and social supports can usually overcome the stresses of caring for a preterm infant. Under these circumstances, even sick preterm babies have a good chance of catching up in development by middle childhood (Liaw & Brooks-Gunn, 1993).

These findings suggest that how well preterm babies develop has a great deal to do with the parent–child relationship. Consequently, interventions directed at supporting both sides of this tie should help these infants recover.

INTERVENING WITH PRETERM INFANTS

A preterm baby is cared for in a special plexiglass-enclosed bed called an *isolette.* Temperature is carefully controlled, since these babies cannot yet regulate their own body temperature effectively. Air is filtered before it enters the isolette to help protect the baby from infection.

Infants born more than 6 weeks early commonly have a disorder called **respiratory distress syndrome** (otherwise known as *hyaline membrane disease)*. Their tiny lungs are so poorly developed that the air sacs collapse, causing serious breathing difficulties. When a preterm infant breathes with the aid of a respirator, is fed through a stomach tube, and receives medication through an intravenous needle, the isolette can be very isolating indeed! Physical needs that otherwise would lead to close contact and other human stimulation are met mechanically. At one time, doctors believed that stimulating such a fragile baby could be harmful. Now we know that in proper doses, certain kinds of stimulation can help preterm infants develop.

■ SPECIAL INFANT STIMULATION. In some intensive care nurseries, preterm babies can be seen rocking in suspended hammocks or lying on waterbeds designed to replace the gentle motion they would have received while still in the mother's uterus. Other forms of stimulation have also been used—for example, an attractive mobile or a tape recording of a heartbeat, soft music, or the mother's voice. These experiences promote faster weight gain, more predictable sleep patterns, and increased alertness during the weeks after birth (Cornell & Gottfried, 1976; Schaefer, Hatcher, & Bargelow, 1980).

Touch is an especially important form of stimulation. In baby animals, touching the skin releases certain brain chemicals that support physical growth—effects believed to occur in humans as well (Schanberg & Field, 1987). In one study, preterm infants who were gently massaged several times each day in the hospital gained weight faster and, at the end of the first year, were advanced in mental and motor development over preterm babies not given this stimulation (Field et al., 1986). In developing countries, where hospitalization is not always possible, skin-to-skin "kangaroo baby care," in which the preterm infant is tucked between the mother's breasts and peers over the top of her clothing, is being encouraged. It fosters oxygenation of the baby's body, temperature regulation, improved feeding, and infant survival (Anderson, 1991; Hamelin & Ramachandran, 1993).

■ TRAINING PARENTS IN INFANT CAREGIVING SKILLS. When preterm infants develop more quickly, parents are likely to feel encouraged and to interact with their baby more effectively. Interventions that support the parenting side of this relationship generally teach parents about the infant's characteristics and promote caregiving skills. For parents with the economic and personal resources to care for a low-birth-weight infant, just a few sessions of coaching in recognizing and responding to the baby's needs leads to steady gains in mental test performance, until scores are similar to those of full-term children (Achenbach et al., 1990).

When preterm infants live in stressed, low-income households, long-term, intensive intervention is necessary. In a recent study, preterm babies born into poverty received a comprehensive intervention that combined medical follow-up, weekly parent training sessions, and enrollment in cognitively stimulating day care from 1 to 3 years of age. More than four times as many intervention children as controls (39 versus 9 percent) were within normal range at age 3 in intelligence, psychological adjustment, and physical growth. However, differences between the two groups declined by age 5. To sustain developmental gains in these vulnerable children, high-quality intervention is needed well beyond age 3—even into the school years (Bradley et al., 1994; Brooks-Gunn et al., 1994).

Unfortunately, even with the best of care, from 30 to 70 percent of babies born at or shortly after the age of viability either die or end up with serious disabilities. Some premature births are unavoidable, but many others can be prevented by improving prenatal nutrition and health care. Today we can save many preterm babies, but an even better course of action would be to combat this serious threat to infant survival and development before it happens.

Are there any general principles that might help us understand how infants who experience a traumatic birth are likely to develop? A long-term study carried out in Hawaii, described in the Lifespan Vista box on page 102, provides answers to this question.

■ **BRIEF REVIEW**

Preterm and small-for-date babies are at risk for many problems. Providing these infants with special stimulation and teaching parents how to care for and interact with them helps restore their growth. Preterm infants living in stressed, low-income households require long-term, intensive intervention to foster and sustain development within normal range. Overall, when newborns with serious complications have access to favorable social environments, they have a good chance of catching up in development.

ASK YOURSELF . . .

■ *Sensitive care can help preterm infants recover, but unfortunately they are less likely to receive this kind of care than are full-term newborns. Explain why.*

■ *List all the factors discussed in this chapter that increase the chances that an infant will be born underweight. How many of these factors could be prevented by better health care for mothers and babies?*

LONG-TERM CONSEQUENCES OF BIRTH COMPLICATIONS: THE KAUAI STUDY

In 1955, Emmy Werner began to follow nearly 700 infants on the island of Kauai who experienced either mild, moderate, or severe birth complications. Each was matched, on the basis of social class and ethnicity, with a healthy newborn. The study had two goals: (1) to discover the long-term effects of birth complications; and (2) to find out how family environments affect the child's chances for recovery.

Results revealed that the likelihood of long-term difficulties increased if birth trauma was severe. But among mild to moderately stressed children, the best predictor of how well they did in later years was the quality of their home environments. Children growing up in stable families developed almost as well as those with no birth difficulties. Those exposed to poverty and family disorganization often did poorly (Werner & Smith, 1992).

This finding has been confirmed in recent research. When low birth weight and poverty combine, children who bounce back during the early years tend to receive more responsive, stimulating, and organized care and live in less crowded homes. Without such protective experiences, the odds that a premature, low-birth-weight

baby will show signs of resilience are low (Bradley et al., 1994).

The Kauai study tells us that as long as birth injuries are not overwhelming, a supportive home environment can restore children's growth. But the most intriguing cases were a handful of exceptions to this rule. A few children with fairly serious birth complications and very troubled families grew into competent adults who fared as well as controls in career attainment and psychological adjustment:

(Consider) Mary, born after 20 hours of labor to an overweight mother who had experienced several miscarriages before that pregnancy. Her father was an unskilled farm laborer with four years of formal education. Between Mary's fifth and tenth birthdays her mother was hospitalized several times for repeated bouts of mental illness, after having inflicted both physical and emotional abuse on her daughter. Surprisingly, by the age of 18, . . . Mary [was an individual] with high self-esteem and sound values who cared about others and [was] liked by [her] peers. (Werner, 1989, p. 108D)

Werner found that resilient children like Mary relied on factors outside the family and within themselves to cope with stress. Some had especially attractive personalities that evoked positive responses from relatives, neighbors, and peers. In other cases, a grandparent, aunt, uncle, or baby-sitter established a warm relationship with the child and provided the needed emotional support.

As we will see in later chapters, child-rearing experiences that permit children to surmount birth complications are the same ones that support healthy development in general—responsive adults and a rich physical and social world. These factors ensure a warm emotional bond with a significant adult; promote cognitive, language, and moral development; and foster high self-esteem and socially skilled behavior. The majority of children without such care continue to show problems into young adulthood. A very few (like Mary) manage to evoke the necessary care elsewhere, thereby "overcoming the odds" (Werner & Smith, 1992).

PRECIOUS MOMENTS AFTER BIRTH

Yolanda and Jay's account of Joshua's birth revealed that the time spent holding and touching him right after delivery was a memorable period filled with intense emotion. A mother given her infant at this time will usually stroke the baby gently, look into the infant's eyes, and talk softly (Klaus & Kennell, 1982). Fathers respond similarly. Regardless of their social class or whether they participated in childbirth classes, they touch, look at, talk to, and kiss their newborn infants as much as and sometimes more than do mothers (Nichols, 1993; Parke & Tinsley, 1981).

Many nonhuman animals engage in specific caregiving behaviors immediately after birth that are essential for the survival of their young. For example, a mother cat licks her newborn kittens and then encircles them with her body (Schneirla, Rosenblatt, & Tobach, 1963). Rats, sheep, and goats engage in similar licking behaviors. But if the mother is separated from the young during the period after delivery, her responsiveness declines until she finally rejects the infant (Poindron & Le Neindre, 1980; Rosenblatt & Lehrman, 1963).

Do human parents also require close physical contact with their babies in the hours after birth for **bonding,** or feelings of affection and concern for the infant, to develop? A few investigators used to think so, but now we know that bonding does not depend on a precise period of togetherness for humans (Eyer, 1992; Lamb, 1994). Some parents report sudden, deep feelings of affection on first holding their babies. For others, these emotions emerge gradually over the first few weeks (MacFarlane, Smith, & Garrow, 1978). In adoptive parents, a warm, affectionate relationship can develop quite successfully even if the child enters the family months or years after birth (Dontas et al., 1985; Hodges & Tizard, 1989). Taken together, these findings reveal that human bonding is a complex process that depends on many factors.

Still, contact with the infant after birth may help build a good relationship between parent and baby. Recent research shows that mothers learn to discriminate their own newborn from other babies on the basis of touch, smell, and sight after as little as one hour of contact (Kaitz et al., 1987, 1988; Kaitz et al., 1993). This early recognition probably contributes to the development of a warm relationship. Realizing this, today hospitals offer an arrangement called *rooming in,* in which the baby stays in the mother's room all or most of the time. If parents choose not to take advantage of this option or cannot for medical reasons, there is no evidence that the baby will suffer emotionally (Lamb, 1994).

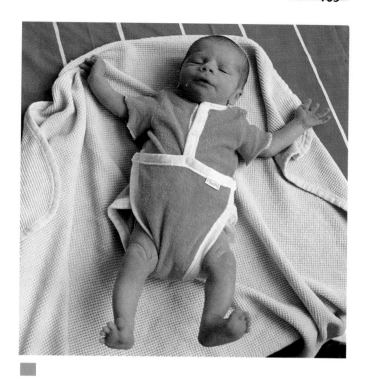

In the Moro reflex, loss of support or a sudden loud sound causes this baby to arch his back, extend his arms outward, and then bring them in toward his body. (Elizabeth Crews)

THE NEWBORN BABY'S CAPACITIES

As recently as 50 years ago, scientists considered the newborn baby to be a passive, disorganized being who could see, hear, feel, and do very little. Today we know that this image is wrong. Newborn babies have a remarkable set of capacities that are crucial for survival and for evoking attention and care from parents. In relating to the physical world and building their first social relationships, they are active from the very start.

NEWBORN REFLEXES

A **reflex** is an inborn, automatic response to a particular form of stimulation. Reflexes are the newborn baby's most obvious organized behavior patterns. As Jay placed Joshua on a table in my classroom, we saw several. When Jay bumped the side of the table, Joshua reacted by flinging his arms wide and bringing them back toward his body. As Yolanda stroked Joshua's cheek, he turned his head in her direction. When she put her finger in Joshua's palm, he grabbed on tightly. Look at Table 3.4 on page 104, and see if you can name the newborn reflexes that Joshua displayed.

Some reflexes have survival value. For example, the rooting reflex helps the infant locate the mother's nipple. And if sucking were not automatic, our species probably would not have survived for a single generation!

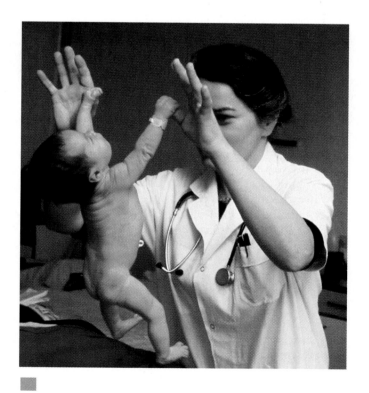

The palmar grasp reflex is so strong during the first week after birth that many infants can use it to support their entire weight. (J. da Cunha/Petit Format/ Photo Researchers, Inc.)

TABLE 3.4

Some Newborn Reflexes

REFLEX	STIMULATION	RESPONSE	AGE OF DISAPPEARANCE	FUNCTION
Eye blink	Shine bright light at eyes or clap hand near head.	Infant quickly closes eyelids.	Permanent	Protects infant from strong stimulation
Rooting	Stroke cheek near corner of mouth.	Head turns toward source of stimulation.	3 weeks (becomes voluntary head turning at this time)	Helps infant find the nipple
Sucking	Place finger in infant's mouth.	Infant sucks finger rhythmically.	Permanent	Permits feeding
Swimming	Place infant face down in pool of water.	Baby paddles and kicks in swimming motion.	4–6 months	Helps infant survive if dropped into body of water
Moro	Hold infant horizontally on back and let head drop slightly, or produce a sudden loud sound against surface supporting infant.	Infant makes an "embracing" motion by arching back, extending legs, throwing arms outward, and then bringing them in toward the body.	6 months	In human evolutionary past, may have helped infant cling to mother
Palmar grasp	Place finger in infant's hand and press against palm.	Spontaneous grasp of adult's finger.	3–4 months	Prepares infant for voluntary grasping
Tonic neck	Turn baby's head to one side while lying awake on back.	Infant lies in a "fencing position." One arm is extended in front of eyes on side to which head is turned, other arm is flexed.	4 months	May prepare infant for voluntary reaching
Stepping	Hold infant under arms and permit bare feet to touch a flat surface.	Infant lifts one foot after another in stepping response.	2 months	Prepares infant for voluntary walking
Babinski	Stroke sole of foot from toe toward heel.	Toes fan out and curl as foot twists in.	8–12 months	Unknown

Sources: Knobloch & Pasamanick, 1974; Prechtl & Beintema, 1965.

A few reflexes form the basis for motor skills that will develop later. The stepping reflex looks like a primitive walking response. In infants who gain weight quickly in the weeks after birth, the stepping reflex drops out because thigh and calf muscles are not strong enough to lift the baby's chubby legs. But if the infant's lower body is dipped in water, the reflex reappears, since the buoyancy of the water lightens the load on the baby's muscles (Thelen, Fisher, & Ridley-Johnson, 1984). When the stepping reflex is exercised regularly, babies are likely to walk several weeks earlier than if it is not practiced (Zelazo, 1983; Zelazo et al., 1993). However, there is no special need for infants to practice the stepping reflex, since all normal babies walk in due time.

Some reflexes contribute powerfully to the infant's early social relationships. A baby who searches for and successfully finds the nipple, sucks easily during feedings, and grasps when the hand is touched encourages parents to respond lovingly and strengthens their sense of competence. Reflexes can also assist caregivers in comforting the baby. For example, on short trips with Joshua to the grocery store, Yolanda brought along a pacifier. If he became fussy, sucking helped quiet him until she could feed, change, or hold and rock him.

Look again at Table 3.4, and note that most newborn reflexes disappear during the first 6 months. Scientists believe that this is due to a gradual increase in voluntary control over behavior as the cortex of the brain matures. Pediatricians test reflexes carefully, especially if a newborn has experienced birth trauma, since reflexes provide one way of assessing the health of the baby's nervous system. Weak or absent reflexes, reflexes that are overly rigid or exaggerated, and reflexes that persist beyond the point in development when they should normally disappear can signal brain damage (Touwen, 1984).

SENSORY CAPACITIES

On his visit to my class, Joshua looked wide-eyed at my bright pink blouse and turned to the sound of his mother's voice. During feedings, he lets Yolanda know by the way he sucks that he prefers the taste of breast milk to a bottle of plain water. Clearly, Joshua has some well-developed sensory capacities. In the following sections, we explore the newborn baby's responsiveness to touch, taste, smell, sound, and visual stimulation.

■ TOUCH. In our discussion of preterm infants, we saw that touch helps stimulate early physical growth. And as we will see in Chapter 6, it is important for emotional development as well. Therefore, it is not surprising that sensitivity to touch is well developed at birth. The reflexes listed in Table 3.4 reveal that the newborn baby responds to touch, especially around the mouth, palms, and soles of the feet. During the prenatal period, these areas, along with the genitals, are the first to become sensitive to touch, followed by other regions of the body (Humphrey, 1978).

Newborn babies also react to temperature change, especially to stimuli that are colder than body temperature (Humphrey, 1978). When Yolanda and Jay undress Joshua, he often expresses his discomfort by crying and becoming more active.

At birth, infants are quite sensitive to pain. When male newborns are circumcised, anesthesia is usually not used because of the risk of giving drugs to a very young infant. Babies often respond with an intense, high-pitched, stressful cry (Hadjistavropoulos et al., 1994). In addition, heart rate and blood pressure rise, irritability increases, and sleep may be disturbed for hours afterward (Anand, 1990). Recent research aimed at developing safe pain-relieving techniques for newborns promises to ease the severe stress of these procedures. One helpful approach is to offer a nipple that delivers a sugar solution. This quickly reduces crying and discomfort, even in preterm newborns (Smith & Blass, 1996).

■ TASTE AND SMELL. All babies come into the world with the ability to communicate their taste preferences to caregivers. When given a sweet liquid instead of water, Joshua uses longer sucks with fewer pauses to savor the taste of his favorite food (Crook & Lipsitt, 1976). In contrast, he is either indifferent to or rejects salty water (Beauchamp et al., 1994). (Not until 4 months, when Joshua is ready for solid foods, will he respond positively to a salty taste.) Newborns also reveal their taste preferences through facial expressions. Like adults, they relax their facial muscles in response to sweetness, purse their lips when the taste is sour, and show a distinct archlike mouth opening when it is bitter (Steiner, 1979). These reactions are important for survival, since (as we will see in Chapter 4) the food that is ideally suited to support the infant's early growth is the sweet-tasting milk of the mother's breast.

Like taste, the newborn's response to the smell of certain foods is similar to that of adults, suggesting that some odor preferences are innate. For example, the smell of bananas or chocolate causes a relaxed, pleasant facial expression, whereas the odor of rotten eggs makes the infant frown (Steiner, 1979). Newborns can also identify the location of an odor and, if it is unpleasant, defend themselves by turning their heads in the other direction (Reiser, Yonas, & Wikner, 1976).

In many mammals, the sense of smell plays an important role in eating and protecting the young from predators by helping mothers and babies recognize each other. Although smell is less well developed in humans, traces of its survival value remain. When breast-fed newborns are exposed to the odor of their own mother and that of a strange mother, within the first week of life they turn more often to their own mother's odor (Cernoch & Porter, 1985). Although bottle-fed infants do not show this response, they prefer the smell of any lactating (milk-producing) woman to that of a nonlactating woman or to their familiar formula (Makin & Porter, 1989; Porter et al., 1992). Newborn infants' attraction to the odor of breast milk probably helps them locate an appropriate food source and, in the process, learn to identify their mother.

■ HEARING. Newborn infants can hear a wide variety of sounds, but they prefer complex sounds, such as noises and voices, to pure tones (Bench et al., 1976). In the first few days, infants can already tell the difference between a few sound patterns, such as a series of tones arranged in ascending versus descending order and utterances with two as opposed to three syllables (Bijeljac-Babic, Bertoncini, & Mehler, 1993; Morrongiello, 1986).

Newborn babies are especially sensitive to sounds within the frequency range of the human voice, and they are biologically prepared to respond to the sounds of any language. Tiny infants can make fine-grained distinctions among a wide variety of speech sounds. For example, when given a nipple that turns on a recording of the "ba" sound, babies suck vigorously and then slow down as the novelty wears off. When the sound switches to "ga," sucking picks up, indicating that infants detect this subtle difference. Using this method, researchers have found only a few speech sounds that infants cannot discriminate (Jusczyk, 1995). These capacities reveal that the baby is marvelously prepared for the awesome task of acquiring language.

Responsiveness to sound supports the newborn's visual exploration of the environment. Infants as young as 3 days turn their eyes and head in the general direction of a sound. The ability to identify its precise location will improve greatly over the first 6 months (Ashmead et al., 1991; Hillier, Hewitt, & Morrongiello, 1992).

Listen carefully to yourself the next time you talk to a young baby. You will probably speak in a high-pitched, expressive voice and use a rising tone at the ends of phrases and sentences. Adults probably communicate with infants in this way because babies are more attentive when they do so. Indeed, newborns prefer human speech with these characteristics. And they are especially attentive when the voice is their mother's and speaks their native tongue rather than a foreign language (Cooper & Aslin, 1990; Spence & DeCasper, 1987). These preferences probably developed from hearing the muffled sounds of the mother's voice before birth.

■ VISION. Vision is the least developed of the senses at birth. Visual centers in the brain are not yet fully formed, nor are structures in the eye. For example, cells in the *retina,* a membrane lining the inside of the eye that captures light and transforms it into messages that are sent to the brain, are not as mature or densely packed as they will be in several months. Also, muscles of the *lens,* which permits us to adjust our visual focus to varying distances, are weak at birth (Banks & Bennett, 1988).

As a result, newborn babies cannot focus their eyes as well as an adult can, and **visual acuity,** or fineness of discrimination, is limited. Research indicates that infants perceive objects at a distance of 20 feet about as clearly as adults do at 600 feet (Courage & Adams, 1990). In addition, unlike adults (who see nearby objects most clearly), newborn babies see equally unclearly across a wide range of distances (Banks, 1980). As a result, images such as the parent's face, even from close up, look quite blurred.

Although newborn babies cannot see well, they actively explore their environment by scanning it for interesting sights and tracking moving objects. However, their eye movements are slow and inaccurate (Aslin, 1993). Joshua's captivation with my pink blouse reveals that he is attracted to bright objects. Although newborns prefer to look at colored rather than gray stimuli, they are not yet good at discriminating colors. It will take a month or two for color vision to improve (Adams, 1987; Brown, 1990).

NEWBORN STATES

Throughout the day and night, newborn infants move in and out of five different **states of arousal,** or degrees of sleep and wakefulness, which are described in Table 3.5. During the first month, these states alternate frequently. Quiet alertness is the most fleeting. It usually moves quickly toward fussing and crying. Much to the relief of their fatigued parents, newborns spend the greatest amount of time asleep—on the average, about 16 to 18 hours a day.

However, striking individual differences in daily rhythms exist that affect parents' attitudes toward and interaction with the baby. A few newborns sleep for long periods, and their well-rested parents have the energy they need for sensitive, responsive care. Other babies cry a great deal, and their parents must exert great effort to soothe them. If they do not succeed, parents may feel less competent and positive toward their infant.

Of the five states listed here, the two extremes of sleep and crying have been of greatest interest to researchers. Each tells us something about normal and abnormal early development.

■ SLEEP. One day, Yolanda and Jay watched Joshua while he slept and wondered why his eyelids and body twitched and his rate of breathing varied. Sleep is made up of at least two states. Yolanda and Jay happened to observe irregular, or **rapid-eye-movement (REM),** sleep. The expression "sleeping like a baby" was probably not meant to describe this state! During REM sleep, electrical brain wave activity is remarkably similar to that of the waking state. The eyes dart beneath the lids; heart rate, blood pressure, and breathing are uneven; and slight body movements occur. In contrast, during **non-rapid-eye-movement (NREM)**

TABLE 3.5

Infant States of Arousal

STATE	DESCRIPTION	DAILY DURATION IN NEWBORN
Regular sleep	The infant is at full rest and shows little or no body activity. The eyelids are closed, no eye movements occur, the face is relaxed, and breathing is slow and regular.	8–9 hours
Irregular sleep	Gentle limb movements, occasional stirring, and facial grimacing occur. Although the eyelids are closed, occasional rapid eye movements can be seen beneath them. Breathing is irregular.	8–9 hours
Drowsiness	The infant is either falling asleep or waking up. Body is less active than in irregular sleep but more active than in regular sleep. The eyes open and close; when open, they have a glazed look. Breathing is even but somewhat faster than in regular sleep.	Varies
Quiet alertness	The infant's body is relatively inactive, with eyes open and attentive. Breathing is even.	2–3 hours
Waking activity and crying	The infant shows frequent bursts of uncoordinated body activity. Breathing is very irregular. Face may be relaxed or tense and wrinkled. Crying may occur.	1–4 hours

Source: Wolff, 1966.

sleep, the body is almost motionless, and heart rate, breathing, and brain wave activity are slow and regular (Dittrichova et al., 1982).

Like children and adults, newborns alternate between REM and NREM sleep. However, they spend far more time in the REM state than they ever will again in their lives. REM sleep accounts for 50 percent of a newborn baby's sleep time. By 3 to 5 years, it has declined to an adultlike level of 20 percent (Roffwarg, Muzio, & Dement, 1966).

Why do young infants spend so much time in REM sleep? In older children and adults, the REM state is associated with dreaming. Babies probably do not dream, at least not in the same way we do. Young infants are believed to have a special need for the stimulation of REM sleep because they spend little time in an alert state, when they can get input from the environment. REM sleep seems to be a way in which the brain stimulates itself. Sleep researchers believe that this stimulation is vital for growth of the central nervous system.

Because the normal sleep behavior of a newborn baby is organized and patterned, observations of sleep states can help identify central nervous system abnormalities. In infants who are brain damaged or have experienced serious birth trauma, disturbed REM–NREM sleep cycles are often present (Theorell, Prechtl, & Vos, 1974; Whitney & Thoman, 1994).

■ CRYING. Crying is the first way that babies communicate, letting parents know that they need food, comfort, and stimulation. During the weeks after birth, all infants have some fussy periods when they are difficult to console. But most of the time, the nature of the cry helps guide parents toward its cause. The baby's cry is actually a complex stimulus that varies in intensity, from a whimper to a message of all-out distress (Gustafson & Harris, 1990).

Newborn infants usually cry because of physical needs. Hunger is the most common cause, but babies may also cry in response to temperature change when undressed, a sudden noise, or a painful stimulus. The next time you hear an infant cry, note your own mental and physical reaction. The sound stimulates strong feelings of arousal and discomfort in just about anyone (Boukydis & Burgess, 1982; Murray, 1985). This powerful response is probably innately programmed in humans to make sure that babies receive the care and protection they need to survive.

Although parents do not always interpret their baby's cry correctly, experience quickly improves their accuracy (Green, Jones, & Gustafson, 1987). Fortunately, there are many ways to soothe a crying baby when feeding and diaper changing do not work. The technique parents usually try first, lifting the baby to the shoulder, is most effective. Being held upright against the parent's gently moving body not only encourages infants to stop crying, but also causes them to become quietly alert and attentive to the environ-

To soothe her crying infant, this mother holds her baby upright against her gently moving body. Besides encouraging infants to stop crying, this technique causes them to become quietly alert and attentive to the environment. (Rosanne Olson/ Tony Stone Images)

ment (Reisman, 1987). The Caregiving Concerns table on page 108 lists other common soothing methods.

Like reflexes and sleep patterns, an infant's cry offers a clue to central nervous system distress. The cries of brain-damaged babies and those who have experienced prenatal and birth complications are often shrill and piercing (Lester, 1987). Most parents respond with extra care and attention, but sometimes the cry is so unpleasant and the infant so difficult to soothe that parents become frustrated, resentful, and angry. Occasionally abusing parents mention a high-pitched, grating cry as one factor that caused them to lose control and harm the baby (Boukydis, 1985; Frodi, 1985).

NEONATAL BEHAVIORAL ASSESSMENT

The many capacities we have just described have been put together into tests that permit doctors, nurses, and researchers to assess a newborn infant's behavior during the first month of life. The most widely used of these tests,

CAREGIVING CONCERNS

Ways of Soothing a Crying Newborn

METHOD	EXPLANATION
Lift the baby to the shoulder and physically rock or walk.	This provides a combination of contact, upright posture, and motion. It is the most effective soothing technique.
Swaddle the baby.	Restricting movement and increasing warmth often soothes a young infant.
Offer a pacifier.	Sucking helps babies control their own level of arousal.
Talk softly or play rhythmic sounds.	Continuous, monotonous, rhythmic sounds, such as a clock ticking, a fan whirring, or peaceful music, are more effective than intermittent sounds.
Take the baby for a short car ride or walk in a baby carriage; swing the baby in a cradle.	Gentle, rhythmic motion of any kind helps lull the baby to sleep.
Massage the baby's body.	Stroke the baby's torso and limbs with continuous, gentle motions. This technique is used in some non-Western cultures to relax the baby's muscles.
Combine several of the methods listed above.	Stimulating several of the baby's senses at once is often more effective than stimulating only one.
If these methods do not work, permit the baby to cry for a short period of time.	Occasionally, a baby responds well to just being put down and will, after a few minutes, fall asleep.

Sources: Campos, 1989; Heinl, 1983; Lester, 1985; Reisman, 1987.

T. Berry Brazelton's (1984) **Neonatal Behavioral Assessment Scale (NBAS),** evaluates the baby's reflexes, state changes, responsiveness to physical and social stimuli, and other reactions.

The NBAS has been given to many infants around the world. As a result, researchers have learned about individual and cultural differences in newborn behavior and how certain child-rearing practices can maintain or change a baby's reactions. For example, NBAS scores of Asian and Native American babies reveal that they are less irritable than Caucasian infants. Mothers in these cultures often encourage their babies' calm dispositions through swaddling, close physical contact, and nursing at the first signs of discomfort (Chisholm, 1989; Freedman & Freedman, 1969; Murett-Wagstaff & Moore, 1989). In contrast, the poor NBAS scores of undernourished infants in Zambia, Africa,

are quickly improved by the way their mothers care for them. The Zambian mother carries her baby about on her hip all day, providing a rich variety of sensory stimulation. By 1 week of age, a once unresponsive newborn has been transformed into an alert, contented baby (Brazelton, Koslowski, & Tronick, 1976).

Can you tell from these examples why a single NBAS score is not a good predictor of later development? Since newborn behavior and parenting styles combine to shape development, changes in NBAS scores over the first week or two of life (rather than a single score) provide the best estimate of the baby's ability to recover from the stress of birth. NBAS "recovery curves" predict intelligence with moderate success well into the preschool years (Brazelton, Nugent, & Lester, 1987).

The NBAS has also been used to help parents get to know their infants. In some hospitals, the examination is given in the presence of parents to teach them about their newborn baby's capacities. Parents of both preterm and full-term newborns who participate in these programs interact more confidently and effectively with their babies (Tedder, 1991). Although lasting effects on development have not been demonstrated, NBAS-based interventions are clearly useful in helping the parent–infant relationship get off to a good start.

Similar to women in the Zambian culture, this Inuit mother of Northern Canada carries her baby about all day, providing close physical contact and a rich variety of stimulation. (Eastcott/Momatiak/Woodfin Camp & Associates)

ASK YOURSELF . . .

■ *How do the capacities of newborn babies contribute to their first social relationships? In answering this question, provide as many examples as you can.*

■ *Jackie, who had a difficult delivery, observes her 2-day-old daughter Kelly being given the NBAS. Kelly scores poorly on many items. Jackie wonders if this means that Kelly will not develop normally. How would you respond to Jackie's concern?*

ADJUSTING TO THE NEW FAMILY UNIT

The early weeks after a new baby enters the family are full of profound changes. The mother needs to recover from childbirth and adjust to massive hormone changes in her body. If she is breast-feeding, energies must be devoted to working out this intimate relationship. The father also needs to become a part of this new threesome while supporting the mother in her recovery. At times, he may feel ambivalent about the baby, who constantly demands and gets his wife's attention (Berman & Pedersen, 1987; Jordan, 1990). And as we will see in Chapter 6, siblings—especially those who are young and previously had their parents all to themselves—understandably feel displaced and often react with jealousy and anger.

While all this is going on, the tiny infant is very assertive about his urgent physical needs, demanding to be fed, changed, and comforted at odd times of the day and night. A family schedule that was once routine and predictable is now irregular and uncertain. Yolanda spoke candidly about the changes that she and Jay experienced:

Our emotions were running high the first few days after Joshua's birth. When we brought him home, we had to come down to earth and deal with the realities of our new responsibility. Both of us were struck by how small and helpless Joshua was, and we worried about whether we would be able to take proper care of him. It took us twenty minutes to change the first diaper. I rarely feel well rested because I'm up two to four times every night, and I spend a good part of my waking hours trying to anticipate Joshua's rhythms and needs. If Jay weren't so willing to help by holding and walking Joshua, I think I'd find it much harder.

How long does this time of adjustment to parenthood last? In Chapter 14, we will see that when husband and wife set aside time to listen to one another and try to support each other's needs, family relationships and routine care of the baby are worked out after a few months. Nevertheless, as one pair of counselors who have worked with many new parents pointed out,

As long as children are dependent on their parents, those parents find themselves preoccupied with thoughts of their children. This does not keep them from enjoying other aspects of their lives, but it does mean that they never return to being quite the same people they were before they became parents (Colman & Colman, 1991).

ASK YOURSELF . . .

■ *Suggest several ways in which parents can help each other make an effective adjustment to the arrival of a new baby.*

PHYSICAL DEVELOPMENT IN EARLY CHILDHOOD

PRENATAL DEVELOPMENT

List the three phases of prenatal development, and describe the major milestones of each.

- Prenatal development is usually divided into three phases. The period of the zygote lasts about 2 weeks, from fertilization until the **blastocyst** becomes deeply **implanted** in the uterine lining. During this time, structures that will support prenatal growth begin to form. The **embryonic disk** is surrounded by the **amnion,** which is filled with **amniotic fluid.** From the **chorion,** villi emerge that burrow into the uterine wall, and the **placenta** starts to develop. The developing organism is connected to the placenta by the **umbilical cord.**

- In the period of the **embryo,** which lasts from 2 to 8 weeks, the foundations for all body structures are laid down. In the first week of this period, the **neural tube** forms, and the nervous system starts to develop. Other organs also grow rapidly.

- The period of the **fetus,** lasting from the ninth week until the end of pregnancy, involves a dramatic increase in body size and completion of physical structures. By the middle of the second **trimester,** the mother can feel movement. The fetus becomes covered with **vernix,** which protects the skin from chapping. Downy white hair called **lanugo** helps the vernix stick to the skin. At the end of the second trimester, production of neurons in the brain is complete.

- **Age of viability** occurs at the beginning of the third trimester, sometime between 22 and 26 weeks. The brain continues to develop, and new sensory and behavioral capacities

emerge. Gradually the lungs mature, the fetus fills the uterus, and birth is near.

PRENATAL ENVIRONMENTAL INFLUENCES

What factors influence the impact of teratogens on the developing organism?

- **Teratogens** are environmental agents that cause damage during the prenatal period. Their effects conform to the sensitive period concept. The developing organism is especially vulnerable during the embryonic period. The impact of teratogens differs from one case to the next, due to amount and length of exposure, the genetic makeup of mother and fetus, and the presence or absence of other harmful agents.

List agents known to be or suspected of being teratogens, and discuss evidence on the harmful impact of each.

- The prenatal impact of many commonly used medications, such as aspirin and caffeine, is hard to separate from other factors correlated with drug taking. Babies whose mothers took heroine, methadone, or cocaine during pregnancy have withdrawal symptoms after birth and are jittery and inattentive. Cocaine is especially risky, since it is associated with physical defects and central nervous system damage.

- Infants of parents who smoke cigarettes are often born underweight and may display inattentiveness and learning problems in childhood. When mothers consume alcohol in large quantities, **fetal alcohol syndrome (FAS),** a disorder involving mental retardation, poor attention, overactivity, slow physical growth, and facial abnormalities, often results. Smaller amounts of alcohol may lead to some of these

problems—a condition known as **fetal alcohol effects (FAE).**

- Radiation, mercury, and lead can result in a wide variety of defects, including physical malformations and severe brain damage. PCBs have been linked to decreased responsiveness to the environment and impairments in memory and intellectual development.

- Among infectious diseases, rubella (German measles) causes a wide variety of abnormalities. Acquired immune deficiency syndrome (AIDS) is linked to brain damage, delayed development, and early death. In the first trimester, toxoplasmosis (caused by a parasite found in many animals) may lead to eye and brain damage.

Describe the impact of additional maternal factors on prenatal development.

- In healthy, physically fit women, regular exercise is related to increased birth weight. When the mother's diet is inadequate, low birth weight and brain damage are major concerns.

- Severe emotional stress is linked to many pregnancy complications, although its impact can be reduced by providing the mother with emotional support. If the **Rh factor** of the mother's blood is negative and the fetus's is positive, special precautions must be taken to ensure that antibodies to the Rh protein do not pass from mother to fetus.

- Aside from the risk of chromosomal abnormalities in older women, maternal age and number of previous births are not major causes of prenatal problems. Instead, poor health and environmental risks associated with poverty are the strongest predictors of pregnancy complications.

Why is early and regular health care vital during the prenatal period?

■ Early and regular prenatal health care is important for all pregnant women. It is especially critical for women unlikely to seek it—in particular, those who are young, single, and poor.

CHILDBIRTH

Describe the three stages of childbirth, the baby's adaptation to labor and delivery, and the newborn baby's appearance.

■ Childbirth takes place in three stages. In the first stage, **dilation and effacement of the cervix** occur as uterine contractions increase in strength and frequency. This stage culminates in **transition,** when the cervix opens completely. In the second stage, the mother pushes with her abdominal muscles, and the baby is born. In the final stage, the placenta is delivered.

■ During labor, infants produce high levels of stress hormones, which help them withstand oxygen deprivation and arouse them into alertness at birth. Newborn infants have large heads, small bodies, and facial features that make adults feel like cuddling them. The **Apgar Scale** assesses the baby's physical condition at birth.

APPROACHES TO CHILDBIRTH

Describe natural childbirth and home delivery, noting any benefits and concerns associated with each.

■ **Natural, or prepared, childbirth** involves classes in which prospective parents learn about labor and delivery, instruction of the mother in relaxation and breathing techniques, and a companion who serves as a coach during childbirth. The method helps reduce the stress and pain of contractions. As long as mothers are healthy and assisted by a well-trained doctor or midwife, it is just as safe to give birth at home as in a hospital.

MEDICAL INTERVENTIONS

List common medical interventions during childbirth, circumstances that justify their use, and any dangers associated with each.

■ Medical interventions during childbirth are more common in the United States than anywhere in the world. When pregnancy and birth complications make **anoxia** likely, **fetal monitors** help save the lives of many babies. However, when used routinely, they may identify infants as in danger who, in fact, are not.

■ Medication to relieve pain is necessary in complicated deliveries. When given in large doses, it produces a depressed state in the newborn that affects the early mother–infant relationship.

■ **Cesarean deliveries** are justified in cases of medical emergency and sometimes when babies are in **breech position.** Many unnecessary cesareans are performed in the United States.

PRETERM AND LOW-BIRTH-WEIGHT INFANTS

What are the risks of preterm and low birth weight, and what factors can help infants who survive a traumatic birth develop?

■ Premature births are high among low-income pregnant women and mothers of twins. Compared to **preterm** babies whose weight is appropriate for time spent in the uterus, **small-for-date** infants are more likely to develop poorly. **Respiratory distress syndrome,** common in babies born more than 6 weeks early, causes serious breathing difficulties.

■ The fragile appearance and unresponsive, irritable behavior of preterm infants affect the care they receive. Some interventions provide special stimulation in the intensive

care nursery. Others teach parents how to care for and interact with their babies. Even infants with serious birth complications can recover with the help of favorable social environments.

PRECIOUS MOMENTS AFTER BIRTH

Is close parent–infant contact shortly after birth necessary for bonding to occur?

■ Human parents do not require close physical contact with the baby immediately after birth for **bonding** to occur. Nevertheless, most parents find early contact with the infant especially meaningful, and it may help them build a good relationship with the baby.

THE NEWBORN BABY'S CAPACITIES

Describe the newborn baby's reflexes and sensory capacities.

■ Infants begin life with remarkable skills for relating to their physical and social worlds. **Reflexes** are the newborn baby's most obvious organized patterns of behavior. Some have survival value, others provide the foundation for voluntary motor skills, and still others contribute to early social relationships.

■ The senses of touch, taste, smell, and sound are well developed at birth. Newborns are especially responsive to high-pitched, expressive voices, and they prefer the sound of their mother's voice. They can distinguish almost all speech sounds in human languages.

■ Vision is the least mature of the newborn's senses. At birth, focusing ability and **visual acuity** are limited. In exploring the visual field, newborn babies are attracted to bright objects but have difficulty discriminating colors.

Describe newborn states of arousal, including sleep characteristics and ways to soothe a crying baby.

■ Although newborns alternate frequently among five different **states of arousal,** they spend most of their time asleep. Sleep consists of at least two states: **rapid-eye-movement (REM) sleep** and **non-rapid-eye-movement (NREM) sleep.** REM provides young infants with stimulation essential for central nervous system development.

■ A crying baby stimulates strong feelings of discomfort in nearby adults. The intensity of the cry and the experiences that led up to it help parents tell what is wrong. Once feeding and diaper changing have been tried, lifting the baby to the shoulder is the most effective soothing technique.

Why is neonatal behavioral assessment useful?

■ The most widely used instrument for assessing the behavior of the newborn infant is Brazelton's **Neonatal Behavioral Assessment Scale (NBAS).** The NBAS has helped researchers understand individual and cultural differences in newborn behavior. Sometimes it is used to teach parents about their baby's capacities.

ADJUSTING TO THE NEW FAMILY UNIT

Describe typical changes in the family after the birth of a new baby.

■ The new baby's arrival is exciting but stressful. The demands of new parenthood often lead to a slight drop in marital happiness, and family roles become more traditional. When husband and wife are sensitive to each other's needs, adjustment problems are usually temporary, and the transition to parenthood goes well.

IMPORTANT TERMS AND CONCEPTS

blastocyst (p. 78)

embryonic disk (p. 78)

implantation (p. 78)

amnion (p. 78)

amniotic fluid (p. 78)

chorion (p. 79)

placenta (p. 79)

umbilical cord (p. 79)

embryo (p. 80)

neural tube (p. 80)

fetus (p. 80)

trimesters (p. 81)

vernix (p. 81)

lanugo (p. 81)

age of viability (p. 81)

teratogen (p. 82)

fetal alcohol syndrome (FAS) (p. 85)

fetal alcohol effects (FAE) (p. 86)

Rh factor (p. 91)

infant mortality rate (p. 92)

dilation and effacement of the cervix (p. 95)

transition (p. 95)

Apgar Scale (p. 95)

natural, or prepared, childbirth (p. 96)

anoxia (p. 98)

breech position (p. 98)

fetal monitors (p. 98)

cesarean delivery (p. 99)

preterm (p. 100)

small for date (p. 100)

respiratory distress syndrome (p. 101)

bonding (p. 102)

reflex (p. 103)

visual acuity (p.106)

states of arousal (p. 106)

rapid-eye-movement (REM) sleep (p. 106)

non-rapid-eye-movement (NREM) sleep (p. 106)

Neonatal Behavioral Assessment Scale (NBAS) (p. 108)

GENERAL CHILDBIRTH INFORMATION

National Association of Parents and Professionals for Safe Alternatives in Childbirth
Route 1, Box 646
Marble Hill, MO 63764-9725
(573) 238-2010
Provides information on all aspects of childbirth. Places special emphasis on choosing safe childbirth alternatives.

American Foundation for Maternal and Child Health
439 East 51st Street, 4th Floor
New York, NY 10022
(212) 759-5510
Provides information on maternal and child health during the birth period.

PRENATAL HEALTH

National Center for Education in Maternal and Child Health
2000 15th Street, N., Suite 701
Arlington, VA 22201-2617
(703) 524-7802
Web site: *www.ncemch.
georgetown.edu*
Government-sponsored agency that provides information on all aspects of maternal and child health.

NATURAL CHILDBIRTH

American Society for Psychoprophylaxis in Obstetrics (ASPO)
1101 Connecticut Avenue, N.W.
Suite 700
Washington, DC 20036
(202) 857-1128
Web site: *www.lamaze-
childbirth.com*
Trains and certifies instructors in Lamaze method of natural childbirth. Provides information to expectant parents.

American Academy of Husband-Coached Childbirth
P.O. Box 5224
Sherman Oaks, CA 91413
(818) 788-6662
(800) 423-2397
Web site: *www.bradleybirth.com*
Certifies instructors in the Bradley method of natural childbirth, which emphasizes coaching by the husband.

HOME BIRTH

Informed Homebirth/Informed Birth and Parenting
P.O. Box 3675
Ann Arbor, MI 48106
(313) 662-6857
Trains and certifies home birth attendants. Offers information to couples interested in home birth.

CESAREAN DELIVERY

Cesareans/Support, Education, and Concern, Inc. (C/Sec, Inc.)
22 Forest Road
Framingham, MA 01701
(508) 877-8266
Provides information on and support for cesarean mothers, including vaginal birth after cesarean.

BIRTH DEFECTS

National Easter Seal Society for Crippled Children and Adults
70 E. Lake Street
Chicago, IL 60601
(312) 726-6200
Web site: *www.seals.com*
Provides information on birth defects. Works with other agencies to help the disabled.

March of Dimes Birth Defects Foundation
1275 Mamaroneck Avenue
White Plains, NY 10605
(914) 428-7100
Web site: *www.modimes.org*
Works to prevent birth defects through public education and community service programs.

Physical Development in Infancy and Toddlerhood

ithin a 2-day period, Lisa, Beth, and Felicia each gave birth to their first child at the same hospital. Over the next 2 years, the mothers permitted me to sit in as they met once a month to talk about their infants—Byron, Rachel, and April. I watched and listened as the three lap babies changed into cruising 1-year-olds and then into walking, talking toddlers.

As the infants grew, the mothers' conversations changed. At first, they worried about physical care—breastfeeding, when to introduce solid foods, and the babies' erratic sleep–waking schedules. Then, between 2 and 3 months, all three youngsters became more alert. Each mother noticed that her baby's daily rhythms were more organized and patterned. "April's much more interested in the world around her," commented Felicia. "Life is easier now that I can anticipate April's feedings and naptimes."

The home setting in which the mothers gathered changed as the babies' motor skills developed. By the second half of the first year, the floor was covered with toys, and all three infants crawled about. Soon crawling became walking. This marked the beginning of *toddlerhood*—a period that spans the second year of life. At first, the children did, indeed, "toddle" with an awkward gait, rocking from side to side and tipping over frequently. But their faces showed the thrill of being upright as they explored enthusiastically. When their 2-year-old birthdays approached, Lisa reflected, "Byron is nearly twice as tall and four times as heavy as he was when he was born. He's starting to look more like a little boy than a baby."

This chapter traces physical growth during the first 2 years—one of the most remarkable and busiest times of development. We will see how rapid changes in the infant's body and brain support new motor skills, ways of learning, and perceptual capacities. Byron, Rachel, and April will join us along the way, to illustrate individual differences and environmental influences on physical development.

BODY GROWTH IN THE FIRST TWO YEARS

uring the first 2 years, the body enlarges more rapidly than at any time after birth. As Figure 4.1 shows, by the end of the first year an infant is about 50 percent longer than at birth, and by 2 years, 75 percent longer. Weight shows similar dramatic gains. By 5 months of age, birth weight has doubled, at 1 year it has tripled, and at 2 years it has quadrupled.

Rather than steady gains, infants and toddlers grow in little spurts. In one study, children followed over the first 21 months of life went for periods of 7 to 63 days with no growth. Then they added as much as a half-inch in a 24-hour period! Almost always, parents described their babies as irritable and unusually hungry on the day before the spurt (Lampl, 1993; Lampl, Veldhuis, & Johnson, 1992).

One of the most obvious changes in Byron, Rachel, and April was their transformation into round, plump babies by the middle of the first year. The early rise in "baby fat," which peaks around 9 months, helps the small infant keep a constant body temperature (Tanner, 1990). During the second year, toddlers become more slender, a trend that continues into middle childhood. In contrast, muscle tissue increases very slowly during infancy and will not reach a peak until adolescence. Babies are not very muscular creatures, and their strength and physical coordination are limited.

INDIVIDUAL AND GROUP DIFFERENCES

As in all aspects of development, differences among children in body size and muscle–fat makeup exist. In infancy, girls are slightly shorter and lighter and have a higher ratio of fat to muscle than do boys. These small sex differences remain throughout early and middle childhood and will be greatly magnified at adolescence. Ethnic differences in body size are apparent as well. Look again at Figure 4.1, and you will see that Rachel, a Japanese-American child, is below the growth norms (height and weight averages) for children her age. In contrast, April is somewhat taller and heavier, as African-American children tend to be (Tanner, 1990).

Children of the same age also differ in *rate* of physical growth. In other words, some make faster progress toward

Rapid body growth supports new motor skills, which have a dramatic impact on the baby's approach to the world. As infants sit up and begin to crawl, their whole view of the environment and capacity to explore it changes. (Tom McCarthy/PhotoEdit)

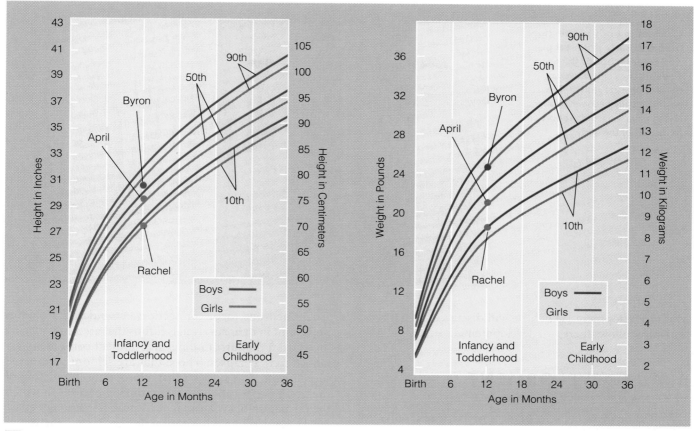

FIGURE 4.1

Gains in height and weight from birth to 2 years among North American children. The steep rise in these growth curves shows that children grow rapidly during this period. At the same time, wide individual differences in body size exist. Infants and toddlers who fall at the 50th percentile are average in height and weight. Those who fall at the 90th percentile are taller and heavier than 90 percent of their agemates. Those who fall at the 10th percentile are taller and heavier than only 10 percent of their peers. Notice that girls are slightly shorter and lighter than boys.

a mature body size than others. We cannot tell how quickly a child's physical growth is moving along just by looking at current body size, since children grow to different heights and weights in adulthood. For example, Byron is slightly larger and heavier than Rachel and April, but he is not physically more mature. In a moment, you will see why.

The best way of estimating a child's physical maturity is to use *skeletal age,* a measure of bone development determined through X-rays. When the skeletal ages of infants and children are examined, girls are considerably ahead of boys. At birth, this sex difference amounts to about 4 to 6 weeks, a gap that widens over infancy and childhood with girls reaching their full body size several years before boys. Girls' greater physical maturity may be partly responsible for the fact that they are more resistant to harmful environmental influences, show fewer developmental problems, and have a lower infant and childhood mortality rate. Besides sex, ethnicity is related to physical maturity: black children tend to be slightly ahead of white children at all ages.

PATTERNS OF BODY GROWTH

As the child's overall size increases, different parts of the body grow at different rates. Two growth patterns describe these changes in body proportions. The first, called the **cephalocaudal trend,** is depicted in Figure 4.2. Translated from Latin, it means "head to tail." As you can see, at birth the head takes up one-fourth of the body, the legs only one-third. Notice how the lower portion of the body catches up. By age 2, the head accounts for only one-fifth and the legs for nearly one-half of total body length. Perhaps this is one reason Lisa thought 2-year-old Byron was starting to look more like a little boy than a baby.

The second pattern is the **proximodistal trend**, meaning growth proceeds from the center of the body outward. In the prenatal period, the head, chest, and trunk grew first, followed by the arms and legs, and finally by the hands and feet. During infancy and childhood, the arms and legs continue to grow somewhat ahead of the hands and feet.

FIGURE 4.2

Changes in body proportions from the early prenatal period to adulthood. This figure illustrates the cephalocaudal trend of physical growth. The head gradually becomes smaller and the legs longer in proportion to the rest of the body.

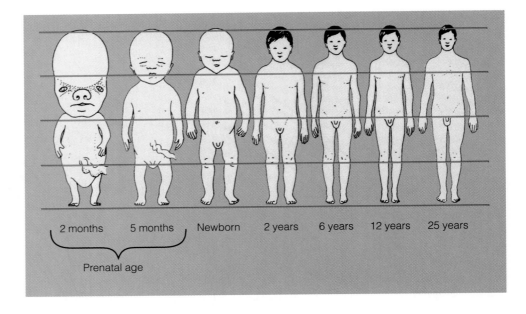

2 months   5 months   Newborn   2 years   6 years   12 years   25 years

Prenatal age

Motor development (which we will discuss in a later section) also follows these two developmental patterns.

BRAIN DEVELOPMENT

Although at birth the brain is nearer to its adult size than any other physical structure, it continues to develop at an astounding pace throughout infancy and toddlerhood. By age 2, the brain is already about 70 percent of its adult weight. To understand brain growth, we need to look at it from two vantage points: (1) the microscopic level of individual brain cells and (2) the larger structural level of the cerebral cortex, responsible for the highly developed intelligence of our species.

DEVELOPMENT OF NEURONS

The basic story of brain growth concerns how **neurons,** or nerve cells that store and transmit information, develop and form an elaborate communication system. The human brain has 100 to 200 billion neurons, many of which have thousands of direct connections with other neurons. There are tiny gaps, or **synapses,** between them, where fibers from different neurons come close together but do not touch. Neurons release chemicals that cross the synapse, sending messages to one another.

Recall that by the end of the second trimester of pregnancy, no more neurons will ever again be produced. After birth, the neurons form complex networks of synaptic connections. As Figure 4.3 shows, during infancy and toddlerhood, growth of neural fibers and synapses increases at an astounding pace (Huttenlocher, 1994; Moore & Persaud, 1993).

As neurons form connections, *stimulation* becomes important in their survival. Neurons that are stimulated by input from the surrounding environment continue to establish new synapses. Those that are seldom stimulated soon die off. This suggests that appropriate stimulation of the brain is critically important during periods in which the formation of synapses is at its peak (Greenough et al., 1993). Research in which animals have been deprived of stimulation during the early weeks and months of life supports this idea. And as the Lifespan Vista box on pages 120–121 indicates, there seem to be periods in which rich and varied stimulation is essential for the human brain to reach its potential as well.

Perhaps you are wondering: If no more neurons are produced after the prenatal period, what causes the dramatic increase in brain size during the first 2 years? About half the brain's volume is made up of **glial cells,** which do not carry messages. Instead, they are responsible for **myelinization,** the coating of neural fibers with an insulating fatty sheath (called *myelin*) that improves the efficiency of message transfer. Glial cells multiply dramatically from the fourth month of pregnancy through the second year of life (Casaer, 1993). Myelinization is responsible for the rapid gain in overall size of the brain.

DEVELOPMENT OF THE CEREBRAL CORTEX

The **cerebral cortex** is the largest, most complex structure of the human brain, accounting for 85 percent of the brain's weight and containing the greatest number of neurons and synapses. It surrounds the rest of the brain, much like a half-shelled walnut. The cerebral cortex is the last part of the brain to stop growing. For this reason, it is be-

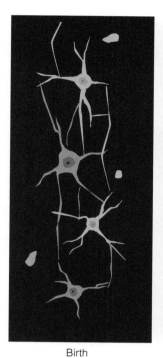 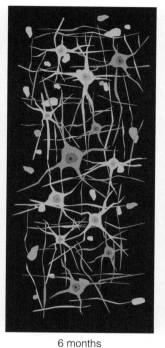 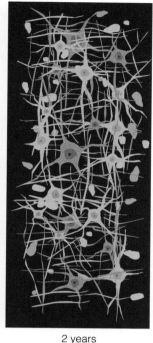

Birth   6 months   2 years

FIGURE 4.3

Development of neurons. Growth of neural fibers takes place rapidly from birth to 2 years. During this time, new synapses form at an astounding pace, supporting the emergence of many new capacities. Stimulation is vitally important for maintaining and increasing this complex communication network. *(From J. L. Conel, The postnatal development of the human cerebral cortex. Cambridge, MA: Harvard University Press. Copyright © 1959 by the President and Fellows of Harvard College. All rights reserved.)*

lieved to be much more sensitive to environmental influences than any other part of the brain.

As Figure 4.4 shows, different regions of the cerebral cortex have specific functions, such as receiving information from the senses, instructing the body to move, and thinking. Not surprisingly, the order in which areas of the cortex develop corresponds to the order in which various capacities emerge in the infant and growing child. The last portion of the cortex to develop and myelinate is the frontal lobe, responsible for thought and consciousness. From age 2 months onward, this area functions more effectively, and it continues to grow well into the second and third decades of life (Fischer & Rose, 1995).

The cortex has two halves, or *hemispheres:* left and right. The hemispheres do not have precisely the same functions. Some tasks are done mostly by one hemisphere and some by the other. For example, each hemisphere receives sensory information from and controls only one side of the body—the one opposite to it.[1] For most of us, the left hemisphere is responsible for verbal abilities (such as spoken and written language) and positive emotion (for example, joy). The right hemisphere handles spatial abilities (judging distances, reading maps, and recognizing geometric shapes) and negative emotion (such as distress). This pattern may be reversed in left-handed people, but

more often, the cortex of left-handers is less clearly specialized than that of right-handers.

Specialization of the two hemispheres is called **lateralization.** Scientists are interested in when brain lateralization occurs because they want to know more about **brain plasticity.** A highly *plastic* cortex is still adaptable because many areas are not yet committed to specific functions. If a

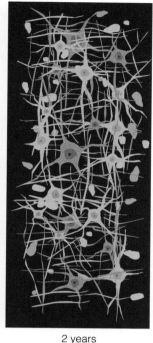

FIGURE 4.4

The left side of the human brain, showing the cerebral cortex. The cortex is divided into different lobes, each of which contains a variety of regions with specific functions. Some major ones are labeled here.

---

[1] The eyes are an exception. Messages from the right halves of each retina go to the right hemisphere; messages from the left halves of each retina go to the left hemisphere. Thus, visual information from both eyes is received by both hemispheres.

A LIFESPAN VISTA

SENSITIVE PERIODS IN BRAIN DEVELOPMENT

The existence of sensitive periods in development of the cortex has been amply demonstrated in studies of animals exposed to extreme forms of sensory deprivation. For example, there seems to be a time when visual experiences must occur for the visual centers of the brain to develop normally. If a month-old kitten is deprived of light for as brief a time as 3 or 4 days, these areas start to degenerate. If the kitten is kept in the dark for as long as 2 months, the damage is permanent. Severe stimulus deprivation also affects overall brain growth. When animals reared as pets are compared to animals reared in isolation, the brains of the pets are heavier and thicker (Greenough & Black, 1992).

Because we cannot ethically expose children to such experiments, researchers interested in identifying sensitive periods for human brain development must rely on naturally occurring circumstances or less direct evidence. They have found some close parallels with the animal findings just described. For example, without early corrective surgery, babies born with *strabismus* (a condition in which one eye does not focus because of muscle weakness) show permanent impairments in depth perception (which depends on blending visual images from

From infancy to early adulthood, several brain growth spurts occur that coinside with peaks in intelligence test performance and major gains in cognitive competence. How brain development can best be supported through stimulation during each of these periods is a challenging question for future research. (John Yurka/The Picture Cube)

both eyes), visual acuity, and perception of the spatial layout of the environment (Birch, 1993).

Focusing on the cortex as a whole, several investigators have identified intermittent brain growth spurts from infancy to early adulthood, based on gains in brain weight and skull size as well as changes in electrical activity of the cortex, as measured by the electroencephalogram (EEG)[2] (Epstein, 1980; Hudspeth & Pribram, 1992; Thatcher, 1991, 1994). These spurts coincide with peaks in intelligence

part of the brain is damaged, other parts can take over tasks that would have been handled by the damaged region. But once the hemispheres lateralize, damage to a particular region means that the abilities controlled by it will be lost forever.

At birth, the hemispheres have already begun to specialize. Most newborns show greater electrical brain-wave activity in the left hemisphere while listening to speech sounds. In contrast, the right hemisphere reacts more strongly to nonspeech sounds as well as stimuli (such as a sour-tasting fluid) that cause infants to display negative emotion (Fox & Davidson, 1986; Hahn, 1987). Nevertheless, dramatic evidence for early plasticity comes from re-

search on infants who had part or all of one hemisphere removed to control violent brain seizures. The remaining hemisphere, whether right or left, took over language and spatial functions. But because lateralization was already underway, full recovery did not take place. In middle childhood and adolescence, these youngsters had difficulty with very complex verbal and spatial tasks (Goodman & Whitaker, 1985).

Before 1 year of age, the brain is more plastic than at any later time of life, perhaps because many of its synapses have not yet been established (Huttenlocher, 1994). Still, the cortex seems to be programmed from the start for hemispheric specialization. A lateralized brain

test performance and major gains in cognitive competence. For example, Figure 4.5 shows the findings of a Swedish study, in which EEGs were measured during a quiet, alert state in individuals ranging from 1 to 21 years of age. The first EEG energy spurt occurred around age 1½ to 2, a period in which representation and language flourish. The next three spurts, at ages 9, 12, and 15, probably reflect the emergence and refinement of abstract thinking. Another spurt, around age 18 to 20, may signal the capacity for mature, reflective thought (Fischer & Rose, 1995; Kitchener et al., 1993).

Researchers speculate that massive production of synapses may underlie the earliest brain growth spurt. Development of more complex and efficient neural networks, due to myelinization and long-distance connections between the frontal lobe and other cortical regions, may account for the later ones. Exactly how brain development might be supported or disrupted by experience during each of these periods is still a question for future research. Once we have such information, it will have major implications for child-rearing and educational practices.

[2] In an EEG, researchers tape electrodes to the scalp, which permit them to measure the electrical activity of the brain.

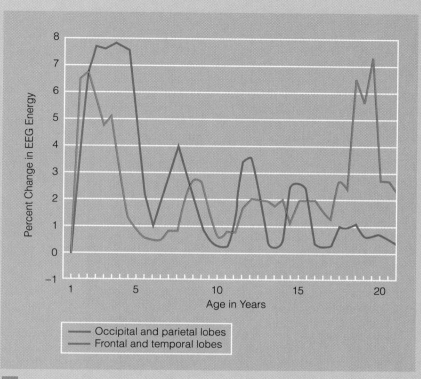

FIGURE 4.5

Brain growth spurts, based on findings of a Swedish cross-sectional study in which EEGs were measured in individuals 1 to 21 years of age. EEG energy peaks indicate periods of rapid growth. These occurred around 1½ to 2, 9, 12, 15, and 18 to 20 years of age in all lobes of the cortex (frontal, temporal, occipital, and parietal). The spurts coincide with peaks in children's intelligence test performance and major gains in cognitive competence. *(From K. W. Fischer & S. P. Rose, 1995, Concurrent Cycles in the Dynamic Development of Brain and Behavior. Newsletter of the Society for Research in Child Development, p. 16. Reprinted by permission of the authors.)*

has the advantage of permitting a much greater variety of talents than if both sides of the cortex served exactly the same functions.

CHANGING STATES OF AROUSAL

Between birth and 2 years, the organization of sleep and wakefulness changes substantially, and fussiness and crying decline. Recall from Chapter 3 that the newborn baby takes round-the-clock naps that total about 16 hours. The average 2-year-old still needs 12 to 13 hours of sleep. The greatest change in sleep and wakefulness is that short periods of each are put together and coincide with a night

and day schedule (Berg & Berg, 1987). By the second year, children generally need only one or two naps a day.

Changes in infants' patterns of arousal are largely due to brain maturation, but they are affected by the social environment as well. In the United States and most Western nations, night waking is regarded as inconvenient. Parents usually succeed in getting their babies to sleep through the night around 4 months of age by offering an evening feeding before putting them down in a separate, quiet room. In this way, they push young infants to the limits of their neurological capacities. Yet as the Cultural Influences box on page 122 shows, isolating infants to promote sleep is rare elsewhere in the world. Influenced by Japanese customs,

CULTURAL INFLUENCES

CULTURAL VARIATION IN INFANT SLEEPING ARRANGEMENTS

While awaiting the birth of a new baby, American middle-class parents typically furnish a special room as the infant's sleeping quarters. Most adults in the United States regard this nighttime separation of baby from parent as perfectly natural. Throughout this century, child-rearing advice from experts has strongly encouraged it. For example, Benjamin Spock, in each edition of *Baby and Child Care* from 1945 to the present, states with authority, "I think it is a sensible rule not to take a child into the parents' bed for any reason" (Spock & Rothenberg, 1992, p. 113).

Yet parent–infant "cosleeping" is common around the globe, in industrialized and nonindustrialized countries alike. Japanese children usually lie next to their mothers throughout infancy and early childhood and continue to sleep with a parent or other family member until adolescence (Takahashi, 1990). Among the Mayans of rural Guatemala, mother–infant cosleeping is interrupted only by the birth of a new baby, at which time the older child is moved beside the father or to another bed in the same room (Morelli et al., 1992). Cosleeping is also frequent in some American subcultures. African-American children are more likely than Caucasian-American children to fall asleep with parents and to remain with them for part or all of the night (Lozoff et al., 1995). Appalachian children of eastern Kentucky typically sleep with their parents for the first 2 years of life (Abbott, 1992).

Available household space plays a minor role in infant sleeping arrangements. Dominant child-rearing beliefs are much more important. In one study, researchers interviewed middle-class American mothers and Guatemalan Mayan mothers about their sleeping practices. American mothers frequently mentioned the importance of early independence training, preventing bad habits, and protecting their own privacy. In contrast, Mayan mothers explained that cosleeping helps build a close parent–child bond, which is necessary for children to learn the ways of people around them. When told that American infants sleep by themselves, Mayan mothers reacted with shock and disbelief, stating that it would be painful for them to leave their babies alone at night (Morelli et al., 1992).

Infant sleeping practices affect other aspects of family life. Sleep problems are not an issue for Mayan parents. Babies doze off in the midst of ongoing social activities and are carried to bed by their mothers. In the United States, getting young children ready for bed often requires an elaborate ritual that takes a good part of the evening. Perhaps bedtime struggles, so common in American middle-class homes but rare elsewhere in the world, are related to the stress young children feel when they are required to fall asleep without assistance.

Infant sleeping arrangements, like other parenting practices, are meant to foster culturally valued characteristics in the young. American middle-class parents view babies as dependent beings who must be urged toward independence, and so they usually require them to sleep alone. In contrast, Japanese, Mayan, and Appalachian parents regard young infants as separate beings who need to establish an interdependent relationship with the community to survive.

Although rare in American middle-class families, parent-infant cosleeping is common around the globe. When children fall asleep with their parents, sleep problems are rare during the early years. And many parents who practice cosleeping believe that it helps build a close parent-child bond. (LaLeche League)

Beth held Rachel close for much of the day. At night, Rachel slept in her mother's bed, waking to nurse at will. For infants experiencing this type of care, the average sleep period remains constant at 3 hours, from 1 to 8 months of age. Only at the end of the first year do these babies move in the direction of an adultlike sleep–waking schedule (Super & Harkness, 1982).

BRIEF REVIEW

Overall size of the body increases rapidly during infancy and toddlerhood. During the first year, body fat increases much faster than muscle. Different parts of the body grow at different rates, following cephalocaudal and proximodistal trends. The human brain grows faster early in development than any other organ. During the first 2 years, synapses, or connections between neurons, are rapidly laid down. Myelinization is responsible for efficient communication among neurons and a dramatic increase in brain weight. The cerebral cortex is the last part of the brain to stop growing. It is also the structure most affected by stimulation. The cortex has already begun to lateralize at birth, but it retains considerable plasticity during the first year of life. Although brain maturation is largely responsible for changes in sleep and wakefulness that gradually coincide with a night and day schedule, the social environment also contributes.

ASK YOURSELF . . .

- *Felicia commented that at 2 months, April's daily schedule seemed more predictable, and she was much more alert. What aspects of brain development might be responsible for this change?*

INFLUENCES ON EARLY PHYSICAL GROWTH

Physical growth, like other aspects of development, results from the interplay between genetic and environmental factors. Heredity, nutrition, and emotional well-being influence early physical growth.

HEREDITY

Since identical twins are much more alike in body size than are fraternal twins, we know that heredity is important in physical growth. When diet and health are adequate, height and rate of physical growth are largely determined by heredity (Tanner, 1990). In fact, as long as negative environmental influences, such as poor nutrition or illness, are not severe, children and adolescents typically show *catch-up growth*—a return to a genetically determined growth path.

Genetic makeup also affects body weight, since the weights of adopted children correlate more strongly with those of their biological than adoptive parents (Stunkard et al., 1986). However, as far as weight is concerned, environment—in particular, nutrition—plays an especially important role.

NUTRITION

Good nutrition is important at any time of development, but it is especially critical in infancy because the baby's brain and body are growing so rapidly. Pound for pound, a young baby's energy needs are twice those of an adult. Twenty-five percent of the infant's total caloric intake is devoted to growth, and extra calories are needed to keep rapidly developing organs functioning properly (Pipes, 1989).

Babies do not just need enough food. They need the right kind of food. In early infancy, breast milk is especially

Breast-feeding is especially important in developing countries, where infants are at risk for malnutrition and early death due to widespread poverty. This baby of Rajasthan, India, is likely to grow normally during the first year because his mother decided to breast-feed. (Jane Schreirman/Photo Researchers)

C A R E G I V I N G   C O N C E R N S

Nutritional and Health Advantages of Breastfeeding

ADVANTAGE	DESCRIPTION
Correct balance of fat and protein	Compared to the milk of other mammals, human milk is higher in fat and lower in protein. This balance, as well as the unique proteins and fats contained in human milk, is ideal for a rapidly myelinating nervous system.
Nutritional completeness	A mother who breast-feeds need not add other foods to her infant's diet until the baby is 6 months old. The milks of all mammals are low in iron, but the iron contained in breast milk is much more easily absorbed by the baby's system. Consequently, bottle-fed infants need iron-fortified formula.
Protection against disease	Through breast-feeding, antibodies are transferred from mother to child. As a result, breast-fed babies have far fewer respiratory and intestinal illnesses and allergic reactions than do bottle-fed infants.
Digestibility	Since breast-fed babies have a different kind of bacteria growing in their intestines than do bottle-fed infants, they rarely become constipated or have diarrhea.
Smoother transition to solid foods	Breast-fed infants accept new solid foods more easily than do bottle-fed infants, perhaps because of their greater experience with a variety of flavors, which pass from the maternal diet into the mother's milk.

Sources: Ford & Labbok, 1993; Räihä & Axelsson, 1995; Sullivan & Birch, 1994.

suited to their needs, and bottled formulas try to imitate it. Later, infants require well-balanced solid foods. If a baby's diet is deficient in either quantity or quality, growth can be permanently stunted.

■ BREAST- VERSUS BOTTLE-FEEDING. For thousands of years, all babies were fed the ultimate human health food: breast milk. Only within the last hundred years has bottle-feeding been available. As formulas became easier to prepare, breast-feeding declined from the 1940s into the 1970s, when over 75 percent of American infants were bottle fed. Partly as a result of the natural childbirth movement (see Chapter 3), breast-feeding gradually became more common, especially among well-educated, middle-class women. Today, over 60 percent of American mothers breast-feed their babies (U.S. Department of Health and Human Services, 1996a).

The Caregiving Concerns table above summarizes the major nutritional and health advantages of breast-feeding. Because of these benefits, breast-fed babies in poverty-stricken regions of the world are much less likely to be malnourished and 6 to 14 times more likely to survive the first year of life. Too often, bottle-fed infants in developing countries get low-grade nutrients, such as rice water or highly dilute cow's and goat's milk. When formula is available, it is generally contaminated due to poor sanitation. Also, because a mother is less likely to get pregnant while she is nursing, breast-feeding helps increase spacing among siblings, a major factor in reducing infant and childhood deaths in economically depressed populations (Grant, 1995). (Note, however, that breast-feeding is not a reliable method of birth control.)

In industrialized nations, most women who choose breast-feeding find it emotionally satisfying, but it is not for everyone. Some mothers simply do not like it or feel embarrassed by it. A few others, for physiological reasons, do not produce enough milk. Occasionally, medical reasons—such as therapy with certain drugs or a viral or bacterial disease that could infect the baby—prevent a mother from nursing (Seltzer & Benjamin, 1990; Van de Perre et al., 1993).

Breast milk is so digestible that an infant becomes hungry quite often—every 1½ to 2 hours in comparison to every 3 or 4 hours for a bottle-fed baby. This makes breast-feeding inconvenient for many employed women. A mother who is not always with her baby can still breast-feed or combine it with bottle-feeding. For example, Lisa returned to her job part-time when Byron was 2 months old. Before she left for work, she pumped her milk into a bottle for later feeding by his caregiver (Hills-Banczyk et al., 1993).

Some women who cannot or do not want to breast-feed worry that they are depriving their baby of an experience essential for emotional development. Yet breast- and bottle-fed youngsters do not differ in psychological adjustment (Fergusson, Horwood, & Shanon, 1987). Regardless of feeding method, a mother can respond promptly to her hungry baby and hold and stroke the infant gently—aspects of feeding that contribute to emotional well-being.

■ ARE CHUBBY BABIES AT RISK FOR LATER OVERWEIGHT AND OBESITY? Byron was an enthusiastic eater from early infancy. He nursed vigorously and gained weight quickly. By 5 months, he began reaching for food from his parents' plates. Lisa wondered: Was she overfeed-

Building a Foundation for Good Eating Habits

SUGGESTION	DESCRIPTION
Introduce solid foods patiently and gradually.	At first, use a small spoon that permits the baby to suck off pureed food, gradually helping the infant to adapt to a solid diet. Patiently offer foods the infant rejects, gradually adding new ones. Food preferences are heavily influenced by experience.
Offer a varied, healthy diet.	Ensure that the older infant and toddler's diet is healthy by offering a wide range of nutritious foods. Limit sweets and avoid "junk" foods, since toddlers will learn to prefer them.
Permit the toddler to choose among healthy foods.	Support the early desire for independence by allowing the toddler to engage in self-feeding (even if messy) and to select from available foods. Over several meals, nutritional needs will be met, and mealtimes will also be emotionally satisfying.
Accept a toddler's idiosyncratic diet.	Relax when a toddler refuses meat and vegetables in favor of peanuts, apples, and cheese slices. These "good foods" may be suited to the child's momentary growth needs. When a varied diet is available, many selections fulfill nutritional requirements, and children eventually modify food jags on their own.
Avoid turning mealtime into an early battleground.	Do not force an infant or toddler to eat. When a young child is consistently unhappy at mealtimes, parents are probably unduly anxious about the child's eating behaviors—a circumstance that can create feeding problems.

ing Byron and increasing his chances of being permanently overweight?

Only a slight correlation exists between fatness in infancy and obesity at older ages (Roche, 1981; Shapiro et al., 1984). Most chubby babies thin out during toddlerhood and the preschool years, as weight gain slows and they become more active. Infants and toddlers can eat nutritious foods freely, without risk of becoming too fat.

How can concerned parents prevent their infants from becoming overweight children and adults? As the Caregiving Concerns table above indicates, one way is to encourage good eating habits. Candy, soft drinks, french fries, and other high-calorie foods loaded with sugar, salt, and saturated fats should be avoided. When young children are given such foods regularly, they start to prefer them (Birch & Fisher, 1995). Physical exercise also guards against excessive weight gain. Once toddlers learn to walk, climb, and run, parents should encourage their natural delight in being able to control their bodies by providing opportunities for energetic play.

MALNUTRITION

Osita is an Ethiopian 2-year-old whose mother has never had to worry about his gaining too much weight. When she weaned him at 1 year, there was little for him to eat besides starchy rice flour cakes. Soon his belly enlarged, his feet swelled, his hair began to fall out, and a rash appeared on his skin. His bright-eyed, curious behavior vanished, and he became irritable and listless.

In developing countries and war-torn areas where food resources are limited, malnutrition is widespread. Recent evidence indicates that 40 to 60 percent of the world's children do not get enough to eat (Bread for the World Institute, 1994). Among the 4 to 7 percent severely affected, malnutrition leads to two dietary diseases: marasmus and kwashiorkor.

Marasmus is a wasted condition of the body caused by a diet low in all essential nutrients. It usually appears in the first year of life, when a baby's mother is too malnourished to produce enough breast milk, and bottle-feeding is also inadequate. Her starving baby becomes painfully thin and is in danger of dying.

Osita has **kwashiorkor**, caused by an unbalanced diet very low in protein. Kwashiorkor usually strikes after weaning, between 1 and 3 years of age. It is common in areas of the world where children get just enough calories from starchy foods, but protein resources are scarce. The child's body responds by breaking down its own protein reserves, causing the swelling and other symptoms that Osita experienced.

Children who manage to survive these extreme forms of malnutrition grow to be smaller in all body dimensions (Galler, Ramsey, & Solimano, 1985a). In addition, their brains are seriously affected. One long-term study of marasmic children revealed that an improved diet led to some catch-up growth in height, but little improvement in head size (Stoch et al., 1982). The malnutrition probably interfered with myelinization, causing a permanent loss in brain weight. By middle childhood, these youngsters score

The swollen abdomen and listless behavior of this child are classic symptoms of kwashiorkor, a nutritional illness that results from a diet very low in protien. (CNRI/Phototake)

low on intelligence tests, show poor fine motor coordination, and have difficulty paying attention in school (Galler et al., 1984, 1990; Galler, Ramsey, & Solimano, 1985b).

Malnutrition is not confined to developing countries. More than 12 percent of American children go to bed hungry at night. Although few of these children have marasmus or kwashiorkor, their physical growth and ability to learn are still affected (Food Research & Action Center, 1991; Wachs, 1995). Recall that poverty and stressful living conditions make the impact of poor diet even worse. Malnutrition is clearly a national and international crisis—one of the most serious problems confronting the human species today.

EMOTIONAL WELL-BEING

We are not used to thinking of affection and stimulation as necessary for healthy physical growth, but they are just as vital as food. **Nonorganic failure to thrive** is a growth disorder, usually present by 18 months of age, resulting from lack of parental love. Infants who have it show all the signs of marasmus: their bodies look wasted, and they are withdrawn and apathetic. But no organic (or biological) cause for the baby's failure to grow can be found. Enough food is offered, and the infant does not have a serious illness.

Lana, an observant nurse at a public health clinic, became concerned about 8-month-old Melanie, who was 3 pounds lighter than she had been at her last checkup. Her mother claimed to feed her often and could not understand why she did not grow. Lana noted Melanie's behavior. Unlike most infants her age, she did not mind separating from her mother. Lana tried offering Melanie a toy, but she showed little interest. Instead, she anxiously kept her eyes on adults in the room. When Lana smiled and tried to look into Melanie's eyes, she turned her head away (Leonard, Rhymes, & Solnit, 1986; Oates, 1984).

The family circumstances surrounding failure to thrive help explain these typical reactions. During feeding and diaper changing, Melanie's mother sometimes acted cold and distant, at other times impatient and hostile. Melanie tried to protect herself by keeping track of her mother's whereabouts and, when her mother approached, avoiding her gaze. Often an unhappy marriage or other family pressures contribute to these serious caregiving problems (Gorman, Leifer, & Grossman, 1993). Melanie's alcoholic father was out of work, and her parents argued constantly. Melanie's mother had little energy to meet the psychological needs of Melanie and her other three children. Without early treatment, by helping parents or arranging for a caring foster home, failure-to-thrive infants remain small and show lasting cognitive and emotional difficulties (Drotar, 1992; Drotar & Sturm, 1988).

BRIEF REVIEW

Heredity, nutrition, affection, and stimulation all contribute to early physical growth. Studies of twins show that body size and rate of maturation are affected by genetic makeup. Breast milk provides babies with the ideal nutrition between birth and 6 months of age. Although bottle- and breast-fed babies do not differ in psychological development, breast-feeding protects many poverty-stricken infants against malnutrition, disease, and early death. Malnutrition is a serious global problem. When marasmus and kwashiorkor are allowed to persist, physical size, brain growth, and ability to learn are permanently affected. Nonorganic failure to thrive reminds us of the close connection between sensitive, loving care and children's growth.

ASK YOURSELF . . .

■ *Explain why breast-feeding offers babies protection against disease and early death in poverty-stricken regions of the world.*

■ *Ten-month-old Shaun is below average in height and painfully thin. He has one of two serious growth disorders. Name them, and indicate what clues you would look for to tell which one Shaun has.*

MOTOR DEVELOPMENT DURING THE FIRST TWO YEARS

Lisa, Beth, and Felicia each kept baby books, proudly noting when their three children held up their heads, reached for objects, sat by themselves, and walked alone. Parents' enthusiasm for these achievements makes perfect sense. With each new motor skill, babies master their bodies and the environment in a new way. For example, sitting alone grants infants an entirely different perspective on the world. Voluntary reaching permits babies to find out about objects by acting on them. And when infants can move on their own, their opportunities for exploration are multiplied.

Babies' motor achievements have a powerful effect on their social relationships. April was the first of the three babies to master crawling. Suddenly Felicia had to "child-proof" the household and restrict April's movements in ways that were unnecessary when, placed on a blanket, she would stay there! At the same time, playful activities expanded. At the end of the first year, April and her parents played a gleeful game of hide-and-seek around the living room sofa. Soon after, April could turn the pages of a picture book and point while Felicia named the objects. April's expressions of delight as she worked on new motor competencies triggered pleasurable reactions in others, which encouraged her efforts further (Mayes & Zigler, 1992). Motor skills, social competencies, cognition, and language were developing together and supporting one another.

THE SEQUENCE OF MOTOR DEVELOPMENT

Gross motor development refers to control over actions that help infants get around in the environment, such as crawling, standing, and walking. In contrast, *fine motor development* has to do with smaller movements, such as reaching and grasping. Table 4.1 shows the average age at which infants and toddlers achieve a variety of gross and fine motor skills. Most children follow this sequence fairly closely.

Notice that the table also presents the age ranges during which the majority of babies accomplish each skill. These indicate that although the *sequence* of motor development is fairly uniform across children, there are large individual differences in *rate* of motor progress. We would be concerned about a child's development only if many motor skills were seriously delayed.

TABLE 4.1

Gross and Fine Motor Development in the First Two Years

MOTOR SKILL	AVERAGE AGE ACHIEVED	AGE RANGE IN WHICH 90 PERCENT OF INFANTS ACHIEVE THE SKILL
When held upright, head erect and steady	6 weeks	3 weeks–4 months
When prone, lifts self by arms	2 months	3 weeks–4 months
Rolls from side to back	2 months	3 weeks–5 months
Grasps cube	3 months, 3 weeks	2–7 months
Rolls from back to side	4½ months	2–7 months
Sits alone	7 months	5–9 months
Crawls	7 months	5–11 months
Pulls to stand	8 months	5–12 months
Plays pat-a-cake	9 months, 3 weeks	7–15 months
Stands alone	11 months	9–16 months
Walks alone	11 months, 3 weeks	9–17 months
Builds tower of two cubes	13 months, 3 weeks	10–19 months
Scribbles vigorously	14 months	10–21 months
Walks up stairs with help	16 months	12–23 months
Jumps in place	23 months, 2 weeks	17–30 months

Source: Bayley, 1969.

New research reveals that motor development is a matter of acquiring increasingly complex, dynamic systems of action. With plenty of movement opportunities and a rich, stimulating environment, this baby has begun to combine previously mastered motor skills—crawling, reaching, pulling to a stand, and stepping—into walking. Her new accomplishment will grant her a more effective way of exploring and controlling the environment. (Menim/Monkmeyer Press)

Look at Table 4.1 once more, and you will see that there is organization and direction to the infant's motor achievements. The *cephalocaudal trend* is evident. Motor control of the head comes before control of the arms and trunk, and control of the arms and trunk before control of the legs. You can also see the *proximodistal trend:* head, trunk, and arm control is advanced over coordination of the hands and fingers. Because physical and motor development follow the same general sequence, early theorists regarded the cephalocaudal and proximodistal trends as genetically determined, maturational patterns. Yet as we will see in the next section, infant motor skills are not isolated, unrelated accomplishments that follow a fixed maturational timetable.

MOTOR SKILLS AS DYNAMIC SYSTEMS OF ACTION

New research reveals that motor development is a matter of acquiring increasingly complex, **dynamic systems of action.** When motor skills work as a system, separate abilities blend together, each cooperating with others to produce more effective ways of exploring and controlling the environment. For example, control of the head and upper chest are combined into sitting with support. Kicking, rocking on all fours, and reaching are gradually

put together into crawling. Then crawling, standing, and stepping are united into walking alone (Pick, 1989; Thelen, 1989).

Each new skill is a joint product of central nervous system maturation, movement possibilities of the body, environmental supports for the skill, and the goal the child has in mind, such as getting a toy or crossing the room. Change in any one of these elements leads to loss of stability in the system, and the child explores and selects new motor patterns. Once a new skill is discovered, the child tries hard to perfect it. Gradually, it becomes accurate, smooth, and efficient. Because motor development is motivated by exploration and the desire to master new tasks, it can only be mapped out by heredity at a very general level (Thelen, 1995; Thelen & Smith, 1997). And since each new skill is learned by revising and combining earlier accomplishments to fit a new goal, infants achieve motor milestones in unique ways.

CULTURAL VARIATIONS IN MOTOR DEVELOPMENT

Cross-cultural research shows how early movement opportunities and a stimulating environment contribute to motor development. Several decades ago, Wayne Dennis (1960) observed infants in Iranian orphanages who were deprived of the tantalizing surroundings that induce infants to acquire motor skills. The Iranian babies spent their days lying on their backs in cribs, without toys to play with. As a result, most did not move on their own until after 2 years of age. When they finally did move, the constant experience of lying on their backs led them to scoot in a sitting position rather than crawl on their hands and knees. As a result, walking was delayed. Since babies who scoot come up against furniture with their feet, not their hands, they are far less likely to pull themselves to a standing position in preparation for walking.

Cultural variations in infant-rearing practices also affect motor development. Take a quick survey of several parents you know, asking these questions: Can babies profit from training? Should sitting, crawling, and walking be deliberately encouraged? Answers vary widely from culture to culture. Japanese mothers believe such efforts are unnecessary and unimportant (Caudill, 1973). Among the Zinacanteco Indians of Southern Mexico, rapid motor progress is actively discouraged. Babies who walk before they know enough to keep away from cooking fires and weaving looms are viewed as dangerous to themselves and disruptive to others (Greenfield, 1992).

In contrast, among the Kipsigis of Kenya and the West Indians of Jamaica, babies hold their heads up, sit alone, and walk considerably earlier than North American infants. Kipsigi parents deliberately teach these motor skills. In the first few months, babies are seated in holes dug in the ground, and rolled blankets are used to keep them up-

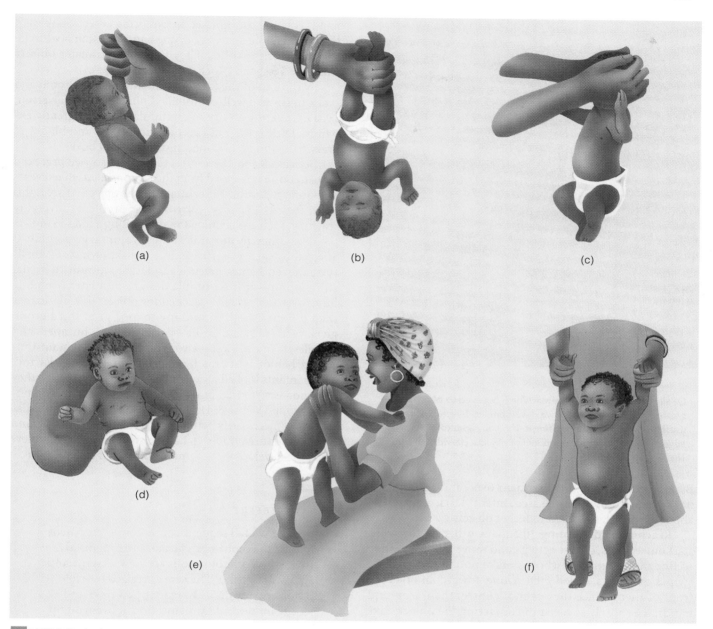

(a) (b) (c)

(d) (e) (f)

FIGURE 4.6

West Indians of Jamaica use a formal handling routine with their babies. Exercises practiced in the first few months include stretching each arm while suspending the baby (a); holding the infant upside-down by the ankles (b); grasping the baby's head on both sides, lifting upward, and stretching the neck (c); and propping the infant with cushions that are gradually removed as the baby begins to sit independently (d). Later in the first year, the baby is "walked" up the mother's body (e) and encouraged to take steps on the floor while supported (f). *(Adapted from B. Hopkins & T. Westra, 1988, "Maternal handling and motor development: An intracultural study," Genetic, Social and General Psychology Monographs, 14, pp. 385, 388, 389. Reprinted by permission of the Helen Dwight Reid Educational Foundation. Published by Heldref Publications, 1319 Eighteenth St., N.W., Washington, DC 20036-1802.)*

right. Walking is promoted by frequently bouncing babies on their feet (Super, 1981). As Figure 4.6 shows, the West Indian mothers use a highly stimulating formal handling routine with their babies, explaining that exercise helps infants grow up strong, healthy, and physically attractive (Hopkins & Westra, 1988).

FINE MOTOR DEVELOPMENT: THE SPECIAL CASE OF VOLUNTARY REACHING

Of all motor skills, voluntary reaching is believed to play the greatest role in infant cognitive development. It opens up a whole new way of exploring the environment.

Of all motor skills, voluntary reaching is believed to play the greatest role in infant cognitive development. By grasping this stuffed animal and manipulating it, this baby adds to his knowledge of the sights, sounds, and feel of objects. (David C. Bitters/The Picture Cube)

By grasping things, turning them over, and seeing what happens when they are released, infants learn a great deal about the sights, sounds, and feel of objects.

Reaching and grasping, like many other motor skills, start out as gross, diffuse activity and move toward mastery of fine movements. Newborns make poorly coordinated swipes or swings, called *prereaching*, toward an object in front of them. Since they cannot control their arms and hands, they rarely succeed in contacting the object. Like newborn reflexes, prereaching eventually drops out, around 7 weeks of age.

At about 3 months, voluntary reaching appears and gradually improves (Bushnell, 1985). Infants of this age reach just as effectively for a sounding object in the dark as for an object in the light (Clifton et al., 1993). As a result, vision is freed from the basic act of reaching so it can focus on more complex adjustments. By 5 months, babies successfully reach for moving objects and reduce their reaching when an object is moved just beyond their reach (Robin, Berthier, & Clifton, 1996; Yonas & Hartman, 1993). And at 9 months, they can obtain a moving object that changes direction (Ashmead et al., 1993).

Once infants can reach, they modify their grasp. The newborn's grasp reflex is replaced by the *ulnar grasp*, a clumsy motion in which the fingers close against the palm. Around 4 to 5 months, when infants begin to sit up, both hands become coordinated in exploring objects. Babies of this age can hold an object in one hand and scan it with the fingertips of the other. They frequently transfer objects from hand to hand (Rochat & Goubet, 1995).

By the end of the first year, infants use the thumb and index finger in a well-coordinated *pincer grasp* (Halverson, 1931). Then the ability to manipulate objects greatly expands. The 1-year-old can pick up raisins and blades of grass, turn knobs, and open and close small boxes.

Between 8 and 11 months, reaching is so well practiced that attention is released from the motor skill itself to events that occur before and after attaining the object. Around this time, infants begin to solve simple problems involving reaching, such as searching for and finding a hidden toy.

Like other motor milestones, voluntary reaching is affected by early experience. In a well-known study, institutionalized babies given a moderate amount of visual stimulation—at first, simple designs and later, a mobile hung over their cribs—reached for objects 6 weeks earlier than did infants given nothing to look at. A third group given massive stimulation—patterned crib bumpers and mobiles at an early age—also reached sooner than unstimulated babies. But this heavy enrichment took its toll. These infants looked away and cried a great deal, and they were not as advanced in reaching as the moderately stimulated group (White & Held, 1966). Clearly, more stimulation is not necessarily better. Trying to push infants beyond their current readiness to handle stimulation can undermine the development of important motor skills.

BRIEF REVIEW

The overall sequence of motor development follows the cephalocaudal and proximodistal trends. Today, motor skills are viewed as dynamic systems of action. They are energized by exploration and the desire to master new tasks and jointly influenced by maturation, movement opportunities, a generally stimulating environment, and infant-rearing practices. Large individual and cultural differences in motor development exist. Voluntary reaching plays a vital role in infant cognitive development. It begins with the newborn baby's prereaching and gradually evolves into a refined pincer grasp around 1 year of age.

ASK YOURSELF . . .

■ *Rosanne read in a magazine that infant motor development could be speeded up through exercise and visual stimulation. She hung mobiles and pictures all over her newborn baby's crib, and she massages and manipulates his body daily. Is Rosanne doing the right thing? Why or why not?*

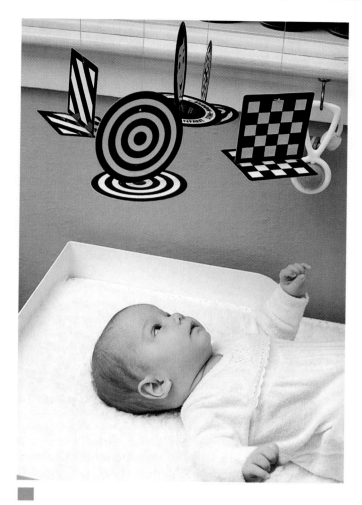

This 3-month-old baby looks at patterns hung over his crib that match his level of visual development. Research shows that a moderate amount of stimulation, tailored to the young baby's needs, results in earlier development of reaching. Either very little or excessive stimulation yields slower motor progress. (Julie O'Neil/The Picture Cube)

LEARNING CAPACITIES

Learning refers to changes in behavior as the result of experience. At birth, the human brain is set up to profit from experience immediately. Infants are capable of two basic forms of learning: classical and operant conditioning. They also learn through their natural preference for novel stimulation. Finally, newborn babies have a remarkable ability to imitate the facial expressions and gestures of adults.

CLASSICAL CONDITIONING

In Chapter 3, we discussed a variety of newborn reflexes. These make **classical conditioning** possible. In this form of learning, a new stimulus is paired with a stimulus that leads to a reflexive response. Once the baby's nervous system makes the connection between the two stimuli, then the new stimulus produces the behavior by itself.

Classical conditioning is of great value to infants because it helps them recognize which events usually occur together in the everyday world. As a result, they can anticipate what is about to happen next, and the environment becomes more orderly and predictable (Rovee-Collier, 1987). Let's take a closer look at the steps of classical conditioning.

As Beth settled down in the rocking chair to nurse Rachel, she often stroked Rachel's forehead. Soon Beth noticed that whenever Rachel's forehead was stroked, she made active sucking movements. Rachel had been classically conditioned. Here is how it happened (see Figure 4.7 on page 132):

1. Before learning takes place, an **unconditioned stimulus (UCS)** must consistently produce a reflexive or **unconditioned response (UCR).** In Rachel's case, sweet breast milk (UCS) resulted in sucking (UCR).

2. Next, a *neutral stimulus* that does not lead to the reflex is presented just before, or at about the same time as, the UCS. Beth stroked Rachel's forehead as each nursing period began. The stroking (neutral stimulus) was paired with the taste of milk (UCS).

3. If learning has occurred, the neutral stimulus by itself produces the reflexive response. The neutral stimulus is then called a **conditioned stimulus (CS),** and the response it elicits is called a **conditioned response (CR).** We know that Rachel has been classically conditioned because stroking her forehead outside the feeding situation (CS) results in sucking (CR).

If the CS is presented alone enough times, without being paired with the UCS, the CR will no longer occur. In other words, if Beth strokes Rachel's forehead again and again without feeding her, Rachel will gradually stop sucking in response to stroking. This is referred to as *extinction*. The occurrence of responses to the CS during the extinction phase of classical conditioning shows that learning has taken place.

Young infants can be classically conditioned most easily when the association between two stimuli has survival value. Rachel learned quickly in the feeding situation, since learning the stimuli that accompany feeding improves the infant's ability to get food and survive (Blass, Ganchrow, & Steiner, 1984). In contrast, some responses are very difficult to condition in young babies. Fear is one of them. Until infants develop the motor skills to escape from unpleasant events, they do not have a biological need to form these associations. But between 8 and 12 months, fear is easy to condition. In Chapter 6, we will discuss the development of fear, as well as other emotional reactions.

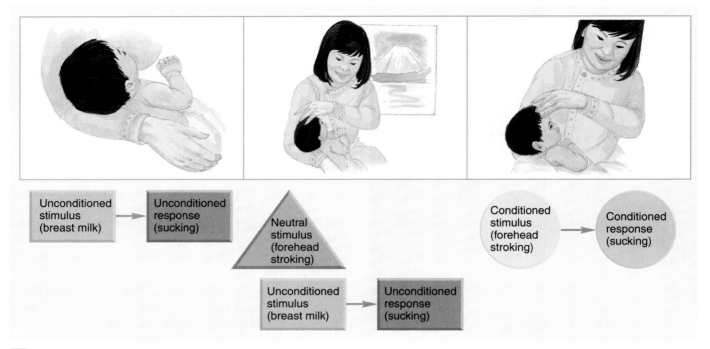

F IGURE 4.7

The steps of classical conditioning. The example here shows how Rachel was classically conditioned to make sucking movements when her forehead was stroked.

OPERANT CONDITIONING

In classical conditioning, babies build expectations about stimulus events in the environment, but they do not influence the stimuli that occur. **Operant conditioning** is quite different. In this form of learning, infants act (or operate) on the environment, and stimuli that follow their behavior change the probability that the behavior will occur again. A stimulus that increases the occurrence of a response is called a **reinforcer**. For example, sweet liquid *reinforces* the sucking response in newborn babies. Removing a desirable stimulus or presenting an unpleasant one to decrease the occurrence of a response is called **punishment.** A sour-tasting fluid *punishes* newborn babies' sucking response. It causes them to purse their lips and stop sucking entirely.

Because the young infant can control only a few behaviors, successful operant conditioning is limited to sucking and head-turning responses. However, many stimuli besides food can serve as reinforcers. For example, researchers have created special laboratory conditions in which the baby's rate of sucking on a nipple produces a variety of interesting sights and sounds. Newborns will suck faster to see visual designs or hear music and human voices (Rovee-Collier, 1987). As these findings suggest, operant conditioning has become a powerful tool for finding out what babies can perceive.

As infants get older, operant conditioning includes a wider range of responses and stimuli. For example, special mobiles have been hung over the cribs of 2- to 6-month-olds. When the baby's foot is attached to the mobile with a long cord, the infant can, by kicking, make the mobile turn. Under these conditions, it takes only a few minutes for infants to start kicking vigorously (Rovee-Collier, 1987).

Operant conditioning soon modifies parents' and babies' reactions to each other. As the infant gazes into the adult's eyes, the adult looks and smiles back, and then the infant looks and smiles again. The behavior of each partner reinforces the other, and as a result, both parent and baby continue their pleasurable interaction. In Chapter 6, we will see that this kind of contingent responsiveness plays an important role in the development of infant–caregiver attachment.

Recall from Chapter 1 that classical and operant conditioning originated with behaviorism, an approach that views the individual as a passive responder to environmental stimuli. Look carefully at the findings just described, and you will see that young babies are not passive. Instead, they use any means they can to explore and control their surroundings. In fact, when infants' environments are so disorganized that their behavior does not lead to predictable outcomes, serious difficulties ranging from intellectual retardation to apathy and depression can result (Cicchetti & Aber, 1986; Seligman, 1975). In addition, as the

Social Issues box on page 134 reveals, problems in brain functioning may prevent some babies from learning certain life-saving responses, leading to sudden infant death syndrome, a major cause of infant mortality.

HABITUATION AND DISHABITUATION

Take a moment to walk through the rooms of the library, your home, or wherever you happen to be reading this book. What did you notice? Probably those things that are new and different, such as a recently purchased picture on the wall or a piece of furniture that has been moved. From birth, the human brain is set up to be attracted to novelty. **Habituation** refers to a gradual reduction in the strength of a response due to repetitive stimulation. Looking, heart rate, and respiration may all decline, indicating a loss of interest. Once this has occurred, a new stimulus—some kind of change in the environment—causes responsiveness to return to a high level. This recovery is called **dishabituation**.

Habituation and dishabituation enable us to focus our attention on those aspects of the environment we know least about. As a result, learning is more efficient. By studying the stimuli that infants of different ages habituate and dishabituate to, scientists can tell much about the infant's understanding of the world. For example, a baby who first habituates to a visual pattern (for example, a 2 x 2 checkerboard) and then dishabituates to a new one (a 4 x 4 checkerboard) clearly remembers the first stimulus and perceives the second one as new and different from it. This method of studying infant perception and cognition, illus-

trated in Figure 4.8, can be used with newborn babies—even those who are 5 weeks preterm (Rose, 1980).

IMITATION

Newborn babies come into the world with a primitive ability to learn through **imitation**—by copying the behavior of another person. For example, Figure 4.9 on page 135 shows infants from 2 days to several weeks old imitating a wide variety of adult facial expressions (Field et al., 1982; Meltzoff & Moore, 1977). The newborn's capacity to imitate extends to certain gestures, such as head movements, and has been demonstrated in many ethnic groups and cultures (Meltzoff & Kuhl, 1994).

Explanations of newborn imitation are more controversial. Some researchers regard the capacity as little more than an automatic response to particular stimuli, much like a reflex. But newborns imitate many facial expressions, and they do so even after short delays—when the adult is no longer demonstrating the behavior. These observations suggest that the capacity is flexible and voluntary (Meltzoff & Moore, 1992).

As we will see in Chapter 5, infants' capacity to imitate changes greatly over the first 2 years. But however limited it is at birth, imitation provides the young baby with a powerful means of learning. Using imitation, newborns begin to get to know people by sharing behavioral states with them. Adults can get babies to express desirable behaviors, and once they do, adults can encourage these further. In addition, caregivers take great pleasure in a baby who imitates their facial gestures and actions. Clearly, imitation is

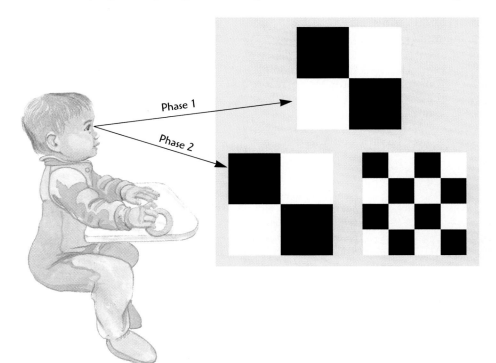

FIGURE 4.8

Example of how the habituation–dishabituation sequence can be used to study infant perception and cognition. In Phase 1, an infant is permitted to look at (habituate to) a 2 x 2 checkerboard. In phase 2, the baby is again shown the 2 x 2 checkerboard, but this time it appears alongside a new, 4 x 4 checkerboard. If the infant dishabituates to (spends more time looking at) the 4 x 4 checkerboard, then we know the baby remembers the first stimulus and can tell that the second one is different.

THE MYSTERIOUS TRAGEDY OF SUDDEN INFANT DEATH SYNDROME

Millie awoke with a start one morning and looked at the clock. It was 7:30, and 3-month-old Sasha had missed her night waking and early morning feeding. Wondering if she was all right, Millie and her husband Stuart tiptoed into the room. Sasha lay still, curled up under her blanket. She had died silently during her sleep.

Sasha was a victim of **sudden infant death syndrome (SIDS).** In industrialized nations, SIDS is the leading cause of infant mortality between 1 and 12 months of age. It accounts for over one-third of these deaths in the United States (Cadoff, 1995).

Although the precise cause of SIDS is not known, its victims show physical abnormalities from the very beginning. Early medical records of SIDS babies reveal higher rates of prematurity and low birth weight, poor Apgar scores, and limp muscle tone (Buck et al., 1989; Malloy & Hoffman, 1995). Abnormal heart rate and respiration, disturbances in sleep–waking activity, and delayed central nervous system development are also involved (Corwin et al., 1995; Froggatt et al., 1988). At the time of death, over half of SIDS babies have a mild respiratory infection. This seems to increase the chances of respiratory failure in an already vulnerable baby (Cotton, 1990).

One hypothesis about the cause of SIDS is that problems in brain functioning prevent these infants from learning how to respond when their survival is threatened—as it is when respiration is suddenly interrupted (Lipsitt, 1990). Between 2 and 4 months of age, when SIDS is most likely, reflexes decline and are replaced by voluntary, learned responses. Respiratory and muscular weaknesses may stop SIDS babies from acquiring behaviors that replace defensive reflexes. As a result, when breathing difficulties occur during sleep, they do not wake up, shift their position, or cry out for help. Instead, they simply give in to oxygen deprivation and death.

In an effort to reduce the occurrence of SIDS, researchers are studying environmental factors related to it. Maternal cigarette smoking, both during and after pregnancy, as well as smoking by other caregivers, is strongly predictive of the disorder. Babies exposed to cigarette smoke are two to three times more likely to die of SIDS than are nonexposed infants (Klonoffcohen et al., 1995). Prenatal abuse of drugs that depress central nervous system functioning (opiates and barbiturates) increases the risk of SIDS 10-fold (Kandall & Gaines, 1991).

SIDS babies are also more likely to sleep on their stomachs than on their backs, and often they are wrapped very warmly in clothing and blankets (Irgens et al., 1995). Scientists think that smoke, depressant drugs, and excessive body warmth (which can be encouraged by putting babies down on their stomachs) place a strain on the respiratory control system in the brain. In an at-risk baby, the respiratory center may stop functioning. In other cases, healthy babies sleeping face down in soft bedding may simply die from continually breathing their own exhaled breath.

Can simple procedures like quitting smoking, avoiding drugs, changing an infant's sleeping position, and removing a few bedclothes prevent SIDS? Research suggests so. For example, public education campaigns that discourage parents from putting babies down on their stomachs were followed by dramatic reductions in SIDS in England, Tasmania, and New Zealand (Taylor, 1991; Wigfield et al., 1992; Willinger, 1995).

When SIDS does occur, surviving family members require a great deal of emotional support. As Millie and Stuart commented 6 months after Sasha's death, "It's the worst crisis we've ever been through. What's helped us most are the comforting words of parents in our support group who've experienced the same tragedy."

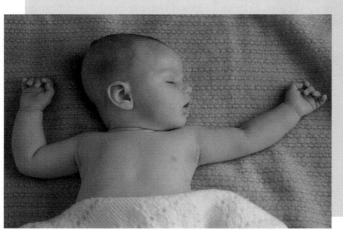

Parents of this sleeping baby put him down on his back, lightly covered. Taking these precautions dramatically reduces the incidence of sudden infant death syndrome. (Charles Compton/Stock Boston)

TRY THIS...

■ Contact a local pediatrician or hospital to find out what advice is offered to parents to help prevent SIDS. How is that advice related to research findings on sids?

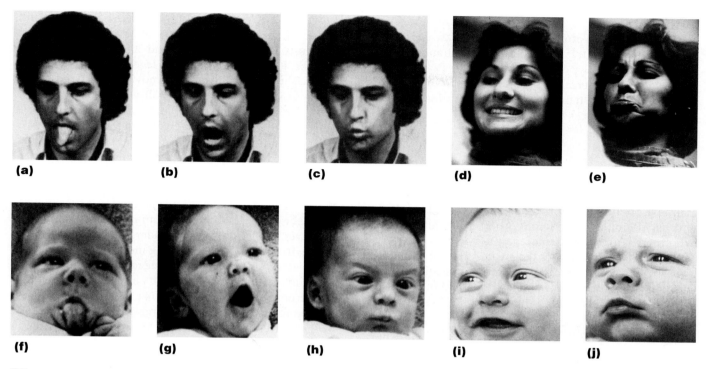

(a)   **(b)**   **(c)**   **(d)**   **(e)**

(f)   **(g)**   **(h)**   **(i)**   **(j)**

FIGURE 4.9

Photographs from two of the first studies of newborn imitation. Those on the left show 2- to 3-week-old infants imitating tongue protrusion (a), mouth opening (b), and lip protrusion (c) of an adult experimenter. Those on the right show 2-day-old infants imitating happy (d) and sad (e) adult facial expressions. *(From A. N. Meltzoff & M. K. Moore, 1977, "Imitation of Facial and Manual Gestures by Human Neonates," Science, 198, p. 75; and T. M. Field et al., 1982, "Discrimination and Imitation of Facial Expressions by Neonates," Science, 218, p. 180. Copyright 1977 and 1982 by the AAAS.)*

one of those behaviors that helps get the baby's relationship with parents off to a good start.

BRIEF REVIEW

Infants are marvelously equipped to learn immediately after birth. Through classical conditioning, infants acquire stimulus associations that have survival value. Operant conditioning permits them to control events in the surrounding world. Habituation and dishabituation reveal that infants, much like adults, are naturally attracted to novel stimulation. Finally, newborns' amazing ability to imitate the facial expressions and gestures of adults helps them get to know their social world.

ASK YOURSELF . . .

■ *Byron has a music box hung on the side of his crib. Each time he pulls a lever, the music box plays a nursery tune. Which learning mechanism is the manufacturer of this toy taking advantage of?*

■ *Earlier in this chapter, we indicated that infants with nonorganic failure to thrive are unlikely to smile at a friendly adult. Also, they keep track of nearby adults in an anxious and fearful way. Explain these reactions using the learning mechanisms discussed in the preceding sections.*

PERCEPTUAL DEVELOPMENT IN INFANCY

I n Chapter 3, you learned that touch, taste, smell, and hearing—but not vision—are remarkably well developed at birth. Now let's turn to a related question: How does perception change over the first year of life?

Our discussion will focus on hearing and vision because almost all research addresses these two aspects of perceptual development. Unfortunately, we know little about how touch, taste, and smell develop after birth. Also, in Chapter 3, we used the word *sensation* to talk about these capacities. Now we are using the word *perception*. The reason is that sensation suggests a fairly passive process—what the baby's receptors detect when they are exposed to stimulation. In contrast, perception is much more active. When we perceive, we organize and interpret what we see. As we look at the perceptual achievements of infancy, you will probably find it hard to tell where perception leaves off and thinking begins.

HEARING

On Byron's first birthday, Lisa bought several tapes of nursery songs, and she turned one on each afternoon at naptime. Soon Byron let her know his favorite tune. If she put on "Twinkle, Twinkle," he stood up in his crib and whimpered until she replaced it with "Jack and Jill." Byron's behavior illustrates the greatest change in hearing over the first year of life: babies start to organize sounds into complex patterns. If two slightly different melodies are played, 1-year-olds can tell they are not the same (Morrongiello, 1986).

As we will see in the next chapter, throughout the first year babies are preparing to acquire language. Recall from Chapter 3 that newborns can detect almost all sounds in human languages. As infants continue to listen actively to the talk of people around them, they learn to focus on meaningful sound variations in their own language. By 6 months of age, they "screen out" sounds that are not useful in their language community (Kuhl et al., 1992; Polka & Werker, 1994).

In the second half of the first year, infants focus on larger speech units that are critical to figuring out the meaning of what they hear. For example, in one study, 7- to 10-month-olds clearly preferred speech with natural breaks between clauses to speech with pauses in unnatural places (Hirsh-Pasek et al., 1987). By 9 months, this rhythmic sensitivity is extended to individual words. Babies of this age listen much longer to speech with stress patterns common in their own language, and they perceive it in wordlike segments (Jusczyk, Cutler, & Redanz, 1993; Morgan & Saffran, 1995).

VISION

If you had to choose between hearing and vision, which would you select? Most people pick vision, for good reason. More than any other sense, humans depend on vision for exploring their environment. Although at first a baby's visual world is fragmented, it undergoes extraordinary changes during the first 7 to 8 months of life.

Visual development is supported by rapid maturation of the eye and visual centers in the brain. In Chapter 3, we saw that the newborn baby focuses and perceives color poorly. By 2 months, infants can focus on objects and discriminate colors about as well as adults can (Banks, 1980;

Brown, 1990). Visual acuity (fineness of discrimination) improves steadily throughout the first year. In the first 6 months, it changes from around 20/600 to 20/100. At 11 months, it reaches a near-adult level (Courage & Adams, 1990). The ability to fixate on a moving object and track it improves steadily over the first half-year (Hainline, 1993).

As babies see more clearly and explore their visual field more adeptly, they work on sorting out features of the environment and how they are arranged in space. We can best understand how they do so by examining the development of two aspects of vision: depth and pattern perception.

■ DEPTH PERCEPTION. *Depth perception* is the ability to judge the distance of objects from one another and from ourselves. It is important for understanding the layout of the environment and for guiding motor activity. To reach for objects, babies must have some idea about depth. Later, when infants learn to crawl, depth perception helps prevent them from bumping into furniture and falling down stairs.

Figure 4.10 shows the well-known *visual cliff*, designed by Eleanor Gibson and Richard Walk (1960) and used in the earliest studies of depth perception. It consists of a glass-covered table with a platform at the center, a "shallow" side with a checkerboard pattern just under the glass, and a "deep" side with a checkerboard several feet below the glass. The researchers found that crawling babies readily crossed the shallow side, but most reacted with fear to the deep side. They concluded that around the time that infants crawl, most distinguish deep and shallow surfaces and avoid dropoffs that look dangerous (Walk & Gibson, 1961).

Gibson and Walk's research shows that crawling and avoidance of dropoffs are linked, but it does not tell us

FIGURE 4.10

The visual cliff. By refusing to cross the deep side and showing a preference for the shallow surface, this infant demonstrates the ability to perceive depth. *(William Vandivert/Scientific American.)*

how they are related. Also, from studies of crawling infants, we cannot tell when sensitivity to depth first appears. To better understand the development of depth perception, recent research has looked at babies' ability to detect particular depth cues, using methods that do not require that they crawl.

■ THE EMERGENCE OF DEPTH PERCEPTION. How do we know when an object is near rather than far away? Try these exercises to find out. Look toward the far wall while moving your head from side to side. Notice that nearby objects move past your field of vision more quickly than those far away. Next, pick up a small object (such as your cup) and move it toward and away from your face. Did its image grow larger as it approached and smaller as it receded?

Motion provides a great deal of information about depth, and it is the first depth cue to which infants are sensitive. Babies 3 to 4 weeks of age blink their eyes defensively when an object is moved toward their face as if it is going to hit (Nánez & Yonas, 1994). As infants are carried about, motion helps them learn more about depth. By 3 to 4 months, they have figured out that objects are not flat but three-dimensional (Arterberry, Craton, & Yonas, 1993).

Binocular depth cues arise because our two eyes have slightly different views of the visual field. The brain registers both images and combines them. Research in which infants look at images while wearing special goggles, like those for 3-D movies, reveals that sensitivity to binocular cues emerges between 2 and 3 months and gradually improves over the first half-year (Banks & Salapatek, 1983).

Finally, around the middle of the first year, babies develop sensitivity to *pictorial* depth cues—the same ones that artists use to make a painting look three-dimensional. Examples are lines that create the illusion of perspective, changes in texture (nearby textures are more detailed than far-away ones), and overlapping objects (an object partially hidden by another object is perceived to be more distant) (Yonas et al., 1986).

Why does perception of depth cues emerge in the order just described? Researchers speculate that motor development is involved. For example, control of the head during the early weeks of life may help babies notice motion cues. Improved focusing ability at 3 months may permit detection of binocular cues. At around 5 to 6 months, the ability to turn, poke, and feel the surface of objects may promote perception of pictorial cues (Bushnell & Boudreau, 1993). Indeed, as we will see next, research shows that one aspect of motor progress—independent movement—plays a vital role in the refinement of depth perception.

■ INDEPENDENT MOVEMENT AND DEPTH PERCEPTION. At 6 months, April started crawling. "She's like a fearless daredevil," exclaimed Felicia. "If I put her down in the middle of our bed, she crawls right over the edge. Several times I stopped her just before she went overboard."

■ Crawling promotes three-dimensional understanding, such as wariness of dropoffs and memory for object locations. As this baby retrieves his pacifier, he takes note of how to get from place to place, where objects are in relation to himself and to other objects, and what they look like from different points of view. (Bob Daemmrich/Stock Boston)

Will April become more wary of the side of the bed as she becomes a more experienced crawler? Research suggests that she will. In one study, infants with more crawling experience (regardless of when they started to crawl) were far more likely to refuse to cross the deep side of the visual cliff (Bertenthal, Campos, & Barrett, 1984). Avoidance of heights, the investigators concluded, is "made possible by independent locomotion" (Bertenthal & Campos, 1987, p. 563).

Independent movement contributes to other aspects of a baby's three-dimensional understanding as well. For example, crawling infants are better at remembering object locations and finding hidden objects than are their noncrawling agemates. The more crawling experience, the better they perform on these tasks (Bai & Bertenthal, 1992; Campos & Bertenthal, 1989).

Why does crawling make such a difference? Compare your own experience of the environment when you are driven from one place to another as opposed to when you walk or drive yourself. When you move on your own, you are much more aware of landmarks, routes of travel, and what things look like from different points of view. The same is true for infants. In fact, some researchers believe that crawling is so important in structuring babies' experience of the world that it may promote a new level of brain

FIGURE 4.11

The way two checkerboards differing in complexity look to infants in the first few weeks of life. *Because of their poor vision, very young infants cannot resolve the fine detail in the more complex checkerboard. It appears blurred, like a gray field. The large, bold checkerboard appears to have more contrast, so babies prefer to look at it. (Adapted from M. S. Banks & P. Salapatek, 1983, "Infant Visual Perception," in M. M. Haith & J. J. Campos (Eds.), Handbook of Child Psychology: Vol. 2. Infancy and Developmental Psychobiology (4th ed., p. 504), New York: Wiley. Copyright © 1983 by John Wiley & Sons. Reprinted by permission.)*

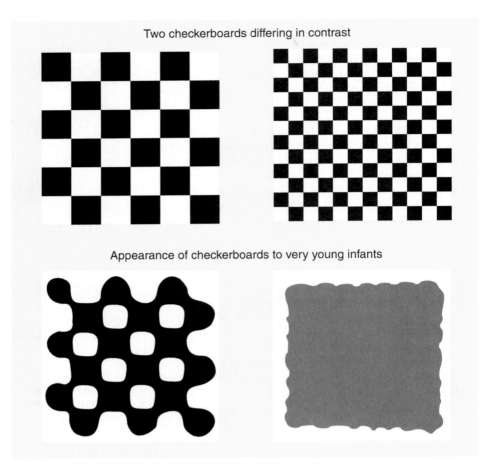

Two checkerboards differing in contrast

Appearance of checkerboards to very young infants

organization by strengthening certain synaptic connections in the cortex (Fox, Calkins, & Bell, 1994).

Many American parents place babies who are not yet crawling in "walkers." Consisting of a seat with a frame on castors, these devices permit babies to move around independently by pushing with their feet. Do walkers help stimulate development in the ways just described? Research suggests they do (Bertenthal, Campos, & Barrett, 1984; Kermoian & Campos, 1988). However, walkers do not have any lasting effects, and they can be dangerous. Infants frequently tip over in them and careen down staircases, perhaps because the frame and seat provide a false sense of security. In the United States, walkers account for almost 28,000 infant injuries each year (Trinkoff & Parks, 1993). For safety's sake, it is best not to put babies in these devices.

■ PATTERN AND FACE PERCEPTION. Even newborns prefer to look at patterned rather than plain stimuli—for example, a drawing of the human face or one with scrambled facial features to a black-and-white oval (Fantz, 1961). As infants get older, they prefer more complex patterns. For example, 3-week-old infants look longest at black-and-white checkerboards with a few large squares, whereas 8- and 14-week-olds prefer those with many squares (Brennan, Ames, & Moore, 1966).

A general principle called **contrast sensitivity** explains these early pattern preferences (Banks & Salapatek, 1981). *Contrast* refers to the difference in the amount of light between adjacent regions in a pattern. If babies can detect the contrast in two or more patterns, they prefer the one with more contrast. To understand this idea, look at the checkerboards in the top row of Figure 4.11. To us, the one with many small squares has more contrasting elements. Now look at the bottom row, which shows how these checkerboards appear to infants in the first few weeks of life. Because of their poor vision, they cannot resolve the features in more complex patterns, so they prefer to look at the large, bold checkerboard. By 2 months of age, when detection of fine-grained detail has improved considerably, infants become sensitive to the greater contrast in complex patterns and spend more time looking at them (Dodwell, Humphrey, & Muir, 1987).

In the early weeks of life, infants respond to the separate parts of a pattern. For example, when shown drawings of human faces, 1-month-olds limit their visual exploration to the border of the stimulus and stare at single, high-contrast features, such as the hairline or chin (see Figure 4.12). At about 2 months, when scanning ability and contrast sensitivity have improved, infants start to thoroughly explore a pattern's internal features by moving their

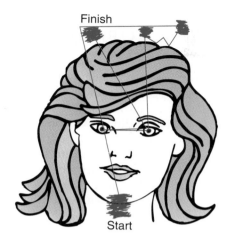

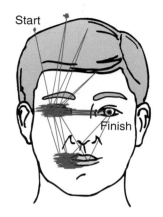

FIGURE 4.12

Visual scanning of the pattern of the human face by 1- and 2-month-old infants. One-month-olds limit their scanning to single features on the border of the stimulus, whereas 2-month-olds explore internal features. *(From P. Salapatek, 1975, "Pattern Perception in Early Infancy," in L. B. Cohen & P. Salapatek (Eds.),* Infant Perception: From Sensation to Cognition, *New York: Academic Press, p. 201. Reprinted by permission.)*

eyes quickly around the figure and pausing briefly to look at each part (Bronson, 1991; Salapatek, 1975).

Once babies can detect all parts of a stimulus, they begin to combine pattern elements. Infants of 6 or 7 months are so good at detecting pattern organization that they even perceive subjective boundaries that are not really present. For example, they perceive a square in the center of Figure 4.13, just as you do (Bertenthal, Campos, & Haith, 1980). Older infants apply their growing knowledge of the world to pattern perception. For example, 9-month-olds show a special preference for an organized series of moving lights that resemble a human being walking in that they look much longer at this display than at upside-down or scrambled versions (Bertenthal, 1993; Bertenthal et al., 1985).

The baby's tendency to search for structure in a patterned stimulus applies to face perception. Infants under 2 months show no special preference for a facial pattern.[3] At 2 to 3 months, they do prefer a face over similar configurations (Dannemiller & Stephens, 1988). By 3 months, infants make fine-grained distinctions among the features of different faces. For example, they can tell the difference between the photos of two moderately similar strangers (Barrera & Maurer, 1981a). Around this time, babies also recognize their mother's face in a photo, since they look longer at it than the face of a stranger (Barrera & Maurer, 1981b). Between 7 and 10 months, infants start to perceive emotional expressions as organized wholes. They treat positive faces (happy and surprised) as different from negative ones (sad and fearful), even when these expressions are demonstrated in slightly varying ways by different people (Ludemann, 1991). As infants recognize and respond to the expressive behavior of others, face perception supports their earliest social relationships.

[3] Perhaps you are wondering how newborns can display the remarkable imitative capacities described earlier if they do not scan the internal features of a face. Recall that the facial expressions in newborn imitation research were not static poses but live demonstrations. Their dynamic quality probably caused infants to notice them.

INTERMODAL PERCEPTION

So far, we have discussed the infant's perceptual abilities one by one. When we take in information from the environment, we often use **intermodal perception.** That is, we combine stimulation from more than one modality, or sensory system. Recent evidence indicates that from the start, babies perceive the world in an intermodal fashion (Meltzoff, 1990; Spelke, 1987).

Recall that newborns turn in the general direction of a sound, and they reach for objects in a primitive way. These behaviors suggest that infants expect sight, sound, and touch to go together. In one study, 1-month-olds were given a pacifier with either a smooth surface or a surface with nubs on it. After exploring it in their mouths, the infants were shown two pacifiers—one smooth and one nubbed. They preferred to look at the one they had sucked, indicating that they could match touch and visual stimulation without much experience seeing and feeling objects

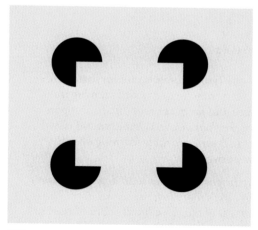

FIGURE 4.13

Subjective boundaries in a visual pattern. Do you perceive a square in the middle of this figure? By 7 months of age, infants do, too. *(Adapted from Bertenthal, Campos, & Haith, 1980.)*

(Meltzoff & Borton, 1979). Within a few months, infants make impressive intermodal matches. Three- and 4-month-olds can relate the shape and tempo of an adult's moving lips to the corresponding sounds in speech. And 7-month-olds can link a happy or angry voice with the appropriate face of a speaking person (Pickens et al., 1994; Soken & Pick, 1992).

Of course, a great many intermodal associations—the way a train sounds or a teddy bear feels—must be based on experience. Infants acquire these relationships remarkably quickly, often after just one exposure to a new situation (Spelke, 1987). In addition, when researchers try to teach intermodal matches by pairing sights and sounds that do not naturally go together, babies will not learn them (Bahrick, 1988, 1992). Intermodal perception is yet another capacity that helps infants build an orderly, predictable perceptual world.

BRIEF REVIEW

During the first year, infants organize sounds into more complex patterns and become sensitive to the sound patterns of their own language. Changes in visual abilities are striking. Depth perception improves as infants detect motion, binocular, and pictorial depth cues. Experience in independent movement plays an important role in avoidance of heights as well as other aspects of three-dimensional understanding. The principle of contrast sensitivity accounts for young babies' pattern preferences. As vision and knowledge of the world improve, infants perceive the parts of patterns, including the human face, as organized wholes. Young infants have a remarkable ability to combine information across different sensory modalities.

ASK YOURSELF . . .

■ *Five-month-old Tyrone sat in his infant seat, passing a teething biscuit from hand to hand, moving it up close to his face and far away, and finally dropping it overboard on the floor below. What aspect of visual development is Tyrone probably learning about? Explain your answer.*

■ *Diane put up bright wallpaper with detailed pictures of animals in Jana's room before she was born. During the first 2 months of life, Jana hardly noticed the wallpaper. Then, around 2 months, she showed keen interest. What new visual abilities probably account for this change?*

UNDERSTANDING PERCEPTUAL DEVELOPMENT

Now that we have reviewed the development of infant perceptual capacities, how can we put together this diverse array of amazing achievements? Eleanor and James Gibson's **differentiation theory** provides widely accepted answers. According to the Gibsons, infants actively search for **invariant features** of the environment—those that remain stable—in a constantly changing perceptual world. For example, in pattern perception, at first babies are confronted with a confusing mass of stimulation. But very quickly, they search for features that stand out along the border of a stimulus. Then they explore its internal features, noticing *stable relationships* among those features. As a result, they detect patterns—crosses, squares, and faces. The development of intermodal perception also reflects this principle. Babies seem to seek out invariant relationships, such as a similar tempo in an object's motion and sound, that unite information across modalities.

The Gibsons use the word *differentiation* (which means analyze or break down) to describe their theory because over time, babies make finer and finer distinctions among stimuli. In addition to pattern perception, differentiation applies to depth perception; recall how sensitivity to overall motion and binocular cues precedes detection of fine-grained pictorial features. So one way of understanding perceptual development is to think of it as a built-in tendency to look for order and consistency, a capacity that becomes more fine-tuned with age (Gibson, 1970; Gibson, 1979).

Acting on the environment plays a major role in perceptual differentiation. Think back to the links between motor milestones and perceptual development discussed in this chapter. Infants constantly look for ways in which the environment *affords* opportunities for action (Gibson, 1988). By moving about and exploring the environment, they figure out which objects can be grasped, squeezed, bounced, or stroked and when a surface is safe to cross or affords the possibility of falling. As a result, they differentiate the world in new ways and act more competently when confronted with diverse perceptual experiences (Adolph, Eppler, & Gibson, 1993).

As we conclude this chapter, it is only fair to note that some researchers believe that babies do not just make sense of experience by searching for invariant features. Instead, they impose *meaning* on what they perceive, constructing categories of objects and events in the surrounding environment. We have already seen the glimmerings of this cognitive point of view in some of the evidence reviewed in this chapter. For example, older babies *interpret* a familiar face as a source of pleasure and affection and a pattern of blinking lights as a moving human being. This cognitive perspective also has merit in understanding the achievements of infancy. In fact, many researchers combine viewpoints, regarding infant development as proceeding from a perceptual to a cognitive emphasis over the first year of life (Mandler, 1992; Salapatek & Cohen, 1987).

SUMMARY

BODY GROWTH IN THE FIRST TWO YEARS

Describe major changes in body growth over the first 2 years.

- Changes in height and weight are rapid during the first 2 years. Body fat is laid down quickly in the first 9 months, whereas muscle development is slow and gradual. Skeletal age is the best way to estimate the child's physical maturity; girls are advanced over boys. The body's growth follows **cephalocaudal** and **proximodistal trends.**

BRAIN DEVELOPMENT

What changes in brain development occur during infancy and toddlerhood, at the level of individual brain cells and the level of the cerebral cortex?

- Early in development the brain grows faster than any other organ of the body. During infancy, **neurons** form **synapses** or connections, at a rapid rate. Stimulation determines which neurons will survive and which will die off. **Glial cells,** which are responsible for **myelinization,** multiply rapidly into the second year, resulting in large gains in brain size. Changes in the electrical activity of the cortex, along with gains in brain weight and skull size, indicate that brain growth spurts occur intermittently from infancy into adolescence. These may be sensitive periods in brain development.
- Different regions of the **cerebral cortex** develop in the same order in which various capacities emerge in the growing child. **Lateralization** refers to specialization of the hemispheres of the cortex. Some brain specialization already exists at birth. In infancy, before many regions of the cortex have specialized, there is high **brain plasticity.**

How does the organization of sleep and wakefulness change over the first 2 years?

- The infant's changing arousal patterns are primarily affected by brain maturation, but the social environment also plays a role. Short periods of sleep and wakefulness are put together and begin to coincide with a night and day schedule. Infants in Western nations sleep through the night much earlier than those in many other cultures.

INFLUENCES ON EARLY PHYSICAL GROWTH

Cite evidence that heredity, nutrition, and affection and stimulation contribute to early physical growth.

- Physical growth results from a continuous and complex interplay between heredity and environment. Heredity contributes to body size and rate of maturation.
- Breast milk is ideally suited to the growth needs of young babies and offers protection against disease. Breast-feeding prevents malnutrition and infant death in poverty-stricken areas of the world. Breast-fed and bottle-fed babies do not differ in psychological development.
- Most chubby babies thin out during toddlerhood and early childhood. Infants and toddlers can eat nutritious foods freely, without risk of becoming overweight.
- **Marasmus** and **kwashiorkor** are dietary diseases caused by malnutrition that affect many children in developing countries. If allowed to continue, body growth and brain development can be permanently stunted. **Nonorganic failure to thrive** illustrates the importance of stimulation and affection for normal physical growth.

MOTOR DEVELOPMENT DURING THE FIRST TWO YEARS

Describe the general course of motor development during the first 2 years along with factors that influence it.

- Like physical development, motor development follows the cephalocaudal and proximodistal trends. New motor skills are a matter of combining existing skills into increasingly complex **dynamic systems of action.** Each new skill is a joint product of central nervous system maturation, movement possibilities of the body, environmental supports for the skill, and the child's motivation to accomplish a task.
- Movement opportunities and a stimulating environment have a profound effect on motor development, as shown by research on infants raised in deprived institutions. Cultural values and child-rearing customs contribute to the emergence and refinement of early motor skills.
- During the first year, infants perfect their reaching and grasping. The

poorly coordinated prereaching of the newborn period drops out. Once voluntary reaching appears, the clumsy ulnar grasp is gradually transformed into a refined pincer grasp.

LEARNING CAPACITIES

Describe four infant learning capacities, the conditions under which they occur, and the unique value of each.

■ Infants can be **classically conditioned** when the pairing of an **unconditioned stimulus (UCS)** and **conditioned stimulus (CS)** has survival value. Young babies are easily conditioned in the feeding situation. Classical conditioning of fear is difficult before 8 to 12 months.

■ **Operant conditioning** enables infants to influence their environment. In addition to food, interesting sights and sounds serve as effective **reinforcers** by increasing the occurrence of a response. **Punishment** involves removing a desirable stimulus or presenting an unpleasant one to decrease the occurrence of a response.

■ **Habituation** and **dishabituation** reveal that at birth, babies are attracted to novelty. Newborn infants also have a primitive ability to **imitate** the facial expressions and gestures of adults.

PERCEPTUAL DEVELOPMENT IN INFANCY

What changes in hearing, depth and pattern perception, and intermodal perception take place during infancy?

■ Over the first year, infants organize sounds into more complex patterns. They also become more sensitive to the sounds and meaningful speech units of their own language.

■ Rapid development of the eye and visual centers in the brain supports the development of focusing, color discrimination, and visual acuity during the first half-year. The ability to track a moving object also improves.

■ Research on depth perception reveals that responsiveness to motion cues develops first, followed by sensitivity to binocular and then pictorial cues. Experience in moving about independently affects babies' three-dimensional understanding.

■ **Contrast sensitivity** accounts for babies' early pattern preferences. At first, infants look at the border of a stimulus and at single features.

Around 2 months, they explore the internal features of a pattern and start to detect pattern organization. Over time, they discriminate increasingly complex, meaningful patterns. Perception of the human face follows the same sequence of development as sensitivity to other patterned stimuli.

■ From the start, infants are capable of **intermodal perception.** During the first year, they quickly combine information across sensory modalities, often after just one exposure to a new situation.

UNDERSTANDING PERCEPTUAL DEVELOPMENT

Explain differentiation theory of perceptual development.

■ According to Eleanor and James Gibson's **differentiation theory,** perceptual development is a matter of searching for **invariant features** in a constantly changing perceptual world. Acting on the world plays a major role in perceptual differentiation. Other researchers take a more cognitive view, suggesting that at an early age, infants impose meaning on what they perceive. Many researchers combine these two ideas.

IMPORTANT TERMS AND CONCEPTS

cephalocaudal trend (p. 117)
proximodistal trend (p. 117)
cerebral cortex (p. 118)
neurons (p. 118)
synapses (p. 118)
glial cells (p. 118)
myelinization (p. 118)
lateralization (p. 119)
brain plasticity (p. 119)
marasmus (p. 125)
kwashiorkor (p. 125)

nonorganic failure to thrive (p. 126)
dynamic systems of action (p. 128)
classical conditioning (p. 131)
unconditioned stimulus (UCS) (p. 131)
unconditioned response (UCR) (p. 131)
conditioned stimulus (CS) (p. 131)
conditioned response (CR) (p. 131)
operant conditioning (p. 132)
reinforcer (p. 132)

punishment (p. 132)
habituation (p. 133)
dishabituation (p. 133)
imitation (p. 133)
sudden infant death syndrome (SIDS) (p. 134)
contrast sensitivity (p. 138)
intermodal perception (p. 139)
differentiation theory (p. 140)
invariant features (p. 140)

FOR FURTHER INFORMATION AND HELP

PHYSICAL GROWTH AND HEALTH

Healthy Mothers, Healthy Babies
409 Twelfth Street, S.W., Suite 309
Washington, DC 20024-2188
(202) 863-2458
A coalition of national and state organizations concerned with maternal and child health. Serves as a network through which information on nutrition, injury prevention, and infant mortality is shared.

United Nations Children's Fund (UNICEF)
3 United Nations Plaza
New York, NY 10017
(212) 326-7000
Website: *www.unicef.org*
International organization dedicated to addressing the problems of children around the world. Develops and implements health and nutrition programs, campaigns to have children vaccinated against disease, and coordinates delivery of food and other aid to disaster-stricken areas.

World Health Organization (WHO)
Avenue Appia
CH–1211 Geneva 27
Switzerland
(22) 791-2111
Website: *www.who.ch*
International health agency of the United Nations that seeks to obtain the highest level of health care for all people. Promotes prevention and treatment of disease and strives to eliminate poverty. Places special emphasis on the health needs of developing countries.

BREAST-FEEDING

La Leche League International
P.O. Box 1209
Franklin Park, IL 60131
(708) 455-7730
Website: *www.lalecheleague.org*
Provides information and support to breast-feeding mothers. Local chapters exist in many cities.

MALNUTRITION

Food Research and Action Center
1875 Connecticut Avenue, N.W., Suite 540
Washington, DC 20009
(202) 986-2200
Provides assistance to community organizations trying to make federal food programs more responsive to the needs of millions of hungry Americans. Seeks to enhance public awareness of the problems of hunger and poverty in the United States.

SUDDEN INFANT DEATH SYNDROME (SIDS)

National Sudden Infant Death Syndrome Clearinghouse
8201 Greensboro Drive, Suite 600
McLean, VA 22102
(703) 821-8955
Provides information to health professionals and the public on SIDS.

National Sudden Infant Death Syndrome Foundation (NSIDSF)
10500 Little Patuxent Pkwy, #420
Columbia, MD 21044
(410) 964-8000
Website: *www.sidsnetwork.org*
Provides assistance to parents who have lost a child to SIDS. Works with families and health professionals in caring for infants at risk due to heart and respiratory problems.

Cognitive Development in Infancy and Toddlerhood

When Byron, Rachel, and April were brought together by their mothers at age 18 months, the room was alive with activity. The three spirited explorers were bent on discovery. Rachel dropped shapes through holes in a plastic box that Beth held and adjusted so the harder ones would fall smoothly into the container. Once a few shapes were inside, Rachel grabbed and shook the box, squealing with delight as the lid fell open and the shapes scattered around her. The clatter attracted Byron, who picked up a shape, carried it to the railing at the top of the basement steps, and dropped it overboard, then followed with a teddy bear, a ball, his shoe, and a spoon. In the meantime, April pulled open a drawer, unloaded a set of wooden bowls, stacked them in a pile, knocked it over, then banged two bowls together.

As the toddlers experimented, I could see the beginnings of language—a whole new way of influencing the world. April was the most vocal of the three youngsters. "All gone baw!" she exclaimed as Byron tossed the bright red ball down the basement steps. "Bye-bye, baw," Rachel chimed in, waving as the ball disappeared from sight. A close look at Rachel revealed that the capacity to represent experience through words and gestures had opened up a new realm of play possibilities. Rachel could pretend. "Night-night," she said as she put her head down and closed her eyes, ever so pleased that in make-believe, she could decide for herself when and where to go to bed.

Over the first 2 years, the small, reflexive newborn baby becomes a self-assertive, purposeful being who solves simple problems and has started to master the most amazing of human abilities—language. "How does all this happen so quickly?" asked Felicia, turning to me. In this chapter, we take up three perspectives on early cognitive development: *Piaget's cognitive-developmental theory, information processing*, and *Vygotsky's sociocultural theory*. We will also consider the usefulness of tests that measure intellectual progress during this earliest phase of development.

Our discussion concludes with the beginnings of language. We will see how toddlers' first words build on early cognitive achievements. Very soon, new words and expressions greatly increase the speed and flexibility of human thinking. Throughout development, cognition and language mutually support one another.

PIAGET'S COGNITIVE-DEVELOPMENTAL THEORY

The Swiss theorist Jean Piaget inspired a vision of children as busy, motivated explorers whose thinking develops as they act directly on the environment. According to Piaget, children move through four stages of development between infancy and adolescence. The first and most elaborate is the **sensorimotor stage**, which spans the first 2 years of life.

KEY PIAGETIAN CONCEPTS

As the name of this stage implies, Piaget believed that infants and toddlers "think" with their eyes, ears, hands, and other sensorimotor equipment. They cannot yet carry out many activities inside their heads. But by the end of toddlerhood, children can solve practical, everyday problems and represent their experiences in speech, gesture, and play. To understand Piaget's view of how these vast changes take place, we need to look at some important concepts, which convey Piaget's ideas about *what changes with development*, and *how cognitive change takes place*.

■ **WHAT CHANGES WITH DEVELOPMENT.** Piaget believed that *psychological structures*—the child's organized ways of making sense of experience—change with age. He referred to specific structures as **schemes.** At first, schemes are motor action patterns. For example, at 6 months, Byron dropped objects in a fairly rigid way, simply by letting go of a rattle or teething ring and watching with interest. By 18 months, his "dropping scheme" had become much more deliberate and creative. Byron tossed all sorts of objects

According to Piaget's theory, at first schemes are motor action patterns. As this 1-year-old takes apart, bangs, and drops these nesting cups, she discovers that her movements have predictable effects on objects and that objects influence one another in regular ways. (Erika Stone)

Test Bank Item 5.1

down the basement stairs, throwing some up in the air, bouncing others off walls, releasing some gently and others with all the force his little body could muster. Soon Byron's schemes will move from an *action-based level* to a *mental level.* Instead of just acting on objects, he will show evidence of thinking before he acts. This change, as we will see later, marks the transition from sensorimotor to preoperational thought.

■ HOW COGNITIVE CHANGE TAKES PLACE. In Piaget's theory, two processes account for changes in schemes: *adaptation* and *organization.*

Adaptation. The next time you have a chance, notice how infants and toddlers tirelessly repeat actions that lead to interesting effects. **Adaptation** involves building schemes through direct interaction with the environment. It consists of two complementary activities: *assimilation* and *accommodation.* During **assimilation,** we use our current schemes to interpret the external world. For example, when Byron dropped objects, he was assimilating them all into his sensorimotor "dropping scheme." In **accommodation,** we create new schemes or adjust old ones after noticing that our current ways of thinking do not fit the environment completely. When Byron dropped objects in different ways, he modified his dropping scheme to take account of the varied properties of objects.

According to Piaget, the balance between assimilation and accommodation varies over time. When children are not changing very much, they assimilate more than they accommodate. Piaget called this a state of cognitive *equilibrium,* implying a steady, comfortable condition. During rapid cognitive change, however, children are in a state of *disequilibrium,* or cognitive discomfort. They realize that new information does not fit their current schemes, so they shift away from assimilation toward accommodation. Once they have modified their schemes, they move back toward assimilation, exercising their newly changed structures until they are ready to be modified again.

Each time this back-and-forth movement between equilibrium and disequilibrium occurs, more effective schemes are produced (Piaget, 1985). Because the times of greatest accommodation are the earliest ones, the sensorimotor stage is Piaget's most complex period of development.

Organization. Schemes change through a second process called **organization.** It takes place internally, apart from direct contact with the environment. Once children form new schemes, they start to rearrange them, linking them with other schemes to create a strongly interconnected cognitive system. For example, eventually Byron will relate "dropping" to "throwing" and to his developing understanding of "nearness" and "farness." According to Piaget, schemes reach a true state of equilibrium when they become part of a broad network of structures that can be jointly applied to the surrounding world (Piaget, 1936/1952).

THE SENSORIMOTOR STAGE

The difference between the newborn baby and the 2-year-old child is so vast that the sensorimotor stage is divided into six substages. Piaget's observations of his own three children served as the basis for this sequence of development. Although this is a very small sample, Piaget watched carefully and also presented his son and two daughters with everyday problems (such as hidden objects) that helped reveal their understanding of the world. In the following sections, we will first describe infant development as Piaget saw it. Then we will consider evidence indicating that the cognitive competence of babies is more advanced than Piaget believed it to be.

■ THE CIRCULAR REACTION. At the beginning of the sensorimotor stage, infants know so little about the world that they cannot purposefully explore it. The **circular reaction** provides them with a special means of adapting their first schemes. It involves stumbling onto a new experience caused by the baby's own motor activity. The reaction is "circular" because the infant tries to repeat the event again and again. As a result, a sensorimotor response that first occurred by chance becomes strengthened into a new scheme. For example, at age 2 months, Rachel accidentally made a smacking sound after a feeding. The sound was new and intriguing, so she tried to repeat it until, after a few days, she became quite expert at smacking her lips.

During the first 2 years, the circular reaction changes in several ways. At first, it centers around the infant's own body. Later, it turns outward, toward manipulation of objects. Finally, it becomes experimental and creative, aimed at producing novel effects in the environment. Piaget considered these changes in the circular reaction so important that he named the sensorimotor substages after them. As you read about each substage, you may find it helpful to refer to the summary in Table 5.1.

■ SUBSTAGE 1: REFLEXIVE SCHEMES (BIRTH TO 1 MONTH). Piaget regarded newborn reflexes as the building blocks of sensorimotor intelligence. At first, babies suck, grasp, and look in much the same way, no matter what experiences they encounter (see Figure 5.1). Beth reported an amusing example of Rachel's indiscriminate sucking at age 2 weeks. She lay next to her father while he took a nap. Suddenly, he awoke with a start. Rachel had latched on and begun to suck on his back!

■ SUBSTAGE 2: PRIMARY CIRCULAR REACTIONS— THE FIRST LEARNED ADAPTATIONS (1 TO 4 MONTHS). Infants start to gain voluntary control over their actions by repeating chance behaviors that lead to satisfying results. This leads to some simple motor habits, such as sucking their fists or thumbs and opening and closing their hands (see Figure 5.2). Babies of this substage also start to vary

TABLE 5.1

Summary of Cognitive Development During the Sensorimotor Stage

SENSORIMOTOR SUBSTAGE	TYPICAL ADAPTIVE BEHAVIORS
1. Reflexive schemes (birth–1 month)	Newborn reflexes (see Chapter 3, page 104).
2. Primary circular reactions (1–4 months)	Simple motor habits centered around the infant's own body; limited anticipation of events.
3. Secondary circular reactions (4–8 months)	Actions aimed at repeating interesting effects in the surrounding world; imitation of familiar behaviors.
4. Coordination of secondary circular reactions (8–12 months)	Intentional, or goal-directed, action sequences; improved anticipation of events; imitation of behaviors slightly different from those the infant usually performs; ability to find a hidden object in the first location in which it is hidden (object permanence).
5. Tertiary circular reactions (12–18 months)	Exploration of the properties of objects by acting on them in novel ways; imitation of unfamiliar behaviors; ability to search in several locations for a hidden object (AB search).
6. Mental combinations (18 months–2 years)	Internal representation of objects and events, as indicated by sudden solutions to sensorimotor problems, ability to find an object that has been moved while out of sight, deferred imitation, and make-believe play.

their behavior in response to environmental demands. For example, they open their mouths differently to a nipple than a spoon. They also show a limited ability to anticipate events. For example, at 3 months, when Byron awoke from his nap, he cried out with hunger. But as soon as Lisa entered the room, Byron's crying stopped. He knew that feeding time was near.

Piaget called the first circular reactions *primary*, and he regarded them as quite limited. Notice how, in the preceding examples, infants' adaptations are oriented toward their own bodies and motivated by basic needs. According to Piaget, babies are not yet very concerned with the effects of their actions on the external world.

■ SUBSTAGE 3: SECONDARY CIRCULAR REACTIONS— MAKING INTERESTING SIGHTS LAST (4 TO 8 MONTHS). Between 4 and 8 months, infants sit up and become skilled at reaching for, grasping, and manipulating objects. These

motor achievements play a major role in turning their attention outward toward the environment. Using the *secondary* circular reaction, infants try to repeat interesting events caused by their own actions. Notice how Piaget's 4-month-old son Laurent gradually builds the sensorimotor scheme of "hitting" over a 10-day period (see Figure 5.3):

At 4 months 7 days [Laurent] looks at a letter opener tangled in the strings of a doll hung in front of him. He tries to grasp (a scheme he already knows) the doll or the letter opener but each time, his attempts only result in his knocking the objects (so they swing out of his reach). . . . At 4 months 15 days, with another doll hung in front of him, Laurent tries to grasp it, then shakes himself to make it swing, knocks it accidentally, and then tries simply to hit it. . . . At 4 months 18 days, . . . Laurent [starts] by simply waving his arms around . . . [and then] hits my hands without trying to grasp

FIGURE 5.1

The newborn baby's schemes consist of reflexes, which will gradually be modified as they are applied to the surrounding environment.

FIGURE 5.2

At 2 months, Byron sees his hand open and close and tries to repeat this action, in a primary circular reaction.

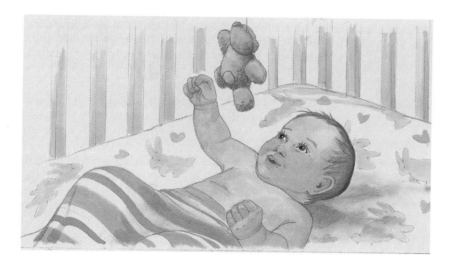

FIGURE 5.3

At 4 months, Piaget's son Laurent accidentally hits a doll hung in front of him. He tries to recapture the interesting effect of the swinging doll. In doing so, he builds a new "hitting scheme" through the secondary circular reaction.

them. The next day, finally, Laurent immediately hits a doll hung in front of him. The [hitting] scheme is now completely differentiated [from grasping]. (Piaget, 1936/1952, pp. 167–168)

Improved control over their own behavior permits infants of this substage to imitate the behavior of others more effectively. However, babies under 8 months only imitate actions they have practiced many times. They cannot adapt flexibly and quickly, copying novel behaviors (Kaye & Marcus, 1981).

■ SUBSTAGE 4: COORDINATION OF SECONDARY CIRCULAR REACTIONS (8 TO 12 MONTHS). Now infants start to organize schemes, combining secondary circular reactions into new, more complex action sequences. Two landmark cognitive changes result.

First, babies can engage in **intentional,** or **goal-directed, behavior.** By 8 months, infants have had enough practice with a variety of schemes that they combine them deliberately in solving sensorimotor problems. The clearest example is provided by Piaget's famous *object-hiding tasks,* in which he shows an attractive toy to a baby and then hides it behind his hand or under a cover. Infants of this substage can find the object. In doing so, they coordinate two schemes—"pushing" aside the obstacle and "grasping" the toy.

Second, the fact that infants can retrieve hidden objects reveals they have begun to attain **object permanence,** the understanding that objects continue to exist when out of sight (see Figure 5.4). But awareness of object permanence is not yet complete. If an object is moved from one hiding place (A) to another (B), 8- to 12-month-olds will search for it only in the first hiding place (A). Because babies make this *AB search error,* Piaget concluded that they do not have a clear image of the object as persisting when not in view.

Substage 4 brings additional advances. First, infants are better at anticipating events, so they sometimes use

their new capacity for intentional behavior to try to change those events. At 10 months, Byron crawled after Lisa when she put on her coat, whimpering to keep her from leaving. Second, babies can imitate behaviors slightly different from those they usually perform. After watching someone else, they try to stir with a spoon, push a toy car, or drop raisins in a cup. Once again, they draw on their capacity for intentional behavior, purposefully modifying schemes to fit an observed action (Piaget, 1945/1951).

■ SUBSTAGE 5: TERTIARY CIRCULAR REACTIONS— DISCOVERING NEW MEANS THROUGH ACTIVE EXPERIMENTATION (12 TO 18 MONTHS). At this substage, the circular reaction—now called *tertiary*—becomes experimental and creative. Toddlers repeat behaviors *with variation,* provoking new effects. Recall how Byron dropped objects over the basement steps, trying this, then that, and then another action (see Figure 5.5). Because they approach the world in this deliberately exploratory way,

FIGURE 5.4

Around 8 months, infants combine schemes deliberately in the solution of sensorimotor problems. They show the beginnings of object permanence, since they can find an object in the first place in which it is hidden.

FIGURE 5.5

At 18 months, Byron dropped a variety of objects down the basement stairs, throwing some up in the air, bouncing others off the wall, releasing some gently and others forcefully, in a deliberately experimental approach. Byron displayed a tertiary circular reaction.

12- to 18-month-olds are far better sensorimotor problem solvers than they were before. For example, Rachel figured out how to fit a shape through a hole in a container by turning and twisting it until it fell through, and she discovered how to use a stick to get toys that were out of reach.

According to Piaget, this new capacity to experiment leads to a more advanced understanding of object permanence. Now toddlers look in several locations to find a hidden toy; they no longer make the AB search error. Their more flexible action patterns also permit them to imitate many more behaviors, such as stacking blocks, scribbling on paper, and making funny faces.

■ SUBSTAGE 6: MENTAL REPRESENTATION—INVENTING NEW MEANS THROUGH MENTAL COMBINATIONS (18 MONTHS TO 2 YEARS). Substage 5 is the last truly *sensorimotor* stage, since Substage 6 brings with it the ability to create **mental representations**—internal images of absent objects and past events. As a result, the older toddler can solve problems symbolically instead of by trial-and-error behavior. One sign of this new capacity is that children arrive at solutions to sensorimotor problems suddenly, suggesting that they experiment with actions inside their heads. For example, at 19 months April played with a toy shopping cart for the first time, rolling it over the carpet

and into the sofa. She paused for a moment, as if to "think," and then immediately turned the toy in a new direction. Had she been in Substage 5, she would have pushed, pulled, and bumped it in a random fashion until it was free to move again.

With the capacity to represent, toddlers arrive at a more advanced understanding of object permanence—that objects can move or be moved when out of sight. Try this object-hiding task with an 18- to 24-month-old as well as a younger child: Put a small toy inside a box and the box under a cover. Then, while the box is out of sight, dump the toy out and show the toddler the empty box. The Substage 6 child finds the hidden toy easily. Younger toddlers are baffled by this situation.

Representation also brings with it the capacity for **deferred imitation**—the ability to remember and copy the behavior of models who are not immediately present. A famous and amusing example is Piaget's daughter Jacqueline's imitation of another child's temper tantrum:

> Jacqueline had a visit from a little boy . . . who, in the course of the afternoon, got into a terrible temper. He screamed as he tried to get out of a playpen and pushed it backwards, stamping his feet. Jacqueline stood watching him in amazement . . . . The next day, she herself screamed in her playpen and tried to move it, stamping her foot lightly several times in succession. (Piaget, 1936/1952, p. 63)

Finally, Substage 6 leads to a major change in the nature of play. Throughout the first year and a half, infants and toddlers engage in **functional play**—pleasurable motor activity with or without objects through which they practice sensorimotor schemes. At the end of the second year, children's capacity to represent experience permits them to engage in **make-believe play,** in which they act out familiar activities. Like Rachel's pretending to go to sleep at the beginning of this chapter, the toddler's make-believe is very simple (see Figure 5.6). Make-believe expands greatly in early childhood and is so important for psychological development that we will return to it again.

■ **R**ECENT RESEARCH ON SENSORIMOTOR DEVELOPMENT

Over the past 20 years, new studies have shown that infants display certain cognitive capacities earlier than Piaget believed. Piaget may have underestimated infant capacities because he did not have the sophisticated experimental techniques that we have today. As we consider recent research on sensorimotor development as well as information processing (covered in a later section), we will see that operant conditioning and the habituation–dishabituation response have been used ingeniously to find out what the young baby knows.

FIGURE 5.6

When Rachel engaged in make-believe by pretending to go to sleep, she created a mental representation of reality. With the capacity for mental representation, the sensorimotor stage draws to a close.

■ OBJECT PERMANENCE. Before 8 months, do babies really believe that an object spirited out of sight no longer exists? In a series of studies in which babies did not have to engage in active search, Renée Baillargeon (1987; Baillargeon & DeVos, 1991) found evidence for object permanence as early as 3½ months of age! One of these studies is illustrated in Figure 5.7 on page 152. Some investigators have challenged Baillargeon, claiming that infants are simply reacting to subtle perceptual differences between her visual displays (Bogartz, Shinsky, & Speaker, 1997). Should future research confirm her findings and those of others, young infants will have been shown to have a remarkably sophisticated understanding of the properties of objects.

If 3½-month-olds grasp the idea of object permanence, then what explains Piaget's finding that much older infants (who are quite capable of voluntary reaching) do not try to search for hidden objects? One explanation is that, just as Piaget's theory suggests, young babies cannot yet put together the separate schemes to retrieve a hidden toy, such as pushing aside the obstacle and grasping the object. In other words, what they *know* about object permanence is not yet *evident* in their searching behavior (Baillargeon et al., 1990).

Once 8- to 12-month-olds actively search for hidden objects, they make the AB search error. Research suggests that one reason they search at A (where they first found the object) instead of B (its most recent location) is that they have trouble inhibiting a previously rewarded response (Diamond, Cruttenden, & Neiderman, 1994). Once again, before 12 months, infants have difficulty integrating object-location knowledge with action. The ability to do so

may depend on rapid maturation of the cortex at the end of the first year (Bell & Fox, 1992; Nelson, 1995).

■ DEFERRED IMITATION. Piaget studied imitation by noting when his own three children demonstrated it in their everyday behavior. Under these conditions, a great deal must be known about the infant's daily life to be sure that deferred imitation has occurred.

Recent research reveals that deferred imitation, a form of representation, is present at 6 weeks of age! Infants who watched an unfamiliar adult's facial expression imitated the facial gesture when exposed to the passive face of the same adult the next day. Perhaps young babies use this way of imitating to identify and communicate with people they have seen before (Meltzoff & Moore, 1994).

As motor capacities improve, infants start to copy actions with objects. In one study, 9-month-olds who saw certain ways of using novel toys were far more likely to engage in these behaviors one day later than were babies exposed to the objects but not shown how they work (Meltzoff, 1988). By 14 months, toddlers use deferred imitation skillfully to enrich their range of sensorimotor

■ Deferred imitation greatly enriches young children's adaptations to their surrounding world. This toddler probably watched an older child or adult sing while playing the keyboard. Later, he sits down and imitates these behaviors, having learned through observation what the keyboard is for. (Joseph Sohm/Stock Boston)

Study in which infants were tested for object permanence using the habituation– dishabituation response.
(a) First, infants were habituated to two events: a short carrot and a tall carrot moving behind a yellow screen, on alternate trials. Then two test events were presented, in which the color of the screen was changed to blue to help the infant notice that now it had a window. (b) In the *possible event,* the short carrot (which was shorter than the window's lower edge) moved behind the blue screen and reappeared on the other side. (c) In the *impossible event,* the tall carrot (which was taller than the window's lower edge) moved behind the screen, did not appear in the window, but then miraculously emerged intact on the other side. Infants as young as 3½ months dishabituated to the impossible event. This suggests that young babies must have some notion of object permanence— that an object continues to exist when it is hidden from view. *(Adapted from R. Baillargeon & J. DeVos, 1991, "Object Permanence in Young Infants: Further Evidence,"* Child Development, 62, *p. 1230. © The Society for Research in Child Development. Reprinted by permission.)*

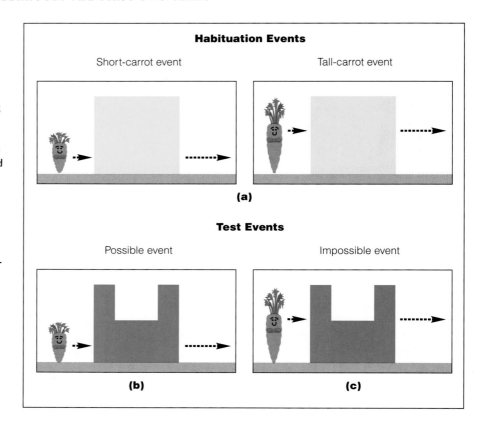

schemes. They retain highly unusual modeled behaviors for several months and can imitate across a change in context—for example, enact a behavior learned at day care in the home (Hanna & Meltzoff, 1993). At the end of the second year, toddlers can imitate actions an adult *tries* to produce, even if these are not fully realized (Meltzoff, 1995).

In sum, deferred imitation is present in early infancy. It does not conclude sensorimotor development as Piaget believed. Nevertheless, deferred imitation becomes far more flexible and complex by the end of toddlerhood, moving beyond specific behaviors to mimicking people's intentions and perspectives. This advance permits young children to better understand and predict others' activities.

EVALUATION OF THE SENSORIMOTOR STAGE

In view of the evidence just described, how should we evaluate the accuracy of Piaget's sensorimotor stage? On the one hand, infants anticipate events, actively search for hidden objects, flexibly vary the circular reaction, and engage in make-believe play within Piaget's time frame. On the other hand, important capacities—among them, secondary circular reactions, understanding of object permanence, deferred imitation, and representation of experience—seem to emerge in preliminary form long before Piaget expected.

Disagreements between Piaget's observations and those of recent research raise controversial questions about how early development takes place. Consistent with Piaget's ideas, infants construct some aspects of experience through motor activity. For example, in Chapter 4, we saw that babies who are experienced crawlers are better at finding hidden objects and perceiving depth on the visual cliff. But infants also comprehend a great deal before they are capable of the motor behaviors Piaget believed were responsible for these understandings. Perhaps some schemes (such as imitation and a basic appreciation of the properties of objects) are prewired into the brain from the start (Spelke, 1991, 1994). Others may be constructed through purely perceptual learning—by looking and listening—rather than through acting directly on the world (Baillargeon, 1994, 1995).

Finally, infants do not develop in the neat, stepwise fashion implied by Piaget's theory, in which a variety of skills change together and abruptly at each new substage. Instead, a baby is likely to be at one level of progress on imitation and quite another on object permanence, depending on the complexity of the task and the child's rate of biological maturation and specific experiences (Rast & Meltzoff, 1995). These ideas—that adultlike capacities are present during infancy in primitive form, that cognitive development is gradual and continuous rather than stagelike, and that it must be described separately for each skill—serve as the basis for a major competing approach to cognitive development: information processing, which we take up next.

But before we turn to this alternative point of view, let's conclude our discussion of the sensorimotor stage by recognizing Piaget's enormous contributions. His work inspired a wealth of new research on infant cognition, including studies that eventually challenged his ideas. In addition, his observations have been useful in practical ways. Teachers and caregivers continue to look to the sensorimotor stage for guidelines on how to create developmentally appropriate environments for infants and toddlers, an issue we take up later in this chapter.

BRIEF REVIEW

According to Piaget, children actively build psychological structures, or schemes, as they manipulate and explore their world. Two processes, adaptation (which combines assimilation and accommodation) and organization, account for the development of schemes. The sensorimotor stage is divided into six substages. The circular reaction, a special means that infants use to adapt schemes, changes from being oriented toward the infant's own body, to being directed outward toward objects, to producing novel effects in the surrounding world. During the last three substages, infants make progress in intentional behavior and understanding object permanence. By the final substage, they start to represent reality and show the beginnings of make-believe play.

Recent research suggests that secondary circular reactions, object permanence, and deferred imitation are present much earlier than Piaget believed. These findings raise questions about his claim that infants must construct all aspects of their cognitive world through motor activity and that sensorimotor development takes place in stages.

ASK YOURSELF . . .

■ *Tony pushed his toy bunny through the slats of his crib onto a nearby table. Using his "pulling scheme," he tried to retrieve it, but it would not fit back through the slats. Next Tony tried jerking, turning, and throwing the bunny. Is Tony in a state of equilibrium or disequilibrium? How do you know?*

■ *Mimi banged her rattle again and again on the tray of her high chair. Then she dropped the rattle, which fell out of sight on her lap, but Mimi did not try to retrieve it. Which sensorimotor substage is Mimi in? Why do you think so?*

INFORMATION PROCESSING IN THE FIRST TWO YEARS

Information-processing theorists agree with Piaget that children are active, inquiring beings, but they do not provide a single, unified theory of cognitive development. Instead, they investigate different aspects of thinking, from attention, memory, and categorization skills to complex problem solving.

Recall that information processing relies on computer-like diagrams and flowcharts to describe the human cognitive system. The computer model of human thinking is very attractive because it is explicit and precise. Information-processing researchers find it useful because they are not satisfied with global concepts, such as assimilation and accommodation, to describe how children think. Instead, they want to know exactly what individuals of different ages do when faced with a task or problem (Klahr, 1992; Siegler, 1998).

A MODEL OF HUMAN INFORMATION PROCESSING

Although many flowcharts of human information processing exist, Richard Atkinson and Richard Shiffrin's (1968) model has inspired more research than any other. As Figure 5.8 on page 154 shows, it divides the mind into three basic parts: *the sensory register; working,* or *short-term, memory;* and *long-term memory.* As information flows through each, **control processes,** or **mental strategies,** operate on and transform it, increasing the efficiency of thinking and the chances information will be retained for later use. Let's take a brief look at each aspect of this model.

First, information enters the **sensory register,** where it is represented directly and very briefly. However, by using mental strategies, we can preserve selected sights and sounds. For example, you can attend to some information more carefully than others, increasing the chances that it will transfer to the next step of the information-processing system.

The second waystation of the mind is **working,** or **short-term, memory.** This is the conscious part of our mental system, where we actively "work" on a limited amount of information to retain it. For example, if you are studying this book effectively, you are constantly applying mental strategies. Perhaps you are taking notes, repeating information to yourself, or grouping pieces of information together—a strategy much like Piaget's notion of organization. Organization is an especially effective way to remember the many new concepts flowing into your working memory at the moment. If you permit input to remain disconnected, you can hold very little in working memory at once, since you must focus on each item separately. But organize it, and you will not just improve your memory.

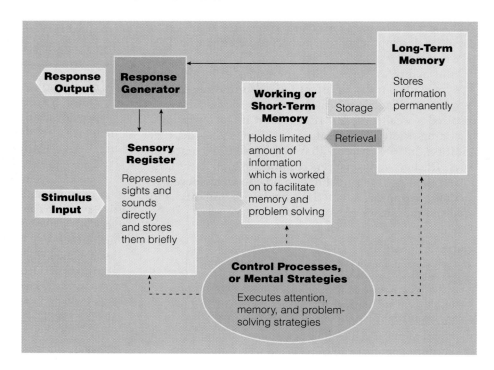

FIGURE 5.8

Atkinson and Shiffrin's model of the human information-processing system. Stimulus input flows through three parts of the mental system: the sensory register; working, or short-term memory; and long-term memory, In each, control processes, or mental strategies, can be used to manipulate information, increasing the efficiency of thinking and the chances that information will be retained. *(Adapted from R. M. Shiffrin & R. C. Atkinson, 1969, "Storage and retrieval processes in long-term memory," Psychological Review, 76, p. 180. Copyright © 1969 by the American Psychological Association. Adapted by permission of the publisher and author.)*

You will increase the chances that information will be transferred to the third, and largest, storage area.

Long-term memory is our permanent knowledge base. Its capacity is limitless. In fact, we store so much in long-term memory that we sometimes have problems with *retrieval,* or getting information back from the system. To aid retrieval, we apply strategies, just as we do in working memory. According to Atkinson and Shiffrin (1968), information in long-term memory is *categorized* according to a master plan, much like a "library shelving system based on the contents of books" (p. 181). As a result, it can be retrieved quite easily by following the same network of assocations used to store it in the first place.

Information-processing researchers believe that the basic structure of the mental system is similar throughout life. However, the *capacity* of the system—the amount of information that can be retained and processed at once—expands, making possible more complex forms of thinking with age (Case, 1992; Halford, 1993). Although gains in information-processing capacity are partly due to brain maturation, they are largely the result of improvements in strategies, such as attending to information and categorizing it effectively. The development of these strategies is already underway in the first year of life.

ATTENTION AND MEMORY

How does attention develop in early infancy? Recall from our discussion of perceptual development in Chapter 4 that between 1 and 2 months, infants shift from attending to a single high-contrast feature of their visual world to exploring objects and patterns more thoroughly. Besides attending to more aspects of the environment, infants gradually become more efficient at managing their attention, taking in information more quickly with age. Habituation–dishabituation research reveals that preterm infants require a long time to habituate and dishabituate to novel stimuli—for example, 5 minutes or more for visual patterns (Werner & Siqueland, 1978). But by 5 months, babies process new information rapidly, requiring as little as 5 to 10 seconds to take in a complex visual stimulus and recognize that it is different from a second one (Fagan, 1977).

Habituation also provides a window into infant memory. For example, infants can be exposed to a stimulus until they habituate. Then they can be shown the same stimulus at a later time. If they habituate more rapidly on the second occasion, this indicates that they recognize they have seen the stimulus before. Using this method, studies show that by 3 months infants remember a visual stimulus for 24 hours (Martin, 1975). By the end of the first year, they remember it for several days and, in the case of some stimuli (such as a photo of the human face), a few weeks (Fagan, 1973).

At this point, let's note that the habituation response tells us about infants' memory in the strange context of the laboratory, but it underestimates their ability to remember real-world events they can actively control. Using operant conditioning, Carolyn Rovee-Collier (1991) studied infant memory in a familiar setting—at home while babies lay in their cribs. First, 2- to 3-month-olds were taught how to move a mobile by kicking a foot tied to it with a long cord.

When the infants were reattached to the mobile after a week's delay, they kicked vigorously, showing that indeed they remembered. And as long as they were reminded of the mobile's dancing motion (the experimenter briefly rotated it for the baby), infants started kicking again up to 2 weeks after they were first trained (Linde, Morrongiello, & Rovee-Collier, 1985; Rovee-Collier, Patterson, & Hayne, 1985). Much like older children and adults, babies seem to remember best when experiences take place in familiar contexts and when they participate actively (Lipsitt, 1990).

So far, we have discussed only **recognition,** the simplest form of memory because all babies have to do is indicate (by looking or kicking) whether a new stimulus is identical or similar to one previously experienced. **Recall** is more challenging, since it involves remembering something not present. Can infants engage in recall? By the middle of the first year, they can. Felicia reported that one day when her husband telephoned, 7-month-old April listened to him speak through the receiver and immediately crawled to the front door. April seemed to be recalling times when she had heard her father's muffled voice as he was about to arrive home from work (Ashmead & Perlmutter, 1980).

By the end of toddlerhood, recall for people, places, and objects is excellent (Bauer, 1996). Yet a puzzling finding is that as adults, we no longer recall our earliest experiences. The Lifespan Vista box on page 156 helps explain why.

CATEGORIZATION

As infants gradually remember more information, they seem to store it in a remarkably orderly fashion. Some creative variations of the operant conditioning research described earlier have been used to find out about infant categorization (see Figure 5.9). In fact, young babies categorize stimuli on the basis of shape, size, number (up to three elements), and other physical properties at such an early age that categorization is among the strongest evidence that babies' brains are set up from the start to structure experience in adultlike ways (Mervis, 1985).

The habituation–dishabituation response has also been used to study infant categorization. Researchers show babies a series of stimuli belonging to one category and then see whether they dishabituate to a picture that is not a member of the category. Findings reveal that 9- to 12-month-olds structure objects into an impressive array of meaningful categories—food items, furniture, birds, animals, vehicles, and more (Mandler & McDonough, 1993; Oakes, Madole, & Cohen, 1991; Ross, 1980; Younger, 1985, 1993). Besides organizing the physical world, infants of this age categorize their emotional and social worlds. They sort people into male and female (Francis & McCroy, 1983; Poulin-DuBois et al., 1994), have begun to distinguish emotional expressions, and can separate the natural movements of people from other motions (see Chapter 4, page 139).

In the second year, children become active categorizers during their play. Try giving a toddler a set of objects that differ in shape and color (such as small blocks). See if the child spontaneously categorizes them. Research shows that around 12 months, toddlers merely touch objects that belong together, without grouping them. A little later, they form single categories. For example, when given four balls and four boxes, a 16-month-old will put all the balls together but not the boxes. And finally, around 18 months, toddlers can sort objects into two classes.

Interestingly, this advanced object-sorting behavior appears at about the same time that toddlers show a "naming explosion," or a sharp spurt in vocabulary in which they label many more objects (Gopnik & Meltzoff, 1987, 1992). Language development seems to facilitate as well as build on improved categorization. In support of this idea, research shows that adult labeling of objects helps direct toddlers' attention to object categories (Waxman & Hall, 1993).

FIGURE 5.9

Study of infant memory using operant conditioning.
Three-month-olds were taught to kick to make a mobile move that was made of small blocks, all with the letter A on them. After a delay, kicking returned to a high level only if the babies were shown a mobile whose elements were labeled with the same form (the letter A). If the form was changed (from As to 2s), infants no longer kicked vigorously. While making the mobile move, the babies had grouped together its features. They associated the kicking response with the category A and, at later testing, distinguished it from the category 2 (Bhatt, Rovee-Collier, & Weiner, 1994; Hayne, Rovee-Collier, & Perris, 1987).

INFANTILE AMNESIA

Groucho: Chicolini, when were you born?
Chico: I don't remember. I was just a little baby at the time.[1]

If toddlers remember many aspects of their everyday lives, then what explains infantile amnesia—the fact that practically none of us can retrieve events that happened to us before age 3? Forgetting cannot be due to the passage of time, since we can recall many events that happened long ago. At present, there are several explanations of infantile amnesia.

One possibility is that brain maturation during early childhood accounts for it. Growth of the frontal lobes of the cortex along with other structures may be necessary before experiences can be stored in ways that permit them to be retrieved many years later (Boyer & Diamond, 1992). Consistent with this view, some researchers believe that two levels of memory exist: one operates unconsciously and automatically, whereas the other is conscious, intentional, and verbal (Newcombe & Fox, 1994).

Yet the idea of vastly different approaches to remembering in younger and older individuals has been questioned, since even toddlers can describe memories verbally and retain them for extensive periods. A growing number of researchers believe that rather than a radical change in the way experience is represented, the offset of infantile amnesia requires the emergence of a special form of recall—*autobiographical memory,* or special, one-time events that are long lasting because they are imbued with personal meaning. For example, perhaps you recall the day a sibling was born, the first time you took an airplane, or a move to a new house.

For memories to become autobiographical, two developments are necessary. First, the child must have a well-developed image of the self. Yet in the first few years, the sense of self is not yet mature enough to serve as an anchor for one-time events (Howe & Courage, 1993). Second, autobiographical memory requires that children integrate personal experiences

into a meaningful life story. Recent evidence reveals that preschoolers learn to structure memories in story-like form by talking about them with adults, who expand on their recollections by explaining what happened when, where, and with whom (Haden, Hayne, & Fivush, 1997; Nelson, 1993). As 3- to 6-year-olds' cognitive and conversational skills improve, their descriptions of special, one-time events become better organized, detailed, and related to the larger context of their lives (Reese, Haden, & Fivush, 1996).

Interestingly, preschool girls' autobiographical memories are more advanced than boys', a difference that fits with findings that parents talk about the past in more detail with daughters than sons (Reese, Haden & Fivush, 1996). Perhaps for this reason, women report an earlier age of first memory and more vivid early memories than do men (Mullen, 1994).

[1] From the Marx Brothers' movie *Duck Soup.*

When he gets older, this toddler won't recall the exciting party that took place on his second birthday. According to recent evidence, the offset of infantile amnesia after age 3 is due to the emergence of autobiographical memory. For this special form of recall to develop, young children must have the language skills to talk about personal experiences and, with the assistance of adults, integrate them into a meaningful life story.(Jeff Greenberg/ The Picture Cube)

EVALUATION OF INFORMATION-PROCESSING FINDINGS

Information-processing research underscores the *continuity* of human thinking from infancy into adult life. In attending to the environment, remembering everyday events, and categorizing objects, Byron, Rachel, and April think in ways that are remarkably similar to our own, even though they are far from being the proficient mental processors we are. Findings on infant memory and categorization join with other research in challenging Piaget's view of early development. If 3-month-olds can remember events for as long as 2 weeks and categorize stimuli, then they must have some ability to mentally represent their experiences. Representation seems to be another capacity that does not have to wait until babies have engaged in many months of sensorimotor activity (even though there is no dispute that representation flourishes during the second year of life).

Information processing has contributed greatly to our view of young babies as sophisticated cognitive beings. Still, it has drawbacks. Perhaps the greatest one stems from its central strength: By breaking cognition down into separate elements (such as perception, attention, and memory), information processing has had difficulty putting all these pieces back together into a broad, comprehensive theory. For this reason, many experts resist abandoning Piaget's ideas in favor of it. But by drawing on the strengths of both perspectives, the field of human development may be moving closer to new, more powerful views of how the mind develops (Case, 1992; Fischer & Farrar, 1987; Halford, 1993).

THE SOCIAL CONTEXT OF EARLY COGNITIVE DEVELOPMENT

Take a moment to review the short episode at the beginning of this chapter in which Rachel dropped shapes into a container. Notice how Rachel actively explored the toy, but she learned about it with her mother's help. Beth held the container, adjusting its position so Rachel could be successful at this challenging task. With Beth's support, Rachel will gradually become better at matching shapes to openings and dropping them into the container. Then she will be able to do the activity (and others like it) on her own.

Vygotsky's (1930–1935/1978) sociocultural theory has brought the field of child development to the realization that children live in rich social contexts that affect the way their cognitive world is structured (Bruner, 1990; Rogoff, 1990). He believed that complex mental activities—those

unique to humans—have their origins in social interaction. Through joint activities with more mature members of their society, children come to master activities and think in ways that have meaning in their culture.

A special Vygotskian concept, the **zone of proximal** (or potential) **development**, explains how this happens. It refers to a range of tasks that the child cannot yet handle alone but can do with the help of more skilled partners. To understand this idea, think of a sensitive teacher or parent (such as Beth) who introduces a child to a new activity. The adult picks a task that the child can master but one challenging enough that the child cannot do it by herself. In this way, the activity is especially suited for fostering development. Then the adult guides and supports, making the task manageable for the child. By joining in the interaction, the child picks up mental strategies, and her competence increases. As this happens, the adult steps back, permitting the child to take over more responsibility for the task.

Although Vygotsky's ideas have mostly been applied to older children, his theory has recently been extended downward to infancy. In earlier parts of this book, we showed how babies are equipped with ways of ensuring

This mother assists her baby in making a music box work. By presenting a task within the child's zone of proximal development and fine-tuning her support to the infant's momentary needs, the mother promotes her son's cognitive development. (Erika Stone)

that caregivers will interact with them. Then adults adjust the environment and their communication in ways that promote learning.

A study by Barbara Rogoff and her collaborators (1984) illustrates this process. The researchers watched how several adults played with Rogoff's son and daughter over the first 2 years, while a jack-in-the-box toy was nearby. In the early months, adults tried to focus the baby's attention by showing the toy and, as the bunny popped out, saying something like "My, what happened?" By the end of the first year (when the baby's cognitive and motor skills had improved), interaction centered on how to use the jack-in-the-box. When the infant reached for the toy, adults guided the baby's hand in turning the crank and putting the bunny back in the box. During the second year, adults helped from a distance. They used verbal instructions and gestures, such as rotating a hand in a turning motion near the crank. Research suggests that this fine-tuned support is related to advanced play, language, and problem solving during the second year (Bornstein et al., 1992; Frankel & Bates, 1990; Tamis-LeMonda & Bornstein, 1989).

By emphasizing that many aspects of cognitive development are socially mediated, Vygotsky adds a new dimension to theories discussed earlier in this chapter. In the Cultural Influences box on the next page, you will find further evidence for this idea. And we will see even more in the next section, as we look at individual differences in mental development during the first 2 years.

BRIEF REVIEW

Information-processing researchers analyze thinking into separate elements and study how each changes with age. Atkinson and Shiffrin's computerlike model divides the mind into three parts: the sensory register; working, or short-term, memory; and long-term memory. As information flows through the system, control processes, or mental strategies, operate on it, increasing the likelihood that information will be retained. Mental strategies gradually improve. With age, infants attend to more aspects of the environment, manage their attention more efficiently, and remember information longer. Findings on categorization support the view that babies structure experience in adultlike ways. Although information processing helps us appreciate infants' remarkable cognitive abilities, it does not offer a comprehensive theory of children's thinking. According to Vygotsky's sociocultural theory, early cognitive development is socially mediated as adults help infants and toddlers master tasks in the zone of proximal development.

ASK YOURSELF . . .

■ *When Rachel was 3 months old, she stared at a little toy dog that Beth dangled in front of her and then looked away. The next day, Beth held up the same toy dog again and Rachel looked at it, but more briefly. What can we conclude about Rachel's processing of the toy?*

■ *At age 18 months, Byron's father stood behind him, helping him throw a large rubber ball into a box. When Byron showed he could throw the ball, his father stepped back and let him try on his own. Using Vygotsky's ideas, explain how Byron's father is supporting his cognitive development.*

INDIVIDUAL DIFFERENCES IN EARLY MENTAL DEVELOPMENT

B yron isn't talking much yet," Lisa remarked as her son neared age 2, "and he rarely stacks things and puts shapes in containers like Rachel and April. He seems so restless and overactive. If he isn't developing as quickly as other children now, will that still be true when he goes to school?" Increasingly concerned, Lisa decided to take Byron to a psychological clinic to be tested.

The testing approach is very different from the cognitive theories we have discussed so far, which try to explain the *process* of development—how children's thinking changes over time. In contrast, designers of mental tests are much more concerned with cognitive *products*. They seek to measure behaviors that reflect mental development and arrive at scores that *predict* future performance, such as later intelligence, school achievement, and adult vocational success. This concern with prediction arose nearly a century ago, when French psychologist Alfred Binet designed the first successful intelligence test, which predicted school achievement. It inspired the design of many new tests, including ones that measure intelligence at very early ages.

INFANT INTELLIGENCE TESTS

Accurately measuring the intelligence of infants is a challenge. Babies cannot answer questions or follow directions, and they often get hungry, distracted, or tired during testing. All we can do is present them with stimuli, coax them to respond, and observe their behavior. As a result, most infant tests consist of perceptual and motor responses along with some tasks that tap early language and problem solving. One commonly used test is the *Bayley Scales of Infant Development*, designed for ages 1 month to 3½ years. It is made up of two scales: (1) the Mental

CULTURAL INFLUENCES

CAREGIVER–TODDLER INTERACTION AND EARLY MAKE-BELIEVE PLAY

One of my husband Ken's activities with our two sons when they were young was to bake pineapple upside-down cake, a favorite treat. At age 4½, David already knew how to arrange the pineapple slices, mix the batter, pour it into the pan, and put it into the oven. One Sunday afternoon when a cake was in the making, 21-month-old Peter stood on a chair at the kitchen sink, busy pouring water from one cup to another.

"He's in the way, Dad!" complained David, trying to pull Peter away from the sink.

"Maybe if we let him help, he'll give us room at the sink," Ken suggested. As David stirred the batter, Ken poured some into a small bowl for Peter, moved his chair to the side of the sink, and handed him a spoon.

"Here's how you do it, Petey," instructed David, with an air of superiority. Peter watched as David stirred, then tried to copy his motion. When it was time to pour the batter, Ken helped Peter hold and tip the small bowl.

"Time to bake it," said Ken.

"Bake it, bake it," repeated Peter, as he watched Ken slip the pan into the oven.

Several hours later, we observed one of Peter's earliest instances of make-believe play. He got his pail from the sandbox, filled it with sand, carried it into the kitchen, and put it down on the floor in front of the oven. "Bake it, bake it," Peter called to Ken. Together, father and son lifted the pretend cake inside the oven.

Until recently, most researchers studied make-believe play apart from the social environment in which it occurs, while children played alone. Probably for this reason, Piaget and his followers concluded that toddlers discover make-believe independently. Vygotsky's theory has challenged this

view. He believed that society provides children with opportunities to represent culturally meaningful activities in play. Make-believe, like other mental functions, is first learned under the guidance of experts (Garvey, 1990). In the example just described, Peter's capacity to represent daily events was extended when Ken drew him into the baking task and helped him act it out in play.

New evidence supports the idea that early make-believe is the combined result of children's readiness to engage in it and social experiences that promote it. In one observational study of middle-class American toddlers, 75 to 80 percent of make-believe during the second year involved mother–child interaction (Haight & Miller, 1993). When adults participate, toddlers' make-believe is more elaborate and advanced (Fiese, 1990; O'Reilly & Bornstein, 1993; Slade, 1987). For example, their play themes are more varied. And they are more likely to combine schemes into complex sequences, as Peter did when he put sand in the bucket ("making the batter"), carried it into the kitchen, and (with Ken's help) put it in the oven ("baking the cake").

In many cultures, adults do not spend much time playing with young children. Instead, older siblings take over this function. For example, in Indonesia and Mexico, where extended family households and sibling caregiving are common, make-believe is more frequent as well as complex with older siblings than with mothers. As early as 3 to 4 years of age, children provide rich, challenging stimulation to their younger brothers and sisters. The fantasy play of these toddlers is just as well developed as that of their middle-class American counterparts (Farver, 1993; Farver & Wimbarti, 1995).

As we will see in Chapter 7, make-believe is a major means through

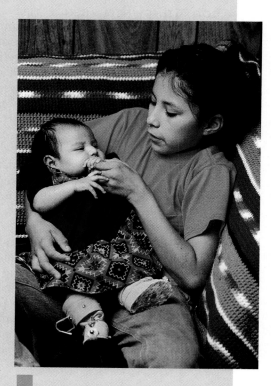

This Mexican girl often cares for her baby sister. When the infant gets older, her sibling caregiver is likely to provide rich, challenging playful stimulation. Opportunities to play with older siblings lead the fantasy play of many ethnic minority toddlers to be just as well developed as that of their middle-class American counterparts, who typically play with adults. (Lawrence Migdale/Stock Boston)

which children extend their cognitive skills and learn about important activities in their culture. Vygotsky's theory, and the findings that support it, tell us that providing a stimulating physical environment is not enough to promote early cognitive development. In addition, toddlers must be invited and encouraged by their elders to participate actively in the social world around them.

Scale, which includes such items as turning to a sound, looking for a fallen object, building a tower of cubes, and naming pictures; and (2) the Motor Scale, which assesses gross and fine motor skills, such as grasping, sitting, drinking from a cup, and jumping (Bayley, 1993).

■ COMPUTING INTELLIGENCE TEST SCORES. Intelligence tests for infants, children, and adults are scored in much the same way. When a test is constructed, it is given to a large, representative sample of individuals. Performances of people at each age level form a *normal,* or *bell-shaped, curve* in which most scores fall near the center (the mean or average) and progressively fewer fall out toward the extremes. On the basis of this distribution, the test designer computes *norms,* or standards against which future test takers can be compared. For example, if Byron does better than 50 percent of his agemates, his score will be 100, an average test score. If he exceeds most children his age, his score will be much higher. If he does better than only a small percentage of 2-year-olds, his score will be much lower.

Scores computed in this way are called **intelligence quotients,** or **IQs,** a term you have undoubtedly heard before. Table 5.2 describes the meaning of a range of IQ scores. Notice how IQ offers a way of finding out whether an individual is ahead, behind, or on time (average) in mental development in relation to others of the same age. The great majority of individuals (96 percent) have IQs between 70 and 130. Only a very few achieve higher or lower scores.

■ PREDICTING LATER PERFORMANCE FROM INFANT TESTS. Many people assume, incorrectly, that IQ is a mea-

sure of inborn ability that does not change with age. Despite careful construction, many infant tests predict later intelligence poorly. In one longitudinal study, most youngsters' scores changed considerably between toddlerhood and adolescence. In fact, the average IQ shift was as great as 28.5 points, and high scorers during the early years were not necessarily high scorers later (McCall, Appelbaum, & Hogarty, 1973).

Both the immaturity of young test takers and the makeup of infant tests limit prediction of later performance. The perceptual and motor items on infant tests are quite different from the tasks given to older children, which emphasize verbal, conceptual, and problem-solving skills. Because of concerns that infant test scores do not tap the same dimensions of intelligence measured at older ages, they are conservatively labeled **developmental quotients,** or **DQs,** rather than IQs.

It is important to note that infant tests do show somewhat better long-term prediction for extremely low-scoring babies. Today, infant tests are largely used for *screening*—helping to identify babies whose very low scores mean that they are likely to have developmental problems in the future (Kopp, 1994).

Because infant tests do not predict later IQ for most children, researchers have turned to information processing for new ways to assess early mental functioning. Their findings show that speed of habituation and dishabituation to visual stimuli are the best available infant predictors of intelligence between 3 and 11 years of age (McCall & Carriger, 1993; Rose & Feldman, 1995). The habituation–dishabituation response may be a more effective predictor of later IQ than traditional infant tests because it assesses quickness of thinking, a characteristic of bright individuals (Colombo, 1995). It also taps basic cognitive processes—attention, memory, and response to novelty—that underlie intelligent behavior at all ages.

Piagetian object permanence tasks also predict later IQ more effectively than traditional infant measures, perhaps because they, too, reflect a basic intellectual process—problem solving (Wachs, 1975). The consistency of these findings prompted designers of the most recent edition of the Bayley test to include several items that tap higher-order skills, such as preference for novel stimuli, ability to find hidden objects, and categorization (Bayley, 1993).

EARLY ENVIRONMENT AND MENTAL DEVELOPMENT

In Chapter 2, we indicated that intelligence is a complex blend of hereditary and environmental influences. Because infant scores are so unstable, researchers have not been able to study genetic contributions to intelligence at early ages. In contrast, many studies have examined the relationship of environmental factors to infant and toddler mental test scores.

TABLE 5.2

Meaning of Different IQ Scores

SCORE	PERCENTILE RANK CHILD DOES BETTER THAN . . . PERCENT OF SAME-AGE CHILDREN	
70	2	
85	16	
100 (average IQ)	50	
115	84	
130	98	

■ HOME ENVIRONMENT. From what you have learned so far, what aspects of young children's home experiences would you expect to influence early mental development? The **Home Observation for Measurement of the Environment (HOME)** is a checklist for gathering information about the quality of children's home lives through observation and parental interview. The Caregiving Concerns table below lists factors measured by HOME during the first 3 years. Each is positively related to toddlers' mental test performance. Also, high HOME scores are associated with IQ gains between 1 and 3 years of age, whereas low HOME scores predict declines. An organized, stimulating physical setting and parental encouragement, involvement, and affection repeatedly predict infant and early childhood IQ, regardless of social class and ethnicity (Bradley & Caldwell, 1982; Bradley et al., 1989).

Can the research summarized so far help us understand Lisa's concern about Byron's development? Indeed, it can. Andrew, the psychologist who tested Byron, found that he scored only slightly below average. Andrew also interviewed Lisa about her child-rearing practices and watched her play with Byron. He noticed that Lisa, anxious about how well Byron was doing, tended to pressure him a great deal. She constantly bombarded him with questions and instructions. Andrew explained that when parents are intrusive in these ways, infants and toddlers are likely to be

distractible, show less mature forms of play, and do poorly on mental tests (Bradley et al., 1989; Fiese, 1990). He coached Lisa in how to establish a sensitive give-and-take while playing with Byron. At the same time, he assured her that Byron's current performance need not forecast his future development. Warm, responsive parenting that builds on toddlers' current capacities is a much better indicator of how they will do later than an early mental test score.

■ INFANT AND TODDLER DAY CARE. Home environments are not the only influential settings in which young children spend their days. Today, over 60 percent of American mothers with a child under age 2 are employed (U.S. Bureau of the Census, 1996). Day care for infants and toddlers has become common, and its quality has a major impact on mental development. One series of studies focused on young children in Bermuda, where 85 percent enter day care before age 2. Verbal stimulation by caregivers and an overall rating of day-care-center quality (based on many characteristics, from physical facilities to daily activities) predicted enhanced cognitive, language, and social skills during the preschool years (McCartney et al., 1985; Phillips, McCartney, & Scarr, 1987). Similarly, in Sweden, entering high-quality day care before age 1 is related to cognitive, emotional, and social competence in middle childhood and adolescence (Andersson, 1989, 1992).

Visit some day care settings in your community, and take notes on what you see. In contrast to most European countries and to Australia and New Zealand, where day care is nationally regulated and funded to ensure its quality, reports on American day care are cause for deep concern. Standards are set by the states and vary greatly across the nation. In some places, caregivers need no special training in child development, and one adult is permitted to care for 8 to 12 babies at once (Helburn, 1995). Large numbers of infants and toddlers everywhere attend unlicensed day-care homes, where no one checks to see that minimum health and safety standards are met (Galinsky et al., 1994). Children who enter low-quality day care during their first year and remain there during the preschool years are rated by teachers as distractible, low in task involvement, and inconsiderate of others when they reach kindergarten age (Howes, 1990).

The Caregiving Concerns table on the page 162 lists signs of high-quality programs that can be used in choosing a day-care setting for an infant or toddler, based on standards for **developmentally appropriate practice** devised by the National Association for the Education of Young Children (Bredekamp & Copple, 1997). These standards specify program characteristics that meet the developmental and individual needs of young children, based on current research and the consensus of experts.

Of course, for parents to make this choice, enough high-quality day care must be available. Unfortunately,

CAREGIVING CONCERNS

Home Observation for Measurement of the Environment (HOME): Infancy and Toddler Subscales

SUBSCALE	SAMPLE ITEM
Emotional and verbal responsiveness of the parent	Parent caresses or kisses child at least once during observer's visit.
Acceptance of the child	Parent does not interfere with child's actions or restrict child's movements more than three times during observer's visit.
Organization of the physical environment	Child's play environment appears safe and free of hazards.
Provision of appropriate play materials	Parent provides toys or interesting activities for child during observer's visit.
Maternal involvement with the child	Parent tends to keep child within visual range and to look at child often during observer's visit.
Variety in daily stimulation	Child eats at least one meal per day with mother or father, according to parental report.

Source: Elardo & Bradley, 1981.

Signs of Developmentally Appropriate Infant and Toddler Day Care

PROGRAM CHARACTERISTIC	SIGNS OF QUALITY
Physical setting	Indoor environment is clean, in good repair, well lighted, and well ventilated. Fenced outdoor play space is available. Setting does not appear overcrowded when children are present.
Toys and equipment	Play materials are appropriate for infants and toddlers and stored on low shelves within easy reach. Cribs, high chairs, infant seats, and child-sized tables and chairs are available. Outdoor equipment includes small riding toys, swings, slide, and sandbox.
Caregiver–child ratio	In day care centers, caregiver–child ratio is no greater than 1 to 3 for infants and 1 to 5 for toddlers. Group size (number of children in one room) is no greater than 6 infants with 2 caregivers and 12 toddlers with 2 caregivers. In day care homes, caregiver is responsible for no more than 6 children; within this group, no more than 2 are infants and toddlers. Staffing is consistent, so infants and toddlers can form relationships with particular caregivers.
Daily activities	Daily schedule includes times for active play, quiet play, naps, snacks, and meals. It is flexible rather than rigid, to meet the needs of individual children. Atmosphere is warm and supportive, and children are never left unsupervised.
Interactions among adults and children	Caregivers respond promptly to infants' and toddlers' distress; hold, talk to, sing, and read to them; and interact with them in a contingent manner that respects the individual child's interests and tolerance for stimulation.
Caregiver qualifications	Caregiver has at least some training in child development, first aid, and safety.
Relationships with parents	Parents are welcome any time. Caregivers talk daily with parents about children's behavior and development.
Licensing and accreditation	Day care setting, whether a center or home, is licensed by the state. Accreditation by the National Academy of Early Childhood Programs or the National Family Day Care Association is evidence of an especially high-quality program.

Source: Bredekamp & Copple, 1997; National Association for the Education of Young Children, 1991.

children most likely to have inadequate day care come from low-income and poverty-stricken families (Phillips et al., 1994). Recognizing that American day care is in crisis, in 1992 Congress allocated additional funds to upgrade its quality and assist parents—especially those with low incomes—in paying for it (Barnett, 1993). This is a hopeful sign, since good day care protects the well-being of all children, and it can serve as effective early intervention for youngsters whose development is at risk, much like the programs we are about to consider in the next section.

EARLY INTERVENTION FOR AT-RISK INFANTS AND TODDLERS

Many studies indicate that children of poverty are likely to show gradual declines in intelligence test scores and to achieve poorly when they reach school age (Brody, 1992). These problems are largely due to stressful home environments that undermine children's ability to learn and increase the chances that they will remain poor throughout their lives. A variety of intervention programs have been developed to break this tragic cycle of poverty.

Although most begin during the preschool years (we will discuss these in Chapter 7), a few start during infancy and continue through early childhood.

Some interventions are center based. The Carolina Abecedarian Project, described in the Social Issues box on the following page, is a well-known example. Others are home based. A skilled adult visits the home and works with parents, teaching them how to stimulate a very young child's development. In most programs, participating youngsters score higher on mental tests than untreated controls by age 2. These gains persist as long as the program lasts and occasionally longer. The more intense the intervention (for example, full-day, year-round high-quality day care plus support services for parents), the greater children's intellectual gains (Bryant & Ramey, 1987).

Without some form of early intervention, many children born into economically disadvantaged families will not reach their potential. Recently, the United States Congress provided limited funding for intervention services directed at infants and toddlers who already have serious developmental problems or who are at risk for them because of poverty (Barnett, 1993). At present, available

SOCIAL ISSUES

THE CAROLINA ABECEDARIAN PROJECT:
A MODEL OF INFANT–TODDLER INTERVENTION

Can educational enrichment at a very early age prevent the declines in mental development known to affect children born into extreme poverty? The findings of the Carolina Abecedarian Project, and others like it, are encouraging. They indicate that intensive intervention beginning in infancy is the most effective way to combat the devastating effects of poverty on children's mental development.

Begun in the 1970s, the Carolina Abecedarian Project identified over a hundred infants at serious risk for school failure, on the basis of parental education and income, poor school achievement among older siblings, and other family problems. Shortly after birth, the babies were randomly assigned to either a treatment or control group. Both groups received nutrition and health services. The primary difference between them was a day-care experience designed to support the treatment group's mental development.

Between 3 weeks and 3 months of age, infants in the treatment group were enrolled in a full-time, year-round day-care program and remained there until they entered kindergarten. During the first 3 years, the children received stimulation aimed at promoting motor, cognitive, language, and social skills. After age 3, the program expanded to include pre-reading and math concepts. At all ages, adult–child communication was emphasized. Teachers were trained to engage in informative, helpful, and nondirective interaction with the children, who were talked and read to daily.

Results were dramatic. Treatment children scored higher than controls on intelligence tests throughout the preschool years (Ramey & Campbell, 1984). Although the high-risk backgrounds of both groups led their IQs to decline over middle childhood, follow-up testing at ages 8 and 12 revealed that treatment children maintained a 5-point advantage over controls. Even more impressive, at ages 8 and 12 treatment youngsters were achieving considerably better in school, especially in reading, writing, and general knowledge. And they were less likely to have been placed in special education or to have repeated a grade (Campbell & Ramey, 1991, 1994; Martin, Ramey, & Ramey, 1990). As we will see in Chapter 7, interventions that start later and last for a shorter time do not have this enduring impact on intelligence and achievement test scores (although they do show other long-term benefits).

The more intensive and extended early intervention is, the greater its impact on mental test scores and school performance. In the Carolina Abecedarian Project, children at risk for school failure entered full-time, year-round day care in early infancy and remained there until kindergarten. A strong emphasis on adult–child communication—talking to and reading to children daily—undoubtedly contributes to the success of this program.

TRY THIS...

■ Evaluate a day-care program serving infants and toddlers on the basis of standards for developmentally appropriate practice given in the Caregiving Concerns table on page 162. Ask the director to indicate the proportion of children enrolled who are low income. In view of its quality, do you think the program could help protect the mental development of poverty-stricken children? How easy is it for parents with limited resources to find good day care in your community?

programs are not nearly enough to meet the need. Nevertheless, those that exist are a promising beginning in a new effort to prevent serious learning difficulties by starting to help children at a very early age.

BRIEF REVIEW

The mental testing approach arrives at IQ scores that compare a child's performance to that of same-age children. Infant intelligence tests consist largely of perceptual and motor responses, and they predict later intelligence poorly. Speed of habituation and dishabituation to visual stimuli and object permanence show better prediction. Factors in the home environment—stimulation provided by the physical setting and parental encouragement, involvement, and affection—are consistently related to early test scores. High-quality day care supports mental development, whereas low-quality care undermines it. Intensive early intervention for poverty-stricken infants and toddlers results in improvements in IQ.

ASK YOURSELF . . .

■ *Fifteen-month-old Joey's DQ is 100. His mother wants to know what this means and what she should do at home to support his mental development. How would you respond to her questions?*

■ *Using what you learned about brain growth in Chapter 4, explain why intensive intervention for poverty-stricken children starting in the first 2 years has a greater impact on IQ scores than intervention beginning at a later age.*

LANGUAGE DEVELOPMENT DURING THE FIRST TWO YEARS

As cognition improves during infancy, it paves the way for an extraordinary human achievement—language. On the average, children say their first word at 12 months of age, with a range of about 8 to 18 months (Whitehurst, 1982). Once words appear, language develops rapidly. Sometime between 1½ and 2 years, toddlers combine two words. By age 6, they have a vocabulary of about 10,000 words, speak in elaborate sentences, and are skilled conversationalists.

How do children acquire language? Two theories developed in the 1950s took opposing views. One, *behaviorism*, regards language development as entirely due to

Infants are communicative beings from the start, as this interchange between a 3-month-old and her grandfather indicates. How will this child accomplish the awesome task of becoming a fluent speaker of her native tongue during the first few years of life? Theorists disagree sharply on answers to this question. (Sotographs/Liason International)

environmental influences. The second, *nativism* (meaning inborn), assumes that children are prewired to master the intricate rules of their native tongue.

THE BEHAVIORIST PERSPECTIVE

The well-known behaviorist B. F. Skinner (1957) proposed that language, like any other behavior, is acquired through *operant conditioning* (see Chapter 4, page 132). As the baby makes sounds, parents reinforce those that are most like words with smiles, hugs, and speech in return. For example, at 12 months, my older son David could often be heard babbling something like this: "book-a-book-a-dook-a-dook-a-book-a-nook-a-book-aaa." One day, I held up his picture book while he babbled away and said, "Book!" Very soon, David was saying "book-aaa" in the presence of books.

Some behaviorists believe that children rapidly acquire complex utterances, such as whole phrases and sentences, through *imitation* (Whitehurst & Vasta, 1975). Imitation can also combine with reinforcement to promote language learning, as when a parent coaxes, "Say, I want a cookie" and delivers praise and a treat after the toddler responds with "wanna cookie!"

Although reinforcement and imitation contribute to early language development, there is wide agreement that they are not the whole story. As Felicia remarked one day, "It's amazing how creative April is with language. She combines words in ways she's never heard before, such as 'needle it' when she wants me to sew up her teddy bear and 'allgone outside' when she has to come in." Felicia's observations reveal that children say many things that are not directly taught. So conditioning and imitation are best viewed as supporting early language learning rather than fully explaining it.

THE NATIVIST PERSPECTIVE

Linguist Noam Chomsky (1957) proposed a nativist theory that regards the young child's amazing language skill as etched into the structure of the human brain. Focusing on grammar, Chomsky reasoned that the rules of sentence organization are much too complex to be directly taught to or independently discovered by a young child. Instead, he argued, all children are born with a **language acquisition device (LAD),** a biologically based innate system that contains a set of rules common to all languages. It permits children, no matter which language they hear, to understand and speak in a rule-oriented fashion as soon as they have picked up enough words.

Are children biologically primed to acquire language? There is evidence that they are. Recall from Chapter 4 that newborn babies are remarkably sensitive to speech sounds and prefer to listen to the human voice. In addition, children the world over reach the major milestones of language development in a similar sequence. This regularity of development certainly fits with Chomsky's idea of a biologically determined language program.

At the same time, challenges to Chomsky's theory suggest that it, too, provides only a partial account of language development. First, researchers have had great difficulty identifying a single underlying system of grammar for all languages (Tomasello, 1995). Second, careful study of children's first word combinations reveals that they often do not follow adult grammatical rules (Maratsos & Chalkley,

1980). Finally, children do not acquire language quite as quickly as nativist theory suggests. Their progress in mastering many sentence constructions is not immediate, but steady and gradual. This suggests that more learning and discovery are involved than Chomsky assumed.

THE INTERACTIONIST PERSPECTIVE

In recent years, new ideas about language development have emerged, emphasizing that innate abilities and environmental influences *interact* to produce children's extraordinary language achievements. Although several interactionist theories exist, all stress the social context of language learning. An active child, well endowed for acquiring language, observes and participates in social exchanges. From these experiences, children gradually discover the functions and regularities of language. According to the interactionist position, native capacity, a strong desire to interact with others, and a rich linguistic and social environment combine to assist children in building a communicative system (Bohannon, 1993).

As we chart the course of early language growth, we will see a great deal of evidence that supports this new view. Table 5.3 provides an overview of early language milestones that we will take up in the next few sections.

GETTING READY TO TALK

Before babies say their first word, they are preparing for language in many ways. They listen attentively to human speech and make speechlike sounds. And as adults, we can hardly help but respond.

Cooing and Babbling. Around 2 months, babies begin to make vowel-like noises, called **cooing** because of their pleasant "oo" quality. Gradually, consonants are added, and

TABLE 5.3

Milestones of Language Development During the First Two Years

APPROXIMATE AGE	MILESTONE
2 months	Infants coo, making pleasurable vowel sounds.
4 months on	Infants and parents establish joint attention, and parents often verbally label what the baby is looking at.
6–14 months	Infants babble, adding consonants to the sounds of the cooing period and repeating syllables. By 7 months, babbling of hearing infants starts to include sounds of mature spoken languages.
6–14 months	Interaction between parent and baby includes simple games, such as pat-a-cake and peekaboo. By 12 months, babies participate actively. These games provide practice in conversational turn-taking and also highlight the meaning and function of spoken words.
8–12 months	Babbling contains sounds and intonation patterns of the infant's language community. Infants begin using preverbal gestures, such as showing and pointing, to influence the behavior of others. Word comprehension first appears.
12 months	Infants say their first recognizable word.
18–24 months	Vocabulary expands from about 50 to 200 words.
20–26 months	Toddlers combine two words.

This 15-month-old delights in playing peekaboo with her mother. As she participates, she practices the *turn-taking* pattern of human conversation. (Tony Freeman/PhotoEdit)

around 6 months **babbling** appears, in which infants repeat consonant–vowel combinations in long strings, such as "babababab" or "nananananana."

The timing of early babbling seems to be due to maturation, since babies everywhere start babbling at about the same age and produce a similar range of early sounds (Stoel-Gammon & Otomo, 1986). But for babbling to develop further, infants must be able to hear human speech. Around 7 months, babbling starts to include the sounds of mature spoken languages. If a baby's hearing is impaired, these speechlike sounds are greatly delayed, and in the case of deaf infants, they are totally absent (Eilers & Oller, 1994).

When a baby coos or babbles and gazes at you, what are you likely to do? One day, as I stood in line at the post office behind a mother and her 7-month-old daughter, the baby babbled, and three adults—myself and two people beside me—started to talk to the infant. We cooed and babbled ourselves, imitating the baby, and also said such things as "My, you're a big girl, aren't you? Out to help Mommy mail letters today?" Then the baby smiled and babbled all the more. As adults interact with infants and they listen to spoken language, babbling increases. By the end of the first year, it contains the consonant–vowel and intonation patterns of the infant's language community (Boysson-Bardies & Vihman, 1991). Through babbling, babies seem to experiment with a great many sounds that can be blended into their first words.

■ BECOMING A COMMUNICATOR. Besides responding to cooing and babbling, adults interact with infants in many other situations. By age 4 months, infants start to gaze in the same direction adults are looking, and adults

follow the baby's line of vision. When this happens, parents often comment on what the infant sees, labeling the environment for the baby. Infants who experience this kind of *joint attention* are likely to talk earlier and show faster vocabulary development (Dunham & Dunham, 1992; Dunham, Dunham, & Curwin, 1993).

Around 6 months, interaction between parent and baby begins to include *turn-taking games,* such as pat-a-cake and peekaboo. At first, the parent starts the game and the baby is an amused observer. By 12 months, babies participate actively, exchanging roles with the parent. As they do so, they practice the turn-taking pattern of human conversation (Ratner & Bruner, 1978).

At the end of the first year, as infants become capable of intentional behavior, they use *preverbal gestures* to influence the behavior of others (Bates, 1979; Fenson et al., 1994). For example, Byron held up a toy to show it and pointed to the cupboard when he wanted a cookie. Lisa responded to his gestures and also labeled them ("Oh, you want a cookie!"). In this way, toddlers learn that using language leads to desired results. Soon they utter words along with their reaching and pointing gestures, the gestures recede, and spoken language is underway (Goldin-Meadow & Morford, 1985).

■ **FIRST WORDS**

Ask several parents to list their toddler's first words. Note how the words build on the sensorimotor foundations that Piaget described. Children's earliest words usually refer to important people ("Mama," "Dada"), objects that move (such as "car," "ball," "cat"), familiar actions ("bye-bye," "up," "more"), or outcomes of familiar act-

Using a *preverbal gesture,* this infant points to show something to his parents. Soon words will be uttered along with these gestures and the gestures will recede as the child makes the transition to spoken language. (Fuji Fotos/The Image Works)

ions ("dirty," "wet," "hot"). In their first 50 words, toddlers rarely name things that just *sit there,* like table or vase (Nelson, 1973).

Some early words are linked to specific cognitive achievements. For example, toddlers begin to use disappearance words, like "all gone," at about the same time they master advanced object permanence problems. And success and failure expressions, such as "There!" and "Uh-oh!", appear when toddlers can solve sensorimotor problems suddenly, in Piaget's Substage 6. According to one pair of researchers, "Children seem to be motivated to acquire words that are relevant to the particular cognitive problems they are working on at the moment" (Gopnik & Meltzoff, 1986, p. 1052).

When young children first learn a new word, they sometimes apply it too narrowly, an error called **underextension.** For example, at 16 months, April used "doll" only to refer to the worn and tattered doll that she carried around with her much of the day. A more common error is **overextension**—applying a word to a wider collection of objects and events than is appropriate. For example, Rachel used "car" for buses, trains, trucks, and fire engines. Toddlers' overextensions reflect their sensitivity to categories. They apply a new word to a group of similar experiences, such as "car" to wheeled objects and "open" to opening a door, peeling fruit, and undoing shoelaces (Behrend, 1988). As their vocabularies enlarge, children make finer distinctions, and overextensions gradually disappear.

Overextensions illustrate another important feature of language development: the distinction between language *production* (the words children use) and language *comprehension* (the words children understand). Children overex-

tend many more words in production than they do in comprehension. That is, a 2-year-old may refer to trucks, trains, and bikes as "car" but look at or point to these objects correctly when given their names (Naigles & Gelman, 1995). At all ages, comprehension develops ahead of production. This tells us that failure to say a word does not mean that toddlers do not understand it. If we rely only on what children say, we will underestimate their knowledge of language.

THE TWO-WORD UTTERANCE PHASE

At first, toddlers add to their vocabularies slowly, at a rate of 1 to 3 words a month. Between 18 and 24 months, a spurt in vocabulary usually takes place. As memory, categorization, and representation improve, many children add 10 to 20 new words a week (Fenson et al. 1994; Reznick & Goldfield, 1992). When vocabulary approaches 200 words, toddlers start to combine two words, such as "Mommy shoe," "go car," and "more cookie." These two-word utterances are called **telegraphic speech** because, like a telegram, they leave out smaller and less important words. Children the world over use them to express an impressive variety of meanings, as Table 5.4 reveals.

Two-word speech contains some simple formulas, such as "want + *X*" and "more + *X*" (with many different words inserted in the *X* position). This does not mean that toddlers grasp the rules of language. Although they rarely make gross grammatical errors (saying "chair my" instead of "my chair"), they can be heard violating the rules. For example, at 20 months, Rachel said "more hot" and "more read,"

TABLE 5.4

Common Meanings Expressed in Toddlers' Two-Word Utterances

MEANING	EXAMPLE
Agent + action	"Tommy hit"
Action + object	"Give cookie"
Agent + object	"Mommy truck" (meaning "Mommy push the truck.")
Action + location	"Put table" (meaning "Put X on the table.")
Entity + location	"Daddy outside"
Entity + attribute	"Big ball"
Possessor + possession	"My truck"
Demonstrative + entity	"That doggie"
Notice + noticed object	"Hi mommy," "Hi truck"
Recurrence	"More milk"
Nonexistent object	"No shirt," "No more milk"

Source: Brown, 1973.

combinations that are not acceptable in English grammar (Braine, 1976). But it does not take long for children to figure out grammatical rules. As we will see in Chapter 7, the beginnings of grammar are in place by age 2½.

INDIVIDUAL AND GROUP DIFFERENCES

Each child's progress in acquiring language results from a complex blend of biological and environmental influences. For example, earlier we saw that Byron's spoken language was delayed, in part because Lisa pressured him a great deal. But Byron is also a boy, and many studies show that girls are slightly ahead of boys in early vocabulary growth (Fenson et al., 1994; Jacklin & Maccoby, 1983). Personality also makes a difference. Reserved, cautious toddlers often wait until they understand a great deal before trying to speak. When, finally, they do speak, their vocabularies grow rapidly (Nelson, 1973).

Young children have unique styles of early language learning. April, like most toddlers, used a **referential style.** Her vocabulary consisted mainly of words that referred to objects. In contrast, Rachel used an **expressive style.** She produced many more pronouns and social formulas, such as "stop it," "thank you," and "I want it," which she uttered as compressed phrases, much like single words (as in "Iwannit"). These styles reflect early ideas about the functions of language. April had an especially active interest in exploring objects, and her parents eagerly labeled them. Consequently, she thought words were for naming things. In contrast, Rachel believed words were for talking about the feelings and needs of herself and other people. Beth

often engaged Rachel in social routines, perhaps because Japanese culture stresses the importance of membership in the social group (Fernald & Morikawa, 1993). April's vocabulary grew faster, since all languages contain many more object labels than social phrases (Bates, Bretherton, & Snyder, 1988; Nelson, 1973). The two toddlers' vocabularies gradually became more similar as they revised their first notions of what language is all about.

At what point should parents be concerned if their child does not talk or says very little? If a toddler's language is greatly delayed when compared to the norms in Table 5.3, then parents should consult the child's doctor or a speech and language therapist. Some toddlers who do not follow simple directions could have a hearing problem. A child over age 2 who has great difficulty putting thoughts into words may have a serious language disorder that requires immediate treatment.

SUPPORTING EARLY LANGUAGE DEVELOPMENT

There is little doubt that children are specially prepared for acquiring language, since no other species can develop as flexible and creative a capacity for communication as we can (Berko Gleason, 1997). At the same time, a great deal of evidence fits the interactionist view that a rich social environment builds on young children's natural readiness to speak their native tongue. The Caregiving Concerns table below summarizes ways that caregivers can consciously support early language learning. They also do so unconsciously—through a special style of speech.

CAREGIVING CONCERNS

Supporting Early Language Learning

SUGGESTION	CONSEQUENCE
Respond to coos and babbles with speech sounds and words	Encourages experimentation with sounds that can later be blended into first words. Provides experience with turn-taking pattern of human conversation.
Establish joint attention and comment on what child sees	Predicts earlier onset of language and faster vocabulary development.
Play social games, such as pat-a-cake and peekaboo	Provides experience with turn-taking pattern of human conversation. Permits pairing of words with actions they represent.
Engage toddlers in joint make-believe play	Promotes all aspects of conversational dialogue.
Engage toddlers in frequent conversations	Predicts faster early language development and academic competence during the school years.
Read to toddlers often, engaging them in dialogues about picture books	Provides exposure to many aspects of language, including vocabulary, grammar, communicative conventions, and information about print and story structures.

Sources: Berk, 1994a; Dunham & Dunham, 1992; Dunham, Dunham, & Curwin, 1993; Hart & Risely, 1995; Ratner & Bruner, 1978; Walker et al., 1994.

This Nepalese mother speaks to her baby daughter in short, clearly pronounced sentences with high-pitched, exaggerated intonation. Adults in many countries use this form of language, called motherese, with infants and toddlers. It eases the task of early language learning. (David Austen/Stock Boston)

Adults in many countries speak to young children in **motherese** (also called **child-directed speech**), a form of language made up of short sentences with high-pitched, exaggerated expression and very clear pronunciation (Fernald et al., 1989). Motherese also contains many simplified words, such as "night-night," "bye-bye," "daddy," and "tummy," that are easy for toddlers to pronounce. Speakers of motherese often repeat phrases, ask questions, and give directions, perhaps as a way of checking to see if their message has been properly received. Parents do not seem to be deliberately trying to teach children to talk when they use motherese, since adults often communicate with foreigners in a similar way. Here is an example of Felicia speaking motherese to 18-month-old April:

Felicia:	"Time to go, April."
April:	"Go car."
Felicia:	"Yes, time to go in the car. Where's your jacket?"
April:	(looks around, walks to the closet) "Dacket!" (pointing to her jacket)

Felicia:	"There's that jacket! Let's put it on. (She helps April into the jacket.) On it goes! Let's zip up. (Zips up the jacket.) Now, say bye-bye to Byron and Rachel."
April:	"Bye-bye, By-on."
Felicia:	"What about Rachel? Bye to Rachel?"
April:	"Bye-bye, Ta-tel (Rachel)."
Felicia:	"Where's your doll? Don't forget your doll."
April:	(looks around)
Felicia:	"Look by the sofa. See? Go get the doll. By the sofa." (April gets the doll.)

From birth on, children prefer to listen to motherese over other kinds of adult talk (Cooper & Aslin, 1994; Fernald, 1993). And parents constantly fine-tune it to fit their children's needs. Notice how Felicia used an utterance length just ahead of April's, creating a sensitive match between her own speech and what April was capable of understanding and producing (Murray, Johnson, & Peters, 1990).

Many features of motherese support early language development. For example, parents who frequently repeat part of their own or the child's previous utterance and use simple questions have 2-year-olds who make faster language progress (Barnes et al., 1983; Hoff-Ginsburg, 1986). But this does not mean that we should deliberately load our speech with repetitions, questions, and other characteristics of motherese! These qualities occur naturally as adults draw young children into dialogues, accepting their attempts to talk as meaningful and worthwhile.

Dialogues about picture books are an especially effective way of stimulating young children's language development. As this father talks about the pictures with his 2-year-old daughter, he exposes her to great breadth of language and literacy knowledge. (Tony Freeman/PhotoEdit)

Conversational give-and-take between parent and toddler is one of the best predictors of early language development and academic competence during the school years. It provides many examples of speech just ahead of the child's current level as well as a sympathetic environment in which children can try out new skills (Hart & Risley, 1995; Walker et al., 1994). Dialogues about picture books are an especially effective way of conversing with young children. They expose toddlers to great breadth of language and literacy knowledge, from vocabulary, grammar, and communicative conventions to information about print and story structures (Crain-Thoreson & Dale, 1992; Whitehurst et al., 1994).

Do social experiences that promote language development remind you of ones discussed earlier in this chapter that strengthen cognitive development in general? Notice how motherese and parent–child conversation create a *zone of proximal development* in which children's language expands. In contrast, impatience and rejection of children's efforts to talk lead them to stop trying and result in immature language skills (Nelson, 1973). In the next chapter, we will see that sensitivity to children's needs and capacities supports their emotional and social development as well.

ASK YOURSELF . . .

■ *Erin's first words included see, give, and thank you, and her vocabulary grew slowly during the second year. What style of early language learning did she display, and what factors might explain it?*

SUMMARY

PIAGET'S COGNITIVE-DEVELOPMENTAL THEORY

According to Piaget, how do schemes change over the course of development?

■ In Piaget's theory, by acting directly on the environment, children move through four stages in which psychological structures, or **schemes,** achieve a better fit with external reality.

■ Schemes change in two ways. The first is through **adaptation,** which is made up of two complementary activities—**assimilation** and **accommodation.** The second is through **organization,** the internal rearrangement of schemes into a strongly interconnected cognitive system.

Describe the major cognitive achievement of the sensorimotor stage.

■ Piaget's **sensorimotor stage** is divided into six substages. Through the **circular reaction,** the newborn baby's reflexes are gradually transformed into the more flexible action patterns of the older infant and finally into the representational schemes of the 2-year-old child.

During Substage 4, infants develop **intentional,** or **goal-directed, behavior** and begin to understand **object permanence.** Throughout the first year and a half, infants and toddlers engage in **functional play.** By Substage 6, they are capable of **mental representation,** as shown by **deferred imitation** and **make-believe play.**

What does recent research have to say about the accuracy of Piaget's sensorimotor stage?

■ Important capacities, such as secondary circular reactions, object permanence, deferred imitation, and representation, emerge earlier than Piaget believed. Infants do not have to construct all aspects of experience through motor activity. Some may be prewired into the brain from the start, and others may be constructed through purely perceptual learning.

INFORMATION PROCESSING IN THE FIRST TWO YEARS

Describe the information-processing view of cognitive development and Atkinson and Shiffrin's model of the information-processing system.

■ Unlike Piaget's stage theory, information processing views development as a continuous process; the cognitive approach of children and adults is assumed to be much the same. Information-processing researchers study many aspects of thinking. They want to know exactly what individuals of different ages do when faced with a task or problem.

■ According to Atkinson and Shiffrin, the human mental system is divided into three parts: the **sensory register; working,** or **short-term, memory;** and **long-term memory.** As information flows through each, **control processes,** or **mental strategies,** operate on it to increase the chances that information will be retained.

What changes in attention, memory, and categorization take place over the first 2 years?

■ With age, infants attend to more aspects of the environment, take information in more quickly, and remember experiences longer. Young infants are capable of **recognition** memory, and by 7 months, they can **recall** past events.

■ Infants remember information in a remarkably orderly fashion. During the first year, they group stimuli into increasingly complex, meaningful categories. By the second year, they become active categorizers, spontaneously sorting objects during their play.

Describe the contributions and limitations of the information-processing approach to our understanding of cognitive development.

■ Information processing has contributed greatly to our appreciation of the baby's cognitive capacities. Its findings join with other research in challenging Piaget's view of babies as purely sensorimotor beings. However, information processing has not yet provided a broad, comprehensive theory of children's thinking.

THE SOCIAL CONTEXT OF EARLY COGNITIVE DEVELOPMENT

How does Vygotsky's concept of the zone of proximal development expand our understanding of early cognitive development?

■ According to Vygotsky's sociocultural theory, complex mental functions originate in social interaction. Through the support and guidance of more skilled partners, infants master tasks within the **zone of proximal development**—ones just ahead of their current capacities.

INDIVIDUAL DIFFERENCES IN EARLY MENTAL DEVELOPMENT

Describe the mental testing approach, the meaning of intelligence test scores, and the extent to which infant tests predict later performance.

■ The mental testing approach measures intellectual development in an effort to predict future performance. **Intelligence quotients,** or **IQs,** are scores on mental tests that compare a child's performance to that of same-age children.

■ Infant tests consist largely of perceptual and motor responses; they predict later intelligence poorly. As a result, scores on infant tests are called **developmental quotients,** or **DQs,** rather than IQs. Speed of habituation and dishabituation to visual stimuli and object permanence, which tap basic cognitive processes, make better predictions.

Discuss environmental influences on early mental development, including home, day care, and early intervention for at-risk infants and toddlers.

■ Research with the **Home Observation for Measurement of the Environment (HOME)** shows that a stimulating home environment and parental encouragement, involvement, and affection repeatedly predict early mental test scores, no matter what the child's social class and ethnic background.

■ The quality of infant and toddler day care also has a major impact on mental development. Standards for **developmentally appropriate practice** specify program characteristics that meet the developmental needs of young children. Intensive early intervention can prevent the gradual declines in intelligence so often experienced by poverty-stricken children.

LANGUAGE DEVELOPMENT DURING THE FIRST TWO YEARS

Describe three theories of language development, and indicate the emphasis each places on innate abilities and environmental influences.

■ Three theories provide different accounts of how young children develop language. According to the behaviorist perspective, parents train children in language skills through operant conditioning and imitation. In contrast, Chomsky's nativist view regards children as naturally endowed with a **language acquisition device (LAD).** New interactionist theories suggest that innate abilities and social contexts combine to promote language development.

Describe major milestones of language development in the first 2 years, individual differences, and ways adults can support infants' and toddlers' emerging capacities.

■ During the first year, a great deal of preparation for language takes place. Infants begin **cooing** at 2 months and **babbling** around 6 months. Adults encourage language progress by responding to infants' coos and babbles, playing turn-taking games with them, and acknowledging their preverbal gestures.

■ Around 12 months, toddlers say their first word. Young children often make errors of **underextension** and **overextension.** Between 18 months and 2 years, a spurt in vocabulary often occurs, and two-word utterances called **telegraphic speech** appear. At all ages, language comprehension is ahead of production.

■ Individual differences in early language development exist. Girls show faster progress than boys, and reserved, cautious toddlers may wait for a period of time before trying to speak. Most toddlers use a **referential style** of language learning, in which most early words are names for objects. A few use an **expressive style,** in which pronouns and social formulas are common and vocabulary grows more slowly.

■ Adults the world over speak to young children in **motherese,** or **child-directed speech,** a simplified form of language that is well suited to their learning needs. Motherese occurs naturally when parents engage toddlers in conversations that accept and encourage their early efforts to talk.

IMPORTANT TERMS AND CONCEPTS

sensorimotor stage (p. 146)

scheme (p. 146)

adaptation (p. 147)

assimilation (p. 147)

accommodation (p. 147)

organization (p. 147)

circular reaction (p. 147)

intentional, or goal-directed, behavior (p. 149)

object permanence (p. 149)

mental representation (p. 150)

deferred imitation (p. 150)

functional play (p. 150)

make-believe play (p. 150)

control processes, or mental strategies (p. 153)

sensory register (p. 153)

working, or short-term, memory (p. 153)

long-term memory (p. 154)

recognition (p. 155)

recall (p. 155)

zone of proximal development (p. 157)

intelligence quotient, or IQ (p. 160)

developmental quotient, or DQ (p. 160)

Home Observation for Measurement of the Environment (HOME) (p. 161)

developmentally appropriate practice (p. 161)

language acquisition device (LAD) (p. 165)

cooing (p. 165)

babbling (p. 166)

underextension (p. 167)

overextension (p. 167)

telegraphic speech (p. 167)

referential style (p. 168)

expressive style (p. 168)

motherese or child-directed speech (p. 169)

FOR FURTHER INFORMATION AND HELP

INFANT AND TODDLER DEVELOPMENT AND EDUCATION

Association for Childhood Education International (ACEI)
11501 Georgia Avenue, Suite 312
Wheaton, MD 20902
(301) 942-2443
Web site: *www.udel.edu/ bateman/acei*
Organization interested in promoting sound educational practice from infancy through early adolescence. Student membership is available and includes a subscription to Childhood Education, *a bimonthly journal covering research, practice, and public policy issues.*

National Association for the Education of Young Children (NAEYC)
1509 16th Street, N.W.
Washington, DC 20036-1426
(202) 232-8777
(800) 424-2460
Web site: *www.naeyc.org/naeyc*

Organization open to all individuals interested in acting on behalf of young children's needs, with primary focus on educational services. Student membership is available and includes a subscription to Young Children, *a bimonthly journal covering theory, research, and practice in infant and early childhood development and education.*

EARLY INTERVENTION

High/Scope Educational Research Foundation
600 N. River Street
Ypsilanti, MI 48198-2898
(313) 485-2000
Web site: *www/highscope.org*
Devoted to improving development and education from infancy through the high school years. Has designed a parent–infant education program. Conducts longitudinal research to determine the effects of early intervention on development.

DAY CARE

Child Care Resource and Referral, Inc.
2116 Campus Drive, S.E.
Rochester, MN 55904
(800) 462-1660
Represents more than 260 local agencies that work for high-quality day care and provide information on available services.

National Association for Family Child Care
206 6th Avenue, Suite 900
Des Moines, IA 50309
(515) 282-8192
Web site: *www.assoc-mgmt.com/users/nafcc*
Organization open to caregivers, parents, and other individuals involved or interested in family day care. Serves as a national voice that promotes high-quality day care.

Emotional and Social Development in Infancy and Toddlerhood

Lisa, Beth, and Felicia's monthly conversations often focused on the emotional and social sides of their infants' development. As the babies reached 8 months of age, Beth reported, "For some reason, Rachel's become more fearful. Recently, I took her to the airport to meet my parents, who were arriving from Japan. When they tried to hug her, she didn't return their enthusiasm, as she would have a month or two ago. Instead, she turned away and buried her head against my shoulder. Several days later, I left Rachel with my parents for several hours—the first time I'd been away from her since she was born. She wailed as soon as she saw me head for the door. And when I returned, Rachel seemed angry. She insisted on being held but also pushed me away and kept crying for 5 or 10 minutes."

Lisa and Felicia also noted that Byron and April showed greater wariness of strangers and wanted to stay close to familiar adults. At the same time, each baby seemed more willful. At 8 months Byron actively resisted when Lisa took away a table knife he had managed to reach. And he could no longer be consoled by a variety of toys that she offered in its place. Taken together, these reactions reflect two related aspects of personality that begin to develop during the first 2 years: *close ties to others* and a *sense of self*—an awareness of one's own separateness and uniqueness.

Our discussion begins with major theories that provide an overall picture of personality development during infancy and toddlerhood. Then we take a look at factors that contribute to these changes. First, we chart the general course of emotional development. As we do so, we will discover why fear and anger became more apparent in Byron, Rachel, and April by the end of the first year. Second, our attention turns to individual differences in temperament and personality. We will examine biological and environmental contributions to these differences and their consequences for future development.

Next, we take up attachment to the caregiver, the child's first affectional tie. We will see how the feelings of security that grow out of this important bond provide a vital source of support for the child's exploration, sense of independence, and expanding social relationships.

Finally, we focus on early self-development. By the end of toddlerhood, April recognized herself in mirrors and photographs, labeled herself as a girl, and showed the beginnings of self-control. "Don't touch!" she instructed herself one day as she resisted the desire to pull a lamp cord out of its socket. Cognitive advances combine with social experiences to produce these changes during the second year.

THEORIES OF INFANT AND TODDLER PERSONALITY

In Chapter 1, we pointed out that psychoanalytic theory is no longer in the mainstream of human development research. But one of its lasting contributions has been its ability to capture the essence of personality development during each phase of life. Sigmund Freud, founder of the psychoanalytic movement, believed that psychological health and maladjustment could be traced to the early years—in particular, to the quality of the child's relationships with parents.

In his *psychosexual theory*, Freud focused on how parents help infants and toddlers discharge instinctual drives originating from the oral and then the anal zone of the body. Freud's limited concern with the channeling of instincts and neglect of important experiences after the early years came to be heavily criticized. But the basic outlines of his theory were elaborated by several noted psychoanalysts who came after him. The leader of these neo-Freudians is Erik Erikson, whose *psychosocial theory* was introduced in Chapter 1.

In the following sections we take a closer look at the tasks of infancy and toddlerhood, as Erikson and a second well-known psychoanalyst—Margaret Mahler—saw them. Together, the two theorists provide us with perceptive insights into early emotional and social development.

According to Erikson, basic trust grows out of the quality of the mother's relationship with the baby. A mother who relieves her infant's discomfort promptly and holds the baby tenderly, especially during feedings, promotes basic trust. *(Jeffrey W. Myers/Stock Boston)*

ERIK ERIKSON: TRUST AND AUTONOMY

Erikson (1950) characterized each Freudian stage as an inner conflict that is resolved positively or negatively, depending on the child's experiences with caregivers. When parenting supports the child's needs, the first year leads to feelings of trust and the next 2 years to a sense of autonomy—early attitudes that provide the foundation for healthy psychological development throughout life.

BASIC TRUST VERSUS MISTRUST.

Freud called the first year the **oral stage** because infants obtain pleasure through the mouth—at first, by sucking, and later, after teeth erupt, by biting and chewing. Gratification of the baby's need for food and pleasurable oral stimulation rests in the hands of the mother, whose task is to provide the right amount of oral satisfaction.

Erikson accepted Freud's emphasis on feeding, but he expanded and enriched Freud's view. A healthy outcome during infancy, Erikson believed, does not depend on the *amount* of food or oral stimulation offered, but rather on the *quality* of the caregiver's behavior. A mother who supports her baby's development holds the infant gently during feedings, patiently waits until the baby has had enough milk, and weans when the infant shows less interest in sucking.

Erikson recognized that no mother can be perfectly in tune with her baby's needs. Many factors affect her responsiveness—her feelings of personal happiness, her momentary life condition (for example, whether she has one or several small children to care for), and child-rearing practices encouraged by her culture. But when the *balance of care* is sympathetic and loving, then the psychological conflict of the first year—**basic trust versus mistrust**—is resolved on the positive side. The trusting infant expects the world to be good and gratifying, so she feels confident about venturing out and exploring it. The mistrustful baby cannot count on the kindness and compassion of others, so she protects herself by withdrawing from people and things around her.

AUTONOMY VERSUS SHAME AND DOUBT.

During Freud's **anal stage,** in the second year, instinctual energies shift to the anal region. Toddlers take pleasure in retaining and releasing urine and feces at will. At the same time, society requires that elimination occur at appropriate times and places. As a result, Freud viewed toilet training, in which children must bring their anal impulses in line with social requirements, as crucial for personality development. If parents insist on training children before they are physically ready or wait too long before expecting self-control, an unresolved battle of wills between parent and child begins.

Erikson agreed that the parent's manner of toilet training is important. But he viewed it as only one of many

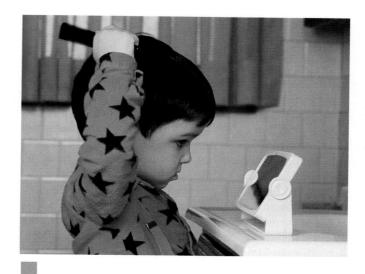

This 2-year-old is intent on combing his hair. Toddlers who are allowed to decide and do things for themselves in appropriate situations develop a sense of autonomy—the feeling that they can control their bodies and act competently on their own. *(Brent Jones/Stock Boston)*

influential experiences for newly walking, talking toddlers. Their familiar refrains—"No!" and "Do it myself!"—reveal that they want to decide for themselves, not just in toileting, but in other situations as well. The great conflict of this stage, **autonomy versus shame and doubt,** is resolved favorably when parents provide young children with suitable guidance and reasonable choices. A self-confident, secure 2-year-old has been encouraged not just to use the toilet, but to eat with a spoon and to help pick up his toys. His parents do not criticize or attack him when he fails at these new skills. And they meet his assertions of independence with tolerance and understanding. For example, they grant him an extra 5 minutes to finish his play before leaving for the grocery store and wait patiently while he tries to zip his jacket.

According to Erikson, the parent who is over- or under-controlling in toileting is likely to be so in other aspects of the toddler's life. The outcome is a child who feels forced and shamed and doubts his ability to control his impulses and act competently on his own.

MARGARET MAHLER: SEPARATION–INDIVIDUATION

Erikson's theory describes how sensitive channeling of the baby's drives leads to positive attitudes toward others (trust) and good feeling about the self (autonomy). Mahler carries this theme further, focusing on how the infant's early relationship with the mother provides the foundation for a sense of self in the second year of life (Mahler, Pine, & Bergman, 1975). According to Mahler, awareness of the self as separate and unique results from events during two phases of development: symbiosis and separation–individuation.

■ SYMBIOSIS. During the first 2 months, babies are barely aware of the surrounding world. They spend most of the day asleep, waking when hunger and other tensions cause them to cry and sinking back into sleep when discomforts are relieved. But around the second month, they enter the phase of **symbiosis** (meaning the blending of two people into an intimate, harmonious relationship). At this time, infants become increasingly alert and interested in sights and sounds around them (see Chapter 4). But unlike the older child and adult, they do not realize these events exist outside themselves. Instead, the self and surrounding world (including the mother) are completely fused. According to Mahler, this oneness is a necessary first step in developing a sense of self. The mother promotes it by responding to the infant's cries, coos, and smiles promptly and with positive emotional tone. The more she does so, the more confidently the infant will separate from her during the next phase. In contrast, infants handled harshly and impatiently are likely to have great difficulty distancing themselves from their mothers.

■ SEPARATION–INDIVIDUATION. In Mahler's second phase, **separation–individuation,** the baby's motor capacity to move away from the mother triggers *individuation,* or self-awareness. The process of separating from the mother begins around 4 to 5 months, when the infant, held in the mother's arms, leans away from her body to scan the environment. But the decisive events of this phase are crawling and then walking.

Newly crawling 8- to 10-month-olds venture only a short distance from the mother. They look back frequently for reassurance and return to the mother's side to reexperience the safety and security of close body contact. Yet as crawling babies come and go, they experience their mothers from a new, distant vantage point and become dimly aware of their own separateness.

Walking brings a dramatic advance in individuation. As toddlers enjoy greater freedom of movement and delight in exploration, they become even more conscious that the mother and the self are distinct beings. Around 18 months, this realization is fullblown, and at first it is frightening. Older toddlers may engage in all kinds of behaviors aimed at resisting and undoing this separateness—following and clinging to the mother, filling her lap with objects from the surrounding environment, and darting away in hopes of being chased, caught, and reminded of her continuing commitment. According to Mahler, the temper tantrums that often occur around this time—called the "terrible twos" by many parents—are signs of the new self's desire to assert itself, mixed with feelings of helpless dependence at not being able to manage all the challenges of the environment.

The mother's patience and reassurance help toddlers surmount this temporary crisis. Between 2 and 3 years, most children develop a sturdy sense of themselves as separate persons. And gains in representation and language (see Chapter 5) enable them to create a positive inner image of the mother that they can rely on in her absence, making separations easier.

SIMILARITIES BETWEEN ERIKSON'S AND MAHLER'S THEORIES

As you read about Erikson's and Mahler's theories, undoubtedly you noticed several common themes. Each regards warm, sensitive parenting as vital for personality development. And each views toddlerhood as a time of budding selfhood. For Erikson, it is a stage in which children achieve autonomous control over basic impulses; for Mahler, it is a period in which they learn to separate confidently from the parent. Both theorists also agree that if children emerge from the first few years without sufficient trust in caregivers and a healthy sense of individuality, the seeds are sown for adjustment problems. Adults who have difficulty establishing intimate ties to others, who are overly dependent on a loved one, or who continually doubt their own ability to meet new challenges may not have fully mastered the tasks of trust, autonomy, and individuation during infancy and toddlerhood.

BRIEF REVIEW

Erikson's and Mahler's psychoanalytic theories provide an overview of the emotional and social tasks of infancy and toddlerhood. According to Erikson, basic trust and autonomy grow out of warm, supportive parenting and reasonable expectations for impulse control during the second year. Mahler's theory suggests that sensitive exchange of emotional signals between mother and baby leads to a symbiotic bond, which provides the foundation for a confident sense of self as infants crawl and then walk on their own. Both theorists agree that the development of trust and individuality during infancy and toddlerhood have lasting consequences for personality development.

ASK YOURSELF . . .

■ *Derek's mother fed him in a warm and loving manner during the first year, but when he became a toddler, she kept him in a playpen for many hours because he got into so much mischief while exploring freely. Use Erikson's theory to evaluate Derek's early experiences.*

■ *Around 18 months, Betina became clingy and dependent. She followed her mother around the house and asked to be held often. How would Mahler account for Betina's behavior? How should Betina's parents respond?*

EMOTIONAL DEVELOPMENT DURING THE FIRST TWO YEARS

In the previous chapter, I suggested that you observe babies' increasingly effective schemes for controlling the environment and ways that adults support cognitive and language development. Now focus on another aspect of infant and caregiver behavior: the exchange of emotions. While you observe, note the various emotions the infant displays, the cues you rely on to interpret the baby's feelings, and how the caregiver responds.

Researchers have conducted many observations like these to find out how babies communicate their feelings and interpret those of others. They have discovered that emotions play a powerful role in organizing the events that Erikson and Mahler regarded as so important—relationships with caregivers, exploration of the environment, and discovery of the self (Barrett & Campos, 1987; Campos et al., 1983; Saarni, Mumme, & Campos, 1997).

DEVELOPMENT OF SOME BASIC EMOTIONS

Do infants come into the world with the ability to express a wide variety of emotions? Controversy surrounds this question. Some investigators believe that all the **basic emotions**—those that can be directly inferred from facial expressions, such as happiness, interest, surprise, fear, anger, sadness, and disgust—are present in the first few weeks of life (Campos et al., 1983; Izard, 1991). Others regard the newborn baby's emotional life as very limited. For example, according to one view, separate emotions gradually emerge over the first year from two global arousal states: the tendency to approach pleasant and withdraw from unpleasant stimulation (Fox, 1991; Sroufe, 1979).

Still, most researchers agree that signs of almost all the basic emotions are present in infancy (Izard et al., 1995; Malatesta-Magai, Izard, & Camras, 1991; Sroufe, 1979). Between 2 and 3 months, as babies inspect the internal features of faces, they start to respond in kind to an adult's facial expressions. By the middle of the first year, face, voice, and posture form distinct patterns that are clearly related to social events (Haviland & Lelwica, 1987; Toda & Fogel, 1993; Weinberg & Tronick, 1994). For example, 6-month-olds typically respond to their mother's playful interaction with a joyful face, pleasant cooing, and excited movements. In contrast, an unresponsive mother is likely to evoke either a sad face and fussy sounds or an angry face, crying, and "pick-me-up" gestures. In sum, by the middle of the first year, emotional expressions are well organized and specific—and therefore able to tell us a great deal about the infant's internal state.

Three basic emotions—happiness, anger, and fear—have received the most research attention. Refer to Table 6.1 for an overview of changes in these emotions as well as others we will take up in this chapter.

■ HAPPINESS. Happiness—first in terms of blissful smiles and later through exuberant laughter—contributes to many aspects of development. Infants smile and laugh when they conquer new skills, expressing their delight in physical and cognitive mastery. The smile also encourages caregivers to be affectionate and stimulating, so the baby will smile even more. Happiness binds parent and baby into

---

TABLE 6.1

Milestones of Emotional Development During the First Two Years

APPROXIMATE AGE	MILESTONE
Birth	Infants show signs of almost all the basic emotions.
2–3 months	Infants engage in social smiling and respond in kind to adults' facial expressions.
3–4 months	Infants begin to laugh at very active stimuli.
6–8 months	Emotional expressions are well organized and clearly related to social events. Infants start to become angry more often and in a wider range of situations. Fear, especially stranger anxiety, begins to rise. Attachment to the familiar caregiver is clearly evident, and separation anxiety appears.
8–10 months	Infants perceive facial expressions as organized patterns, and ability to detect their meaning improves. Social referencing appears.
10–12 months	Infants laugh at subtle elements of surprise.
18–24 months	Toddlers display self-conscious emotions of shame, embarrassment, and pride. A vocabulary for talking about feelings develops rapidly. Emotional self-regulation improves. First signs of empathy appear.

This 1-year-old smiles and laughs as she masters new motor skills. Her exuberance leads caregivers to return her joy and encourage her efforts. As a result, she approaches her surroundings with enthusiasm. (John Coletti/The Picture Cube)

a warm, supportive relationship and fosters the infant's developing competence.

During the early weeks, newborn babies smile when full, during sleep, and in response to gentle touches and sounds, such as stroking the skin, rocking, and the mother's soft, high-pitched voice. By the end of the first month, infants start to smile at interesting sights, but these must be dynamic and eye-catching, such as a bright object jumping suddenly across the baby's field of vision. Between 6 and 10 weeks, the human face evokes a broad grin called the **social smile** (Sroufe & Waters, 1976). Not surprisingly, these changes in smiling parallel the development of infant perceptual capacities—in particular, babies' increasing sensitivity to visual patterns, including the human face (see Chapter 4).

Laughter, which appears around 3 to 4 months, reflects faster processing of information than smiling. But like smiling, the first laughs occur in response to very active stimuli, such as the mother saying playfully, "I'm gonna get you!" and kissing the baby's tummy. Over time, babies laugh at events with subtler elements of surprise. At 10 months, Byron chuckled as Lisa played a silent game of peekaboo. At 1 year, he laughed heartily as she crawled on all fours and then walked like a penguin (Sroufe & Wunsch, 1972).

Around the middle of the first year, infants smile and laugh more often when interacting with familiar people, a preference that strengthens the parent–child bond. During the second year, the smile becomes a deliberate social signal. Toddlers break their play with an interesting toy to communicate their delight to an attentive adult (Jones & Raag, 1989).

■ ANGER AND FEAR. Newborn babies respond with generalized distress to a variety of unpleasant experiences, including hunger, painful medical procedures, changes in body temperature, and too much or too little stimulation. During the first 2 months, fleeting expressions of anger appear as babies cry. These gradually increase in frequency and intensity from 4 to 6 months into the second year. Older babies also show anger in a wider range of situations—for example, when an object is taken away, the caregiver leaves for a brief time, or they are put down for a nap (Camras et. al., 1992; Stenberg & Campos, 1990).

Like anger, fear rises during the second half of the first year. Older infants often hesitate before playing with a new toy, and research with the visual cliff reveals that they start to show fear of heights around this time (see Chapter 4). But the most frequent expression of fear is to unfamiliar adults, a response called **stranger anxiety.** Many infants and toddlers are quite wary of strangers, although the reaction does not always occur. It depends on several factors: the infant's temperament (some babies are generally more fearful), past experiences with strangers, and the situation in which baby and stranger meet (Thompson & Limber, 1991).

To understand these influences, let's return to Rachel's fearful withdrawal from her grandparents, described at the beginning of this chapter. From birth, Rachel was continuously cared for by her mother. She had little opportunity to get to know strange adults. Also, she met her grandparents in an unfamiliar environment (a crowded airport), and they rushed over and tried to hold her. Under these conditions, babies are most likely to display fearful reactions. Later, at home, Rachel watched with interest and approached as her grandmother sat quietly on the sofa, smiled, and held out a teddy bear. A familiar setting, the opportunity to become acquainted from a distance, and warmth and friendliness on the part of the stranger reduced Rachel's fear (Horner, 1980).

What is the significance of this rise in anger and fear after 6 months of age? Researchers believe that these emotions have survival value as babies begin to move on their own. Older infants can use the energy mobilized by anger to defend themselves or overcome obstacles. And fear keeps babies' enthusiasm for exploration in check, increasing the likelihood that they will remain close to the mother

Stranger anxiety rises during the second half of the first year. When an unfamiliar man picks this baby up, the infant bursts into tears. Had the man sat quietly, expressed warmth and friendliness, and offered an interesting toy while permitting the baby to approach from a distance with the support of his mother, the infant might have reacted positively. (Elizabeth Crews/ The Image Works)

and be careful about approaching unfamiliar people and objects. Anger and fear are also strong social signals that motivate caregivers to comfort a suffering infant.

Finally, cognitive development plays an important role in infants' angry and fearful reactions, just as it does in their expressions of happiness. Between 8 and 12 months, when (as Piaget pointed out) babies grasp the notion of intentional behavior, they have a better understanding of the cause of their frustrations. They know whom or what to get angry at. In the case of fear, improved memories permit older infants to distinguish familiar from unfamiliar events better than before.

UNDERSTANDING AND RESPONDING TO THE EMOTIONS OF OTHERS

Infants' emotional expressions are closely tied to their ability to interpret the emotional cues of others. Already we have seen that within the first few months, babies match the feeling tone of the caregiver in face-to-face communication. This suggests that they have begun to discriminate emotional states. Between 7 and 10 months, infants perceive facial expressions as organized patterns (see Chapter 4, page 139)—a clear indication that emotional messages are meaningful to them.

Soon after, babies realize that an emotional expression not only has meaning, but is a meaningful response to a specific object or event. By the end of the first year, they engage in **social referencing,** in which they actively seek emotional information from a trusted person in an uncertain situation. Many studies show that the caregiver's emotional expression (happiness, anger, or fear) influences whether a 1-year-old will be wary of strangers, play with an unfamiliar toy, or cross the deep side of the visual cliff (Rosen, Adamson, & Bakeman, 1992; Sorce et al., 1985; Walden & Ogan, 1988).

Social referencing grants infants and toddlers a powerful means for learning. By responding to caregivers' emotional messages, they can avoid harmful situations, such as a shock from an electric outlet or a fall down a steep staircase. And parents can capitalize on social referencing to teach their youngsters, whose capacity to explore is rapidly expanding, how to react to a great many novel events.

EMERGENCE OF SELF-CONSCIOUS EMOTIONS

Besides basic emotions, humans are capable of a second, higher-order set of feelings, including shame, embarrassment, guilt, envy, and pride. These are called **self-conscious emotions** because each involves injury to or enhancement of our sense of self. For example, when we are ashamed or embarrassed, we feel negatively about ourselves. In contrast, pride reflects delight in the self's achievements (Campos et al., 1983).

Self-conscious emotions appear at the end of the second year, as the sense of self emerges. Between 18 and 24 months, children can be seen feeling ashamed and embarrassed as they lower their eyes, hang their heads, and hide their faces with their hands. Pride also emerges at this time, and envy and guilt are present by age 3 (Lewis et al., 1989; Sroufe, 1979). Besides self-awareness, self-conscious emotions require an additional ingredient: adult instruction in *when* to feel proud, ashamed, or guilty. Parents begin this tutoring early when they say, "My, look at how far you can throw that ball!" or "You should feel ashamed for grabbing that toy from Billy!"

Self-conscious emotions help children acquire socially and culturally valued behaviors and goals. For example, in most of the United States, children are taught from an early age to feel pride over personal achievement—throwing a ball the farthest, winning a game, and (later on) getting good grades. Around age 2, American children are likely to call a parent's attention to an achievement, such as a completed puzzle, by pointing and saying, "Look, Mom!" (Stipek, Recchia, & McClintic, 1992). Among the Zuni Indians, shame and embarrassment are responses to purely personal success, whereas pride is evoked by helpfulness and generosity (Benedict, 1934). In Japan, violating cultural standards of concern for others—a parent, a teacher, or an employer—is cause for intense shame (Lewis, 1992b).

Children learn when to experience self-conscious emotions from adult instruction. Among the !Kung of Botswana, Africa, helping and sharing are encouraged at an early age. Perhaps this toddler already feels a sense of pride as she tries to assist her grandmother with food preparation. (Konner/Anthro-Photo)

BEGINNINGS OF EMOTIONAL SELF-REGULATION

Besides expanding their range of emotional reactions, infants and toddlers begin to find ways to manage their emotional experiences. **Emotional self-regulation** refers to the strategies we use to adjust our emotional state to a comfortable level of intensity so we can accomplish our goals (Eisenberg et al., 1995; Thompson, 1994). If you drank a cup of coffee to wake up this morning, reminded yourself that an anxiety-provoking event would be over soon, or decided not to see a scary horror film, you were engaging in emotional self-regulation.

In the early months of life, infants have only a limited capacity to regulate their emotional states. Although they can turn away from unpleasant stimulation and mouth and suck when their feelings get too intense, they are easily overwhelmed. They depend for help on the soothing interventions of caregivers—lifting the distressed baby to the shoulder, rocking, and talking softly.

Rapid development of the cortex gradually increases the baby's tolerance for stimulation. Between 2 and 4 months, caregivers start to build on this capacity through face-to-face play and attention to objects. In these interactions, parents arouse pleasure in the baby while adjusting the pace of their own behavior so the infant does not become overwhelmed and distressed. As a result, the baby's tolerance for stimulation increases further (Field, 1994). By the end of the first year, infants' ability to move about permits them to regulate feelings more effectively by approaching or retreating from various situations.

As caregivers help infants regulate their emotional states, they also provide lessons in socially approved ways of expressing feelings. Beginning in the first few months, American middle-class mothers match their baby's positive feelings far more often than the negative ones. In this way, they encourage happiness and discourage anger and sadness. Interestingly, infant boys get more training in hiding their unhappiness than do girls. The well-known sex difference—females as emotionally expressive and males as emotionally controlled—is promoted at a very tender age (Malatesta & Haviland, 1982; Malatesta et al., 1986).

By the second year, growth in representation and language leads to new ways of regulating emotions. A vocabulary for talking about feelings, such as "happy," "love," "surprised," "scary," "yucky," and "mad," develops rapidly after 18 months (Bretherton et al., 1986). By describing their emotions, toddlers can guide the caregiver in helping them feel better. For example, while listening to a story about monsters, Rachel whimpered, "Mommy, scary." Beth put the book down and gave Rachel a comforting hug.

BRIEF REVIEW

Changes in infants' ability to express emotion and respond to the emotions of others reflect their developing cognitive capacities and have both social and survival value. The social smile appears between 6 and 10 weeks, laughter around 3 to 4 months. In the middle of the first year, anger and fear start to increase. Young infants match the feeling tone of their caregivers' facial expressions. By 8 to 10 months, they engage in social referencing, actively seeking emotional information from a trusted person in an uncertain situation. Self-conscious emotions, such as shame, embarrassment, and pride, begin to emerge in the second year as toddlers develop self-awareness. Emotional self-regulation is supported by brain maturation, improvements in cognition and language, and sensitive child-rearing practices.

ASK YOURSELF . . .

■ *Dana is planning to meet her 10-month-old niece Laureen for the first time. How should Dana expect Laureen to react? How would you advise Dana to go about establishing a positive relationship with Laureen?*

■ *One of Byron's favorite games was dancing with his mother while she sang "Old MacDonald," clapping his hands and stepping from side to side. At 14 months, Byron danced joyfully as Beth and Felicia watched. At 20 months, he began to show signs of embarrassment—smiling, looking away, and covering his eyes with his hands. What explains this change in Byron's emotional reaction?*

TEMPERAMENT AND DEVELOPMENT

Beginning in early infancy, Byron, Rachel, and April showed distinct patterns of emotional responding. Byron was intense, active, and distractible—qualities that often made parenting quite trying for Lisa. In contrast, Rachel seemed calm, patient, and easily soothed (except on the occasion of her mother's brief absence). And April was an especially joyous baby whose friendliness consistently evoked positive responses from others.

When we describe one person as cheerful and "upbeat," another as active and energetic, and still others as calm, cautious, or prone to angry outbursts, we are referring to **temperament**—stable individual differences in quality and intensity of emotional reaction (Goldsmith, 1987; Kagan, 1994). Researchers have become increasingly interested in temperamental differences among infants and children, since the child's style of emotional responding is believed to form the cornerstone of the adult personality.

The New York Longitudinal Study, initiated in 1956 by Alexander Thomas and Stella Chess, is the longest and most comprehensive study of temperament to date. A total of 141 children were followed from early infancy well into adulthood. Results showed that temperament is a major factor in increasing the chances that a child will experience psychological problems or, alternatively, be protected from the damaging effects of a highly stressful home life. However, Thomas and Chess (1977) also found that temperament is not fixed and unchangeable. Parenting practices can modify children's emotional styles considerably.

These findings inspired a growing body of research on temperament, including its stability, its biological roots, and its interaction with child-rearing experiences. But before we review what is known about these issues, let's look at how temperament is measured.

MEASURING TEMPERAMENT

Temperament is usually assessed in one of three ways: through interviews and questionnaires given to parents; through behavior ratings by doctors, nurses, or caregivers who know the child well; or through direct observation by researchers. Recently, physiological measures have been used to supplement these techniques. For example, highly inhibited, shy children show greater electrical activity in the right than in the left hemisphere of the cortex, whereas their more sociable counterparts show

the reverse pattern (Fox, 1994; Fox, Bell, & Jones, 1992).[1] They also produce more cortisol (a hormone that regulates blood pressure and that is involved in reducing stress) than their uninhibited agemates (Gunnar & Nelson, 1994; Kagan & Snidman, 1991). Researchers hope these measures will shed light on the genetic basis and role of brain structures in temperament.

Most often, parental reports are used to assess temperament because of their convenience and parents' depth of knowledge. Information from parents has been criticized for being biased and subjective, and it is only modestly related to observational measures (Seifer et al., 1994). Nevertheless, parental reports are useful for understanding the way parents view and respond to their child.

In the New York Longitudinal Study, detailed descriptions of each child's behavior were collected regularly from parents. When carefully analyzed, these yielded nine dimensions of temperament, summarized in Table 6.2. The researchers noticed that certain characteristics clustered together, producing three temperamental types that described the majority of their sample:

- The **easy child** (40 percent of the sample): This child quickly establishes regular routines in infancy, is generally cheerful, and adapts easily to new experiences.

- The **difficult child** (10 percent of the sample): This child is irregular in daily routines, is slow to accept new experiences, and tends to react negatively and intensely.

- The **slow-to-warm-up child** (15 percent of the sample): This child is inactive, shows mild, low-key reactions to environmental stimuli, is negative in mood, and adjusts slowly to new experiences.

Notice that 35 percent of the children did not fit any of these patterns. Instead, they showed unique blends of temperamental characteristics. Other systems for classifying temperament do exist (Buss & Plomin, 1984; Goldsmith & Rothbart, 1990). However, Thomas and Chess's nine dimensions and three styles provide a fairly complete picture of the traits most often studied.

STABILITY OF TEMPERAMENT

It would be difficult to claim that temperament really exists if children's emotional styles were not stable over time. Indeed, the findings of many studies provide support for the long-term stability of temperament. An infant who scores low or high on activity level, rhythmicity, attention span, irritability, sociability, or shyness is likely to respond similarly in childhood and, occasionally, even into the adult

---

[1] Recall from Chapter 4 that the right cortical hemisphere mediates the display of negative emotion, whereas the left hemisphere mediates positive emotion (see page 119).

TABLE 6.2

Nine Dimensions of Temperament

DIMENSION	DESCRIPTION AND EXAMPLE
Activity level	Proportion of active periods to inactive ones. Some babies are always in motion. Others move about very little.
Rhythmicity	Regularity of body functions. Some infants fall asleep, wake up, get hungry, and have bowel movements on a regular schedule, whereas others are much less predictable.
Distractibility	Degree to which stimulation from the environment alters behavior. Some hungry babies stop crying temporarily if offered a pacifier or a toy to play with. Others continue to cry until fed.
Approach–withdrawal	Response to a new object or person. Some babies accept new foods and smile and babble at strangers, whereas others pull back and cry on first exposure.
Adaptability	Ease with which the child adapts to changes in the environment. Although some infants withdraw when faced with new experiences, they quickly adapt, accepting the new food or person on the next occasion. Others continue to fuss and cry over an extended period of time.
Attention span	Amount of time and persistence devoted to an activity. Some babies watch a mobile or play with a toy for a long time, whereas others lose interest after a few minutes.
Intensity of reaction	Intensity or energy level of response. Some infants laugh and cry loudly, whereas others react only mildly.
Threshold of responsiveness	Intensity of stimulation required to evoke a response. Some babies startle at the slightest change in sound or lighting. Others take little notice of these changes in stimulation.
Quality of mood	Amount of friendly, joyful behavior as opposed to unpleasant, unfriendly behavior. Some babies smile and laugh frequently when playing and interacting with people. Others fuss and cry.

Source: Thomas, Chess, & Birch, 1970.

years (Caspi & Silva, 1995; Kochanska & Radke-Yarrow, 1992; Pedlow et al., 1993; Riese, 1987; Ruff et al., 1990). The temperamental styles identified in the New York longitudinal study are also fairly stable. Compared to easy children, difficult preschoolers are more likely to have problems concentrating and getting along with peers after school entry. And slow-to-warm-up youngsters are often overwhelmed by demands that they adapt quickly to new experiences in school, a circumstance that may intensify their withdrawal in middle childhood (Chess & Thomas, 1984).

Yet overall, temperament is only moderately stable from one age period to the next. In fact, some characteristics, such as shyness and sociability, are stable over the long term only in children at the extremes—those who are very inhibited or very outgoing to begin with (Kerr et al., 1994; Robinson et al., 1992).

The fact that early in life, children show marked individual differences in temperament, some of which are related to physiological reactions, indicates that biological factors play an important role. At the same time, the changes that occur for many youngsters suggest that temperament can be modified by experience (although children rarely change from one extreme to another—that is, a shy toddler practically never becomes a highly sociable school-age child). Let's take a close look at genetic and environmental contributions to temperament.

GENETIC INFLUENCES

The very word *temperament* implies a genetic foundation for individual differences in emotional style. Research shows that identical twins are more similar than fraternal twins across a wide range of temperamental and personality traits (DiLalla, Kagan, & Reznick, 1994; Emde et al., 1992; Robinson et al., 1992). In Chapter 2, we indicated that twin studies suggest a moderate role for heredity in temperament and personality: About half of the individual differences among us are due to differences in our genetic makeup.

Evidence on shyness provides further support for the importance of heredity. Besides differing from their sociable agemates on the physiological measures mentioned earlier, inhibited, withdrawn babies show high rates of motor activity and crying when faced with new sights and sounds (Kagan & Snidman, 1991). By the end of the first year, highly stimulating, unfamiliar experiences (such as a battery-powered robot) cause their hearts to race, their pupils to dilate, and their muscles to tense up. Under the same conditions, sociable babies remain relaxed and composed. Shy people are also more likely to have certain physical traits—blue eyes, thin faces, and hay fever—known to be influenced by heredity (Arcus & Kagan, 1995). The genes controlling these characteristics may also contribute to a fearful, reactive temperamental style.

Finally, consistent ethnic and sex differences in early temperament imply a role for heredity. Like Rachel, Chinese and Japanese babies tend to be calmer, more easily soothed, and better at quieting themselves than are Caucasian infants (Kagan et al., 1994; Lewis, Ramsay, & Kawakami, 1993). And Byron's high activity level is consistent with a trend for boys to be more active and daring and girls more anxious and timid—a difference reflected in boys' higher accident rates throughout childhood and adolescence (Jacklin & Maccoby, 1983; Richardson, Koller, & Katz, 1986).

ENVIRONMENTAL INFLUENCES

Although genetic influences on temperament are clear, no study has shown that infants maintain their emotional styles in the absence of environmental supports. Instead, heredity and environment often combine to strengthen the stability of temperament, since a child's approach to the world affects the experiences to which she is exposed. To see how this works, let's take a second look at ethnic and sex differences in temperament.

As I watched Beth care for Rachel as a 3-month-old baby, her calm, soothing manner and use of gentle rocking and touching contrasted with Lisa's and Felicia's tendency to stimulate Byron and April through lively facial expressions and talking. These differences in early caregiving appear repeatedly in studies comparing American and Asian infant–mother pairs (Fogel, Toda, & Kawai, 1988). They suggest that some differences in early temperament are encouraged by cultural beliefs and practices. When asked about their approach to child rearing, Japanese mothers say that babies come into the world as independent beings who must learn to rely on their mothers through close physical contact. American mothers are likely to believe just the opposite—that they must wean babies away from dependence into autonomy (Kojima, 1986). As a result, Japanese mothers do more comforting and American mothers more stimulating—behaviors that enhance early temperamental differences between their infants.

A similar process seems to contribute to sex differences in temperament. Within the first 24 hours after birth (before they have had much experience with the baby), parents already perceive boys and girls differently. Sons are rated as larger, better coordinated, more alert, and stronger. Daughters are viewed as softer, more awkward, weaker, and more delicate (Rubin, Provenzano, & Luria, 1974; Stern & Karraker, 1989). These gender-stereotyped beliefs carry over into the way parents treat their infants and toddlers. For example, parents more often encourage sons to be physically active and daughters to seek help and physical closeness. These practices promote temperamental differences between boys and girls (Fagot, 1978; Smith & Lloyd, 1978).

In families with several children, an additional influence on temperament is at work. Parents often look for and emphasize each child's unique characteristics (Plomin, 1994c). You can see this in the comments they make after the birth of a second baby: "He's so much calmer," "She's a lot more active," or "He's more sociable." When one child in a family is perceived as easy, another is likely to be regarded as difficult, even though the second child might not be very difficult when compared to children in general (Schachter & Stone, 1985). Each child, in turn, evokes

Beginning at birth, Chinese infants are calmer, more easily soothed when upset, and better at quieting themselves than are Caucasian infants. These differences are probably hereditary, but cultural variations in child rearing support them. (Alan Oddie/PhotoEdit)

responses from caregivers that are consistent with parental views and with the child's actual temperamental style. In sum, temperament and personality can only be understood in terms of complex interdependencies between genetic and environmental factors.

TEMPERAMENT AND CHILD REARING: THE GOODNESS-OF-FIT MODEL

Children with different temperaments have unique child-rearing needs. If a child's disposition interferes with learning or getting along with others, it is important for adults to gently but consistently counteract the child's maladaptive behavior.

The concept of **goodness-of-fit** describes how temperament and environment can work together to produce favorable outcomes (Thomas & Chess, 1977). When a child's style of responding and environmental demands are in harmony, or achieve a "good fit," then development is at its best. But a "poor fit" between temperament and environment disrupts development.

As the Lifespan Vista box on page 186 shows, parenting that is in tune with the child's temperament is particularly important for difficult youngsters. These children, at least in Western middle-class society, are far less likely to receive sensitive care that makes firm but reasonable demands for mastering new experiences (van den Boom & Hoeksma, 1994). By the second year, their parents often resort to

DIFFICULT CHILDREN: WHEN PARENTS ESTABLISH A GOOD FIT

In the New York Longitudinal Study, 70 percent of children who were temperamentally difficult in the early years developed serious behavior problems by middle childhood. But in no case was this outcome the result of temperament alone. Instead, it occurred because the negative behaviors of difficult children often provoked parental reactions that fit poorly with their basic dispositions. The case of Carl, one of the most difficult youngsters in the New York Longitudinal sample, shows how long-term outcomes for these children are a function of rearing experiences (Thomas & Chess, 1977).

As a baby, Carl rejected almost all new situations, such as his first bath and spoonfuls of solid food. He shrieked, cried, and struggled to get away. Yet his mother and father recognized that his behavior did not mean they were "bad parents." To the contrary, Carl's father viewed his son's emotional intensity as a sign of strength and vigor. And both parents believed that if they were patient, reduced the number of new situations Carl had to deal with at one time, and provided him with opportunities for repeated exposure, he would, in the end, adapt positively.

By the time Carl reached school age, he was doing remarkably well. The energies that he had put into tantrums were now channeled constructively. He was a good student and became enthusiastically involved in several activities. One of these was playing the piano—lessons that he had asked for but (as with other new experiences) at first disliked intensely. Carl's mother had granted the piano instruction on one condition: that he continue for 6 months. Held to this bargain, Carl came to love his introduction to music. His parents'

patience and consistency had helped him reorganize his behavior, benefit from new learning opportunities, and avoid adjustment difficulties. Throughout middle childhood and adolescence, Carl encountered few radical changes in his life. He remained in the same community and went through school with the same group of peers. As a result, he did well academically, made friends easily, and explored a wide range of interests.

On entering college, however, Carl faced many new situations at once—different approaches to learning, a more complex daily schedule, and unfamiliar living arrangements and people. For a time, his temperamental pattern of withdrawal and intensely negative reactions resurfaced. Fortunately, Carl had acquired insights into his emotional style and successful coping strategies. After only one counseling session, he reduced the number of new academic subjects, disciplined himself to study daily, and made a point of attending social events, even if he felt uncomfortable. By the end of his freshman year, Carl was again functioning well (Chess & Thomas, 1990).

Interviewed at age 29, Carl's temperament was quite different from what it had been in infancy and childhood. His mood had changed from negative to positive, and he approached rather than withdrew from new tasks. Nurtured by his parents' patience and consistency, Carl developed high self-esteem and broad interests and was zestful and relaxed. As a young adult, his intensity had become an asset rather than a liability.

Unlike Carl, many difficult children do not receive sensitivity and encouragement. Because of a poor fit between temperament and child rearing, they fail to reach their educa-

tional and occupational potential. And their early social difficulties carry over into unhappy marriages and a high rate of divorce in adulthood (Caspi, Elder, & Herbener, 1990).

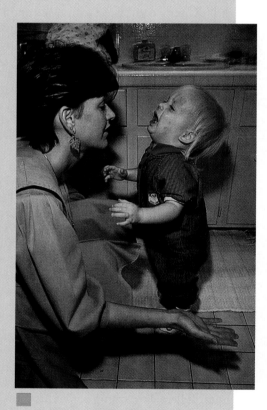

This mother is perplexed because her 1-year-old baby is not responding to her efforts to help him calm down. Difficult children react negatively and intensely to many new experiences. When parents are patient and provide opportunities for gradual, repeated exposure to new situations, difficultness often subsides. (Nubar Alexanian/ Stock Boston)

angry, punitive discipline, and the child reacts with defiance and disobedience. Then parents behave inconsistently, rewarding the child's noncompliant behavior by giving in to it, although they resisted at first (Lee & Bates, 1985). These practices maintain and even increase the child's irritable, conflict-ridden style.

Goodness-of-fit depends in part on cultural values. For example, difficult children in working-class Puerto Rican families are treated with sensitivity and patience; they are not at risk for adjustment problems (Gannon & Korn, 1983). In Western nations, shy, withdrawn children are regarded as socially incompetent—an attitude that discourages parents from helping them approach new social situations. Yet in Chinese culture, adults evaluate such children positively—as advanced in social maturity and understanding (Chen, Rubin, & Li, 1995).

The concept of goodness-of-fit reminds us that babies come into the world with unique styles of responding that adults need to accept. Children cannot be molded in ways that do not blend with their basic styles. This means that parents can neither take full credit for their children's virtues nor be blamed for all their faults. But they can turn an environment that exaggerates a child's difficulties into one that builds on the youngster's strengths, helping each child master the challenges of development.

In the following sections, we will see that goodness-of-fit is at the heart of infant–caregiver attachment. This first intimate relationship grows out of interaction between parent and baby, to which the emotional styles of both partners contribute.

BRIEF REVIEW

Children's unique temperamental styles are apparent in infancy. However, the stability of temperament is only moderate; some children retain their original dispositions, while others change over time. Heredity influences early temperament, but child-rearing experiences determine whether a child's emotional style is sustained or modified over time. A good fit between parenting practices and a child's temperament helps children at risk for adjustment problems develop favorably.

ASK YOURSELF . . .

■ *Rachel, like many other Asian infants, is calm and easily soothed when upset. What factors contribute to her temperamental style?*

■ *At 18 months, highly active Byron climbed out of his high chair long before his meal was finished. Exasperated with Byron's behavior, his father made him sit at the table until he had eaten all his food. Soon Byron's behavior escalated into a full-blown tantrum. Using the concept of goodness-of-fit, suggest another way of handling Byron.*

DEVELOPMENT OF ATTACHMENT

Attachment is the strong, affectional tie we feel for special people in our lives that leads us to feel pleasure and joy when we interact with them and to be comforted by their nearness during times of stress. By the end of the first year, infants have become attached to familiar people who have responded to their needs for physical care and stimulation. Watch babies of this age, and notice how parents are singled out for special attention. A whole range of responses are reserved just for them. For example, when the mother enters the room, the baby breaks into a broad, friendly smile. When she picks him up, he pats her face, explores her hair, and snuggles against her. When he feels anxious or afraid, he crawls into her lap and clings closely.

Freud first suggested that the infant's emotional tie to the mother is the foundation for all later relationships. We will see shortly that research on the consequences of attachment is consistent with Freud's idea. But attachment has also been the subject of intense theoretical debate. Turn back to the description of Freud's and Erikson's theories at the beginning of this chapter, and notice how *psychoanalytic theory* regards feeding as the primary context in which caregivers and babies build this close emotional bond. *Behaviorism*, too, emphasizes the importance of feeding, but for different reasons. According to a well-known behaviorist account, as the mother satisfies the baby's hunger, infants learn to prefer her soft caresses, warm smiles, and tender words of comfort because these events have been paired with tension relief (Sears, Maccoby, & Levin, 1957).

Although feeding is an important context for building a close relationship, attachment does not depend on hunger satisfaction. In the 1950s, a famous experiment showed that rhesus monkeys reared with terry-cloth and wire mesh "surrogate mothers" clung to the soft terry-cloth substitute, even though the wire mesh "mother" held the bottle and infants had to climb on it to be fed (Harlow & Zimmerman, 1959). Observations of human infants also reveal that they become attached to family members who seldom if ever feed them, including fathers, siblings, and grandparents. And perhaps you have noticed that toddlers in Western cultures develop strong emotional ties to cuddly objects, such as blankets and teddy bears. Yet such objects have never played a role in infant feeding!

Baby monkeys reared with "surrogate mothers" from birth preferred to cling to a soft terry cloth "mother" instead of a wire mesh "mother" that held a bottle. These findings reveal that the drive-reduction explanation of attachment, which assumes that the mother-infant relationship is based on feeding, is incorrect. (Martin Rogers/Stock Boston)

ETHOLOGICAL THEORY OF ATTACHMENT

Today, **ethological theory** is the most widely accepted view of the development of attachment. According to ethology, many human behaviors have evolved over the history of our species because they promote survival. John Bowlby (1969), who first applied this idea to the infant–caregiver bond, was inspired by Konrad Lorenz's studies of imprinting in baby geese (see Chapter 1). He believed that the human infant, like the young of other animal species, is endowed with a set of built-in behaviors that help keep the parent nearby, increasing the chances that the infant will be protected from danger. Contact with the parent also ensures that the baby will be fed, but Bowlby was careful to point out that feeding is not the basis for attachment. Instead, the attachment bond itself has strong biological roots. It can best be understood in an evolutionary context where survival of the species is of utmost importance.

According to Bowlby, the infant's relationship with the parent begins as a set of innate signals that call the adult to the baby's side. Over time, a true affectional bond develops, which is supported by new cognitive and emotional capac-ities as well as a history of warm, responsive care. Attachment develops in four phases:

1. *The preattachment phase* (birth to 6 weeks). Built-in behaviors—grasping, smiling, crying, and gazing into the adult's eyes—help bring newborn babies into close contact with other humans. Once an adult responds, infants encourage her to remain nearby, since they are comforted when picked up, stroked, and talked to softly. Babies of this age can recognize their own mother's smell and voice (see Chapter 3). But they are not yet attached to her, since they do not mind being left with an unfamiliar adult.

2. *The "attachment-in-the-making" phase* (6 weeks to 6–8 months). During this phase, infants start to respond differently to a familiar caregiver than to a stranger. But even though they can recognize the parent, babies do not yet protest when separated from her. Therefore, attachment is underway but not yet established.

3. *The phase of "clearcut" attachment* (6–8 months to 18 months–2 years). Now attachment to the familiar caregiver is clearly evident. Babies display **separation anxiety,** becoming upset when the adult whom they have come to rely on leaves. (Recall Rachel's behavior when her mother left her with her grandparents.) Separation anxiety appears universally around the world after 6 months of age, increasing until about 15 months (see Figure 6.1).
   Besides protesting the parent's departure, older infants and toddlers try hard to maintain her presence. Crawling and walking babies approach, follow, and climb on her in preference to others. And they use her as a **secure base** from which to explore, venturing into the environment and then returning for emotional support.

4. *Formation of a reciprocal relationship* (18 months–2 years and on). By the end of the second year, rapid growth in representation and language permits toddlers to begin to understand the parent's coming and going and to predict her return. As a result, separation protest declines. Now children start to negotiate with the caregiver, using requests and persuasion rather than crawling after and clinging to her. For example, at age 2, April asked Felicia to read a story before leaving her with a baby-sitter. The extra time with her mother, along with a better understanding of where Felicia was going ("to a movie with Daddy") and when she would be back ("right after you go to sleep"), helped April tolerate her mother's absence.

According to Bowlby (1980), out of their experiences during these four phases, children construct an inner representation of the parent–child bond that they can use as a

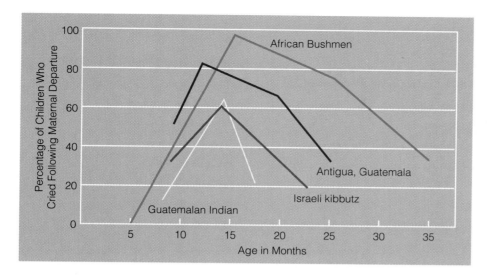

FIGURE 6.1

Development of separation anxiety. In cultures around the world, separation anxiety emerges in the second half of the first year, increasing until about 15 months and then declining. *(From J. Kagan, R. B. Kearsley, & P. R. Zelazo, 1978,* Infancy: Its Place in Human Development, *Cambridge, MA: Harvard University Press, p. 107. Copyright © 1978 by the President and Fellows of Harvard College. All rights reserved. Reprinted by permission.)*

secure base in the parent's absence. This inner representation becomes a vital part of personality. It serves as an **internal working model,** or set of expectations about the availability of attachment figures and their likelihood of providing support during times of stress. This image becomes the model, or guide, for all future close relationships—through childhood and adolescence and into adult life (Bretherton, 1992).

MEASURING THE SECURITY OF ATTACHMENT

Although virtually all family-reared babies become attached to a familiar caregiver by the second year, the quality of this relationship differs from child to child. A widely used technique for assessing the quality of attachment between 1 and 2 years of age is the **Strange Situation.** In designing it, Mary Ainsworth and her colleagues (1978) reasoned that securely attached infants and toddlers should use the parent as a secure base from which to explore an unfamiliar playroom. In addition, when the parent leaves for a brief time, the child should show separation anxiety and find a strange adult less comforting than the parent. As summarized in Table 6.3, the Strange Situation takes the baby through eight short episodes in which brief separations from and reunions with the parent occur.

Observing the responses of infants to these episodes, researchers have identified a secure attachment pattern and three patterns of insecurity (Ainsworth et al., 1978; Main & Solomon, 1990). Which of the following patterns do you think Rachel displayed in the incident described at the beginning of this chapter?

- **Secure attachment.** These infants use the parent as a secure base from which to explore. When separated,

they may or may not cry, but if they do, it is due to the parent's absence, since they prefer her to the stranger. When the parent returns, they actively seek contact, and their crying is reduced immediately.

- **Avoidant attachment.** These babies seem unresponsive to the parent when she is present. When she leaves, they are usually not distressed, and they react to the stranger in much the same way as the parent. During reunion, they avoid or are slow to greet the parent, and when picked up, they often fail to cling.

- **Resistant attachment.** Before separation, these infants seek closeness to the parent and often fail to explore. When she returns, they display angry, resistive behavior, sometimes hitting and pushing. Many continue to cry after being picked up and cannot be comforted easily.

- **Disorganized–disoriented attachment.** This pattern reflects the greatest insecurity. At reunion these infants show a variety of confused, contradictory behaviors. For example, they might look away while being held by the parent or approach her with a flat, depressed gaze. A few cry out unexpectedly after having calmed down or display odd, frozen postures.

Infants' reactions in the Strange Situation resemble their use of the parent as a secure base and their response to separation at home (Pederson & Moran, 1996; Vaughn & Waters, 1990). For this reason, the procedure is a powerful tool for assessing attachment security.

CULTURAL VARIATIONS

Despite the usefulness of the Strange Situation, infants' responses must be interpreted cautiously in other cultures. For example, as Figure 6.2 reveals, German infants show

TABLE 6.3

Episodes in the Strange Situation

EPISODE	EVENTS	ATTACHMENT BEHAVIORS OBSERVED
1	Experimenter introduces parent and baby to playroom and then leaves.	
2	Parent is seated while baby plays with toys.	Parent as a secure base
3	Stranger enters, is seated, and talks to parent.	Reaction to unfamiliar adult
4	Parent leaves room. Stranger responds to baby and offers comfort if upset.	Separation anxiety
5	Parent returns, greets baby, and if necessary offers comfort. Stranger leaves room.	Reaction to reunion
6	Parent leaves room.	Separation anxiety
7	Stranger enters room and offers comfort.	Ability to be soothed by stranger
8	Parent returns, greets baby, if necessary offers comfort, and tries to reinterest baby in toys.	Reaction to reunion

Note. Episode 1 lasts about 30 seconds; the remaining episodes each last about 30 minutes. Separation episodes are cut short if the baby becomes very upset. Reunion episodes are extended if the baby needs more time to calm down and return to play.

Source: Ainsworth et al., 1978.

considerably more avoidant attachment than American babies do. But German parents encourage their infants to be nonclingy and independent, so the baby's behavior may be an intended outcome of cultural beliefs and practices (Grossmann et al., 1985). Did you classify Rachel's attachment behavior as resistant? An unusually high num-

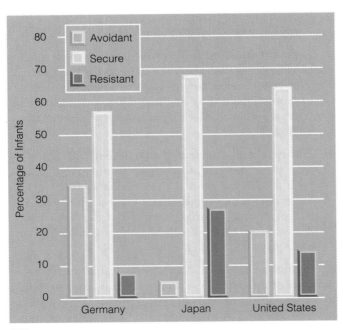

FIGURE 6.2

A cross-cultural comparison of infants' reactions in the Strange Situation. A high percentage of German babies seem avoidantly attached, whereas a substantial number of Japanese infants appear resistantly attached. Note that these responses may not reflect true insecurity. Instead, they are probably due to cultural differences in rearing practices. *(Adapted from van IJzendoorn & Kroonenberg, 1988.)*

ber of Japanese infants display a resistant response, but the reaction may not represent true insecurity. Japanese mothers rarely leave their babies in the care of unfamiliar people, so the Strange Situation probably creates far greater stress for them than it does for infants who frequently experience maternal separations (Miyake, Chen, & Campos, 1985; Takahashi, 1990). Despite these cultural variations, the secure pattern is still the most common pattern of attachment in all societies studied to date (van IJzendoorn & Kroonenberg, 1988).

FACTORS THAT AFFECT ATTACHMENT SECURITY

What factors affect attachment security? Researchers have looked closely at four important influences: (1) opportunity to establish a close relationship; (2) quality of caregiving; (3) the baby's characteristics; and (4) the family context in which infant and parent live.

■ OPPORTUNITY FOR ATTACHMENT. What happens when a baby does not have the opportunity to establish an affectional tie to a caregiver? In a series of landmark studies René Spitz (1945, 1946) observed institutionalized babies who had been given up by their mothers between 3 and 12 months of age. The infants were placed on a large ward where they shared a nurse with at least seven other babies. In contrast to the happy, outgoing behavior they had shown before separation, they wept and withdrew from their surroundings, lost weight, and had difficulty sleeping. If a caregiver whom the baby could get to know did not replace the mother, the depression deepened rapidly.

According to Spitz, institutionalized babies had emotional difficulties not because they were separated from their mothers, but because they were prevented from forming a bond with one or a few adults. A more recent study supports this conclusion. Researchers followed the development of infants in an institution with a good caregiver–child ratio and a rich selection of books and toys. However, staff turnover was so rapid that the average child had 50 different caregivers by age 4½! Many of these children became "late adoptees" who were placed in homes after age 4. Since most developed deep ties with their adoptive parents, this study indicates that a first attachment bond can develop as late as 4 to 6 years of age. But throughout childhood and adolescence, these youngsters were more likely to display emotional and social problems, including an excessive desire for adult attention, "over friendliness" to unfamiliar adults and peers, and difficulties in making friends (Hodges & Tizard, 1989; Tizard & Hodges, 1978; Tizard & Rees, 1975). Although follow-ups into adulthood are necessary to be sure, these findings leave open the possibility that fully normal development may depend on establishing close bonds with caregivers during the first few years of life.

■ QUALITY OF CAREGIVING. What kind of parental behavior promotes secure attachment? To find out, researchers have related various aspects of maternal caregiving to the attachment bond. The findings of many studies reveal that securely attached babies have mothers who respond promptly to infant signals, express positive emotion, and handle their babies tenderly and carefully. In contrast, insecurely attached infants have mothers who dislike physical contact, handle them awkwardly, and behave insensitively when meeting the baby's needs (Ainsworth et al., 1978; Belsky, Rovine, & Taylor, 1984; Isabella, 1993).

Exactly what do mothers of securely attached babies do to support their infant's feelings of trust? In one study, mother–infant interaction was videotaped and carefully coded for each partner's behavior. Findings indicated that a special form of communication called **interactional synchrony** distinguished secure from insecure babies (Isabella & Belsky, 1991). It is best described as a sensitively tuned "emotional dance," in which the caregiver responds to infant signals in a well-timed, appropriate fashion. In addition, both partners match emotional states, especially the positive ones (Stern, 1985).

But more research is needed to document the importance of interactional synchrony. Other research reveals that only 30 percent of the time are exchanges between mothers and their babies so perfectly in tune with one another. The remaining 70 percent of the time, interactive errors occur (Tronick, 1989). Perhaps warm, sensitive caregivers and their infants become especially skilled at repairing these errors and returning to a synchronous state. Nevertheless, finely tuned, coordinated interaction does not characterize mothers and infants everywhere. Among the Gusii of Kenya, mothers rarely cuddle, hug, and interact playfully with their babies, although they are very responsive to their infant's needs (LeVine et al., 1994). This

This mother and baby engage in a sensitively tuned form of communication called interactional synchrony in which they match emotional states, especially the positive ones. Interactional synchrony may support the development of secure attachment, but it does not characterize mother-infant interaction in all cultures. (Julie O'Neil/The Picture Cube)

suggests that secure attachment depends on attentive caregiving, but its association with moment-by-moment contingent interaction is probably culture specific.

Compared to securely attached infants, avoidant infants tend to receive overstimulating, intrusive care. Their mothers might, for example, talk energetically to a baby who is looking away or falling asleep. Resistant infants often experience interaction at the other extreme. Their mothers are minimally involved in caregiving and unresponsive to infant signals. It is as if avoidant babies try to escape from overwhelming, poorly paced interaction, whereas resistant infants react with anger and frustration to a lack of maternal involvement (Cassidy & Berlin, 1994; Isabella & Belsky, 1991).

When caregiving is extremely inadequate, it is a powerful predictor of disruptions in attachment. Child abuse and neglect (topics we will consider in Chapter 8) are associated with all three forms of attachment insecurity. Among maltreated infants, the most worrisome classification—disorganized–disoriented attachment—is especially high (Carlson et al., 1989). Infants of depressed mothers also show the uncertain behaviors of this pattern, mixing closeness, resistance, and avoidance while looking very sad and depressed themselves (Lyons-Ruth et al., 1990; Teti et al., 1995).

■ INFANT CHARACTERISTICS. Since attachment is the result of a *relationship* that builds between two partners, infant characteristics should affect how easily it is established. In Chapter 3, we saw that prematurity, birth complications, and newborn illness make caregiving more taxing for parents. In poverty-stricken, stressed families, these difficulties are linked to attachment insecurity (Willie, 1991). But when parents have the time and patience to care for a baby with special needs and the infant is not very sick, at-risk newborns fare quite well in the development of attachment (Pederson & Moran, 1995).

Infants also vary in temperament, but its role in attachment security has been intensely debated. Some researchers believe that babies who are irritable and fearful may simply react to brief separations with intense anxiety, regardless of the parent's sensitivity to the baby (Kagan, 1989). Consistent with this view, proneness to distress in early infancy is moderately related to later insecure attachment (Seifer et al., 1996; Vaughn et al., 1992).

But other evidence suggests that caregiving may account for this correlation. For example, in one study, distress-prone infants who became insecurely attached tended to have rigid, controlling mothers who probably had trouble comforting a baby who often fussed and cried (Mangelsdorf et al., 1990). In another study, an intervention that taught mothers how to respond sensitively to their irritable 6-month-olds led to gains in maternal responsiveness and in children's attachment security, exploration, cooperativeness, and sociability that were still present at 3½ years of age (van den Boom, 1995).

A major reason that temperament and other infant characteristics are not strongly related to attachment security may be that their influence depends on goodness-of-fit. From this perspective, *many* child attributes can lead to secure attachment as long as the caregiver sensitively adjusts her behavior to fit the baby's needs (Seifer & Schiller, 1995). But when a parent's capacity to do so is strained—for example, by her own personality or stressful living conditions—then difficult babies are at greater risk for attachment problems.

■ FAMILY CIRCUMSTANCES. Around April's first birthday, Felicia's husband Lonnie lost his job, and constant arguments with Felicia over the monthly bills caused him to leave the family for a time. Although Felicia tried not to let these worries affect her caregiving, April sensed the tension. Several times, Felicia left April at Beth's house while she looked for employment. April, who had previously taken separations like these quite well, cried desperately on her mother's departure and clung for a long time after she returned.

April's behavior reflects a repeated finding: When families experience major life changes, such as a shift in employment or marital status or the birth of a younger sibling, the quality of attachment often changes—sometimes positively and at other times negatively (Thompson, Lamb, & Estes, 1982; Touris, Kromelow, & Harding, 1995). This is an expected outcome, since family transitions affect parent–child interaction and, in turn, the attachment bond. As Felicia and Lonnie resolved their difficulties, April's clinginess declined and she felt more secure.

Parents bring to the family context a long history of attachment experiences, out of which they construct internal working models that they apply to the bonds established with their babies. Lisa remembered her mother as tense and preoccupied. She expressed regret at not having had a closer relationship. Felicia recalled her mother as deeply affectionate and caring but much too controlling. Her mother's strictness, she explained, was probably influenced by the dangerous inner-city neighborhood in which the family lived. Do these images of parenthood affect the quality of Byron's and April's attachments to their mothers?

To answer this question, researchers have assessed adults' internal working models by having them evaluate childhood memories of attachment experiences (George, Kaplan, & Main, 1985). Mothers who show objectivity and balance in discussing their childhoods, regardless of whether they were positive or negative, tend to have securely attached infants. In contrast, mothers who dismiss the importance of early relationships or describe them in angry, confused ways usually have insecurely attached babies (van IJzendoorn, 1995a). Fathers' childhood recollections are less clearly related to attachment, perhaps because fathers typically spend less time with their babies than do mothers (van IJzendoorn & Bakermans-Kranenburg, 1995).

This baby boy has formed a secure, deeply affectionate relationship with his father. According to ethological theory, the father's internal working model, or set of expectations about the availability of attachment figures in his own life, is likely to be objective, balanced, and optimistic. (Steve Starr/Stock Boston)

SOCIAL ISSUES

IS INFANT DAY CARE A THREAT TO ATTACHMENT SECURITY?

Recent research suggests that American infants placed in full-time day care (more than 20 hours per week) before 12 months of age are more likely than home-reared babies to display insecure attachment—especially avoidance—in the Strange Situation. Does this mean that infants who experience daily separations from their employed mothers and early day care are at risk for developmental problems? Some researchers think so (Belsky & Braungart, 1991; Sroufe, 1988), whereas others disagree (Clarke-Stewart, 1989; Scarr, Phillips, & McCartney, 1990). Yet a close look at the evidence reveals that we should be extremely cautious about concluding that day care is harmful to babies.

First, the rate of attachment insecurity among day-care infants is only slightly higher than that of non-day-care infants (36 versus 29 percent), and it is similar to the overall figure for children in industrialized countries around the world (Lamb, Sternberg, & Prodromidis, 1992). In fact, most infants of employed mothers are securely attached! Furthermore, not all studies report a difference in attachment quality between day-care and home-reared infants (Roggman et al., 1994). This suggests that early emotional development of day-care children is within normal range.

Second, we have seen that family conditions affect attachment security. Many employed women find the pressures of handling both work and motherhood stressful. Some respond less sensitively to their babies because they are fatigued and harried, thereby risking the infant's security (Owen & Cox, 1988). Others probably value and encourage their infant's independence. In these cases, avoidance in the Strange Situation may represent healthy autonomy rather than insecurity. Finally, poor-quality day care may contribute to the slightly higher rate of insecure attachment among infants of employed mothers. In one study, babies who were insecurely attached to both mother and caregiver tended to be in day-care environments with large numbers of children and few adults, where their bids for attention were frequently ignored (Howes et al., 1988). In contrast, a warm infant–caregiver bond in a high-quality day-care setting can protect development, especially when children's relationship with one or both parents is insecure (Egeland & Hiester, 1995; Howes, Hamilton, & Matheson, 1994).

Taken together, research indicates that a small number of infants may be at risk for attachment insecurity due to inadequate day care and the joint pressures of full-time employment and parenthood. In view of these findings, it makes sense to increase the availability of good day care and to educate parents about the vital role of sensitive, responsive caregiving in early emotional development.

TRY THIS...

■ Return to Chapter 5 and review the signs of developmentally appropriate day care for infants and toddlers on page 161. Which ones are especially important for ensuring early emotional adjustment?

Look carefully at these findings, and you will see that adults with unhappy upbringings are not destined to become insensitive parents. Instead, the way people *view* their childhoods—their ability to look back in an understanding, forgiving way—is much more influential in how they relate to their own children than the actual history of care they received (van IJzendoorn, 1995b).

Felicia eventually took a job; Lisa had returned to work many months before. When mothers divide their time between work and parenting and place their infants in day care, is the quality of attachment affected? The Social Issues box above reviews the controversy over whether infant day care threatens attachment security.

MULTIPLE ATTACHMENTS

We have already indicated that babies develop attachments to a variety of familiar people—not just mothers, but fathers, siblings, grandparents, and substitute caregivers. Although Bowlby (1969) made room for multiple attachments in his theory, he believed that infants are predisposed to direct their attachment behaviors to a single special person, especially when they are distressed. For example, when an anxious, unhappy 1-year-old is permitted to choose between the mother and father for comfort and security, the infant generally chooses the mother (Lamb, 1976). This preference typically declines over the second year of life. An expanding world of attachments enriches the emotional and social lives of many babies.

■ FATHERS. Like mothers', fathers' sensitive caregiving predicts secure attachment—an effect that becomes stronger the more time they spend with babies (Cox et al., 1992). But as infancy progresses, mothers and fathers in many cultures—Australia, India, Israel, Italy, Japan, and the United States—relate to babies in different ways. Mothers

CULTURAL INFLUENCES

FATHER–INFANT RELATIONSHIPS AMONG THE AKA

Among the Aka hunters and gatherers of Central Africa, fathers devote more time to infants than in any other known society. Observations reveal that Aka fathers are within arm's reach of their babies more than half the day. They pick up and cuddle their infants at least five times more often than do fathers in other African hunting-and-gathering societies.

Why are Aka fathers so involved with their babies? Research shows that when husband and wife help each other with many tasks, fathers assist more with infant care. The relationship between Aka husband and wife is unusually cooperative and intimate. Throughout the day, they share hunting, food preparation, and social and leisure activities. Babies are brought along on hunts, and mothers find it hard to carry them long distances.

This explains, in part, why fathers spend so much time holding their infants. But when the Aka return to the campground, fathers continue to devote many hours to infant caregiving. The more Aka parents are together, the greater the father's interaction with his baby (Hewlett, 1992).

This Aka father spends much time in close contact with his baby. In Aka society, husband and wife share many tasks of daily living and have an unusually cooperative and intimate relationship. Infants are usually within arms reach of their fathers, who devote many hours to caregiving. (Barry Hewlett)

devote more time to physical care and expressing affection. Fathers spend more time in playful interaction (Lamb, 1987; Roopnarine et al., 1990). Mothers and fathers also play differently with babies. Mothers more often provide toys, talk to infants, and start conventional games like pat-a-cake and peekaboo. In contrast, fathers tend to engage in more exciting, highly physical bouncing and lifting games, especially with their infant sons (Yogman, 1981).

However, this picture of "mother as caregiver" and "father as playmate" has changed in some families due to the revised work status of women. Employed mothers engage in more playful stimulation of their babies than do unemployed mothers, and their husbands are somewhat more involved in caregiving (Cox et al., 1992). When fathers are the primary caregivers, they retain their highly arousing play style (Lamb & Oppenheim, 1989). Such highly involved fathers are less gender stereotyped in their beliefs, have sympathetic, friendly personalities, and regard parenthood as an especially enriching experience (Lamb, 1987; Levy-Shiff & Israelashvili, 1988). A warm, gratifying marital relationship supports both parents' involvement with babies. But as the Cultural Influences above reveals, it is particularly important for fathers.

■ SIBLINGS. Despite a declining family size, 80 percent of American children still grow up with at least one sibling. In a survey that asked married couples why they wanted more than one child, the most frequent reason was sibling companionship (Bulatao & Arnold, 1977). Yet the arrival of a baby brother or sister is a difficult experience for most preschoolers, who quickly realize that now they must share their parents' attention and affection. They often become demanding and clingy for a period of time and engage in "deliberate naughtiness," especially when the mother is busy caring for the baby (Dunn & Kendrick, 1982).

However, resentment about being displaced is only one dimension of a rich emotional relationship that starts to build between siblings after a baby's birth. An older child can also be seen kissing, patting, and calling out "Mom, he needs you" when the baby cries—signs of growing caring and affection. By the time the baby is about 8 months old, siblings typically spend much time together. Infants of this age are comforted by the presence of their preschool-age brother or sister during their mother's short absences (Stewart, 1983). And in the second year, they often imitate and actively join in play with the older child (Dunn, 1989).

Nevertheless, individual differences in sibling relationships appear shortly after a baby's birth and persist through early childhood (Dunn, 1992). Temperament plays an important role. For example, conflict increases when one sibling is emotionally intense or highly active (Brody, Stoneman, & McCoy, 1994; Dunn, 1994). Parental behavior also makes a difference. Secure infant–mother attachment and warmth toward both children are related to positive sibling ties, whereas coldness is associated with sibling friction (Volling & Belsky, 1992).

Still, when a mother is very positive and playful with a new baby, her preschool-age child is likely to feel slighted and act in a less friendly way toward the infant. Setting

CAREGIVING CONCERNS

Encouraging Affectional Ties Between Infants and Their Preschool Siblings

SUGGESTION	DESCRIPTION
Spend extra time with the older child	To minimize feelings of being deprived of affection and attention, set aside time to spend with the older child. Fathers can be especially helpful in this regard, planning special outings with the preschooler and taking over care of the baby so the mother can be with the older child.
Handle sibling misbehavior with patience.	Respond patiently to the older sibling's misbehavior and demands for attention, recognizing that these reactions are temporary. Give the preschooler opportunities to feel proud of being more grown up than the baby. For example, encourage the older child to assist with feeding, bathing, dressing, and offering toys, and show appreciation for these efforts.
Discuss the baby's wants and needs.	Discuss the baby's feelings and intentions with the preschooler, such as "He's so little that he just can't wait to be fed" or "He's trying to reach his rattle and can't." By helping the older sibling understand the baby's point of view, parents can promote friendly, considerate behavior.

Sources: Dunn & Kendrick, 1982; Howe & Ross, 1990.

aside special times to devote to the older child supports sibling harmony. In addition, mothers who discuss a baby's feelings and intentions have preschoolers who are more likely to comment on the infant as a person with special wants and needs. Such children behave in an especially considerate and friendly manner toward the baby (Dunn & Kendrick, 1982; Howe & Ross, 1990). The Caregiving Concerns table above suggests ways to promote positive relationships between babies and their preschool siblings.

■ ATTACHMENT AND LATER DEVELOPMENT. According to psychoanalytic and ethological theories, the inner feelings of affection and security that result from a healthy attachment relationship support all aspects of psychological development. Consistent with this view, research indicates that quality of attachment to the mother in infancy is related to cognitive and social development. Preschoolers who were securely attached as babies show greater enthusiasm and persistence on problem-solving tasks. And such children also tend to be rated higher in social skills and play maturity and to have more favorable relationships

Although the arrival of a baby brother or sister is a difficult experience for most preschoolers, a rich emotional relationship quickly builds between siblings. This toddler is already actively involved in play with his 4-year-old brother, and both derive great pleasure from the interaction. (Erika Stone/Photo Researchers, Inc.)

with peers, both in early and middle childhood (Elicker, Englund, & Sroufe, 1992; Matas, Arend, & Sroufe, 1978; Shulman, Elicker, & Sroufe, 1994).

However, a close look at insecurely attached infants reveals a less clear picture: In some studies, they developed adjustment problems, but in others they did not (Fagot & Kavanaugh, 1990; Lyons-Ruth, Alpern, & Repacholi, 1993; Rothbaum et al., 1995). Perhaps *continuity of caregiving* determines whether attachment insecurity is linked to later problems (Lamb et al., 1985). In other words, when parents react insensitively for a very long time, children are likely to become maladjusted. But infants and young children are resilient beings. A child who has compensating, affectional ties outside the family or whose parent's caregiving improves is likely to fare well. This suggests that efforts to create warm, responsive environments are not just important in infancy; they are also worthwhile at later ages. Indeed, we will discover this in subsequent chapters.

BRIEF REVIEW

According to ethological theory, infant–caregiver attachment has evolved because it promotes survival. In early infancy, the baby's innate signals help keep the parent nearby. By 6 to 8 months, separation anxiety and use of the mother as a secure base indicate that a true attachment has formed. Representation and language help toddlers tolerate brief separations from the parent. Research on infants deprived of a consistent caregiver suggests that fully normal development depends on establishing a close affectional bond during the first few years of life. Caregiving that is responsive to babies' needs supports secure attachment; insensitive caregiving is linked to attachment insecurity. Family conditions, such as a change in marital or employment status, can affect the quality of attachment. Besides mothers, fathers and siblings are influential attachment figures. Early attachment security predicts cognitive and social competence at later ages, but continuity of caregiving may be largely responsible for this relationship.

ASK YOURSELF . . .

■ *Return to Chapter 5 and list the child-rearing practices that promote early cognitive development. How do they compare with those that foster secure attachment?*

■ *Recall from Chapter 5 that Lisa tended to overwhelm Byron with questions and instructions. How would you expect Byron to respond in the Strange Situation? Explain your answer.*

■ *Maggy works full-time and leaves her 14-month-old son Vincent at a day-care center. When she arrives to pick him up at the end of the day, Vincent keeps on playing and hardly notices her. Maggy wonders whether Vincent is securely attached. Is Maggy's concern warranted?*

SELF-DEVELOPMENT DURING THE FIRST TWO YEARS

Infancy is a rich, formative period for the development of physical and social understanding. In Chapter 5, you learned that infants develop an appreciation of the permanence of objects. And in this chapter, we saw that over the first year, infants recognize and respond appropriately to others' emotions and distinguish familiar people from strangers. The fact that both objects and people achieve an independent, stable existence in infancy implies that knowledge of the self as a separate, permanent entity emerges around this time.

SELF-RECOGNITION

Felicia hung a large plastic mirror on the side of April's crib. As early as the first few months, April smiled and returned friendly behaviors to her own image. At what age did she realize that the charming baby gazing and grinning back was really herself?

To answer this question, researchers have exposed infants and toddlers to images of themselves in mirrors, on videotapes, and in photos. In one study, 9- to 24-month-olds were placed in front of a mirror. Then, under the pretext of wiping the baby's face, each mother rubbed red dye on her infant's nose. Younger infants touched the mirror as if the red mark had nothing to do with themselves. But by 15 months, toddlers began to rub their strange-looking little red noses—a response that indicates the beginnings of self-recognition (Bullock & Lutkenhaus, 1990; Lewis & Brooks-Gunn, 1979). By the end of toddlerhood, self-recognition is well established. Two-year-olds look and smile more at a photo of themselves than one of another child. And almost all use their name or a personal pronoun ("I" or "me") to refer to themselves (Lewis & Brooks-Gunn, 1979).

How do toddlers develop an awareness of the self's existence? Many theorists believe that the beginnings of self lie in infants' *sense of agency*—recognition that their own actions cause objects and people to react in predictable ways. Parents who encourage babies to explore the environment and who respond to their signals consistently and sensitively help them construct a sense of

This infant notices the correspondence between his own movements and the movements of the image in the mirror, a cue that helps him figure out that the grinning baby is really himself. (Paul Damien/Tony Stone Worldwide)

agency. Then, as infants act on the environment, they notice different effects that may help them sort out self from objects and other people (Lewis, 1991; Pipp, Easterbrooks, & Brown, 1993).

Self-awareness quickly becomes a central part of children's emotional and social lives. Recall that self-conscious emotions depend on toddlers' emerging sense of self. Self-recognition also leads to the first signs of **empathy**—the ability to understand and respond sympathetically to the feelings of others. For example, toddlers start to give to others what they themselves find most comforting—a hug, a reassuring comment, or a favorite doll or blanket (Bischof-Köhler, 1991; Zahn-Waxler et al., 1992). At the same time, they demonstrate a clearer awareness of how to upset and frustrate other people. One 18-month-old heard her mother comment to another adult, "Anny (sibling) is really frightened of spiders. In fact, there's a particular toy spider that we've got that she just hates" (Dunn, 1989, p. 107). The innocent-looking toddler ran to get the spider out of the toy box, returned, and pushed it in front of Anny's face!

CATEGORIZING THE SELF

Self-awareness permits toddlers to compare themselves to other people, in much the same way that they group together physical objects (see Chapter 5). Between 18 and 30 months, children categorize themselves and others on the basis of age ("baby," "boy," or "man"), sex ("boy" or "girl"), and even goodness and badness ("I a good girl" and "Tommy mean!") (Stipek, Gralinski, & Kopp, 1990). Toddlers' understanding of these social categories is quite lim-

ited, but they use this knowledge to organize their own behavior. For example, children's ability to label their own gender is associated with a sharp rise in gender-stereotyped responses (Fagot & Leinbach, 1989). As early as 18 months, toddlers select and play in a more involved way

As early as 18 months, children select gender-stereotyped toys and play with them in a more involved way. This toddler can probably label his own gender—an attainment associated with a sharp rise in gender-stereotyped responses. He has begun to use his knowledge of social categories to organize his behavior. (Suzanne Haldane/Stock Boston)

Helping Toddlers Develop Compliance and Self-Control

SUGGESTION	RATIONALE
Respond to the toddler warmly and sensitively.	Toddlers who experience warmth and sensitivity are far more compliant and cooperative than negative and resistant.
Provide advance notice when the toddler must stop an enjoyable activity.	Toddlers find it more difficult to stop a pleasant activity already underway than to wait before engaging in a desired action.
Offer many prompts and reminders.	Toddlers' ability to remember and comply with rules is limited; they need continuous adult oversight.
Respond to self-controlled behavior with verbal and physical approval.	Praise and hugs reinforce appropriate behavior, increasing its likelihood of occurring again.
Support language development.	Early language development is related to self-control. During the second year, children begin to use language to remind themselves about adult expectations.
Gradually increase rules in accord with the toddler's developing capacities.	As cognition and language improve, toddlers can follow more rules related to safety, respect for people and property, family routines, manners, and simple chores.

with toys that are stereotyped for their own gender—dolls and tea sets for girls, trucks and cars for boys. Then parents encourage these preferences by responding positively when toddlers display them (Fagot, Leinbach, & O'Boyle, 1992). As we will see in Chapter 8, gender-typed behavior increases dramatically in early childhood.

EMERGENCE OF SELF-CONTROL

Self-awareness also provides the foundation for **self-control,** the capacity to resist an impulse to engage in socially disapproved behavior. Self-control is essential for morality, another dimension of the self that will flourish during childhood and adolescence. To behave in a self-controlled fashion, children must have some ability to think of themselves as separate, autonomous beings who can direct their own actions. And they must also have the representational and memory capacities to recall a caregiver's directive (such as "April, don't touch that light socket!") and apply it to their own behavior (Kopp, 1987).

The first glimmerings of self-control appear as **compliance.** Between 12 and 18 months, children show clear awareness of caregivers' wishes and expectations and can obey simple requests and commands (Kaler & Kopp, 1990). And as every parent knows, they can also decide to do just the opposite! One way toddlers assert their autonomy is by resisting adult directives. But think back to Erikson's theory, which suggests that parenting practices have much to do with a healthy sense of self. Among toddlers

who experience warm, sensitive caregiving and reasonable expectations for mature behavior, opposition is far less common than compliance and cooperation (Kochanska, Aksan, & Koenig, 1995). Compliance quickly leads to toddlers' first consciencelike verbalizations, as when April said, "No, can't" as she reached out for a light socket (Kochanska, 1993).

Around 18 months, the capacity for self-control appears, and it improves steadily into early childhood. For example, in one study, toddlers were given three tasks that required them to resist temptation. In the first, they were asked not to touch a toy telephone that was within arm's reach. In the second, raisins were hidden under cups, and they were instructed to wait until the experimenter said it was all right to pick up a cup and eat a raisin. In the third, they were told not to open a gift until the experimenter had finished her work. On all three problems, the ability to wait increased steadily between 18 and 30 months. Toddlers who were especially self-controlled were advanced in language development. In fact, some sang and talked to themselves to keep from touching the desired objects (Vaughn, Kopp, & Krakow, 1984).

As self-control improves, mothers require toddlers to follow more rules, from safety and respect for property and people to family routines, manners, and simple chores (Gralinski & Kopp, 1993). Still, toddlers' control over their own actions is fragile. It depends on constant oversight and reminders by parents. To get April to stop playing to go on an errand, Felicia found that several prompts ("Remember, we're going to go in just a minute") and gentle insistence were usually necessary. The Caregiving Concerns table on the opposite page summarizes effective ways of helping toddlers develop compliance and self-control.

As the second year of life drew to a close, Lisa, Beth, and Felicia were delighted at their youngsters' readiness to learn the rules of social life. As we will see in Chapter 8, advances in cognition and language, along with parental warmth and reasonable maturity demands, lead children to make tremendous strides in this area during early childhood.

ASK YOURSELF . . .

■ *Nine-month-old Harry turned his cup upside down and spilled juice all over the tray of his high chair. His mother said sharply, "Harry, put your cup back the right way!" Can Harry comply with his mother's request? Why or why not?*

SUMMARY

THEORIES OF INFANT AND TODDLER PERSONALITY

Explain Erikson's and Mahler's theories of infant and toddler personality.

■ According to Erikson, warm, responsive caregiving leads infants to resolve the psychological conflict of Freud's **oral stage—basic trust versus mistrust**—positively. During the **anal stage, autonomy versus shame and doubt** is resolved favorably when parents provide toddlers with appropriate guidance and reasonable choices.

■ In Mahler's theory, sensitive, loving care fosters **symbiosis** in infancy, which provides the foundation for **separation–individuation** during toddlerhood. The capacity to move away from the mother, by crawling and then walking, leads to self-awareness. Gains in representation and language help create a positive, inner image of the mother that the child can rely on in her absence.

EMOTIONAL DEVELOPMENT DURING THE FIRST TWO YEARS

Describe changes in happiness, anger, and fear over the first year, noting the adaptive function of each.

■ Signs of almost all the **basic emotions** are present in infancy. The **social smile** appears between 6 and 10 weeks, laughter around 3 to 4 months. Happiness strengthens the parent–child bond and reflects as well as supports physical and cognitive mastery. Anger and fear, especially in the form of **stranger anxiety**, increase in the second half of the first year. These reactions have survival value as infants move about on their own.

Summarize changes in understanding others' emotions, expression of self-conscious emotions, and emotional self-regulation during the first 2 years.

■ The ability to understand the feelings of others expands over the first year as babies perceive facial expressions as organized patterns. Around 8 to 10 months, **social referencing** appears. During toddlerhood, self-awareness and adult instruction provide the foundation for **self-conscious emotions,** such as shame, embarrassment, and pride. Caregivers help infants with **emotional self-regulation** by relieving distress, engaging in stimulating play, and discouraging negative emotion. During the second year, growth in representation and language leads to more effective ways of regulating emotion.

TEMPERAMENT AND DEVELOPMENT

What is temperament, and how is it measured?

■ Infants differ greatly in **temperament,** or style of emotional responding. Temperament is most often assessed through parental reports, although behavior ratings and direct observations are also used. Researchers are beginning to identify physiological reactions that are markers of temperament. Through parental descriptions, three temperamental patterns—the **easy child,** the **difficult child,** and the **slow-to-warm-up child**—were identified in the New York Longitudinal Study.

Discuss the role of heredity and environment in the stability of temperament, including the goodness-of-fit model.

■ Many temperamental characteristics are moderately stable over time. Temperament has biological roots, but child rearing has much to do with maintaining or changing it. The term **goodness-of-fit** describes how temperament and a child's environment work together to affect later development. Parenting practices that have a good fit with the child's basic emotional style are important for all children, but especially for those who are difficult or shy and withdrawn.

DEVELOPMENT OF ATTACHMENT

Describe the ethological theory of attachment and the development of attachment during the first 2 years.

■ The development of **attachment,** infants' strong affectional tie to familiar caregivers, has been the subject of intense theoretical debate. The most widely accepted perspective is **ethological theory,** which views attachment as a relationship that evolved because it promotes survival.

■ The baby contributes actively to the formation of the attachment bond. In early infancy, a set of built-in behaviors encourages the parent to care for the baby. Around 6 to 8 months, **separation anxiety** and use of the parent as a **secure base** from which to explore indicate that a true attachment relationship has formed. As representation and language develop, toddlers try to alter the parent's coming and going through requests and persuasion rather than following and clinging. Out of early

caregiving experiences, children construct an **internal working model** that serves as a guide for all future close relationships.

Describe the Strange Situation, the four attachment patterns assessed by it, and factors that affect attachment security.

■ The **Strange Situation** is used between ages 1 and 2 to identify four patterns of attachment: **secure attachment, avoidant attachment, resistant attachment, and disorganized–disoriented attachment.** Cultural conditions must be considered in interpreting reactions in the Strange Situation.

■ Attachment quality is influenced by the infant's opportunity to develop a close affectional tie with one or a few adults, the caregivers' sensitivity and responsiveness, the fit between the baby's temperament and parenting practices, and family circumstances. In some (but not all) cultures, **interactional synchrony** characterizes communication between mothers and securely attached babies. Mothers' internal working models are good predictors of infant attachment patterns.

Discuss infants' attachments to fathers and siblings.

■ Besides attachments to mothers, infants develop strong affectional ties to fathers, usually through stim-

ulating, playful interaction. Early in the first year, infants begin to build rich emotional relationships with siblings. Individual differences in the quality of sibling relationships are influenced by temperament and parenting practices.

Describe and interpret the relationship between secure attachment in infancy and cognitive and social competence in childhood.

■ Secure attachment in infancy is related to more effective cognitive and social development in childhood. However, continuity of parenting may determine whether an infant who is insecurely attached shows later adjustment problems.

SELF-DEVELOPMENT DURING THE FIRST TWO YEARS

When does self-recognition develop, and what emotional and social capacities does it contribute to?

■ By the end of the second year, infants recognize themselves in mirrors, on videotapes, and in photos. Self-awareness permits toddlers to compare themselves to other people and form primitive social categories based on age, sex, and goodness and badness. It also provides the foundation for self-conscious emotions, **empathy, compliance,** and **self-control.**

IMPORTANT TERMS AND CONCEPTS

oral stage (p. 177)

basic trust versus mistrust (p. 177)

anal stage (p. 177)

autonomy versus shame and doubt (p. 177)

symbiosis (p. 178)

separation–individuation (p. 178)

basic emotions (p. 179)

social smile (p. 180)

stranger anxiety (p. 180)

social referencing (p. 181)

self-conscious emotions (p. 181)

emotional self-regulation (p. 182)

temperament (p. 183)

easy child (p. 183)

difficult child (p. 183)

slow-to-warm-up child (p. 183)

goodness-of-fit (p. 185)

attachment (p. 187)

ethological theory of attachment (p. 188)

separation anxiety (p. 188)

secure base (p. 188)

internal working model (p. 189)

Strange Situation (p. 189)

secure attachment (p. 189)

avoidant attachment (p. 189)

resistant attachment (p. 189)

disorganized–disoriented attachment (p. 189)

interactional synchrony (p. 191)

empathy (p. 197)

self-control (p. 198)

compliance (p. 198)

AGE	PHYSICAL	COGNITIVE	LANGUAGE	EMOTIONAL/SOCIAL
Birth–6 months	■ Rapid height and weight gain. ■ Reflexes decline. ■ Sleep organized into a day/night schedule. ■ Holds head up, rolls over, and reaches for objects. ■ Can be classically and operantly conditioned. ■ Habituates to unchanging stimuli. ■ Hearing well developed; by the end of this period, displays greater sensitivity to speech sounds of own language. ■ Depth and pattern perception emerge and improve.	■ Engages in immediate imitation and deferred imitation of adults' facial expressions. ■ Repeats chance behaviors leading to pleasurable and interesting results. ■ Displays object permanence in habituation–dishabituation task. ■ Recognition memory for people, places, and objects improves. ■ Can categorize stimuli on the basis of size, shape, number (up to three elements), and other physical properties.	■ Engages in cooing and, by the end of this period, babbling. ■ Establishes joint attention with caregiver, who labels objects and events. 	■ Shows signs of almost all basic emotions (happiness, interest, surprise, fear, anger, sadness, disgust). ■ Social smile and laughter emerge. ■ Matches adults' emotional expressions during face-to-face interaction. ■ Emotional expressions become better organized and clearly tied to social events. 
7–12 months	■ Sits alone, crawls, and walks. ■ Shows refined pincer grasp. ■ Perceives larger speech units critical to understanding meaning. ■ Depth and pattern perception improve further.	■ Combines sensorimotor schemes. ■ Engages in intentional, or goal-directed, behavior. ■ Finds object hidden in one place. ■ Engages in deferred imitation of adults' actions on objects. ■ Recall memory for people, places, and objects improves. ■ Groups stimuli into wide range of meaningful categories.	■ Babbling expands to include sounds of spoken languages and, by the end of this period, sounds of the child's language community. ■ Uses preverbal gestures (showing, pointing) to communicate. 	■ Anger and fear increase in frequency and intensity. ■ Stranger anxiety and separation anxiety appear. ■ Uses caregiver as a secure base for exploration. ■ Shows "clearcut" attachment to familiar caregivers. ■ Ability to detect the meaning of others' emotional expressions improves. ■ Engages in social referencing.

AGE	PHYSICAL	COGNITIVE	LANGUAGE	EMOTIONAL/SOCIAL
13–18 months	■ Height and weight gain rapid, but not as great as in first year. ■ Walking better coordinated. ■ Scribbles with pencil. ■ Builds tower of 2–3 cubes.	■ Experiments with objects in a trial-and-error fashion. ■ Finds object hidden in more than one place. ■ Actively categorizes objects during play.	■ Actively joins in turn-taking games, such as pat-a-cake and peeka-boo. ■ Says first words. ■ Makes errors of underextension and overextension.	■ Actively joins in play with siblings. ■ Recognizes image of self in mirror and on videotape. ■ Shows signs of empathy. ■ Capable of compliance.

AGE	PHYSICAL	COGNITIVE	LANGUAGE	EMOTIONAL/SOCIAL
19–24 months	■ Jumps, runs, and climbs. ■ Manipulates objects with good coordination. ■ Builds tower of 4–5 cubes.	■ Solves sensorimotor problems suddenly. ■ Finds object moved while out of sight. ■ Engages in deferred imitation of complex action sequences and imitates actions an adult *tries* to produce, even if not fully realized. ■ Engages in make-believe play. ■ Actively categorizes objects more effectively.	■ Vocabulary increases to 200 words. ■ Combines two words; a beginning appreciation of grammar is present.	■ Displays self-conscious emotions (shame, embarrassment, and pride). ■ Acquires a vocabulary of emotional terms. ■ Starts to use language to assist with emotional self-regulation. ■ Begins to tolerate caregiver absences more easily. ■ Uses own name or personal pronoun to label self. ■ Categorizes self and others on the basis of age, sex, and goodness and badness. ■ Shows gender-stereotyped toy choices. ■ Self-control appears.

Physical and Cognitive Development in Early Childhood

For more than a decade, my fourth-floor office window overlooked the preschool and kindergarten play yard of our university laboratory school. On mild fall and spring mornings, the doors of the classrooms swung open, and sand table, woodworking bench, easels, and large blocks spilled out into a small, fenced courtyard. Around the side of the building was a grassy area with jungle gyms, swings, a small playhouse, and a flower garden planted by the children. Beyond it, I could see a circular path lined with tricycles and wagons. Each day, the setting was alive with activity.

The years from 2 to 6 are often called "the play years," and aptly so, since play blossoms during this time and supports every aspect of development. Our discussion of early childhood opens with the physical achievements of this period—growth in body size, improvements in motor coordination, and refinements in perception. We pay special attention to biological and environmental factors that support these changes, as well as to their intimate connection with other aspects of development.

Then we explore the many facets of early childhood cognition, drawing from three theories with which you are already familiar. We begin with Piaget's preoperational stage, which, for the most part, emphasizes young children's deficits rather than strengths. Recent research on preoperational thought, along with Vygotsky's sociocultural theory and information processing, extends our understanding of preschoolers' cognitive competencies. Next, we turn to a variety of factors that contribute to early childhood mental development—the home environment, the quality of preschool and day care, and the many hours children spend watching television. Our chapter concludes with language development, the most awesome achievement of early childhood.

The preschool children whom I came to know well, first by watching from my office window and later by observing at close range in their classrooms, will join us as we trace the changes of early childhood. They will provide us with many examples of developmental trends and individual differences.

PHYSICAL DEVELOPMENT IN EARLY CHILDHOOD

BODY GROWTH

Look at Figure 7.1, and you will see that the rapid increase in body size that took place in infancy tapers off into a slower pattern of growth during early childhood. On the average, children add 2 to 3 inches in height and about 5 pounds in weight each year. Boys

Toddlers and 5-year-olds have very different body shapes. During early childhood, body fat declines, the torso enlarges to better accommodate the internal organs, and the spine straightens. Compared to her younger brother, this girl looks more streamlined. Her body proportions resemble those of an adult. *(Bob Daemmrich/Stock Boston)*

continue to be slightly larger than girls. At the same time, the "baby fat" that began to decline in toddlerhood drops off further. The child gradually becomes thinner, although girls retain somewhat more body fat, and boys are slightly more muscular. As the torso lengthens and widens, internal organs tuck neatly inside, and the spine straightens. By age 5, the top-heavy, bowlegged, potbellied toddler has become a more streamlined, flat-tummied, longer-legged child with body proportions similar to those of adults (Tanner, 1990).

Individual differences in body size are even more apparent during early childhood than in infancy. Looking down at the play yard one day, I watched 5-year-old Darryl speed around the bike path. At 48 inches tall and 55 pounds, he towered over his kindergarten classmates and was, as his mother put it, "off the growth charts" at the doctor's office. (The average American 5-year-old boy is 43 inches tall and weighs 42 pounds.) Priti, an Asian-Indian child, was unusually small because of her ethnic heritage. Lynette and Hallie, two Caucasian children with impoverished home lives, were well below average for reasons we will discuss shortly.

SKELETAL GROWTH

Skeletal changes underway in infancy continue throughout early childhood. Between ages 2 and 6, approximately 45 new *epiphyses,* or growth centers in which cartilage hardens into bone, emerge in various parts of the skeleton and can be seen in X-rays. Other epiphyses will appear in middle childhood. X-rays permit doctors to estimate

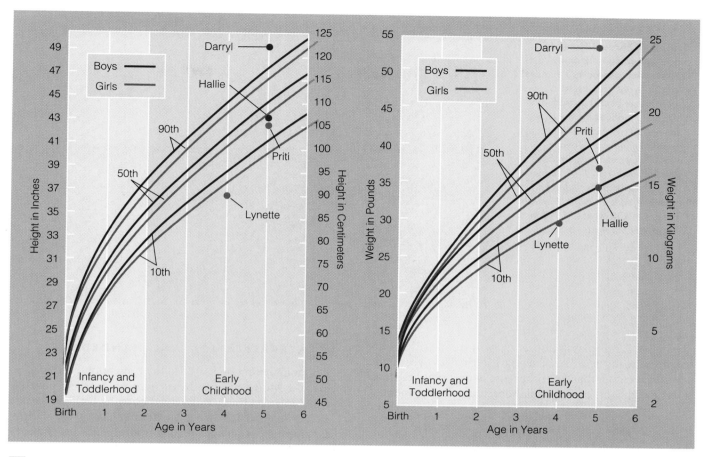

FIGURE 7.1

Gains in height and weight during early childhood among North American children. Compared to the first 2 years of life, growth is slower. Girls continue to be slightly shorter and lighter than boys. Wide individual differences in body size exist, as the percentiles on these charts reveal. Darryl, Priti, Lynette, and Hallie's heights and weights differ greatly.

children's *skeletal age*, the best measure of progress toward physical maturity. During early and middle childhood, information about skeletal age can help diagnose growth disorders.

Parents and children are especially aware of another aspect of skeletal growth: By the end of the preschool years, children start to lose their primary or "baby" teeth. The age at which they do so is heavily influenced by physical maturation. For example, girls, who are ahead of boys, lose their primary teeth sooner. Environmental influences, especially prolonged malnutrition, can delay the appearance of permanent teeth.

Even though primary teeth are temporary, care of them is important. Diseased baby teeth can affect the health of permanent teeth. Consistent brushing, avoiding sugary foods, and regular dental visits prevent tooth cavities. Unfortunately, childhood tooth decay remains high, especially among low-income youngsters in the United States and among children in developing countries. Poor diet, lack of fluoridation, and inadequate health care are responsible (Edelstein & Douglass, 1995).

ASYNCHRONIES IN PHYSICAL GROWTH

From what you have learned so far in this chapter and in Chapter 4, can you come up with a single overall description of physical growth? If you answered no, you are correct. As Figure 7.2 on page 208 shows, physical growth is *asynchronous:* different body systems have unique patterns of maturation. Body size (as measured by height and weight) and a variety of internal organs follow the **general growth curve.** It involves rapid growth during infancy, slower gains in early and middle childhood, and rapid growth again during adolescence. Yet there are exceptions to this trend. The genitals develop slowly from birth to age 4, change little throughout middle childhood, and then grow rapidly during adolescence. In contrast, the lymph tissue (small clusters of glands found throughout the body) grows at an astounding pace in infancy and childhood but declines in adolescence. The lymph system helps fight infection and assists with absorption of nutrients, thereby supporting children's health and survival (Malina & Bouchard, 1991).

FIGURE 7.2

Growth of three different organ systems and tissues contrasted with the body's general growth. Growth is plotted in terms of percentage of change from birth to 20 years. Notice how the lymph tissue rises to twice its adult level by the end of childhood. Then it declines. *(Reprinted by permission of the publisher from J. M. Tanner, 1990, Foetus into Man [2nd ed.], Cambridge, MA: Harvard University Press, p. 16. Copyright © 1990 by J. M. Tanner. All rights reserved.)*

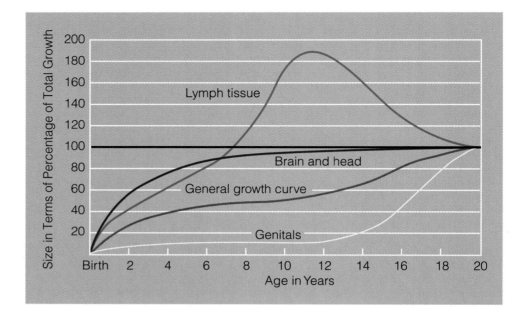

Figure 7.2 displays another growth trend with which you are already familiar: During the first few years, the brain grows faster than any other part of the body. Let's look at some highlights of early childhood brain development.

BRAIN DEVELOPMENT

Between 2 and 6 years, the brain increases from 70 to 90 percent of its adult weight (Tanner, 1990). As neural fibers continue to *myelinate* (see Chapter 4), preschoolers improve in a wide variety of skills—physical coordination, perception, attention, memory, language, logical thinking, and imagination.

Recall that the cortex is made up of two *hemispheres.* Measures of brain electrical activity reveal that the hemispheres develop at different rates. For most children, the left hemisphere shows a dramatic growth spurt between 3 and 6 years and then levels off. In contrast, the right hemisphere matures slowly throughout early and middle childhood, with a slight growth spurt between ages 8 and 10 (Thatcher, Walker, & Giudice, 1987).

These findings fit nicely with what we know about several aspects of cognitive development. Language skills (typically housed in the left hemisphere) increase at an astonishing pace in early childhood. In contrast, spatial skills (such as finding one's way from place to place, drawing pictures, and recognizing geometric shapes) develop gradually over childhood and adolescence. Differences between the two hemispheres in rate of maturation suggest that they are continuing to *lateralize,* or specialize, in cognitive functions.

LATERALIZATION AND HANDEDNESS

One morning on a visit to the preschool, I watched 3-year-old Moira as she drew pictures, worked puzzles, joined in snacktime, and played outside. Unlike most of her classmates, Moira does most things—drawing, eating, and zipping her jacket—with her left hand. But she uses her right hand for a few activities, such as throwing a ball.

Although left-handedness is associated with developmental problems, the large majority of left-handed children are completely normal. (Gael Zucker/Stock Boston)

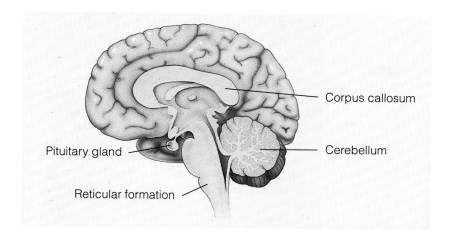

FIGURE 7.3

Cross-section of the human brain, showing the location of the cerebellum, the reticular formation, and the corpus callosum. These structures undergo considerable development during early childhood. Also shown is the pituitary gland, which secretes hormones that control body growth (see page 210).

Handedness shows up early in development. By 5 to 6 months, more infants reach for objects with their right hand than with their left. By age 2, hand preference is stable, and it strengthens during early and middle childhood. As we will see, this indicates that specialization of brain regions increases during this time.

A strong hand preference reflects the greater capacity of one side of the brain—the individual's **dominant cerebral hemisphere**—to carry out skilled motor action. Other abilities located on the dominant side may also be superior. In support of this idea, for right-handed people, who make up 90 percent of the population, language is housed with hand control in the left hemisphere. For the remaining 10 percent, who are left-handed, language is often shared between the hemispheres rather than located in only one. This indicates that the brains of left-handers tend to be less strongly lateralized than those of right-handers (Hiscock & Kinsbourne, 1987). Consistent with this idea, many left-handed individuals (like Moira) are also *ambidextrous*. Although they prefer their left hand, they sometimes use their right hand skillfully as well (McManus et al., 1988).

Is handedness and, along with it, specialization of brain functions hereditary? Although researchers disagree on this issue, certain findings argue against a genetic explanation. Twins—whether identical or fraternal—are more likely than ordinary siblings to differ in handedness. The hand preference of each twin is related to body position during the prenatal period (twins usually lie in opposite orientations) (Perelle & Ehrman, 1994). According to one theory, the way most fetuses lie (turned toward the left) may promote greater postural control by the right side of the body (Previc, 1991).

What about children like Moira whose hand use suggests an unusual organization of brain functions? Do these youngsters develop normally? Perhaps you have heard that left-handedness is more frequent among severely retarded and mentally ill people than it is in the general population. Although this is true, recall that when two variables are correlated, one does not necessarily cause the other. Atypi-

cal lateralization is probably not responsible for the problems of these individuals. Instead, they may have suffered early damage to the left hemisphere, which caused their disabilities and also led to a shift in handedness. In support of this idea, left-handedness shows a slight association with prenatal and birth difficulties that can result in brain damage, including prematurity, prolonged labor, Rh incompatibility, and breech delivery (Coren & Halpern, 1991; Williams, Buss, & Eskenazi, 1992).

Keep in mind, however, that only a small number of left-handers show developmental problems of any kind. The great majority, like Moira, are normal in every respect. In fact, the unusual lateralization of left-handed children may have certain advantages. Although we do not yet know why, left- and mixed-handed youngsters are more likely than their right-handed agemates to develop outstanding verbal and mathematical talents by adolescence (Benbow, 1986).

OTHER ADVANCES IN BRAIN DEVELOPMENT

Besides the cortex, other parts of the brain make strides during early childhood (see Figure 7.3). As we look at these changes, you will see that all involve establishing links between different parts of the brain, increasing the coordinated functioning of the central nervous system.

At the rear of the brain is the **cerebellum,** a structure that aids in balance and control of body movement. Fibers linking the cerebellum to the cerebral cortex begin to myelinate after birth, but they do not complete this process until about age 4 (Tanner, 1990). This change undoubtedly contributes to dramatic gains in motor control, so that by the end of the preschool years children can play a simple game of hopscotch, pump a swing, and throw a ball with a well-organized set of movements.

The **reticular formation,** a structure of the brain that maintains alertness and consciousness, myelinates throughout early childhood and into adolescence. Neurons in the reticular formation send out fibers to the frontal lobe of the cortex, contributing to improvements in sustained, controlled attention (McGuinness & Pribram, 1980).

The **corpus callosum** is a large bundle of fibers that connects the two cortical hemispheres so they can communicate directly. Myelinization of the corpus callosum does not begin until the end of the first year of life. By age 4 to 5, its development is fairly advanced, supporting gains in the efficiency of thinking (Witelson & Kigar, 1988).

BRIEF REVIEW

Compared to infancy, gains in body size take place more slowly during early childhood, and the body becomes more streamlined. By the end of the preschool years, children start to lose their primary teeth. Physical growth is asynchronous; different parts of the body have their own unique patterns of maturation. During early childhood, the brain continues to grow more rapidly than the rest of the body. Hand preference strengthens, a sign of increasing brain lateralization. Although a slight association between left-handedness and developmental abnormalities exists, the large majority of left-handed children show no problems of any kind. The cerebellum, the reticular formation, and the corpus callosum develop significantly during early childhood, contributing to connections between different parts of the brain.

ASK YOURSELF . . .

- *After graduating from dental school, Norm entered the Peace Corps and was assigned to rural India. He found that many Indian children had extensive tooth decay. In contrast, the young patients of his dental school friends in the United States had much less. What factors probably account for this difference?*

- *Crystal has a left-handed cousin who is mentally retarded. Recently, she noticed that her 2-year-old daughter Shana is also left-handed. Crystal has heard that left-handedness is a sign of developmental problems, so she is worried about Shana. How would you respond to Crystal's concern?*

- *In one study, preschoolers were asked to close their eyes and feel a small textured pillow with one hand. Then they had to say whether a pillow rubbed across the other hand was made of the same fabric. (Recall from Chapter 4 that each half of the brain receives information from only one side of the body.) Between 3 and 5 years, performance improved greatly (Galin et al., 1979). Which brain structure is responsible for this change?*

INFLUENCES ON PHYSICAL GROWTH AND HEALTH

In earlier chapters, we considered a wide variety of influences on physical growth during the prenatal period and infancy. As we discuss growth and health in early childhood, you will encounter a variety of familiar themes. Although heredity remains powerfully important, environmental factors continue to play crucial roles. Emotional well-being, good nutrition, relative freedom from disease, and physical safety are essential. And as the Social Issues box on the following page illustrates, environmental pollutants can threaten healthy growth. The extent to which low-level lead—one of the most common—undermines children's physical and mental well-being is the subject of intensive research and heated debate.

HEREDITY AND HORMONES

The impact of heredity on physical growth is evident throughout childhood. Children's physical size and rate of growth are related to their parents' (Malina & Bouchard, 1991). Genes influence growth by controlling the body's production of hormones, especially two that are released by the **pituitary gland,** located at the base of the brain.

The first is **growth hormone (GH),** which is necessary for physical development from birth on. Children who lack it reach an average mature height of only 4 feet, 4 inches. When treated with injections of GH, such children show catch-up growth and then begin to grow at a normal rate. Reaching their genetically expected height depends on starting treatment early, before the epiphyses of the skeleton are very mature (Tanner, 1990).

The second hormone affecting children's growth is **thyroid-stimulating hormone (TSH).** It stimulates the thyroid gland (located in the neck) to release *thyroxine,* which is necessary for normal development of the nerve cells of the brain and for GH to have its full impact on body size. Infants born with a deficiency of thyroxine must receive it at once, or they will be mentally retarded. At later ages, children with too little thyroxine grow at a below-average rate. By then, the central nervous system is no longer affected, since the most rapid period of brain development is complete. With prompt treatment, such children catch up in body growth and eventually reach normal adult size (Tanner, 1990).

EMOTIONAL WELL-BEING

As in infancy, mind and body continue to be closely linked during childhood. Preschoolers with very stressful home lives (due to divorce, financial difficulties, or a change in their parents' employment status) suffer more respiratory and intestinal illnesses and unintentional injuries (Beautrais, Fergusson, & Shannon, 1982).

SOCIAL ISSUES

LEAD POISONING IN CHILDHOOD

Five-year-old Desonia lives in an old, dilapidated tenement in a slum area of a large American city. Layers of paint applied over the years can be seen flaking off the inside walls and the back porch. The oldest paint chips are lead based. As an infant and young preschooler, Desonia picked them up and put them in her mouth. The slightly sweet taste of the leaded flakes encouraged her to nibble more. Soon Desonia became listless, apathetic, and irritable, and her appetite dropped off. When she complained of constant headaches, began to walk with an awkward gait, and experienced repeated convulsions (involuntary muscle contractions), her parents took her to a nearby public health clinic. Blood tests showed that she had severe lead poisoning, a condition that results in permanent brain damage and, if allowed to persist, early death (Friedman & Weinberger, 1990).

Severe lead poisoning like Desonia's has declined over the past twenty years in the United States, following passage of laws restricting use of lead-based paint. But lead already present in homes is difficult to remove. And children can also absorb lead through residues in dust and soil—a danger that is especially great in inner-city industrial areas. The persistence of lead in the environment has sparked a broader concern: Is lead contamination a "silent epidemic"? Do children exposed to even small quantities show impaired intellectual functioning?

To answer this question, seven longitudinal studies—three in the United States, two in Australia, one in Costa Rica, and one in Yugoslavia—were completed during the past decade. In each, lead exposure during the second year of life (when blood levels peak) was used to predict later mental development. Other factors (such as poverty, poor nutrition, and stressful home environments) that might account for the lead–IQ relationship were carefully controlled.

Only one of these investigations found a pervasive influence of low-level lead—not just on mental test scores, but on academic performance and educational attainment as well (Needleman et al., 1990). These associations occurred at very low blood concentra-tions—amounts found in 20 percent of all American children and 60 percent of African-American children in urban areas (Berney, 1993).

Alarmed by these findings, in 1991 the U.S. Public Health Service lowered the official "level of concern" for lead and required all children who met it to be followed up with efforts to reduce their exposure immediately. Although widespread agreement exists about controlling environmental lead, the new policy has sparked controversy. Supporters regard the more stringent standard as a necessary precaution. Critics point out that the majority of longitudinal studies showed no effects of low-level lead exposure! They also worry that expensive, time-consuming monitoring of children for low blood lead will divert attention away from far more powerful causes of intellectual impairment, such as poverty and poor nutrition (Wasserman et al., 1992; Wolf, Jimenez, & Lozoff, 1994).

TRY THIS...

■ Return to Chapter 2, page 66, and review the factors that affect public policy. What might have prompted the U.S. Public Health Service to lower the "level of concern" for lead, despite inconsistent research findings? Do you agree with the new policy? Why or why not?

■ Telephone a pediatrician's office or health clinic. Ask what kind of lead testing is routinely done (by law or otherwise) as part of children's physical exams in your community.

When emotional deprivation is extreme, it can interfere with the production of growth hormone and lead to **deprivation dwarfism,** a growth disorder that appears between 2 and 15 years of age. Lynette, the very small 4-year-old mentioned earlier in this chapter, was diagnosed with this condition. She had been placed in foster care after child welfare authorities discovered that she spent most of the day at home alone, unsupervised. She may also have been physically abused. To help her recover, Lynette was enrolled in our laboratory preschool. She showed the typical characteristics of deprivation dwarfism—very short stature, light weight in proportion to her height, immature skeletal age, and decreased GH secretion. When such chil-dren are removed from their emotionally inadequate environments, their GH levels quickly return to normal, and they grow rapidly. But if treatment is delayed, the dwarfism can be permanent (Oates, Peacock, & Forrest, 1985).

NUTRITION

With the transition to early childhood, many children become unpredictable and picky eaters. One father I know wistfully recalled his son's eager sampling of the cuisine at a Chinese restaurant during toddlerhood. "He ate rice, chicken chow mein, egg rolls, and more. Now, at age 3, the only thing he'll try is the ice cream!"

Unpredictable appetites and picky eating are common during early childhood. At dinnertime, this 3-year-old girl seems more interested in playing than eating. Fortunately, her parents are sensitive to her nutritional needs. They offer a well-balanced meal and put only a small portion on her plate. *(Joel Gordon)*

This decline in appetite is normal. It occurs because growth has slowed. And preschoolers' wariness of new foods is adaptive, since they are still learning which items are safe to eat and which are not. By sticking to familiar foods, they are less likely to swallow dangerous substances when adults are not around to protect them (Rozin, 1990).

 Even though they eat less, preschoolers need a high-quality diet. They require the same foods adults do—only in smaller amounts. Fats, oils, and salt should be kept to a minimum because of their link to high blood pressure and heart disease in adulthood. Foods high in sugar should also be avoided. In addition to causing tooth decay, sugary cereals, cookies, cakes, soft drinks, and candy reduce young children's appetite for healthy foods.

How can preschoolers be encouraged to eat a variety of foods? Repeated exposure to a new food (without any direct pressure to eat it) increases children's acceptance. In one study, preschoolers were given one of three versions of a food they had never eaten (sweet, salty, or plain tofu). After 8 to 15 exposures, they readily ate the food. But they preferred the version they had already tasted. For example, children in the "sweet" condition liked sweet tofu best, and those in the "plain" condition liked plain tofu best (Sullivan & Birch, 1990). These findings reveal that children's tastes are trained by foods served often in their culture. Adding sugar or salt in hopes of increasing a young child's willingness to eat a healthy food simply teaches the child to like a sugary or salty taste.

The emotional climate at mealtimes has a powerful impact on children's eating habits. Many parents worry about how well their preschoolers eat, so meals become unpleasant and stressful. Sometimes parents bribe their children, saying, "Finish your vegetables, and you can have an extra cookie." Unfortunately, this practice only causes children to like the healthy food less and the treat more! (Birch, Johnson, & Fisher, 1995). The Caregiving Concerns table on the following page offers some suggestions for promoting healthy, varied eating in young children.

Finally, as we indicated in earlier chapters, many children in the United States and in developing countries are deprived of diets that support healthy growth. Five-year-old Hallie was bused to our laboratory preschool from a poor neighborhood on the west side of town. His mother's welfare check was barely enough to cover the cost of housing, let alone food. Hallie's diet was deficient in protein as well as essential vitamins and minerals—iron (to prevent anemia), calcium (to support development of bones and teeth), vitamin A (to help maintain eyes, skin, and a variety of internal organs), and vitamin C (to facilitate iron absorption and wound healing). These are the most common dietary deficiencies of the preschool years. Not surprisingly, Hallie was small for his age, and he often seemed tired. By age 7, American low-income children are, on the average, about 1 inch shorter than their middle-class counterparts (Yip, Scanlon, & Trowbridge, 1993).

INFECTIOUS DISEASE

Two weeks into the school year, I looked out my window and noticed that Hallie was absent from the play yard. Several weeks passed; still, I did not see him. When I asked Leslie, his preschool teacher, what had happened, she explained, "Hallie's been hospitalized with the measles. He's had a difficult time recovering—lost weight when there wasn't much to lose in the first place." In well-nourished children, ordinary childhood illnesses have no effect on physical growth. But when children are undernourished, disease interacts with malnutrition in a vicious spiral, and the consequences for physical growth can be severe.

Hallie's reaction to the measles is commonplace among children in developing nations, where a large proportion of the population lives in poverty. In these countries, many children are not immunized against infectious disease. Illnesses such as measles and chicken pox, which typically do not appear until after age 3 in industrialized nations, occur much earlier. This is because poor diet depresses the body's

CAREGIVING CONCERNS

Encouraging Good Nutrition in Early Childhood

SUGGESTION	DESCRIPTION
Offer, a varied, healthy diet.	Provide a well-balanced variety of nutritious foods that are colorful and attractively served. Avoid serving sweets and "junk" foods as a regular part of meals and snacks.
Offer predictable meals as well as several snacks each day.	Preschoolers' stomachs are small, and they may not be able to eat enough in three meals to satisfy energy requirements. They benefit from extra opportunities to eat.
Offer small portions, and permit the child to ask for seconds.	When too much food is put on the plate, preschoolers may be overwhelmed and not even try to eat.
Offer new foods early in the meal and over several meals, and respond with patience if the child rejects the food.	Introduce new foods before the child's appetite is satisfied. Let children see you eating and enjoying the new food. If the child rejects it, accept the refusal and serve it again at another meal. As foods become more familiar, they are more readily accepted.
Keep mealtimes pleasant, and include the child in mealtime conversations.	A pleasant, relaxed eating environment helps children develop positive attitudes about food. Avoid confrontations over disliked foods and table manners.
Avoid using food as a reward.	Saying, "No dessert until you clean your plate," tells children that they must eat regardless of how hungry they are and that dessert is the best part of the meal.

Sources: Birch, Johnson, & Fisher, 1995; Kendrick, Kaufmann, & Messenger, 1991.

immune system, making children far more susceptible to disease. Disease, in turn, is a major cause of malnutrition and, through it, affects physical growth. Illness reduces appetite and limits the body's ability to absorb foods (Grant, 1995).

In industrialized nations, childhood diseases have declined dramatically during the past half century, largely due to widespread immunization of infants and young children. Hallie got the measles because, unlike his classmates from more advantaged homes, he did not receive a full program of immunizations during his first 2 years of life. Although the majority of preschoolers in the United States are immunized, some do not receive full protection until 5 or 6 years of age, when it is required for school entry (Children's Defense Fund, 1997).

To remedy this problem, in 1994 all medically uninsured American children were guaranteed free immunizations. Yet immunization rates in the United States continue to lag behind those in western European nations. How is it that European countries have managed to protect the health of young children more effectively? To explore this question, turn to the Cultural Influences box on page 214.

CHILDHOOD INJURIES

Three-year-old Tory caught my eye as I visited the preschool classroom one day. More than any other child, he had trouble sitting still and paying attention at storytime. Outside, he darted from one place to another, spending little time at a single activity. On a field trip to our campus museum, Tory ignored Leslie's directions and ran

across the street without holding his partner's hand. Later in the year, I read in our local newspaper that Tory had narrowly escaped serious injury when he put his mother's car into gear while she was outside scraping its windows. The vehicle rolled through a guardrail and over the side of a 10-foot concrete underpass. There it hung until rescue

To inform parents about the importance of immunizations, the U.S. Department of Health and Human Services distributes this poster free of charge. In the poster's next printing, chicken pox will be added to the list of preventable childhood diseases.

CHILD HEALTH CARE IN THE UNITED STATES AND EUROPEAN NATIONS

In the United States, economically disadvantaged children are far less likely than well-off youngsters to receive basic health care. Because of the high cost of medical treatment, low-income children see a doctor only half as often as middle-class children with similar illnesses (Newacheck & Starfield, 1988). Estimates indicate that 60 percent of American children under age 5 who come from poverty-stricken families are in less than excellent health (Children's Defense Fund, 1992).

American health insurance is an optional, employment-related fringe benefit. Many American businesses that rely on low-wage and part-time help do not insure their employees. If they do, they often do not cover other family members, including children. Although a variety of public health programs are available in the United States, they reach only the most needy individuals. This leaves nearly 10 million children from poor, low-income, and moderate-income families uninsured and, therefore, without affordable medical care (Children's Defense Fund, 1997).

The inadequacies of American child health care stand in sharp contrast to services provided in other industrialized nations, where medical insurance is government sponsored and available to all citizens. Let's look at two examples.

In the Netherlands, each child receives free medical examinations from birth through adolescence. During the early years, health care also includes parental counseling in nutrition, disease prevention, and child development. The Netherlands achieves its high childhood immunization rate by giving parents of every newborn baby a written schedule that shows exactly when and where the child should be immunized. If the child is not brought in at the specified time, a public health nurse calls the family. In instances of repeated missed appointments, the nurse goes to the home to ensure that the child receives the recommended immunizations (Verbrugge, 1990a, 1990b).

In Norway, federal law requires that well baby and child clinics be established in all communities and

that examinations by doctors take place three times during the first year and at ages 2 and 4. Specialized nurses see children on additional occasions, monitoring their growth and development, providing immunizations, and counseling parents on physical and mental health. Although citizens pay a small fee for routine medical visits, hospital services are free of charge. Parents with a seriously ill hospitalized child are given leave from work with full salary, a benefit financed by the government (Lie, 1990).

In Chapter 2, we noted that Americans have historically been strongly committed to the idea that parents should assume total responsibility for the care and rearing of children. This belief, in addition to powerful economic interests in the medical community, has prevented government-sponsored health services from being offered to all children. In European nations, child health care is regarded as a fundamental human right, no different from the right to education.

Widespread, government-sponsored immunization of infants and young children is a cost-effective means of supporting healthy growth by dramatically reducing the incidence of childhood diseases. Although this boy finds a routine inoculation painful, it will offer him lifelong protection. (Russell D. Curtis/ Photo Researchers)

workers arrived. Tory's mother was charged with failing to use a restraint seat for children under age 5.

Unintentional injuries—auto collisions, pedestrian accidents, drownings, poisonings, firearm wounds, burns, falls, and swallowing foreign objects—are the leading cause of childhood mortality. These largely preventable events account for 40 to 50 percent of deaths in early and middle childhood and as many as 75 percent during adolescence. Approximately 22,000 American youngsters die from these incidents each year. And for each death, thousands of other injured children survive but suffer pain, brain damage, and permanent physical disabilities (Kronenfeld & Glik, 1995).

As Figure 7.4 shows, auto accidents, fires, and drownings are the most common injuries during the preschool years. Motor vehicle collisions are by far the most frequent source of injury from 1 year of age through early adulthood.

■ FACTORS RELATED TO CHILDHOOD INJURIES. We are used to thinking of childhood injuries as "accidental." But a close look reveals that meaningful causes underlie them, and we can, indeed, do something about them.

As Tory's case suggests, individual differences exist in the safety of children's behaviors. Because of their higher activity level and greater willingness to take risks during play, boys are more likely to be injured than girls. (Look again at Figure 7.4.) Temperamental characteristics—irritability, inattentiveness, and negative mood—are also related to childhood injuries. As we saw in Chapter 6, children with these traits present special child-rearing challenges. Where safety is concerned, they are likely to protest when placed in auto seat restraints, refuse to take a companion's hand when crossing the street, and disobey after repeated adult instruction and discipline (Matheny, 1991).

At the same time, families whose preschoolers get injured tend to have characteristics that increase the likelihood of exposure to danger. Poverty and low parental education are strongly associated with injury deaths. Parents who must cope with many daily stresses often have little time and energy to monitor the safety of their youngsters. And their homes and neighborhoods pose further risks. Noise, crowding, and confusion characterize these households, and they tend to be located in run-down, inner-city neighborhoods where children have few places to play other than the streets (Peterson & Brown, 1994).

Broad societal conditions also affect childhood injury. Among Western industrialized nations, the United States ranks among the highest in childhood injury mortality. Although injury deaths have steadily declined in nearly all developed countries during the past 30 years, recent figures suggest that they may be on the rise in the United States (Children's Defense Fund, 1996). Widespread poverty, a shortage of high-quality day care (to supervise children in their parents' absence), and an alarmingly high rate of births to teenagers (who are neither psychologically nor financially ready to raise a child) account for this trend. But

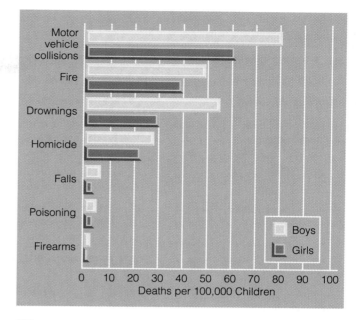

FIGURE 7.4

Rate of injury mortality in the United States for children between 1 and 4 years of age by type of injury and sex. Note that for all kinds of injuries, mortality is higher for boys than girls. (Adapted from Williams & Kotch, 1990.)

children from advantaged families are also at somewhat greater risk for injury in the United States than they are in European nations (Williams & Kotch, 1990). This indicates that besides reducing poverty and teenage pregnancy and upgrading the status of day care, additional steps must be taken to ensure the safety of American children.

■ PREVENTING CHILDHOOD INJURIES. Childhood injuries have many causes, so a variety of approaches are needed to control them. Laws can prevent a great many injuries by requiring car safety seats for young children, child-resistant caps on medicine bottles, flame-proof clothing, and fences around backyard swimming pools (the site of 90 percent of early childhood drownings).

Communities can also help by modifying their physical environments. Inexpensive and widely available public transportation can reduce the time that children spend in cars. Playgrounds, a common site of injury, can be covered with protective surfaces, such as rubber matting, sand, and wood chips. Free, easily installed window guards can be given to families in high-rise apartment buildings to prevent falls. And widespread media campaigns and information distributed in doctors' offices, schools, and day care centers help inform parents and children about safety issues.

Nevertheless, many dangers cannot be eliminated from the environment. And even though they know better, many parents and children behave in ways that compromise safety. How can we change human behavior? A variety of

CAREGIVING CONCERNS

Reducing Unintentional Injuries in Early Childhood

SUGGESTION	DESCRIPTION
Provide age-appropriate supervision and safety instruction.	Despite increasing self-control, preschoolers need nearly constant supervision. Establish and enforce safety rules, and explain the reasons behind them, thereby encouraging the child to remember, understand, and obey.
Know the child's temperament.	Children who are unusually active, distractible, and curious have more than their share of injuries and need extra monitoring.
Eliminate the most serious dangers from the home.	Examine all spaces for safety. For example, in the kitchen, store dangerous products in high cabinets out of sight, and keep sharp implements in a latched drawer. Always accompany young preschoolers to the bathroom, and keep all medicines in containers with safety caps.
During automobile travel, always restrain the child in a car seat.	Use an age-appropriate, properly installed car seat, and strap the child in correctly every time.
Select safe playground equipment and sites.	Make sure sand, wood chips, or rubberized matting has been placed under swings, seesaws, and jungle gyms. Check yards for dangerous plants. Always supervise outdoor play.
Be extra cautious around water.	Constantly observe children during water play; even shallow, inflatable pools can be sites of drownings. While swimming, young children's heads should not be immersed in water; they may swallow so much that they develop water intoxication, which can lead to convulsions and death.
Practice safety around animals.	Wait to get a pet until the child is mature enough to handle and care for it—usually around age 5 or 6. Never leave a young child alone with an animal; bites often occur during playful roughhousing. Model and teach humane pet treatment.

Note. For additional advice on how to keep young children safe, consult American Academy of Pediatrics (1993), *Caring for Your Baby and Young Child: Birth to Age 5,* pp. 375–409. (New York: Bantam).

programs based on applied behavior analysis (modeling and reinforcement) have successfully improved safety practices. In one, counselors helped parents identify dangers in the home—fire hazards, objects that young children might swallow, poisons, firearms, and others. Then they demonstrated specific ways to eliminate them (Tertinger, Greene, & Lutzker, 1984). In several other programs, parents and children were rewarded with prizes if the youngsters arrived at day care or school each morning in car seats (Roberts, Alexander, & Knapp, 1990). The Caregiving Concerns table above lists ways that caregivers can reduce unintentional injuries in early childhood.

BRIEF REVIEW

Genes influence children's growth by regulating the production of hormones. Two pituitary hormones, growth hormone (GH) and thyroid-stimulating hormone (TSH), are important in children's growth. Many environmental influences affect growth and health in early childhood. Extreme emotional deprivation can interfere with the production of GH, resulting in deprivation dwarfism. Although appetites of preschoolers decline and they resist new foods, good nutrition remains important. The emotional climate of mealtimes influences the quality and variety of foods that young children will eat. Disease can interact with malnutrition to seriously undermine children's growth, a common problem in developing countries. Unintentional injuries are the leading cause of death in childhood. Injury rates are related to child and family characteristics as well as to broad societal conditions. Consequently, a variety of approaches are needed to prevent them.

ASK YOURSELF . . .

■ *One day, Leslie prepared a new snack to serve at preschool: celery stuffed with ricotta cheese and pineapple. The first time she served it, few of the children touched it. How can Leslie encourage her pupils to accept the snack? What should she avoid doing?*

■ *Chapter 1 introduced you to ecological systems theory, which shows how development and well-being are affected by several levels of the environment. List ways to reduce childhood injuries by intervening in the microsystem, mesosystem, and macrosystem.*

TABLE 7.1

Changes in Gross and Fine Motor Skills During Early Childhood

AGE	GROSS MOTOR SKILLS	FINE MOTOR SKILLS
2–3 years	Walks more rhythmically; hurried walk changes to run. Jumps, hops, throws, and catches with rigid upper body. Pushes riding toy with feet; little steering.	Puts on and removes simple items of clothing. Zips and unzips large zippers. Uses spoon effectively.
3–4 years	Walks up stairs, alternating feet, and downstairs, leading with one foot. Jumps and hops, flexing upper body. Throws and catches with slight involvement of upper body; still catches by trapping ball against chest. Pedals and steers tricycle.	Fastens and unfastens large buttons. Serves self food without assistance. Uses scissors. Copies vertical line and circle. Draws first picture of person, using tadpole image.
4–5 years	Walks downstairs, alternating feet. Runs more smoothly. Gallops and skips with one foot. Throws ball with increased body rotation and transfer of weight on feet; catches ball with hands. Rides tricycle rapidly, steers smoothly.	Uses fork effectively. Cuts with scissors following line. Copies triangle, cross, and some letters.
5–6 years	Increases running speed. Gallops more smoothly; engages in true skipping. Displays mature throwing and catching pattern. Rides bicycle with training wheels.	Uses knife to cut soft food. Ties shoes. Draws person with six parts. Copies some numbers and simple words.

Source: Furuno et al., 1987; Newborg, Stock, & Wnek, 1984.

MOTOR DEVELOPMENT

Visit a playground at a neighborhood park, preschool, or day-care center, and observe several children between 2 and 6 years of age. You will see that an explosion of new motor skills occurs in early childhood, each of which builds on the simpler movement patterns of toddlerhood.

The same principle that governs motor development during the first 2 years of life continues to operate during the preschool years. Children integrate previously acquired skills into more complex *dynamic systems of action.* (Return to Chapter 4, page 128, if you need to review this concept.) Then they revise each new skill as their bodies grow larger and stronger, their central nervous systems develop, and their environments present new challenges.

GROSS MOTOR DEVELOPMENT

As children's bodies become more streamlined and less top-heavy, their center of gravity shifts downward, toward the trunk. As a result, balance improves greatly, paving the way for new motor skills involving large muscles of the body (Ulrich & Ulrich, 1985). By age 2, the preschooler's gait becomes smooth and rhythmic—secure enough so that soon they leave the ground, at first by running and later by jumping, hopping, galloping, and skipping. As children become steadier on their feet, their arms and torsos are freed to experiment with new skills—throwing and catching balls, steering tricycles, and swinging on horizontal bars and rings. Then upper and lower body skills combine into more refined actions. Five- and 6-year-olds dash across the play yard to catch a ball, steer and pedal a tricycle at the same time, and flexibly move their whole body when hopping and jumping. Table 7.1 provides a closer look at the development of gross motor skills during the preschool years.

FINE MOTOR DEVELOPMENT

Like gross motor development, fine motor skills take a giant leap forward during early childhood. Because control of the hands and fingers improves, young children put puzzles together, build with small blocks, cut and paste, and string beads. To parents, fine motor progress is most apparent in two areas: (1) children's care of their own bodies, and (2) the drawings and paintings that fill the walls at home, day care, and preschool.

SELF-HELP SKILLS. As Table 7.1 shows, young children gradually become self-sufficient at dressing and feeding, although parents need to be patient as preschoolers develop these abilities. When tired and in a hurry, young children often revert to eating with their fingers. And the 3-year-old who dresses himself sometimes ends up with his shirt on inside out, his pants on backward, and his left snow boot on his right foot!

Perhaps the most complex self-help skill of early childhood is shoe tying, mastered around age 6. Success requires a longer attention span, memory for an intricate series of hand movements, and the dexterity to perform them. Shoe

When young children experiment with crayons and paint, they not only develop fine motor skills but acquire the artistic traditions of their culture. This Australian Aboriginal 4–year-old creates a dot painting. To Westerners, it looks abstract. To the child, it expresses a "dreamtime" story about the life and land of his ancestors. If asked about the painting, he might respond, "Here are the boulders on the creek line, the hills with kangaroos and emus, and the campsites." (Laura Berk)

tying, along with drawing and writing, illustrates the close connection between motor and cognitive development.

■ DRAWING AND WRITING. When given crayon and paper, even toddlers scribble in imitation of others. As the young child's ability to represent the world expands, marks on the page take on meaning. At first, children's artful representation occurs through gestures. For example, one 18-month-old hopped her crayon around the page, explaining as she made a series of dots, "Rabbit goes hop-hop." Around age 2, children realize that pictures can depict pretend objects—a basic feature of artistic expression (Kavanaugh & Harris, 1994). By age 3, children's scribbles become pictures. Often this happens after they make a gesture with the crayon, notice that they have drawn a recognizable shape, and then decide to label it (Winner, 1986).

A major milestone in children's drawing occurs when they use lines to represent the boundaries of objects. This permits them to draw their first picture of a person by age 3 or 4. Look at the tadpole image on the left in Figure 7.5. It is a universal one in which the limits of the preschooler's fine motor skills reduce the figure to the simplest form that still looks human (Gardner, 1980; Winner, 1986).

Unlike many adults, young children do not demand that a drawing be realistic. But as they get older and their

fine motor control improves, they learn to desire greater realism. As a result, they create more complex drawings, like the one on the right in Figure 7.5 by a 6-year-old child. Still, there are perceptual distortions, since only gradually over the school years do children figure out how to represent depth in their drawings (Braine et al., 1993).

As young children experiment with lines and shapes, notice print in storybooks, and observe people writing, they try to print letters and, later on, words. Often the first word printed is the child's name. Initially, it may be represented by a single letter. "How do you make a *D?*" my older son David asked at age 3. When I printed a large uppercase *D*, he tried to copy. "*D* for David," he said as he wrote, quite satisfied with his backward, imperfect creation. A year later, David added several letters, and around age 5, he wrote his name clearly enough for others to read. Like many children, David continued to reverse some letters in his printing until well into second grade. Not until they learn to read do children find it useful to distinguish between mirror-image forms, such as *b* and *d* and *p* and *q* (Bornstein, 1992).

▮ FACTORS THAT AFFECT EARLY CHILDHOOD MOTOR SKILLS

We have discussed motor milestones in terms of the average age at which children reach them, but, of course, there are wide individual differences. Body build, ethnicity, sex, and opportunity for physical play are among the many factors that affect young children's motor progress.

A child with a tall, muscular body tends to move more quickly and acquire certain skills earlier than a short, stocky youngster. Researchers believe that body build contributes to the superior performance of African-American over Caucasian children in running and jumping. African-American youngsters tend to have longer limbs, so they have better leverage (Lee, 1980; Wakat, 1978).

Sex differences in motor skills are evident in early childhood. Boys are slightly ahead of girls in skills that emphasize force and power. By age 5, they can jump slightly farther, run slightly faster, and throw a ball much farther (about 5 feet beyond the distance covered by girls). Girls have an edge in fine motor skills and in certain gross motor skills that require a combination of good balance and foot movement, such as hopping and skipping. Boys' greater muscle mass and (in the case of throwing) slightly longer forearms may contribute to their skill advantages. And girls' greater overall physical maturity may be partly responsible for their better balance and precision of movement.

From an early age, boys and girls are usually encouraged into different physical activities. For example, fathers often play catch with their sons but seldom do so with their daughters. Baseballs and footballs are purchased for boys, jump ropes, hula hoops, and jacks for girls. As children get older, sex differences in motor skills get larger, yet differences in physical capacity remain small until adolescence.

FIGURE 7.5

Examples of young children's drawings. The universal tadpolelike shape that children use to draw their first picture of a person is shown on the left. The tadpole soon becomes an anchor for greater detail as arms, fingers, toes, and facial features sprout from the basic shape. By the end of the preschool years, children produce more complex, differentiated pictures like the one on the right drawn by a 6-year-old child. *(Tadpole drawings from H. Gardner, 1980,* Artful Scribbles: The Significance of Children's Drawings, *New York: Basic Books, p. 64. Reprinted by permission of Basic Books, a division of HarperCollins Publishers, Inc. Six-year-old's picture from E. Winner, August 1986, "Where Pelicans Kiss Seals,"* Psychology Today, *20[8], p. 35. Reprinted by permission of the author.)*

These trends suggest that social pressures for boys to be active and physically skilled and for girls to play quietly at fine motor activities may exaggerate small, genetically based sex differences (Coakley, 1990).

Children seem to master the motor skills of early childhood naturally, as part of their everyday play. Aside from throwing (where direct instruction seems to make some difference), there is no evidence that preschoolers exposed to formal lessons are ahead in motor development (Roberton, 1984). Therefore, the appropiate goal of adult involvement in young children's motor activities should be "fun" rather than learning the "correct" technique (Kutner, 1993).

When children have play spaces appropriate for running, climbing, jumping, and throwing and are encouraged to use them, they respond eagerly to these challenges. Preschools, day-care centers, and city playgrounds need to accommodate a wide range of physical abilities by offering pieces of equipment that differ in size or that can be adjusted to fit the needs of individual children.

BRIEF REVIEW

Motor development proceeds rapidly during early childhood. Improvements in balance contribute to a variety of gross motor skills, including

As early as the preschool years, boys are slightly ahead of girls in motor skills that emphasize force and power. Although only a small difference in physical capacity exists, this gap in performance widens as boys are given baseballs and footballs and encouraged to use them. (David W. Trozzo)

running, jumping, hopping, galloping, skipping, throwing, and catching. Advances in fine motor development can be seen in children's ability to dress themselves and eat efficiently with utensils. Preschoolers also start to draw and print, using fine motor skills to represent the world through pictures and written symbols. Many factors are related to early childhood motor development, including body build, ethnicity, sex, and the richness and appropriateness of the child's play environment.

ASK YOURSELF . . .

■ *Mabel and Chad want to do everything they can to support their 3-year-old daughter's athletic development. What advice would you give them?*

COGNITIVE DEVELOPMENT IN EARLY CHILDHOOD

One rainy morning, as I observed in our laboratory preschool, Leslie, the children's teacher, joined me at the back of the room to watch for a moment herself. "Preschoolers' minds are such a blend of logic, fantasy, and faulty reasoning," Leslie reflected. "Every day, I'm startled by the maturity and originality of what they say and do. Yet at other times, their thinking seems limited and inflexible."

Leslie's comments sum up the puzzling contradictions of early childhood cognition. Over the previous week, I had seen many examples as I followed the activities of 3-year-old Sammy. That day, I found him at the puzzle table, moments after a loud clash of thunder occurred outside. Sammy looked up, startled, then turned to Leslie and pronounced, "The man turned on the thunder!" Leslie patiently explained that people can't turn thunder on or off. "Then a lady did it," Sammy stated with certainty.

In other respects, Sammy's cognitive skills seemed surprisingly advanced. At snack time, he accurately counted, "One, two, three, four!" and then got four cartons of milk, giving one to each child at his table. But when more than four children joined his snack group, Sammy's counting broke down. And some of his notions about quantity seemed as fantastic as his understanding of thunder. Across the snack table, Priti dumped out her raisins, and they scattered in front of her. "How come you got lots, and I only got this little bit?" asked Sammy, failing to realize that he had just as many; they were simply all bunched up in a tiny red box. Piaget's theory helps us understand Sammy's curious reasoning.

PIAGET'S THEORY: THE PREOPERATIONAL STAGE

As children move from the sensorimotor to the **preoperational stage,** the most obvious change is an extraordinary increase in representational, or symbolic, activity. Recall that infants and toddlers have some ability to mentally represent the world. During early childhood, this capacity blossoms.

ADVANCES IN MENTAL REPRESENTATION

Piaget acknowledged that language is our most flexible means of mental representation. By detaching thought from action, it permits cognition to be far more efficient than it was during the sensorimotor stage. When we think in words, we overcome the limits of our momentary perceptions. We can deal with the past, present, and future all at once, combining images of the world in unique ways (Miller, 1993).

Yet despite the power of language, Piaget did not believe that it plays a major role in cognitive development. According to Piaget, language does not give rise to representational thought. Instead, sensorimotor activity provides the foundation for language, just as it underlies deferred imitation and make-believe play. Can you think of evidence that supports Piaget's view? Recall from Chapter 5 that the first words toddlers use have a strong sensory basis. In addition, toddlers acquire an impressive range of cognitive categories long before they use words to label them (see page 155). Still, other theorists regard Piaget's account of the link between language and thought as incomplete, as we will see later in this chapter.

MAKE-BELIEVE PLAY

Make-believe play is another example of representation. Like language, it increases dramatically during early childhood. Piaget believed that through pretending, young children practice and strengthen newly acquired representational schemes. Drawing on his ideas, several investigators have traced the development of make-believe during the preschool years.

■ DEVELOPMENT OF MAKE-BELIEVE. One day, Sammy's 18-month-old brother Dwayne came to visit the classroom. Dwayne wandered around, picked up the receiver of a toy telephone, said, "Hi, Mommy," and then dropped it. In the housekeeping area, he found a cup, pretended to drink, and then toddled off again.

In the meantime, Sammy joined a group of children in the block area for a space shuttle launch. "That can be our control tower," he suggested to Vance, pointing to a corner by a bookshelf.

"Wait, I gotta get it all ready," said Lynette, who was still arranging the astronauts (two dolls and a teddy bear) inside a circle of large blocks, which represented the rocket.

"Countdown!" Sammy announced, speaking into a small wooden block, his pretend walkie-talkie.

"Five, six, two, four, one, blastoff!" responded Vance, commander of the control tower.

Lynette made one of the dolls push a pretend button and reported, "Brrrm, brrrm, they're going up!"

A comparison of Dwayne's pretend with that of Sammy and his classmates illustrates three important changes in make-believe. Each reflects the preschool child's growing symbolic mastery.

First, over time, play becomes increasingly detached from the real-life conditions associated with it. In early pretending, toddlers use only realistic objects—for example, a toy telephone to talk into or a cup to drink from. Around age 2, use of less realistic toys, such as a block for a telephone receiver, becomes more frequent. Between 3 and 5 years, children become better at imagining objects and events without support from the real world, as when Sammy invented the control tower in a corner of the room. We can see that children's representations are becoming more flexible, since a play symbol no longer has to resemble the object for which it stands (Corrigan, 1987; O'Reilly, 1995).

Second, the way in which the "child as self" participates in play changes with age. At first, make-believe is directed toward the self—for example, Dwayne pretends to feed only himself. A short time later, children direct pretend actions toward other objects, as when the child feeds a doll. And early in the third year, they use objects as active agents. The child becomes a detached participant who makes a doll feed itself or (in Lynette's case) push a button to launch a rocket. Make-believe gradually becomes less self-centered as children realize that agents and recipients of pretend actions can be independent of themselves (Corrigan, 1987; McCune, 1993).

Finally, over time, make-believe includes more complex scheme combinations. For example, Dwayne can pretend to drink from a cup but he does not yet combine pouring and drinking. Later on, children combine schemes, especially in **sociodramatic play,** the make-believe with peers that appears around age 2½ and increases rapidly until 4 to 5 years (Haight & Miller, 1993; Howes & Matheson, 1992). Already, Sammy and his classmates create and coordinate several roles in an elaborate plot. By the end of early childhood, children have a sophisticated understanding of role relationships and story lines (Göncü, 1993).

■ ADVANTAGES OF MAKE-BELIEVE. Today, Piaget's view of make-believe as mere practice of representational schemes is regarded as too limited. Research indicates that make-believe play not only reflects but contributes to chil-

During the preschool years, make-believe play blossoms. This child uses objects as active agents in a complex play scene. (M. Siluk/The Image Works)

dren's cognitive and social skills (Nicolopoulou, 1993; Singer & Singer, 1990). In comparison to other social activities (such as drawing or putting puzzles together), during sociodramatic play preschoolers' interactions last longer, show more involvement, draw larger numbers of children into the activity, and are more cooperative (Connolly, Doyle, & Reznick, 1988). When we consider these findings, it is not surprising that preschoolers who spend more time at sociodramatic play are advanced in general intellectual development and are seen as more socially competent by their teachers (Burns & Brainerd, 1979; Connolly & Doyle, 1984). And many studies reveal that make-believe strengthens a wide variety of mental abilities, including memory, language, logical reasoning, and imagination and creativity (Berk, 1994a). We will return to the topic of early childhood play in Chapter 8.

LIMITATIONS OF PREOPERATIONAL THOUGHT

Aside from gains in representation, Piaget described preschoolers in terms of what they *cannot,* rather than *can,* understand (Beilin, 1992). He compared them to older, more competent children in the concrete operational stage, as the term "*pre*operational" suggests. According to Piaget, young children are not capable of *operations*—mental actions that obey logical rules. Instead, their thinking is rigid, limited to one aspect of a situation at a time, and strongly influenced by the way things appear at the moment.

FIGURE 7.6

Piaget's three mountains problem. Each mountain is distinguished by its color and by its summit. One has a red cross, another a small house, and the third a snow-capped peak. Children at the preoperational stage are egocentric. They cannot select a picture that shows the mountains from the doll's perspective. Instead, they simply choose the photo that shows their own vantage point.

■ EGOCENTRISM. According to Piaget, the most serious deficiency of the preoperational stage, the one that underlies all others, is **egocentrism.** By this, Piaget did not mean selfishness or inconsiderateness. Instead, he believed that when children begin to mentally represent the world, they are egocentric with respect to their symbolic viewpoints. They are unaware of any perspectives other than their own, and they believe everyone else perceives, thinks, and feels the same way they do (Piaget, 1950).

Piaget's most convincing demonstration of egocentrism involves a task called the *three-mountains problem* (see Figure 7.6). A child is permitted to walk around a display of three mountains arranged on a table. Then the child stands on one side, and a doll is placed at various locations around the display. The child must choose a photograph that shows what the display looks like from the doll's perspective. Before age 6 or 7, most children simply select the photo that shows the mountains from their own point of view (Piaget & Inhelder, 1948/1956).

Egocentrism, Piaget pointed out, shows up in other aspects of children's reasoning. Recall Sammy's firm insistence that someone must have turned on the thunder, in much the same way that he uses a switch to turn on a light or radio. Similarly, Piaget regarded egocentrism as responsible for preoperational children's **animistic thinking**— their belief that inanimate objects have lifelike qualities, such as thoughts, wishes, feelings, and intentions, just like

themselves. The 3-year-old who charmingly explains that the sun is angry at the clouds and has chased them away is demonstrating this kind of reasoning. According to Piaget, because young children egocentrically assign human purposes to physical events, magical thinking is especially common during the preschool years.

Piaget argued that egocentrism leads to the rigidity and illogical nature of young children's thinking. Thought proceeds so strongly from a single point of view that children do not *accommodate,* or revise their faulty reasoning, in response to their physical and social worlds. But to fully appreciate this shortcoming, let's consider some additional tasks that Piaget gave to children.

■ INABILITY TO CONSERVE. Piaget's most important tasks are the conservation problems. **Conservation** refers to the idea that certain physical characteristics of objects remain the same, even when their outward appearance changes. At snack time, Priti and Sammy each had identical boxes of raisins, but after Priti spread hers out on the table, Sammy was convinced that she had more.

Another type of conservation task involves liquid. The child is shown two identical tall glasses of water and asked if they contain equal amounts. Once the child agrees, the water in one glass is poured into a short, wide container, changing the appearance of the water but not its amount. Then the child is asked whether the amount of water is still the same or whether it has changed. Preoperational children think the quantity of water is no longer the same. They explain, "There is less now because the water is way down here" (that is, its level is so low in the short, wide container) or "There is more now because it is all spread out." In Figure 7.7, you will find other conservation tasks that you can try with children.

Preoperational children's inability to conserve highlights several related aspects of their thinking. First, their understanding is **perception-bound.** They are easily distracted by the appearance of objects (It *looks* like there is less water in the short, wide container, so there *must be* less water). Second, their thinking is *centered,* or characterized by **centration.** They focus only on one aspect of a situation, neglecting other important features. In the case of conservation of liquid, the child centers on the height of the water in the two containers, failing to realize that all changes in height of the liquid are compensated by changes in width. Third, children of this stage emphasize **states versus transformations.** They tend to focus on momentary conditions rather than on dynamic changes between them. For example, in the conservation of liquid problem, they treat the initial and final states of the water as completely unrelated events.

The most important illogical feature of preoperational thought is its **irreversibility.** Children of this stage cannot mentally go through a series of steps in a problem and then reverse direction, returning to the starting point.

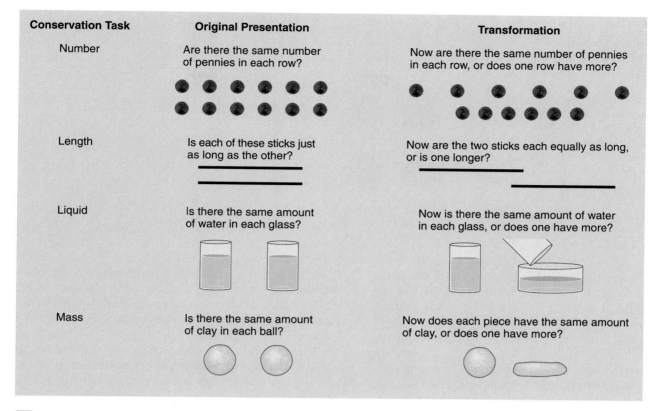

| Conservation Task | Original Presentation | Transformation |

FIGURE 7.7

Some Piagetian conservation tasks. Children at the preoperational stage cannot yet conserve.

Reversibility, the opposite of this concept, is part of every logical operation. After Priti spills her box of raisins, Sammy cannot reverse by thinking to himself, "I know that Priti doesn't have more raisins than I do. She just poured them out of that little red box, and if we put them back in again, her raisins and my raisins would look just the same."

■ TRANSDUCTIVE REASONING. Because preoperational children are not capable of reversible thinking, their explanations often seem like collections of disconnected facts and contradictions. Piaget called this feature of young children's thought **transductive reasoning,** which means reasoning from one particular event to another particular event. Preschoolers simply link together two events that occur close in time and space in a cause-and-effect fashion. For example, one day Leslie asked the children, "Why does it get dark at night?" "Because we go to bed!" responded 4-year-old Lynette.

■ LACK OF HIERARCHICAL CLASSIFICATION. Lack of logical operations leads preschoolers to have difficulty with **hierarchical classification.** That is, they cannot yet organize objects into classes and subclasses on the basis of similarities and differences. Piaget's famous *class inclusion problem,* illustrated in Figure 7.8 on page 224, demon-strates this limitation. Show a child a set of common objects, such as 16 flowers, most of which are yellow and a few of which are blue. Then ask whether there are more yellow flowers or more flowers. Preoperational children respond confidently, "More yellow flowers!" They tend to center on the overriding perceptual feature of yellow. They do not think rever-sibly by moving from the whole class (flowers) to the parts (yellow and blue) and back again.

RECENT RESEARCH ON PREOPERATIONAL THOUGHT

Over the past two decades, researchers have challenged Piaget's notion of the preschool child as cognitively deficient. They have found that many Piagetian problems contain confusing or unfamiliar elements or too many pieces of information for young children to handle at once. As a result, preschoolers' responses do not reflect their true abilities. Piaget also missed many naturally occurring instances of preschoolers' effective reasoning. Let's look at some examples that illustrate these points.

■ EGOCENTRIC, ANIMISTIC, AND MAGICAL THINKING. Are young children really so egocentric that they believe a person standing elsewhere in a room sees the same thing

FIGURE 7.8

A Piagetian class inclusion problem. Children are shown 16 flowers, 4 of which are blue and 12 of which are yellow. Asked whether there are more yellow flowers or more flowers, the preoperational child responds, "More yellow flowers," failing to realize that both yellow and blue flowers are included in the category of "flowers."

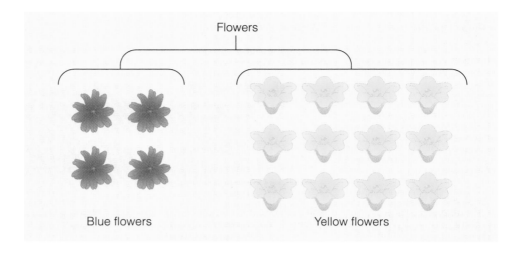

Flowers

Blue flowers   Yellow flowers

they see? When researchers change the nature of the three-mountains problem to include familiar objects and use methods other than picture selection (which is difficult even for 10-year-olds), 4-year-olds show clear awareness of others' vantage points (Borke, 1975; Newcombe & Huttenlocher, 1992).

Nonegocentric responses also appear in young children's everyday interactions with people. For example, preschoolers adapt their speech to fit the needs of their listeners. Sammy uses shorter, simpler expressions when talking to his little brother Dwayne than to agemates or adults (Gelman & Shatz, 1978). Also, in describing objects, children do not use such words as "big" and "little" in a rigid, egocentric fashion. Instead, they *adjust* their descriptions, taking account of context. By age 3, children judge a 2-inch shoe as small when seen by itself (because it is much smaller than most shoes) but as big for a very tiny 5-inch doll (Ebeling & Gelman, 1994).

Recent studies also indicate that Piaget overestimated children's animistic beliefs because he asked children about objects with which they have little direct experience, such as the clouds, sun, and moon. Children as young as 3 rarely think that very familiar inanimate objects, such as rocks and crayons, are alive. They do make errors about certain vehicles, such as trains and airplanes, that can move and have headlights that look like eyes (Dolgin & Behrend, 1984; Richards & Siegler, 1986). Preschoolers' responses result from incomplete knowledge about some objects, not from a rigid belief that inanimate objects are alive.

The same is true for other fantastic beliefs of the preschool years. Most 3- and 4-year-olds believe in the supernatural powers of fairies, goblins, and other enchanted creatures that appear in storybooks and movies. But they deny that magic can alter their everyday experiences—for example, turn a picture into a real object (Subbotsky, 1994). Between 4 and 8 years, as familiarity with physical events and principles increases, children's magical beliefs decline (Phelps & Woolley, 1994). Most figure out who is really behind the activities of Santa Claus and the Tooth Fairy!

■ **ILLOGICAL THOUGHT.** Many studies have reexamined the illogical characteristics of the preoperational stage. Results show that when tasks are simplified and made relevant to their everyday lives, preschoolers do better than Piaget might have expected.

For example, when a conservation-of-number task is scaled down to include only three items instead of six or seven, preschoolers respond correctly (Gelman, 1972). And when preschoolers are asked carefully worded questions about what happens to substances (such as sugar) after they are dissolved in water, they give accurate explanations. Most 3- to 5-year-olds know that the substance is conserved—that it continues to exist, can be tasted, and makes the liquid heavier, even though it is invisible in the water (Au, Sidle, & Rollins, 1993; Rosen & Rozin, 1993). These findings suggest that preschoolers notice transformations, reverse their thinking, and understand causality in familiar contexts.

Indeed, 3- and 4-year-olds' descriptions of their world reveal that they use causal terms, such as "if–then" and "because," with the same degree of accuracy as adults do (McCabe & Peterson, 1988). Transductive reasoning seems to occur only when preschoolers grapple with topics they know little about. Although young children cannot consider the complex interplay of forces that adolescents and adults can, they often analyze their experiences accurately in terms of basic cause-and-effect relations.

■ **HIERARCHICAL CLASSIFICATION.** Even though preschoolers have difficulty with Piagetian class inclusion tasks, their everyday knowledge is organized into nested categories at an early age. By the second year, children have formed a variety of global categories, such as kitchen utensils, bathroom objects, animals, vehicles, plants, and furniture (Bauer & Mandler, 1989; Mandler, Bauer, & McDonough, 1991). Notice that each of these categories includes objects that differ widely in appearance. This challenges Piaget's assumption that preschoolers' thinking is always perception bound.

Over the preschool years, these global categories differentiate. Children form many *basic-level categories*—ones at an intermediate level of generality, such as "chairs," "tables," "dressers," and "beds." Performance on object-sorting tasks indicates that 3- and 4-year-olds can easily move back and forth between basic-level categories and *general categories,* such as "furniture." They also break down the basic-level categories into subcategories, such as "rocking chairs" and "desk chairs" (Mervis, 1987). Preschoolers' category systems are not yet very complex, but the capacity to classify hierarchically is present in early childhood.

■ APPEARANCE VERSUS REALITY. So far, we have seen that when presented with familiar situations and simplified problems, young children show some remarkably advanced reasoning. Yet in certain situations, they are easily tricked by the outward appearance of things, just as Piaget suggested.

In a series of studies, John Flavell and his colleagues presented children with objects that were disguised in various ways and asked what the objects were, "really and truly." Preschoolers were easily tricked by sights and sounds. When asked whether a white piece of paper placed behind a blue filter is "really and truly blue" or whether a can that sounds like a baby crying when turned over is

Dressed as a pumpkin for Halloween, this 3-year-old is confused and upset when he looks in the mirror. Because he has not yet sorted out the distinction between appearance and reality, he may be wondering, "Do I just look like a pumpkin, or have I really become one?" (Rita Nannini/Photo Researchers)

"really and truly a baby," they often responded, "Yes!" Not until 6 to 7 years did children do well on these tasks (Flavell, 1993; Flavell, Green, & Flavell, 1987). These findings help explain why Sammy, on seeing his teacher Leslie dressed as a witch for Halloween, became concerned that she had really become one!

How do children master distinctions between appearance and reality? Make-believe play may be important. Preschoolers can tell the difference between pretend play and real experiences long before they answer many appearance–reality problems correctly (Golomb & Galasso, 1995; Woolley & Wellman, 1990). Often we hear them say such things as "Pretend this block is a telephone," or "Let's use this box for a house." Experiencing the contrast between everyday and playful use of objects may help children refine their understanding of what is real and what is unreal in the surrounding world.

■ EVALUATION OF THE
PREOPERATIONAL STAGE

The evidence as a whole indicates that Piaget was partly wrong and partly right about young children's cognitive capacities. When given scaled-down tasks based on familiar experiences, preschoolers show the beginnings of logical operations. But their reasoning is not as well developed as that of school-age children, since they fail Piaget's three-mountains, conservation, and class inclusion problems and have difficulty separating appearance from reality.

The fact that preschoolers have some logical understanding suggests that operational thought is not absent during one age period and suddenly present at another. Instead, children demonstrate mature reasoning at an early age, although it is fragile and incomplete.

The idea that logical operations develop gradually poses a serious challenge to Piaget's stage concept, which assumes sudden and abrupt change toward logical reasoning around 6 or 7 years of age. Does a preoperational stage of development really exist? Some researchers no longer think so. Instead, they believe that children work out their understanding of each type of task separately. Their thought processes are regarded as basically the same at all ages—just present to a greater or lesser extent. Recall from earlier chapters that this view of cognitive development as continuous forms the basis for an alternative approach: *information processing,* which we take up shortly.

Other experts think the stage concept is still valid, but it must be modified. For example, some theorists combine Piaget's stage approach with the information-processing emphasis on task-specific change (Case, 1992; Fischer & Farrar, 1987; Halford, 1992). They believe that Piaget's strict stage definition needs to be transformed into a less tightly knit concept, one in which a related set of competencies develops over an extended time period, depending on biological maturity and specific experiences. This flexible stage notion recognizes the unique qualities of early childhood thinking. At the same time, it provides a better account of why, to use Leslie's words, "preschoolers' minds are such a blend of logic, fantasy, and faulty reasoning."

PIAGET AND EDUCATION

Over the past 30 years, Piaget's theory has had a major impact on education, especially during early childhood. Leslie was greatly influenced by Piaget's work, which she studied in college. Three educational principles derived from his theory have become realities in her classroom:

1. *An emphasis on discovery learning.* In a Piagetian classroom, children are encouraged to discover for themselves through spontaneous interaction with the environment. Leslie's classroom has a rich variety of materials and play areas designed to promote exploration and discovery—art, puzzles, table games, dress-up clothing, building blocks, reading corner, woodworking, and more. For most of the morning, children choose freely from among these activities.

2. *Sensitivity to children's readiness to learn.* A Piagetian classroom does not try to speed up development. Instead, Piaget believed that appropriate learning experiences build on children's current thinking. Leslie watches and listens to her pupils, introducing experiences that permit them to practice newly discovered cognitive schemes and that are likely to challenge their incorrect ways of viewing the world. But she does not impose new skills before children indicate they are interested and ready (Johnson & Hooper, 1982).

3. *Acceptance of individual differences.* Piaget's theory assumes that all children go through the same sequence of development, but at different rates. Leslie makes a special effort to plan activities for individual children and small groups rather than for the total class (Ginsburg & Opper, 1988). In addition, Leslie evaluates educational progress by comparing each child to his or her own previous development. She is less interested in how children measure up to normative standards, or the average performance of same-age peers.

Educational applications of Piaget's theory, like his stages, have met with criticism. Perhaps the greatest challenge has to do with his insistence that young children learn mainly through acting on the environment. In the next section, we will see that they also use language-based routes to knowledge, which Piaget de-emphasized.

BRIEF REVIEW

During Piaget's preoperational stage, mental representation flourishes, as indicated by growth in language and make-believe play. Aside from repre-

sentation, Piaget's theory emphasizes the young child's cognitive limitations. Egocentrism underlies a variety of illogical features of preoperational thought, including animism, an inability to pass conservation tasks, transductive reasoning, and lack of hierarchical classification. Recent research reveals that when tasks are simplified and made relevant to children's everyday experiences, preschoolers show the beginnings of logical reasoning. These findings indicate that operational thought is not absent during early childhood, and they challenge Piaget's notion of stage. Piaget's theory has had a powerful influence on education, promoting discovery learning, sensitivity to children's readiness to learn, and acceptance of individual differences.

ASK YOURSELF . . .

■ *Recently, 2-year-old Brooke's father shaved off his thick beard and mustache. When Brooke saw him, she was very upset. Using Piaget's theory, explain why Brooke was distressed by her father's new appearance.*

■ *One weekend, 4-year-old Will went fishing with his family. When his father asked, "Why do you think the river is flowing along?" Will responded, "Because it's alive and wants to." Yet at home, Will understands very well that his tricycle isn't alive and can't move by itself. What explains this contradiction in Will's reasoning?*

■ *Sammy returned to preschool after several days of illness due to the flu. Leslie said, "Sammy, I'm so glad you're back. Why were you sick?" "I was sick because I threw up," replied Sammy. What kind of reasoning is Sammy demonstrating, and why did he reason this way in response to Leslie's question?*

VYGOTSKY'S SOCIOCULTURAL THEORY

Piaget's de-emphasis on language brought on yet another challenge, this time from Vygotsky's sociocultural theory, which stresses the social context of cognitive development. During early childhood, rapid growth in language enhances young children's social communication. Soon they start to talk to themselves in much the same way that they converse with others, and this greatly enhances cognitive development. Let's see how this happens.

CHILDREN'S PRIVATE SPEECH

Watch preschoolers as they go about their daily activities, and you will see that they frequently talk out loud to themselves. For example, as Sammy worked a puzzle one day, I heard him say, "Where's the red piece? I need the red one. Now, a blue one. No, it doesn't fit. Try it here." On another occasion, Sammy sat next to Mark on the rug and blurted out, "It broke," without explaining what or when.

Piaget (1923/1926) called these utterances *egocentric speech,* reflecting his belief that young children are unable to take the perspectives of others. He believed that their talk is often "talk for self" in which they run off thoughts in whatever form they happen to occur, regardless of whether a listener can understand. According to Piaget, cognitive maturity and certain social experiences—namely, disagreements with peers—eventually bring an end to egocentric speech. Through arguments with agemates, children repeatedly see that others hold viewpoints different from their own. As a result, egocentric speech declines.

Vygotsky (1934/1987) voiced a powerful objection to Piaget's conclusion that young children's language is egocentric and nonsocial. He reasoned that children speak to themselves for self-guidance and self-direction. Because language helps children think about their own behavior and plan courses of action, Vygotsky viewed it as the foun-

dation for all higher cognitive processes, including sustained attention, deliberate memory, planning, problem solving, and self-reflection. As children get older and find tasks easier, their self-directed speech is internalized as silent, *inner speech*—the verbal dialogues we carry on with ourselves while thinking and acting in everyday situations.

Over the past two decades, almost all studies have supported Vygotsky's perspective. As a result, children's speech-to-self is now referred to as **private speech** instead of egocentric speech. Research shows that children use more self-talk when tasks are difficult and they are confused about how to proceed (Berk, 1992a). Also, just as Vygotsky predicted, private speech goes underground with age, changing into whispers and silent lip movements (Berk, 1994b; Berk & Landau, 1993). Finally, children who use private speech freely during a challenging activity are more attentive and involved and do better than their less talkative agemates (Berk & Spuhl, 1995; Bivens & Berk, 1990).

SOCIAL ORIGINS OF EARLY CHILDHOOD COGNITION

Where does private speech come from? Vygotsky's answer highlights the social origins of cognition, his main difference of opinion with Piaget. Recall from Chapter 5 that Vygotsky believed children's learning takes place within a *zone of proximal development*—a range of tasks too difficult for the child to do alone but that can be accomplished with the help of others. During infancy, communication in the zone of proximal development is largely nonverbal. In early childhood, verbal dialogues are added as adults and more skilled peers help children master challenging activities. Consider the way Sammy's mother helped him put a difficult puzzle together:

Sammy: "I can't get this one in." (tries to insert a piece in the wrong place)

Mother: "Which piece might go down here?" (points to the bottom of the puzzle)

Sammy: "His shoes." (looks for a piece resembling the clown's shoes but tries the wrong one)

Mother: "Well, what piece looks like this shape?" (pointing again to the bottom of the puzzle)

Sammy: "The brown one." (tries it and it fits. Then attempts another piece and looks at his mother)

Mother: "Try turning it just a little." (gestures to show him)

Sammy: "There!" (puts in several more pieces. His mother watches.)

Eventually, children take the language of these dialogues, make it part of their private speech, and use this speech to organize their independent efforts in the same way.

To be effective, the adult's communication must be carefully coordinated with the child's current abilities

During the preschool years, children frequently talk to themselves as they play and explore the environment. Research supports Vygotsky's theory that children use private speech to guide their behavior when faced with challenging tasks. With age, private speech is transformed into silent, inner speech, or verbal thought. (Kopstein/Monkmeyer Press)

(Wood, 1989). Notice how Sammy's mother offers help at just the right moment and in the right amount. As Sammy becomes better at handling the task himself, his mother permits him to take over her guiding role and apply it to his own activity.

Is there evidence to support Vygotsky's ideas on the social origins of private speech and cognitive development? In two studies, mothers who used patient, sensitive communication in teaching their preschoolers a challenging puzzle had children who used more private speech and were more successful when working by themselves (Behrend, Rosengren, & Perlmutter, 1992; Berk & Spuhl, 1995). Other research shows that although children benefit from working on tasks with same-age peers, their planning and problem solving show more improvement when their partner is either an "expert" peer (especially capable at the task) or an adult (Azmitia, 1988; Radziszewska & Rogoff, 1988). And peer conflict and disagreement (emphasized by Piaget) does not seem to be as important in fostering cognitive development as the extent to which children resolve differences of opinion and cooperate (Cannella, 1993; Tudge, 1990).

VYGOTSKY AND EDUCATION

Today, educators are eager to use Vygotsky's ideas to enhance children's learning. Piagetian and Vygotskian classrooms clearly have features in common. Both emphasize active participation and acceptance of individual differences. But a Vygotskian classroom goes beyond learning through independent discovery. Instead, it promotes *assisted discovery*. Teachers guide children's learning with explanations, demonstrations, and verbal prompts, carefully tailoring their efforts to each child's zone of proximal development. Assisted discovery is also helped along by peer collaboration. Teachers arrange *cooperative learning* experiences, grouping together classmates of differing abilities and encouraging them to teach and help one another.

Note, however, that some of Vygotsky's ideas, like Piaget's, have been challenged. Verbal communication may not be the only means through which children's thinking develops, or even the most important one in some cultures. For example, the young child learning to sail a canoe in Micronesia or weave a garment on a foot loom in Guatemala may gain more from direct observation and practice than from joint participation with and verbal guidance by adults (Rogoff & Chavajay, 1995; Rogoff et al., 1993). Thus, we are reminded once again that children learn in a great many ways, and as yet, no single theory provides a complete account of cognitive development.

BRIEF REVIEW

Piaget regarded preschoolers' self-directed speech as egocentric and nonsocial. In contrast, Vygotsky viewed private speech as a means of self-guidance and self-direction. According to Vygotsky, language is the foundation for all higher cognitive processes. As adults and more skilled peers provide children with verbal guidance in the zone of proximal development, children integrate these dialogues into their private speech and use them to guide their own behavior. Research supports Vygotsky's ideas. A Vygotskian classroom emphasizes assisted discovery. Verbal support from teachers and peer collaboration are important.

ASK YOURSELF . . .

■ *Tanisha sees her 5-year-old son Toby talking out loud to himself while he plays. She wonders whether she should discourage this behavior. Use Vygotsky's theory to explain why Toby talks to himself. How would you advise Tanisha?*

A Vygotskian classroom promotes assisted discovery. Teachers guide children's learning, carefully tailoring their efforts to each child's zone of proximal development. They also arrange cooperative learning experiences, grouping together classmates of differing abilities and encouraging them to guide and help one another. (Paul Conklin/Monkmeyer Press)

INFORMATION PROCESSING

Return for a moment to the model of information processing discussed on page 154 of Chapter 5. Recall that information processing focuses on *control processes,* or *mental strategies,* that children use to transform stimuli flowing into their mental systems. During early childhood, advances in representation and children's ability to guide their own behavior lead to more efficient ways of attending, manipulating information, and solving problems. Preschoolers also begin to acquire academically relevant knowledge important for school success.

ATTENTION

Parents and teachers are quick to notice that preschoolers spend only short times involved in tasks, have difficulty focusing on details, and are easily distracted. The capacity to sustain attention does improve during early childhood, and fortunately so, since children will rely on it greatly once they enter school (Ruff & Lawson, 1990). But even 5- and 6-year-olds do not remain attentive for very long. When observed during free play at preschool, the average time they spend in a single activity is about 7 minutes (Stodolsky, 1974).

By the end of early childhood, attention also becomes more *planful.* In one study, researchers had 3- to 5-year-olds look for a lost object in their preschool play yard. Older preschoolers searched more systematically and exhaustively than 3-year-olds (Wellman, Somerville, & Haake, 1979). Still, children's attentional strategies have a long way to go. When given detailed pictures or written materials, preschoolers fail to search thoroughly (Ruff & Rothbart, 1996).

MEMORY

Try showing a young child a set of 10 pictures or toys. Then mix them up with some unfamiliar items and ask the child to point to the ones in the original set. You will find that preschoolers' *recognition* memory (ability to tell whether a stimulus is the same as or similar to one they have seen before) is remarkably good. In fact, 4- and 5-year-olds perform nearly perfectly.

Now give the child a more demanding task. Keep the items out of view and ask the child to name the ones she saw. This requires *recall*—that the child generate a mental image of an absent stimulus. Young children's recall is much poorer than their recognition. At age 2, they can recall no more than 1 or 2 of the items, at age 4 only about 3 or 4 (Perlmutter, 1984).

Of course, recognition is much easier than recall for adults as well, but in comparison to adults, children's recall is quite deficient. The reason is that young children are less

effective at using **memory strategies,** deliberate mental activities that improve our chances of remembering. For example, when you want to retain information, you might *rehearse,* or repeat the items over and over again. Or you might *organize,* grouping together items that are alike so you can easily retrieve them by thinking of their similar characteristics.

Preschoolers do show the beginnings of memory strategies. For example, when circumstances permit, they arrange items in space to aid their memories. In a study of 2- to 5-year-olds, an adult placed either an M&M or a wooden peg in each of 12 identical containers and handed them one by one to the child, who was asked to remember where the candy was hidden. By age 4, children put the candy containers in one place on the table and the peg containers in another. This strategy almost always led to perfect recall (DeLoache & Todd, 1988). But preschoolers do not yet rehearse or organize items into categories (for example, all the vehicles together, all the animals together) when asked to remember. Even when trained to do so, their memory performance may not improve, and they rarely apply these strategies in new situations (Gathercole, Adams, & Hitch, 1994; Miller & Seier, 1994).

Perhaps young children are not very strategic memorizers because they see little need to remember information for its own sake. In support of this explanation, the memory strategies that preschoolers do use are most effective when recall leads to a desired goal—for example, an M&M to eat, as in the research just described (Wellman, 1988).

■ MEMORY FOR EVERYDAY EXPERIENCES. Think about the difference in your recall of the listlike information discussed in the previous section and your memory for common, everyday experiences. Like adults, preschoolers remember familiar events in terms of **scripts,** general descriptions of what occurs and when it occurs in a particular situation.

Young children's scripts begin as a structure of main acts. For example, when asked to tell what happens when you go to a restaurant, a 3-year-old might say, "You go in, get the food, eat, and then pay." Although children's first scripts contain only a few acts, they are almost always recalled in correct sequence (Fivush, Kuebli, & Clubb, 1992). With age, scripts become more elaborate, as in the following restaurant account given by a 5-year-old child: "You go in. You can sit in the booths or at a table. Then you tell the waitress what you want. You eat. If you want dessert, you can have some. Then you pay and go home" (Hudson, Fivush, & Kuebli, 1992).

Scripts seem to be a basic means through which children organize and interpret their everyday worlds. Once familiar events become scripted in memory, any specific instance of a scripted experience becomes difficult to recall. For this reason, both children and adults remember

Like adults, young children remember familiar experiences in terms of scripts. After washing her hands before lunch at preschool many times, this girl is unlikely to recall the details of what she did on a particular day. But she will be able to describe what typically happens when you have lunch at preschool, and her account will become more elaborate with age. Scripts help us organize and interpret our everyday experiences. (Will Faller)

novel experiences much better than recurring ones (Hudson, 1988).

Parents and teachers can enhance preschoolers' memory for everyday events by the way they talk about them. Children whose mothers converse often about the past, ask many questions, and provide a great deal of elaborative information build a more complex *autobiographical memory* (see Chapter 5, page 156). They recount special, one-time occurrences in more complete and personally meaningful ways (Fivush, 1995). In line with Vygotsky's ideas, these findings indicate that social experiences are important in the development of early memory skills.

THE YOUNG CHILD'S THEORY OF MIND

As representation of the world, memory, and problem solving improve, children start to reflect on their own thought processes. They begin to construct a *theory of mind*, or set of beliefs about mental activities. This understanding is often called **metacognition**. The prefix "meta-," meaning beyond or higher, is applied to the term because metacognition means "thinking about thought." As adults, we have a complex appreciation of our inner mental worlds. For example, you can tell the difference between a wide variety of cognitive activities, such as knowing, remembering, guessing, forgetting, and imagining, and you are aware of a great many factors that influence them.

We rely on these understandings to interpret our own and others' behavior and to improve our performance on various tasks.

■ AWARENESS OF AN INNER MENTAL LIFE. Children's conversations reveal that awareness of mental activity emerges remarkably early. "Think," "remember," and "pretend" are among the first verbs to appear in children's vocabularies, and after age 2 ½ they use them appropriately. By age 3, children distinguish thinking from other mental activities. They realize that it takes place inside their head and that a person can think about something without seeing it, talking about it, or touching it (Estes, 1994; Flavell, Green, & Flavell, 1995).

More convincing evidence that preschoolers grasp the difference between an inner mental and outer physical world comes from games that involve tricking another person about the location of an object by moving it, laying false tracks, or giving incorrect information. By age 4, children understand that belief and reality can differ—in other words, that people can hold *false beliefs* (Robinson & Mitchell, 1994; Sodian et al., 1991).

Preschoolers are also conscious of some factors that affect the mind's functioning. For example, 3- and 4-year-olds realize that noise, lack of interest, and thinking about other things can interfere with attention to a task (Miller & Zalenski, 1982). And by age 5, most children know that information briefly presented or that must be held for a long time is more likely to be forgotten (Kreutzer, Leonard, & Flavell, 1975; Lyon & Flavell, 1993).

■ LIMITATIONS OF THE YOUNG CHILD'S THEORY OF MIND. Although surprisingly advanced, preschoolers' awareness of inner cognitive activities is far from complete. When questioned about subtle distinctions between mental states, such as "know" and "forget," children below age 5 often express confusion (Lyon & Flavell, 1994). Furthermore, young children believe that all events must be directly observed to be known. They do not understand that *mental inferences* can be a source of knowledge (Sodian & Wimmer, 1987). Finally, preschoolers are unaware that people continue to think while they are waiting or otherwise not doing something. They conclude that mental activity stops when there are no obvious external cues to indicate a person is thinking (Flavell, Green, & Flavell, 1993, 1995).

In sum, young children seem to view the mind as a passive container of information. They believe that physical experience with the environment determines mental experience. In contrast, older children view the mind as an active, constructive agent that selects and transforms information and affects how the world is perceived (Pillow, 1988; Wellman, 1990). We will consider this change further in Chapter 12 when we take up metacognition in middle childhood.

Preschoolers acquire a great deal of knowledge about literacy informally as they participate in everyday activities involving written symbols. They try to figure out how print conveys meaningful information, just as they strive to make sense of other aspects of their world. (Will Faller)

EARLY CHILDHOOD LITERACY

One week Leslie's pupils brought empty food boxes from home to place on special shelves in the classroom. Soon a make-believe grocery store opened. Children labeled items with prices, made shopping lists, and wrote checks at the cash register. A sign at the entrance announced the daily specials: *APLS BNS 5¢* ("apples bananas 5¢").

As their grocery store play reveals, preschoolers understand a great deal about written language long before they learn to read or write in conventional ways. This is not surprising when we consider that children in industrialized nations live in a world filled with written symbols. Each day, they observe and participate in activities involving storybooks, calendars, lists, and signs. As part of these experiences, they try to figure out how written symbols convey meaningful information, just as they strive to make sense of other aspects of their world.

Young preschoolers search for units of written language as they "read" memorized versions of stories and recognize familiar signs, such as *ON* and *OFF* on light switches and *PIZZA* at their favorite fast-food counter. But their early ideas about how written language is related to meaning are quite different from our own. For example, many preschoolers think that a single letter stands for a whole word or that each letter in a person's signature represents a separate name. Often they believe that letters (just like pictures) look like what they mean. One child explained that the word deer begins with the letter *O* because it is shaped like a deer. Then he drew an *O* and added antlers to it (Dyson, 1984; Sulzby, 1985).

As their perceptual and cognitive capacities improve and as they encounter writing in many different contexts, preschoolers gradually revise these ideas. Soon they become aware of some general characteristics of written language and create their own print-like symbols. Figure 7.9 shows a story and grocery list by a 4-year-old. This child understands that stories are written from left to right, that print appears in rows, that letters have certain features, and that stories look different from shopping lists (McGee & Richgels, 1996).

By the end of early childhood, children combine letters in their writing. However, their first ideas about how letters contribute to larger units are usually incorrect. Children often think that each letter represents a syllable. For example, a child named Santiago wrote his name with three letters (*SIO*), which he read, "San-tia-go." Soon children realize that letters and sounds are linked in systematic ways, as you can see in the invented spellings that are typical between ages 5 and 7. At first children rely heavily on the names of letters, as in *ADE LAFWTS KRMD NTU A LAVATR* ("eighty elephants crammed into a[n] elevator"). Over time, they will switch to standard spellings (Gentry, 1981; McGee & Richgels, 1996).

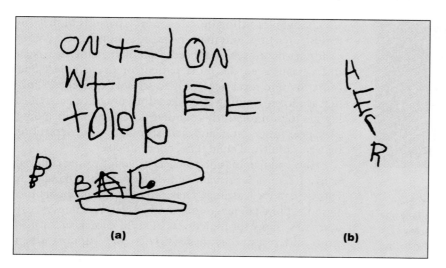

(a) (b)

FIGURE 7.9

A story (a) and a grocery list (b) written by a 4-year-old. This child's writing has many features of real print. It also reveals an awareness of different kinds of written expression. *(From L. M. McGee & D. J. Richgels, 1990, Literacy's Beginnings, Boston: Allyn and Bacon, p. 166. Reprinted by permission.)*

Literacy development builds on a broad foundation of spoken language and knowledge about the world. The more literacy-related experiences young children have in their everyday lives, the better prepared they are to tackle the complex tasks involved in reading and writing. Adults can provide literacy-rich physical environments, encourage literacy-related play, and read to children often (Roskos & Neuman, 1993; Snow, 1993). In early childhood, adults need not be overly concerned about the correctness of children's interpretations of written language. Instead, they can help most by accepting preschoolers' ideas and supporting their active efforts to revise and extend their knowledge.

YOUNG CHILDREN'S MATHEMATICAL REASONING

Mathematical reasoning, like literacy, builds on a foundation of informally acquired knowledge. In the early preschool period, children start to attach verbal labels (such as "lots," "little," "big," "small") to different amounts and sizes. And between ages 2 and 3, they also begin to count. At first, counting is little more than a memorized

Counting on fingers is an early, spontaneous strategy that young children rely on to master basic math facts. As they repeatedly use this technique and others, answers become more strongly associated with problems, and children retrieve the correct answer automatically. (Elizabeth Zuckerman/PhotoEdit)

routine, as in "Onetwothreefourfivesix!" Or children simply repeat a few number words while vaguely pointing in the direction of objects (Fuson, 1988).

Very soon counting becomes more precise. Most 3- to 4-year-olds have established an accurate one-to-one correspondence between a short sequence of number words and the items they represent. Sometime between ages 4 and 5, they grasp the vital principle of **cardinality**. They understand that the last number in a counting sequence indicates the quantity of items in a set (Bermejo, 1996). By the late preschool years, children no longer need to start a counting sequence with the number "one." Instead, knowing that there are six items in one pile and some additional ones in another, they begin with the number "six" and *count on* to determine the total quantity. Eventually, they generalize this strategy and *count down* to find out how many items remain after some are taken away. Once they master these procedures, children start to manipulate numbers without countable objects (Fuson, 1988). At this point, counting on fingers becomes an intermediate step on the way to automatically doing simple addition and subtraction.

Cross-cultural research suggests this basic arithmetic knowledge emerges universally around the world. However, children acquire it at different rates, depending on counting experiences available in their everyday lives. When adults provide many informal occasions and requests for counting, children probably come to these basic understandings sooner (Geary, 1995). Then they are solidly available as supports for the mathematical skills they will be taught in school.

A NOTE ON ACADEMICS IN EARLY CHILDHOOD

Does preschoolers' developing grasp of literacy and mathematical concepts mean that we should expose them to school-like instruction aimed at accelerating these skills? Experts in early childhood education agree that it would be a serious mistake to do so. Premature academic training, which involves sitting quietly for long periods of drill on reading, writing, and math facts, does not fit with young children's developmental needs. These demands are likely to frustrate even the most patient of preschoolers and cause them to react negatively to school experiences. Formal academic training also takes time away from activities known to promote early cognitive development, including play, peer interaction, and rich, stimulating conversations between adult and child (Greenberg, 1990).

How can parents and teachers acquaint young children with academically relevant knowledge without risking their enthusiasm for learning? They can offer many opportunities to explore language and number concepts in everyday activities, allow for high levels of child choice and initiation, and respond with sensitivity and support when preschoolers show interest in mastering new skills.

With age, preschoolers sustain attention for longer periods of time and search planfully for missing objects in familiar environments. By the end of early childhood, recognition memory is highly accurate. In contrast, recall improves slowly because preschoolers use memory strategies less effectively than do older children and adults. Like adults, young children remember familiar experiences in terms of scripts, which become more elaborate with age. Talking with children about the past enhances their memory for special, one-time occurrences. Preschoolers begin to construct a theory of mind, but they view the mind as a passive container of information. They also develop a basic understanding of written symbols and arithmetic concepts through informal experiences.

ASK YOURSELF . . .

■ *Lena notices that her 4-year-old son Gregor can recognize his name in print and count to 20. She wonders why Gregor's preschool teacher permits him to spend so much time playing instead of teaching him academic skills. Gregor's teacher responds, "I am teaching him academics—through play." Explain what she means.*

INDIVIDUAL DIFFERENCES IN MENTAL DEVELOPMENT

Five-year-old Hallie sat in a testing room while Sarah gave him an intelligence test. Some of the questions Sarah asked were *verbal*. For example, she held out a picture of a shovel and said, "Tell me what this shows"—an item measuring vocabulary. Then she tested his memory by asking him to repeat sentences and lists of numbers back to her. Other *nonverbal* tasks largely assessed spatial reasoning. Hallie copied designs with special blocks, figured out the pattern in a series of shapes, and indicated what a piece of paper folded and cut would look like when unfolded (Thorndike, Hagen, & Sattler, 1986).

Preschoolers differ markedly in cognitive progress, just as they vary in physical growth and motor skills. Psychologists and educators typically measure these differences with intelligence tests, such as the one Hallie took. Scores are computed in the same way as they are for infants and toddlers. But instead of emphasizing perceptual and motor responses, tests for preschoolers sample a wide range of verbal and nonverbal cognitive abilities. By age 5 to 6, intel-

ligence tests become good predictors of later intelligence and academic achievement (Hayslip, 1994).

Note that the questions Sarah asked Hallie tap knowledge and skills that not all children have had equal opportunity to learn. The issue of *cultural bias* in mental testing is a hotly debated topic that we will take up in Chapter 9. For now, keep in mind that intelligence tests do not sample all human abilities, and performance is affected by cultural and situational factors. At the same time, test scores remain important because they predict school achievement, which, in turn, is strongly related to vocational success in industrialized societies. Let's see how the environments in which young children spend their days—home, preschool, and day care—affect mental test performance.

HOME ENVIRONMENT AND MENTAL DEVELOPMENT

A special version of the *Home Observation for Measurement of the Environment (HOME)*, covered in Chapter 5, assesses aspects of 3- to 6-year-olds' home lives that support intellectual growth (see the Caregiving Concerns table below). As Piaget's and Vygotsky's theories suggest, physical surroundings and child-rearing practices play

CAREGIVING CONCERNS

Home Observation for the Measurement of the Environment (HOME): Early Childhood Subscales

SUBSCALE	SAMPLE ITEM
Stimulation through toys, games, and reading material	Home includes toys to learn colors, sizes, and shapes.
Language stimulation	Parent teaches child about animals through books, games, and puzzles.
Organization of the physical environment	All visible rooms are reasonably clean and minimally cluttered.
Pride, affection, and warmth	Parent spontaneously praises child's qualities once or twice during observer's visit.
Stimulation of academic behavior	Child is encouraged to learn colors.
Modeling and encouragement of social maturity	Parent introduces interviewer to child.
Variety in daily stimulation	Family member takes child on one outing at least every other week (picnic, shopping).
Avoidance of physical punishment	Parent neither slaps nor spanks child during observer's visit.

Source: Bradley & Caldwell, 1979.

important roles. Preschoolers who develop well intellectually have a home rich in toys and books. Their parents are warm and affectionate, stimulate language and academic knowledge, and arrange outings to places with interesting things to see and do. Such parents also make reasonable demands for socially mature behavior—for example, that the child do simple chores and behave courteously toward others. And when conflicts arise, these parents use reason to resolve them instead of physical force and punishment (Bradley & Caldwell, 1979, 1982).

As we saw in Chapter 2, these characteristics are less likely to be found in poverty-stricken families where parents lead highly stressful lives (Garrett, Ng'andu, & Ferron, 1994). When low-income parents manage, despite daily pressures, to obtain high HOME scores, their preschoolers do substantially better on intelligence tests (Bradley & Caldwell, 1979, 1981). These findings (as well as others we will discuss in Chapter 9) suggest that the home plays a major role in the generally poorer intellectual performance of low-income children in comparison to their middle-class peers.

PRESCHOOL AND DAY CARE

Children between 2 and 6 spend even more time away from their homes and parents than infants and toddlers do. Over the last 30 years, the number of young children enrolled in preschool or day care has steadily increased. This trend is largely due to the dramatic rise in women's participation in the labor force. Currently, 64 percent of American preschool-age children have mothers who are employed (U.S. Bureau of the Census, 1996). Figure 7.10 shows where preschoolers spend their days while their parents are at work.

A *preschool* is a half-day program with planned educational experiences aimed at enhancing the development of 2- to 5-year-olds. In contrast, *day care* identifies a variety of arrangements for supervising children of employed parents, ranging from care in someone else's or the child's own home to some type of center-based program. The line between preschool and day care is a fuzzy one. As Figure 7.10 indicates, parents often select a preschool as a child care option. Many preschools (and public school kindergartens as well) have increased their hours to full days in response to the needs of employed parents (U.S. Department of Education, 1996). At the same time, today we know that good day care is not simply a matter of keeping children safe and adequately fed in their parents' absence. Day care should provide the same high-quality educational experiences that an effective preschool does, the only difference being that children attend for an extended day.

▧ TYPES OF PRESCHOOL. Preschool programs come in great variety, ranging along a continuum from child-centered to teacher-directed. In **child-centered preschools,** teachers provide a wide variety of activities from which children select, and most of the day is devoted to spontaneous play. In contrast, in **academic preschools,** teachers structure the program. Children are taught letters, numbers, colors, shapes, and other academic skills through repetition and drill, and play is de-emphasized. Despite grave concern about the appropriateness of this approach, preschool teachers have felt increased pressure to provide formal academic training. The trend is motivated by a widespread belief that academic instruction at earlier ages will improve the ultimate achievement of American youths. However, in countries that outperform the United States in math and science achievement, such as

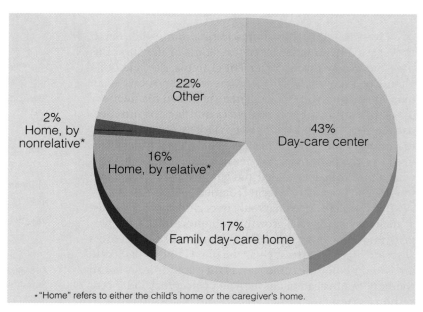

FIGURE 7.10

Who's minding America's preschoolers?
The chart refers to settings in which 3- and 4-year-olds spend most time while their mothers are at work. The "other" category consists mostly of children cared for by their mothers during working hours. Over one-fourth of 3- and 4-year-olds actually experience more than one type of child care, a fact not reflected in the chart. *(Adapted from B. Willer, S. L. Hofferth, E. E. Kisker, P. Divine-Hawkins, E. Farquhar, & F. B. Glantz, 1991,* The Demand and Supply of Child Care in 1990: Joint Findings from the National Child Care Survey 1990 and A Profile of Child Care Settings, *Washington, DC: National Association for the Education of Young Children. Reprinted by permission.)*

2%
Home, by nonrelative*

22%
Other

43%
Day-care center

16%
Home, by relative*

17%
Family day-care home

*"Home" refers to either the child's home or the caregiver's home.

Japan and Korea, children have a relaxed early childhood and are not hurried into academic work (Song & Ginsburg, 1987; Tobin, Wu, & Davidson, 1989).

■ EARLY INTERVENTION FOR AT-RISK PRESCHOOLERS. For low-income children, the intellectual benefits of preschool are considerable—much more than for middle-class children, whose home lives already provide many advantages. In the 1960s, when the United States launched a "War on Poverty," a wide variety of preschool intervention programs for disadvantaged children were initiated. Their underlying assumption was that the learning problems of these youngsters were best treated early, before formal schooling begins. **Project Head Start,** begun by the federal government in 1965, is the most extensive of these experiments.

A typical Head Start program provides children with a year or two of preschool, along with nutritional and medical services. Parent involvement is a central part of the Head Start philosophy. Parents serve on policy councils and contribute to program planning. They also work directly with children in classrooms, attend special programs on parenting and child development, and receive services directed at their own emotional, social, and vocational needs. Currently, over 1,300 Head Start centers located around the country serve about 720,000 children each year (Kassebaum, 1994).

Over two decades of research establishing the long-term benefits of preschool intervention have helped Head Start survive. The most important of these studies combined data from seven university-based interventions. Results showed that children who attended the programs scored higher in IQ and school achievement than controls during the first 2 to 3 years of elementary school. After that time, these gains declined. Nevertheless, children who received intervention remained ahead on measures of real-life school adjustment into adolescence. They were less likely to be placed in special education or retained in grade, and a greater number graduated from high school. There were also lasting benefits in attitudes and motivation. Children who attended the programs were more likely to give achievement-related reasons (such as school or job accomplishments) for being proud of themselves, and their mothers held higher vocational aspirations for them (Lazar & Darlington, 1982).

Do these findings on the impact of outstanding university-based programs generalize to Head Start centers located in American communities? As long as programs are of high quality, the outcomes are much the same (Zigler & Styfco, 1994a). In fact, as the Lifespan Vista box on page 237 indicates, the Head Start model, which combines early childhood education with support for parents in improving their own lives, is especially likely to induce lasting changes in both generations.

■ DAY CARE. We have seen that high-quality early intervention can enhance the development of economically disadvantaged children. However, as we noted in Chapter 5, much day care in the United States is not of this high quality. Regardless of whether they come from middle- or low-income homes, preschoolers exposed to inadequate day care score lower on measures of cognitive and social skills (Howes, 1988, 1990; Vandell & Powers, 1983). In contrast, good day care can reduce the negative impact of an underprivileged home life, and it sustains the benefits of growing up in an advantaged family (Phillips et al., 1994).

What are the ingredients of high-quality day care in early childhood? Large-scale studies of center- and home-based settings reveal that the following factors are related to sensitive, stimulating caregiver behavior and children's cognitive language, and social development: group size (number of children in a single space), caregiver–child ratio, caregiver's educational preparation, and caregiver's personal commitment to learning about and taking care of children (Galinsky et al., 1994; Helburn, 1995; Howes, Phillips, & Whitebook, 1992). Other research shows that spacious, well-equipped environments and activities that meet the needs and interests of preschool-age children also contribute to positive outcomes (Howes, 1988; Scarr et al., 1993).

The Caregiving Concerns table on page 236 summarizes characteristics of high-quality early childhood programs, based on standards for developmentally appropriate practice devised by the National Association for the Education of Young Children. Together, they offer a set of worthy goals as we strive to expand and upgrade day care and educational services for young children.

EDUCATIONAL TELEVISION

Besides home and preschool, young children spend much time in another learning environment: television. The average 2- to 6-year old watches TV from 2 to 3 hours a day—a very long time in the life of a young child (Comstock, 1993). Each afternoon, Sammy looked forward to watching certain educational programs. The well-known "Sesame Street" was his favorite.

Find a time to watch an episode of "Sesame Street" yourself. The program was originally designed for the same population served by Head Start—low-income children who enter school academically behind their middle-class peers. It uses fast-paced action, lively sound effects, and humorous puppet characters to stress letter and number recognition, counting, vocabulary, and basic concepts. Today, half of America's 2- to 5-year-olds regularly watch "Sesame Street," and it is broadcast in more than 40 countries around the world.

Research shows that "Sesame Street" works well as an academic tutor. The more children watch, the higher they

Signs of Developmentally Appropriate Early Childhood Programs

PROGRAM CHARACTERISTIC	SIGNS OF QUALITY
Physical setting	Indoor environment is clean, in good repair, and well ventilated. Classroom space is divided into richly equipped activity areas, including make-believe play, blocks, science, math, games, puzzles, books, art, and music. Fenced outdoor play space is equipped with swings, climbing equipment, tricycles, and sandbox.
Group size	In preschools and day care centers, group size is no greater than 18 to 20 children with 2 teachers.
Caregiver–child ratio	In day care centers, teacher is responsible for no more than 8 to 10 children. In family day care homes, caregiver is responsible for no more than 6 children.
Daily activities	Most of the time, children work individually or in small groups. Children select many of their own activities and learn through experiences relevant to their own lives. Teachers facilitate children's involvement, accept individual differences, and adjust expectations to children's developing capacities.
Interactions among adults and children	Teachers move among groups and individuals, asking questions, offering suggestions, and adding more complex ideas. They use positive guidance techniques, such as modeling and encouraging expected behavior and redirecting children to more acceptable activities.
Teacher qualifications	Teachers have college-level specialized preparation in early childhood development, early childhood education, or a related field.
Relationships with parents	Parents are encouraged to observe and participate. Teachers talk frequently with parents about children's behavior and development.
Licensing and accreditation	Program is licensed by the state. If a preschool or day care center, accreditation by the National Academy of Early Childhood Programs is evidence of an especially high quality program. If a day care home, accreditation by the National Association for Family Day Care is evidence of high-quality experiences for children.

Sources: Bredekamp & Copple, 1997; National Association for the Education of Young Children, 1991.

score on tests designed to measure the program's learning goals (Bogatz & Ball, 1972; Rice et al., 1990). In other respects, however, the rapid-paced format of "Sesame Street" has been criticized. When different types of programs are compared, those with slow-paced action and easy-to-follow story lines lead to more elaborate make-believe play. Those presenting quick, disconnected bits of information do not (Tower et al., 1979).

Some experts argue that because television presents such complete data to the senses, in heavy doses it encourages passive thinking. Too much television also takes time away from reading, playing, and interacting with others (Singer & Singer, 1990). But television can support cognitive development as long as children's viewing is not excessive and programs meet their developmental needs. We will look at the impact of television on young children's emotional and social development in the next chapter.

BRIEF REVIEW

By 5 to 6 years of age, IQ scores become good predictors of school achievement. A stimulating home environment, warm parenting, and reason-

able demands for mature behavior are positively related to mental test scores. Project Head Start, an intervention for low-income children that combines preschool education with parent involvement, results in immediate gains in IQ and achievement. Although these decline over time, children show lasting benefits in real-life indicators of school adjustment. High-quality day care can serve as effective early intervention, whereas poor day care undermines the development of children from all social classes. Preschoolers who watch "Sesame Street" score higher on tests of academic knowledge, but too much TV watching takes time away from many cognitively stimulating, worthwhile activities.

ASK YOURSELF . . .

■ *Senator Smith heard that IQ gains resulting from Head Start do not last, so he plans to vote against funding for the program. Write a letter to Senator Smith explaining why he should support Head Start.*

THE FUTURE OF EARLY INTERVENTION

The impact of preschool intervention may be considerably greater than previous evidence suggests. For example, studies reporting only minimal benefits of Head Start are often biased by one very important factor: Because not all poor children can be served, Head Start typically enrolls the most economically disadvantaged. Controls to whom they are compared often do not come from such extremely impoverished families. In a series of studies that took this into account, Head Start children showed greater gains on mental tests than did "other preschool" and "no preschool" comparison groups. Two years later, this advantage remained, although it was no longer as great as it had been immediately after Head Start (Lee, Brooks-Gunn, & Schnur, 1988; Lee et al. 1990). However, Head Start graduates entered poorer quality public schools, which explains, at least in part, why test score effects often fade out (Lee & Loeb, 1994).

Despite declining mental test performance, Head Start graduates (like those of university-based programs) show an improved ability to meet basic educational requirements during elementary and secondary school. And in one experimental program, high-quality preschool education plus parent involvement was associated with a reduction in delinquency and teenage pregnancy and a greater likelihood of employment in early adulthood (Schweinhart, Barnes, & Weikart, 1993).

Although researchers are not sure how these lasting effects are achieved, one possibility is that they are the result of changes in the attitudes, behaviors, and life circumstances of parents, who create better environments for both themselves and their children. Consequently, new interventions are being conceived as *two-generation models*. A typical parent component of early intervention emphasizes teaching child-rearing skills and providing other supports that encourage parents to act as supplementary interveners for their child. Researchers believe these efforts may not be enough (White, Taylor, & Moss, 1992). By expanding intervention to include developmental goals for both parents and children, program benefits might be extended. A parent helped to move out of poverty with education, vocational training, and other social services is likely to show improved psychological well-being and greater interest in planning for the future. When combined with child-centered intervention, these gains may translate into lasting changes in parents' lives and exceptionally strong benefits for children (Zigler & Styfco, 1994a).

Currently, a variety of two-generation models are being tried—several in conjunction with Head Start. Although the approach is too new to have yielded much research, several pioneering efforts provide cause for optimism (Smith, 1995). In one, called Project Redirection, teenage mothers received services for themselves as well as their babies, including education; health care; workshops on life management, employment, and parenting; and group and individual counseling. In addition, each teenager was paired with a community woman, who served as mentor, role model, and friend. A 5-year follow-up revealed that participants obtained higher HOME scores, were less likely to be on welfare, and had higher family earnings than controls receiving less intensive intervention. In addition, program children were more likely to have enrolled in Head Start, and they had higher verbal IQs and fewer behavior problems (Quint, 1991).

Head Start and other programs like it are highly cost-effective. Program expenses are far less than the funds required to provide special education, treat delinquency, and support welfare (Zigler & Styfco, 1994b). More intensive, longer-lasting efforts that focus on the development of both parents and children offer hope of increasing its positive outcomes.

LANGUAGE DEVELOPMENT

Language is intimately related to virtually all the cognitive changes discussed in this chapter. Between ages 2 and 6, children make awesome and momentous advances in language. Preschoolers' remarkable achievements, as well as their mistakes along the way, indicate that they master their native tongue in an active, rule-oriented fashion.

VOCABULARY

At age 2, Sammy had a vocabulary of 200 words. By age 6, he will have acquired around 10,000 words. To do so, Sammy will learn about 5 new words each day (Anglin, 1993). How do children build their vocabularies so quickly? Researchers have discovered that they can connect a new word with an underlying concept after only a brief encounter, a process called **fast-mapping.**

Young preschoolers fast-map some words more accurately and easily than others. For example, they learn labels for objects especially rapidly because these refer to concrete items they already know much about (Gentner, 1982). Soon they add large numbers of words for actions ("go," "run," "broke") as well as modifiers referring to noticeable features of objects and people ("red," "round," "sad"). If modifiers are related to one another in meaning, they take somewhat longer to learn. For example, 2-year-olds grasp the general distinction between "big" and "small," but not until age 3 to 5 are more refined differences between "tall" and "short," "high" and "low," and "long" and "short" understood. Similarly, children acquire "now"–"then" before "yesterday"–"today"–"tomorrow" (Clark, 1983; Stevenson & Pollitt, 1987).

Preschoolers figure out the meanings of new words by contrasting them with ones they already know. But exactly how they discover which concept each word picks out is not yet fully understood. Ellen Markman (1989, 1992) believes that in the early phases of vocabulary growth, children adopt a **principle of mutual exclusivity.** They assume that words refer to entirely separate (nonoverlapping) categories. Consistent with this idea, when 2-year-olds are told the names of two very different novel objects (a clip and a horn), they assign each label correctly, to the whole object and not a part of it (Waxman & Senghas, 1992).

However, sometimes adults call a single object by more than one name. Under these conditions, preschoolers are just as systematic in inferring word meanings. They assume the new word refers either to a higher- or lower-order category or to particular features, such as a part of the object, its shape, its color, or its proper name (Hall, 1996; Tomasello & Barton, 1994; Waxman & Hatch, 1992). When no such cues are available, children as young as 2 demonstrate remarkable flexibility in their word learning strategies. They abandon the mutual exclusivity principle and treat the new word as a second name for the object.

These findings tell us something about how children master labels for objects, but far less is known about the principles they use for other types of words. Children seem to draw on other aspects of language to help in these instances. By observing how words are used in the structure of sentences, they can distinguish the meanings of certain verbs, such as "give" and "take" or "drop" and "move" (Gleitman, 1990). And children also rely on social cues. For example, when an adult uses a new label while looking back and forth between the child and an object in an inviting way, the label probably refers to an action ("come," "help," or "play"), not the object (Tomasello & Akhtar, 1995).

Once preschoolers have a sufficient vocabulary, they use words creatively to fill in for ones they have not learned. As early as age 2, children coin new words in systematic ways. For example, Sammy said "plant-man" for a gardener (created a compound word) and "crayoner" for a child using crayons (added the ending -er) (Clark, 1995). Preschoolers also extend language meanings through

metaphor. For example, one 3-year-old used the expression "fire engine in my tummy" to describe a stomachache (Winner, 1988). Not surprisingly, the metaphors preschoolers use and understand involve concrete, sensory comparisons, such as "clouds are pillows" and "leaves are dancers." Once their vocabulary and knowledge of the world expand, they appreciate metaphors based on nonsensory comparisons as well, such as, "Friends are like magnets" (Karadsheh, 1991; Keil, 1986). Metaphors permit young children to communicate in especially vivid and memorable ways.

GRAMMAR

Grammar refers to the way we combine words into meaningful phrases and sentences. Between ages 2 and 3, English-speaking children use simple sentences that follow a subject–verb–object word order. This shows that they have a beginning grasp of the grammar of their language (de Villiers & de Villiers, 1992).

As young children conform to rules about word order, they also begin to make the small additions and changes in words that enable us to express meanings flexibly and efficiently. For example, they add -s for plural ("cats"), use prepositions ("in" and "on"), and form various tenses from the verb *to be* ("is," "are," "were," "has been," "will"). All English-speaking children master these grammatical markers in a regular sequence, starting with the ones that involve the simplest meanings and structures (Brown, 1973; de Villiers & de Villiers, 1973).

By age 3½, children have acquired a great many of these rules, and they apply them so consistently that once in a while they overextend the rules to words that are exceptions, a type of error called **overregularization.** "My toy car *breaked*," "I *runned* faster than you," and "We each have two *feets*" are expressions that start to appear between 2 and 3 years of age (Marcus et al., 1992; Marcus, 1995).

Between 3 and 6, children master even more complex grammatical forms, although they make predictable errors along the way. In asking questions, preschoolers are reluctant to let go of the subject–verb–object structure of the English language. At first, they use only a rising intonation and fail to invert the subject and verb, as in "Mommy baking cookies?" and "What you are doing, Daddy?" Because they cling to a consistent word order, they also have trouble with some passive sentences. When told, "The car was pushed by the truck," young preschoolers often make a toy car push a truck. By age 5, they understand expressions like these, but full mastery of the passive form is not complete until the end of middle childhood (Horgan, 1978; Lempert, 1989).

Even though they make errors, preschoolers' grasp of grammar is impressive. By age 4 to 5, children use embedded sentences ("I think *he will come*"), tag questions ("Dad's going to be home soon, *isn't he?*"), and indirect objects ("He showed *his friend* the present"). As the preschool years draw to a close, children use most of the grammatical constructions of their language competently (Tager-Flusberg, 1993).

CONVERSATION

Besides acquiring vocabulary and grammar, children must learn to engage in effective and appropriate communication with others. This practical side of language is called **pragmatics,** and preschoolers make considerable headway in mastering it.

At the beginning of early childhood, children are already skilled conversationalists. In face-to-face interaction with peers, they take turns, respond appropriately to their partner's remarks, and maintain a topic over time (Garvey, 1975; Podrouzek & Furrow, 1988). The number of turns over which children can sustain interaction increases with age. These surprisingly advanced abilities probably grow out of early interactive experiences with adults and siblings (see Chapter 6).

By age 4, children already know a great deal about culturally accepted ways of adjusting their speech to fit the age, sex, and social status of their listeners. For example, in acting out different roles with hand puppets, 4- to 7-year-olds use more commands when playing both high status and male roles, such as doctor, teacher, and father. In con-trast, they speak more politely and use more indirect requests when acting out lower status and female roles, such as patient, pupil, and mother (Anderson, 1984). Older preschoolers also adjust their speech on the basis of how well they know their conversational partner. They give fuller explanations to a stranger than to someone with whom they share common experiences, such as a family member or friend (Sonnenschein, 1986).

Preschoolers' conversational skills occasionally do break down. For example, have you tried talking on the telephone with a preschooler lately? Here is an excerpt of one 4-year-old's phone conversation with his grandfather:

Grandfather : "How old will you be?"

John: "Dis many." (Holding up four fingers)

Grandfather: "Huh?"

John: "Dis many." (Again holding up four fingers)

(Warren & Tate, 1992, pp. 259–260)

Young children's conversations appear less mature in highly demanding situations in which they cannot see their listeners' reactions or rely on typical conversational aids, such as gestures and objects to talk about. Conversational skills provide yet another example of how preschoolers' competencies depend on the demands of the situation.

SUPPORTING LANGUAGE LEARNING IN EARLY CHILDHOOD

How can adults foster preschoolers' language development? Interaction with more skilled speakers, which is so important in toddlerhood, remains critical during early childhood. Conversational give-and-take with adults, either at home or in preschool, is consistently related to general measures of language progress (Byrne & Hayden, 1980; McCartney, 1984).

Sensitive, caring adults use additional techniques that promote early language skills. When children use words incorrectly or communicate unclearly, they give helpful, explicit feedback, as in "There are several balls over there, and I can't tell exactly which one you want. Do you mean a large or small one or a red or green one?" At the same time, such adults do not overcorrect, especially when children make grammatical mistakes. Criticism discourages children from actively experimenting with language rules.

Instead, adults provide subtle, indirect feedback about grammar by using two strategies, often in combination: **expansions** and **recasts** (Bohannon & Stanowicz, 1988). For example, if a child says, "I gotted new red shoes," the parent might respond, "Yes, you got a pair of new red shoes," *expanding* the complexity of the child's statement and *recasting* it into appropriate form. Parents who reformulate children's utterances in these ways have preschoolers who make especially rapid language progress (Nelson et al., 1984). Nevertheless, some researchers question

In highly demanding situations, preschool children's conversational skills can break down. This 3-year-old is likely to have trouble communicating clearly on the telephone because he lacks the supports available in face-to-face interaction, such as visual access to his partner's reaction and to objects that are topics of conversation. *(Tony Freeman/PhotoEdit)*

Conversational give-and-take with adults, either at home or in preschool, promotes young children's language development. (R. Sidney/The Image Works)

whether expansions and recasts are as important in children's mastery of grammar as mere exposure to a rich language environment containing abundant examples of correct forms. Adults do not use these techniques often, and they are not provided to children in all cultures (Marcus, 1993; Valian, 1993).

Do the findings just described remind you once again of Vygotsky's theory? In language as in other aspects of intellectual growth, effective parents and teachers seem to interact with young children in ways that gently prompt them to take the next developmental step forward. In the next chapter, we will see that this special combination of warmth and encouragement of mature behavior is at the heart of early childhood emotional and social development as well.

ASK YOURSELF . . .

- *One day, Sammy's mother explained to him that the family would take a vacation in Miami. The next morning, Sammy emerged from his room with belongings spilling out of a suitcase and remarked, "I gotted my bag packed. When are we going to Your-ami?" What do Sammy's errors reveal about his approach to mastering language?*

SUMMARY

PHYSICAL DEVELOPMENT IN EARLY CHILDHOOD

BODY GROWTH

Describe major trends in body growth during early childhood.

- Compared to infancy, children grow more slowly in early childhood. Body fat declines, and children become longer and leaner. New epiphyses appear in the skeleton, and by the end of early childhood, children start to lose their primary teeth.

- Different parts of the body grow at different rates. Overall body size (as measured by height and weight) increases rapidly in infancy, more slowly in early and middle childhood, and rapidly again in adolescence—a pattern represented by the **general growth curve.**

BRAIN DEVELOPMENT

Describe brain development during early childhood, including lateralization (as reflected in handedness) and myelinization of specific structures.

- During early childhood, the left cerebral hemisphere grows more rapidly than the right, supporting young children's rapidly expanding language skills.

- Hand preference is stable by age 2 and increases during early and middle childhood, indicating that lateralization strengthens during this time. Handedness indicates an individual's **dominant cerebral hemisphere.** Although left-handedness is associated with developmental problems, the great majority of left-handed children show no abnormalities of any kind.

- During early childhood, connections are established among different brain structures. Fibers linking the **cerebellum** to the cerebral cortex myelinate, enhancing children's balance and motor control. The **reticular formation,** responsible for alertness and consciousness, and the **corpus callosum,** which connects the two cerebral hemispheres, also myelinate rapidly.

INFLUENCES ON PHYSICAL GROWTH AND HEALTH

Explain how heredity influences physical growth.

- Heredity influences physical growth by controlling the release of hormones from the **pituitary gland.** Two hormones are especially influential: **growth hormone (GH)** and **thyroid-stimulating hormone (TSH).**

What are the effects of emotional well-being, nutrition, and infectious disease on physical growth in early childhood?

■ Emotional well-being continues to influence body growth in middle childhood. An emotionally inadequate home life can lead to **deprivation dwarfism.**

■ Preschoolers' slower growth rate causes their appetite to decline, and often they become picky eaters. Repeated exposure to new foods and a positive emotional climate at mealtimes can promote healthy, varied eating in young children.

■ Malnutrition can combine with infectious disease to undermine healthy growth. In developing countries, many children do not receive early immunizations that protect against childhood illnesses. Immunization rates are lower in the United States than in other industrialized nations because many economically disadvantaged children do not have access to health insurance and good medical care.

What factors increase the risk of unintentional injuries, and how can childhood injuries be prevented?

■ Unintentional injuries are the leading cause of childhood mortality. Injury victims are more likely to be boys; to be temperamentally irritable, inattentive, and negative; and to grow up in poor inner-city families. Laws that promote child safety, modifications in home and play environments, public education, and interventions designed to change parent and child behavior can help prevent childhood injuries.

MOTOR DEVELOPMENT

Cite major milestones of gross and fine motor development in early childhood.

■ During early childhood, the child's center of gravity shifts toward the trunk, and balance improves, paving the way for many gross motor achievements. Preschoolers run, jump, hop, gallop, eventually skip, throw and catch, and generally become better coordinated.

■ Increasing control of the hands and fingers leads to dramatic improvements in fine motor skills. Preschoolers gradually become self-sufficient at dressing and using a knife and fork. By age 3, children's scribbles become pictures. Their drawings increase in complexity with age, and they try to print letters of the alphabet.

What factors influence early childhood motor skills?

■ Body build, race, sex, and opportunity for physical play affect early childhood motor development. Differences in motor skills between boys and girls are partly genetic, but environmental pressures exaggerate them. Children master the motor skills of early childhood through informal play experiences.

COGNITIVE DEVELOPMENT IN EARLY CHILDHOOD

PIAGET'S THEORY: THE PREOPERATIONAL STAGE

Describe advances in mental representation and limitations of thinking during the preoperational stage.

■ Rapid advances in mental representation, notably language and make-believe play, mark the beginning of Piaget's **preoperational stage.** With age, make-believe becomes increasingly complex, evolving into **sociodramatic play** with others.

■ Aside from representation, Piaget described the young child in terms of deficits rather than strengths. Preoperational children are **egocentric**—unable to imagine the perspectives of others and reflect on their own thinking. Egocentrism leads to a variety of illogical features of thought. Preschoolers engage in **animistic thinking,** and their cognitions are **perception-bound, centered,** focused on **states rather than transformations,** and **irreversible.** In addition, preoperational children display **transductive reasoning** rather than truly causal reasoning. Because of these difficulties, they fail Piaget's **conservation** and **hierarchical classification** tasks.

What are the implications of recent research for the accuracy of the preoperational stage?

■ When young children are given simplified problems relevant to their everyday lives, their performance appears more mature than Piaget assumed. Operational thinking develops gradually over the preschool years, a finding that challenges Piaget's concept of stage.

What educational principles can be derived from Piaget's theory?

■ Piaget's theory has had a lasting impact on education for young children. A Piagetian classroom promotes discovery learning, sensitivity to children's readiness to learn, and acceptance of individual differences.

VYGOTSKY'S SOCIOCULTURAL THEORY

Explain Vygotsky's perspective on the origins and significance of children's private speech, and describe applications of his theory to education.

■ In contrast to Piaget, Vygotsky regarded language as the foundation for all higher cognitive processes. According to Vygotsky, **private speech,** or language used for self-guidance, emerges out of social communication as adults and more skilled peers help children master challenging tasks. Eventually private speech is internalized as inner, verbal thought.

■ A Vygotskian classroom emphasizes assisted discovery. Verbal guidance from teachers and peer collaboration are vitally important.

INFORMATION PROCESSING

How do attention and memory change during early childhood?

- Preschoolers spend only short times involved in tasks, have difficulty focusing on details, and are easily distracted. Attention gradually becomes more sustained and planful during early childhood.

- Young children's recognition memory is very accurate. Their recall for listlike information is much poorer than that of older children and adults because preschoolers use **memory strategies** less effectively. Like adults, preschoolers remember recurring everyday experiences in terms of **scripts,** which become more elaborate with age. Conversing with children about special, one-time events helps them build a more complex autobiographical memory.

Describe the young child's theory of mind.

- Preschoolers begin to construct a theory of mind, indicating that they are capable of **metacognition,** or thinking about thought. Between ages 2 and 3, they distinguish between an inner mental and outer physical world, and by age 4 they understand that people can hold false beliefs. However, young children regard the mind as a passive container of information.

Summarize children's literacy and mathematical knowledge during early childhood.

- Children understand a great deal about literacy long before they read or write in conventional ways. Preschoolers gradually revise incorrect ideas about the meaning of written symbols as their perceptual and cognitive capacities improve and as they encounter writing in many different contexts. They also experi-

ment with counting strategies and discover basic mathematical principles, including **cardinality.**

INDIVIDUAL DIFFERENCES IN MENTAL DEVELOPMENT

Describe the impact of testing conditions, home, preschool and day care, and educational television on mental development in early childhood.

- Intelligence tests in early childhood sample a wide variety of verbal and nonverbal skills. During such testing, low-income and ethnic minority children, especially, benefit from time to get to know the examiner and generous praise and encouragement.

- Children growing up in warm, stimulating homes with parents who make reasonable demands for mature behavior score higher on mental tests.

- **Child-centered preschools** emphasize spontaneous play. In **academic preschools,** teachers train academic skills through repetition and drill. Formal academic instruction, however, is inconsistent with young children's developmental needs.

- **Project Head Start** is the largest federally funded preschool program for low-income children in the United States. High-quality preschool intervention results in immediate test score gains and long-term improvements in school adjustment. The Head Start model, which combines early childhood education with parental support, is especially effective. Good day care can also serve as early intervention. In contrast, poor-quality day care undermines the development of children from all social classes.

- Children pick up many cognitive skills from educational television programs like "Sesame Street." Programs with slow-paced action and easy-to-follow story lines foster make-believe play.

LANGUAGE DEVELOPMENT

Trace the development of vocabulary, grammar, and conversational skills in early childhood.

- Supported by **fast mapping,** children's vocabularies grow dramatically during early childhood. Preschoolers figure out the meaning of new words by contrasting them with ones they already know. The **principle of mutual exclusivity** explains their acquisition of some, but not all, early words. Once preschoolers have sufficient vocabulary, they extend language meanings, coining new words and creating metaphors.

- Between ages 2 and 3, children adopt the basic word order of their language. As they master grammatical rules, they occasionally **overregularize,** or apply the rules to words that are exceptions. By the end of early childhood, children have acquired complex grammatical forms.

- **Pragmatics** refers to the practical, social side of language. In face-to-face interaction with peers, young preschoolers are already skilled conversationalists. By age 4, they adapt their speech to their listeners in culturally accepted ways. Preschoolers' communicative skills appear less mature in highly demanding contexts.

Cite factors that support language learning in early childhood.

- Conversational give-and-take with more skilled speakers fosters preschoolers' language skills. When adults provide explicit feedback on the clarity of children's utterances and use **expansions** and **recasts,** preschoolers show especially rapid language progress. However, the impact of these strategies on grammatical development has been challenged. For this aspect of language, exposure to a rich language environment may be sufficient.

IMPORTANT TERMS AND CONCEPTS

general growth curve (p. 207)

dominant cerebral hemisphere (p. 209)

cerebellum (p. 209)

reticular formation (p. 209)

corpus callosum (p. 210)

pituitary gland (p. 210)

growth hormone (GH) (p. 210)

thyroid-stimulating hormone (TSH) (p. 210)

deprivation dwarfism (p. 211)

preoperational stage (p. 220)

sociodramatic play (p. 221)

egocentrism (p. 222)

animistic thinking (p. 222)

conservation (p. 222)

perception-bound (p. 222)

centration (p. 222)

states versus transformations (p. 222)

irreversibility (p. 222)

transductive reasoning (p. 223)

hierarchical classification (p. 223)

private speech (p. 227)

memory strategies (p. 229)

scripts (p. 229)

metacognition (p. 230)

cardinality (p. 232)

child-centered preschools (p. 234)

academic preschools (p. 234)

Project Head Start (p. 235)

fast-mapping (p. 237)

principle of mutual exclusivity (p. 238)

overregularization (p. 238)

pragmatics (p. 239)

expansions (p. 239)

recasts (p. 239)

FYI FOR FURTHER INFORMATION AND HELP

INFECTIOUS DISEASE IN CHILDHOOD

American Academy of Pediatrics
P.O. Box 927
Elk Grove Village, IL 60009-0927
(847) 228-5005
Web site: *www.kidsdoc@aap.org*
Provides public education on a variety of childhood health issues. Among the many pamphlets and publications available are written guidelines for effective control of infectious disease.

Child Care Information Exchange
Box 2890
Redmond, WA 98052
(206) 883-9394
Web site: *www.wolfenet.com/~ccie*
A bimonthly publication written especially for day care directors that addresses the practical issues of running a center. Articles discussing health and safety are often included.

CHILDHOOD INJURY CONTROL

U.S. Consumer Product Safety Commission
5401 Westbard Avenue
Bethesda, MD 20207
(800) 638-2772
Web site: *www.cpsc.gov*
Establishes and enforces product safety standards. Operates a hot line providing information on safety issues and recall of dangerous consumer products.

EARLY CHILDHOOD DEVELOPMENT AND EDUCATION

Association for Childhood Education International (ACEI)
11561 Georgia Avenue, Suite 312
Wheaton, MD 20902
(301) 942-2443
Web site: *www.udel.edu/ bateman/acei*
Organization interested in promoting sound educational practice from infancy through early adolescence. Student membership is available and includes a subscription to Childhood Education, *a bimonthly journal covering research, practice, and public policy issues.*

National Association for the Education of Young Children (NAEYC)
1509 16th Street, N.W.
Washington, DC 20036-1426
(202) 232-8777; (800) 424-2460
Web site: *www.naeyc.org/naeyc*
Organization open to all individuals interested in acting on behalf of young children's needs, with primary focus on educational services. Student membership is available and includes a subscription to Young Children, *a bimonthly journal covering theory, research, and practice in infant and early childhood development and education.*

PRESCHOOL INTERVENTION

National Head Start Association
201 N. Union Street, Suite 320
Alexandria, VA 22314
(703) 739-0875
Web site: *www.nhsa.com*
Association of Head Start directors, parents, staff, and others interested in the Head Start program. Works to upgrade the quantity and quality of Head Start services.

High/Scope Educational Research Foundation
600 N. River Street
Ypsilanti, MI 48198
(313) 485-2000
Web site: *www.highscope.org*
Devoted to improving development and education from infancy through adolescence. Conducts longitudinal research to determine the effects of early intervention on development.

DAY CARE

National Association for Family Child Care
206 6th Avenue, Suite 900
Des Moines, IA 50309
(515) 282-8192
Organization open to caregivers, parents, and other individuals involved or interested in family day care. Serves as a national voice that promotes high-quality day care.

8

Emotional and Social Development in Early Childhood

As the children in Leslie's classroom moved through the preschool years, their personalities took on clearer definition. By age 3, they voiced firm likes and dislikes as well as new ideas about themselves. "Stop bothering me," 3-year-old Sammy said to his classmate, Mark, as he aimed a beanbag toward the mouth of a large clown face. "See, I'm great at this game," Sammy announced with confidence, an attitude that kept him trying, even though he missed most of the throws.

The children's conversations also revealed their first notions about morality. Often they combined statements about right and wrong they had heard from adults with forceful attempts to defend their desires. "You're 'posed to share," stated Mark, grabbing a beanbag out of Sammy's hand.

"I was here first! Gimme it back," demanded Sammy, who pushed Mark while reaching for the beanbag. The two boys struggled until Leslie intervened, provided an extra set of beanbags, and showed them how they could both play at once.

As Sammy and Mark's interaction reveals, preschoolers are quickly becoming complex social beings. Although arguments and aggression take place among all young children, cooperative exchanges are far more frequent. Between the years of 2 and 6, first friendships emerge in which children converse, act out complementary roles, and learn that their own desires for companionship and toys are best met when they consider the needs and interests of others.

Children's developing understanding of their social world was especially evident in the attention they gave to the dividing line between male and female. While Lynette and Karen cared for a sick baby doll in the housekeeping area, Sammy, Vance, and Mark transformed the block corner into a busy intersection. "Green light, go!" shouted police officer Sammy as Vance and Mark pushed large wooden cars and trucks across the floor. Already, the children preferred to interact with same-sex peers, and their play themes mirrored the gender stereotypes of their cultural community.

This chapter is devoted to the many facets of emotional and social development in early childhood. We begin with Erik Erikson's overview of personality change during the preschool years. Then we consider children's concepts of themselves, their insights into their social and moral worlds, their increasing ability to manage their emotional and social behaviors, and the many factors that support these competencies. In the final sections of this chapter, we answer the question "What is effective child rearing?" We also consider the complex conditions that support good parenting or lead it to break down. Today, child abuse and neglect rank among America's most serious national problems.

ERIKSON'S THEORY: INITIATIVE VERSUS GUILT

Erikson (1950) described early childhood as a period of "vigorous unfolding" (p. 255). Once children have a sense of autonomy and feel secure about separating from parents, they become more relaxed and less contrary than they were as toddlers. Their energies are freed for tackling the critical psychological conflict of the preschool years: **initiative versus guilt.**

The word *initiative* means spirited, enterprising, and ambitious. It suggests that young children have a new sense of purposefulness. They are eager to tackle new tasks, join in activities with peers, and discover what they can do with the help of adults.

Erikson regarded play as a central means through which young children find out about themselves and their social world. Play permits preschoolers to try out new skills with little risk of criticism and failure. It also creates a small social organization of children who must cooperate to achieve common goals. Make-believe, especially, offers unique opportunities for developing initiative. In cultures around the world, children act out family scenes and highly visible occupations—police officer, doctor, nurse, and teacher in Western societies, rabbit hunter and potter among the Hopi Indians, and hut builder and spear maker among the Baka of West Africa (Garvey, 1990). In this way, make-believe provides children with important insights into the link between self and wider society.

Recall that Erikson's theory builds on Freud's psychosexual stages. Freud's **phallic stage** of early childhood is a time when sexual impulses transfer to the genital region of the body, and the well-known *Oedipus conflict* arises. The young boy wishes to have his mother all to himself and feels hostile and jealous of his father. Freud described a similar *Electra conflict* for girls, who want to possess their fathers and envy their mothers. These feelings soon lead to intense anxiety, since children fear they will lose their parents' love and be punished for their unacceptable wishes. To master the anxiety, avoid punishment, and maintain the affection of parents, children form a *superego*, or conscience, through **identification** with the same-sex parent. They take the parent's characteristics into their personality and, as a result, adopt the moral and gender-role standards of their society. Each time the child disobeys standards of conscience, painful feelings of guilt occur.

For Erikson, the negative outcome of early childhood is an overly strict superego, one that causes children to feel too much guilt because they have been threatened, criticized, and punished excessively by adults. When this happens, preschoolers' exuberant play and bold efforts to master new

According to Freud and Erikson, preschoolers form a superego, or conscience, by identifying with the same-sex parent and, thereby, adopting the moral and gender-role standards of their society. This young boy displays a sense of initiative when he joins his father in making home repairs. (Ann Hagen Griffiths/OPC)

SELF-DEVELOPMENT IN EARLY CHILDHOOD

During the preschool years, new powers of representation permit children to reflect on themselves. They start to develop a **self-concept,** the sum total of attributes, abilities, attitudes, and values that an individual believes defines who he or she is.

FOUNDATIONS OF SELF-CONCEPT

Ask a 3- to 5-year-old to tell you about him- or herself, and you will probably hear something like this:

> I'm Tommy. See, I got this new red T-shirt. I'm 4 years old. I can brush my teeth, and I can wash my hair all by myself. I have a new Tinkertoy set, and I made this big, big tower.

As these statements indicate, preschoolers' self-concepts, like other aspects of their thinking, are very concrete. Usually, they mention observable characteristics, such as their name, physical appearance, possessions, and everyday behaviors (Keller, Ford, & Meacham, 1978).

However, young children's understanding of themselves is not limited to observable attributes. By age 2½, they also describe themselves in terms of typical beliefs, emotions, and attitudes, as in "I'm happy when I play with my friends" or "I don't like being with grown-ups" (Eder, 1989). This indicates that they have a beginning understanding of their unique psychological characteristics. Furthermore, a child who says that she "doesn't push in front of other people in line" is also likely to indicate that she "feels like being quiet when angry," as if she recognizes that she is high in self-control (Eder, 1990). But preschoolers do not yet make explicit reference to internal dispositions. This capacity must wait for the greater cognitive maturity of middle childhood.

In fact, very young preschoolers' concepts of themselves are so bound up with concrete possessions and actions that they spend much time asserting their rights to objects, as Sammy did in the beanbag incident at the beginning of this chapter. The stronger the child's self-definition, the more possessive he or she tends to be about objects, claiming them as "Mine!" (Levine, 1983). These findings indicate that rather than a sign of selfishness, early struggles over objects seem to be a sign of developing selfhood, an effort to clarify boundaries between self and others.

The ability to distinguish self from others also permits children to cooperate for the first time in playing games, solving simple problems, and resolving disputes over objects (Brownell & Carriger, 1990; Caplan et al., 1991). Adults might take both of these capacities into account when trying to promote friendly peer interaction. For

tasks break down. Their self-confidence is shattered, and they approach the world timidly and fearfully.

At this point, it is important to note that Freud's Oedipus and Electra conflicts are no longer regarded as satisfactory explanations of children's emotional, moral, and gender-role development. Later, when we discuss these topics in detail, we will critically evaluate Freud's psychosexual ideas. At the same time, Erikson's image of initiative captures the diverse changes that take place in young children's emotional and social lives. The preschool years are, indeed, a time when children develop a confident self-image, more effective control over emotions, new social skills, the foundations of morality, and a clear sense of self as boy or girl. Now let's look closely at each of these aspects of development.

example, teachers and parents can accept the young child's possessiveness as a sign of self-assertion ("Yes, that's your toy") and then encourage compromise ("but in a little while, can you give someone else a turn?"), rather than simply insisting on sharing.

EMERGENCE OF SELF-ESTEEM

A special aspect of self-concept emerges in early childhood: **self-esteem,** the judgments we make about our own worth and the feelings associated with those judgments. Self-esteem ranks among the most important aspects of self-development, since evaluations of our own competencies affect emotional experiences, future behavior, and long-term psychological adjustment. Take a moment to think about your own self-esteem. Besides a global appraisal of your worth as a person, you have a variety of separate self-judgments. For example, you may regard yourself as well liked by others, very good at schoolwork, but only so-so at sports.

Preschoolers' self-esteem is not as well defined as that of older children and adults. And usually they rate their own ability as extremely high and underestimate the difficulty of a task (Harter, 1990). Sammy's announcement that he was great at beanbag throwing despite his many misses of the target is a typical self-evaluation during the preschool years. A high sense of self-esteem contributes

Acting on the world is an especially important source of self-definition to preschoolers. When questioned about her self-concept, this child is likely to describe typically behaviors, such as "I can dress myself." She may also mention commonly experienced emotions and attitudes, indicating that she has a beginning appreciation of her unique psychological characteristics. (Mary Kate/PhotoEdit)

greatly to preschoolers' initiative during a period in which they must master many new skills. Most young children know they are growing bigger and stronger, and they see that failure on one occasion often translates into success on another. Their belief in their own capacities is supported by the patience and encouragement of adults, who realize that a child who has trouble riding a tricycle or cutting with scissors will be able to do so a short time later.

Nevertheless, by age 4, some children give up easily when faced with a challenge, such as working a hard puzzle or building a tall block tower. They conclude that they cannot do the task and are discouraged after failure (Cain & Dweck, 1995; Smiley & Dweck, 1994). When these young nonpersisters are asked to act out with dolls an adult's reaction to failure, they often respond, "He's punished because he can't do [the puzzle]" or "Daddy's gonna be very mad and spank her" (Burhans & Dweck, 1995, p. 1727). They are also likely to report that their parents would berate them for making small mistakes (Heyman, Dweck, & Cain, 1992). The Caregiving Concerns table on the following page suggests ways to avoid these self-defeating reactions and foster a healthy self-image in young children.

EMOTIONAL DEVELOPMENT IN EARLY CHILDHOOD

Gains in representation, language, and self-concept support emotional development in early childhood. Between ages 2 and 6, children achieve a better understanding of their own and others' feelings, and their ability to regulate the expression of emotion improves. Self-development also contributes to a rise in *self-conscious emotions,* such as shame, embarrassment, guilt, envy, and pride.

UNDERSTANDING EMOTION

Young children's vocabulary for talking about emotion expands rapidly. Early in the preschool years, they refer to causes, consequences, and behavioral signs of emotion. Over time, their understanding improves in accuracy and complexity. By age 4 to 5, children correctly judge the causes of many basic emotions, as in "He's happy because he's swinging very high," "I'm mad because he wouldn't share the toy," or "He's sad because he misses his mother." However, they are likely to emphasize external factors over internal states as explanations—a balance that changes with age (Levine, 1995). Preschoolers are also good at predicting what a playmate expressing a certain emotion might do next. For example, they know that an angry child might hit someone or grab a toy back and that a happy child is more likely to share (Russell, 1990). They are even aware that a lingering mood can affect a person's behavior for some time in the future (Bretherton et al., 1986).

Fostering a Healthy Self-Image in Young Children

SUGGESTION	DESCRIPTION
Build a positive relationship.	Indicate that you want to be with the child by arranging times to be fully available. Listen without being judgmental, and express some of your own thoughts and feelings. Mutual sharing helps children feel valued.
Nurture success.	Adjust expectations appropriately, and provide assistance when asking the child to do something beyond his or her current limits. Accentuate the positive in the child's work or behavior. Promote self-motivation by emphasizing praise over concrete rewards. Instead of simply saying, "That's good," mention effort and specific accomplishments. Display the child's artwork and other products, pointing out increasing skill.
Foster the freedom to choose.	Choosing gives children a sense of responsibility and control over their own lives. Where children are not yet capable of deciding on their own, involve them in some aspect of the choice, such as when and in what order a task will be done.
Acknowledge the child's emotions.	Accept the child's strong feelings, and suggest constructive ways to handle them. When a child's negative emotion results from an affront to his or her self-esteem, offer sympathy and comfort along with a realistic appraisal of the situation so the child feels supported and secure.
Use a rational approach to discipline.	The discipline strategies discussed on pages 254 and 266–267 (induction and authoritative child rearing) promote self-confidence and self-control.

Source: Berne & Savary, 1993.

Young children also use emotional language to guide and influence a companion's behavior. One day, when Karen's eyes filled with tears because her mother was late picking her up, Lynette said soothingly, "Aw, that's all right. She'll come. Don't worry. Want to sit by me?" As this example suggests, preschoolers come up with effective ways to relieve others' negative feelings. For example, they suggest physical comfort to reduce sadness and giving a desired object to a playmate to reduce anger (Fabes et al., 1988). Overall, preschoolers have an impressive ability to interpret, predict, and change others' feelings—knowledge that is of great help in getting along with peers and adults.

At the same time, there are limits to young children's emotional understanding. In situations in which there are conflicting cues about how a person is feeling, preschoolers have difficulty making sense of what is going on. For example, when asked what might be happening in a picture showing a happy-faced child with a broken bicycle, 4- and 5-year-olds tended to rely on the emotional expression: "He's happy because he likes to ride his bike." Older children more often reconciled the two cues: "He's happy because his father promised to help fix his broken bike" (Gnepp, 1983). As in their approach to Piagetian tasks, preschoolers focus on the most obvious aspect of a complex emotional situation to the neglect of other relevant information.

IMPROVEMENTS IN EMOTIONAL SELF-REGULATION

Language contributes to preschoolers' improved *emotional self-regulation,* or ability to control the expression of emotion. By age 3 to 4, children verbalize a variety of

strategies for adjusting their emotional arousal to a more comfortable level. For example, they know that emotions can be blunted by restricting sensory input (covering your eyes or ears to block out a scary sight or sound), talking to yourself ("Mommy said she'll be back soon"), or changing your goals (deciding that you don't want to play anyway after being excluded from a game) (Thompson, 1990).

Children's awareness and use of these strategies means that intense emotional outbursts become less frequent over the preschool years. Adult instruction also promotes this change. Most cultures encourage their members to communicate positive feelings and inhibit unpleasant ones as a way of fostering good interpersonal relations, and young children try hard to conform to this rule. As early as the preschool years, children who have trouble coping with negative emotion get along poorly with adults and peers (Eisenberg et al., 1995).

By watching adults handle their own feelings, children pick up strategies for regulating emotion. Discussing feelings can also be helpful. When parents prepare children for difficult experiences by describing what to expect and ways to handle anxiety, they offer coping strategies that children can apply to themselves. Preschoolers' vivid imaginations combined with their difficulty in separating appearance from reality make fears common in early childhood. The Caregiving Concerns table on page 250 lists ways that parents can help young children manage them.

CHANGES IN SELF-CONSCIOUS EMOTIONS

One morning in Leslie's classroom, a group of children crowded around for a bread-baking activity. Leslie asked

CAREGIVING CONCERNS

Helping Children Manage Common Fears of Early Childhood

FEAR	SUGGESTION
Monsters, ghosts, and darkness	Reduce exposure to frightening stories in books and on TV until the child is better able to sort out appearance from reality. Make a thorough "search" of the child's room for monsters, showing him or her that none are there. Leave a night-light burning, sit by the child's bed until he or she falls asleep, and tuck in a favorite toy for protection.
Preschool or day care	If the child resists going to preschool but seems content once there, then the fear is probably separation. Under these circumstances, provide a sense of warmth and caring while gently encouraging independence. If the child fears being at preschool, try to find out what is frightening—the teacher, the children, or perhaps a crowded, noisy environment. Provide extra support by accompanying the child at the beginning and lessening the amount of time you are present.
Animals	Do not force the child to approach a dog, cat, or other animal that arouses fear. Let the child move at his or her own pace. Demonstrate how to hold and pet the animal, showing the child that when treated gently, the animal reacts in a friendly way. If the child is bigger than the animal, emphasize this: "You're so big. That kitty is probably afraid of you!"
Very intense fears	If a child's fear is very intense, persists for a long time, interferes with daily activities, and cannot be reduced in any of the ways just suggested, it has reached the level of a phobia. Sometimes phobias are linked to family problems, and special counseling is needed to reduce them. At other times, phobias simply diminish without treatment.

them to wait patiently while she got a baking pan. In the meantime, Sammy reached to feel the dough, but the bowl tumbled over the side of the table. When Leslie returned, Sammy looked at her for a moment, covered his eyes with his hands, and said, "I did something bad." He was feeling ashamed and guilty.

As children's self-concepts become better developed, they experience *self-conscious emotions* more often—feelings that involve injury to or enhancement of their sense of self (see Chapter 6). Yet as Sammy's reaction indicates, preschoolers do not yet label these emotions precisely. And they experience them under somewhat different conditions than do older children and adults. For example, young children are likely to feel guilty for any act that can be described as wrongdoing, even if it was accidental (Graham, Doubleday, & Guarino, 1984). Also, the presence of an audience seems to be necessary for preschoolers to experience self-conscious emotions. In the case of pride, children depend on external recognition, such as a parent or teacher saying, "That's a great picture you drew." And they are likely to experience guilt and shame only if their misdeeds are observed or detected by others (Harter & Whitesell, 1989).

Self-conscious emotions play an important role in children's achievement-related and moral behavior. Since preschoolers are still developing standards of excellence and conduct, they depend on instruction, feedback, and examples from adults to know when to feel pride, guilt, and shame. As children develop guidelines for good behavior, the presence of others will no longer be necessary to evoke these emotions. In addition, these feelings will be

limited to situations in which children feel personally responsible for an outcome (Lewis, 1992; Stipek, Recchia, & McClintic, 1992).

DEVELOPMENT OF EMPATHY

Another self-conscious emotion—*empathy*—becomes more common in early childhood. The ability to recognize and respond sympathetically to the feelings of others is important in motivating positive social behavior. Young children who react with empathy are more likely to share

As young children's language skills expand and their ability to take the perspective of others improves, expressions of empathy become more common. *(Lora E. Askinazi/The Picture Cube)*

and help when they notice another person in distress (Eisenberg & Miller, 1987).

Compared to toddlers (see Chapter 6), preschoolers rely more on words to console others, a change that indicates a more reflective level of empathy. And as the ability to take another's perspective improves over early and middle childhood, empathic responding increases.

The development of empathy depends on cognitive and language development, but it is also supported by early experience. Parents who are warm and encouraging and show a sensitive, empathic concern for their children have preschoolers who are likely to react in a concerned way to the distress of others (Radke-Yarrow & Zahn-Waxler, 1984). This is not true for children who are repeatedly scolded and punished. In one study, researchers observed physically abused preschoolers at a day-care center. Compared to nonabused youngsters, they rarely showed signs of empathy. Instead, they responded to a peer's unhappiness with fear, anger, and physical attacks (Klimes-Dougan & Kistner, 1990). Harsh, punitive parenting interferes with the development of empathy at an early age.

PEER RELATIONS IN EARLY CHILDHOOD

As children become increasingly self-aware, more effective at communicating, and better at understanding the thoughts and feelings of others, their skill at interacting with peers improves rapidly. Peers provide young children with learning experiences they can get in no other way. Because peers relate to one another on an equal footing, they must assume greater responsibility

These children are engaged in parallel play. Although they sit side by side and use similar materials, they do not try to influence one another's behavior. Parallel play remains frequent and stable over the preschool years, accounting for about one-fifth of children's play time. (George Doedwin/Monkmeyer Press)

As these preschoolers trade toys and comment on each other's activities at the sand table, they engage in a form of true social interaction called associative play. (Mary Kate Denny/PhotoEdit)

for keeping a conversation going, cooperating, and setting goals in play than when they associate with adults or older siblings. With peers, children form friendships—special relationships marked by attachment and common interests.

ADVANCES IN PEER SOCIABILITY

In the early part of this century, Mildred Parten (1932) noticed a dramatic rise over the preschool years in children's ability to engage in joint, interactive play. Observing 2- to 5-year-olds, she concluded that social development proceeds in a three-step sequence. It begins with **nonsocial activity**—unoccupied, onlooker behavior and solitary play. Then it shifts to a limited form of social participation called **parallel play,** in which a child plays near other children with similar materials but does not try to influence their behavior. At the highest level, preschoolers engage in two forms of true social interaction. One is **associative play,** in which children engage in separate activities, but they interact by exchanging toys and commenting on one another's behavior. The other is **cooperative play**—a more advanced type of interaction in which children orient toward a common goal, such as acting out a make-believe theme or working on the same product, such as a sand castle or a painting.

Find a time to observe young children of varying ages, and note how long they spend in each of these types of play. You will probably discover that these play forms emerge in the order suggested by Parten, but they do not form a neat developmental sequence in which later-appearing ones replace earlier ones (Howes & Matheson, 1992). Instead, all types coexist during the preschool years. Furthermore, although nonsocial activity declines with age, it is still the most frequent form of behavior among 3-

In cooperative play, the most advanced form of social participation, children orient toward a common goal. These 4-year-olds develop an imaginative scene with the assistance of a very cooperative family pet. By the end of early childhood, associative and cooperative play account for nearly half of children's play time. (Tom McCarthy/Stock South)

to 4-year-olds. Even among kindergartners it continues to take up as much as a third of children's free-play time. Also, solitary and parallel play remain fairly stable from 3 to 6 years, and together, these categories account for as much of the young child's play as highly social, cooperative interaction. Social development during the preschool years is not just a matter of eliminating nonsocial and partially social activity from the child's behavior.

We now understand that it is the *type*, rather than the amount, of solitary and parallel play that changes during early childhood. In studies of preschoolers' play in Taiwan and the United States, researchers rated the *cognitive maturity* of nonsocial, parallel, and cooperative play by applying the categories shown in Table 8.1. Within each of Parten's play types, older children displayed more cognitively mature behavior than did younger children (Pan, 1994; Rubin, Watson, & Jambor, 1978).

Often parents wonder if a preschooler who spends large amounts of time playing alone is developing nor-

mally. Only *certain types* of nonsocial activity—aimless wandering, hovering near peers, and functional play involving immature, repetitive motor action—are cause for concern during the preschool years (Coplan et al., 1994). Most nonsocial play of preschoolers is positive and constructive, and teachers encourage it by setting out art materials, puzzles, and building toys during free play. Children who spend much time in these activities are not maladjusted. Instead, they tend to be very bright youngsters who, when they do play with peers, show socially skilled behavior (Rubin, 1982).

As we noted in Chapter 7, *sociodramatic play* (or make-believe with peers) becomes especially common during the preschool years. It supports both cognitive and social development. In joint make-believe, preschoolers act out and respond to one another's pretend feelings. Their play is rich in references to emotional states. Young children also explore and gain control over fear-arousing experiences when they play doctor or dentist or pretend to search for monsters in a magical forest. As a result, they are better able to understand the feelings of others and regulate their own. Finally, to create and manage complex plots, preschoolers must resolve their disputes through negotiation and compromise. These experiences contribute greatly to their ability to get along with others (Howes, 1992).

FIRST FRIENDSHIPS

As preschoolers interact, first friendships form that serve as important contexts for emotional and social development. Take a moment to jot down what the word *friendship* means to you. You probably thought of a mutual relationship involving companionship, sharing, understanding of thoughts and feelings, and caring for and comforting one another in times of need. In addition, mature friendships endure over time and survive occasional conflicts.

Preschoolers understand something about the uniqueness of friendship. They know that a friend is someone "who likes you" and with whom you spend a lot of time playing (Youniss, 1980). Yet their ideas about friendship are

TABLE 8.1

Developmental Sequence of Cognitive Play Categories

PLAY CATEGORY	DESCRIPTION	EXAMPLES
Functional play	Simple, repetitive motor movements with or without objects. Especially common during the first 2 years of life.	Running around a room, rolling a car back and forth, kneading clay with no intent to make something
Constructive play	Creating or constructing something. Especially common between 3 and 6 years.	Making a house out of toy blocks, drawing a picture, putting together a puzzle
Make-believe play	Acting out everyday and imaginary roles. Especially common between 2 and 6 years.	Playing house, school, or police officer; acting out fairy tales or television characters.

Sources: Rubin, Fein, & Vandenberg, 1983; Smilansky, 1968.

far from mature. Four- to 7-year-olds regard friendship as pleasurable play and sharing of toys. As yet, friendship does not have a long-term, enduring quality based on mutual trust (Damon, 1977; Selman, 1980). Indeed, Sammy could be heard declaring, "Mark's my best friend" on days when the boys got along well. But he would state just the opposite—"Mark, you're not my friend!"—when a dispute arose that was not quickly settled.

Nevertheless, interactions between friends already have unique features. Preschoolers give twice as much reinforcement, in the form of greetings, praise, and compliance, to children whom they identify as friends, and they also receive more from them. Friends are also more emotionally expressive, talking, laughing, and looking at each other more often than nonfriends (Hartup, 1996). Apparently, sensitivity, spontaneity, and intimacy characterize friendships very early, although children are not able to say that these qualities are essential to a good friendship until much later.

As early as the preschool years, some children have difficulty making friends. In Leslie's classroom, Robbie was one of them. His demanding, aggressive behavior caused other children to dislike him. Wherever he happened to be, such comments as "Robbie ruined our block tower" and "Robbie hit me for no reason" could be heard. You will learn more about Robbie's problems as we take up moral development in the next section.

BRIEF REVIEW

Erikson's stage of initiative versus guilt provides an overview of the personality changes that take place in early childhood. During the preschool years, children develop a self-concept made up of observable characteristics and typical beliefs, emotions, and attitudes. Most preschoolers have a high sense of self-esteem, which supports their enthusiasm for mastering new skills. Language for talking about emotion grows rapidly. Young children understand the causes and consequences of basic emotions, and they verbalize a variety of strategies for regulating the expression of emotion. As self-awareness increases, children experience self-conscious emotions more often. Cognition, language, and warm, sensitive parenting support the development of empathy in early childhood.

Beginning in early childhood, peer interaction is important for developing social skills. Over the preschool years, cooperative play becomes common, although solitary and parallel play are also frequent. First friendships form in early childhood. Although preschoolers do not have a mature understanding of friendship, interactions between friends

are already more positive, emotionally expressive, and rewarding.

ASK YOURSELF . . .

- *Reread the description of Sammy and Mark's argument at the beginning of this chapter. On the basis of what you know about self-development, why was it a good idea for Leslie to resolve the dispute by providing an extra set of beanbags so that both boys could play at once?*

- *Four-year-old Tia had just gotten her face painted at a carnival. Soon the heat of the afternoon caused her balloon to pop. When Tia started to cry, her mother said, "Oh, Tia, balloons aren't such a good idea when it's hot outside. We'll get another on a cooler day. If you cry, you'll mess up your beautiful face painting." What aspect of emotional development is Tia's mother trying to promote, and why is her intervention likely to be helpful to Tia?*

- *Three-year-old Bart lives in the country where there are no other preschoolers nearby. His parents wonder whether it is worth driving Bart into town once a week to play with his 3-year-old cousin. What advice would you give Bart's parents, and why?*

FOUNDATIONS OF MORALITY IN EARLY CHILDHOOD

If you watch children's behavior and listen in on their conversations, you will find many examples of their developing moral sense. By age 2, they show great concern with deviations from the way objects should be and people should act. They point to destroyed property, such as broken toys, with an expression of discomfort, often exclaiming, "Uh-oh!" In addition, they typically react with alarm to behaviors that are aggressive or that might otherwise harm someone. Soon they comment directly on their own and others' actions: "I naughty. I wrote on wall," or (after having been hit by another child) "Connie not nice" (Kochanska, Casey, & Fukumoto, 1995).

Throughout the world, adults take note of this budding capacity to distinguish right from wrong. In some cultures, special terms are used to describe it. The Utku Indians of Hudson Bay say the child develops *ihuma* (reason). The Fijians believe that *vakayalo* (sense) appears. In response, parents hold children more responsible for their behavior (Kagan, 1989). By the end of early childhood, children can state a great many moral rules, such as "You're not supposed to take things without asking" or "Tell the truth!" In addition, they argue over matters of justice, as when they say, "You sat there last time, so it's my turn" or "It's not fair. He got more!"

All theories of moral development recognize that conscience begins to take shape in early childhood. And most agree that at first, the child's morality is *externally controlled* by adults. Gradually it becomes regulated by *inner standards*. Truly moral individuals do not just do the right thing when authority figures are around. Instead, they have developed principles of good conduct, which they follow in a wide variety of situations.

Although there are points of agreement among major theories, each emphasizes a different aspect of moral functioning. Psychoanalytic theory stresses the *emotional side* of conscience development—in particular, identification and guilt as motivators of good conduct. Social learning theory focuses on *moral behavior* and how it is learned through reinforcement and modeling. And the cognitive-developmental perspective emphasizes *thinking*—children's ability to reason about justice and fairness. These theories also differ in the extent to which they view children as actively contributing to their own moral development.

THE PSYCHOANALYTIC PERSPECTIVE

Recall from our discussion earlier in this chapter that in Freud's view, young children form a conscience by *identifying* with the same-sex parent, whose moral standards they take into their own personalities. Children obey their conscience to avoid *guilt*, a painful emotion that arises each time they are tempted to misbehave. According to Freud, moral development is largely complete by 5 to 6 years of age, at the end of the phallic stage.

Although Freud's theory of conscience development is accepted by psychoanalysts, most child development researchers disagree with it. First, if you look carefully at the Oedipus and Electra conflicts (see page 246), you will see that discipline promoting fear of punishment and loss of parental love should motivate young children to behave morally (Kochanska, 1993). Yet children whose parents frequently use threats, commands, or physical force usually feel little guilt after harming others, and they show poor self-control. In the case of love withdrawal—for example, when a parent refuses to speak to or actually states a dislike for the child—children often respond with high levels of self-blame. They might think to themselves, "I'm no good, and nobody loves me." Eventually, these youngsters may protect themselves from overwhelming feelings of guilt by denying the emotion when they do something wrong. So they, too, develop a weak conscience (Zahn-Waxler et al., 1990).

In contrast, a special type of discipline called **induction** does support conscience formation. It involves pointing out the effects of the child's misbehavior on others. For example, a parent might say, "If you keep pushing him he'll fall down and cry" or "She feels so sad because you won't

Induction, a form of discipline that involves pointing out the effects of the child's misbehavior on others, supports conscience formation. For example, this teacher might say, "If you grab Susie's tricycle without asking for a turn, you'll make her feel very unhappy." (Will Faller)

give back her doll" (Hoffman, 1988). As long as the explanation matches the child's capacity to understand, induction is effective with children as young as 2 years of age. Preschoolers whose parents use it are more likely to make up for their misdeeds. And they also show more **prosocial, or altruistic, behavior**—actions that benefit another person without any expected reward for the self (Zahn-Waxler, Radke-Yarrow, & King, 1979).

Why is induction so effective? First, it tells children how to behave so they can use this information in future situations. Second, by pointing out the impact of the child's actions on others, parents encourage empathy, which promotes prosocial behavior (Krevans & Gibbs, 1996). In contrast, discipline that relies too heavily on threats of punishment or love withdrawal makes children so anxious and afraid that they cannot think clearly enough to figure out what they should do. These practices may stop unacceptable behavior temporarily, but in the long run they do not get children to internalize moral rules.

Although there is little support for the Freudian view of conscience development, Freud was correct that guilt is an important motivator of moral action. Around age 3, guilt reactions are clearly evident, and temperament affects their intensity. Highly inhibited, nonimpulsive preschoolers are more likely to confess and make amends after a transgression and (if they are girls) to experience emotional discomfort and be concerned about restoring parental approval (Kochanska, 1995, 1997).

But contrary to what Freud believed, guilt is not the only force that compels us to act morally. And moral development is not complete by the end of early childhood. Instead, it is a much more gradual process, beginning in the preschool years and extending into adulthood.

BEHAVIORISM AND SOCIAL LEARNING THEORY

According to the traditional behaviorist view, operant conditioning is an important way in which children pick up new responses. From this perspective, children start to behave in ways consistent with adult moral standards because parents and teachers follow up "good behavior" with *positive reinforcement* in the form of approval, affection, and other rewards.

■ THE IMPORTANCE OF MODELING. Operant conditioning is not enough for children to acquire moral responses. Recall from Chapter 4 that for a behavior to be reinforced, it must first occur spontaneously. Yet many prosocial behaviors, such as sharing, helping, or comforting an unhappy playmate, do not occur often enough at first for reinforcement to explain their rapid development in early childhood. Instead, social learning theorists believe that children largely learn to act morally through *modeling*—by observing and imitating people who demonstrate appropriate behavior (Bandura, 1977; Grusec, 1988). Once children acquire a moral response, such as sharing, helping, or telling the truth, reinforcement in the form of praise, along with adult reminders of the rules for moral behavior, increases its frequency (Mills & Grusec, 1989).

Many studies show that exposure to models who behave helpfully or generously increases young children's prosocial responses. In fact, models exert their most powerful effect during the preschool years. At the end of early childhood, children with a history of consistent exposure to caring adults tend to behave prosocially regardless of whether a model is present. By that time, they have internalized prosocial rules from repeated observations of and encouragement by others (Mussen & Eisenberg-Berg, 1977).

A model's characteristics affect children's willingness to imitate their behavior. First, preschoolers are more likely to copy the prosocial actions of an adult who is warm and responsive than one who is cold and distant (Yarrow, Scott, & Waxler, 1973). Second, children tend to select competent, powerful models to imitate—the reason they are especially willing to copy the behavior of older peers and adults (Bandura, 1977). A final characteristic that affects children's willingness to imitate is whether adults "practice what they preach." When models say one thing and do another—for example, announce that "It's important to help others" rarely engage in helpful acts—children generally choose the most lenient standard of behavior that adults demonstrate (Mischel & Liebert, 1966).

■ EFFECTS OF PUNISHMENT. Most parents are aware of the limited usefulness of *punishment,* such as scolding, criticism, and spankings, and apply it sparingly. The use of sharp reprimands or physical force to restrain or move a child from one place to another may be justified when immediate obedience is necessary—for example, when a 3-year-old is about to run into the street or it is time to leave the house for an appointment and the child protests. In fact, parents are most likely to use forceful techniques under these conditions. When they are interested in fostering long-term goals, such as acting kindly toward others, they tend to rely on warmth and reasoning (Kuczynski, 1984).

Indeed, a great deal of research shows that punishment only promotes momentary compliance, not lasting changes in children's behavior. For example, Robbie's parents punished often, spanking, shouting, and criticizing when he did something wrong. Robbie usually stopped misbehaving when his mother and father were around, but he engaged in the behavior again as soon as they were out of sight and he thought he could get away with it. As a result, Robbie was especially unmanageable in settings away from home, such as preschool (Strassberg et al., 1994).

Harsh punishment has undesirable side effects. First, it provides children with adult models of aggression. Second, children who are frequently punished soon learn to avoid the punishing adult. When Robbie's parents came into the room, Robbie braced himself for something unpleasant and kept his distance. Consequently, they had little opportunity to teach Robbie desirable behaviors to replace unacceptable responses. Finally, as punishment "works" to stop children's misbehavior temporarily, it offers immediate relief to adults, and they are reinforced for using coercive discipline. For this reason, a punitive adult is likely to punish with greater frequency over time, a course of action that can spiral into serious abuse.

■ ALTERNATIVES TO HARSH PUNISHMENT. Alternatives to criticism, slaps, and spankings can reduce the undesirable side effects of punishment. A technique called **time out** involves removing children from the immediate setting—for example, by sending them to their rooms—until they are ready to act appropriately. Time out is useful when a child is out of control and other effective methods of discipline cannot be applied at the moment (Betz, 1994). It usually requires only a few minutes to change behavior, and it also offers a "cooling off" period for angry parents. Another approach is *withdrawal of privileges,* such as playing outside or going to the movies. Removing privileges may generate some resentment in children, but parents avoid harsh techniques that could easily intensify into violence.

When parents do decide to use punishment, they can increase its effectiveness in several ways:

1. *Consistency.* Punishment that is unpredictable is related to especially high rates of disobedience in children. When parents permit children to act inappropriately on some occasions but scold them on others, children are confused about how to behave, and the unacceptable act persists.

2. *A warm parent–child relationship.* Children of involved and caring parents find the interruption in parental

One alternative to harsh punishment is time out, in which children are removed from the immediate setting until they are ready to act appropriately. Time out is useful when a child is out of control and other effective methods of discipline cannot be applied at the moment. But the best way to motivate good conduct is to let children know ahead of time how to act, serve as good examples, and praise them when they behave well. Then time out will seldom be necessary. (Photo courtesy: Goodman/Monkmeyer Press)

affection that accompanies punishment to be especially unpleasant. As a result, they want to regain the warmth and approval of parents as quickly as possible.

3. *Explanations.* Explanations help children recall the misdeed and relate it to expectations for future behavior.

Finally, parenting practices that encourage and reward good conduct are the most effective forms of discipline. Instead of waiting for children to misbehave, parents can let children know ahead of time how to act, serve as good examples, and praise children when they behave well (Zahn-Waxler & Robinson, 1995). Adults can also reduce opportunities for misbehavior. For example, on a long car trip, parents can bring along back-seat activities that relieve children's restlessness and boredom. At the supermarket, they can engage preschoolers in conversation and encourage them to assist with shopping (Holden, 1983;

Holden & West, 1989). When adults help children acquire acceptable behaviors to replace forbidden acts, the need for punishment is reduced.

THE COGNITIVE-DEVELOPMENTAL PERSPECTIVE

Both the psychoanalytic and behaviorist approaches to morality focus on how children acquire ready-made standards of good conduct from adults. The cognitive-developmental perspective is different. It regards children as *active thinkers* about social rules. As early as the preschool years, children make moral judgments, deciding what is right or wrong on the basis of concepts they construct about justice and fairness (Gibbs, 1991).

Young children already have some well-developed ideas about morality. For example, 3-year-olds think that a child who intentionally knocks a playmate off a swing is worse than one who does so accidentally (Yuill & Perner, 1988). They are also aware that disobeying *moral rules,* such as being kind to others and not taking someone else's possessions, is much more serious than violating *social conventions,* such as not saying "please" or "thank you" or eating messy food with fingers (Smetana & Braeges, 1990; Turiel, 1983).

How do young children come to make these distinctions? According to cognitive-developmental theorists, not through direct teaching, modeling, and reinforcement, since adults insist that children conform to social conventions just as often as they press for obedience to moral rules. Instead, children *actively make sense* of their experiences. They observe that people respond differently to violations of moral rules than to breaks with social convention. When a moral offense occurs, children react emotionally, describe their own injury or loss, tell another child to stop, or retaliate. An adult who intervenes is likely to call attention to the rights and feelings of the victim. In contrast, children are less likely to react to violations of social convention. And in these situations, adults tend to demand obedience without explanation, as when they state, "Say the magic word!" or "Don't eat with your fingers" (Smetana, 1989; Turiel, Smetana, & Killen, 1991).

Young children are clearly off to a good start in appreciating that moral rules are important because they protect people's welfare. Disputes with peers and siblings over rights, possessions, and property give preschoolers an opportunity to work out their first ideas about justice and fairness. The way adults handle violations of rules and discuss moral issues with children also helps them reason about morality. Children who are advanced in moral thinking have parents who adapt their communications about fighting, honesty, and ownership to what their children can understand, respect the child's opinion, and gently stimulate the child to think further, without being hostile or critical (Walker & Taylor, 1991).

Preschoolers who are disliked by peers because of their aggressive approach to resolving conflict have trouble distinguishing between moral rules and social conventions, and they violate both often (Sanderson & Siegal, 1988). Without special help, such children show long-term disruptions in moral development.

THE OTHER SIDE OF MORALITY: DEVELOPMENT OF AGGRESSION

Beginning in late infancy, all children display aggression from time to time as they become better at identifying sources of anger and frustration. By the early preschool years, two forms of aggression emerge. The most common is **instrumental aggression.** In this form, children are not deliberately hostile. Instead, they want an object, privilege, or space and, in trying to get it, they push, shout at, or otherwise attack a person who is in the way. The other type, **hostile aggression,** is meant to hurt, as when the child hits, insults, or tattles on a playmate to injure the other person.

For most preschoolers, instrumental aggression declines with age as they learn to compromise over possessions. In contrast, hostile aggression increases between 4 and 7, although it is rare compared to children's friendly interactions (Shantz, 1987). This slight rise in hostile encounters occurs as children become better at detecting others' intentions. Older preschoolers are more likely to recognize when another child is being deliberately malicious and to try to get even by attacking in return.

Although children of both sexes show this general pattern of development, on the average boys are more aggressive than girls, a trend that appears in many cultures (Whiting & Edwards, 1988). The sex difference is, in part, due to biology—in particular, to male sex hormones, or androgens. Androgens contribute to boys' greater physical activity, which may increase their opportunities for aggressive encounters (Hines & Green, 1991). At the same time, the development of gender typing (a topic we will take up shortly) is also important. As soon as 2-year-olds become dimly aware of gender stereotypes—that males and females are expected to behave differently—aggression drops off in girls but is maintained in boys (Fagot & Leinbach, 1989). Then (as we will see in a moment) parents' tendency to discipline boys more harshly magnifies this effect.

An occasional aggressive exchange between young children is normal and to be expected. But some preschoolers, like Robbie, show abnormally high rates of aggression. Researchers have traced their problems to strife-ridden families, poor parenting practices, and exposure to television violence—factors that often can be found together.

■ **THE FAMILY AS TRAINING GROUND FOR AGGRESSIVE BEHAVIOR.** "I can't control him, he's impossible," complained Nadine, Robbie's mother, at a conference with Leslie. When Leslie asked what might be going on at home

An occasional expression of aggression is normal in early childhood. This preschooler displays instrumental aggression as he pushes a playmate to get an attractive toy. Instrumental aggression declines with age as children learn how to compromise and share. (Crews/The Image Works)

that made it hard to handle Robbie, she discovered that Robbie's parents fought constantly. Their conflict led to family stress and a "spillover" of hostile communication into child rearing that stimulated and perpetuated Robbie's aggression (Miller et al., 1993).

Observations in families like Robbie's reveal that anger and punitiveness can quickly spread from one family member to another, creating a conflict-ridden family atmosphere and an "out-of-control" child. The pattern begins with forceful discipline, which is made more likely by stressful life experiences, a parent's own personality, or a temperamentally difficult child. Once the parent threatens, criticizes, and punishes, then the child whines, yells, and refuses until the parent finds the child's behavior to be too much and "gives in." As these cycles become more frequent, they generate anxiety and irritability among other family members, who soon join in the hostile interactions (Dodge, Pettit, & Bates, 1994; Patterson, Reid, & Dishion, 1992). Aggressive children who are products of these family processes soon learn to view the world from a violent perspective. Because they expect others to react with anger and physical force, they see hostile intent where it does not exist (Dodge & Somberg, 1987). As a result, they make many unprovoked attacks, which contribute to aggressive cycles.

For at least two reasons, boys are more likely than girls to become involved in family interactions that promote aggressive behavior. First, parents more often use commands and physical punishment with sons, which encourages them to adopt the same tactics (Lytton & Romney,

1991). Second, parents are less likely to interpret fighting among boys as aggressive, so they may overlook it more than they do with girls (Condry & Ross, 1985). By middle childhood, boys expect less parental disapproval and report feeling less guilty over aggression than do girls (Perry, Perry, & Weiss, 1989).

Unfortunately, highly aggressive children often develop serious adjustment problems. Because of their hostile style of responding and their poor self-control, they tend to be rejected by peers, to fail in school, and (by adolescence) to seek out deviant peer groups, which lead them toward delinquency and adult criminality (see Chapter 12). These children need treatment early, before their antisocial behavior becomes so well practiced that it is difficult to change.

■ HELPING CHILDREN AND PARENTS CONTROL AGGRESSION. Help for aggressive children must break the cycle of hostilities between family members, replacing it with effective interaction styles. Leslie suggested that Robbie's parents see a family therapist, who observed their inept practices and coached them in alternatives. They learned not to give in to Robbie, to pair commands with reasons, and to replace verbal insults and spankings with more effective punishments, such as time out and withdrawal of privileges (Patterson, 1982).

At the same time, Leslie began teaching Robbie more successful ways of relating to peers. As opportunities arose, she encouraged Robbie to talk about a playmate's feelings and express his own. This helped Robbie take the perspective of others and empathize. Soon he showed greater willingness to share and cooperate (Feshbach & Feshbach, 1982). Robbie also participated in *social problem-solving training*. Over several months, he met with Leslie and a small group of preschoolers. The children used puppets to act out common conflicts, discussed effective and ineffective ways of resolving them, and tried out successful strategies. Children who receive such training show gains in social competence that are still present a year later (Spivack & Shure, 1974). Effective social problem solving also provides children with a sense of mastery in the face of stressful life events. It reduces the risk of adjustment difficulties in children from low-income and troubled families (Goodman, Gravitt, & Kaslow, 1995).

Finally, Robbie's parents got help with their marital problems. This, in addition to their improved ability to manage Robbie's behavior, greatly reduced tension and conflict in the household.

■ TELEVISION, AGGRESSION, AND OTHER ASPECTS OF SOCIAL LEARNING. In the United States, 57 percent of television programs between 6 A.M. and 11 P.M. contain violent scenes, often in the form of repeated aggressive acts that go unpunished. In fact, most TV violence does not show victims experiencing any serious harm, and few programs condemn violence or depict other ways of solving

problems. To the contrary, over one-third of violent TV scenes are embedded in humor, a figure that rises to two-thirds for children's shows. Of all TV programs, children's cartoons are the most violent (Mediascope, 1996).

Young children are especially likely to be influenced by television. One reason is that below age 8, children do not understand a great deal of what they see on TV. Because they have difficulty connecting separate scenes into a meaningful story line, they do not relate the actions of a TV character to motives or consequences (Collins et al., 1978). Young children also find it hard to separate true-to-life from fantasized television content. Not until age 7 do they fully realize that fictional characters do not retain the same roles in real life (Wright et al., 1994). These misunderstandings increase young children's willingness to uncritically accept and imitate what they see on TV.

Reviewers of thousands of studies have concluded that TV violence provides children with an extensive "how-to course in aggression" (Slaby et al., 1995, p. 163). Violent programming not only creates short-term difficulties in parent and peer relations, but has long-term effects as well. Longitudinal research reveals that highly aggressive children have a greater appetite for violent TV. As they watch more, they become increasingly likely to resort to hostile ways of solving problems. This spiraling pattern of learning contributes to serious antisocial acts by adolescence and young adulthood (Huesmann, 1986). Television violence also "hardens" children to aggression, making them more willing to tolerate it in others. Heavy TV viewers begin to see the world as a mean and scary place where aggressive acts are a normal and acceptable means for solving problems (Donnerstein, Slaby, & Eron, 1994).

TV promotes conflict in another way: through its advertising. Preschoolers (as well as many older children) innocently believe that the promises of TV ads are true, so they ask for products that they see. In the aisles of grocery and toy stores, adult refusals often lead to arguments between parents and children (Atkin, 1978). Commercials for sugary foods make up about 80 percent of advertising aimed at children. When parents give in to children's demands, young TV viewers come to prefer these snacks and are convinced by TV messages that they are healthy (Gorn & Goldberg, 1982).

Finally, television conveys ethnic and gender stereotypes that are common in American society. Although African Americans are better represented on TV than they once were, too often they are segregated from whites. Other ethnic minorities rarely appear, and when they do, they tend to be depicted negatively, as villains or victims of violence. Similarly, women appear less often than men as main characters, and they are usually cast in "feminine" roles, such as wife, mother, nurse, teacher, or secretary, and as victims (Signorielli, 1993; Zillman, Bryant, & Huston, 1994).

The ease with which television can manipulate the beliefs and behavior of children has resulted in strong public pressures to improve its content. Unfortunately, as the

Social Issues box on page 260 indicates, these efforts have not been very successful. At present, it is up to parents to regulate their children's exposure—a heavy burden, given that children find TV so attractive.

BRIEF REVIEW

The young child's morality gradually shifts from externally controlled responses to internalized standards. Contrary to predictions from Freudian theory, power assertion and love withdrawal do not promote the development of conscience. Instead, induction is far more successful. Behaviorism and social learning theory have shown that modeling combined with reinforcement in the form of praise is very successful in encouraging prosocial acts. In contrast, harsh punishment only promotes temporary compliance, not lasting changes in children's behavior. Cognitive-developmental theory views children as active thinkers about justice and fairness. During the preschool years, children recognize that intentionally hurting someone is worse than doing so accidentally, and they distinguish moral rules and social conventions. Hostile family atmospheres, poor parenting practices, and heavy television viewing promote childhood aggression, which can spiral into serious antisocial activity. TV also fosters a naive belief in the truthfulness of advertising and ethnic and gender stereotypes.

ASK YOURSELF . . .

■ *Alice and Wayne want their two young children to develop a strong, internalized conscience and to become generous, caring individuals. List as many parenting practices as you can that would foster these goals.*

■ *Nanette told her 3-year-old son Darren not to go into the front yard without asking, since the house faces a very busy street. Darren disobeyed several times, and now Nanette thinks it's time to punish him. How would you recommend that Nanette discipline Darren, and why?*

GENDER TYPING IN EARLY CHILDHOOD

Gender typing refers to the process of developing gender roles, or gender-linked preferences and behaviors valued by the larger society. Early in the preschool years, gender typing is well underway. In Leslie's classroom, children tended to play and form friendships with peers of their own sex. Girls spent more time in the housekeeping, art, and reading corners, whereas boys gathered more often in blocks, woodworking, and active play spaces.

The same three theories that provide accounts of morality have been used to explain gender-role development. According to *psychoanalytic theory,* gender-stereotyped beliefs and behaviors are adopted in the same way as other social standards—through identification with the same-sex parent. But as in the area of morality, Freud's ideas have difficulty accounting for gender typing. Research shows that the same-sex parent is only one of many influences in gender-role development. Opposite-sex parents, peers, teachers, and the broader social environment are important as well.

Social learning theory, with its emphasis on modeling and reinforcement, and *cognitive-developmental theory,* with its focus on children as active thinkers about their social world, are major current approaches to understanding gender typing. We will see that neither has proven adequate by itself. Consequently, a new perspective that combines elements of both, called *gender schema theory,* has recently arisen. In the following sections, we consider the early development of gender typing, along with genetic and environmental factors that contribute to it.

PRESCHOOLERS' GENDER-STEREOTYPED BELIEFS AND BEHAVIOR

Recall from Chapter 6 that around age 2, children begin to label their own sex and that of other people. As soon as basic gender categories are established, children start to sort out what they mean in terms of activities and behavior. A wide variety of gender stereotypes are quickly mastered.

Preschoolers associate many toys, articles of clothing, tools, household items, games, occupations, and even colors (pink and blue) with one sex as opposed to the other (Huston, 1983; Picariello, Greenberg, & Pillemer, 1990). And their actions fall in line with their beliefs—not only in play preferences, but in personality traits as well. We have already seen that boys tend to be more active, assertive, and aggressive, whereas girls tend to be more fearful, dependent, compliant, and emotionally sensitive (Feingold, 1994).

Over the preschool years, children's gender-stereotyped beliefs become stronger—so much so that they operate like blanket rules rather than flexible guidelines (Biernat, 1991; Martin, 1989). Once, when Leslie showed the children a picture of a Scottish bagpiper wearing a kilt, they insisted, "Men don't wear skirts!" During free play, they often exclaimed that girls don't drive fire engines and can't be police officers and boys don't take care of babies and can't be the teacher. These rigid ideas are a joint product of gender stereotyping in the environment and young children's cognitive immaturity. As we will see later, most preschoolers do not yet realize that

SOCIAL ISSUES

REGULATING CHILDREN'S TELEVISION

Exposure to television is almost universal in the United States and other Western industrialized nations. Ninety-eight percent of American homes have at least one television set, and a TV is switched on in a typical household for a total of 7.1 hours a day. TV enters the lives of children at an early age, becoming a major teacher of undesirable attitudes and behavior. Yet television has as much potential for good as it does for ill. If the content of television were changed, it could promote prosocial attitudes and behavior and convey information about nonviolent aspects of the world, such as history, science, literature, fine arts, and other cultures (Graves, 1993).

Unfortunately, recent changes in the media world have increased children's exposure to harmful messages. Over 60 percent of American families are cable subscribers, and nearly half own VCRs—rates that are rapidly increasing. Children with cable or VCR access can see films with far more graphic violence and sexual content than are shown on commercial TV (Huston et al., 1992).

Many organizations concerned with the well-being of children and families have recommended government regulation of TV. But the First Amendment right to free speech has made the federal government reluc-tant to place limits on television content. Today, there are fewer restrictions than there once were on pro-gram content and advertising for children. For example, 15 years ago, characters in children's programs were not permitted to sell products. Now they commonly do—a strategy that greatly increases children's desire to buy (Kunkel, 1993).

Consequently, professionals and committed public officials have sought ways to counteract the nega-tive impact of television that are con-sistent with the First Amendment—for example, requiring a minimum amount of educational programming for children and insist-ing on antiviolence public service announcements. Broadcasters, whose profits are at risk, are against any gov-ernment restrictions (Donnerstein, Slaby, & Eron, 1994).

Parents face an awesome task in monitoring and controlling children's TV viewing, given the extent of harm-ful messages on the screen. Here are some strategies they can use:

- Avoid using TV as a baby-sitter. Provide children with clear rules that limit the amount of time they can watch—for example, an hour a day and only certain programs— and stick to the rules.

- Do not use television to reward or punish children, a practice that increases its attractiveness.

- Encourage children to watch pro-grams that are child-appropriate, informative, and prosocial.

- As much as possible, watch with children, helping them understand what they see. When adults express disapproval of on-screen behavior, raise questions about the realism of televised information, and encourage children to discuss it, they teach children to evaluate TV content rather than accept it uncritically.

- Build on TV programs in con-structive ways, encouraging chil-dren to move away from the set into active engagement with their surroundings. For example, a pro-gram on animals might spark a trip to the zoo, a visit to the library for books about animals, or new ways of observing and caring for the family pet.

- Avoid excess television viewing, especially violent programs, your-self. Parental viewing patterns influence children's viewing pat-terns.

- Respond to children with warmth and reasonable demands for mature behavior. Children who experience these practices prefer programs with prosocial content and are less attracted to violent TV.

In this home, the television is turned off, and an older child engages in an art activity while her younger sister looks on and learns from her. Using televi-sion to promote children's prosocial behavior and active engagement with the sur-rounding world is a great chal-lenge for parents in the United States, given the antisocial con-tent of many programs. (Dusty Willison/International Stock)

TRY THIS...

- Watch several Saturday morning cartoons, late-afternoon children's shows, and prime-time adult television programs. Count the number of prosocial and aggressive acts you see. Also note the roles that male and female characters play. How do the programs compare in terms of the social messages they send to children?

characteristics *associated* with sex—activities, toys, occupations, hairstyle, and clothing—do not *determine* whether a person is male or female.

GENETIC INFLUENCES ON GENDER TYPING

The sex differences we have just described appear in many cultures around the world (Whiting & Edwards, 1988). Certain of them—the preference for same-sex playmates as well as male activity level and aggression—are also widespread among animal species (Beatty, 1992). To what extent might gender typing be influenced by genetic factors?

Eleanor Maccoby (1990) argues that hormonal differences between males and females have important consequences for gender typing. We have already seen that androgens promote active play among boys, making aggression more likely. Early on, hormones affect play styles, leading to rough, noisy movements among boys and calm, gentle actions among girls. Then, as children begin to interact with peers, they naturally choose same-sex partners whose interests and behaviors are compatible with their own. Over the preschool years, girls increasingly seek out other girls because of their common preference for quieter activities. And boys come to prefer other boys, who share each other's desire to run, climb, play-fight, and build up and knock down. At age 4, children already spend three times as much time with same-sex as other-sex playmates. By age 6, this ratio has climbed to 11 to 1 (Benenson, 1993; Maccoby & Jacklin, 1987).

However, we must be careful not to overemphasize the role of heredity in gender typing. As we will see in the next section, a wide variety of environmental forces build on genetic influences to promote children's awareness of and conformity to gender roles.

ENVIRONMENTAL INFLUENCES ON GENDER TYPING

A wealth of evidence reveals that family influences, encouragement by teachers and peers, and examples in the broader social environment combine to promote the vigorous gender typing of early childhood.

■ THE FAMILY. Beginning at birth, parents hold different perceptions and expectations of their sons and daughters. Many parents state that they want their children to play with "gender-appropriate" toys, and they also believe that boys and girls should be reared differently. Parents are likely to describe achievement, competition, and control of emotion as important for sons and warmth, "ladylike" behavior, and close supervision of activities as important for daughters (Brooks-Gunn, 1986; Turner & Gervai, 1995).

These beliefs carry over into actual parenting practices. Mothers and fathers are far more likely to purchase guns, cars, and footballs for sons and dolls, tea sets, and jump ropes for daughters—toys that promote very different play styles. In addition, parents actively reinforce many gender-typed behaviors. For example, they react more positively when a young son as opposed to a daughter plays with cars and trucks, demands attention, or tries to take toys from others, thereby rewarding his active and assertive behavior (Fagot & Hagan, 1991). In contrast, they more often direct play activities, provide help, and discuss emotions with a daughter, encouraging her dependency and emotional sensitivity (Kuebli & Fivush, 1992; Lytton & Romney, 1991).

These factors are major influences in children's gender-role learning, since parents who consciously avoid behaving in these ways have less gender-typed children (Weisner & Wilson-Mitchell, 1990). Other family members also contribute to gender typing. For example,

During the preschool years, children seek out playmates of their own sex. Sex hormones affect children's play styles, leading to rough, noisy movements among boys and calm, gentle actions among girls. Then preschoolers naturally choose same-sex partners who share their interests and behavior. Social pressures for "gender-appropriate" play are also believed to promote gender segregation. (Left, Michael Newman/ PhotoEdit; right, Merritt Vincent/PhotoEdit)

Test Bank Items 8.71 through 8.72

preschoolers with older, other-sex siblings have many more opportunities to imitate and participate in "cross-gender" play (Stoneman, Brody, & MacKinnon, 1986). In any case, of the two sexes, boys are clearly the more gender typed. One reason is that parents—particularly fathers—are far less tolerant of "cross-gender" behavior in their sons than in their daughters. They are more concerned if a boy acts like a sissy than if a girl acts like a tomboy (Lytton & Romney, 1991; Maccoby, 1980).

■ TEACHERS. Besides parents, teachers encourage children's gender typing. Several times, Leslie caught herself promoting gender segregation and stereotyping in her classroom. One day when the class was preparing to leave for a field trip, she called out, "Will the girls line up on one side and the boys on the other?" Then, as the class became noisy with excitement, she pleaded, "Boys, I wish you'd quiet down like the girls!"

As at home, girls get more encouragement to participate in adult-structured activities at preschool. They can frequently be seen clustered around the teacher, following directions in an activity. In contrast, boys more often choose areas of the classroom where teachers are minimally involved or entirely absent. As a result, boys and girls practice very different social behaviors. Compliance and bids for help occur more often in adult-structured contexts, whereas assertiveness, leadership, and creative use of materials appear more often in unstructured pursuits (Carpenter, 1983).

■ PEERS. Children's same-sex peer groups strengthen gender-stereotyped beliefs and behavior. By age 3, same-sex peers positively reinforce one another for gender-typed play by praising, imitating, or joining in the activity of an agemate who shows a "gender-appropriate" response (Fagot & Patterson, 1969). In contrast, when preschoolers engage in "gender-inappropriate" activities—for example, when boys play with dolls or girls with woodworking tools—they receive criticism from peers. Boys are especially intolerant of "cross-gender" play in their male companions (Fagot, 1977). A boy who frequently crosses gender lines is likely to be ignored by other boys even when he does engage in "masculine" activities!

Children also develop different styles of social influence in gender-segregated peer groups. To get their way with male peers, boys more often rely on commands, threats, and physical force, whereas girls learn to use polite requests and persuasion. These tactics succeed with other girls but not with boys, who pay little attention to girls' gentle tactics (Borja-Alvarez, Zarbatany, & Pepper, 1991; Leaper, 1991). Consequently, an additional reason that girls may stop interacting with boys is that they do not find it very rewarding to communicate with an unresponsive social partner.

■ THE BROADER SOCIAL ENVIRONMENT. Although American society has changed to some degree, children's everyday environments contain many examples of gender-typed behavior—in occupations, leisure activities, and achievements of men and women. As we will see in the next section, young children do not just imitate the many gender-linked responses they observe. They also start to view themselves and their environment in gender-biased ways, a perspective that can seriously limit their interests and skills.

GENDER-ROLE IDENTITY

As adults, each of us has a **gender-role identity**—an image of oneself as relatively masculine or feminine in characteristics. By middle childhood, researchers can measure gender-role identity by asking children to rate themselves on personality traits. Individuals differ considerably in their responses. A child or adult with a "masculine" identity scores high on traditionally masculine items (such as self-sufficient, ambitious, and forceful) and low on traditionally feminine ones (such as affectionate, soft-spoken, and cheerful). Someone with a "feminine" identity does just the reverse. Although most individuals view themselves in gender-typed terms, a substantial minority (especially females) have a kind of gender-role identity called **androgyny.** They score high on *both* masculine and feminine characteristics (Bem, 1974; Boldizar, 1991).

Gender-role identity is a good predictor of psychological adjustment. Masculine and androgynous children and adults have higher self-esteem, whereas feminine individuals often think poorly of themselves, perhaps because many feminine traits are not highly valued in our society (Alpert-Gillis & Connell, 1989; Boldizar, 1991). Androgynous individuals are more adaptable in behavior—for example, able to show masculine independence or feminine sensitivity, depending on the situation (Taylor & Hall, 1982). Research on androgyny shows that it is possible for children to acquire a mixture of positive qualities traditionally associated with each gender—an orientation that may best help them realize their potential.

■ EMERGENCE OF GENDER-ROLE IDENTITY. How do children develop a gender-role identity? Both social learning and cognitive-developmental answers to this question exist. According to *social learning theory,* behavior comes before self-perceptions. Preschoolers first acquire gender-typed responses through modeling and reinforcement. Only then do they organize these behaviors into gender-linked ideas about themselves.

In contrast, *cognitive-developmental theory* regards the direction of development as the other way around. Over the preschool years, children first acquire a cognitive appreciation of their own gender. They develop **gender**

constancy, the understanding that their sex is a permanent characteristic that remains the same even if clothing, hairstyle, and play activities change. Then children use this idea to guide their behavior, and a preference for gender-typed activities appears (Kohlberg, 1966).

Research indicates that gender constancy is not present in most children until the end of the preschool years, when they pass Piagetian conservation tasks (De Lisi & Gallagher, 1991). Shown a doll whose hairstyle and clothing are transformed before their eyes, a child younger than age 6 is likely to insist that the doll's sex has changed as well (McConaghy, 1979). And when asked such questions as "When you (a girl) grow up, could you ever be a daddy?" or "Could you be a boy if you wanted to?" young children freely answer yes (Slaby & Frey, 1975).

Yet cognitive immaturity is not the only reason for preschoolers' poor performance on gender constancy tasks. It also results from limited social experience—in particular, lack of opportunity to learn about genital differences between the sexes. In many households in Western cultures, young children do not see members of the opposite sex naked. Therefore, they distinguish males and females using the only information they do have—the way each gender dresses and behaves. Children as young as 3 who are aware of genital differences usually answer gender constancy questions correctly (Bem, 1989).

Is cognitive-developmental theory correct that gender constancy is responsible for children's gender-typed behavior? Evidence for this assumption is weak. "Gender-appropriate" behavior appears so early in the preschool years that modeling and reinforcement must account for its initial appearance, as social learning theory suggests. At present, researchers disagree on just how gender constancy

contributes to gender-role development (Bussey & Bandura, 1992; Lutz & Ruble, 1995). But they do know that once children begin to reflect on gender roles, they form basic gender categories that strengthen gender-typed self-images and behavior. Yet another theoretical perspective shows how this happens.

■ GENDER SCHEMA THEORY. **Gender schema theory** is an information-processing approach to gender typing that combines social learning and cognitive-developmental features. It emphasizes that both environmental pressures and children's cognitions work together to shape gender-role development (Bem, 1984; Martin & Halverson, 1981, 1987). Beginning at an early age, children respond to instruction from others, picking up gender-stereotyped preferences and behavior. At the same time, they start to organize their experiences into *gender schemas,* or masculine and feminine categories, that they use to interpret their world. A young child who says, "Only boys can be doctors" or "Cooking is a girl's job" already has some well-formed gender schemas. As soon as preschoolers can label their own sex, they begin to select gender schemas consistent with it, applying those categories to themselves (Fagot & Leinbach, 1989; Martin & Little, 1990). As a result, their self-perceptions become gender typed and serve as additional schemas that they use to process information and guide their own behavior.

Let's look at the example in Figure 8.1 to see exactly how this network of gender schemas strengthens gender-typed preferences and behavior. Three-year-old Mandy has been taught that "dolls are for girls" and "trucks are for boys." She also knows that she is a girl. Mandy uses this information to make decisions about how to behave.

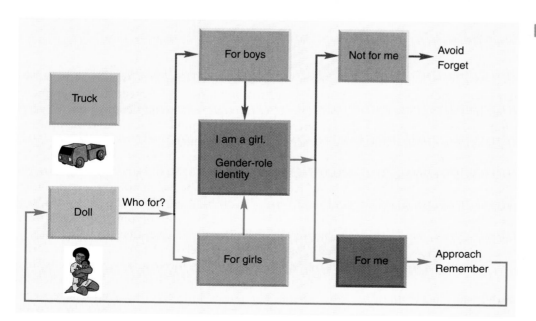

FIGURE 8.1

Effect of gender schemas on gender-stereotyped preferences and behavior. Mandy's network of gender schemas leads her to approach and explore "feminine" toys, such as dolls, and to avoid "masculine" ones, such as trucks. *(From C. L. Martin & C. F. Halverson, 1981, "A Schematic Processing Model of Sex Typing and Stereotyping in Children," Child Development, 52, p. 1121. © The Society for Research in Child Development, Inc. Adapted by permission.)*

A LIFESPAN VISTA

SWEDEN'S COMMITMENT TO GENDER EQUALITY: IMPACT ACROSS THE LIFESPAN

Of all nations in the world, Sweden is unique in its valuing of gender equality and its social programs that translate this commitment into action. Over a century ago, Sweden's ruling political party adopted equality as a central tenet. One social class was not to exploit another, nor one gender another. In the 1960s, Sweden's expanding economy required that women enter the labor force in large numbers. When the question arose as to who would help sustain family life, the Swedish populace called on the principle of equality and answered: fathers, just like mothers. Sweden is the only country in the world where increasing men's involvement in child care is an explicit government objective (Dahlberg, 1994).

The Swedish "equal roles family model" maintains that husband and wife should have the same opportunity to pursue a career and be equally responsible for housework and child care. To support this goal, day-care centers had to be made available out-side the home. Otherwise, a class of less privileged women might be exploited for caregiving and domestic work—an outcome that would contradict the principle of equality. And since full-time employment for both parents often strains a family with young children, Sweden mandated that mothers and fathers with children under age 8 could reduce the length of their working day to 6 hours, with a corresponding reduction in pay but not in benefits (Sandqvist, 1992).

According to several indicators, Sweden's family model has affected the experiences, beliefs, and behaviors of both adults and young people. Maternal employment is extremely high; over 80 percent of mothers with infants and preschoolers work outside the home. Although Swedish fathers do not yet share housework and child care equally with mothers, they are more involved than fathers in North America and other western European nations. In one international study of fourty countries, Swedish men rated themselves lowest in "masculine" traits. On most, they fell at about the same level as Swedish women, and on some, below them (Hofstede, 1980). These findings provide further evidence that gender roles in Sweden are less differentiated than elsewhere.

Parenting of young children has also changed. In interviews with three generations of Swedish women about their childhoods, a shift toward more democratic child rearing and greater freedom and independence appeared (Aström, 1986). Another study showed that Swedish parents try especially hard to be sensitive and empathic, to ensure participation in decision making, and to spend most of their free time with their children. These parenting characteristics, as we will see shortly, are consistently related to positive psychological outcomes for children. Furthermore, Swedish day-care centers are numerous, of high quality, and heavily subsidized by the government. Children enrolled at an early age are advantaged in cognitive, emotional, and social development (Andersson, 1989; 1992).

Because her schemas lead her to conclude that "dolls are for me," when given a doll, she approaches it, explores it, and learns more about it. In contrast, on seeing a truck, she uses her gender schemas to conclude that "trucks are not for me" and responds by avoiding the "gender-inappropriate" toy (Martin & Halverson, 1981). Gender schemas are so powerful that when children see others behaving in "gender-inconsistent" ways, they often cannot remember the information or distort it to make it "gender-consistent" (Liben & Signorella, 1993; Signorella & Liben, 1984).

REDUCING GENDER STEREOTYPING IN YOUNG CHILDREN

How can we help young children avoid developing rigid gender schemas that restrict their behavior and learn-ing opportunities? As the Lifespan Vista box above illustrates, when a nation makes a concerted effort to promote gender equality through its child and family policies, gender stereotyping is substantially reduced and life options across generations are enhanced.

In cultures where values are slower to change, gender-linked associations are usually so pervasive that parents and teachers must work especially hard to prevent young children from absorbing them (Bem, 1984). Adults can begin by eliminating gender stereotyping from their own behavior and from the alternatives they provide for children. For example, mothers and fathers can take turns making dinner, bathing children, and driving the family car. They can provide sons and daughters with both trucks and dolls and pink and blue clothing. And teachers can make sure that all children spend some time each day in

Finally, a study of adolescent gender-related attitudes revealed that valuing the "masculine" over the "feminine" role was less pronounced in Sweden than in the United States. Swedish young people regarded each gender as a blend of instrumental and expressive traits. And although Swedish adolescents more often aspired to stereotyped occupations than did their American counterparts, Swedish girls felt considerably better about their gender. The difference may be due to greater equalization in men's and women's pay scales and a widespread attitude in Sweden that "feminine" work is important to society. Compared to American adolescents, Swedish young people more often viewed gender roles as a matter of learned tasks than inborn personality traits or sets of rights and duties (Intons-Peterson, 1988).

Traditional gender typing is not completely eradicated in Sweden. But great progress has been made as a result of steadfastly pursuing a program of equal opportunity for males and females for several decades—an approach that has had a significant impact on the lives of children, adolescents, and adults.

Sweden places a high value on gender equality. Swedish parents try especially hard to be sensitive and empathic, to ensure participation in decision making, and to spend most of their free time with their children. Compared to fathers in North America and other Western European nations, Swedish fathers are more involved in housework and child care. (Bo Zaunders/The Stock Market)

both adult-structured and unstructured activities. Also, efforts can be made to shield children from television and other media presentations that indicate males and females differ in what they can do.

Once children notice the vast array of gender stereotypes in their society, parents and teachers can point out exceptions. For example, they can arrange for children to see men and women pursuing nontraditional careers. And they can reason with children, explaining that interests and skills, not gender, should determine a person's occupation and activities. Recent evidence shows that such reasoning is very effective in reducing children's tendency to view the world in a gender-biased fashion (Bigler & Liben, 1990, 1992). And as we will see in the next section, a rational approach to child rearing promotes healthy, adaptable functioning in many other areas as well.

BRIEF REVIEW

During the preschool years, children develop a wide variety of gender-typed beliefs, personality traits, and behaviors. Although heredity contributes to several aspects of gender typing, environmental forces play an especially powerful role. Parents view and treat boys and girls differently, and traditional gender-role learning receives further support from teachers, same-sex peers, and the wider social environment. Children gradually develop a gender-role identity, a view of themselves as masculine, feminine, or androgynous in characteristics. Gender schema theory provides a more complete account of the development of gender-role identity than does social learning or cognitive-developmental theory. It shows how environmental pressures and children's cognitions combine to sustain gender-typed self-perceptions, preferences, and behavior.

ASK YOURSELF . . .

■ *Geraldine cut her 3-year-old daughter Fern's hair very short for the summer. When Fern looked in the mirror, she said, "I don't wanna be a boy," and began to cry. Why is Fern upset about her short hairstyle, and what can Geraldine do to help?*

■ *When 4-year-old Roger was in the hospital, he was cared for by a male nurse named Jared. After Roger recovered, he told his friends about Dr. Jared. Using gender schema theory, explain why Roger remembered Jared as a doctor, not a nurse.*

CHILD REARING AND EMOTIONAL AND SOCIAL DEVELOPMENT IN EARLY CHILDHOOD

Throughout this chapter and the previous one, we have seen how parents can foster their children's development—by serving as warm models and reinforcers of mature behavior, by using reasoning, explanation, and inductive discipline, by avoiding harsh punishment, and by encouraging children to master new skills. Now let's put these elements together into an overall view of effective parenting.

CHILD-REARING STYLES

In a series of landmark studies, Diana Baumrind gathered information on child-rearing practices by watching parents interact with their preschoolers. Two broad dimensions of child rearing emerged. The first is *demandingness*. Some parents establish high standards for their children and insist that their youngsters meet those standards. Other parents demand very little and rarely try to influence their children's behavior. The second dimension is *responsiveness*. Some parents are accepting of and responsive to their children. They frequently engage in open discussion and verbal give-and-take. Others are rejecting and unresponsive. As Table 8.2 shows, the various combinations of demandingness and responsiveness yield four styles of child rearing. Baumrind's research focused on three of them: authoritative, authoritarian, and permissive.

TABLE 8.2

A Two-Dimensional Classification of Parenting Styles

	RESPONSIVE	UNRESPONSIVE
DEMANDING	Authoritative parent	Authoritarian parent
UNDEMANDING	Permissive parent	Uninvolved parent

Source: Adapted from E. E. Maccoby & J. A. Martin, 1983, "Socialization in the Context of the Family: Parent–Child Interaction," in E. M. Hetherington (Ed.), *Handbook of Child Psychology: Vol. 4. Socialization, Personality, and Social Development* (4th ed., p. 39), New York: Wiley. Copyright © 1983 by John Wiley & Sons. Reprinted by permission.

■ AUTHORITATIVE CHILD REARING. The **authoritative style** is the most adaptive approach to child rearing. Authoritative parents make reasonable demands for maturity and enforce them by setting limits and insisting that the child obey. At the same time, they express warmth and affection, listen patiently to their child's point of view, and encourage participation in family decision making. Authoritative child rearing is a rational, democratic approach that recognizes and respects the rights of both parents and children.

Baumrind's findings revealed that children of authoritative parents were developing especially well. They were lively and happy, self-confident in their mastery of new tasks, and self-controlled (Baumrind, 1967). These children also seemed less gender typed. Girls scored particularly high in independence and desire to master new tasks and boys in friendly, cooperative behavior (Baumrind & Black, 1967).

■ AUTHORITARIAN CHILD REARING. Parents who use an **authoritarian style** are also demanding, but they place such a high value on conformity that they are unresponsive—even outright rejecting—when children are unwilling to obey. "Do it because I said so!" is the attitude of these parents. As a result, they engage in very little give-and-take with children, who are expected to accept the adult's word for what is right in an unquestioning manner. If they do not, authoritarian parents resort to force and punishment.

Baumrind found that preschoolers with authoritarian parents were anxious, withdrawn, and unhappy. When interacting with peers, they tended to react with hostility when frustrated (Baumrind, 1967). Boys, especially, showed high rates of anger and defiance. Girls were dependent and lacking in exploration, and they retreated from challenging tasks (Baumrind, 1971).

■ PERMISSIVE CHILD REARING. The **permissive style** of child rearing is nurturant and accepting, but it avoids making demands or imposing controls of any kind. Permissive parents allow children to make many of their own decisions at an age when they are not yet capable of doing so.

They can eat meals and go to bed when they feel like it and watch as much television as they want. They do not have to learn good manners or do any household chores. Although some permissive parents truly believe that this style of child rearing is best, many others lack confidence in their ability to influence their child's behavior and are disorganized and ineffective in running their households.

Baumrind found that children of permissive parents were very immature. They had difficulty controlling their impulses and were disobedient and rebellious when asked to do something that conflicted with their momentary desires. They were also overly demanding and dependent on adults, and they showed less persistence on tasks at preschool than children whose parents exerted more control. The link between permissive parenting and dependent, nonachieving behavior was especially strong for boys (Baumrind, 1971).

WHAT MAKES AUTHORITATIVE CHILD REARING SO EFFECTIVE?

Since Baumrind's early work, a great many studies have confirmed her findings. Throughout childhood and adolescence, authoritative parenting is associated with task persistence, social maturity, high self-esteem, internalized moral standards, and superior academic achievement (Denham, Renwick, & Holt, 1991; Lamborn et al., 1991; Steinberg et al., 1992, 1994).

Why does authoritative parenting work so well? First, control that appears fair and reasonable to the child, not abrupt and arbitrary, is far more likely to be complied with and internalized. Second, nurturant parents who are secure in the standards they hold for their children provide models of caring concern as well as confident, assertive behav-

ior. Finally, authoritative parents make demands that are reasonable in terms of their child's developing capacities. By adjusting expectations to fit children's ability to take responsibility for their own behavior, these parents let children know that they are competent individuals who can do things successfully for themselves. As a result, high self-esteem and mature, independent behavior are fostered (Kuczynski et al., 1987).

CULTURAL AND SITUATIONAL INFLUENCES ON CHILD-REARING STYLES

Some parents show variations in child-rearing styles that are adaptive when viewed in light of cultural values and family living conditions. For example, Chinese adults describe their parenting techniques as highly demanding—an emphasis that reflects deeply ingrained Confucian beliefs in the importance of strict discipline, respect for elders, and teaching socially desirable behavior (Berndt et al., 1993; Chao, 1994; Lin & Fu, 1990). In Hispanic and Asian Pacific Island families, high parental control (particularly by the father) is paired with unusually high maternal warmth. This combination is believed to promote compliance and strong family commitment in children (Fracasso & Busch-Rossnagel, 1992). Although wide variation among African Americans exists, some research suggests that black mothers (especially those who are younger, less educated, and single) often rely on an adult-centered approach in which they expect immediate obedience from children (Kelley, Power, & Wimbush, 1992). But when parents have few social supports and live in dangerous neighborhoods, forceful discipline may be necessary to protect children from becoming victims of crime or involved in antisocial activities (Ogbu, 1985).

Asked about their child-rearing practices, the grandparents and mother in this Chinese family would probably describe themselves as quite demanding. Yet clearly, the control they exert occurs in the context of high responsiveness—expressions of warmth and acceptance toward the child. (Ira Kirschenbaum/Stock Boston)

Turn back to Table 8.2, and you will see that we have not yet considered one pattern of parenting: the *uninvolved style*, which combines undemanding with indifferent, rejecting behavior. Uninvolved parents show little commitment to their role as caregivers beyond the minimum effort required to feed and clothe the child. Often they are emotionally detached and depressed and so overwhelmed by the many stresses in their lives that they have little time and energy to spare for children (Maccoby & Martin, 1983).

At its extreme, uninvolved parenting is a form of child maltreatment called neglect. Especially when it begins early, it disrupts virtually all aspects of development, including attachment, cognition, play, and social and emotional skills (Egeland & Sroufe, 1981; Radke-Yarrow et al., 1985). As we turn now to the topic of child maltreatment, we will see that effective child rearing is sustained not just by the desire of mothers and fathers to be good parents. Almost all want to be. A great many factors, both within and outside the family, contribute to parents' capacity to be warm, consistent, and appropriately demanding. Unfortunately, when these vital supports for good parenting break down, children as well as their parents can suffer terribly.

CHILD MALTREATMENT

Child maltreatment is as old as human history, but only recently has there been widespread acceptance that the problem exists and research aimed at understanding it. Perhaps the increase in public concern is due to the fact that child maltreatment is especially common in large, industrialized nations. It occurs so often in the United States that a recent government committee called it "a national emergency." A total of 3.1 million cases were reported to juvenile authorities in 1995, an increase of 132 percent over the previous decade (Children's Defense Fund, 1997). The true figure is surely much higher, since most cases go unreported.

Child maltreatment takes the following forms:

1. *Physical abuse:* assaults on children that produce pain, cuts, welts, bruises, burns, broken bones, and other injuries

2. *Sexual abuse:* sexual comments, fondling, intercourse, and other forms of exploitation

3. *Physical neglect:* living conditions in which children do not receive enough food, clothing, medical attention, or supervision

4. *Emotional neglect:* failure of caregivers to meet children's needs for affection and emotional support

5. *Psychological abuse:* actions that seriously damage children's emotional, social, or cognitive functioning

Although all experts recognize that these five types exist, they differ on how frequent and intense an adult's actions must be to be called maltreatment. While all of us can agree that broken bones, cigarette burns, and bite marks are abusive, the decision is harder to make in instances in which an adult touches or makes degrading comments to a child (Barnett, Manly, & Cicchetti, 1993). Some investigators regard psychological and sexual abuse as the most destructive forms. The rate of psychological abuse may be the highest, since it accompanies most other types. Over 200,000 cases of child sexual abuse are reported each year. Yet this statistic greatly underestimates the actual number, since affected children feel frightened, confused, and guilty and are usually pressured into silence. Although children of all ages are affected, the largest number of sexual abuse victims are identified in middle childhood. Therefore, we will pay special attention to this form of maltreatment in Chapter 10.

■ THE ORIGINS OF CHILD MALTREATMENT. Early findings suggested that child maltreatment was rooted in adult psychological disturbance (Kempe et al., 1962; Spinetta & Rigler, 1972). But it soon became clear that although child abuse was more common among disturbed parents, a single "abusive personality type" did not exist. Sometimes even "normal" parents harmed their children! Also, parents who had been abused as children did not always repeat the cycle with their own youngsters (Kaufman & Zigler, 1989; Simons et al, 1991).

Recent research reveals that many interacting variables—at the family, community, and cultural levels—promote child abuse and neglect. Table 8.3 summarizes factors associated with child maltreatment. The more of these risks that are present, the greater the likelihood that it will occur. Let's examine each set of influences in turn.

The Family. Within the family, certain children—those whose characteristics make them more of a challenge to rear—have an increased likelihood of becoming targets of abuse. These include premature or very sick babies and children who are temperamentally difficult, inattentive and overactive, or who have other developmental problems. But whether such children actually are maltreated depends on characteristics of parents (Belsky 1993). In one study, temperamentally difficult youngsters who were physically abused had mothers who believed that they could do little to control the child's behavior. Instead, they attributed the child's unruliness to a stubborn or bad disposition, a perspective that led them to move quickly toward physical force when the child misbehaved (Bugental, Blue, & Cruzcosa, 1989).

Once child abuse gets started, it quickly becomes part of a self-sustaining family relationship. The small irritations to which abusive parents react—a fussy baby, a preschooler who knocks over a glass of milk, or a child who will not mind immediately—soon become bigger ones.

TABLE 8.3

Factors Related to Child Maltreatment

FACTOR	DESCRIPTION
Parent Characteristics	Psychological disturbance; substance abuse; history of abuse as a child; belief in harsh, physical discipline; desire to satisfy unmet emotional needs through the child; unreasonable expectations for child behavior; young age (most under 30); low educational level
Child characteristics	Premature or very sick baby; difficult temperament; inattentiveness and overactivity; other developmental problems
Family characteristics	Low income; poverty; homelessness; marital instability; social isolation; physical abuse of mother by husband or boyfriend; frequent moves; large, closely spaced families; overcrowded living conditions; disorganized household; lack of steady employment; other signs of high life stress.
Community	Characterized by social isolation; few parks, day care centers, preschool programs, recreation centers, and churches to serve as family supports
Culture	Approval of physical force and violence as ways to solve problems

Sources: Belsky, 1993; Pianta, Egeland, & Erickson, 1989; Simons et al., 1991.

Then the harshness of parental behavior increases. By the preschool years, abusive and neglectful parents seldom interact with their children. When they do, they rarely express pleasure and affection; the communication is almost always negative (Trickett et al., 1991).

Most parents, however, have enough self-control not to respond to their child's misbehavior with abuse, and not all children with developmental problems are mistreated. Other factors must combine with these conditions to prompt an extreme parental response. Research reveals that unmanageable parental stress is strongly associated with all forms of maltreatment. Low income, unemployment, marital conflict, overcrowded living conditions, frequent moves, and extreme household disorganization are common in abusive homes. These conditions increase the chances that parents will be too overwhelmed to meet basic child-rearing responsibilities or will vent their frustrations at their children (Pianta, Egeland, & Erickson, 1989).

The Community. The majority of abusive parents are isolated from both formal and informal social supports in their communities. There are at least two causes of this social isolation. First, because of their own life histories, many of these parents have learned to mistrust and avoid others. They do not have the skills necessary for establishing and maintaining positive relationships with friends and relatives (Polansky et al., 1985). Second, abusive parents are more likely to live in neighborhoods that provide few links between family and community, such as parks, day-care centers, preschool programs, recreation centers, and churches (Coulton et al., 1995). For these reasons, they lack "lifelines" to others and have no one to turn to for help during particularly stressful times.

The Larger Culture. Societal values, laws, and customs profoundly affect the chances that parents who feel overburdened will maltreat their children. Cultures that view force and violence as appropriate ways to solve problems

set the stage for child abuse. These conditions exist in the United States. Although all fifty states have laws designed to protect children from maltreatment, strong support still exists for the use of physical force in parent–child relations. For example, during the past quarter century, the United States Supreme Court has twice upheld the right of school officials to use corporal punishment. Crime rates have risen in American cities, and television sets beam graphic displays of violence into family living rooms. In view of the widespread acceptance of violent behavior in American culture, it is not surprising that 90 percent of American parents report using slaps and spankings to discipline their children (Staub, 1996). In countries where physical punishment is not accepted, such as China, Japan, Luxembourg, and Sweden, child abuse is rare (Zigler & Hall, 1989).

■ CONSEQUENCES OF CHILD MALTREATMENT. The family circumstances of maltreated children impair the development of emotional self-regulation, self-concept, and social skills. Over time, these youngsters show serious learning and adjustment problems, including academic failure, difficulties with peers, severe depression, substance abuse, and delinquency (Simons, Conger, & Whitbeck, 1988).

What explains these damaging consequences? Think back to our earlier discussion of hostile cycles of parent–child interaction, which are especially severe for abused children. Clearly, their home lives overflow with opportunities to learn to use aggression as a way of solving problems. The low warmth and control to which neglected children are exposed also promote aggressive, acting-out behavior (Miller et al., 1993).

Furthermore, demeaning parental messages, in which children are ridiculed, humiliated, rejected, or terrorized, result in low self-esteem, high anxiety, self-blame, and efforts to escape from extreme psychological pain—at

you idiot!
you idiot!
you idiot!
you idiot!
you idiot!
you idiot!
you idiot!
you idiot!
you idiot!

Children learn by repetition.

You don't have to hit to hurt.

San Francisco
Child Abuse Council
(415) 668-0494

Public service announcements help prevent child abuse by educating people about the problem and informing them of where to seek help. This poster reminds adults that degrading remarks can hit as hard as a fist. (Courtesy San Francisco Child Abuse Council)

times severe enough to lead to attempted suicide in adolescence (Briere, 1992; Sternberg et al., 1993). At school, maltreated children are serious discipline problems. Their noncompliance, poor motivation, and cognitive immaturity interfere with academic achievement—an outcome that further undermines their chances for life success (Eckenrode, Laird, & Doris, 1993).

■ PREVENTING CHILD MALTREATMENT. Since child maltreatment is embedded in families, communities, and society as a whole, efforts to prevent it must be directed at each of these levels. Many approaches have been suggested.

These include interventions that teach high-risk parents effective child-rearing and disciplinary strategies, high school child development courses that include direct experience with children, and broad social programs aimed at bettering economic conditions for low-income families.

In earlier parts of this book, we saw that providing social supports to families is very effective in easing parental stress. It sharply reduces child maltreatment as well. Research indicates that a trusting relationship with another person is the most important factor in preventing mothers with childhood histories of abuse from repeating the cycle with their own youngsters (Caliso & Milner, 1992; Egeland, Jacobvitz, & Sroufe, 1988). Parents Anonymous, a national organization that has as its main goal helping child-abusing parents learn constructive parenting practices, does so largely through social supports. Its local chapters offer self-help group meetings, daily phone calls, and regular home visits to relieve social isolation and teach alternative child-rearing skills.

Even with intensive treatment, some adults persist in their abusive acts. An estimated 1,500 American children die from maltreatment each year (Children's Defense Fund, 1997). In cases in which parents are unlikely to change their behavior, taking the drastic step of separating parent from child and legally terminating parental rights is the only reasonable course of action.

Child maltreatment is a distressing and horrifying topic—a sad note on which to end our discussion of a period of childhood that is so full of excitement, awakening, and discovery. But there is reason to be optimistic. Great strides have been made in understanding and preventing child maltreatment over the last several decades.

ASK YOURSELF . . .

- *Earlier in this chapter, we discussed induction as an especially effective form of discipline. Of Baumrind's three child-rearing styles, which is most likely to be associated with use of induction, and why?*

- *Chandra heard a news report that ten severely neglected children, living in squalor in an inner-city tenement, were discovered by Chicago police. Chandra thought to herself, "What could possibly lead parents to mistreat their children so badly?" How would you answer Chandra's question?*

SUMMARY

ERIKSON'S THEORY: INITIATIVE VERSUS GUILT

What personality changes take place during Erikson's stage of initiative versus guilt?

■ According to Erikson, during early childhood, children grapple with the psychological conflict of **initiative versus guilt.** Erikson's stage corresponds to Freud's **phallic stage,** in which conscience is formed through **identification** with the same-sex parent. Although Freud's ideas are no longer widely accepted, Erikson's image of initiative captures the emotional and social changes of the preschool years.

SELF-DEVELOPMENT IN EARLY CHILDHOOD

Describe preschoolers' self-concepts and self-esteem.

■ Preschoolers' **self-concepts** largely consist of observable characteristics and typical beliefs, emotions, and attitudes. Their increasing self-awareness underlies struggles over objects as well as first efforts to cooperate.

■ Preschoolers' high **self-esteem** contributes to their mastery-oriented approach to the environment. However, even a little adult disapproval can undermine a young child's self-esteem and enthusiasm for learning.

EMOTIONAL DEVELOPMENT IN EARLY CHILDHOOD

Cite changes in understanding and expression of emotion during early childhood.

■ Young children have an impressive understanding of the causes and consequences of basic emotional reactions. By age 3 to 4, they are also aware of various strategies for emotional self-regulation. As a result, intense emotional outbursts become less frequent.

■ Preschoolers experience self-conscious emotions more often as their self-concepts become better developed and they become increasingly sensitive to the praise and criticism of others. Empathy becomes more common over the preschool years.

PEER RELATIONS IN EARLY CHILDHOOD

Trace the development of peer sociability in early childhood.

■ During early childhood, interactive play with peers increases. According to Parten, it begins with **nonsocial activity,** shifts to **parallel play,** and then moves to **associative** and **cooperative play.** However, preschoolers do not follow this straightforward developmental sequence. Solitary play and parallel play remain common throughout early childhood. Sociodramatic play becomes especially frequent and supports many aspects of emotional and social development.

Describe the quality of preschoolers' friendships.

■ Preschoolers view friendship in concrete, activity-based terms. Their interactions with friends are especially positive and cooperative.

FOUNDATIONS OF MORALITY IN EARLY CHILDHOOD

What are the central features of psychoanalytic, behaviorist and social learning, and cognitive-developmental approaches to moral development?

■ The psychoanalytic and behaviorist and social learning approaches to morality focus on how children acquire ready-made standards held by adults. In contrast to Freud's theory, discipline using fear of punishment and loss of parental love does not foster conscience development. Instead, **induction** is far more effective in encouraging self-control and **prosocial,** or **altruistic, behavior.**

■ Behaviorism and social learning theory regard reinforcement and modeling as the basis for moral action. Effective adult models of morality are warm, powerful, and practice what they preach. Harsh punishment does not promote moral internalization and socially desirable behavior. Alternatives, such as **time out** and withdrawal of privileges, can reduce the undesirable side effects of punishment.

■ The cognitive-developmental perspective views children as active thinkers about social rules. Preschoolers understand that disobeying moral rules is more serious than violating social conventions. Parents who discuss moral issues with their children help them reason about morality.

Describe the development of aggression in early childhood, including family and television as major influences.

■ All children display aggression from time to time. During early childhood, **instrumental aggression** declines while **hostile aggression** increases. Boys tend to be more aggressive than girls, a difference that may be linked to boys' higher activity level.

■ Ineffective discipline and a conflict-ridden family atmosphere promote and sustain aggression in children. Teaching parents effective child-rearing practices and providing children with social problem-solving training can reduce aggressive behavior. Television promotes childhood aggression, belief in the truthfulness of advertising, and ethnic and gender stereotypes.

GENDER TYPING IN EARLY CHILDHOOD

Discuss genetic and environmental influences on preschoolers' gender-stereotyped beliefs and behavior.

■ **Gender typing** is well underway in the preschool years. Genetic factors are believed to play a role in boys' higher activity level and aggression and children's preference for same-sex playmates. At the same time, the environment provides powerful support for gender typing. Parents, teachers, peers, and the broader social environment encourage many gender-typed responses.

Describe and evaluate the accuracy of major theories of the emergence of gender-role identity.

■ Although most people have a traditional **gender-role identity,** some are **androgynous,** combining both "masculine" and "feminine" characteristics.

■ According to social learning theory, preschoolers first acquire gender-typed responses through modeling and reinforcement and then organize them into gender-linked ideas about themselves. Cognitive-developmental theory suggests that **gender constancy** must be mastered before children develop gender-typed behavior.

■ In contrast to cognitive-developmental predictions, gender-role behavior is acquired long before gender constancy. **Gender schema theory** is an information-processing approach to gender typing that combines social learning and cognitive-developmental features. As children acquire gender-stereotyped preferences and behaviors, they form masculine and feminine categories, or gender schemas, that they apply to themselves and use to interpret their world.

CHILD REARING AND EMOTIONAL AND SOCIAL DEVELOPMENT IN EARLY CHILDHOOD

Describe the impact of child-rearing styles on children's development, and explain why authoritative parenting is so effective.

■ Compared to the **authoritarian** and **permissive styles,** the **authoritative style** of child rearing promotes cognitive and social competence. Warmth, explanations, and reasonable demands for mature behavior account for the effectiveness of the authoritative style.

Discuss the multiple origins of child maltreatment and its consequences for development.

■ Child maltreatment is related to factors within the family, community, and larger culture. Child and parent characteristics often feed on one another to produce abusive behavior. Unmanageable parental stress and social isolation greatly increase the chances that abuse and neglect will occur. When a society approves of force and violence as means of solving problems, child abuse is promoted.

IMPORTANT TERMS AND CONCEPTS

initiative versus guilt (p. 246)
phallic stage (p. 246)
identification (p. 246)
self-concept (p. 247)
self-esteem (p. 248)
nonsocial activity (p. 251)
parallel play (p. 251)
associative play (p. 251)

cooperative play (p. 251)
induction (p. 254)
prosocial, or altruistic, behavior (p. 254)
time out (p. 255)
instrumental aggression (p. 257)
hostile aggression (p. 257)
gender typing (p. 259)

gender-role identity (p. 262)
androgyny (p. 262)
gender constancy (p. 262)
gender schema theory (p. 263)
authoritative style (p. 266)
authoritarian style (p. 266)
permissive style (p. 266)

FOR FURTHER INFORMATION AND HELP

CHILD ABUSE AND NEGLECT

Child Help USA, Inc.
6463 Independence Avenue
Woodland Hills, CA 91370
(800) 422-4453
Web site: *www.childhelpusa.org*
Promotes public awareness of child abuse through publications, media campaigns, and a speakers' bureau. Supports the National Child Abuse Hotline, (800) 4-A-CHILD. Callers may request information about child abuse or speak with a crisis counselor.

Parents Anonymous
2701 N. 16th Street, Suite 316
Phoenix, AZ 85006
(602) 248-0428
Web site: *www.parentsanon.org*
Dedicated to prevention and treatment of child abuse. Local groups provide support to child-abusing parents and training in nonviolent child-rearing techniques.

National Clearinghouse
on Child Abuse and
Neglect Information
P.O. Box 1182
Washington, DC 20013
(800) 394-3366
Web site: *www.calib.com/nccanch*
Provides information to states and communities wishing to develop programs and activities that identify, prevent, and treat child abuse and neglect.

MILESTONES
OF DEVELOPMENT IN EARLY CHILDHOOD

AGE	PHYSICAL	COGNITIVE	LANGUAGE	EMOTIONAL/SOCIAL
2 years	■ Slower gains in height and weight than in toddlerhood. ■ Balances improves, walking becomes better coordinated. ■ Running, jumping, hopping, throwing, and catching appear. ■ Puts on and removes some item of clothing. ■ Uses spoon effectively.	■ Make-believe becomes less dependent on realistic toys, less self-centered, and more complex. ■ Can take the perspective of others in simple situations. ■ Recognition memory well developed. ■ Aware of the difference between inner mental and outer physical events.	■ Vocabulary increases rapidly. ■ Sentences follow basic word order of native language; grammatical markers are added. ■ Displays effective conversational skills, such as turn taking and topic maintenance. 	■ Begins to develop self-concept and self-esteem. ■ Cooperation and instrumental aggression appear. ■ Understands causes and consequences of basic emotions. ■ Empathy increases. ■ Gender-stereotyped beliefs and behavior increases.
3–4 years	■ Running, jumping, hopping, throwing, and catching become better coordinated. ■ Galloping and one-foot skipping appear. ■ Rides tricycle. ■ Uses scissors, draws first picture of a person. ■ Can tell the difference between writing and nonwriting. 	■ Notices transformations, reverses thinking, and has a basic understanding of causality in familiar situations. ■ Classifies familiar objects hierarchically. ■ Uses private speech to guide behavior when engaged in challenging tasks. ■ Remembers familiar experiences in terms of scripts. ■ Can generalize remembered information from one situation to another. ■ Understands that people can hold false beliefs. ■ Aware of some meaningful features of written language. ª Counts small numbers of objects and grasps the principle of cardinality.	■ Occasionally overextends grammatical rules to exceptions. ■ Understands many culturally accepted ways of adjusting speech to fit the age, sex, and social status of speakers and listeners.  "Nose your touch!" "That's backwards!"	■ Emotional self-regulation improves. ■ Self-conscious emotions (shame, embarrassment, guilt, envy, and pride) become more common. ■ Nonsocial activity declines and joint, interactive play increases. ■ Instrumental aggression declines and hostile aggression increases. ■ Forms first friendships. ■ Distinguishes moral rules and social conventions. ■ Preference for same-sex playmates increases. 

AGE	PHYSICAL	COGNITIVE	LANGUAGE	EMOTIONAL/SOCIAL

5–6 years

PHYSICAL
- Body is streamlined and longer-legged with proportions similar to that of an adult.
- First permanent tooth erupts.
- Skipping appears.
- Ties shoes, draws more elaborate pictures, writes name.

COGNITIVE
- Ability to distinguish appearance from reality improves.
- Attention becomes more sustained and planful.
- Recall and scripted memory, and memory for unique everyday events improve.
- Understands that letters and sounds are linked in systematic ways.
- Counts on and counts down, engaging in simple addition and subtraction.

LANGUAGE
- Vocabulary reaches about 10,000 words.
- Has mastered many complex grammatical forms.

EMOTIONAL/SOCIAL
- Ability to interpret, predict, and influence other's emotional reactions improves.
- Relies on language to express empathy.
- Has acquired many morally relevant rules and behaviors.
- Grasps the genital basis of sex differences and shows gender constancy.

Physical and Cognitive Development in Middle Childhood

"I'm on my way, Mom!" hollered 10-year-old Joey as he stuffed the last bite of toast into his mouth, slung his book bag over his shoulder, dashed out the door, jumped on his bike, and headed down the street for school. Joey's 8-year-old sister Lizzie followed, quickly kissing her mother goodbye and hurrying to catch up with Joey. Off she raced, peddling furiously, until soon she was side-by-side with her older brother. Rena, the children's mother and one of my colleagues at the university, watched from the front porch and her son and daughter disappeared in the distance.

"They're branching out," Rena remarked to me over lunch that day as she described the children's expanding activities and relationships. Homework, household chores, soccer teams, music lessons, scouting, friends at school and in the neighborhood, and Joey's new paper route were all part of the children's routine. "It seems as if the basics are all there; I don't have to monitor Joey and Lizzie so constantly anymore. But being a parent is still very challenging. Now it's more a matter of refinements—helping them become independent, competent, and productive individuals."

Joey and Lizzie have entered the phase of development called middle childhood, which spans the years from 6 to 11. Around the world, children of this age are assigned new responsibilities. Joey and Lizzie, like other youngsters in industrialized nations, spend many long hours in school. Indeed, middle childhood is often called the "school years," since its onset is marked by the start of formal schooling. In village and tribal cultures, the school may be a field or a jungle. But universally, mature members of society guide children of this age toward more realistic tasks that increasingly resemble those they will perform as adults.

This chapter focuses on physical and cognitive development in middle childhood—changes that are less spectacular than those of the earlier years. By age 6, the brain has reached 95 percent of its adult size, and the body continues to grow slowly. In this way, nature grants school-age children the mental powers to master challenging activities as well as added time to learn before reaching physical maturity.

We begin by reviewing typical growth trends, special health concerns, and gains in motor abilities. Then we return to Piaget's theory and the information-processing approach, which provide an overview of cognitive change during the school years. Next, we examine the genetic and environmental roots of IQ scores, which often enter into important educational decisions. Our discussion continues with language, which blossoms further during middle childhood. Finally, we consider the importance of schools in children's learning and development.

PHYSICAL DEVELOPMENT IN MIDDLE CHILDHOOD

BODY GROWTH

The rate of physical growth during the school years is an extension of the slow, regular pattern that characterized early childhood. At age 6, the average child weighs about 45 pounds and is 3½ feet tall. As Figure 9.1 shows, children add about 2 to 3 inches in height and 5 pounds in weight each year.

Look closely at Figure 9.1, and you will see that at ages 6 to 8 girls are slightly shorter and lighter than boys. By age 9, this trend reverses. Already, Rena noticed, Lizzie was starting to catch up with Joey in size. For many girls, the 10-year-old height spurt overlaps with the much more dramatic adolescent growth spurt, which takes place 2 years earlier in girls than boys.

During middle childhood, the lower portion of the body is growing fastest. These 8-year-old girls are taller and longer legged than they were as preschoolers. *(Arnie Katz/Stock South)*

Test Bank Items 9.1 through 9.2

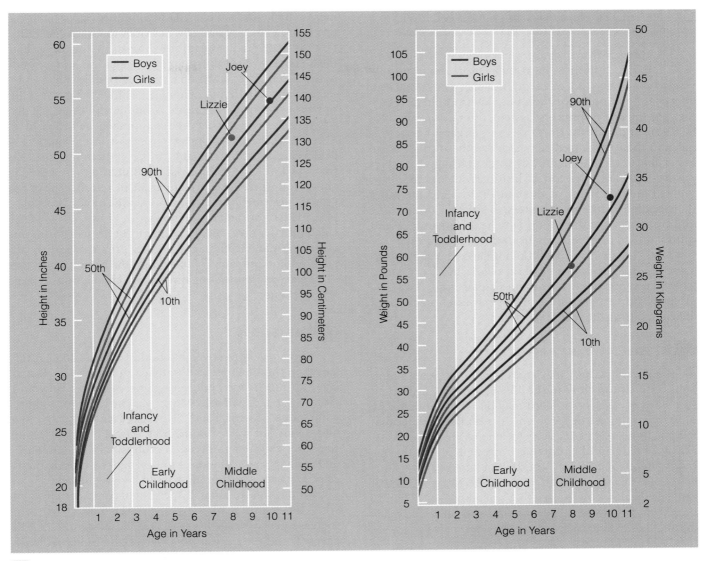

FIGURE 9.1

Gains in height and weight during middle childhood among North American children. The slow rate of growth established in early childhood extends into the school years. Girls are slightly shorter and lighter than boys until age 9, at which time this trend is reversed as girls approach the adolescent growth spurt. Eight-year-old Lizzie is beginning to catch up with 10-year-old Joey in physical size. Wide individual differences in body size continue to exist, as the percentiles on these charts reveal.

Because the lower portion of the body is growing fastest, Joey and Lizzie appeared longer-legged than they had in early childhood. Rena discovered that they grew out of their jeans more quickly than their jackets and frequently needed larger shoes. As in early childhood, girls have slightly more body fat and boys more muscle. After age 8, girls begin accumulating fat at a faster rate, and they will add even more during adolescence (Tanner, 1990).

During middle childhood, the bones of the body lengthen and broaden. However, ligaments are not yet firmly attached to bones. This, combined with increasing muscle strength, grants children unusual flexibility of movement as they turn cartwheels, perform handstands,

and practice fancy break-dance routines. As their bodies become stronger, many children experience a greater desire for physical exercise. Nighttime "growing pains"—aches and muscle pulls—are common as muscles adapt to an enlarging skeleton (Sheiman & Slomin, 1988).

Between the ages of 6 and 12, all twenty primary teeth are replaced by permanent ones, with girls losing their teeth slightly earlier than boys. The first teeth to go are upper front teeth, giving many first and second graders a "toothless" smile. For a while, permanent teeth seem much too large. Growth of facial bones, especially the jaw and chin, gradually causes the child's face to lengthen and mouth to widen, accommodating the newly erupting teeth.

COMMON HEALTH PROBLEMS

Children from advantaged homes, like Joey and Lizzie, appear to be at their healthiest in middle childhood, full of energy and play. The cumulative effects of good nutrition, combined with rapid development of the body's immune system, offer greater protection against disease. At the same time, larger lungs permit more air to be exchanged with each breath, so children are better able to exercise vigorously without tiring.

Nevertheless, a variety of health problems do occur. We will see that many of them affect low-income more than middle-income youngsters. Because economically disadvantaged families often lack health insurance and cannot afford to pay for medical visits on their own (see Chapter 7), many youngsters do not have regular access to a doctor. And growing numbers also lack such basic necessities as a comfortable home and regular meals. Not surprisingly, poverty continues to be a powerful predictor of ill health during the school years.

VISION AND HEARING

Rena often warned Joey and Lizzie not to read in dim light or to sit too close to the TV set, exclaiming, "You'll ruin your eyes!" Her concern may be well founded. The most common vision problem in middle childhood is *myopia,* or nearsightedness. By the end of the school years, nearly 25 percent of children are affected. Heredity contributes to myopia, but the condition is also related to how people use their eyes (Teikari et al., 1991). Myopia is one of the few health conditions that increases with family income and education. The more time people spend reading and doing other close-up work, the more likely they are to be myopic (Angle & Wissmann, 1980). Fortunately, for those youngsters who develop nearsightedness because they love reading, drawing, or model building, the condition can be corrected easily with glasses.

During middle childhood, the Eustachian tube (canal that runs from the inner ear to the throat) becomes longer, narrower, and more slanted, preventing fluid and bacteria from traveling so easily from the mouth to the ear. As a result, ear infections become less frequent. Still, about 3 to 4 percent of the school-age population, and as many as 18 to 20 percent of low-income youngsters, develop permanent hearing loss from repeated ear infections (Mott, James, & Sperhac, 1990). Regular screening for both vision and hearing problems is important so that defects can be corrected before they lead to serious learning difficulties.

MALNUTRITION

School-age children need a well-balanced, plentiful diet to provide energy for successful learning in school and increased physical activity. Many youngsters are so focused on play, friendships, and new activities that they spend little time at the table. Readily available, healthy between-meal snacks—cheese, fruit, raw vegetables, and peanut butter—help meet nutritional needs in middle childhood.

As long as parents encourage healthy eating, the mild nutritional deficits that result from the child's busy daily schedule have no impact on development. But as we have seen in earlier chapters, many poverty-stricken youngsters in developing countries and in the United States suffer from serious and prolonged malnutrition. By middle childhood, the effects are apparent in retarded physical growth, low intelligence test scores, poor motor coordination, inattention, and distractibility.

Unfortunately, when malnutrition persists for many years, permanent damage is done. Prevention through government-sponsored food programs beginning in the early years and continuing throughout childhood and adolescence is necessary.

OBESITY

Mona, a very overweight child in Lizzie's class, often stood on the sidelines and watched during recess. When she did join in the children's games, she was slow and clumsy. On a daily basis, Mona was the target of unkind comments: "Move it, Tubs!" "No fatsoes allowed!" On most afternoons, she walked home from school by herself while the other children gathered in groups, talking, laughing, and chasing. Once at home, Mona sought comfort in high-calorie snacks, which promoted further weight gain.

Mona is one of about 27 percent of American children who suffer from **obesity,** a greater-than-20-percent increase over average body weight, based on the child's age, sex, and physical build. Overweight and obesity are growing problems in affluent nations like the United States. Childhood obesity has climbed steadily since the 1960s, with over 80 percent of youngsters like Mona remaining overweight as adults (Dietz, Bandini, & Gortmaker, 1990; Muecke et al., 1992).

Obese children have serious emotional and social difficulties and are at risk for life long health problems. High blood pressure and cholesterol levels along with respiratory abnormalities begin to appear in the early school years, symptoms that are powerful predictors of heart disease, other serious illnesses, and early death.

CAUSES OF OBESITY. Childhood obesity is a complex physical disorder with multiple causes. Not all children are equally at risk for becoming overweight. Fat children tend to have fat parents, and identical twins are more likely to share the disorder than are fraternal twins. But similarity among family members is not strong enough for genetics to account for any more than a tendency to gain weight

Overweight children tend to have overweight parents. But genetics accounts for no more than a tendency to gain weight. A family environment where high-calorie, high-fat foods are readily available, food takes on powerful emotional meaning, and much time is spent in sedentary activities affects the health of his father and his son. (Dick Luria/FPG International)

(Bouchard, 1994). One indication that environment is powerfully important is the consistent relation between social class and obesity. Low-income youngsters in industrialized nations are not just at greater risk for malnutrition. They are also more likely to be overweight (Stunkard & Sørensen, 1993). Among factors responsible are lack of knowledge about healthy diet; a tendency to buy high-fat, low-cost foods; and family stress, which prompts overeating in some individuals.

Parental feeding practices contribute to childhood obesity. In infancy and early childhood, some parents anxiously overfeed their children, interpreting almost all their discomforts as a desire for food. Others are overly controlling, constantly monitoring what their children eat. In either case, they fail to help children regulate their own energy intake. Furthermore, parents of obese children often use food as a reward. When food is used to reinforce other behaviors, children start to value the treat itself as well as other similar foods (Birch & Fisher, 1995). Gradually high-calorie snacks and desserts come to symbolize warmth, comfort, and relief of tension.

Perhaps because of these experiences, obese children soon develop maladaptive eating habits. They are more responsive to external stimuli associated with food—taste, sight, smell, and time of day—and less responsive to internal hunger cues than individuals of normal weight (Ballard et al., 1980; Constanzo & Woody, 1979). Overweight individuals also eat faster and chew their food less thoroughly,

a behavior pattern that appears as early as 18 months of age (Drabman et al., 1979).

Furthermore, fat children are less physically active than their normal-weight peers. This inactivity is both cause and consequence of their overweight condition. Recent evidence indicates that the rise in childhood obesity in the United States over the past 30 years is in part due to television viewing. Next to already existing obesity, time spent in front of the TV set is the best predictor of future obesity among school-age children (Gortmaker, Dietz, & Cheung, 1990). Television greatly reduces the time that children devote to physical exercise, and TV ads encourage them to eat fattening, unhealthy snacks (see Chapter 8).

■ CONSEQUENCES OF OBESITY. Unfortunately, physical attractiveness is a powerful predictor of social acceptance in our culture. Both children and adults rate obese youngsters as less likable than children with a wide range of physical disabilities (Brenner & Hinsdale, 1978; Lerner & Schroeder, 1971). By middle childhood, obese children have a low sense of self-esteem, report feeling more depressed, and display more behavior problems than their peers. A vicious cycle emerges in which unhappiness and overeating contribute to one another, and the child remains overweight. As we will see in Chapter 13, these psychological consequences combine with continuing discrimination to result in reduced life chances in close relationships and employment.

■ TREATING OBESITY. Childhood obesity is difficult to treat because it is a family disorder. In Mona's case, the school nurse suggested that Mona and her obese mother enter a weight loss program together. But Mona's mother, unhappily married for many years, had her own reasons for overeating. She rejected this idea, claiming that Mona would eventually lose weight on her own. Although many obese youngsters do try to slim down in adolescence, often they go on crash diets that deprive them of essential nutrients during a period of rapid growth. These efforts can make matters worse. Temporary starvation leads to physical stress and fatigue. Soon the young person returns to old eating patterns, and weight rebounds to a higher level. Then, to protect itself, the body burns calories more slowly and becomes more resistant to weight loss (Pinel, 1997).

When parents decide to seek treatment for an obese child, long-term changes in body weight do occur. A recent study found that the most effective interventions were family based and focused on changing behaviors. Both parent and child revised eating patterns, exercised daily, and reinforced each other with praise and points for progress, which they exchanged for special activities and times together (Epstein et al., 1987). A follow-up after 5 years showed that children maintained their weight loss more effectively than did adults (Epstein et al., 1990). This finding underscores the importance of intervening with obese children at an early age, before harmful eating behaviors become well established.

BEDWETTING

One Friday afternoon, Terry called up Joey to see if he could sleep over, but Joey refused. "I can't," said Joey anxiously, without giving an explanation.

"Why not? We can take our sleeping bags out in the backyard. Come on, it'll be super!"

"My mom won't let me," Joey responded, unconvincingly. "I mean, well, I think we're busy, we're doing something tonight."

"Gosh, Joey, this is the third time you've said no. See if I'll ask *you* again!" snapped Terry as he hung up the phone.

Joey is one of 8 percent of American school-age children who suffer from **nocturnal enuresis,** or bedwetting during the night (Rappaport, 1993). Enuresis evokes considerable distress in children and parents alike. In the overwhelming majority of cases, the problem has biological roots. Heredity is a major contributing factor. Parents with a history of bedwetting are far more likely to have a child with the problem (McGuire & Savashino, 1984). Enuresis is unrelated to the depth of a child's sleep, and only rarely is it due to abnormalities in the urinary tract. Most often, it is caused by a failure of muscular responses that inhibit urination or a hormonal imbalance that permits too much urine to accumulate during the night (Houts, 1991). Punishing a school-age child for wetting is only likely to make matters worse.

The most effective treatment for enuresis is a urine alarm that wakes the child at the first sign of dampness and works according to conditioning principles. Success rates of about 70 percent occur after 4 to 6 months of treatment. Most children who relapse achieve dryness after trying the alarm a second time (Rushton, 1989). Although many children outgrow enuresis without any form of intervention, it generally takes years for them to do so (Houts, 1991).

ILLNESSES

Children experience a somewhat higher rate of illness during the first 2 years of elementary school than they will later, due to exposure to sick children, an immune system that is not yet mature, and greater physical activity, which sometimes leads to physical injury. Allergies, colds, influenza, muscle sprains, and bone fractures are common reasons for missing school.

But the most frequent cause of school absence and childhood hospitalization is *asthma,* a condition in which the bronchial tubes (passages that connect the throat and lungs) are highly sensitive. In response to a variety of stimuli, such as cold weather, infection, exercise, or allergies, they fill with mucus and contract, leading to coughing, wheezing, and serious breathing difficulties. The number of children with asthma has increased by 50 percent over the last decade. Today, 6 to 12 percent of American youngsters are affected, and asthma-related deaths have risen in recent years (Celano & Geller, 1993). Although heredity contributes to asthma, researchers believe that environmental factors are necessary to spark the illness. Boys, African-American children, and children who had low birth weights, whose parents smoke, and who live in poverty are at greatest risk (Chilmonczyk et al., 1993; Weitzman, Gortmaker, & Sobol, 1990). Perhaps black and poverty-stricken youngsters experience a higher rate of asthma because of pollution in inner-city areas (which triggers allergic reactions), stressful home lives, and lack of access to good health care.

UNINTENTIONAL INJURIES

As we conclude our discussion of threats to children's health during the school years, let's return for a moment to the topic of unintentional injuries (discussed in detail in Chapter 7). As Figure 9.2 shows, the frequency of injuries increases steadily over middle childhood into adolescence, with boys continuing to show a higher rate than girls. Auto and bicycle collisions account for most of this rise (Brooks & Roberts, 1990).

As children range farther from home, safety education becomes especially important. By middle childhood, the greatest risk-takers tend to be children whose parents do not act as safety-conscious models or who try to enforce rules with punitive or inconsistent discipline (Roberts, Elkins, & Royal, 1984). These child-rearing techniques, as we saw in Chapter 8, spark defiance in children and may actually promote high-risk behavior. The greatest challenge for injury control programs is how to reach these "more difficult to reach" youngsters, alter their family contexts, and reduce the dangers to which they are exposed (Brooks & Roberts, 1990).

HEALTH EDUCATION

Researchers and educators have become intensely interested in finding ways to help school-age children understand how their bodies work, acquire mature conceptions of health and illness, and develop patterns of behavior that foster good health throughout life. Successfully targeting children for intervention on any health issue requires information on their current health-related knowledge. The Social Issues box on page 284 sum-

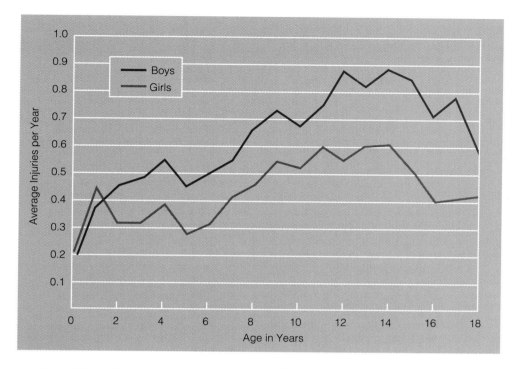

FIGURE 9.2

Rates of unintentional injury by age and sex of child in a sample drawn from nearly 700 American families. Injuries increase steadily from 5 to 14 years, after which they decline. Boys experience more injuries than girls throughout childhood and adolescence. *(From E. L. Schor, 1987, "Unintentional Injuries,"* American Journal of Diseases of Children, 194, *p. 1281. Reprinted by permission.)*

marizes children's concepts of health and illness during middle childhood.

The school-age period may be an especially important time for fostering healthy life-styles because of the child's growing independence, increasing cognitive capacities, and rapidly developing self-concept, which includes perceptions of physical well-being (Harter, 1990). Yet in virtually every effort to teach health concepts to school-age children, researchers have found that health habits show little change.

Why are health practices hard to modify in middle childhood? There are several reasons. First, health is not an important goal to children, who feel good most of the time and are far more concerned about schoolwork, friends, and play. Second, children do not yet have an adultlike time perspective, which relates past, present, and future. Engaging in preventive behaviors is difficult when so much time intervenes between what children do now and its health consequences. Finally, much health information that children get is contradicted by other sources, such as television advertising and the examples of adults and peers.

Consequently, teaching school-age children health-related facts must be supplemented by other efforts. As we saw in earlier chapters, one effective way to foster children's health is to reduce hazards, such as pollution, inadequate medical care, and nonnutritious foods widely available in homes and school cafeterias. At the same time, since environments will never be totally free of health risks, parents and teachers need to model and reinforce good health practices as much as possible (Friedman, Greene, & Stokes, 1991).

This 10-year-old stops in the midst of her busy day to grab a hot dog. The school years are an especially important time for fostering a healthy life-style. Yet health practices are hard to modify in middle childhood, especially when healthy behaviors are contradicted by television advertising and the examples of adults and peers. (Clay/Monkmeyer Press)

SOCIAL ISSUES

CHILDREN'S UNDERSTANDING OF HEALTH AND ILLNESS

Lizzie lay on the living room sofa with a stuffy nose and sore throat, disappointed that she was missing her soccer team's final game and pizza party. "How'd I get this dumb cold anyhow?" she wondered aloud to Joey. "I probably did it to myself by playing outside without a hat when it was freezing cold."

"No, no," Joey contradicted. "You can't get sick that way. Some creepy little viruses got into your bloodstream and attacked, just like an army."

"What're viruses? I didn't eat any viruses," answered Lizzie, puzzled.

"You don't eat them, silly, you breathe them in. Somebody probably sneezed them all over you at school. That's how you got sick!"

Lizzie and Joey are at different developmental levels in their understanding of health and illness—ideas that are influenced by cognitive development, exposure to biological knowledge, and attitudes in their social world. Researchers have asked 4- to 14-year-olds in Western cultures questions designed to tap what they know about the causes of health and certain illnesses, such as colds, AIDS, and cancer.

During the preschool and early school years, children do not have much biological knowledge. For example, if you ask 4- to 8-year-olds to tell you what is inside their bodies, you will find that they know little about their internal organs and how they work. As a result, young children fall back on their rich knowledge of people's behavior to account for health and illness (Carey, 1985). Children of this age regard health as a matter of engaging in specific practices (eating the right foods, getting enough exercise, and wearing warm clothing on cold days) and illness as a matter of failing to follow these rules or coming too close to a sick person (Bibace & Walsh, 1980).

During middle childhood, children acquire more knowledge about their bodies and are cognitively better able to make sense of it. By age 9 or 10, they name a wide variety of internal organs and view them as interconnected, working as a system (Carey, 1985). Around this time, concepts of health and illness shift to biological explanations. Joey understands that illness can be caused by contagion—breathing in a harmful substance (a virus), which affects the operation of the body in some way.

By early adolescence, explanations become more elaborate and precise. Eleven- to 14-year-olds recognize health as a long-term condition that depends on the interaction of body, mind, and environmental influences (Hergenrather & Rabinowitz, 1991). And their notions of illness involve clearly stated ideas about interference in normal body processes: "You get a cold when your sinuses fill up with mucus. Sometimes your lungs do, too, and you get a cough. Colds come from viruses. They get into the bloodstream and make your platelet count go down" (Bibace & Walsh, 1980).

School-age children everywhere are capable of grasping basic biological ideas, but whether or not they do so depends on information in their everyday environments. Research reveals that children are likely to generalize their knowledge of familiar diseases to less familiar ones. As a result, they often conclude that risk factors for colds (sharing a Coke or sneezing on someone) can cause AIDS. And lacking much understanding of cancer, they assume that it (like colds and AIDS) is communicable (Sigelman et al., 1993). These incorrect ideas can lead to unnecessary anxiety about getting a serious disease as well as negative attitudes toward its victims.

The extent to which children comprehend illness from a biological rather than magical point of view can be undermined by social attitudes.

When certain diseases take on powerful symbolic meanings—for example, cancer as a malignant, destructive evil and AIDS as a sign of moral decay—even adults who have an accurate biological understanding irrationally expect bad things to happen from associating with affected people. These beliefs are quickly picked up by children, and they help explain the severe social rejection experienced by some youngsters with chronic diseases, such as AIDS and cancer (Pryor & Reeder, 1993; Whalen et al., 1995).

This child comforts her grandmother, who is dying of cancer. Helping school-age children understand that cancer is not communicable can prevent them from developing negative attitudes toward its victims. (Spencer Grant/The Picture Cube)

TRY THIS...

■ Ask a school-age child and an early adolescent about the causes of an ordinary, short-term illness (such as the common cold or flu) and a chronic disease (such as AIDS or cancer). Are differences in their responses consistent with recent research? What factors might account for those differences?

MOTOR DEVELOPMENT AND PLAY

Gains in body size and muscle strength support improved motor coordination during middle childhood. In addition, greater cognitive and social maturity permit older children to use their new motor skills in more complex ways. A major change in children's play takes place at this time.

GROSS MOTOR DEVELOPMENT

During middle childhood, running, jumping, hopping, and ball skills become more refined. Third to sixth graders burst into sprints as they race across the playground, jump quickly over rotating ropes, engage in intricate patterns of hopscotch, kick and dribble soccer balls, swing bats at balls pitched by their classmates, and balance adeptly as they walk toe-to-toe across narrow ledges. Table 9.1 summarizes gross motor achievements between ages 6 and 12.

These diverse skills reflect gains in flexibility, balance, agility (quickness and accuracy), and force of movement. More efficient information processing also plays an important role (Roberton, 1984). For example, *reaction time* improves steadily during middle childhood, with 14-year-olds responding almost twice as quickly to a stimulus as 6-year-olds (Southard, 1985). As a result, younger children often have difficulty with skills that require immediate responses, such as batting and dribbling. Since 6- and 7-year-olds are seldom successful at batting a thrown ball, T-ball is more appropriate for them than baseball. And handball, four-square, and kickball should precede instruction in tennis, basketball, and football (Thomas, 1984).

Sex differences in gross motor skills that appeared during the preschool years extend into middle childhood and, in some instances, become more pronounced. Girls continue to have an edge in skipping, jumping, and hopping, which depend on balance and agility. But boys outperform girls on all other skills in Table 9.1, especially in throwing and kicking (Cratty, 1986).

School-age boys' genetic advantage in muscle mass is not great enough to account for their superiority in so many gross motor skills. Instead, environment is believed to play a much larger role. In a study of over 800 elementary school pupils, children of both sexes viewed sports in a gender-stereotyped fashion—as much more important for boys. And boys more often stated that it was vital to their parents that they participate in athletics. These attitudes affected children's self-confidence and behavior. Girls saw themselves as having less talent at sports, and by sixth grade they devoted less time to athletics than did their male classmates (Eccles & Harold, 1991; Eccles, Jacobs, & Harold, 1990).

Special measures need to be taken to raise girls' confidence that they can do well at sports. Sensitizing parents to unfair biases against girls' athletic ability may prove helpful. In addition, greater emphasis on skill training for girls along with increased attention to their athletic achievements in schools and communities is likely to improve

TABLE 9.1

Changes in Gross Motor Skills During Middle Childhood

SKILL	DEVELOPMENTAL CHANGE
Running	Running speed increases from 12 feet per second at age 6 to over 18 feet per second at age 12.
Other gait variations	Skipping improves. Sideways stepping appears around age 6 and becomes more continuous and fluid with age.
Vertical jump	Height jumped increases from 4 inches at age 6 to 12 inches at age 12.
Standing broad jump	Distance jumped increases from 3 feet at age 6 to over 5 feet at age 12.
Precision jumping and hopping (on a mat divided into squares)	By age 7, children can accurately move from square to square, a performance that improves until age 9 and then levels off.
Throwing	Throwing speed, distance, and accuracy increase for both sexes, but much more for boys than girls. At age 6, a ball thrown by a boy travels 39 feet per second, one by a girl 29 feet per second. At age 12, a ball thrown by a boy travels 78 feet per second, one by a girl 56 feet per second.
Catching	Ability to catch small balls thrown over greater distances improves with age.
Kicking	Kicking speed and accuracy improve, with boys considerably ahead of girls. At age 6, a ball kicked by a boy travels 21 feet per second, one by a girl 13 feet per second. At age 12, a ball kicked by a boy travels 34 feet per second, one by a girl 26 feet per second.
Batting	Batting motions become more effective with age, increasing in speed and accuracy and involving the entire body.
Dribbling	Style of hand dribbling gradually changes, from awkward slapping of the ball to continuous, relaxed, even stroking.

Sources: Cratty, 1986; Malina & Bouchard, 1991; Roberton, 1984.

Ted's Picture with Label

Ted's Space Trip Story

The rocket is blasting off! They are going through 100 galaxies! People are climbing a ladder to see them. It has a special shield all around it. In case there is an attack on the rocket. Because you never know. They are on TV also.

Ted's War of 1812 Essay

There were several reasons that Congress declared war with Britain. First, the British attacked American ships to keep France from obtaining supplies. Second, Britain kidnapped American sailors. But Britain claimed that there were deserters, and Britain forced sailors to serve the British Navy. Third, settlers feared the British in Canada because they wanted to claim land in Canada. Fourth, all peaceful solutions failed. So that's what caused the War of 1812.

FIGURE 9.3

Fine motor coordination improves over middle childhood, as these writing samples reveal. At age 5, Ted printed in large uppercase letters, asking for help in spelling the words. By age 7, he had mastered the lowercase alphabet, and his printing was small and evenly spaced. At age 9, he used cursive writing. Notice, also, how letter reversals and invented spellings decline with age. *(Ted's picture with label and space trip story from L. M. McGee & D. J. Richgels, 1990,* Literacy's Beginnings: Supporting Young Readers and Writers, *Boston: Allyn and Bacon, p. 312. Reprinted by permission of the authors. Ted's War of 1812 essay from D. J. Richgels, L. M. McGee, & E. A. Slaton, 1989, "Teaching Expository Text Structure in Reading and Writing," in K. D. Muth, Ed.,* Children's Comprehension of Text, *Newark, DE: International Reading Association, p. 180. Reprinted by permission.)*

their performance. Middle childhood is a crucial time to take these steps, since school-age children are starting to discover what they are good at and to make some definite skill commitments.

FINE MOTOR DEVELOPMENT

Fine motor development also improves steadily over the school years. On rainy afternoons, Joey and Lizzie could be found experimenting with yo-yos, building model airplanes, and weaving potholders on small looms. Like many children, they took up musical instruments, which demand considerable fine motor control. Girls remain ahead of boys in the fine motor area, including handwriting and drawing.

By age 6, most children can print the alphabet, their first and last names, and the numbers from 1 to 10 with reasonable clarity. However, as Figure 9.3 reveals, their

FIGURE 9.4

Integration of depth cues into children's drawings increases dramatically over the school years. Compare this 10-year-old's drawing to the one by a 6-year old on page 219. Here, depth is indicated by overlapping objects, diagonal placement, and converging lines, as well as by making distant objects smaller than near ones. Although the young artist from Singapore who created this picture is highly talented, by the late elementary school years many children who attend school make good use of depth cues in their artwork. Experience with writing and instruction in artistic representation probably foster children's competence at depicting the third dimension. (Laura Berk, left; International Museum of Children's Art, right)

writing tends to be quite large because they use the entire arm to make strokes rather than just the wrist and fingers. Uppercase letters are usually mastered first because their horizontal and vertical motions are easier to control than the small curves of the lowercase alphabet. Legibility of writing gradually increases as children produce more accurate letters with uniform height and spacing. These improvements prepare children for mastering cursive writing by third grade.

Children's drawings show dramatic gains in organization, detail, and representation of depth. By the end of the preschool years, children can accurately copy many two-dimensional shapes, and they integrate these into their drawings. Some depth cues have also begun to appear, such as making distant objects smaller than near ones (Braine et al., 1993). Around 9 to 10 years, the third dimension is clearly evident in children's drawings through overlapping objects, diagonal placement, and converging lines (see Figure 9.4).

ORGANIZED GAMES WITH RULES

The physical activities of school-age children reflect an important advance in the quality of their play: Organized games with rules become common. In cultures around the world, the variety of children's spontaneous, rule-based games is enormous. Some are variants on popular sports, such as soccer, baseball, and basketball. Others are well-known childhood games like tag, jacks, and hopscotch.

Gains in perspective taking—in particular, children's ability to understand the roles of several players in a game—permit this transition to rule-oriented games.

Their contribution to development is great. Child-invented games usually involve simple physical skills and a sizable element of luck. As a result, they rarely become contests of individual ability. Instead, they permit children to try out different styles of competing, winning, and losing with little personal risk. Also, in their efforts to organize a game, children discover why rules are necessary and which ones work well. In fact, children often spend as much time working out the details of how a game should proceed as they do playing the game itself! As we will see in Chapter 10, these experiences help children form more mature concepts of fairness and justice.

Today, school-age youngsters spend less time gathering on sidewalks and playgrounds than they did in generations past. Television, video games, and adult-organized sports (such as Little League Baseball and city soccer and hockey leagues) fill many hours that children used to devote to spontaneous play. Some researchers worry that adult-structured athletics, which mirror professional sports, are robbing children of critical learning experiences and endangering their development. Perhaps you participated in one of these sports during your own childhood. If so, what were your experiences like? The Caregiving Concerns table on page 288 summarizes the pros and cons of adult-organized athletics and suggests ways to make them positive experiences in the lives of children.

PHYSICAL EDUCATION

Physical activity supports many aspects of children's development—the health of their bodies, their sense of self-worth as active and capable beings, and the cognitive

Is this coach careful to encourage rather than criticize? To what extent does he emphasize teamwork, fair play, courtesy, and skill development over winning? These factors determine whether or not adult-organized sports are pleasurable, constructive experiences for children. (Lawrence Migdale/Photo Researchers)

and social skills necessary for getting along with others. Yet only one-third of elementary school pupils have a daily physical education class. The average school-age child gets only 20 minutes of physical education a week (Steinhardt, 1992).

The lives of modern children are sedentary. Many ride to and from school in buses and cars, sit in classrooms most of the day, and watch TV for 3 to 4 hours after they return home. The National Children and Youth Fitness Study, which tested thousands of schoolchildren on a variety of fitness items (such as pull-ups, sit-ups, and one-mile run), found that only two-thirds of 10- to 12-year-old boys and half of 10- to 12-year-old girls met basic fitness standards for children their age (Looney & Plowman, 1990).

Besides offering more frequent physical education classes, many experts believe that schools need to change their physical education programs. Training in competitive sports is a high priority, but it is unlikely to reach the least physically fit youngsters. Instead, programs should emphasize informal games that most children can perform well and individual exercise—walking, running, jumping, tumbling, and climbing. These activities are most likely to last into later years and lead to an active and healthy lifestyle.

BRIEF REVIEW

Children's bodies grow slowly in middle childhood, at a pace similar to that of the preschool

CAREGIVING CONCERNS

Pros and Cons of Adult-Organized Sports in Middle Childhood

PROS	CONS	RECOMMENDATIONS FOR COACHES AND PARENTS
Adult-structured athletics prepares children for realistic competition—the kind they may face as adults.	Adult involvement leads games to become overly competitive, placing too much pressure on children.	Permit children to select from among appropriate activities the ones that suit them best. Do not push children into sports they do not enjoy.
Regularly scheduled games and practices ensure that children get plenty of exercise and fill free time that might otherwise be devoted to less constructive pursuits.	When adults control the game, children learn little about leadership, followership, and fair play. When adults assign children to specific roles (such as catcher, first base), children lose the opportunity to experiment with rules and strategies.	Before age 9, emphasize basic skills, such as kicking, throwing, and batting, and simplified games that grant all participants adequate playing time. Permit children to progress at their own pace and to play for the fun of it, whether or not they become expert athletes.
Children get instruction in physical skills necessary for future success in athletics. Parents and children share an activity that both enjoy.	Highly structured, competitive sports are less fun than child-organized games; they resemble "work" more than "play."	Adjust practice time to children's attention spans and need for time with peers, with family, and for homework. Two practices a week, each no longer than 30 minutes for younger school-age children and 60 minutes for older school-age children, are sufficient. Emphasize effort, skill gains, and teamwork. Avoid criticism for errors and defeat, which promote anxiety and avoidance of athletics.

Sources: Horn, 1987; Kolata, 1992.

Physical activity supports many aspects of children's development—the health of their bodies, their sense of self-worth as active and capable beings, and the cognitive and social skills necessary for getting along with others. These Chinese elementary school pupils play relay races during their physical education class. (Ann & Myran Sutton/ FPI International)

years. Gains in height occur in slight spurts followed by lulls; growth of the legs accounts for most of the increase. Between ages 6 and 12, the bones of the body lengthen and broaden, and all primary teeth are replaced by permanent ones.

Although many children are at their healthiest in middle childhood, a variety of health problems do occur. Most are more common among low-income youngsters, who are exposed to more health risks throughout development. Vision and hearing difficulties, malnutrition, overweight and obesity, nighttime bedwetting, respiratory illnesses, and unintentional injuries are among the most frequent health concerns during the school years. Genetic and environmental factors combine to increase children's vulnerability to certain health problems, such as nearsightedness, obesity, and asthma. Besides teaching health concepts, interventions for school-age children must provide healthier environments and directly promote good health practices.

In middle childhood, gains in size and strength combine with more efficient information processing to produce refinements in many gross motor skills. Sex differences present in early childhood persist into the school years, with boys receiving more encouragement for athletic accomplishment than girls. Fine motor development is evident in the legibility of children's writing and in the organization, detail, and representation of depth in their drawings. Organized games with rules become common in school-age children's play. Besides contributing to children's health, physical activity fosters a wide variety of cognitive and social skills.

ASK YOURSELF . . .

■ *How is body growth during the school years consistent with the cephalocaudal trend of development that you studied in Chapter 4?*

■ *Joey complained to his mother one evening that it wasn't fair that his younger sister Lizzie was almost as tall as he was. He worried that he wasn't growing fast enough. How should Rena respond to Joey's concern?*

■ *On Saturdays, 8-year-old Gina gathers with friends at a city park to play kickball. Besides improved ball skills, what is she learning?*

COGNITIVE DEVELOPMENT IN MIDDLE CHILDHOOD

Finally!" Lizzie exclaimed the day she entered first grade. "Now I get to go to real school just like Joey!" Rena remembered how 6-year-old Lizzie had walked confidently into her classroom, pencils, crayons, and writing pad in hand, ready for a more disciplined approach to learning than she had experienced in early childhood.

Lizzie entered a whole new world of challenging mental activities. In a single morning, she and her classmates wrote in journals, met in reading groups, worked on addition and subtraction, and sorted leaves gathered on the playground for a science project. As Lizzie and Joey moved through the elementary school grades, they tackled increasingly complex tasks and became more accomplished at reading, writing, math skills, and general knowledge of the world. Cognitive development had prepared them for this new phase.

PIAGET'S THEORY: THE CONCRETE OPERATIONAL STAGE

When Lizzie visited my child development class as a 4-year-old, she was easily confused by Piaget's conservation problems (see Chapter 7, page 223). For example, in conservation of liquid, she insisted that the amount of water had changed after it had been poured from a tall, narrow container into a short, wide one. At age 8, when Lizzie returned, these tasks were easy. "Of course it's the same," she exclaimed. "The water's shorter but it's also wider. Pour it back," she instructed the college student who was interviewing her. "You'll see, it's the same amount!"

Lizzie has entered Piaget's **concrete operational stage,** which spans the years from 7 to 11. During this period, thought is far more logical, flexible, and organized than it was during early childhood.

CONSERVATION

Piaget regarded *conservation* as the single most important achievement of the concrete operational stage. It provides clear evidence of *operations*—mental actions that obey logical rules. Notice how Lizzie coordinates several aspects of the task rather than *centering* on only one, as a preschooler would do. In other words, Lizzie is capable of **decentration.** She recognizes that a change in one aspect of the water (its height) is compensated for by a change

An improved ability to categorize underlies children's interest in collecting objects during middle childhood. These fourth graders can sort this shell collection into an elaborate structure of classes and subclasses. (Brian Smith)

in another aspect (its width). Lizzie also demonstrates **reversibility,** the capacity to think through a series of steps in a problem and then mentally reverse direction, returning to the starting point. Recall from Chapter 7 that reversibility is part of every logical operation. It is solidly achieved in middle childhood.

CLASSIFICATION

By the end of middle childhood, children pass Piaget's *class inclusion problem* (see page 000). This indicates that they can categorize more effectively (Hodges & French, 1988). You can see this in children's play activities. Collections—stamps, coins, baseball cards, rocks, bottle caps, and more—become common in middle childhood. At age 10, Joey spent hours sorting and resorting his large box of baseball cards. At times he grouped them by league and team, at other times by playing position and batting average. He could separate the players into a variety of classes and subclasses and flexibly move back and forth between them. This understanding is beyond preschoolers, who usually insist that objects can be sorted in only one way.

SERIATION

Seriation refers to the ability to order items along a quantitative dimension, such as length or weight. To test for it, Piaget asked children to arrange sticks of different lengths from shortest to longest. Older preschoolers can create the series, but they do so haphazardly. They put the sticks in a row but make many errors and take a long time to correct them. In contrast, 6- to 7-year-olds are guided by an orderly plan. They create the series efficiently by beginning with the smallest stick, then moving to the next smallest, and so on, until the ordering is complete.

The concrete operational child's improved grasp of quantitative arrangements is also evident in a more challenging seriation problem—one that requires children to seriate mentally. This ability is called **transitive inference.** In a well-known transitive inference problem, Piaget showed children pairings of differently colored sticks. From observing that Stick A is longer than Stick B and Stick B is longer than Stick C, children must make the mental inference that A is longer than C. Not until age 7 or 8 do children perform well on this task (Chapman & Lindenberger, 1988; Piaget, 1967).

SPATIAL REASONING

Piaget found that school-age youngsters have a more accurate understanding of space than they did earlier. For example, their comprehension of distance improves, as a special conservation task reveals. To give this problem, make two small trees out of modeling clay and place them apart on a table. Next, put a block or thick piece of cardboard between the trees. Then ask the child whether the trees are nearer together, farther apart, or still the same distance apart. Preschoolers say that the distance has become smaller. Four-year-olds can conserve distance when questioned about very familiar objects, but their understanding is not as solid and complete as that of the school-age child (Fabricius & Wellman, 1993; Miller & Baillargeon, 1990).

School-age children's more advanced understanding of space can also be seen in their ability to give directions. When asked to name an object to the left or right of another person, 5- and 6-year-olds answer incorrectly; they simply apply their own frame of reference. Between 7 and 8 years, children start to perform *mental rotations,* in which they align the self's frame to match that of a person in a different orientation. As a result, they can identify left and right for positions they do not occupy (Roberts & Aman, 1993). Around 8 to 10 years, children can give clear, well-organized directions for how to get from one place to another by using a "mental walk" strategy in which they imagine another person's movements along a route. Without special prompting, 6-year-olds focus on the end point and do not say exactly how to get there (Gauvain & Rogoff, 1989; Plumert et al., 1994).

LIMITATIONS OF CONCRETE OPERATIONAL THOUGHT

Clearly, school-age children are far more capable problem solvers than they were during the preschool years. But concrete operational thinking suffers from one important limitation. Children think in an organized, logical fashion only when dealing with concrete information that they can perceive directly. Their mental operations work poorly with abstract ideas—ones not apparent in the real world.

Children's solutions to transitive inference problems provide a good illustration. When shown pairs of sticks of unequal length, Lizzie easily figured out that if Stick A is longer than Stick B and Stick B is longer than Stick C, then Stick A is longer than Stick C. But she had great difficulty with a hypothetical version of this task, such as "Susan is taller than Sally and Sally is taller than Mary. Who is the tallest?" Not until age 11 or 12 can children solve this problem easily.

That logical thought is at first tied to immediate situations helps account for a special feature of concrete operational reasoning. Perhaps you noticed that school-age children master Piaget's concrete operational tasks step by step, not all at once. For example, they usually grasp conservation problems in a certain order: first number, followed by length, mass, and liquid (Brainerd, 1978). Piaget used the term **horizontal décalage** (meaning development within a stage) to describe this gradual mastery of logical concepts. The horizontal décalage is another indication of the concrete operational child's difficulty with abstractions. School-age youngsters do not come up with the gen-

eral principle of conservation and then apply it to all relevant situations. Rather, they seem to work out the logic of each problem separately.

RECENT RESEARCH ON CONCRETE OPERATIONAL THOUGHT

From researchers' attempts to verify Piaget's assumptions about concrete operations, two themes emerge. The first has to do with the impact of culture and schooling on children's thinking. The second has to do with how best to explain children's sequential mastery of logical problems during middle childhood.

■ THE IMPACT OF CULTURE AND SCHOOLING. According to Piaget, brain maturation combined with experience in a rich and varied external world should lead children everywhere to reach the concrete operational stage. Yet a large body of evidence reveals that in non-Western societies, conservation is often greatly delayed. For example, among the Hausa of Nigeria, who live in small agricultural settlements and rarely send their children to school, children do not understand even the most basic conservation tasks—number, length, and liquid—until age 11 or later (Fahrmeier, 1978).

These findings suggests that for children to master conservation and other Piagetian concepts, they must take part in everyday activities that promote this way of thinking (Light & Perrett-Clermont, 1989). Joey and Lizzie, for example, think of fairness in terms of equal distribution—a value emphasized in their culture. They frequently divide materials, such as crayons, Halloween treats, and lemonade, equally among themselves and their friends. Because they often see the same quantity arranged in different ways, they grasp conservation early. But in societies where equal sharing of goods is not common, conservation is unlikely to appear at the expected age.

The very experience of going to school is related to certain Piagetian attainments, such as transitive inference. This makes perfect sense, when we consider the many opportunities that schooling affords for seriating objects and learning about order relations (Artman & Cahan, 1993). On the basis of findings like these, some investigators have concluded that concrete operational reasoning is not a natural form of logic that emerges spontaneously in all children. Instead, it may be socially generated—an outcome of practical activities in particular cultures. Does this view remind you of Vygotsky's sociocultural theory, which we discussed in earlier chapters?

■ AN INFORMATION-PROCESSING VIEW OF SCHOOL-AGE CHILDREN'S THINKING. Piaget's notion of the horizontal décalage raises a familiar question about his theory: Is an abrupt stagewise transition to logical thought the best way to describe cognitive development in middle childhood? Recall from Chapter 7 that preschoolers show the beginnings of logical thinking on simplified and familiar tasks. The horizontal décalage suggests that logical understanding continues to improve over the school years.

Some theorists argue that the development of operational thinking can best be understood in terms of gains in information-processing capacity rather than a sudden shift to a new stage. For example, Robbie Case (1985, 1992) proposes that as children use cognitive schemes, they demand less attention and become more automatic. This frees up space in *working memory* (see page 153) so that children can focus on combining old schemes and generating new ones. For instance, the child confronted with water poured from one container to another recognizes that the height of the liquid changes. As this understanding becomes routine, the child notices that the width of the water changes as well. Soon the child coordinates these observations, and conservation of liquid is achieved. Then, as this logical idea becomes well practiced, the child transfers it to more demanding situations, such as area and weight. A similar explanation of the horizontal décalage has been suggested by Kurt Fischer (1980; Fischer & Farrar, 1987). He believes that school-age children eventually coordinate several task-specific skills into a general logical principle. At this point, thinking becomes highly efficient and abstract—the kind of change we will see when we discuss formal operational thought in Chapter 11.

EVALUATION OF THE CONCRETE OPERATIONAL STAGE

Piaget was indeed correct that school-age youngsters approach a great many problems in systematic and rational ways that were not possible just a few years before. But still at issue is whether this period is best viewed in terms of *continuous* improvement in logical skills or *discontinuous* restructuring of children's thinking as they enter a new stage. A growing number of researchers think that both types of change may be involved (Carey, 1985; Case, 1992; Fischer, 1980; Sternberg & Odagaki, 1989). From early to middle childhood, children apply logical schemes to a much wider range of tasks. Yet in the process, their thought seems to undergo qualitative change—toward a more comprehensive grasp of the underlying principles of logical thought. Piaget himself seems to have recognized this possibility in the very concept of the horizontal décalage. So perhaps some blend of Piagetian and information-processing ideas holds greatest promise for understanding cognitive development in middle childhood.

BRIEF REVIEW

During the concrete operational stage, thought is more logical, flexible, and organized than it was during the preschool years. The ability to conserve indicates that children can decenter and reverse their thinking. School-age children also have an improved grasp of classification, seriation, and spatial concepts, but they cannot yet think abstractly. Cross-cultural findings raise questions about Piaget's assumption that all children spontaneously master concrete operational tasks. In addition, the gradual development of operational reasoning challenges Piaget's notion of an abrupt stagewise transition to logical thought. A blend of Piagetian and information-processing views may be the best way to understand cognitive change in middle childhood.

ASK YOURSELF . . .

- *Nine-year-old Adrienne spends many hours helping her father build furniture in his woodworking shop. Explain how this experience may have contributed to Adrienne's advanced performance on Piagetian seriation problems.*

INFORMATION PROCESSING

In contrast to Piaget's focus on overall cognitive change, the information-processing perspective examines separate aspects of thinking. Attention and memory, which underlie every act of cognition, are central concerns in middle childhood, just as they were during infancy and the preschool years. Also, increased understanding of how school-age children process information is being applied to their academic learning—in particular, to reading and mathematics.

ATTENTION

During middle childhood, attention changes in three ways. It becomes more controlled, adaptable, and planful.

First, children become better at deliberately attending to just those aspects of a situation that are relevant to their goals, ignoring other information. Researchers study this increasing control of attention by introducing irrelevant stimuli into a task and seeing how well children attend to its central elements. Performance improves sharply between 6 and 9 years of age as children apply strategies for selecting information more effectively (Miller & Seier, 1994).

Second, older children flexibly adapt their attention to momentary requirements of situations. For example, when studying for a spelling test, 10-year-old Joey devoted most attention to words he knew least well. Lizzie was much less likely do so (Masur, McIntyre, & Flavell, 1973).

Finally, greater planfulness can be seen in school-age children's ability to scan detailed pictures and written materials for similarities and differences more thoroughly than do preschoolers. And on tasks with many parts, school-age children make decisions about what to do first and what to do next in an orderly fashion. In one study, 5- to 9-year-olds were given lists of 25 items to obtain from a play grocery store. Before starting on a shopping trip, older children more often took time to scan the store, and they also followed shorter routes through the aisles (Szepkouski, Gauvain, & Carberry, 1994).

Why does attention improve from early to middle childhood? Brain maturation and the demands of school tasks are probably jointly responsible. Unfortunately, some children have great difficulty focusing and sustaining attention, a problem that often remains with them into adulthood. See the Lifespan Vista box on pages 294–295 for a discussion of the serious learning and behavior problems of children with attention-deficit hyperactivity disorder.

MEMORY STRATEGIES

As attention improves with age, so do *memory strategies,* the deliberate mental activities we use to store and retain information. When Lizzie had a list of things to learn, such as a phone number, the state capitals of the United States, or the names of geometric shapes, she immediately used **rehearsal,** repeating the information to herself over and over again. This memory strategy first appears in the early grade school years. Soon after, a second strategy becomes common: **organization.** Children group related items together (for example, all state capitals in the same part of the country), an approach that improves recall dramatically (Bjorklund & Muir, 1988).

Memory strategies require time and effort to perfect. At first, school-age children do not use them very effectively. For example, 8-year-old Lizzie rehearsed in a piece-meal fashion. After being given the word *cat* in a list of items, she said, "Cat, cat, cat." In contrast, 10-year-old Joey combined previous words with each new item, saying, "Desk, man, yard, cat, cat" (Kunzinger, 1985; Ornstein, Naus, & Liberty, 1975). Joey also organized more skillfully, grouping items into fewer categories. In addition, he used organization in a wide range of memory tasks, whereas Lizzie used it only when categorical relations among items were obvious. Not surprisingly, Joey retained much more information (Bjorklund et al., 1994).

By the end of middle childhood, children start to use a third memory strategy, **elaboration.** It involves creating a relationship, or shared meaning, between two or more pieces of information that are not members of the same category. For example, suppose the words *fish* and *pipe* are among those you want to learn. In using elaboration, you might generate a mental image of a fish smoking a pipe or

recite a sentence expressing this relationship ("The fish puffed the pipe"). Once children discover this memory technique, they find it so effective that it tends to replace other strategies. Elaboration develops late because it requires a great deal of mental effort and considerable working memory capacity. It becomes increasingly common during adolescence and young adulthood (Schneider & Pressley, 1989).

Because the strategies of organization and elaboration combine items into *meaningful chunks,* they permit children to hold on to much more information at once. As a result, these strategies contribute to expansion of working memory. In addition, when children link a new item to information they already know, they can *retrieve* it easily by thinking of other items associated with it. As we will see in the next section, this is one reason that memory improves steadily during the school years.

THE KNOWLEDGE BASE AND MEMORY PERFORMANCE

During middle childhood, the long-term knowledge base grows larger and becomes organized into elaborate, hierarchically structured networks. Many researchers believe this rapid growth of knowledge helps children use strategies and remember. In other words, knowing more about a particular topic makes new information more meaningful and familiar so it is easier to store and retrieve (Bjorklund & Muir, 1988).

To test this idea, Michelene Chi (1978) looked at how well third- through eighth-grade chess experts could remember complex chessboard arrangements. The children recalled the configurations considerably better than adults, who knew how to play chess but were not especially knowledgeable. Yet the adults did better than the children at recalling lists of numbers on a standard memory test. The children showed superior memory performance only in the domain of knowledge in which they were expert.

Although knowledge clearly plays an important role in memory development, it must be quite broad and well structured before it can facilitate strategy use and recall (DeMarie-Dreblow, 1991). Furthermore, knowledge is not the only important factor in children's strategic memory processing. Children who are expert in an area, whether it be chess, math, social studies, or spelling, are usually highly motivated. As a result, they not only acquire knowledge more quickly, but they *actively use what they know* to add more. Academically successful and unsuccessful children differ in just this way. Poor students fail to approach memory tasks by asking how previously stored information can clarify new material. This, in turn, interferes with the development of a broad knowledge base (Brown et al., 1983). So at least by the end of the school years, knowledge acquisition and use of memory strategies are intimately related and support one another.

ATTENTION-DEFICIT HYPERACTIVITY DISORDER

While the other fifth graders worked quietly at their desks, Calvin squirmed in his seat, dropped his pencil, looked out the window, fiddled with his shoelaces, and talked out. "Hey, Joey," he yelled across the room, "wanna play ball after school?" Joey didn't answer. He wasn't eager to play with Calvin. Out on the playground, Calvin was a poor listener and failed to follow the rules of the game. When up at bat, he had difficulty taking turns. In the outfield, he tossed his mitt up in the air and looked elsewhere when the ball came his way. Calvin's desk at school and his room at home were a chaotic mess. He often lost pencils, books, and other materials necessary for completing assignments.

Calvin is one of 3 to 5 percent of school-age children with **attention-deficit hyperactivity disorder,** or **ADHD** (American Psychiatric Association, 1994). Although boys are diagnosed five to ten times more often than girls, recent evidence suggests that just as many girls may suffer from the disorder. Girls are less likely to be identified because their symptoms are usually not as flagrant (Hynd et al. , 1991).

Children with ADHD have great difficulty staying on task for more than a few minutes. They often act impulsively, ignoring social rules and lashing out with hostility when frustrated. Many (but not all) are *hyperactive.* They charge through their days with excessive motor activity, leaving parents and teachers frazzled and other children annoyed. These youngsters have few friends and are soundly rejected by their classmates.

The intelligence of children with ADHD is normal, and they show no signs of serious emotional disturbance. According to one view, their diverse symptoms are due to an impaired ability to postpone action in favor of thought (Barkley, 1997). Consequently, they do poorly on tasks requiring sustained attention, and they find it hard to ignore irrelevant information.

Heredity plays a major role in ADHD, since the disorder runs in families, and identical twins share it more often than do fraternal twins (Zametkin, 1995). At the same time, these children are somewhat more likely to come from homes in which marriages are unhappy and family stress is high (Bernier & Siegel, 1994). But researchers agree that a stressful home life rarely causes ADHD. Instead, the behaviors of these children can contribute to family problems, which (in turn) may intensify the child's preexisting difficulties.

Calvin's doctor eventually prescribed stimulant medication, the most common treatment for ADHD. As long as dosage is carefully regulated, these drugs reduce activity level and improve attention, academic performance, and peer relations for 70 percent of children who take them (Barkley, DuPaul, & Costello, 1993). Researchers do not know precisely why stimulants are helpful. Some speculate that they change the chemical balance in brain regions that inhibit impulsivity and hyperactivity, thereby decreasing the child's need to engage in off-task and self-stimulating behavior.

Although stimulant medication is relatively safe, its impact is short

CULTURE AND MEMORY STRATEGIES

Think, for a moment, about the kinds of situations in which the strategies of rehearsal, organization, and elaboration are useful. People usually employ these techniques when they need to remember information for its own sake. On many other occasions, they participate in daily activities that produce excellent memory as a natural by-product of the activity itself.

 A repeated finding of cross-cultural research is that people in non-Western cultures who have no formal schooling do not use or benefit from instruction in memory strategies (Rogoff, 1990). Tasks that require children to recall isolated bits of information are common in classrooms, and they provide children with a great deal of motivation to come up with ways to memorize. In fact, Western children get so much practice with this type of learning that they do not refine other techniques for remembering that rely on spatial location and arrangement of objects, cues that are readily available in everyday life. Australian Aboriginal and Guatemalan Mayan children are considerably better at these memory skills (Kearins, 1981; Rogoff, 1986). Looked at in this way, the development of memory strategies is not just a matter of a more competent information-processing system. It is also a product of task demands and cultural circumstances.

THE SCHOOL-AGE CHILD'S THEORY OF MIND

During middle childhood, children's *theory of mind,* or beliefs about mental activities, becomes more elaborate and refined. Recall from Chapter 7 that this awareness of cognitive processes is often referred to as *metacognition.*

term. Drugs cannot teach children ways of compensating for inattention and impulsivity (Whalen & Henker, 1991). Combining medication with interventions that model and reinforce appropriate academic and social behavior seems to be the most effective approach to treatment (Barkley, 1994). Family intervention is also important. Inattentive, overactive children strain the patience of parents, who are likely to react punitively and inconsistently—a child-rearing style that strengthens inappropriate behavior. Breaking this cycle is as important for ADHD children as it is for the defiant, aggressive youngsters we discussed in Chapter 8. In fact, 35 percent of the time, these two sets of behavior problems occur together (Nottelmann & Jensen, 1995).

Although some children with ADHD outgrow their difficulties, about half continue to have problems concentrating and finding friends in adolescence. Furthermore, low self-esteem, impaired social skills, antisocial behavior, difficulties with intimate relationships, and poor job performance are present in about 30 percent of young adults who were diagnosed with ADHD in childhood (Denckla, 1991). Severity of symp-

toms, presence of other psychological disorders, and family stress increase the likelihood that ADHD will persist.

Because ADHD can be a lifelong disorder, it often requires long-term therapy. Adults with ADHD need help structuring their environment, regulating negative emotion, listening to others, planning for an appropriate career, and understanding their condition as a biological deficit rather

than a character flaw. At present, findings on the effectiveness of stimulant medication for adults with ADHD are inconclusive. But a caring and involved parent, spouse, friend, or counselor does make a difference. When adults with ADHD were asked what factors had helped them most, they mentioned people who had been particularly supportive (Weiss & Hechtman, 1993).

While his classmates try to work, this boy constantly looks up from his assignment and fires paper airplanes across the room. Children with ADHD have great difficulty staying on task and often act impulsively, ignoring social rules. (David Young-Wolff/PhotoEdit)

School-age children's improved ability to reflect on their own mental life is another reason for the advances in thinking and problem solving that take place at this time.

Unlike preschoolers, who view the mind as a passive container, older children regard it as an active, constructive agent capable of selecting and transforming information (Wellman & Hickling, 1994). Consequently, they have a much better understanding of the impact of psychological factors on cognitive performance. They know, for example, that doing well on a task depends on focusing attention—concentrating on it, wanting to do it, and not being tempted by anything else (Miller & Bigi, 1979; Sodian & Wimmer, 1987). They are also aware that in studying material for later recall, it is helpful to devote most effort to items that you know least well. And when asked, elementary school pupils show that they know quite a bit about effective memory strategies. Witness this 8-year-old's

response to the question of what she would do to remember a phone number:

> Say the number is 663-8854. Then what I'd do is—say that my number is 663, so I won't have to remember that, really. And then I would think now I've got to remember 88. Now I'm 8 years old, so I can remember, say, my age two times. Then I say how old my brother is, and how old he was last year. And that's how I'd usually remember that phone number. [Is that how you would most often remember a phone number?] Well, usually I write it down. (Kreutzer, Leonard, & Flavell, 1975, p. 11)

This child clearly understands the importance of establishing connections between new information and existing knowledge. And she also recognizes that she

School-age children have an improved ability to reflect on their own mental life. This child is aware that external aids to memory are often necessary to ensure that information will be retained. (Will Faller)

can use external aids to enhance memory—in this case, writing down the phone number.

Once children are aware of the many factors that influence mental activity, they combine these into an integrated understanding. School-age children take account of *interactions* among variables—how age and motivation of the learner, effective use of strategies, and nature and difficulty of the task work together to affect cognitive performance (Wellman, 1990). In this way, metacognition truly becomes a comprehensive theory in middle childhood.

SELF-REGULATION

Although metacognition expands, school-age youngsters often have difficulty putting what they know about thinking into action. They are not yet good at **self-regulation,** the process of continuously monitoring progress toward a goal, checking outcomes, and redirecting unsuccessful efforts. For example, Lizzie is aware that she should group items together in a memory task and read a complicated paragraph more than once to make sure she understands it. But she does not always do these things when working on an assignment (Beal, 1990; Brown et al., 1983).

Self-regulation does not become well developed until adolescence. By then, it is a strong predictor of academic success. Students who do well in school know when they possess a skill and when they do not. If they run up against obstacles, such as poor study conditions, a confusing text passage, or a class presentation that is unclear, they take steps to organize the learning environment, review the material, or seek other sources of support. This active, purposeful approach contrasts sharply with the passive orien-

tation of students who do poorly (Borkowski et al., 1990; Zimmerman, 1990).

Parents and teachers can foster children's self-regulation by pointing out the special demands of tasks, indicating how use of strategies will improve performance, and emphasizing the value of self-correction. Many studies show that providing children with instructions to check and monitor their progress toward a goal has a substantial impact on how well they do (Pressley, 1995).

Children who acquire effective self-regulatory skills develop confidence in their own ability—a belief that supports the use of self-regulation in the future (Paris & Newman, 1990; Schunk, 1990). Unfortunately, some children receive messages from parents and teachers that seriously undermine their academic self-esteem and self-regulatory skills. We will consider the special problems of these *learned-helpless* youngsters, along with ways to help them, in Chapter 10.

APPLICATIONS OF INFORMATION PROCESSING TO ACADEMIC LEARNING

Over the past decade, fundamental discoveries about the development of information processing have been applied to children's learning of reading and mathematics. Researchers have begun to identify the cognitive ingredients of high-level performance and to pinpoint ways in which poor learners are deficient. They hope, as a result, to design teaching methods that will help school-age children master these essential skills.

■ READING. While reading, we use a large number of skills at once, taxing all aspects of our information-processing systems. Joey and Lizzie must perceive single letters and letter combinations, translate them into speech sounds, hold chunks of text in working memory while interpreting their meaning, and combine the meanings of various parts of a text passage into an understandable whole. In fact, reading is such a demanding process that most or all of these skills must be done automatically. If one or more are poorly developed, they will compete for space in our limited working memories, and reading performance will decline (Perfetti, 1988).

Researchers do not yet know how children manage to acquire and combine all these varied skills into fluent reading. Currently, psychologists and educators are engaged in a "great debate" about how to teach beginning reading. On one side are those who take a **whole language approach** to reading instruction. They argue that reading should be taught in a way that parallels natural language learning. From the very beginning, children should be exposed to text in its complete form—stories, poems, letters, posters, and lists—so they can appreciate the communicative function of written language. According to these experts, as long as reading is kept whole and meaningful, children will be motivated to discover the specific skills they need as they gain experience with the printed word (Goodman, 1986; Watson, 1989). On the other side of the debate are those who advocate a **basic skills approach.** According to this view, children should be given simplified text materials. At first, they should be coached on *phonics*—the basic rules for translating written symbols into sounds. Only later, after they have mastered these skills, should they get complex reading material (Rayner & Pollatsek, 1989).

As yet, research does not show clearcut superiority for either of these approaches (Stahl, McKenna, & Pagnucco, 1994). In fact, a third group of experts believes that children may learn best when they receive a balanced mixture of both (Biemiller, 1994; Pressley, 1994; Stahl, 1992). Learning the basics—relationships between letters and sounds—enables children to decipher words they have never seen before. As this process becomes more automatic, it releases children's attention to the higher-level activities involved in comprehending the text's meaning. But if practice in basic skills is overemphasized, children may lose sight of the goal of reading—understanding. Many teachers report cases of pupils who can read aloud fluently but who register little or no meaning. These children might have been spared serious reading problems if they had been exposed to meaning-based instruction that included attention to basic skills.

■ MATHEMATICS. Once children enter elementary school, they apply their rich informal knowledge of number concepts and counting to more complex mathematical skills (Geary, 1994). For example, children first understand multiplication as a kind of repeated addition. When given the following problem, "Sue has 5 books. Joe has 3 times as many. How many books does Joe have?" Lizzie thought to herself, "What's 5 x 3?" When she had difficulty remembering the answer, she said, "Okay, it's got to be 5 books + 5 books + 5 books. I know, it's 15!" Lizzie's use of addition strengthened her understanding of multiplication. It also helped her recall a multiplication fact that she had been trying to memorize.

Mathematics as taught in many classrooms, however, does not make good use of children's basic grasp of number concepts. Children are given procedures for solving problems without linking these to their informally acquired understandings. Consequently, they often apply a rule that is close to what they have been taught but that yields a wrong answer. Their mistakes indicate that they have tried to memorize a method, but they do not comprehend the basis for it (Geary, 1994).

Arguments about how to teach early mathematics closely resemble those in the area of reading. Drill in computational skills is pitted against "number sense" or understanding. Yet once again, a blend of the two is probably

Culture and language–based factors contribute to Asian children's skill at mathematics. The abacus supports these Japanese pupils' understanding of place value. Ones, tens, hundreds, and thousands are each represented by a different column of beads, and calculations are performed by moving the beads to different positions. As children become skilled at using the abacus, they learn to think in ways that facilitate solving complex arithmetic problems. (Fuji Fotos/The Image Works)

best. Research indicates that conceptual knowledge serves as a vital base for the development of accurate, efficient computation in middle childhood (Byrnes & Wasik, 1991).

Cross-cultural evidence suggests that in Asian countries, pupils receive supports for learning mathematics that are not broadly available in the United States. For example, use of the metric system helps Asian children think in ways that foster a grasp of place value. The consistent structure of number words in Asian languages ("ten two" for twelve, "ten three" for thirteen) also makes this idea clear (Fuson & Kwon, 1992). Furthermore, Asian number words are shorter and more quickly pronounced. Therefore, more digits can be held in working memory at once, increasing the speed of arithmetic computation (Geary et al., 1993, 1996). Finally, as we will see later in this chapter, in Asian classrooms, much more time is spent exploring underlying math concepts and much less on drill and repetition.

■ **BRIEF REVIEW**

Over the school years, attention becomes more controlled, adaptable, and planful, and memory strategies become more effective. An expanding knowledge base contributes to improved memory. However, children's motivation to use what they know in learning new information is also important. Metacognition moves from a passive to an active view of mental functioning. School-age youngsters have difficulty with self-regulation, but they can be taught to improve their self-regulatory skills. Information processing has been applied to children's academic learning in school. Instruction that attends to both basic skills and understanding seems to be most effective in reading and mathematics.

ASK YOURSELF . . .

■ *One day, the children in Lizzie and Joey's school saw a slide show about endangered species. They were told to remember as many animal names as they could. Fifth and sixth graders recalled considerably more than second and third graders. What factors might account for this difference?*

■ *Lizzie knows that if you have difficulty learning part of a task, you should devote most of your attention to that aspect. But she plays each of her piano pieces from beginning to end instead of picking out the hard parts for extra practice. What explains Lizzie's failure to apply what she knows?*

INDIVIDUAL DIFFERENCES IN MENTAL DEVELOPMENT

During middle childhood, intelligence tests become increasingly important in assessing individual differences in mental development. Around age 6, IQ becomes more stable than it was at earlier ages, and it correlates well with academic achievement. Because IQ predicts school performance, it plays an important role in educational decisions. Do intelligence tests provide an accurate indication of the school-age child's ability to profit from instruction? In the following sections, we take a close look at this controversial issue.

▪ DEFINING AND MEASURING INTELLIGENCE

Virtually all intelligence tests provide an overall score (the IQ), which is taken to represent *general intelligence* or reasoning ability. Yet a diverse array of tasks appear on most tests for children. Today, there is widespread agreement that intelligence is a collection of many mental capacities, not all of which are included on currently available tests.

Test designers use a complicated statistical technique called *factor analysis* to identify the various abilities assessed by intelligence tests. This procedure determines which sets of items on the test correlate strongly with one another. Those that do are assumed to measure a similar ability and, therefore, are designated as a separate factor. To understand the types of intellectual factors measured in middle childhood, let's look at some representative intelligence tests and how they are administered.

■ **SOME REPRESENTATIVE INTELLIGENCE TESTS.** The intelligence tests that Joey and Lizzie take every so often in school are *group administered tests*. They permit large numbers of pupils to be tested at once and require very little training of teachers who give them. Group tests are useful for instructional planning and identifying children who require more extensive evaluation with *individually administered tests*. Unlike group tests, individually administered ones demand considerable training and experience to give well. The examiner not only considers the child's answers, but also carefully observes the child's behavior, noting such things as attentiveness to and interest in the tasks and wariness of the adult. These reactions provide insight into whether the test score is accurate or underestimates the child's abilities.

Two individual tests—the Stanford-Binet and the Wechsler—are often used to identify highly intelligent children and diagnose those with learning problems. As we look at them, refer to Figure 9.5, which illustrates items that typically appear on intelligence tests for children.

The Stanford-Binet Intelligence Scale. The modern descendent of Alfred Binet's first successful intelligence test is the **Stanford-Binet Intelligence Scale,** for individuals between 2 years of age and adulthood. Its latest version measures general intelligence and four intellectual factors: verbal reasoning, quantitative reasoning, spatial reasoning, and short-term memory (Thorndike, Hagen, & Sattler, 1986). Within these factors are 15 subtests that permit a detailed analysis of each individual's mental abilities. The verbal and quantitative factors emphasize culturally loaded, fact-oriented information, such as vocabulary and sentence comprehension. In contrast, the spatial reasoning factor is believed to be less culturally biased because it does not demand specific information. Instead, it tests children's ability to see complex relationships, as in the spatial visualization item shown in Figure 9.5.

Like many current tests, the Stanford-Binet is designed to be sensitive to diversity among children and to reduce gender bias. Pictures of children from different ethnic groups, a child in a wheelchair, and "unisex" figures that can be interpreted as male or female are included.

The Wechsler Intelligence Scale for Children–III. The **Wechsler Intelligence Scale for Children–III (WISC–III)** is the third edition of a widely used test for 6- through 16-

Item Type	Typical Verbal Items
Vocabulary	Tell me what "carpet" means.
General Information	How many ounces make a pound? What day of the week comes right after Thursday?
Verbal Comprehension	Why are police officers needed?
Verbal Analogies	A rock is hard; a pillow is _____ .
Logical Reasoning	Five girls are sitting side by side on a bench. Jane is in the middle and Betty sits next to her on the right. Alice is beside Betty, and Dale is beside Ellen, who sits next to Jane. Who are sitting on the ends?
Number Series	Which number comes next in the series? **4 8 6 12 10 ___**

Typical Nonverbal Items

Picture Oddities	Which picture does not belong with the others?

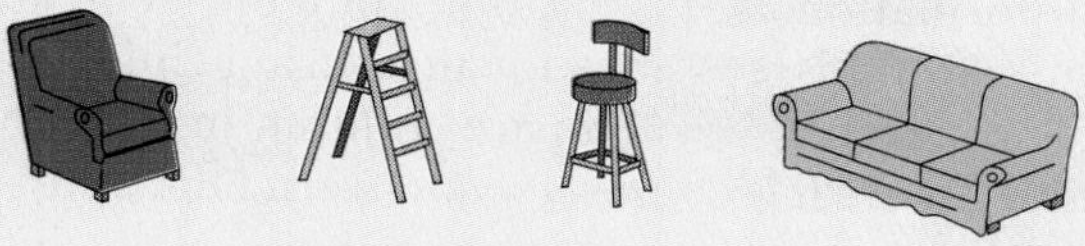

Spatial Visualization	Which of the boxes on the right can be made from the pattern shown on the left?

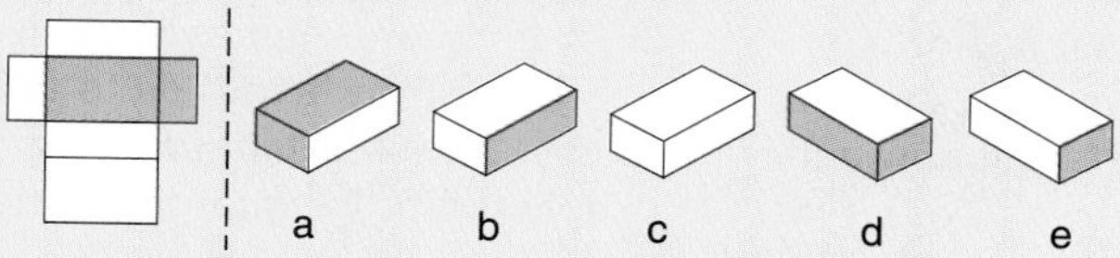

Typical Performance Items

Picture Series	Put the pictures in the right order so that what is happening makes sense.

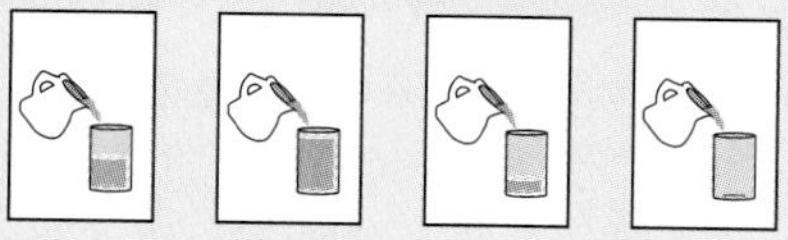

Puzzles	Put these pieces together so they make a wagon.

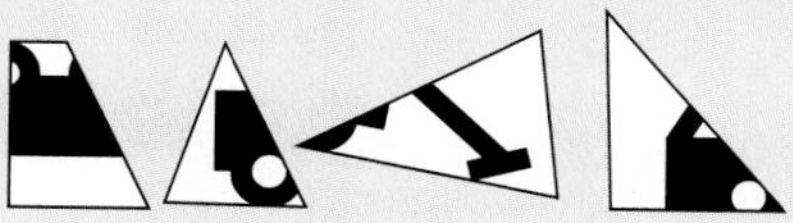

F IGURE 9.5

Test items like those on common intelligence tests for children. In contrast to verbal items, nonverbal items do not require reading or direct use of language. Performance items are also nonverbal, but they require the child to draw or construct something rather than merely give a correct answer. As a result, they appear only on individually administered intelligence tests. (Logical reasoning, picture oddities, and spatial visualization examples are adapted with permission of *The Free Press, a Division of Macmillan, Inc., from A. R. Jensen, 1980, Bias in Mental Testing, New York: Free Press, pp. 150, 154, 157.)*

year-olds. A downward extension of it—the *Wechsler Preschool and Primary Scale of Intelligence–Revised (WPPSI–R)*—is appropriate for children 3 through 8 (Wechsler, 1989, 1991). The Wechsler tests offered both a measure of general intelligence and a variety of factor scores long before the Stanford-Binet. As a result, over the past two decades, psychologists and educators have come to prefer the WISC and WPPSI.

Both the WISC–III and the WPPSI–R assess two broad intellectual factors: verbal and performance. Each contains 6 subtests, yielding 12 separate scores in all. Performance items (see examples in Figure 9.5) require the child to arrange materials rather than talk to the examiner. Consequently, these tests provided one of the first means through which non-English-speaking children and children with speech and language disorders could demonstrate their intellectual strengths. The Wechsler tests were also the first to be standardized on samples representing the total population of the United States, including ethnic minorities.

RECENT DEVELOPMENTS IN DEFINING INTELLIGENCE

In recent years, some researchers have started to combine the factor analytic approach to defining intelligence with information processing. They believe that factors on intelligence tests are of limited use unless we can identify the cognitive processes responsible for them. Once we understand the underlying basis of IQ, we will know much more about why a particular child does well or poorly and what capacities must be worked on to improve performance. These researchers conduct *componential analyses* of children's IQ scores. This means that they look for relationships between aspects (or components) of information processing and intelligence test scores. Preliminary findings reveal that the speed with which individuals perceive and manipulate information and the effectiveness with which they apply strategies to remember and solve problems are related to IQ (Deary, 1995; Geary & Burlingham-Dubree, 1989; Vernon, 1993).

The componential approach has one major shortcoming: It regards intelligence as entirely due to causes within the child. Yet throughout this book, we have seen how cultural and situational factors affect children's cognitive skills. Robert Sternberg has expanded the componential approach into a comprehensive theory that regards intelligence as a product of both inner and outer forces.

■ STERNBERG'S TRIARCHIC THEORY OF INTELLIGENCE. As Figure 9.6 shows, Sternberg's (1985, 1988b) **triarchic theory of intelligence** is made up of three interacting subtheories. The first, the *componential subtheory,* spells out the information-processing skills that underlie intelligent behavior. You are already familiar with its main elements—strategy application, knowledge acquisition, metacognition, and self-regulation.

According to Sternberg, children's use of these components is not just a matter of internal capacity. It is also a function of the conditions under which intelligence is assessed. The *experiential subtheory* states that highly intelligent individuals, compared to less intelligent ones, process information more skillfully in novel situations. When given a new task, the bright person learns rapidly, making strategies automatic so working memory is freed for more complex aspects of the situation.

Think, for a moment, about the implications of this idea for measuring children's intelligence. To accurately compare children in brightness—in ability to deal with novelty and learn efficiently—all children would need to be presented with equally unfamiliar test items. Otherwise some children will appear more intelligent than others simply become of their past experiences, not because they are really more cognitively skilled.

This point brings us to the third part of Sternberg's model, the *contextual subtheory.* It proposes that intelligent people skillfully *adapt* their information-processing skills

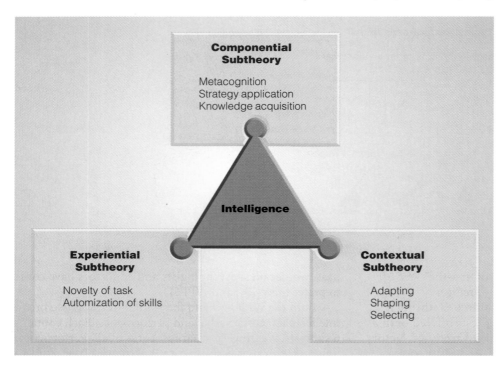

FIGURE 9.6

Sternberg's triarchic theory of intelligence.

to fit with their personal desires and their everyday worlds. When they cannot adapt to a situation, they try to *shape,* or change, it to meet their needs. If they cannot shape it, they *select* new contexts that are consistent with their goals. The contextual subtheory emphasizes that intelligent behavior is never culture-free. Because of their backgrounds, some children come to value behaviors required for success on intelligence tests, and they easily adapt to the tasks and testing conditions. Others with different life histories misinterpret the testing context or reject it entirely because it does not suit their needs.

Sternberg's theory emphasizes the complexity of intelligent behavior. As you can already see, his ideas are relevant to the controversy surrounding cultural bias in intelligence testing, which we will address shortly.

■ GARDNER'S THEORY OF MULTIPLE INTELLIGENCES. Howard Gardner's (1983) **theory of multiple intelligences** provides yet another view of how information-processing skills underlie intelligent behavior. But unlike the componential approach, it does not begin with existing mental tests and try to isolate the processing elements required to succeed on them. Instead, Gardner believes that intelligence should be defined in terms of distinct sets of processing operations that permit individuals to engage in a wide range of culturally valued activities. Therefore, Gardner dismisses the idea of general intelligence and proposes seven independent intelligences (see Table 9.2).

Gardner argues that each intelligence has a unique biological basis, a distinct course of development, and different expert, or "end-state," performances. At the same time, he emphasizes that a lengthy process of education is required to transform any raw potential into a mature

social role. This means that cultural values and learning opportunities have a great deal to do with the extent to which a child's intellectual strengths are realized and how they are expressed.

Gardner's list of abilities has yet to be firmly grounded in research. For example, biological evidence for the independence of his abilities is weak. Furthermore, current mental tests do tap several of Gardner's intelligences (linguistic, logico-mathematical, and spatial), and research suggests that they have at least some common features. Nevertheless, Gardner's theory has been especially helpful in efforts to understand and nurture children's special talents, a topic we will discuss at the end of this chapter.

EXPLAINING INDIVIDUAL AND GROUP DIFFERENCES IN IQ

When we compare individuals in terms of academic achievement, years of education, and the status of their occupations, it quickly becomes clear that certain sectors of the population are advantaged over others. In trying to explain these differences, researchers have examined the intelligence test performance of children from different ethnic and social-class backgrounds. Many studies show that American black children score, on the average, 15 IQ points below American white children (Brody, 1992). Social class differences in IQ also exist. The gap between middle-income and low-income children is about 9 points (Jensen & Figueroa, 1975). These figures are, of course, averages. There is considerable variation *within* each ethnic and social-class group. Still, group differences in IQ are large enough and of serious enough consequence that they cannot be ignored.

TABLE 9.2

Gardner's Multiple Intelligences

INTELLIGENCE	PROCESSING OPERATIONS POSSIBILITIES	END-STATE PERFORMANCE
Linguistic	Sensitivity to the sounds, rhythms, and meanings of words and the different functions of language	Poet, journalist
Logico-mathematical	Sensitivity to, and capacity to detect, logical or numerical patterns; ability to handle long chains of logical reasoning	Mathematician, scientist
Musical	Ability to produce and appreciate pitch, rhythm (or melody), and aesthetic-sounding tones; understanding of the forms of musical expressiveness	Violinist, composer
Spatial	Ability to perceive the visual–spatial world accurately, to perform transformations on those perceptions, and to re-create aspects of visual experience in the absence of relevant stimuli	Sculptor, navigator
Bodily–kinesthetic	Ability to use the body skillfully for expressive as well as goal-directed purposes; ability to handle objects skillfully	Dancer, athlete
Interpersonal	Ability to detect and respond appropriately to the moods, temperaments, motivations, and intentions of others	Therapist, salesperson
Intrapersonal	Ability to discriminate complex inner feelings and to use them to guide one's own behavior; knowledge of one's own strengths, weaknesses, desires, and intelligences	Person with detailed, accurate self-knowledge

Sources: Gardner, 1983.

In 1969, psychologist Arthur Jensen published a controversial article entitled, "How Much Can We Boost IQ and Scholastic Achievement?" Jensen's answer to this question was "not much." He argued that heredity is largely responsible for individual, social-class, and ethnic variations in intelligence, a position he continues to maintain (Jensen, 1980, 1985). Jensen's work was followed by an outpouring of responses and research studies, leading to a heated nature–nurture debate on the origins of IQ. Recently, the controversy was rekindled in Richard Herrnstein and Charles Murray's (1994) book, *The Bell Curve*. Like Jensen, these authors concluded that the contribution of heredity to individual and social-class differences in IQ is substantial. At the same time, they stated that the relative role of heredity and environment in the black–white IQ gap remains unresolved. Let's look closely at some important evidence.

■ NATURE VERSUS NURTURE. In Chapter 2, we introduced the *heritability estimate*. Recall that heritabilities are obtained from *kinship studies,* which compare family members. The most powerful evidence on the role of heredity in IQ involves twin comparisons. The IQ scores of identical twins (who share all their genes) are more similar than those of fraternal twins (who are genetically no more alike than ordinary siblings). On the basis of this and other kinship evidence, researchers estimate that about half the differences in IQ among children can be traced to their genetic makeup (Plomin 1994a). However, in Chapter 2 we noted that heritabilities risk overestimating genetic influences and underestimating the importance of environment. Although research offers convincing evidence that genes contribute to IQ, disagreement persists over just how large the role of heredity really is (Ceci, 1990).

Furthermore, a widespread misconception exists that if a characteristic is heritable, then the environment can do little to affect it. A special type of kinship study, involving adopted children, shows that this assumption is incorrect. In one investigation of this kind, children of two extreme groups of biological mothers, those with IQs below 95 and those with IQs above 120, were studied. All the children were adopted at birth by parents who were well above average in income and education. During the school years, children of the low-IQ biological mothers scored above average in IQ, indicating that test performance can be greatly improved by an advantaged home life! At the same time, they did not do as well as children of high-IQ biological mothers placed in similar adoptive families (Horn, 1983). Adoption research confirms the balanced position that both heredity and environment affect IQ scores.

Some intriguing adoption evidence sheds light on the black–white IQ gap. Black children placed into well-to-do white homes early in life also score high on intelligence tests. In two such studies, they attained mean IQs of 110 and 117 by middle childhood, well above average and 20 to 30 points higher than the typical scores of children growing up in low-income black communities (Moore, 1986; Scarr & Weinberg, 1983). A follow-up of one sample revealed that although the test scores of black adoptees declined by adolescence, they remained above the IQ average for low-income African Americans (Weinberg, Scarr, & Waldman, 1992).

Adoption findings do not completely resolve questions about ethnic differences in IQ. Nevertheless, the IQ gains of adopted black children "reared in the culture of the tests and schools" are consistent with a wealth of evidence indicating that poverty severely depresses the intelligence of large numbers of ethnic minority youngsters (Scarr & Weinberg, 1983, p. 261). And in many other cases, unique cultural values and practices do not prepare these children for the intellectual tasks sampled by intelligence tests and valued in school.

■ CULTURAL INFLUENCES. Jermaine, an African-American child in Lizzie's third-grade class, participated actively in class discussion and wrote complex, imaginative stories. Two years earlier as a first grader, Jermaine had responded, "I don't know," to the simplest of questions, including "What's your name?" Fortunately, Jermaine's teacher understood his uneasiness. Slowly and gently, she helped him build a bridge between the learning style fostered by his cultural background and the style necessary for academic success. A growing body of evidence reveals that IQ scores are affected by specific learning experiences, including exposure to certain language customs and knowledge.

Language Customs. Ethnic minority subcultures often foster unique language skills that do not fit the expectations of most classrooms and testing situations. Shirley Brice Heath (1982, 1989), an anthropologist who spent many hours observing in low-income black homes in a southeastern American city, found that adults asked black children very different kinds of questions than is typical in white middle-class families. From an early age, white parents ask knowledge-training questions, such as "What color is it?" and "What's this story about?" that resemble the questioning style of tests and classrooms. In contrast, the black parents asked only "real" questions—ones that they themselves did not know the answer to. Often these were analogy questions ("What's that like?") or story-starter questions ("Didja hear Miss Sally this morning?") that called for elaborate responses about whole events and had no single right answer. The black children developed complex verbal skills at home, such as storytelling and exchanging quick-witted remarks. Unfortunately, these worked poorly in school. The children were confused by questions in classrooms and often withdrew into silence.

When faced with the strangeness of the testing situation, the minority child may look to the examiner for cues about how to respond. Yet most intelligence tests permit

When black children grow up "in the culture of the tests and schools," they perform well above the population mean in IQ. Transracial adoption research supports the view that rearing environment is responsible for the black–white IQ gap. (Bob Daemmrich/Stock Boston)

tasks to be presented in only one way, and they allow no feedback to children. Consequently, minority children may simply give the first answer that comes to mind, not one that truly represents what they know, as in this example:

Tester: "How are wood and coal alike? How are they the same?"

Child: "They're hard."

Tester: "An apple and a peach?"

Child: "They taste good."

Tester: "A ship and an automobile?"

Child: "They're hard."

Tester: "Iron and silver?"

Child: "They're hard." (Miller-Jones, 1989, p. 362)

Earlier in the testing session, this child asked whether she was doing all right but got no reply. She probably repeated her first answer because she had trouble figuring out the task's meaning, not because she was unable to classify objects. Had the tester prompted her to look at the questions in a different way, she might have done better.

Familiarity with Test Content. Many researchers argue that IQ scores are affected by specific information acquired as part of middle-class upbringing. Unfortunately, attempts to change tests by eliminating fact-oriented verbal tasks and relying only on spatial reasoning and performance items (believed to be less culturally loaded) have not raised the scores of low-income minority children very much (Kaplan, 1985).

Nevertheless, even these nonverbal test items depend on learning opportunities. In one study, children's performance on a spatial reasoning task was related to their experience playing a popular but expensive game that (like the test items) required them to arrange blocks to duplicate a design as quickly as possible (Dirks, 1982). Low-income minority children, who often grow up in more "people oriented" than "object oriented" homes, may lack experience with games and objects that promote certain intellectual skills. In line with this possibility, when ethnically diverse parents were asked to describe their view of an intelligent first grader, Anglo-Americans emphasized cognitive traits. In contrast, ethnic minorities (Cambodian, Filipino, Vietnamese, and Mexican immigrants) saw noncognitive characteristics— motivation, self-management, and social skills—as equally or more important (Okagaki & Sternberg, 1993).

That specific experiences affect performance on intelligence tests is also supported by evidence indicating that the amount of time a child spends in school is a strong predictor of IQ. In comparisons of children who are the same age but in different grades, those who have been in school longer score higher on intelligence tests (Ceci, 1990, 1991). Taken together, these findings indicate that teaching children the factual knowledge and ways of thinking valued in classrooms has a sizable impact on their intelligence test performance.

■ REDUCING CULTURAL BIAS IN INTELLIGENCE TESTS. Although not all experts agree, today there is greater acknowledgment that IQ scores can underestimate the intelligence of culturally different children. A special concern exists about incorrectly labeling minority children as slow learners and assigning them to remedial classes, which are far less stimulating than regular school experiences. Because of this danger, test scores need to be combined with assessments of adaptive behavior—children's ability to cope with the demands of their everyday environments (Landesman & Ramey, 1989). The child who does poorly on an IQ test yet plays a complex game on the playground or figures out how to rewire a broken TV is unlikely to be mentally deficient.

In addition, culturally relevant testing procedures enhance minority children's test performance. In one approach, called *dynamic assessment,* the adult introduces purposeful teaching into the testing situation to see what the child can attain with social support. Many minority children perform more competently after adult assistance. And the approach helps identify forms of intervention likely to help children who are learning poorly in classrooms (Lidz, 1991).

In sum, intelligence tests are useful when interpreted carefully by examiners who are sensitive to cultural influences on test performance. And despite their limitations, IQ scores continue to be fairly accurate measures of school learning potential for the majority of Western children.

BRIEF REVIEW

Intelligence tests for children measure overall IQ as well as a variety of separate intellectual factors. The Stanford-Binet Intelligence Scale and the Wechsler Intelligence Scale for Children–III (WISC–III) are commonly used tests in middle childhood. Sternberg's triarchic theory of intelligence states that information-processing skills, prior experience with the tasks, and contextual factors (the child's cultural background and interpretation of the testing situation) interact to determine IQ. According to Gardner's theory of multiple intelligences, seven distinct abilities, each defined by unique processing operations, represent the diversity of human intelligence.

Heritability and adoption research shows that both genetic and environmental factors contribute to individual differences in intelligence. Because of different language customs and unfamiliar test content, the IQ scores of low-income minority children often do not reflect their true abilities.

ASK YOURSELF . . .

■ *Desiree, a low-income African-American child, was quiet and withdrawn while taking an intelligence test. Later she remarked to her mother, "I can't understand why that lady asked me all those questions, like what a ball and stove are for. She's a grownup. She* must *know what a ball and stove are for!" Using Sternberg's triarchic theory, explain Desiree's reaction to the testing situation. Why is Desiree's score likely to underestimate her intelligence?*

LANGUAGE DEVELOPMENT

Vocabulary, grammar, and pragmatics continue to develop in middle childhood, although the changes are less obvious than those of earlier ages. In addition, school-age children's attitude toward language undergoes a fundamental shift. They develop *language awareness.*

VOCABULARY

The average 6-year-old's vocabulary is already quite large (about 10,000 words). By the end of the school years, recognition vocabulary reaches about 40,000 words, a rate of growth that exceeds that of early childhood. School-age children enlarge their vocabularies by analyzing the structure of complex words. From "happy" and "decide," they quickly derive the meanings of "happiness" and "decision"

(Anglin, 1993). In addition, many new words are picked up from context, especially while reading (Miller, 1991).

As their conceptual knowledge becomes better organized, school-age children can think about and use words more precisely. Word definitions offer examples of this change. Five- and 6-year-olds give very concrete descriptions that refer to functions or appearance—for example, knife: "when you're cutting carrots"; bicycle: "it's got wheels, a chain, and handlebars." By the end of elementary school, synonyms and explanations of categorical relationships appear—for example, knife: "something you could cut with. A saw is like a knife. It could also be a weapon" (Wehren, DeLisi, & Arnold, 1981). This advance reflects the older child's ability to deal with word meanings on an entirely verbal plane. Fifth and sixth graders no longer need to be shown what a word refers to in order to understand it. They can add new words to their vocabulary simply by being given a definition (Dickinson, 1984).

School-age children's more reflective and analytical approach to language permits them to appreciate the multiple meanings of words. For example, they recognize that a great many words, such as "cool" or "neat," have psychological as well as physical meanings: "What a *cool* shirt!" or "That movie was really *neat!*" This grasp of double meanings permits 8- to 10-year-olds to comprehend subtle metaphors, such as "sharp as a tack" and "spilling the beans" (Wellman & Hickling, 1994; Winner, 1988). It also leads to a change in children's humor. In middle childhood, riddles and puns requiring children to go back and forth between different meanings of the same key word are common.

GRAMMAR

Although children have mastered most of the grammar of their language by the time they enter school, use of complex grammatical constructions improves. For example, during middle childhood, children use the passive voice more frequently, and it expands from an abbreviated structure ("It broke") into full statements ("The glass was broken by Mary") (Horgan, 1978; Lempert, 1989; Pinker, Lebeaux, & Frost, 1987).

Another grammatical achievement of middle childhood is the understanding of infinitive phrases, such as the difference between "John is eager to please" and "John is easy to please" (Chomsky, 1969). Like gains in vocabulary, appreciation of these subtle grammatical distinctions is supported by an improved ability to analyze and reflect on language.

PRAGMATICS

Improvements in *pragmatics,* the communicative side of language, take place in middle childhood. School-age children become better at adapting to the needs of listeners in challenging communicative situations. In one study, 3-

to 10-year-olds were asked which of eight similar objects they liked best as a birthday present for an imaginary friend. Preschoolers gave ambiguous descriptions, such as "the red one." In contrast, school-age youngsters were much more precise—for example, "the round red one with stripes on it" (Deutsch & Pechmann, 1982).

Conversational strategies also become more refined. For example, older children are better at phrasing things to get their way. When faced with an adult who refuses to hand over a desired object, 9-year-olds, but not 5-year-olds, state their second requests more politely (Axia & Baroni, 1985). School-age children are also sensitive to subtle distinctions between what people say and what they really mean. Lizzie, for example, knew that when her mother said, "The garbage is beginning to smell," she really meant, "Take that garbage out!" (Ackerman, 1978).

LEARNING TWO LANGUAGES AT A TIME

Like most American children, Joey and Lizzie speak only one language, their native tongue of English. Yet throughout the world, a great many children grow up *bilingual.* They learn two languages, and sometimes more than two, during childhood. An estimated 6 million American school-age children speak a language other than English at home (U. S. Bureau of the Census, 1996).

■ BILINGUAL DEVELOPMENT. Children can become bilingual in two ways: (1) by acquiring both languages at the same time in early childhood, or (2) by learning a second language after mastering the first. Children of bilingual parents who teach them both languages in early childhood show no special problems with language development. They acquire normal native ability in the language of their surrounding community and good to native ability in the second language, depending on their exposure to it. When children acquire a second language after they already speak a first language, it generally takes them about a year to become as fluent as native-speaking agemates (Reich, 1986).

Research shows that bilingualism has a positive impact on cognitive development. Children who are fluent in two languages do better than others on tests of analytical reasoning, concept formation, and cognitive flexibility (Hakuta, Ferdman, & Diaz, 1987). In addition, bilingual children are advanced in ability to reflect on language. They are more aware that words are arbitrary symbols, more conscious of language structure and detail, and better at noticing errors of grammar and meaning in spoken and written prose (Galambos & Goldin-Meadow, 1990; Ricciardelli, 1992).

■ BILINGUAL EDUCATION. The advantages of bilingualism provide strong justification for bilingual education programs in American schools. Yet the question of how ethnic minority children with limited English proficiency should be educated is hotly debated. On one side of the controversy are those who believe that time spent communicating in the child's native tongue detracts from English language achievement, which is crucial for success in the worlds of school and work. On the other side are educators committed to truly *bilingual* education: developing minority children's native language while fostering mastery of English. Providing instruction in the native tongue lets children know that their heritage is respected (McGroarty, 1992). In addition, it prevents *semilingualism,* or inadequate proficiency in both languages. When minority children gradually lose their first language as a result of being

When bilingual education programs provide instruction in both the child's native language and in English, children are more involved in learning and are advanced in language development. (Will Faller)

taught the second, they end up limited in both languages for a period of time, a circumstance that leads to serious academic difficulties (August & Garcia, 1988).

At present, public opinion sides with the first of these two viewpoints. Many states have passed laws declaring English to be their official language, creating conditions in which schools have no obligation to teach minority pupils in languages other than English. Yet in classrooms where both languages are integrated into the curriculum, minority children are more involved in learning, participate more actively in class discussions, and acquire the second language more easily. In contrast, when teachers speak only in a language their pupils can barely understand, children display frustration, boredom, and withdrawal (Cazden, 1984).

Bilingualism offers one of the best examples of how language, once learned, becomes an important tool of the mind and fosters cognitive growth. From this perspective, the goals of schooling could reasonably be broadened to include helping all children become bilingual, thereby fostering the cognitive, language, and cultural enrichment of the entire nation.

BRIEF REVIEW

During the school years, vocabulary increases rapidly, and children develop a more precise and flexible understanding of word meanings. Mastery of complex grammatical constructions becomes more refined. School-age children express themselves well in challenging communicative situations, and they acquire more subtle conversational strategies. Bilingualism has a positive impact on cognitive development and language awareness.

ASK YOURSELF . . .

■ *Ten-year-old Shana arrived home from school after a long day, sank into the living room sofa, and commented, "I'm totally wiped out!" Megan, her 5-year-old sister, looked puzzled and asked, "What did'ya wipe out, Shana?" Explain Shana and Megan's different understanding of the meaning of this expression.*

CHILDREN'S LEARNING IN SCHOOL

Throughout this chapter, we have touched on evidence indicating that schools are vital forces in children's cognitive development. How do schools exert such a powerful impact? Research looking at schools as complex social systems—their educational philosophies,

teacher–pupil interaction patterns, and the larger cultural context in which they are embedded—provides important insights into this question.

THE EDUCATIONAL PHILOSOPHY

Each teacher brings to the classroom an educational philosophy that plays a major role in children's learning experiences. Two philosophical approaches have been studied in American education. They differ in what children are taught, the way they are believed to learn, and how their progress is evaluated.

■ TRADITIONAL VERSUS OPEN CLASSROOMS. In a **traditional classroom,** children are relatively passive in the learning process. The teacher is the sole authority and does most of the talking. Pupils spend most of their time at their desks—listening, responding when called on, and completing assigned tasks. Their progress is evaluated by how well they keep pace with a uniform set of standards for all pupils in their grade.

In an **open classroom,** children are viewed as active agents in their own development. The teacher assumes a flexible authority role, sharing decision making with pupils, who learn at their own pace. Pupils are evaluated in relation to their own prior development; comparisons with same-age pupils are less important. A glance inside an open classroom reveals richly equipped learning centers, small groups of pupils working on tasks they choose themselves, and a teacher who moves from one area to another, guiding and supporting in response to children's individual needs.

During the past few decades, the pendulum in American education has swung back and forth between these two views. In the 1960s and early 1970s, open education gained in popularity, inspired by Piaget's vision of the child as an active, motivated learner. Then, as high school students' scores on the Scholastic Aptitude Test (SAT) declined over the 1970s, a "back to basics" movement arose. Classrooms returned to traditional, teacher-directed instruction, which remains the dominant approach today.

The combined results of many studies reveal that children in traditional classrooms have a slight edge in terms of academic achievement. At the same time, open settings are associated with other benefits. Open-classroom pupils are more independent, and they value and respect individual differences in their classmates more. Pupils in open environments also like school better than those in traditional classrooms, where lack of pupil autonomy contributes to a general decline in motivation throughout the school years (Eccles et al., 1993; Skinner & Belmont, 1993).

■ NEW PHILOSOPHICAL DIRECTIONS. The philosophies of some teachers fall somewhere in between traditional and open. They want to foster high achievement as well as independence, positive social relationships, and excitement about learning. New experiments in elementary education, grounded in Vygotsky's sociocultural theory, represent this intermediate point of view (Forman, Minick, & Stone, 1993). One that has received widespread attention is the Kamehameha Elementary Education Program (KEEP), applied in public schools in Hawaii and Los Angeles and on a Navajo reservation in Arizona. Vygotsky's concept of the *zone of proximal development*—a range of challenging tasks that the child is ready to master with the help of a more skilled partner (see Chapter 5)—serves as the foundation for KEEP's theory of instruction.

KEEP classrooms are rich in activity settings specially designed to enhance teacher–child and child–child dialogues. In each setting, children work on a project that ensures that their learning will be active and directed toward a meaningful goal. For example, they might read a story and discuss its meaning or draw a map of the playground to promote an understanding of geography. Sometimes activity settings include the whole class. More often, they involve small groups that foster cooperative learning and permit teachers to stay in touch with how well each child is doing. The precise organization of each KEEP classroom is adjusted to fit the unique learning styles of its pupils, creating culturally responsive environments (Tharp, 1993; Tharp & Gallimore, 1988).

So far, research suggests that the KEEP approach is highly effective in promoting learning among children who typically achieve poorly. In KEEP schools, minority pupils performed at their expected grade level in reading, much better than children of the same background enrolled in traditional schools. KEEP pupils also participated actively in class discussion, used elaborate language structures, frequently supported one another's learning, and were more attentive and involved than were non-KEEP controls. As the KEEP model is more widely applied, perhaps it will prove successful with all types of children because of its comprehensive goals and effort to meet the learning needs of a wide range of pupils.

TEACHER–PUPIL INTERACTION

In all classrooms, teachers vary in the way they interact with children—differences that are consistently related to academic achievement. Lizzie's third-grade teacher, for example, organized the learning environment so that activities ran smoothly and transitions were brief and orderly. Teachers who are effective classroom managers have pupils who spend more time learning, and this is reflected in higher achievement test scores (Good & Brophy, 1994).

Although Lizzie's teacher was well organized and efficient, she usually emphasized factual knowledge. Down the hall, Joey's teacher encouraged the children to grapple with ideas and apply their knowledge to new situations. He asked, "Why is the main character in this story a hero?" and "Now that you are good at division, how many teams should we have at recess? How many children on each team?" In an observational study of fifth-grade social studies and math lessons, students were far more attentive when teachers encouraged higher-level thinking rather than limiting instruction to simple memory exercises (Stodolsky, 1988). The Caregiving Concerns table on the following page suggests ways to encourage children to think critically—to consider the deeper meaning of problems by reflecting, evaluating, and expressing their ideas in clear and useful ways.

Teachers do not interact in the same way with all children. Some children get more attention and praise than others. Well-behaved, high-achieving pupils experience positive interactions with their teachers. In contrast, teachers especially dislike children who achieve poorly and are also disruptive. These unruly pupils are often criticized and are rarely called on in class discussions. When they seek special help or permission, their requests are usually denied (Good & Brophy, 1994).

Unfortunately, once teachers' attitudes toward pupils are established, they are in danger of becoming more extreme than is warranted by children's behavior. A special concern is that **educational self-fulfilling prophecies** can be set in motion: Children may adopt teachers' positive or negative views and start to live up to them. Many studies show that school-age pupils become increasingly aware of teacher opinion, and it can influence their performance (Harris & Rosenthal, 1985; Skinner & Belmont, 1993). This effect is especially strong in classrooms where teachers emphasize competition and frequently make public comparisons among children (Weinstein et al., 1987).

In many schools, pupils are assigned to groups or classes in which children of similar achievement levels are taught together. Teachers' treatment of different ability groups can be an especially powerful source of self-fulfilling prophecies. In "low" groups, pupils get more drill on basic facts and skills, a slower learning pace, and less time on academic work. Gradually, these children lose self-esteem and are viewed by themselves and others as "not smart." Not surprisingly, ability grouping widens the gap between high and low achievers (Dornbusch, Glasgow, & Lin, 1996).

TEACHING CHILDREN WITH SPECIAL NEEDS

We have seen that effective teachers flexibly adjust their teaching strategies to accommodate pupils with a wide range of abilities and characteristics. But such adjustments are increasingly difficult to make at the very low and high ends of the ability distribution. How do schools serve children with special learning needs?

CAREGIVING CONCERNS

Encouraging Critical Thinking in School-Age Children

SUGGESTION	DESCRIPTION
Promote open-ended questioning and interaction.	Engaging in open-ended questioning ("why" and "how") and discussing answers to those questions encourage children to be adventurous and inquiring and to prompt each other's thinking.
Ask children to clarify and draw inferences.	Asking "What is the main point?", "What does it mean?", or "What might happen next?" encourages children to be active, precise thinkers and to consider the implications of their ideas.
Help children view situations from multiple perspectives.	Once children give their own opinion on an issue, ask them to state and try to support opposing perspectives. Seeing things from other viewpoints makes children more aware of their own thinking patterns and encourages them to be open-minded and flexible.
Provide opportunities for children to use what they know to solve problems.	Effective problem solving involves stating a plausible hypothesis, identifying what evidence would support it and what would refute it, and evaluating the hypothesis against evidence. As early as age 6, children have some understanding of the relation between hypotheses and evidence. Many opportunities to set aside personal beliefs in favor of conclusions that follow from evidence promote effective problem solving.
Model critical thinking.	When children observe adults engaged in critical thinking, they are more likely to adopt it as a valued activity for themselves.
Use a precise "language of thinking."	When adults use a specific "thinking vocabulary" that includes such terms as as *hypothesis, reasons, conclusions, evidence,* and *opinion,* their verbal cues evoke a richer variety of thought processes in children.
Explain thinking possibilities.	Explaining thinking possibilities makes children conscious of the ingredients of critical thinking and, therefore, more likely to apply this knowledge. For example, an adult might discuss with children what the word *hypothesis* means and how a hypothesis in science differs from one in history or everyday life.
Give feedback that supports critical thinking.	Positive feedback and suggestions for how to do better rather than criticism establish a safe haven for critical thinking that permits children to take risks.

Sources: Byrnes, 1993; Ruffman et al., 1993; Tishman, Perkins, & Jay, 1995.

■ MAINSTREAMING CHILDREN WITH LEARNING DIFFICULTIES. The Individuals with Disabilities Education Act (Public Law 101–475) mandates that schools place children requiring special supports for learning in the "least restrictive" environments that meet their educational needs. The law led to a rapid increase in **mainstreaming** of many pupils who otherwise would have been served in special education classes. Instead, they were integrated into regular classrooms for part or all of the school day, a practice designed to better prepare them for participation in society.

Characteristics of Mainstreamed Children. Some mainstreamed pupils are **mildly mentally retarded**—children whose IQs fall between 55 and 70 and who also show problems in adaptive behavior (social and self-help skills in everyday life). Typically, a mildly mentally retarded child can be educated to the level of an average sixth grader. In adulthood, most live independently and hold routine jobs, although they require extra guidance and support during times of stress (American Psychiatric Association, 1994).

The largest number of mainstreamed children have **learning disabilities.** About 5 to 10 percent of school-age children are affected. These youngsters obtain average or above-average IQ scores. But they have a specific learning disorder (for example, in reading, writing, or math computation) that results in poor school achievement. Their problems cannot be traced to any obvious physical or emotional difficulty or environmental disadvantage. Instead, faulty brain functioning is believed to underlie their difficulties (Hammill, 1990). Some of the disorders run in families, suggesting that they are at least partly genetic (Pennington & Smith, 1988). In most instances, the cause is unknown.

The learning problems of these children are so frustrating that they can lead to serious emotional, social, and family difficulties (Silver, 1989). Yet those who receive appropriate educational intervention have a good chance of making a satisfactory adjustment in adulthood. Although the disability usually persists, these individuals find ways to compensate for it. Often they select college majors and careers that do not rely heavily on the skill in which they are deficient. The majority of learning-disabled adults manage to equal or exceed the average educational and occupational attainment of the general population (Horn, O'Donnell, & Vitulano, 1983).

How Effective Is Mainstreaming? Does mainstreaming succeed in providing more appropriate academic experiences as well as integrated participation in classroom life?

The pupil on the left, who has a learning disability, has been mainstreamed into a regular classroom. Because his teacher takes special steps to encourage peer acceptance, individualizes instruction, minimizes comparisons with classmates, and promotes cooperative learning, this boy looks forward to school and is doing well. (Will Hart)

At present, research findings are not positive on either of these points. Achievement differences between mainstreamed pupils and those taught in self-contained classrooms are not great (MacMillan, Keogh, & Jones, 1986). Furthermore, mainstreamed children are often rejected by peers. Retarded pupils are overwhelmed by the social skills of their classmates; they cannot interact quickly or adeptly in a conversation or game. And the processing deficits of some learning-disabled children lead to problems in social awareness and responsiveness (Rourke, 1988; Taylor, Asher, & Williams, 1987).

Does this mean that mainstreaming is not a good way to serve children with special learning needs? This extreme conclusion is not warranted. Often these children do best when they receive instruction in a *resource room* for part of the day and in the regular classroom for the remainder. In the resource room, a special education teacher works with pupils on an individual and small-group basis. Then, depending on their abilities, children are mainstreamed for different subjects and amounts of time. This flexible approach makes it more likely that the unique academic needs of each child will be served (Keogh, 1988; Lerner, 1989).

Once children enter the regular classroom, special steps must to be taken to promote their acceptance by normal peers. When instruction is carefully individualized and teachers minimize comparisons with higher-achieving classmates, mainstreamed pupils show gains in self-esteem and achievement (Madden & Slavin, 1983). Also, cooperative learning experiences in which a mainstreamed child and several normal peers work together on the same task have been found to promote friendly interaction and social acceptance (Scruggs & Mastropieri, 1994).

■ GIFTED CHILDREN. In Joey and Lizzie's school, some children were **gifted.** They displayed exceptional intellectual strengths. Like mainstreamed pupils, their characteristics were diverse. In every grade were one or two pupils

with IQ scores above 130, the standard definition of giftedness based on intelligence test performance (Horowitz & O'Brien, 1986). High-IQ children, as we have seen, are particularly quick at academic work. They have keen memories and an exceptional capacity to analyze a challenging problem and efficiently move toward a correct solution. Yet earlier in this chapter, we noted that intelligence tests do not sample the entire range of human mental skills. Recognition of this fact has led to an expanded conception of giftedness in schools.

Creativity and Talent. High *creativity* can result in a child being designated as gifted. Tests of creativity tap **divergent thinking**—the generation of multiple and unusual possibilities when faced with a task or problem. Divergent thinking contrasts sharply with **convergent thinking,** which involves arriving at a single correct answer and is emphasized on intelligence tests (Guilford, 1985).

Recognizing that highly creative children (like high-IQ children) are often better at some tasks than others, researchers have devised verbal, figural, and "real-world-problem" tests of divergent thinking (Runco, 1992; Torrance, 1988). A verbal measure might ask children to name as many uses for common objects (such as a newspaper) as they can. A figural measure might ask them to come up with as many drawings based on a circular motif as possible (see Figure 9.7 on page 310). A "real-world problem" measure either gives children everyday problems or requires them to think of such problems and then suggest solutions. Responses to all these tests can be scored for the number of ideas generated as well as for their originality.

Yet critics of these measures point out that at best, they are imperfect predictors of creative accomplishment in everyday life, since they tap only one of the complex, cognitive contributions to creativity (Cramond, 1994). Also involved are the ability to define a new and important problem, to evaluate divergent ideas and choose the most promising, and to call on relevant knowledge to understand and solve the problem (Sternberg & Lubart, 1995).

FIGURE 9.7

Responses of a highly creative 8-year-old to a figural measure of creativity. This child was asked to make as many pictures as she could from the circles on the page. The titles she gave her drawings, from left to right, are as follows: "dracula," "one-eyed monster," "pumpkin," "hula-hoop," "poster," "wheelchair," "earth," "moon," "planet," "movie camera," "sad face," "picture," "stoplight," "beach ball," "the letter O," "car," "glasses." *(Test form copyright © 1980 by Scholastic Testing Service, Inc. Reprinted by permission of Scholastic Testing Service, Inc., from* The Torrance Tests of Creative Thinking *by E. P. Torrance.)*

Consider these additional ingredients, and you will see why both children and adults usually demonstrate creativity in only one or a few related areas. Partly for this reason, definitions of giftedness have been extended to include *specialized talent.* There is clear evidence that outstanding performances in particular fields, such as mathematics, science, music, art, athletics, and leadership, have roots in specialized skills that first appear in childhood.

At the same time, talent must be nurtured in a favorable environment. Studies of the backgrounds of talented children and highly accomplished adults often reveal homes rich in stimulating activities and parents who emphasize curiosity (Albert, 1994; Perleth & Heller, 1994). In addition, such parents recognize their child's creative potential and (as the talent develops) arrange for apprenticeship under inspiring teachers (Bloom, 1985; Feldman, 1991). Classrooms in which children can take risks, challenge the instructor, and reflect on ideas without being rushed to the next assignment are also important (Sternberg & Lubart, 1991).

Educating the Gifted. Although programs for the gifted exist in many schools, debate about their effectiveness usually focuses on factors irrelevant to giftedness—whether to offer enrichment in regular classrooms, to pull children out for special instruction (the most common practice), or to advance brighter pupils to a higher grade. Children of all ages fare well academically and socially within each of these models (Moon & Feldhusen, 1994;

Southern, Jones, & Stanley, 1994). Yet the extent to which they foster creativity and talent depends on opportunities to acquire relevant skills.

Recently, Gardner's theory of multiple intelligences has inspired several model programs that include all pupils. A wide variety of meaningful activities, each tapping a specific intelligence or set of intelligences, serve as contexts for assessing strengths and weaknesses and, on that basis, teaching new knowledge and original thinking. For example, linguistic intelligence might be fostered through storytelling or playwriting, spatial intelligence through drawing, sculpting, or taking apart and reassembling objects (Gardner, 1993).

Evidence is still needed on how effectively these programs nurture children's talents. But so far they have succeeded in one way—by highlighting the strengths of some pupils who previously had been considered ordinary or even at risk for school failure. Consequently, they may be especially useful in identifying talented minority children, who are underrepresented in school programs for the gifted (Frazier, 1994).

HOW WELL EDUCATED ARE AMERICA'S CHILDREN?

Our discussion of schooling has largely focused on what teachers can do in classrooms to support the education of children. Yet a great many factors, both within and

CULTURAL INFLUENCES

EDUCATION IN JAPAN, TAIWAN, AND THE UNITED STATES

Why do Asian children perform so well academically? Research into societal, school, and family conditions in Japan, Taiwan, and the United States provides some answers:

CULTURAL VALUING OF ACADEMIC ACHIEVEMENT. In Japan and Taiwan, natural resources are limited. Progress in science and technology is essential for economic well-being. Because a well-educated work force is necessary to meet this goal, children's mastery of academic skills is vital. In the United States, attitudes toward academic achievement are far less unified. Many Americans believe that it is more important for children to feel good about themselves and to explore various areas of knowledge than to perform well in school.

EMPHASIS ON EFFORT. Japanese and Taiwanese parents and teachers believe that all children have the potential to master challenging academic tasks if they work hard enough. In contrast, many American parents and teachers regard native ability as the key to academic success (Stevenson, 1992). These differences in attitude may contribute to the fact that American parents are less likely to encourage activities at home that might improve school performance. Japanese and Taiwanese children spend more free time reading and playing academic-related games than do children in the United States (Stevenson & Lee, 1990).

INVOLVEMENT OF PARENTS IN EDUCATION. Asian parents devote many hours to helping their children with homework. American parents spend very little and do not regard homework in elementary school as especially important. Overall, American parents hold much lower standards for their children's academic performance and are far less

concerned about how well their youngsters are doing in school (Stevenson, Chen, & Lee, 1993; Stevenson & Lee, 1990).

HIGH-QUALITY EDUCATION FOR ALL. Unlike teachers in the United States, Japanese and Taiwanese teachers do not make early educational decisions on the basis of achievement. There are no separate ability groups in elementary schools. Instead, all pupils receive the same high-quality instruction. Academic lessons are particularly well organized and presented in ways that capture children's attention. Topics in mathematics are treated in greater depth, and there is less repetition of material taught the previous year (Stevenson & Lee, 1990).

MORE TIME DEVOTED TO INSTRUCTION. In Japan and Taiwan, the school year is over 50 days longer than in the United States. Much more of the day is devoted to academic pursuits, especially mathematics. However, Asian schools are not regimented places, as many Americans believe. An 8-hour school day means extra recesses and a longer lunch period, with plenty of time for play, social interaction, and extracurricular activities (Stevenson, 1994).

COMMUNICATION BETWEEN TEACHERS AND PARENTS. Japanese and Taiwanese teachers come to know their pupils especially well. They teach the same children for 2 or 3 years and make visits to the home once or twice a year. Teachers and parents communicate by writing messages about assignments, academic performance, and behavior in small notebooks that children carry back and forth every day. No such formalized system of frequent teacher–parent communication exists in the United States (Stevenson & Lee, 1990).

Do Japanese and Taiwanese children pay a price for the pressure

placed on them to succeed? By high school, academic work often displaces other experiences, since Asian adolescents must pass a highly competitive entrance exam to gain admission to college. Yet the highest performing Asian students are very well adjusted (Crystal et al., 1994). Awareness of the ingredients of Asian success has prompted Americans to rethink current educational practices.

Japanese children achieve considerably better than their American counterparts for a variety of reasons. Their culture stresses the importance of working hard to master academic skills. Their parents help more with homework and communicate more often with teachers. And a longer school day permits frequent alternation of academic instruction with pleasurable activity. This approach makes learning easier and more enjoyable. (Eiji Miyazawa/Black Star)

outside schools, affect children's learning. Societal values, school resources, quality of teaching, and parental encouragement all play important roles. Nowhere are these multiple influences more apparent than when schooling is examined in cross-cultural perspective.

Perhaps you are aware from recent news reports that American children fare poorly when their achievement is compared to that of children in other industrialized nations. In international studies of mathematics and science achievement, young people from Hong Kong, Japan, Korea, and Taiwan have consistently been among the top performers, whereas Americans have scored no better than at the mean (International Education Association, 1988; Lapointe, Askew, & Mead, 1992; Lapointe, Mead, & Askew, 1992). These trends emerge early in development. In a comparison of elementary school children in Japan, Taiwan, and the United States, large differences in mathematics achievement were present in kindergarten and became greater with increasing grade. Less extreme gaps occurred in reading, in which Taiwanese children scored highest, Japanese children lowest, and American children in between (Stevenson, 1992; Stevenson & Lee, 1990).

Why do American children fall behind in academic accomplishments? A common assumption is that Asian children are "smarter," but this is not true. They do not score better on intelligence tests than their American agemates (Stevenson et al., 1985). Instead, as the Cultural Influences box on page 311 indicates, a variety of social forces combine to foster a strong commitment to learning in Asian families and schools.

The Japanese and Taiwanese examples underscore that families, schools, and the larger society must work together to upgrade education. The current educational reform movement in the United States is an encouraging sign. Throughout the country, academic standards and teacher certification requirements are being strengthened. In addition, many schools are working to increase parent involvement in children's education. Parents who create stimulating learning environments at home, monitor their youngster's academic progress, help with homework, and communicate often with teachers have children who consistently show superior academic progress (Bradley, Caldwell, & Rock, 1988; Stevenson & Baker, 1987).

ASK YOURSELF . . .

■ *Saul, a third-grade teacher, places stars by the names of children who get A's on assignments. Children who earn at least 10 stars each week are called "All-Stars." What effect is this practice likely to have on children who achieve poorly, and why?*

■ *Carrie is a mainstreamed first grader with a learning disability. What steps can her teacher take to make sure her academic and social needs are met in a regular classroom?*

SUMMARY

PHYSICAL DEVELOPMENT IN MIDDLE CHILDHOOD

BODY GROWTH

Describe major trends in body growth during middle childhood.

■ Gains in body size during middle childhood extend the pattern of slow, regular growth established during the preschool years. Bones continue to lengthen and broaden, and all twenty primary teeth are replaced by permanent ones. By age 9, girls overtake boys in physical size.

COMMON HEALTH PROBLEMS

What vision and hearing problems are common in middle childhood?

■ During middle childhood, children from advantaged homes are at their healthiest, due to good nutrition and rapid development of the immune system. At the same time, a variety of health problems do occur, many of which are more prevalent among low-income children.

■ The most common vision problem in middle childhood is myopia, or nearsightedness. It is influenced by heredity and the way children use their eyes. Myopia is one of the few health conditions that increases with family education and income. Because of untreated ear infections, many low-income children experience some hearing loss during the school years.

Describe the causes and consequences of serious nutritional problems in middle

childhood, granting special attention to obesity.

■ When malnutrition is allowed to persist for many years, its negative impact on physical growth, intelligence, and motor performance is permanent.

■ Overweight and **obesity** are growing problems in affluent nations like the United States. Although heredity contributes to obesity, parental feeding practices, maladaptive eating habits, and lack of exercise also play important roles. Obese children are often rejected by peers and have serious adjustment problems. Family-based interventions are the most effective approaches to treatment.

What factors contribute to nocturnal enuresis and asthma, and how can these health problems be reduced?

■ In the majority of cases, **nocturnal enuresis,** or bedwetting during the night, has biological roots. The most effective treatment is a urine alarm that works according to conditioning principles.

■ Asthma is the most common cause of school absence and childhood hospitalization. Although heredity contributes to asthma, environmental factors—pollution, stressful home lives, and lack of access to good health care—have led the illness to increase.

Describe changes in unintentional injuries in middle childhood.

■ Unintentional injuries increase over middle childhood and adolescence, with auto and bicycle collisions accounting for most of the rise. Safety education is especially important during the school years.

HEALTH EDUCATION

What can parents and teachers do to encourage good health practices in school-age children?

■ Learning health-related information seldom changes school-age children's behavior. Besides educating children about good health, adults need to reduce health hazards in children's environments and model and reinforce good health practices.

MOTOR DEVELOPMENT AND PLAY

Cite major changes in motor development and play during middle childhood.

■ Improvements in flexibility, balance, agility, force, and reaction time contribute to school-age children's gross motor development. Adult encouragement of boys' athletic participa-

tion largely accounts for their superior performance on a wide range of skills.

■ Fine motor development also improves during the school years. Children's writing becomes more legible, and their drawings improve in organization, detail, and representation of depth.

■ Organized games with rules become common during the school years. Children's spontaneous games support cognitive and social development as well as health.

■ Many school-age youngsters are not physically fit. Frequent, high-quality physical education classes help ensure that all children have access to the benefits of regular exercise and play.

COGNITIVE DEVELOPMENT IN MIDDLE CHILDHOOD

PIAGET'S THEORY: THE CONCRETE OPERATIONAL STAGE

What are the major characteristics of concrete operational thought?

■ During the **concrete operational stage,** children can reason logically about concrete, tangible information. Attainment of conservation indicates that they can **decenter** and **reverse** their thinking. They are also better at classification, **seriation,** and **transitive inference.** School-age youngsters' spatial reasoning improves, as their understanding of distance and ability to give directions reveal.

■ Piaget used the term **horizontal décalage** to describe the school-age child's gradual mastery of logical concepts. Concrete operational thought is limited in that children have difficulty with abstractions.

Discuss recent research on concrete operational thought.

■ Recent evidence indicates that specific cultural practices, especially those associated with schooling, affect children's mastery of Piagetian tasks. Some researchers believe that the gradual development of operational thought can best be understood within an information processing framework.

INFORMATION PROCESSING

How do attention and memory change in middle childhood?

■ During the school years, attention becomes more controlled, adaptable, and planful, and memory strategies improve. **Rehearsal** appears first, followed by **organization** and then **elaboration.** The inattention, impulsivity, and excessive motor activity of children with **attention-deficit hyperactivity disorder** lead to serious learning and behavior problems.

■ Development of the long-term knowledge base facilitates memory by making new information easier to store and retrieve. Children's motivation to use what they know also contributes to memory development. Memory strategies are promoted by learning activities in school.

Describe the school-age child's theory of mind and capacity to engage in self-regulation.

■ Metacognition expands over middle childhood. School-age children regard the mind as an active, constructive agent, and they develop an integrated theory of mind. However, they are not yet good at **self-regulation**—putting what they know about thinking into action. Self-regulation improves with instruction in how to check and monitor progress toward a goal.

Discuss current controversies in teaching reading and mathematics to elementary school children.

- Skilled reading draws on all aspects of the information-processing system. Experts disagree on whether a **whole language approach** or **basic skills approach** should be used to teach beginning reading. A balanced mixture of both is probably most effective. Teaching that blends practice in basic skills with conceptual understanding also seems best in mathematics.

INDIVIDUAL DIFFERENCES IN MENTAL DEVELOPMENT

Cite commonly used intelligence tests in middle childhood, and describe major approaches to defining intelligence.

- During the school years, IQ becomes more stable, and it correlates well with academic achievement. Most intelligence tests yield an overall score as well as scores for separate intellectual factors. The **Stanford-Binet Intelligence Scale** and **Wechsler Intelligence Scale for Children–III (WISC–III)** are widely used, individually administered intelligence tests for children.

- Sternberg's **triarchic theory** views intelligence as a complex interaction of information-processing skills, specific experiences, and contextual (or cultural) influences. According to Gardner's **theory of multiple intelligences,** seven mental abilities exist, each of which has a unique biological basis and distinct course of development. Gardner's theory has been especially helpful in understanding and nurturing children's talents.

Describe evidence indicating that both heredity and environment contribute to intelligence.

- Heritability estimates and adoption research indicate that intelligence is a product of both heredity and environment. Studies of black children adopted into white middle-class homes indicate that the black–white IQ gap is substantially determined by environment.

- IQ scores are affected by specific learning experiences, including exposure to certain language customs and knowledge sampled by the test. Special precautions need to be taken when evaluating the mental abilities of low-income ethnic minority children.

LANGUAGE DEVELOPMENT

Describe changes in vocabulary, grammar, and pragmatics during middle childhood, along with the advantages of bilingualism.

- During middle childhood, vocabulary continues to grow rapidly, and children have a more precise and flexible understanding of word meanings. Children's grasp of complex grammatical constructions and use of conversational strategies also become more refined. Research shows that bilingual children are advanced in cognitive development and language awareness.

CHILDREN'S LEARNING IN SCHOOL

Describe the impact of educational philosophies on children's motivation and academic achievement.

- Pupils in **traditional classrooms** are advantaged in terms of academic achievement. Those in **open classrooms** tend to be independent learners who respect individual differences and have more positive attitudes toward school.

- The Kamehameha Elementary Education Program (KEEP) uses a balanced philosophical viewpoint based on Vygotsky's theory. It is highly effective in promoting learning among children who typically achieve poorly.

Discuss the role of teacher–pupil interaction in academic achievement.

- Effective classroom management and instruction that encourages high-level thinking have a positive impact on children's academic progress. **Educational self-fulfilling prophecies,** in which children start to live up to the opinions of their teachers, are most likely to occur in classrooms that emphasize competition and public evaluation.

How can schools best meet the learning needs of children at the very low and high ends of the ability distribution?

- Pupils with **mild mental retardation** and **learning disabilities** are often integrated into regular classrooms. The success of **mainstreaming** depends on individualized instruction and positive peer relations.

- **Giftedness** includes high IQ, high creativity, and exceptional talent. Tests of creativity tap **divergent thinking,** which contrasts with the **convergent thinking** typically emphasized on intelligence tests. Gifted children are best served by educational programs that build on their special strengths.

Why do American children fall behind children in other industrialized nations in academic accomplishments?

- American children fare poorly when their achievement is compared to that of children in other industrialized nations. In contrast, children from Japan and Taiwan are consistently among the top performers. A strong commitment to learning in families and schools underlies the greater academic success of Asian over American pupils.

IMPORTANT TERMS AND CONCEPTS

obesity (p. 280)

nocturnal enuresis (p. 282)

concrete operational stage (p. 290)

decentration (p. 290)

reversibility (p. 290)

seriation (p. 290)

transitive inference (p. 290)

horizontal décalage (p. 291)

rehearsal (p. 293)

organization (p. 293)

elaboration (p. 293)

attention-deficit hyperactivity
  disorder (ADHD) (p. 294)

self-regulation (p. 296)

whole language approach (p. 296)

basic skills approach (p. 297)

Stanford-Binet Intelligence Scale
  (p. 298)

Wechsler Intelligence Scale for
  Children–III (WISC–III) (p. 298)

triarchic theory of intelligence
  (p. 300)

theory of multiple intelligences
  (p. 301)

traditional classroom (p. 306)

open classroom (p. 306)

educational self-fulfilling prophecy
  (p. 307)

mainstreaming (p. 308)

mild mental retardation (p. 308)

learning disabilities (p. 308)

gifted (p. 309)

divergent thinking (p. 309)

convergent thinking (p. 309)

FOR FURTHER INFORMATION AND HELP

ADULT-ORGANIZED SPORTS FOR CHILDREN

Little League Baseball
P.O. Box 3485
Williamsport, PA 17701
(717) 326-1921
Web site: *www.littleleague.org*
*Organizes baseball and softball
programs for children 6 to 18 years of
age. Operates a special division for
handicapped children and sponsors
an annual world series.*

Soccer Association for Youth
4903 Vine Street
Cincinnati, OH 45217
(513) 769-3800
*Supports soccer programs for children
between 6 and 18 years of age throughout
the United States. Seeks to encourage
widespread participation and offer equal
opportunity regardless of ability or sex.
Distributes supplies and support necessary to form teams.*

ATTENTION-DEFICIT HYPERACTIVITY DISORDER

Children with Attention Deficit
Disorders
499 N.W. 70th Avenue, Suite 308
Plantation, FL 33317
(954) 587-3700
Web site: *www.chadd.org*

*Provides support and education to fami-
lies of children with attention-deficit
hyperactivity disorder. Encourages
schools and health care professionals
to be responsive to their needs.*

BILINGUAL EDUCATION

National Association for
Bilingual Education
Union Center Plaza
810 First Street, N.E., Third Floor
Washington, DC 20002
(202) 898-1829
Web site: *www.nabe.org*
*An organization of educators, public
citizens, and students aimed at increasing
public understanding of the importance
of bilingual education.*

LEARNING DISABILITIES

Learning Disabilities Association
of America
4156 Library Road
Pittsburgh, PA 15234
(412) 341-1515
Web site: *www.ldanatl.org*
*A 60,000-member organization of
parents of learning-disabled children
and interested professionals. Local groups
provide parent support and education
and sponsor recreational programs and
summer camps for children.*

GIFTEDNESS

Association for the Gifted of the
Council for Exceptional Children
1920 Association Drive
Reston, VA 22091
(703) 620-3660
Web site: *www.cec.sped.org*
*An organization of educators and parents
aimed at stimulating interest in program
development for gifted children. Publishes
Journal for the Education of the Gifted.*

National Association for
Gifted Children
1155 15th Street N.W., No. 1002
Washington, DC 20005
(202) 785-4268
Web site: *www.nagc.org*
*Association of scholars, educators, and
librarians devoted to advancing education for gifted children. Distributes information and sponsors institutes. Publishes
the journal* Gifted Child Quarterly.

Emotional and Social Development in Middle Childhood

One late afternoon, Rena heard Joey dash through the front door, run upstairs, and call up his best friend Terry. "Terry, gotta talk to you," pleaded Joey, out of breath from running home. "Everything was going great until that word I got—*porcupine*," remarked Joey, referring to the fifth-grade spelling bee at school that day. "Just my luck! P-o-r-k, that's how I spelled it! I can't believe it. Maybe I'm not so good at social studies," Joey confided, "but I *know* I'm better at spelling than that stuck-up Belinda Brown. Gosh, I knocked myself out studying those spelling lists. Then *she* got all the easy words. Did'ya see how snooty she acted after she won? If I *had* to lose, why couldn't it be to a nice person, anyhow!"

Joey's conversation reflects a whole new constellation of emotional and social capacities. First, Joey shows evidence of *industriousness*. By entering the spelling bee, he energetically pursued meaningful achievement in his culture—a major change of the middle childhood years. At the same time Joey's social understanding has expanded. He can size up himself and others in terms of strengths, weaknesses, and personality characteristics. Furthermore, friendship means something quite different to Joey than it did at younger ages. Terry is a best friend whom Joey counts on for understanding and emotional support.

We begin this chapter by returning to Erikson's theory for an overview of the personality changes of middle childhood. Then we take a close look at a variety of aspects of emotional and social development. We will see how, as children reason more effectively and spend more time in school and with peers, their views of themselves, of others, and of social relationships become more complex.

Although school-age children spend less time with parents than they did at younger ages, the family remains a powerful context for development. Joey and Lizzie are growing up in homes profoundly affected by social change. Unlike her own mother, Rena has been employed since her children were preschoolers. In addition, Joey and Lizzie's home life has been disrupted by family discord: Rena is divorced. The children's personal experiences in adjusting to these departures from traditional family arrangements will help us appreciate the vital role of family relationships in all aspects of development. Our chapter concludes with a discussion of some common emotional problems of middle childhood.

ERIKSON'S THEORY: INDUSTRY VERSUS INFERIORITY

According to Erikson (1950), the personality changes of the school years build on Freud's **latency stage**, a period in which the sexual instincts lie dormant. In Chapter 8, we noted that Freud's theory is no longer widely accepted. Yet when early experiences have been positive, children enter middle childhood with the calm confidence that the word *latency* suggests. Their energies are redirected from the make-believe of early childhood into realistic accomplishment.

Erikson believed that the combination of adult expectations and children's drive toward mastery sets the stage for the psychological conflict of middle childhood: **industry versus inferiority.** Industry means developing competence at useful skills and tasks. In cultures everywhere, improved physical and cognitive capacities mean that adults impose new demands. Children, in turn, are ready to meet these challenges and benefit from them.

In industrialized nations, the transition to middle childhood is marked by the beginning of formal schooling. With it comes literacy training, which prepares children for the vast array of specialized careers in complex societies. In school, children become aware of their own and others' unique capacities, learn the value of division of labor, and develop a sense of moral commitment and responsibility. The danger at this stage is *inferiority*, reflected in the sad

When Joey and Belinda Brown competed in the spelling bee, they showed evidence of industriousness by pursuing meaningful achievement in their culture. According to Erikson, developing a sense of industry is the critical psychosocial task of middle childhood. (Charles Gupton/Stock Boston)

Test Bank Items 10.1 through 10.4

pessimism of children who have come to believe they will never be good at anything. This profound sense of inadequacy can develop when family life has not prepared children for school life or when experiences with teachers and peers are so negative that they destroy children's feelings of competence and mastery.

Erikson's sense of industry combines several critical developments of middle childhood: a positive but realistic self-concept, pride in doing things well, moral responsibility, and cooperative participation with agemates. Let's see how these aspects of self and social relationships change over the school years.

SELF-DEVELOPMENT IN MIDDLE CHILDHOOD

Several transformations in self-understanding take place in middle childhood. First, children can describe themselves in terms of psychological traits. Second, they start to compare their own characteristics to those of their peers. Finally, they speculate about the causes of their strengths and weaknesses. These new ways of thinking about themselves have a major impact on children's developing self-esteem.

CHANGES IN SELF-CONCEPT

During the school years, children organize their observations of behaviors and internal states into general dispositions that they can verbalize to others, with a major change taking place between ages 8 and 11. The following self-descriptions reflect this change:

A boy age 7: I am 7 and I have hazel brown hair and my hobby is stamp collecting. I am good at football and I am quite good at sums and my favourite game is football and I love school and I like reading books and my favourite car is an Austin. (Livesley & Bromley, 1973, p. 237)

A girl age 11½: My name is A. I'm a human being. I'm a girl. I'm a truthful person. I'm not pretty. I do so-so in my studies. I'm a very good cellist. I'm a very good pianist. I'm a little bit tall for my age. I like several boys. I like several girls. I'm old-fashioned. I play tennis. I am a very good swimmer. I try to be helpful. I'm always ready to be friends with anybody. Mostly I'm good, but I lose my temper. I'm not well-liked by some girls and boys. I don't know if I'm liked by boys or not. (Montemayor & Eisen, 1977, pp. 317–318)

Notice that instead of specific behaviors, school-age children emphasize competencies, as in "I am quite good at sums" or "I'm a very good cellist" (Damon & Hart, 1988). Also, the older child clearly describes her personality. For example, she comments that she is truthful, old-fashioned, helpful, friendly, and short-tempered.

Another change in self-concept takes place in middle childhood: Around age 7, children begin to make *social comparisons.* They judge their appearance, abilities, and behavior in relation to those of others. Return to the beginning of this chapter, and note that Joey expressed some thoughts about how good he was compared to Belinda Brown—better at spelling but not so great at social studies (Stipek & Mac Iver, 1989).

What factors are responsible for these revisions in self-concept? Cognitive development affects the changing *structure* of the self—children's ability to combine typical experiences and behaviors into psychological dispositions. The *content* of self-concept, however, comes largely from interactions with others. Early in this century, sociologist George Herbert Mead (1934) proposed that a well-organized psychological self emerges when children can imagine the attitude that others take toward them. In other words, *perspective-taking* skills—in particular, the ability to imagine what other people are thinking—play a crucial role in the development of a psychological self. As we will see later in this chapter, school-age children become better at reading the messages they receive from others and incorporating these into their self-definitions.

During middle childhood, children look to more people for information about themselves as they enter a wider range of settings in school and community. This is reflected in children's frequent reference to social groups in their self-descriptions (Livesley & Bromley, 1973). "I'm a Boy Scout, a paper boy, and a Prairie City soccer player," Joey remarked when asked to describe himself. Gradually, as children move toward adolescence, their sources of self-definition become more selective. Although parents remain influential, between ages 8 and 15 peers become more important. And over time, self-concept becomes increasingly vested in feedback from close friends (Rosenberg, 1979).

DEVELOPMENT OF SELF-ESTEEM

Recall that most preschoolers have very high self-esteem. As children move into middle childhood, they receive much more feedback about their performance in different activities compared to that of their peers. As a result, self-esteem differentiates, and it also adjusts to a more realistic level.

■ A HIERARCHICALLY STRUCTURED SELF-ESTEEM. Susan Harter (1982, 1990) asked children to indicate the extent to which a variety of statements, such as "I am good at homework" or "I'm usually the one chosen for games," are true of themselves. Her findings, and those of other researchers, reveal that classrooms, peer groups, and playgrounds are key contexts in which children evaluate their

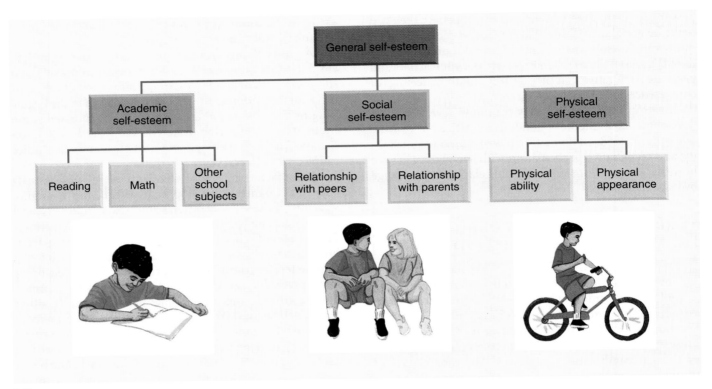

FIGURE **10.1**

Hierarchical structure of self-esteem in the mid-elementary school years. From their experiences in different settings, children form at least three separate self-esteems—academic, social, and physical. These differentiate into additional self-evaluations and combine to form an overall sense of self-worth.

own competence. By age 7 to 8, children have formed at least three separate self-esteems—academic, social, and physical—that become more refined with age. For example, academic self-worth divides into performance in different school subjects, social self-worth into peer and parental relationships (Marsh, 1990). As a result, self-esteem takes on the hierarchical structure shown in Figure 10.1.

■ CHANGES IN LEVEL OF SELF-ESTEEM. As children adjust their self-judgments to fit the opinions of others and their objective performance, self-esteem drops during the first few years of elementary school (Stipek & Mac Iver, 1989). Typically, this decline is not great enough to be harmful. Most (but not all) children appraise their characteristics and competencies realistically while maintaining an attitude of self-acceptance and self-respect. Then, from fourth to sixth grade, self-esteem rises for the majority of youngsters, who feel especially good about their peer relationships and athletic capabilities (Nottelmann, 1987).

INFLUENCES ON SELF-ESTEEM

From middle childhood on, strong relationships exist between self-esteem and everyday behavior. Academic self-esteem predicts children's school achievement (Marsh,

Smith, & Barnes, 1985). Children with high social self-esteem are consistently better liked by their peers (Harter, 1982). And as we saw in Chapter 9, boys come to believe they have more athletic talent than do girls, and they are also more advanced in a variety of physical skills. What social influences might lead self-esteem to be high for some children and low for others?

■ CHILD-REARING PRACTICES. Children whose parents use an *authoritative* child-rearing style—combining warmth and responsiveness with firm but reasonable expectations for behavior (see Chapter 8)—feel especially good about themselves (Bishop & Ingersoll, 1989; Lord, Eccles, & McCarthy, 1994). Warm, positive parenting lets children know that they are accepted as competent and worthwhile. And firm but appropriate expectations, backed up with explanations, seem to help children make sensible choices and evaluate their own behavior against reasonable standards.

Yet note that these findings are correlational. We cannot actually tell whether child-rearing styles are causes of or reactions to children's characteristics and behavior. Research on the precise content of adults' messages to children has been far more successful at isolating factors that affect children's sense of self-worth. Let's see how these communicative forces mold children's evaluations of themselves in achievement contexts.

■ MAKING ACHIEVEMENT-RELATED ATTRIBUTIONS. *Attributions* are our common, everyday explanations for the causes of behavior—our answers to the question "Why did I (or another person) do that?" Look back at Joey's conversation about the spelling bee at the beginning of this chapter. Notice how he attributes his disappointing performance to *luck* (Belinda got all the easy words) and his usual success at spelling to *ability* (he *knows* he's a better speller than Belinda). Joey also appreciates that *effort* makes a difference; he "knocked himself out studying those spelling words."

Cognitive development permits school-age children to separate all these variables in explaining performance (Chapman & Skinner, 1989). Yet they differ greatly in how they account for their successes and failures. Children who are high in academic self-esteem make **mastery-oriented attributions.** They believe their successes are due to ability—a characteristic they can count on when faced with new challenges. And they attribute failure to factors that can be changed and controlled, such as insufficient effort or a very difficult task. So regardless of whether these children succeed or fail, they take an industrious, persistent, and enthusiastic approach to learning.

Unfortunately, children who develop **learned helplessness** hold very discouraging explanations for their performance. They attribute their failures, not their successes, to ability. And on occasions when they do succeed, they are likely to conclude that external factors, such as luck, are responsible. Furthermore, unlike their mastery-oriented counterparts, they have come to believe that ability is fixed and cannot be changed. They do not think competence can be improved by trying hard. So when a task is difficult, these children experience an anxious loss of control—in Erikson's terms, a pervasive sense of inferiority. They quickly give up, saying "I can't do this," before they have really tried (Elliott & Dweck, 1988).

Over time, the ability of learned-helpless children no longer predicts their performance. Because they fail to make the connection between effort and success, learned-helpless children do not develop metacognitive and self-regulatory skills necessary for high achievement (see Chapter 9). Lack of effective learning strategies, reduced persistence, and a sense of being controlled by external forces sustain one another in a vicious cycle (Heyman & Dweck, 1992).

■ INFLUENCES ON ACHIEVEMENT-RELATED ATTRIBUTIONS. What accounts for the very different attributions of mastery-oriented and learned-helpless children? The messages they receive from parents and teachers play a key role. Children with a learned-helpless style tend to have

■

Repeated negative evaluations of their competence can cause children to develop learned helplessness—the belief that trying hard will not help them succeed. This learned-helpless boy is overwhelmed by a poor grade. He seems to have concluded that there is little he can do to improve his performance and is likely to display anxious loss of control when faced with a challenging task. *(MacDonald Photography/Envision)*

parents who set unusually high standards yet believe their child is not very capable and has to work harder to succeed (Parsons, Adler, & Kaczala, 1982; Phillips, 1987). In one study, researchers manipulated the feedback that fourth and fifth graders received after they failed at a task. Those receiving negative messages about their competence more often attributed failure to lack of ability than did children who were told that they had not tried hard enough (Dweck et al., 1978).

Some children are especially likely to have their performance undermined by the feedback they receive from adults. Girls more often than boys blame their ability for poor performance. Girls also tend to receive messages from teachers and parents that their ability is at fault when they do not do well (Phillips & Zimmerman, 1990). Low-income ethnic minority children are also vulnerable to learned helplessness. In several studies, African-American and Mexican-American children received less favorable feedback from teachers (Aaron & Powell, 1982; Irvine, 1986; Losey, 1995). Also, when ethnic minority children observe that adults in their own family are not rewarded by society for their achievement efforts, they may give up themselves. Many African-American children come to believe that even if they do try in school, social prejudice will prevent them from succeeding in the end (Ogbu, 1988).

■ SUPPORTING CHILDREN'S SELF-ESTEEM. Attribution research suggests that even adults who are, on the whole, warm and supportive may send subtle messages to children that undermine their competence. *Attribution retraining* is an approach to intervention that encourages learned-helpless children to believe that they can overcome failure by exerting more effort. Most often, children are asked to work on tasks that are hard enough so they will experience some failure. Then they get repeated feedback such as "You can do it if you try harder." Children are also taught to view their successes as due to both ability and effort rather than to chance factors, by giving them additional feedback after they succeed, such as "You're really good at this" or "You really tried hard on that one." Another approach is to encourage low-effort children to focus less on grades and more on mastering a task for its own sake (Stipek & Kowalski, 1989). Finally, learned-helpless children may need instruction in metacognition and self-regulation to make up for learning lost in this area because of their attributional styles (Borkowski et al., 1990).

To work well, attribution retraining is best begun in middle childhood, before children's views of themselves become hard to change. An even better approach is to prevent low self-esteem—by minimizing comparisons among children, helping them overcome failures, and designing school environments that accommodate individual differences in development and styles of learning. Finally, extra measures should be taken to support the self-esteem of girls and ethnic minority children—by providing role models of adult success, fostering ethnic pride, and ensuring equality of opportunity in society at large.

BRIEF REVIEW

Erikson's stage of industry versus inferiority indicates that when family, school, and peer experiences are positive, school-age children develop an industrious approach to productive work and feelings of competence and mastery. During middle

childhood, psychological traits and social comparisons appear in children's self-descriptions. A differentiated, hierarchically organized self-esteem emerges, and children's sense of self-worth declines as they adjust their self-judgments to fit the opinions of others and objective performance. Parental warmth and reasonable demands are related to high self-esteem. Attribution research has identified adult communication styles that affect children's explanations for success and failure and, in turn, their self-esteem and task performance.

ASK YOURSELF . . .

■ *Return to page 285 of Chapter 9 and review the messages that parents send to girls about their athletic talent. On the basis of what you know about children's attributions for success and failure, why do school-age girls like Lizzie perform more poorly and spend less time at sports than do boys?*

■ *In view of Joey's attributions for his spelling bee performance, is he likely to enter the next spelling bee and try hard to do well? Why or why not?*

EMOTIONAL DEVELOPMENT IN MIDDLE CHILDHOOD

Greater self-awareness and social sensitivity support emotional development in middle childhood. Gains take place in children's experiencing of self-conscious emotions, awareness of emotional states, and emotional self-regulation.

In middle childhood, the self-conscious emotions of pride and guilt become clearly integrated with personal responsibility. A teacher or parent no longer has to be present for a new accomplishment to spark a glowing sense of pride or for a transgression to arouse painful pangs of guilt (Harter, Wright, & Bresnick, 1987). Also, children do not report guilt for any mishap, as they did at younger ages, but only for intentional wrongdoing, such as ignoring responsibilities, cheating, or lying (Graham, Doubleday, & Guarino, 1984).

School-age children's understanding of psychological dispositions means that they are more likely to explain emotion by making reference to internal states than external events (Strayer, 1993). They are also much more aware of the diversity of emotional experiences. By age 8, children recognize that they can experience more than one emotion at a time (Harter & Buddin, 1987; Wintre & Vallance, 1994). For example, recalling the birthday present he received from his grandmother, Joey reflected, "I was happy that I got something but a little sad that I didn't get just what I wanted." Similarly, Joey appreciates that emotional reactions need not reflect a person's true feelings. Consequently, he is much better at hiding his emotions when it is socially appropriate to do so. "I got all excited and told grandma I liked that dumb plastic toy train," Joey said to Rena one day, "but I really don't. It's too babyish for a 10-year-old" (Saarni, 1989).

Rapid gains in emotional self-regulation occur as children find many more ways to handle emotionally arousing situations. In several studies, 5- to 11-year-olds were told stories about positive and negative events, such as having to wait before receiving an attractive prize or getting a bad grade on a test. Then they were asked what could be done to control emotions under these conditions. Although children of all ages were aware that they could distract themselves with alternative behaviors, such as reading or watching TV, older children more often mentioned cognitive strategies for handling feelings. When an event could not be changed, they came up with ways of reinterpreting it (Altshuler & Ruble, 1989; Band & Weisz, 1988). For example, to a bad grade, one child said, "Things could be worse. There'll be another test."

School-age children are not just better at recognizing and responding to their own feelings. They are also more aware of the thoughts and feelings of others, as we will see in the next section.

UNDERSTANDING OTHERS

By the mid-elementary school years, children discover consistencies in the behavior of people they know. As with their self descriptions, they begin to describe other people in terms of psychological traits (Droege & Stipek, 1993; Eder, 1989). This increases their awareness that others may react differently than they do to social situations. Middle childhood brings major advances in **perspective taking**—the capacity to imagine what other people may be thinking and feeling.

SELMAN'S STAGES OF PERSPECTIVE TAKING

Robert Selman's five-stage model describes changes in children's perspective-taking skill. He asked preschool through adolescent youngsters to respond to social dilemmas in which the characters have differing information and opinions about an event. Here is one example:

Holly is an 8-year-old girl who likes to climb trees. She is the best tree climber in the neighborhood. One day while climbing down from a tall tree she falls off the bottom branch but does not hurt herself. Her father sees her fall. He is upset and asks her to promise not to climb trees anymore. Holly promises.

Later that day, Holly and her friends meet Sean. Sean's kitten is caught up in a tree and cannot get down. Something has to be done right away or the kitten may fall. Holly is the only one who climbs trees well enough to reach the kitten and get it down, but she remembers her promise to her father. (Selman & Byrne, 1974, p. 805)

After the dilemma is presented, children answer questions that highlight their ability to interpret the story from varying points of view, such as

Does Sean know why Holly cannot decide whether or not to climb the tree?

What will Holly's father think? Will he understand if she climbs the tree?

Does Holly think she will be punished for climbing the tree? Should she be punished for doing so?

Table 10.1 on page 324 summarizes Selman's five stages of perspective taking. At first, children have only a limited idea of what other people might be thinking and feeling. Over time, they become more aware that people can interpret the same event in quite different ways. Soon, they can "step in another person's shoes" and reflect on how that person might regard their own thoughts, feelings, and behavior. Finally, they can examine the relationship between two people's perspectives simultaneously, at first from the vantage point of a disinterested spectator and, finally, by making reference to societal values.

PERSPECTIVE TAKING AND SOCIAL SKILLS

Perspective taking is related to a wide variety of social skills. Good perspective takers are more likely to display empathy and compassion (Eisenberg et al., 1987). In addition, they are better at *social problem solving*, or thinking of effective ways to handle difficult social situations (Marsh, Serafica, & Barenboim, 1981). In fact, once children are capable of self-reflective perspective-taking (see Table 10.1), they often rely on it to clear up everyday misunderstandings. For example, one day when Joey teased Terry in a friendly way, Terry took offense. Joey made use of advanced perspective taking to patch things up. "Terry, I didn't mean it," he explained, "I *thought you would think* I was just kidding when I said that."

Perspective taking varies greatly among children of the same age. Individual differences are due to cognitive maturity as well as experiences in which adults and peers explain their viewpoints, encouraging children to notice another's perspective (Dixon & Moore, 1990). Children with very poor social skills—in particular, the angry, aggressive styles that we discussed in Chapter 8—have great difficulty imagining the thoughts and feelings of

TABLE 10.1

Selman's Stages of Perspective Taking

STAGE	APPROXIMATE AGE RANGE	DESCRIPTION	TYPICAL RESPONSE TO "HOLLY" DILEMMA
Level 0: Undifferentiated perspective taking	3–6	Children recognize that self and other can have different thoughts and feelings, but they frequently confuse the two.	The child predicts that Holly will save the kitten because she does not want it to get hurt and believes that Holly's father will feel just as she does about her climbing the tree: "Happy, he likes kittens."
Level 1: Social-informational perspective taking	4–9	Children understand that different perspectives may result because people have access to different information.	When asked how Holly's father will react when he finds out that she climbed the tree, the child responds, "If he didn't know anything about the kitten, he would be angry. But if Holly shows him the kitten, he might change his mind."
Level 2: Self-reflective perspective taking	7–12	Children can "step in another person's shoes" and view their own thoughts, feelings, and behavior from the other person's perspective. They also recognize that others can do the same.	When asked whether Holly thinks she will be punished, the child says, "No. Holly knows that her father will understand why she climbed the tree." This response assumes that Holly's point of view is influenced by her father being able to "step in her shoes" and understand why she saved the kitten.
Level 3: Third-party perspective taking	10–15	Children can step outside a two-person situation and imagine how the self and other are viewed from the point of view of a third, impartial party.	When asked whether Holly should be punished, the child says, "No, because Holly thought it was important to save the kitten. But she also knows that her father told her not to climb the tree. So she'd only think she shouldn't be punished if she could get her father to understand why she had to climb the tree." This response steps outside the immediate situation to view both Holly's and her father's perspectives simultaneously.
Level 4: Societal perspective taking	14–adult	Individuals understand that third-party perspective taking can be influenced by one or more systems of larger societal values.	When asked if Holly should be punished, the individual responds, "No. The value of humane treatment of animals justifies Holly's action. Her father's appreciation of this value will lead him not to punish her."

Sources: Selman, 1976; Selman & Byrne, 1974.

others. They often mistreat adults and peers without feeling the guilt and remorse prompted by awareness of another's point of view. Interventions that provide coaching and practice in perspective taking are helpful in reducing antisocial behavior and increasing empathy and prosocial responding (Chalmers & Townsend, 1990; Chandler, 1973).

MORAL DEVELOPMENT IN MIDDLE CHILDHOOD

Recall from Chapter 8 that preschoolers pick up a great many morally relevant behaviors through modeling and reinforcement. By middle childhood, they have had time to reflect on these experiences and internalize rules for good conduct, such as "It's good to help others in trouble" or "It's wrong to take something that doesn't belong to you." This change leads children to become considerably more independent and trustworthy. They can take on many more responsibilities, from running an errand at the supermarket to making sure a younger sibling does not wander into the street. Of course, these advances take place only when children have had much time to profit from the consistent guidance and example of caring adults in their lives.

In Chapter 8, we also saw that children do just copy their morality from others. As the cognitive-developmental approach emphasizes, they actively think about right and wrong. Children's expanding social world and their increasing ability to take the perspective of others leads moral understanding to improve greatly in middle childhood.

LEARNING ABOUT JUSTICE THROUGH SHARING

In everyday life, children frequently experience situations that involve **distributive justice**—beliefs about how to divide up material goods fairly. Heated discussions often take place over how much weekly allowance is to be given to siblings of different ages, who has to sit where in the family car on a long trip, and in what way an eight-slice pizza is to be shared by six hungry playmates. William Damon (1977,

These five school-age children have figured out how to divide up two pizzas fairly among themselves. Already, they have a well-developed sense of distributive justice. (Will Faller)

1988) has studied children's changing concepts of distributive justice over early and middle childhood.

Even 4-year-olds recognize the importance of sharing, but their reasons for doing so often seem contradictory and self-serving: "I shared because if I didn't, she wouldn't play with me" or "I let her have some, but most are for me because I'm older." As children enter middle childhood, they express more mature notions of distributive justice (see Table 10.2). At first, these ideas of fairness are based on *equality*. Children in the early school grades are intent on making sure that each person gets the same amount of a treasured resource, such as money, turns in a game, or a delicious treat.

A short time later, children start to view fairness in terms of *merit*. Extra rewards should go to someone who has worked especially hard or otherwise performed in an exceptional way. Finally, around age 8, children can reason on the basis of *benevolence*. They recognize that special consideration should be given to those who are at a disadvantage, like the needy or the disabled. Older children indicate that an extra amount might be given to a child who cannot produce as much or who does not get any allowance from his parents. They also adapt their basis of fairness to fit the situation—for example, relying more on

equality when interacting with strangers and more on benevolence when interacting with friends (McGillicuddy-De Lisi, Watkins, & Vinchur, 1994).

According to Damon (1988), the give-and-take of peer interaction makes children more sensitive to others' perspectives, and this, in turn, supports their developing ideas of justice. Indeed, mature distributive justice reasoning shows many of the same relationships to everyday social behavior as perspective taking. For example, it is associated with more effective social problem solving and a greater willingness to help and share with others (Blotner & Bearison, 1984; McNamee & Peterson, 1986).

CHANGES IN MORAL AND SOCIAL-CONVENTIONAL UNDERSTANDING

As their ideas about justice advance, children clarify and create links between moral rules and social conventions. During middle childhood, they realize that moral rules and social conventions sometimes overlap. For example, saying "Thank you" after receiving a present is an arbitrary practice arrived at by social agreement. At the same time, not doing so can injure others by hurting their feelings (Turiel, 1983).

Culture influences the extent to which children separate moral rules from social conventions. American children of diverse religious backgrounds regard morality as distinct from religious rituals and rules for everyday living. In contrast, Hindu children view violating certain customs, such as eating chicken the day after a father's death or calling parents by their first names, as serious moral transgressions (Nucci & Turiel, 1993; Shweder, Mahapatra, & Miller, 1990). In India, as in many developing countries, practices that strike Western outsiders as arbitrary often have profound moral and religious significance.

BRIEF REVIEW

In middle childhood, self-conscious emotions become clearly linked to personal responsibility. Children's awareness of emotional experience expands, and they become better at emotional

TABLE 10.2

Damon's Sequence of Distributive Justice Reasoning

BASIS OF REASONING	AGE	DESCRIPTION
Equality	5–6	Fairness involves strictly equal distribution of goods. Special considerations like merit and need are not taken into account.
Merit	6–7	Fairness is based on deservingness. Children recognize that some people should get more because they have worked harder.
Benevolence	8	Fairness includes giving special consideration to those who are disadvantaged. More should be given to those who are in need.

self-regulation. Perspective taking undergoes major advances and is related to a wide variety of social skills. Moral understanding also improves. Children develop more advanced notions of distributive justice, and they appreciate the overlap between moral rules and social conventions.

ASK YOURSELF . . .

- *Return to Joey's description of Belinda Brown at the beginning of this chapter. How does it reflect changes in children's understanding of others during middle childhood?*

- *When given the Holly dilemma and asked, "Does Holly think she will be punished for climbing the tree?" Lizzie responded, "No, Holly knows that her father will understand how sad she would feel if she let that kitten fall out of the tree." Which of Selman's perspective-taking stages is Lizzie at?*

Peer groups first form in middle childhood. These boys have probably established a social; structure of leaders and followers as they gather often for joint activities, such as bike riding and basketball. Their body language suggests that they feel a strong sense of group belonging. (R. Sidney/The Image Works)

PEER RELATIONS IN MIDDLE CHILDHOOD

In middle childhood, the society of peers becomes an increasingly important context for development. Peer contact, as we have seen, plays an important role in perspective taking and understanding of self and others. These developments, in turn, contribute to the quality of peer interaction, which becomes more prosocial over the school years. In line with this change, aggression declines, but the drop is greatest for physical attacks. As we will see, verbal insults among boys and social ostracism among girls occur as school-age children form peer groups and start to distinguish "insiders" from "outsiders."

PEER GROUPS

Watch children in the school yard or neighborhood, and notice how groups of three to a dozen or more often gather. The organization of these collectives changes greatly with age. By the end of middle childhood, children display a strong desire for group belonging. Together, they generate unique values and standards for behavior. They also create a social structure of leaders and followers that ensures group goals will be met. When these characteristics are present, a **peer group** is formed.

The practices of these informal groups lead to a "peer culture" that typically consists of a specialized vocabulary, dress code, and place to "hang out" during leisure hours. For example, Joey formed a club with three other boys.

They met in the treehouse in Joey's backyard, used nicknames, and wore a "uniform" consisting of T-shirts, jeans, and tennis shoes. Calling themselves "the pack," the boys developed a secret handshake and chose Joey as their leader. Their activities included improving the clubhouse, trading baseball cards, making trips to the video arcade, and—just as important—keeping girls and adults out!

As children develop these exclusive associations, the codes of dress and behavior that grow out of them become more broadly influential. At school, children who deviate are often rebuffed by their peers. "Kissing up" to teachers, wearing the wrong kind of shirt or shoes, and tattling on classmates are grounds for critical glances and comments.

These special customs bind peers together, creating a sense of group identity. Within the group, children acquire many valuable social skills—cooperation, leadership, followership, and loyalty to collective goals. Through these experiences, children experiment with and learn about social organizations.

The beginning of peer group ties is also a time when some of the "nicest children begin to behave in the most awful way" (Redl, 1966, p. 395). From third grade on, gossip, rumor spreading, and exclusion rise among girls, who (because of gender-role expectations) express aggression in subtle, indirect ways (Crick & Grotpeter, 1995). Boys are more straightforward in their hostility toward the "outgroup." Prank playing, such as egging a house, making a funny phone call, or ringing a doorbell and running away, often occurs among small groups of boys, who provide one another with temporary social support for these mildly antisocial behaviors (Fine, 1980).

The school-age child's desire for group belonging can also be satisfied through formal group ties—Girl Scouts, Boy Scouts, 4-H, church groups, and other associations. Adult involvement holds in check the negative behaviors associated with children's informal peer groups. In addition, children gain in social and moral understanding as they work on joint projects and help in their communities (Keasey, 1971).

FRIENDSHIPS

While peer groups provide children with insight into larger social structures, one-to-one friendships contribute to the development of trust and sensitivity. During the school years, friendship becomes more complex and psychologically based. Look at the following 8-year-old's answers to questions about what makes a best friend:

> Why is Shelly your best friend? *Because she helps me when I'm sad, and she shares . . . .* What makes Shelly so special? *I've known her longer, I sit next to her and got to know her better. . . .* How come you like Shelly better than anyone else? *She's done the most for me. She never disagrees, she never eats in front of me, she never walks away when I'm crying, and she helps me with my homework. . . .* How do you get someone to like you? *. . . . If you're nice to [your friends], they'll be nice to you.* (Damon, 1988, pp. 80–81)

As these responses show, friendship is no longer just a matter of engaging in the same activities. Instead, it is a mutually agreed-on relationship in which children like each other's personal qualities and respond to one another's needs and desires. And once a friendship forms, *trust* becomes its defining feature. School-age children state that a good friendship is based on acts of kindness that signify each person can be counted on to support the other. Consequently, events that break up a friendship are quite different than they were during the preschool years. Older children regard violations of trust, such as not helping when others need help, breaking promises, and gossiping behind the other's back, as serious breaches of friendship (Damon, 1977; Selman, 1980).

Because of these features, school-age children's friendships are more selective. Preschoolers say they have lots of friends—sometimes everyone in their class! By age 8 or 9, children have only a handful of people they call friends and, very often, only one best friend. Girls, especially, are likely to be exclusive in their friendships because they demand greater closeness than do boys. In addition,

During middle childhood, concepts of friendship become more psychologically based. These girls want to spend time together because they like each other's personal qualities. Mutual trust is a defining feature of their friendship. Each counts on the other to provide support and assistance. In integrated schools, many children form close other-race friendships. (Will Faller)

friends tend to be of the same age, sex, race, ethnicity, and social class. They also resemble one another in personality (sociability) and academic achievement (Hartup, 1996). Note, however, that characteristics of schools and neighborhoods affect friendships. For example, in integrated schools, as many as 50 percent of pupils report at least one close other-race friend (DuBois & Hirsch, 1990).

Through friendship, children learn the importance of emotional commitment. They come to realize that close relationships can survive disagreements if both parties are secure in their liking for one another (Hartup et al., 1993). Most school-age friendships last for several years. But as children approach adolescence, their varying rates of development and changing interests cause many friendships to break up and new ones to be established.

PEER ACCEPTANCE

As we all know from our own childhoods, some children make friends and enter peer groups far more easily than others. In Chapter 9, we saw that obese children and children with serious learning problems often have great difficulty with peer acceptance. Yet there are other children whose appearance and intellectual abilities are quite normal; still, their classmates despise them.

Researchers assess peer acceptance with self-report measures that ask classmates to evaluate one another's likability. Children's responses reveal four different categories: **popular children,** who get many positive votes; **rejected children,** who are actively disliked; **controversial children,** who get a large number of positive and negative votes; and **neglected children,** who are seldom chosen, either positively or negatively. About two-thirds of pupils in a typical elementary school classroom fit one of these categories. The rest are *average* in peer acceptance; they do not receive extreme scores (Coie, Dodge, & Coppotelli, 1982).

Peer acceptance is a powerful predictor of psychological adjustment. Rejected children, especially, are unhappy, alienated, poorly achieving children with low self-esteem. Both teachers and parents rate them as having a wide range of emotional and social problems. Peer rejection in middle childhood is also strongly associated with poor school performance, dropping out, antisocial behavior, and delinquency in young adulthood (DeRosier, Kupersmidt, & Patterson, 1994; Parker & Asher, 1987).

■ DETERMINANTS OF PEER ACCEPTANCE. What causes one child to be liked and another to be rejected? A wealth of research reveals that social behavior plays a powerful role. Popular children have very positive social skills. They communicate with peers in sensitive, friendly, and cooperative ways. When they do not understand another child's reaction, they ask for an explanation. If they disagree with a play partner in a game, they go beyond voicing their displeasure and suggest what the other child could do instead. When they want to enter an ongoing play group, they adapt their behavior to the flow of the activity (Gottman, Gonso, & Rasmussen, 1975; Ladd & Price, 1987; Newcomb, Bukowski, & Pattee, 1993).

Rejected children, in contrast, display a wide range of negative social behaviors. At least two subtypes exist. **Rejected-aggressive children,** the largest subgroup, show high rates of conflict, hostility, and hyperactive, inattentive, and impulsive behavior. They are also deficient in social understanding. For example, they are more likely than other children to misinterpret the innocent behaviors of peers as hostile and to blame others for their social difficulties (Crick & Ladd, 1993; Dekovic & Gerris, 1994). In contrast, **rejected-withdrawn children,** a smaller subgroup, are passive and socially awkward. These children, especially, feel lonely, hold negative expectations for how peers will treat them, and are very concerned about being scorned and attacked (Rabiner, Keane, & MacKinnon-Lewis, 1993; Stewart & Rubin, 1995). Because of their inept, submissive style of interaction, they are at risk for abuse by bullies, who look for victims unlikely to retaliate.

Consistent with the mixed peer opinion they engender, controversial children display a blend of positive and negative social behaviors. Like rejected-aggressive youngsters, they are hostile and disruptive, but they also engage in positive, prosocial acts. Even though some peers dislike them, they have qualities that protect them from social exclusion. As a result, they appear to be relatively happy and comfortable with their peer relationships (Newcomb, Bukowski, & Pattee, 1993).

Finally, perhaps the most surprising finding is that neglected children, once thought to be in need of treatment, are usually well adjusted. Although these youngsters engage in low rates of interaction and are considered shy, the majority are just as socially skilled as average children. They do not report feeling especially lonely or unhappy, and when they want to, they can break away from their usual pattern of playing by themselves (Crick & Ladd, 1993; Wentzel & Asher, 1995). Neglected children remind us that there are other paths to emotional well-being besides the outgoing, gregarious personality style so highly valued in our culture.

■ HELPING REJECTED CHILDREN. A variety of interventions aimed at improving the rejected child's peer relations and psychological adjustment have been developed. Most involve coaching, modeling, and reinforcing positive social skills, such as how to begin interacting with a peer, cooperate in play, and respond to another child with friendly emotion and approval. Several of these programs have produced lasting gains in social competence and peer acceptance (Lochman et al., 1993; Mize & Ladd, 1990).

Some researchers believe that these interventions might be even more effective when combined with other treatments. Often rejected children are poor students. Intensive academic tutoring improves both their school achievement and social acceptance (Coie & Kreihbel, 1984). Other interventions focus on training in perspective taking and social problem solving (Ladd & Mize, 1983). Still another approach is to increase rejected children's expectations for social success. Many conclude, after repeated rebuffs from peers, that no matter how hard they try, they will never be liked. Rejected youngsters make better use of the social skills they do have when they believe peers will accept them (Rabiner & Coie, 1989).

GENDER TYPING IN MIDDLE CHILDHOOD

Children's understanding of gender roles broadens in middle childhood, and their gender-role identities (views of themselves as relatively masculine or feminine) change as well. We will see that development is different for boys and girls, and it can vary considerably across cultures.

GENDER-STEREOTYPED BELIEFS

During the school years, children extend the gender-stereotyped beliefs they acquired in early childhood. As they think more about people as personalities, they label some traits as more typical of one gender than the other. For example, they regard "tough," "aggressive," "rational," and "dominant" as masculine and "gentle," "sympathetic," "excitable," and "affectionate" as feminine, in much the same way adults do (Best et al., 1977; Serbin, Powlishta, & Gulko, 1993).

Shortly after entering elementary school, children figure out which academic subjects and skill areas are "masculine" and which are "feminine." Throughout the school years, they regard reading, art, and music as more for girls and mathematics, athletics, and mechanical skills as more for boys (Eccles, Jacobs, & Harold, 1990; Huston, 1983). These stereotypes influence children's preferences for certain subjects and, in turn, how well they do at them. In a study in which children in Japan, Taiwan, and the United States were asked to name the school subject they liked best, girls were more likely to choose reading and boys mathematics in all three cultures. Asked to predict their future performance, boys thought they would do better in math than did girls. In contrast, no sex difference in favor of girls emerged in predictions about reading (Lummis & Stevenson, 1990). As we will see in Chapter 12, these beliefs become realities in adolescence.

Although school-age children are aware of many stereotypes, they have a more open-minded view of what males and females *can do* than they did at younger ages. The ability to classify flexibly underlies this change. School-age children realize that a person can belong to more than one social category—for example, be a "boy" yet "like to play house" (Serbin, Powlishta, & Gulko, 1993). But acknowledging that people *can* cross gender lines does not mean that children always *approve* of doing so. A recent study showed that both children and adults are fairly tolerant of girls' violations of gender roles. But they judge boys' violations ("playing with dolls" or "wearing a dress") very harshly—as just as bad as violating a moral rule! (Levy, Taylor, & Gelman, 1995).

GENDER-ROLE IDENTITY AND BEHAVIOR

Boys' and girls' gender-role identities follow different paths in middle childhood. Self-ratings on personality traits reveal that from third to sixth grade, boys strengthen their identification with the masculine role. In contrast, girls' identification with feminine attributes declines. Although girls still lean toward the feminine side, they begin to describe themselves as having some "other-gender" characteristics (Boldizar, 1991; Serbin, Powlishta, & Gulko, 1993). This difference is also evident in children's activities. Although boys usually stick to "masculine" pursuits, girls feel free to experiment with a wider range of options. Besides cooking, sewing, and baby-sitting, they join organized sports teams, take up science projects, and build forts in the backyard.

In Chapter 8, we saw that parents are especially concerned about boys' gender-role conformity. Pressure from peers is similar. A tomboyish girl can make her way into

boys' activities without losing status with her female peers, but a boy who hangs out with girls is likely to be ridiculed and rejected. Finally, perhaps school-age girls realize that society attaches greater prestige to "masculine" characteristics. As a result, they want to try some of the activities and behaviors associated with the more highly valued gender role.

CULTURAL INFLUENCES ON GENDER TYPING

Although the sex differences just described are typical in Western nations, they do not apply to children everywhere. Girls are less likely to experiment with "masculine" activities in cultures and subcultures where the gap between male and female roles is especially wide. And when social and economic conditions make it necessary for boys to take over "feminine" tasks, their personalities and behavior are less stereotyped.

For example, in Nyansongo, a small agricultural settlement in Kenya, mothers work 4 to 5 hours a day in the gardens. They assign the care of young children, the tending of the cooking fire, and the washing of dishes to older siblings. Half the boys between ages 5 and 8 take care of infants, and

This Kenyan boy is often assigned "feminine" tasks, such as caring for infants and helping with household chores. Compared to boys in other cultures, he is likely to be less gender-stereotyped in personality characteristics. (Paul Conklin/ Monkmeyer Press)

half help with household chores. As a result, girls are relieved of total responsibility for "feminine" tasks and have more idle time to interact with agemates. Their greater freedom and independence leads them to score higher than girls of other village and tribal cultures in dominance, assertiveness, and playful roughhousing. In contrast, boys' caregiving responsibilities mean that they frequently engage in help giving and emotional support (Whiting & Edwards, 1988a, 1988b).

Should these findings be taken to suggest that boys in Western cultures be assigned more "cross-gender" tasks? The consequences of doing so are not straightforward. Recent evidence shows that when fathers hold traditional gender-role beliefs and their sons engage in "feminine" housework, boys experience strain in the father–child relationship, feel stressed by their responsibilities, and judge themselves as less competent (McHale et al., 1990). So parental values may need to be consistent with task assignments for children to benefit from those tasks.

BRIEF REVIEW

During middle childhood, children become members of peer groups, through which they learn about the functioning of larger social structures. Friendships change, emphasizing mutual trust and assistance. Peer acceptance is a powerful predictor of psychological adjustment. Popular children interact positively; rejected children behave antisocially and ineptly; and controversial children display a mixture of positive and negative social behaviors. Neglected children tend to be socially competent youngsters who prefer solitary activities. Interventions that train social skills, improve academic performance, and increase social understanding lead to gains in peer acceptance of rejected children.

Over the school years, children extend their gender-typed beliefs to personality characteristics and achievement areas. Boys' "masculine" gender-role identities strengthen, whereas girls' identities become more flexible. However, cultural values and practices can modify these trends.

ASK YOURSELF . . .

■ *Apply your understanding of attributions to rejected children's social self-esteem. How are rejected children likely to explain their failure to gain peer acceptance? What impact on future efforts to get along with agemates are these attributions likely to have?*

■ *Return to Chapter 8, page 262, and review the concept of androgyny. Which of the two sexes is more androgynous in middle childhood, and why?*

FAMILY INFLUENCES IN MIDDLE CHILDHOOD

As children move into school, peer, and community contexts, the parent–child relationship changes. At the same time, the child's developing sense of competence continues to depend on the quality of family interaction. We will see that recent changes in the American family—high rates of divorce, remarriage, and maternal employment—can have positive as well as negative effects on children.

PARENT–CHILD RELATIONSHIPS

In middle childhood, the amount of time children spend with parents declines dramatically. The child's growing independence means that parents must deal with new issues. "I've struggled with how many chores to assign, how much allowance to give, and whether their friends are good influences," noted Rena. "And then there's the problem of how to keep track of them when they're out of the house or even when they're home and I'm not there to see what's going on."

Although parents face new concerns, child rearing actually becomes easier for those who established an authoritative style during the early years (Maccoby, 1984b). Reasoning works more effectively with school-age children because of their greater capacity for logical thinking. Of course, older children sometimes use their cognitive powers to bargain and negotiate. "Mom," Joey pleaded for the third time, "if you let Terry and me go to the mall tonight, I'll rake all the leaves in the yard, I promise."

Fortunately, parents can appeal to the child's better developed sense of self-esteem, humor, and morality to resolve these difficulties. "Joey, you know it's a school night, and you have a test tomorrow," Rena responded. "You'll be unhappy at the results if you stay out late and don't study. Come on, no more wheeler-dealering!" Perhaps because parents and children have, over time, learned how to resolve conflicts, coercive discipline declines over the school years (Maccoby, 1984a).

As children demonstrate that they can manage daily activities and responsibilities, effective parents gradually shift control from adult to child. This does not mean they let go entirely. Instead, they engage in **coregulation,** a transitional form of supervision in which they exercise general oversight, while permitting children to be in charge of moment-by-moment decision making. Coregulation supports and protects children, who are not yet ready for total independence. At the same time, it prepares them for adolescence, when they will need to make many important decisions themselves.

Coregulation grows out of a cooperative relationship between parent and child—one based on give-and-take and mutual respect. Here is a summary of its critical ingredients:

The parents' tasks . . . are threefold: First, they must monitor, guide, and support their children at a distance—that is, when their children are out of their presence; second, they must effectively use the times when direct contact does occur; and third, they must strengthen in their children the abilities that will allow them to monitor their own behavior, to adopt acceptable standards of good [conduct], to avoid undue risks, and to know when they need parental support and guidance. Children must be willing to inform parents of their whereabouts, activities, and problems so that parents can mediate and guide when necessary. (Maccoby, 1984a, pp. 191–192)

Although school-age children often press for greater independence, they know how much they need their parents' continuing support. In one study, fifth and sixth graders described parents as the most influential people in their lives. They often turned to mothers and fathers for affection, advice, enhancement of self-worth, and assistance with everyday problems (Furman & Buhrmester, 1992).

SIBLINGS

In addition to parents and friends, siblings are important sources of support to school-age youngsters. Siblings provide one another with companionship, help with difficult tasks, and comfort during times of emotional stress. Yet sibling rivalry tends to increase in middle childhood. As children participate in a wider range of activities, parents often compare siblings' traits and accomplishments. The child who gets less parental attention, more disapproval, and fewer material resources is likely to resent a sibling who receives more favorable treatment (Boer, Goedhart, & Treffers, 1992; Brody, Stoneman, & McCoy, 1994).

When siblings are close in age and the same sex, parental comparisons are more frequent, resulting in more quarreling and antagonism. This effect is particularly strong when fathers prefer one child. Perhaps because fathers spend less time with children, their favoritism is more noticeable, thereby triggering greater anger during sibling interactions (Brody, Stoneman, & McCoy, 1992; Brody et al., 1992).

Siblings often take steps to reduce this rivalry by striving to be different from one another (Huston, 1983). For example, two brothers I know deliberately selected different school subjects, athletic pursuits, and music lessons. If the older one did especially well at an activity, the younger one did not want to try it. Of course, parents can reduce these effects by making a special effort not to compare children. But some feedback about their competencies is inevitable, and as siblings strive to win recognition for their own uniqueness, they shape important aspects of each other's development.

Although sibling rivalry tends to increase in middle childhood, siblings also provide one another with emotional support and help with difficult tasks. (Erika Stone)

Birth order plays an important role in sibling experiences. For a time, oldest children have their parents' attention all to themselves. Even after brothers and sisters are born, they receive greater pressure for mature behavior from parents. For this reason, oldest children are slightly advantaged in IQ and school achievement (Zajonc & Mullally, 1997). Younger siblings, in contrast, tend to be more popular with agemates, perhaps as the result of becoming skilled at negotiating and compromising from interacting with more powerful brothers and sisters (Miller & Maruyama, 1976).

ONLY CHILDREN

Although sibling relationships bring many benefits, they are not essential for normal development. Contrary to popular belief, only children are not spoiled, self-centered, and selfish. Instead, they are just as well adjusted as other children and advantaged in some respects. Children growing up in one-child families score higher in self-esteem and achievement motivation. Consequently, they do better in school and attain higher levels of education (Falbo, 1992). One reason for these trends may be that only children have somewhat closer relationships with their parents, who exert more pressure for mastery and accomplishment (Falbo & Polit, 1986).

Are these findings restricted to Western industrialized countries, where family size is a matter of individual choice? The People's Republic of China has a rigid family policy in which urban couples are given strong economic incentives to have no more than one child. Because they are limited to a single offspring, are Chinese parents rearing the new generation with indulgence? To find out, refer to the Cultural Influences box on the following page.

DIVORCE

Sibling relationships are affected by other aspects of family life. Joey and Lizzie's interaction, Rena told me, had been particularly negative only a few years before. Joey pushed, hit, taunted, and called Lizzie names. Although she tried to retaliate, she was little match for Joey's larger size. The arguments usually ended with Lizzie running in tears to her mother. Joey and Lizzie's fighting coincided with Rena and her husband's growing marital unhappiness. When Joey was 8 and Lizzie 5, their father, Drake, moved out.

The children were not alone in having to weather this traumatic event. Between 1960 and 1980, the divorce rate in the United States tripled and then stabilized. Currently, it is the highest in the world, nearly doubling that of the second-ranked country, Sweden. Over one million American children experience the separation and divorce of their parents each year. At any given time, about one-fourth of American youngsters live in single-parent households. Although the large majority (85 percent) reside with their mothers, father custody has increased over the past decade, from 9 to 15 percent (Meyer & Garasky, 1993).

Children spend an average of 5 years in a single-parent home, or almost a third of their total childhood. For many, divorce eventually leads to new family relationships. About two-thirds of divorced parents marry a second time. Half of these children eventually experience a third major change—the end of their parent's second marriage (Hetherington & Jodl, 1994).

These figures reveal that divorce is not a single event in the lives of parents and children. Instead, it is a transition that leads to a variety of new living arrangements, accompanied by changes in housing, income, and family roles and responsibilities (Hetherington, 1995). Since the 1960s, many studies have reported that marital breakup is quite stressful for children. But the research also reveals great individual differences in how children respond. The custodial parent's psychological health, the child's characteristics, and social supports within the family and surrounding community contribute to children's adjustment.

■ IMMEDIATE CONSEQUENCES. "Things were worst during the period Drake and I decided to separate," Rena

reflected. "We fought over everything—from custody of the children to the living room furniture, and the kids really suffered. Once, sobbing, Lizzy told me she was 'sorry she made Daddy go away.' Joey kicked and threw things at home. At school, he was distracted and often didn't do his work. In the midst of everything, I could hardly deal with their problems. We had to sell the house; I couldn't afford it alone. And I needed a better paying job. I had to change from teaching part-time to full-time at the university or start looking."

Rena's description captures conditions in many newly divorced households. Family conflict often rises for a time as parents try to settle disputes over children and personal belongings. Once one parent moves out, additional events threaten supportive interactions between parents and children. Mother-headed households typically experience a sharp drop in income. Three-fourths of divorced mothers get less than the full amount of child support from the absent father or none at all (Children's Defense Fund, 1997). They often have to move to new housing for economic reasons, reducing supportive ties to neighbors and friends.

These life circumstances often lead to a highly disorganized family situation called "minimal parenting" (Wallerstein & Kelly, 1980). "Meals and bedtimes were at all hours, the house didn't get cleaned, and I stopped taking Joey and Lizzie on weekend outings," said Rena. As children react with distress and anger to their less secure home lives, discipline may become harsh and inconsistent as mothers try to recapture control of their upset youngsters. Fathers usually spend more time with children immediately after divorce, but often this contact decreases over time. When fathers see their children only occasionally, they are inclined to be permissive and indulgent. This often conflicts with the mother's style of parenting and makes her task of managing the child on a day-to-day basis even more difficult (Furstenberg & Nord, 1985).

In view of these changes, it is not surprising that children experience painful emotional reactions. But the intensity of their feelings and the way these are expressed vary with the child's age, temperament, and sex.

Children's Age. Five-year-old Lizzie's fear that she had caused her father to leave is not unusual. The cognitive immaturity of preschool and early elementary school children makes it difficult for them to grasp the reasons behind their parents' separation. Younger children often blame themselves and take the marital breakup as a sign that they could be abandoned by both parents. They may whine and cling, displaying intense separation anxiety. Preschoolers are especially likely to fantasize that their parents will get back together (Wallerstein, Corbin, & Lewis, 1988).

Older children are better able to understand the reasons behind their parents' divorce. They recognize that strong differences of opinion, incompatible personalities, and lack of caring for one another are responsible (Mazur, 1993). Still, many school-age and adolescent youngsters

CULTURAL INFLUENCES

ONLY CHILDREN IN THE PEOPLE'S REPUBLIC OF CHINA

The People's Republic of China has 21 percent of the world's population but only 7 percent of its fertile land. In the late 1970s, the Central Committee of the Communist Party concluded that quality of life for China's citizens would not improve without drastic efforts to control its swelling population. In 1979, it implemented a one-child family policy, which was strictly enforced in urban areas and strongly encouraged in rural regions. As a result, a profound generational change occurred. Although the majority of adults of child-bearing age had siblings, most of their children did not. By 1985, 80 to 90 percent of babies born in urban areas and 50 to 60 percent born in rural regions were only children (Tseng et al., 1988).

Critics of the policy complained that it might ruin the character of the Chinese people. In a culture where kinship relations form the basis of daily life, media reports predicted that parents and grandparents would engage in pampering and spoiling, causing children to behave like "little emperors" (Falbo, 1992).

Toni Falbo and Dudly Poston (1993) have carried out the most comprehensive study of the characteristics of Chinese only and non-only children to date. They selected a representative sample of 4,000 third and sixth graders, 1,000 from each of four provinces, each geographically distinct and containing a highly diverse population. The older pupils had been born before implementation of the one-child policy, the younger pupils afterward. All took verbal and mathematical achievement tests. Parent, teacher, peer, and self-ratings on a variety of attributes provided a measure of personality, which was specially designed to tap the "little emperor" syndrome—whether children were selfish, disrespectful of elders, dependent, and uncooperative. Finally, children's height and weight were measured to see if the new policy was associated with improved physical development.

Much like findings in Western nations, Chinese only children were advanced in academic achievement. In 3 of the 4 provinces, they outscored non-only children in verbal skills. Although differences were less clear in math, only children never scored lower than first and last borns, and they usually did better than at least one of these groups. Only children had few distinguishing personality characteristics. And in two of the provinces, they were physically larger than their classmates. The economic benefits parents received from limiting family size may have resulted in better nutrition and health care for children.

Other research reveals that only children in China feel more emotionally secure than children with siblings (Yang et al., 1995). Perhaps government disapproval promotes tension and unhappiness in families with more than one child.

Chinese only children seem to be developing as well as or better than their counterparts with siblings. Why are so many Chinese adults convinced that only children are "little emperors"? When rapid social change takes place, people used to different conditions may regard the new lifestyle as a threat to the social order. In the case of the one-child family policy, they may incorrectly assume that lack of siblings is the cause of children's difficulties, failing to consider that the large number of only children in China today guarantees that they will dominate *both* success and failure groups.

This mural reminds citizens that limiting family size is a basic national policy in the People's Republic of China. In urban areas, the majority of couples have no more than one child. (Owen Franken/Stock Boston)

react strongly to the end of their parents' marriage, particularly when family conflict is high (Borrine et al., 1991; Forehand et al., 1991). Some escape into undesirable peer activities, such as running away, truancy, and delinquent behavior (Doherty & Needle, 1991; Dornbusch et al., 1985).

However, not all older children react this way. For some—especially the oldest child in the family—divorce can trigger more mature behavior. These youngsters may willingly take on extra burdens, such as household tasks, care and protection of younger siblings, and emotional support of a depressed, anxious mother. But if these demands are too great, older children may eventually become resentful and withdraw into some of the destructive behavior patterns just described (Hetherington, 1995).

Children's Temperament and Sex. When temperamentally difficult children are exposed to stressful life events and inadequate parenting, their problems are magnified. In contrast, easy children are less often targets of parental anger and are also better at coping with adversity when it hits.

These findings help us understand sex differences in response to divorce. Girls sometimes respond as Lizzie did, with internalizing reactions, such as crying, self-criticism, and withdrawal. More often, they show some demanding, attention-getting behavior. But in mother-custody families, boys experience more serious adjustment problems. Recall from Chapter 8 that boys are more active and noncompliant—behaviors that increase with exposure to parental conflict and inconsistent discipline. Coercive mother–child interaction and impulsive, defiant behavior on the part of sons are common in divorcing households (Hetherington, 1991).

Perhaps because their behavior is so unruly, boys receive less emotional support from mothers, teachers, and peers. And as Joey's behavior toward Lizzie illustrates, the coercive cycles of interaction between boys and their divorced mothers soon spread to sibling relations (MacKinnon, 1989). These outcomes compound boys' difficulties. Children of both sexes show declines in school achievement during the aftermath of divorce, but school problems are greater for boys (Guidubaldi & Cleminshaw, 1985).

■ LONG-TERM CONSEQUENCES. Rena eventually found full-time work at the university and gained control over the daily operation of the household. Her own feelings of anger and rejection also declined. And after several meetings with a counselor, Rena and Drake realized the harmful impact of their quarreling on Joey and Lizzie. They resolved to keep the children out of future disagreements. Drake visited regularly and handled Joey's unruliness with firmness and consistency. Soon Joey's school performance improved, his behavior problems subsided, and both children seemed calmer and happier.

Like Joey and Lizzie, most children show improved adjustment by 2 years after divorce. Yet a few continue to have serious difficulties into early adulthood (Chase-Lans-

dale, Cherlin, & Kiernan, 1995). Boys and children with difficult temperaments are especially likely to experience lasting emotional problems. Among girls, the major long-term effects involve heterosexual behavior—a rise in sexual activity at adolescence, short-lived sexual relationships in early adulthood, and lack of self-confidence with men (Cherlin, Kiernan, & Chase-Lansdale, 1995; Hetherington & Clingempeel, 1992).

The overriding factor in positive adjustment following divorce is effective parenting—in particular, how well the custodial parent handles stress, shields the child from family conflict, and uses authoritative parenting (Hetherington, 1991). Contact with fathers is also important. For girls, a good father–child relationship appears to contribute to heterosexual development. For boys, it seems to play a crucial role in overall psychological well-being. In fact, several studies indicate that outcomes for sons are better when the father is the custodial parent (Camara & Resnick, 1988; Santrock & Warshak, 1986). Fathers are more likely than mothers to praise a boy's good behavior and less likely to ignore his disruptiveness. The father's image of greater power and authority may also help him obtain more compliance from a son.

Although divorce is painful for children, remaining in a stressed intact family is much worse than making the transition to a low-conflict, single-parent household (Block, Block, & Gjerde, 1988). When divorcing parents put aside their disagreements and support one another in their child-rearing roles, children have the best chance of adapting well to a single-parent household and growing up

■ After parents divorce, most children reside with their mothers. Those who also stay involved with their fathers fare best in development. For girls, a good father–daughter relationship appears to contribute to heterosexual development. For boys, it seems to play a crucial role in overall psychological well-being. (Will Hart)

competent, stable, and happy. Caring extended family members, teachers, and friends are also important in reducing the likelihood that divorce will result in long-term disruption (Hetherington, 1995).

■ DIVORCE MEDIATION, JOINT CUSTODY, AND CHILD SUPPORT. Awareness that divorce is highly stressful for children and families has led to community-based services aimed at helping families through this difficult time. One is **divorce mediation.** It consists of a series of meetings between divorcing adults and a trained professional, who tries to help them settle disputes, such as property division and child custody. Its purpose is to avoid legal battles that intensify family conflict. Research reveals that it increases out-of-court settlements, compliance with these agreements, and feelings of well-being among divorcing parents. By reducing family hostilities, it probably has great benefits for children (Emery, Mathews, & Kitzmann, 1994).

A relatively new child custody option tries to keep both parents involved with children. In **joint custody,** the court grants the mother and father equal say in important decisions about the child's upbringing. Yet many experts have raised questions about the practice. Joint custody results in a variety of living arrangements. In most instances, children reside with one parent and see the other on a fixed schedule, much like the typical sole custody situation. But in other cases, parents share physical custody, and children must move between homes and sometimes

schools and peer groups. These transitions introduce a new kind of instability that is especially hard on some children (Johnston, Kline, & Tschann, 1989). The success of joint custody requires a cooperative relationship between divorcing parents. If they continue to quarrel, it prolongs children's exposure to a hostile family atmosphere (Furstenberg & Cherlin, 1991).

Finally, many single-parent families depend on child support from the absent parent to relieve financial strain. In response to a recent federal law, all states have established procedures for withholding wages from parents who fail to make these court-ordered payments. Although child support is usually not enough to lift a single-parent family out of poverty, it can ease its burdens substantially. An added benefit is that children are more likely to maintain contact with a noncustodial father if he pays child support (Stephen, Freedman, & Hess, 1993). The Caregiving Concerns table below summarizes ways to help children adjust to their parents' divorce.

REMARRIAGE

"If you get married to Wendell, and Daddy gets married to Carol," Lizzie wondered aloud to Rena, "then I'll have two sisters and one more brother. And let's see, how many grandmothers and grandfathers? Gosh, a lot!" exclaimed Lizzie. "But what will I call them all?" she asked, looking worried.

CAREGIVING CONCERNS
Helping Children Adjust to Their Parents' Divorce

SUGGESTION	EXPLANATION
Shield children from conflict.	Witnessing intense parental conflict is very damaging to children. If one parent insists on expressing hostility, children fare better if the other parent does not respond in kind.
Provide children with as much continuity, familiarity, and predictability as possible.	Children adjust better during the period surrounding divorce when their lives have some stability—for example, the same school, bedroom, baby-sitter, and playmates and a dependable daily schedule.
Explain the divorce and tell children what to expect.	Children are more likely to develop fears of abandonment if they are not prepared for their parents' separation. They should be told that their mother and father will not be living together any more, which parent will be moving out, and when they will be able to see that parent. If possible, mother and father should explain the divorce together. Parents should provide a reason for the divorce that the child can understand and assure the child that he or she is not to blame.
Emphasize the permanence of the divorce.	Fantasies of parents getting back together can prevent children from accepting the reality of their current life. Children should be told that the divorce is final and that there is nothing they can do to change that fact.
Respond sympathetically to children's feelings.	Children need a supportive and understanding response to their feelings of sadness, fear, and anger. For children to adjust well, their painful emotions must be acknowledged, not denied or avoided.
Promote a continuing relationship with both parents.	When parents disentangle their lingering hostility toward the former spouse from the child's need for a continuing relationship with the other parent, children adjust well. Grandparents and other extended family members can help by not taking sides.

Source: Teyber, 1992.

For many children, life in a single-parent family is temporary. Their parents find a new partner within a few years. As Lizzie's comments indicate, entry into these *blended*, or *reconstituted*, *families* leads to a complex set of new relationships. For some children, this expanded family network is a positive turn of events that brings greater adult attention. But for most, it presents difficult adjustments. Stepparents often use different child-rearing practices than the child was used to, and having to switch to new rules and expectations can be stressful. In addition, children often regard steprelatives as "intruders" into the family. But how well children adapt is, once again, related to the overall quality of family functioning. This often depends on which parent remarries as well as the age and sex of the child. As we will see, older children and girls seem to have the hardest time.

■ MOTHER–STEPFATHER FAMILIES. The most frequent form of blended family is a mother–stepfather arrangement, since mothers generally retain custody of the child. Boys usually adjust quickly. They welcome a stepfather who is warm and responsive and who offers relief from the coercive cycles of interaction that tend to build with their divorced mothers. Mothers' friction with sons also declines due to greater economic security, another adult to share household tasks, and an end to loneliness (Hetherington, Cox, & Cox, 1985). In contrast, girls adapt less favorably when custodial mothers remarry. Stepfathers disrupt the close ties many girls established with mothers in a single-parent family, and girls often react to the new arrangement with sulky, resistant behavior (Vuchinich et al., 1991).

Note, however, that the child's age affects these findings. Older school-age and adolescent youngsters of both sexes find it harder to adjust to blended families (Hetherington, 1993). Perhaps because they are more aware of the impact of remarriage on their own lives, they challenge some aspects of it that younger children simply accept, creating more relationship issues with their steprelatives.

■ FATHER–STEPMOTHER FAMILIES. Research reveals more confusion for children in father–stepmother families. In the case of noncustodial fathers, remarriage often leads to reduced contact. They tend to withdraw from their "previous" families, more so if they have daughters than sons. When fathers have custody, children typically react negatively to remarriage. One reason is that children living with fathers often start out with more problems. Perhaps the biological mother could no longer handle the unruly child (usually a boy), so the father and his new wife are faced with a youngster who has serious behavior problems. In other instances, the father is granted custody because of a very close relationship with the child, and his remarriage disrupts this bond (Brand, Clingempeel, & Bowen-Woodward, 1988).

Girls, especially, have a hard time getting along with their stepmothers (Hobart & Brown, 1988). Sometimes (as just mentioned) this occurs because the girl's relationship with her father is threatened by the remarriage. In addition, girls often become entangled in loyalty conflicts between their two mother figures. But the longer girls live in father–stepmother households, the more positive their interaction with stepmothers becomes. With time and patience they do adjust, and eventually girls benefit from the support of a second mother figure (Brand, Clingempeel, & Bowen-Woodward, 1988).

■ MATERNAL EMPLOYMENT

Today, single and married mothers are in the labor market in nearly equal proportions, and over 70 percent of those with school-age children are employed (U.S. Bureau of the Census, 1996). In Chapter 6, we saw that the impact of maternal employment on infant development depends on the quality of substitute care and the continuing parent–child relationship. This same conclusion applies during later years. In addition, a host of factors—the mother's work satisfaction, the support she receives from her husband, the child's sex, and the family's social class—have a bearing on how well children do in an employed-mother family.

■ MATERNAL EMPLOYMENT AND CHILD DEVELOPMENT. Children of mothers who enjoy their work and remain committed to parenting show especially positive adjustment—a higher sense of self-esteem, more positive family and peer relations, less gender-stereotyped beliefs, and better grades in school (Hoffman, 1989; Williams & Radin, 1993). These benefits undoubtedly result from parenting practices. Employed mothers who value their parenting role are more likely to use authoritative child rearing (Greenberger & Goldberg, 1989). They schedule special times to devote to their children and also encourage greater responsibility and independence. A modest increase in fathers' involvement in child care and household duties also accompanies maternal employment. More contact with the father is related to higher intelligence and achievement, mature social behavior, and flexible gender-role attitudes (Gottfried, 1991; Williams, Radin, & Allegro, 1992).

But there are some qualifiers to these encouraging findings. Outcomes are more favorable for daughters than sons. Girls, especially, profit from the image of female competence. Daughters of employed mothers have higher educational aspirations and, in college, are more likely to choose nontraditional careers, such as law, medicine, and physics. In contrast, boys in low-income families are sometimes adversely affected. They tend to admire their fathers less and interact more negatively with them. These findings are probably due to a lingering belief in many lower-class

As long as this employed mother enjoys her job, remains committed to parenting, and finds satisfactory child care arrangements, her children are likely to display high self-esteem, positive family and peer relations, flexible beliefs about gender, and good school performance. (Michael Newman/PhotoEdit)

homes that when a mother works, the father has failed in his provider role (Hoffman, 1989).

Furthermore, when employment places heavy demands on the mother's schedule, children are at risk for ineffective parenting. Working long hours and spending little time with school-age children are associated with less favorable adjustment (Moorehouse, 1991). In contrast, part-time employment seems to have benefits for children of all ages, probably because it permits mothers to meet the needs of children with a wide range of characteristics (Lerner & Abrams, 1994; Williams & Radin, 1993).

■ SUPPORT FOR EMPLOYED MOTHERS AND THEIR FAMILIES. As long as mothers have the necessary supports to engage in effective parenting, maternal employment offers children many advantages. In dual-earner families, the husband's willingness to share responsibilities is crucial. If the father helps very little or not at all, the mother carries a double load, at home and at work, leading to fatigue, distress, and reduced time and energy for children.

Besides fathers, work settings and government policies can help employed mothers. Part-time employment and liberal paid maternity and paternity leaves (including time off when children are ill) make juggling the demands of work and child rearing easier. Although these workplace supports are available in Canada and Western Europe, at present only unpaid employment leave is mandated by U.S. federal law.

■ CHILD CARE FOR SCHOOL-AGE CHILDREN. High-quality child care is vital for parents' peace of mind and children's well-being, even during middle childhood. In recent years, much public concern has been voiced about the estimated 2.4 million 5- to 13-year-olds in the United States who regularly look after themselves during after-school hours.

Research on these **self-care children** reveals inconsistent findings. Some studies report that they suffer from low self-esteem, antisocial behavior, poor academic achievement, and fearfulness, whereas others show no such effects (Padilla & Landreth, 1989). What explains these contradictions? The way self-care children spend their time seems to be the crucial factor. Children who have a history of authoritative child rearing, are monitored from a distance by parental telephone calls, and have regular after-school chores appear responsible and well-adjusted. In contrast, those left to their own devices are more likely to bend to peer pressures and engage in antisocial behavior (Steinberg, 1986, 1988).

The Caregiving Concerns table on the page 338 lists signs of readiness for self-care along with ways to help

Child care for school-age children is widely available in Australia but rare in the United States. These Australian children have access to special lessons as part of a government-supported out-of-school-hours program. Low-income children, who otherwise would have few opportunities for such enrichment activities, benefit greatly—in school performance, peer relations, and psychological adjustment. (Carly Wolinsky/Stock Boston)

Signs of Readiness for Self-Care and Ways to Help Children Manage on Their Own

SIGNS OF READINESS	HELPING CHILDREN MANAGE
At least 9 or 10 years old[1]	Establish a telephone check-in procedure with the parent, a relative, or a friend.
Can follow important rules and directions	
Can recognize dangerous situations and respond appropriately	Leave emergency numbers, as well as the numbers of friends and neighbors, by the telephone.
Can make phone calls and take messages in an emergency	Teach safety skills, including a fire escape plan and basic first aid.
Can use household appliances safely	
Can respond to strangers properly (not opening the door, not saying he or she is alone)	Structure the child's after-school time by assigning regular responsibilities.
Can keep track of keys and lock and unlock doors	Establish rules about having friends over, going out, how much television, and which appliances can be used.
Can resolve sibling conflicts independently	Select a safe, well-traveled route home from school, and do not let the child wear a house key on a chain that advertises his or her self-care status.
Does not feel frightened or unhappy	

Source: Peterson, 1989.

[1] Before age 9 or 10, children should not be left unsupervised because they do not yet have the cognitive and social skills to deal with emergencies.

children manage on their own. Unfortunately, when children are not mature enough to handle the self-care arrangement, many employed parents have no alternative. After-school programs for 6- to 13-year-olds are rare in American communities. Where high-quality programs do exist, low-income children who otherwise would have few opportunities for enrichment activities (scouting, music lessons, organized sports) show improved school performance, peer relations, and psychological adjustment (Posner & Vandell, 1994).

BRIEF REVIEW

During the school years, child rearing shifts toward coregulation, a transitional form of supervision in which parents exercise general oversight while granting children more decision-making power. Sibling rivalry tends to increase, and children often take steps to reduce it by striving to be different from one another. Only children are just as well adjusted as children with siblings, and they are advantaged in academic achievement and educational attainment.

Large numbers of American children experience the divorce of their parents. Although many adjust well by 2 years after the divorce, boys and temperamentally difficult children are likely to experience lasting emotional problems. Effective child rearing is the most important factor in help-

ing children adapt to a single-parent family. When parents remarry, children living in father-stepmother families, and daughters especially, display more adjustment difficulties.

Maternal employment is related to high self-esteem, reduced gender stereotyping, and mature social behavior. However, these outcomes vary with children's sex and social class, the demands of the mother's job, and the father's participation in child rearing. The impact of self-care on school-age children varies with parenting practices and how children spend their time.

ASK YOURSELF . . .

■ *"How come you don't study hard and get good grades like your sister?" a mother exclaimed in exasperation after seeing her son's poor report card. What impact do remarks like this have on sibling interaction, and why?*

■ *What advice would you give a divorcing couple with two school-age sons about how to help their children adapt to life in a single-parent family?*

■ *Nine-year-old Bobby's mother has just found employment, so Bobby takes care of himself after school. What factors are likely to affect Bobby's adjustment to this arrangement?*

SOME COMMON PROBLEMS OF DEVELOPMENT

Throughout our discussion, we have considered a variety of stressful experiences that place children at risk for future problems. In the following sections, we touch on two more areas of concern: school-age children's fears and anxieties and the consequences of child sexual abuse. Finally, we sum up factors that help children cope effectively with stress.

FEARS AND ANXIETIES

Although fears of the dark, thunder and lightning, and supernatural beings (stimulated by movies and television) occasionally persist into middle childhood, children's anxieties are largely directed toward new concerns. As children begin to understand the realities of the wider world, the possibility of personal harm (being robbed, stabbed, or shot) and media events (war and disasters) often trouble them. Other common worries include academic performance; parents' health; physical injuries; and rejection by classmates (Silverman, La Greca, & Wasserstein, 1995). Most children handle their fears constructively, by talking about them with parents, teachers, and friends and relying on the more sophisticated emotional self-regulation strategies that develop in middle childhood.

About 20 percent of school-age youngsters develop an intense, unmanageable anxiety of some kind (Beidel, 1991). **School phobia** is an example. Typically, children with this disorder are middle-class youngsters whose achievement is average or above. Still, they feel severe apprehension about attending school, often accompanied by physical complaints (dizziness, nausea, stomachaches, and vomiting) that disappear once they are allowed to remain home. About one-third are 5- to 7-year-olds, most of whom do not fear school so much as separation from their mother. The difficulty can often be traced to a troubled parent–child relationship in which the mother encourages dependency. Intensive family therapy is necessary to help these children (Pilkington & Piersel, 1991).

Most cases of school phobia appear later, around 11 to 13, during the transition from middle childhood to adolescence. These children usually find a particular aspect of school frightening—an overcritical teacher, a school bully, a threatening gang, or too much parental pressure for school success. Treating this form of school phobia may require a change in school environment or parenting practices. Firm insistence that the child return to school along with training in how to cope with difficult situations is also helpful (Klungness, 1990).

Severe childhood anxieties may arise from harsh living conditions. A great many children live in the midst of constant violence. In inner-city ghettos and war-torn areas of the world, they learn to drop to the floor at the sound of gunfire and witness the wounding and killing of friends and relatives. As the Lifespan Vista box on page 340 reveals, these youngsters often suffer from long-term emotional distress. Finally, as we saw in our discussion of child abuse in Chapter 8, too often violence and other destructive acts become part of adult–child relationships. During middle childhood, child sexual abuse increases.

CHILD SEXUAL ABUSE

Until very recently, child sexual abuse was viewed as a rare occurrence. When children came forward with it, adults usually did not take their claims seriously. In the 1970s, efforts by professionals along with widespread media attention caused child sexual abuse to be recognized as a serious national problem. In the United States, several hundred thousand cases are reported each year (see Chapter 8).

■ CHARACTERISTICS OF ABUSERS AND VICTIMS. Sexual abuse is committed against children of both sexes, but more often against girls. The most likely victims are between the ages of 9 and 11. However, sexual abuse also occurs at younger and older ages, and few children experience only a single incident. For some, the abuse begins early in life and continues for many years (Burkhardt & Rotatori, 1995).

Generally, the abuser is male—a parent or someone the parent knows well. Often it is a father, stepfather, or live-in boyfriend, somewhat less often an uncle or older brother. In a few instances, mothers are the offenders, more often with sons than daughters. If it is a nonrelative, it is usually someone the child has come to know and trust (Alter-Reid et al., 1986).

In the overwhelming majority of cases, the abuse is serious—vaginal or anal intercourse, oral-genital contact, fondling, and forced stimulation of the adult. Abusers make the child comply in a variety of distasteful ways, including deception, bribery, verbal intimidation, and physical force (Gomez-Schwartz, Horowitz, & Cardarelli, 1990).

You may be wondering how any adult—especially, a parent or close relative—could possibly violate a child sexually. Many offenders deny their own responsibility. They blame the abuse on the willing participation of a seductive youngster. Yet children are not capable of making a deliberate, informed decision to enter into a sexual relationship! Even at older ages, they are not free to say yes or no. Instead, abusers tend to have characteristics that predispose them toward sexual exploitation. They have great difficulty controlling their impulses, may suffer from psychological disorders, and are often addicted to alcohol or drugs. Often they pick children who are unlikely to defend themselves—those who are physically weak, emotionally deprived, and socially isolated.

A LIFESPAN VISTA

CHILDREN OF WAR

On May 27, 1992, Zlata Filipovic, a 10-year-old Bosnian girl, recorded the following reactions to the intensifying Serb attack on the city of Sarajevo in her diary:

> SLAUGHTER! MASSACRE! HORROR! CRIME! BLOOD! SCREAMS! TEARS! DESPAIR!

That's what Vaso Miskin Street looks like today. Two shells exploded in the street and one in the market. Mommy was nearby at the time. She ran to Grandma and Granddad's. Daddy and I were beside ourselves because she hadn't come home. I saw some of it on TV but I still can't believe what I actually saw. It's unbelievable. I've got a lump in my throat and a knot in my tummy. HORRIBLE. They're taking the wounded to the hospital. It's a madhouse. We kept going to the window hoping to see Mommy, but she wasn't back. . . . Daddy and I were tearing our hair out. . . . I looked out the window one more time and . . . I SAW MOMMY RUNNING ACROSS THE BRIDGE. As she came into the house she started shaking and crying. Through her tears she told us how she had seen dismembered bodies. . . . Thank God, Mommy is with us. Thank God. (Filipovic, 1994, p. 55)

Violence stemming from ethnic and political tensions is being felt increasingly around the world. Since World War II, almost all the hundreds of conflicts around the globe have been internal civil wars. Besides being armed encounters, modern wars are usually social upheavals in which well-established ways of life are threatened or destroyed and women and children are frequent victims (Ressler, 1993).

Children's experiences under conditions of armed conflict are diverse. Some may participate in the fighting, either because they are forced or because they want to please adults. Others are kidnapped, terrorized, or tortured. Those who are bystanders often come under direct fire and may be killed or physically maimed for life. And as Zlata's diary entry illustrates, many children of war watch in horror as family members, friends, and neighbors flee, are wounded, or die (Ladd & Cairns, 1996).

When war and social crises are temporary, most children are comforted by caregivers' reassuring messages and do not show long-term emotional difficulties. But chronic danger requires children to make substantial adjustments, and their psychological functioning can be seriously impaired. Many children of war lose their sense of safety, acquire a high tolerance for violence, are haunted by terrifying memories, become suspicious of others, and build a pessimistic view of the future that can last for decades (Cairns, 1996).

The extent to which children are negatively affected by war depends on other factors. Closeness to wartime events increases the chances of maladjustment. For example, an estimated 50 percent of traumatized 6- to 12-year-old Cambodian war refugees continued to show intense stress reactions when they reached young adulthood (Kinzie et al., 1989). The support and affection of parents is the best safeguard against lasting problems. Unfortunately, many children of war are separated from family members. Sometimes, the child's community can offer protection. For example, Israeli children who lost a parent in battle fared best when they lived in kibbutzim, cohesive agricultural settlements where many adults knew the child well and felt responsible for his or her welfare (Lifschitz et al., 1977).

When wartime drains families and communities of resources, international organizations need to step in and help children. Until we know how to prevent war, efforts to preserve children's physical, psychological, and educational well-being may be the best way to stop transmission of violence to the next generation in many parts of the world.

This Rwandan refugee child has experienced the trauma of civil war. Here he watches as his home burns, and he has probably witnessed the wounding and death of family members, friends, and neighbors. If he survives, he is likely to show lasting emotional problems without special support from caring adults. (Michael Simpson/ The Picture Cube)

Reported cases of child sexual abuse are strongly linked to poverty, marital instability, and resulting weakening of family ties. Children who live in homes with a history of constantly changing characters—repeated marriages, separations, and new partners—are especially vulnerable. But community surveys reveal that middle-class children in relatively stable homes are also victims. Economically advantaged families are simply more likely to escape detection (Gomez-Schwartz, Horowitz, & Cardarelli, 1990).

■ CONSEQUENCES FOR CHILDREN. The adjustment problems of child sexual abuse victims are often severe. Depression, low self-esteem, mistrust of adults, feelings of anger and hostility, and difficulties in getting along with peers are common. Younger children often react with sleep difficulties, loss of appetite, and generalized fearfulness. Adolescents sometimes show runaway and suicidal reactions, substance abuse, and delinquency (Kendall-Tacket, Williams, & Finkelhor, 1993).

Sexually abused children frequently display sexual knowledge and behavior beyond their years. They have learned from their abusers that sexual overtures are acceptable ways to get attention and rewards. Abused girls are likely to enter into unhealthy relationships as they move toward young adulthood. Many become promiscuous, believing that their bodies are for the use of others. When they marry, they are likely to choose husbands who abuse them and their children (Faller, 1990). And as mothers, they often show poor parenting skills, abusing and neglecting their youngsters (Pianta, Egeland, & Erickson, 1989). In these ways, the harmful impact of sexual abuse is transmitted to the next generation.

■ PREVENTION AND TREATMENT. Treating child sexual abuse is difficult. The reactions of family members—anxiety about harm to the child, anger toward the abuser, and sometimes hostility toward the victim for telling—can increase children's distress. Since sexual abuse typically appears in the midst of other serious family problems, long-term therapy with children and families is usually necessary (Doyle, 1994).

The best way to reduce the suffering of victims is to prevent child sexual abuse from continuing. Today, courts are prosecuting abusers (especially nonrelatives) more rigorously. As the Social Issues box on the page 342 indicates, children's testimony is being taken more seriously. In schools, sex education programs can help children recognize inappropriate sexual advances and encourage them to report these actions. Finally, educating teachers, parents, and other adults who work with children about the signs and symptoms of sexual abuse can help ensure that victimized children are identified early and receive the help they need.

So there really was a monster in her bedroom.

For many kids, there's a real reason to be afraid of the dark.

Last year in Indiana, there were 6,912 substantiated cases of sexual abuse. The trauma can be devastating for the child and for the family. So listen closely to the children around you.

If you hear something you don't want to believe, perhaps you should. For helpful information on child abuse prevention, contact the LaPorte County Child Abuse Prevention Council, 7451 Johnson Road, Michigan City, IN 46360. (219) 874-0007

LaPorte County Child Abuse Prevention Council

As this poster points out, child sexual abuse, until recently regarded as a product of children's vivid imaginations, is a devastating reality. Victims are in urgent need of protection and treatment. (La Porte County Child Abuse Protection Council)

STRESS AND COPING: THE RESILIENT CHILD

Throughout middle childhood—and other phases of development as well—children are confronted with challenging and sometimes threatening situations that require them to cope with psychological stress. In this and the previous chapter, we have considered such topics as chronic health problems, learning disabilities, divorce, and child sexual abuse. Each taxes children's coping resources, creating serious risks for development.

At the same time, many studies show only a modest relationship between stressful life experiences and psychological disturbance in childhood (Garmezy, 1993; Rutter, 1979). Recall our discussion in Chapter 3 of the long-term consequences of birth complications. We noted that some

SOCIAL ISSUES

CHILDREN'S EYEWITNESS TESTIMONY

Increasingly, children are being called on to testify in court cases involving child abuse and neglect, child custody, and other matters. Having to provide such information can be difficult and traumatic. Almost always, children must report on highly stressful events. In doing so, they may have to speak against a parent or other relative to whom they feel loyal. In some family disputes, they may fear punishment for telling the truth. In addition, child witnesses are faced with an unfamiliar situation—at the very least an interview in the judge's chambers, and at most an open courtroom with judge, jury, spectators, and the possibility of unsympathetic cross-examination. Not surprisingly, there is considerable debate about the accuracy of children's recall under these conditions.

In most states, it is rare for children under age 5 to be asked to testify, whereas those age 6 and older often are. Children between ages 10 and 14 are generally assumed competent to testify. These guidelines make good sense in terms of what we know about memory development. Compared to preschoolers, school-age children are better able to give detailed descriptions of past experiences and make accurate inferences about others' motives and intentions. Also, older children are more resistant to misleading questions of the sort attorneys ask when, in cross-examination, they try to influence the content of the child's response (Ceci & Bruck, 1993; Goodman & Tobey, 1994).

Nevertheless, when properly questioned, even 3-year-olds can recall recent events accurately—including ones that were highly stressful (Baker-Ward et al., 1993; Goodman et al., 1991). But court testimony often involves repeated interviews in which children are asked suggestive questions. These circumstances increase the likelihood of incorrect reporting—even among school-age children, whose descriptions are usually elaborate and dependable (Leichtman & Ceci, 1995). By the time children come to court, it is weeks, months, or even years after the occurrence of the target events. When a long delay is combined with suggestions about what happened, children can easily be misled into giving false information (Ceci, Leichtman, & Bruck, 1994).

When children are interviewed in a frightening legal setting, their ability to report past events accurately is reduced further (Saywitz & Nathanson, 1993). To ease the task of testifying, special interviewing methods have been devised for children. Sometimes professionals have puppets ask questions and the child respond through them. In child sexual abuse cases, anatomically correct dolls have been used to prompt children's recall. However, serious concerns have been raised about this method. Research indicates that it does not improve the accuracy of young children's answers. And it can encourage them to report physical and sexual contact that never happened (Bruck et al., 1995; Wolfner, Faust, & Dawes, 1993).

Child witnesses need to be prepared so that they understand the courtroom process and know what to expect. In some places, "court schools" exist in which children are taken through the setting and given an opportunity to role-play court activities. As part of this process, children can be encouraged to admit not knowing an answer rather than guessing or going along with what an adult expects. At the same time, legal professionals need to take steps to lessen the risk of suggestibility—by limiting the number of times children are interviewed and asking questions in nonleading ways.

If a child is likely to experience emotional trauma or later punishment (in a family dispute), then courtroom procedures can be adapted to protect them. For example, they can testify over closed-circuit TV so they do not have to face an abuser. When it is not wise for a child to participate directly, expert witnesses can provide testimony that reports on the child's psychological condition and includes important elements of the child's story (Ceci & Bruck, 1995).

Will this 6-year-old boy recount events accurately and completely on the witness stand? The answer to this question depends on many factors—his cognitive maturity, the way he is questioned, how long ago the events occurred, whether adults in his life have tried to influence his responses, how the doll is used to prompt his recall, and his understanding of the courtroom process. (Stacy Pick/Stock Boston)

TRY THIS...

■ Contact a local therapist, lawyer, or judge who has had experience interviewing children in court proceedings. Find out what procedures he or she uses to question children. Do they match the recommendations of research on how to increase the chances of accurate reporting?

children manage to overcome the combined effects of birth trauma, poverty, and a deeply troubled family life. The same is true when we look at findings on school difficulties, family transitions, and child maltreatment. What promotes such remarkable resilience in the face of adversity? Research on stress-resistant children highlights three broad factors that consistently protect against maladjustment:

■ Personal characteristics of children—an easy temperament, high self-esteem, and a mastery-oriented approach to new situations.

■ A family environment that provides warmth, closeness, and order and organization to the child's life.

■ A person outside the immediate family—perhaps a grandparent, teacher, or close friend—who develops a special relationship with the child, offering a support system and a positive coping model.

Any one of these ingredients can account for why one child fares well and another poorly when exposed to extreme hardship. Yet most of the time, personal and environmental resources are interconnected (Smith & Prior, 1995; Sorenson, 1993). Throughout this book, we have seen many examples of how unfavorable life experiences increase the chances that parents and children will act in ways that expose them to further hardship. Children can usually handle one stressor in their lives, even if it is chronic. But when negative conditions pile up, such as marital discord, poverty, crowded living conditions, and parental psychological disorder, the rate of maladjustment is multiplied (Capaldi & Patterson, 1991; Sameroff et al., 1993).

Social supports are especially important during periods of developmental transition—when children are more vulnerable because they are faced with many new tasks (Rutter, 1987). One such turning point is the beginning of middle childhood, a time of new challenges in academic work and peer relations. As the next two chapters will reveal, another major turning point is the transition to adolescence. Especially at these times, families, schools, communities, and society as a whole can do much to enhance development.

SUMMARY

ERIKSON'S THEORY: INDUSTRY VERSUS INFERIORITY

What personality changes take place during Erikson's stage of industry versus inferiority?

■ According to Erikson, the personality changes of the school years build on Freud's **latency** stage. Children who successfully resolve the psychological conflict of **industry versus inferiority** develop a sense of competence at useful skills and tasks.

SELF-DEVELOPMENT IN MIDDLE CHILDHOOD

Describe school-age children's self-concept and self-esteem, and discuss factors that affect their achievement-related attributions.

■ During middle childhood, children's self-concepts include personality traits and social comparisons. Self-esteem becomes hierarchically organized and declines over the early school years as children adjust their self-judgments to feedback from the environment.

■ Children with **mastery-oriented attributions** credit their successes to high ability and their failures to insufficient effort. In contrast, children who receive negative feedback about their ability develop **learned helplessness,** attributing their failures to low ability.

EMOTIONAL DEVELOPMENT IN MIDDLE CHILDHOOD

Cite changes in understanding and expression of emotion in middle childhood.

■ In middle childhood, the self-conscious emotions of pride and guilt become integrated with personal responsibility. School-age children also recognize that people can experience more than one emotion at a time. Emotional self-regulation improves as children use cognitive strategies for controlling feelings.

UNDERSTANDING OTHERS

How does perspective taking change in middle childhood?

■ **Perspective taking** improves greatly during the school years, as Selman's five-stage sequence indicates. Cognitive maturity and experiences in which adults and peers encourage children to take note of another's viewpoint support changes in perspective-taking skill. Good perspective takers are more socially skilled.

MORAL DEVELOPMENT IN MIDDLE CHILDHOOD

Describe changes in moral understanding during middle childhood.

■ By middle childhood, children have internalized a wide variety of moral rules. Their concepts of **distributive justice** change, from equality to merit to benevolence. School-age children also begin to grasp the relationship between moral rules and social conventions.

PEER RELATIONS IN MIDDLE CHILDHOOD

How do peer sociability and friendship change in middle childhood?

■ In middle childhood, peer interaction becomes more prosocial, and aggression declines. By the end of the school years, children organize themselves into **peer groups.** Friendships develop into mutual relationships based on trust.

Describe major categories of peer acceptance and ways to help rejected children.

■ On measures of peer acceptance, **popular children** are liked by many agemates, **rejected children** are actively disliked, **controversial children** are both liked and disliked, and **neglected children** are seldom chosen, either positively or negatively.

■ At least two subtypes of peer rejection exist: **rejected-aggressive children,** who show severe conduct problems, and **rejected-withdrawn children,** who are passive and socially awkward. Both subgroups often experience lasting adjustment difficulties. Interventions that provide coaching in social skills, academic tutoring, and training in social understanding have been used to help rejected youngsters.

GENDER TYPING IN MIDDLE CHILDHOOD

What changes in gender-stereotyped beliefs and gender-role identity take place during middle childhood?

■ School-age children extend their awareness of gender stereotypes to personality characteristics and academic subjects. Boys strengthen their identification with the masculine role, whereas girls often experiment with "opposite-gender" activities. Cultures shape gender-typed behavior through daily activities assigned to children.

FAMILY INFLUENCES IN MIDDLE CHILDHOOD

How do parent–child communication and sibling relationships change in middle childhood?

■ Effective parents of school-age youngsters engage in **coregulation,** exerting general oversight while permitting children to be in charge of moment-by-moment decision making. Coregulation depends on a cooperative relationship between parent and child.

■ During middle childhood, sibling rivalry increases as children participate in a wider range of activities and parents compare their abilities. Oldest children are slightly advantaged in IQ and school achievement. Younger siblings are more popular. Only children are as well adjusted as children with siblings, and they do better in school and attain higher levels of education.

What factors influence children's adjustment to divorce and remarriage?

■ Divorce is common in the lives of American children. Although most experience painful emotional reactions, younger children and boys in mother-custody homes tend to react more strongly. Boys and children with difficult temperaments are more likely to show lasting psychological problems.

■ The most important factor in a positive adjustment following divorce is effective parenting. Contact with noncustodial fathers is also important. Because **divorce mediation** helps parents resolve disputes, it can reduce children's exposure to conflict. **Joint custody** is a controversial practice that may create additional strains for children.

■ Many divorced parents remarry, a transition that also creates difficulties for children. Children in father–stepmother families, and

girls especially, display the greatest adjustment problems.

What factors influence the impact of maternal employment on school-age children?

■ As long as mothers enjoy their work and remain committed to parenting, maternal employment is associated with a higher sense of self-esteem in children, more positive family and peer relations, less gender-stereotyped beliefs, and better grades in school. However, outcomes are more positive for daughters than sons, and boys in low-income homes sometimes show adverse effects.

■ **Self-care children** who are monitored from a distance and experience authoritative parenting appear responsible and well adjusted. High-quality after-school programs, fathers' involvement in family responsibilities, and opportunities for part-time employment help mothers balance the multiple demands of work and child rearing.

SOME COMMON PROBLEMS OF DEVELOPMENT

Cite common fears and anxieties in middle childhood.

■ School-age children's fears are directed toward new concerns having to do with personal harm, media events, academic performance, parents' health, and peer relations. Some children develop intense, unmanageable fears, such as **school phobia.** Severe anxiety can also result from harsh living conditions.

Discuss factors related to child sexual abuse and its consequences for children's development.

■ Child sexual abuse is generally committed by male family members, more often against girls than boys. Abusers have characteristics that predispose them toward sexual exploitation of children. Reported cases are strongly associated with poverty and marital instability. Abused children often have severe adjustment problems.

STRESS AND COPING: THE RESILIENT CHILD

Cite factors that help children cope with stress and reduce the chances of maladjustment.

■ Overall, only a modest relationship exists between stressful life experi-

ences and psychological disturbance in childhood. Personal characteristics of children, a warm, well-organized home life, and social supports outside the family are related to childhood resilience in the face of stress.

IMPORTANT TERMS AND CONCEPTS

latency stage (p. 318)

industry versus inferiority (p. 318)

mastery-oriented attributions (p. 321)

learned helplessness (p. 321)

perspective taking (p. 323)

distributive justice (p. 324)

popular children (p. 327)

rejected children (p. 327)

controversial children (p. 327)

neglected children (p. 328)

rejected-aggressive children (p. 328)

rejected-withdrawn children (p. 328)

coregulation (p. 330)

divorce mediation (p. 335)

joint custody (p. 335)

peer group (p. 336)

self-care children (p. 337)

school phobia (p. 339)

FOR FURTHER INFORMATION AND HELP

DIVORCE

Parents Without Partners
8807 Colesville Road
Silver Spring, MD 20910
(800) 637-7974
Organization of custodial and noncustodial single parents that provides support in the upbringing of children. Many local groups exist throughout the United States.

REMARRIAGE

Stepfamily Association of America
215 Centennial Mall South,
Suite 212
Lincoln, NE 68508
(402) 477-7837
Association of families interested in stepfamily relationships. Organizes support groups and offers education and children's services.

Stepfamily Foundation
333 West End Avenue
New York, NY 10023
(212) 877-3244
Web site: *www.stepfamily.org*
Organization of remarried parents, interested professionals, and divorced individuals. Arranges group counseling sessions for stepfamilies and provides training for professionals.

CHILD SEXUAL ABUSE

Parents United International
232 Gish Road
San Jose, CA 95112
(408) 453-7616
Organization of individuals who have experienced child sexual abuse. Assists families affected by incest and other types of sexual abuse by providing information and arranging for medical and legal counseling.

Society's League Against Molestation
c/o Women Against Rape/ Childwatch
P.O. Box 346
Collingswood, NJ 08108
(609) 858-7800
A 100,000-member organization that works to prevent child sexual abuse through public education. Offers counseling and assistance to victims and their families.

AGE	PHYSICAL	COGNITIVE	LANGUAGE	EMOTIONAL/SOCIAL
6–8 years	■ Slow gains in height and weight continue until adolescent growth spurt.	■ Thought becomes more logical, as shown by the ability to pass Piagetian conservation, class inclusion, and seriation problems.	■ Vocabulary increases rapidly throughout middle childhood.	■ Self-esteem differentiates, hierarchically organized, and declines to a more realistic level.
	■ Gradual replacement of primary teeth by permanent teeth throughout middle childhood.	■ Understanding of spatial concepts, including the ability to give directions, improves.	■ Word definitions are concrete, referring to functions and appearance.	■ Distinguishes ability, effort, and luck in attributions for success and failure.
	■ Writing becomes smaller and more legible. Letter reversals decline.	■ Attention becomes more controlled, adaptable, and planful.	■ Language awareness improves over middle childhood.	■ Understands that access to different information often causes people to have different perspectives.
	■ Drawings become more organized and detailed and include some depth cues.	■ Uses memory strategies of rehearsal and organization.		■ Becomes more responsible and independent.
	■ Organized games with rules become common.	■ Awareness of the importance of memory strategies and the impact of psychological factors (attention, motivation) in task performance improves.		■ Distributive justice reasoning changes from equality to merit to benevolence.
				■ Peer interaction becomes more prosocial, and physical aggression declines.
				■ Self-conscious emotions of pride and guilt are integrated with personal responsibility.

AGE	PHYSICAL	COGNITIVE	LANGUAGE	EMOTIONAL/SOCIAL
9–11 years	■ Adolescent growth spurt begins 2 years earlier for girls than boys.	■ Logical thought remains tied to concrete situations.	■ Word definitions emphasize synonyms and categorical relations.	■ Self-concept includes personality traits and social comparisons.
	■ Gross motor skills of running, jumping, throwing, catching, kicking, batting, and dribbling are executed more quickly and with better coordination.	■ Piagetian tasks continue to be mastered in a step-by-step fashion.	■ Understanding of complex grammatical forms improves.	■ Self-esteem tends to rise.
		■ Memory strategies of rehearsal and organization become more effective. Begins to use memory strategy of elaboration.	■ Grasps double meanings of words, as reflected in comprehension of metaphors and humor.	■ Explains emotion by referring to internal states; recognizes that individuals can experience more than one emotion at a time.
	■ Reaction time improves, contributing to motor skill development.		■ Adapts messages to the needs of listeners in complex communicative situations.	■ Emotional self-regulation includes cognitive strategies.
	■ Representation of depth in drawings expands.	■ Long-term knowledge base grows larger and becomes better organized.	■ Conversational strategies become more refined.	■ Can "step in another's shoes" and view the self from that person's perspective.
		■ Self-regulation improves.		■ Later, can view the relationship between self and other from the perspective of a third, impartial party.
				■ Appreciates the linkage between moral rules and social conventions.
				■ Peer groups emerge.
				■ Friendships are based on mutual trust.
				■ Personality traits and academic subjects become gender stereotyped, but school-age children (especially girls) view the capacities of males and females more flexibly.
				■ Sibling rivalry tends to increase.

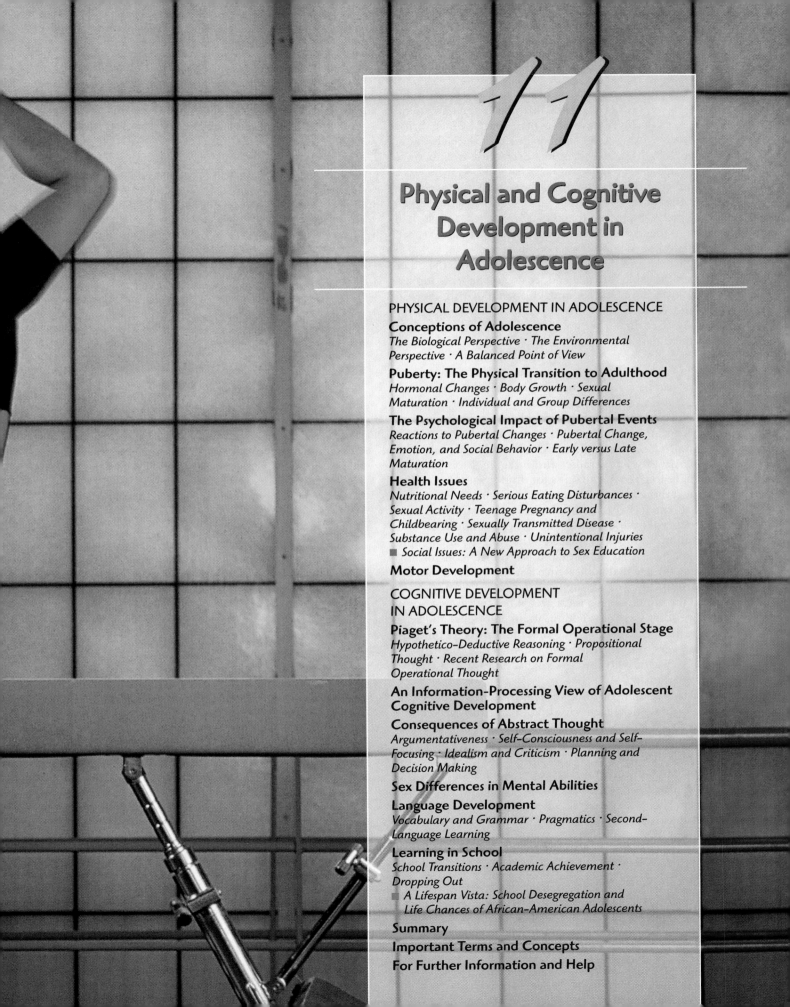

11

Physical and Cognitive Development in Adolescence

On her eleventh birthday, Sabrina's friend Joyce gave a surprise party, but Sabrina appeared somber during the celebration. Although Sabrina and Joyce had been close friends since third grade, their relationship was faltering. Sabrina was a head taller and some 20 pounds heavier than most of the other girls in her sixth-grade class. Her breasts were well developed, her hips and thighs had broadened, and she had begun to menstruate. In contrast, Joyce still had the short, lean, flat-chested body of a school-age child.

Ducking into the bathroom while Joyce and the other girls set the table for cake and ice cream, Sabrina looked herself over in the mirror and whispered, "Gosh, I feel so big and heavy." At church youth group on Sunday evenings, Sabrina broke away from Joyce and spent time with the eighth-grade girls, around whom she didn't feel so large and awkward.

Once every 2 weeks, parents gathered at Sabrina and Joyce's school for discussions about child-rearing concerns. Sabrina's Italian-American parents, Franca and Antonio, came whenever they could. "How you know they are becoming teenagers is this," Antonio volunteered one evening. "The bedroom door is closed, and they want to be alone. Also, they contradict and disagree. I tell Sabrina, 'You have to go to Aunt Gina's for dinner with the family.' The next thing I know, she is arguing with me."

Sabrina has entered adolescence, a period in which she will cross the dividing line between childhood and adulthood. In modern societies, the skills young people must master are so complex and the choices confronting them so diverse that adolescence lasts for nearly a decade. But around the world, the basic tasks of this phase are much the same. Sabrina must accept her full-grown body, acquire adult ways of thinking, attain emotional and economic independence, develop more mature ways of relating to peers of both sexes, and construct an identity—a secure sense of who she is, sexually, morally, politically, and vocationally.

The beginning of adolescence is marked by **puberty,** a flood of biological events leading to an adult-sized body and sexual maturity. As Sabrina's reactions suggest, entry into adolescence can be a trying time, more so for some youngsters than for others. In this chapter, we trace the events of puberty and take up a variety of health concerns. We also discuss motor skills, which highlight the large sex differences in physical development at this time.

Adolescence brings with it the capacity for abstract thinking, which opens up new realms of learning. Teenagers can grasp complex scientific principles, grapple with social and political issues, detect the hidden meaning of a poem or story, and deal with language in increasingly flexible, creative ways. The second part of this chapter traces these extraordinary changes from both Piaget's and the information-processing perspective. Then we take a close look at some findings that have attracted a great deal of public attention: sex differences in mental abilities. The final portion of this chapter is devoted to the primary setting in which adolescent thought takes shape: the school.

PHYSICAL DEVELOPMENT IN ADOLESCENCE

CONCEPTIONS OF ADOLESCENCE

THE BIOLOGICAL PERSPECTIVE

Ask several parents of young children what they expect their sons and daughters to be like as teenagers. You will probably get answers like these: "Rebellious and uncontrollable." "Full of rages and tempers." This view, widespread in contemporary American society, dates back to the writings of eighteenth-century philosopher Jean-Jacques Rousseau, whom we introduced in Chapter 1. Rousseau (1762/1955) believed that a natural outgrowth of the biological upheaval of puberty was heightened emotionality, conflict, and defiance of adults.

In the twentieth century, this perspective was picked up by major theorists. The most influential was G. Stanley Hall, whose view of development was grounded in Darwin's theory of evolution. Hall (1904) described adolescence as a cascade of instinctual passions, a phase of growth so turbulent that it resembled the period in which humans evolved from savages into civilized beings.

Sigmund Freud, as well, emphasized the emotional storminess of the teenage years. He called adolescence the **genital stage,** a period in which instinctual drives reawaken and shift to the genital region of the body. The struggle of the earlier phallic period is renewed, resulting in psychological conflict and volatile, unpredictable behavior. But unlike preschool children, adolescents can find romantic partners outside the family. As they do so, inner forces gradually achieve a new, more mature harmony, and the stage concludes with marriage, birth, and rearing of children. In this way, young people fulfill their biological destiny: sexual reproduction and survival of the species.

THE ENVIRONMENTAL PERSPECTIVE

Recent research on large numbers of teenagers suggests that the notion of adolescence as a biologically determined period of storm and stress is greatly exaggerated. A number of problems, such as eating disorders, depression, suicide, and lawbreaking, occur more often in adolescence than earlier. But the overall rate of severe psychological disturbance rises only slightly (by 2 percent) from childhood to adolescence, when it is the same as in the adult population—about 15 to 20 percent (Powers, Hauser, & Kilner, 1989). Although some teenagers encounter serious difficulties, emotional turbulence is not a routine feature of adolescence.

The first researcher to point out the wide variability in adolescent adjustment was anthropologist Margaret Mead (1928). She traveled to the Pacific islands of Samoa and returned with a startling conclusion: Because of the culture's relaxed social relationships and openness toward sexuality, adolescence "is perhaps the pleasantest time the Samoan girl (or boy) will ever know" (p. 308).

Mead offered an alternative view in which the social environment is entirely responsible for the range of teenage experiences, from erratic and agitated to calm and stress-free. Yet this conclusion is just as extreme as the biological perspective it tried to replace! Later researchers found that Samoan adolescence was not as untroubled as Mead had assumed (Freeman, 1983). Still, Mead convinced researchers that greater attention must be paid to social and cultural influences for adolescent development to be understood.

A BALANCED POINT OF VIEW

Today, we know that adolescence is a product of *both* biological and social forces. Biological changes are universal—found in all primates and all cultures. These internal stresses and the social expectations accompanying them—that the young person give up childish ways, develop new interpersonal relationships, and take on greater responsibility—are likely to prompt moments of uncertainty, self-doubt, and disappointment in all teenagers.

At the same time, the length of adolescence varies greatly from one culture to the next. Although simpler societies have a shorter transition to adulthood, usually adolescence is not absent. A recent study of 186 tribal and village cultures revealed that almost all had an intervening phase, however brief, between childhood and full assumption of adult roles (Schlegel & Barry, 1991).

In industrialized societies, where successful participation in economic life requires many years of education, adolescence is greatly extended. Young people face extra years of dependence on parents and postponement of sexual gratification as they prepare for a productive work life. We will see that the extent to which the social environment supports young people in achieving adult responsibilities has much to do with how well they fare. For all the biological tensions and uncertainties about the future that modern teenagers feel, most are surprisingly good at negotiating this period of life. With this idea in mind, let's look closely at puberty, the dawning of adolescent development.

PUBERTY: THE PHYSICAL TRANSITION TO ADULTHOOD

The changes of puberty are dramatic and momentous. Within a few years, the body of the school-age child is transformed into that of a full-grown adult. Pubertal growth is regulated by genetically influenced hormonal processes. Girls, who have been advanced in physical maturity since the prenatal period, reach puberty, on the average, 2 years earlier than do boys. Heredity also contributes to wide individual differences. Identical twins are more similar in timing and rate of pubertal changes than are fraternal twins and siblings (Tanner, 1990). Yet environment remains important. We will see that nutrition, exercise, and physical health play important roles.

H ORMONAL CHANGES

The complex hormonal changes that underlie puberty take place gradually and are underway by age 8 or 9. Secretion of *growth hormone (GH)* and *thyroxine* (see Chapter 7, page 210) increase, leading to tremendous gains in body size and attainment of skeletal maturity.

Sexual maturation is controlled by the sex hormones. Although *estrogens* are thought of as female hormones and *androgens* as male hormones, both types are present in each sex, but in different amounts. The boy's testes release large quantities of the androgen *testosterone,* which leads to muscle growth, body and facial hair, and other male sex characteristics. Testosterone also contributes to gains in body size. Estrogens released by the girl's ovaries cause the breasts, uterus, and vagina to mature, the body to take on feminine proportions, and fat to accumulate. In addition, estrogens help regulate the menstrual cycle. *Adrenal androgens,* released from the adrenal glands on top of each kidney, influence the girl's height spurt and stimulate growth of underarm and pubic hair. They have little impact on boys, whose physical characteristics are mainly affected by androgen secretions from the testes.

As you can already tell, pubertal changes can be divided into two broad types: (1) overall body growth; and (2) maturation of sex characteristics. Although we will discuss these changes separately, they are interrelated. We have seen that hormones responsible for sexual maturity also affect body growth; boys and girls differ in both aspects.

B ODY GROWTH

The first outward sign of puberty is the rapid gain in height and weight known as the **growth spurt.** On the average, it is underway for American girls shortly after age 10, for boys around age 12½ (Malina, 1990). The girl is taller and heavier during early adolescence, but this advantage is short-lived. At age 14, she is surpassed by the typical

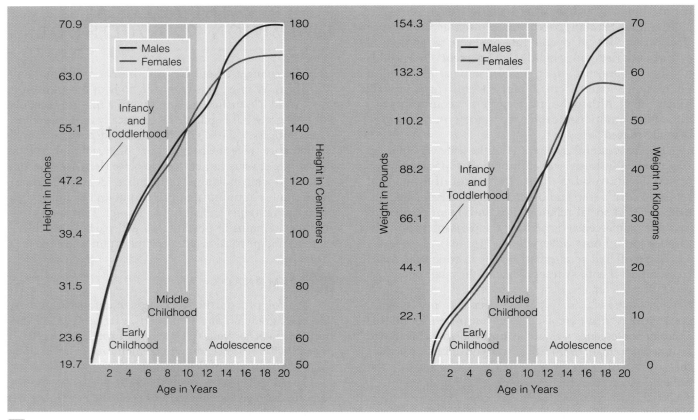

FIGURE 11.1

Average height and weight from infancy through adolescence among North American young people. Note that the adolescent growth spurt takes place earlier for girls than boys. (*From R. M. Malina, 1975,* Growth and Development: The First Twenty Years in Man, *Minneapolis: Burgess Publishing Company, p. 19. Adapted by permission.*)

boy, whose adolescent growth spurt has started, whereas hers is almost finished. Growth in body size is complete for most girls by age 16 and for boys by age 17½, when the epiphyses at the ends of the long bones close completely (see Chapter 7, page 206). Altogether, adolescents add almost 10 inches in height and about 40 pounds in weight during puberty. Figure 11.1 provides an overview of general body growth from infancy through adolescence.

■ BODY PROPORTIONS. In adolescence, the cephalocaudal growth trend of infancy and childhood reverses. At first, the hands, legs, and feet accelerate, and then the torso, which accounts for most of the adolescent height gain (Wheeler, 1991). This pattern of development helps us understand why early adolescents often appear awkward and out of proportion—long-legged, with giant feet and hands.

Large sex differences in body proportions also appear, caused by the action of sex hormones on the skeleton. Boys' shoulders broaden relative to the hips, whereas girls' hips broaden relative to the shoulders and waist. Of course, boys also end up considerably larger than girls, and their legs are longer in relation to the rest of the body. The major

The pubertal growth spurt takes place, on the average, 2 years earlier for girls than boys. Although the boy and girl standing next to each other are both sixth graders, the girl is much taller and more mature looking. (Will Hart)

reason is that boys have 2 extra years of preadolescent growth, when the legs are growing fastest (Tanner, 1990).

■ MUSCLE–FAT MAKEUP AND OTHER INTERNAL CHANGES. One reason 11-year-old Sabrina became concerned about her weight is that compared to her later developing girlfriends, her body had accumulated much more fat. Around age 8, girls start to add more fat than boys on their arms, legs, and trunk, and they continue to do so throughout puberty. In contrast, arm and leg fat of adolescent boys decreases. Although both sexes gain in muscle, this increase is much greater for boys, who develop larger skeletal muscles, hearts, and lung capacity. Also, the number of red blood cells, and therefore the ability to carry oxygen from the lungs to the muscles, increases in boys but not in girls (Katchadourian, 1977). Altogether, boys gain far more muscle strength than do girls, a difference that contributes to boys' superior athletic performance during the teenage years.

SEXUAL MATURATION

Accompanying the rapid increase in body size are changes in physical features related to sexual functioning. Some, called **primary sexual characteristics,** involve the reproductive organs directly (ovaries, uterus, and vagina in females; penis, scrotum, and testes in males). Others, called **secondary sexual characteristics,** are visible on the outside of the body and serve as additional signs of sexual maturity (for example, breast development in females, underarm and pubic hair in both sexes). As Table 11.1 shows, these characteristics develop in a fairly standard sequence, but the age at which each begins and is completed varies greatly (Dubas & Petersen, 1993).

■ SEXUAL MATURATION IN GIRLS. Female puberty usually begins with the budding of the breasts and the growth spurt. **Menarche,** or first menstruation, typically happens around 12 ½ years for North American girls, around 13 for Europeans. But the age range is wide, extending from 10 ½ to 15½ years. Following menarche, pubic hair and breast development are completed, and underarm hair appears. Most girls take 3 to 4 years to complete this sequence, although this, too, can vary greatly.

Notice in Table 11.1 how nature delays sexual maturity until the girl's body is large enough for childbearing; menarche occurs after the peak of the height spurt. As an extra measure of security, for 12 to 18 months following menarche, the girl's ovaries usually do not produce mature ova. However, this temporary period of sterility does not apply to all girls, and it cannot be counted on for protection against pregnancy (Tanner, 1990; Wheeler, 1991).

■ SEXUAL MATURATION IN BOYS. The first sign of puberty in boys is the enlargement of the testes (glands that manufacture sperm), accompanied by changes in the texture and color of the scrotum. Pubic hair emerges a short time later, about the same time that the penis begins to enlarge (Wheeler, 1991).

Refer again to Table 11.1, and you will see that the growth spurt occurs much later in the sequence of pubertal events for boys than girls. When it reaches its peak (around age 14), enlargement of the testes and penis is nearly complete, and underarm hair appears soon after.

TABLE 11.1

Average Age and Age Range of Major Pubertal Changes in North American Girls and Boys

GIRLS	AVERAGE AGE	AGE RANGE	BOYS	AVERAGE AGE	AGE RANGE
Breasts begin to "bud"	10	(8–13)	Testes begin to enlarge	11.5	(9.5–13.5)
Height spurt begins	10	(8–13)	Pubic hair appears	12	(10–15)
Pubic hair appears	10.5	(8–14)	Penis begins to enlarge	12	(10.5–14.5)
Peak of strength spurt	11.6	(9.5–14)	Height spurt begins	12.5	(10.5–16)
Peak of height spurt	11.7	(10–13.5)	Spermarche (first ejaculation) occurs	13	(12–16)
Menarche (first menstruation) occurs	12.8	(10.5–15.5)	Peak of height spurt	14	(12.5–15.5)
Adult stature reached	13	(10–16)	Facial hair begins to grow	14	(12.5–15.5)
Breast growth completed	14	(10–16)	Voice begins to deepen	14	(12.5–15.5)
Pubic hair growth completed	14.5	(14–15)	Penis growth completed	14.5	(12.5–16)
			Peak of strength spurt	15.3	(13–17)
			Adult stature reached	15.5	(13.5–17.5)
			Pubic hair growth completed	15	(14–17)

Sources: Malina & Bouchard, 1991; Tanner, 1990.

Facial and body hair also emerges just after the peak in body growth and develops for several years. Another landmark of male physical maturity is the deepening of the voice as the larynx enlarges and the vocal cords lengthen. (Girls' voices also deepen slightly.) Voice change usually takes place at the peak of the male growth spurt and is often not complete until puberty is over.

While the penis is growing, the prostate gland and seminal vesicles (which together produce semen, the fluid in which sperm are bathed) enlarge. (To see where these organs are located, return to Chapter 2, page 47). Then, around age 13, **spermarche,** or first ejaculation, occurs (Jorgensen & Keiding, 1991). For a while, the semen contains few living sperm. So, like girls, boys have an initial period of reduced fertility.

INDIVIDUAL AND GROUP DIFFERENCES

Heredity is partly responsible for the timing of pubertal maturation, since identical twins generally reach menarche within a month or two of each other, whereas fraternal twins differ by about 12 months (Tanner, 1990). Nutrition and exercise also contribute. In females, a sharp rise in body weight and fat may trigger sexual maturation. Girls who begin serious athletic training at young ages or who eat very little (both of which reduce the percentage of body fat) often show greatly delayed menstruation. In contrast, overweight girls typically start menstruating early (Rees, 1993).

Variations in pubertal growth can also be found among regions of the world and social classes. Heredity probably plays little role, since groups with very different genetic origins living under similarly advantaged conditions resemble one another in pubertal timing (Eveleth & Tanner, 1976). Instead, physical health is largely responsible. In poverty-stricken regions where malnutrition and infectious disease are widespread, menarche is greatly delayed. In many parts of Africa, it does not occur until age 14 to 17. And within countries, girls from higher-income families reach menarche 6 to 18 months earlier than those living in economically disadvantaged homes.

A **secular trend,** or generational change, in pubertal timing, lends added support to the role of physical well-being in adolescent growth. In industrialized nations, age of menarche declined steadily from 1860 to 1970, by about 3 to 4 months per decade. Nutrition, health care, sanitation, and control of infectious disease improved greatly during this time. Of course, humans cannot keep maturing earlier indefinitely, since we cannot exceed the genetic limitations of our species. Secular gains have slowed or stopped in some developed countries, such as Canada, England, Sweden, Norway, Japan, and the United States (McAnarney et al., 1992; Roche, 1979). Consequently, modern young people reared under good nutritional circumstances are likely to resemble their parents in physical growth more than at any time during the previous 130 years.

Because of improved health and nutrition, secular trends in physical growth have taken place in industrialized nations. The adolescent girl on the left is taller than her grandmother, mother, and aunt, and she probably reached menarche at an earlier age. Improved nutrition and health are responsible for gains in body size and faster physical maturation and from one generation to the next. (Bob Daemmrich/The Image Works)

THE PSYCHOLOGICAL IMPACT OF PUBERTAL EVENTS

Think back to your late elementary and junior high school days. As you reached puberty, how did your feelings about yourself and your relationships with others change? As we will see in the following sections, pubertal events affect the adolescent's self-image, mood, and interaction with parents and peers. Some of these outcomes are a response to dramatic physical change, regardless of when it occurs. Others have to do with the timing of pubertal maturation.

REACTIONS TO PUBERTAL CHANGES

A generation or two ago, menarche was often traumatic. Today, girls commonly react with "surprise," undoubtedly due to the sudden nature of the event. Otherwise, they typically report a mixture of positive and negative emotions— "excited and pleased" as well as "scared and upset." Yet wide individual differences exist that depend on prior knowledge and support from family members. Both are influenced by cultural attitudes toward puberty and sexuality.

For girls who have no advance information, menarche can be shocking and disturbing. In the 1950s, up to 50 percent were given no prior warning (Shainess, 1961). Today, no more than 10 to 15 percent are uninformed (Brooks-Gunn, 1988b). This shift is probably due to modern parents' greater willingness to discuss sexual matters with their youngsters. Currently, almost all girls get some information from their mothers. And girls whose fathers are told about pubertal changes adjust especially well. Perhaps a father's involvement reflects a family atmosphere that is highly understanding and accepting of physical and sexual matters (Brooks-Gunn & Ruble, 1980, 1983).

Like girls' reactions to menarche, boys' responses to spermarche reflect mixed feelings. Virtually all boys know about ejaculation ahead of time, but few get any information from parents. Usually they obtain it from reading material (Gaddis & Brooks-Gunn, 1985). In addition, whereas almost all girls eventually tell a friend that they are menstruating, far fewer boys tell anyone about spermarche (Brooks-Gunn et al., 1986). Overall, boys get much less social support for the physical changes of puberty than do girls. This suggests that boys might benefit, especially, from opportunities to ask questions and discuss feelings with a sympathetic male teacher at school.

The experience of puberty is affected by the larger culture in which boys and girls live. Many tribal and village societies celebrate puberty with a *rite of passage*—a community-wide event that marks an important change in privilege and responsibility. Consequently, all young people know that pubertal changes are honored and valued in their culture. In contrast, Western societies grant little formal recognition to movement from childhood to adolescence or from adolescence to adulthood. Certain religious ceremonies, such as confirmation and the Jewish bar or bat mitzvah, do resemble a rite of passage. But they usually do not lead to any meaningful change in social status.

Instead, modern adolescents are confronted with many ages at which they are granted partial adult status—for example, an age for starting employment, for driving, for leaving high school, for voting, and for drinking. In some contexts (on the highway and at work), they may be treated like adults. In others (at school and at home), they

may still be regarded as children. The absence of a widely accepted marker of physical and social maturity makes the process of becoming an adult especially confusing.

PUBERTAL CHANGE, EMOTION, AND SOCIAL BEHAVIOR

In the preceding section, we considered adolescents' reactions to their sexually maturing bodies. Puberty can also affect emotional state and social behavior. A common belief is that puberty has something to do with adolescent moodiness and the desire for greater physical and psychological separation from parents.

■ ADOLESCENT MOODINESS. Although research reveals that higher pubertal hormone levels are linked to greater moodiness, these relationships are not strong. And we cannot be sure that a rise in pubertal hormones actually causes adolescent moodiness (Buchanan, Eccles, & Becker, 1992).

What else might contribute to the common observation that adolescents are moody creatures? In several studies, the mood fluctuations of children, adolescents, and adults were tracked over a week by having them carry electronic pagers. At random intervals, they were beeped and asked to write down what they were doing, who they were with, and how they felt.

As expected, adolescents reported somewhat lower moods than did school-age children or adults (Csikszentmihalyi & Larson, 1984; Larson & Lampman-Petraitis, 1989). But young people whose moods were especially negative were experiencing a greater number of negative life events, such as difficulties with parents and disciplinary actions at school. Negative events increased steadily from childhood to adolescence, and teenagers also seemed to react to them with greater emotion than did children (Larson & Ham, 1993).

Furthermore, compared to the moods of adults, adolescents' feelings were less stable. They often varied from cheerful to sad and back again. But teenagers also moved from one situation to another more often, and their mood swings were strongly related to these changes. High points of their days were times spent with friends and in self-chosen leisure and hobby activities. Low points tended to occur in adult-structured settings—class, job, school halls, school library, and church. Taken together, these findings suggest that situational factors may combine with hormonal influences to affect teenagers' moodiness—an explanation consistent with the balanced view of adolescence described earlier in this chapter.

■ PARENT–CHILD RELATIONSHIPS. Sabrina's father noticed that as his children entered adolescence, their bedroom doors started to close, they resisted spending time with the family, and they became more argumentative. Within a 2-day period, Sabrina and her mother squabbled over Sabrina's messy room ("Mom, it's *my* room. You don't

In this adolescent initiation ceremony, N'Jembe women of Gabon in west-central Africa celebrate the arrival of puberty in two young girls (located just behind the leader, wearing elaborate headdresses) with a special ritual. (Sylvain Grandadam/Photo Researchers, Inc.)

have to live in it!") and her clothing purchases ("Sabrina, if you *buy* it, then *wear* it. Otherwise, you are wasting money!"). And Sabrina resisted the family's regular weekend visit to Aunt Gina's ("Why do I have to go *every* week?"). Many studies show that puberty is related to a rise in parent–child conflict. Bickering and standoffs increase as adolescents move toward the peak of pubertal growth. During this time, both parents and teenagers report feeling less close to one another (Holmbeck & Hill, 1991; Paikoff & Brooks-Gunn, 1991).

Why should a youngster's new, more adultlike appearance trigger these petty disputes between parent and child? Researchers believe the association may have some adaptive value. Among nonhuman primates, the young typically leave the family group around puberty. The same is true in many nonindustrialized cultures (Caine, 1986; Schlegel & Barry, 1991). Departure of young people from the family discourages sexual relations among close blood relatives. But because children in industrialized societies remain economically dependent on parents long after puberty, they cannot leave the family. Consequently, a modern substitute for physical departure seems to have emerged—psychological distancing between parents and children (Steinberg, 1987).

As we will see later, adolescents' new powers of reasoning may also contribute to a rise in family tensions. Also, the need for families to redefine relationships as children demand to be treated in adultlike ways may produce a temporary period of conflict. The quarreling that does take place is generally mild. In reality, parents and adolescents display both conflict and affection. Although separation from parents is adaptive, both generations benefit from warm, protective family bonds throughout the lifespan (Steinberg, 1990).

EARLY VERSUS LATE MATURATION

"All our children were early developers," said Franca during the parents' discussion group. "The three boys were tall by age 12 or 13, but it was easier for them. They felt big and important. Sabrina was skinny as a little girl, but now she says she is too fat and wants to diet. She thinks about boys and doesn't concentrate on her schoolwork."

Findings of several studies match the experiences of Sabrina and her brothers. Early maturing boys were seen as relaxed, independent, self-confident, and physically attractive by both adults and peers. Popular with agemates, they held many leadership positions in school and tended to be athletic stars. In contrast, late maturing boys were not well liked. Adults and peers viewed them as anxious, overly talkative, and attention seeking (Brooks-Gunn, 1988a; Clausen, 1975; Jones, 1965; Jones & Bayley, 1950).

Among girls, the impact of maturational timing was just the reverse. Early maturing girls were below average in popularity; appeared withdrawn, lacking in self-confidence, and psychologically stressed; and held few positions of leader-

These boys are all 13 years old, yet they differ sharply in pubertal maturation. The two early maturing boys are probably popular, self-confident, athletic stars with a positive body image. The three late maturing boys are likely to have a low sense of self-esteem and a negative body image. (Kopstein/Monkmeyer Press)

ship. In contrast, their late maturing counterparts were especially well off. They were regarded as physically attractive, lively, sociable, and leaders at school (Ge, Conger, & Elder, 1996; Jones & Mussen, 1958).

Two factors seem to account for these trends: (1) how closely the adolescent's body matches cultural ideals of physical attractiveness, and (2) how well young people "fit in" physically with their peers.

■ THE ROLE OF PHYSICAL ATTRACTIVENESS. Flip through the pages of your favorite popular magazine. You will see evidence for our society's view of an attractive female as thin and long-legged and a good-looking male as tall, broad-shouldered, and muscular. The female image is a girlish shape that favors the late developer. The male image fits the early maturing boy.

A consistent finding is that early-maturing girls have a less positive **body image**—conception of and attitude toward their physical appearance—than do their on-time and late-maturing agemates. Among boys, the opposite is true: early maturation is linked to a positive body image, whereas late maturation predicts dissatisfaction with the physical self (Alsaker, 1995). Both boys and girls with physical characteristics that are commonly regarded as less attractive have a lower sense of self-esteem and are less well liked by agemates (Langlois & Stephan, 1981).

■ THE IMPORTANCE OF FITTING IN WITH PEERS. A second way of explaining differences in adjustment between early and late maturers is in terms of their physical status in relation to peers. From this perspective, early maturing girls and late maturing boys have difficulty because they fall at the extremes of physical development and are self-conscious about not "fitting in." Not surprisingly, adolescents feel most comfortable with peers who match their own level of biological maturity (Brooks-Gunn et al., 1986; Stattin & Magnusson, 1990).

Because few agemates of the same biological status are available, early maturing adolescents of both sexes seek older companions—sometimes with unfavorable consequences. Older peers often encourage them into activities that they are not yet ready to handle emotionally, including sexual activity, drug and alcohol use, and minor delinquent acts. Perhaps because of involvements like these, the school performance of early maturers tends to suffer (Caspi et al., 1993; Stattin & Magnusson, 1990).

■ LONG-TERM CONSEQUENCES. Do the effects of early and late maturation persist into adulthood? Long-term follow-ups show striking turnabouts in overall well-being. Many early maturing boys and late maturing girls, who had been so admired in adolescence, became rigid, conforming, and somewhat discontented adults. In contrast, late maturing boys and early maturing girls, who were stress-ridden as teenagers, often developed into adults who were independent, flexible, cognitively competent, and satisfied with the direction of their lives (Macfarlane, 1971). What explains these remarkable reversals? Perhaps the confidence-inducing adolescence of early maturing boys and late maturing girls does not promote the coping skills needed to solve life's later problems. In contrast, painful experiences associated with off-time pubertal growth may, in time, contribute to sharpened awareness, clarified goals, and greater stability.

It is important to note, however, that these outcomes may not hold completely in all cultures. In a Swedish study, achievement difficulties of early maturing girls persisted into young adulthood, in the form of lower educational attainment than their on-time and later maturing counterparts (Stattin & Magnusson, 1990). In countries with highly selective college entrance systems, perhaps it is harder for early maturers to recover from declines in school performance. Clearly, the effects of maturational timing involve a complex blend of biological, immediate social setting, and cultural factors.

The physical changes of adolescence are regulated by growth and sex hormones. On the average, girls reach puberty 2 years earlier than do boys. However, wide individual differences exist, to which both heredity and environment contribute. Nutrition and health account for regional and social-class differences in pubertal timing and the secular trend in industrialized nations.

Puberty has important psychological and social consequences. Typically, girls' reactions to menarche and boys' reactions to spermarche are mixed, although prior knowledge and social support affect their responses. Adolescent moodiness is related to both sex hormones and changes in the social environment. Puberty prompts increased conflict and psychological distancing between parent and child. Standards and expectations of the culture and peer group lead early maturing boys and late maturing girls to be advantaged in emotional and social adjustment. In contrast, late maturing boys and early maturing girls have adjustment difficulties. However, the stresses associated with off-time pubertal growth may eventually spark more effective coping skills.

ASK YOURSELF . . .

■ *Millie, mother of an 11-year-old son, is convinced that the rising sexual passions of puberty cause teenage rebelliousness. Where did this belief originate? Explain why it is incorrect.*

■ *Sasha remembers menarche as a traumatic experience. When she discovered she was bleeding, she thought she had a deadly illness and didn't tell anyone for two days. What are the likely causes of Sasha's negative reaction?*

■ *How might adolescent moodiness contribute to the psychological distancing between parents and children that accompanies puberty? (Hint: Think about bidirectional influences in parent–child relationships discussed in previous chapters.)*

BRIEF REVIEW

Adolescence is a time of dramatic physical change leading to an adult-sized body and sexual maturity. Early biologically oriented theories viewed puberty as a period of inevitable storm and stress. Modern researchers recognize that adolescent development and adjustment are a product of both biological and social forces. Adolescence is greatly extended in complex societies that require a long period of education for a productive work life.

HEALTH ISSUES

As young people move into adolescence, they begin to view physical health in a broader way—as more than just the absence of illness. To teenagers, being healthy means functioning physically, mentally, and socially at their best. Consistent with this new view, the arrival of puberty is accompanied by new health concerns related to the young person's striving to meet physical and psychological needs. As adolescents are granted greater autonomy, their personal decision making becomes important, in health as well as other areas (Millstein & Litt, 1990). Yet none of the health difficulties we are about to discuss can be traced to a single cause within the individual. Throughout development, physical, psychological, family, and cultural factors jointly contribute to health and well-being.

NUTRITIONAL NEEDS

When their sons reached puberty, Franca and Antonio noticed a kind of "vacuum cleaner effect" in the kitchen as the boys routinely emptied the refrigerator. Rapid body growth leads to a dramatic rise in food intake. During the growth spurt, boys require about 2,700 calories a day and much more protein, girls about 2,200 calories and somewhat less protein than boys because of their smaller size and muscle mass (Mott, James, & Sperhac, 1990).

This increase in nutritional requirements comes at a time when the diets of many young people are the poorest. Of all age groups, adolescents are the most likely to consume empty calories and eat on the run. These eating habits are particularly harmful if they extend a lifelong pattern of poor nutrition, less serious if they are just a temporary response to peer influences and a busy schedule.

The most common nutritional problem of adolescence is iron deficiency, affecting about 75 percent of American teenagers (McWilliams, 1986). A tired, irritable teenager may be suffering from anemia rather than unhappiness and should have a medical checkup. Most adolescents do not get enough calcium, and they are also deficient in riboflavin (vitamin B₂) and magnesium, both of which support metabolism.

Adolescents, especially girls concerned about their weight, tend to be attracted to fad diets. Unfortunately, most are too limited in nutrients and calories to be healthy for fast-growing, active teenagers. When a youngster insists on trying a special diet, parents should, in turn, insist that he or she first consult a doctor or dietitian.

SERIOUS EATING DISTURBANCES

Franca worried about Sabrina's desire to lose weight, explained to her that she was really quite average in build for an adolescent girl, and reminded Sabrina that her Italian ancestors thought a plump female body was more beautiful than a thin one. Girls who reach puberty early, who are very dissatisfied with their body image, and who grow up in economically advantaged homes where a cultural concern with weight and thinness is especially strong are at risk for serious eating problems.

■ ANOREXIA NERVOSA. **Anorexia nervosa** is a tragic eating disturbance in which young people starve themselves because of a compulsive fear of getting fat. About 1 in every 50 teenage girls in the United States is affected (Garner, 1993; Seligmann, 1994). Caucasian Americans are at

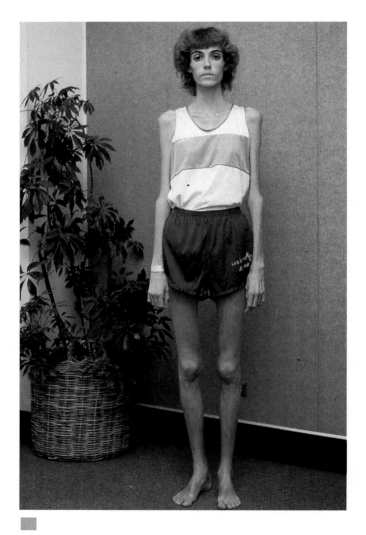

This anorexic girl's strict, self-imposed diet and obsession with strenuous physical exercise has led her to become painfully thin. Even so, her body image is so distorted that she probably regards herself as fat. (Wm. Thompson/The Picture Cube)

greater risk than African Americans, who are more satisfied with their size and shape (Story et al., 1995).

Anorexics have an extremely distorted body image. Even after they have become severely underweight, they conclude that they are fat. Most go on a self-imposed diet so strict that they struggle to avoid eating in response to hunger. To enhance weight loss, they exercise strenuously.

In their attempt to reach "perfect" slimness, anorexics lose between 25 and 50 percent of their body weight and appear painfully thin. Because a normal menstrual cycle requires about 15 percent body fat, either menarche does not occur or menstrual periods stop. Malnutrition causes pale skin, brittle discolored nails, fine dark hairs all over the body, and extreme sensitivity to cold. If allowed to continue, anorexia nervosa can result in shrinking of the heart muscle and kidney failure. About 5 percent die of the disorder (Harris, 1991).

Anorexia nervosa is the combined result of forces within the individual, the family, and the larger culture. We have already seen that the societal image of "thin is beautiful" contributes to the poorer body image of early maturing girls, who are at greatest risk for anorexia (Graber et al., 1994). But though almost all adolescents go on diets at one time or another, anorexics persist in weight loss to an extreme. Many are perfectionists with high standards for their own behavior and performance. Typically, these girls are excellent students who are responsible and well behaved—ideal daughters in many respects.

Yet family interactions of parents and anorexic daughters reveal problems related to adolescent autonomy. Often these parents have high expectations for achievement and social acceptance and are overprotective and controlling. Although the daughter tries to meet these demands, inside she is angry at not being recognized as an individual in her own right. Instead of rebelling openly, the anorexic girl indirectly tells her parents, "I am a separate person from you, and I can do what I want with my own body!" At the same time, this youngster, who has been so used to having parents make decisions for her, meets the challenges of adolescence with little self-confidence. Starving herself is also a way of avoiding new expectations by returning to a much younger, preadolescent image (Halmi, 1987; Maloney & Kranz, 1991).

Because anorexic girls typically deny that any problem exists, treating the disorder is difficult. Hospitalization is often a first step. Since malnutrition affects brain and cognitive functioning, usually the girl's diet must be improved before she can gain insight into her problems. Family therapy, aimed at changing parent–child interaction and expectations, is the most successful treatment. About 30 percent of treated anorexics fully recover. For many others, eating problems continue in less extreme form. One-third show signs of a less severe disorder—bulimia.

■ BULIMIA. When Sabrina's 16-year-old brother Louis brought his girlfriend Cassie to the house, Sabrina admired her good figure. "What willpower! Cassie hardly touches food," Sabrina thought to herself. "But what in the world is wrong with Cassie's teeth?"

Willpower was not the secret of Cassie's slender shape. When it came to food, she actually had great difficulty controlling herself. Cassie suffered from **bulimia,** an eating disorder in which young people (again, mainly girls) engage in binge eating followed by deliberate vomiting, purging with laxatives, and strict dieting. When by herself, Cassie often felt lonely, unhappy, and anxious. She responded with eating rampages, consuming thousands of calories in an hour or two. The vomiting that followed eroded the enamel on Cassie's teeth. In some cases, life-threatening damage to the throat and stomach occurs (Halmi, 1987).

Bulimia is much more common than anorexia nervosa. About 1 to 3 percent of teenage girls are affected; only 5 percent have previously been anorexic (Fairburn & Belgin, 1990). Bulimics share with anorexics a pathological

fear of getting fat, a middle-class family background, and good school performance. But they lack self-control in other areas of their lives. Many engage in petty shoplifting and alcohol abuse. They also differ from anorexics in that they are aware of their abnormal eating habits and feel depressed and guilty about them. As a result, bulimia is usually easier to treat through individual and family therapy, support groups, and nutrition education (Harris, 1991; Thackwray et al., 1993).

SEXUAL ACTIVITY

Louis and Cassie hadn't planned to have intercourse; it "just happened." But before and after, a lot of things passed through their minds. Cassie had been dating Louis for 3 months, and she began to wonder, "Will he think I'm normal if I don't have sex with him? If he wants to and I say no, will I lose him?" Both young people knew their parents wouldn't approve. In fact, when Franca and Antonio noticed how attached Louis was to Cassie, they talked to him about the importance of waiting and the dangers of pregnancy. But that Friday evening, Louis and Cassie's feelings for each other seemed overwhelming. As things went farther and farther, Louis thought, "If I don't make a move, will she think I'm a wimp?" And Cassie had heard from one of her girlfriends that you couldn't get pregnant the first time.

With the arrival of puberty, hormonal changes—in particular, the production of androgens in young people of both sexes—lead to an increase in the sex drive (Udry, 1990). As Louis and Cassie's inner thoughts reveal, adolescents become very concerned about how to manage sexuality in social relationships. New cognitive capacities involving perspective taking and self-reflection affect their efforts to do so. Yet like the eating behaviors we have just discussed, adolescent sexuality is heavily influenced by the social context in which the young person is growing up.

■ THE IMPACT OF CULTURE. Think, for a moment, about when you first learned the facts of life and how you found out about them. In your family, was sex discussed openly or treated with secrecy? Exposure to sex, education about it, and efforts to limit the sexual curiosity of children and adolescents vary widely around the world.

Despite publicity granted to the image of a sexually free modern adolescent, sexual attitudes in the United States are relatively restrictive. Typically, American parents give children little information about sex, discourage them from engaging in sex play, and rarely talk about sex in their presence. When young people become interested in sex, they seek information from friends, books, magazines, movies, and television. On prime-time TV, which adolescents watch the most, premarital sex occurs two to three times each hour and is spontaneous and passionate. Characters are rarely shown taking steps to avoid pregnancy or sexually transmitted disease (Braverman & Strasburger, 1994; Strasburger, 1989).

Adolescence is an especially important time for the development of sexuality. American teenagers receive contradictory and confusing messages from the social environment about the appropriateness of sex. Although the rate of premarital sex has risen among adolescents, most engage in low levels of sexual activity and have only a single partner. (Richard Hutchings/PhotoEdit)

The messages delivered by these two sets of sources are contradictory and confusing. On one hand, adults emphasize that sex at a young age and before marriage is bad. On the other hand, adolescents encounter much in the broader social environment that extols the excitement and romanticism of sex. American teenagers are left bewildered, poorly informed about sexual facts, and with little sound advice on how to conduct their sex lives responsibly.

■ ADOLESCENT SEXUAL ATTITUDES AND BEHAVIOR. Although differences among subcultural groups exist, over the past 30 years the sexual attitudes of American adolescents and adults have become more liberal. Compared to a generation ago, more people believe that sexual intercourse before marriage is all right, as long as two people are emotionally committed to one another (Michael et al., 1994). Recently a slight swing back in the direction of conservative sexual beliefs has occurred, largely due to the risk of sexually transmitted disease, especially AIDS (Roper Starch Worldwide, 1994).

Trends in the sexual behavior of adolescents are quite consistent with their beliefs. The rate of premarital sex among young people has risen. For example, high school girls claiming to have had sexual intercourse grew from 28 percent in 1971 to 48 percent in 1990 (Braverman & Strasburger, 1993). As Table 11.2 reveals, a substantial minority of boys and girls are sexually active quite early, by ninth grade. Males tend to have their first intercourse earlier than females, and sexual activity is especially high among African-American adolescents—particularly boys.

Yet timing of first intercourse provides only a limited picture of adolescent sexual behavior. Most teenagers engage in relatively low levels of sexual activity. The typical 15- to 19-year-old sexually active male—white, black, or Hispanic—has relations with only one girl at a time and spends half the year with no partner at all (Sonenstein, Pleck, & Ku, 1991). Contrary to popular belief, a runaway sexual revolution does not characterize American young people. In fact, the rate of teenage sexual activity is about the same in the United States as in Western European nations (Creatsas et al., 1995).

■ CHARACTERISTICS OF SEXUALLY ACTIVE ADOLESCENTS. Teenage sexual activity is linked to a wide range of personal, family, peer, and educational characteristics. These include early physical maturation, parental separation and divorce, large family size, sexually active friends and older siblings, poor school performance, and lower educational aspirations (Braverman & Strasburger, 1994; Ku, Sonenstein, & Pleck, 1993). Since many of these factors are associated with growing up in a low-income family, it is not surprising that early sexual activity is more common among young people from economically disadvantaged homes. In fact, the high rate of premarital intercourse among black teenagers can largely be accounted for by widespread poverty in the black population (Sullivan, 1993).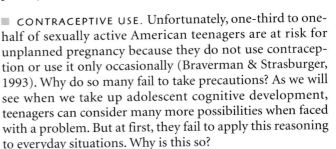

■ CONTRACEPTIVE USE. Unfortunately, one-third to one-half of sexually active American teenagers are at risk for unplanned pregnancy because they do not use contraception or use it only occasionally (Braverman & Strasburger, 1993). Why do so many fail to take precautions? As we will see when we take up adolescent cognitive development, teenagers can consider many more possibilities when faced with a problem. But at first, they fail to apply this reasoning to everyday situations. Why is this so?

One reason is that advances in perspective taking lead teenagers, for a time, to be extremely concerned about others' opinion of them. Recall how Cassie and Louis each worried about what the other would think if they did not have sex. Furthermore, in the midst of everyday social pressures, adolescents often overlook the consequences of engaging in risky behaviors (Jaskiewicz & McAnarney, 1994).

The social environment also contributes to teenagers' reluctance to use contraception. Adolescents who talk openly about sex with their parents are not less sexually active, but they are more likely to use birth control (Brooks-Gunn, 1988b). Unfortunately, many adolescents are too scared or embarrassed to ask parents questions. When a sexual encounter occurs, they are ambivalent about whether to go ahead, and many leap into the experience without forethought. Although most get some sex education at school, teenagers' knowledge about sex and contraception is often incomplete or just plain wrong. Many do not know where to get birth control counseling and devices (Winn, Roker, & Coleman, 1995).

TABLE 11.2

Teenage Sexual Activity Rates by Sex, Ethnic Group, and Grade

| SEX | ETHNIC GROUP | | | GRADE | | | | TOTAL |
	WHITE	BLACK	HISPANIC	9	10	11	12	
Male	56.4	87.8	63.0	48.7	52.5	62.6	76.3	60.8
Female	47.0	60.0	45.0	31.9	42.9	52.7	66.6	48.0
Total	51.6	72.3	53.4	39.6	47.6	57.3	71.9	54.2

Note: Data reflect the percentage of high school students who report ever having had sexual intercourse.
Source: U.S. Centers for Disease Control, 1992.

■ SEXUAL ORIENTATION. About 3 to 6 percent of teenagers discover that they are lesbian or gay (Patterson, 1995). Adolescence is an equally crucial time for the sexual development of these young people, and societal attitudes, once again, loom large in how well they fare.

New evidence indicates that heredity makes an important contribution to homosexuality. Identical twins of both sexes are much more likely than fraternal twins to share a homosexual orientation. The same is true for biological as opposed to adoptive relatives (Bailey & Pillard, 1991; Bailey et al., 1993). Furthermore, male homosexuality tends to be more common on the maternal than paternal side of families. Indeed, a recent gene mapping study indicated that one or several genes on the X chromosome (which males inherit from the mother) might predispose males toward homosexuality (Hamer et al., 1993). Yet these findings do not apply to all homosexuals, since some do not have the genetic markers. Homosexuality probably results from combinations of biological and environmental factors that are not yet well understood (Horton, 1995).

Homosexual adolescents in industrialized nations often experience intense inner conflict, isolation, and loneliness. Family rejection and social stigma contribute to high rates of problem behaviors, including depression, suicide, substance abuse, and high-risk sexual behavior (Baumrind, 1995; Hershberger & D'Augelli, 1995). Gay and lesbian adolescents have a special need for caring adults and peers who can help them establish a positive sexual identity and find self- and social acceptance.

Sexually active adolescents, both heterosexual and homosexual, face serious health risks. In the following sections, we examine the high rates of pregnancy, childbirth, and sexually transmitted disease among American teenagers.

TEENAGE PREGNANCY AND CHILDBEARING

Cassie was lucky not to get pregnant after having sex with Louis, but some of her high school classmates weren't so fortunate. She'd heard about Veronica, who missed several periods, pretended nothing was wrong, and didn't go to a doctor until a month before she gave birth. Veronica lived at home until she became pregnant a second time. At that point, her parents told her there wasn't room in the house for a second baby. So Veronica dropped out of school and moved in with her 17-year-old boyfriend, Todd, who worked in a fast-food restaurant. A few months later, Todd left Veronica because he couldn't stand being tied down with the babies. Veronica had to apply for public aid to support herself and the two children.

Each year, more than a million American teenagers become pregnant, 30,000 under the age of 15. As Figure 11.2 shows, the adolescent pregnancy rate in the United States is twice that of Australia, England, Canada, and France, three times that of Sweden, and six times that of the Netherlands. The United States differs from these nations in three important ways: (1) effective sex education reaches fewer teenagers; (2) convenient, low-cost contraceptive services for adolescents are scarce; and (3) many more families live in poverty, which encourages young people to take risks without considering the future (Jones et al., 1988).

About 40 percent of teenage pregnancies end in abortion, 13 percent in miscarriage (Chase-Lansdale & Brooks-Gunn, 1994; Jaskiewicz & McAnarney, 1994). Because the United States has one of the highest adolescent abortion rates of any developed country, the number of teenage births is actually lower than it was 30 years ago. But teenage parenthood is a much greater problem today because modern adolescents are far less likely to marry before childbirth. In 1960, only 15 percent of teenage births were to unmarried females, whereas today, nearly 70 percent are (Children's Defense Fund, 1997). Increased social acceptance of single motherhood, along with the belief that a baby might fill a void in their lives, has meant that only a small number of girls give up their infants for adoption. Each year, about 320,000 unmarried adolescent girls take on the responsibilities of parenthood before they are psychologically mature.

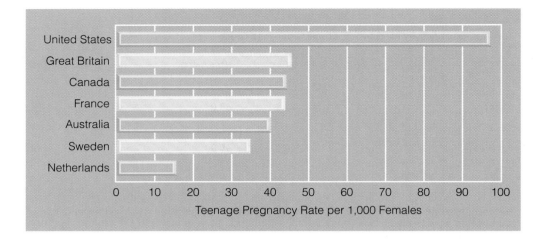

FIGURE 11.2

Teenage pregnancy rate in seven industrialized nations. (Adapted from United Nations, 1991b.)

■ CORRELATES AND CONSEQUENCES OF TEENAGE PARENTHOOD. Becoming a parent is challenging and stressful for anyone, but it is especially difficult for adolescents. Teenage parents have not yet established a clear sense of direction for their own lives. Also, teenage mothers are many times more likely to be poor than older mothers. A high percentage of out-of-wedlock births are to members of low-income minorities, especially African-American, Native-American, and Hispanic teenagers. Many of these young people seem to turn to early parenthood as a way to move into adulthood when educational and career avenues are unavailable (Caldas, 1993; Murry, 1992).

Early childbearing imposes lasting hardships on two generations—adolescent and newborn baby. The lives of pregnant teenagers are often troubled in many ways, and after the baby is born, their circumstances worsen. Only 50 percent of girls who give birth before age 18 finish high school, compared to 96 percent of those who wait to become parents. Both teenage mothers and fathers are likely to be on welfare. If they are employed, their limited education restricts them to unsatisfying, low-paid jobs (Furstenberg, Brooks-Gunn, & Chase-Landsdale, 1989).

Because many pregnant girls do not receive early prenatal care, their babies often experience prenatal and birth complications, especially low birth weight (Scholl, Heidiger, & Belsky, 1996). Children of teenagers are also at risk for poor parenting. Compared to adult mothers, adolescent mothers know less about child development, feel less positively about the parenting role, and interact less effectively with their infants (Sommer et al., 1993). Many of their children score low on intelligence tests, achieve poorly in school, and engage in disruptive social behavior. Too often, the cycle of adolescent pregnancy is repeated in the next generation. About one-third of girls who have a baby before age 19 were born to teenage mothers (Furstenberg, Levine, & Brooks-Gunn, 1990).

Outcomes are more favorable when teenage mothers return to school after giving birth and continue to live in their parents' homes, sharing child care with experienced adults. If the adolescent finishes high school, avoids additional births, and finds a stable marriage partner, long-term disruptions in her own and her child's development are less severe. The small minority of young mothers who fail in all three of these ways face a life of continuing misfortune (Furstenberg, Brooks-Gunn, & Morgan, 1987).

■ PREVENTION STRATEGIES. Preventing teenage pregnancy means addressing the many factors underlying early sexual activity and lack of contraceptive use. Informing adolescents about sex and contraception is crucial. Too often, sex education courses in public schools are given too late (after sexual activity has begun), last only a few sessions, and are limited to a catalogue of facts about anatomy and reproduction. Sex education that goes beyond this minimum does not encourage early sex, as some opponents claim. It does improve awareness of sexual facts—knowledge that is necessary for responsible sexual behavior (Katchadourian, 1990).

Knowledge, however, is not sufficient to influence teenagers' behavior. Sex education must also help them build a bridge between what they know and what they do. As the Social Issues box on the following page illustrates, a new wave of sex education programs has emerged that uses creative techniques to teach adolescents the skills they need to resist pressures to engage in early and unprotected sex (Kirby et al., 1994).

The most controversial aspect of adolescent pregnancy prevention involves providing easy access to contraceptives. Some large cities have school-based health clinics that offer this service. Many Americans argue that placing birth control pills or condoms in the hands of teenagers is equivalent to saying that early sex is okay. Yet in western Europe, where these clinics are common, teenage sexual activity is no higher than in the United States, but pregnancy and abortion rates are much lower (Zabin & Hayward, 1993).

Finally, teenagers who look forward to a promising future are far less likely to engage in early and irresponsible sex. By expanding educational, vocational, and employment

SOCIAL ISSUES

A NEW APPROACH TO SEX EDUCATION

In Atlanta, Georgia, a select group of high school seniors begins an intensive, 20-hour series of training sessions that will prepare them to serve as sex educators of younger students. They are part of Postponing Sexual Involvement, a program designed to promote attitudes and skills that early adolescents can use to delay intercourse until they are mature enough to manage their sexuality responsibly. Postponing Sexual Involvement is based on three assumptions:

1. Most young teenagers do not yet have the cognitive maturity to consider the future implications of their sexual behavior.

2. The needs that early adolescents try to meet through sexual activity—desire for social acceptance and affection, confirmation of masculinity or femininity, and escape from loneliness and boredom—can be satisfied in other ways.

3. Young people often do not want to engage in sexual intercourse but are pressured into it by peers and glamorous media images.

Recognizing that merely conveying facts about sexuality will not change teenagers' behavior, Postponing Sexual Involvement translates the facts into active learning experiences. A "social inoculation" approach is used in which eighth graders, through discussion and role playing, confront sexual situations similar to those they will encounter in everyday life. Specific activities help students identify the sources of pressures to have sex, analyze why they are tempted to give in, and develop skills to resist. Delaying sex became a major goal of the program after one survey of

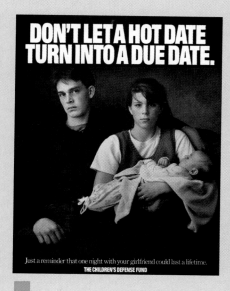

DON'T LET A HOT DATE TURN INTO A DUE DATE.

Just a reminder that one night with your girlfriend could last a lifetime.
THE CHILDREN'S DEFENSE FUND

Convincing teenagers of the consequences of early sexual activity and helping them acquire social skills to resist can help reduce the high rate of teenage pregnancy and child bearing in the United States. This public service poster is distributed to American communities by the Children's Defense Fund as part of its adolescent pregnancy prevention program. (See Chapter 2, page 75, for the address, phone number, and web site of the Children's Defense Fund.) (The Children's Defense Fund)

more than 1,000 sexually active teenage girls revealed that the overwhelming majority wanted more information on how to "say no" without hurting another person's feelings. Older peers serve as teachers because they are effective models for younger students. They can also demonstrate that there are many ways to win social admiration other than through sexual activity.

Postponing Sexual Involvement has produced remarkable results among teenagers who were not yet sexually active. A follow-up more than a year after the program ended revealed that many more nonparticipants than participants had begun sexual activity (61 versus 39 percent for boys; 27 versus 17 percent for girls). Program students who did engage in intercourse were more likely than controls to use contraceptives and to experiment with sex only once or twice (Howard & McCabe, 1990).

Perhaps because of its theme of abstinence, Postponing Sexual Involvement did not change the behavior of participants who were already sexually experienced. But other research shows that learning activities relevant to the situations of these young people are effective. Sexually active teenagers who are encouraged to relate sexual information to their own values, to personalize that information ("Each time Ann and I have unprotected intercourse, we risk pregnancy"), and to use it in decision making ("Since neither of us wants her to get pregnant, we had better start using birth control") do practice more effective contraception (Barth, Petro, & Leland, 1992; Schinke, Blythe, & Gilchrist, 1981).

TRY THIS...

■ Using what you know about influences on adolescent sexual activity, design your own course in sex education, considering the following issues: At what age should sex education begin? What topics are most important to discuss? How can sex education be made less threatening and embarrassing? Should boys and girls be taught separately or together? Can peers be used to teach sex education, and if so, how?

TABLE 11.3

Most Common Sexually Transmitted Diseases of Adolescence

DISEASE	INCIDENCE (RATE PER 100,000)	CAUSE	SYMPTOMS AND CONSEQUENCES	TREATMENT
AIDS	20[a]	Virus	Fever, weight loss, severe fatigue, swollen glands, and diarrhea. As the immune system weakens, severe pneumonias and cancers, especially on the skin, appear. Death due to other diseases usually occurs.	No cure; experimental drugs prolong life.
Chlamydia	215	Bacteria	Discharge from the penis in males; painful itching and, burning vaginal discharge, and dull pelvic pain in females. Often no symptoms. If left untreated, can lead to inflammation of the pelvic region, infertility, and sterility.	Antibiotic drugs.
Cytomegalovirus	Unknown[b]	Virus of the herpes family	No symptoms in most cases. Sometimes, a mild flu-like reaction. In a pregnant woman, can spread to the embryo or fetus and cause miscarriage or serious birth defects (see page 87).	None. Usually disappears on its own.
Genital warts	451	Virus	Warts that grow near the vaginal opening in females, on the penis or scrotum in males. Can cause severe itching. Related to cancer of the cervix.	Removal of warts.
Gonorrhea	300	Bacteria	Discharge from the penis or vagina, painful urination. Sometimes no symptoms. If left untreated, can spread to other regions of the body, resulting in such complications as infertility, sterility, blood poisoning, arthritis, and inflammation of the heart.	Antibiotic drugs.
Herpes simplex 2 (genital herpes)	167	Virus	Fluid-filled blisters on the genitals, high fever, severe headache, and muscle aches and tenderness. No symptoms in a few people. In a pregnant woman, can spread to the embryo or fetus and cause birth defects (see page 87).	No cure. Can be controlled with drug treatment.
Syphillis	75	Bacteria	Painless chancre (sore) at site of entry of germ and swollen glands, followed by rash, patchy hair loss, and sore throat within 1 week to 6 months. These symptoms disappear without treatment. Latent syphilis varies from no symptoms to damage to the brain, heart, and other organs after 5 to 20 years. In pregnant women, can spread to the embryo and fetus and cause birth defects (see page 87).	Antibiotic drugs.

[a] This figure includes both adolescents and young adults. For most of these cases, the virus is contracted in adolescence, and symptoms appear in early adulthood.

[b] Cytomegalovirus is the most common STD. Because there are no symptoms in most cases, its precise rate of occurrence is unknown. Half of the population or more may have had the virus sometime during their lives.

Source: U.S. Department of Health and Human Services, 1996b.

opportunities, society can provide young people with good reasons to postpone child bearing.

■ INTERVENING WITH TEENAGE PARENTS. The most difficult and costly way to deal with adolescent parenthood is to wait until after it has happened. Young single mothers need health care for themselves and their children, encouragement to stay in school, job training, instruction in parenting and life management techniques, and high-quality, affordable day care. Schools that provide these services reduce the incidence of low-birth-weight babies, increase mothers' educational success, and prevent additional child bearing (Seitz & Apfel, 1993, 1994).

Programs that focus on fathers are attempting to increase their financial and emotional commitment to the baby (Children's Defense Fund, 1997). But fathers are difficult to reach because most either do not admit their paternity or abandon the young mother and infant after a short time. At present, the majority of adolescent parents of both sexes do not receive the help they need.

CAREGIVING CONCERNS

Preventing Sexually Transmitted Disease

STRATEGY	DESCRIPTION
Know your partner well.	Take time to get to get to know your partner. Find out whether your partner has had sex with many people or has used injectable drugs.
Maintain mutual faithfulness.	For this strategy to work, neither partner can have an STD at the start of the relationship.
Do not use drugs.	Using a needle, syringe, or drug liquid previously used by others can spread STD. Alcohol, marijuana, or other illegal substances impair judgment, reducing your capacity to think clearly about the consequences of behavior.
Always use a latex condom and vaginal contraceptive when having sex with a nonmarital partner.	Latex condoms give good (but not perfect) protection against STD by reducing the passage of bacteria and viruses. Vaginal contraceptives containing nonoxynol-9 can kill several kinds of STD microbes. They increase protection when combined with condom use.
Do not have sex with a person you know has an STD.	Even if you are protected by a condom, the risk of contracting the disease is still there. In the case of the AIDS virus, you are risking your life.
If you get an STD, inform all recent sexual partners.	Notifying people you may have exposed to STD permits them to get treatment before spreading the disease to others.

Source: Daugirdas, 1992.

SEXUALLY TRANSMITTED DISEASE

Early sexual activity and inconsistent contraceptive use lead to another widespread health problem: sexually transmitted disease (STD) (see Table 11.3). Adolescents have the highest rates of STD of any age group. One out of six sexually active teenagers contract one of these illnesses each year. If left untreated, sterility and life-threatening complications can result (Braverman & Strasburger, 1994). Teenagers in greatest danger of STD are the same ones who tend to engage in irresponsible sexual behavior—poverty-stricken young people who feel a sense of hopelessness about their lives (Holmbeck, Waters, & Brookman, 1990).

By far the most serious STD is AIDS. One-fifth of cases in the United States occur between ages 20 and 29. Nearly all of these originate in adolescence, since AIDS symptoms typically take 8 to 10 years to develop. Drug-abusing and homosexual adolescents account for most cases, but heterosexual spread of the disease is increasing, especially among females (U.S. Centers for Disease Control, 1996).

Besides helping teenagers understand sex, pregnancy, and contraception, another important goal of sex education is to help them avoid STD. Over 90 percent of high school students are aware of basic facts about AIDS. But some hold false beliefs that put them at risk—for example, that birth control pills provide some protection or that it is possible to tell whether people have AIDS by looking at them (DiClemente, 1993). The Caregiving Concerns table above lists strategies for preventing STD.

SUBSTANCE USE AND ABUSE

At age 14, Louis took some cigarettes out of his uncle's pack and smoked them when he was alone in the house. At an unchaperoned party, he and Cassie drank several cans of beer, largely because everyone else was doing it. Louis got little physical charge out of these experiences. He was a good student and well liked by peers and got along well with his parents. He had no need for drugs as an escape valve from daily life. But he knew of others at school for whom things were different—students who started with alcohol and cigarettes, moved to harder substances, and eventually were hooked.

In the United States, teenage alcohol and drug use is pervasive—higher than in any other industrialized nation. By age 14, 56 percent of young people have already tried cigarette smoking, 81 percent drinking, and 39 percent at least one illegal drug (usually marijuana). By the end of high school, 15 percent are regular cigarette users, 32 percent have engaged in heavy drinking at least once, and over 43 percent have experimented with illegal drugs. Of these, about one-third have tried at least one highly addictive and toxic substance, such as amphetamines, cocaine, phencyclidine (PCP), or heroin (U.S. Department of Health and Human Services, 1995).

Surprisingly, these high figures actually represent a decline in adolescent alcohol and drug use during the past decade (see Figure 11.3). Why do so many young people subject themselves to the health risks of these substances?

FIGURE 11.3

Percentage of high school seniors reporting use of alcohol, cigarettes, and illegal drugs in the past month, 1976–1994. Substance use continues to be widespread among adolescents, although overall it has declined during the past decade. However, the sharp increase from 1992 to 1994—largely accounted for by marijuana and cigarette smoking—has sounded a note of alarm. (*From U.S. Department of Health and Human Services, 1995.*)

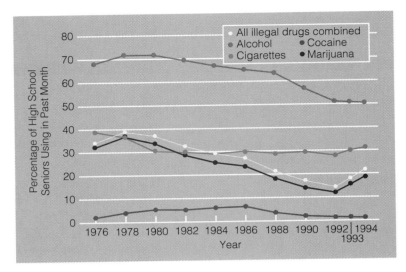

Part of the reason is cultural. Modern adolescents live in a drug-dependent society. They see adults using caffeine to wake up in the morning, cigarettes to cope with daily hassles, a drink to calm down in the evening, and other remedies to relieve stress, headaches, depression, and physical illness.

For most young people, substance use simply reflects their intense curiosity about "adultlike" behaviors. A recent study revealed that the majority of teenagers dabbled in alcohol, tobacco, and marijuana. These *experimenters* were not headed for a life of decadence and addiction, as many adults believe. Instead, they were psychologically healthy, sociable, curious young people (Shedler & Block, 1990).

Yet adolescent drug experimentation should not be taken lightly. Because most drugs impair perception and thought processes, a single heavy dose can lead to permanent injury or death. And a worrisome minority of teenagers move from substance *use* to *abuse*—taking drugs regularly, requiring increasing amounts to achieve the same effect, and finding themselves unable to stop. Four percent of high school seniors are daily drinkers, and almost as many take an illegal drug on a daily basis (U.S. Department of Health and Human Services, 1995).

■ CORRELATES AND CONSEQUENCES OF ADOLESCENT SUBSTANCE ABUSE. In contrast to experimenters, drug abusers are seriously troubled adolescents who express their unhappiness through antisocial acts. Longitudinal research shows that their impulsive, disruptive style is often present in early childhood and is promoted by other factors, including low income, family mental health problems, parental drug use, and lack of parental involvement. By early adolescence, peer encouragement—friends who use and provide access to illegal substances—is a strong predictor of substance abuse (Chassin et al., 1996; Dobkin et al., 1995; Stice & Barrera, 1995).

Teenage substance abuse is a devastating turn of events that often has lifelong consequences. When adolescents depend on alcohol and hard drugs to deal with daily stresses, they fail to learn responsible decision-making skills and alternative coping techniques—crucial lessons during this time of transition to adulthood. Over time, these young people enter into marriage, childbearing, and the work world prematurely and fail at them readily. Adolescent drug addiction is associated with high rates of divorce and job loss—painful outcomes that encourage further addictive behavior (Newcomb & Bentler, 1988).

Modern adolescents observe adults using drugs to help themselves through their daily lives and relieve many common symptoms. So it is not surprising that most teenagers experiment with drugs at one time or another. Those who make the transition from use to abuse are seriously troubled young people who are inclined to express their unhappiness through impulsive, antisocial behavior. (Michaud/Photo Researchers)

■ PREVENTION STRATEGIES. School-based programs that promote effective parenting (including monitoring of teenagers' activities), educate students about the dangers of drugs and alcohol, and teach them how to resist peer pressure help reduce experimentation to some degree (O'Malley, Johnston, & Bachman, 1995; Steinberg, Fletcher, & Darling, 1994). But some drug taking seems inevitable. Weekend on-call transportation services that any young person can contact for a safe ride home, with no questions asked, can prevent teenagers from endangering themselves and others when they do experiment.

Drug abuse, as we have seen, occurs for quite different reasons than does occasional use. Therefore, different strategies are required to deal with it. Hospitalization is often a necessary and even lifesaving first step. Once the young person is weaned from the drug, long-term therapy to treat low self-esteem, anxiety, and impulsiveness and academic and vocational training to improve life success and satisfaction are generally needed. Not much is known about the best way to treat adolescent drug abuse. Even the most comprehensive programs have alarmingly high relapse rates—from 35 to 70 percent (Newcomb & Bentler, 1989).

∪NINTENTIONAL INJURIES

Adolescent risk taking, fueled by a tendency to act without forethought, results in a rise in certain kinds of unintentional injuries. Motor vehicle collisions are the leading killer of adolescents, accounting for almost 40 percent of deaths between ages 15 and 24 (U.S. Bureau of the Census, 1996). Parents need to set firm limits on their teenager's car use, particularly with respect to drinking and driving. These efforts work best when there is a history of good parent–child communication—a powerful preventive of adolescent injury (Millstein & Irwin, 1988).

Other unintentional injuries account for an additional 11 percent of adolescent deaths. Most are caused by firearms. The rate of disability and death from firearms is especially high among low-income minority youths, particularly African-American males in inner-city ghettos (Children's Defense Fund, 1997). Where gun sales are banned (Canada, all of Western Europe, and Japan), such needless deaths almost never happen.

MOTOR DEVELOPMENT

Puberty is accompanied by steady improvement in motor performance, but the pattern of change is quite different for boys and girls. Girls' gains are slow and gradual, leveling off by age 14. In contrast, boys show a dramatic spurt in strength, speed, and endurance that continues through the end of the teenage years. Figure 11.4 illustrates this sex difference for running speed, broad jump, and vertical jump. Notice how the gender gap in physical skill widens over time. By midadolescence, for example, very few girls perform as well as the average boy, and practically no boys score as low as the average girl (Malina & Bouchard, 1991).

Because girls and boys are no longer well matched physically, gender-segregated physical education usually

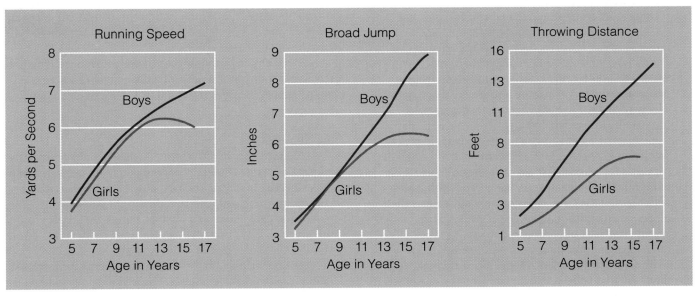

FIGURE 11.4

Age changes in running speed, broad jump, and throwing distance for boys and girls. The gender gap in athletic performance widens during adolescence. (*From A. Espenschade & H. Eckert, 1974, "Motor Development," in W. R. Warren & E. R. Buskirk, Eds., Science and Medicine of Exercise and Sport, pp. 329–330, New York: Harper & Row. Adapted by permission of HarperCollins Publishers, Inc.*)

begins in junior high school. At the same time, athletic options for both sexes expand. Many new sports are added to the curriculum—track and field, wrestling, tackle football, weight lifting, floor hockey, archery, tennis, and golf, to name just a few.

In 1972, the federal government required schools receiving public funds to provide equal opportunities for males and females in all education programs, including athletics. The law sparked a dramatic rise in girls' participation in sports. Still, it falls far short of boys'. About 64 percent of males but only 41 percent of females are active in high school sports (Berk, 1992b). In Chapter 9, we saw that girls get less encouragement and recognition for athletic achievement. This pattern continues into the teenage years.

Besides improving motor performance, sports influence cognitive and social development. Interschool and intramural athletics provide important lessons in competition, assertiveness, problem solving, and teamwork. These experiences are less available to females because of the lower status of girls' sports. A positive sign is that the sex difference in high school athletic participation is becoming smaller.

BRIEF REVIEW

Puberty brings new health concerns. Greater nutritional requirements of a rapidly growing body come at a time when the eating habits of many young people are the poorest. For some teenagers, the cultural ideal of thinness combines with family and psychological problems to produce the serious eating disorders of anorexia nervosa and bulimia.

Hormonal changes of puberty lead to an increase in the sex drive, but social experience affects how young people manage their sexuality. American adolescents get mixed messages from adults and the larger culture about sexual activity. The percentage of sexually active teenagers has increased. Homosexual young people face special challenges in establishing a positive sexual identity. Adolescent cognitive processes and lack of social supports for responsible sexual behavior contribute to high rates of teenage pregnancy, abortion, and childbirth in the United States. Sexually transmitted disease is an additional danger of early sexual activity and lack of contraceptive use.

Although most adolescents experiment with alcohol and drugs, a worrisome minority make the transition from use to abuse. Teenage substance abuse is linked to a variety of psychological, family, and peer problems, and treatment is difficult. Motor vehicle and firearm injuries increase during adolescence.

Adolescents improve greatly in motor performance, although boys' gains are much greater than girls'. Boys continue to receive more recognition and encouragement for athletic accomplishments.

ASK YOURSELF . . .

- *Fourteen-year-old Lindsay says she couldn't possibly get pregnant because her boyfriend told her he would never do anything to "mess her up." What factors might account for Lindsay's unrealistic reasoning?*

- *Return to page 361 and review Veronica's life circumstances after becoming a teenage mother. Why is it likely that Veronica and her children will experience long-term hardships?*

- *Explain how adolescent substance abuse follows the pattern of other teenage health problems in being a product of both psychological and social forces.*

COGNITIVE DEVELOPMENT IN ADOLESCENCE

One mid-December evening, a knock at the front door announced the arrival of Franca and Antonio's oldest son, Jules, home for vacation after the fall semester of his sophomore year at college. Moments later, the family gathered around the kitchen table. "How did it all go, Jules?" inquired Antonio while passing out pieces of apple pie.

"Well, physics and philosophy were awesome. The last few weeks, our physics prof introduced us to Einstein's relativity theory. Boggles my mind, it's so incredibly counterintuitive."

"Counter-what?" asked Sabrina, trying hard to follow the conversation.

"Counterintuitive. Unlike what you'd normally expect," explained Jules. "Imagine you're on a train, going unbelievably fast, like 160,000 miles a second. The faster you go approaching the speed of light, the slower time passes and the denser and heavier things get relative to on the ground. The theory revolutionized the way we think about time, space, matter—the entire universe."

Sabrina wrinkled her forehead, unable to comprehend Jules's otherworldly reasoning. "Time slows down when I'm bored, like right now, not on a train when I'm going somewhere exciting. No speeding train ever made me denser and heavier, but this apple pie will if I eat any more of it," Sabrina announced, getting up and leaving the table.

Sixteen-year-old Louis reacted differently. "Totally cool, Jules. So what'd you do in philosophy?"

"It was a course in philosophy of technology. We studied the ethics of futuristic methods in human reproduction. For example, we argued the pros and cons of a world in which all embryos develop in artificial wombs."

"What do you mean? asked Louis. "You order your kid at the lab?"

"That's right. I wrote my paper on it. I had to evaluate it in terms of principles of justice and freedom. I can see some advantages but also lots of dangers...."

As this conversation illustrates, adolescence brings with it vastly expanded powers of reasoning. At age 11, Sabrina finds it difficult to move beyond her firsthand experiences to a world of possibilities. Over the next few years, her thinking will take on the abstract qualities that characterize the cognition of her older brothers. Jules juggles variables in complex combinations and thinks about situations not easily detected in the real world or that do not exist at all. Compared to school-age children's thinking, adolescent thought is more enlightened, imaginative, and rational.

PIAGET'S THEORY: THE FORMAL OPERATIONAL STAGE

According to Piaget, the capacity for abstract thinking begins around age 11. In the **formal operational stage**, the adolescent reasons much like a scientist searching for solutions in the laboratory. Concrete operational children can only "operate on reality," but formal operational adolescents can "operate on operations." In other words, concrete things and events are no longer required as objects of thought. Instead, adolescents can come up with new, more general logical rules through internal reflection (Inhelder & Piaget, 1955/1958). Let's look at two major features of the formal operational stage.

HYPOTHETICO-DEDUCTIVE REASONING

At adolescence, young people become capable of **hypothetico-deductive reasoning.** When faced with a problem, they start with a *general theory* of all possible factors that might affect the outcome and *deduce* from it specific *hypotheses* (or predictions) about what might happen. Then they test these hypotheses in an orderly fashion to see which ones work in the real world. Notice how this form of problem solving begins with possibility and proceeds to reality. In contrast, concrete operational children start with reality—with the most obvious predictions about a situation. When these are not confirmed, they cannot think of alternatives and fail to solve the problem.

Adolescents' performance on Piaget's famous *pendulum problem* illustrates this new approach. Suppose we present several school-age children and adolescents with strings of different lengths, objects of different weights to attach to the strings, and a bar from which to hang the strings. Then we ask each of them to figure out what influences the speed with which a pendulum swings through its arc.

Formal operational adolescents come up with four hypotheses: (1) the length of the string; (2) the weight of the object hung on it; (3) how high the object is raised before it is released; and (4) how forcefully the object is pushed. Then, by varying one factor at a time while holding all the others constant, they try out each of these possibilities. Eventually they discover that only string length makes a difference.

In contrast, concrete operational children experiment unsystematically. They cannot separate out the effects of each variable. For example, they may test for the effect of string length without holding weight constant by comparing a short, light pendulum with a long, heavy one. Also, school-age youngsters fail to notice variables that are not immediately suggested by the concrete materials of the task—the height and forcefulness with which the pendulum is released.

PROPOSITIONAL THOUGHT

A second important characteristic of the formal operational stage is **propositional thought.** Adolescents can evaluate the logic of propositions (verbal statements)

In Piaget's formal operational stage, adolescents engage in propositional thought. As these students discuss problems in a precalculus class, they show that they can reason with symbols that do not necessarily represent objects in the real world. (Will Hart)

without referring to real-world circumstances. In contrast, concrete operational children can evaluate the logic of statements only by considering them against concrete evidence in the real world.

In a study of propositional reasoning, an experimenter showed children and adolescents a pile of poker chips and asked whether statements about the chips were true, false, or uncertain. In one condition, the experimenter hid a chip in her hand and presented the following propositions:

- "*Either* the chip in my hand is green *or* it is not green."

- "The chip in my hand is green *and* it is not green."

In another condition, the experimenter held either a red or a green chip in full view and made the same statements.

School-age children focused on the concrete properties of the poker chips. When the chip was hidden from view, they replied that they were uncertain about both statements. When it was visible, they judged both statements to be true if the chip was green and false if it was red. In contrast, adolescents analyzed the logic of the statements. They understood that the "either–or" statement is always true and the "and" statement is always false, regardless of the poker chip's color (Osherson & Markman, 1975).

Although Piaget believed that language does not play a central role in children's cognitive development (see Chapter 7), he acknowledged it is more important in adolescence. Abstract thought requires language-based systems of representation that do not stand for real things, such as those in higher mathematics. Around age 14 or 15, high school students start to use these systems in algebra and geometry. Formal operational thought also involves verbal reasoning about abstract concepts. Jules showed that he could think in this way when he pondered the relation among time, space, and matter in physics and wondered about justice and freedom in philosophy.

RECENT RESEARCH ON FORMAL OPERATIONAL THOUGHT

Recent research on formal operational thought poses questions similar to those we discussed with respect to Piaget's earlier stages: Is there evidence that abstract thinking appears earlier than Piaget expected? Do all individuals reach formal operations during their teenage years?

■ ARE CHILDREN CAPABLE OF ABSTRACT THINKING? Adolescents are capable of a much deeper grasp of scientific principles than they were at younger ages. For example, in simplified situations—ones involving no more than two possible causal variables—6-year-olds understand that hypotheses must be confirmed by appropriate evidence (Ruffman et al., 1993). But school-age children cannot sort

out evidence that bears on three or more variables at once. And they have difficulty explaining why a pattern of observations supports a hypothesis, even when they recognize the connection between the two (Kuhn, Amsel, & O'Loughlin, 1988; Schauble, 1990).

School-age children's capacity for propositional thought is also limited. For example, they have great difficulty reasoning from premises that contradict reality or their own beliefs. Consider the following set of statements: "If dogs are bigger than elephants and elephants are bigger than mice, then dogs are bigger than mice." Children younger than 10 judge this reasoning to be false, since all the relations specified do not occur in real life (Moshman & Franks, 1986). As Piaget's theory indicates, around age 11 young people in Western nations are much better at analyzing the logic of propositions irrespective of their content. Propositional thought improves steadily over the adolescent years (Markovits & Vachon, 1989, 1990).

■ DO ALL INDIVIDUALS REACH THE FORMAL OPERATIONAL STAGE? Try giving one or two of the formal operational tasks just described to your friends, and see how well they do. Even well-educated adults have difficulty with abstract reasoning! About 40 to 60 percent of college students fail Piaget's formal operational problems (Keating, 1979).

Why is it that so many college students, and adults in general, are not fully formal operational? The reason is that people are most likely to think abstractly in situations in which they have had extensive experience. This is supported by evidence that taking college courses leads to improvements in formal reasoning related to course content (Lehman & Nisbett, 1990). The physics student grasps Piaget's pendulum problem with ease. The English enthusiast excels at analyzing the themes of a Shakespeare play, whereas the history buff skillfully evaluates the causes and consequences of the Vietnam War. Because of differences in training and experience, the person who does well at one of these tasks may not be especially good at the others.

Finally, recall from Chapter 9 that the development of concrete operational thought is greatly delayed in some village and tribal societies. In many of these cultures, formal operational reasoning does not appear at all (Cole, 1990; Gellatly, 1987). Piaget acknowledged that because of lack of opportunity to solve hypothetical problems, abstract thought might not appear in some societies. Still, these findings raise further questions about the universal nature of Piaget's stage sequence. Is the highest stage really an outgrowth of children's independent efforts to make sense of their world? Or is it a culturally transmitted way of reasoning that is specific to literate societies and taught in school? These issues have prompted many investigators to turn toward an information-processing view.

AN INFORMATION-PROCESSING VIEW OF ADOLESCENT COGNITIVE DEVELOPMENT

Information-processing theorists agree with the broad outlines of Piaget's description of adolescent cognition. Compared to children, teenagers are much better at abstract reasoning because of gains in an underlying capacity to attend to information, hold it in memory, and combine it into more efficient and effective representations (Case, 1985, 1992; Demetriou et al., 1993; Fischer, 1980). But unlike Piaget, who thought that formal operations emerge out of the young person's efforts to act on and interpret complex situations, information-processing researchers believe that abstract reasoning can and often needs to be directly taught.

One influential information-processing view is that cognitive development involves acquiring increasingly powerful rules, or cognitive procedures, for solving problems. Consider Sabrina, who, during a free moment in science class, experimented with a balance scale. "If I put equal weights in the same place on both sides," she thought to herself, "it balances. If I put them in different places, it doesn't. If I add more weight to one side, I can make it balance again. So both weight and distance make a difference. But how do they work together?" Sabrina has a far more sophisticated understanding of this formal operational problem than does a younger child. Yet her grasp of the balance scale principle is still imprecise. Research by Robert Siegler (1978, 1983, 1998) shows that giving Sabrina information that addresses specific gaps in her reasoning will help her master the abstract rule—that only when the product of weight times distance on both sides of the scale is equal will it balance.

Consistent with these findings, recent evidence indicates that adolescents grasp formal operations in a step-by-step fashion on different kinds of problems. In one series of studies, 10- to 20-year-olds were given sets of tasks graded in difficulty. For example, one set consisted of quantitative-relational tasks, like the balance scale problem. Another set contained verbal-propositional tasks, like the poker chip problem on page 370. And in still another set were causal-experimental tasks, like the pendulum problem on page 369. For each type of task, adolescents mastered essential skills in sequential order. For example, on causal-experimental problems, they first became aware of the many possibilities that could influence an outcome. This enabled them to formulate and test hypotheses. Over time, adolescents combined separate skills into a smoothly functioning system. They constructed a general model that could be applied to many instances of a given type of problem (Demetriou, Efklides, & Platsidou, 1993; Demetriou et al., 1993).

What factors support adolescents' progress through this sequence? Greater information-processing capacity, exposure to increasingly complex problems, and instruction that highlights critical features of tasks and effective strategies are believed to be important (Kuhn et al., 1995; Schauble, 1996). Since adolescents often have more experience with certain kinds of problems than others, their level of development can vary considerably across tasks.

Today, information processing is offering the field a more precise description of cognitive change than Piaget's broad concepts of assimilation and accommodation (Klahr, 1992; Kuhn, 1992). In doing so, it is stimulating new methods of instruction that enhance the thinking of children and adolescents.

CONSEQUENCES OF ABSTRACT THOUGHT

The development of formal operations leads to dramatic revisions in the way adolescents see themselves, others, and the world in general. But just as adolescents are occasionally awkward in the use of their transformed bodies, they are sometimes faltering and clumsy in their abstract thinking. Parents and teachers must be careful not to mistake the many typical reactions of the teenage years—argumentativeness, self-concern, insensitive remarks, and indecisiveness—for anything other than inexperience with new reasoning powers. The Caregiving Concerns table on the following page suggests ways to handle the everyday consequences of teenagers' newfound capacity for abstraction.

ARGUMENTATIVENESS

As adolescents acquire formal operations, they are motivated to use them. The once pliable school-age child becomes a feisty, argumentative teenager who can marshal facts and ideas to build a case (Elkind, 1994). "A simple, straightforward explanation used to be good enough to get Louis to obey," complained Antonio. "Now, he wants a thousand reasons. And worse yet, he finds a way to contradict them all!"

As long as parent–child disagreements remain focused on principles and do not deteriorate into meaningless battles, they can promote development. Through discussions of family rules and practices, adolescents become more aware of their parents' values and the reasons behind them. Gradually, they come to see the validity of parental beliefs and adopt many as their own (Alessandri & Wozniak, 1987). Teenagers' capacity for effective argument also opens the door to intellectually stimulating pastimes, such as debate teams and endless bull-sessions with friends over moral, ethical, and political concerns.

CAREGIVING CONCERNS

Ways to Handle the Consequences of Teenagers' New Capacity for Abstraction

CONSEQUENCE	SUGGESTION
Argumentativeness	During disagreements, remain calm, rational, and focused on principles. Express your point of view and the reasons behind it. Although adolescents may continue to challenge, explanations permit them to consider the validity of your beliefs at a later time.
Sensitivity to public criticism	Refrain from finding fault with the adolescent in front of others. If the matter is important, wait until you can speak to the teenager alone.
Exaggerated sense of personal uniqueness	Acknowledge the adolescent's unique characteristics. At opportune times, point out how you felt similarly as a young teenager, encouraging a more balanced perspective.
Idealism and criticism	Respond patiently to the adolescent's grand expectations and critical remarks. Point out positive features of targets, helping them see that all worlds and people are blends of virtues and imperfections.
Difficulty making everyday decisions	Refrain from deciding for the adolescent. Offer patient reminders and everyday diplomatic suggestions until he or she can make choices more confidently.

SELF-CONSCIOUSNESS AND SELF-FOCUSING

Adolescents' ability to reflect on their own thoughts, combined with the physical and psychological changes they are undergoing, means that they start to think more about themselves. Piaget believed that the arrival of formal operations is accompanied by a new form of egocentrism: the inability to distinguish the abstract perspectives of self and other (Inhelder & Piaget, 1955/1958). For a time, teenagers become very wrapped up in the importance of their own thoughts, appearance, and behavior. As they imagine what others must be thinking, two distorted images of the relation between self and other appear.

The first is called the **imaginary audience**. Young teenagers regard themselves as always on stage. They are convinced that they are the focus of everyone else's attention and concern (Elkind & Bowen, 1979). As a result, they become extremely self-conscious, often going to great lengths to avoid embarrassment. Sabrina, for example, woke up one Sunday morning with a large pimple on her chin. "I can't possibly go to church!" she cried. "*Everyone* will notice how ugly I look." The imaginary audience helps us understand the long hours adolescents spend inspecting every detail of their appearance. It also accounts for their extreme sensitivity to public criticism. To teenagers, who believe that everyone is monitoring their performance, a critical remark from a parent or teacher can be mortifying.

A second cognitive distortion is the **personal fable**. Because teenagers are so sure that others are observing and thinking about them, they develop an inflated opinion of their own importance. They start to feel that they are special and unique. Many adolescents view themselves as

reaching great heights of glory as well as sinking to unusual depths of despair—experiences that others could not possibly understand (Elkind, 1994). On one occasion, for example, Sabrina had a crush on a boy who failed to return her affections. As she lay on the sofa feeling depressed, Franca tried to assure her that there would be other boys. "Mom," Sabrina snapped, "you don't know what it's like to be in love!" The personal fable may contribute to adolescent risk taking. Teenagers who have sex without contraceptives or weave in and out of traffic at 80 miles an hour

These seventh- and eighth-grade drama students really are on stage. But the imaginary audience leads them to think that everyone is monitoring their performance at other times as well. Consequently, young teenagers are extremely self-conscious and go to great lengths to avoid embarrassment. (Will Hart)

seem, at least for the moment, convinced of their uniqueness and invulnerability.

The imaginary audience and personal fable are strongest during the transition to formal operations. They gradually decline as abstract reasoning becomes better established (Enright, Lapsley, & Shukla, 1979; Lapsley et al., 1988). Some experts believe these ways of thinking do not represent a return to egocentrism, as Piaget's theory suggests. Instead, they may be an outgrowth of gains in perspective taking, which cause young teenagers to be very concerned with what others think (Lapsley et al., 1986). Adolescents may also cling to the idea that others are preoccupied with their appearance and behavior for emotional reasons. Doing so helps them maintain a hold on important relationships as they struggle to separate from parents and establish an independent sense of self (Lapsley, 1993).

IDEALISM AND CRITICISM

Because abstract thinking permits adolescents to go beyond the real to the possible, it opens up the world of the ideal and of perfection. Teenagers can imagine alternative family, religious, political, and moral systems, and they want to explore them. Doing so is part of investigating new realms of experience, developing larger social commitments, and defining their own values and preferences.

The idealism of teenagers leads them to construct grand visions of a perfect world—with no injustice, discrimination, or tasteless behavior. Then they insist that reality submit itself completely to their ideals. They do not make room for the shortcomings of everyday life. Adults, with their longer life experience, have a more realistic outlook. The disparity between adults' and teenagers' world views is often called the "generation gap," and it creates tension between parent and child. Aware of the perfect family against which their real parents and siblings do not measure up, adolescents may become fault-finding critics.

Teenage idealism and criticism have benefits in the long run. Once adolescents learn to see others as having both strengths and weaknesses, they are better able to work constructively for social change and to form positive and lasting relationships (Elkind, 1994). Parents can help teenagers forge a better balance between the ideal and the real by being tolerant of their criticism while reminding the young person that all people, including adolescents themselves, are blends of virtues and imperfections.

PLANNING AND DECISION MAKING

Adolescents, who think more analytically, handle cognitive tasks more effectively than they did at younger ages. Given a homework assignment, they are far better at *self-regulation*—planning what to do first and what to do next, monitoring progress toward a goal, and redirecting unsuc-

cessful efforts. For this reason, study skills improve from middle childhood into adolescence.

But when it comes to planning and decision making in everyday life, teenagers (especially young ones) often feel overwhelmed by the possibilities before them. Their efforts to choose among alternatives frequently break down, and they may resort to habit, act on impulse, or not make a decision at all (Elkind, 1994). On many mornings, for example, Sabrina tried on five or six outfits before leaving for school. Often she shouted from the bedroom, "Mom, what shall I wear?" Then, when Franca made a suggestion, Sabrina rejected it, opting for one of the two or three sweaters she had worn for weeks. Similarly, Louis procrastinated about registering for college entrance tests. When Franca mentioned that he was about to miss the deadline, Louis sat over the forms, unable to decide when or where he wanted to take the test. Parents may have to help with patient reminders and diplomatic suggestions until the young person gathers more experience and can make choices with greater confidence.

BRIEF REVIEW

In Piaget's formal operational stage, adolescents become capable of abstraction, as indicated by hypothetico-deductive reasoning and propositional thought. Recent research shows that abstract reasoning is not well developed until after age 11. Adolescents and adults are most likely to think abstractly in areas in which they have had extensive experience. In village and tribal cultures, formal operations often do not appear at all. According to the information-processing perspective, abstract reasoning develops gradually, is situation-specific, and often must be directly taught.

The dramatic cognitive changes of adolescence are reflected in everyday behavior. The ability to think in more sophisticated ways leads teenagers to become more argumentative, self-conscious and self-focused, and idealistic and critical. Because of gains in self-regulation, study skills improve. Yet in everyday life, adolescents are often overwhelmed by possibilities and may react impulsively and indecisively.

ASK YOURSELF . . .

■ *Cassie insisted that she had to have high heels to go with her prom dress. "No way I can wear those low heels, Mom. They'll make me look way too short next to Louis, and the whole evening will be ruined!" Why is Cassie so concerned about a detail of her appearance that most people would be unlikely to notice?*

SEX DIFFERENCES IN MENTAL ABILITIES

Sex differences in intellectual performance have been studied since the beginning of this century, and they have sparked almost as much controversy as the ethnic and social-class differences in IQ considered in Chapter 9. Although boys and girls do not differ in general intelligence, they do vary in specific mental abilities. Throughout the school years, girls attain higher scores on reading achievement tests and account for a lower percentage of children referred for remedial reading instruction (Halpern, 1992; Mullis et al., 1994). Girls' advantage on tests of general verbal ability is still present in adolescence. However, it is so slight that it is not really meaningful (Hyde & Linn, 1988).

By adolescence, boys start to do better than girls in mathematics, especially in abstract problem solving (Feingold, 1993; Hyde, Fenema, & Lamon, 1990). The gender gap is largest among gifted youngsters. In some widely publicized research on bright seventh and eighth graders invited to take the Scholastic Assessment Test (SAT), boys outscored girls on the mathematics subtest year after year. Twice as many boys as girls had scores above 500; 13 times as many scored over 700 (Benbow & Stanley, 1980, 1983).

Some researchers believe that the gender gap in mathematics is genetically based. One assumption is that it is due to sex differences in spatial skills and that the same biological factors underlie both spatial and mathematical performance. In support of this view, EEG measures of brain electrical activity show that mathematically talented junior high school boys have a unique capacity to activate the right cortical hemisphere while working on spatial tasks (Benbow & Lubinski, 1993). Also, adolescent girls and college women who score high on the SAT mathematics subtest are particularly good at spatial reasoning (Casey et al., 1995).

At the same time, social pressures contribute to girls' underrepresentation among the mathematically talented. Beginning in elementary school, before sex differences in math achievement appear, both boys and girls view math as a masculine subject. In addition, girls regard math as less useful for their future lives and more often blame their errors on lack of ability. These beliefs, in turn, lead girls to take fewer math courses in high school and college. The end result is that girls are less likely to acquire abstract mathematical concepts and reasoning skills (Byrnes & Takahira, 1993; Marsh, 1989).

A positive sign is that sex differences in cognitive abilities of all kinds have declined steadily over the past several decades. Paralleling this change is an increase in girls' enrollment in advanced high school math and science courses, a critical factor in reducing sex differences in knowledge and skill (Mullis et al., 1994). Still, extra steps are needed to promote girls' interest in and confidence at doing math and science. Teachers need to do a better job of demonstrating the relevance of math and science to everyday life. In addition, all students benefit from encouragement and constructive feedback rather than criticism when engaged in problem solving. Finally, exposure to role models of successful women is likely to improve girls' belief in their capacity to do well at math and science.

LANGUAGE DEVELOPMENT

Although language development is largely complete by the end of childhood, subtle but important changes take place in adolescence. These gains are related to teenagers' improved capacity for reflective thought and abstraction.

VOCABULARY AND GRAMMAR

Adolescents add a wide variety of abstract words to their vocabulary and can define them easily and accurately. In the conversation described on pages 368–369, note Jules's use of "counterintuitive," "incredible," "revolutionized," "philosophy," "reproduction," and "justice." As a 9- or 10-year-old, he rarely used such words and had difficulty grasping their meaning.

Formal operations permit adolescents to become masters of irony and sarcasm (Winner, 1988). "Don't have a major brain explosion," Louis commented to Sabrina when she complained about her homework one evening. And he quipped, "Oh boy, my favorite!" when Franca served a dinner that he disliked. Young children sometimes realize that a sarcastic remark is insincere when it is said in a very exaggerated way. But adolescents and adults need only notice the discrepancy between the statement and its context to grasp the intended meaning (Capelli, Nakagawa, & Madden, 1990). Increased sensitivity to the nuances of language enables teenagers to read and understand adult literary works.

Adolescents also use more elaborate grammatical constructions. Their sentences are longer and have a greater number of subordinate clauses than those of children. In addition, they are much better at analyzing and reflecting on the grammar of their language. Not surprisingly, diagramming sentences is a skill reserved for the junior high and high school years.

PRAGMATICS

Perhaps the most obvious change in language at adolescence is an improved ability to vary language style according to the situation (Obler, 1993). This change is partly the result of entering many more situations in which there is pressure to adjust speech style. To be successful on the debate team, Louis had to speak in a rapid-fire, well-organized manner. In theater class, he worked on reciting memorized lines as if they were natural. At work, his boss insisted that he respond to customers cheerfully and courteously. The ability to reflect on language and engage in self-regulation also helps adolescents use language styles effectively. They are far more likely than school-age children to practice what they want to say in an expected situation, review what they did say, and figure out how they could say it better (Romaine, 1984).

SECOND-LANGUAGE LEARNING

Many people assume that the best time to become bilingual is in childhood—that picking up a second language is harder during adolescence and adulthood. Compared to children, adolescents and adults make faster initial progress when learning a new language, but their ultimate attainment is not as high. For example, in a study of Chinese and Korean adults who had immigrated to the United States at varying ages, those who began mastering English between 3 and 7 scored as well as native speakers on a test of grammatical knowledge. As age of arrival in the United States increased through adolescence, test scores gradually declined (Johnson & Newport, 1989). The ability to pronounce a second language without an accent also decreases with age—gradually during childhood and sharply at adolescence (Anderson & Graham, 1994).

Research on children deprived of early language stimulation shows that there is a *sensitive period* for first-language development. That is, mastery of a native tongue must begin sometime in childhood for full development to occur (Mayberry, 1993). This same principle applies to a second language. Because biological readiness for language is greatest in childhood, bilingual development is best begun at an early age. Still, adolescents can become quite competent speakers of a second language. In fact, most aspects of language continue to improve with use, well into mature adulthood (Obler, 1993).

BRIEF REVIEW

Although adolescent girls score better than boys on tests of general verbal abilities, the difference is so slight that it is not meaningful. Boys are ahead of girls in mathematics performance, a difference related to spatial skills and social pressures that reduce girls' chances of doing well at math.

During adolescence, language continues to develop in subtle but important ways. Teenagers add many abstract words to their vocabularies. The ability to move beyond the literal meaning of words improves, and the grammatical structure of speech becomes more complex. Adolescents are better than children at modifying their language style to fit different situations. Because a sensitive period for acquiring language has passed, second-language learning is harder in adolescence and adulthood than in childhood.

ASK YOURSELF . . .

■ *Research shows that girls perform more poorly than boys on certain formal operational tasks, such as the pendulum and balance scale problems (Meehan, 1984). On the basis of what you know about the development of formal operational thought and sex differences in cognitive abilities, how would you account for this finding?*

■ *At home, Louis often created humorous parodies of the way his teachers spoke at school. What might account for his skill in mimicking the speech mannerisms of people?*

LEARNING IN SCHOOL

In complex societies, adolescence coincides with entry into secondary school. Most young people move into either a middle or junior high school and then into a high school. With each change, academic achievement becomes more serious, affecting college choices and job opportunities. In the following sections, we take up a variety of aspects of secondary school life.

SCHOOL TRANSITIONS

When Sabrina started junior high, she left a small, intimate, self-contained sixth-grade classroom for a much larger, impersonal school. "I don't know most of the kids in my classes," Sabrina complained to her mother at the end of the first week. "Besides, there's just too much homework. I get assignments in all my classes at once. I can't do all this!" she shouted, bursting into tears.

Moving from a small, self-contained elementary school classroom to a large, impersonal secondary school is stressful for adolescents. School grades, extracurricular participation, and self-esteem decline, especially for girls. (M. Antman/The Image Works)

■ IMPACT OF SCHOOL TRANSITIONS. As Sabrina's reactions suggest, school transitions can drastically alter academic and social experiences, creating new adjustment problems. With each school change—from elementary to middle or junior high and then to high school—adolescents' grades decline. The drop is partly due to tighter academic standards. At the same time, the transition to secondary school often brings with it less personal attention, more whole-class instruction, and less chance to participate in classroom decision making (Eccles et al., 1993a).

Inevitably, students must readjust their feelings of self-confidence and self-worth as environments become more impersonal and academic expectations are revised. A large-scale study showed that the earlier the school transition, the more powerful and longer-lasting its consequences, especially for girls (Simmons & Blyth, 1987). The researchers followed over 300 adolescents living in a large midwestern city from sixth to tenth grade. Some were enrolled in schools with a 6–3–3 grade organization (a K–6 elementary school, a 3-year junior high, and a 3-year high school). These students made two school changes, one to junior high and one to high school. A comparison group attended schools with an 8–4 grade organization. They made only one school transition, from a K–8 elementary school to high school.

For the sample as a whole, grade point average dropped and feelings of anonymity increased after each school transition. Participation in extracurricular activities declined more in the 6–3–3 than in the 8–4 arrangement, although the drop was greater for girls. Sex differences in self-esteem were even more striking. As Figure 11.5 shows, boys' self-esteem increased steadily, except in 6–3–3 schools, where it leveled off after entry into high school. Girls in the 6–3–3 arrangement fared especially poorly. Their self-esteem declined with each school change. In contrast, their 8–4 counterparts gained steadily in feelings of self-worth throughout secondary school.

These findings show that any school transition is likely to temporarily depress adolescents' psychological well-being, but the earlier it occurs, the more dramatic and long-lasting its impact. Girls in 6–3–3 schools fared poorest, the researchers argued, because movement to junior high tended to coincide with other life changes—namely, the onset of puberty and dating. Adolescents who must cope with added life stresses, such as family disruption or a shift in residence, around the time they change schools are at greatest risk for academic and emotional difficulties. Poorly achieving and poverty-stricken young people show an especially sharp drop in school perfor-

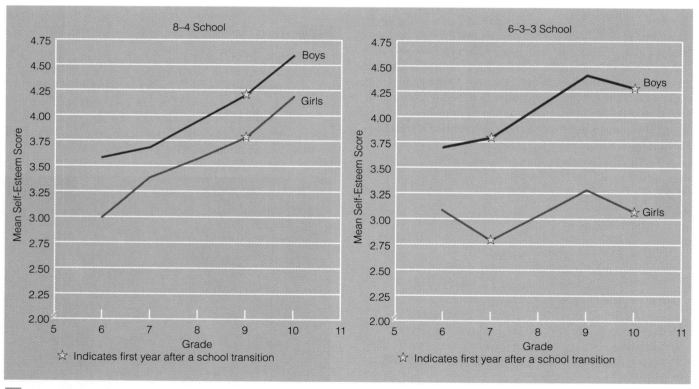

FIGURE 11.5

Self-esteem from sixth to tenth grade by school type for boys and girls. In this longitudinal study of over 300 adolescents, self-esteem increased steadily for both sexes in the 8–4 school arrangement. Girls in 6–3–3 schools fared especially poorly. Their self-esteem dropped sharply after each school change. (*Adapted from Simmons and Blyth, 1987.*)

Factors that Support High Achievement During Adolescence

FACTOR	DESCRIPTION
Child-rearing practices	Authoritative parenting Joint parent–adolescent decision making Parent involvement in the adolescent's education
Peer influences	Peer valuing of and support for high achievement
School characteristics	Teachers who are warm and supportive Learning activities that encourage high-level thinking
Employment schedule	Job commitment limited to under 15 hours a week High-quality vocational education for non-college-bound adolescents

mance after the junior high transition (Seidman et al., 1994). For some, it initiates a downward spiral in academic performance and school involvement that eventually leads to failure and dropping out (Simmons, Black, & Zhou, 1991).

■ **HELPING ADOLESCENTS ADJUST TO SCHOOL TRANSITIONS.** Fortunately, there are ways to ease the strain of going from elementary to secondary school. Since most students do better in an 8–4 school arrangement, school districts considering reorganization should seriously consider this plan. When this is not possible, smaller social units can be formed within large schools to relieve students' feelings of anonymity. For example, homerooms can be provided in which teachers offer academic and personal counseling and work closely with parents to promote favorable school adjustment. Students can also be assigned to classes with several familiar peers or a constant group of new peers. These interventions lead to better academic performance and psychological adjustment and reduce the rate of school dropout (Felner & Adan, 1988).

Finally, successful transitions are most likely to occur in schools that foster adolescents' growing capacity for autonomy, responsibility, and control over their own lives. Rigid school rules that strike young people as unfair and punitive frustrate their developmental needs, contributing to long-term dissatisfaction with school life (Eccles et al., 1993a).

ACADEMIC ACHIEVEMENT

Adolescent achievement is the result of a long history of cumulative effects. Early on, positive educational environments, both family and school, lead to personal traits that support achievement—intelligence, confidence in one's own abilities, the desire to succeed, and high educational aspirations. Nevertheless, improving an unfavorable environment can help a poorly performing young person bounce back, opening the door to a more satisfying adult life. The Caregiving Concerns table to the left summarizes environmental factors that enhance achievement during the teenage years.

■ **CHILD-REARING PRACTICES.** Authoritative parenting (which combines warmth with reasonable demands for maturity) is linked to achievement in adolescence, just as it predicts mastery-oriented behavior during the childhood years. In research involving thousands of adolescents, authoritative parenting predicted higher grades, whereas authoritarian and permissive styles were associated with lower grades (Dornbusch et al., 1987; Steinberg et al., 1992). Parents who engage in joint decision making with adolescents, gradually permitting more autonomy with age, have youngsters who achieve especially well (Dornbusch et al., 1990).

Parent involvement in the adolescent's secondary school career also fosters academic success. High-achieving young people typically have parents who keep tabs on their

This parent is involved with her adolescent's school career. Besides keeping tabs on his progress, she is probably in frequent contact with the school. She sends a message to her son about the importance of education and teaches him how to solve academic problems and make wise educational decisions. *(Erika Stone)*

child's progress, communicate with teachers, and make sure that their child is enrolled in challenging, well-taught classes. These efforts are just as important during junior and senior high school as they were earlier (Grolnick & Slowiaczek, 1994).

■ PEER INFLUENCES. Peers play an important role in adolescent achievement, in a way that is related to both family and school. Teenagers whose parents value achievement are likely to choose friends who share those values (Berndt & Keefe, 1995). For example, when Sabrina began to make new friends in junior high, she often studied with her girlfriends. Each wanted to do well in school and reinforced the same desire in the others.

Peer support for high achievement also depends on the overall climate of the peer culture. In some schools, ethnic minority students react against working hard, convinced that getting good grades will have little payoff in the future and regarding it as a threat to their ethnic identity. In one case study of an inner-city high school, African-American students who did achieve were labeled as "brainiacs" and had to cope with the "burden of acting white." Many capable adolescents felt caught between achievement and peer approval. They resolved the dilemma by "putting the brakes" on academic effort (Fordham & Ogbu, 1986).

■ CLASSROOM AND SCHOOL ENVIRONMENTS. To realize their potential for abstract thought, adolescents need classroom environments that are responsive to their expanding powers of reasoning. Yet a large, departmentalized school organization leads many adolescents to report that their classes lack warmth and supportiveness—a circumstance that dampens their motivation (Eccles et al., 1993b). Adolescents (not just children) need to form close relationships with teachers. As they begin to develop an identity beyond the family, they seek out adult models other than their parents (Eccles et al., 1993a). Of course, one important reason for separate classes in each subject is so adolescents can be taught by experts, who are more likely to encourage high-level thinking. But the classroom experiences of many secondary school students do not work out this way. In a study of English, social studies, and science teachers with reputations for excellence, students were not equally or consistently given assignments that stimulated abstract thought (Sanford, 1985).

Because of the uneven quality of instruction in American schools, a great many seniors graduate from high school deficient in basic academic skills. Although the achievement gap separating African-American and Hispanic students from white students has declined since the 1970s, mastery of reading, writing, mathematics, and science by low-income ethnic minority students remains disappointing (Mullis et al., 1994; Wolf, 1993). Many attend underfunded schools with run-down buildings,

outdated equipment, and textbook shortages. In some, attention to crime and discipline problems has taken the lead over learning and instruction. By junior high, large numbers of poverty-stricken minority students have been placed in low academic tracks, compounding their learning difficulties.

■ TRACKING. Ability grouping, as we saw in Chapter 9, is detrimental during the elementary school years. Students in low groups generally get poor quality instruction. Soon they view themselves as failures, and their peers label them this way as well. At least into the early years of secondary school, mixed-ability classes are desirable. Research shows that they do not stifle the more able students, and they have intellectual and social benefits for poorly performing youngsters (Oakes, Gamoran, & Page, 1992).

By high school, some grouping is unavoidable because certain aspects of education must dovetail with the young person's future educational and career plans. In the United States, high school students are counseled into college preparatory, vocational, or general education tracks. Unfortunately, this sorting tends to perpetuate educational inequalities of earlier years. Low-income minority students are assigned in large numbers to noncollege tracks. One study found that a good student from an economically disadvantaged family had only half as much chance of ending up in an academically oriented program as a student of equal ability from a middle-class background (Vanfossen, Jones, & Spade, 1987).

High school students are separated into academic and vocational tracks in virtually all industrialized nations. But in China, Japan, and most Western European nations, students take a national examination to determine their placement in high school. The outcome usually fixes future possibilities for the young person. In the United States, educational decisions are more fluid. Students who are not assigned to a college preparatory track or who do poorly in high school can still get a college education. But by the adolescent years, social-class differences in quality of education and academic achievement have already sorted American students more drastically than is the case in other countries. In the end, many young people do not benefit from this more open system. Compared to other developed nations, the United States has a higher percentage of high school dropouts and adolescents with very limited educational skills (Hamilton, 1990; McAdams, 1993). As the Lifespan Vista box on the following page indicates, school desegregation is an important means for counteracting minority students' reduced educational opportunities and life chances.

■ PART-TIME WORK. Almost half of American high school students work part-time during the school year—a greater percentage than in any other developed country.

SCHOOL DESEGREGATION AND LIFE CHANCES OF AFRICAN-AMERICAN ADOLESCENTS

In 1954, a historic U.S. Supreme Court decision proclaimed the injustice of separate schools for racially and ethnically different students. The law gradually led to desegregation of many American schools, but not without community polarization in many areas of the country. Prejudice, fierce defense of the neighborhood school, and strong objections to busing led many parents to oppose the practice. Where desegregation did occur, teachers had to take steps to promote cross-race contact and interracial acceptance. Only then did gains in black students' self-esteem and achievement appear (Johnson, Johnson, & Maruyama, 1984). School desegregation efforts began to decline in the 1970s because many studies reported little or no impact on intergroup relations and test scores.

Yet research reveals that although short-term benefits of desegregation are mixed, long-term effects are striking. Attending desegregated schools is related to higher occupational aspirations among African-American students (Dawkins, 1983; Gable, Thompson, & Iwanicki, 1983). Desegregation also increases the likelihood that black young people will make educational and employment decisions appropriate to their goals. Black graduates of ethnically diverse high schools are more likely to enroll in desegregated colleges and to enter careers in which African Americans have been underrepresented (Braddock & McPartland, 1982, 1987). And once in the work world, they can more often be found in professional jobs and in ethnically mixed settings, where they report feeling more comfortable than do their counterparts with a segregated education (Trent, 1991).

How does school desegregation achieve these effects? According to several theorists, attending a mixed high school grants all groups equal access to information about educational and occupational opportunities and methods of taking advantage of them. College fairs, reminders about college possibilities and application deadlines by teachers and peers, help with the college application process, and counselors with strong ties to college admissions offices are far more abundant in racially mixed schools (Wells, Crain, & Uchetelle, 1995). Furthermore, black graduates of interracial high schools are more likely to use desegregated networks to find jobs,

including informal referrals and unsolicited approaches to white employers. These are popular techniques for securing entry-level positions that lead to higher occupational status and income in later life (Braddock & McPartland, 1987).

In sum, desegregated schools grant black students access to better educational and occupational opportunities, which help break the cycle of perpetual segregation by contributing to success in the labor market. An even more extended outcome could be improved academic achievement for future generations of African-American students (Wells & Crain, 1994).

Desegregated schools grant African-American students access to better educational and occupational opportunities. Through such supports as college fairs and reminders about college possibilities and application deadlines, ethnically diverse high schools help break the cycle of perpetual segregation by contributing to success in the labor market. (R. Sidney/The Image Works)

Most are middle-class adolescents in pursuit of spending money rather than vocational training. Low-income teenagers who need to contribute to family income find it harder to get jobs (Children's Defense Fund, 1997).

Furthermore, the jobs adolescents hold are limited to low-level repetitive tasks that provide little contact with adult supervisors. A heavy commitment to such jobs is harmful. Students who work more than 15 hours per week have poorer school attendance, lower grades, and less time for extracurricular activities. They also report more drug and alcohol use and feel more distant from their parents. And perhaps because of the menial nature of their jobs, employed teenagers tend to be cynical about work life. Many admit to having stolen from their employers (Greenberger & Steinberg, 1986; Steinberg & Dornbusch, 1991.)

When work experiences are specially designed to meet educational and vocational goals, outcomes are different. Work–study programs are related to positive school and work attitudes and improved achievement among teenagers whose low-income backgrounds and weak academic skills make them especially vulnerable to school dropout and unemployment (Owens, 1982; Steinberg, 1984). Yet high-quality vocational preparation for non-college-bound American adolescents is scarce. Unlike Western European nations, the United States has no widespread training system to prepare its youths for skilled business and industrial occupations and manual trades. The federal government does support some job training programs. But most are too short to make a difference in the lives of poorly skilled adolescents, who need intensive training and academic remediation before they are ready to enter the job market. And at present, these programs serve only a small minority of young people who need assistance (Children's Defense Fund, 1997).

DROPPING OUT

Across the aisle from Louis in math class sat Norman, who daydreamed, crumpled his notes into his pocket after class, and rarely did his homework. On test days, he twirled a rabbit's foot for good luck but left most of the questions blank. Louis had been in school with Norman since fourth grade, but the two boys had little to do with one another. To Louis, who was quick at schoolwork, Norman seemed to live in another world. Once or twice each week, Norman cut class, and one spring day, he stopped coming altogether.

Norman is one of 14 percent of American young people who, by 18 years of age, leave high school without a diploma (Children's Defense Fund, 1997). The dropout rate is particularly high among low-income ethnic minority youths, especially Hispanic teenagers (see Figure 11.6). The decision to leave school has dire conse-

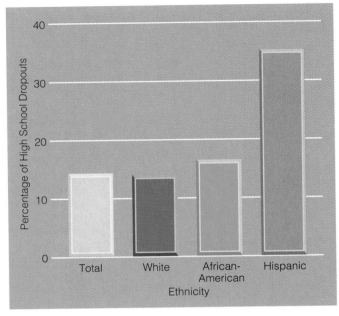

FIGURE 11.6

Percentage of high school dropouts by ethnicity. Because African-American and Hispanic teenagers are more likely to come from low-income and poverty-stricken families, their dropout rates are above the national average. The rate for Hispanic young people is especially high. (*From U.S. Department of Education, 1996.*)

quences. As Figure 11.7 shows, dropouts are far less likely to be employed than are high school graduates who do not go to college. And even when they are employed, they have a much greater chance of remaining in menial, low-paying jobs and of being out of work from time to time (Eccles, 1990).

■ FACTORS RELATED TO DROPPING OUT. Long before he stopped coming, Norman showed signs that he would drop out. Because of a history of poor school performance, his perception of his own ability was extremely low. He gave up on tasks that presented the least bit of challenge and counted on luck—his rabbit's foot—to get him by. As Norman got older, he attended class less regularly, failed to pay attention when he was there, rarely did his homework, and was a discipline problem. He didn't participate in any school clubs or sports but, instead, often used drugs and engaged in lawbreaking acts. Because he was so uninvolved, few teachers or students got to know him well. The day Norman left, he felt alienated from all aspects of school life.

Like other dropouts, Norman's family background contributed to his problems. Compared to other students, even those with the same social-class and grade profile, dropouts are more likely to have parents who are less involved in their youngster's education. Many did not finish high school themselves and are unemployed, on welfare, or coping with the aftermath of divorce. When their youngsters bring home poor report cards, these parents are more likely to respond with punishment and anger—reactions that cause adolescents to rebel further against academic work (Garnier, Stein, & Jacobs, 1997; Rumberger et al., 1990).

Academically marginal students who drop out often have school experiences that undermine their chances for success. Recent reports indicate that over 60 percent of adolescents in some crime-ridden inner-city high schools do not graduate. Students in general education and vocational tracks, where teaching tends to be the least stimulating, are three times more likely to drop out as those in an academic program (Office of Educational Research and Improvement, 1993).

■ PREVENTION STRATEGIES. The strategies for helping teenagers at risk for dropping out are diverse, but several common themes are related to success:

■ *High-quality vocational training.* For many marginal students, the real-life nature of vocational education is more comfortable and effective than purely academic work. But to work well, it must carefully integrate academic and job-related instruction so students see the relevance of what happens in the classroom to their future goals (Hamilton, 1993).

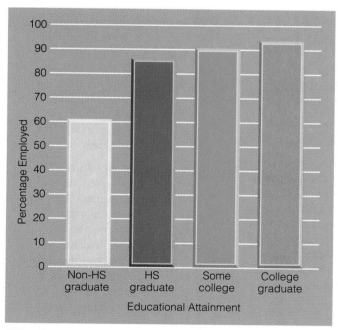

FIGURE 11.7

Employment rates of 16- to 24-year-olds by educational attainment. Non–high school graduates are much less likely to get jobs than are their counterparts with high school diplomas. Employment rates increase with years of schooling completed. (*From U.S. Department of Education, 1996.*)

■ *Remedial instruction and counseling that offer personalized attention.* Most potential dropouts need intensive remedial instruction in small classes that permit warm, caring teacher–student relationships to form. To overcome the negative psychological effects of repeated school failure, good academic assistance must be combined with social support and special counseling (Rumberger, 1990).

■ *Efforts to address the many factors in students' lives related to leaving school early.* Programs that strengthen parent involvement, offer flexible work–study arrangements, and provide on-site child care for teenage mothers can make staying in school easier for at-risk adolescents.

■ *Participation in extracurricular activities.* Another way of helping marginal students is to draw them into the community life of the school (Mahoney & Cairns, 1997). The most powerful influence on extracurricular involvement is small school size. In smaller high schools (500 to 700 students or less), a greater proportion of the student body is needed to staff and operate activities. Potential dropouts are far more likely to participate, feel needed, gain recognition for their abilities, and remain until graduation. Note that "house" plans, which create smaller units within large schools, can have the same effect (Berk, 1992b).

As we conclude our discussion of academic achievement, let's place the school dropout problem in historical perspective. Over the last half century, the percentage of American adolescents completing high school rose dramatically—from 39 percent in 1940 to 86 percent in the 1990s. During this same period, college attendance also increased. Today, nearly 40 percent of 18- to 24-year-olds are working toward college degrees—the highest rate in the world. Finally, about one-third of all high school dropouts return to finish their education within a few years, and some extend their schooling further (Children's Defense Fund, 1997). As the end of adolescence approaches, many young people realize how essential education is for a rewarding job and career.

ASK YOURSELF . . .

■ *Tanisha is finishing sixth grade. She could either continue in her current school through eighth grade or switch to a much larger junior high school. What would you suggest she do, and why?*

■ *In a workshop for parents of adolescents, one father asks what he might do to encourage his teenage children to do well in school. Provide a list of suggestions along with reasons each is effective.*

PHYSICAL DEVELOPMENT IN ADOLESCENCE

CONCEPTIONS OF ADOLESCENCE

How have conceptions of adolescence changed over the twentieth century?

- Early biologically oriented theories viewed **puberty** as an inevitable period of storm and stress. For example, in Freud's **genital stage**, reawakening of instinctual drives results in psychological conflict and volatile behavior. An alternative perspective regarded the social environment as entirely responsible for the wide variability in adolescent adjustment. Modern research shows that adolescence is a product of both biological and social forces.

PUBERTY: THE PHYSICAL TRANSITION TO ADULTHOOD

Describe pubertal changes in body size, proportions, and sexual maturity.

- Hormonal changes beginning in middle childhood initiate puberty, on the average, 2 years earlier for girls than boys. The first outward sign of puberty is the **growth spurt**. As the body enlarges, girls' hips and boys' shoulders broaden. Girls add more fat, boys more muscle.

- Sex hormones regulate changes in **primary** and **secondary sexual characteristics. Menarche** occurs late in the girl's sequence of pubertal events, following the rapid increase in body size. Among boys, as the sex organs and body enlarge and pubic and underarm hair appear, **spermarche** (first ejaculation) takes place.

What factors influence the timing of puberty?

- In addition to heredity, nutrition and overall physical health contribute to the timing of puberty.

A **secular trend** toward an earlier age of menarche has occurred in industrialized nations.

THE PSYCHOLOGICAL IMPACT OF PUBERTAL EVENTS

What factors influence adolescents' reactions to the physical changes of puberty?

- Girls generally react to menarche with surprise and mixed emotions, but whether their feelings lean in a positive or negative direction depends on advance information and support from family members. Boys usually know ahead of time about spermarche, but they receive less support for the physical changes of puberty than do girls.

- Besides higher hormone levels, negative life events and situational changes are associated with adolescent moodiness. Puberty is accompanied by psychological distancing between parent and child. The reaction may be a modern substitute for physical departure from the family, which typically occurs at sexual maturity in primate species.

Describe the impact of maturational timing on adolescent adjustment, noting sex differences.

- Early maturing boys and late maturing girls, whose appearance closely matches cultural standards of physical attractiveness, have a more positive **body image** and usually adjust well in adolescence. In contrast, early maturing girls and late maturing boys, who fit in least well physically with peers, experience emotional and social difficulties.

HEALTH ISSUES

Describe nutritional needs, and cite factors related to serious eating disturbances during adolescence.

- As the adolescent's body grows, nutritional requirements increase. Because of poor eating habits, many adolescents suffer from iron, vitamin, and mineral deficiencies.

- Girls who reach puberty early, who are very dissatisfied with their body images, and who grow up in economically advantaged homes where thinness is idealized are at increased risk for eating disorders. **Anorexia nervosa** tends to appear in girls with perfectionist personalities and overprotective, controlling parents. The impulsive eating of **bulimia** is associated with lack of self-control in other areas of life.

Discuss social and cultural influences on adolescent sexual attitudes and behavior.

- The hormonal changes of puberty lead to an increase in sex drive, but social factors affect how teenagers manage their sexuality. Compared to most other cultures, the United States is fairly restrictive in its attitude toward adolescent sex. Young people receive contradictory messages from the social environment. Sexual attitudes of adults and adolescents have become more liberal in recent decades, and the rate of teenage sexual activity has risen.

- Early sexual activity is linked to a variety of factors associated with economic disadvantage. About half of sexually active American teenagers do not practice contraception regularly. Adolescent cognitive processes and a lack of social support for responsible sexual behavior underlie the failure of many young people to protect themselves against pregnancy.

Describe factors related to the development of homosexuality.

- About 3 to 6 percent of adolescents discover that they are lesbian or gay. New evidence indicates that genetic factors play an important role in homosexuality. Lesbian and gay teenagers face special problems in developing a positive sexual identity.

Discuss factors related to teenage pregnancy, child bearing, and sexually transmitted disease.

■ Adolescent pregnancy, abortion, and child bearing are higher in the United States than in many industrialized nations. Teenage parenthood is often associated with school dropout and poverty, which threaten the well-being of both adolescent and child. Improved sex education and contraceptive services reduce teenage pregnancy and child bearing.

■ A high rate of sexually transmitted disease (STD) among American adolescents is another consequence of early sexual activity and inconsistent contraceptive use. Many individuals contract the AIDS virus as teenagers.

What personal and social factors are related to adolescent substance use and abuse?

■ Teenage alcohol and drug use is widespread in the United States. Adolescents who experiment with drugs are well-adjusted, curious young people. The minority who go from use to abuse have psychological, family, and school problems.

Cite common unintentional injuries in adolescence.

■ Motor vehicle collisions are the leading cause of adolescent injury and death. The rate of disability and death caused by firearms is also high.

MOTOR DEVELOPMENT

Describe sex differences in motor development during adolescence.

■ Adolescents improve in gross motor performance, although boys show much larger gains than girls. Girls continue to receive less encouragement and recognition for developing athletic skill, although their involve-

ment in high school sports has increased in recent years.

COGNITIVE DEVELOPMENT IN ADOLESCENCE

PIAGET'S THEORY: THE FORMAL OPERATIONAL STAGE

What are the major characteristics of formal operational thought?

■ During Piaget's **formal operational stage,** abstract thinking appears. Adolescents engage in **hypothetico-deductive reasoning.** When faced with a problem, they think of all possibilities and test them systematically. **Propositional thought** also develops. Young people can evaluate the logic of verbal statements without considering them against real-world circumstances.

Discuss recent research on formal operational thought and its implications for the accuracy of Piaget's formal operational stage.

■ Adolescents are capable of a much deeper grasp of scientific principles than are school-age children. However, college students and adults think abstractly only in situations in which they have had extensive experience, and formal thought does not appear in many village and tribal cultures. These findings indicate that Piaget's highest stage is affected by specific learning opportunities.

AN INFORMATION-PROCESSING VIEW OF ADOLESCENT COGNITIVE DEVELOPMENT

How do information-processing researchers account for the development of abstract reasoning?

■ Information-processing researchers believe that gains in information-processing capacity and opportunities to acquire knowledge and cognitive strategies account for the development of abstract reasoning.

On different kinds of problems, adolescents grasp formal operational abilities in a similar, step-by-step fashion.

CONSEQUENCES OF ABSTRACT THOUGHT

Describe typical reactions of adolescents that result from new abstract reasoning powers.

■ Using their new cognitive powers, teenagers become more argumentative, idealistic, and critical. As they think more about themselves, two distorted images of the relation between self and other appear—the **imaginary audience** and **personal fable.** Adolescents show gains in self-regulation, but they often have difficulty making decisions in everyday life.

SEX DIFFERENCES IN MENTAL ABILITIES

Describe sex differences in mental abilities at adolescence.

■ During adolescence, the female advantage in verbal ability is very small. Boys start to do better in mathematics, especially abstract problem solving. Biological factors common to spatial and mathematical reasoning may contribute to the gender gap. At the same time, social pressures related to stereotyping of math as a masculine subject are involved.

LANGUAGE DEVELOPMENT

Describe changes in vocabulary, grammar, and pragmatics during adolescence, and compare adolescents' capacity for second-language learning to that of children.

■ Teenagers add many abstract words to their vocabulary, use more elaborate grammatical constructions, and become masters of irony and sarcasm. The most obvious change in language at adolescence is the ability to make subtle adjustments in language style, depending on the situation.

■ Compared to children, teenagers make faster initial progress when learning a second language, but their ultimate level of attainment is not as high. Biological readiness for language learning is greatest in childhood.

LEARNING IN SCHOOL

Discuss the impact of school transitions on adolescent adjustment.

■ School transitions in adolescence can be stressful. Teenagers who must cope with added stresses around the time they change schools are at greatest risk for academic and emotional difficulties.

Discuss family, peer, and school influences on academic achievement during adolescence.

■ Authoritative parenting and parent involvement in the adolescent's secondary school career promote high achievement. Teenagers with parents who encourage achievement are likely to choose friends who do the same. Warm, supportive learning environments with activities that emphasize high-level thinking enable adolescents to reach their cognitive potential.

■ By high school, separate educational tracks that dovetail with students' future plans are necessary. Unfortunately, high school tracking in the United States usually extends the educational inequalities of earlier years.

What factors are related to dropping out of school?

■ Fourteen percent of American young people leave high school without a diploma. Many are low-income, minority youths. Dropping out is the result of a slow, cumulative process of disengagement from school. Family and school influences combine to undermine the young person's chances for success.

IMPORTANT TERMS AND CONCEPTS

puberty (p. 350)

genital stage (p. 350)

growth spurt (p. 351)

primary sexual characteristics (p. 353)

secondary sexual characteristics (p. 353)

menarche (p. 353)

spermarche (p. 354)

secular trends (p. 354)

body image (p. 356)

anorexia nervosa (p. 358)

bulimia (p. 359)

formal operational stage (p. 369)

hypothetico-deductive reasoning (p. 369)

propositional thought (p. 369)

imaginary audience (p. 372)

personal fable (p. 372)

EATING DISORDERS

Anorexia Nervosa and Related
Eating Disorders, Inc.
P.O. Box 5102
Eugene, OR 97405
(541) 344-1144
Web site: *www.anred.com*
*An association of anorexics and bulimics,
their families and friends, and concerned
professionals that provides information
about the disorders.*

SEXUAL BEHAVIOR

Alan Guttmacher Institute
111 Fifth Avenue
New York, NY 10003
(212) 248-1111
*Compiles statistics on sexual behavior
and fertility and promotes public policy
related to birth control. Publishes the
journal* Family Planning Perspectives,
*which includes articles on teenage sexual
behavior, pregnancy, and childbearing.*

SEXUALLY TRANSMITTED DISEASE (STD)

American Social Health Association
P.O. Box 13827
Research Triangle Park, NC 27709
(919) 361-8400
Web site: *www.sunsite.unc.edu/asha*
*A national health agency that works to
expand research, provide information to
communities, and improve public health
policy related to STD. Operates the Herpes Resource Center, a support program
for sufferers of incurable genital herpes.*

Teens Teaching AIDS Prevention
3030 Walnut
Kansas City, MO 64108
(816) 561-8784
*Provides teenagers with information on
AIDS through peer counseling. Operates
a toll-free hotline staffed by trained high
school students and adult advisers, (800)
234-TEEN.*

SUBSTANCE ABUSE

Do It Now Foundation
P.O. Box 27568
Tempe, AZ 85285
(602) 491-0393
*Works to provide factual information to
adolescents and adults about alcohol,
drugs, and related health issues. Assists
organizations engaged in drug education.*

(See page 449 for additional organizations related to alcohol abuse.)

TEENAGE PREGNANCY

National Organization
of Adolescent Pregnancy
and Parenting
4421A East–West Highway
Bethesda, MD 20814
(202) 783-5770
*An association that promotes community
services designed to treat problems associated with adolescent pregnancy and
childbearing.*

DROPOUT PREVENTION

National Dropout Prevention
Center
Clemson University
205 Martin Street
Clemson, SC 29634-5111
(864) 656-2599
Web site: *www.dropoutprevention.org*
*A center offering information on school
dropout prevention and identifying high-risk youth. Provides consultation and
referral services to school systems, agencies, and associations dealing with the
dropout problem.*

Emotional and Social Development in Adolescence

ouis sat on the grassy hillside overlooking the high school, waiting for his best friend, Darryl, to arrive from his fourth-period class. The two boys often met at noontime and then crossed the street to have lunch at a nearby hamburger stand.

Watching as hundreds of students poured onto the school grounds, Louis reflected on what Mrs. Kemp had said in government class that day. "Suppose by chance I *had* been born in the People's Republic of China. I'd be sitting here, speaking a different language, being called by a different name, and thinking about the world in different ways. Gosh, I am who I am through some quirk of fate," Louis pondered.

Louis awoke from his thoughts with a start. Darryl was standing in front of him. "Hey, dreamer! I've been shouting and waving from the bottom of the hill for 5 minutes. How come you're so spaced out lately, Louis?" Darryl asked as they walked off.

"Oh, just wondering about stuff—like what I want, what I believe in. My older brother Jules—I envy him. He seems to know just where he's going. Most of the time, I'm up in the air about it. You ever feel that way?"

"Yeah, a lot," admitted Darryl, looking at Louis seriously as they approached the hamburger stand. "I often think—What am I really like? Who will I become?"

Louis and Darryl's introspective remarks are signs of a major reorganization of the self that takes place at adolescence: the development of identity. Both young people are attempting to formulate who they are—their personal values and the directions they will pursue in life. We begin this chapter with Erikson's account of identity development and the research it has stimulated on teenagers' thoughts and feelings about themselves. The quest for identity

extends to many aspects of development. We will see how moral understanding, sense of cultural belonging, and masculine and feminine self-images are refined during the teenage years. And as parent–child relationships are revised and young people become increasingly independent of the family, friendships and peer networks become crucial contexts for bridging the gap between childhood and adulthood. Our chapter concludes with a discussion of several serious adjustment problems of adolescence: depression, suicide, and delinquency.

ERIKSON'S THEORY: IDENTITY VERSUS IDENTITY DIFFUSION

rikson (1950, 1968) was the first to recognize **identity** as the major personality achievement of adolescence and as a vital step toward becoming a productive, happy adult. Constructing an identity involves defining who you are, what you value, and the directions you choose to pursue in life. It is the driving force behind many new commitments—to a sexual orientation, to a vocation, and to ethical, political, religious, and cultural ideals.

Erikson called the psychological conflict of adolescence **identity versus identity diffusion.** He believed that successful outcomes of earlier stages pave the way toward its positive resolution. For example, young people who reach adolescence with a weak sense of *trust* have trouble finding ideals to have faith in. Those with little *autonomy* or *initiative* do not engage in the active exploration required to choose among alternatives. And those who lack a sense of *industry* fail to select a vocation that matches their interests and skills.

Although the seeds of identity formation are planted early, not until adolescence do young people become absorbed in this task. According to Erikson, in complex societies, teenagers experience an *identity crisis*—a temporary period of confusion and distress as they experiment with alternatives before settling on values and goals. During this period, adolescents question what they once took for granted. "I've gone to church every Sunday morning since I was a little kid," Louis confided in Darryl. "Now I'm not so sure I can accept my parents' way of thinking about God." Teenagers who go through a process of inner soul-searching eventually arrive at a mature identity. They sift

"Exploration" describes the typical adolescent's gradual, uneventful approach to identity formation. This teenager's eager experimentation with photography in his high school journalism class may help crystallize a vocational path or lead to an absorbing, long-term hobby. (Bob Daemmrich/The Image Works)

through characteristics that defined the self in childhood and combine them with new commitments. Then they mold these into a solid inner core that provides a sense of sameness as they move through different roles in daily life. Once formed, identity continues to be refined in adulthood as people reevaluate earlier commitments and choices.

Although current theorists agree with Erikson that questioning of the self's values, plans, and priorities is necessary for a mature identity, they no longer refer to this process as a "crisis" (Baumeister, 1990). For some young people, identity development is traumatic and disturbing, but for most it is not. "Exploration" better describes the typical adolescent's gradual, uneventful approach to identity formation. The many daily choices that teenagers make—"whom to date, whether or not to break up, having intercourse, taking drugs, going to college or working, which college, what major, studying or playing, being politically active"—and the reasons for them are gradually put together into an organized self-structure (Marcia, 1980, p. 161).

The negative outcome of Erikson's fifth stage is *identity diffusion.* Some adolescents appear shallow and directionless, either because earlier conflicts have been resolved negatively or society restricts their choices to ones that do not match their abilities and desires. As a result, they are unprepared for the psychological challenges of adulthood. For example, individuals find it difficult to risk the self-sharing involved in Erikson's young adult stage—intimacy—if they do not have a firm sense of self (an identity) to which they can return.

Is there research to support Erikson's ideas about identity development? In the following sections, we will see that adolescents go about the task of defining themselves in ways that closely match Erikson's description.

SELF-DEVELOPMENT IN ADOLESCENCE

During adolescence, cognitive advances transform the young person's vision of the self into a more complex, well-organized, and consistent picture. Changes in self-concept and self-esteem set the stage for development of a unified personal identity.

CHANGES IN SELF-CONCEPT

Recall that by the end of middle childhood, children describe themselves in terms of personality traits, such as "I'm smart," "I'm shy," or "I'm honest." This change permits young people to link their past, present, and future selves. But the self-statements of early adolescents are not interconnected, and sometimes they even include contradictory descriptions (Damon & Hart, 1988).

By middle to late adolescence, teenagers combine their various traits into an organized system. They add integrating principles, which make sense out of apparent contradictions. For example, one young person remarked, "I'm very adaptable. When I'm around my friends, who think that what I say is important, I'm very talkative; but around my family I'm quiet because they're never interested enough to really listen to me" (Damon, 1990, p. 88).

Teenagers' self-statements also reflect their concern about being liked and viewed in a positive light by others. Compared to school-age children, they place more emphasis on social virtues, such as being friendly, considerate, kind, and cooperative (Rosenberg, 1979). In addition, personal and moral values appear as key themes in older adolescents' self-concepts. As adolescents' views of themselves include enduring beliefs and plans, they move toward the unity of self involved in building a mature identity.

CHANGES IN SELF-ESTEEM

Self-esteem, the evaluative side of self-concept, continues to differentiate during adolescence. Besides academic competence, social self-worth, and physical ability, new dimensions reflecting important concerns of this period are added: close friendship, romantic appeal, and job competence (Harter, 1990).

Level of self-esteem also changes. Turn back to Figure 11.5 on page 376, and you will see that except for a temporary decline associated with school transitions, self-esteem rises for most adolescents (Nottelmann, 1987). This steady increase is yet another reason that modern researchers question the assumption that adolescence is a time of emotional turmoil. To the contrary, for most young people,

This teenager is a volunteer at a day camp for Native-American children on a Wyoming reservation. Here, she assists the children in t-shirt tie dyeing. Community service activities like this one undoubtedly contribute to the rise in self-esteem that characterizes most young people during adolescence. (John Eastcott/ The Image Works)

becoming an adolescent leads to feelings of pride and self-confidence (Powers, Hauser, & Kilner, 1989). This is true not just in the United States, but around the world. A study of self-esteem in ten industrialized countries showed that the majority of teenagers had an optimistic outlook on life, a positive attitude toward school and work, and faith in their ability to cope with life's problems (Offer, 1988).

Of course, as we saw in Chapter 11, adolescents vary widely in self-esteem. Those who are off-time in pubertal development, who are heavy drug users, and who fail in school feel poorly about themselves. In addition, of those young people whose self-esteem drops in adolescence, most are girls (Block & Robins, 1994). Recall that girls worry more about their physical appearance and feel more insecure about their abilities. Finally, economically advantaged teenagers evaluate themselves more positively than do low-income adolescents for several reasons—because they are more likely to experience authoritative parenting, receive positive feedback from teachers, and perform well academically (Rosenberg, Schooler, & Schoenbach, 1989).

PATHS TO IDENTITY

Adolescents' well-organized self-descriptions and expanded sense of self-esteem provide the cognitive foundation for identity development. Using a clinical interviewing procedure, researchers have grouped adolescents into four categories, which show their progress in formulating a mature identity (Marcia, 1980). Table 12.1 summarizes these identity statuses: **identity achievement, moratorium, identity foreclosure,** and **identity diffusion.**

Adolescents often shift from one status to another until identity is achieved. For example, in junior high school, Louis accepted his parents' religious beliefs (foreclosure) and gave only passing thought to a vocational direction (diffusion). In high school, he began to explore these identity issues. Many young people start out as identity-foreclosed and diffused, but by late adolescence they have moved toward moratorium and identity achievement (Archer, 1982; Meilman, 1979). College triggers increased exploration, exposing young people to new career options and lifestyles. Most teenagers who go to work after high school graduation settle on a self-definition earlier than college-bound youths (Munro & Adams, 1977). But those who find it difficult to realize their occupational goals because of lack of training or vocational choices are at risk for identity diffusion (Archer, 1989).

At one time, researchers thought that adolescent girls postponed establishing an identity and focused instead on Erikson's sixth stage, intimacy development. Girls do show more sophisticated reasoning in identity areas related to intimacy, such as sexuality and family versus career priorities. In this respect, they are actually ahead of boys in identity development. Otherwise, the process and timing of

TABLE 12.1

The Four Identity Statuses

IDENTITY STATUS	DESCRIPTION	EXAMPLE
Identity achievement	Having already explored alternatives, identity-achieved individuals are committed to a clearly formulated set of self-chosen values and goals. They feel a sense of psychological well-being, of sameness through time, and of knowing where they are going.	When asked how willing she would be to give up going into her chosen occupation if something better came along, Darla responded, "Well, I might, but I doubt it. I've thought long and hard about law as a career. I'm pretty certain it's for me."
Moratorium	The word *moratorium* means "delay or holding pattern." These individuals have not yet made definite commitments. They are in the process of exploration—gathering information and trying out activities, with the desire to find values and goals to guide their life.	When asked if he had ever had doubts about his religious beliefs, Ramon said, "Yes, I guess I'm going through that right now. I just don't see how there can be a God and yet so much evil in the world."
Identity foreclosure	Identity-foreclosed individuals have committed themselves to values and goals without taking time to explore alternatives. Instead, they accept a ready-made identity that authority figures (usually parents but sometimes teachers, religious leaders, or romantic partners) have chosen for them.	When asked if she had ever reconsidered her political beliefs, Hillary answered, "No, not really, our family is pretty much in agreement on these things."
Identity diffusion	Identity-diffused individuals lack clear direction. They are not committed to values and goals, nor are they actively trying to reach them. They may have never explored alternatives, or they may have tried to do so but found the task too threatening and overwhelming.	When asked about his attitude toward nontraditional gender roles, Joel responded, "Oh, I don't know. It doesn't make much difference to me. I can take it or leave it."

identity formation are the same for boys and girls (Archer & Waterman, 1994; Streitmatter, 1993).

IDENTITY STATUS AND PERSONALITY CHARACTERISTICS

According to Erikson, identity achievement and moratorium are psychologically healthy routes to a mature self-definition, whereas foreclosure and diffusion are maladaptive. Research supports this conclusion. Young people who are identity achieved or actively exploring have a higher sense of self-esteem, are more likely to think abstractly and critically, report greater similarity between their ideal self (what they hoped to become) and their real self, and are more advanced in moral reasoning (Josselson, 1994; Marcia et al., 1993). Also, identity-achieved individuals are less self-conscious and self-focused and more secure about revealing their true selves to others (O'Connor, 1995).

Adolescents who get stuck in either foreclosure or diffusion have adjustment difficulties. Foreclosed individuals tend to be dogmatic, inflexible, and intolerant. Some use their commitments defensively, regarding any difference of opinion as a threat (Kroger, 1995). Most are afraid of rejection by people on whom they depend for affection and self-esteem. A few foreclosed teenagers who are alienated from their families and society may join cults or other extremist groups, uncritically adopting a way of life that is different from their past.

Long-term diffused teenagers are the least mature in identity development. They typically entrust themselves to luck or fate, have an "I don't care" attitude, and tend to go along with whatever the "crowd" is doing at the moment. As a result, they are most likely to use and abuse drugs. At the heart of their apathy is often a sense of hopelessness about the future (Archer & Waterman, 1990). Ethnic and religious prejudices are typical of both foreclosed and diffused young people. The foreclosed teenager tends to pick them up from authority figures, the diffused young person from peers (Streitmatter & Pate, 1989).

FACTORS THAT AFFECT IDENTITY DEVELOPMENT

Adolescent identity is the beginning of a lifelong process of refinement in personal commitments. In a fast-paced, changing world, individuals need to retain the capacity to engage in moratorium–achievement cycles throughout life (Archer, 1989). A wide variety of factors influence identity development.

Cognitive processes—in particular, the way adolescents grapple with competing beliefs and values—play an important role. Those who assume that absolute truth is always attainable tend to be foreclosed, whereas those who lack confidence about ever knowing anything with cer-

tainty are more often identity diffused or in a state of moratorium. Adolescents who appreciate that rational criteria can be used to choose among alternative visions are likely to be identity achieved (Boyes & Chandler, 1992).

Recall from Chapter 6 that toddlers with a healthy sense of self have parents who provide both emotional support and freedom to explore. A similar link between parenting and identity exists at adolescence. When the family serves as a "secure base" from which teenagers can confidently move out into the wider world, identity development is enhanced. Adolescents who feel attached to their parents but also free to voice their own opinions tend to be identity achieved or in a state of moratorium (Grotevant & Cooper, 1985; Lapsley, Rice, & FitzGerald, 1990). Foreclosed teenagers usually have close bonds with parents, but they lack opportunities for healthy separation. And diffused young people report the lowest levels of warm, open communication at home (Papini, 1994).

Identity development also depends on schools and communities offering rich and varied opportunities for exploration. Erikson (1968) noted that it is "the inability to settle on an occupational identity which most disturbs young people" (p. 132). Classrooms that promote high-level thinking, extracurricular and community activities that enable teenagers to take on responsible roles, and vocational training programs that immerse young people in the real world of adult work foster identity achievement. A chance to talk with adults and older peers who have worked through identity questions can also be helpful (Waterman, 1989).

Finally, the larger cultural context and historical time period affect identity development. Among modern adolescents, exploration and commitment take place earlier in vocational choice and gender-role preference than in religious and political values. Yet a generation ago, when the Vietnam War divided Americans, political beliefs took shape sooner (Archer, 1989; Waterman, 1985). Societal forces are also responsible for the special problems that ethnic minority young people face in forming a secure personal identity, as the Cultural Influences box on the following page describes. The Caregiving Concerns table on page 393 summarizes ways that adults can support identity development in adolescence.

BRIEF REVIEW

Erikson's stage of identity versus identity diffusion recognizes the formation of a coherent set of values and life plans as the major personality achievement of adolescence. An organized self-concept and more differentiated sense of self-esteem prepare the young person for constructing an identity. For most teenagers, self-worth rises,

IDENTITY DEVELOPMENT AMONG ETHNIC MINORITY ADOLESCENTS

Most Caucasian-American adolescents are aware of their cultural ancestry but are not intensely concerned about it (Phinney, 1993). But for teenagers who are members of minority groups, ethnicity is central to the quest for identity, and it presents difficult, sometimes overwhelming challenges. Different skin colors, native languages, and neighborhoods set minority youths apart and increase the prejudices to which they are exposed. As they develop cognitively and become more sensitive to feedback from the social environment, they become painfully aware that they are targets of discrimination and inequality. This discovery complicates their efforts to develop a sense of cultural belonging and a set of personally meaningful life goals. One African-American journalist, looking back on his own adolescence, remarked, "[Y]ou didn't see black people doing certain things, and you couldn't rationalize it. I mean, you don't think it out but you say, 'Well, it must mean that white people are better than we are. Smarter, brighter—whatever'" (Monroe, Goldman, & Smith, 1988, pp. 98–99).

Minority youths often feel caught between the standards of the larger society and the traditions of their culture of origin. Some respond by rejecting aspects of their ethnic background. In one study, Asian-American 15- to 17-year-olds were more likely than blacks and Hispanics to hold negative attitudes toward their subcultural group. Perhaps the absence of a social movement stressing ethnic pride of the kind available to black and Hispanic teenagers underlies this finding. Asian-American young people in this study had trouble naming well-known personalities who might serve as ethnic role models (Phinney, 1989). Some Asian parents are overly restrictive of their teenagers out of fear that assimilation into the larger society will undermine their cultural traditions, and their youngsters rebel. One Southeast-Asian refugee described his daughter's behavior: "She complains about going to the Lao temple on the weekend and instead joined a youth group in a neighborhood Christian Church. She refused to wear traditional dress on the Lao New Year. The girl is setting a very bad example for her younger sisters and brothers" (Nidorf, 1985, pp. 422–423).

Other minority teenagers react to years of shattered self-esteem, school failure, and barriers to success in the American mainstream by defining themselves in contrast to majority values. A Mexican-American teenager who had given up on school commented, "Mexicans don't have a chance to go on to college and make something of themselves." Another, responding to the question of what it takes to be a successful adult, mentioned "being on the streets" and "knowing what's happening." He pointed to his uncle, leader of a local gang, as an example (Matute-Bianchi, 1986, p. 250–251).

Because it is painful and confusing, minority high school students often dodge the task of forming an ethnic identity. Many are diffused or foreclosed on ethnic identity issues (Markstrom-Adams & Adams, 1995). How can society help them resolve identity conflicts constructively? A variety of efforts are relevant, including reducing poverty and ensuring that schools respect minority youths' ethnic heritage and unique learning styles. Minority adolescents who are ethnic-identity achieved—who have explored and adopted values from both their primary culture and the dominant culture—tend to be achieved in other areas of identity as well. They also have a higher sense of self-esteem, a greater sense of mastery over the environment, and more positive family and peer relations (Phinney, 1989; Phinney & Alipuria, 1990).

Finally, lack of concern by many white adolescents with their own ethnic origins (other than American) implies a view of the social world that is out of touch with the pluralistic nature of American society. The ethnic distinctions with which most of us are familiar (African-American, Asian-American, Caucasian, Hispanic, and Native-American) oversimplify the rich cultural diversity of the American populace (Spencer & Dornbusch, 1990). Interventions that increase the multicultural sensitivity of white teenagers lead to greater awareness of their own ethnic heritage. And majority adolescents who are secure in their own ethnic identity are less likely to hold negative stereotypes of their minority peers (Rotheram-Borus, 1993).

These Native-American adolescents dress in traditional costume in preparation for demonstrating a ceremonial dance to citizens of a small Wyoming town. When minority youths encounter respect for their cultural heritage in schools and communities, they are more likely to retain ethnic values and customs as an important part of their identities. (John Eastcott/Yva Momatiuk/The Image Works)

Test Bank Items 12.26 through 12.27

Ways Adults Can Support Healthy Identity
Development in Adolescence

STRATEGY	RATIONALE
Warm, open communication	Provides both emotional support and freedom to explore values and goals
Discussions at home and school that promote high-level thinking	Encourages rational and deliberate selection among competing beliefs and values
Opportunities to participate in extracurricular activities and vocational training programs	Permits young people to explore the real world of adult work
Opportunities to talk with adults and peers who have worked through identity questions	Offers models of identity achievement and advice on how to resolve identity concerns
Opportunities to explore ethnic heritage and learn about other cultures in an atmosphere of respect	Fosters identity achievement in all areas and ethnic tolerance, which supports the identity explorations of others

and the identity task does not spark a serious emotional crisis. Four identity statuses describe progress toward forming a mature identity. Identity achievement and moratorium are adaptive statuses associated with positive personality characteristics. Teenagers who are identity foreclosed or diffused tend to have adjustment difficulties. Identity development is fostered by the realization that rational procedures can be used to choose among competing beliefs and values, by parents who provide emotional support and freedom to explore, by schools and communities rich in opportunities, and by societies that permit young people from all backgrounds to realize their personal goals.

ASK YOURSELF . . .

- *Review the conversation between Louis and Darryl at the opening of this chapter. What identity status best characterizes the two boys? Explain your answer.*

- *Jules is an identity-achieved young person, secure in his self-chosen values and future goals. In previous chapters, what did you learn about Franca and Antonio's parenting style that helps explain Jules's adaptive approach to identity formation?*

MORAL DEVELOPMENT IN ADOLESCENCE

Sabrina sat at the kitchen table reading the Sunday newspaper, her face wide-eyed with interest. "You gotta see this," she said to Louis, who sat munching cereal. Sabrina held up a page of large photos showing a 70-year-old woman standing in her home. The floor and furniture were piled with stacks of newspapers, cardboard boxes, tin cans, glass containers, food, and clothing. The plaster on the walls was crumbling, the pipes in the house were frozen, and the sinks, toilet, and furnace no longer worked. The headline read: "Loretta Perry: My Life Is None of Their Business."

"Look what they're trying to do to this poor lady," exclaimed Sabrina. "They wanna throw her out of her house and tear it down! Those city inspectors must not care about anyone. Here it says, 'Mrs. Perry has devoted much of her life to doing favors for people.' Why doesn't someone help *her?*"

"Sabrina, you missed the point," Louis responded. "Mrs. Perry is in violation of thirty building code standards. The law says you're supposed to keep your house clean and in good repair."

"But Louis, she's old and she needs help. She says her life will be over if they destroy her home."

"The building inspectors aren't being mean, Sabrina. Mrs. Perry is stubborn. She refuses to obey the law. By not taking care of her house, she's not just a threat to herself. She's a danger to her neighbors, too. Suppose her house caught on fire. You can't live around other people and say your life is nobody's business."

"You don't just knock someone's home down," Sabrina replied angrily. "Where're her friends and neighbors in all this? Why aren't they over there fixing up that house? You're like those building inspectors, Louis. You've got no feeling!"

Louis and Sabrina's disagreement over Mrs. Perry's plight illustrates tremendous advances in moral understanding. Changes in cognition and social experience permit adolescents to better understand larger social structures—societal institutions and lawmaking systems—that govern moral responsibilities. As their grasp of social arrangements expands, adolescents' ideas about what ought to be done when the needs and desires of people are in conflict also change, toward increasingly just, fair, and balanced solutions to moral problems (Gibbs, 1991).

PIAGET'S THEORY OF MORAL DEVELOPMENT

The most influential approach to moral development is Lawrence Kohlberg's cognitive-developmental perspective, which was inspired by Piaget's early work on the moral judgment of the child. Piaget (1932/1965) saw children as moving through two broad stages of moral understanding.

The first stage is **heteronomous morality,** which extends from about 5 to 10 years of age. The word *heteronomous* means under the authority of another. As the term suggests, children of this stage view rules as handed down by authorities (God, parents, and teachers), as having a permanent existence, as unchangeable, and as requiring strict obedience. Also, in judging an act's wrongness, they focus on objective consequences rather than intent to do harm. When asked to decide which child is naughtier— John, who accidentally breaks 15 cups while on his way to dinner, or Henry, who breaks 1 cup while stealing some jam—a 6- or 7-year-old chooses John.

According to Piaget, around age 10 children make the transition to the stage of **autonomous morality.** They realize that people can have different perspectives on moral matters and that intentions, not just outcomes, should serve as the basis for judging behavior. Piaget believed that improvements in perspective taking, which result from cognitive development and opportunities to interact with peers, are responsible for this change. Autonomous individuals no longer view rules as fixed. Instead, they regard them as socially agreed-on principles that can be revised when there is a need to do so. In creating and changing rules, older children and adolescents use a standard of fairness called *reciprocity.* They express the same concern for the welfare of others as they do for themselves. Most of us are familiar with reciprocity in the form of the Golden Rule: "Do unto others as you would have them do unto you."

Take a moment to consider Piaget's theory in light of what you learned about moral development in earlier chapters (return to pages 253–257 and 324–325). You will see that his account of young children as rigid, external, and focused on physical consequences underestimates their moral capacities. Nevertheless Piaget's account of morality, like his cognitive theory, does describe the general direction of moral development. Although children are more sophisticated moral thinkers than Piaget made them out to be, they are not as advanced as adolescents and adults. Over the past two decades, Piaget's groundbreaking work has been replaced by Kohlberg's more comprehensive theory, which regards moral development as extending beyond childhood into adolescence and adulthood in a six-stage sequence.

KOHLBERG'S EXTENSION OF PIAGET'S THEORY

Kohlberg used a clinical interviewing procedure to study the development of moral understanding. He gave children and adolescents *moral dilemmas*—stories that present a conflict between two moral values—and asked them what the main actor should do and why. The best known of these is the "Heinz dilemma," which asks subjects to choose between the value of obeying the law (not stealing) and the value of human life (saving a dying person):

In Europe a woman was near death from cancer. There was one drug that the doctors thought might save her. A druggist in the same town had discovered it, but he was charging ten times what the drug cost him to make. The sick woman's husband, Heinz, went to everyone he knew to borrow the money, but he could only get together half of what it cost. The druggist refused to sell it cheaper or let Heinz pay later. So Heinz got desperate and broke into the man's store to steal the drug for his wife. Should Heinz have done that? Why? (paraphrased from Colby et al., 1983, p. 77)

Kohlberg emphasized that it is *the way an individual reasons* about the dilemma, not *the content of the response* (whether to steal or not to steal), that determines moral maturity. Individuals who believe Heinz should take the drug and those who think he should not can be found at each of Kohlberg's first four stages. Only at the two highest stages are moral reasoning and content integrated into a coherent ethical system (Kohlberg, Levine, & Hewer, 1983). Given a choice between obeying the law and preserving individual rights, the most advanced moral thinkers support individual rights (in the Heinz dilemma, stealing the drug to save a life). Does this remind you of adolescents' effort to formulate a sound, well-organized set of personal values in identity development? According to some theorists, the development of identity and moral understanding are part of the same process.

■ KOHLBERG'S STAGES OF MORAL UNDERSTANDING. Kohlberg organized his six stages into three general levels of moral development. He believed that moral understanding is promoted by the same factors Piaget thought were important for cognitive growth: (1) actively grappling with moral issues and noticing weaknesses in one's current reasoning; and (2) advances in perspective taking, which permit individuals to resolve moral conflicts in more effective ways. As Table 12.2 shows, Kohlberg's moral stages are related to Piaget's cognitive and Selman's perspective-taking stages. As we examine Kohlberg's developmental sequence and illustrate it with responses to the Heinz dilemma, look for changes in perspective taking that each stage assumes:

The Preconventional Level. At the **preconventional level,** morality is externally controlled. As in Piaget's heteronomous stage, children accept the rules of authority figures and judge actions by their consequences. Behaviors that result in punishment are viewed as bad and those that lead to rewards are seen as good.

Stage 1: The punishment and obedience orientation. Children at this stage find it difficult to consider two points of view in a moral dilemma. As a result, they ignore peoples' intentions and, instead, focus on fear of authority and avoidance of punishment as reasons for behaving morally.

Prostealing: "If you let your wife die, you will get in trouble. You'll be blamed for not spending the money to

TABLE 12.2

Relations Among Kohlberg's Moral, Piaget's Cognitive, and Selman's Perspective-Taking Stages

KOHLBERG'S MORAL STAGE	DESCRIPTION	PIAGET'S COGNITIVE STAGE	SELMAN'S PERSPECTIVE-TAKING STAGE[a]
Punishment and obedience orientation	Fear of authority and avoidance of punishment are reasons for behaving morally.	Preoperational, early concrete operational	Social-informational
Instrumental purpose orientation	Satisfying personal needs determines moral choice.	Concrete operational	Self-reflective
"Good boy–good girl" orientation	Maintaining the affection and approval of friends and relatives motivates good behavior.	Early formal operational	Third-party
Social-order-maintaining orientation	A duty to uphold laws and rules for their own sake justifies moral conformity.	Formal operational	Societal
Social contract orientation	Fair procedures for changing laws to protect individual rights and the needs of the majority are emphasized.		
Universal ethical principle orientation	Abstract universal principles that are valid for all humanity guide moral decision making.		

[a] To review these stages, return to Chapter 10, page 324.

help her and there'll be an investigation of you and the druggist for your wife's death." (Kohlberg, 1969, p. 381)

Antistealing: "You shouldn't steal the drug because you'll be caught and sent to jail if you do. If you do get away, your conscience would bother you thinking how the police would catch up with you any minute." (Kohlberg, 1969, p. 381)

Stage 2: The instrumental purpose orientation. Children become aware that people can have different perspectives in a moral dilemma, but this understanding is, at first, very concrete. Individuals view right action as what satisfies their personal needs, and they believe others also act out of self-interest. Reciprocity is understood as equal exchange of favors—"you do this for me and I'll do that for you."

Prostealing: "The druggist can do what he wants and Heinz can do what he wants to do. . . . But if Heinz decides to risk jail to save his wife, it's his life he's risking; he can do what he wants with it. And the same goes for the druggist; it's up to him to decide what he wants to do." (Rest, 1979, p. 26)

Antistealing: "[Heinz] is running more risk than it's worth unless he's so crazy about her he can't live without her. Neither of them will enjoy life if she's an invalid." (Rest, 1979, p. 27)

The Conventional Level. At the **conventional level**, individuals continue to regard conformity to social rules as necessary, but not for reasons of self-interest. They believe that actively maintaining the current social system is important for ensuring positive human relationships and societal order.

Stage 3: The "good boy–good girl" orientation, or the morality of interpersonal cooperation. The desire to obey rules because they promote social harmony first appears in the context of close personal ties. Stage 3 individuals want to maintain the affection and approval of friends and relatives by being a "good person"—trustworthy, loyal, respectful, helpful, and nice. The capacity to view a two-person relationship from the vantage point of an impartial, outside observer supports this new approach to morality. At this stage, the individual understands reciprocity in terms of the Golden Rule.

Prostealing: "No one will think you're bad if you steal the drug, but your family will think you're an inhuman husband if you don't. If you let your wife die, you'll never be able to look anyone in the face again." (Kohlberg, 1969, p. 381)

Antistealing: "It isn't just the druggist who will think you're a criminal, everyone else will too. After you steal it, you'll feel bad thinking how you've brought dishonor on your family and yourself; you won't be able to face anyone again." (Kohlberg, 1969, p. 381)

Stage 4: The social-order-maintaining orientation. At this stage, the individual takes into account a larger perspective—that of societal laws. Moral choices no longer depend on close ties to others. Instead, rules must be enforced in the same evenhanded fashion for everyone, and each member of society has a personal duty to uphold them. The Stage 4 individual believes that laws cannot be disobeyed under any circumstances because they are vital for ensuring societal order.

Prostealing: "He should steal it. Heinz has a duty to protect his wife's life; it's a vow he took in marriage. But it's wrong to steal, so he would have to take the drug with the idea of paying the druggist for it and accepting the penalty for breaking the law later."

Antistealing: "It's a natural thing for Heinz to want to save his wife, but . . . . You have to follow the rules regardless of how you feel or regardless of the special circumstances. Even if his wife is dying, it's still his duty as a citizen to obey the law. No one else is allowed to steal, why should he be? If everyone starts breaking the law in a jam, there'd be no civilization, just crime and violence." (Rest, 1979, p. 30)

The Postconventional or Principled Level. Individuals at the **postconventional level** move beyond unquestioning support for the laws and rules of their own society. They define morality in terms of abstract principles and values that apply to all situations and societies.

Stage 5: The social contract orientation. At Stage 5, individuals regard laws and rules as flexible instruments for furthering human purposes. They can imagine alternatives to their social order, and they emphasize fair procedures for interpreting and changing the law when there is a good reason to do so. When laws are consistent with individual rights and the interests of the majority, each person follows them because of a *social contract orientation*—free and willing participation in the system because it brings about more good for people than if it did not exist.

Prostealing: "Although there is a law against stealing, the law wasn't meant to violate a person's right to life. Taking the drug does violate the law, but Heinz is justified in stealing in this instance. If Heinz is prosecuted for stealing, the law needs to be reinterpreted to take into account situations in which it goes against people's natural right to keep on living."

Stage 6: The universal ethical principle orientation. At this highest stage, right action is defined by self-chosen ethical principles of conscience that are valid for all humanity, regardless of law and social agreement. These values are abstract, not concrete moral rules like the Ten Commandments. Stage 6 individuals typically mention such principles as equal consideration of the claims of all human beings and respect for the worth and dignity of each person.

Prostealing: "If Heinz does not do everything he can to save his wife, then he is putting some value higher than the value of life. It doesn't make sense to put respect for property above respect for life itself. [People] could live together without private property at all. Respect for human life and personality is absolute and accordingly [people] have a mutual duty to save one another from dying. (Rest, 1979, p. 37)

■ RESEARCH ON KOHLBERG'S STAGE SEQUENCE. Longitudinal studies provide the most convincing evidence for Kohlberg's stage sequence. With few exceptions, individuals move through the stages in the order that Kohlberg expected (Colby et al., 1983; Walker, 1989; Walker & Taylor, 1991b). A striking finding is that moral development is very slow and gradual. Stages 1 and 2 reasoning decrease in early adolescence, whereas Stage 3 increases through mid-adolescence and then declines. Stage 4 reasoning rises over the teenage years until, by early adulthood, it is the typical response. Few people move beyond it to Stage 5. In fact, postconventional morality is so rare that there is no clear evidence that Kohlberg's Stage 6 actually follows Stage 5. The highest stage of moral development is still a matter of speculation.

As you read the Heinz dilemma, you probably came up with your own solution to it. Now, try to think of a moral dilemma you recently faced in everyday life. How did you solve it, and did your reasoning fall at the same stage as your thinking about Heinz and his dying wife? Real-life conflicts often elicit moral reasoning below a person's actual capacity because they bring out the many practical considerations involved in an actual moral conflict (Trevethan & Walker, 1989). The influence of situational factors on moral judgments suggests that like Piaget's cognitive stages, Kohlberg's moral stages are best viewed in terms of a loose rather than strict concept of stage. Each individual draws on a range of moral responses, which vary with context. With age, this range shifts upward as less mature moral reasoning is gradually replaced by more advanced moral thought.

■ **ENVIRONMENTAL INFLUENCES ON MORAL REASONING**

Many environmental factors promote moral stage change, including child-rearing practices, schooling, peer interaction, and aspects of culture. In line with Kohlberg's belief, growing evidence suggests that the way these experiences work is to present young people with cognitive challenges, which stimulate them to think about moral problems in more complex ways.

■ CHILD-REARING PRACTICES. As in childhood, moral understanding in adolescence is fostered by warm, democratic child rearing and discussions of moral concerns. Teenagers who gain most in moral development have parents who create a supportive atmosphere by listening sensitively, asking clarifying questions, and presenting higher-level reasoning. In contrast, parents who lecture,

Intense, animated discussions about moral issues in which peers confront and criticize each other's statements lead to gains in moral understanding. (Stephen Marks)

use threats, or make sarcastic remarks have youngsters who change little or not at all (Boyes & Allen, 1993; Walker & Taylor, 1991a).

■ SCHOOLING. Years of schooling completed is one of the most powerful predictors of moral development. Longitudinal research reveals that moral reasoning advances in late adolescence and early adulthood only as long as a person enters and remains in college (Rest & Narvaez, 1991; Speicher, 1994). Perhaps higher education has a strong impact on moral development because it introduces young people to social issues that extend beyond personal relationships to entire political and social groups. Consistent with this idea, college students who report more academic perspective-taking opportunities (for example, classes that emphasize open discussion of opinions) and who indicate that they have become more aware of social diversity tend to be advanced in moral reasoning (Mason & Gibbs, 1993).

■ PEER INTERACTION. Many studies confirm Piaget's belief that interaction among peers, who confront one another with differing viewpoints, promotes moral understanding. But peer interaction must have certain features to be effective. Look back at Sabrina and Louis's argument over the plight of Loretta Perry on page 393. Each teenager directly confronts and criticizes the other's statements, and emotionally intense expressions of disagreement occur. Peer discussions like this lead to much greater stage change than those in which adolescents state their opinions in a disorganized, uninvolved way (Berkowitz & Gibbs, 1983; Haan, Aerts, & Cooper, 1985). Also, note that Sabrina and Louis do not revise their thinking after just one discussion. Because moral development is a gradual process, it takes many peer interaction sessions over weeks or months to produce moral change.

■ CULTURE. Young people in industrialized nations move through Kohlberg's stages more quickly and advance to higher levels than do individuals in simpler societies, who rarely move beyond Stage 3. In tribal and village cultures, moral cooperation is based on direct relations between people. Laws and government institutions do not exist to regulate it. Yet Stage 4 to 6 reasoning depends on an understanding of the role of larger societal structures in resolving moral conflict (Snarey, 1995).

In cultures where young people begin to participate in the institutions of their society at early ages, moral development is advanced. For example, on *kibbutzim*, small but technologically advanced agricultural settlements in Israel, children receive training in the governance of their community in middle childhood. By third grade, they mention more concerns about societal laws and rules when discussing moral conflicts than do Israeli city-reared or American children (Fuchs et al., 1986). During adolescence and young adulthood, a greater percentage of kibbutz than American individuals reach Kohlberg's Stages 4 and 5 (Snarey, Reimer, & Kohlberg, 1985).

These cultural differences raise an important question about Kohlberg's theory: Are the highest stages culturally universal, or do they emerge only in Western societies? Researchers have tried to answer this question by studying cultures just as complex as Western nations but guided by very different religious and philosophical traditions. In one investigation, adolescents and adults in India showed the same pattern of movement through Kohlberg's stages as their American agemates, and just as many or more

Young people growing up on Israeli kibbutzim receive training in the governance of their community at an early age. As a result, they understand the role of societal laws in resolving moral conflict and are advanced in moral reasoning. (Louis Goldman/Photo Researchers)

reached the postconventional level. Still, the Indian participants often dealt with moral conflicts in ways that did not fit neatly into Kohlberg's scheme. For example, many stated that moral problems (such as Heinz's) cannot be solved at the level of a single individual's conscience. Instead, their resolution is the responsibility of the entire society (Vasudev & Hummel, 1987). Kohlberg's theory does seem to tap an important universal dimension of morality. At the same time, it does not capture all aspects of moral thinking in every culture.

ARE THERE SEX DIFFERENCES IN MORAL REASONING?

The debate over the universality of Kohlberg's stages has also been extended to gender. Return once again to Sabrina and Louis's moral discussion at the beginning of this section. Sabrina's argument focuses on caring and commitment to others. Louis's approach is more impersonal. He looks at the dilemma of Loretta Perry in terms of competing rights and justice.

Carol Gilligan (1982) is the most well-known figure among those who have argued that Kohlberg's theory does not adequately represent the morality of girls and women. She believes that feminine morality emphasizes an "ethic of care" that is devalued in Kohlberg's system. For example, Sabrina's reasoning falls at Stage 3 (because it focuses on the importance of mutual trust and affection between people), whereas Louis's is at Stage 4 (because he emphasizes the value of obeying the law to ensure societal order). According to Gilligan, a concern for others is a *different*, not less valid, basis for moral judgment than a focus on impersonal rights.

Many studies have tested Gilligan's claim that Kohlberg's approach underestimates the moral maturity of females, and most do not support it. On hypothetical dilemmas as well as everyday moral problems, adolescent and adult females do not fall behind males in development. Also, themes of justice and caring appear in the responses of both sexes, and when girls do raise interpersonal concerns, they are not downscored in Kohlberg's system (Jadack et al., 1995; Kahn, 1992; Walker, 1995). These findings suggest that although Kohlberg emphasized justice rather than caring as the highest of moral ideals, his theory taps both sets of values.

Still, Gilligan's claim that the study of moral development has been limited by too much attention to rights and justice (a "masculine" ideal) and too little attention to care and responsiveness (a "feminine" ideal) is a powerful one. Some evidence shows that although the morality of males and females taps both orientations, females do tend to stress care, whereas males either stress justice or use justice and care equally (Galotti, Kozberg, & Farmer, 1991; Garmon et al., 1996). The difference appears most often on real-life rather than hypothetical dilemmas. Consequently,

it may be largely a function of women's greater involvement in daily activities involving care and concern for others.

MORAL REASONING AND BEHAVIOR

A final question about moral development concerns the relation between moral reasoning and behavior. If people do not act in accord with their principles, then their morality must be questioned. Kohlberg believed that moral thought and action come closer together at the higher levels of moral understanding. Mature moral thinkers realize that behaving in line with their beliefs is an important part of creating a just social world (Gibbs, 1995). Consistent with this idea, higher-stage individuals more often engage in prosocial acts, such as helping, sharing, and defending victims of injustice. They are also more honest. For example, they are less likely to cheat on assignments and tests in school (Harris, Mussen, & Rutherford, 1976).

Yet even though a clear connection between advanced moral reasoning and action exists, it is only moderate. As we saw in earlier chapters, moral behavior is influenced by many factors besides cognition, including the emotions of empathy and guilt and a long history of experiences that affect moral decision making. Researchers have yet to discover how all these complex facets of morality work together (Blasi, 1990).

BRIEF REVIEW

According to Piaget, children move from an authority-focused, heteronomous morality to an autonomous morality based on reciprocity by the end of childhood. Lawrence Kohlberg's three-level, six-stage theory was inspired by Piaget's work. From late childhood into adulthood, morality changes from concrete, externally controlled reasoning to more abstract, principled justifications for moral choices. Like Piaget's cognitive stages, Kohlberg's moral stages fit a loose rather than strict concept of stage. A broad range of experiences fosters moral development, including moral discussions with parents and peers, years of schooling, and contact with larger social structures in complex societies. Although Kohlberg's theory emphasizes a "masculine" morality of justice rather than a "feminine" morality of care, it does not underestimate the moral maturity of females. As individuals advance through Kohlberg's stages, moral reasoning becomes better related to behavior.

ASK YOURSELF . . .

■ *In our discussion of Kohlberg's theory, why were examples of both pro stealing and anti stealing responses to the Heinz dilemma presented for Stages 1 through 4 but only pro-stealing responses for Stages 5 and 6?*

■ *Tam grew up in a small, isolated village culture. Lydia was raised in a large industrial city. At age 15, Tam reasons at Kohlberg's Stage 2, Lydia at Stage 4. What factors might account for the difference?*

GENDER TYPING IN ADOLESCENCE

As Sabrina entered adolescence, some of her thinking and behavior became more gender typed. For example, she began to place more emphasis on excelling in traditionally feminine subjects of language, art, and music than in math and science. And when with peers, Sabrina worried about how she should walk, talk, eat, dress, laugh, and compete, judged according to accepted social standards for maleness and femaleness.

Research suggests that early adolescence is a period of **gender intensification**—increased gender stereotyping of attitudes and behavior (Galambos, Almeida, & Petersen, 1990). Although it occurs in both sexes, gender intensification is stronger for girls, who feel less free to experiment with "opposite-gender" activities and behavior than they did in middle childhood (Huston & Alvarez, 1990).

What accounts for gender intensification? Biological, social, and cognitive factors are involved. Puberty magnifies gender differences in appearance, causing teenagers to spend more time thinking about themselves in gender-linked ways. Pubertal changes also prompt gender-typed pressures from others. Parents (especially those with traditional gender-role beliefs) may encourage "gender-appropriate" activities and behavior to a greater extent than they did in middle childhood (Crouter, Manke, & McHale, 1995). And when adolescents start to date, they often become more gender typed as a way of increasing their attractiveness to the other sex (Crockett, 1990). Finally, cognitive changes—in particular, greater concern with what others think—make young teenagers more responsive to gender-typed expectations.

Gender intensification seems to decline by middle to late adolescence, but not all young people move beyond it to the same degree. Teenagers who are encouraged to explore non-gender-typed options and to question the value of gender stereotypes for themselves and society at large are more likely to build an androgynous gender-role identity, selecting "masculine" and "feminine" traits that suit their personally chosen goals (Eccles, 1987). Overall, androgynous adolescents tend to be psychologically healthier—more self-confident, better liked by peers, and

Early adolescence is a period of gender intensification. Puberty magnifies gender differences in appearance, causing teenagers to begin thinking about themselves in gender-linked ways. And when adolescents start to date, they often become more gender typed as a way of increasing their attractiveness to the other sex. (Bob Daemmrich/Stock Boston)

identity achieved (Dusek, 1987; Massad, 1981; Ziegler, Dusek, & Carter, 1984).

Now let's turn to two aspects of adolescent social experience that play powerful roles in the identity, moral, and gender-role processes we have considered: family and peer relations.

THE FAMILY IN ADOLESCENCE

Franca works as a chemist in a large drug company. Antonio owns and operates a downtown hardware store. Both parents remember Louis's freshman year of high school as a difficult time. Because of a demanding project at work, Franca was away from home many evenings and weekends. Antonio took over in her absence, but when two of his employees quit, he, too, had less time for the family. That year, Louis and two friends used their computer know-how to crack the code of a long-distance telephone service. From the family basement, they made calls around the country. Louis's grades fell, and he often left the house without saying where he was going. Franca and Antonio began to feel uncomfortable about the long hours Louis spent in the basement and their lack of contact with him. Finally, when the telephone company traced the illegal calls to the family's phone number, Franca and Antonio knew they had cause for concern.

Development at adolescence involves striving for **autonomy** on a much higher plane than during the second year of life, when independence first became a major issue for the child. Teenagers seek to establish themselves as separate, self-governing individuals. This means relying more on oneself and less on parents for guidance and decision making (Hill & Holmbeck, 1986; Steinberg & Silverberg, 1986). A major way that teenagers seek greater self-directedness is to shift away from family to peers, with whom they explore courses of action that depart from earlier, more secure and stable patterns. Nevertheless, parent–child relationships remain vital for assisting adolescents in becoming autonomous, responsible individuals.

PARENT–CHILD RELATIONSHIPS

Adolescents require freedom to experiment. Yet as Franca and Antonio's episode with Louis reveals, they also need guidance and, at times, protection from dangerous situations. Think back to what we said earlier about parent–child relationships that foster academic achievement (Chapter 11), identity formation, and moral maturity. Effective

Adolescent autonomy is effectively achieved in the context of warm parenting. This mother supports her daughter's desire to try new experiences, relaxing control in accord with the adolescent's readiness to take on new responsibilities without threatening the parent–child bond. (Bob Daemmrich/The Image Works)

parenting of adolescents strikes a balance between connection and separation. Parental warmth and acceptance combined with firm (but not overly restrictive) monitoring of the teenagers' activities is strongly related to many aspects of adolescent competence (Baumrind, 1991; Kurdek & Fine, 1994; Steinberg et al., 1994). Note that these features make up the authoritative style that was so adaptive in childhood as well.

Maintaining an authoritative style during adolescence involves special challenges and adjustments. In Chapter 11, we showed that puberty brings increased parent–child conflict, for both biological and cognitive reasons. Teenagers' improved ability to reason about social relationships adds to family tensions as well. Perhaps you can recall a time when you stopped viewing your parents as all-knowing and perfect and saw them as "just people" (Steinberg & Silverberg, 1986). Once teenagers *de-idealize* their parents, they no longer bend as easily to parental authority as they did at earlier ages. They begin to regard many matters (such as cleaning their rooms, coming and going from the household, and doing their schoolwork) as their own personal business, whereas parents continue to think of these as shared concerns (Smetana, 1988; Smetana & Asquith, 1994). Disagreements are harder to settle when parents and adolescents approach situations from such different perspectives.

In Chapter 2, we described the family as a *system* that must adapt to changes in its members. But when development is very rapid, the process of adjustment is harder. Adolescents are not the only family members undergoing a major life transition. Many parents are in their forties and are reassessing their own lives. While teenagers face a boundless future and a wide array of choices, their parents must come to terms with the fact that half their life is over and possibilities are narrowing. The pressures experienced by each generation act in opposition to one another (Hill & Holmbeck, 1987). For example, parents often can't understand why the adolescent wants to skip family activities to be with peers. And teenagers fail to appreciate that parents want the family to be together as often as possible because an important stage in adult life—parenthood—will soon be over.

Finally, as Franca and Antonio's experience with Louis reminds us, the family is embedded in a larger context. Parents' difficulties at work as well as other life stresses can affect parent–child relationships. However, we must keep in mind that maternal employment by itself does not reduce parental time with teenagers, nor is it harmful to adolescent development (Richards & Duckett, 1994). In fact, parents who are financially secure, invested in their work, and content with their marriages usually have fewer midlife difficulties and find it easier to grant their teenagers an appropriate degree of autonomy (Silverberg & Steinberg, 1990). When Franca and Antonio realized Louis's need for more support and guidance, their problems with him subsided.

As teenagers move closer to adulthood, the task for parents and children is not one of just separating. They must establish a blend of togetherness and independence. In healthy families, teenagers remain attached to parents and continue to seek their advice, but they do so in a context of greater freedom. By middle to late adolescence, most parents and children achieve this more mature, mutual relationship (Allen et al., 1994; Larson et al., 1996).

SIBLINGS

Like parent–child relationships, sibling interactions adapt to change at adolescence. As younger siblings mature and become more self-sufficient, they accept less direction from their older brothers and sisters. Consequently, teenage siblings relate to one another on a more equal footing than they did earlier. Furthermore, as teenagers become more involved in friendships and romantic relationships, they invest less time and energy in their siblings (Furman & Buhrmester, 1992; Stocker & Dunn, 1994). As a result, sibling relationships become less intense during adolescence, in both positive and negative feelings.

Despite a drop in companionship, attachment between siblings, like closeness to parents, remains strong for most young people. Quality of sibling relationships is quite stable over time. Brothers and sisters who established a positive bond in early childhood continue to display greater affection and caring during the teenage years (Dunn, Slomkowski, & Beardsall, 1994).

PEER RELATIONS IN ADOLESCENCE

As adolescents spend less time with family members, peers become increasingly important. In industrialized nations, young people spend the majority of each weekday with agemates in school. Teenagers also spend much out-of-class time together, especially in the United States. American teenagers average 18 nonschool hours with peers, compared to 12 hours for Japanese and 9 hours for Taiwanese adolescents (Fuligni & Stevenson, 1995). Higher rates of maternal employment and less demanding academic standards probably account for this difference.

Is the large amount of time American teenagers spend together beneficial or harmful? We will see that adolescent peer relations can be both positive and negative. At their best, peers serve as critical bridges between the family and adult social roles.

ADOLESCENT FRIENDSHIPS

Adolescents report that they enjoy being with friends more than any other activity (Csikszentmihalyi & Larson,

1984). The number of "best friends" declines from about 4 to 6 in early adolescence to 1 or 2 in adulthood. At the same time, the nature of the relationship changes. When asked about the meaning of friendship, teenagers stress two characteristics. The first, and most important, is *intimacy*. Adolescents seek psychological closeness, trust, and mutual understanding from their friends. Second, more than younger children, teenagers want their friends to be *loyal*— to stick up for them and not to leave them for somebody else (Berndt & Perry, 1990).

As frankness and faithfulness increase in friendships, teenagers get to know each other better as personalities. Besides sex, race, ethnicity, and social class, adolescent friends tend to be alike in educational aspirations, political beliefs, and willingness to try drugs and engage in law-breaking acts. And they become more similar in these ways over time (Berndt & Keefe, 1995). Cooperation, generosity, and mutual affirmation between friends also rise at adolescence. This change may reflect greater effort and skill at preserving the relationship as well as increased sensitivity to a friend's needs and desires (Windle, 1994).

Ask several adolescent girls and boys to describe their close friendships. You are likely to find a consistent sex difference. Although boys and girls have just as many friends, emotional closeness and trust are more common in girls' talk about friends than boys'. Girls also rate their friendships as more intimate (Buhrmester & Furman, 1987; Parker & Asher, 1993). This does not mean that boys rarely form close friendship ties. They often do, but the quality of their friendships is more variable. The intimacy of boys' friendships is related to gender-role identity. Androgynous boys are just as likely as girls to form intimate same-sex ties, whereas boys who identify strongly with the traditional masculine role are less likely to do so (Jones & Dembo, 1989).

What benefits do adolescents derive from their friendships? Although young people who are well adjusted to begin with are better able to form and sustain close peer ties, friendships further their emotional and social development. The reasons are several:

1. *Close friendships provide opportunities to explore the self and develop a deep understanding of another.* Through open, honest communication, adolescent friends become sensitive to each other's strengths and weaknesses, needs and desires. This process supports the development of self-concept, perspective taking, identity, and intimate ties beyond the family.

2. *Close friendships help young people deal with the stresses of adolescence.* Teenagers with supportive friendships report fewer daily hassles and more "uplifts" than do others (Kanner et al., 1987). As a result, anxiety and loneliness are reduced while self-esteem and sense of well-being are fostered.

3. *Close friendships promote good school adjustment in both middle- and low-income students.* When teenagers enjoy interacting with friends at school, perhaps they begin to view all aspects of school life more positively (Berndt & Keefe, 1995).

CLIQUES AND CROWDS

Friends also gather in *peer groups* (see Chapter 10), which become increasingly common and more tightly structured and exclusive than they were in middle childhood. Adolescent peer groups are organized around **cliques,** small groups of about five to seven members who are good friends and, therefore, resemble one another in family background, attitudes, and values. In early adolescence, cliques are limited to members of the same sex, but by mid-adolescence mixed groups become common. The cliques within a typical high school can be identified by their interests and social status, as the well-known "popular" and "unpopular" groups reveal (Cairns et al., 1995; Gillmore et al., 1997). Cliques develop dress codes, ways of speaking, and behaviors that separate them from one another and from the adult world.

Sometimes several cliques with similar values form a larger, more loosely organized group called a **crowd.** Unlike the more intimate clique, membership in a crowd is based on reputation and stereotype. Whereas the clique serves as the main context for direct interaction, the crowd grants the adolescent an identity within the larger social structure of the school. For example, Louis and Darryl hung out with the debate team, who wore identical sweatshirts to practices, tournaments, and informal weekend gatherings. Other crowds included the "jocks," who were very involved in athletics; the "brains," who worried about their grades; and the "workers," who had part-time jobs and lots of spending money. The "druggies" used drugs on more than a one-time basis, while the "greasers" wore leather jackets, crossed the street to smoke cigarettes, and felt alienated from most aspects of school life (Urberg et al., 1995).

What influences the assortment of teenagers into cliques and crowds? Research reveals that family influences are important. In a study of 8,000 ninth to twelfth graders, adolescents who described their parents as authoritative were members of "brain," "jock," and "popular" groups that accepted both the adult and peer reward systems of the school. In contrast, boys whose parents were permissive valued interpersonal relationships and aligned themselves with the "fun culture," or "partyer" crowd. And teenagers who viewed their parents as uninvolved more often affiliated with "partyer" and "druggie" crowds, suggesting lack of identification with adult reward systems (Durbin et al., 1993).

In early adolescence, as interest in dating increases, boys' and girls' cliques come together. As mixed-sex cliques form and "hang out," they provide a supportive context for boys and girls to get to know each other. Cliques offer models for how to interact with the opposite sex and a chance to do so without having to be intimate. Gradually, the larger group divides into couples, several of whom spend time together, going to parties and movies. By late adolescence, boys and girls feel comfortable enough about approaching each other directly that the mixed-sex clique is no longer needed and disappears (Padgham & Blyth, 1990).

Crowds also decline in importance. As adolescents formulate their own personal values and goals, they no longer feel a need to broadcast, through dress, language, and preferred activities, who they are. Nevertheless, both cliques and crowds serve vital functions. The clique provides a context for acquiring new social skills and for experimenting with values and roles in the absence of adult monitoring. The crowd offers adolescents the security of a temporary identity as they separate from the family and begin to construct a coherent sense of self (Brown, 1990).

DATING

Although sexual interest is affected by the hormonal changes of puberty, the beginning of dating is regulated by social expectations of the peer group (Dornbusch et al., 1981). In one study, adolescents were asked about their reasons for dating. Younger teenagers were more likely to date for recreation and to achieve status with agemates. By late adolescence, as young people became ready for greater psychological intimacy in a dating relationship, they looked for someone who shares their interests, who has clear goals for the future, and who is likely to make a good permanent partner (Roscoe, Diana, & Brooks, 1987).

As long as dating does not begin too soon, it extends the benefits of adolescent friendships. Besides fun and enjoyment, dating promotes sensitivity, empathy, and identity development as teenagers relate to someone whose needs are different from their own. (Jeff Greenberg/The Picture Cube)

The achievement of intimacy in dating relationships typically lags behind that of same-sex friendships (Buhrmester & Furman, 1987). Perhaps because communication between boys and girls remains stereotyped and shallow through midadolescence, early dating has a negative rather than positive impact on social maturity (Douvan & Adelson, 1966). It also carries with it a higher risk of premarital pregnancy. Sticking with activities in mixed groups, such as parties and dances, before becoming involved with a steady boyfriend or girlfriend is best for young teenagers.

As long as dating does not begin too soon, romantic involvements extend the benefits of adolescent friendships. Besides fun and enjoyment, dating provides lessons in cooperation, etiquette, and how to deal with people in a wider range of situations. As teenagers relate to someone whose needs are different from their own, sensitivity, empathy, and identity development are enhanced (Zani, 1993). About half of first romances do not survive high school graduation, and those that do become less satisfying (Shaver, Furman, & Buhrmester, 1985). Because young people are still forming their identities, those who like each other at one point in time often find that they do not have much in common later.

These high school international club members form a crowd. Unlike the more intimate clique, the larger, more loosely organized crowd grants adolescents an identity within the larger social structure of the school. (Will Faller)

PEER PRESSURE AND CONFORMITY

When Franca and Antonio discovered Louis's lawbreaking during his freshman year of high school, they began to worry (as many parents do) about the negative side of adolescent peer networks. Conformity to peer pressure is greater during adolescence than in childhood or young adulthood—a finding that is not surprising, when we consider how much time teenagers spend together. But contrary to popular belief, adolescence is not a time when young people blindly do what their peers ask. Peer conformity is actually a complex process that varies with the adolescent's age and need for social approval and with the situation.

In one study of nearly 400 junior and senior high school students, adolescents felt greatest pressure to conform to the most obvious aspects of the peer culture—dressing and grooming like everyone else and participating in social activities, such as dating and going to parties and school dances. Peer pressure to engage in proadult behavior, such as getting good grades and cooperating with parents, was also strong. Although pressure toward misconduct rose in early adolescence, compared to other areas it was low. Many teenagers said that their friends actively discouraged antisocial acts. These findings show that peers and parents often act in concert, toward desirable ends! Finally, peer pressures were only modestly related to teenagers' values and behaviors. Clearly, these young people did not always follow the dictates of peers (Brown, Lohr, & McClenahan, 1986).

Perhaps because of greater concern with what their friends think of them, early adolescents are more likely than younger or older individuals to give in to peer pressure (Brown, Clasen, & Eicher, 1986). Yet when parents and peers disagree, even young teenagers do not consistently rebel against the family. Instead, parents and peers differ in their spheres of greatest influence. Parents have more impact on teenagers' basic life values and educational plans (Sebald, 1986). Peers are more influential in short-term day-to-day matters, such as dress, music, and choice of friends. Adolescents' personal characteristics also make a difference. Young people who feel competent and worthwhile are less likely to fall in line behind peers.

Finally, authoritative child rearing, which fosters high self-esteem, social and moral maturity, and a positive view of parents, is related to greater resistance to unfavorable peer pressure among both white and black teenagers (Fletcher et al., 1995; Mason et al., 1996). In contrast, adolescents who experience extremes of parental behavior—either too much restrictiveness or too little monitoring—tend to be highly peer oriented. They more often rely on friends for advice about their personal lives and future and are more willing to break their parents' rules, ignore their schoolwork, and hide their talents to be popular with agemates (Fuligni & Eccles, 1993).

BRIEF REVIEW

Biological, social, and cognitive factors make early adolescence a period of gender intensification. Within the family, parents who strike a balance between connection and separation through authoritative child rearing assist teenagers in achieving autonomy. Sibling relationships become less intense as adolescents increase their independence from the family and spend more time with peers.

Intimacy and loyalty become central to friendship during the teenage years. Close friendship ties are related to many aspects of psychological health and competence. Adolescent peer groups are organized around cliques and crowds. The clique provides a setting in which teenagers learn social skills and try out new values and roles. The crowd offers a temporary identity as adolescents work on constructing their own. As teenagers become interested in dating, mixed-sex cliques form, which divide into couples. Although dating relationships increase in intimacy with age, they lag behind same-sex friendships. Conformity to peer pressure rises in adolescence, but teenagers do not mindlessly "follow the crowd." Peers are more powerful influences on dress, music, and social activities, parents on life values and educational plans.

ASK YOURSELF . . .

■ *How does Louis and Darryl's lunchtime conversation at the beginning of this chapter reflect the special characteristics of adolescent friendship?*

■ *Phyllis likes her 14-year-old daughter Farrah's friends, but she wonders what Farrah gets out of hanging out at Jake's Pizza Parlor with them on Friday and Saturday evenings. Explain to Phyllis what Farrah is learning.*

■ *How might gender intensification contribute to the shallow quality of early adolescent dating relationships?*

PROBLEMS OF DEVELOPMENT

Although most young people move through adolescence with little difficulty, we have seen that some experience major disruptions in development, such as premature parenthood, substance abuse, and school failure. Our discussion has also shown that psychological and behavior problems cannot be explained by any single factor alone. Instead, biological and psychological change, families, schools, peers, communities, and culture act together to produce a particular outcome. This theme is apparent in three additional problems of the teenage years: depression, suicide, and delinquency.

DEPRESSION

Depression—feeling sad, frustrated, and hopeless about life—is the most common psychological problem of the teenage years. Such feelings are not absent in childhood, but they increase dramatically around puberty. This change is understandable if we think about the many challenges adolescents face and their greater capacity for focusing on themselves. About 20 to 35 percent of teenagers experience mild feelings of depression, but they bounce back after a short period. Others display a more worrisome picture. About 12 to 15 percent are moderately depressed, and 5 percent severely so (Brooks-Gunn & Petersen, 1991). These teenagers are gloomy and self-critical for weeks at a time. They withdraw from pleasurable activities and show a loss of energy, a change in appetite, and disturbed sleep.

Depression prevents young people from mastering critical developmental tasks. Without treatment, depressed teenagers have a high likelihood of becoming depressed adults. Adolescent depression is also associated with drug abuse, lawbreaking, and car accidents, and it predicts future problems in school performance, employment, and marriage (Lewinsohn et al., 1993). Unfortunately, the symptoms tend to be overlooked by parents and teachers alike. Because of the popular stereotype of adolescence as a period of storm and stress, many adults interpret depressive reactions as "normal" and just a passing phase (Strober, McCracken, & Hanna, 1990).

Both biological and environmental factors are related to severe depression. Kinship studies reveal that heredity plays an important role. Perhaps depression rises at adoles-

Depression in teenagers should not be dismissed as a temporary side effect of puberty. Because adolescent depression can lead to long-term emotional problems, it deserves to be taken seriously. Without treatment, depressed teenagers have a high likelihood of becoming depressed adults. (Nancy Richmond/The Image Works)

cence because the hormonal changes of puberty trigger it in susceptible young people (Kennedy, 1993). But environmental events can also activate depression. Sometimes it follows a profound loss, such as parental divorce or the end of a close friendship or dating relationship. At other times, failing at something important sets it off. Still other cases are linked to stressful life circumstances. Poverty and ethnic minority status place teenagers at higher risk for depressive symptoms (Garrison et al., 1989; Sadler, 1991). At the same time, parental warmth can shield young people from the mood-lowering effects of stress. In contrast, parental hostility and lack of support are consistently associated with adolescent depression (Ge et al., 1996; Greenberger & Chen, 1996).

Another finding is that depression occurs twice as often in girls as boys (Culbertson, 1997). On the basis of what you know about sex differences in adolescent development, perhaps you can think of several explanations. In Chapter 11, we noted that adolescents who experience puberty and school transition at the same time (most of whom are girls) are more vulnerable to depression (Peterson, Sarigiani, & Kennedy, 1991). But gender-typed coping styles seem to be a crucial link in these relationships. The gender intensification that girls experience in early adolescence promotes passivity and dependency—an approach to the world that results in anxiety and helplessness in the face of stress and challenge. Consistent with this idea, one study found that girls with either an androgynous or masculine gender-role identity showed a much lower rate of depressive symptoms—one no different from that of masculine-identified boys (Wilson & Cairns, 1988).

SUICIDE

Profound depression can lead to suicidal thoughts, which all too often are translated into action. Suicidal adolescents usually show signs of extreme despondency during the period before the suicidal act. Many verbalize the wish to die, lose interest in school and friends, neglect their personal appearance, and give away treasured possessions. When a teenager tries or succeeds at taking his or her own life, depression is one of the factors that precedes it.

■ FACTORS RELATED TO ADOLESCENT SUICIDE. The suicide rate increases over the lifespan. As Figure 12.1 shows, it is lowest in childhood and highest in old age, but it jumps sharply at adolescence. Currently, suicide is the third leading cause of death among American young people, after motor vehicle collisions and homicides. It is a growing national problem, having tripled over the past 30 years, perhaps because modern teenagers face more stresses and have fewer supports than in decades past. Adolescent suicide has risen throughout Europe, although not as much as it has in the United States (U.S. Department of Health and Human Services, 1996c).

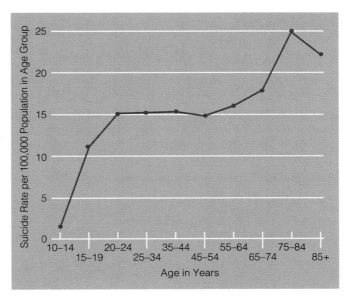

FIGURE 12.1

Suicide rate over the lifespan in the United States. Although teenagers do not commit suicide as often as adults and the aged, suicide rises sharply from childhood to adolescence, a trend also apparent in other industrialized nations. (*From U.S. Department of Health and Human Services, 1996c.*)

Striking sex differences in suicidal behavior exist. The number of boys who kill themselves exceeds the number of girls by 4 or 5 to 1. This may seem surprising, since girls show a higher rate of depression. Yet the findings are not inconsistent. Girls make more unsuccessful suicide attempts and use methods with a greater likelihood of revival, such as a sleeping pill overdose. In contrast, boys tend to select more active techniques that lead to instant death, such as firearms or hanging. Once again, gender-role expectations may be responsible. There is less tolerance for feelings of helplessness and failed efforts in males than females (Garland & Zigler, 1993).

Suicide tends to occur in two types of young people. In the first group are adolescents who are highly intelligent but solitary, withdrawn, and unable to meet their own standards or those of important people in their lives. A second, larger group shows antisocial tendencies. These young people express their despondency through bullying, fighting, stealing, and increased risk taking and drug use (Kandel, Raveis, & Davies, 1991; Lehnert, Overholser, & Spirito, 1994). Besides turning their anger and disappointment inward, they are hostile and destructive toward others.

Family turmoil, parental emotional problems, and marital breakup are common in the backgrounds of suicidal teenagers, who typically feel distant from parents and peers (Shagle & Barber, 1993). Their fragile self-esteem disintegrates in the face of stressful life events. Common circumstances just before a suicide include conflict or

TABLE 12.3

Warning Signs of Suicide

Efforts to put personal affairs in order—smoothing over troubled relationships, giving away treasured possessions

Verbal cues—saying goodbye to family members and friends, making direct or indirect references to suicide ("I won't have to worry about these problems much longer"; "I wish I were dead"; "I wonder what dying is like")

Feelings of sadness, despondency, "not caring" anymore

Extreme fatigue, lack of energy, and boredom

No desire to socialize; withdrawal from friends

Easily frustrated

Emotional outbursts—spells of crying or laughing, bursts of energy

Inability to concentrate, distractible

Decline in grades, absence from school, discipline problems

Neglect of personal appearance

Sleep change—loss of sleep or excessive sleepiness

Appetite change—eating more or less than usual

Physical complaints—stomachaches, backaches, headaches

Source: Capuzzi, 1989.

rejection by a romantic partner or the humiliation of having been caught engaging in irresponsible, antisocial acts.

Why is suicide rare in childhood but on the rise in adolescence? Teenagers' improved ability to plan ahead seems to be involved. Few successful suicides are sudden and impulsive. Instead, young people at risk usually take purposeful steps toward killing themselves. Warning signs, some of which are intended as calls for help, are listed in Table 12.3. Other cognitive changes also contribute to adolescent suicide. Belief in the personal fable leads many depressed young people to conclude that no one could possibly understand the intense pain they feel. As a result, their despair, hopelessness, and isolation deepen.

■ PREVENTION AND TREATMENT. Picking up on the signals that a troubled teenager sends is a critical first step in suicide prevention. Parents and teachers need to be trained in warning signs. Schools can help by providing sympathetic counselors, peer support groups, and information about telephone hot lines that adolescents can call in an emergency. Once a teenager takes steps toward suicide, staying with the young person, listening, and expressing sympathy and concern until professional help can be obtained is essential. The Cargiving Concerns table below suggests specific ways to respond to a young person who might be suicidal.

Intervention with depressed and suicidal adolescents ranges from antidepressant medication to individual, family, and group therapy. Sometimes hospitalization is neces-

CAREGIVING CONCERNS

Ways to Respond to a Young Person Who Might Be Suicidal

STRATEGY	DESCRIPTION
Be psychologically and physically available.	Grant the young person your full attention; indicate when and where you can be located, and emphasize that you are always willing to talk.
Communicate a caring, capable attitude.	Such statements as "I'm concerned. I care about you . . ." encourage the adolescent to discuss feelings of despair. Conveying a capable attitude helps redirect the young person's world of confusion toward psychological order.
Assess the immediacy of risk.	Gently inquire into the young person's motives with such questions as "Do you want to harm yourself? Do you want to die or kill yourself?" If the answer is yes, inquire into the adolescent's plan. If it is specific (involves a method and a time), the risk of suicide is very high.
Empathize with the young person's feelings.	Empathy, through such statements as "I understand your confusion and pain," increases your persuasive power and defuses the adolescent's negative emotion.
Oppose the suicidal intent.	Communicate sensitively but firmly that suicide is not an acceptable solution and that you want to help the adolescent explore other options.
Offer a plan for help.	Offer to assist the young person in finding professional help and in telling others, such as parents and school officials, who need to know about the problem.
Obtain a commitment.	Ask the adolescent to agree to the plan. If he or she refuses, negotiate a promise to contact you or another supportive person if and when suicidal thoughts return.

Source: Kirk, 1993.

sary to ensure the teenager's safety and swift entry into treatment. Until the adolescent improves, parents are usually advised to remove weapons, knives, razors, scissors, and drugs from the home. On a broader scale, gun control legislation that limits adolescents' access to the most frequent and deadly suicide method would greatly reduce both the number of suicides and the high teenage homicide rate (Clark & Mokros, 1993).

After a suicide, family and peer survivors need support to assist them in coping with grief, anger, and guilt for not having been able to help the victim. Teenage suicides often take place in clusters. When one occurs, it increases the likelihood of others among peers who knew the young person or heard about the death through the media. In view of this trend, an especially watchful eye needs to be kept on vulnerable adolescents after a suicide happens. Restraint by journalists in reporting teenage suicides on television or in newspapers can also aid in preventing them (Diekstra, Kienhorst, & de Wilde, 1995).

DELINQUENCY

Juvenile delinquents are children or adolescents who engage in illegal acts. Young people under age 21 account for a large proportion of police arrests in the United States—about 30 percent (U.S. Department of Justice, 1996). Yet when teenagers are asked directly and confidentially about lawbreaking, almost all admit that they are guilty of an offense of one sort or another (Farrington, 1987). Most of the time, they do not commit major crimes. Instead, they engage in petty stealing, disorderly conduct, and acts that are illegal only for minors, such as underage drinking, violating curfews, and running away from home.

Both police arrests and self-reports show that delinquency rises over the early teenage years, remains high during middle adolescence, and then declines into young adulthood (see Figure 12.2 on page 408). What is responsible for this trend? Recall that the desire for peer approval increases antisocial behavior among young teenagers. Over time, peers become less influential, moral reasoning matures, and young people enter social contexts (such as marriage, work, and career) that are less conducive to lawbreaking.

For most adolescents, a brush with the law does not forecast long-term antisocial behavior. But teenagers who have many encounters with the police are usually serious offenders. About 12 percent of violent crimes (homicide, rape, robbery, and assault) and 22 percent of property crimes (burglary and theft) are committed by adolescents (U.S. Department of Justice, 1996). A small percentage of young people are responsible.

■ FACTORS RELATED TO DELINQUENCY. Depending on the estimate, about three to seven times as many boys as girls commit major crimes. Although social class and ethnicity are strong predictors of arrest records, they are only

■

Most of the time, juvenile delinquency involves petty stealing, disorderly conduct, and acts that are illegal only for minors, such as underage drinking. Usually two or more peers commit these acts together. In early adolescence, the desire for peer approval increases antisocial behavior among young people. (Smith/Monkmeyer Press)

mildly related to teenagers' self-reports of antisocial acts. This is probably due to biases in the juvenile justice system—in particular, the tendency to arrest, charge, and punish low-income ethnic minority youths more often than their middle-class white and Asian counterparts (Fagan, Slaughter, & Hartstone, 1987).

Low verbal intelligence, poor school performance, peer rejection in childhood, and entry into antisocial peer groups are also linked to delinquency. How do these factors fit together? One of the most consistent findings about delinquent youths is that their family environments are low in warmth, high in conflict, and characterized by inconsistent discipline. Beginning in early childhood, these forms of child rearing breed antisocial behavior (Feldman & Weinberger, 1994; Miller et al., 1993). Research suggests that the path to chronic delinquency unfolds through the series of steps shown in Figure 12.3 on the following page.

Factors beyond the family and peer group also contribute to delinquency. Students in schools that fail to meet their developmental needs—those with large classes, poor-quality instruction, and rigid rules—show higher rates of lawbreaking, even after other influences are controlled (Hawkins & Lam, 1987). And in poverty-stricken neighborhoods with fragmented community ties and adult criminal subcultures, teenagers have few constructive alternatives to antisocial behavior. Youth gangs often originate in these environments.

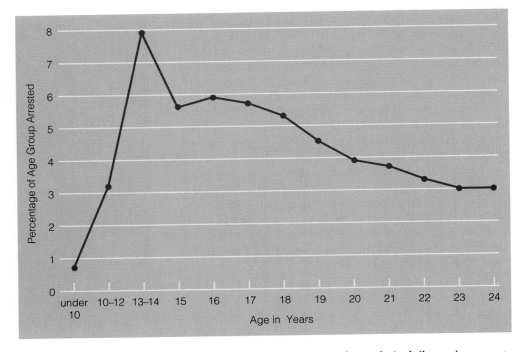

■ PREVENTION AND TREATMENT. Because delinquency has roots in childhood and results from events in several contexts, prevention must start early and take place at multiple levels. Helping parents to use authoritative child rearing, schools to teach more effectively, and communities to provide the economic and social conditions necessary for healthy development would go a long way toward reducing adolescent criminality.

Treating serious offenders also requires an approach that recognizes the multiple determinants of delinquency. So far as possible, adolescents are best kept in their own homes and communities to increase the possibility that treatment changes will transfer to their daily environment. Many treatment models exist, including individual therapies and community-based interventions involving halfway houses, day treatment centers, special classrooms, work experience programs, and summer camps. Those that work best are lengthy and intensive and use problem-focused methods that teach cognitive and social skills needed to overcome family, peer, and school difficulties (Guerra, Tolan, & Hammond, 1994).

The hard-core delinquent for whom other efforts have failed may have to be removed from the community and placed in an institution. Overall, the success of institu-

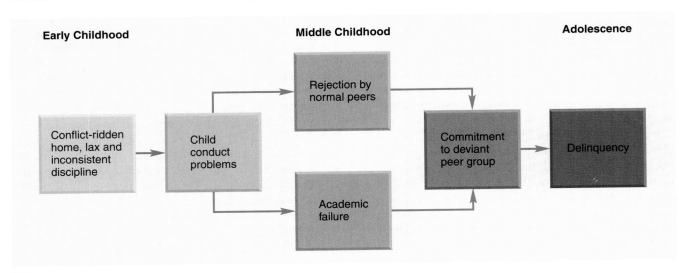

FIGURE 12.3

Developmental path to chronic delinquency. *(From G. R. Patterson, B. D. DeBaryshe, & E. Ramsey, 1989, "A Developmental Perspective on Antisocial Behavior," American Psychologist, 44, p. 331. Copyright 1989 by the American Psychological Association. Reprinted by permission.)*

A LIFESPAN VISTA

LIVES OF CRIME: PATHWAYS AND TURNING POINTS

Antisocial behavior and criminal activity are relatively stable across the life course. Disruptiveness in childhood often translates into persistent adolescent delinquency, which, in turn, predicts adult criminality. Nevertheless, these relationships are far from perfect. Many defiant children and lawbreaking teenagers do not become hardened criminals. To find out what might interrupt a life of crime, Robert Samson and John Laub (1993) examined the most comprehensive records ever compiled on juvenile delinquents in the United States. Between 1940 and 1965, extensive information was gathered on the life experiences of 500 10- to 17-year-old delinquent boys and 500 carefully matched nondelinquent controls. They were restudied at ages 25 and 32, with special attention granted to their family, employment, and criminal histories.

Quality of adult ties to family and work predicted behavior in both groups, often overriding the influence of childhood dispositions. For example, 74 percent of delinquents with low job stability in late adolescence and early adulthood were arrested again between ages 25 and 32, compared to an arrest rate of only 32 percent among those with high job stability. A good marriage led to an equally large reduction in repeat offending. Having a family to support and a means of doing so seemed to induce responsible behavior and a turn from criminality. As one former delinquent told the researchers, "Marriage settled me down—a good wife and fine healthy sons." Another commented, "I worked steadily to support [my family] and take care of my responsibilities.

I never had any time to get into trouble."

Although all the delinquent boys had been imprisoned at one time or another, the more time they spent in correctional institutions, the more likely they were to sustain a life of crime. Imprisonment denied them opportunities to obtain jobs and become stable family members who turned from lawbreaking. According to the researchers, these findings suggest a need for a fresh look at policies aimed at stopping crime. Contrary to popular belief, severe prison sentences do not reduce criminality. Despite a rapidly rising prison population, violence has increased in many American cities. Keeping adolescent and early adult offenders locked up for many years disrupts their marital and vocational lives during a crucial period of development, committing them to a future that is bleak.

tional programs has not been encouraging. Positive behavior changes typically do not last once young people return to the everyday settings that contributed to their difficulties (Quay, 1987). As the Lifespan Vista box above reveals, spending years in juvenile detention centers or prisons usually extends into a life of crime. These disappointing outcomes remind us, once again, that the best way to combat adjustment problems in adolescence is through prevention, beginning early in life.

ASK YOURSELF . . .

■ *Return to Chapter 11 and reread the sections on teenage pregnancy and substance abuse. What factors do these problems have in common with adolescent suicide and delinquency? How would you explain the finding that teenagers who experience one of these difficulties are likely to display others?*

SUMMARY

ERIKSON'S THEORY: IDENTITY VERSUS IDENTITY DIFFUSION

According to Erikson, what is the major personality achievement of adolescence?

■ Erikson's theory emphasizes **identity** as the major personality achievement of adolescence. Young people who successfully resolve the psychological conflict of **identity versus**

identity diffusion construct a solid self-definition consisting of self-chosen values and goals.

SELF-DEVELOPMENT IN ADOLESCENCE

Describe changes in self-concept and self-esteem during adolescence.

■ Changes in self-concept and self-esteem set the stage for identity for-

mation. Adolescents' self-descriptions become more organized and consistent, and personal and moral values appear as key themes. New dimensions of self-esteem are also added. For most young people, self-esteem rises over the teenage years.

Describe the four identity statuses, along with factors that promote identity development.

Test Bank Items 12.96 through 12.98

- In complex societies, a period of exploration is necessary to form a personally meaningful identity. **Identity achievement** and **moratorium** are psychologically healthy identity statuses. **Identity foreclosure** and **identity diffusion** are related to adjustment difficulties.

- Adolescents who recognize that rational criteria can be used to choose among beliefs and values and who feel attached to parents but free to disagree are likely to be advanced in identity development. Schools and communities that provide young people with options for exploration support the search for identity.

MORAL DEVELOPMENT IN ADOLESCENCE

Describe Piaget's theory of moral development and Kohlberg's extension of it, and evaluate the accuracy of each.

- Piaget identified two stages of moral understanding: (1) **heteronomous morality,** in which moral rules are viewed as fixed dictates of authority figures, and (2) **autonomous morality,** in which rules are seen as flexible, socially agreed-on principles. Although Piaget's theory describes the general direction of moral development, it underestimates young children's moral capacities.

- According to Kohlberg, moral development is a gradual process that extends into adulthood. Moral reasoning advances through three levels, each of which contains two stages: (1) the **preconventional level,** in which morality is viewed as controlled by rewards, punishments, and the power of authority figures, (2) the **conventional level,** in which conformity to laws and rules is regarded as necessary to preserve positive human relationships and societal order, and (3) the **postconventional level,** in which individuals develop abstract, universal principles

of justice. Because situational factors affect moral judgments, Kohlberg's moral stages are best viewed in terms of a loose rather than strict concept of stage.

What environmental factors affect moral reasoning?

- Many experiences contribute to moral maturity, including warm, rational child-rearing practices, schooling, peer discussions of moral issues, and the complexity of society. Kohlberg's theory does not capture all aspects of moral thinking in every culture.

Evaluate claims that Kohlberg's theory does not adequately represent the morality of females, and describe the relationship of moral reasoning to behavior.

- Although Kohlberg's theory does not underestimate the moral maturity of females, it emphasizes justice rather than caring as a moral ideal. As individuals advance to higher stages, moral reasoning and behavior come closer together.

GENDER TYPING IN ADOLESCENCE

Why is early adolescence a period of gender intensification?

- **Gender intensification** occurs in early adolescence for several reasons. Physical and cognitive changes prompt young teenagers to view themselves in gender-linked ways, and gender-typed pressures from parents and peers increase. Teenagers who eventually build an androgynous gender-role identity show better psychological adjustment.

THE FAMILY IN ADOLESCENCE

Discuss changes in parent–child and sibling relationships during adolescence.

- Effective parenting of adolescents requires an authoritative style that strikes a balance between connection

and separation. Adapting family interaction to meet adolescents' need for **autonomy** is especially challenging. As teenagers de-idealize their parents, they often question parental authority. Because both adolescents and parents are undergoing major life transitions, they approach situations from different perspectives.

- Sibling relationships become less intense as adolescents separate from the family and turn toward peers. Still, attachment to siblings remains strong for most young people.

PEER RELATIONS IN ADOLESCENCE

Describe adolescent friendships, peer groups, and dating relationships and their consequences for development.

- During adolescence, the nature of friendship changes, toward greater intimacy and loyalty. Intimate sharing is more common in girls' than boys' friendships. Close friendships promote perspective taking, self-concept, identity, the capacity for romantic involvements, and improved academic performance.

- Adolescent peer groups are organized into **cliques,** small groups of friends with common interests, dress styles, and behavior. Sometimes several cliques form a larger, more loosely organized group called a **crowd** that grants the adolescent an identity within the larger social structure of the school. Mixed-sex cliques provide a supportive social context for boys and girls to get to know one another.

- Intimacy in dating relationships lags behind that of same-sex friendships. First romances serve as practice for later, more mature bonds. They generally dissolve or become less satisfying after graduation from high school.

Discuss conformity to peer pressure in adolescence.

- Peer conformity is greater during adolescence than at younger or older ages. Young teenagers are most likely to give in to peer pressure for antisocial behavior. Yet most peer pressures are not in conflict with important adult values. Peers have greatest influence on short-term, day-to-day matters. Adults have more impact on long-term values and educational plans.

PROBLEMS OF DEVELOPMENT

What factors are related to adolescent depression and suicide?

- Depression is the most common psychological problem of the teenage years. Adolescents who are severely depressed are likely to remain so as adults. Heredity contributes to depression, but stressful life events are necessary to trigger it. Depression is more common in girls than boys.

- Profound depression often leads to suicidal thoughts. The suicide rate increases dramatically at adolescence. Boys account for most teenage suicides, whereas girls make more unsuccessful attempts. Family stress is common in the backgrounds of suicidal adolescents, who react intensely to loss, failure, or humiliation.

Discuss factors related to delinquency.

- Almost all teenagers become involved in some delinquent activity, but only a few are serious offenders. Most of these are boys with a childhood history of antisocial behavior. Although many factors are related to delinquency, one of the most consistent is a family environment low in warmth, high in conflict, and characterized by inconsistent discipline. Schools that fail to meet adolescents' developmental needs and poverty-stricken neighborhoods with high crime rates also contribute to delinquency.

MPORTANT TERMS AND CONCEPTS

identity (p. 388)
identity versus identity diffusion (p. 388)
identity achievement (p. 390)
moratorium (p. 390)
identity foreclosure (p. 390)

identity diffusion (p. 390)
heteronomous morality (p. 394)
autonomous morality (p. 394)
preconventional level (p. 394)
conventional level (p. 395)

postconventional level (p. 396)
gender intensification (p. 399)
autonomy (p. 400)
clique (p. 402)
crowd (p. 402)

FYI FOR FURTHER INFORMATION AND HELP

SUICIDE

Youth Suicide Prevention
11 Parkman Way
Needham, MA 02194
(617) 738-0700
A volunteer network of parents and professionals that works to increase public awareness of youth suicide, publicize warning signs, and develop prevention programs in schools and communities.

DELINQUENCY

National Council on Crime and Delinquency
685 Market Street, No. 620
San Francisco, CA 94105
(415) 896-6223
Web site: *www.cascomm/users/nccd*
An 11,000-member organization of professionals and other concerned individuals interested in the development of programs that prevent and treat crime and delinquency. Publishes the journal Crime and Delinquency.

MILESTONES
OF DEVELOPMENT IN ADOLESCENCE

AGE	PHYSICAL	COGNITIVE	LANGUAGE	EMOTIONAL/SOCIAL
11–14 years	■ If a girl, reaches peak of growth spurt. ■ If a girl, adds more body fat than muscle. ■ If a girl, starts to menstruate. ■ If a boy, starts to ejaculate seminal fluid. ■ If a girl, motor performance gradually increases and then levels off. ■ If a boy, begins growth spurt.	■ Becomes capable of formal operational thought. ■ Can argue more effectively. ■ Becomes more self-conscious and self-focused. ■ Becomes more idealistic and critical. ■ Self-regulation of cognitive performance continues to improve.	■ Vocabulary continues to increase as abstract words are added. ■ Grasps irony and sarcasm. ■ Understanding of complex grammatical forms continues to improve. ■ Can make subtle adjustments in speech style, depending on the situation.	■ Moodiness and parent–child conflict increase. ■ Spends less time with parents and siblings. ■ Spends more time with peers. ■ Friendships are defined by intimacy and loyalty. ■ Peer groups become organized around cliques. ■ Several cliques form a crowd on the basis of common values. ■ Peer pressure to conform increases.

AGE	PHYSICAL	COGNITIVE	LANGUAGE	EMOTIONAL/SOCIAL
15–20 years	■ If a girl, completes growth spurt.	■ Is likely to show formal operational reasoning on familiar tasks.	■ Can read and interpret adult literary works.	■ Combines features of the self into an organized self-concept.
	■ If a boy, reaches peak and then completes growth spurt.	■ Long-term knowledge base continues to expand.		■ Self-esteem differentiates further.
	■ If a boy, voice deepens.	■ Develops more complex rules for solving problems.		■ Self-esteem tends to rise.
	■ If a boy, adds muscle while body fat declines.	■ Masters the components of formal operational reasoning in sequential order.		■ Is likely to be searching for an identity.
	■ May have sexual intercourse.			■ Is likely to engage in societal perspective taking.
	■ If a boy, motor performance increases dramatically.	■ Becomes less self-conscious and self-focused.		■ Is likely to have a conventional moral orientation.
		■ Becomes better at everyday planning and decision making.		■ Has probably started dating.

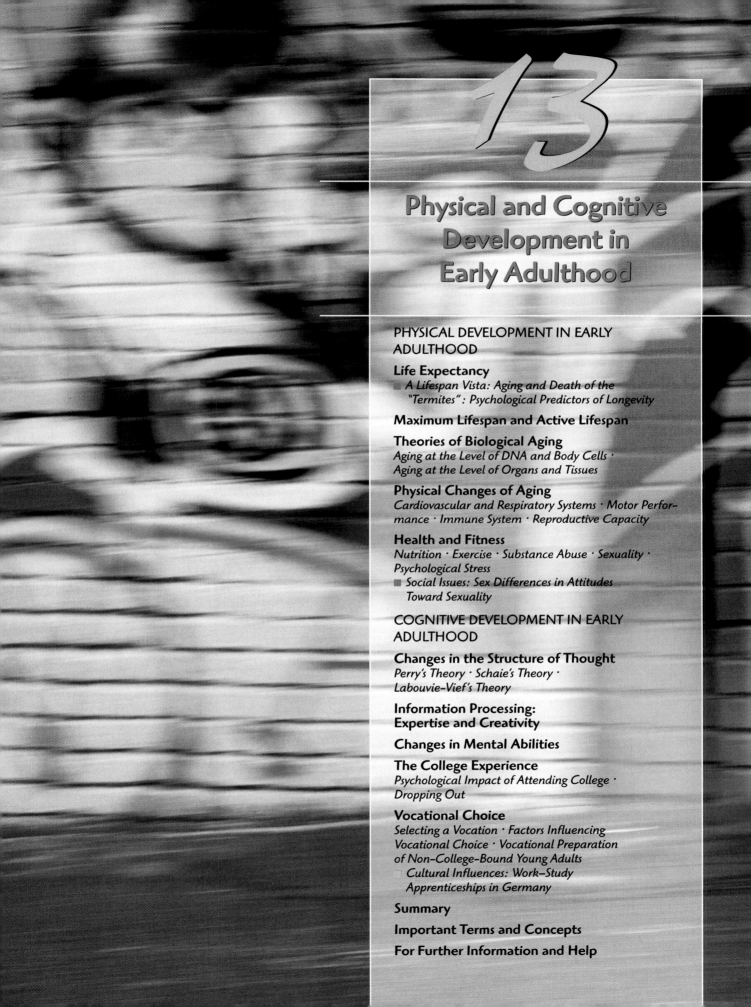

Physical and Cognitive Development in Early Adulthood

The back seat and trunk piled high with belongings, 22-year-old Sharese hugged her mother and brother good-bye, jumped in the car, and headed toward the interstate with a sense of newfound independence mixed with apprehension. Three months earlier, the family had watched proudly as Sharese received her bachelor's degree in chemistry from a small university 40 miles from her home. Her college years had been ones of gradual release from economic and psychological dependence on her family. She returned home on weekends as often as she desired and lived there during the summer months. Her mother supplemented Sharese's loans with a monthly allowance. But this day marked a turning point. She was moving to her own apartment in a new city 800 miles away, with plans to begin working on a master's degree the following week. In charge of all her educational and living expenses, Sharese felt more self-sufficient than at any previous time in her life.

The college years had been ones in which Sharese made important lifestyle changes and settled on a vocational direction. Overweight throughout high school, she lost 20 pounds during her freshman year, revised her diet, and began a regimen of exercise by joining the university's Ultimate Frisbee team. The sport helped her acquire healthier habits and leadership skills as team captain. A summer spent as a counselor at a camp for chronically ill children, combined with personal events we will take up later, convinced Sharese to apply her background in science to a career in public health.

Still, she wondered whether her choice was right. Two weeks before she was scheduled to leave, Sharese confided to her mother that she had doubts and might not go. Her mother advised, "Sharese, we never know ahead of time whether the things we choose are going to suit us just right, and most times they aren't perfect. It's what we make of them—how we view and mold them—that turns a choice into a success." And so Sharese embarked on her journey and found herself face to face with a multitude of exciting challenges and opportunities.

In this chapter, we take up the physical and cognitive sides of early adulthood—the period of the twenties and thirties. In Chapter 1, we emphasized that the adult years are difficult to divide into discrete periods, since the timing of important milestones varies greatly among individuals—much more so than in childhood and adolescence. But for most people, this first phase of adult life involves a common set of tasks: leaving home, completing education, beginning full-time work, attaining economic independence, establishing a long-term sexually intimate relationship, and starting a family. As Sharese's conversation with her mother reveals, the momentous decisions of early adulthood inevitably lead to second thoughts and disappointments. But with the help of family, community, and societal contexts, most young adults make the best of wrong turns and solve problems successfully. These are energetic decades that, more than any phase, offer the potential for living to the fullest.

PHYSICAL DEVELOPMENT IN EARLY ADULTHOOD

In earlier chapters, we saw that the body grows larger and stronger, coordination improves, and sensory systems gather information more effectively during childhood and adolescence. Once body structures reach maximum capacity and efficiency, **biological aging** begins —genetically influenced declines in the functioning of organs and systems that are universal in all members of our species (Williams, 1992). However, like physical growth, biological aging is *asynchronous* (see Chapter 7, pages 207–208). Change varies widely across parts of the body, with some structures not affected at all. In addition, individual differences are great—variation that the lifespan perspective helps us understand. Biological aging is influenced by a host of contextual factors, each of which can accelerate or slow age-related declines. These include the person's unique genetic makeup, lifestyle, living environment, and historical period (Arking, 1991). As a result, the physical changes of the adult years are, indeed, multidimensional and multidirectional.

In the following sections, we examine the process of biological aging, first at the larger level of human life expectancy and limits of the human lifespan. Then we turn to theories of aging and specific physical and motor changes,

The decades of early adulthood—the twenties and thirties—are an energetic time that, more than any other period of the lifespan, offer the potential for living to the fullest. (B. Bachmann/The Image Works)

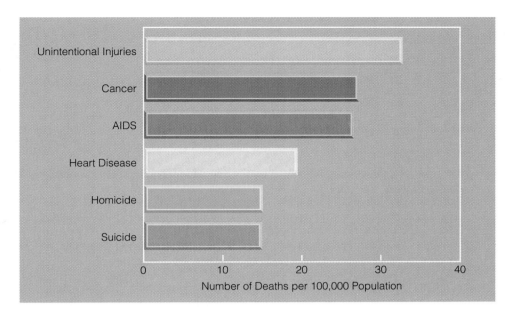

FIGURE 13.1

Leading causes of death between 25 and 44 years of age in the United States. (*Adapted from U.S. Bureau of the Census, 1996.*)

some of which are already underway in early adulthood. Our discussion will show that biological aging is not fixed and immutable. Instead, it can be modified substantially through behavioral and environmental interventions.

LIFE EXPECTANCY

Dramatic gains in **average life expectancy**—the number of years an individual born in a particular year can expect to live—provide powerful support for the malleability of biological aging. An American baby born in 1900 had an average life expectancy of about 45 years. In 1995, the figure reached 76 years. Twentieth-century gains in life expectancy are so extraordinary that they equal those of the previous 5,000 years! Improved nutrition, medical treatment, sanitation, and safety seem to be responsible. Steady declines in infant mortality (see Chapter 3) are a major contributor to longer life expectancy. But death rates among adults have decreased as well. For example, heart disease, the leading cause of overall adult death in the United States, has dropped more than 30 percent in the past 30 years. If all mortality due to heart disease could be eliminated, 12 years would be added to life expectancy.

As Figure 13.1 shows, in early adulthood, unintentional injuries (two-thirds of which are due to motor vehicle collisions) continue to be the leading cause of death, followed by cancer and AIDS. During the past 15 years, young adult deaths from various forms of cancer have dropped between 20 and 43 percent, extending life expectancy by several years (McGinnis & Lee, 1995; U.S. Bureau of the Census, 1996). Aspects of personality and a good marriage can also add to life, as the Lifespan Vista box on pages 418–419 reveals.

Consistent group differences in life expectancy underscore the joint contribution of heredity and environment to biological aging. On the average, women can look forward to 7 to 8 more years of life than men—a sex difference found in almost all cultures (see Table 13.1). The female life expectancy advantage also characterizes several animal species, including rats, mice, and dogs (Shock, 1977). The protective value of the female's extra X chromosome (see Chapter 2) is believed to be responsible.

TABLE 13.1

Average Life Expectancy by Sex in Selected Nations, 1995

NATION	TOTAL POPULATION	AVERAGE LIFE EXPECTANCY MALE	FEMALE
Japan	79.4	76.6	82.4
Iceland	79.0	76.7	81.4
Sweden	78.4	75.6	81.4
France	78.4	74.5	82.4
Canada	78.3	74.9	80.8
Netherlands	78.0	74.9	81.2
Italy	77.9	74.7	81.2
Spain	77.9	74.7	81.4
Australia	77.8	74.7	81.0
Norway	77.6	73.3	81.2
Belgium	77.2	73.9	80.7
Great Britain	77.0	74.2	90.0
New Zealand	76.7	73.1	80.4
United States	76.0	72.8	79.7

Source: Central Intelligence Agency, 1996.

A LIFESPAN VISTA

AGING AND DEATH OF THE "TERMITES": PSYCHOLOGICAL PREDICTORS OF LONGEVITY

In 1921, Louis Terman, who adapted Binet's original intelligence test for use with American children (see Chapter 1), recruited 1,500 highly intelligent California boys and girls (IQs of at least 135) into one of the most comprehensive longitudinal studies ever conducted. The clever participants, who were observed, tested, and interviewed throughout their lives, nicknamed themselves the "Termites." Today, an enormous data base charting all aspects of their development exists. Since half the Termites are now dead, the lifespan archives offered an unusual opportunity to examine the effects of personality, stressful life events, and health-related behaviors on length of life (Friedman et al., 1995).

Researchers have long believed that people with resilient personalities—able to withstand major disruptions in their lives—are less prone to disease and premature death. What kind of personality makes a difference, and just how do life stress and personal resources combine to affect longevity?

To find out, investigators focused on two sources of stress: parental divorce in childhood and marital instability in adulthood. Each was a powerful predictor of age of death. Average life expectancy among the Termites was 76 for men whose parents divorced, 80 for men whose parents stayed married. For women, the corresponding figures were 82 and 86 years (Schwartz et al., 1995). Marital instability in adulthood had similar effects. At age 40, Termites who were "steadily married" (had one lasting marriage) or who had never married were at lowest risk for premature death. Risk increased considerably for those who were "inconsistently married" (remarried after divorce), and it was highest for those who were separated, divorced, or widowed.

Furthermore, stressful life events tended to go together. Termites whose parents divorced were themselves more likely to divorce—and they more often reported a highly stressful early home environment and adjustment difficulties in adulthood. Still, each stressful marital event contributed independently to a shorter lifespan. And the longer life expectancy for steadily married than inconsistently married Termites suggests lasting negative effects of divorce that are not completely overcome by remarriage.

Finally, childhood personality characteristics contributed to length of life. In previous research, outgoing, cheerful children were found to be especially resilient in the face of stress (see Chapter 6, page 183). In this study, however, conscientiousness and dependability predicted survival; impulsive, untrustworthy, and selfish individuals tended to have shorter lives. And sociability—expected to foster longevity—did just the opposite! Cheerful, "happy-go-lucky" children (as well as children whose parents divorced) grew into adults more likely to smoke, drink, and take risks—behaviors known to shorten length of life. As the researchers

Life expectancy varies substantially among social-class and ethnic groups and nations. For example, a white child born in the United States in the 1990s is likely to live 6 to 8 years longer than a black child—a gap accounted for by higher rates of infant mortality, unintentional injuries, violent death, life-threatening disease, and poverty-linked stress in the African-American population (Kochanek, Maurer, & Rosenberg, 1994; McGinnis & Lee, 1995). Length of life can be predicted by the quality of health care, housing, and social services a country provides. As Table 13.1 shows, among the world's nations, the United States ranks only fourteenth in life expectancy. In developing nations with widespread poverty, malnutrition, disease, and armed conflict, average life expectancy hovers around 50 years and is sometimes lower (Grant, 1995; Sivard, 1993).

MAXIMUM LIFESPAN AND ACTIVE LIFESPAN

Perhaps you are wondering: What is the **maximum lifespan**, or genetic limit to length of life for a person free of external risk factors? According to current estimates, it varies between 70 and 110 years for most people, with 85 or 90 about average (Olshansky, Carnes, & Cassel, 1990). The oldest verified age to which an individual has lived is 122 years.

Do these figures actually reflect the upper bound of human longevity, or can our lifespans be extended further? At present, scientists disagree on answers to this question (Barinaga, 1991). Some believe that about 85 or 90 years is as much as most humans can expect, since gains in average life expectancy are largely the result of reducing health risks in the first 20 or 30 years; expected life for persons age

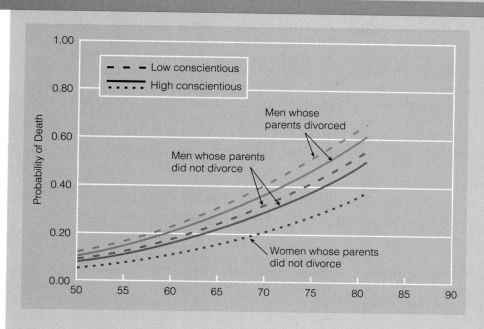

FIGURE 13.2

Age-related likelihood of death for a 20-year-old Termite by conscientiousness and parental divorce. The bottom curve shows the longest-lived women. The difference between this curve and the lowest curve for men highlights the well-known gender difference in longevity. But note that the difference between these two curves at age 70 *is smaller than* the difference between the highest and lowest men's curves. This indicates that the *combined effects* of personality and parental divorce on length are life are greater than the effect of gender. (*Adapted from H. S. Friedman et al., 1995, "Psychosocial and Behavioral Predictors of Longevity: The Aging and Death of the 'Termites'," American Psychologist, 50, p. 73. Copyright © by the American Psychological Association. Adapted by permission.*)

explained, whereas sociability may be helpful in the midst of difficult times, it can be harmful in the long run if it leads a person to be careless and irresponsible throughout life (Friedman et al., 1995).

In view of the many variables that affect longevity, the role of psychological factors in this study, summarized in Figure 13.2, is striking. The bright participants faced unique life challenges and developed in a particular historical context, so we should be cautious about generalizing the results. Nevertheless, we will see that they fit quite well with recent evidence on the relationship of psychological stress to mortality.

65 and older has increased very little (about 5 months) over the past decade (U.S. Department of Health and Human Services, 1994b). Others think we have not yet identified the human genetic limit, since the lifespans of several species have been stretched in the laboratory. For example, biologists have doubled the lifespan of fruit flies through selective breeding, increased the lifespan of roundworms 70 percent through genetic engineering, and raised rodents that live 30 percent longer by feeding them very low-calorie diets (Langreth, 1993).

The possibility of similar achievements in humans raises another issue: *Should* the lifespan be increased as far as possible? In considering this question, most people respond that *quality of life,* not just quantity, is the truly important goal. When researchers estimate **active lifespan,** they find that Americans can expect an average of 64 years of vigorous, healthy life, a level that has actually declined slightly over the past few years (U.S. Department

of Health and Human Services, 1994b). This disheartening trend is largely due to continued high rates of preventable illness, death, and disability among low-income and ethnic minority sectors of the population. Many experts argue that only after reducing health risks among these groups and wiping out age-related diseases does it make sense to invest in extending the maximum lifespan.

THEORIES OF BIOLOGICAL AGING

A t an intercollegiate tournament, Sharese dashed across the playing field for hours, leaping into the air to catch Frisbees sailing her way. In her early twenties, she is at her peak in strength, endurance, sensory acuteness, and immune system responsiveness. Why will she begin to age over the next two decades, gradually showing more noticeable declines in physical functioning as she moves into middle and late adulthood?

Biological aging is the sum of many causes, some operating at the level of DNA, others at the level of cells, and still others at the level of tissues, organs, and whole organisms. Dozens of theories exist, indicating that our understanding is still in an early stage. One popular but inaccurate idea is that the body wears out from use. Unlike parts of a machine, worn-out parts of the body usually replace or repair themselves. Furthermore, no relationship exists between hard physical work and early death. To the contrary, vigorous exercise predicts a longer life (Schneider, 1992). We now know that this "wear and tear" theory is an oversimplification.

AGING AT THE LEVEL OF DNA AND BODY CELLS

Current explanations of biological aging are of two types: those that emphasize the *programmed effects of specific genes* and those that emphasize the *cumulative effects of random events,* both internal and external, that damage genetic and cellular material. Support for both views exists, and it is likely that a combination will eventually prove to be correct.

One "programmed" theory proposes the existence of "aging genes" that control certain biological changes, such as menopause, gray hair, and deterioration of body cells. The strongest evidence for this view comes from research showing that human cells allowed to divide in the laboratory have a lifespan of 50 divisions plus or minus 10, a limit controlled by a small number of genes (Schneider, 1992). As further support for genetically programmed aging, longevity is a family trait. People whose parents had long lives tend to live longer themselves, and there is greater similarity in lifespans for identical than fraternal twins (Kanugo, 1994).

According to an alternative, "random" view, DNA in body cells is gradually damaged due to spontaneous or externally caused mutations. As these accumulate, cell replacement and repair is less efficient, or abnormal cancerous cells are produced. Studies of various animal species confirm an increase in DNA breaks and deletions and damage to other cellular material with age. Among humans, similar evidence has begun to appear, although it is not yet conclusive (Kanugo, 1994; Mazin, 1993; Vijg & Gossen, 1993).

One probable cause of age-related DNA and cellular abnormalities is the release of **free radicals,** naturally occurring, highly reactive chemicals that form in the presence of oxygen. (Radiation and certain pollutants and drugs can trigger similar effects.) When oxygen atoms break down within the cell, the reaction strips away an electron, creating a free radical. As it seeks a replacement from its surroundings, it destroys nearby cellular material, including DNA, proteins, and fats essential for cell functioning. Free radicals are believed to be involved in more than sixty disorders of aging, including heart disease, cancer, cataracts, and arthritis (Levine & Stadtman, 1992). Although our bodies produce substances that neutralize free radicals, some harm occurs and accumulates.

Some researchers believe that genes for longevity work by defending against free radicals. In this way, a "programmed" genetic response may limit "random" DNA and cellular deterioration. Foods rich in vitamins C and E and beta-carotene forestall free-radical damage as well (Arking, 1991).

AGING AT THE LEVEL OF ORGANS AND TISSUES

What consequences might the genetic and cellular effects just described have for the structure and functioning of organs and tissues? There are many possibilities. Among those with clear support is the **cross-linkage theory** of aging. Over time, protein fibers that make up the body's connective tissue form bonds, or links, with one another. When these normally separate fibers cross link, tissue becomes less elastic, leading to many negative outcomes, including loss of flexibility in the skin and other organs, clouding of the lens of the eye, clogging of arteries, and damage to the kidneys. Like other aspects of aging, cross-linking can be reduced by external factors, including regular exercise and a vitamin-rich, low-fat diet (Arking, 1991; Schneider, 1992).

Gradual failure of the endocrine system, which is responsible for production and regulation of hormones, is yet another route to aging. An obvious example is decreased estrogen production in women, which culminates in menopause. Since hormones affect so many body functions, disruptions in the endocrine system can have widespread effects on health and survival. At present, scientists are studying the impact of key hormones on aging. Recent evidence indicates that a gradual drop in growth hormone (GH) is associated with loss of muscle, addition of body fat, and thinning of the skin. In adults with abnormally low levels of GH, hormone therapy can slow these symptoms, but it has serious side effects for most people. So far, diet and physical activity are safer ways to control these outcomes (Corpas, Harman, & Blackman, 1993).

Finally, declines in immune system functioning contribute to many conditions of aging. Among these are increased susceptibility to infectious disease, heightened risk of cancer, and changes in blood vessel walls associated with cardiovascular disease (Schneider, 1992). Decreased vigor of the immune response seems to be genetically programmed, but other aging processes we have considered can intensify it. Let's look closely at its physical signs, along with other characteristics of aging, in the following sections.

PHYSICAL CHANGES OF AGING

During the twenties and thirties, changes in physical appearance and declines in body functioning are so gradual that many are hardly noticeable. Later, they will accelerate. The physical changes of aging are summarized in Table 13.2 on page 422. We will examine several in detail, taking up others in later chapters. But before we begin, it is important to note that the data forming the basis for these trends are largely cross-sectional. Since younger cohorts have experienced better health care and nutrition, cross-sectional studies can exaggerate impairments associated with aging (Arking, 1991). Fortunately, longitudinal evidence is accumulating, helping to correct this picture.

CARDIOVASCULAR AND RESPIRATORY SYSTEMS

During her first month in graduate school, Sharese pored over research articles on cardiovascular functioning. In her African-American extended family, her father, an uncle, and three aunts had died of heart attacks in their forties and fifties. The tragedies had prompted Sharese to worry about her own lifespan, reconsider her health-related behaviors, and enter the field of public health in hopes of finding ways to relieve the health problems of black Americans. *Hypertension,* or high blood pressure, occurs 12 percent more often in the American black than white population; deaths from heart disease are over 50 percent greater (U.S. Department of Health and Human Services, 1994a).

Sharese was surprised to learn that there are fewer age-related changes in the heart than we might expect, in view of the fact that heart disease rises with age and is a leading cause of adult death. In healthy individuals, the heart's ability to meet the body's oxygen requirements under typical conditions (as measured by heart rate in relation to volume of blood pumped) does not change during adulthood. Only when people are stressed by exercise does performance of the heart decline with age. The change is due to a decrease in maximum heart rate and greater rigidity of the heart muscle. Consequently, the heart has difficulty delivering enough oxygen to the body during high activity and bouncing back from strain (Arking, 1991).

One of the most serious diseases of the cardiovascular system is *atherosclerosis,* in which heavy deposits of plaque containing cholesterol and fats collect on the walls of the main arteries. If it is present, it usually begins early in life, progresses during middle adulthood, and culminates in serious illness. Atherosclerosis is multiply determined, making it hard to separate the contributions of biological aging from individual genetic and environmental influences. The complexity of causes is illustrated by animal research indicating that before puberty, a high-fat diet produces only fatty streaks on the artery walls. In sexually mature adults, it leads to serious plaque deposits (Arking, 1991). These findings suggest that sex hormones may heighten the insults of a high-fat diet.

As we mentioned earlier, heart disease has decreased considerably since the middle of this century. An especially large drop occurred over the last 10 years due to cultural changes in diet, exercise, cigarette smoking, and medical detection and treatment of high blood pressure and cholesterol (McGinnis & Lee, 1995). Later, when we consider health and fitness in early adulthood, we will see why heart attacks were so common in Sharese's family—and occur at especially high rates in the African-American population.

Like the heart, functioning of the lungs shows few age-related changes at rest, but during physical exertion, respiratory volume decreases and breathing rate increases with age. Maximum vital capacity (amount of air that can be forced in and out of the lungs) declines by 50 percent from the 20s to the 80s. Connective tissue in the lungs, chest muscles, and ribs stiffens with age, making it more difficult for the lungs to expand to full volume. Fortunately, under normal conditions, we use less than half our vital capacity; most of it is held in reserve (Arking, 1991). Nevertheless, aging of the lungs contributes to the fact that during heavy exercise, older adults find it more difficult to meet the body's oxygen needs.

MOTOR PERFORMANCE

Declines in heart and lung functioning under conditions of exertion, combined with gradual muscle loss, lead to changes in motor performance. In ordinary people, the impact of biological aging on motor skills is difficult to

Athletic skill peaks between ages 20 and 30 and then declines. However, as long as practice continues, performance drops only slightly—about 2 percent per decade—into the sixties and seventies. Here John Kelley, age 78, smiles as he approaches the finish line of the Boston Marathon, his fifty-fifth completion of the race. Although his two first-place finishes occurred in his twenties and thirties, he continued to train and run for many more years. (AP/Wide World Photos)

TABLE 13.2

Physical Changes of Aging

ORGAN OR SYSTEM	TIMING OF CHANGE	DESCRIPTION
Sensory		
Vision	From age 30	As the lens stiffens and thickens, ability to focus on close objects declines. Yellowing of the lens, weakening of muscles controlling the pupil, and clouding of the vitreous (gelatin-like substance that fills the eye) reduce light reaching the retina, impairing color discrimination and night vision. Visual acuity, or fineness of discrimination, decreases, with a sharp drop between ages 70 and 80.
Hearing	From age 30	Sensitivity to sound declines, especially at high frequencies but gradually extending to all frequencies. Change is more than twice as rapid for men than women.
Taste	From age 60	Sensitivity to the four basic tastes—sweet, salty, sour, and bitter—is reduced. May be due to factors other than aging, since number and distribution of taste buds do not change.
Smell	From age 60	Loss of smell receptors reduces ability to detect and identify odors.
Touch	Gradual	Loss of touch receptors reduces sensitivity on the hands, particularly the fingertips.
Cardiovascular	Gradual	As the heart muscle becomes more rigid, maximum heart rate decreases, reducing the heart's ability to meet the body's oxygen requirements when stressed by exercise. As artery walls stiffen and accumulate plaque, blood flow to body cells is reduced.
Respiratory	Gradual	Under physical exertion, respiratory capacity decreases and breathing rate increases. Stiffening of connective tissue in the lungs and chest muscles makes it more difficult for the lungs to expand to full volume.
Immune	Gradual	Shrinking of the thymus limits maturation of T cells and disease-fighting capacity of B cells, impairing the immune response.
Muscular	Gradual	As nerves stimulating them die, fast-twitch muscle fibers (responsible for speed and explosive strength) decline in number and size to a greater extent than do slow-twitch fibers (which support endurance). Tendons and ligaments (which transmit muscle action) stiffen, reducing speed and flexibility of movement.
Skeletal	Begins in the late 30s, accelerates in the 50s, slows in the 70s	Cartilage in the joints thins and cracks, leading bone ends beneath it to erode. New cells continue to be deposited on the outer layer of the bones, and mineral content of bone declines. The resulting broader but more porous bones weaken the skeleton and make it more vulnerable to fracture. Change is more rapid in women than men.
Reproductive	In women, accelerates after age 35; in men, begins after age 40	Fertility problems (including difficulty conceiving and carrying a pregnancy to term) and risk of having a baby with a chromosomal disorder increase.
Nervous	From age 60	Brain weight declines as neurons die, mostly in the cerebral cortex, and as ventricles (spaces) within the brain enlarge. Development of new synapses can, in part, compensate for the decline in number of neurons.
Skin	Gradual	Epidermis (outer layer) is held less tightly to the dermis (middle layer); fibers in the dermis and hypodermis (inner layer) thin; fat cells in the hypodermis decline. As a result, the skin becomes looser, less elastic, and wrinkled. Change is more rapid in women than men.
Hair	From age 35	Grays and thins.
Height	From age 50	Loss of bone strength leads to collapse of disks in the spinal column, leading to a height loss of as much as 2 inches by the seventies and eighties.
Weight	Increases to age 50; declines from age 60	Weight change reflects a rise in fat and a decline in muscle and bone mineral. Since muscle and bone are heavier than fat, the resulting pattern is weight gain followed by loss. Body fat accumulates on the torso and decreases on the extremities.

Sources: Abrass, 1990; Arking, 1991; Fabsitz, Sholinsky, & Carmelli, 1994; Pearson et al., 1995; Receputo et al., 1994; Thornbury & Mistretta, 1981.

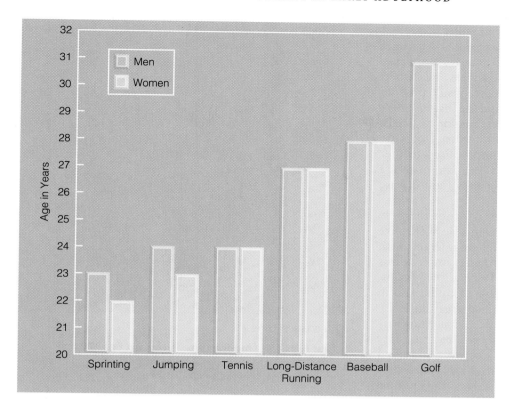

FIGURE 13.3

Age of peak performance of Olympic and professional athletes by type of sport for men and women. *Sprinting, jumping, and tennis, which require speed of limb movement, explosive strength, and gross body coordination, peak in the early twenties. Long–distance running, baseball, and golf, which depend on endurance, arm–hand steadiness, and aiming, peak in the late twenties and early thirties. (Adapted from Schulz & Curnow, 1988.)*

separate from decreases in motivation and practice. Therefore, researchers study outstanding athletes, who try to attain their very best performance in real life. As long as athletes continue intensive training, their attainments at each age approach the limits of what is biologically possible (Ericsson, 1990).

Many studies show that athletic skill peaks between ages 20 and 30 and then declines. In several investigations, the mean ages for best performance of Olympic and professional athletes in a variety of sports were charted over time. Absolute performance in many events improved over the past century. Athletes continually set new world records, suggesting improved training methods. Nevertheless, ages of best performance remained relatively constant. As Figure 13.3 reveals, athletic tasks that require speed of limb movement, explosive strength, and gross body coordination—sprinting, jumping, and tennis—peak in the early twenties. Those that depend on endurance, arm–hand steadiness, and aiming—long-distance running, baseball, and golf—peak in the late twenties and early thirties. Since these tasks have to do with either stamina or precise motor control, they take longer to develop (Schulz & Curnow, 1988).

Research on outstanding athletes tells us that the upper biological limit of motor capacity is reached in the first part of early adulthood. How quickly do athletic skills weaken in later years? Longitudinal research on master runners reveals that as long as practice continues, performance drops only slightly—about 2 percent per decade—into the sixties and

seventies (Hagberg et al., 1985). Indeed, sustained training leads to adaptations in body structures that minimize motor decline. For example, vital capacity is almost twice as great in older people actively participating in sports as in healthy controls without a history of exercise (Hagberg, 1987). Training also slows muscle loss and leads fast-twitch muscle fibers to be converted into slow-twitch fibers, which support excellent long-distance running performance as well as other endurance skills (Ericsson, 1990).

In sum, although athletic skill is at its best in early adulthood, only a small part of age-related decline is due to biological aging. Lower levels of performance in older healthy people largely reflect reduced capacities that result from adaptation to a less physically demanding lifestyle.

IMMUNE SYSTEM

The immune response is the combined work of specialized cells that neutralize or destroy antigens (foreign substances) in the body. Two types of white blood cells play vital roles. *T cells,* which originate in the bone marrow and mature in the thymus (a small gland located in the upper part of the chest), attack antigens directly. *B cells,* manufactured in the bone marrow, secrete antibodies into the bloodstream that multiply, capture antigens, and permit the blood system to destroy them. Since receptors on their surfaces recognize only a single antigen, T and B cells come in great variety. They join forces with additional cells to produce immunity.

The capacity of the immune system to offer protection against disease increases through adolescence and declines after age 20. The trend is partly due to changes in the thymus, which is largest during the teenage years and shrinks to being barely detectable after age 50. As a result, production of thymic hormones is reduced, and the thymus is less able to promote full maturity and differentiation of T cells. Since B cells release far more antibodies when T cells are present, the immune response is compromised further (Schneider, 1992). Research indicates that administering thymic hormones can help the aging body fight disease (Licastro et al., 1990).

Withering of the thymus, however, is not the only reason that the body gradually becomes less effective in warding off illness. The immune system interacts with the nervous and endocrine systems. For example, stress can weaken the immune response. During final exams, Sharese found herself more susceptible to colds. And in the month after her father died, she had great difficulty recovering from the flu. Divorce, caring for an ill aging parent, and chronic depression can also reduce immunity (Cohen & Williamson, 1991). The link between stress and illness makes perfect sense when we consider that stress hormones mobilize the body for action, whereas the immune response is fostered by reduced activity (Maier, Watkins, & Fleshner, 1994). But this also means that increased difficulty coping with physical and psychological stress can contribute to age-related declines in immune system functioning.

REPRODUCTIVE CAPACITY

 Sharese was born when her mother was in her early twenties—an age at which Sharese was still single and entering graduate school a generation later. Many people believe that pregnancy during the twenties is ideal, not only because the risks of miscarriage and chromosomal disorders are reduced (see Chapter 2), but also because younger parents have more energy to keep up with active children. However, as Figure 13.4 reveals, first births to women in their thirties have increased greatly over the past two decades. Many people are delaying child bearing until their education is complete, their careers are well established, and they know they can support a child.

Nevertheless, reproductive capacity does decline with age. Fertility problems among women increase from age 15 to 50, with a sharp rise in the mid-thirties. Between ages 30 and 34, nearly 15 percent are affected, a figure that climbs to 28 percent for 35- to 39-year-olds (McFalls, 1990). Age also affects male reproductive capacity. Amount of semen and concentration of sperm in each ejaculation gradually decline after age 40 (Murray & Meacham, 1993). Although there is no best time during adulthood to begin parenthood, individuals who decide to put off childbirth until their late thirties or their forties risk having fewer children than they desired or none at all.

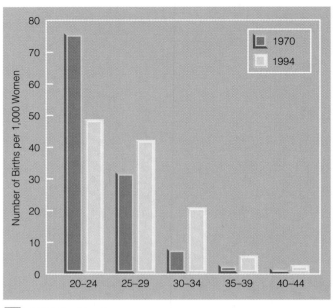

FIGURE 13.4

First births to American women by age in 1970 and 1994. The birth rate decreased over this period for women 20 to 24 years of age, whereas it increased for women 25 years and older. For women in their thirties, the birth rate more than doubled. (*U.S. Department of Health and Human Services, 1996; Adapted from Ventura, 1989.*)

BRIEF REVIEW

Biological aging varies widely across parts of the body and can be slowed by environmental factors that promote health and safety, as revealed by dramatic gains in average life expectancy over this century. Consistent sex and ethnic differences in length of life underscore the joint contribution of heredity and environment to age-related physical decline. Scientists disagree on whether human longevity can or should be extended. Far greater consensus exists about lengthening active lifespan.

Some theories of biological aging emphasize the programmed effects of specific genes, whereas others stress the cumulative effects of random events that damage genetic and cellular material. The combined effects of both processes are probably involved. Some physical changes of aging are already underway in early adulthood. Under conditions of high activity, the heart and lungs show a reduced capacity to meet the body's oxygen requirements. Declines in motor performance occur from early to late adulthood, but they can be minimized by regular exercise and practice. The immune response gradually weakens, although stress can magnify this change. Reproductive capacity decreases, especially after age 35 for women and age 40 for men.

ASK YOURSELF . . .

■ *Explain why the United States ranks only fourteenth in life expectancy among the world's nations. (For help in answering this question, return to the section on public policies and lifespan development in Chapter 2, pages 64–67.)*

■ *Thirty-year-old Len noticed the term* free radicals *in an article on adult health in his local newspaper, but he doesn't understand how they contribute to biological aging. Explain the role of free radicals to Len.*

■ *Penny is a long-distance runner for her college's track team. She wonders what her running performance will be like 10 or 20 years from now. Describe physical changes and environmental factors that will affect Penny's long-term athletic skill.*

HEALTH AND FITNESS

Look back at Figure 13.1 on page 417, which shows the leading causes of death in early adulthood. In later chapters, we will see that the number of adults whose lives are cut short by unintentional injury or homicide declines with age, whereas the number affected by disease and physical disability rises. Biological aging clearly contributes to this trend. But we have already noted wide individual and group differences in physical changes that are linked to environmental risks and health-related behaviors.

Social-class variations in health over the lifespan reflect these influences. Income, education, and occupational status show a strong and continuous relationship to almost every disease and health indicator (Adler et al., 1994). Furthermore, when a representative sample of 3,600 Americans were asked about chronic illnesses and health-related limitations on their daily lives, social-class differences were relatively small at 25 to 35 years, widened between ages 35 and 65, and contracted during old age (see Figure 13.5). Longitudinal findings confirm that economically advantaged individuals tend to sustain good health over most of their adult lives, whereas the health of lower-income individuals steadily declines (House et al., 1990a, 1990b). Social-class differences in health-related circumstances and habits—stressful life events, crowding, pollution, health care, diet, exercise, and access to supportive social relationships—are largely responsible (House et al., 1991).

These findings reveal, once again, that the living conditions that surround us and those we create for ourselves play powerful roles in how well we age. Since the incidence of health problems is much lower during the twenties and thirties than in succeeding decades, early adulthood is an excellent time to prevent later problems by adopting a healthier lifestyle. In the following sections, we take up a variety of major health concerns—nutrition, exercise, substance abuse, sexuality, and psychological stress.

FIGURE 13.5

Age differences in self-reported health status by social class. Variations among income groups are small at 25 to 34 years, increase from 35 to 64 years, and then contract. Environmental risks and poor health habits contribute to the earlier decline in health status among low-income and poverty-stricken adults. *(From J. S. House et al., 1990, "Age, Socioeconomic Status, and Health," Milbank Quarterly, 68, p. 396. Reprinted by permission.)*

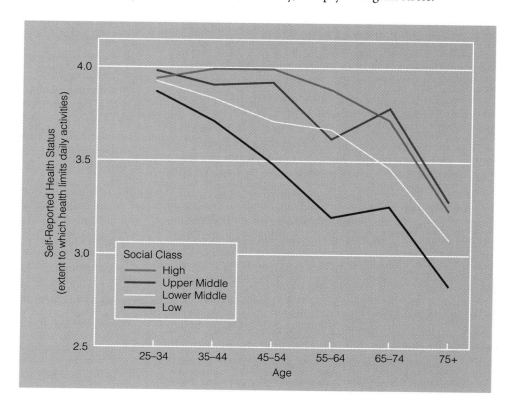

NUTRITION

Bombarded with advertising claims and an extraordinary variety of food choices, adults in industrialized nations find it increasingly difficult to make wise dietary decisions. An abundance of food combined with a heavily scheduled life means that most Americans eat because they feel like it or because it is time to do so rather than to maintain the body's functions (Donatelle & Davis, 1997). As a result, many eat the wrong types and amounts of food. Overweight and obesity and a high-fat diet are widespread nutritional problems with long-term consequences for health in adulthood.

■ OVERWEIGHT AND OBESITY. Obesity (a 20 percent increase over average body weight, based on age, sex, and physical build) is increasing in the United States, having doubled since 1900. Currently, more than a quarter of the population is affected. The incidence is particularly high for low-income ethnic minorities, especially females. Sharese learned that obesity rises to 60 percent among African-American women over age 40—an important reason (as we will see in a moment) that several of Sharese's family members succumbed to heart attacks in middle age. Overweight is even more prevalent among Hispanic and Native-American groups (Ernst & Harlan, 1991; U.S. Department of Health and Human Services, 1994a). Recall from Chapter 9 that fat children are very likely to become fat adults. But a substantial number of people (4 percent of men and 8 percent of women) show large weight gains in adulthood, most often between ages 25 and 34. And young adults who were already obese (especially women) typically get heavier (Williamson et al., 1990).

Causes and Consequences. As we noted in Chapter 9, heredity plays a role in body weight; some people are more vulnerable to obesity than others. But environmental pressures underlie the rising rates of obesity in the United States and other industrialized nations. With the decline in need for physical labor in the home and workplace, our lives have become more sedentary. And the average number of calories and amount of fat Americans consume have increased over most of this century (Brownell & Wadden, 1992).

Adding some weight between 25 and 50 is a normal part of aging, since **basal metabolic rate (BMR),** the amount of energy the body uses at complete rest, gradually declines as the number of active muscle cells (which create the greatest energy demand) drops off (see Table 13.2). But excess fat is strongly associated with serious health problems. These include high blood pressure, circulatory difficulties, atherosclerosis, adult-onset diabetes,[1] arthritis, sleep and digestive disorders, several forms of cancer, and early death (Robison et al., 1993). Furthermore, fat adults suffer enormous social discrimination. They are less likely

[1] A mild form of diabetes first appearing in adulthood that is usually treated through dietary management and exercise.

to find mates and be rented apartments, given financial aid for college, and offered jobs (Allison & Pi-Sunyer, 1994; Gortmaker et al., 1993).

Treatment. Because obesity climbs in early adulthood, treatment for adults should begin as soon as possible—preferably in the early 20s. Even moderate weight loss reduces health problems substantially (Robison et al., 1993). But successfully intervening in obesity is difficult. At present, 95 percent of individuals who start a weight-loss program return to their original weight within 5 years (Allison & Pi-Sunyer, 1994). The high rate of failure is partly due to limited knowledge of just how obesity disrupts the complex neural, hormonal, and metabolic factors that maintain a normal body-weight setpoint. Until more information is available, researchers are examining the characteristics of treatments and participants associated with greater success. The following elements promote lasting behavior change:

■ *Low-fat diet plus exercise.* To lose weight, Sharese sharply reduced calories and fat in her diet and exercised regularly. This combination is essential for limiting the impact of a genetic tendency to overweight. In addition (as we will see shortly), exercise offers physical and psychological benefits that help prevent overeating.

■ *Training participants to keep an accurate record of what they eat.* About 30 to 35 percent of obese people sincerely believe they eat less than they do, and from 25 to 45 percent report problems with binge eating. When Sharese became aware of how often she ate when she was not hungry, she was better able to limit her food intake (Martin, White, & Hulsey, 1991).

■ *Social support.* Group or individual counseling and encouragement of friends and relatives help sustain weight-loss efforts by fostering self-esteem (Prochaska, DiClemente, & Norcross, 1992). Once Sharese decided to act with the support of her family and a weight-loss counselor, she began to feel better about herself—walking and holding herself differently—even before the first pounds were shed.

■ *Teaching problem-solving skills.* Acquiring cognitive and behavioral strategies for coping with tempting situations and periods of slowed progress is associated with long-term change. Weight-loss maintainers are more likely than individuals who relapse to be conscious of their behavior, use social support, confront problems directly, and use personally developed strategies (Kayman, Bruvold, & Stern, 1990).

■ *Extended intervention.* Longer treatments (from 25 to 40 weeks) that include the components just listed grant people time to develop new habits.

Finally, it is important to note that nearly 50 percent of American women and 25 percent of American men report that they are trying to lose weight—statistics that far exceed the number of obese people (Robison et al., 1993). Recall from Chapter 11 that the high value placed on thinness in American society creates unrealistic expectations about desirable body weight and contributes to dangerous eating disorders of anorexia nervosa and bulimia, which remain common in early adulthood (see pages 358–359). Overweight people in weight-loss programs who set too stringent weight goals are more likely to regain their lost pounds. And normal-weight people who diet often display "starvation syndrome" symptoms, including depression, anxiety, weakness, and preoccupation with food. A *sensible* body weight—neither too low nor too high—predicts physical and psychological health and longer life (Manson et al., 1995).

■ DIETARY FAT. After entering college, Sharese altered the diet of her childhood and adolescent years, sharply limiting red meat, eggs, butter, and fried foods. Public service announcements about the health risks of a high-fat diet have led the fat consumption of American adults to drop slightly—from an average of 36 to 34 percent of caloric intake—during the past 10 years (McGinnis & Lee, 1995). The nation's health goals include reducing dietary fat to 30 percent, with no more than 10 percent made up of saturated fat, by the year 2000—and for good reason. Fat consumption may play a role in breast cancer and (when it includes large amounts of red meat) is clearly linked to colon cancer (Clifford & Kramer, 1993). But the main reason people need to watch how much and what kind of fat they eat is the strong connection between saturated fat and cardiovascular disease.

Fat is a type of molecule made up of carbon atoms dotted with hydrogen atoms. *Saturated fat* has as many hydrogen atoms as it can chemically hold, generally comes from meat and dairy products, and is solid at room temperature. *Unsaturated fat* has less than its full complement of hydrogen atoms and includes most vegetable oils that are liquid at room temperature.

Moderate fat consumption is essential for normal body functioning. But when we consume too much, especially of the saturated variety, some is converted to cholesterol, which accumulates as plaque on the arterial walls in atherosclerosis. Earlier in this chapter, we noted that atherosclerosis is determined by multiple biological and environmental factors. But excess fat consumption (along with other societal conditions) is largely responsible for the high incidence of heart disease in the American black population, since black Africans have among the lowest rates of heart disease in the world (Begley, 1995).

The best rule of thumb is to eat less fat of all kinds and use unsaturated instead of saturated fat whenever possible. Furthermore (as we will see next), regular exercise can reduce the harmful influence of dietary fat, since it creates chemical byproducts that help eliminate cholesterol from the body.

EXERCISE

Three times a week, over the noon hour, Sharese delighted in running, making her way to a wooded trail that cut through a picturesque area of the city. Regular exercise kept her fit and slim. Compared to earlier days when she had been sedentary and overweight, it also limited the number of respiratory illnesses she caught. And as Sharese explained to a friend one day, "Exercise gives me a positive outlook. It calms me down—takes the edge off. Afterward, I feel a burst of energy that gets me through the day. If I don't do it, I get tired in the afternoon." Although most Americans are aware of the health benefits of exercise, only 24 percent engage in regular physical activity (McGinnis & Lee, 1995).

Besides reducing body fat and building muscle, exercise fosters resistance to disease. Frequent bouts of moderate-intensity exercise enhance the immune response, thereby lowering the risk of colds or flu and, when these illnesses do strike, promoting faster recovery (Nieman, 1994). Furthermore, in several longitudinal studies extending over 10 to 20 years, physical activity was linked to reduced incidence of cancer at all body sites except the skin, with the strongest findings for cancer of the rectum and colon (Albanes, Blair, & Taylor, 1989; Wannamethee, Shaper, & Macfarlane, 1993). Exercise also helps prevent adult-onset diabetes

Regular exercise, through such activities as basketball, running, swimming, cycling, rowing, or brisk walking, can lead to a healthier, longer life. About 20 minutes, 3 to 5 times a week, of relatively vigorous use of large muscles of the body can help protect against obesity, enhance the immune response, and foster a positive outlook on life. (Spencer Grant/The Picture Cube)

(Kriska, Blair, & Pereira, 1994). Finally, physically active people are less likely to develop cardiovascular disease. If they do, it typically occurs at a later age and is less severe than among their inactive agemates (Bokovoy & Blair, 1994).

Just how does exercise help prevent the serious illnesses just mentioned? It may do so by reducing the incidence of obesity—a risk factor for heart disease, diabetes, and several forms of cancer. In addition, people who exercise probably adopt other healthful behaviors, thereby lowering the risk of diseases associated with high-fat diets, alcohol consumption, and smoking. Exercise can also have a direct impact on disease prevention. For example, in animal research, it inhibits growth of cancerous tumors beyond the impact of diet, body fat, and the immune response (Mackinnon, 1992). And it promotes cardiovascular functioning by strengthening the heart muscle, producing a form of "good cholesterol" (high-density lipoproteins) that helps remove "bad cholesterol" (low-density lipoproteins) from the artery walls, and decreasing blood pressure (Donatelle & Davis, 1997).

Yet another way that exercise may guard against illness is through its mental health benefits. Many studies show that physical activity reduces anxiety and depression, improves mood, and enhances alertness and energy (Weyerer & Kupfer, 1994). The impact of exercise on a "positive outlook," as Sharese expressed it, is most obvious just after a workout and can last for several hours (Chollar, 1995). The stress-reducing properties of exercise undoubtedly strengthen immunity to disease. And as physical activity enhances psychological well-being, it promotes self-esteem, ability to cope with stress, on-the-job productivity, and life satisfaction.

When we consider the evidence as a whole, it is not surprising that physical activity is associated with substantially lower death rates from all causes (Blair et al., 1989). The contribution of exercise to longevity cannot be accounted for by preexisting illness in inactive people, since sedentary individuals who start to exercise live longer than those who remain inactive (Paffenbarger et al., 1993).

How much exercise is recommended for a healthier, happier, and longer life? The usual prescription is 20 minutes, 3 to 5 times a week, of relatively vigorous use of the large muscles of the body in which heart rate is elevated to 60 to 90 percent of its maximum—through rowing, cycling, dancing, cross-country skiing, swimming, running, or brisk walking (American College of Sports Medicine, 1991). At the same time, exercise should be pleasurable, not a chore. If the exercise program is too stressful, its health benefits will be reduced.

SUBSTANCE ABUSE

Despite an overall decrease in use of both legal and illegal substances in recent years, alcohol and drug dependency remains a serious problem in early adulthood. It impairs cognitive processes, intensifies psychological problems that underlie the addiction, and increases the risk of unintentional injury and death. Return to Chapter 11, page 366, and review the personal and situational factors that lead individuals to become substance abusers in adolescence. The same characteristics are predictive in the adult years. Cigarette smoking and alcohol consumption are the two most common substance disorders (Lowinson, Ruiz, & Millman, 1992).

■ CIGARETTE SMOKING. Dissemination of information on the harmful effects of cigarette smoking has helped reduce its prevalence from 40 percent of American adults in 1965 to 25 percent in 1994. Still, smoking has declined very slowly. Most of the drop is among college graduates; those who did not finish high school have changed very little. More than 90 percent of men and 85 percent of women who smoke started before age 21. The average number of cigarettes smoked per day is high—about 20. And the earlier people start smoking, the greater their daily cigarette consumption, an important reason that preventive efforts in adolescence are vital (Breslau, 1993; Jarvik & Schneider, 1992; U.S. Bureau of the Census, 1996).

The ingredients of cigarette smoke—nicotine, tar, carbon monoxide, and other chemicals—leave their damaging mark throughout the body. As the person inhales, delivery of oxygen to tissues is reduced, and heart rate and blood pressure rise. Over time, insufficient oxygen results in limited night vision, more rapid wrinkling of the skin, and a lower sperm count and higher rate of male sexual impotence (Margolin, Morrison, & Hulka, 1994; Tur, Yosipovitch, & Oren-Vulfs, 1992). More deadly outcomes include increased risk of heart attack, stroke, and cancer of the mouth, throat, larynx, esophagus, lungs, pancreas, kidneys, and bladder (Allen, Piccone, & D'Amanda, 1993).

The link between smoking and mortality is dose related. The more cigarettes consumed, the greater the chance of premature death. At the same time, the benefits of quitting are great, including return of disease risks to nonsmoker levels within 3 to 8 years and a healthier living environment. Although millions of people have stopped smoking without help, those in treatment programs or using cessation aids (for example, nicotine gum, nasal spray, or patches, designed to reduce dependency gradually) often fail. After 1 year, 70 to 80 percent start smoking again.

Unfortunately, too few treatments are long enough and teach skills for avoiding relapse (Jarvik & Schneider, 1992). Sometimes a personal experience or the advice of a doctor is enough to energize quitting. My father, a heavy smoker in his 20s and 30s, sought treatment for chronic bronchitis (inflammation of the bronchial tubes, common among smokers). His doctor told him that if he did not stop smoking, he would not live to see his children grow up. That day marked his last cigarette. He died at age 80, having lived nearly 30 years longer than the average life expectancy for his year of birth.

■ ALCOHOL. National surveys reveal that about 13 percent of men and 3 percent of women in the United States are heavy drinkers (average two or more drinks per day). About one-third of these are *alcoholics*—people who are unable to limit their alcohol use and are often unaware that it is the cause of their problems. In men, alcoholism usually begins in the teens and early twenties and worsens over the following decade. In women, its onset is typically later, in the twenties and thirties, and its course more variable. Many alcoholics are also addicted to other drugs. About 80 percent are heavy cigarette smokers (Goodwin, 1992).

Twin and adoption studies support a genetic contribution to alcoholism (Anthenelli & Schuckit, 1992). But whether or not a person comes to deal with life's problems through drinking is greatly affected by personal characteristics and circumstances, since half of hospitalized alcoholics do not have a family history of problem drinking (Hawkins, Catalano, & Miller, 1992). Alcoholism crosses social-class and ethnic lines, but it is higher in some groups than others. For example, in cultures where alcohol is a traditional part of religious or ceremonial activities, people are less likely to abuse it. Where access to alcohol is carefully controlled and a sign of adulthood, dependency is more likely. Poverty and hopelessness also promote it (Donatelle & Davis, 1997).

Alcohol acts as a depressant, impairing the ability of the brain to control thought and action. In a heavy drinker, it relieves guilt and anxiety, then induces these states as its effects wear off, so the alcoholic drinks again. Chronic alcohol use does widespread damage to the body. Its most

well-known complication is liver disease, but it is also linked to cardiovascular disease, inflammation of the pancreas, irritation of the intestinal tract, bone marrow problems, disorders of the blood and joints, and some forms of cancer. Over time, alcohol damages the central nervous system, leading to confusion, apathy, inability to learn, and impaired memory. The costs to society are enormous. About 40 percent of highway fatalities in the United States involve drivers who have been drinking. Nearly half of convicted felons are alcoholics, and about half of police activities in large cities involve alcohol-related offenses. Alcohol frequently plays a part in date rape and other forms of sexual coercion (Goodwin, 1992).

The most successful treatments are comprehensive, combining personal and family counseling, group support, and aversion therapy (in which medication is used to produce a physically unpleasant reaction, such as nausea and vomiting, to alcohol). Alcoholics Anonymous, a community support approach, helps many people exert greater control over their lives through the encouragement of others with similar problems.Nevertheless, breaking an addiction that has dominated a person's life is difficult; about 60 percent of alcoholics relapse within 3 months after treatment (Goodwin, 1992).

SEXUALITY

By the time their teenage years are over, more than 80 percent of young people have had sexual intercourse; by age 22, the figure rises to 90 percent (Michael et al., 1994). Compared to earlier generations, a wider range of sexual choices and lifestyles can be observed in modern adults, including cohabitation, marriage, extramarital experiences, and orientation toward a heterosexual or homosexual partner. In this chapter, we explore the attitudes, behaviors, and health concerns related to sexual activity becoming a regular event in young people's lives. In Chapter 14, we focus on the emotional side of close relationships.

■ HETEROSEXUAL ATTITUDES AND BEHAVIOR. One Friday evening, Sharese accompanied her roommate Heather to a young singles bar. Shortly after they arrived, two young men joined them. Faithful to her boyfriend Ernie (whom she met in college and who worked in another city), Sharese remained aloof for the next hour. In contrast, Heather was talkative and friendly and gave one of the men (Rich) her phone number. The next weekend, Heather went out with Rich. On the second date, they had intercourse, but the romance was short-lived. Within a few weeks, each went in search of a new partner. Aware of Heather's varied sex life, Sharese couldn't help but wonder whether her own was normal. Only after nearly a year of dating exclusively had she and Ernie slept together.

Since the 1950s, public display of sexuality in movies, newspapers, magazines, and books has steadily increased,

In cultures where alcohol is a traditional part of religious or ceremonial activities, people are less likely to abuse it. For example, alcoholism is very low among Jews. Here, following a Friday evening Sabbath service, Jewish families say the kiddush, or blessing over the fruit of the vine. (David Austen/Stock Boston)

fostering the impression that Americans are more sexually active than ever before. This escalation in sexual openness has prompted people whose sexual history falls short of popular images to question whether they might be missing something. What are modern adults' sexual attitudes and behaviors really like? Answers to this question have been difficult to obtain, until the completion of the National Health and Social Life Survey, the first in-depth study of Americans' sex lives based on a nationally representative sample. Nearly 4 out of 5 randomly chosen 18- to 59-year-olds agreed to participate—3,400 in all. Findings were remarkably similar to recent survey results in France, Great Britain, and Finland (Laumann et al., 1994; Michael et al., 1994).

Recall from Chapter 11 that the sex lives of most teenagers do not dovetail with exciting media images. A similar picture emerges for adults. Although their sexual practices are diverse, Americans are far less sexually active than we have come to believe. Couples like Sharese and Ernie are more typical (and satisfied) than are those like Heather and Rich.

Sexual partners, whether dating, cohabiting, or married, usually do not select each other arbitrarily. Instead, they tend to be alike in age (within 5 years), education, ethnicity, and (to a lesser extent) religion. In addition, most people who establish lasting relationships meet in conventional ways: They are introduced by family members or friends or get to know each other at work, school, or social events where people similar to themselves congregate. Only 10 percent of adults who marry and 14 percent who decide to live together meet at bars, through personal ads, or on vacations. The powerful influence of social networks on sexual choice is adaptive, since sustaining an intimate relationship is easier when adults share interests and values and people they know approve of the match.

Consistent with popular belief, Americans today have more sexual partners than they did a decade or two ago. For example, one-third of adults over age 50 have had five or more partners in their lives, whereas half of 30- to 50-year-olds have accumulated that many in much less time. But when adults—either younger or older—are asked how many partners they have had in the past year, the usual reply (for 71 percent) is one.

What explains this trend toward more relationships in the context of sexual commitment? In the past, dating several partners was followed by marriage. Today, dating more often gives way to cohabitation, which leads either to marriage or to breakup. In addition, people are marrying later, and the divorce rate remains high. Together, these factors create more opportunities for new sexual partners. Still, most people spend the majority of their lives with one partner. And only a small minority of Americans (3 percent), most of whom are men, report five or more partners in a single year. (See the Social Issues box on the following page for a discussion of sex differences in sexual attitudes and behavior.)

How often do Americans have sex? Not nearly as frequently as the media would have us believe. Only one-third of 18- to 59-year-olds have intercourse as often as twice a week, another third have it a few times a month, and the remaining third have it a few times a year or not at all. Three factors affect frequency of sexual activity: age, whether people are married or cohabiting, and how long the couple has been together. Younger, dating people have more partners, but this does not translate into more sex! Sexual activity increases through the twenties as people either marry or cohabit. Then, around age 30, it declines, even though hormone levels have not changed much. The demands of daily life—working, commuting, taking care of home and children—are probably responsible. Despite the common assumption that sexual practices vary greatly across social groups, the patterns just described are unaffected by education, social class, or ethnicity.

Although most Americans have only modest amounts of sex, they are happy with their sex lives. For those in committed relationships, feeling "extremely physically and emotionally satisfied" exceeds 80 percent; it rises to 88 percent for married couples. As number of sex partners increases, satisfaction declines sharply. These findings challenge the stereotype of marriage as sexually dull and people with many partners as having the "hottest" sex.

Furthermore, only a minority of adults—women more often than men—report persistent sexual problems. The two most frequent difficulties for women are lack of interest in sex (33 percent) and inability to achieve orgasm (24 percent); for men, climaxing too early (29 percent) and anxiety about performance (16 percent). But a completely trouble-free physical experience is not essential for sexual happiness. Satisfying sex involves more than technique; it is attained in the context of love, affection, and fidelity. In the words of the survey researchers, "happiness with partnered sex is linked to happiness with life" (Michael et al., 1994, pp. 130).

■ HOMOSEXUAL ATTITUDES AND BEHAVIOR. The tragic spread of AIDS has increased the visibility of homosexuals in the United States, reenergizing the gay liberation movement of the 1970s. Efforts at AIDS prevention, securing health care, and combating discrimination against people with AIDS have resulted in renewed attention to the civil rights of homosexuals. During the past decade, attitudes toward homosexuals, although still quite negative, have begun to change, particularly in the area of equal employment opportunities (Hugick, 1992; Seltzer, 1993). Homosexuals' greater openness about their sexual orientation has contributed to slow gains in acceptance, since prior exposure and interpersonal contact with gay men and lesbians reduce negative attitudes (Ellis & Vasseur, 1993). As in the past, men judge homosexuals more harshly than do

SOCIAL ISSUES

SEX DIFFERENCES IN ATTITUDES TOWARD SEXUALITY

Differences between men and women in sexual attitudes and behavior are widely assumed, and modern theories offer diverse explanations for them. For example, Nancy Chodorow, a feminist psychoanalytic theorist, and Carol Gilligan, a feminist theorist of moral development (see Chapter 12, page 398), believe that the emotional intensity of the infant–caregiver relationship is carried over into future intimate ties for girls. However, it is disrupted for boys as they form a "masculine" gender-role identity stressing individuation and independence (Chodorow, 1978; Gilligan, 1982). According to ethological theory, sexuality is powerfully shaped by the desire to have children and ensure their survival. Since sperm are plentiful and ova far less numerous, women must be more careful than men about selecting a partner with the commitment and resources to protect children's development (Symons, 1987). Finally, social learning theory views sex differences as due to modeling and reinforcement of gender-role expectations (see Chapter 8, page 262). Women receive more disapproval for having numerous partners and engaging in casual sex than do men, who are sometimes rewarded with admiration and social status.

Both small-scale studies and large-scale surveys confirm that women are more opposed to premarital and extramarital sex than are men. However, the difference, although substantial, has decreased over the past three decades (Laumann et al., 1994; Oliver & Hyde, 1993). Furthermore, when a small number of men with a great many sexual partners are excluded, modern men and women differ very little in average number of lifetime sexual partners (Chatterjee, Handcock, & Simonoff, 1995). Why is

In spite of media images, for most Americans, satisfying sex is attained in the context of love, affection, and fidelity. People tend to seek out partners similar in age, education, and background to themselves, and partners in committed relationships have sex more often and are more satisfied. (Walter Bibikow/The Image Bank)

this so? From an ethological perspective, more effective contraception has permitted sexual activity without risk of reproduction. Under these circumstances, women can have as many partners as men without risking the welfare of their offspring.

Still, when women complain that the men they meet are not interested in long-term commitments, their laments have a ring of truth. Many more men than women report that they are looking for sexual play and pleasure, with marriage and love not necessarily a part of it. These conflicting goals are greatest among young adults. With age, both sexes seem to become more traditional, regarding a loving relationship as more central to sexuality (Michael et al., 1994).

TRY THIS...

- Ask several middle-aged people for their impressions of how sexual attitudes and behavior of young people today differ from those of their own youth. On the basis of evidence reviewed in this chapter, to what extent do the views expressed represent myth and to what extent do they reflect reality?

women, perhaps because heterosexual men are more concerned than women with conforming to prescribed gender roles (Herek & Glunt, 1993; Whitley & Kite, 1995).

In the National Health and Social Life Survey, 2.8 percent of men and 1.4 percent of women regarded themselves as homosexual or bisexual—figures that are similar to those of other recent national surveys conducted in the United States, France, and Great Britain (Billy et al., 1993; Spira, 1992; Wellings et al., 1994). These low numbers—and the possible unwillingness of homosexuals to answer questions about their sexual practices in a climate of persecution—have limited what researchers have been able to find out about the sex lives of gay men and lesbians. The little evidence available indicates that homosexual sex follows many of the same rules as heterosexual sex: People tend to seek out partners similar in education and background to themselves; partners in committed relationships have sex more often and are more satisfied; and the overall frequency of sex is modest (Laumann et al., 1994; Michael et al., 1994).

Although homosexuals are few in number, they tend to live in large cities where many others share their sexual orientation, rather than in small towns and rural areas. Gay pride has filtered down to some nonurban communities, but it is still rare. Living outside of big cities, without a social network in which to find compatible homosexual partners and where prejudices are more intense, can be isolating and lonely (Miller, 1989). People who identify themselves as gay or lesbian also tend to be well educated. In the National Health and Social Life Survey, twice as many college-educated as high-school-educated men and eight times as many college-educated as high-school-educated women stated they were homosexual. Although the reasons for these findings are not clear, they probably reflect greater social and sexual liberalism among the more highly educated and therefore greater willingness to disclose homosexuality.

■ SEXUALLY TRANSMITTED DISEASE. A large number of Americans—one in four—are likely to contract a sexually transmitted disease (STD) sometime in their lives; 1.5 percent report having had one in the past year (Barringer, 1993; Laumann et al., 1994). (See Chapter 11, page 364, for a description of the most common STDs.) Although the incidence is highest in adolescence, STD continues to be prevalent in early adulthood. During the teens and twenties, people accumulate most of their sexual partners, and they often do not take appropriate precautions to prevent the spread of STD (see page 000). The overall rate of STD is higher among women than men, since it is at least twice as easy for a man to infect a woman with any STD, including AIDS, as it is for a woman to infect a man (Michael et al., 1994).

Although AIDS, the most deadly STD, remains concentrated among gay men and intravenous drug abusers, many homosexuals have responded to its spread by changing their sexual practices—limiting the number of partners with whom they have sex and using latex condoms consistently and correctly (Johnson et al., 1992; Laumann et al., 1994). As a result, the number of infections is lower among gay men today than it was in the early 1980s. However, AIDS has not declined among drug abusers and their sexual partners. Consequently, some experts fear that it will spread most rapidly in marginal groups where intravenous drug abuse is high and combines with poverty, inadequate education, poor health, high life stress, and hopelessness. People overwhelmed by these problems are least likely to take preventive measures (Jonsen & Stryker, 1992).

Yet AIDS can be contained and reduced—through sex education extending from childhood into adulthood and access to health services, condoms, and clean needles and syringes for high-risk individuals (Stryker et al., 1995). In view of the rise in AIDS cases among women over the past decade, a special need exists for female-controlled preventive measures. The recently developed female condom offers some promise, but its effectiveness and acceptability are not yet established (Amaro, 1995).

■ SEXUAL COERCION. STD is not the only serious health threat to accompany sexual activity. One afternoon after a long day of classes, Sharese flipped on the TV and caught a talk show on sex without consent. Karen, a 25-year-old woman, described her husband Mike pushing, slapping, verbally insulting, and forcing her to have sex. "It was a control thing," Karen explained tearfully. "He complained that I wouldn't always do what he wanted. I was confused and blamed myself. I didn't leave because I was sure he'd come after me and get more violent."

One day, as Karen was speaking long distance to her mother on the phone, Mike grabbed the receiver and shouted, "She's not the woman I married! I'll kill her if she doesn't shape up!" Karen's parents realized the seriousness of her situation, arrived by plane to rescue her the next day, and helped her start divorce proceedings and get treatment.

An estimated 14 to 25 percent of women have endured **rape,** legally defined as intercourse by force, by threat of harm, or when the victim is incapable of giving consent (due to mental illness, mental retardation, or intoxication). The majority of victims (8 out of 10) are under age 30 (Koss, 1993; National Victims Center, 1992). Women are vulnerable in the company of partners, acquaintances, and strangers, but most of the time their abusers are men they know well. One study suggested that up to one-third of American women will be assaulted by an intimate partner (Hamberger, Saunders, & Hovey, 1992). Sexual coercion crosses social-class and ethnic lines; people of all walks of life are offenders and victims. For example, in a national sample of nearly 3,000 college men, nearly 8 percent admitted to attempting or committing date rape (Koss, Gidycz, & Wisniewski, 1987).

Psychological reactions to rape are severe enough that therapy is vital. In this rape crisis center in Bergen County, New Jersey, a professional counselor and volunteer assists a victim in recovering from trauma through social support, validation of her experience, and safety planning. (Sidney/Monkmeyer Press)

Personal characteristics of the man with whom a woman is involved are far better predictors of her chances of becoming a victim than her own characteristics. Men who engage in sexual assault tend to believe in traditional gender roles, approve of violence against women, and accept rape myths ("Women want to be raped" or "Any healthy woman can resist if she really wants to"). Many have an intense desire to dominate their victim but deny this, reasoning that "she brought it on herself" (Scully, 1990). Perpetrators also have difficulty interpreting women's social behavior accurately. They are likely to view friendliness as seductiveness, assertiveness as hostility, resistance as desire (Allison & Wrightsman, 1993).

Cultural forces—in particular, strong gender typing—contribute to sexual coercion. When men are taught from an early age to be dominant, competitive, and aggressive and women to be submissive, cooperative, and passive, the themes of rape are reinforced. Under these conditions, men may view a date not in terms of getting to know a partner, but in terms of sexual conquest. And a husband may regard satisfaction of his sexual needs as his wife's duty, even if she is uninterested (Amaro, 1995). The widespread acceptance of violence in American society (see Chapter 8, page 269) also sets the stage for rape. It typically occurs in relationships where other forms of aggression are commonplace (Browne, 1993).

Consequences. Women's psychological reactions to rape resemble those of survivors of extreme trauma. Immediate responses include shock, confusion, withdrawal, and psychological numbing. These eventually give way to chronic fatigue, tension, disturbed sleep, depression, and suicidal thoughts (Resnick & Newton, 1992). When sexual coercion is ongoing, taking any action may seem dangerous (as it did to Karen), so the victim falls into a pattern of extreme passivity, helplessness, and fear (Goodman, Koss, & Russo, 1993; Herman, 1992).

One-third to one-half of rape victims are physically injured. Some contract sexually transmitted diseases, and pregnancy results in about 5 percent of cases (Beebe, 1991; Koss, Koss, & Woodruff, 1991). Furthermore, women victimized by rape (and other crimes) report more symptoms of illness across almost all body systems. And they are more likely to engage in negative health behaviors, including smoking and alcohol use (Koss, Koss, & Woodruff, 1991).

Prevention and Treatment. Many rape victims are not as fortunate as Karen, since their anxiety about provoking another attack keeps them from telling anyone, even trusted family members and friends (Browne, 1991). If they seek help for other problems, conflict over issues surrounding sexuality may lead a sensitive health professional to detect a possible rape. A variety of community services exist to assist women in taking refuge from abusive partners, including safe houses, crisis hotlines, support groups, and legal assistance. Most, however, are underfunded and cannot reach out to everyone in need.

The trauma induced by rape is severe enough that therapy is important. Many experts advocate group intervention, since contact with other survivors is especially helpful in countering isolation and self-blame (Koss & Harvey, 1991). Other critical features that foster recovery include

- *Routine screening for victimization* when women seek health care to ensure referral to community services and protection from future harm.

- *Validation of the experience,* by acknowledging that many other women have been physically and sexually assaulted by intimate partners; that such assaults lead to a wide range of persisting symptoms, are illegal and inappropriate, and should not be tolerated; and that the trauma can be overcome.

Preventing Sexual Coercion

SUGGESTION	DESCRIPTION
Reduce gender stereotyping and gender inequalities.	The roots of sexual coercion lie in the historically subordinate role of women. Restricted educational and employment opportunities keep many women economically dependent on men and therefore poorly equipped to avoid partner violence.
Mandate treatment for men who physically or sexually assault women.	Ingredients of effective intervention include inducing personal responsibility for violent behavior; teaching social awareness, social skills, and anger management; and developing a support system to prevent future attacks.
Expand interventions for children and adolescents who have witnessed violence between adult caregivers.	Although most child witnesses do not become involved in abusive relationships as adults, boys are at increased risk of assaulting their female partners and girls are at increased risk of becoming victims if they observed violence between their parents.
Teach women to take precautions that lower the risk of sexual assault.	Risk of sexual assault can be reduced by communicating sexual limits clearly to a date; developing neighborhood ties to other women; increasing the safety of the immediate environment (for example, installing deadbolt locks, checking the back seat of the car before entering); avoiding deserted areas; and not walking alone after dark. (For additional suggestions, consult *The New* Our Bodies Ourselves: *A Book by and for Women.*)

Sources: Boston Women's Health Book Collective, 1992; Browne, 1993.

- *Safety planning* even when the abuser is no longer present, to prevent recontact and reassault. This includes how to obtain police protection, legal intervention, a safe shelter, and other aid should a rape survivor find herself at risk again.

Finally, many steps can be taken at the level of the individual, the community, and society to prevent sexual coercion. Some ways of doing so are listed in the Caregiving Concerns table above.

 ■ MENSTRUAL CYCLE. The menstrual cycle is central to women's lives, and it presents unique health concerns. Although almost all women experience some discomfort during menstruation, others have more severe difficulties. **Premenstrual syndrome (PMS)** refers to an array of physical and psychological symptoms that usually appear 6 to 10 days prior to menstruation. The most common are abdominal cramps, fluid retention, diarrhea, tender breasts, backache, headache, fatigue, tension, irritability, and depression; the precise combination varies from person to person. PMS is usually experienced for the first time after age 20. Nearly three out of four women have some form of it, usually mild. For 10 percent, PMS is severe enough to interfere with academic, occupational, and social functioning. PMS is a worldwide phenomenon—just as common in Italy and the Islamic nation of Bahrain as it is in the United States (Brody, 1992; Monagle et al., 1993).

The causes of PMS are not well established, but evidence for a biological basis is accumulating (Asso &

Magos, 1992). Researchers believe that PMS may be related to the rise in estrogens that precedes menstruation. But since hormone treatment is not consistently effective, sensitivity of brain centers to these hormones, rather than the hormones themselves, may be responsible (Lewis, 1992). Besides hormones, current treatments include analgesics for pain, diuretics for fluid buildup, limiting caffeine intake (which often intensifies symptoms), vitamin therapy, exercise, and other strategies for reducing stress. As yet, no method has been successful in curing PMS.

PSYCHOLOGICAL STRESS

A final health concern, threaded throughout previous sections, has such a broad impact that it merits a comment of its own. Psychological stress, measured in terms of adverse social conditions, negative life events, or daily hassles, is related to a wide variety of health outcomes (Cooper, Cooper, & Eaker, 1988; Lazarus, 1985). In addition to its association with many unhealthy behaviors, stress has clear physical consequences. For example, as it mobilizes the body for action, it elevates blood pressure. Chronic stress resulting from economic hardship and inner-city living is consistently linked to hypertension, a relationship that contributes to the high incidence of heart disease in low-income groups, especially African Americans (Calhoun, 1992). Earlier we mentioned that psychological stress also interferes with immune system functioning, a link that may underlie its relationship to several forms of cancer. And by reducing digestive activity as blood flows to the brain, heart, and extremities, stress can cause gastrointestinal difficulties, including constipation, diarrhea, colitis, and ulcers (Donatelle & Davis, 1997).

In previous chapters, we repeatedly noted the stress-buffering effect of social support, which continues throughout adulthood (Markides & Cooper, 1989). Consequently, helping people who are isolated develop and maintain satisfying social ties is as important a health intervention as any we have mentioned. Before we turn to the cognitive side of early adulthood, you may find it helpful to examine the Caregiving Concerns table below, which summarizes the many ways we can foster a healthy adult life.

BRIEF REVIEW

Environmental conditions and behavior have a major impact on health and longevity. Overweight and obesity lead to serious physical, psychological, and social problems. Dietary fat is linked to cardiovascular disease and colon cancer. Exercise protects against illness and mortality by reducing body fat, building muscle, and enhancing the immune response, cardiovascular functioning, and psychological well-being. Cigarettes and alcohol are the most commonly abused substances. They do damage throughout the body and are difficult addictions to break.

Americans are less sexually active and more likely to be committed to a single partner than media images would have us believe. Both heterosexuals and homosexuals select partners similar in education and background to themselves and are most satisfied in committed relationships. Sexually transmitted disease (STD) is widespread in early adulthood. Although the incidence of AIDS has declined among gay men, it remains high among drug abusers and their sexual partners and has risen among women. Many American women have been traumatized by sexual coercion, a form of abuse promoted by strong gender typing and acceptance of violence in American society. The menstrual cycle presents women with unique health concerns. About 10 percent experience premenstrual syndrome (PMS) severe enough to interfere with their everyday lives.

Psychological stress leads to unhealthy behaviors and unfavorable physical consequences. Social support is a vital health intervention in early adulthood and at other times of life.

CAREGIVING CONCERNS

Fostering a Healthy Adult Life

SUGGESTION	DESCRIPTION
Engage in healthy eating behavior.	Educate yourself and those with whom you live about the makeup of a healthy diet. Eat in moderation, and learn to separate true hunger from eating due to boredom or stress.
Maintain a reasonable body weight.	If you need to lose weight, make a commitment to a lifelong change in the way you eat, not just a temporary diet. Select a sensible, well-balanced dietary plan, eat small meals spaced over the day, and exercise regularly.
Keep physically fit.	Choose a specific time to exercise, and stick with it. To help sustain physical activity and make it more enjoyable, exercise with your partner or a friend and encourage one another. Set reasonable expectations, and allow enough time to reach your fitness goals; many people become exercise dropouts because their expectations were too high.
Control alcohol intake, and do not smoke cigarettes.	Drink moderately or not at all. Do not allow yourself to feel you must drink to be accepted or to enjoy a social event. If you smoke, choose a time that is relatively stress-free to quit. Seek the support of your partner or a friend.
Engage in responsible sexual behavior.	Identify attitudes and behaviors that you need to change to develop a healthy intimate relationship. Educate yourself about sexual anatomy and functioning so you can make sound decisions about contraception and protect yourself against sexually transmitted disease.
Manage stress.	Seek a reasonable balance among work, family, and leisure. Become more aware of stressors, and identify ways of coping with them so you are better prepared when they arise. Engage in regular exercise, and find time each day for relaxation and quiet reflection.

Source: Donatelle & Davis, 1997.

- *List as many factors as you can that may have contributed to heart attacks and early death among Sharese's relatives.*

- *Tom began going to a health club three days a week after work. Soon the pressures of his job and his desire to spend more time with his girlfriend convinced him that he no longer had time for so much exercise. Explain to Tom why he should keep up his exercise regimen, and suggest ways to fit it into his busy life.*

- *Why are people in committed relationships likely to be more sexually active and satisfied than those who are dating several partners?*

COGNITIVE DEVELOPMENT IN EARLY ADULTHOOD

What happens to cognitive development with the transition to adulthood? Lifespan theorists have examined this question from three familiar vantage points. First, they have proposed transformations in the structure of thought—new, qualitatively distinct ways of thinking that extend the cognitive-developmental changes of childhood and adolescence. Second, adulthood is a time of attaining advanced knowledge in a particular area, an accomplishment that has important implications for information processing and creativity. Finally, researchers have been interested in the extent to which the diverse mental abilities assessed by intelligence tests remain stable or change during the adult years. Let's examine each of these perspectives on adult cognition in turn.

CHANGES IN THE STRUCTURE OF THOUGHT

Sharese described her first year in graduate school as a "cognitive turning point." As part of her internship in a public health clinic, she observed firsthand the many factors that affect human health-related behaviors. For a time, she was intensely uncomfortable about the fact that clear-cut solutions to dilemmas in the everyday world were so hard to come by. "Working in this messy reality is so different from the problem solving I did in my undergraduate classes," she commented to her mother in a phone conversation one day.

Sharese's reflections agree with those of a variety of researchers who have studied **postformal thought**—

cognitive development beyond Piaget's formal operational stage. Even Piaget (1967) acknowledged the possibility that important advances in thinking follow the attainment of formal operations. He observed that adolescents place excessive faith in abstract systems, preferring a logical, internally consistent but inaccurate perspective on the world to one that is vague, contradictory, and adapted to particular circumstances (see Chapter 11, page 373). To clarify how thinking is restructured in adulthood, let's look at some influential theories.

PERRY'S THEORY

In a well-known study, William Perry (1970, 1981) interviewed students at the end of each of their four years of college, asking "what stood out" during the previous year. Responses indicated that students' cognitive perspectives changed as they were exposed to the complexities of university life and moved closer to adult roles.

Younger students regarded knowledge as made up of separate units (beliefs and propositions) whose truth could be determined by comparing them to abstract standards—ones that exist apart from the thinking person and his or her situation. As a result, they engaged in **dualistic thinking,** dividing information, values, and authority into right and wrong, good and bad, we and they. As one college freshman stated, "When I went to my first lecture, what the man said was just like God's word. I believe everything he said because he is a professor . . . and this is a respected position" (Perry, 1981, p. 81). In contrast, older students were aware of a diversity of opinions on almost any topic. They moved toward **relativistic thinking,** viewing all knowledge as embedded in a framework of thought. As a result, they gave up the possibility of absolute truth in favor of multiple truths, each relative to its context. A college senior reasoned, "Just seeing how [famous philosophers] fell short of an all-encompassing answer, [you realize] that ideas are really individualized. And you begin to have respect for how great their thought could be, without its being absolute" (p. 90).

Perry's (1970, 1981) theory is based on a highly educated sample. He acknowledges that movement from dualism to relativism is probably limited to young people confronted with the multiplicity of viewpoints typical of a college education. But the underlying theme of *adaptive cognition*—thought less constrained by the need to find one answer to a question and more responsive to its context—is also evident in two additional views of the restructuring of thinking in adulthood.

SCHAIE'S THEORY

According to K. Warner Schaie (1977/1978), it would be difficult for human cognitive capacities to exceed the complexity of Piaget's formal operational stage. But with

Compared to adolescents, young adults focus less on acquiring knowledge and more on applying it to everyday life. Problems encountered at work, unlike those posed in classrooms or on intelligence tests, often do not have a single correct solution. (John Coletti)

entry into adulthood, the situations in which people must reason become more diverse. As a result, the goals of mental activity shift from acquiring knowledge to using it, as the following stage sequence reveals:

1. The **acquisitive stage** (childhood and adolescence): The first two decades of life are largely devoted to *knowledge acquisition*. As young people move from concrete to formal operational thought, they develop more powerful procedures for storing information, combining it, and drawing conclusions.

2. The **achieving stage** (early adulthood): In early adulthood, people must adapt their cognitive skills to situations (such as marriage and employment) that have profound implications for *achieving long-term goals*. As a result, they focus less on acquiring knowledge and more on applying it to everyday life. The problems encountered in work, intimate relationships, and child rearing, unlike those posed in classrooms or on intelligence tests, often do not have a single correct solution. Yet how the individual handles them can affect the entire life course. Consequently, young adults must attend to both the problem and its context, not just the problem, to be successful.

3. The **responsibility stage** (middle adulthood): In middle adulthood, as families and work lives become well established, *expansion of responsibilities to others* takes place on the job, in the community, and at home. As a result, cognition extends to situations involving social obligations. Maintaining a gratifying relationship with an intimate partner, staying involved in children's lives, assuming greater leadership on the job, and fulfilling

community roles must be juggled simultaneously and effectively. The most advanced form of this type of thinking, called the **executive stage,** characterizes individuals whose responsibilities have become highly complex. People at the helm of large organizations—businesses, academic institutions, and churches—must monitor organizational progress and the activities of many people. This demands that they understand the dynamic forces that affect an elaborate social structure and combine information from many sources to make decisions.

4. The **reintegrative stage** (late adulthood): As the future becomes shorter, the need to acquire knowledge and monitor decisions in terms of later consequences declines. In addition, executive thought becomes less important with retirement. The cognitive challenges of late adulthood move beyond "What should I know?" and "How should I use what I know?" to "Why should I know?" (Schaie & Willis, 1992, pp. 390–391). In this stage, people reexamine and *reintegrate their interests, attitudes, and values,* using them as a guide for what knowledge to acquire and apply. Consequently, the elderly are more selective about the circumstances in which they expend their cognitive energies. They are less likely than younger people to waste time on tasks that appear meaningless to them or that rarely occur in their daily lives.

LABOUVIE-VIEF'S THEORY

Gisella Labouvie-Vief's (1980, 1985) portrait of adult cognition echoes features of Schaie's theory. Adolescence, she points out, permits the individual to operate within a world of possibility. Adulthood brings a return to pragmatic concerns—thinking focused on concrete situations. But rather than a regression to less mature thinking, movement from hypothetical to **pragmatic thought** is a structural advance in which logic becomes a tool to solve real-world problems.

According to Labouvie-Vief, the need to specialize motivates this change. As adults select and pursue one path out of many alternatives, they become more aware of the constraints of everyday life. And in the course of balancing various roles, they give up their earlier need to resolve contradictions. Instead, they accept inconsistencies as part of life and develop ways of thinking that thrive on imperfection and compromise. Sharese's friend Christy, a student and mother of two young children, illustrates:

> I've always been a feminist, and I wanted to remain true to my beliefs in family and career. But this is Gary's first year of teaching high school, and he's saddled with four preparations and coaching the school's basketball team. At least for now, I've had to settle for 'give-and-take feminism'— going to school part-time and shouldering most of the responsibility for the kids while he gets used to his new job. Otherwise, we'd never make it financially.

Awareness of multiple truths, integration of logic with reality, and tolerance for the gap between the ideal and real sum up qualitative transformations in thinking during the adult years. As we will see in the next section, adults' increasingly specialized and context-bound thought, although closing off options, also opens new cognitive doors to higher levels of competence.

INFORMATION PROCESSING: EXPERTISE AND CREATIVITY

In Chapter 9, we noted that children's expanding knowledge of the world improves their ability to remember new information related to what they already know. **Expertise**—acquisition of extensive knowledge in a field or endeavor—is supported by the specialization that begins with selecting a college major or an occupation, since it takes many years for a person to master any complex domain. Once expertise is attained, it has a profound impact on information processing.

Compared to novices, experts remember and reason more quickly and effectively. This makes sense, since the expert knows more domain-specific concepts and represents them in richer ways—at a deeper and more abstract level and as having more features that can be linked to other concepts. As a result, experts approach problems with underlying principles in mind, whereas the understanding of novices is superficial. For example, a highly trained physicist notices when several problems deal with conservation of energy and can therefore be solved similarly. In contrast, a beginning physics student focuses only on surface features—whether the problem contains a disk, a pulley, or a coiled spring (Chi, Glaser, & Farr, 1988). Experts appear more cognitively mature because they can use what they know to arrive at solutions automatically—through quick and easy remembering. And when a problem is especially challenging, they are more likely to plan ahead, analyzing and categorizing it before attempting to solve it.

Besides enhancing problem solving, expertise is necessary for creativity. The creative products of adulthood differ

Expertise is supported by specialization. This furniture maker's thorough knowledge of his craft permits him to move from problem to solution quickly and effectively. (John Millar/Tony Stone Images)

from those of childhood in that they are not just original, but directed at a social or aesthetic need. Consequently, mature creativity requires a unique cognitive capacity—the ability to formulate new, culturally meaningful problems, to ask significant questions that have not been posed before. According to Patricia Arlin (1984, 1989), movement from *problem solving* to *problem finding* is a core feature of postformal thought evident in highly accomplished artists and scientists.

A century of research reveals that creative accomplishment rises in early adulthood, peaks in the late thirties or early forties, and gradually declines. However, important exceptions to this trend exist. People off to an early start in creativity tend to peak and drop off sooner, whereas "late bloomers" show their full stride at older ages. This suggests that creativity is more a function of "career age" than chronological age. The course of creativity also varies across disciplines. For example, artists and musicians typically show an early rise in creativity, perhaps because they do not need to undergo extensive formal education before they begin to produce. Academic scholars and scientists usually display their achievements later and over a longer time, since they must earn a higher degree and spend years doing research to make worthwhile contributions (Simonton, 1988, 1991).

Creativity is rooted in expertise, but not all experts are creative. Other qualities are needed, including an innovative thinking style, tolerance of ambiguity, a special drive to succeed, and a willingness to experiment and try again after failure (Sternberg & Lubart, 1995). In addition, creativity requires time and energy. For women especially, it can be postponed or disrupted by child rearing, divorce, or an unsupportive partner (Valliant & Valliant, 1990). In sum, creativity is multiply determined. When personal and situational factors that promote it are present, it can continue for many decades, well into old age.

CHANGES IN MENTAL ABILITIES

Intelligence tests are a convenient tool for assessing changes in a variety of mental abilities during the adult years. Factors included on tests for adults are similar to those for school-age children and adolescents (see page 299). Think back to our discussion of postformal thought. It suggests that intelligence tests, well suited to measuring skills needed for success in school, are less adequate for assessing competencies relevant to many adults' everyday lives (Schaie & Willis, 1992). Nevertheless, research with mental tests sheds light on the widely held belief that intelligence declines in adulthood as structures in the brain deteriorate.

Many cross-sectional studies show this pattern—a peak in test performance at age 35 followed by a steep drop

into old age. But widespread testing of young adults in college and in the armed services during the 1920s offered researchers a convenient opportunity to conduct longitudinal research, retesting people in middle adulthood. Results revealed an age-related increase! What explains this contradiction? To find out, K. Warner Schaie (1994) applied the *longitudinal-sequential design,* which combines longitudinal and cross-sectional approaches (see Chapter 1, pages 35–36), in the Seattle Longitudinal Study.

In 1956, people ranging in age from 22 to 70 were tested cross-sectionally. Then, at regular intervals, longitudinal follow-ups were conducted and new samples added, yielding a total of 5,000 participants, five cross-sectional comparisons, and longitudinal data spanning 35 years. Findings on five mental abilities showed the typical cross-sectional drop after the mid-thirties. But longitudinal trends revealed modest gains sustained into the fifties and early sixties, after which performance decreased very gradually.

Figure 13.6 illustrates Schaie's cross-sectional and longitudinal outcomes for just one mental ability. *Cohort effects*—improvements in education and health from one generation to the next—are responsible for the difference. Since each new generation performed better than the one before it, the early cross-sectional decline is an illusion. Instead, the picture of development, accurately portrayed in longitudinal research, is one of steady improvement in diverse intellectual performances from early to middle adulthood, followed by a modest drop that is not underway until late in life. In later chapters, we will return to factors that affect this pattern as well as vast individual differences present throughout the adult years.

BRIEF REVIEW

Lifespan theorists studying postformal thought have identified structural changes in adult cognition. According to Perry, during the college years, young adults give up the search for a single truth in favor of multiple truths, each relative to its context. In Schaie's stage sequence, adulthood is accompanied by a shift from acquiring knowledge to using it in everyday life. Similarly, Labouvie-Vief stresses pragmatic thought, in which logic becomes a tool to solve real-world problems.

As adults specialize, they develop expertise, which enhances information processing and provides the foundation for creativity. Longitudinal research reveals steady gains in a variety of mental abilities well into middle adulthood.

ASK YOURSELF . . .

■ *As a college sophomore, Steven took a course in decision making but didn't find it very interesting. At age*

FIGURE 13.6

Cross-sectional and longitudinal trends in verbal ability (ability to understand ideas expressed in words). The cross-sectional decline is due to cohort effects—improvements in education and health in younger generations. When adults are followed longitudinally, their performance rises over early and middle adulthood and declines very gradually in the later years. A similar pattern holds for four other mental abilities: spatial orientation, inductive reasoning, numeric ability, and verbal memory. *(Adapted from K. W. Schaie, 1988, "Variability in Cognitive Functioning in the Elderly." In M. A. Bender, R. C. Leonard, & A. D. Woodhead, Eds., Phenotypic Variation in Populations, New York: Plenum, p. 201. Reprinted by permission.)*

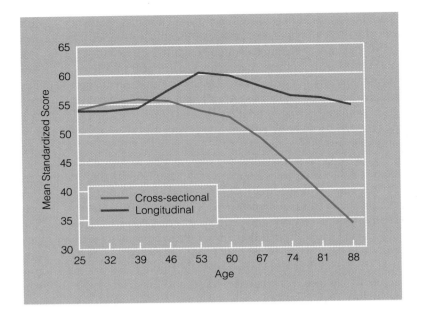

25, newly married and employed, he realized the value of what the professor had taught him. How do changes in the structure of thought help explain Steven's revised attitude?

■ *How does the development of creativity in adulthood illustrate assumptions of the lifespan perspective, discussed in Chapter 1?*

THE COLLEGE EXPERIENCE

ooking back at the trajectory of their lives, many people view the college years as formative—more influential than any other period of adulthood. This is not surprising, since college serves as a "developmental testing ground," a time when full attention can be devoted to exploring alternative values, roles, and behaviors. To facilitate this exploration, college exposes young people to a form of "culture shock"—encounters with new ideas, new teachers and friends with varied beliefs, new freedoms and opportunities, and new academic and social demands (Pascarella & Terenzini, 1991). In the United States, 75 percent of high school graduates—more than in any other nation in the world—enroll in an institution of higher education. Besides offering a route to a high-status career and its extrinsic rewards (salary and job security), Americans expect colleges and universities to have a transforming impact on the coming generation. A wealth of evidence suggests that they do.

PSYCHOLOGICAL IMPACT OF ATTENDING COLLEGE

A comprehensive review of thousands of studies revealed broad psychological changes from the freshman to senior year of college (Pascarella & Terenzini, 1991). In addition to knowledge of their major field of study, students gain in ways less obviously tied to their academic programs. As Perry's theory of postformal thought predicts, they become better at applying reason and evidence to problems for which there is no clear answer and identifying the strengths and weaknesses of different sides of complex issues. These cognitive changes are complemented by revisions in attitudes and values—increased interest in literature, the performing arts, and philosophical and historical issues and greater tolerance for ethnic and cultural diversity. As we noted in Chapter 12, college also leaves its mark on moral reasoning by fostering concern with individual rights and human welfare. Finally, gains in reflective thought combined with exposure to multiple world views encourage young people to look more closely at themselves. During the college years, students develop greater self-understanding, enhanced self-esteem, and a firmer sense of identity.

How do these interrelated changes come about? Type of college attended makes little difference, since cognitive growth is just as great for students who initially enroll in a two-year community college as a four-year institution (Bohr et al., 1994). Instead, the impact of college is jointly determined by the person's involvement in academic and nonacademic activities and the richness and diversity of the campus setting. Residential living is one of the most consis-

The impact of attending college on cognition, attitudes, and values is enhanced by residential living. When students live on campus, they are more likely to experience the richness and diversity of the campus setting. (Richard Pasley/Stock Boston)

tent determinants of change because it maximizes involvement in the educational and social systems of the institution (Pascarella et al., 1993). Since over half of American college students commute, these findings underscore the importance of programs that integrate commuting students into campus life. Quality of academic experiences also affects college outcomes. Students' effort and willingness to participate in class combined with instruction that is challenging, that integrates learning in separate courses around a central theme, and that provides extensive contact with faculty are linked to a wide range of psychological benefits (Franklin, 1995; Pascarella & Terenzini, 1991).

DROPPING OUT

A college degree channels people's postcollege lives, affecting their interests and opportunities in enduring ways. Yet 40 percent of freshmen drop out, most within the first year and many within the first 6 weeks. The price paid is high, in lifelong earnings and personal development (Edwards, Cangemi, & Kowalski, 1990).

Academic factors are not the sole reason for leaving, since young people who withdraw usually have the ability to succeed in the college to which they were admitted. Other student and institutional characteristics play a larger role. Most entering freshmen have high hopes for college life but find the transition difficult. Those who find it very hard to adapt—due to lack of motivation, poor study skills, financial pressures, or emotional dependence on parents—quickly develop negative attitudes toward the college environment, which leads them to abandon it. Often these exit-prone students do not meet with their advisors or professors. At the same time, colleges that do little to help high-risk students through encouragement, understanding, and assistance with decision making have a higher percentage of students who do not graduate (Cangemi, 1983).

Factors involved in the decision to withdraw from college are usually not catastrophic; they are typical problems of early adulthood. Most dropouts either find it more difficult to deal with these problems or are unable to get assistance in doing so. Reaching out to students, especially during the early weeks and throughout the first year, is crucial. Young people who sense that they have entered a college community concerned about them as individuals are far more likely to persist to graduation (Billson & Terry, 1987).

VOCATIONAL CHOICE

Young adults—college- and non-college-bound alike —face a major life decision: the choice of a suitable work role. Being a productive worker calls for many of the same qualities needed to be an active citizen and nurturant family member—good judgment, responsibility, dedication, and cooperation. An individual well prepared for work is better able to fulfill other adult roles. How do young people make decisions about careers, and what influences their choices? What is the transition from school to work like, and what factors make it easy or difficult?

SELECTING A VOCATION

In societies with an abundance of career possibilities, occupational choice is a gradual process, beginning long before adolescence. Major theorists view the young person as moving through several phases of vocational development (Ginzberg, 1972, 1988; Super, 1980, 1984):

1. The **fantasy period** (early and middle childhood). As we saw in Chapter 7, young children gain insight into career options by fantasizing about them. However, their preferences are largely guided by familiarity,

glamour, and excitement and bear little relation to the decisions they will eventually make.

2. The **tentative period** (early and middle adolescence). Between ages 11 and 17, adolescents start to think about careers in more complex ways. At first, they evaluate vocational options in terms of their *interests*. Later, as they become more aware of personal and educational requirements for different vocations, they take into account their *abilities and values*. "I like science and the process of discovery," Sharese thought to herself as she neared high school graduation. "But I'm also good with people, and I'd like to do something to help others. So maybe teaching or medicine would suit my needs."

3. The **realistic period** (late adolescence and early adulthood). By the end of the teenage years, the economic and practical realities of adulthood are just around the corner, and young people start to narrow their options. At first, many do so through further *exploration*, gathering more information about a set of possibilities that blends with their personal characteristics. Then they enter a final phase of *crystallization* in which they focus on a general vocational category. Within it, they experiment for a period of time before settling on a single occupation. As a college sophomore, Sharese pursued her interest in science, but she had not yet selected a major. Once she decided on chemistry, she considered whether to pursue teaching, medicine, or public health.

FACTORS INFLUENCING VOCATIONAL CHOICE

Although most people follow this general pattern of vocational development, there are exceptions in both timing and sequence. A few know from an early age just what they want to be and follow a direct path to a career goal. Others keep their options open for an extended period. College students are granted ample time to explore and decide. In contrast, the life conditions of low-income youths prevent them from having the choices that economically advantaged young people do.

Consider for a moment how a vocational choice is made, and you will see that it is not just a rational process in which young people weigh abilities, interests, and values against career options. Like other developmental milestones, it is the result of a dynamic interaction between person and environment. A great many influences feed into the decision.

■ PERSONALITY. People are attracted to occupations that complement their personalities. John Holland (1966, 1985) has identified six personality types that affect vocational choice. The *investigative person* enjoys working with ideas and is likely to select a scientific occupation (for example, anthropologist, physicist, or engineer). The *social person* likes interacting with people and gravitates toward human services (counseling, social work, or teaching). The *realistic person* prefers real-world problems and work with objects and tends to choose a mechanical occupation (construction, plumbing, or surveying). The *artistic person* is emotional and high in need for individual expression and looks toward an artistic field (writing, music, or the visual arts). The *conventional person* likes well-structured tasks and values material possessions and social status—traits well suited to certain business fields (accounting, banking, or quality control). Finally, the *enterprising person* is adventurous, persuasive, and a strong leader and is drawn to sales and supervisory positions.

Research reveals a clear relationship between personality and vocational choice, but it is only moderate. Personality is not a stronger predictor because many people are blends of several personality types and can do well at more than one kind of occupation. Furthermore, career decisions are made in the context of family background, educational opportunities, and current life circumstances. For example, Sharese's friend Christy scored high on Holland's "investigative" dimension. But when she married and had children early, she postponed her dream of becoming a college professor and chose a human services career that required fewer years of educational preparation. As Christy's case illustrates, personality takes us only part way in understanding vocational choice.

■ FAMILY INFLUENCES. Young peoples' vocational aspirations are strongly correlated with the jobs of their parents. Individuals who grew up in middle-class homes are more likely to select high-status, white-collar occupations, such as doctor, lawyer, scientist, and engineer. In contrast, those with low-income backgrounds tend to choose lower-status, blue-collar careers—for example, plumber, construction worker, food service employee, and secretary. Parent–child similarity is partly a function of educational attainment. The single best predictor of occupational status is number of years of schooling completed (Featherman, 1980).

Family resemblance in occupational choice also comes about for other reasons. Middle-class parents are more likely to give their children important information about the world of work and to have connections with people who can help the young person obtain a high-status position (Grotevant & Cooper, 1988). Parenting practices also shape work-related values. Recall from Chapter 2 that middle-class parents tend to promote curiosity and independence, which are required for success in many high-status careers. Lower-class parents, in contrast, are more likely to emphasize conformity and obedience. At work, they are used to following the directives of others. Eventually, young people choose careers that are compatible with these values. The jobs that appeal to them tend to be like those of their parents (Mortimer & Borman, 1988).

■ TEACHERS. Teachers also play a powerful role in career decisions. In one study, college freshmen were asked who had the greatest impact on their choice of a field of study. The people most often mentioned (by 39 percent of the sample) were high school teachers (Johnson, 1967, as cited by Rice, 1993). Young people who attend college have more opportunity to develop close bonds with teachers than do their non-college-educated counterparts, whose parents are more influential. These findings provide yet another reason to promote positive teacher–student relations, especially for low-income high school students. The power of teachers as role models could serve as an important source of upward mobility for these young people.

■ GENDER STEREOTYPES. Over the past two decades, young men's career preferences have remained strongly gender stereotyped, whereas young women have expressed increasing interest in occupations largely held by men (Sandberg et al., 1991). Changes in gender-role attitudes along with the dramatic rise in employed mothers, who serve as career-oriented models for their daughters (see Chapter 13), are common explanations for women's growing attraction to nontraditional careers.

At the same time, women's progress in entering and excelling at male-dominated professions has been slow. As Table 13.3 shows, the percentage of women engineers, lawyers, and doctors increased between 1972 and 1995 in

TABLE 13.3

Percentage of Women in Various Professions, 1972, 1983, 1995

PROFESSION	1972	1983	1995
Engineering	0.8	5.8	8.5
Law	3.8	15.8	21.4
Medicine	9.3	15.8	20.4
Business—executive and managerial	17.6	32.4	43.0[a]
Writing, art, entertainment	31.7	42.7	47.2
Social work	55.1	64.3	68.9
Elementary and secondary education	70.0	70.9	74.8
Higher education	28.0	36.3	40.9
Library, museum curatorship	81.6	84.4	83.6
Nursing	92.6	95.8	94.3

Source: U.S. Bureau of the Census, 1996.
[a]This percentage includes executives and managers at all levels. Women make up only 9 percent of senior management at big firms, although that figure represents a three-fold increase in the past decade.

the United States, but it falls far short of equal representation. Women remain heavily concentrated in the less well-paid, traditionally feminine professions of literature, social work, education, and nursing (U.S. Bureau of the Census, 1996). In virtually all fields, their achievements lag behind those of men, who write more books, make more discoveries, hold more positions of leadership, and produce more works of art.

Ability cannot account for these dramatic sex differences. Recall from Chapter 11 that the gender gap in cognitive performance of all kinds is small and is declining. Instead, gender-stereotyped messages from the social environment play a key role. Although girls' grades are higher than boys', girls reach secondary school less confident of their ability and more likely to underestimate their achievement (Bornholt, Goodnow, & Cooney, 1994). Between tenth and twelfth grade, the proportion of girls in gifted programs decreases. When asked what discouraged them from continuing in gifted classes, parental and peer pressures and attitudes of teachers and counselors ranked high on girls' lists (Read, 1991).

During college, the career aspirations of academically talented females decline further. In one longitudinal study, high school valedictorians were followed over a 10-year period—through college and into the work world. By their sophomore year, young women shifted their expectations toward less demanding careers because of concerns about combining work with child rearing and unresolved questions about their ability. Even though female valedictorians outperformed their male counterparts in college courses, they achieved at lower levels after career entry (Arnold, 1994). Another study reported similar results. Educational aspirations of mathematically talented women declined considerably during college, as did the number majoring in the sciences (Benbow & Arjmand, 1990).

These findings reveal a pressing need for programs that sensitize high school and college personnel to the special problems women face in developing and maintaining high career aspirations. Research shows that the aspirations of young women rise in response to career guidance that encourages them to set goals that match their abilities, interests, and values (Kerr, 1983). Models of accomplished women who have successfully dealt with family–career role conflict are also important (Schroeder, Blood, & Maluso, 1993).

VOCATIONAL PREPARATION OF NON-COLLEGE-BOUND YOUNG ADULTS

Sharese's younger brother Leon graduated from high school in a vocational track. Like 25 percent of young people with a high school diploma, he had no plans to go to college. While in school, Leon held a part-time job selling candy at the local shopping mall. He hoped to work in data

WORK—STUDY APPRENTICESHIPS IN GERMANY

Rolf, an 18-year-old German vocational student, is an apprentice at Brandt, a large industrial firm known worldwide for its high-quality products. Like many German companies, Brandt has a well-developed apprenticeship program that includes a full-time professional training staff, a suite of classrooms, and a lab equipped with the latest learning aids. Apprentices move through more than ten major divisions in the company that are carefully selected to meet their learning needs. Rolf has worked in purchasing, inventory, production, personnel, marketing, sales, and finance. Now in cost accounting, he assists Herr Stein, his supervisor, in designing a computerized inventory control system. Rolf draws a flowchart of the new system under the direction of Herr Stein, who explains that each part of the diagram will contain a set of procedures to be built into a computer program.

Rolf is involved in complex and challenging projects, guided by caring mentors who love their work and want to teach it to others. Two days a

week, he attends a Berufsschule, a part-time vocational school. On the job, Rolf applies a wide range of academic skills, including reading, writing, problem solving, and logical thinking. His classroom learning is directly relevant to his daily life (Hamilton, 1990).

Germany has the most successful apprenticeship system in the world for preparing young people to enter modern business and industry. More than 60 percent of adolescents participate in it, making it the most common form of secondary education. German adolescents who do not go to a *Gymnasium* (college preparatory high school) usually complete full-time schooling by age 15 or 16, but education remains compulsory until age 18. They fill the 2-year gap with part-time vocational schooling combined with apprenticeship. Students are trained for a wide range of occupations—more than 400, leading to over 20,000 specialized careers. Each apprenticeship is jointly planned by educators and employers. Apprentices who complete training and pass a qualifying examination are certified

as skilled workers and earn union-set wages for that occupation. Businesses provide financial support for the program because they know it guarantees a competent, dedicated work force (Hamilton, 1990, 1993).

The German apprenticeship system offers a smooth and rewarding path from school to career for young people who do not enter higher education. Many apprentices are hired by the firms in which they were trained. Most others find jobs in the same occupation. For those who change careers, the apprentice certificate is a powerful credential. Employers view successful apprentices as responsible and capable workers. They are willing to invest in further training to adapt the individual's skills to other occupations. As a result, German young people establish themselves in well-paid careers with security and advancement possibilities between the ages of 18 and 20 (Hamilton, 1990).

The success of the German system suggests that some kind of national apprenticeship program would improve the transition from school to work for young people in the United States. Nevertheless, implementing an American apprenticeship system poses major challenges. Among these are overcoming the reluctance of employers to assume part of the responsibility for vocational training; creating institutional structures that ensure cooperation between schools and businesses; and finding ways to prevent low-income youths from being concentrated in the lowest-skilled apprenticeship placements, which would perpetuate current social inequalities (Bailey, 1993; Hamilton, 1993). Pilot apprenticeship projects are currently underway, in an effort to solve these problems and build a bridge between learning and working in the United States.

High-quality vocational training combined with apprenticeship enables German youths who do not go to college to enter well-paid careers. This electronics trainee works at a hydraulics system factory. His education involves integrating academic skills with on-the-job experience to ensure that he becomes competent at both. (M. Granitsas/The Image Works)

processing after graduation, but 6 months later he was still a part-time sales clerk at the candy store. Although Leon had filled out many job applications, he got no interviews or offers. He soon despaired of any relationship between his schooling and a career.

Leon's inability to find a job other than the one he held as a student is typical for American non-college-bound high school graduates. His sister Sharese will have a much easier time. As a college student, Sharese profited from the advice of faculty in her major field of study, access to a wide variety of career services, and applied experiences related to her vocational goals.

Although high school graduates are more likely to find employment than those who drop out, they have fewer work opportunities than they did several decades ago. More than one-fourth of recent high school graduates who do not continue their education are unemployed (U.S. Bureau of the Census, 1996). When they do find work, most are limited to low-paid, unskilled jobs. In addition, they have few alternatives to turn to for vocational counseling, career information, and job placement as they make the transition from school to work (Bailey, 1993; Hamilton, 1990).

Unlike Western European nations, the United States has no widespread training system to prepare its youths for skilled business and industrial occupations and manual trades (Hamilton & Hurrelmann, 1994). The federal government does support some job training programs, and funding for them has increased. But most are too short to make a difference in the lives of poorly skilled individuals, who need intensive training and academic remediation before they are ready to enter the job market. And at present, these programs serve only a small minority of young people who need assistance.

Inspired by successful programs in Western Europe, youth apprenticeship strategies that coordinate on-the-job training with classroom instruction are being considered as an important dimension of educational reform in the United States. The Cultural Influences box on the opposite page describes Germany's highly successful apprenticeship system. Bringing together the worlds of schooling and work offers many benefits. These include helping non-college-bound young people establish productive lives right after graduation, motivating at-risk youths to stay in school, and contributing to the nation's economic growth (Bailey, 1993; Hamilton, 1993).

Although vocational development is a lifelong process, the most important periods for launching a career are adolescence and early adulthood. Young people well prepared for an economically and personally satisfying work life are much more likely to become productive citizens, devoted family members, and contented adults. The support of families, schools, businesses, communities, and society as a whole can contribute greatly to a positive outcome. We will take up the challenges of establishing a career and integrating it with other life tasks in Chapter 14.

ASK YOURSELF . . .

■ *During her freshman year of college, Sharese participated in a program in which a professor and five students met once a month to talk about common themes in three courses that the students took together. Why is this program likely to increase student retention?*

UMMARY

PHYSICAL DEVELOPMENT IN EARLY ADULTHOOD

LIFE EXPECTANCY

How has average life expectancy changed in the twentieth century?

■ Both heredity and environment affect **average life expectancy.** Twentieth-century gains in length of life, amounting to about 30 years in the United States, are largely due to improved nutrition, medicine, sanitation, and safety. These changes clearly indicate that **biological aging** can be modified by behavior and environmental interventions.

MAXIMUM LIFESPAN AND ACTIVE LIFESPAN

Discuss the extent to which maximum lifespan and active lifespan can be improved.

■ Scientists disagree about whether **maximum lifespan** can be extended beyond an average of 85 or 90 years. Reducing health problems among low-income and ethnic minority groups would most likely add to Americans' **active lifespan** of 64 years.

THEORIES OF BIOLOGICAL AGING

Describe current theories of biological aging, including those at the level of DNA and body cells and those at the level of tissues and organs.

■ Biological aging may eventually prove to be the combined result of the programmed effects of specific genes and the cumulative effects of random events. Aging genes may control certain biological changes in DNA and body cells. DNA may also be damaged as mutations accumulate, leading to less efficient cellular repair and replacement and to abnormal cells. Release of highly reactive **free radicals** is a probable cause of age-related DNA and cellular damage.

- Genetic and cellular deterioration affects organs and tissues. The **cross-linkage theory** of aging suggests that over time, protein fibers form links and become less elastic, producing negative changes in many organs. Declines in the endocrine and immune systems may also contribute to aging.

PHYSICAL CHANGES OF AGING

Describe the physical changes of aging, paying special attention to the cardiovascular and respiratory systems, motor performance, the immune system, and reproductive capacity.

- The physical changes of aging are gradual in early adulthood but accelerate later. Declines in heart and lung performance show up only during exercise. Heart disease is a leading cause of death in adults, although it has decreased considerably in the past 50 years due to lifestyle changes and medical advances. Atherosclerosis is a serious, multiply determined cardiovascular disease involving fatty deposits on artery walls.

- Athletic skill peaks in the twenties. Less active lifestyles, rather than biological aging, account for most of the subsequent decline in athletic skill and motor performance.

- The immune response strengthens in adolescence and declines after age 20. This trend is partly due to shrinking of the thymus gland. Increased difficulty coping with physical and psychological stress also contributes.

- Reproductive capacity declines with age, especially after age 35 in women and age 40 in men.

HEALTH AND FITNESS

Describe the impact of nutrition and exercise on health, and discuss obesity in adulthood.

- Many Americans eat the wrong kinds and amounts of food. Obesity has doubled in the United States in the twentieth century, currently affecting more than a quarter of the population, especially low-income ethnic minority females. Heredity, sedentary lifestyles, and high-fat diets contribute to obesity, which is associated with serious health problems and social discrimination.

- Adding some weight between ages 25 and 50 is normal due to a decrease in **basal metabolic rate (BMR)**, but many young adults show large gains. Successful treatments for obesity typically involve a low-fat diet plus exercise, accurate recording of food consumption, social support, teaching problem-solving skills, and extended intervention.

- Regular exercise reduces body fat, builds muscle, helps prevent illness (including cardiovascular disease), and enhances psychological well-being. Nevertheless, only 24 percent of Americans engage in regular physical activity.

Name the two most common substance disorders, and discuss the health risks they entail.

- Cigarette smoking and alcohol consumption, the two most common substance disorders, are problems in early adulthood and can be difficult to overcome. Most adults who smoke began before age 21. They are at increased risk for heart attack, stroke, numerous cancers, and premature death.

- About one-third of heavy drinkers suffer from alcoholism, to which both heredity and environment contribute. Alcohol is implicated in liver and cardiovascular disease, certain cancers, numerous other physical

disorders, and social problems such as highway fatalities, crime, and sexual coercion.

Describe sexual attitudes and behavior of young adults, and discuss sexually transmitted disease, sexual coercion, and premenstrual syndrome (PMS).

- American adults are less sexually active than popular media images suggest, but compared to earlier generations, they display a wider range of sexual choices and lifestyles and have had more sexual partners. Nevertheless, most people spend the majority of their lives with one partner. Adults in committed relationships report high satisfaction with their sex lives.

- Over the past decade, negative attitudes toward homosexual men and women have begun to change, particularly in the area of equal employment opportunities. Homosexual relationships share many characteristics with heterosexual relationships, including similarity in partners' backgrounds, greater satisfaction in committed relationships, and modest frequency of sexual activity.

- Twenty-five percent of Americans are likely to contract a sexually transmitted disease (STD) during their lifetime; women are more vulnerable than men. AIDS remains concentrated among gay men and intravenous drug abusers, but changes in sexual practices have reduced the number of infections.

- Fourteen to twenty-five percent of women have been **raped,** typically by men they know well; they suffer psychological trauma that warrants therapy. Cultural acceptance of strong gender typing and violence contributes to sexual coercion.

- Nearly three out of four women experience **premenstrual syndrome (PMS).** usually in mild form. For some, PMS is severe enough to interfere with daily life.

How does psychological stress affect health?

■ Chronic psychological stress induces physical responses that contribute to heart disease, several types of cancer, and gastrointestinal problems. Supportive social ties can help alleviate stress.

COGNITIVE DEVELOPMENT IN EARLY ADULTHOOD

CHANGES IN THE STRUCTURE OF THOUGHT

What are the characteristics of adult thought, and how does thinking change in adulthood?

■ Cognitive development beyond Piaget's formal operational stage is known as **postformal thought.** Adult cognition typically reflects an awareness of multiple truths, integrates logic with reality, and tolerates the gap between the ideal and the real.

■ According to Perry's theory, college students move from **dualistic thinking,** dividing information into right and wrong, to **relativistic thinking,** or awareness of multiple truths.

■ In Schaie's theory, the emphasis shifts from acquiring to using knowledge as a person goes through the **acquisitive stage** in childhood and adolescence; the **achieving stage** in early adulthood; the **responsibility stage** and, for some, the **executive stage** in middle adulthood; and finally, the **reintegrative stage** in late adulthood.

■ According to Labouvie-Vief's theory, the need to specialize motivates adults to progress from the adolescent's ideal world of possibilities to **pragmatic thought,** which uses logic as a tool to solve real-world problems.

INFORMATION PROCESSING: EXPERTISE AND CREATIVITY

What roles do expertise and creativity play in adult thought?

■ Specialization in college and in an occupation leads to **expertise,** which enhances problem solving and is necessary for creativity. Mature creativity involves formulating meaningful new problems and questions. Although creativity tends to rise in early adulthood and peak in the late thirties or early forties, its development varies considerably and is multiply determined.

CHANGES IN MENTAL ABILITIES

What do intelligence tests reveal about changes in adults' mental abilities over the course of a lifetime?

■ Cross-sectional studies show a peak in intelligence test performance around age 35 followed by a steep decline, whereas longitudinal studies reveal an age-related increase. Applying the longitudinal-sequential design in the Seattle Longitudinal Study, Schaie discovered that the cross-sectional decline is due to cohort effects. Intellectual performance improves steadily into middle adulthood and does not decline until late in life.

THE COLLEGE EXPERIENCE

Describe the impact of a college education on young people's lives, and discuss the problem of dropping out.

■ Through involvement in academic programs and campus life, college students engage in exploration that produces gains in knowledge and reasoning ability, revised attitudes and values, enhanced self-esteem and self-knowledge, and preparation for a high-status career. The 40 percent of freshmen who find the transition to college life so difficult that they drop out are usually struggling with typical problems of early adult-

hood and could benefit from supportive interventions.

VOCATIONAL CHOICE

Trace the development of vocational choice, and cite factors that influence it.

■ Vocational development moves through three phases: a **fantasy period** in which children explore career options through play; a **tentative period,** in which teenagers weigh different careers against their interests, abilities, and values; and a **realistic period,** in which young people settle on a vocational category and then a specific career. Vocational choices are influenced by personality, parents' occupations, teachers, and gender stereotypes. The single best predictor of occupational status is number of years of schooling completed.

Discuss recent findings on sex differences in occupational preferences and achievement.

■ Young men's career preferences have remained strongly gender stereotyped, whereas young women's interest in male-dominated occupations has increased. Yet women's progress in male-dominated professions has been slow, and their achievements lag behind those of men in virtually all fields. Gender-stereotyped messages prevent many women from reaching their career potential.

What career options are open to non-college-bound high school graduates?

■ American non-college-bound high school graduates have fewer work opportunities today than they did several decades ago. More than one-fourth are unemployed, and most are limited to low-paid, unskilled jobs. To address their need for vocational training, youth apprenticeships inspired by those in Western Europe are being considered as an important dimension of educational reform in the United States.

IMPORTANT TERMS AND CONCEPTS

biological aging (p. 416)

average life expectancy (p. 417)

maximum lifespan (p. 418)

active lifespan (p. 419)

free radicals (p. 420)

cross-linkage theory (p. 420)

basal metabolism rate (BMI)
  (p. 426)

rape (p. 432)

premenstrual syndrome (PMS)
  (p. 434)

postformal thought (p. 436)

dualistic thinking (p. 436)

relativistic thinking (p. 436)

acquisitive stage (p. 437)

achieving stage (p. 437)

responsibility stage (p. 437)

executive stage (p. 437)

reintegrative stage (p. 437)

pragmatic thought (p. 438)

expertise (p. 438)

fantasy period (p. 441)

tentative period (p. 442)

realistic period (p. 442)

INFERTILITY

(See page 75 for organizations related to infertility.)

CARDIOVASCULAR DISEASE

American Heart Association
7272 Greenville Avenue
Dallas, TX 75231-4596
(800) 242-8721
Web site: *www.amhrt.org*
Supports research, public education, and community service programs to reduce disability and premature death from cardiovascular disease.

OVERWEIGHT AND OBESITY

Overeaters Anonymous
383 Van Ness Avenue, Suite 1601
Torrance, CA 90501
(310) 618-8835

An international organization aimed at helping people who have a desire to stop eating compulsively. Patterned after the treatment approach of Alcoholics Anonymous.

ALCOHOL ABUSE

Alcoholics Anonymous
475 Riverside Drive
New York, NY 10163
(212) 870-3400
Web site: *www. alcoholicanonymous.org*
An international organization aimed at helping people recover from alcoholism. Local chapters exist in many countries.

Mothers Against Drunk Driving (MADD)
511 E. John Carpenter Freeway, Suite 700
Irving, TX 75062
(214) 744-6233
(800) GET-MADD
Web site: *www.grannet.com/ mad.htm*
Encourages citizens to work toward reducing the drunk driving problem. Engages in public education, lobbies for more stringent laws, and provides assistance to victims and their families.

(See page 385 for additional organizations related to substance abuse.)

GAY AND LESBIAN CIVIL RIGHTS

International Gay and Lesbian Human Rights Commission
1360 Mission Street, Suite 200
San Francisco, CA 94103
(415) 255-8680
Web site: *www.iglhrc.org*
Monitors and responds to human rights violations against lesbians, gays, bisexuals, and people with AIDS worldwide. Promotes AIDS education and prevention.

National Gay and Lesbian Task Force
1734 14th Street, N.W.
Washington, DC 20009-4309
(202) 332-6483
Web site: *www.ngltf.org*
Dedicated to the elimination of prejudice based on sexual orientation. Lobbies Congress, engages in community organiz

ing, and disseminates information to the public.

SEXUAL COERCION

Women Against Rape (WAR)
Box 02084
Columbus, OH 43202
(614) 566-5153
An organization devoted to rape prevention. Sponsors crisis information services, rape survivor support groups, and rape prevention training including self-defense classes for women.

SEXUALLY TRANSMITTED DISEASE (STD)

U.S. Centers for Disease Control and Prevention
Mail Stop E06
1600 Clifton Road, N.E.
Atlanta, GA 30333
(404) 639-3311
Web site: *www.cdc.gov*
Government-sponsored agency that monitors the incidence and spread of diseases in the United States and recommends preventive measures. Sponsors the National STD Hotline, (800) 227-8922, and the National AIDS Hotline, (800) 342-AIDS (English) and (800) 344-SIDA (Spanish).

(See pages 000 and 000 for additional organizations related to AIDS and other sexually transmitted diseases.)

14

Emotional and Social Development in Early Adulthood

After completing her master's degree, Sharese returned to her home town, where her marriage to Ernie was soon to take place. A year-long engagement had preceded the wedding, during which Sharese had vacillated about whether to follow through. At times, she looked with envy at Heather, still unattached and free to pursue the full range of career options before her.

Sharese also pondered the life circumstances of Christy and her husband Gary—married their junior year in college and two children born within the next few years. Despite his good teaching performance, Gary's relationship with the high school principal deteriorated, and he quit his teaching position by the end of his first year. Financial pressures and the demands of parenthood had put Christy's education and career plans on hold. Sharese began to wonder if it was possible for a woman to have both family and career.

Sharese's ambivalence intensified as her wedding day approached. When Ernie asked why she was so agitated, she blurted out that she had doubts about getting married. Ernie's admiration and respect for Sharese had strengthened over their courtship, and he reassured her of his love. His career was launched, and at age 27, he felt ready to start a family. Uncertain and conflicted, Sharese felt swept toward the altar. Relatives, friends, and presents began to arrive. On the appointed day, she walked down the aisle.

In this chapter, we take up the emotional and social sides of early adulthood. Having achieved independence from the family, young people find that they still want and need close, affectionate ties. Yet like Sharese, they often fear losing their freedom. Once this struggle is resolved, the years from 20 to 40 lead to new family units and parenthood, accomplished in the context of diverse lifestyles. At the same time, young adults must learn to perform the skills and tasks of their chosen occupation. We will see that love and work are inevitably intertwined, and society expects success at both. In negotiating both arenas, young adults do more choosing, planning, and changing course than any other age group. When their decisions are in tune with themselves and their social and cultural worlds, they acquire many new competencies, and life is full and rewarding.

ERIKSON'S THEORY: INTIMACY VERSUS ISOLATION

Erikson's contributions have energized the study of adult personality development. All modern theories have been influenced by his vision (McCrae & Costa, 1990). According to Erikson (1964), adults move through three stages, each of which confronts the individual with both opportunity and risk— "a turning point for better or worse" (p. 139). The critical psychological conflict of early adulthood is **intimacy versus isolation.** It is reflected in the young person's thoughts and feelings about making a permanent commitment to an intimate partner.

As Sharese's inner turmoil reveals, establishing a mutually gratifying close relationship is a challenging task. Most young adults have only recently attained economic independence from parents, and many are still involved in the quest for identity. Yet intimacy requires that they give up some of their newfound independence and redefine their identity in terms of the values and interests of two people, not just themselves. During their first year of marriage, Sharese separated from Ernie twice as she tried to reconcile her needs for independence and intimacy. Maturity involves balancing these two forces. Without a sense of independence, people define themselves only in terms of their partner and sacrifice self-respect and initiative. Without intimacy, they face the negative outcome of Erikson's stage of early adulthood: loneliness and self-absorption. Ernie's patience and stability helped Sharese realize that marriage requires generosity and compromise but not total surrender of the self.

A secure sense of intimacy is also evident in the quality of other close relationships. For example, in friendships and work ties, young people who have achieved intimacy are cooperative, tolerant, and accepting of differences in background and values. Although they enjoy being

Establishing a mutually gratifying close relationship is a challenging task for young adults. Intimacy demands that they give up some of their newfound independence and redefine their identity in terms of two people. This husband and wife, both police officers, share an occupation—a basis for common interests. But to achieve intimacy, they must also be tolerant and accepting of differences in background and values. (Dick Blume/The Image Works)

Test Bank Items 14.1 through 14.3

with others, they are also comfortable when alone. People who have a sense of isolation hesitate to form close ties because they fear loss of their own identity, tend to compete rather than cooperate, are not accepting of differences, and are easily threatened when others get too close (Hamachek, 1990). Erikson believed that successful resolution of intimacy versus isolation prepares the individual for the middle adulthood stage, which focuses on *generativity*—caring for the next generation and helping to improve society.

Research based on self-reports confirms that intimacy is a central concern of early adulthood (Ryff & Migdall, 1984; Whitbourne et al., 1992). But it also reveals (as we have noted before) that a fixed series of tasks tied neatly to age does not describe the life course of many adults. Childbearing and child rearing (aspects of generativity) usually occur in the twenties and thirties, and contributions to society through work are also underway at this time. Furthermore, as we will see shortly, many combinations of marriage, children, and career exist, each with a unique pattern of timing and commitment (McAdams, de St. Aubin, & Logan, 1993; Weiland, 1993).

In sum, both intimacy and generativity seem to emerge in early adulthood, with shifts in emphasis that differ from one person to the next. Recognizing that Erikson's theory provides only a broad sketch of adult personality development, other theorists have expanded and modified his stage approach, adding detail and flexibility.

OTHER THEORIES OF ADULT PSYCHOSOCIAL DEVELOPMENT

In the 1970s, growing interest in adult development, sparked by people's personal experiences with change, led to the publication of several widely read books on the topic. Two of these volumes—Daniel Levinson's (1978) *The Seasons of a Man's Life* and George Vaillant's (1977) *Adaptation to Life*—present psychosocial theories in the tradition of Erikson that have gathered considerable research support. Each is summarized in Table 14.1.

LEVINSON'S SEASONS OF LIFE

Conducting extensive biographical interviews with forty 35- to 45-year-old men from four occupational subgroups (hourly workers in industry, business executives, university biologists, and novelists), Levinson (1978) looked for an underlying order in the adult life course. Later he investigated the lives of 45 women, also 35 to 45 years of age, from three subgroups (homemakers, business executives, and university professors). His results, and those of others, reveal a common path of change, within which men and women approach developmental tasks in somewhat different ways (Levinson, 1996; Roberts & Newton, 1987).

Like Erikson, Levinson (1978, 1996) conceives of development as a sequence of qualitatively distinct *eras*

TABLE 14.1

Stages of Adult Psychosocial Development

PERIOD OF DEVELOPMENT	ERIKSON	LEVINSON	VAILLANT
Early Adulthood (20–40 years)	Intimacy versus isolation	Early adult transition: 17–22 years	Intimacy
		Entry life structure for early adulthood: 22–28 years	
		Age 30 transition: 28–33 years	Career consolidation
		Culminating life structure for early adulthood: 33–40 years	
Middle Adulthood (40–65 years)	Generativity versus stagnation	Midlife transition: 40–45 years	Generativity
		Entry life structure for middle adulthood: 45–50 years	
		Age 50 transition (50–55 years)	Keeper of meanings
		Culminating life structure for middle adulthood (55–60 years)	
Late Adulthood (65 years–death)	Ego integrity versus despair	Late adult transition (60–65 years)	Ego integrity
		Late adulthood (65 years–death)	

(stages or seasons), each of which has its own time and brings certain psychological challenges to the forefront of the person's life. Within each era, important changes occur in a predictable age-related sequence jointly determined by biological and social forces. As Table 14.1 shows, each era begins with a *transition*, lasting about 5 years, which concludes the previous era and prepares the person for the next one. Between transitions, people move into stable periods in which they concentrate on building a *life structure* that harmonizes the inner personal and outer societal demands of that phase.

The **life structure,** a key concept in Levinson's theory, is the underlying pattern or design of a person's life at a given time. Its components are the person's relationships with significant others—individuals, groups, or institutions. The life structure can have many components, but usually only a few (having to do with marriage/family and occupation) are central to the person's life. However, wide individual differences in the weight of central and peripheral components exist.

Look again at Table 14.1, and notice that each of Levinson's eras consists of alternating transitional and structure-building periods. The primary task of a structure-building period is to select and integrate components in ways that enhance one's life. A structure-building period ordinarily lasts about 5 to 7 years—at most, 10 years. Then the life structure that formed the basis for stability comes into question, and a new transition arises in preparation for modifying it. Take a moment to list the components of your own life structure. Are you in midst of building—choosing and blending components and pursuing values and goals within that structure? Or are you in transition—terminating an existing structure to create possibilities for a new one?

Biographical reports of many individuals confirm Levinson's description of the life course. They also reveal that early adulthood is the era of "greatest energy and abundance, contradiction and stress" (Levinson, 1986, p. 5). These years can bring rich satisfactions in love, sexuality, family life, occupational advancement, and realization of major life goals. But they also carry great burdens—serious decisions about marriage, children, work, and lifestyle before many people have the life experience to choose wisely.

■ DREAMS AND MENTORS. How do young adults cope with the opportunities and hazards of this period? Levinson found that during the early adult transition (17 to 22 years), most construct a *dream,* an image of the self in the adult world that guides their decision making. The more specific the dream, the more purposeful the individual's structure building. For men, the dream usually emphasizes an independent achiever in an occupational role. In contrast, only a minority of women report dreams in which career dominates. Instead, most career-oriented women display "split dreams" in which both marriage and career

To realize their dream, young adults form a relationship with a mentor, who fosters their occupational skills and knowledge of the values, customs, and characters of the workplace. This senior executive advises a young employee who is new to the banking business. (Sepp Seitz/Woodfin Camp & Associates)

are prominent. Also, women's dreams tend to define the self in terms of relationships with husband, children, and colleagues. Men's dreams are more individualistic: They view significant others, especially wives, as vital supporters of their goals. Less often do they see themselves as supporters of the goals of others.

Young adults also form a relationship with a mentor who facilitates the realization of their dream. The mentor is generally several years older and experienced in the world the person seeks to enter. Most of the time, a senior colleague at work fills this role, occasionally a friend, neighbor, or relative. Mentors may act as teachers who enhance the person's occupational skills, guides who acquaint the person with the values, customs, and characters of the occupational setting, and sponsors who foster the person's advancement in the workplace. As we will see when we take up vocational development, finding a supportive mentor is easier for men than women.

According to Levinson (1978), men oriented toward high-status careers (doctor, lawyer, scientist) spend their twenties acquiring the skills, values, and credentials of their profession. Although some women follow this path, for many others the process of developing a career extends well into middle age (Levinson, 1996; Roberts & Newton, 1987).

■ THE AGE 30 TRANSITION. During the age 30 transition, young people reevaluate their life structure and try to change components they find inadequate. Those who were preoccupied with their career and are still single usually become concerned with finding a life partner. However, men rarely reverse the relative priority of career and family, whereas career-oriented women sometimes do.

Women who stressed marriage and motherhood early often develop more individualistic goals at this time. Recall that Christy had dreamed of becoming a college professor. In her mid-thirties, she finally got her doctoral degree and undertook the task of making it professionally. During the age 30 transition, women become conscious of aspects of their marriage that threaten to inhibit further development of the independent side of their dream. Married women tend to demand that their husbands recognize and accommodate their interests and aspirations beyond the home.

For men and women who are satisfied with neither their relational nor occupational accomplishments, the age 30 transition can be a crisis. For many others who question whether they will be able to create a meaningful life structure, it is a period of considerable conflict and instability.

■ **SETTLING DOWN FOR MEN, CONTINUED INSTABILITY FOR WOMEN.** Levinson (1978, 1996) describes age 33 to 40 as a period of settling down. To create the culminating life structure of early adulthood, men emphasize certain relationships and aspirations and make others secondary, or set them aside. In doing so, they try to establish a stable niche in society by anchoring themselves more firmly in family, occupation, and community. At the same time, they advance within the structure—by improving their skills and contributing to society in ways consistent with their values, whether those be wealth, power, prestige, artistic or scientific achievement, or particular forms of family or community participation. During these years, Sharese's husband Ernie expanded his knowledge of real estate accounting, became a partner in his firm, coached his son's soccer team, and was elected treasurer of his church. He paid less attention to golf, travel, and playing the guitar than he had in his twenties.

"Settling down," however, does not accurately describe women's experiences during their thirties. Many remain highly unsettled because of the addition of an occupational or relationship commitment that must be integrated into their life structure. When her two children were born, Sharese felt torn between her research position in the state health department and spending time with her family. She took 6 months off after the arrival of each baby. When she returned to work, she did not pursue attractive administrative openings because they required travel and time away from home. And shortly after Christy got her Ph.D., she and Gary divorced. Becoming a single parent while starting her professional life introduced new strains. Not until middle adulthood do many women attain the stability typical of men in their thirties—reaching professional maturity and taking on more authority in the community.

VAILLANT'S ADAPTATION TO LIFE

Vaillant (1977) examined the development of 94 men born in the 1920s, selected for study while they were students at a highly competitive liberal arts college and followed over the next 30 years. In college, the participants underwent extensive interviews. During each succeeding decade, they answered lengthy questionnaires about their current lives. Then Vaillant interviewed each man at age 47 about work, family, and physical and psychological health.

Other than denying a strict age-related schedule of change, Vaillant's theory is compatible with Levinson's. Both agree that quality of relationships with important people shape the life course. In examining the ways that men altered themselves and their social world to adapt to life, Vaillant confirmed Erikson's stages but filled in gaps between them. Following a period in their twenties devoted to intimacy concerns, the men focused on career consolidation in their thirties—working hard and making the grade in their occupations. During their forties, they pulled back from individual achievement and became more generative—giving to and guiding others. In their fifties, they became "keepers of meaning," or guardians of their culture. Many men became more philosophical—concerned about the values of the new generation and the state of their society. They wanted to teach others what they had learned from life experience and sought to perpetuate a capacity to care (Vaillant & Koury, 1994).

Unlike Levinson, Vaillant did not study women. But longitudinal research that includes women born around the same time as his participants suggests a similar series of changes (Block, 1971; Oden & Terman, 1968).

LIMITATIONS OF LEVINSON'S AND VAILLANT'S THEORIES

Although there is substantial consensus among psychosocial theorists about adult development, we must keep in mind that their conclusions are largely based on interviews with people born in the 1920s to 1940s. The patterns identified by Levinson and Vaillant fit the life paths of Sharese, Ernie, Christy, and Gary, but it is still possible that they do not apply as broadly to modern young people as they do to the previous generation (Rossi, 1980).

Two other factors limit the conclusions of these theorists. First, although non-college-educated, low-income adults were included in Levinson's sample, they were few in number, and low-income women remain almost entirely uninvestigated. Yet social class can profoundly affect the life course. For example, Levinson's blue-collar workers rarely implemented an occupational dream. Perhaps because career advancement is less salient for them, lower-class men perceive "early adulthood" to end and "maturity" to arrive at a younger age than do their middle-class counterparts (Neugarten, 1979). Finally, Levinson's participants were middle aged when interviewed, and they might not have remembered all aspects of their early adult lives accurately. In sum, studies of new generations are needed before we can definitely conclude that the developmental sequences just described apply to most or all young people.

THE SOCIAL CLOCK

In the previous section and in earlier parts of this book, we emphasized that changes in society from one generation to the next can affect the life course. Bernice Neugarten (1968, 1979) points out that an important cultural and generational influence on adult development is the **social clock**—age-graded expectations for life events, such as beginning a first job, getting married, birth of the first child, buying a home, and retiring. All societies have timetables for accomplishing major developmental tasks. Being on time or off time can profoundly affect self-esteem, since adults (like children and adolescents) make social comparisons, measuring the progress of their lives against their friends', siblings', and colleagues'. Especially when evaluating family and occupational attainments, people often ask, "How am I doing for my age?"

A major source of personality change in adulthood is conformity to or departure from the social clock. In a study of college women born in the 1930s who were followed up at ages 27 and 43, researchers determined how closely participants followed a "feminine" social clock (marriage and parenthood in the early or mid-twenties) or a "masculine" social clock (entry into a high-status career and advancement by the late twenties). Those who started families on time became more responsible, self-controlled, tolerant, and nurturant but declined in self-esteem and felt more vulnerable as their lives progressed. Those who followed an occupational timetable typical for men became more dominant, sociable, independent, and intellectually effective. Women not on a social clock—who had neither married nor begun a career by age 30—were doing especially poorly. They suffered from self-doubt, feelings of incompetence, and loneliness. One stated, "My future is a giant question mark" (Helson, Mitchell, & Moane, 1984, p. 1090).

As we noted in Chapter 1, expectations for appropriate behavior during early, middle, and late adulthood are no longer as definite as they once were. Still, many adults experience some psychological distress when they are substantially behind in timing of life events (Rook, Catalano, & Dooley, 1989). Following a social clock of some kind seems to foster confidence during early adulthood because it guarantees that the young person will engage in the work of society, develop skills, and increase in understanding of the self and others (Helson & Moane, 1987). As Neugarten (1979) suggests, the stability of society depends on having people committed to social clock patterns. With this in mind, let's take a closer look at how young men and women traverse the major tasks of young adulthood.

All societies have social clocks, or timetables for accomplishing major developmental tasks. Yet today, expectations for appropriate behavior are not as definite as they once were. This first-time mother has taken time to establish herself as a research scientist before having children. Is she departing from the social clock or establishing a new pattern? (Bob Daemmrich/The Image Works)

BRIEF REVIEW

Erikson describes the critical psychological conflict of early adulthood as intimacy versus isolation. It involves establishing a mutually gratifying close relationship with an intimate partner. Other psychosocial theorists have expanded and refined Erikson's stages. According to Levinson, adults move through an age-related sequence of alternating transitional and stable periods in which they build and revise their life structure in response to personal and societal demands. In the twenties, young people construct a dream and form a relationship with a mentor who facilitates its realization. During the age 30 transition, they reevaluate relational and occupational commitments in preparation for readjusting their life structure over the following decade. Vaillant fills in gaps between Erikson's stages. In his theory, the twenties are largely devoted to intimacy concerns, the thirties to career consolidation. Conformity to or departure from the social clock, a societal timetable for major life events, can profoundly affect self-esteem and personality development.

ASK YOURSELF . . .

■ *Return to Chapter 1 and review commonly used methods of studying human development. Which method did Vaillant use to chart the course of adult life? What are its strengths and limitations?*

■ *In Levinson's theory, during which periods are people likely to be moving within a social clock timetable? When are they likely to be moving out of one and establishing a new schedule?*

■ *Using the concept of the social clock, explain why Sharese was so conflicted about getting married to Ernie after she finished graduate school.*

CLOSE RELATIONSHIPS

To establish an intimate tie to another person, people must find a partner, build an emotional bond, and sustain it over time. Although young adults are especially concerned with romantic love, intimacy is also reflected in other relationships in which there is a mutual commitment—friends, siblings, and co-workers. Let's examine the multiple faces of intimacy.

ROMANTIC LOVE

During her junior year of college, Sharese glanced around the room in government class, her eyes often settling on Ernie, a senior and one of the top students. One weekend, Sharese and Ernie were invited to a party given by a mutual friend, and they struck up a conversation. Within a short time, Sharese discovered that Ernie was as warm and interesting as he had seemed from a distance. And Ernie found Sharese to be lively, intelligent, and attractive. By the end of the evening, the couple realized they had similar opinions on important social issues and liked the same leisure activities. They began to date steadily. Four years later, they married.

Finding a partner with whom to share one's life is a major milestone of adult development, with profound consequences for self-concept and psychological well-being. It is also a complex process that unfolds over time and is affected by a variety of events, as Sharese and Ernie's relationship reveals.

■ SELECTING A MATE. Recall from Chapter 13 that intimate partners tend to meet in places where there are people of their own age, ethnicity, social class, and religion. Once in physical proximity, people usually select partners who resemble themselves in other ways—attitudes, personality, educational plans, intelligence, physical attractiveness, and even height (Keith & Schafer, 1991; Simpson & Harris, 1994). Although romantic partners sometimes have abilities that complement one another, there is little support for the idea that "opposites attract." Instead, compatibility is the most powerful force in transforming strangers into lovers. The more similar two people are, the more satisfied they tend to be with their relationship and the more likely they are to stay together (Caspi & Herbener, 1990).

Sex differences exist in the importance placed on certain characteristics. Women assign greater weight to intelligence, ambitiousness, financial status, and character, men to physical attractiveness. Women prefer a same-age or slightly older partner, men a younger partner (Feingold, 1992; Kenrick & Keefe, 1992). Ethological theory helps us understand why. In Chapter 13, we noted that because their capacity to reproduce is limited, women seek a mate with traits that help ensure children's survival and well-being. In contrast, men look for a mate with characteristics that signal youth, health, and ability to bear offspring. Consistent with this explanation, men often want the relationship to move quickly toward physical intimacy. Women prefer a longer time for partners to get to know each other before sexual intercourse occurs.

As the Lifespan Vista box on the following page reveals, yet another influence on the partner we choose and the quality of the relationship we build is our early parent–child bonds. Finally, for romance to lead to marriage, it must happen at the right time. Two people may be right for each other, but if one or both are not ready to marry in terms of their own development or social clock, then the relationship is likely to dissolve.

■ THE COMPONENTS OF LOVE. What feelings and behaviors tell us that we are in love? According to one well-known theory, love has three components: intimacy, passion, and commitment. *Intimacy* is the emotional component. It involves warm, tender communication, expressions of concern about the other's well-being, and a desire for the partner to reciprocate. *Passion,* the desire for sexual activity and romance, is the physical and psychological arousal component. *Commitment* is the cognitive component. It leads partners to decide that they are in love and to maintain that love (Sternberg, 1987, 1988a).

The balance among these components changes as romantic relationships develop. At the beginning, **passionate love**—intense sexual attraction—is strong. Gradually, passion declines in favor of intimacy and commitment, which form the basis for **companionate love**—warm, trusting affection and caregiving (Berscheid, 1988; Hatfield, 1988). Each aspect of love, however, helps sustain

CHILDHOOD ATTACHMENT PATTERNS
AND ADULT ROMANTIC RELATIONSHIPS

Recall from Chapter 6 (page 000) that according to Bowlby's ethological theory, patterns of attachment originating in the infant–caregiver relationship are crucial for later emotional and social development. Early attachment bonds lead to the construction of an *internal working model,* or set of expectations about attachment figures, that serves as a guide for close relationships throughout life. In Chapter 6, we saw that adults' evaluations of their early attachment experiences are related to their parenting behaviors—specifically, to the quality of attachments they build with their own babies (see page 192). Additional evidence indicates that recollections of childhood attachment patterns are also strong predictors of romantic relationships in adulthood.

In studies carried out in Australia, Israel, and the United States, researchers asked people to describe their early parental bonds (attachment history), their attitudes toward intimate relationships (internal working model), and their actual experiences with romantic partners. Consistent with Bowlby's theory, childhood attachment patterns[2] served as remarkably good indicators of adult internal working models and relationship experiences:

SECURE ATTACHMENT. Adults who described their attachment history as *secure* (warm, loving, and supportive) had internal working models that reflected this security. They viewed themselves as likable and easy to get to know, were comfortable with intimacy, and rarely worried about

[2] To review patterns of infant–caregiver attachment, return to page 189.

abandonment or someone getting too close to them. In line with these attitudes, they characterized their most important love relationship in terms of trust, happiness, and friendship.

AVOIDANT ATTACHMENT. Adults with an avoidant attachment history (demanding, disrespectful, and critical parents) displayed internal working models that stressed independence, mistrust of love partners, and anxiety about people getting too close. They were convinced that others disliked them and that romantic love is hard to find and rarely lasts. Jealousy, emotional distance, and lack of acceptance pervaded their most important love relationship.

RESISTANT ATTACHMENT. Adults who described a resistant attachment history (parents who responded unpredictably and unfairly) presented internal working models in which they wanted to merge completely with another person and fell in love quickly. At the same time, they worried that their intense feelings would overwhelm others, who really did not love them and would not want to stay with them. Their most important love relationship was riddled with jealousy, emotional highs and lows, and desperation about whether the partner would return their affection (Bartholomew & Horowitz, 1991; Hazan & Shaver, 1987; Kirkpatrick & Davis, 1994; Mikulincer & Erev, 1991; Shaver & Brennan, 1992; Simpson, Rholes, & Nelligan, 1992).

Bowlby's predictions about the long-term consequences of early attachment are reflected not only in adults' self-reports, but also in the reports of their acquaintances and by their social behavior. Peers describe young adults with a secure attach-

ment history as more competent, charming, cheerful, and likable than their insecure counterparts (Kobak & Sceery, 1988). Perhaps because secure adults tend to choose partners who also have a childhood history of attachment security, their dating relationships last twice as long and their marriages are less likely to end in divorce than those of insecure people (Collins & Read, 1990; Hill, Young, & Nord, 1994). Furthermore, an insecure attachment history is associated with a variety of maladaptive social behaviors. For example, avoidant adults tend to deny attachment needs through excessive work and brief sexual encounters and affairs. Resistant adults are quick to express fear and anger, and they disclose information about themselves at inappropriate times. Both groups drink excessive amounts of alcohol to reduce tension and anxiety (Brennan & Shaver, 1995).

Because current psychological state might bias recall of parental behavior, we cannot know for sure if adults' descriptions of their childhood attachment experiences are accurate (Fox, 1995). Also, we must keep in mind that quality of attachment to parents is not the only factor that influences later intimate ties. Characteristics of the partner and current life conditions can powerfully affect relationships. When adults with unhappy love lives have opportunities to form new, more satisfying intimate ties, they may revise their internal working models, thereby weakening the continuity between early and later relationships. Still, a wealth of evidence suggests that unfavorable childhood attachment experiences predispose people to conclude that they are undeserving of love or that their intimate partners cannot be trusted.

modern relationships. Early passionate love is a strong predictor of whether partners stay together. But without the quiet intimacy, predictability, and shared attitudes and values of companionate love, most romances eventually break up (Hendrick & Hendrick, 1992).

An ongoing relationship with a mate requires effort from both partners, as a study of newlyweds' feelings and behavior over the first year of marriage reveals. Husbands and wives gradually felt less "in love" and pleased with married life. A variety of factors contributed to this change. A sharp drop in time spent talking to one another and in doing things that brought each other pleasure (for example, saying "I love you" or making the other person laugh) occurred. In addition, although couples engaged in just as many joint activities at the beginning and end of the year, leisure pursuits gave way to household tasks and chores. Less pleasurable activities may have contributed to the decline in satisfaction (Huston, McHale, & Crouter, 1986).

In the transformation of romantic involvements from passionate to companionate, commitment may be the aspect of love that determines whether a relationship survives. Communicating that commitment—through warmth, sensitivity, caring, acceptance, and respect—can be of great benefit (Knapp & Taylor, 1994). For example, Sharese's doubts about getting married subsided largely because of Ernie's expressions of commitment. In the most dramatic of these, he painted a large sign on her birthday and placed it in their front yard. It read, "I LOVE SHARESE." Sharese returned Ernie's sentiments, and the intimacy of their bond deepened.

Couples who consistently express their commitment to each other report higher-quality relationships (Duck, 1994; Hecht, Marston, & Larkey, 1994). The Caregiving Concerns table below lists ways to help keep the embers of love aglow in a romantic partnership.

Finally, we have seen that love is multidimensional and can be experienced in different ways. Passion and intimacy, which form the basis of romantic love, do not figure as heavily into mate selection in all societies, as the Cultural Influences box on page 460 reveals.

FRIENDSHIPS

Like romantic partners and childhood friends, adult friends are usually similar in age, sex, and social class—factors that contribute to common interests, experiences, and needs and therefore to the pleasure derived from the relationship. Friends offer many of the same benefits in adulthood that they did in earlier years. They enhance self-esteem through affirmation and acceptance and provide social support during times of stress. Friends also make life more interesting by expanding social opportunities and access to knowledge and points of view.

Trust, intimacy, and loyalty continue to be important in adult friendships, as they were in middle childhood and adolescence. Sharing thoughts and feelings is sometimes greater in friendship than in marriage, although commitment is less strong as friends come and go over the life course. Even so, some adult friendships continue for many years. In one study of people age 60 and older, the majority reported having at least one friendship that lasted through-

CAREGIVING CONCERNS
Keeping Love Alive in a Romantic Partnership

SUGGESTION	DESCRIPTION
Make time for your relationship.	To foster relationship satisfaction and a sense of being "in love," plan regular times to be together during enjoyable activities.
Tell your partner of your love.	Express affection and caring, including the powerful words "I love you," at appropriate times. These messages increase perceptions of commitment and encourage your partner to respond in kind.
Be available to your partner in times of need.	Provide emotional support, giving of yourself when your partner is distressed.
Communicate constructively and positively about relationship problems.	When you or your partner are dissatisfied, suggest ways of overcoming difficulties, and ask your partner to collaborate in choosing and implementing a course of action.
Show an interest in important aspects of your partner's life.	Ask about your partner's work, friends, family, and hobbies, and express appreciation for his or her special abilities and achievements. In doing so, you grant your partner a sense of being valued.
Confide in your partner.	Share innermost feelings, keeping intimacy alive.
Forgive minor offenses, and try to understand major offenses.	Whenever possible, overcome feelings of anger through forgiveness. In this way, you acknowledge unjust behavior but avoid becoming preoccupied with it.

Sources: Donatelle & Davis, 1997; Enright, Gassin, & Wu, 1992; Knapp & Taylor, 1994.

Test Bank Items 14.38 through 14.41, 14.46 through 14.47

CULTURAL INFLUENCES

A CROSS-CULTURAL PERSPECTIVE ON LOVE

Romantic love did not become a primary basis for marriage until the eighteenth century. It became the dominant factor in Western nations during the twentieth century, as the value of individualism strengthened (Hatfield, 1993). In the United States, mature love is based on autonomy, appreciation of the partner's unique qualities, and intense emotional experience. When a person tries to satisfy dependency needs through an intimate bond, the relationship is usually regarded as immature (Dion & Dion, 1988).

This Western view of love contrasts sharply with the perspectives of Eastern cultures, such as China and Japan. In Japanese, *amae*, or love, means "to depend on another's benevolence." Dependency throughout life is recognized and viewed positively (Doi, 1973). The traditional Chinese collectivist view defines the self in terms of role relationships. A Chinese man considers himself a son, a brother, a husband, and a father; he rarely thinks in terms of an independent self (Chu, 1985).

In Chinese society, acceptance of dependency on others lessens the emotional intensity of any one relationship because such feelings are dis-tributed across a broader social network. In choosing a mate, a Chinese adult is expected to consider obligations to others, especially parents, in addition to personal feelings. As one writer summarized, "An American asks, 'How does my heart feel?' A Chinese asks, 'What will other people say?'" (Hsu, 1981, p. 50). Consistent with this difference, college students of Asian ethnicity are less likely than those of Caucasian-American, Cana-dian, or European descent to endorse a view of love based on physical attraction and intensity of emotion and more likely to stress companionship and practical considerations, such as similarity of background, career promise, and likelihood of being a good parent (Dion & Dion, 1993; Hendrick & Hendrick, 1986). Clearly, the interpersonal standards of our culture shape the way we view love.

This young Japanese man and woman exchange vows in a traditional Shinto ceremony. Their view of love and marriage is likely to be very different from that of most Western couples. (David Ball/The Image Cube)

out life (Roberto & Kimboko, 1989). Friendship continuity is greater for women, who also see their friends more often—a factor that helps maintain the relationship (Field & Minkler, 1988).

■ SAME-SEX FRIENDSHIPS. Throughout life, women continue to have more intimate same-sex friendships than do men. When together, female friends say they prefer to "just talk," whereas male friends say they like to "do something," such as play sports. Consequently, female friendships have been described as "face to face," male friendships as "side by side" (Wright, 1982). Men report barriers to intimacy with other men. For example, they indicate that they sometimes feel in competition with male friends and are therefore unwilling to disclose any weaknesses. And they also worry that if they tell about themselves, their friends may not reciprocate (Reid & Fine, 1992).

Of course, individual differences in friendship quality exist, to which gender-role identity and marital status contribute. Compared to traditionally oriented adults, androgynous men and women report disclosing more intimate information to their friends (Fischer & Narus, 1981; Lombardo & Lavine, 1981). And marriage reduces personal sharing between men, who seem to redirect their disclosures toward their wives.

■ OTHER-SEX FRIENDSHIPS. Other-sex friendships are also important in adulthood, although they occur less often and do not last as long as same-sex friendships. These bonds decline after marriage for men, but they increase with age for women, who tend to form them in the workplace. Highly educated, employed women have the largest number of other-sex friends. Through these relationships, young adults learn a great deal about masculine and feminine styles of intimacy. And because males confide especially easily in their female friends, men with other-sex friends are granted a unique opportunity to broaden their expressive capacity (Swain, 1992).

■ SIBLINGS AS FRIENDS. Whereas intimacy is essential to friendship, commitment—in terms of willingness to maintain a relationship and care about the other—is the defining characteristic of family ties. As young people marry and need to invest less time in developing a romantic partnership, siblings become more frequent companions than they were in adolescence. Often there is spillover between friend and sibling roles. For example, Sharese described Heather's practical assistance—helping with moving and running errands during an illness—in kin terms: "She's like a sister to me. I can always turn to her." And sibling ties in adulthood are often like friendships, where the main concern is keeping in contact, offering social support, and enjoying being together (O'Connor, 1992). Relationships between same-sex siblings can be especially close. Despite rivalries and differences in interests that emerged in childhood, a shared background of experiences within the family promotes similarity in values and perspectives and the possibility for deep mutual understanding.

Sibling relationships are among the longest we have in life. As we grow older, they become increasingly important sources of well-being. In Vaillant's (1977) study of well-educated men, the single best predictor of emotional health at age 65 was having had a close tie with a sibling in early adulthood. In another investigation, one-fifth of married women identified a sister as a best friend (Oliker, 1989).

LONELINESS

Because it is a time when people expect to form intimate ties, early adulthood is a vulnerable period for **loneliness**—unhappiness that results from a gap between social relationships we currently have and those we desire. Adults may feel lonely because they do not have an intimate partner or because they lack gratifying friendships. Both situations give rise to similar emotions, but they are not interchangeable (Brehm, 1992). For example, even though she had several enjoyable friendships, Heather felt lonely from time to time because she was not dating someone she cared about. And although Sharese and Ernie were happily married, they felt lonely after moving to a new town where they did not know anyone.

Loneliness is at its peak during the late teens and early twenties, after which it declines steadily into the seventies (Brehm, 1992; Liefbroer & de Jong-Gierveld, 1990). This is understandable, since young people must constantly develop new relationships as they move through school

and employment settings. Also, younger adults may expect much more from their intimate ties than do older adults, who have learned to live with imperfections.

Who is most likely to experience loneliness, and under what conditions? Separated, divorced, or widowed adults are lonelier than their married, cohabiting, or single counterparts, suggesting that loneliness is especially intense after loss of an intimate tie (Liefbroer & de Jong-Gierveld, 1990; Rubenstein & Shaver, 1982). When not involved in a romantic relationship, men feel lonelier than women, perhaps because they have fewer alternatives for satisfying intimacy needs. In marriages, wives report greater loneliness than husbands, especially if they are unemployed, have recently moved, or have young children—circumstances that limit their access to a wider social network (Fischer & Phillips, 1982).

Personal characteristics also contribute to the likelihood of being lonely. Temperament is involved, since shy, socially anxious people report more loneliness (Bruch et al., 1989; Cheek & Busch, 1981). When loneliness persists, it is associated with a wide variety of self-defeating attitudes and behaviors. Lonely people evaluate themselves and others more negatively, tend to be socially unresponsive and insensitive, and are slow to develop intimacy because they are reluctant to tell others about themselves. The extent to which these responses are cause or consequence of loneliness is unclear, but once in place, they certainly promote further isolation (Jones, 1990).

■ Loneliness is at its peak during the late teens and early twenties. It usually stems from lack of an intimate partner or gratifying friendships. When loneliness persists, it is associated with negative self-evaluations and socially unresponsive and insensitive behavior—responses that promote further isolation. (Ron Coppock/Liaison International)

Although loneliness is extreme for some people, most young adults encounter it from time to time as they struggle with unfulfilled relationship desires. As long as it is not overwhelming, loneliness can motivate young people to take social risks and reach out to others. It can also encourage them to find ways to be comfortably alone and use this time to deepen self-understanding. Much of healthy personality development involves striking this balance—between "[developing] satisfying relationships with other people and [creating] a secure, internal base of satisfaction within ourselves" (Brehm, 1992, p. 345).

BRIEF REVIEW

Finding an intimate partner to share one's life is a major task of early adulthood. The greater the resemblance in background and psychological characteristics, the more likely two people are to fall in love and sustain their relationship. As romantic ties develop, passion gives way to warm, affectionate companionship.

Like friendships in middle childhood and adolescence, adult friendships enhance self-esteem, provide social support, and are based on trust, intimacy, and loyalty. Women continue to have more intimate same-sex friendships than men, although an androgynous gender-role identity enhances friendship closeness in both sexes. Cross-sex friendships decline after marriage for men, but increase with age for women. Sibling companionship strengthens and is a powerful predictor of psychological well-being. As young people strive to develop gratifying intimate ties, they are vulnerable to loneliness.

ASK YOURSELF . . .

- *After living together for a year, Mindy and Graham wondered why their relationship seemed less passionate and satisfying. What factors probably contributed to this change? Suggest ways that Mindy and Graham can breathe new life into their romantic partnership.*

- *Claire and Tom, both married to other partners, got to know each other at work and occasionally have lunch together. What is each likely to gain from this other-sex friendship?*

THE FAMILY LIFE CYCLE

For the majority of young people, the quest for intimacy leads to marriage. Their life course takes shape within the **family life cycle**—a sequence of phases that characterizes the development of most families

around the world (Framo, 1994). In early adulthood, people typically live on their own, marry, and bear and rear children. As they become middle aged and their children leave home, their parenting responsibilities diminish. Late adulthood brings retirement, growing old, and (mostly for women) death of one's spouse (Duvall, 1977; McGoldrick, Heiman, & Carter, 1993).

However, we must be careful not to think of the family life cycle as a fixed progression. Recall from Chapter 2 that the family is a dynamic system of interdependent relationships embedded in community, cultural, and historical contexts. All of these factors affect the way it changes over time. Today, wide variations in sequence and timing of these phases exist. High rates of out-of-wedlock births, delayed childbearing, divorce, and remarriage are but a few illustrations. And some people—either voluntarily or nonvoluntarily—do not go through some or all of the family life cycle.

Still, the family life cycle approach is useful. It provides us with an organized way of thinking about how the family system changes over time and the impact of these changes on both the family unit and the individuals within it. Each phase requires that roles be modified and needs be met in new ways.

LEAVING HOME

After Sharese had been away at college for 6 months, she noticed a change in the way she related to her mother. She found it more enjoyable to discuss daily experiences and life goals, sought advice and listened with greater openness, and expressed affection more freely. Over the next few years, Sharese's bedroom gradually began to seem more like a guest room. Looking around before she moved out permanently, Sharese felt some pangs of nostalgia for the warmth and security of her childhood days coupled with a sense of pride at being on her own.

Departure from the parental home is a major step in assuming adult responsibilities. The average age of leaving has decreased in recent years as more young people live independently before marriage. In 1940, over 80 percent of American 18- to 24-year-olds resided with their parents. Today, only about 50 percent do. Most industrialized nations show this trend (Kerckhoff & Macrae, 1992; White, 1994).

Timing of departure varies with the reason for leaving. Departures for education tend to be at younger ages, those for full-time work and marriage later. Since a larger percentage of young adults go to college in the United States than other nations, more Americans leave home early, around age 18. Still, studies of nationally representative samples in Australia, Great Britain, and the United States show that marriage is the most common reason for departure in all three nations. And because women usually marry at a younger age than men, they are likely to leave home sooner. Many young people also leave to be independent or to get away from friction at home (Kerckhoff & Macrae, 1992).

Nearly half of young adults return home for a brief time after initial leaving. Those who departed to marry are least likely to return. But premarital independent living is a fragile arrangement. As people encounter unexpected twists and turns on the road to independence, the parental home serves as a safety net and base of operation for launching adult life. Failures in work or marriage can prompt a move back home. Also, young people who left because of family conflict usually return—largely because they were not ready for independent living. But most of the time, role transitions, such as the end of college or military service and the beginning of full-time work, bring people back. Contrary to popular belief, returning home is usually not a sign of weakness. Instead, it is a common and ordinary event among unmarried adults (DaVanzo & Goldscheider, 1990; White, 1994).

Although most high school seniors expect to live on their own before marriage, the extent to which they do so varies with social class and ethnicity. Economically well off young people are more likely to establish their own residence. Among African Americans and Hispanics, poverty and a cultural tradition of extended family living lead to low rates of home leaving. Unmarried Asian Americans also tend to live with their parents. But the longer Asian families have been in the United States and are exposed to individualistic values, the more likely their children are to move out after finishing high school (Goldscheider & Goldscheider, 1993).

When young people are prepared for independence, departure from the home is linked to more satisfying parent–child interaction and successful transition to adult roles (Bloom, 1987). However, leaving home very early may contribute to long-term disadvantage, since it is associated with an emphasis on employment rather than education and lack of parental financial assistance, advice, and emotional support. Not surprisingly, those who depart at a young age have less successful marriages and work lives (White, 1994).

Besides timing of departure, personal characteristics make a difference. For example, young adults who perceive themselves as adaptable to change and in control of their environments and who find a middle ground between closeness and distance in family relationships adjust especially well to the first year of college. For men, attaining this middle ground often means developing themselves emotionally and expressively. For women, it usually means acquiring a firmer sense of autonomy (Holmbeck & Wandrei, 1993). Interventions directed at helping young people reevaluate traditional gender roles may be especially valuable at this time.

JOINING OF FAMILIES IN MARRIAGE

The United States remains a culture strongly committed to marriage. Nearly 90 percent of Americans marry at least once in their lives. Young adults wait considerably longer before marrying today than they did at mid-century. In 1950, the average age of first marriage was 20.3 for women and 22.8 for men; in 1994, it was 24.5 and 26.5, respectively (U.S. Bureau of the Census, 1996). As Figure

14.1 on page 464 shows, the number of first and second marriages has declined over the last few decades as more people remain single, cohabit, or do not remarry after divorce. Still, both marriage and divorce rates remain high.

Marriage is often thought of as the joining of two individuals. In actuality, it requires that two entire systems—the husband's and wife's families—adapt and overlap to create a new subsystem. Consequently, marriage presents couples with complex challenges, especially today when husband–wife roles have only begun to move in the direction of a true partnership—educationally, occupationally, and in emotional connectedness (Eisler, 1987; McGoldrick, Heiman, & Carter, 1993).

■ MARITAL ROLES. Their wedding and honeymoon over, Sharese and Ernie turned to a myriad of issues that they had previously decided individually or that their families of origin had prescribed. They had to consider everyday matters—when and how to eat, sleep, talk, work, relax, have sex, and spend money. They also had to decide which family traditions and rituals to retain and which to develop for themselves. And relationships with parents, siblings, extended family, friends, and co-workers had to be renegotiated.

Recent alterations in the context of marriage, including changing gender roles and living farther away from family members, mean that modern couples must do more work to define their relationship. Although husbands and wives are usually similar in religious and ethnic background, "mixed" marriages occur more often today than they did in the past. Since 1970, other-race unions have quadrupled, affecting over 1 million American couples, or 2 percent of the married population (U.S. Bureau of the Census, 1996). More young people also choose a mate of a different religion than their own. For example, between one-third and one-half of American Jews who marry today select a non-Jewish spouse (Greenstein, Carlson, & Howell, 1993). When their backgrounds are very distinct, these couples face extra challenges in achieving a successful transition to married life.

Many modern couples live together before marriage, making it less of a turning point in the family life cycle than in the past. Still, the burden of defining marital roles can be great. American women who marry before age 20 (25 percent) are twice as likely to divorce as those who marry in their twenties. Women who marry after age 30 (20 percent) are least likely to divorce, but if they do, they end the marriage sooner (McGoldrick, Heiman, & Carter, 1993).

These findings suggest that it is better to marry later than earlier. They also indicate that people who fall outside the normative age range for marriage often face stresses that make the transition more difficult. Those who marry early may be running away from their own family or seeking the family they never had. Most have not developed a secure enough identity or sufficient independence to be ready for a mature marital bond. Early marriage followed by childbirth and reversals of family life cycle events

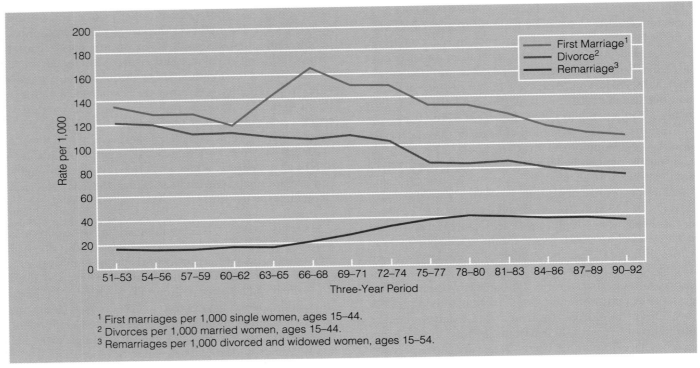

FIGURE 14.1

Rates of first marriage, divorce, and remarriage for women, 1951 to 1992. The number of first marriages has dropped steadily since mid–century. After increasing between 1960 and 1980, the divorce rate stabilized and then declined slightly. Remarriages rose sharply in the 1960s as divorce climbed. Since then, they have decreased. Still, both marriage and divorce rates are high in the United States. *(From U.S. Bureau of the Census, 1992a.)*

(childbirth before marriage) are more common among low-income young people. This acceleration of family formation complicates adjustment to life as a couple (Fulmer, 1989). People who marry very late—most of whom are economically advantaged adults—are sometimes ambivalent about compromising their independence or in conflict about marriage and career.

Despite progress in the area of women's rights, **traditional marriages** still exist in Western nations. In this form of marriage, there is a clear division of husband's and wife's roles. The man is the head of household; his primary responsibility is to provide for the economic well-being of his family. The woman devotes herself to caring for her husband and children and to creating a nurturant, comfortable home. However, traditional marriages have changed in recent decades. Motherhood remains a full-time job—or at least the top priority—while children are young, but many women return to their occupations at a later date. Unlike previous generations, their role is not fixed throughout adult life.

Egalitarian marriages reflect the values of the women's movement. Husband and wife relate as equals, and power and authority are shared. Both partners try to balance the time and energy they devote to the workplace, the children, and their relationship. Well-educated, career-oriented couples tend to adopt this form of marriage—especially before

they have children. But at least so far, women's employment has not had a dramatic effect on household division of labor. Men in these families participate more than do those in single-earner families. But their wives continue to do the bulk of the housework, averaging almost three times as many hours as their husbands (see Figure 14.2). True equality in marriage is still rare, and couples who strive for it more often attain a form of marriage in between traditional and egalitarian (Starrels, 1994).

■ MARITAL SATISFACTION. Despite its rocky beginnings, Sharese and Ernie's marriage grew to be especially happy. In contrast, Christy and Gary became increasingly discontent. What distinguishes marriages high in satisfaction from less successful partnerships? Differences between the two couples mirror the findings of a large body of research, summarized in Table 14.2.

Christy and Gary had a brief courtship, married at a young age, had children early, and struggled financially. Gary's negative, critical personality led him to get along poorly with Christy's parents and to feel threatened when he and Christy disagreed. Christy tried her best to offer Gary encouragement and support, but her own needs for nurturance and individuality were not being met. Gary felt threatened by Christy's career aspirations. As she came closer to attaining them, the couple grew further apart. In contrast,

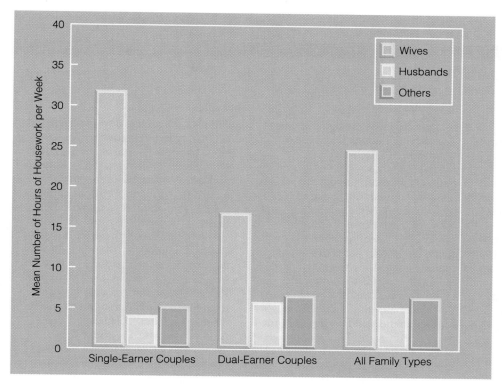

FIGURE 14.2

Hours of housework done by wives, husbands, and other family members for single-earner couples (only the husband is employed) and dual-earner couples (both husband and wife are employed). Wives in dual-earner households continue to do most of the work, averaging almost three times as many hours as their husbands. Total number of hours devoted to domestic work is lowest in these families because they have fewer children and tend to hire household help. In addition, many employed women handle work overload by reducing time devoted to housework. *(Adapted from Berardo, Shehan, & Leslie, 1987; Starrels, 1994.)*

Sharese and Ernie married later, after their education was complete. They postponed having children until their careers were underway and they had built a sense of togetherness that allowed each to thrive as an individual. Patience, caring, shared values, enjoyment of each other's company, and good conflict resolution skills contributed to their compatibility.

Although the factors just described differentiate troubled from gratifying marital relationships, research also reveals clear sex differences in marital satisfaction. Many more men than women report being happily married (Holahan, 1984; Kaslow, Hansson, & Lundblad, 1994; Levenson, Carstensen, & Gottman, 1993). And marriage is associated with gains in men's mental and physical health, whereas married women are less healthy than single women in every respect—emotionally, physically, even in crime statistics (see the Social Issues Box on pages 466-467). Marriage can take a heavy toll on women who are overwhelmed by the demands of husband, children, household, and career. And when a marriage is not going well, women are more willing to evaluate the relationship as problematic and try to work on it. Men often withdraw from conflict, magnifying the stress on their wives (McGoldrick, 1989).

TABLE 14.2

Factors Related to Marital Satisfaction

FACTOR	HAPPY MARRIAGE	UNHAPPY MARRIAGE
Family backgrounds	Similar in social class, education, religion, and ages of partners	Very different in social class, education, religion, or ages of partners
Age of marriage	After age 20	Before age 20
Length of courtship	At least 6 months	Less than 6 months
Timing of first pregnancy	After first year of marriage	Before or within first year of marriage
Relationship to extended family	Warm and positive	Negative, wish to maintain distance
Marital patterns in extended family	Stable	Unstable; frequent separations and divorces
Financial and employment status	Secure	Insecure
Personality characteristics	Emotionally positive; good conflict resolution skills	Emotionally negative and impulsive; poor conflict resolution skills

Note: The more factors present, the greater the likelihood of marital happiness or unhappiness.

Sources: Levenson, Carstensen, & Gottman, 1993; McGoldrick, Heiman, & Carter, 1993; Russell & Wells, 1994; Skolnick, 1981.

SOCIAL ISSUES

SPOUSE ABUSE

Violence has reached epidemic proportions in the United States, permeating many families. In previous chapters, we discussed domestic violence several times—child maltreatment in Chapter 8, child sexual abuse in Chapter 10, and marital sexual assault in Chapter 13. Here we focus on spouse abuse, estimated to affect over 1.8 million women annually (Carden, 1994).

Within a given family, violence in one form predicts its occurrence in another. Recall the story of Karen in Chapter 13. Her husband Mike not only assaulted her sexually and physically, but abused her psychologically—isolating, humiliating, and demeaning her. Property destruction can occur as well. A violent husband may break his wife's favorite possessions, punch holes in walls, or throw things. Tactics used to force the woman into submission are diverse. If children are present, they may also be victims of abuse as well as pawns in the husband's effort to intimidate his wife (Carden, 1994; Jouriles & Le Compte, 1991).

FACTORS RELATED TO SPOUSE ABUSE. Karen's experience helps us understand how wife battering may emerge and escalate. During their courtship, Mike felt suspicious of Karen's friends, so she began to distance herself from them. Soon he was her only companion. Shortly after their wedding, Mike's abuse began. First, he spewed insults about her family and co-workers. Then he started to restrict her activities, forbidding her to leave the house, controlling her access to money, and complaining so much about her job that she finally quit. Anxious and lonely, Karen spent her days at home and gained weight. The first time Mike struck her, he yelled about her careless housekeeping and her unkempt appearance. The next morning, Mike apologized and promised not to hurt her again. But his violence continued, becoming more frequent and extreme. Most of the time, it was followed by a brief period in which he went out of his way to atone for his loss of control.

These abuse–remorse cycles in which aggression escalates characterize many batterers (Walker, 1979). Why do some men behave this way? As with other forms of domestic violence, the reasons are multiple and complex. Recall that a woman's characteristics do not predict sexual coercion. The same is true of spouse abuse (Hotaling & Sugarman, 1986). Rather, personality and developmental history of the husband, family circumstances, and cultural factors combine to make spouse abuse more likely.

Batterers usually have serious psychological problems. Many are overly dependent on their wives as well as jealous and possessive. The thought of Karen leaving induced such high anxiety in Mike that he resorted to violence to keep her (Allen et al., 1989). At the same time, these men are ambivalent about intimacy; they want but fear the intensity and sharing of a close relationship. An excessive need to be in control can be seen in the husband's desire to make all family decisions and monitor everything his wife does. Dissatisfaction with life, both at home and at work, is manifested in depression, anxiety, and low self-esteem. Trivial events, such as an unironed shirt or a late meal, can set off violent episodes, indicating a tendency to react to frustration with hostility and great difficulty managing anger (Else et al., 1993; Vaselle-Augenstein & Ehrlich, 1992).

■ MARITAL EXPECTATIONS AND MYTHS. In a recent study in which 50 happily married couples were interviewed about their marriages, each participant reported good times and bad; none was happy all the time. Many admitted that there were moments when they wanted out, when they felt they had made a mistake. Clearly, marital happiness did not signify a "rose garden." Instead, it was grounded in mutual respect, pleasure and comfort in each other's company, and joint problem solving. All couples emphasized the need to reshape their relationship in response to new circumstances and each partner's changing needs and desires (Wallerstein & Blakeslee, 1995).

Yet cultural expectations work against this view of marriage as an ongoing project requiring both partners' involvement and cooperation. Historically, women had little power in marriage and society; a wife's status came from her husband. This gender gap has such deep cultural roots that it continues to influence marital expectations today. In a study of college students, more women than men said their partners should be superior to themselves, and more men than women said their partners should be inferior to themselves—in intelligence, education, vocational success, and income (Ganong & Coleman, 1992). Under these circumstances, women are likely to play down their abilities, sacrificing part of themselves. And men tend to limit themselves to the provider role rather than participating fully in family life.

Furthermore, many young people have a mythical image of marital bliss—one that is a far cry from reality. In a recent survey, a substantial number of college students endorsed the following beliefs not supported by facts:

A high proportion of spouse abusers experienced or witnessed abuse in their homes as children (Hotaling & Sugarman, 1986; Tolman & Bennett, 1990). Men who were exposed to domestic violence are not doomed to repeat it. But their childhoods provided them with expectations and behaviors to model in their future close relationships. Stressful life events, such as job loss or financial difficulties, increase the likelihood of battering, although it occurs in couples of all social-class and ethnic backgrounds. Alcohol abuse is also related to it, but rather than triggering violence, drinking probably helps the perpetrator avoid responsibility for his behavior (Vaselle-Augenstein & Ehrlich, 1992).

At a societal level, cultural norms that endorse male dominance and female submissiveness and use of physical force to preserve this inequality promote spouse abuse. Battering is far more likely to occur when a marriage is husband controlled, when the man holds traditional gender-role beliefs, and when he approves of violence as a way to solve family problems (Stith & Farley, 1993).

Why don't women leave these destructive relationships before the abuse escalates? A variety of situational factors discourage them from

doing so, including dependence on the husband's greater earning power; fear of retaliation against herself or her children; hope, based on promises after each explosive episode, that he will change; and the shame and embarrassment associated with going to the police (Carden, 1994). Furthermore, anxiety, depression, and physical injury can prevent an abused woman from thinking clearly about how to get help.

INTERVENTION AND TREATMENT. Community services available to victims of sexual coercion also provide shelter, protection, and advocacy for battered women (see pages 433–434). Because many return to their abusive partners several times before making their final move, community agencies usually offer treatment for batterers. The therapies available are diverse, and no single approach has been shown to be superior. Most rely on group sessions that confront rigid gender stereotyping; teach communication, problem solving, and anger control; and use social support to foster behavior change. Sometimes couple therapy is used, but only after the abuse has stopped (Harway & Hansen, 1994).

Although existing treatments are far better than no treatment, almost

all are too brief to pay sufficient attention to alcohol problems and marital difficulties. Consequently, of the small number of men who agree to participate, at least half continue their violent behavior with either the same or a new partner. At present, we have only a beginning understanding of how best to intervene in spouse abuse (Carden, 1994; Vaselle-Augenstein & Ehrlich, 1992).

TRY THIS...

- Return to Chapter 8, pages 268–269, and reread the section on child maltreatment. What predisposing factors do child abuse and spouse abuse have in common?

- What services are available in your community for battered women? Contact one of them and ask for a description. Is treatment for abusers provided? If so, how comprehensive is it? Does it address influences at the level of the individual, the family, and society?

- A couple's satisfaction increases through the first year of marriage.

- The best single predictor of marital satisfaction is the quality of a couple's sex life.

- If my spouse loves me, he or she should instinctively know what I want and need to be happy.

- No matter how I behave, my spouse should love me simply because he or she is my spouse. (Larson, 1988, p. 5)

As these myths are overturned, couples react with disappointment, and marriage becomes less satisfying and more conflictual.

In view of its long-term implications, it is surprising that most couples spend little time reflecting on the decision to marry before their wedding day (McGoldrick, Heiman, &

Carter, 1993). Courses in family life education in high schools and colleges can help dispel marital myths. More realistic expectations, in turn, can promote better mate selection and ease adjustment to marriage (Honeycutt, 1991).

PARENTHOOD

In the past, the issue of whether to have children was, for many adults, "a biological given or an unavoidable cultural demand" (Michaels, 1988, p. 23). Today, in Western industrialized nations, it is a matter of true individual choice. Effective birth control techniques enable adults who do not want to become parents to avoid having children in most instances. And changing cultural values allow people to remain childless with less fear of social criticism and rejection than was the case a generation or two ago.

In 1950, 78 percent of married couples were parents. Today, 72 percent of couples bear children, and they tend to have their first child at a later age. Consistent with this pattern of delayed childbearing, family size in industrialized nations has declined. In 1950, the average number of children per couple was 3.1. Currently, it is 2.1, a downward trend that is expected to continue into the twenty-first century (U.S. Bureau of the Census, 1996).

Nevertheless, the vast majority of married people continue to embrace parenthood as one of life's most meaningful experiences. Why do they do so, and how do the challenges of child rearing affect the adult life course?

■ THE DECISION TO HAVE CHILDREN. The choice of parenthood is affected by a complex array of factors, including financial circumstances, personal and religious values, and biological and medical conditions. Overall, women with traditional gender-role orientations are more likely to have children. Whether a woman is employed has less impact on her decision than her occupation. Women who work in managerial positions are less likely to become parents than women in less demanding careers (White & Kim, 1987).

When American couples are asked about their desire to have children, they mention a variety of advantages and disadvantages, which are listed in Table 14.3. Take a moment to consider which ones are most important to you. Although some ethnic and regional differences exist, reasons for having children that are most important to all groups include the desire for a warm, affectionate relationship and the stimulation and fun that children provide. Also frequently mentioned are growth and learning experiences that children bring into the lives of adults, the desire to have someone carry on after one's own death, and the feelings of accomplishment and creativity that come from helping children develop (Hoffman, Thornton, & Manis, 1978; Michaels, 1988).

Most young adults are also aware that having children means years of extra burdens and responsibilities. When asked about the disadvantages of parenthood, they mention "loss of freedom" most often, followed by "financial strain." Indeed, the cost of child rearing is a major factor in modern family planning. According to a conservative estimate, parents will spend about $230,000[1] to rear a child from birth through four years of college. Finally, many adults worry greatly about bringing children into a troubled world—one filled with crime, war, and pollution (Michaels, 1988).

Careful weighing of the pros and cons of having children is increasingly common today. This means that many more couples are making informed and personally meaningful choices about becoming parents—a trend that should increase the chances that they are ready to have children and that their own lives will be enriched by their decision.

■ ADJUSTMENT TO PARENTHOOD. Childbirth profoundly alters the lives of husband and wife. Disrupted sleep schedules, less time to devote to each other and to leisure activities, and new financial responsibilities often lead to a mild decline in marital happiness. In addition, entry of children into the family usually causes the roles of husband and wife to become more traditional (Cowan & Cowan, 1988, 1992; Klinnert et al., 1992; Palkovitz & Copes, 1988). This is true even for couples like Sharese and Ernie, who were strongly committed to gender-role equality and used to sharing household tasks. Movement toward traditional roles is hardest on new mothers who have been involved in a career. The larger the difference in men's and women's responsibilities, the more conflict increases and

[1] This figure is based on a 1988 estimate, corrected for later inflation (Glick, 1990; U.S. Department of Labor, 1996). It includes basic expenses related to food, housing, clothing, medical care, and education.

TABLE 14.3

Advantages and Disadvantages of Parenthood Mentioned by Modern American Couples

ADVANTAGES	DISADVANTAGES
Giving and receiving warmth and affection	Loss of freedom, being tied down
Experiencing the stimulation and fun that children add to life	Financial strain
Being accepted as a responsible and mature member of the community	Worries over children's health, safety, and well-being
Experiencing new growth and learning opportunities that add meaning to life	Interference with mother's employment opportunities
Having someone carry on after one's own death	Risks of bringing up children in a world plagued by crime, war, and pollution
Gaining a sense of accomplishment and creativity from helping children grow	Reduced time to spend with spouse
Learning to become less selfish and to sacrifice	Loss of privacy
Having offspring who help with parents' work or add their own income to the family's resources	Fear that children will turn out badly, through no fault of one's own

Sources: Hoffman, Thornton, & Manis, 1978; Michaels, 1988.

marital satisfaction and mental health decrease after child-birth, especially for women (Belsky et al., 1991; Hawkins et al., 1993; Levy-Shiff, 1994).

Violated expectations about jointly caring for a new baby contribute to the decline in marital happiness just mentioned. Women, especially, count on far more help from their husbands than usually occurs (Hackel & Ruble, 1992). Postponing childbearing until the late twenties or thirties, as more couples are doing today (see Chapter 13, page 424), eases the transition to parenthood. Waiting permits couples to pursue occupational goals and gain life experience. Under these circumstances, men are more enthusiastic about becoming fathers and therefore more willing to participate actively. And women whose careers are underway are more likely to encourage their husbands to share housework and child care (Coltrane, 1990).

Men who view themselves as especially nurturant and caring show less decline in marital satisfaction after the birth of a baby, probably because they are better at meeting the needs of their wives and infants. The father's involvement may increase his understanding of the challenges a mother faces in coping with a new baby, reduce his feelings of being an "outsider" as the infant demands his wife's attention, and free up time for partners to spend together. Also, men's participation enhances the marital relationship because women tend to see it as a loving act toward themselves (Levy-Shiff, 1994).

In many non-Western cultures, the birth of children is less likely to threaten marital satisfaction. In these societies, parenthood is highly valued, family life is central for women, and traditional gender roles are widely accepted. Consequently, becoming a mother grants a woman considerable status, and husband–wife division of labor is not questioned. In addition, the extended family typically assists with household and child care tasks (Levy-Shiff, 1994; Lozoff, Jordan, & Malone, 1988). In Western industrialized nations, however, the trend toward gender equality and isolation of the nuclear family unit leads marital and parenting roles to be closely linked; happiness in one profoundly affects happiness in the other.

■ ADDITIONAL BIRTHS. How many children a couple chooses to have is affected by the same array of factors that influenced their decision to become parents in the first place. Besides more effective birth control, a major reason that family size has declined in industrialized nations is the increased career orientation of many women. Also, the high divorce rate means that many couples do not complete their childbearing plans.

Research indicates that adults and children benefit from small family size. Because parents are less economically and emotionally stressed, they are more patient with each other and have more time to devote to each child's development. Furthermore, in smaller families, siblings are more likely to be widely spaced (born more than 2 years apart), which adds to the attention and resources husband

and wife can invest in one another and in each child. Together, these findings may account for the fact that marital satisfaction tends to be greater and children tend to have higher IQs, do better in school, and attain higher levels of education in smaller families (Anderson, Russell, & Schumm, 1983; Blake, 1989; Powell & Steelman, 1993).

■ FAMILIES WITH YOUNG CHILDREN. With the entry of children, the family system becomes a permanent one for the first time. Before parenthood, if a spouse leaves, the system dissolves; afterward, it continues. In this way, the arrival of children constitutes a key change in the family life cycle.

A year after the birth of their first child, Sharese and Ernie received a phone call from Heather, who asked how they liked being parents: "Is it a joy, a dilemma, a stressful experience—how would you describe it?"

Chuckling, Sharese and Ernie responded in unison, "All of the above!"

Child rearing is an enormous task, and seldom are young people fully prepared for it. In today's complex

Despite its many challenges, rearing young children is a powerful source of adult development. Parents report that it expands their emotional capacities and enriches their lives. Involved parents say that child rearing helped them become more sensitive, tolerant, self-confident, and responsible. (James Wilson/Woodfin Camp & Associates)

world, men and women are not as certain about how to rear children as they were in previous generations. Clarifying child-rearing values and implementing them in warm, supportive, and appropriately demanding ways are crucial for the welfare of the next generation and society. Yet cultures do not always place a high priority on parenting, as indicated by the lack of many societal supports for children and families in the United States (see Chapter 2, pages 64–66). Furthermore, changing family forms mean that the lives of modern parents differ substantially from those of past generations.

In previous chapters, we discussed a wide variety of influences on child-rearing styles—personal characteristics of children and parents, family economic conditions, social class, ethnicity, and more. The marital relationship is also important. Husbands and wives who cooperate and respond to each other's needs are more likely to be sensitive to their children. Support from a spouse can even reduce the disruptive impact of stressful life events, economic strain, and parental depression on child rearing (Simons et al., 1990, 1992, 1993).

For employed parents, a major struggle during this phase of family life is finding good day care. The younger the child, the greater the parents' sense of risk and difficulty finding the help they need (Pleck, 1985). When competent, convenient, affordable day care is not available, it usually leads to additional pressures on the woman. Either she must curtail or give up her career, or she must endure unhappy children, missed workdays, and constant searches for new arrangements.

Despite its many challenges, rearing young children is a powerful source of adult development. Parents report that it expands their emotional capacities and enriches their lives. For example, Ernie remarked that he was very "goal oriented" before his children were born but felt "rounded out" by sharing in their care. Other involved parents say that child rearing helped them tune into others' feelings and needs and become more tolerant, self-confident, and responsible (Coltrane, 1990).

■ FAMILIES WITH ADOLESCENTS. Adolescence brings sharp changes in parental roles. In Chapters 11 and 12, we noted that parents of adolescents must a establish a new relationship with their children—blending guidance with freedom and gradually relinquishing control. As adolescents gain in autonomy and explore values and goals in their search for identity, parents often complain that their teenager is too focused on peers and no longer seems to care about the family.

Flexibility is the key to family success during this period. As adolescents move from childhood to adulthood, they alternately become dependent when they cannot handle things alone and experiment with greater independence. Their behavior seems unpredictable and ever-changing. When parents try to tighten the reins by disciplining in ways

appropriate for younger children or withdraw to avoid conflict, they heighten family tensions. Although most couples handle these difficulties, others reach impasses. More people seek or are referred for family therapy during this phase of the family life cycle than any other (Young, 1991).

■ PARENT EDUCATION. In the past, when there was greater continuity across generations in family life, adults learned what they needed to know about parenting through modeling and direct experience. Today's world confronts husbands and wives with a host of societal factors that impinge on their ability to succeed as parents. In addition, the scientific literature on child development has mushroomed, much of it offering practical implications that can help parents do a better job of rearing children.

Modern adults eagerly seek information on child rearing through popular books. New mothers regard these sources as particularly valuable, second in importance only to their doctors (Deutsch et al., 1988). Special courses have also emerged, designed to help parents understand how children develop, clarify child-rearing values, explore family communication, and apply more effective parenting strategies. Day care centers, schools, churches, community health centers, and colleges often sponsor these parent education programs. A variety of positive outcomes have been demonstrated, including improved parent–child interaction, more open and flexible parent attitudes, and heightened awareness by parents of their role as educators of their children (Powell, 1986). Although these courses take a variety of forms, at present there is no convincing evidence that any one approach is best (Todres & Bunston, 1993). Perhaps the most effective programs adapt their goals and procedures to fit the specific needs and characteristics of their participants.

BRIEF REVIEW

In modern industrialized nations, the sequence and timing of phases of the family life cycle varies widely. Departure from the parental home usually marks the beginning of adult responsibilities, although many young people return temporarily as they launch their adult lives. Today, marriage requires that young adults work harder to define their relationship. Traditional marriages have changed as many full-time mothers return to their occupations when their children are older. Well-educated, career-oriented couples are more likely to have egalitarian marriages. However, true equality, especially in household division of labor, is rare. Consequently, men are more satisfied with and benefit more from marriage than do women.

The demands of new parenthood often lead to a mild drop in marital happiness, and family roles become more traditional. A positive marital relationship promotes satisfaction with parenthood and effective rearing of young children. When children become adolescents, parents must cope with their unpredictable behavior and growing independence. Marital happiness is lowest at this time.

ASK YOURSELF . . .

■ *After her wedding, Sharese was convinced she had made a mistake and never should have gotten married. Cite factors that helped sustain her marriage to Ernie and that led it to become especially happy.*

■ *Suggest several ways in which a new mother and father can help each other make an effective adjustment to parenthood.*

THE DIVERSITY OF ADULT LIFESTYLES

The modern array of adult lifestyles arose in the 1960s, a decade of rapid social change in which American young people began to question the conventional wisdom of previous generations. Many asked, How can I find happiness? What kinds of commitments should I make to live a full and rewarding life? As the public became more accepting of diverse lifestyles, choices seemed more available than in the past—among them, staying single, cohabiting, remaining childless, and divorcing.

Today, nontraditional family options have penetrated the American mainstream. Many adults experience not just one, but several. As we consider these variations in the following sections, we will see that the factors that induce people to enter them are complex. Some adults make a deliberate decision to adopt a particular lifestyle, whereas others drift into it. The lifestyle may be imposed by society, as is the case for cohabiting homosexual couples, who cannot marry legally. Or people may decide on a certain lifestyle because they feel pushed away from another, such as a marriage gone sour. Thus, the adoption of a lifestyle can be within or beyond the person's control.

SINGLEHOOD

On finishing her education, Heather joined the Peace Corps and spent 5 years in Africa. Although open to a long-term relationship with a man whose sense of adventure equaled her own, she had only fleeting romances. When she returned to the United States, she accepted an executive position with an insurance company. Professional challenge and travel preoccupied her. At age 35, she reflected on her life circumstances over lunch with Sharese: "I was open to marriage, but after my career took off, it would have interfered. Now I'm so used to independence that it would take a lot of adjustment to live with another person. I like being able to pick up and go where I want, when I want, without having to ask anybody or think about having to take care of anybody. But there's a tradeoff: I sleep alone, eat most of my meals alone, and spend a lot of my leisure time alone."

Singlehood characterizes individuals not living with an intimate partner. It has increased in recent years, especially among young adults. For example, never-married 30- to 34-year-olds have tripled since 1970, rising to 19 percent among males and 13 percent among females. Besides more people marrying later or not at all, divorce has added to the rate of singlehood. In view of these trends, it is likely that most Americans will spend a substantial part of their adult lives single, and a growing minority—about 10 to 12 percent—will stay that way (U.S. Bureau of the Census, 1996).

Because they marry later, more men than women are single in early adulthood. But women are far more likely than men to remain single for many years or their entire life. As women get older, there are fewer men available with characteristics that most look for in a mate—the same age or older, better educated, and more professionally successful. Men—whether never married, divorced, or widowed—find partners more easily, since they can select from a large pool of younger unmarried women. Because of the tendency for women to "marry up" and men to "marry down," men in blue-collar occupations and women in highly demanding, prestigious careers are overrepresented among singles after age 30.

Ethnic differences also exist. For example, the percentage of never-married African Americans is nearly twice as great as that of Caucasian Americans in early adulthood. As we will see later, high unemployment among black men interferes with marriage. But many African Americans eventually marry in their late thirties and forties, a period in which the black and white marriage rates come closer together (Cherlin, 1992; U.S. Bureau of the Census, 1996).

Singlehood is a multifaceted experience with different meanings. At one extreme are people who choose it deliberately, at the other people who regard themselves as single because of circumstances beyond their control. Most, like Heather, are in the middle—adults who wanted to marry but made choices that took them in a different direction (Shostak, 1987). In a study in which never-married, childless women were interviewed about their lives, some said they focused on occupational goals instead of marriage. Others reported that they found singlehood preferable to the disappointing relationships they had with men. And still others commented that they just did not meet "the right person" (Dalton, 1992).

Of the various advantages of singlehood, those mentioned most often are freedom and mobility. But singles also recognize drawbacks—loneliness, the dating grind, limited sexual and social life, reduced sense of security, and feelings of exclusion from the world of married couples (Chasteen, 1994). Single men have more physical and mental health problems than do single women, who usually come to terms with their lifestyle and fare well. The greater social support available to women through intimate same-sex friendships is partly responsible. In addition, never-married men are more likely to have problematic family backgrounds and personal characteristics that contribute to both their singlehood and their adjustment difficulties (Buunk & Driel, 1989).

Many single people go through a stressful period in their late twenties, when more of their friends get married and they become an exception. The mid-thirties is another trying time for single women, due to the approaching biological deadline for bearing children. A few decide to become parents through adoption, artificial insemination, or an extramarital affair.

COHABITATION

Cohabitation refers to the lifestyle of unmarried couples who have an intimate, sexual relationship and share a residence. Until the 1960s, cohabitation in Western nations was largely limited to low-income people. Since then, it has increased in all groups, with an especially dramatic rise among well-educated and economically advantaged young adults. As Figure 14.3 shows, seven times as many American

couples are cohabiting in the 1990s as did in 1970. About one-third of these households include children, since in half of cohabiting relationships, one or both partners are separated or divorced (U.S. Bureau of the Census, 1996).

Like singlehood, cohabitation has different meanings. For some, it serves as *preparation for marriage*—a time to test the relationship and get used to the realities of living together. For others, it is an *alternative to marriage*—an arrangement that offers the rewards of sexual intimacy and companionship along with the possibility of easy departure if there is a decline in satisfaction. In view of this variation, it is not surprising that great range exists in the extent to which cohabitors share money and possessions and take responsibility for each other's children.

Americans are more open to cohabitation than they were in the past, although attitudes are not yet as positive as they are in Western Europe. In the Netherlands, Norway, and Sweden, cohabitation is thoroughly integrated into society. As a result, cohabitors are nearly as devoted to one another as are married people (Buunk & Driel, 1989; Kaslow, Hansson, & Lundblad, 1994; Ramsøy, 1994). When they decide to marry, Dutch, Norwegian, and Swedish cohabitors more often do so to legalize their relationship, especially for the sake of children. American cohabitors typically marry to confirm their love and commitment—sentiments that their Western European counterparts attach to cohabitation.

Largely as a result of cultural ambivalence about cohabitation, American cohabitors differ from people who are

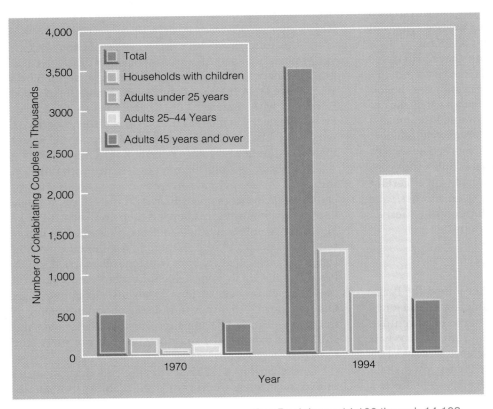

FIGURE 14.3

Rates of cohabitation in 1970 and 1994 in the United States. The figure shows total number of cohabiting couples, number of cohabiting couples with children, and cohabiting rates by age. Notice the dramatic rise in cohabitation for all groups. However, postponement of marriage and the high rate of divorce (see Figure 14.1) have caused the age distribution of cohabitors to change. In 1970, the majority were middle aged and elderly. Today, most are young adults. *(From U.S. Bureau of the Census, 1996.)*

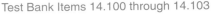
Test Bank Items 14.100 through 14.103

married or living in separate residences. They are less religious, more politically liberal, more androgynous, and have had more sexual partners. Overall, they are more likely to endorse and engage in nonconventional behavior. In addition, a larger number have parents who are divorced and say they get along poorly with them (Cunningham & Antill, 1994; Thornton, 1991; Thornton, Axinn, & Hill, 1992).

These personal characteristics are responsible for outcomes associated with cohabitation. Compared to married people, American cohabitors are less likely to pool finances or jointly own a house. (This difference is much smaller in Western Europe.) Furthermore, American couples who live together before marriage are more prone to divorce—an association not present in other Western countries (Buunk & Driel, 1989; Cherlin, 1992). But we must keep in mind that American cohabitors have experienced more conflict in their own homes, and they tend to emphasize self-fulfillment over obligations to others in their relationships. As they carry this individualistic outlook into marriage, they are more likely to dissolve a union when it becomes less satisfying.

Certain couples are exceptions to the trends just described. People who cohabit after separation or divorce often test a new relationship carefully to prevent another painful failure, especially when children are involved. As a result, they cohabit longer and are less likely to move toward marriage (Cherlin, 1992). In contrast to American heterosexual cohabitors, homosexual couples report strong commitment—as intense as that of married people. When their relationships become problematic, they end more often than marriages do only because there are fewer barriers to separating, including children in common, financial dependence on a partner, and concerns about the costs of divorce (Peplau, 1991).

Clearly, cohabitation has advantages and drawbacks. For people not ready for marriage, it combines the rewards of a close relationship with the opportunity to remain at least partially uncommitted. Although most couples cohabit to avoid legal obligations, they can encounter difficulties because they do not have them. Bitter fights over property, money, rental contracts, and responsibility for children are the rule rather than the exception when unmarried couples split up (Buunk & Driel, 1989).

CHILDLESSNESS

At work, Sharese got to know Beatrice and Daniel. Married for 7 years and in their mid-thirties, they did not have children and were not planning to have any. To Sharese, their relationship seemed especially caring and affectionate. "At first, we were open to becoming parents," Beatrice explained, "but eventually we decided to focus on our marriage."

In our discussion of reproductive technologies in Chapter 2, we noted that about 20 percent of couples have fertility problems. Because treatment is not always successful, some remain *involuntarily* childless. Others who are childless by circumstance simply did not find a partner with whom to share parenthood. Beatrice and Daniel are in another category—men and women who are *voluntarily* childless.

How many American couples choose not to have children is a matter of dispute. Some researchers claim the rate has been low for several decades—between 3 and 6 percent (Houseknecht, 1987; Jacobson & Heaton, 1989). Others say it rose in the 1980s and early 1990s and is currently higher than ever before—from 10 to 15 percent (Ambry, 1992; Morell, 1994). These differing reports may be due to the fact that voluntary childlessness is not always a permanent condition. A few people decide very early that they do not want to be parents and stick to these plans. But most, like Beatrice and Daniel, make their decision after they are married and have developed a lifestyle they do not want to give up. Later, some change their minds.

Besides marital satisfaction and freedom from childcare responsibilities, common reasons for not having children include the woman's career and economic security. Consistent with these motives, the voluntarily childless are usually college educated, have prestigious occupations, and are highly committed to their work. Many were only or first-born children whose parents encouraged achievement and independence. In a culture that negatively stereotypes childlessness, it is not surprising that voluntarily childless women are more self-reliant and assertive. Typically the wife is responsible for the decision not to have children (Houseknecht, 1987; Morell, 1994).

Voluntarily childless adults are just as content with their lives as are parents who have rewarding relationships with their children. In contrast, adults who cannot overcome their infertility and parents whose children have serious psychological or physical problems are likely to be dissatisfied and depressed (Connidis & McMullin, 1993). Think about these findings, and you will see that they do not support the prevailing stereotype of childless people as lonely and unfulfilled. Instead, they indicate that parenthood enhances well-being only when the parent–child relationship is warm and affectionate. And childlessness interferes with adjustment and life satisfaction only when it is beyond a person's control.

DIVORCE AND REMARRIAGE

If current rates continue (see Figure 14.1 on page 464), about half of all marriages in the United States will end in separation or divorce. Most people who divorce do so within 5 to 10 years of their marriage, so many divorces happen when children are still at home. Furthermore, 61 percent of divorced men and 54 percent of divorced women remarry (Hetherington, Law, & O'Connor, 1994). Divorce occurs at an especially high rate during the first few years of second marriages—7 percent above that for first marriages. Afterward, the divorce rate for first and second marriages is about the same (Cherlin, 1992; Ganong & Coleman, 1994).

Why do so many marriages fail? As Christy and Gary's divorce illustrates, the most obvious reason is a disrupted husband–wife relationship. Christy and Gary did not argue more often than Sharese and Ernie did. But their style of resolving conflict was markedly different. When Christy raised concerns, Gary responded with resentment, anger, and retreat—a conflict-confronting–conflict-avoiding pattern that can be found in many partners who split up. Another typical style involves little conflict, but partners increasingly lead separate lives because they have different expectations of marriage and family life and few shared interests, activities, or friends (Hetherington, Law, & O'Connor, 1994).

A few decades ago, unhappy couples like Christy and Gary often stayed together. Shifting societal conditions, including widespread family poverty and the changing status of women, provide the larger context for the high divorce rate. Economically disadvantaged couples who suffer multiple life stresses are especially likely to split up. But Christy's case represents another trend—rising marital breakup among well-educated, economically independent women. Women are twice as likely as men to initiate divorce (Rice, 1994).

When Sharese heard that Christy and Gary's marriage had dissolved, she remarked that it seemed as if "someone had died." Her description is fitting, since divorce involves the loss of a way of life and therefore part of the self sustained by that way of life. As a result, it carries with it opportunities for both positive and negative change.

Immediately after separation, both men and women are depressed and anxious and display impulsive, antisocial behavior. For most, these reactions subside within 2 years. Women who were in traditional marriages and who organized their identities around their husband have an especially hard time. As one divorcee remarked, "I used to be Mrs. John Jones, the bank manager's wife. Now I'm Mary Jones. Who is Mary Jones?" (Hetherington, Law, & O'Connor, 1994, p. 216). Many noncustodial fathers feel disoriented and rootless as a result of decreased contact with their children. Others distract themselves through a frenzy of social activity (Cherlin, 1992).

Finding a new partner contributes most to the life satisfaction of divorced adults. But it is more crucial for men, who show more positive adjustment in the context of marriage than on their own. Despite a drop in income, inadequate child care, and loneliness (see Chapter 10), most divorced women prefer their new life to an unhappy marriage. For example, Christy developed new skills and a sense of self-reliance that might not have emerged had she remained married to Gary. However, a few women—especially those who are anxious and fearful or who remain strongly attached to their ex-spouse—show a drop in self-esteem, become depressed, and tend to form repeated unsuccessful relationships (Ganong & Coleman, 1994). Job training, continued education, and career advancement play vital roles in the economic and psychological well-being of divorced women. The availability of high-quality child care is essential for these activities (Hetherington, 1995).

On the average, people remarry within 4 years of divorce, men somewhat faster than women. Why do many remarriages break up? There are several reasons. First, although people often remarry for love, practical matters—financial security, help in rearing children, relief from loneliness, and social acceptance—figure more heavily into a second marriage than a first. These concerns do not provide a sound footing for a lasting partnership. Second, some people transfer the negative patterns of interaction and problem solving learned in their first marriage to the second. Third, people who have already had a failed marriage are more likely to view divorce as an acceptable solution when marital difficulties resurface. And finally, remarried couples experience more stress from stepfamily situations (Ganong & Coleman, 1994). Recall from Chapter 10 that adults in blended families have few societal guidelines for how to relate to their steprelatives, including stepchildren. As we will see in the next section, stepparent–stepchildren ties are powerful predictors of marital happiness.

In divorce and remarriage, as in other adult lifestyles, there are multiple pathways leading to diverse outcomes. It generally takes about 2 years for blended families to develop the connectedness and comfort of intact biological families. Family life education and couples and group therapy can help divorced and remarried adults adapt to the complexities of their new circumstances (Forgatch, Patterson, & Ray, 1996).

VARIANT STYLES OF PARENTHOOD

Diverse family forms result in varied styles of parenthood. Among these are a growing number of remarried parents, never-married parents, and gay and lesbian parents. Each type of family presents unique challenges to parenting competence and adult psychological well-being.

■ **REMARRIED PARENTS.** Whether stepchildren live in the household or visit only occasionally, stepparents are in a difficult position. Since the parent–child tie predates the remarriage, the stepparent enters as an outsider. All too often, stepparents try to move into their new role too quickly. Because they do not have a warm attachment bond to build on, their discipline is usually ineffective. Stepparents frequently criticize the biological parent for being too lenient. The parent, in turn, tends to view the stepparent as too harsh. These differences can become major issues that divide the remarried couple (Ganong, Coleman, & Fine, 1995; Papernow, 1993).

Stepmothers, especially, are likely to experience conflict and poor adjustment. Expected to be in charge of family relationships, they quickly find that stepparent–stepchild ties do not develop instantly. Even when their husbands do not have custody, stepmothers feel stressed. As stepchildren go in and out of the home, stepmothers compare life with and without resistant children, and many prefer life without! No matter what a stepmother does to build a close parent–child bond, her efforts are probably doomed to failure in the short run.

Stepfathers with children of their own have an easier time. They establish positive ties with stepchildren relatively quickly, perhaps because they feel less pressure to plunge into parenting than do stepmothers. At the same time, they have had enough experience to know how to build a warm parent–child relationship. Stepfathers without biological children are new to child rearing. When they have unrealistic expectations or their wives push them into the father role, their interactions with stepchildren can be troublesome. After making several overtures that are ignored or rebuffed, they often withdraw from parenting (Bray, 1992; Hetherington & Clingempeel, 1992).

A caring husband–wife relationship, the cooperation of the absent biological parent, and the willingness of children to accept their parents' new spouse are crucial for stepparent adjustment. Because stepparent–stepchild bonds are hard to establish, the divorce rate is higher for couples with stepchildren than for those without (Ganong & Coleman, 1994).

■ NEVER-MARRIED PARENTS. Earlier we mentioned that single adults occasionally decide to become parents and rear children on their own. Births to women in high-status occupations who have not married by their thirties have increased in recent years. However, they are still few in number, and little is known about how these mothers and their children fare.

During the past several decades, never-married parenthood has been especially high among African-American young women. For example, in 1993, over 60 percent of births to black women in their twenties were out of wedlock, whereas only 18 percent to white women were. A sharp difference between blacks and whites in the timing of family life cycle events underlies these trends. African-American women postpone marriage more and childbirth less than do their Caucasian-American counterparts (U.S. Bureau of the Census, 1996).

Loss of manufacturing jobs, rising unemployment, and consequent inability of many black men to support a family have contributed to never-married parenthood among African Americans (Rice, 1994). African Americans have also responded to changing family values—the shift toward cohabitation, postponement of marriage, and separation of marriage from childbearing that has spread throughout American culture (Cherlin, 1992).

Recall from Chapter 2 that never-married black mothers tap a traditional source of strength in their culture by relying heavily on their own mothers and other extended family members for help in caring for their children. For most, marriage follows birth of the first child by several years and is not necessarily to the child's father. Nevertheless, these couples function much like other first-marriage parents.

For most never-married African-American mothers, marriage follows birth of the first child by several years and is not necessarily to the child's father. Nevertheless, these couples function much like other first-marriage parents. The children are often unaware that the father is a stepfather, and parents do not report the relationship problems typical of blended families.
(Bernard Boutrit/Woodfin Camp & Associates)

Their children are often unaware that the father is a stepfather, and parents do not report the relationship problems typical of blended families (Ganong & Coleman, 1994).

Still, never-married parenthood among low-income women is costly, since living in a female-headed household makes it more difficult to overcome poverty. Furthermore, the extreme deprivation and danger of crime-ridden, inner-city neighborhoods hinders extended family networks from providing assistance, thereby isolating many young mothers. (For an example, return to the story of Zinnia Mae in Chapter 2, pages 61 and 63.) Strengthening vocational education and employment opportunities for African Americans would encourage marriage as well as help unmarried-mother families.

■ GAY AND LESBIAN PARENTS. Several million American gay men and lesbians are parents, most through previous heterosexual marriages, a few through adoption or reproductive technologies (Hare, 1994). In the past, laws assuming that homosexuals could not be adequate parents led those who divorced a heterosexual partner to lose custody of children. Today, several states hold that sexual orientation is irrelevant to custody, but in others, fierce prejudice against homosexual parents still prevails.

Families headed by a homosexual parent or a gay or lesbian couple are very similar to those of heterosexuals. Gay and lesbian parents are as committed to and effective at the parental role, and sometimes more so. Indeed, some research indicates that gay fathers are more consistent in setting limits and more responsive to their children's needs than are heterosexual fathers, perhaps because gay men's

Homosexual parents are as committed to and effective at child rearing as heterosexual parents. Their children are well adjusted, and the large majority develop a heterosexual orientation. (Mark Richards/PhotoEdit)

nontraditional gender-role identity fosters involvement with children (Bigner & Jacobsen, 1989a, 1989b). Children of gay and lesbian parents are as well adjusted as children of heterosexual parents, and the large majority are heterosexual (Bailey et al., 1995; Golombok & Tasker, 1996).

When extended family members have difficulty accepting them, homosexual mothers and fathers often build "families of choice" through friends who assume functions traditionally expected of relatives. But most of the time, parents of gays and lesbians cannot endure a permanent rift. With the passage of time, interactions of homosexual parents with their families of origin resemble those of heterosexuals (Hare, 1994; Lewin, 1993).

Partners of homosexual parents usually take on some caregiving responsibilities and are attached to the children. The extent of involvement varies with the way children were brought into the relationship. When children were adopted or conceived through reproductive technologies, partners tend to be more involved than when children originated in a previous heterosexual relationship (Hare & Richards, 1993). In a few instances, homosexual partners have become the joint legal parents of children, an arrangement that enhances family stability (Green & Bozett, 1991).

Overall, families headed by homosexuals can be distinguished from other families only by issues related to living in a nonsupportive society. The greatest concern of gay and lesbian parents is that their children will be stigmatized by their parents' sexual orientation (Hare, 1994; Lewin, 1993).

BRIEF REVIEW

Since the 1960s, the diversity of adult lifestyles has expanded. Long-term singlehood occurs most often among women in high-status careers and men in blue-collar occupations or who are unemployed. Depending on their life circumstances, the adjustment of single people varies greatly. Cohabitation is common in Western industrialized nations. Since living together before marriage is not yet broadly accepted in the United States, American cohabitors tend to be less conventional than other young adults. For this reason, the marriages of former cohabitors are more likely to break up. Voluntarily childless adults are usually college educated, highly involved in their careers, and content with their lives. Widespread family poverty and the changing status of women have contributed to the high divorce rate in the United States. Divorced adults usually adjust to their new circumstances within 2 years. Finding a new partner contributes most to their life satisfaction. Unfortunately, the divorce rate for second marriages is high.

Modern families also include varied styles of parenthood. Stepparent–stepchild bonds are difficult to establish and depend on the support of the spouse and the cooperation of the absent parent and the child. Never-married parenthood is widespread among African-American women, who often rely on extended family members (particularly their own mothers) for child-rearing assistance. Gay and lesbian parents are just as effective at child rearing as are heterosexual parents.

ASK YOURSELF . . .

■ *Return to Chapter 10, pages 332–335 and review the impact of divorce and remarriage on children and adolescents. How do those findings resemble outcomes for adults? What might account for the similarities?*

■ *After dating for a year, Wanda and Scott decided to live together. Wanda's mother heard that couples who cohabit before marriage are more likely to get divorced. She worried that cohabitation would reduce Wanda and Scott's chances for a successful life together. Is her fear justified? Why or why not?*

VOCATIONAL DEVELOPMENT

Besides family life, vocational life is a vital domain of development in early adulthood. After choosing an occupation, young people must learn how to perform its tasks competently, get along with co-workers, respond to authority, and protect their own interests. When work experiences go well, adults develop new competencies, feel a sense of personal accomplishment, make new friends, and become financially independent and secure.

ESTABLISHING A VOCATION

In our discussion of Levinson and Vaillant's theories earlier in this chapter, we noted divergent paths and timetables for vocational development in contemporary adulthood. Consider, once again, the wide variation in establishing a career among Sharese, Ernie, Christy, and Gary. As is typical for men, Ernie and Gary's vocational lives were long and *continuous,* beginning after completion of formal education and ending with retirement. Like many women, Sharese and Christy had *discontinuous* career paths—ones that were interrupted or deferred by childbearing and child rearing (Betz, 1993; Ornstein & Isabella, 1990). Furthermore, not all people embark on the vocation of their dreams. For example, although half of young people aspire to professional occupations, only 20 percent of the work force attains them (U.S. Bureau of the Census, 1996).

Even for young people who enter their chosen field, initial experiences can be discouraging. At the health department, Sharese discovered that committee meetings and paper work consumed much of her day. Since each grant proposal and research project had a deadline, the pressure of productivity weighed heavily on her. Adjusting to unanticipated disappointments in salary, supervisors, and co-workers is difficult (Hatcher & Crook, 1988). As new workers become aware of the gap between their expectations and reality, resignations are common. On the average, people in their twenties move to a new job every 2 years; five or six changes are not unusual (Seligman, 1994).

After a period of evaluation and adjustment, young adults generally settle into an occupation. In careers with opportunities for promotion, high aspirations must often be revised downward, since the structure of most work settings resembles a pyramid. In businesses, there are few high-level executive positions; in factories, there are a limited number of supervisory jobs. In a longitudinal study of over 400 AT&T lower-level male managers, the importance of work in men's lives varied with career advancement and age. For men who advanced very little, "work disengagement" occurred early; family, recreation, and community service assumed greater importance by the early thirties. Men with average levels of career success emphasized nonwork roles at a later age. In contrast, men who were highly successful became more involved in their jobs over time.

Wide variation exists today among young adults' career paths and timetables. Men's vocational lives are typically long and continuous. In contrast, women often have discontinuous career paths—ones that are interrupted or deferred by childbearing and child rearing. (Jonathan Nourok/Tony Stone Images)

Although the desire for advancement declines with age for many, most workers still seek challenges and find satisfaction in their work roles (Howard & Bray, 1988).

Besides opportunity, personal characteristics affect career progress. As we will see in the next section, a sense of *self-efficacy*—belief in one's own ability to be successful—affects career choice and development. (Return to Chapter 1, pages 18-19, if you need to review this idea.) Young people who are very anxious about the possibility of failing or making mistakes tend to set career aspirations that are either too high or too low. As a result, they achieve far less than their abilities would permit (Lopez, 1989).

Recall from our discussion of Levinson's theory that success in a career often depends on the quality of a mentoring relationship. Access to an effective mentor is jointly influenced by the availability of willing people and the worker's capacity to select an appropriate individual. Interestingly, the best mentors are usually not top executives, who tend to be preoccupied and therefore less helpful and sympathetic. Most of the time, it is better for a young person to choose a mentor lower on the corporate ladder (Seligman, 1994).

Test Bank Items 14.128 through 14.131

WOMEN AND ETHNIC MINORITIES

Although women and ethnic minorities have penetrated nearly all professions, they attain less than they otherwise would if their talents were developed to the fullest. Women in general—and African-American, Hispanic, and Native-American women in particular—remain concentrated in occupations that offer little opportunity for advancement, and they are underrepresented in executive and managerial roles (see Chapter 13, page 443). Despite the massive influx of women into the labor force, their earnings relative to men's have changed little during the past 50 years. For every dollar earned by a man, the average woman earns only 70 cents, a gap that is larger in the United States than in Western European nations (Adelman, 1991).

Women seem to go through a different, often more complex process of vocational development than do men. Especially for those who pursue traditionally feminine occupations, career planning is often short term and subject to considerable change. Low self-efficacy with respect to male-dominated fields limits not only women's occupational choices, but their progress once their vocational lives have begun. Women who pursue nontraditional careers usually have "masculine" qualities—high achievement orientation, an emphasis on individualism, and the expectation that they will have to make a life for themselves through their own efforts (Seligman, 1994).

Even when women enter high-status professions, very few move into high-level management positions. Gender-stereotyped images of women as followers rather than leaders and role conflict between work and family slow their advancement. Singlehood or late marriage and few or no children are strongly associated with career achievement in women—a relationship that does not hold for men (Betz & Fitzgerald, 1987). Furthermore, since men dominate high-status fields, few women are available to serve as mentors. (A similar situation exists for ethnic minorities.) Some evidence suggests that women with female mentors are more productive than those with male mentors, perhaps because female mentors can provide guidance on the unique problems women encounter in the workplace (Goldstein, 1979).

Despite obstacles to success, women benefit greatly from achievement in the outside world. Recall the study of the highly gifted "Termites" who were followed over their life course, described in Chapter 13 (see page 418). At age 60, women who reported the highest levels of life satisfaction had developed rewarding careers. The least satisfied women had been housewives all their lives (Sears & Barbie, 1977). These findings suggest that at least some of the problems experienced by married women may not be due to marriage per se, but rather to lack of a gratifying vocation. Consistent with this idea, most young women express a preference for blending work and family in their adult lives (Barnett & Rivers, 1996). For women in financially stressed families, this is usually not a choice; it is a necessity.

COMBINING WORK AND FAMILY

Whether women work because they want to or have to (or both), the dominant family form today is the **dual-earner marriage,** in which both husband and wife are employed. Most dual-earner couples are also parents, since the majority of women with children are in the work force (see Chapter 10). In about one-third of these families, moderate to severe conflict occurs over trying to meet both work and family responsibilities (Pleck, 1985).

What are the main sources of strain in dual-earner marriages? When Sharese returned to her job after her children were born, she felt a sense of *work overload.* Not only did she have a demanding career, but (like most employed women) she shouldered most of the household and child care tasks. Furthermore, Sharese and Ernie felt torn between the desire to excel at their jobs and the desire to spend more time with each other and their children. At times, their lives became so busy that they had little energy left over for visiting or entertaining friends, relatives, and work associates (Pleck, 1985).

Work–family role conflict is greater for women, and it negatively affects quality of life in both settings (Higgins, Duxbury, & Irving, 1992). It is especially intense for wives

After a full day at the office, this mother picks up her baby from day care and runs errands, attending to her second full-time job. Like most women in dual-earner marriages, she shoulders most of the child care and household tasks. Work-family role conflict is greater for women. (Tom Stewart/The Stock Market).

in low-status occupations with rigid schedules and little worker autonomy. Couples in prestigious careers have more control over both work and family domains. For example, Sharese and Ernie devised ways to spend more time with their children. They picked them up at day care early one day a week, compensating by doing certain occupational tasks on evenings and weekends. Like other career-oriented mothers, Sharese coped with role pressures by setting priorities. She decreased the amount of time she spent on household chores, not child rearing (Duxbury & Higgins, 1994).

Having two careers in one family usually means that certain career decisions become more complex. A move to a new job can mean vocational sacrifices for one partner. Usually this is the wife, since a decision in favor of the husband's career (typically further along and better paid) is more likely to maximize family income. One solution to the geographical limitations of the dual-earner marriage is to live apart. Although more couples are doing this, the strain of separation and risk of divorce are high.

Clearly, dual-earner marriages pose difficulties, but when couples cooperate to surmount these, they gain in many ways. Besides higher earnings and a better standard of living, the greatest advantage for college-educated couples is self-fulfillment of the wife. Ernie took great pride in Sharese's accomplishments, and he reaped other benefits. Sharese's career orientation contributed to his view of her as an interesting, self-confident, and capable helpmate in life.

In sum, a challenging, rewarding occupation in the context of a supportive spouse can strengthen a marriage and foster adult development. Under other circumstances —for example, when a woman tries to combine a low-status, low-paying job with marriage to a disapproving man—the physical and psychological toll can be severe. The Caregiving Concerns table below lists strategies that help dual-earner couples combine work and family roles in ways that promote mastery and pleasure in both spheres of life.

ASK YOURSELF . . .

■ *Heather climbed the career ladder of her company quickly, reaching a top-level executive position by her early thirties. In contrast, Sharese and Christy did not attain managerial roles in early adulthood. What accounts for the disparity in career progress of the three women?*

■ *Work life and family life are inseparably intertwined. Explain how this is so in early adulthood.*

CAREGIVING CONCERNS
Strategies That Help Dual-Earner Couples Combine Work and Family Roles

STRATEGY	DESCRIPTION
Devise a plan for sharing household tasks.	As soon as possible in the relationship, talk about division of household responsibilities. Decide who does a particular chore on the basis of who has the needed skill and time, not gender. Schedule regular times to rediscuss your plan to fit changing family circumstances.
Begin sharing child care right after the baby's arrival.	For fathers, strive to spend equal time with the baby early. For mothers, refrain from imposing your standards on your partner. Instead, share the role of "child-rearing expert" by discussing parenting values and concerns often. Attend a parent education course together.
Talk over conflicts about decision making and responsibilities.	Face conflict through communication. Clarify your feelings and needs and express them to your partner. Listen and try to understand your partner's point of view. Then be willing to negotiate and compromise.
Establish a balance between work and family.	Critically evaluate the time you devote to work in view of your values and priorities. If it is too much, cut back.
Make sure your relationship receives regular nurturance and attention.	See the Caregiving Concerns table on page 459.
Press for workplace and public policies that assist dual-earner couples.	Difficulties faced by dual-earner couples are partly due to lack of workplace and societal supports. Encourage your employer to provide benefits that help combine work and family roles, such as flexible work hours; parental leave with pay; and onsite high-quality, affordable day care. Communicate with lawmakers and other citizens about improving public policies for children and families.

S UMMARY

ERIKSON'S THEORY: INTIMACY VERSUS ISOLATION

According to Erikson, what personality changes take place during early adulthood?

- In Erikson's theory, young adults must resolve the conflict of **intimacy versus isolation,** balancing independence and intimacy as they form a close relationship with a partner. Research indicates that young people deal with intimacy concerns in the context of a variety of lifestyles. It also shows that aspects of generativity, including work and career as well as childbearing and child rearing, are prime concerns in the twenties and thirties.

OTHER THEORIES OF ADULT PSYCHOSOCIAL DEVELOPMENT

Describe Levinson's and Vaillant's psychosocial theories of adult personality development.

- Building on Erikson's stage approach, both Levinson and Vaillant outlined patterns in the adult life course that have been supported by research.

- Levinson described a predictable series of eras, each consisting of a transition and a stable period, in which the **life structure** is revised and elaborated. Young adults usually construct a dream, typically involving career for men and both marriage and career for women, and find a mentor to help them in their career. In their thirties, they focus on aspects of their lives that have received less attention. Men settle down, whereas women continue in an unsettled phase into middle adulthood.

- Vaillant refined Erikson's stages, portraying the twenties as devoted to intimacy, the thirties to career consolidation, the forties to guiding others, and the fifties to cultural and philosophical values.

What is the social clock, and how does it affect personality in adulthood?

- Although societal expectations have become less rigid, conformity to or departure from the **social clock,** the culturally determined timetable for major life events, can be a major source of personality change in adulthood. Following a social clock grants confidence to young adults, whereas departures can bring psychological distress.

CLOSE RELATIONSHIPS

Describe the role of romantic love in the young adult's quest for intimacy.

- Finding a partner is a major milestone of adult development. Intimate partners tend to resemble one another in age, ethnicity, social class, education, and various personal and physical attributes. The emphasis placed on romantic love as the basis for mate selection in Western nations does not characterize all cultures.

- The balance among passion, intimacy, and commitment changes as romantic relationships move from the intense sexual attraction of **passionate love** toward more settled **companionate love.** Effort and commitment are key in enduring relationships.

Describe adult friendships and sibling relationships.

- Adult friendships have many of the same characteristics and benefits as earlier friendships and are based on trust, intimacy, and loyalty. Women's same-sex friendships tend to be more intimate than men's. Other-sex friendships are important in adulthood but less frequent and enduring than same-sex friendships.

- Siblings become more frequent companions in early adulthood than they were in adolescence, often taking on the characteristics of friendship, especially between same-sex siblings.

Describe the role of loneliness in adult development.

- Young adults are vulnerable to **loneliness,** but it declines with age as they form satisfying intimate relationships and learn to be comfortably alone.

THE FAMILY LIFE CYCLE

Trace phases of the family life cycle that are prominent in early adulthood, and cite factors that influence these phases today.

- Although the majority of young people marry, wide variations in the sequence and timing of phases of the **family life cycle** reflect the diversity of modern life.

- Leaving home is a major step in assuming adult responsibilities. A large percentage of American teenagers depart relatively early when they go to college. Social class and ethnicity influence the likelihood that a young person will live independently before marriage. Returning to live at home for a period of time is common among unmarried young adults.

- Nearly 90 percent of Americans marry, although most do so later than in the past. Both **traditional marriages** and **egalitarian marriages** are affected by women's employment and changing gender roles. Husbands and wives today must work harder to define their marital roles.

- Many young people enter marriage with unrealistic expectations. Even happy marriages have their ups and downs and require adaptability on the part of both partners. Men tend to be happier and healthier in marriage than women.

- Effective birth control techniques and changing cultural values make childbearing a matter of choice in Western industrialized nations. Although most American couples bear children, they are doing so later, having fewer children, and weighing the pros and cons more carefully than in the past.

- The arrival of children is a key change in the family life cycle, making the family system permanent for the first time. New parents must adjust to increased responsibilities, less time for each other, and more traditional roles. Marital satisfaction typically declines somewhat with these adjustments.

- Challenges facing families with young children include inadequate preparation for child rearing, lack of societal supports for parenting, the need for cooperation in the marital relationship, and difficulties in finding good day care.

- In families with adolescents, parents must establish new relationships with their increasingly autonomous teenagers, blending guidance with freedom and gradually relinquishing control. Marital satisfaction is often lowest in this phase, but flexibility and parent education are helpful.

THE DIVERSITY OF ADULT LIFESTYLES

Discuss the diversity of adult lifestyles, focusing on singlehood, cohabitation, and childlessness.

- Diverse nontraditional family options are now part of the American mainstream. Adults may deliberately choose a lifestyle, drift into it, or be forced into it by circumstances.

- Singlehood has risen in recent years and includes both never-married and divorced individuals. Women with high-status careers and men who are unemployed or in blue-collar occupations are most likely to remain single. Single women tend to be better adjusted than single men.

- **Cohabitation** has risen dramatically, especially among well-educated, economically advantaged young adults. It serves either as preparation for marriage or as an alternative to marriage, and it often includes divorced partners and their children. Because American cohabitors tend to be less conventional than other people, their subsequent marriages are more likely to fail.

- Voluntarily childless adults tend to be well educated and career oriented and are just as satisfied with their lives as are parents who have good relationships with their children.

Discuss today's high rates of divorce and remarriage, and cite factors that contribute to them.

- Half of all marriages in the United States will end in separation or divorce, often while children are at home. More than half of divorced people will remarry, and many will divorce again. Unhappy marital relationships, family poverty, and the changing status of women contribute to the divorce rate.

- Finding a new partner is important to many divorced adults, especially men. Remarriages break up for several reasons, including the rominence of practical concerns rather than love in the decision to remarry, the persistence of negative patterns of interaction, the acceptance of divorce as a solution to marital difficulties, and problems adjusting to a stepfamily.

Discuss the challenges associated with variant styles of parenthood, including remarried parents, never-married parents, and gay and lesbian parents.

- Establishing stepparent–stepchild ties is difficult, especially for stepmothers and for stepfathers without children of their own. A caring husband–wife relationship, the cooperation of the absent biological parent, and children's acceptance are crucial for stepparent adjustment.

- Never-married parenthood is especially high among low-income African-American women in their twenties. Unemployment among black men and changing American family values contribute to this trend. Although these young mothers often receive help from extended family members, they find it difficult to overcome poverty.

- Families headed by homosexuals face difficulties related to living in an unsupportive society. Gay and lesbian parents are just as loving and effective as heterosexual parents.

VOCATIONAL DEVELOPMENT

Discuss men's and women's patterns of vocational development, and cite difficulties faced by women, ethnic minorities, and couples seeking to combine work and family.

- In addition to intimacy and family concerns, vocational development is a key task for young adults. Men's career paths are usually continuous, whereas women's are often discontinuous due to childbearing and child rearing. After adjusting to the realities of the work world, young adults settle into an occupation. Their progress is affected by opportunities for promotion in their chosen occupation, personal characteristics such as self-efficacy, and access to an effective mentor.

- Women and ethnic minorities have penetrated nearly all professions but have made limited progress in advancement and earnings. Women tend to be hampered by low self-efficacy with respect to traditionally male-dominated fields, gender stereotypes, role conflict between work and family, and difficulties in finding a suitable mentor.

- Couples, and particularly women, in **dual-earner marriages** experience stresses from work overload, work–family role conflict, and the need to make vocational sacrifices to further their partner's career. Benefits include higher earnings, a better standard of living, and self-fulfillment for the wife.

IMPORTANT TERMS AND CONCEPTS

intimacy versus isolation (p. 452)

life structure (p. 454)

social clock (p. 456)

passionate love (p. 457)

companionate love (p. 457)

loneliness (p. 461)

family life cycle (p. 462)

traditional marriage (p. 464)

egalitarian marriage (p. 464)

cohabitation (p. 472)

dual-earner marriage (p. 478)

FOR FURTHER INFORMATION AND HELP

PARENT EDUCATION

Family Resource Coalition
200 S. Michigan Avenue, Suite 1600
Chicago, IL 60604
(312) 341-0900
A national organization that helps families develop support systems to strengthen family life and children's development. Educates public, government, and business leaders about the needs of families and how family resource programs can meet those needs.

SINGLEHOOD AND DIVORCE

Single Mothers by Choice
P.O. Box 1642,
Gracie Square Station
New York, NY 10028
(212) 988-0993
Web site: *www.parentsplace.com/ readroom/smc*
Organization of primarily single professional women in their thirties and forties who have either decided to have or are considering having children outside of marriage.

Parents Without Partners
8807 Colesville Road
Silver Spring, MD 20910
(301) 588-9354
Web site: *www.fwst.net/interact/ pwp.htm*
Organization of custodial and noncustodial single parents that provides support in the upbringing of children. Many local groups exist throughout the United States.

REMARRIAGE

Stepfamily Association of America
215 Centennial Mall South,
Suite 212
Lincoln, NE 68508
(402) 477-7837
Web site: *www.stepfam.org*
Association of families interested in stepfamily relationships. Organizes support groups and offers education and children's services.

Stepfamily Foundation
333 West End Avenue
New York, NY 10023
(212) 877-3244
Organization of remarried parents, interested professionals, and divorced individuals. Arranges group counseling sessions for stepfamilies and provides training for professionals.

GAY AND LESBIAN PARENTS

Gay and Lesbian Parents
Coalition International
Box 50360
Washington, DC 20091
(202) 583-8029
Strives to educate society about the compatibility of homosexuality and parenting. Supports efforts to eliminate discrimination due to sexual orientation, coordinates support groups for parents and children, and conducts public education programs.

DOMESTIC VIOLENCE

National Council on Child Abuse
and Family Violence
1155 Connecticut Avenue, N.W.,
Suite 400
Washington, DC 20036
(202) 429-6695
Supports community prevention and treatment programs for women and children who are victims of abuse. Seeks to increase public awareness of domestic violence.

AGE	PHYSICAL	COGNITIVE	EMOTIONAL/SOCIAL

20–30 years

PHYSICAL

- Athletic skills that require speed of limb movement, explosive strength, and gross body coordination peak early in this decade and then decline.

- Athletic skills that depend on endurance, arm–hand steadiness, and aiming peak at the end of this decade and then decline.

- Declines in touch sensitivity; respiratory, cardiovascular, and immune system functioning; and elasticity of the skin begin and continue throughout adulthood.

- As basal metabolism declines, gradual weight gain begins in the middle of this decade and continues through middle adulthood.

- Sexual activity increases.

COGNITIVE

- If college educated, dualistic thinking (dividing information, values, and authority into right and wrong) declines in favor of relativistic thinking (viewing all knowledge as embedded in a framework of thought).

- Narrows vocational options and settles on a specific career.

- With entry into marriage and employment situations, focuses less on acquiring knowledge and more on applying it to everyday life.

- Develops expertise (acquisition of extensive knowledge in a field or endeavor), which enhances problem solving.

- Creativity (generating unusual products) increases.

- Modest gains in a variety of mental abilities assessed by intelligence tests occur during this and the following decade.

EMOTIONAL/SOCIAL

- Leaves home permanently.

- Strives to make a permanent commitment to an intimate partner.

- Usually constructs a dream, an image of the self in the adult world that guides decision making.

- Usually forms a relationship with a mentor, who facilitates realization of the dream.

- If in a high-status career, acquires professional skills, values, and credentials (for women, may be delayed and take a longer time).

- Begins to develop mutually gratifying adult friendships and work ties.

- May cohabit, marry, and bear children.

- Sibling relationships become more companionate.

- As people move in and out of relationships, loneliness peaks early in this decade and then declines steadily throughout adulthood.

AGE	PHYSICAL	COGNITIVE	EMOTIONAL/SOCIAL

30–40 years

PHYSICAL

- Declines in vision, hearing, and the skeletal system begin and continue throughout adulthood.

- In women, fertility problems increase sharply in the middle of this decade.

- Hair begins to gray and thin in the middle of this decade.

- Sexual activity declines, probably due to the demands of daily life.

COGNITIVE

- As family and work lives expand, the cognitive capacity to juggle many responsibilities simultaneously improves.

- Creativity (generating unusual products) often peaks.

EMOTIONAL/SOCIAL

- Reevaluates life structure and tries to change components that are inadequate.

- Establishes a more stable niche within society through family, occupation, and community activities (for women, career consolidation may be delayed).

Physical and Cognitive Development in Middle Adulthood

On a snowy December evening, Devin and Trisha sat down to read the holiday cards that were piled high on the kitchen counter. Devin's fifty-fifth birthday had just passed; Trisha would reach 48 within 2 months. During the past year, they had celebrated their twenty-fourth wedding anniversary. These milestones, along with greetings from friends who sent annual updates about their lives, brought the changes of midlife into bold relief.

Instead of new births, children starting school, or a first promotion at work, notations on the cards sounded new themes. Jewel's recap of the past year reflected a growing awareness of a finite lifespan, one in which time had become more precious. She wrote,

> My mood has been lighter ever since my birthday. There was some burden I laid down by turning 49. My mother passed away when she was 48, so it all feels like a gift now. Blessed be!

George and Anya reported on their son's graduation from law school and their daughter Michelle's first year of university. The house empty of children, George wrote,

> Anya is filling the gap created by the children's departure by returning to college for a nursing degree. After enrolling this fall, she was surprised to find herself in the same psychology class as Michelle. At first, Anya was worried about whether she'd be able to handle the academic work, but after a semester of success, she's feeling much more confident.

Tim's message reflected continuing robust health, acceptance of physical change, and a new burden: caring for aging parents—a certain reminder of the limits of the lifespan:

> I used to be a good basketball player in college, but I've recently noticed that my 20-year-old nephew Brent can dribble and shoot circles around me. It must be my age! I ran our city marathon in September, coming in seventh in the over-50 division. Brent ran, too, but he opted out a few miles short of the finish line to get some pizza while I pressed on. This must be my age, too!

> The saddest news this year is that my dad had a bad stroke. His mind is clear, but his body is partially paralyzed. It's really upsetting because he was getting to enjoy the computer I gave him, and it was so wonderfully upbeat to talk to him in the few months before the stroke.

Middle age begins around age 40 and ends at about 60. This phase of adulthood is marked by a narrowing of life options and sense of a shrinking future as children leave home and career pathways become more determined. In other ways, middle adulthood is difficult to define, since

wide variations in attitudes and behaviors exist. Some individuals seem physically and mentally young at age 60—active and optimistic, having attained a sense of serenity and stability. Others feel old at age 35 or 40—as if their lives had peaked and were on a downhill course.

Yet another reason that middle age escapes clear definition is that it is a product of modern times. In centuries past, only a brief interval existed between the tasks of early adulthood and those of old age. For example, women often became widows by their mid-fifties, before their youngest child left home. And harsh living conditions led people to accept a ravaged body as a natural part of life. As average life expectancy—and with it, health and vigor—increased over this century, adults became more aware of their own aging and mortality.

Societal forces also affect whether a middle-aged person feels stagnant or in the prime of life. In a depressed job market, employers can tell 40-year-olds that they are "too old." Hearing this often enough, they come to believe what they are told. When people have highly valued skills and are financially well off, middle age can be an exciting and liberating time of life.

In this chapter, we trace physical and cognitive development during the fifth and sixth decades of the lifespan. In both domains, we will encounter not just progressive declines, but sustained performance and compensating gains. As in earlier chapters, we will be reminded that change occurs in manifold ways. Besides heredity and biological aging, our personal approach to passing years combines with family, community, and cultural contexts to profoundly affect the way we age.

Middle adulthood is difficult to define, since it encompasses wide variations in attitudes and behaviors. These middle-aged individuals seem physically and mentally young—active and optimistic, having attained a sense of serenity and stability. (Dick Luria/FPG International)

PHYSICAL DEVELOPMENT IN MIDDLE ADULTHOOD

Physical development in midlife is a continuation of the gradual changes already underway in early adulthood. Yet enough time has passed that a look in the mirror or at family photos prompts awareness of an older body, even in the most vigorous adults. Hair grays and thins, new lines appear on the face, and a less youthful, fuller body shape is evident. In addition to these obvious signs, it is during midlife that most individuals begin to experience life-threatening health episodes personally—if not in themselves, then in their partners and friends. And a change in time orientation, from "years since birth" to "years left to live," adds to consciousness of aging (Neugarten, 1968).

These factors lead to a revised physical self-image, which often emphasizes fewer hoped-for gains and more feared declines. Prominent concerns among 40- to 60-year-olds are getting a fatal disease, being too ill to maintain independence, and losing mental capacities. Yet notice that these outcomes are common stereotypes of aging. Unfortunately, many middle-aged adults fail to mention realistic alternatives as central life goals—becoming a more physically fit person and developing into a healthy, energetic older adult (Hooker & Kaus, 1994).

As we examine physical changes and health issues of middle adulthood, we will see that certain aspects of aging cannot be controlled. Yet many positive outcomes can be attained and feared outcomes avoided. There is much we can do to promote physical vigor and good health during midlife.

PHYSICAL CHANGES

As they dressed one morning before work, Trisha commented half-jokingly to Devin, "I just let the mirror get dusty; then I can't see the wrinkles and gray hairs very well." As she turned and took a closer look, her tone became more serious. "I'm certainly not happy about my weight. Look at this fat pad on my abdomen. It just doesn't want to go! I need to get back to some regular exercise—adjust my life to fit it in." Devin responded by glancing down soberly at his own enlarged midriff.

At breakfast, Devin took his glasses on and off and squinted while trying to read the paper. "Trish—what's the eye doctor's phone number? I need to get these bifocals adjusted again." When conversing between the kitchen and adjoining den, Devin sometimes asked Trisha to repeat herself. And he turned the radio and TV volume up high

enough so that Trisha frequently remarked, "Does it need to be so loud?" Devin, it seemed, couldn't hear quite as clearly as before.

Let's look at the major physical changes of middle adulthood—those especially salient to Devin and Trisha along with others occurring in the reproductive system and skeleton. As we do so, you may find it helpful to refer back to Table 13.2 on page 422, which provides a summary.

VISION

By the forties, difficulty reading small print is a common experience. It is due to growth in size of the *lens* of the eye combined with weakening of the muscle that enables it to adjust its focus (accommodate) to nearby objects. As new fibers appear on the surface of the lens, they compress older fibers toward the center, creating a thicker, denser, less pliable structure that eventually cannot be transformed at all. By age 50, the accommodative ability of the lens is one-sixth of what it was at age 20. Around age 60, the lens loses its capacity to adjust to objects at varying

Because the lens of the eye gradually loses its ability to adjust its focus to nearby objects, difficulty reading small print is a common experience by the forties. Corrective lenses for reading, which for nearsighted people take the form of bifocals, ease this problem. (G. S. Zimbel/Monkmeyer Press)

distances entirely, a condition called **presbyopia** (meaning, literally, "old eyes"). Corrective lenses for reading, which for nearsighted people take the form of bifocals, ease this problem. Because of the enlarging lens, the eye gradually becomes more farsighted between ages 30 and 60 (Whitbourne, 1996).

A second set of changes limits ability to see clearly in dim light. Throughout adulthood, the size of the pupil shrinks and the lens yellows. In addition, starting at age 40, the *vitreous* (transparent gelatin like substance that fills the eye) develops opaque areas. Consequently, the amount of light reaching the retina is reduced. Changes in the lens and vitreous also cause light to scatter within the eye, increasing sensitivity to glare. Devin, who as a college student had enjoyed driving at night, now found it more challenging to do so. He sometimes had trouble making out signs and moving objects. And his vision was more disrupted by bright light sources, such as the headlights of oncoming cars (Kline et al., 1992).

Finally, yellowing of the lens limits color discrimination, especially at the green-blue-violet end of the spectrum (Weale, 1992). Devin noticed that from time to time, he had to ask whether his sport coat, tie, and socks matched.

HEARING

An estimated 14 percent of American adults between 45 and 64 years of age have a hearing loss. Adult-onset hearing impairments account for many of these cases. Although some conditions run in families and may be hereditary, most are age-related. These are called **presbycusis** (meaning "old hearing") and are caused by several factors (Sill et al., 1994).

As we age, inner-ear structures that transform mechanical sound waves into neural impulses deteriorate due to natural cell death or reduced blood supply as a result of atherosclerosis. The first sign is a sharp hearing loss at high frequencies, around 50 years of age. Gradually, the impairment extends to all frequencies so that after age 60 human speech is more difficult to make out, although hearing loss remains greatest for high tones.

Men's hearing declines earlier and at a faster rate than women's, a difference thought to be due to exposure to environmental noise in male-dominated occupations. Over 9 million American production workers are being or have been exposed to noise capable of doing damage—the most common cause of hearing loss in middle-aged adults (Brechtelsbauer, 1990).

Most middle-aged and elderly people with hearing difficulties benefit from sound amplification with hearing aids. Once perception of the human voice is affected, speaking to the person patiently, clearly, and with good eye contact aids understanding. Finally, many severe cases of adult hearing loss can be prevented—by avoiding intense noise or using ear plugs during noise exposure.

SKIN

Our skin consists of three layers: (1) the *epidermis,* or outer protective layer, where new skin cells are constantly produced; (2) the *dermis,* or middle supportive layer, which consists of connective tissue that stretches and bounces back, granting the skin flexibility; and (3) the *hypodermis,* an inner fatty layer that adds to the soft lines and shape of the skin. As we age, the epidermis becomes less firmly attached to the dermis, fibers in the dermis thin, and fat in the hypodermis diminishes.

These changes lead the skin to wrinkle and loosen. In the thirties, lines develop on the forehead as a result of smiling, furrowing the brow, and other facial expressions. In the forties, these become more pronounced, and "crow's-feet" appear around the eyes. Gradually, the skin loses elasticity and begins to sag, especially on the face, arms, and legs. After age 50, "age spots," collections of pigment under the skin, increase.

Because sun exposure hastens wrinkling and spotting, individuals who have spent much time outdoors without proper skin protection look older than their contemporaries. And partly because the dermis of women is not as thick as that of men, women's skin ages more quickly (Whitbourne, 1985, 1996).

MUSCLE–FAT MAKEUP

As Trisha and Devin make clear, weight gain—referred to as "middle-age spread"—is a concern to both men and women. A common pattern of change is an increase in body fat and loss of lean body mass (muscle and bone). The rise in fat largely affects the torso and occurs as fatty deposits within the body cavity; as we noted earlier, fat beneath the skin on the limbs declines. On the average, size of the abdomen increases 6 to 16 percent in men, 25 to 35 percent in women from early through middle adulthood (Whitbourne, 1996). Sex differences in fat distribution also appear. Men accumulate more on the back and upper abdomen, women around the waist and upper arms.

Muscle mass declines very gradually in the forties and fifties, largely due to atrophy of fast-twitch fibers, responsible for speed and explosive strength. Yet as we indicated in Chapter 13, large weight gain and loss of muscle power are not inevitable, since continued exercise offsets these changes. Within the same individual, strength varies between often-used and little-used muscles (Arking, 1991). And consider Devin's 57-year-old friend Tim, who has ridden his bike to and from work for years and jogged on weekends, averaging 1 hour of vigorous activity per day. Like many endurance athletes, he weighed the same and had the same muscular physique at age 57 as at 25 (Suominen et al., 1980).

REPRODUCTIVE SYSTEM

The midlife transition in which fertility declines is called the **climacteric.** It differs markedly between the two sexes, since it brings an end to reproductive capacity in women, whereas fertility in men diminishes but is retained.

■ REPRODUCTIVE CHANGES IN WOMEN. The changes involved in women's climacteric occur gradually over a 10-year period, during which the production of estrogen drops. As a result, the number of days in a woman's monthly cycle shortens, from about 28 in her twenties and thirties to perhaps 23 by her late forties. Her cycles also become more irregular. In some, ova are not released; when they are, more are defective (see Chapter 2, page 52–53). The climacteric concludes with **menopause,** the end of menstruation and reproductive capacity. This usually occurs between 50 and 55, although the age range is large—from 40 to 59.

Following menopause, estrogen declines further. This causes the reproductive organs to shrink in size, the genitals to be less easily stimulated, and the vagina to lubricate more slowly during arousal. The drop in estrogen also contributes to other physical changes of aging. For example, it promotes decreased elasticity of the skin and loss in bone mass. And estrogen's ability to slow accumulation of plaque on the walls of the arteries is lost. Consequently, doctors often recommend **estrogen replacement therapy (ERT)** for postmenopausal women. Its advantages are clear: protection against bone deterioration and cardiovascular disease and a reduction in certain discomforts of the climacteric, such as hot flashes due to fluctuating hormones that cause sleeplessness and irritability. Nevertheless, ERT is controversial. It is associated with an increase in cancer of the endometrium (lining of the uterus) and, possibly, risk of breast cancer. Therefore, women and their doctors should make decisions about ERT carefully, on an individual basis (Bergkvist & Persson, 1996; Richedwards & Hennekens, 1996; Smith et al., 1996).

■ WOMEN'S PSYCHOLOGICAL REACTIONS TO MENOPAUSE. How do women react to menopause—a clear-cut signal that their childbearing years are over? The answer lies in how they interpret the event in relation to their past and future lives.

For Jewel, who had wanted marriage and family but never attained these goals, menopause was traumatic. Her sense of physical competence was still bound up with the ability to have children. Discomfort from physical symptoms—hot flashes, headaches, and loss of sleep—can also turn menopause into a difficult time. And as the climacteric evokes concern about an aging body in a society that values a youthful appearance, some women respond with disappointment to a loss of sex appeal.

At the same time, many women find menopause to be little or no trouble at all. They do not want more children, and they are relieved to see their menstrual periods end and to no longer have to worry about birth control. More highly educated, career-oriented women with fulfilling lives outside the home usually have more positive attitudes than do women with less education (Theisen et al., 1995). And the little evidence we have suggests that African-American and Mexican-American women have a generally favorable view, a trend that is stronger for Mexican Americans who have not yet adopted the language (and perhaps certain beliefs) of the larger society (Bell, 1995; Standing & Glazer, 1992).

The wide variation in physical symptoms and attitudes indicates that menopause is more than just a hormonal event; it is profoundly affected by societal beliefs and practices. For a cross-cultural look at menopause, turn to the Cultural Influences box on page 492.

■ REPRODUCTIVE CHANGES IN MEN. Men also experience a climacteric, but the change is limited to a decrease in quantity of semen and sperm after age 40 (Murray & Meacham, 1993). However, sperm continue to be produced throughout life, and men in their nineties have fathered children.

Testosterone production, along with its effects on male sex organs and sexual interest, remains stable throughout adulthood in healthy men who continue to engage in sexual activity (which stimulates cells that release testosterone). Consequently, no male counterpart to menopause exists.

Nevertheless, because of reduced blood flow to and changes in connective tissue in the penis, more stimulation is required for an erection, and it may be harder to maintain (Whitbourne, 1996). The inability to attain an erection when desired can occur at any age, but it becomes more common in midlife, affecting 20 percent of men by age 60. An episode or two of impotence is not serious, but frequent bouts can lead some men to fear that their sex life is over and undermine their self-image. Stress, alcohol abuse, and illness can contribute to the problem. But whatever the cause, the chances are excellent that with medical or psychological treatment, impotence can be cured (Larson, 1996).

SKELETON

As new cells accumulate on their outer layers, the bones broaden, but their mineral content declines, so they become more porous. This leads to a gradual loss in bone mass that begins in the late thirties and accelerates in the fifties, especially among women. Women's reserve of bone minerals is lower than men's to begin with. And following menopause, the favorable impact of estrogen on bone mineral absorption is lost. Reduction in bone density during adulthood is substantial—about 8 to 12 percent for men, 20 to 30 percent for women (Arking, 1991).

CULTURAL INFLUENCES

MENOPAUSE AS A BIOCULTURAL EVENT

Biology and culture join forces to influence women's response to menopause, making it a *biocultural event*. In the United States and other industrialized nations, menopause is "medicalized"—assumed to be a syndrome requiring treatment. Many women experience physical and emotional symptoms, and the more symptoms they report, the more negative their attitude toward menopause tends to be (Theisen et al., 1995).

Yet change the circumstances in which menopause is evaluated, and attitudes toward it change as well. In one study, nearly 600 men and women between ages 19 and 85 described their view of menopause in one of three contexts—when discussing it as a medical problem, as a life transition, or as a symbol of aging. The medical context evoked many more negative statements than did the other contexts (Gannon & Ekstrom, 1993).

Research in non-Western cultures reveals that changes in women's roles also affect the experience of menopause. In tribal and village societies where social power rests in the hands of men and non-childbearing women have low social status, women report the greatest distress during the climacteric. In cultures where older women are respected and the mother-in-law and grandmother roles bring new privileges and responsibilities, complaints about menopausal symptoms are rare (Patterson & Lynch, 1988).

A recent comparison of rural Mayan women of the Yucatan with rural Greek women on the island of Evia reveals additional biocultural influences on the menopausal experience (Beyene, 1992). In both cultures, old age is a time of increased status, and menopause brings freedom from child rearing and more time to participate in leisure activities. Otherwise, the experiences of the Mayan and Greek women, before and during the climacteric, differ greatly.

Mayan women marry at age 13 or 14. By 35 to 40, they have had many children but few menstrual periods (due to repeated pregnancies and breast-feeding). Because they are eager for childbearing to end, they welcome menopause, describing it with such phrases as "being happy" and "free like a young girl again." None report hot flashes or any other symptoms of discomfort.

Like Americans, rural Greek women use birth control to limit family size. And most report hot flashes and cold sweats during the climacteric. But they regard these as temporary discomforts that will stop on their own, not as medical symptoms requiring treatment. "Pay no attention," "Go outside for fresh air," and "Throw off the covers at night," the Greek women reply when asked what they do when they have hot flashes.

Does frequency of childbearing affect menopausal symptoms, as this contrast between Mayan and Greek women suggests? Much more research is needed to find out for sure. At the same time, the difference between American and Greek women in attitudes toward and management of hot flashes is striking. This—along with other cross-cultural findings—highlights the impact of culture on the menopausal experience.

For these rural Mayan women of the Yucatan, old age is a time of increased status, and menopause brings freedom. After decades of childbearing, Mayan women welcome menopause, describing it as "being happy" and "free like a young girl again." (Michael Gallagher/Liaison International)

Loss of bone strength causes the disks in the spinal column to collapse. Consequently, height may drop by as much as 1 inch by age 60, a change that will hasten thereafter. In addition, the weakened bones cannot support as much load, fracture more easily, and heal more slowly. A healthy lifestyle, including exercise and adequate calcium intake, can slow bone loss—in postmenopausal women by as much as 30 to 50 percent. As noted earlier, estrogen replacement therapy also helps preserve bone density and, consequently, prevents fractures (Reid, 1996).

When bone loss is very great, it leads to a debilitating disorder called *osteoporosis*. We will take up this condition when we consider illness and disability in the next section.

BRIEF REVIEW

Outward signs of aging—graying and thinning hair, wrinkling and sagging skin, and a weight gain leading to "middle-age spread"—are evident in midlife. In addition, the lens of the eye loses its capacity to accommodate, and color discrimination and ability to see clearly in dim light decline. Changes in the inner ear lead to hearing loss at high frequencies that gradually extends to the full range of sounds. Women's physical symptoms during the climacteric and psychological reactions to menopause vary widely among individuals and cultural groups. Although quantity of semen and sperm decline, men remain fertile throughout adulthood. Loss of bone mass occurs in both sexes, but it is more rapid in postmenopausal women.

ASK YOURSELF . . .

■ *At age 42, Stan's optician told him that he needed bifocals, and over the next 10 years, Stan required an adjustment to his corrective lenses almost every year. What physical changes account for Stan's repeated need for new eye wear?*

■ *Nancy noticed that between ages 40 and 50, she gained 20 pounds, and her arm and leg muscles seemed weaker. She had trouble opening tightly closed jars, and her legs ached after she climbed a flight of stairs. Nancy thought to herself, "Exchanging muscle for fat must be an inevitable part of aging." Is Nancy correct? Why or why not?*

HEALTH AND FITNESS

I n midlife, nearly 60 percent of people rate their health as either "excellent" or "good"—still a majority, but nevertheless far fewer than in early adulthood, when the figure is nearly 75 percent. Whereas younger people usually attribute health complaints to temporary infections, middle-aged adults more often point to chronic diseases. Visits to the doctor's office and hospital stays become more frequent. Middle-aged men are slightly more likely to consider their health as either "excellent" or "poor," whereas women more often place themselves in the middle

categories of "good" and "fair." The reason, as we will see shortly, is that men are more likely to suffer from fatal illnesses, whereas women more often have nonfatal, limiting health problems (Markides, 1994).

Our discussion takes up sexuality as a positive indicator of health, in addition to typical negative indicators—major diseases and disabling conditions. Before we begin, it is important to note that our understanding of health in middle and late adulthood is limited by lack of research on women and ethnic minorities. Most studies of illness risk factors, prevention, and treatment have been carried out on men. Fortunately, this situation is changing. For example, the Women's Health Initiative represents a 15-year commitment by the U.S. federal government to study the impact of prevention strategies—diet, exercise, estrogen replacement, and stress reduction—on the health of over 164,000 postmenopausal women of all ethnic groups and social classes (Finnigan, 1996).

SEXUALITY

Sexual activity among married couples tends to decline in middle adulthood, but only slightly. Longitudinal research reveals that stability of sexual activity is far more typical than dramatic change. Couples having sex often in early adulthood continue to do so in midlife. And the best predictor of sexual frequency is marital happiness (Edwards & Booth, 1994). Apparently, sexual pleasure and emotional intimacy remain strongly linked for most people.

Having sex is also associated with psychological well-being in the middle years of marriage. In a study of 3,000 married men and women, people not having sex with reasonable frequency were usually dissatisfied with the entire relationship (Blumenstein & Schwartz, 1983). The sexual activity–well-being association is probably bidirectional. Sex is more likely to occur in the context of a good marriage, and couples having sex often probably view their relationship more positively.

Nevertheless, intensity of sexual response declines somewhat in midlife, due to the physical changes of the climacteric. Both men and women take longer to feel aroused and to reach orgasm (Edwards & Booth, 1994). If partners perceive each other as less attractive, this may contribute to a drop in sexual desire. Yet in the context of a positive outlook, sexual activity can become more satisfying. Devin and Trisha, for example, viewed each other's aging bodies with acceptance and affection—as a sign of their enduring and deepening relationship. And with greater freedom from the demands of work and family, their sex life became more spontaneous.

Finally, when surveys include both married and unmarried people, a striking difference between men and women in age-related sexual activity appears. The proportion of men with no sexual partners in the previous year increases only slightly, from 8 percent in the thirties to 11 percent in the late fifties. In contrast, the rise for women is dramatic, from 9 to 30 percent, a gender gap that will become even greater in late adulthood (Michael et al., 1994). Opportunity, not desire, is responsible. A higher male mortality rate and the value women place on affection and continuity in sexual relations make partners less available to them. When we look at the evidence as a whole, sexual activity in midlife, as in earlier periods, is the combined result of biological, psychological, and social forces.

ILLNESS AND DISABILITY

As Table 15.1 shows, cancer and cardiovascular disease are the leading causes of death in middle age, with cancer by far the leading killer of women. Third in line is stroke, but it is not yet very common. Since it increases sharply from middle to late adulthood, we will consider it in Chapter 17 (U.S. Bureau of the Census, 1996). Despite a rise in vision problems, the unintentional injury rate due to motor vehicle collisions declines in midlife, perhaps because older people compensate with many years of driving experience or are more cautious. Not evident in the table is that certain injuries increase, especially bone fractures.

Extending trends established in earlier decades, men continue to be more vulnerable than women to most health problems, and economic disadvantage remains a strong predictor of poor health and premature death (see Chapter 13, page 425). As we look closely at major causes of illness, disability, and death, we will encounter yet another familiar theme: the close connection between emotional and physical well-being. Personality traits that magnify stress—especially hostility and anger—are serious threats to health in midlife.

■ CANCER. The death rate due to cancer multiplies tenfold from early to middle adulthood (see Table 15.1). In midlife, it is responsible for approximately 130,000 deaths in the United States each year—36 percent of all deaths. Although the incidence of many types of cancer is currently leveling off or declining, cancer mortality was on the rise for many decades, largely because of a dramatic increase in lung cancer due to cigarette smoking. In the last 5 years, lung cancer dropped in men, since 50 percent fewer smoke today than did in the 1950s. In contrast, lung cancer has increased in women, many of whom took up smoking after World War II (American Cancer Society, 1996).

Cancer occurs when the genetic programming of a normal cell is disrupted, leading to uncontrolled growth and spread of abnormal cells that crowd out normal tissues and organs. Why does this happen? Recall from Chapter 13 that according to one theory, random error in duplication of body cells increases with age, either due to release of free radicals or breakdown of the immune system. External agents may also initiate or intensify this process. Scientists recently discovered that damage to a gene called p53, which keeps cells with defective DNA from multiplying, is involved in 60 percent of cancers, including bladder, breast, cervix, liver, lung, prostate, and skin. New cancer therapies are beginning to target the p53 gene (Hager, 1996). Furthermore, an inherited proneness to certain cancers exists. For example, many patients with familial breast and ovarian cancer who respond poorly to treatment lack a particular tumor-suppressing gene. Genetic screening for this mutation is now available so prevention can begin as early as possible (Berchuck et al., 1996; Vandenberg et al., 1996).

Overall, a complex interaction of heredity, biological aging, and environment contributes to cancer. Figure 15.1 shows the most frequent types and the incidence of each. Illness and mortality rates are higher for certain groups of people than others. For cancers that affect both sexes, men are generally more vulnerable than women. The difference may be due to genetic makeup, exposure to cancer-causing agents as a result of lifestyle or occupation, and a tendency to delay going to the doctor. In Chapter 13, we noted that heart disease occurs more often among African Americans

TABLE 15.1

Death Rates[a] by Leading Causes in the United States

	25–44 YEARS	45–64 YEARS		
	TOTAL	TOTAL	MALES	FEMALES
Cancer	27	275	307	245
Cardiovascular Disease	20	215	314	123
Stroke	2	30	34	27
Unintentional Injury	31	29	43	16
Motor Vehicle	17	14	19	9

[a]Number of deaths per 100,000 population.
Source: U.S. Bureau of the Census, 1996.

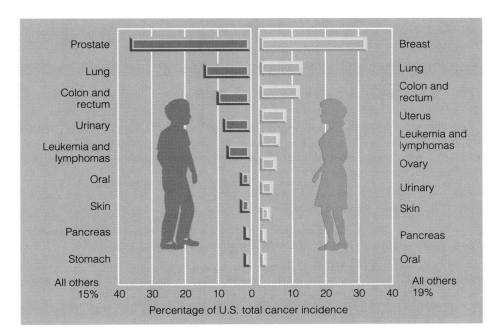

F IGURE 15.1

Cancer incidence in the United States by site and sex. (Adapted from American Cancer Society, 1996.)

than Caucasian Americans. The same is true for cancer. Researchers believe that unhealthy living conditions and behaviors linked to poverty, along with limited access to health care, are responsible (Flack et al., 1995).

Many people fear cancer because they believe it is incurable. Yet 40 percent of people with cancer are cured (free of the disease for 5 or more years). Breast cancer is the leading type for women, prostate cancer for men. Lung cancer—largely preventable through avoiding tobacco— ranks second for both sexes, followed closely by colon and rectal cancer. Scheduling annual medical checkups that screen for these and other forms and taking the additional steps listed in the Caregiving Concerns table on page 496 can reduce the incidence of and mortality from cancer considerably.

Surviving cancer is a triumph, but it also brings emotional challenges. During cancer treatment, relationships focus on the illness. Afterward, they must refocus on health and full participation in daily life. Unfortunately, stigmas associated with cancer exist. Friends, family, and co-workers may need to be reminded that cancer is not contagious and that research shows cancer survivors to be just as productive on the job as other people (American Cancer Society, 1996).

■ CARDIOVASCULAR DISEASE. Despite a decline in heart disease during the last few decades (see Chapter 13), about one hundred thousand 45- to 64-year-old Americans die of it annually, accounting for nearly 30 percent of deaths in this age group (U.S. Bureau of the Census, 1996). We associate cardiovascular disease with heart attacks, but Devin (like many middle-aged and older adults) first discovered it during an annual checkup. The doctor detected high blood pressure, high blood cholesterol, and *atherosclerosis*

—a buildup of plaque in his coronary arteries, which encircle the heart and provide its muscles with oxygen and nutrients. These signs of cardiovascular disease are known as "silent killers," since they often have no symptoms.

When symptoms are evident, they take different forms. The most extreme is a *heart attack*—blockage of normal blood supply to an area of the heart. It is usually brought on by a blood clot in one or more plaque-filled coronary arteries. Intense pain results as muscle in the affected region dies. A heart attack is a medical emergency; over 50 percent of victims die before reaching the hospital, another 15 percent during treatment, and an additional 7 to 10 percent over the next few years (Manson et al., 1992).

Among other less extreme symptoms of cardiovascular disease are *arrhythmia,* or irregular heartbeat. When it persists, it can prevent the heart from pumping enough blood and result in faintness. It can also allow clots to form within the heart's chambers, which may break loose and travel to the brain. In some individuals, indigestionlike or crushing chest pains, called *angina pectoris,* reveal an oxygen-deprived heart.

Today, cardiovascular disease can be treated in many ways—through coronary bypass surgery, medication, pacemakers to regulate heart rhythm, and more. To relieve arterial blockage, Devin had angioplasty, a procedure in which a surgeon threaded a needle-thin catheter into his arteries and inflated a balloon at its tip, which flattened fatty deposits to allow blood to flow more freely. At the same time, his doctor warned that unless Devin took other measures to reduce his risk, the arteries would clog again within a year. As the Caregiving Concerns table on page 497 indicates, there is much we can do to prevent heart disease or slow its progress.

Reducing the Incidence of Cancer and Cancer Deaths

INTERVENTION	DESCRIPTION
Know the seven warning signs of cancer.	Change in bowel or bladder habits; sore that does not heal; unusual bleeding or discharge; thickening or lump in a breast or elsewhere in your body; indigestion or swallowing difficulty; obvious change in a wart or mole; nagging cough or hoarseness. If you have any of these signs, consult your doctor immediately.
Engage in self-examination.	Women should self-examine the breasts and men the testicles for lumps and other changes once a month. If detected early, breast and testicular cancers can usually be cured.
Schedule regular medical checkups and cancer-screening tests.	Regular medical checkups, mammograms every 1 to 2 years and Pap tests every year for women, and other screening tests for both sexes increase early detection and cure.
Avoid tobacco.	Cigarette smoking causes 90 percent of lung cancer deaths and 30 percent of all cancer deaths. Smokeless (chewing) tobacco increases risk of cancer of the mouth, larynx, throat, and esophagus.
Avoid sun exposure.	Sun exposure causes many cases of skin cancer. When in the sun for an extended time, use a sun blocker and cover exposed skin.
Avoid unnecessary X-ray exposure.	Excessive exposure to X-rays increases cancer risk. Most medical X-rays are adjusted to deliver the lowest possible dose but should not be used unnecessarily.
Avoid exposure to industrial chemicals and other pollutants.	Exposure to nickel, chromate, asbestos, vinyl chloride, and other industrial agents increases risk of various cancers.
Weigh the benefits versus risks of estrogen replacement therapy (ERT).	Because ERT increases risk of cancer of the endometrium, carefully consider it with your doctor.
Maintain a healthy diet.	Avoid too much dietary fat and salt-cured, smoked, and nitrite-cured foods; eat vegetables and foods rich in fiber and vitamins A and C.

Source: Larson, 1996.

Of course, some risks, such as heredity, advanced age, and being male, cannot be changed. But cardiovascular disease is so disabling and deadly that we must be alert for it even where it is least expected. A special concern is accurate diagnosis in women. Since men account for over 70 percent of cases in middle adulthood, doctors may come to view a heart condition as a "male problem." Consequently, they may overlook symptoms in women. Adding to this possibility is that the first sign of an ailing heart in men is often a full-blown heart attack; in women, it is usually mild angina. Finally, most research on cardiovascular disease involves middle-aged men. Less is carried out on all people over age 65, when women are at greatest risk (Wichmann & Martin, 1992).

■ OSTEOPOROSIS. When age-related bone loss (described earlier in this chapter) is severe, a condition called **osteoporosis** is present. About 1 in every 4 postmenopausal women and the majority of people of both sexes over age 70 are affected (Donatelle & Davis, 1997; Larson, 1996). We associate osteoporosis with a slumped-over posture, a shuffling gait, and a "dowager's hump" in the upper back, but this extreme is rare. Because the bones gradually become more porous over many years, osteoporosis may not be evident until fractures—typically in the spine, hips, and wrist—occur or are discovered through X-rays.

Besides age and sex, factors related to osteoporosis include early menopause (and therefore, early decline in estrogen); a calcium-deficient diet; lack of physical activity (which reduces muscle and bone); a thin, small-framed body (which usually signifies a low peak bone mass); a family history of the disorder (heredity); cigarette smoking; and alcohol and caffeine intake (how these substances exert their effects is unknown). Much remains to be learned about osteoporosis, since some people with few of these risk factors still develop the disorder. Because of their higher bone density, African Americans are less likely than Caucasian and Asian Americans to be affected (Donatelle & Davis, 1997).

In the United States, osteoporosis affects over 25 million people and causes more than 1.5 million fractures annually. When the hip breaks, 12 to 20 percent of patients die within a year (Office of Technology Assessment, 1994). To treat the disorder, doctors recommend a diet enriched with calcium and vitamin D (which promotes calcium absorption), weight-bearing exercise (walking rather than

Reducing the Risk of Heart Attack

INTERVENTION	RISK REDUCTION
Quit smoking.	────────────────► 70% 5 years after quitting, up to 70 percent lower risk compared to current smokers
Reduce blood cholesterol level.	────────────────► 60% 2 to 3 percent decline in risk for each 1 percent reduction in blood cholesterol. Reductions in cholesterol average 10 percent with diet therapy and can exceed 20 percent with drug therapy.
Treat high blood pressure.	────────────────► 60% Combined diet and drug therapy can lower blood pressure substantially, leading to as much as a 60 percent risk reduction
Maintain ideal weight.	────────────────► 55% Up to 55 percent lower risk for those who maintain ideal body weight compared to those who are obese
Exercise regularly.	──────────────► 45% 45 percent lower risk for people who maintain an active rather than sedentary lifestyle
Drink a occasional glass of wine or beer.ᵃ	──────────────► 45% Up to 45 percent lower risk for people who consume small-to-moderate amounts of alcohol; believed to promote high-density lipoproteins, a form of "good cholesterol" that reduces "bad cholesterol"
Begin estrogen replacement therapy (ERT) after menopause.	─────────────► 44% 44 percent lower risk in users compared with nonusers (should be carefully considered; ERT increases risk of cancer of the endometrium)
Take low-dose aspirin.	────────────► 33% Up to 33 percent lower risk for people who take 325 mg (1 tablet) daily or every other day; reduces the likelihood of blood clots (should be medically recommended; long-term aspirin use can have serious side effects)
Reduce hostility and other forms of psychological stress.	Extent of risk reduction not yet known

ᵃRecall from Chapter 13 that heavy alcohol use increases the risk of cardiovascular disease.

Sources: Harpaz et al., 1996; Larson, 1996; Manson et al., 1992.

swimming), and estrogen replacement therapy (for women not at high risk for certain forms of cancer). Each of these interventions helps the bones regain mineral content (Birdwood, 1996; Larson, 1996). Prevention early in life is an even better course of action. Increasing calcium and vitamin D intake and engaging in regular exercise in childhood, adolescence, and early adulthood reduce lifelong risk by maximizing peak bone density (Anderson, Rondano, & Holmes, 1996; Sabatier et al., 1996).

HOSTILITY AND ANGER

Each time Trisha's sister Dottie called, she seemed like a powder keg ready to explode. Dottie was critical of her boss at work and dissatisfied with the way Trisha, a lawyer, had handled the family's affairs after their father died. All conversations ended the same way—with Dottie making demeaning, hurtful remarks as her anger rose to the surface. "Any lawyer knows that, Trisha; how could you be so stupid! I should have called a *real* lawyer." "You and Devin are so stuck in your privileged lives that you can't think of anyone else. You don't know what work *is*."

Trisha listened as long as she could bear. Then she warned, "Dottie, if you continue, I'm going to hang up. . . . Dottie, I'm ending this right now!"

Off the telephone, Dottie's life was full of health-related complaints and problems. At age 53, she had high blood pressure, difficulty sleeping, and back pain. During the past 5 years, she had been hospitalized five times—twice for treatment of digestive problems, twice for an irregular heartbeat, and once for a benign tumor on her thyroid gland. Trisha often wondered whether Dottie's personal style was partly responsible for her physical condition.

That hostility and anger might have negative effects on health is a centuries-old idea. Several decades ago, researchers first tested it by identifying 35- to 59-year-old men who displayed the **Type A behavior pattern**—extreme competitiveness, ambition, impatience, hostility, angry outbursts, and a sense of time pressure. They found that within the next 8 years, Type A's were more than twice as likely as Type B's (people with a more relaxed disposition) to develop heart disease (Rosenman et al., 1975).

Later studies, however, often failed to confirm these results, perhaps because Type A is a mix of behaviors, only one or two of which affect health. Indeed, recent evidence pinpoints hostility as the "toxic" ingredient of Type A, since isolating it from global Type A consistently predicts heart disease—and other health problems as well (Adams, 1994; Dembroski et al., 1989; Helmers, Posluszny, & Krantz, 1994). And new findings suggest that *expressed hostility,* especially, in the form of frequent angry outbursts and rude, disagreeable behavior, leads to greater cardiovascular arousal, health complaints, and illness. Why is this so? As people get angry, heart rate, blood pressure, and stress hormones rise—physical reactions that escalate anger, until the body's response is extreme (Siegman, 1994).

Of course, people who are repeatedly enraged are more likely to be dissatisfied with their lives and engage in unhealthy behaviors. But hostility predicts health problems even when such factors as smoking, alcohol consumption, overweight, general unhappiness, and negative life events are controlled. And since men score higher in hostility than do women (Dottie is an exception), emo-

tional style may contribute to the sex differences in heart disease described earlier (Stoney & Engebretson, 1994).

To preserve her health, should Dottie bottle up her hostility instead of expressing it? New findings suggest that inhibiting expression of emotion is also associated with increased mortality from heart disease, and it may pose other as yet unknown health risks (Denollet et al., 1996; Denollet, Sys, & Brutsaert, 1995). As we will see shortly, a better alternative is to develop effective ways of handling stress and conflict.

ADAPTING TO THE PHYSICAL CHALLENGES OF MIDLIFE

Middle adulthood is often a productive period in which people attain their greatest accomplishments and life satisfactions. Nevertheless, it takes considerable stamina to cope with the full array of changes this phase can bring. Devin responded to his enlarged waistline and cardiovascular symptoms by reducing job-related stress through daily 10-minute meditation sessions and leaving his desk twice a week to attend a low-impact aerobics class. Aware of her sister Dottie's difficulties, Trisha resolved to handle her own hostile feelings more adaptively. And her generally optimistic outlook enabled her to cope successfully with the physical changes of midlife, the pressures of her legal career, and Devin's chronic illness.

STRESS MANAGEMENT

Turn back to Chapter 13, pages 434–435, and review the negative consequences of psychological stress on the cardiovascular, immune, and gastrointestinal systems. As adults encounter problems at home and at work, daily hassles can add up to a serious stress load. Stress management is important at any age for a more satisfying life. In middle adulthood, it can limit the age-related rise in illness and, when disease strikes, reduce its severity.

The Caregiving Concerns table on the following page summarizes effective ways to reduce stress. Take a moment to list significant stressors in your life. Notice that many cannot be eliminated, but you can change the way you view them. For example, Trisha learned to distinguish normal emotional reactions from unreasonable self-blame. She stopped interpreting Dottie's anger as a sign of her own incompetence and, instead, reminded herself of Dottie's difficult temperament and hard life. At work, Trisha focused on problems she could control—not her boss's irritability, but how to delegate routine tasks to her staff so she could spend more time on problems that required her knowledge and skills. When pressures mounted, she

Is this angry man heading for an early death? Recent evidence pinpoints hostility as the "toxic" ingredient in the Type A behavior pattern. Expressed hostility, especially, in the form of frequent angry outbursts and rude, disagreeable behavior, leads to greater cardiovascular arousal, health complaints, and illness. (Tom Wurl/Stock Boston)

avoided rushing into action and carefully weighed alternatives. And greater life experience helped her accept change as inevitable. Consequently, she was better equipped to deal with the jolt of sudden change, such as Devin's hospitalization for treatment of heart disease.

Although communities often provide social supports aimed at easing the challenges of young adulthood and old age, they are less likely to do so for the middle adult years. Jewel had little knowledge of what to expect during the climacteric. "It would have helped to have a support group at church or the community center so I could have learned about menopause," she commented in a phone call to Trisha. Support services addressing other midlife concerns, such as adult learners returning to college and caregivers of elderly parents, would also reduce stress during this period (Patterson & Lynch, 1988).

Finally, constructive approaches to anger reduction are a vital health intervention (refer again to Table 15.4). Teaching people to be assertive rather than hostile and to negotiate rather than explode interrupts the intense physical response that intervenes between psychological stress and illness. If reasonable communication is not possible, it is may be best to delay responding by leaving a provocative situation, as Trisha did when she told Dottie that one more insult would cause her to hang up the phone (Deffenbacher, 1994).

EXERCISE

Regular exercise, as we showed in Chapter 13, has a wide variety of physical and psychological benefits. On the way to his first aerobics class, Devin wondered, "Can starting to exercise at age 50 counteract my many years of physical inactivity?" Devin's question is important, since 40 to 50 percent of adults are sedentary. Of those who begin an exercise program, 50 percent discontinue within the first 6 months. Among those who stay active, less than 20 percent exercise at levels that lead to health benefits (McAuley & Jacobson, 1991).

A person beginning to exercise in midlife must overcome initial barriers and obstacles along the way, such as lack of time and energy, inconvenience, and work conflicts. What helps people start exercising and stay with it? *Self-efficacy*—belief in one's ability to succeed—is just as vital in adopting, maintaining, and exerting oneself in an exercise regimen as it is in career progress (see Chapter 14). An important outcome of starting an exercise program is that it leads sedentary adults to gain in self-efficacy, which promotes physical activity all the more (McAule, Bane, & Mihalko, 1995; McAuley & Jacobson, 1991).

Most beginners prefer to exercise in groups, perhaps because they recognize that camaraderie will help keep them going. As Jewel wrote to Trisha:

CAREGIVING CONCERNS

Managing Stress

TECHNIQUE	DESCRIPTION
Reevaluate the situation.	Learn to tell the difference between normal emotional reactions and those based on irrational beliefs.
Focus on events you can control.	Don't worry about things you cannot change or that may never happen; focus on strategies for handling events under your control.
View life as fluid.	Expect change and accept it as inevitable; then many unanticipated changes will have less emotional impact.
Consider alternatives.	Don't rush into action; think before you act.
Set reasonable goals for yourself.	Aim high, but be realistic in terms of your capacities, motivation, and the situation.
Exercise regularly.	A physically fit person can handle stress better, both physically and emotionally.
Master relaxation techniques.	Relaxation helps refocus energies and reduce the physical discomfort of stress. Classes and self-help books can teach these techniques.
Use constructive approaches to anger reduction.	In addition to the techniques listed above, seek a delay in responding ("Let me check into that and get back to you"); use mentally distracting behaviors (counting to 10 backward) and self-instruction (a covert "Stop!") to control anger arousal; then engage in calm, self-controlled problem solving ("I should call him rather than confront him personally").
Seek social support.	Friends, family members, co-workers, and organized support groups can offer information, assistance, and suggestions for how to handle stressful situations.

What helps people start exercising and stay with it? Like these women who meet twice a week to walk, most beginners prefer to exercise in groups, perhaps because they recognize that camaraderie will help keep them going. (D. Greco/The Image Works)

I generally get up between 6:20 and 7A.M., depending on whether I swim or do step classes. Believe me, getting up and dragging yourself out in subzero weather to exercise is a testament to internal fortitude. However, having a "best friend" in the class frequently is a major motivating force—and warms the heart, especially on arctic days.

Indeed, group exercisers adhere to an exercise plan at twice the rate of individual exercisers. Encouragement, feedback, and praise from an enthusiastic group leader also sustain involvement. Combining these program characteristics with flexibility in goal setting (permitting participants to modify their exercise goals on the basis of how they feel that day) and cognitive distraction techniques (teaching participants to attend to pleasant, distracting stimuli when exercise gets uncomfortable) is particularly effective. Under these conditions, it is not unusual for up to 80 to 85 percent of enrollees to stick with a fitness program (Martin et al., 1984).

AN OPTIMISTIC OUTLOOK

Our ability to handle the inevitable changes of life depends in part on personality strengths. What type of individual is likely cope with stress adaptively, thereby reducing its impact on illness and mortality? Searching for answers to this question, researchers have studied a set of three personal qualities—control, commitment, and challenge—that, together, they called **hardiness** (Kobasa, 1979; Maddi & Kobasa, 1984).

Trisha fit the pattern of a hardy individual. First, she regarded most experiences as *controllable.* "You can't stop all bad things from happening," she advised Jewel after hearing about her menopausal symptoms, "but you can try to do something about them." Second, Trisha displayed a

committed, involved approach to daily activities, finding interest and meaning in almost all of them. Finally, she viewed change as a *challenge*—a normal part of life and a chance for personal growth.

Do these characteristics help people remain healthy in the face of stress? There is evidence that they do. Hardiness influences the extent to which people appraise stressful situations as manageable, interesting, and enjoyable. These positive appraisals, in turn, predict health-promoting behaviors, tendency to seek social support, and fewer physical symptoms (Florian, Mikulincer, & Taubman, 1995; Wiebe & Williams, 1992). Furthermore, high-hardy individuals are likely to use active, problem-centered coping strategies in situations they can control. In contrast, low-hardy people more often use emotion-centered and avoidant coping strategies—for example, saying "I wish I could change how I feel," denying that the stressful event occurred, or eating and drinking to forget about it (Carver, Scheier, & Weintraub, 1989; Williams, Wiebe, & Smith, 1992).

In some studies, hardiness-related positive appraisals were associated with lower physiological arousal to stress—a major means by which hardiness is believed to protect against illness. The link between hardiness and reduced physiological reactivity does not always appear, perhaps because a hardy person's active coping sometimes leads to greater arousal. But over time, this increase may be offset by a calm that comes with effective stress management (Wiebe & Williams, 1992). In support of this idea, a recent study found that hardiness was associated with higher blood levels of cortisol, a hormone that regulates blood pressure and is involved in resistance to stress (Zorilla, DeRubeis, & Redei, 1995).

In this and earlier chapters, we have seen that many factors act as stress-resistant resources—heredity, diet, exercise, social support, coping strategies, and more.

Research on hardiness adds yet another ingredient: a generally optimistic outlook and zest for life.

GENDER AND AGING: A DOUBLE STANDARD

Earlier in this chapter, we mentioned that negative stereotypes of aging lead many middle-aged adults to fear what is happening to their bodies. These unfavorable stereotypes are more likely to be applied to women than men. Despite the fact that many woman in midlife say they have "hit their stride"—feel assertive, confident, versatile, and capable of resolving life's problems—the broader cultural image of an older woman is one who is unattractive, incompetent, and passive (Sontag, 1979).

This double standard is evident when people are asked to evaluate older adults. Middle-aged women are rated as less attractive and as having more negative characteristics than are middle-aged men, a gender gap that widens with age. In some studies, aging men actually gain slightly in positive judgments of appearance, maturity, and power, whereas aging women show a sharp decline. Furthermore, women are often viewed as having reached middle and old age earlier (see Figure 15.2). And in some studies, the sex of the person doing the rating makes a difference: Compared to women, men judge an aging female much more harshly (Kogan & Mills, 1992).

These effects appear more often when people rate photos as opposed to verbal descriptions of men and women. Consequently, researchers believe that the ideal of a sexually attractive woman—smooth skin, good muscle tone, and lustrous hair—is at the heart of the double standard of aging. In Chapter 14, we noted that women prefer same-age or slightly older sexual partners, whereas men prefer younger partners. To explain why, ethological theory points to sex differences in reproductive capacity (see page 424). Does the end of a woman's ability to bear children contribute to negative judgments, especially by men? Some investigators think so (Kogan & Mills, 1992). Yet societal forces can exaggerate this view. For example, when media ads include middle-aged people, they are usually males—executives, fathers, and grandfathers who are images of competence and security. And the cosmetic industry offers many products designed to hide signs of aging for women, but far fewer for men (Harris, 1994).

At one time in our evolutionary history, a double standard of aging may have been adaptive. Today, as many couples limit childbearing and devote more time to career and leisure pursuits, it has become irrelevant. Role models of older women, whose lives are full of intimacy, hope, imagination, and accomplishment, can promote acceptance of physical aging and a new vision of growing older—one that emphasizes gracefulness, fulfillment, and inner strength.

BRIEF REVIEW

Despite more health concerns, the majority of middle-aged adults rate their health as either "excellent" or "good." Although intensity of the sexual response declines, sexual activity among married couples is fairly stable from early to middle adulthood and predicts psychological well-being. The proportion of women with no sexual partner increases substantially in midlife.

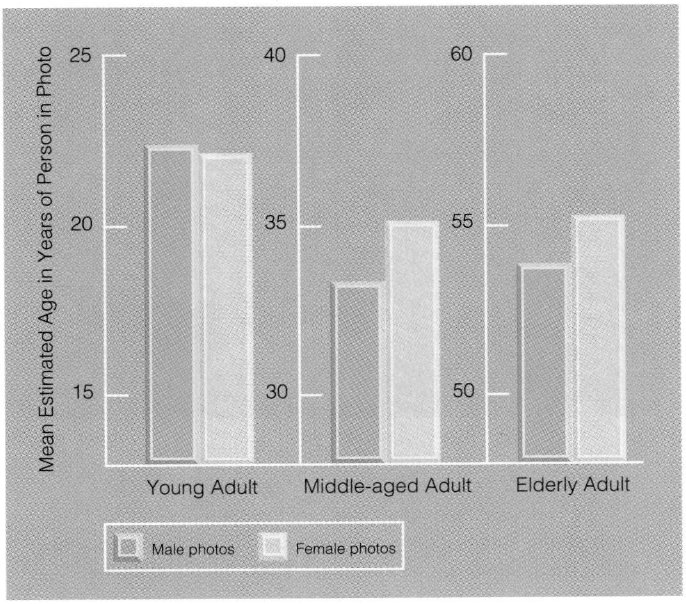

FIGURE 15.2

Findings of a study in which adults were asked to classify photos as young, middle-aged, and elderly and estimate the age of the person in each. In the middle-aged and elderly categories, age assignments were considerably younger for male than female photos. Many people believe that middle and old age arrive earlier for women than men. *(From N. Kogan, 1979, "A Study of Categorization," Journal of Gerontology, 34, p. 363. Copyright 1979 by the Gerontological Society of America. Reprinted by permission.)*

During middle adulthood, most people have their first personal encounter with life-threatening illness, either in themselves or their partners and friends. Death rates due to cancer and cardiovascular disease rise substantially. The overall incidence of these illnesses is higher in economically disadvantaged adults and in men. In contrast, far more women than men develop osteoporosis due to loss of estrogen after menopause. Although heredity and biological aging contribute to illness, much can be done to prevent disease, disability, and early death. Besides avoiding harmful substances (such as tobacco), improving diet, and exercising regularly, limiting hostility and other forms of stress through effective stress management techniques promotes healthy aging. Personal qualities that make up hardiness—control, commitment, and challenge—help people remain healthy in the face of stress.

ASK YOURSELF . . .

■ *When Cara complained of chest pains to Dr. Furrow, he decided to "wait and see" before conducting additional tests. In contrast, Cara's husband, Bill, received a battery of tests aimed at detecting cardiovascular disease, even though he did not complain of symptoms. What might account for Dr. Furrow's different approach to Cara than Bill?*

■ *Because his assistant misplaced some important files, Tom lost a client to a competitor. Tom felt his anger building to the breaking point. Explain why Tom's response is unhealthy, and suggest effective ways for dealing with it.*

COGNITIVE DEVELOPMENT IN MIDDLE ADULTHOOD

In middle adulthood, the cognitive demands of everyday life extend to new and, at times, more challenging situations. Consider a typical day in the lives of Devin and Trisha. Recently appointed dean of faculty at a small college, Devin was at his desk by 7 A.M. In between strategic planning meetings, he reviewed files of applicants for new positions, worked on the coming year's budget, and spoke at an alumni luncheon. In the meantime, Trisha prepared for a civil trial, participated in jury selection at the courthouse, then returned to her firm to join the other top lawyers for a conference about management issues. That evening, Trisha and Devin advised their 20-year-old son, Mark, who had dropped by to discuss his uncertainty over

whether to change his college major. By 7:30 P.M., Trisha was off to an evening meeting of the local school board. And Devin left for a biweekly gathering of an amateur quartet, in which he played the cello.

Recall from Chapter 13 that Schaie characterized middle adulthood as the *responsibility stage*—a time when expansion of responsibilities takes place on the job, in the community, and at home. To juggle diverse roles and perform effectively, Devin and Trisha called on a wide array of intellectual abilities, including accumulated knowledge, verbal fluency, memory, rapid analysis of information, reasoning, problem solving, and expertise in their areas of specialization. What changes in thinking take place in middle adulthood? How does vocational life—a major arena in which cognition is expressed—influence intellectual skills? And what can we do to support the rising tide of adults who are returning to college in hopes of enhancing their knowledge and quality of life? Let's see what research has to say about these topics.

This police officer brings his 12-year-old daughter with him on "Take our daughters to work day." He exemplifies Schaie's responsibility stage—a time when responsibilities expand on the job, in the community, and at home. (Bob Daemmrich/The Image Works)

CHANGES IN MENTAL ABILITIES

At age 50, when he occasionally couldn't recall a name or had to pause in the midst of a lecture or speech to think about what to say next, Devin wondered to himself, "Are these the first signs of an aging mind?" Twenty years earlier, Devin took little notice of the very same mental events. His current questioning stems from a widely held stereotype—a view of the aging mind as on a path of inevitable decline. The majority of aging research has focused on deficits because they are cause for concern, while neglecting cognitive stability and gains (Salthouse, 1991a).

In Chapter 13, we noted that some apparent decrements in cognitive aging result from weaknesses in the research itself. In cross-sectional studies, *cohort effects,* in which older participants are less educated and in poorer health, can create an inaccurate picture of both the timing and extent of decline. Also, the tests given may tap abilities less often used by older individuals, whose lives no longer require that they learn information for its own sake but, instead, build knowledge and skills that help them tackle current, real-world problems.

As we examine changes in thinking in middle adulthood, we will revisit the theme of diversity in development. Different aspects of cognitive functioning show different patterns of change. Although declines occur in some areas, most people display cognitive competence, especially in familiar contexts, and some attain outstanding accomplishment. Overall, the evidence supports an optimistic view of adult cognitive potential.

The changes we are about to consider bring into bold relief core assumptions of the lifespan perspective: development as *multidimensional,* or the combined result of biological, personal, and social forces; development as *multidirectional,* or the joint expression of growth and decline, with the precise mix varying across abilities and individuals; and development as *plastic,* or open to change, depending on how a person's biological and environmental history combines with current life conditions. Before we begin, you may find it helpful to return to Chapter 1, pages 9–10, to review these important ideas.

CRYSTALLIZED AND FLUID INTELLIGENCE

Many studies report consistent age-related trends in two broad-ranging mental abilities. Each includes a diverse array of specific intellectual factors tapped by intelligence tests.

The first of these broad abilities is called **crystallized intelligence.** It refers to skills that depend on accumulated knowledge and experience, good judgment, and mastery of social conventions. Together, these capacities represent abilities acquired because they are valued by the individual's culture. Devin made use of crystallized intelligence when he expressed himself articulately at the alumni luncheon and suggested effective ways to save money in budget planning. On intelligence tests, vocabulary, general information, verbal analogy, and logical reasoning items measure crystallized intelligence. (Turn back to Chapter 9, page 299, to see examples.)

In contrast, **fluid intelligence** depends more heavily on basic information-processing skills—the speed with which we can analyze information, the capacity of working memory, and the ability to detect relationships among stimuli. Fluid intelligence often works with crystallized intelligence to support effective reasoning, abstraction, and problem solving. But fluid intelligence is believed to be influenced more by conditions in the brain and learning unique to the individual, less by culture (Horn, 1982). Intelligence test items that reflect fluid abilities include number series, spatial visualization, figure matrices, and picture series. (Refer again to page 299 for examples.)

Research shows that crystallized intelligence increases steadily with age into late adulthood, whereas fluid intelligence starts to decline in the late twenties or early thirties (see Figure 15.3). These trends are found repeatedly in

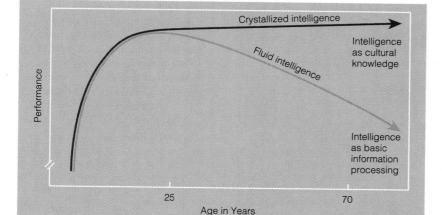

FIGURE 15.3

Changes in crystallized and fluid intelligence over the lifespan. Many cross-sectional studies reveal that crystallized intelligence increases into late adulthood, whereas fluid intelligence starts to decline in the late twenties or early thirties. Longitudinal research presents a somewhat more optimistic picture for fluid abilities. Nevertheless, it confirms that they decline earlier and more rapidly than crystallized abilities. *(Adapted from Baltes, 1987.)*

cross-sectional comparisons, even when education and health status of younger and older participants are equivalent (Horn & Donaldson, 1980; Horn, Donaldson, & Engstrom, 1981). The rise in crystallized abilities makes sense, since adults are constantly adding to their knowledge and skills at work, at home, and in leisure activities. In addition, many crystallized skills are practiced almost daily. But perhaps you are questioning, as others have, the early decline in fluid intelligence on the basis of longitudinal evidence you read about in Chapter 13. Now let's take a closer look.

Recall the Seattle Longitudinal Study, which revealed modest gains in five mental abilities from early to middle adulthood that were sustained until the fifties and early sixties, after which performance gradually declined. Figure 15.4 shows these trends. The five abilities—verbal ability, inductive reasoning, verbal memory, spatial orientation, and numeric ability—tap both crystallized and fluid skills. But a sixth ability is also shown: perceptual speed, a fluid skill in which participants must, for example, identify during a time limit which of five shapes is identical to a model or whether pairs of multidigit numbers are the same or different (Schaie, 1994, 1996). Perceptual speed dropped steadily from the twenties to the late eighties, a finding that fits with a wealth of research indicating that cognitive processing slows as people get older. And notice in Figure 15.4 how, late in life, the fluid factors (spatial orientation, numeric ability, and perceptual speed) show greater decrements than the crystallized factors (verbal ability, inductive reasoning, and verbal memory).

Some theorists believe that a general slowing of central nervous system functioning underlies nearly all age-related declines in cognitive performance (Salthouse, 1985). Many

researchers have tested this idea and found at least partial support. They have also discovered other important changes in information processing, some of which may be triggered by the decline in speed (Kausler, 1994).

But before we turn to this evidence, let's clarify why longitudinal research reveals stability in a variety of crystallized and fluid abilities during middle adulthood, despite declines in basic information-processing skills. There are several reasons. First, the decrease in basic processing is modest until late in life—not great enough to affect many well-practiced performances. For the most part, it shows up only on complex, unfamiliar tasks. Second, as we will see shortly, adults often find ways to compensate for cognitive weaknesses by drawing on their cognitive strengths. Finally, as people discover they are no longer as good as they once were at certain tasks, they accommodate, shifting to activities that depend less on cognitive efficiency and more on accumulated knowledge. The basketball player becomes a coach, the quick-witted salesperson a manager (Salthouse, 1991b).

INDIVIDUAL DIFFERENCES

Hidden beneath the age trends just described are large individual differences. Some people, because of disease or very unfavorable environments, begin to decline intellectually in their forties. A few maintain full functioning at a very advanced age (Schaie, 1989).

Using intellectual skills seems to affect the degree to which they are retained. In the Seattle Longitudinal Study, declines were delayed for people with above-average education, highly complex occupations, and stimulating leisure pursuits that included reading, traveling, attending

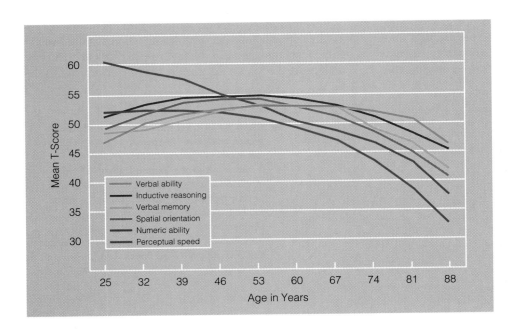

FIGURE 15.4

Longitudinal trends in six mental abilities. In five abilities, modest gains were sustained until the fifties and early sixties, followed by gradual declines. The sixth ability—perceptual speed—dropped steadily from the twenties to the late eighties. Late in life, fluid factors (spatial orientation, perceptual speed, and numeric ability) showed greater decrements than crystallized factors (verbal ability, inductive reasoning, and verbal memory).

cultural events, and participating in clubs and professional organizations. People with flexible personalities, lasting marriages (especially to a cognitively high-functioning partner), and absence of cardiovascular and other chronic diseases were also likely to maintain mental abilities well into late adulthood. And being economically well off was linked to favorable cognitive development, undoubtedly because social class is associated with many of the factors just mentioned (Schaie, 1996).

Finally, maintaining high levels of perceptual speed predicted an advantage on other mental abilities. As we turn now to information processing in midlife, we will see why a decrement in speed of processing affects other aspects of cognitive functioning.

INFORMATION PROCESSING

nformation-processing researchers interested in adult development usually use the same model of the mental system introduced in Chapter 5 (see page 154) to guide their exploration of different aspects of thinking. As processing speed slows, certain aspects of attention and memory decline as well. Yet midlife is also a time of cognitive gains as adults apply their vast knowledge and life experience to problem solving in the everyday world.

SPEED OF PROCESSING

One day, Devin looked over the shoulder of his 20-year-old son Mark, who was playing a computer game. Mark responded to multiple cues on the screen in rapid-fire fashion. "Let me try it," suggested Devin, who practiced over several days but remained well behind Mark in performance. And on a family holiday in Australia, Mark adjusted quickly to driving on the left side of the road. Trisha and Devin took longer. After several days, they still felt confused at intersections, where they had to respond rapidly to lights, cross traffic, and oncoming vehicles.

These real-life experiences fit with laboratory findings. On simple reaction time tasks (pushing a button in response to a light) and complex reaction time tasks (pushing a left-hand button to a blue light, a right-hand button to a yellow light), response time increases steadily from the early twenties into the nineties. Older adults are increasingly disadvantaged as situations requiring rapid responding become more complex. The decline is small, under 1 second in most studies, but it is nevertheless of practical significance. And for unknown reasons, reaction time slows at a faster rate for women than men (Fozard et al., 1994; Salthouse, 1985).

What causes this age-related slowing of cognitive processing? Although researchers agree that changes in the

How do older adults adapt to age-related slowing of cognitive processing? They make up for declines in processing speed through practice, experience, and a compensatory approach that permits them to prepare responses in advance. This older musician's many years of playing and familiarity with complex music help him perform swiftly and fluidly. (Andy Sacks/Tony Stone Images)

brain are probably responsible, they disagree on the precise explanation. According to the **neural network view,** as neurons in the brain die, breaks in neural networks occur. The brain adapts by forming bypasses—new synaptic connections that go around the breaks but are less efficient (Cerella, 1990). A second hypothesis, the **information-loss view,** suggests that older adults differ from younger adults in the rate at which information is lost as it moves through the system. Imagine making a photocopy and using it to make another copy. Each time we do this, the copy becomes less clear. Similarly, with each step of thinking, information degrades. The older the adult, the more exaggerated this effect. As a result, the whole system must slow down to inspect and interpret the information. Since complex tasks have more processing steps, they are more affected by information loss (Myerson et al., 1990).

Yet many older adults perform complex, familiar tasks with considerable efficiency. Devin, for example, played a Mozart quartet on his cello with great speed and dexterity, keeping up with three other players who were 10 years his junior. How did he manage? Compared to the other players, he more often looked ahead in the score, taking a greater chunk of music into working memory. This compensatory approach permitted him to prepare a response in advance, thereby minimizing the importance of speed. In one study, researchers asked 19- to 72-year-olds to do transcription typing and also tested their reaction time. Although reaction time slowed with age, speed of typing did not change. Like Devin, older individuals looked further ahead in the transcription, processing more characters at once (Salthouse, 1984).

Practice and experience can also compensate for declines in processing speed. Devin's many years of playing the cello and his familiarity with the Mozart quartet undoubtedly supported his ability to play swiftly and fluidly.

ATTENTION

Studies of attention focus on how much information adults can take into their mental systems at once and the extent to which they can control their attention by ignoring irrelevant information. When Dottie telephoned, Trisha sometimes tried to prepare dinner or continue working on a legal brief while talking on the phone. With age, she found it harder to divide her attention between the two activities. Consistent with Trisha's experience, laboratory research reveals that when tasks are complex, sustaining two at once becomes harder with age (Kausler, 1991; Salthouse, Rogan, & Prill, 1984). An age-related decrement also occurs in the ability to keep attention focused on relevant information in the presence of many surrounding irrelevant stimuli (Hartley & McKenzie, 1991).

These declines in attention might be due to the slowdown in information processing described earlier, which limits the amount of information a person can handle at once (Plude & Hoyer, 1985). Reduced speed of processing may also contribute to a related finding: a decrement with age in the ability to combine many pieces of visual information into a meaningful pattern. When the mind inspects stimuli slowly, they are more likely to remain disconnected. This problem, in turn, can intensify attentional difficulties (Plude & Doussard-Roosevelt, 1989). Without a coherent pattern to serve as a guide, adults may have difficulty inhibiting the processing of irrelevant information. Consequently, attention is more disorganized.

But once again, adults can compensate for these changes. People highly experienced in attending to certain information and ignoring others, such as air traffic controllers, golfers, and birdwatchers, know exactly what to look for. As a result, they may be less affected by attentional difficulties. Furthermore, practice can improve the ability to divide and control attention. When older adults receive training in these skills, their performance improves as much as that of younger adults, although training does not close the gap between age groups (Kausler, 1994).

MEMORY

Memory is crucially important for all aspects of information processing—an important reason that we place great value on a good memory in middle and late adult-

hood. From the twenties into the sixties, the amount of information people can retain in working memory diminishes. Whether given lists of words or digits or meaningful prose passages to learn, middle-aged and older adults recall less than young adults (Salthouse, 1991b; Salthouse & Skovronek, 1992).

This change is largely due to a decline in use of memory strategies on these tasks. Older individuals rehearse less than do younger individuals—a difference believed to be due to a slower rate of thinking. Older people cannot repeat new information to themselves as quickly as younger people (Salthouse & Babcock, 1991).

Memory strategies of organization and elaboration are also applied less often and less effectively with age. (See Chapter 9, page 293, if you need to review these strategies.) Both require people to link incoming information with already stored information. One reason older adults are less likely to use organization and elaboration is that they have more difficulty retrieving information from long-term memory that would help them recall. For example, given a list of words containing "robin," "parrot," and "blue jay," they find it harder to access the category "bird," even though they know it well (Hultsch, 1975). Why does this happen? Greater difficulty keeping attention on relevant as opposed to irrelevant information seems to be involved. As irrelevant stimuli take up space in working memory, less is available for the memory task at hand (Hartman & Hasher, 1991; Salthouse & Meinz, 1995).

But we must keep in mind that the memory tasks given by researchers require strategies that many adults seldom use and may not be motivated to use, since most are not in school (see Chapter 9, page 294). When instructed to organize or elaborate, middle-aged and older people willingly do so, and their performance improves. And there are other ways to compensate for age-related declines in working memory. For example, we can slow the pace at which information is presented so adults have enough time to process it or cue the link between new and previously stored information ("To learn these words, try thinking of the category 'bird'") (Kausler, 1994).

Furthermore, when we consider the variety of memory skills we call on in daily life, the decrements just described are limited in scope. General *factual knowledge* (such as historical events), *procedural knowledge* (such as how to drive a car, ride a bike, or solve a math problem), and knowledge related to one's occupation either remain unchanged or increase into midlife (Baltes, Dittmann-Kohli, & Dixon, 1984). Middle-aged people who have trouble recalling something often draw on decades of accumulated *metacognitive knowledge* about how to maximize performance—reviewing major points before an important presentation, organizing notes and files so information can be found quickly, and parking the car in the same area of the parking lot each day.

In sum, age changes in memory vary widely across tasks and individuals as people use their cognitive capacities to meet the requirements of their everyday worlds. This may remind you of Sternberg's *triarchic theory of intelligence*, described in Chapter 9—in particular, his *contextual subtheory* (see page 300). It emphasizes that intelligent people adapt their information-processing skills to fit with their personal desires and the demands of their environments. To understand memory development (and other aspects of cognition) in adulthood, we must view it in context. As we turn to problem solving, expertise, and creativity, we will see this theme again.

PRACTICAL PROBLEM SOLVING AND EXPERTISE

One evening, as Devin and Trisha sat in the balcony of the Chicago Opera House awaiting curtain time, a small figure appeared on the stage to announce that 67-year-old Ardis Krainik, general director and "life force" of the opera company, had died. A hush fell over the theater. Soon, members of the audience turned to one another, wanting to know more about the woman who had made the opera company into one of the world's greatest.

Starting as a chorus singer and clerk typist, Ardis rose rapidly through the ranks, becoming assistant to the director and developing a reputation for tireless work and unmatched organizational skill. When the opera company fell deeply into debt, the board of directors turned to Ardis to save it from disaster. As newly appointed general director, she erased the deficit within a year and began to restore the company's sagging reputation. She charmed executives who knew nothing about opera into making big contributions, attracted world-class singers, and filled the house to

As its general director, Ardis Krainik (right) saved the Lyric Opera of Chicago from disaster, erasing the opera's deficit and restoring its sagging reputation. Her creative achievements depended on her ability to reconcile the company's subjective, artistic needs with its objective, financial needs. Drawing on her life experience, she dealt with a complex practical problem in a unique way. (She is shown here, during the 1980s, with Patricia Ryan of Lyric Opera's Executive Committee and world renown tenor Plácido Domingo.) (Lyric Opera of Chicago Archives)

98 percent capacity. On her office wall hung a sign she received as a gift after one year as head of the company. It read, "Wonder Woman" (Rhein, 1997).

As Ardis's story illustrates, middle-aged adults have special opportunities to display continued cognitive growth in the realm of **practical problem solving,** which requires people to size up real-world situations and analyze how best to achieve goals that have a high degree of uncertainty. Gains in *expertise*—an extensive, highly organized, and integrated knowledge base that can be used to support a high level of performance—help us understand why practical problem solving takes a leap forward.

The development of expertise is underway in early adulthood. But it reaches its height in midlife, leading to highly efficient and effective approaches to solving problems that are organized around abstract principles and intuitive judgments. Saturated with experience, the expert intuitively feels when an approach to a problem will work and when it will not. This rapid, implicit application of knowledge is the result of years of learning and experience. It is unlikely to show up on laboratory tasks that assess basic processing but do not call on this knowledge (Smith & Baltes, 1992; Wagner & Sternberg, 1985).

Expertise is not just the province of the highly educated and those who rise to the top of administrative ladders. It can emerge in any field of endeavor. In a study of food service workers, researchers identified the diverse ingredients of expert performance in terms of physical skills (strength and dexterity), technical knowledge (of menu items, ordering, and food presentation); organizational skills (a sense of priority, anticipating customer needs); and social skills (confident presentation; pleasant, polished manner). Next, 20- to 60-year olds with less than 2 to more than 10 years experience were evaluated on these qualities. Although physical strength and dexterity declined with age, job knowledge and behavior increased. Among younger and older adults with similar years of experience, middle-aged employees performed more competently, serving customers in an especially adept, attentive way (Perlmutter, Kaplan, & Nyquist, 1990).

Age-related advantages in problem solving are also evident in leisure activities and everyday dilemmas. A study of middle-aged men who regularly attended harness races and were expert in predicting winners revealed that they used a highly complex reasoning process in which they integrated a large number of variables, such as a horse's speed, its lifetime earnings, the driver's ability, and track size. This permitted the men to compare eight or more horses at once. The ability to engage in high-level, integrative thinking in the harness-racing context was unrelated to IQ (Ceci, 1990).

Similarly, research on everyday problem solving reveals that it peaks between 40 and 59 years of age and may be sustained even longer (Denney, 1990; Denney & Pearce, 1989). Consider the following problem:

What would you do if you had a landlord who refused to make some expensive repairs you want done because he or she thinks they are too costly?

(a) Try to make the repairs yourself.

(b) Try to understand your landlord's view and decide whether they are necessary repairs.

(c) Try to get someone to settle the dispute between you and your landlord.

(d) Accept the situation and don't dwell on it.
    (Cornelius & Caspi, 1987, p. 146)

On tasks like these, middle-aged and older adults select better solutions (as rated by independent judges) than do young adults. In the preceding example, the preferred response is (b), a problem-centered approach that involves seeking information and using it to guide action (Cornelius & Caspi, 1987).

CREATIVITY

Like problem solving, creativity may change with advancing age. Some researchers believe it shifts from a focus on generating unusual products (divergent thinking) to an emphasis on integrating experience and knowledge into unique ways of thinking and acting (Abra, 1989). For example, Ardis's creative achievements depended on her ability to reconcile the opera company's subjective, artistic needs with its objective, financial needs. Fueled by life experience, she dealt in a unique way with a complex practical problem (Sasser-Coen, 1993).

Creativity in middle adulthood may also reflect a transition from a largely egocentric concern with self-expression to more altruistic goals. As the middle-aged person overcomes the illusion of youth that "life can last forever," the desire to contribute to humanity and enrich the lives of others increases (Dacey, 1989). This change may be partly responsible for the midlife decline in overall creative output described in Chapter 13. In reality, however, creativity in emerges in new forms.

INFORMATION PROCESSING IN CONTEXT

In sum, cognitive gains in midlife are especially likely in areas involving experience-based buildup and transformation of knowledge and skills. Consider the evidence we have just reviewed, and you will see that processing speed varies with the situation. When given a challenging real-world problem related to his or her expertise, the middle-aged adult is likely win out not just in efficiency, but in quality and originality of thinking. And as the Lifespan Vista box on the following page illustrates, when researchers' laboratory tasks are relevant to the real-life endeavors of intelligent, cognitively active adults, older people respond as quickly and competently as their younger counterparts do!

By midlife, people's past and current experiences vary enormously—more so than in previous decades. Therefore, thinking in middle adulthood is characterized by an increase in specialization; it branches out in a multitude of directions. Yet for middle-aged adults to realize their cognitive potential, they must have opportunities for continued growth. Let's see how vocational and educational environments can support cognition in middle adulthood.

BRIEF REVIEW

Cognitive development in midlife reflects core assumptions of the lifespan perspective: development as multidimensional, multidirectional, and plastic. Longitudinal research reveals that mental abilities reflecting crystallized intelligence remain stable or increase, whereas those depending on fluid intelligence start to decline. At the same time, large individual differences exist that are related to opportunities to use intellectual skills, personality, physical health, and social class.

With age, speed of information processing declines, a trend believed to be due to changes in the brain. The slowdown, although small, may contribute to decrements in attention, capacity of working memory, and use of memory strategies on laboratory memory tasks. Yet middle-aged and older adults have ways of compensating, and performance improves with training and practice.

Aspects of memory used often in daily life remain unchanged or increase during middle adulthood. Greater life experience and expertise in specialized fields permit large gains in practical problem solving and a shift in creativity toward integrating experience and knowledge into original ways of handling complex everyday problems.

ASK YOURSELF . . .

■ *In what aspects of cognition did Devin decline, and in what aspects did he gain? How do changes in Devin's thinking reflect assumptions of the lifespan perspective*

■ *Asked about hiring older adults as waitpeople, one restaurant manager replied, "I cannot hire enough older workers . . . they are my best employees . . . I do not even know what you mean by slowness. They are the fastest ones on the floor" (Perlmutter, Kaplan, & Nyquist, 1990, p. 189). Why does this manager find older employees desirable, despite the age-related decline in speed of processing?*

HOW GOOD IS YOUR PROFESSOR'S MEMORY? THE TASK MAKES A DIFFERENCE

Compare the performance of a typical-age college student and a professor on reaction time tasks and memory for unrelated information. Although the professor has many more years of education, the college student will probably do better. But when a task requires general or metacognitive knowledge used often in academic life, the professor is likely to display considerable competence, even at a mature age.

Research on professors' cognitive processing is especially revealing, since factors that ordinarily make age-related change difficult to interpret are controlled. For example, professors of different ages share similar cultural experiences, in that their years of education, work environments, and social class are alike. And as long as they are still employed, there is a good chance that all have been mentally active throughout their lives. These similarities reduce cohort effects.

In a recent study, three groups of university professors at the University of California, Berkeley—young (age 30 to 44), middle-aged (age 45 to 59), and senior (age 60 to 71)—were administered a battery of speed-of-

processing and memory tasks. All had 21 to 22 years of education and were actively engaged in academic work. Control groups of very young adults (college students age 18 to 23) and older adults (age 60 to 71) with 14 to 16 years of education were also tested (Shimamura et al., 1995).

How did the professors and the controls perform? When asked to push a button as rapidly as possible after a signal, both showed the typical age-related decline in speed of processing. Both groups also declined on tasks requiring them to remember arbitrary associations. For example, in one task, they were given six pairs of faces, each consisting of a female and a male. Then they were shown each female face and asked to select its male partner from a display.

Older professors' performance on a task designed to tax the capacity of working memory, however, deviated from the researchers' expectations of age-related intellectual decrements. An array of 16 visual patterns was presented sixteen times on a computer screen, randomly arranged each time. On each trial, the adult had to point to a different pattern, monitoring previous responses so none would be repeated (see Figure 15.5). Then, a second block of sixteen trials was presented, adding the chal-

lenge of remembering responses not just within a block but also between blocks. In previous investigations and among controls in this study, older individuals typically did much worse on the second than on the first block of trials. The reason is believed to be their greater difficulty ignoring irrelevant information. Yet senior professors did not show this increasing error rate from one block of trials to the next!.

Finally, when asked to recall meaningful prose passages, once again the professors showed no decline in performance with age. In contrast, the recall of controls dropped off in a predictable pattern.

What are the implications of these results? The similar declines among professors and controls in cognitive speed and memory for unrelated information suggest that these trends are strongly affected by biological aging. But what about the two performances in which professors showed no age decrements? Both require metacognitive skills of planning, organization, and manipulating information in memory. And the prose recall task depends on integrating new knowledge with a large fund of existing knowledge. Perhaps years of applying these strategies in work life counteracted the effects of aging.

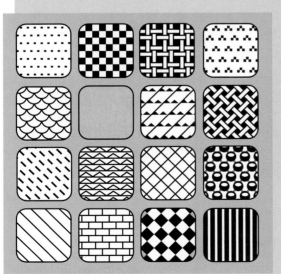

FIGURE 15.5

Array of visual patterns in a task designed to tax the capacity of working memory. On each trial, the adult had to point to a different pattern while monitoring previous responses so none would be repeated. The task was presented twice, yielding two blocks of sixteen trials. Senior professors did not show the expected age-related increase in errors from the first to the second block of trials, a pattern of change strongly evident in the performance of older controls. *(From A. P. Shimamura, J. M. Berry, J. A. Mangels, C. L. Rusting, & P. J. Jurica, 1995, Memory and cognitive abilities in university professors: Evidence for successful aging. Psychological Science, 6, p. 272. Copyright © 1995 by the American Psychological Society. Reprinted by permission.)*

VOCATIONAL LIFE AND COGNITIVE DEVELOPMENT

Vocational settings are vital contexts for encouraging people to maintain, restore, and apply old skills and learn new ones. Yet they vary in the extent to which they are cognitively stimulating and promote autonomy. At times, work environments carry unfavorable stereotypes about age-related problem-solving and decision-making skills that may lead older employees to be assigned less challenging work or encouraged to retire before they would otherwise do so.

In Chapter 13, we indicated that intellectual and personality characteristics affect the occupations people select. Devin, for example, chose college teaching and administration because he enjoyed reading and writing, framing new ideas, and helping others learn. Once a person is immersed in a job, it influences cognition. In a study of over 600 American men representing a wide range of occupations, researchers asked participants about the task complexity and self-direction of their jobs. During this interview, they also assessed cognitive flexibility, based on logical reasoning, seeing both sides of an issue, and making judgments independently. A decade later, the job and cognitive variables were remeasured, permitting a look at their effects on each other. As expected, findings indicated that cognitively flexible men sought work that offered challenge and autonomy. But complex work also led to gains in cognitive flexibility! In other words, the relation between vocational life and cognition was reciprocal (Kohn & Schooler, 1978).

These same findings emerged in large-scale studies carried out in Japan and Poland—cultures quite different from the United States (Kohn et al., 1990; Kohn & Slomczynski, 1990). In each nation, having a stimulating, nonroutine job helped explain the relationship between social class and flexible, abstract thinking. Furthermore, learning on the job generalizes to other realms of life. People who do intellectually demanding work seek out stimulating leisure pursuits (Miller & Kohn, 1983). And they come to value self-direction, both for themselves and for their children. Consequently, they are likely to pass their cognitive preferences to the next generation.

Is the impact of a challenging job on cognitive growth greater for young adults, who are in the early phase of vocational development? Research shows that it is not. People in their fifties and early sixties gain as much as do those in their twenties and thirties. The relationship also holds for people of different generations and, therefore, widely varying life experiences (Miller, Slomczynski, & Kohn, 1985). Once again, we are reminded of the plasticity of development. Cognitive flexibility is responsive to vocational experience well into middle adulthood, and perhaps beyond.

ADULT LEARNERS: BECOMING A COLLEGE STUDENT IN MIDLIFE

Adults are returning to college in large numbers. Over the past quarter century, students over age 25 (the definition of adult students) in American colleges and universities increased from 28 to 44 percent (U.S. Department of Education, 1996). Their reasons for going back to school are diverse—a career change, a better

The reciprocal relationship between vocational life and cognition is likely to benefit this teacher of English literature in Kyoto, Japan. Large-scale studies in cultures as diverse as the United States, Japan, and Poland show that cognitively flexible individuals seek work that offers challenge and autonomy. In return, complex work leads to gains in cognitive flexibility. (Cameramann/The Image Works)

This mature woman's return to college may have been sparked by a life transition, such as divorce, widowhood, or job layoff. At first, adult learners often question their ability to be successful at academic work. When family members, friends, and educational institutions are supportive, returning students reap great personal benefits and do well academically. (Yvonne Henesy/ Gamma Liaison)

income, self-enrichment, a sense of personal achievement, or simply a degree. Life transitions often trigger a return to formal education, as was the case for Devin and Trisha's friend Anya, who entered a nursing program after her last child left home. A divorce, widowhood, job layoff, or a youngest child going to school for the first time are other transitions that commonly precede reentry (Bradburn, Moen, & Dempster-McClain, 1995).

CHARACTERISTICS OF RETURNING STUDENTS

Women are the majority of adult learners—about 57 percent. A recent large rise in students over age 35 is due almost exclusively to women, who make up nearly 70 percent of this age group. As Anya's fear of not being able to handle classwork suggests (see page 488), first-year reentry women report feeling more self-conscious, inadequate, and scared to talk in class than do either returning men or traditional-age students (under age 25). Their anxiety is partly due to not having practiced academic learning for many years. It is also promoted by stereotypes of aging, since returning women are often convinced that traditional-age students are smarter (Wilke & Thompson, 1993).

Role demands outside the educational setting—from spouses, children, other family members, friends, and employers—pull many returning women in several, often conflicting directions. Those reporting high psychological stress typically have vocational rather than enrichment goals, young children, limited financial resources, and non-supportive husbands (Novak & Thacker, 1991). As a classmate told Anya one day, "I tried keeping the book open and

reading, cooking, and talking to the kids. It didn't work. They felt I was ignoring them." Because of multiple demands on their time, mature-age women tend to take fewer credits, experience more interruptions in their academic programs, and progress at a slower rate than do mature-age men (Robertson, 1991).

SUPPORTING RETURNING STUDENTS

As these findings suggest, social supports for returning students can make the difference between continuing in school and dropping out. Adult students need family members and friends who encourage their efforts and help them find time for uninterrupted study. Anya's classmate explained, "My doubts subsided when my husband saw me with a book in one hand and a pile of dirty clothes in the other and volunteered, 'I can cook dinner and do the laundry. You take your books and do what you need to do.'" Institutional services for returning students are also essential. Personal relationships with faculty; peer networks enabling adults to get to know one another; conveniently scheduled evening, Saturday, and off-campus classes; and financial aid for part-time students (many of whom are returning adults) increase the chances of academic success (Novak & Thacker, 1991).

The Caregiving Concerns table below suggests a variety of ways to facilitate adult reentry to college. Low-income and ethnic minority students need special assistance. Without it, the gap between the educationally advantaged and disadvantaged can widen in adulthood. Some programs offer field trips to colleges, testing to identify academic strengths and weaknesses, and opportunities to speak with

CAREGIVING CONCERNS

Facilitating Adult Reentry to College

SOURCE OF SUPPORT	DESCRIPTION
Partner and children	Value and encourage educational efforts. Assist with household tasks to permit time for uninterrupted study.
Extended family and friends	Value and encourage educational efforts.
Educational institution	Provide orientation programs and literature that inform adult students about services and social supports. Provide counseling and intervention addressing academic weaknesses and self-doubts about success. Facilitate peer networks through regular meetings or phone contact. Promote personal relationships with faculty. Encourage active engagement and discussion in classes and integration of course content with real-life experiences. Offering evening, Saturday, and off-campus classes. Provide financial aid for part-time students. Initiate campaigns to recruit returning students, including low-income and ethnic minority groups. Assist students with young children in finding child care arrangements, and provide on-campus child care.
Workplace	Value and encouraging educational efforts. Accommodate work time to class schedules.

financial aid officers, academic advisers, and students before enrollment. After entry, academic tutoring and sessions in confidence building and assertiveness are provided (Safman, 1988). Minority adults may need help in adjusting to styles of learning that are at odds with their cultural background. One Chinese returning student noted that she found criticizing ideas and arguing with her professors very difficult, since Chinese students are taught to respect, not disagree with, their teachers.

BENEFITS OF RETURNING TO COLLEGE

When support systems are in place, most returning students reap great personal benefits and do well academically. They especially value new personal relationships, sharing opinions and experiences, and relating subject matter to their own lives. An improved ability to integrate knowledge in midlife leads to a changed appreciation of classroom experiences and assignments, as these comments reveal:

> [As a traditional-age student,] I felt disaffiliated with most of my courses. Now courses are relevant. I'm taking them because I want to take them. I can see how everything relates. Before, they were disconnected.

> Most things used to be external to me.... It makes learning a lot more interesting and memorable when you can bring your experiences into it. It makes you want to learn and want to keep learning more and more. (Wilke & Thompson, 1993, p. 86)

Another benefit of large numbers of adult students in college classes is intergenerational contact and communication. As young students see firsthand the capacities and talents of older individuals, unfavorable stereotypes of aging decline.

In previous chapters, we underscored the power of education to shape the life course. Returning to school continues to alter life paths in middle adulthood. For Anya, it led to a position as a parish nurse with creative opportunities to inform and counsel members of a large congregation about health concerns. Higher education granted Anya new options, paid labor force rewards, and a higher sense of self-esteem as she reevaluated her own competencies (Redding & Dowling, 1992). At times, revised values and greater self-reliance sparked by education prompt changes in other spheres of life, such as a divorce or a new intimate partnership (Esterberg, Moen, & Dempster-McClain, 1994). In this way, reentering school is not only initiated by life change but precipitates it, leading middle-aged adults to reshape the course of their development.

ASK YOURSELF . . .

- *Consider the famous saying "You can't teach an old dog new tricks." Evaluate its accuracy in terms of evidence on the impact of vocational and educational experiences on cognitive development in midlife.*

- *Why do most high-level government and corporate positions go to middle-aged and older adults rather than to young adults? What cognitive capacities enable mature adults to perform well in these jobs?*

- *Marcella completed only one year of college, in her twenties. Now, at age 42, she has returned to earn a degree. Plan a set of experiences for the month before Marcella enrolls and her first semester that will increase her chances of success.*

SUMMARY

PHYSICAL DEVELOPMENT IN MIDDLE ADULTHOOD

PHYSICAL CHANGES

Describe the physical changes of middle adulthood, paying special attention to vision, hearing, the skin, muscle–fat makeup, and the skeleton.

- The gradual physical changes begun in early adulthood continue in midlife, contributing to a revised physical self-image, which often emphasizes fewer hoped-for gains and more feared declines.

- Vision is affected by **presbyopia,** or loss of the accommodative ability of the lens, a reduced ability to see in dim light, and diminished color discrimination. Middle-aged people, particularly men, experience hearing loss called **presbycusis,** which begins at high frequencies and spreads to other tones.

- The skin wrinkles, loosens, and starts to develop age spots, especially in women and people exposed to the sun. Muscle mass declines and fat deposits increase, with men and women developing different patterns of fat distribution. Continued exercise can offset changes in strength and weight.

- Bone density declines in both sexes, but to a greater extent in women, especially after menopause. Loss in height and bone fractures can result.

Describe reproductive changes in women during middle adulthood, and discuss women's psychological reactions to menopause.

- The **climacteric** in women occurs gradually over a 10-year period as estrogen production drops. Doctors often recommend **estrogen replacement therapy (ERT)** to protect post-menopausal women from bone deterioration, cardiovascular disease, skin changes, and discomforts such as hot flashes, sleeplessness, and -

irritability. ERT remains controversial because of an increased risk of certain cancers.

- **Menopause** is a biocultural event— affected by hormonal changes as well as societal beliefs and practices. Physical symptoms and psychological reactions vary widely. Whether women find menopause to be traumatic or liberating depends on how they interpret it in relation to their past and future lives.

Describe reproductive changes in men during middle adulthood.

- Men experience a climacteric, but their reproductive capacity declines without ending. Therefore, no male counterpart to menopause exists. Occasional episodes of impotence are more common in midlife but can usually be treated successfully.

HEALTH AND FITNESS

Discuss sexuality in middle adulthood and its association with psychological well-being.

- Sexual activity among married couples remains fairly stable in midlife, declining only slightly, and is associated with psychological well-being. Many more women than men are without sexual partners, a trend that continues into later life.

Discuss cancer, cardiovascular disease, and osteoporosis, noting risk factors and interventions.

- The death rate from cancer increases tenfold from early to middle adulthood. Cancer is the leading killer of middle-aged women. A complex interaction of heredity, biological aging, and environment contributes to cancer. Today 40 percent of people with cancer are cured. Annual screenings and various preventive steps (such as not smoking) can reduce the incidence of cancer and cancer deaths.

- Cardiovascular disease has declined in recent decades, but it remains a major cause of death in middle adulthood, especially among men. Symptoms include high blood pressure, high cholesterol, atherosclerosis, heart attack, arrhythmia, and angina pectoris. Diet, exercise, drug therapy, and stress reduction can reduce risks and aid in treatment.

- **Osteoporosis** affects one in four postmenopausal women and the majority of people of both sexes over age 70. Weight-bearing exercise, calcium and vitamin D, and ERT (for women) can help prevent and treat osteoporosis.

Discuss the association of hostility and anger with heart disease and other health problems.

- Hostility is the component of the **Type A behavior pattern** that predicts heart disease and other health problems, largely due to physiological arousal associated with anger. Since inhibiting the expression of emotion is also related to health problems, a better alternative is to develop effective ways of handling stress and conflict.

ADAPTING TO THE PHYSICAL CHANGES OF MIDLIFE

Discuss the benefits of stress management, exercise, and an optimistic outlook in dealing effectively with the changes of midlife.

- The changes and responsibilities of middle adulthood can cause psychological stress, with negative consequences for the cardiovascular, immune, and gastrointestinal systems. Stress management techniques, social support, and constructive approaches to anger reduction can alleviate stress.

- Regular exercise confers many physical and psychological benefits, making it worthwhile for sedentary middle-aged people to begin exercising. Developing a sense of self-efficacy and exercising in a group increase the chances that a beginner will stick with an exercise regimen.

- **Hardiness** is made up of three personal qualities: control, commitment, and challenge. By inducing a generally optimistic outlook on life, hardiness helps people cope with stress adaptively.

Explain the double standard of aging.

- Negative stereotypes of aging discourage older adults of both sexes. Yet middle-aged women are more likely to be viewed unfavorably, especially by men. Although this double standard of aging may have been adaptive at an earlier time in our evolutionary history, it is irrelevant in an era of limited childbearing and greater involvement in career and leisure pursuits.

COGNITIVE DEVELOPMENT IN MIDDLE ADULTHOOD

CHANGES IN MENTAL ABILITIES

Describe changes in crystallized and fluid intelligence in middle adulthood, and discuss individual differences in intellectual functioning.

- Longitudinal research reveals gains from early to middle adulthood in skills that tap both **crystallized intelligence** (which depends on accumulated knowledge and experience) and **fluid intelligence** (which depends more on information-processing skills). An exception is perceptual speed, a fluid skill that drops steadily from the twenties to the eighties.

- Large individual differences among middle-aged adults remind us that intellectual development is multidimensional, multidirectional, and plastic. Some people decline intellectually in their forties due to disease or very unfavorable environments. Individuals who use their intellectual skills are likely to retain them. Stimulating occupations and leisure pursuits, good health, lasting marriages, flexible personalities, and economic advantage are linked to favorable cognitive development

INFORMATION PROCESSING

How does information processing change in midlife?

- Speed of cognitive processing slows with age, a change that researchers explain with either the **neural network view** or the **information-loss view.** Slower processing speed makes it harder for middle-aged people to divide their attention, focus on relevant stimuli, and combine many pieces of visual information into a meaningful pattern.

- Adults in midlife retain less information in working memory, largely due to a decline in use of memory strategies. Training, practice, experience, and metacognitive knowledge enable middle-aged and older adults to compensate for decrements in processing speed, attention, and memory.

Discuss the development of practical problem solving, expertise, and creativity in middle adulthood.

- Middle-aged adults in all walks of life often become good at **practical problem solving,** largely due to development of expertise, which peaks in midlife. Creativity in middle adulthood shifts from generating unusual products and expressing oneself to integrating experience and knowledge in unique ways and fulfilling altruistic goals.

VOCATIONAL LIFE

Identify the relationship between vocational life and cognitive development.

- At all ages and in very different cultures, the relationship between vocational life and cognitive development is reciprocal. Stimulating, complex work and flexible, abstract, autonomous thinking support one another.

ADULT LEARNERS: BECOMING A COLLEGE STUDENT IN MIDLIFE

Discuss the challenges facing adults returning to college, ways to support returning students, and benefits of earning a degree in midlife.

- Often motivated to return to college by life transitions, women make up the majority of the growing number of adult students, especially those over age 35. Returning students must cope with a lack of recent practice at academic work, stereotypes of aging, and multiple role demands. Low-income and ethnic minority students need special assistance.

- Social support from family and friends and institutional services suited to their needs can help returning students succeed. Further education brings personal rewards, new relationships, intergenerational communication, and reshaped life paths.

IMPORTANT TERMS AND CONCEPTS

presbyopia, (p. 490)
presbycusis, (p. 490)
climacteric, (p. 491)
menopause, (p. 491)

estrogen replacement therapy (ERT), (p. 491)
osteoporosis, (p. 496)
Type A behavior pattern, (p. 498)
hardiness, (p. 500)

crystallized intelligence, (p. 503)
fluid intelligence, (p. 503)
neural network view, (p. 505)
information-loss view, (p. 505)
practical problem solving, (p. 507)

FOR FURTHER INFORMATION AND HELP

VISION

Vision Foundation
818 Auburn Street
Watertown, MA 02172
(800) 852-3029
Provides help to those with total or partial vision loss as well as those living with progressive eye disease. Offers referral services, a "buddy" telephone network to share information, and self-help support groups.

HEARING

Better Hearing Institute
5021 Backlick Road
Annandale, VA 22003
(800) 327-9355
Dedicated to helping people with impaired hearing. Disseminates information about hearing loss and available medical and amplification interventions.

MENOPAUSE

Menopause News
2074 Union Street, Suite 10
San Francisco, CA 94123
(800) 241-6366
Web site: *www.well.comp/mnews*
A newsletter that aims to provide an in-depth view of the physical and emotional aspects of menopause.

HEALTH AND FITNESS

Midlife Women's Network
5129 Logan Avenue, S.
Minneapolis, MN 55419
(800) 886-4354
Web site: *www.users.aol.com/ mdlfwoman/info.htm*
An organization that informs and empowers women who want to improve their health and quality of life during middle adulthood.

National Women's Health
Resource Center
2440 M. Street, N.W., Suite 325
Washington, DC 20037
(202) 293-6045
Web site: *www.womenshealth.com*
Disseminates information about women's health.

SEXUALITY

Sexuality Information and
Education Council of the
United States
130 W. 42nd Street, Suite 2500
New York, NY 10036-7901
(212) 819-9770
Web site: *www.siecus.org*
Provides information about and advocacy of sexuality as a healthy part of life.

CANCER

American Cancer Society
1599 Clifton Road, N.E.
Atlanta, GA 30329
(404) 320-3333
Supports research and public education on cancer. Local chapters have lists of community resources, home care items for loan, transportation services, and public education programs.

Y-Me National Organization for
Breast Cancer Information and
Support
212 W. Van Buren
Chicago, IL 60607
(800) 221-2141
Web site: *www.y-me.org*
Offers presurgical counseling, support, and self-help meetings for women who have or suspect they have breast cancer, their families, and friends.

CARDIOVASCULAR DISEASE

(See page 449 for information on the American Heart Association.)

OSTEOPOROSIS

National Osteoporosis Foundation
1150 17th Street, N.W., Suite 500
Washington, DC 20036-4603
(800) 464-6700
Web site: *www.nof.org*
Disseminates information about osteoporosis and supports research.

STRESS MANAGEMENT

American Institute of Stress
124 Park Avenue
Yonkers, NY 10703
(800) 247-3529
Web site: *http://www.stress.org*
Disseminates information on the personal and social consequences of stress and on stress management.

RETURNING STUDENTS

American Association for Adult
and Continuing Education
1201 16th Street, N.W., Suite 230
Washington, DC 90036
(202) 463-6333
Provides leadership in advancing adult education as a lifelong learning process. Works to stimulate continuing education efforts at local, state, and regional levels.

American Association
of University Women
Educational Foundations Program
2401 Virginia Avenue, N.W.
Washington, DC 20037
(202) 785-7700
Organization of graduates of community colleges, colleges, and universities that conducts research and advocates educational equity for women and girls.

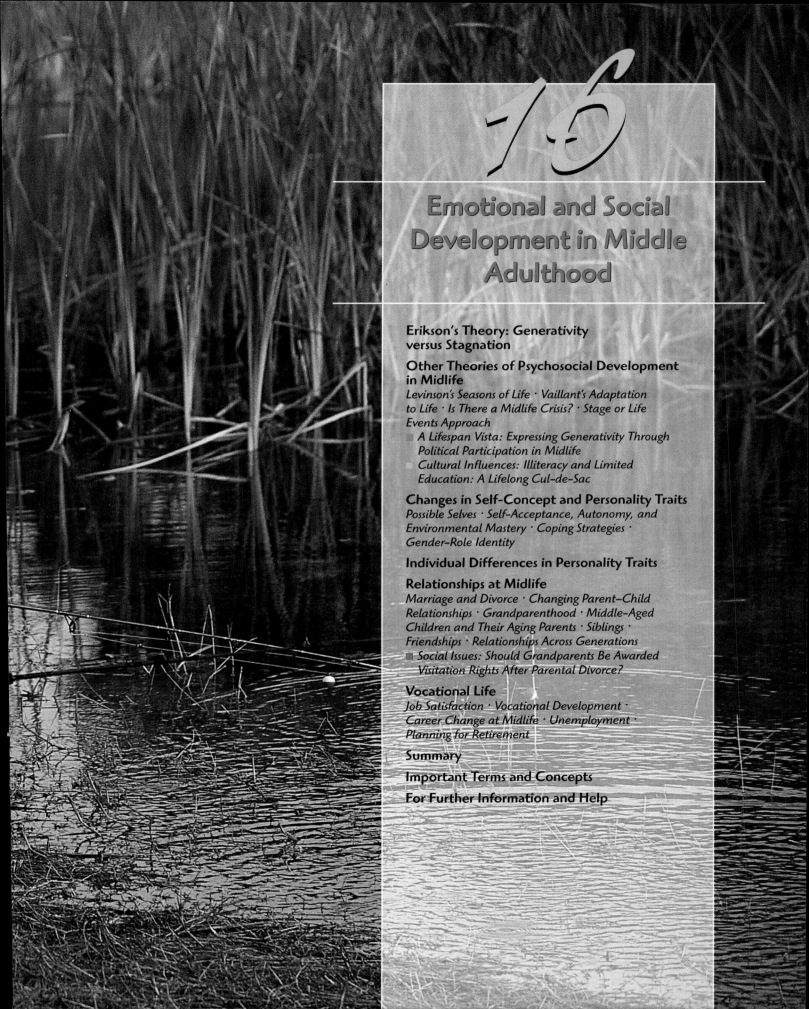

16

Emotional and Social Development in Middle Adulthood

Trisha and Devin passed through middle adulthood uneventfully, building on earlier strengths and intensifying commitment to values and concerns relevant to their desire to leave a legacy for those who would come after them. As their son Mark graduated from college, took his first job, fell in love, and married, they felt a sense of pride at having escorted a member of the next generation into responsible adult roles. Family activities, which had declined during Mark's adolescent and college years, now increased as Trisha and Devin related to their son and daughter-in-law not just as kin with whom they felt a close emotional bond, but as enjoyable adult companions. Challenging work and more time for community involvement, leisure pursuits, and each other contributed to a richly diverse and gratifying time of life.

The years of midlife were not as smooth for two of Trisha and Devin's friends. Fearing that she might grow old alone, Jewel frantically pursued her quest for an intimate partner by attending singles events, registering with dating services, and traveling in hopes of meeting a like-minded companion. "I can't stand the thought of turning 50. I look like an old bag with big circles under my eyes," she lamented in a letter to Trisha. In other ways, Jewel's life had compensating satisfactions—friendships that had grown more meaningful, a warm relationship with a nephew and niece, and a successful consulting business.

Tim, Devin's best friend from graduate school, had been divorced for over 15 years. Recently, he had met Elena, for whom he cared deeply. But Elena was in the midst of major life changes—not just a divorce of her own, but also dealing with a troubled daughter, a change in careers, and a move from the city that was a constant reminder of her unhappy past. Whereas Tim had reached the peak of his vocation and was ready to enjoy life, Elena wanted to recapture much of what she had missed in earlier decades—not just a satisfying close relationship, but opportunities to real-

ize her talents. "I don't know where I fit into Elena's plans," Tim wondered aloud in a phone conversation with Trisha.

With the arrival of middle adulthood, half or more of the lifespan is over. Increasing awareness of limited time ahead prompts adults to reevaluate the meaning of their lives and reach out to future generations. As we will see in this chapter, most middle-aged people make modest adjustments in their outlook and daily lives. A few experience profound inner turbulence and initiate major changes, often in an effort to make up for lost time. Besides advancing years, family and vocational transitions are intimately involved in emotional and social development during the middle years of life.

ERIKSON'S THEORY: GENERATIVITY VERSUS STAGNATION

Erikson's critical psychological conflict of midlife is called **generativity versus stagnation**. In Chapter 14, we noted that generativity is underway in early adulthood, typically through childbearing and child rearing and establishing a niche in the occupational world. It expands greatly in midlife.

Generativity involves reaching out to others in ways that give to and guide the next generation. At this stage, commitment extends beyond oneself (identity) and one's life partner (intimacy) to a larger group—family, community, or society in general. The generative adult combines his or her need for self-expression with a need for communion, integrating personal goals with the welfare of the larger social environment (Wrightsman, 1994). The resulting strength is the capacity to care for others in a broader way than in previous stages.

Erikson (1950) selected the term *generativity* to encompass everything generated that can outlive the self and ensure society's continuity and improvement: children, ideas, products, and works of art. Although parenting is a major means of realizing generativity, some people, because of misfortune or special gifts, do not express it through

Generativity expands greatly in midlife. It involves reaching out to others in ways that give to and guide the next generation. Dressed as Santa Claus, this middle-aged man brightens the holidays for neighborhood children. He combines his need for self-expression with the welfare of the larger social environment. (Joel Gordon)

Test Bank Items 16.1 through 16.3

EXPRESSING GENERATIVITY THROUGH POLITICAL PARTICIPATION IN MIDLIFE

After age 40, adults show greater engagement with political issues that may affect future generations. Yet events in a person's life history can have a profound impact on midlife political consciousness. Specifically, politically relevant experiences in childhood, adolescence, and early adulthood are likely to mold identity development. Those occurring later have less opportunity to lead to a radical transformation of the self (Stewart & Healy, 1989).

Research on political activists of the 1960s and early 1970s reveals a strong link between early political convictions sparked by historical events and dedication to political causes in midlife. For example, follow-ups of male protesters in student movements of the 1960s showed that they became "ideal middle-aged citizens" who were more socially and politically involved than others of their generation (Fendrich, 1993). Similarly, women who as late adolescents and young adults were either concerned about or active participants in the civil rights and women's movements seemed empowered by their youthful experiences. When they reached their forties, they were highly involved in effecting political change in their local com-

munities (Cole & Stewart, 1996; Stewart & Healy, 1989).

Intensive case studies illustrate this early identity–midlife political participation connection (Stewart & Gold-Steinberg, 1990). Consider Sarah, whose family and school upbringing instilled a profound sense of social responsibility. Her mother had devoted many hours to charitable causes; her father had often stressed the importance of good citizenship. As a young pupil, Sarah recalled a teacher saying, "[Y]ou are very privileged people and you absolutely have to make a difference in this world, you can't just sit back and enjoy it." Sarah frequently thought about her teacher's message, and her parents underscored it. During her college years, involvement in the civil rights, antiwar, and women's movements of the 1960s strengthened Sarah's political values. She became a socially committed college history teacher, an occupational identity that served her well after she married and had children.

In the 1980s, when her children were grown, Sarah experienced an intense desire to renew her political activism. Searching for a cause in which to invest her midlife consciousness, she became involved in gender-equity issues on campus and in antinuclear education efforts in her

community. Sarah recognized the link between her early experiences and her generative impulses in middle adulthood. She commented,

> In my twenties, when we were trying to end the war in Vietnam, we simultaneously thought that we had all the power in the world and that we had no power at all. Neither was true. We did actually accomplish some things . . . but the 25-year-old consciousness . . . is not the 45-year-old consciousness. I now think that I have a much more limited, accurate, and in some important ways energizing and empowering sense of my own powers and limitations. (Stewart & Gold-Steinberg, 1990, p. 557)

Political awareness and student activism in the early decades of life seem to produce a sense of self-efficacy and commitment, which in turn promotes active political participation later in life. As the researchers comment, "In this light, student protesters may be seen not as society's malcontents but as tomorrow's exemplars of social responsibility" (Cole & Stewart, 1996, p. 138).

their own children. Adults can be generative in other family relationships (as Jewel was with her nephew and niece), as mentors in the workplace, in volunteer endeavors, and through many forms of productivity and creativity.

Look closely at what we have said so far, and you will see that generativity brings together inner desires and cultural demands. Middle-aged adults feel a need to be needed; they want to make a contribution that will survive their death (Kotre, 1984; McAdams, 1985). According to Erikson, "belief in the species"—a conviction that life is basically good and worthwhile, even in the face of human destructiveness and deprivation—is the underlying motivation for generative action. Without this optimistic world

view, people would not have any hope of improving humanity. At the same time, society requires adults to take responsibility for the next generation in their roles as parents, teachers, mentors, leaders, organizers, and guardians of the culture. As the Lifespan Vista box above illustrates, cultural opportunities and injustices shape adults' generative activities.

The negative outcome of this stage is stagnation. Erikson recognized that once a person has attained certain life goals—marriage, children, and vocation—there is a temptation to become self-centered and self-indulgent. People with a sense of stagnation are unable to contribute to society's welfare because they place their own comfort and security above challenge and sacrifice. Their self-absorption is expressed in many ways—through lack of involvement with and concern for young people (including their own children); through a focus on what they can get from others rather than what they can give; and through little interest in being productive at work, developing their talents, or bettering the world in other ways (Hamachek, 1990).

Many studies confirm that generativity is a prominent concern for middle-aged adults. Some researchers assess personality traits believed to be components of generativity, such as assertiveness, nurturance, responsibility, and breadth of interests (Ryff & Migdal, 1984). Others ask people to rate themselves on generative characteristics and activities—for example, "I try to pass along the knowledge I have gained through my experiences" and "I have made and created things that have had an impact on other people" (Oches & Plug, 1986; Ryff & Heincke, 1983). Still others look for generative themes in people's narrative descriptions of themselves. These indicate that generativity has become part of an expanding identity that gives purpose and meaning to life (McAdams, 1993). Each approach reveals that generativity increases in midlife (McAdams & de St. Aubin, 1992; McAdams, de St. Aubin, & Logan, 1993).

Just as Erikson's theory suggests, highly generative people tend to be psychologically fulfilled and healthy. They are open to differing viewpoints, possess leadership qualities, want more from work than just financial rewards, and care greatly about the welfare of their children, partner, and the wider society (Peterson & Klohnen, 1995). Having children seems to foster men's generative development more than women's. In two studies, fathers scored higher in generativity than did childless men (McAdams & de St. Aubin, 1992; Snarey et al., 1987). Perhaps parenthood awakens in men a tender, caring attitude toward the next generation that women have opportunities to develop in many other ways.

OTHER THEORIES OF PSYCHOSOCIAL DEVELOPMENT IN MIDLIFE

Recall that Erikson's theory provides only a broad sketch of adult personality development. For a closer look at psychosocial change in midlife, let's revisit Levinson's and Vaillant's theories, to which you were introduced in Chapter 14.

LEVINSON'S SEASONS OF LIFE

Return to page 453 to review Levinson's eras (stages or seasons), and notice that like early adulthood, middle adulthood begins with a transitional period (age 40 to 45), followed by the building of an entry life structure (age 45

to 50). Then this structure is evaluated and revised (age 50 to 55), resulting in a culminating life structure (age 55 to 60). Among the men and women Levinson (1978, 1996) interviewed, the majority displayed the first two of these phases. Nevertheless, because of gender stereotypes and differences in opportunity, male and female experiences were somewhat different.

■ MIDLIFE TRANSITION. Around age 40, people evaluate their success in meeting early adulthood goals. Realizing that from now on, more time will lie behind than ahead, they regard the remaining years as increasingly precious. According to Levinson, virtually all people experience considerable confusion as they start to modify marriage/family and occupational components of the life structure. Some make drastic changes, divorcing, remarrying, shifting occupations and lifestyle, or displaying great progress in creativity. Others make smaller changes while staying in the same marriage, surroundings, occupation, and workplace.

Whether these years bring a gust of wind or a storm, most people turn inward for a time, focusing on personally meaningful living (Neugarten, 1968). Part of the reason is that for many middle-aged adults, only limited advancement and personal growth at work remain possible. Others are disappointed in not having fully realized their early adulthood dream and feel an inner pressure to find a more satisfying path before it is too late. They ask, Can I still achieve what I wanted? If not, can I accept what I have attained? What are my alternatives, and can I build a better life? Even people who have reached their goals ask, What good are these accomplishments to others, to society, and to myself?

According to Levinson, for middle-aged adults to reassess their relation to themselves and the external world, they must confront four developmental tasks, summarized in Table 16.1. Each requires the person to reconcile two opposing tendencies within the self. The midlife transition brings both an opportunity and a need to achieve a better balance than before. Let's see how this happens.

■ MODIFYING THE LIFE STRUCTURE: GENDER SIMILARITIES AND DIFFERENCES. At midlife, adults must give up certain youthful qualities, find age-appropriate ways to express others, and accept being older, thereby creating a youth–age balance more in tune with their time of life. Physical changes, personal encounters with illness, and aging parents intensify this task, and it often triggers reassessment of what is important. Due to the double standard of aging (see Chapter 15), women find it harder than men to accept being older. For Jewel, the stereotypical image of an older woman prompted a desperate fear of becoming unattractive and unlovable. She tried numerous remedies, from skin creams to a face lift, to maintain her youth. Indeed, women are more likely than men to perceive themselves as younger than their chronological age.

TABLE 16.1

Levinson's Four Developmental Tasks of Middle Adulthood

TASK	DESCRIPTION
Young–Old	The middle-aged person must seek new ways of being both young and old. This means giving up certain youthful qualities, retaining and transforming others, and finding positive meaning in being older.
Destruction–Creation	With greater awareness of mortality, the middle-aged person focuses on ways he or she has acted destructively and others have done the same. Past hurtful acts toward parents, intimate partners, children, friends, and rivals are countered by a strong desire to become more creative—by making products of value to the self and others and participating in activities that advance human welfare.
Masculinity–Femininity	The middle-aged person must come to terms with masculine and feminine parts of the self, creating a better balance. For men, this means becoming more empathic and caring; for women, it often means becoming more autonomous, dominant, and assertive.
Engagement–Separateness	The middle-aged person must create a better balance between engagement with the external world and separateness. For men, this generally means pulling back from ambition and achievement and becoming more in touch with the self. Women who have devoted themselves to child rearing or who have unfulfilling jobs typically need to move in the other direction—toward greater involvement in the work world and wider community.

Sources: Levinson, 1978, 1996.

And as Figure 16.1 indicates, especially for women, the gap between subjective and objective age widens over time (Montepare & Lachman, 1989).

As middle-aged adults confront their own mortality and the actual or impending death of agemates, they become more aware of ways people can act destructively—to parents, intimate partners, children, friends, and co-workers. Countering this force is a desire to be creative through endeavors that advance human welfare. For example, Devin and Trisha became more involved in improving quality of life in their community, desiring to leave a legacy for future generations. The image of a legacy, which flourishes in midlife, can be satisfied in many ways—through charitable gifts, creating products valued by the self and others, volunteer service, or mentoring young people. According to Levinson, creativity in midlife is stimulated in part by recognition that destructiveness must be overcome by loving, life-affirming aspects of the self.

Middle age is also a time when people must reconcile masculine and feminine parts of the self. For men, this means being more accepting of "feminine" traits of nurturance and caring, which enhance love relationships and lead to a more compassionate exercise of authority in the workplace. For women, it generally means being more open to "masculine" characteristics of autonomy, dominance, and assertiveness (Gilligan, 1982; Harris, Ellicott, & Holmes,

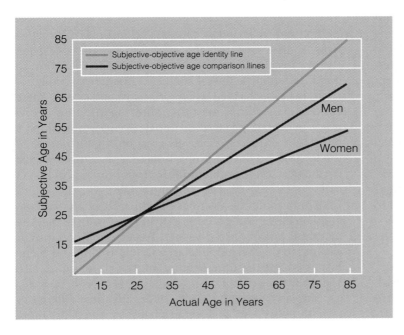

F IGURE 16.1

Relationship between subjective and objective age across the lifespan. After age 25, men and women report perceiving themselves as younger than they actually are. Especially for women, the gap between subjective and objective age widens over time. Due to the double standard of aging, women find it harder to accept being older. *(Adapted from J. Montepare & M. Lachman, 1989, "'You're Only as Old as You Feel': Self-Perceptions of Age, Fears of Aging, and Life Satisfaction from Adolescence to Old Age,"* Psychology and Aging, 4, *p. 75. Copyright © 1989 by the American Psychological Association. Reprinted by permission.)*

1986; Reinke, Holmes, & Harris, 1985). Recall from Chapter 8 that people who combine masculine and feminine traits have an androgynous gender-role identity. Later we will see that androgyny is associated with many favorable personality traits.

Finally, midlife requires a middle ground between engagement with the external world and separateness. Men generally need to reduce their concern with ambition and achievement and attend more fully to the self. A few women, who have had active, successful careers may need to do so as well. But those who devoted their early adulthood to child rearing or an unfulfilling job, after a period of self-reflection, often feel compelled to move in the other direction (Levinson, 1996). At age 48, for example, Elena left her position as a reporter for a small-town newspaper, pursued an advanced degree in creative writing with passionate commitment, eventually accepted a college teaching position, and began work on two novels. As Tim looked inward, he recognized his overwhelming desire for a gratifying intimate partnership. He realized that by scaling back his own vocational pursuits, he could grant Elena the time and space she needed to build a rewarding career—and that doing so might deepen their attachment to one another.

■ THE LIFE STRUCTURE IN SOCIAL AND CULTURAL CONTEXT. Rebuilding the life structure depends on contexts that support people's reassessment of what they want to give to the world and to themselves. When poverty, unemployment, and lack of a respected place in society dominate the life course, energies are directed toward survival rather than pursuit of a satisfying life structure (Levinson, 1978).

Even adults whose jobs are secure and who live in pleasant neighborhoods may find that employment conditions place too much emphasis on productivity and profit and too little on the meaning of work, thereby restricting possibilities for growth. In her early forties, Trisha left a large law firm for a small practice because of constant pressures to bring in high-fee clients and little acknowledgment of the quality of her efforts by the managing partners.

Opportunities for advancement permit realization of the early adulthood dream, thereby easing the transition to middle adulthood. Yet they are far less available to women than men. Individuals of both sexes in blue-collar jobs also have few possibilities for promotion. The industrial workers in Levinson's (1978) sample made whatever adjustments they could in middle age—becoming active union members, shop stewards, or mentors of younger workers. Many found compensating rewards in moving from the junior to senior generation in their families.

Education is crucial for personally enriching work, limiting family size so women can blend child rearing with vocational life, acquiring the abstract verbal skills needed to reflect on the inner self, and attaining the economic security that grants time and energy for rebuilding the life structure. Yet on a worldwide basis, 26.5 percent of the adult population is illiterate—nearly twice as many women as men. Many more have only limited schooling (United Nations, 1991a). As the Cultural Influences box on the following page reveals, without education, there is no dream to challenge in middle adulthood—and no hope of taking control of one's life.

■ VAILLANT'S ADAPTATION TO LIFE

Because the men and women interviewed by Levinson were between ages 35 and 45, they cannot tell us anything certain about psychosocial change in the fifties or beyond. In Vaillant's (1977) longitudinal study of men who attended a highly competitive liberal arts college, participants were followed past the half-century mark. Recall from Chapter 14 that they became "keepers of meaning," or guardians of their culture (see page 455). Adults in their late forties and fifties carry peak responsibility for the functioning of society. Vaillant reported that the most successful and best adjusted individuals entered a calmer, quieter time of life, an outcome verified for both men and women in other research (Whitbourne & Weinstock, 1979). "Passing the torch"—concern that the positive aspects of their culture survive—became a major preoccupation.

Older people in societies around the world are guardians of traditions, laws, and cultural values. This stabilizing force holds in check too rapid change sparked by the questioning and challenging of adolescents and young adults. As people move toward the end of middle age, they focus on longer term, less personal goals, such as the state of human relations in their society. And they become more philosophical, accepting that not all problems can be solved in their lifetime.

■ IS THERE A MIDLIFE CRISIS? Levinson (1978, 1996) reported that most men and women in his samples experienced substantial inner turmoil during the transition to middle adulthood. Yet Vaillant (1977) saw few examples of crisis. Instead, change was typically slow and steady. These contrasting findings raise the following questions: How much personal upheaval accompanies entry to midlife? Is self-doubt and stress especially great during the decade of the forties, and does it prompt major restructuring of the personality, as the phrase **midlife crisis** implies? Are men's and women's experiences similar or different?

Think about the reactions of Trisha, Devin, Jewel, Tim, and Elena to middle adulthood, and you will see that the picture is one of great diversity. Trisha and Devin moved easily into this period. In contrast, Jewel, Tim, and Elena displayed greater questioning of their situations and sought alternate life paths.

Similarly, research suggests wide individual differences in response to midlife. Overall, changes for men are more

CULTURAL INFLUENCES

ILLITERACY AND LIMITED EDUCATION: A LIFELONG CUL-DE-SAC

Among the 948 million people around the globe who cannot read, 63 percent are women. Almost 35 percent of the world's female population is illiterate, a rate that rises to 45 percent when only developing countries are considered (United Nations, 1995).

Illiteracy compounds other problems of women—in the family, workplace, and community. Education determines a woman's access to paid employment, earning capacity, overall health, and ability to control fertility. Women in nonindustrialized nations with 7 or more years of education marry, on the average, 4 years later and have 2.2 fewer children than do women with no schooling. An educated woman with a smaller family size is better equipped to overcome social prejudice and develop an identity and status beyond childbearing—factors that lead to fuller participation in public life.

When illiteracy among women is allowed to persist into adulthood, it carries over to the next generation. Poverty-stricken, illiterate girls marry early and give birth throughout their adult lives. They and their children take on all manner of low-paid jobs to survive—circumstances that limit their children's education. In households where women cannot read, gender typing is high. Even mothers underinvest in girls, giving them less food and clothing than they give to boys and more often taking them out of school to work or help with household tasks. Poorly educated women are more likely to have children out of wedlock, be divorced or abandoned by their husbands, or (due to poor family health) widowed at an early age. To ease economic burdens in these families, girls are given away in marriage as soon as possible. Then they repeat the cycle of illiteracy, early childbearing, large family size, and poverty (United Nations, 1991a, 1995).

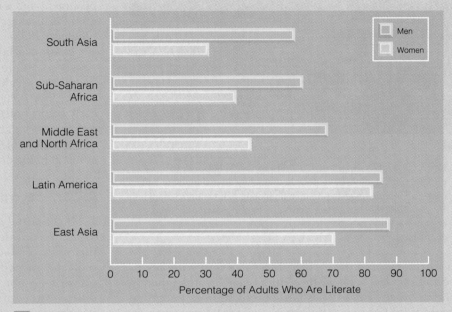

FIGURE 16.2

Literacy rates in developing regions of the world. Although literacy has improved considerably during the past several decades, women continue to fall far behind men. In South Asia, sub-Saharan Africa, the Middle East, and North Africa more than half of women are still illiterate. *(Adapted from United Nations, 1995.)*

Great strides have been made in increasing literacy in developing nations during the past few decades. Nevertheless, as Figure 16.2 shows, the majority of adult women remain illiterate in South Asia, Africa, and the Middle East. The industrialized world must help governments in these regions extend compulsory education from primary school through grade 12, prevent parents from taking girls out of school for child labor or early marriage, and teach illiterate adults to read. When a developing country educates a woman, it reaches out to an entire family and, therefore, to the next generation—a high-return investment in the economic development of a society.

Educating girls like these primary school students in Bardera, India, prevents the cycle of illiteracy, early childbearing, large family size, and poverty from carrying over to the next generation. (Betty Press/Woodfin Camp & Associates)

Research suggests wide individual differences in response to midlife. Only a minority of people drastically alter their lives. This man and woman might appear to be having a flirtation sparked by a midlife crisis, but in fact they are a happily married couple, comparing busy schedules so they can plan an evening out. (Will Hart)

likely to occur in the early forties (in accord with Levinson's timetable). Those for women may be postponed to the late forties and fifties until a reduction in parenting responsibilities grants time and freedom to confront personal issues (Harris, Ellicott, & Holmes, 1986; Mercer, Nichols, & Doyle, 1989; Tamir, 1980). In addition, compared to men, the direction of change for women is more variable. It depends in part on whether as young adults they focused on a "feminine" social clock (marriage and motherhood), a "masculine" social clock (high-status career), or a combination of the two (Helson & Roberts, 1994; Stewart & Vandewater, 1992).

But sharp disruption and agitation are more the exception than the rule. For example, Elena had considered both a divorce and a new career long before she initiated these changes. In her thirties, she separated from her husband, later reconciled, and told him of her desire to return to school, which he firmly opposed. She put her own life on hold because of her daughter's academic and emotional difficulties and her husband's resistance. In an intensive study of ten people who changed careers between ages 35 and 55 (representing a minority of midlifers), only three were judged to be in a state of crisis. Like Elena, the remaining seven had planned the change for a long time. Although entry into the second career seemed sudden to outsiders, the career changers did not perceive it as a radical shift. Instead, the kind of work they did and enjoyed in their first careers was related to what they chose to do in their second (Lawrence, 1980).

Furthermore, midlife is sometimes experienced in a way that resembles relief rather than crisis. In Chapter 15, we noted that women often welcome menopause, as it liberates them from unwanted pregnancy and the need to use birth control. In a study of well-known contemporary novelists, many expressed "feeling safe at last in the middle years," free of anxieties about the adequacy of their performance. Midlife brought a sense of confidence and accomplishment and a more insightful view of difficulties—that even the bleakest event is an episode, that it too would pass (Gullette, 1988).

In sum, although concern with mortality and lasting values is common during middle age, only a minority of people drastically alter their life structure (Wrightsman, 1988, 1994). The few who are in crisis typically have had early adulthoods in which gender roles, family pressures, or low-income and poverty severely limited their ability to attain reasonable satisfaction of personal needs and goals, either at home or in the wider world (McAdams, 1988).

STAGE OR LIFE EVENTS APPROACH

If crisis and major restructuring are rare, is it appropriate to consider middle adulthood a *stage* of development, as Erikson, Levinson, and Vaillant's theories indicate? According to some researchers, the mid-adult transition is not stagelike; it does not lead to qualitatively different ways of thinking, feeling, and behaving. Instead, it is simply an adaptation to life events—external pressures that affect many people during the middle years, such as children growing up, reaching the crest of a career, and impending retirement (McCrae & Costa, 1990).

Yet recall from earlier chapters that life events are no longer as age graded as they were in the past. Their timing varies enough that they cannot be the single cause of midlife change. Furthermore, in several studies, people were asked to trace their thoughts, feelings, attitudes, and hopes during early and middle adulthood. Psychosocial change, in terms of personal disruption followed by reassessment, coincided with both family life cycle events and chronological age (Ellicott, 1985). For this reason, most experts regard adaptation during midlife as the combined result of growing older and social experiences (Schroots & Birren, 1990; Whitbourne et al., 1992). Return to our discussion of generativity and the midlife transition, and notice how both factors are involved.

Finally, in describing their lives, the large majority of middle-aged people report troubling moments that prompt new understandings and goals—a finding consistent with the stage approach. As we take a closer look at emotional and social development in middle adulthood, we will see that this period, like others, is characterized by *both continuity and stagewise change.* With this in mind, let's turn to the diverse inner concerns and outer experiences that contribute to psychological well-being and decision making during this phase of life.

BRIEF REVIEW

The critical psychological conflict of Erikson's middle adulthood stage is generativity versus stagnation. It involves giving to and guiding the next generation through children, ideas, products, and works of art. Compared to Erikson, Levinson presents a more detailed picture of psychosocial change. During the midlife transition, awareness of limited time ahead and reassessment of early adulthood goals prompt people to modify their life structure. To do so effectively, people must reconcile four opposing tendencies within the self. Men and women do so in somewhat different ways. According to Vaillant, middle adulthood is a time when people become "keepers of meaning"—concerned with survival of the traditions, laws, and values of their culture.

Building a better relationship with the self and surrounding world depends on supportive family, work, and cultural contexts. At the same time, only rarely does desire for change bring about a midlife crisis. Adaptation during midlife is the combined result of growing older and life events.

ASK YOURSELF . . .

- *Explain how Elena's decision to change careers is a joint product of inner needs and external pressures.*

- *After years of experiencing little personal growth at work, 42-year-old Mel started to look for a new job. When he received an attractive offer in another city, he felt torn between leaving friendships built over many years and a long-awaited career opportunity. After several weeks of soul searching, he took the new job. Was Mel's dilemma a midlife crisis? Why or why not?*

CHANGES IN SELF-CONCEPT AND PERSONALITY TRAITS

In middle adulthood, changes in self-concept and personality traits reflect growing awareness of a finite lifespan, longer life experience, and generative concerns.

POSSIBLE SELVES

On a business trip, Jewel found a spare afternoon to visit Trisha. The two women sat in a coffee shop, reminisc-ing about the past and thinking about the future. "It's been tough living on my own and building the business," Jewel commented. "What I hope for is to become better at my work, to be more community oriented, and to stay healthy and available to my friends. Of course, I don't want to grow old alone, but if I don't find that special person, I suppose I can take comfort in the fact that I'll never have to face divorce or widowhood."

Jewel is discussing **possible selves,** future-oriented representations of what she hopes to become and what she is afraid of becoming. Possible selves are the temporal dimension of self-concept—what the individual is striving for and attempting to avoid. Lifespan researchers regard these hopes and fears as just as vital for explaining behavior as people's views of their current characteristics. Indeed, possible selves may be an especially strong motivator of action in midlife, as more meaning becomes attached to time. Some researchers speculate that as we age, we rely less on social comparisons in judging our self-worth and more on temporal comparisons—how well we are doing in relation to what we had planned for ourselves (Suls & Mullen, 1982).

Throughout adulthood, the way people describe their current selves is quite stable. A 30-year-old who says he is cooperative, competent, outgoing, or successful is likely to report a similar picture at a later age. But possible selves show great change. Adults in their early twenties mention many possible selves, and they are lofty and idealistic—being "perfectly happy," "rich and famous," "healthy throughout life," and not being "down and out" or "a person who does nothing important." With age, possible selves become fewer in number and more modest and concrete. Most middle-aged people no longer desire to be the best and the most successful. Instead, they are largely concerned with performance of roles and responsibilities already begun—being "competent at work," "a good husband and father," "able to put my children through the colleges of their choice," and not being "in poor health" or "without enough money to meet my daily needs" (Cross & Markus, 1991; Ryff, 1991).

What explains these shifts in possible selves? As the future no longer holds limitless opportunities, adults adjust their hopes and fears so they can preserve mental health. They must maintain a sense of unachieved possibility to stay motivated, yet they must still manage to feel good about themselves and their lives in spite of disappointments (Baltes & Carstensen, 1991). For example, Jewel no longer desired to be an executive in a large company, as she had in her twenties. Instead, she wanted to grow in her current occupation. And although she feared loneliness in old age, she reminded herself that marriage could also lead to equally negative outcomes, such as divorce and widowhood—possibilities that made not having attained an important interpersonal goal easier to bear.

Unlike current self-concept, which is constantly responsive to others' feedback, possible selves (although influenced by others) can be defined and redefined by the individual, as needed. Consequently, they permit affirmation of the self, even when things are not going well (Cross & Markus, 1991). Researchers believe that possible selves may be the key to continued well-being in adulthood, as people revise these future images to achieve a better match between desired and achieved goals. Many studies reveal that the self-esteem of middle-aged and older individuals equals or surpasses that of younger individuals, perhaps because of the protective role of possible selves (Bengston, Reedy, & Gordon, 1985).

SELF-ACCEPTANCE, AUTONOMY, AND ENVIRONMENTAL MASTERY

An evolving mix of competencies and experiences leads certain personality traits to change in middle adulthood. One of the most consistent findings is a rise in introspection as people contemplate the second half of life (Neugarten & Datan, 1973). Middle-aged adults tend to be in better touch with themselves than younger individuals,

Is middle age "the prime of life" for this man and woman? Compared to younger people, middle-aged adults tend to be more at ease with themselves, acknowledging and accepting both their good and their bad qualities. (Ronnie Kaufman/The Stock Market)

and many have reshaped contexts to suit their personal needs and values.

These developments undoubtedly contribute to other gains in personal functioning. In a study of well-educated individuals ranging in age from the late teens into the seventies, three traits increased from early to middle adulthood and then leveled off. The first was *self-acceptance.* More than young adults, middle-aged people acknowledged and accepted both their good and bad qualities and felt positively about themselves and life. Second, they saw themselves as more *autonomous*—less concerned about expectations and evaluations of others and more concerned with following self-chosen standards. Third, they regarded themselves as high in *environmental mastery*—capable of managing a complex array of tasks easily and effectively (Ryff, 1991).

In Chapter 15, we noted that midlife brings gains in expertise and practical problem solving. These cognitive changes may support the confidence, initiative, and decisiveness of this period. Overall, midlife is a time of increased comfort with the self, independence, and commitment to personal values—outcomes not just apparent in cross-sectional research, but in longitudinal studies as well (Block, 1982; Helson & Wink, 1992; Mitchell & Helson, 1990). Perhaps because of these personal attributes, people sometimes refer to middle age as "the prime of life."

COPING STRATEGIES

In Chapter 15, we discussed the importance of stress management in the prevention of illness. It is also vital for psychological well-being. Three longitudinal studies found that midlife brought an increase in effective **coping strategies**—thoughts and behaviors used to manage stress and the emotions associated with stress (Haan, 1972; Helson & Wink, 1992; Vaillant, 1976). For example, in Vaillant's sample, middle-aged men were more likely to look for the "silver lining" or positive side of a difficult situation, postpone action to permit evaluation of alternative courses of action, anticipate and plan ways to handle future discomforts, and use humor to express ideas and feelings without negative effects on others. Younger individuals more often engaged in denial of troubling emotions, acting out (temper outbursts), avoidance (sleep, substance use), and blaming others.

People use two general strategies to cope with stress. In *problem-centered coping,* they appraise the situation as changeable, identify the difficulty, and decide what to do about it. If problem solving does not work, people engage in *emotion-centered coping,* which is internal, private, and aimed at controlling distress when there is little we can do about a situation (Lazarus & Lazarus, 1994). Adults who cope effectively use a mixture of problem- and emotion-centered techniques, depending on the situation. And their approach is deliberate, thoughtful, and respectful of both

Test Bank Items 16.31 through 16.39

themselves and others. In contrast, ineffective coping is largely emotion centered and either impulsive or escapist (Lazarus, 1991).

Why might effective coping increase in middle adulthood? Other personality changes probably support it, such as greater self-acceptance and confidence at handling life's problems. Return to Chapter 11, page 357, to review evidence indicating that many individuals who were stress ridden as adolescents developed into contented, well-adjusted adults. In contrast, teenagers who rarely faced troubling situations had more difficulty coping with later problems. Similarly, in a longitudinal study of well-educated women, taking initiative to overcome difficult times in early adulthood predicted advanced self-understanding, social and moral maturity, and high life satisfaction at age 43 (Helson & Roberts, 1994). These findings suggest that years of experience in overcoming stress may contribute to more sophisticated, flexible coping during middle age.

GENDER-ROLE IDENTITY

In her forties and early fifties, Trisha appeared more assertive at work, speaking out more freely at meetings and taking the lead in devising strategy when a team of lawyers worked on an especially complex case. She was also more dominant in family relationships, expressing her preferences and points of view to her husband and son more readily than she had 10 or 15 years earlier. In contrast, Devin's sense of empathy and caring became more apparent, and he was less assertive and more accommodating to Trisha's wishes than he had been in early adulthood.

Gender-role identity becomes more androgynous in midlife—a mixture of both "masculine" and "feminine" characteristics. Like this father cooking with his children, men typically become more emotionally sensitive and nurturant. At the same time, women become more confident, self-sufficient, and forceful. (Richard Smith/Monkmeyer Press)

Many studies report an increase in traditionally masculine traits in women and feminine traits in men across middle age (Huyck, 1990; James et al., 1995). Women become more confident, self-sufficient, and forceful; men more emotionally sensitive, nurturing, considerate, and dependent. These trends appear in cross-sectional and longitudinal research, in people of different social classes, and in highly diverse cultures—not just Western industrialized nations, but small village societies such as the Mayans of Guatemala, the Navajo of the United States, and the Druze of the Middle East (Fry, 1985; Guttmann, 1977; Turner, 1982). They are evident in impressions of everyday behavior as well as the personality traits people say are true of themselves. This indicates that gender-role identity becomes more androgynous in midlife—a mixture of both "masculine" and "feminine" characteristics. It also reveals a tendency for men and women to become more similar in gender-related self-descriptions as they age.

Although there is little argument that these changes occur, the reasons for them are controversial. According to a well-known ethological view called **parental imperative theory,** traditional gender roles are maintained during the active parenting years to help ensure the survival of children. Men become more goal oriented, whereas women emphasize nurturance of husband and children. After children reach adulthood, parents are free to express the "other-gender" side of their personalities, which they had to restrain while rearing children (Gutmann, 1987; Gutmann & Huyck, 1994). A related explanation is that the decline in sex hormones associated with aging may contribute to androgyny in later life (Rossi, 1980).

These biological accounts have been criticized, and evidence for them is mixed. Think back to what you learned in earlier chapters, and you will see that both warmth and assertiveness (in the form of firmness and consistency) are necessary for parents to rear children effectively. Furthermore, although children's departure from the home is related to men's openness to the "feminine" side of their personalities, it is less clearly linked to a rise in "masculine" traits among women (Huyck, 1996). In longitudinal research, college-educated women in the labor force showed a rise in independence by their early forties, regardless of whether they had children; those who were homemakers did not. Women attaining high status at work gained most in dominance, assertiveness, and outspokenness by their early fifties (Helson & Picano, 1990; Wink & Helson, 1993). Finally, increased androgyny is not associated with menopause—a finding at odds with a hormonal explanation of gender-role change (Helson & Wink, 1992).

Besides reduced parenting responsibilities, other demands and experiences of midlife may prompt a more androgynous orientation. For example, among men, a need to enrich a marital relationship after children have departed, along with reduced career advancement opportunities, may be involved in the awakening of emotionally sensitive traits. And men's greater number of health problems may prompt a rise in dependency and passivity. Compared to men, women are far more likely to face economic and social disadvantages. A greater number remain divorced, are widowed, and encounter discrimination in the workplace. Self-reliance and assertiveness are vital for coping with these circumstances.

In sum, androgyny in midlife is the product of a complex combination of social roles and life conditions. In Chapter 8, we noted that androgyny predicts high self-esteem. It is also associated with advanced moral reasoning and psychosocial maturity (Prager & Bailey, 1985; Waterman & Whitbourne, 1982). Indeed, men and women who do not integrate the masculine and feminine sides of their personalities tend to have mental health problems, perhaps because they are unable to adapt flexibly to the challenges of aging (Huyck, 1996; Turner, 1982).

INDIVIDUAL DIFFERENCES IN PERSONALITY TRAITS

Although Trisha and Jewel both became more self-assured and assertive in midlife, in other respects they seemed different. Trisha had always been more organized and hardworking, whereas Jewel was more gregarious and fun loving. Once the two women had traveled together. At the end of each day, Trisha was disappointed if she had not kept to a schedule and visited every tourist attraction. Jewel liked to "play it by ear"—wandering through streets and stopping to talk with shopkeepers and residents.

In previous sections, we reviewed personality changes common to many people in middle adulthood, but stable individual differences also exist. The hundreds of personality traits on which people differ have been organized into five basic factors, called the **"big five" personality traits:**

neuroticism, extroversion, openness to experience, agreeableness, and conscientiousness. Table 16.2 provides a description of each. Notice that Trisha is high in conscientiousness, whereas Jewel is high in extroversion (Costa & McCrae, 1994; McCrae & Costa, 1990).

Cross-sectional studies show that neuroticism and extroversion decline from the teenage years to the end of the twenties, whereas agreeableness and conscientiousness increase—changes that reflect "settling down" and greater maturity. After that time, little change takes place, a finding that has led some theorists to conclude that personality development ends somewhere between age 25 and 30. As further support for this view, longitudinal research shows that over intervals ranging from 3 to 30 years, adults score similarly on the "big five" traits (Costa & McCrae, 1994).

How can there be stability on these characteristics, yet significant change in aspects of personality discussed earlier in this chapter? Studies of the "big five" traits examine development by taking age-related averages on very large samples. Typically, they do not examine the impact of life experiences or the social clock, which shape aspirations, goals, and expectations for appropriate behavior. Furthermore, look closely at the traits in Table 16.2 and you will see that they are very different from the attributes we considered in previous sections. They do not take into account motivations, values, preferred tasks, and coping styles. And they do not consider how certain aspects of personality, such as masculinity and femininity, are integrated (Block, 1995; Helson & Stewart, 1994).

Theorists concerned with change focus on life stories—the way personal needs and life events induce new strategies, plans, and goals. In contrast, theorists who emphasize stability measure traits on which individuals can easily be compared and that are present at any time of life (McAdams, 1994). Perhaps we can resolve the apparent contradiction by thinking of people as changing in overall organization and integration of personality, but doing so on a foundation of basic, enduring dispositions.

TABLE 16.2

The "Big Five" Personality Traits

TRAIT	DESCRIPTION
Neuroticism	Individuals who are high on this trait are worrying, temperamental, self-pitying, self-conscious, emotional, and vulnerable. Individuals who are low are calm, even-tempered, self-content, comfortable, unemotional, and hardy.
Extroversion	Individuals who are high on this trait are affectionate, talkative, active, funloving, and passionate. Individuals who are low are reserved, quiet, passive, sober, and emotionally unreactive.
Openness to experience	Individuals who are high on this trait are imaginative, creative, original, curious, and liberal. Individuals who are low are down-to-earth, uncreative, conventional, uncurious, and conservative.
Agreeableness	Individuals who are high on this trait are soft-hearted, trusting, generous, acquiescent, lenient, and good-natured. Individuals who are low are ruthless, suspicious, stingy, antagonistic, critical, and irritable.
Conscientiousness	Individuals who are high on this trait are conscientious, hardworking, well organized, punctual, ambitious, and persevering. Individuals who are low are negligent, lazy, disorganized, late, aimless, and nonpersistent.

Source: McCrae & Costa, 1990.

Test Bank Items 16.44 through 16.51

BRIEF REVIEW

As middle-aged adults adjust their life structures, personality changes in important ways. Possible selves become fewer in number and are largely concerned with roles and responsibilities already begun. People become more introspective as they work on readjusting their goals and ambitions. At the same time, self-acceptance, autonomy, sense of environmental mastery, and effective coping strategies increase. A variety of demands and experiences prompt gains in traditionally masculine traits in women and feminine traits in men. Consequently, adults become more androgynous in gender-role identity in midlife. Despite changes in attributes that tap motivations, values, preferred tasks, and coping styles, stable individual differences on the "big five" personality traits persist.

ASK YOURSELF . . .

■ *Around age 40, Luellen no longer thought about becoming a performing pianist. Instead, she decided to concentrate on accompanying other musicians in her community and expanding her studio of young pupils. How do Luellen's plans reflect changes in possible selves at midlife?*

■ *On his fifty-second birthday, Tom was asked how he felt now that another year had gone by. He replied, "I feel calmer and more content than at any time in my life." What personality changes might have contributed to Tom's response?*

■ *Jeff, age 46, suggested to his wife, Julia, that they set aside a special time each year to discuss their relationship—both positive aspects and ways to improve. Julia reacted to Jeff's suggestion with surprise, since he had never before seemed interested in working on the quality of their marriage. What developments at midlife probably fostered this new concern?*

RELATIONSHIPS AT MIDLIFE

The emotional and social changes of midlife take place within a complex web of family relationships as well as friendships. Although a few middle-aged people (like Jewel and Tim) live alone, the vast majority (9 out of every 10 in the United States) live in families—most with a spouse (U.S. Bureau of the Census, 1996).

The middle adulthood phase of the family life cycle is often referred to as "launching children and moving on." At one time it was called the "empty nest," but this phrase

President Bill Clinton and First Lady Hillary Rodham Clinton exemplify the middle adulthood phase of the family life cycle called "launching the children and moving on." Here they accompany their daughter Chelsea to Stanford University, where she began college in the fall of 1997. *(San Francisco Chronicle)*

implies a negative transition, especially for women (McCullough & Rutenberg, 1989). When adults devote themselves entirely to their children, the end of active parenting can trigger feelings of emptiness and regret. But for many people, middle adulthood is a liberating time, offering a sense of completion and an opportunity to strengthen existing ties and build new ones.

This phase is the newest and longest of the family life cycle. Early in this century, most parents reared children for almost all of their active adulthood. Due to a declining birthrate and longer life expectancy, modern parents launch children about 20 years before retirement and then seek other rewarding activities. Because of the lengthening of this period, it is marked by the greatest number of exits and entries of family members (McGoldrick, Heiman, & Carter, 1993). As adult children leave home and marry, middle-aged people must adapt to new roles of parent-in-law and grandparent. At the same time, they must establish a different type of relationship with their aging parents, who may become ill or infirm and die.

These changes reveal that middle adulthood is much more than the label "empty nest" implies. Let's see how ties within and beyond the family are modified during this time of life.

MARRIAGE AND DIVORCE

Although not all couples are financially comfortable, compared to other age groups middle-aged households are well off economically. Americans between 45 and 54 have the highest average annual income (U.S. Bureau of the Census, 1996). Partly for this reason and because the time between departure of the last child and retirement is so long, the contemporary social view of marriage in midlife has become one of expansion and new horizons.

These forces strengthen the need to review and adjust the marital relationship. For couples like Devin and Trisha, this shift was gradual. By middle age, their marriage had permitted gratification of both family and individual needs, endured many changes, and culminated in a period in which more love was expressed than at any other time. Elena's marriage, in contrast, became more conflict ridden as her teenage daughter's problems introduced added strains and as departure of children made marital difficulties more obvious. Tim's failed marriage revealed yet another pattern. With passing years, both love expressed and number of problems declined. As less happened in the relationship, either good or bad, the couple had little to keep them together (McCullough & Rutenberg, 1989).

Satisfaction with family life continues to be a strong predictor of psychological well-being in midlife (Ishii-Kuntz, 1990). Middle-aged men who have focused only on career often realize the limited nature of their pursuits. At the same time, their wives may insist on a different and deeper relationship. In addition, children fully engaged in adult roles remind middle-aged parents that they are in the latter part of their lives. Consequently, many decide that the time for improving their marriages is now.

What does the future hold for this divorcing couple? For the woman, it is likely to bring a sharp decline in standard of living. Yet if she weathers divorce successfully, she will probably experience personal growth, becoming more tolerant, comfortable with uncertainty, and nonconforming in personality. The man is likely to enter a new relationship and remarry within a short time. (L. D. Gordon/The Image Bank)

As in early adulthood, divorce is one way of resolving an unsatisfactory marriage in midlife. Although most divorces occur within 5 to 10 years of marriage, in the United States 11 percent take place after 20 years or more (U.S. Public Health Service, 1996). Because the current midlife generation was married at a time when divorce rates were lower, marital breakup is often more difficult than it is for young adults. Middle-aged people are more likely to view it as a personal failure. And they typically find less peer support because fewer divorces occur in their age group. Women in traditional marriages, who devoted themselves to caring for husband and children, have the hardest time adjusting to divorce.

A substantial number of midlife divorces occur to people who have experienced one or more previous unsuccessful marriages, since the divorce rate is more than twice as great among remarried couples than couples in a first marriage. For women, the negative impact of marital breakup—especially when it is repeated—on standard of living can be devastating. In a study of over five thousand 30- to 44-year-old women followed over 15 years, divorce led to a 39 percent drop in average income. Many African-American women opted for separation rather than divorce, probably because of the high cost of divorce proceedings. Women who separated had the highest poverty rate—before the marital transition (27 percent) and afterward (57 percent). Neither separated nor divorced women escaped economic disadvantage after adjusting to their new life circumstances. Even 7 or 8 years after their marriages ended, the high rate of poverty persisted (Morgan, 1991).

As these findings reveal, marital breakup, in midlife and earlier, is a strong contributor to the **feminization of poverty**—a trend in which women who support themselves or their families have become the majority of the adult poverty population, regardless of age and ethnic group. Because of weak public policies safeguarding families (see Chapter 2), the gender gap in poverty is much higher in the United States than in other Western industrialized nations (Goldberg & Kremen, 1990).[1]

What do recently divorced middle-aged people say about why their marriages ended? In one study, women mentioned communication problems most often, followed by husband's substance abuse, husband's physical and verbal abuse, and their own desire for autonomy. Notice how these responses dwell on neutral circumstances (communication) and the husband's faults. When women attribute divorce to a cause that is more self-accusing ("my husband's lack of interest in me"), they show poorer adjustment (Davis & Aron, 1989).

Longitudinal evidence reveals that middle-aged women who weather divorce successfully tend to become more tolerant, comfortable with uncertainty, and nonconforming in personality—factors believed to be fostered by

[1] The feminization of poverty has become a global problem. Return to the Cultural Influences box on page 523 to find out why it is widespread in developing regions of the world.

divorce-forced independence. As in earlier periods, divorce represents "both a time of trauma and a time of growth" (Rockwell, Elder, & Ross, 1979, p. 403). Unfortunately, little is known about divorce adjustment among middle-aged men, perhaps because most enter new relationships and remarry within a short time.

CHANGING PARENT–CHILD RELATIONSHIP

Parents' positive relationships with their grown children are the result of a gradual process of "letting go," starting in childhood, gaining momentum in adolescence, and culminating in children's independent living. As we mentioned earlier, most middle-aged parents adjust well to the launching phase of the family life cycle; only a minority have difficulty. Investment in nonparental relationships and roles, children's characteristics, parents' marital and economic circumstances, and cultural forces affect the extent to which this transition is expansive and rewarding or sad and distressing.

After moving their son Mark into his college dormitory at the start of his freshman year, Devin and Trisha felt a twinge of nostalgia. On the way home, they recalled his birth, first day of school, and high school graduation and commented on their suddenly tranquil household. Beyond this, they returned to rewarding careers and community participation and delighted in having more time for each other. Parents who have developed gratifying alternative activities typically welcome their children's adult status. A strong work orientation, especially, predicts gains in life satisfaction after children depart from the home (Seltzer & Ryff, 1994; Silverberg & Steinberg, 1990).

Regardless of whether they reside with parents, adolescent and young adult children who are "off time" in development—not showing expected signs of independence and accomplishment—can lead to parental strain (Ryff et al., 1994). Consider Elena, whose daughter was frequently truant from high school and in danger of not graduating. The need for greater parental oversight and guidance caused anxiety and unhappiness for Elena, who was ready to complete the active parenting phase and focus on her own personal and vocational development (Raup & Myers, 1989).

The end of parent–child coresidence is accompanied by a substantial decline in parental authority. Devin and Trisha no longer knew of Mark's daily comings and goings nor expected that he inform them. Nevertheless, Mark telephoned at regular intervals to report on events in his life and seek advice about major decisions. Although the parental role changes, its continuation is important to middle-aged adults. Departure of children is a relatively minor event when parent–child contact and affection are sustained. When it results in little or no communication, parents' life satisfaction declines (White, 1994; White & Edwards, 1990).

Throughout middle adulthood, parents continue to give more assistance to children than they receive, espe-

cially while children are young adults and unmarried or when they face difficulties, such as marital breakup or unemployment (Rossi & Rossi, 1990). Providing emotional and financial support while children get their lives underway is related to midlife psychological well-being. Due to disrupted relationships and economic need, divorced and remarried parents are less likely than parents in first marriages to offer adult children support, and they are also less content (Marks, 1995). Overall, favorable adjustment to the launching phase depends on feeling successful as a parent and not being estranged from one's children.

Recall from Chapter 13 that many young people from low-income homes and with cultural traditions of extended family living do not leave home early. Consequently, the timing of the launching phase varies greatly with social class and ethnicity. In traditional families with strong investment of authority in the father, departure of an adult child can threaten parental status, undermining well-being in midlife (Hartley, 1995).

When children marry, parents face additional challenges in enlarging the family network to include in-laws. Difficulties occur when parents do not approve of their child's partner or when the young couple adopts a way of life inconsistent with the values of the family of origin. When warm, supportive relationships endure, intimacy between parents and children increases over the adult years, with great benefits for parents' life satisfaction (Rossi & Rossi, 1990). Once young adults strike out on their own, members of the middle generation, especially mothers, usually take on the role of **kinkeeper,** gathering the family for celebrations and making sure everyone stays in touch.

As children become adults, parents expect a mature relationship with them, marked by tranquility and contentment. Yet many factors—on the child's and adult's side—affect whether that goal is achieved. The Caregiving Concerns table on page 532 suggests ways middle-aged parents can increase the chances that bonds with adult children will be loving and rewarding and serve as contexts for personal growth.

GRANDPARENTHOOD

Two years after Mark married, Devin and Trisha were thrilled to learn that a granddaughter was on the way. Although the stereotypical image of grandparents as very elderly persists, today the average age of grandparenthood in the United States is 49 to 51 for women, 51 to 53 for men. A longer life expectancy means that adults will spend as much as one–third of their lifespan in the grandparent role (Smith, 1991).

■ MEANINGS OF GRANDPARENTHOOD. Why did Trisha and Devin, like many people their age, greet the announcement of a grandchild with such enthusiasm? Grandparent-

CAREGIVING CONCERNS

Ways Middle-Age Parents Can Promote Positive Ties with Their Adult Children

SUGGESTION	DESCRIPTION
Emphasize positive communication.	Let adult children know of your respect, support, and interest. This not only communicates affection, but permits conflict to be handled in a constructive context.
Avoid unnecessary comments that are a holdover from childhood.	Adult children, like younger children, appreciate an age-appropriate relationship. Comments that have to do with safety, eating, and cleanliness (for example, "Be careful on the freeway," "Don't eat those foods," and "Make sure you wear a sweater on a cold day") annoy adult children and can stifle communication.
Accept the possibility that some cultural values and practices and aspects of lifestyle will be modified in the next generation.	In constructing a personal identity, most adult children have gone through a process of evaluating the meaning of cultural values and practices for their own lives. Traditions and lifestyles cannot be imposed on adult children; parents' efforts to do so can be a source of great conflict and stress. Recognize that relationships with grown children can outlive each family member's beliefs and lifestyle preferences. Rather than focusing on conflicts, look for common ground.
When an adult child encounters difficulties, resist the urge to "fix" things.	Although you may have difficulty tolerating the situation, accept the fact that no meaningful change can take place without the willing cooperation of the adult child. Stepping in and taking over communicates a lack of confidence and respect. Find out whether the adult child wants your help, advice, and decision-making skills.
Be clear about your own needs and preferences.	When it is difficult to arrange for a visit, baby-sit, or provide other assistance, say so and negotiate a reasonable compromise rather than letting resentment build.

Source: Toder, 1994.

hood is a highly significant milestone to most who experience it. When asked about its meaning, people generally mention one or more of the following gratifications:

■ Valued elder—being perceived as a wise, helpful person

■ Immortality through descendants—leaving behind not just one but two generations after death

■ Reinvolvement with personal past—being able to pass family history and values to a new generation

■ Indulgence—having fun with children without major child-rearing responsibilities (Kivnick, 1983; Miller & Cavanaugh, 1990)

■ GRANDPARENT–GRANDCHILD RELATIONSHIPS. Grandparents' styles of relating to grandchildren vary as widely as the meanings they derive from their new role. The grandparent and grandchild's age and sex make a difference. When their granddaughter was young, Trisha and Devin enjoyed an affectionate, playful relationship with her. As she got older, she looked to them for information and advice in addition to warmth and caring. By the time their granddaughter reached adolescence, Trisha and Devin had become role models, family historians, and conveyers of social, vocational, and religious values (Hurme, 1991; Kornhaber & Woodward, 1981). Typically, relationships are closer between grandparents and grandchildren of the same sex and, especially, between maternal grandmothers

and granddaughters—a pattern found in many countries, including Canada, Finland, Poland, and the United States (Smith, 1991). Grandmothers also report higher satisfaction with the grandparent role than do grandfathers, perhaps because it is an important means through which middle-aged women satisfy their kinkeeping function (Thomas, 1986a, 1986b, 1989).

Nearness of residence made Trisha and Devin's pleasurable interaction with their granddaughter possible. Grandparents who live far from their grandchildren usually have more distant relationships with them, appearing mainly on holidays, birthdays, and other formal occasions but otherwise having little contact (Bengtson & Robertson, 1985). Despite high family mobility in Western industrialized nations, most grandparents live within 30 minutes of at least one grandchild, permitting regular visits (Werner, 1991).

Social class and ethnicity are additional influences on grandparent–grandchild ties. In middle-class families, the grandparent role is not central to family maintenance and survival. For this reason, it is fairly unstructured and takes many forms. In contrast, grandparents perform essential activities in low-income families. For example, many single parents live with their family of origin, where grandparents' financial and caregiving assistance reduce the impact of poverty. Grandchildren in single-parent and stepparent families rate the quality and range of activities with their grandparents higher than do grandchildren in intact families. As children experience the stress of family

In some subcultures, grandparents are absorbed into an extended family household and become actively involved in child rearing. When a Chinese-American, Japanese-American, or Mexican-American maternal grandmother is a homemaker, she is the preferred caregiver while parents of young children are at work. (David W. Hamilton/The Image Bank)

transition, bonds with grandparents take on increasing importance (Kennedy & Kennedy, 1993).

In some subcultures, grandparents are absorbed into an extended family household and become actively involved in child rearing. When a Chinese-American, Japanese-American, or Mexican-American maternal grandmother is a homemaker, she is the preferred caregiver while parents of young children are at work. Similarly, grandparent involvement in child care is high among Native Americans. In the absence of a biological grandparent, Native-American parents may adopt an unrelated elder into the family to serve as a mentor and disciplinarian for their children (Werner, 1991). (See Chapter 2, page 65, for a description of the grandmother's role in the African-American extended family.)

Increasingly, grandparents have stepped in as primary caregivers in the face of serious family problems. Today, 2 million American children live apart from their parents in grandparent-headed households. Despite their willingness to help, grandparents have considerable difficulty obtaining legal custody of grandchildren. Although they are the relatives most often called on in foster placement of children, grandparents receive a lower monthly stipend to support child care than do other foster caregivers. The majority of these grandparents have limited financial means and express a need for additional assistance from community and government agencies (Robertson, 1995).

In most families, parents serve as gatekeepers of grandparents' contact with grandchildren. When grandparents and parents get along poorly, the grandparent–grandchild tie usually suffers. After marital breakup, for example, grandparents on the custodial parent's side have more frequent contact with grandchildren than do grandparents on the noncustodial side. A growing concern among grandparents is maintaining contact with grandchildren after parental divorce. As the Social Issues box page 534 explains, American grandparents barred from seeing their grandchildren can petition the courts for visitation rights. Yet this is a controversial practice, since it can harm rather than help children if it intensifies family hostilities.

When family relationships are positive, grandparenthood provides an important means of fulfilling personal and societal needs in midlife and beyond. Typically, grandparents are a frequent source of pleasure, support, and knowledge for children, adolescents, and young adults. They also provide the young with firsthand experience in how older people think and function. In return, grandchildren become deeply attached to grandparents and keep them abreast of social change. Clearly, grandparenthood is a vital context for sharing between generations.

MIDDLE-AGED CHILDREN AND THEIR AGING PARENTS

Compared to earlier generations, today's adults not only spend more years as parents and grandparents, but also as children of aging parents. The percentage of middle-aged people with living parents has risen dramatically — from 10 percent in 1900 to nearly 50 percent in the 1990s (U.S. Bureau of the Census, 1996). A longer life expectancy means that adult children and their parents are increasingly likely to grow old together. What are middle-aged children's relationships with their aging parents like? And how does life change for adult children when an aging parent's health declines?

■ **FREQUENCY AND QUALITY OF CONTACT.** A widespread myth is that adults of past generations were more devoted to their aging parents than are adults of the present generation. Although adult children spend less time in close proximity to their parents, the reason is not neglect or isolation. Fewer aging adults live with younger generations now than in the past because of a desire to be independent, made possible by gains in health and financial security. Nevertheless, two-thirds of older adults in the United States live close to at least one of their children, and frequency of contact is high through both visits and

SHOULD GRANDPARENTS BE AWARDED VISITATION RIGHTS AFTER PARENTAL DIVORCE?

Before the 1970s, grandparents did not have the right to petition the courts for visitation privileges with their grandchildren after parental separation and divorce. Legally mandated visitation was reserved for parents, who regulated grandparents' access to children. Although grandparent visitation cases occasionally reached the courts, judges were wary about granting these requests. They recognized that intense conflict between parents and grandparents lay behind most petitions and that it was not in the best interests of children to embroil them in intergenerational disputes (Derdeyn, 1985).

In recent years, a rising population of middle-aged and older Americans has led to a broadening of grandparents' rights. Interest groups representing senior citizens have convinced state legislators to support the grandparent–grandchild relationship during an era in which a high rate of marital breakup has threatened extended family ties. Today, all 50 states permit grandparents to seek legal visitation judgments (Bostock,

1994). The new policy is also motivated by a well-intentioned desire on the part of lawmakers to foster children's access to social supports within the family and widespread belief in the specialness of the grandparent–grandchild relationship.

Nevertheless, serious questions have been raised about legalizing children's ties to their grandparents. The most significant factor in how children's development is affected by interaction with grandparents is the quality of the grandparents' relationship with the children's parents. If it is positive, children are likely to benefit. But courtroom battles that turn parents and grandparents into adversaries may close the door to gains that would otherwise result from frequent grandparent–grandchild contact (Thompson et al., 1989). As one therapist observed, in families where parents are divorcing, the behavior of grandparents varies greatly, from constructive help to entanglement in parental battles and efforts to undermine the child's relationship with one and sometimes both parents (Derdeyn, 1985).

In sum, research suggests that the courts should exercise considerable restraint in awarding grandparent visitation rights in divorce cases. Yet statutes providing for grandparent visitation are expanding to include a broader range of circumstances, largely as a result of senior citizen pressure. More laws take the position that grandparents have rights of access to their grandchildren simply as a result of their status as grandparents. As evidence for this trend, in half the states, it is now possible for grandparents to petition for visitation with a child in an intact home (Bostock, 1994).

TRY THIS...

■ Suppose you are an expert witness in a grandparent visitation petition. Would you advise the judge to consider grandparents' previous and current relationships with family members? How about the reason the grandparents filed the petition? Why or why not?

telephone calls. Proximity increases with age. Elders who move usually do so in the direction of kin, and younger people tend to move in the direction of their aging parents. The majority of older people are satisfied with the amount of face-to-face contact they have with their children (Jerrome, 1990b).

Middle age is a time when adults reassess relationships with their parents, just as they rethink other close ties (Helson & Moane, 1987). Many adult children become more appreciative of their parents' strengths and generosity. Trisha, for example, marveled at her parents' fortitude in rearing three college-educated children despite limited income. And she recalled her mother's sound advice just before her marriage to Devin nearly three decades earlier: "Build a life together but also forge your own life. You'll be happier." Trisha had called on that advice at several turning points, and it influenced her decisions.

In the non-Western world, older adults most often live with their married children. Chinese and Japanese elderly, for example, generally reside with the eldest son and his wife and children (United Nations, 1991c). Regardless of whether coresidence and daily contact are typical, relationship quality varies widely. Usually, patterns established earlier persist; positive parent–child ties remain so, as do conflict-ridden interactions.

Help exchanged between adult children and their aging parents is responsive to past and current family circumstances. The closer family ties were when children were growing up, the more help given and received (Whitbeck, Hoyt, & Huck, 1994). Also, parents give more to single than married children, and children give more to widowed parents and parents in poor health. At the same time, a shift in helping occurs over the adult years. Parent-to-child advice, household aid, gift giving, and financial assistance decline,

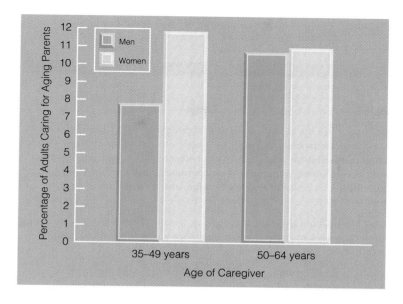

FIGURE 16.3

Percentage of middle-aged men and women involved in caring for an aging parent in the United States, based on interviews with a nationally representative sample of over 13,000 adults. Although women are usually the primary caregivers, men make substantial contributions to the care of ill and frail parents. With age, the sex difference in parental caregiving declines. (Adapted from Marks, 1996.)

whereas child-to-parent help of various kinds increases (Rossi & Rossi, 1991).

■ CARING FOR AGING PARENTS. The burden of caring for aging parents can be great. In Chapter 2, we noted that the family structure has become more "top-heavy," with more generations alive, but fewer younger members as birthrates have declined. This means that more than one older family member is likely to need assistance, with fewer younger adults available to provide it. Adults with ill or frail parents are caught in a **middle-generation squeeze,** faced with competing demands of children (some of whom are under age 18 and still at home) and employment (Gatz, Bengtson, & Blum, 1990). About 1 in every 10 middle-aged adults in the United States is involved in caring for a parent with a disability or chronic illness (Marks, 1996).

When an aging parent's spouse cannot provide care, adult daughters are the next most likely relatives to do so. Even when the spouse is available, adult children—again, usually daughters—often pitch in as needed. Why are women more often elected as caregivers? Families turn to the person who seems most available—living nearby and with fewer commitments regarded as interfering with the ability to assist. These unstated rules, in addition to parents' preference for same-sex caregivers (aging mothers live longer), lead more women to fill the role. About 50 percent of women caregivers are employed; another 9 to 28 percent quit their jobs to provide care (Brody et al., 1987; Stone, Cafferata, & Sangl, 1987).

Nevertheless, men make a substantial contribution to care of aging parents—one that should not be overlooked. Tim, for example, looked in on his father, a recent stroke victim, every evening, reading to him, running errands, making household repairs, and taking care of finances. His sister, however, provided more hands-on care in her own home—cooking, feeding, and bathing. The care sons and

daughters provide tends to be divided along gender-role lines. However, no sex differences exist in sense of obligation or affection toward aging parents (Kahana, Biegel, & Wykle, 1994).

As adults move from early to later middle age, the sex difference in parental caregiving declines (see Figure 16.3). Perhaps as men reduce their vocational commitments and feel less need to conform to a "masculine" gender role, they are more able and willing to provide basic care (Marks, 1996). At the same time, parental caregiving may contribute to men's greater openness to the "feminine" side of their personalities. A man who cared for his mother, severely impaired by Alzheimer's disease (see Chapter 17, page 569), commented on how the experience altered his outlook:

> Having to do personal care, becoming a male nurse, was a great adjustment. It was so difficult to do these tasks; things a man, a son, is not supposed to do. But, I had to alter, since charity must come before maintaining a selfish, conventional view. I have definitely modified my views on conventional expectations. (Hirsch, 1996, p. 112)

Although most adult children help willingly, caring for a chronically ill or disabled parent is highly stressful. Some people regard it as similar to caring for a young child, but it is very different. The need for parental care typically occurs suddenly, after a heart attack, fall, stroke, or diagnosis of cancer, leaving little time for preparation. Whereas children become increasingly independent, the parent usually gets worse and the caregiving task as well as its cost escalates. "One of the most difficult aspects is the emotional strain of being such a close observer of my father's physical and mental decline," Tim explained to Devin and Trisha. Tim also felt a sense of grief over the loss of a cherished relationship, as his father no longer seemed to be his

CAREGIVING CONCERNS

Relieving the Stress of Caring for an Aging Parent

STRATEGY	DESCRIPTION
Use effective coping strategies.	Use problem-centered coping to manage the parent's behavior and caregiving tasks. Delegate responsibilities to other family members, seek assistance from friends and neighbors, and recognize the parent's limits while calling on capacities the parent does have. Use emotion-centered coping to reinterpret the situation in a positive way, such as emphasizing its opportunity for personal growth and for giving to parents in the last years of their life. Avoid denial of anger, depression, and anxiety in response to the caregiver work burden, which heightens stress.
Seek social support.	Confide in family members and friends about the stress of caregiving, seeking their encouragement and help. So far as possible, avoid quitting work to care for an ill parent because it leads to social isolation and loss of financial resources.
Make use of community resources.	Contact your Area Agency on Aging (see Chapter 2, page 66), church, synagogue, and other community organizations to seek information and assistance, in the form of in-home respite help, home-delivered meals, transportation, and adult day care.
Press for workplace and public policies that relieve the emotional and financial burdens of caring for an aging parent.	Difficulties faced by caregivers are partly due to lack of workplace and societal supports. Encourage your employer to provide elder care benefits, such as flexible work hours and caregiver leave without pay. Communicate with lawmakers and other citizens about the need for additional government funding, in the form of tax relief and other reimbursements, to help pay for elder care. Emphasize the need for improved health insurance plans that reduce the financial strain of elder care on middle- and low-income families.

former self. Because duration of caregiving is uncertain, caregivers often feel as if they no longer have control over their lives (Gatz, Bengtson, & Blum, 1990).

Adults who share a household with ill parents—about 10 percent of caregivers—experience the most stress (Marks, 1996). In addition to the factors just mentioned, a parent and child who have lived separately for years usually dislike moving in together. Conflicts are likely to arise over routines and lifestyles. But the greatest source of stress is problem behavior, especially for caregivers of parents who have deteriorated mentally. Tim's sister reported that their father would wake during the night, ask repetitive questions, follow her around the house, and become agitated and combative (Gatz, Bengtson, & Blum, 1990). Parental caregiving often has emotional and physical health consequences, with a rate of depression as high as 30 to 50 percent (Schulz, Visintainer, & Williamson, 1990).

Social support is highly effective in reducing caregiver stress. Tim's encouragement, assistance, and willingness to listen helped his sister cope with in-home care of their father (Birkel, 1987). Despite more time to care for an ill parent, women who quit work generally fare poorly, probably because of social isolation and financial strain (Pohl et al., 1994). Surrogate in-home care (by a nonfamily member) is often not a solution because of its high cost and low availability outside large cities. And unless they must, few people want to place their parents in formal care, such as nursing homes, which are also expensive. At present, adult

children have limited options in how to provide parental care. The Caregiving Concerns table above summarizes ways to relieve the stress of caring for an aging parent—at the individual, family, community, and societal levels.

SIBLINGS

As Tim's relationship with his sister reveals, siblings are ideally suited to provide social support. The majority of adult siblings report frequent contact, seeing or talking on the phone at least monthly. As in early adulthood, sister–sister relationships are closer than sister–brother and brother–brother ties, a trend apparent in many industrialized nations (Cicirelli, 1995; White & Riedmann, 1992).

Middle adulthood is often a time when sibling bonds strengthen in response to major life events. Parenting and employment pressures may have caused the sibling relationship to be set aside temporarily. Launching and marriage of children seem to prompt sisters and brothers to think more about each other. As Tim commented, "It helped our relationship when my sister's children were out of the house and married. I'm sure she cared about me. I think she just didn't have time!" The impact of parental illness can have a profound impact on sibling closeness. Brothers and sisters who previously had little to do with one another find themselves in frequent contact as they cooperate to provide care. When parents die, adult children realize they have become the oldest generation in the family and must look to each other to sustain family ties (Gold, 1996; Moyer, 1992).

Although the trend is toward closer relationships, not all sibling bonds improve in midlife. Recall Trisha's negative encounters with her sister, Dottie (see Chapter 15, page 498). Yet Dottie's difficult temperament had made her hard to get along with since childhood. Her temper flared when their father died and problems arose over family finances. As one expert expressed it, "As siblings grow older, good relationships [often] become better and rotten relationships get worse" (Moyer, 1992, p. 57). Clearly, sibling ties are affected by a combination of sibling personalities and events in siblings' lives.

In Western nations, adult siblings usually live independently, and the relationship is voluntary. Frequency of interaction depends on both psychological closeness and nearness of residence. Ethnic background also makes a difference. In one study, Italian-American siblings (especially sisters) had much greater contact and warmer relationships than did white Protestant siblings. The researchers speculated that strong parental authority in Italian immigrant families led siblings to turn to one another for support early in life. These bonds persisted into adulthood (Johnson, 1985).

Closer ties between sisters do not necessarily characterize nonindustrialized societies, where sibling relationships are basic to family functioning. For example, among the Asian Pacific Islanders, brother–sister attachments are strong, and family social life is organized around them. A brother–sister pair is often treated as a special unit in exchange marriages with another family. After marriage, brothers are expected to protect sisters, and sisters serve as spiritual mentors for brothers. Families not only include a large number of biological siblings, but also grant other relatives, such as cousins, the status of brother or sister. This leads to an unusually large network of sibling support throughout life (Cicirelli, 1995).

In village and tribal societies, cultural norms prevent sibling conflict, which could threaten family cooperation (Weisner, 1993). In industrialized nations, promoting positive sibling interaction in childhood is vital for ensuring that siblings will support one another in later years. With each life event in middle adulthood, many siblings grow closer together, whereas others drift further apart.

FRIENDSHIPS

In his forties and fifties, Devin had more time to spend with friends than he did in the previous decade, when family responsibilities took away from friendship. Late Friday afternoons, he met several male friends at a coffee house, and they chatted for a couple of hours. However, the majority of Devin's friends were couple based—ones in which he and Trisha gathered with other husband–wife pairs. Compared to Devin, Trisha more often got together with friends on her own (Blieszner & Adams, 1992).

Characteristics of middle-age friendships are a continuation of trends we discussed in Chapter 14. In a study of 18- to 75-year-olds, men were less expressive with friends than were women across all age groups. Whereas men tended to talk about sports, politics, and business, women focused on feelings and life problems. For this reason, when Trisha and Devin gathered with friends, men often congregated in one area, women in another (Fox, Gibbs, & Auerbach, 1985; Wellman, 1992). Women report a greater number of close friends and say they both receive and provide their friends with more emotional support. Not surprisingly, women are more satisfied with their friendships than are men (Antonucci, 1994).

Nevertheless, for both sexes, number of friends declines with age, probably because people become less willing to invest in nonfamily ties that are unimportant to them. As selectivity of friendship increases, middle-aged and older people express deeper and more complex ideas about friendship. They also try harder to reduce disagreements with friends than do young adults (Antonucci & Akiyama, 1995; Fox, Gibbs, & Auerbach, 1985). Having chosen a friend, adults in midlife attach great value to the relationship and take extra steps to protect it.

By midlife, family relationships and friendships support different aspects of psychological well-being. Family ties protect the individual against serious threats and losses. Because such ties are based on obligation, they promote security within a long-term time frame. In contrast, friendships are voluntary ties that serve as current sources of pleasure and satisfaction (Antonucci & Akiyama, 1995). As middle-aged husbands and wives renew their sense of partnership and companionship, they may combine the best of family and friendship. Indeed, research indicates that viewing a spouse as a best friend contributes greatly to marital happiness (Bengtson, Rosenthal, & Burton, 1990).

RELATIONSHIPS ACROSS GENERATIONS

A widely recognized goal of families, communities, and society as a whole is that each generation of adults leave the next better off than they themselves were. For more than two centuries, this aspiration has been met in the United States. For example, Trisha's high-school-educated parents did everything possible to support her desire to earn a professional degree. And Devin and Trisha reared their son under financially more comfortable circumstances than they had experienced as children.

Yet Trisha and Devin often wondered, as many Americans do, whether the current generation would ever be as well off as the previous one. And on the basis of high rates of marital breakup, single-parent families, youth crime rates, and geographical distance among extended family members, they questioned whether younger people would continue to respect their elders and look after their well-being. Trisha and Devin also worried that improved financial conditions among the elderly in comparison to the high poverty rate among families with young children (see Chapter 2, page 61) might fuel intergenerational anger and resentment.

Is solidarity between generations eroding in the United States? A recent survey of a nationally representative sample of 1,500 people between ages 18 and 90 addressed this question. Findings showed that despite the social changes just mentioned, supportive ties among younger and older citizens have endured (Bengtson & Harootyan, 1994).

At the family level, the large majority (90 percent) of adult children reported feeling close to an aging parent. In fact, adult children described closer ties to their aging parents than aging parents felt toward their adult children (although 80 percent of these ties were also strong). In line with the high rate of help giving described in earlier in this chapter, younger adults expressed a deep sense of obligation toward their parents—more than their parents expected from them! For example, only 19 percent of 18- to 34-year-olds and 28 percent of 35- to 44-year-olds agreed with the statement "Grown children should not be expected to support their parents." In contrast, 51 percent of 45- to 64-year-olds and 55 percent of people age 65 and over agreed with this statement (Lawton, Silverstein, & Bengtson, 1994).

Young, middle-aged, and older survey respondents reported a wide range of volunteer activities in their communities. Middle-aged adults frequently devoted time to children and adolescents—usually through church or synagogue activities and academic tutoring programs. Older adults, by contrast, more often provided volunteer assistance to the elderly. But they were more likely to do so if they had warm relationships with their adult children—a finding that reflects an important link between family relationships and giving to the community. More than 70 percent of the sample described some kind of informal assistance to relatives, friends, neighbors, and other community residents (Harootyan & Vorek, 1994).

Finally, most participants expressed no resentment about government benefits to other age groups. Instead, a *norm of equity* characterized their responses. When people believed the needs of one generation were unmet, they expressed dissatisfaction about government benefits to another age group. Only rarely did they say that their own age group was neediest and should get more (Schlesinger & Kronebusch, 1994).

In sum, despite public concern over intergenerational conflicts, strong connections between age groups within the family and the community exist, and most people are not guided by self-interest in how they think government resources should be distributed. The values, opinions, and behaviors of ordinary citizens reveal "hidden bridges" among people of all ages. They suggest that intergenerational responsibility is likely to guide the activities of Americans for many years to come.

BRIEF REVIEW

"Launching children and moving on" is the newest and longest phase of the family life cycle. As years without parenting responsibilities stretch ahead, many couples review and adjust their marital relationships. Although divorce remains a common solution to an unhappy marriage, middle-aged adults find marital breakup more difficult than do younger individuals. For women, the impact on standard of living can be severe. Nevertheless, women who adapt successfully to divorce often show signs of personal growth.

Adults who have gratifying alternative activities and feel successful as parents typically adjust well to departure of children from the home. When children marry, problems accepting in-laws occur when parents do not approve of the child's partner or lifestyle. As children establish their own families, mothers usually become kinkeepers, taking responsibility for sustaining family bonds. Grandparents derive many meanings from their role. Their relationships with grandchildren vary widely, depending on sex, age, nearness of residence, social class, ethnicity, and quality of grandparent–parent ties.

Caring for aging parents can cause considerable stress, especially among women (who assume most of the burden) and when parents have deteriorated mentally. With age, the sex difference in parental caregiving declines. Sibling bonds that were positive in earlier years often strengthen in response to midlife events. Sex differences in friendship that characterized earlier years extend into middle adulthood. At the same time, number of friends declines, and middle-aged adults take extra steps to preserve friendships ties. Contrary to widespread belief, intergenerational commitment—in terms of family relationships, community involvement, and beliefs about the distribution of government resources—remains strong in American society.

ASK YOURSELF . . .

■ *Freda divorced her husband after 25 years of marriage. Compared to her daughter Salena, who got divorced at age 30, Freda had a harder time adjusting. Explain why adapting to divorce is more difficult for middle-aged than for young adults and is particularly stressful for women.*

■ *Raylene and her brother Walter live in the same city as their aging mother, Elsie. When Elsie could no longer live independently, Raylene took primary responsibility for her care. What factors probably contributed to Raylene's involvement in caregiving and Walter's lesser role?*

■ *As a young adult, Daniel maintained close friendship ties with six college classmates. At age 45, he continued to see only two of them. What explains Daniel's reduced circle of friends in midlife?*

VOCATIONAL LIFE

We have already seen that the midlife transition typically involves vocational readjustments. For Devin, it resulted in a move up the career ladder to a demanding administrative post as college dean. Trisha reoriented her career from a large to a small law firm, where she could pursue cases that interested her and felt her efforts were appreciated. Recall from Chapter 15 that after her oldest child left home, Anya earned a college degree and entered the work force for the first time. Jewel strengthened her commitment to an already successful business, whereas Elena changed careers. Finally, Tim reduced his vocational obligations as he prepared for retirement.

Work continues to be a salient aspect of identity and self-esteem in middle adulthood. More so than in earlier or later years, middle-aged workers attempt to increase the personal meaning and self-direction of their vocational lives (Levinson, 1978, 1996). At the same time, certain aspects of job performance improve. Older employees have lower rates of absenteeism, turnover, and accidents and show no change in work productivity (Warr, 1994). Consequently, the value of an older employee ought to be equal to, and possibly even greater than, that of a younger employee.

Understanding the midvocational years is especially important. The post–World War II baby boom, along with the elimination of mandatory retirement age in many industrialized nations, means that the number of older workers will rise dramatically over the next few decades (Salthouse & Maurer, 1996). Yet a favorable transition from adult worker to older worker is hindered by negative stereotypes of aging—incorrect beliefs about limited learning capacity, slower decision making, and resistance to change and supervision. Furthermore, gender discrimination continues to restrict the vocational attainments of many women. Let's take a close look at middle-aged work life.

JOB SATISFACTION

Job satisfaction has both psychological and economic significance. If people are dissatisfied at work, the consequences include strikes, grievances, absenteeism, and turnover, all of which are costly to employers.

Research shows that overall job satisfaction increases with age at all occupational levels, from executives to hourly

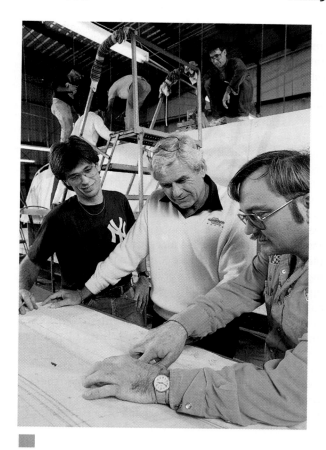

These middle-aged boat designers share their thoughts with a younger co-worker as they look over new plans. Job satisfaction rises in midlife. Intrinsic satisfaction—contentment with the work itself—is largely responsible for the overall gain. (Bob Daemmrich/The Image Works)

workers. The relationship is weaker for women than men, probably because of women's reduced chances for advancement, which result in a sense of unfairness (Stagner, 1985). When different aspects of jobs are considered, intrinsic satisfaction—contentment with the work itself—shows a strong age-related gain. Extrinsic satisfaction—for example, with supervision, pay, and promotions—changes very little (Warr, 1994).

What explains the rise in job satisfaction over the adult years? A broader time perspective probably contributes. "I recall complaining about how much I had to do when I first started teaching," remarked Devin. "Since then, I've seen a lot of hard times. From my current vantage point, I can tell a big problem from a trivial one." Moving out of unrewarding work roles, as Trisha did, can also boost morale. Key characteristics that predict job well-being include involvement in decision making, reasonable workloads, and good physical working conditions. Older people may have greater access to jobs that are attractive in these ways.

Finally, having fewer alternative positions into which they can move, older workers generally reduce their vocational aspirations. As the perceived gap between actual and possible achievements declines, work involvement and commitment increase (Kalleberg & Loscocco, 1983; Warr, 1992).

Although emotional engagement with work is usually seen as psychologically healthy, it can also result in **burnout**—a condition in which long-term job stress leads to emotional exhaustion, a sense of loss of personal control, and feelings of reduced accomplishment (Lee & Ashforth, 1990). Burnout occurs more often in the helping professions, including health care, human services, and teaching. Managers with substantial responsibility for others are also susceptible. Although people in interpersonally demanding jobs are as psychologically healthy as other people, sometimes a worker's dedication exceeds his or her coping skills and the social supports available in the work setting. High intrinsic and low extrinsic job satisfaction is associated with the emotional exhaustion and loss of personal control aspects of burnout. Low intrinsic and low extrinsic job satisfaction is linked to feelings of reduced accomplishment (Zedeck et al., 1988).

Burnout is a serious occupational hazard, since it is linked to absenteeism, turnover, poor job performance, and impaired health (Kahill, 1988). It is more common today than several decades ago, perhaps because extraordinarily stressful conditions are more frequent in jobs involving work with people (Jenkins & Maslach, 1994). Employers can prevent burnout by making sure workloads are reasonable, providing opportunities for workers to take time out from stressful situations, and limiting hours of stressful work (Pines & Aronson, 1988).

A widespread movement to increase worker participation in production planning and interaction through job teams is likely to boost job satisfaction at all ages. Other proposals include provisions for work at home, which would accommodate the desire of some people for less noise and time pressure and more opportunities to take breaks without disturbing others.

VOCATIONAL DEVELOPMENT

After several years as a parish nurse, Anya felt a need for additional training to find ways to do her job better. Trisha especially appreciated her current work setting because of its extensive support of workshop and course attendance, which helped her keep abreast of new legal developments. As college dean, Devin attended several seminars each year on management effectiveness and innovations in instructional technology. As Anya, Trisha, and Devin reveal, vocational development is vital during all phases of work life.

■ JOB TRAINING. Anya's 35-year-old supervisor Roy was surprised when she asked for time off to take several semi-nars to upgrade her skills. "You're in your fifties," he commented insensitively. "What're you going to do with so much new information at this point in your life?"

Although Roy's narrow-minded attitude is usually left unspoken, it is all too common among managers—even some who are older themselves! Research suggests that less training and on-the-job career counseling are available to older workers. And when vocational development activities are offered, older employees may be less likely to volunteer for them (Cleveland & Shore, 1992; Salthouse & Maurer, 1996). What influences willingness to engage in job training and updating? Characteristics of the person and the work environment make a difference.

On the person side, the degree to which an individual wants to change is important. With age, growth needs decline somewhat in favor of security needs. Consequently, learning and challenge may have less intrinsic value to many older workers. Perhaps for this reason, older employees depend more on co-worker and supervisor encouragement for vocational development. Yet we have just seen that they are less likely to have supportive supervisors. Furthermore, stereotypes of aging reduce older workers' self-efficacy, or confidence that they can renew and expand their skills—another reason they may not volunteer for training experiences (Maurer & Tarulli, 1994; Noe & Wilk, 1993).

Challenging tasks facilitate updating. An employee given work that requires new learning has to obtain that learning to complete the assignment. Unfortunately, older workers sometimes receive more routine tasks than do younger workers. Therefore, some of their reduced participation in vocational development may be due to the type of assignments they receive. Interaction among co-workers can also have a profound impact. Inside project teams, people similar in age communicate more often. Age-balanced work groups (with more than one person in each age range) foster on-the-job learning because communication is a source of support as well as a means of acquiring job-relevant information (Zenger & Lawrence, 1989).

■ GENDER AND ETHNICITY: THE GLASS CEILING. In her thirties, Jewel became a company president by starting her own business. As a woman, she decided that her chances of rising to a top-flight executive position in a large corporation were so slim that she didn't even try. Recall from Chapter 14 that women and ethnic minorities rarely move into high-level management jobs. At major firms in the United States, only 9 percent of senior executives are women. Far less than 1 percent are Asian-American, African-American, or Hispanic (Barr, 1996; Morrison, 1992).

Women and ethnic minorities face a **glass ceiling**, or invisible barrier to advancement up the corporate ladder. Contrary to popular belief, their low numbers cannot be attributed to poor management skills. In a recent survey of employees at six large American firms, women managers were rated as more effective and satisfying to work for and

Women and ethnic minorities face a glass ceiling, or invisible barrier to advancement up the career ladder, in major corporate and government organizations. A few, such as Secretary of State Madeleine Albright, succeed in shattering the glass ceiling. Many others go around it by starting their own businesses. *(Librado Romero/The New York Times)*

more likely to generate extra effort from people than were men. Characteristics that distinguished them from their male counterparts were charisma, inspiration, and considerateness (Bass & Avolio, 1994). Modern businesses realize that the best managers must not only display "masculine" authority and decisiveness, but build consensus through wider participation in decision making—an approach that requires "feminine" qualities of caring and collaboration.

Why is there a glass ceiling? Management is an art and skill that must be taught. Yet women and members of ethnic minorities have less access to mentors, role models, and informal networks that serve as training routes. Also, the majority of chief executives acknowledge that they spend less money on formal training programs for their female employees. Reasons given are stereotyped uncertainties about women's commitment to their jobs and their capacity to become strong managers. Furthermore, challenging, high-risk, high-visibility assignments that require leader-

ship and open the door to advancement, such as startup ventures, international experience, and troubleshooting, are rarely granted to women and minorities (Barr, 1996; Bily & Manoochehri, 1995).

Like Jewel, many women have shattered the glass ceiling by going around it. Nearly twice as many female as male middle managers quit their jobs in large corporations, largely because of lack of career opportunities (Stroh, Brett, & Reilly, 1996). Today, there are 4 million American women who, like Jewel, have started their own businesses—a rate six times that of men (Mergenhagen, 1996).

When women and ethnic minorities leave the corporate world to further their careers, companies not only lose valuable talent, but fail to address the leadership needs of an increasingly diversified work force. The Caregiving Concerns table below summarizes strategies for breaking the glass ceiling, enabling women and ethnic minorities to reach the top in major organizations.

CAREGIVING CONCERNS

Strategies for Breaking the Glass Ceiling

STRATEGY	DESCRIPTION
Reduce gender and ethnic stereotypes among executives.	Workshops that teach executives about diversity through training in self-awareness, sensitivity to differences, effective mentoring, and leadership skills
Recruit women and ethnic minorities.	Procedures that seek out women and minorities as job candidates and that hold managers accountable for equal employment opportunities
Equalize access to management training.	The same variety of training experiences for men and women and ethnic majority and minority employees; should include access to mentoring, informal networking, and high-risk, high-visibility assignments
Establish equitable promotion and salary policies.	Policies assuring that everyone is judged by common criteria and that employees doing the same job are paid the same, regardless of gender and ethnicity

CAREER CHANGE AT MIDLIFE

Although most people remain in the same vocation through middle age, career change does occur, as Elena's shift from journalism to teaching and creative writing illustrates. Recall that circumstances at home and at work motivated Elena's decision to pursue a new vocation. Like other career changers, she sought a more satisfying life—a goal attained by ending an unhappy marriage and initiating a long-awaited vocational move at the same time.

As we noted earlier, midlife career changes are usually not radical; they typically involve leaving one line of work for a related one. Elena sought a more stimulating, involving job. But other people move in the reverse direction—to careers that are more relaxing, free of painful decisions, and less demanding in terms of responsibility for others. The decision to change is often difficult. The individual must weigh years invested in one set of skills, current income, and job security against present frustrations and hoped-for gains from a new vocation (Stagner, 1985).

When an extreme career shift occurs, it usually signals a personal crisis. In a study of professionals who abandoned their well-paid, prestigious positions for routine, poorly paid, semiskilled work, nonwork problems influenced the break with an established career. For example, an eminent 55-year-old TV producer became a school bus driver; a New York banker became a waiter in a ski resort. Each responded to feelings of personal meaninglessness by escaping from family conflict, difficult relationships with colleagues, and work that had become unsatisfying into a freer, more independent lifestyle (Sarason, 1977).

UNEMPLOYMENT

Devin and Trisha's friend George worked in a corporate retirement-planning office, counseling retirees on how to enjoy leisure time or find new work. When he lost his job at age 54, he had to apply his counseling skills to himself. George found unemployment to be a profound culture shock. For the first two weeks, he spent most of the day in bed, didn't shave, and drank heavily. Even after this initial phase, George was depressed and had frequent bouts of illness.

As companies downsize and jobs are eliminated, the majority of people affected are middle-aged and older. Although unemployment is difficult at any time, middle-aged workers show much greater psychological distress than do their younger counterparts. Older workers affected by layoffs remain without work for a longer time. In addition, people over age 40 who must reestablish occupational security find themselves "off time" in terms of the social clock. Consequently, job loss can disrupt major tasks of midlife, including generativity and reappraisal of life goals

and accomplishments. Finally, having become more involved in and committed to an occupation, the older employed worker has also lost something of greater value (Broomhall & Winefield, 1990; Warr, 1987).

After a despondent period, George began to follow the advice he had given his clients. He made a list of what he really liked to do, what he didn't want to do again, and the risks he could take given his current financial and life circumstances. He formed his own small business and continued to advise retirees, write articles, and give speeches on all aspects of retirement, working from home in a T-shirt instead of a business suit. Effective problem-centered coping strategies enabled George to adjust (Beck, 1992).

Social support is vital for reducing stress and reassuring middle-aged job seekers of their worth. However, not all forms of social support work equally well. Recognition of the person's abilities and communication with others who share interests and values help the most. Both are social experiences that occur often in relations with co-workers (Mallinckrodt & Fretz, 1988).

People who have lost their jobs in midlife usually do not duplicate the status and pay of their previous positions. As they search, they encounter age discrimination and find that they are overqualified for many openings. Counseling that focuses on financial planning, reducing unhappiness related to deprivation of the family provider role, and encouraging personal flexibility can help people find alternative, gratifying work roles.

PLANNING FOR RETIREMENT

One evening, Devin and Trisha met George and Anya for dinner. Halfway through the meal, Devin inquired, "George, you're an expert on the topic. Tell us what you and Anya are going to do about retirement. Are you planning to close down your business or work part-time? Do you think you'll stay here or move out of town?"

Three or four generations ago, the two couples would not have had this conversation. In 1900, 69 percent of men age 65 and over were in the labor force. By 1965, the figure had dropped to 30 percent. In 1995, it was 12 percent (U.S. Bureau of the Census, 1996). Retirement is no longer a privilege reserved for the wealthy. In 1935, the U.S. Social Security Act initiated government-sponsored retirement benefits based on employment history. Today, the federal government pays Social Security benefits to more than 82 percent of the aged. Others are covered by employer-based private pension plans (Meyer & Bellas, 1995).

Most American workers report looking forward to retirement, and an increasing number are leaving full-time work in midlife. The average age of retirement in the United States declined over the past two decades. Currently, it is age 62. One-third of all household heads retire before age 55,

Ingredients of Effective Retirement Planning

ISSUE	DESCRIPTION
Finances	Ideally, financial planning for retirement should start with the first paycheck; at a minimum, it should begin 10 to 15 years before retirement, since most people spend more than 20 years retired.
Fitness	Starting a fitness program in middle age is important, since good health is crucial for well-being in retirement.
Role adjustment	Retirement is harder for people accustomed to giving orders, delegating work, and being identified by their professions. Preparing for a radical role adjustment reduces stress.
Where to live	The pros and cons of moving should be considered carefully, since where one lives affects access to health care, friends, family, recreation, entertainment, and part-time employment.
Leisure activities	A retiree typically benefits from an additional 50 hours of time a week. Careful planning of what to do with that time has a major impact on psychological well-being.
Health insurance	Finding out about Medicare and Medicaid benefits and other health insurance options is important in ensuring quality of life after retirement.
Legal affairs	The preretirement period is an excellent time to finalize a will and begin estate planning.

Source: Pery, 1995.

one-half before age 60 (Atchley, 1991; Turner, Bailey, & Scott, 1994). This means that increasing numbers of people spend up to one-fourth of their lives in retirement.

Retirement is a lengthy, complex process beginning as soon as the middle-aged person first thinks about it (Atchley, 1985). Planning is important, since retirement leads to a loss of two important work-related rewards—income and status—and to a change in many other aspects of life. Like other life transitions, retirement is often stressful.

"Retirement planning helps you evaluate your options, learn about the availability of resources, and prepare emotionally for the changes ahead," Devin and Trisha heard George explain when they attended one of his retirement seminars. Yet the majority of middle-aged people spend more time planning a two-week vacation than preparing for retirement. Research consistently shows that planning leads to better retirement adjustment and satisfaction (Atchley, 1991; Pery, 1995).

The Caregiving Concerns table above lists the variety of issues addressed in a typical retirement preparation program. Financial planning is vital, since income typically drops by 50 percent. Although more people are involved in financial planning than other forms of preparation, as many as 25 percent fail to look closely at their financial well-being (Turner, Bailey, & Scott, 1994).

Retirement leads to ways of spending time that are largely guided by what one wants to do, not by what one has to do. Individuals who have not thought carefully about how to fill this time may find their sense of purpose in life seriously threatened. Research in Canada and the United States reveals that planning for an active life has an even greater impact on happiness after retirement than does financial planning. Undoubtedly this is because participation in activities promotes many factors essential for psychological well-being, including a structured time schedule, social contact, and self-esteem (MacEwen et al., 1995; Ostling & Kelloway, 1992). Carefully considering whether or not to relocate at retirement is related to an active life, since it affects access to family, friends, recreation, entertainment, and part-time work.

Devin retired at age 62, George at age 66. Although several years younger, Trisha and Anya—like many married women—coordinated their retirements with those of their husbands (Ruhm, 1996). In good health but without an intimate partner to share her life, Jewel kept her consulting business going until age 75. Tim took early retirement and moved to be near Elena, where he devoted himself to public service—tutoring second graders in a public school, transporting inner-city children to museums, and coaching after-school and weekend youth sports. For Tim, like many executives, retirement offered the first opportunity to pay attention to the world around him.

Unfortunately, people with lower lifetime earnings and less education are least likely to attend retirement preparation programs, yet they stand to benefit the most. Employers need to take extra steps to encourage their participation (Gibson & Burns, 1991). In addition, improving retirement among the economically disadvantaged depends on access to better health care, vocational training, and jobs at early ages. Clearly, a lifetime of opportunities and experiences affect the transition to retirement. In Chapter 18, we will consider the decision to retire and retirement adjustment in greater detail.

ASK YOURSELF . . .

■ *Ira recalls complaining about his job far more in early than middle adulthood. At present, he finds his work interesting and gratifying. Describe factors that probably caused Ira's job satisfaction to increase with age.*

■ *Trevor assigned the older members of his work team routine tasks because he was certain they could no longer handle complex assignments. Cite evidence presented in this and the previous chapter that shows Trevor is wrong.*

■ *An executive asks you what his large corporation ought to do to promote advancement of women and ethnic minorities to upper management positions. What would you recommend?*

UMMARY

ERIKSON'S THEORY: GENERATIVITY VERSUS STAGNATION

According to Erikson, how does personality change in middle age?

■ Generativity begins in early adulthood but expands greatly as middleaged adults face Erikson's psychological conflict of **generativity versus stagnation.** Highly generative people find fulfillment as they make contributions to society through parenthood, other family relationships, the workplace, volunteer activities, or intellectual and creative endeavors.

OTHER THEORIES OF PSYCHOSOCIAL DEVELOPMENT IN MIDLIFE

Describe Levinson's and Vaillant's views of psychosocial development in middle adulthood, and discuss similarities and differences between men and women.

■ According to Levinson, middle-aged adults reassess their relation to themselves and the external world. They must confront four developmental tasks, each of which requires them to reconcile two opposing

tendencies within the self: young–old, destruction–creation, masculinity–femininity, and engagement–separateness.

■ Due to the double standard of aging, women have more trouble than men accepting being older. Men adopt "feminine" traits of nurturance and caring, whereas women take on "masculine" traits of autonomy, dominance, and assertiveness. Men and successful career-oriented women often reduce their focus on ambition and achievement. Women who have devoted themselves to child rearing or an unfulfilling job typically increase their involvement in work and the community.

■ According to Vaillant, middle-aged adults become guardians of their culture. In their late forties and fifties, they carry peak responsibility for the functioning of society.

Does the phrase midlife crisis *fit most people's experience of middle adulthood?*

■ Wide individual differences exist in response to midlife. Only a minority of people experience a **midlife crisis** in which intense self-doubt and inner turmoil lead them to make

drastic changes in their personal lives and careers.

Characterize middle adulthood using a life events approach and a stage approach.

■ Both continuity and stagewise change characterize emotional and social development in middle adulthood. Some changes are adaptations to external events (such as the family life cycle), but these events are less age graded than in the past. At the same time, the majority of middle-aged adults report stagelike development, involving troubling moments that prompt new understandings and goals.

CHANGES IN SELF-CONCEPT AND PERSONALITY TRAITS

Describe changes in self-concept and personality traits in middle adulthood.

■ **Possible selves** become fewer in number as well as more modest and concrete as middle-aged individuals adjust their hopes and fears to the circumstances of their lives. Revising possible selves enables adults to maintain self-esteem and stay motivated.

■ Adults become more introspective and in touch with themselves in midlife. Self-acceptance, autonomy, and environmental mastery increase, leading some people to consider middle age the "prime of life." In addition, **coping strategies** become more effective.

Describe changes in gender-role identity in midlife.

■ Both men and women become more androgynous in middle adulthood. Biological explanations, such as **parental imperative theory,** have been criticized, and evidence for them is mixed. A complex combination of social roles and life conditions is most likely responsible for midlife changes in gender-role identity.

INDIVIDUAL DIFFERENCES IN PERSONALITY TRAITS

Discuss individual differences in personality traits in adulthood.

■ After ages 25 to 30, little change takes place in the **"big five" personality traits** of neuroticism, extroversion, openness to experience, agreeableness, and conscientiousness. Although adults change in overall organization and integration of personality, they do so on a foundation of basic, enduring dispositions.

RELATIONSHIPS AT MIDLIFE

Describe the middle adulthood phase of the family life cycle, and discuss midlife relationships with a marriage partner, adult children, grandchildren, and aging parents.

■ "Launching children and moving on" is the newest and longest phase of the family life cycle. Middle-aged adults must adapt to many entries and exits of family members as their own children leave, marry, and produce grandchildren, and as their own parents age and die.

■ The changes of midlife prompt many adults to focus on improving their marriages. When divorce occurs, middle-aged adults typically have a harder time adjusting than do young adults. For women of all ages, marital breakup usually brings significant economic disadvantage, contributing to the **femininization of poverty** in the United States.

■ Most middle-aged parents adjust well to the launching phase of the family life cycle, especially if they have a strong work orientation and parent–child contact and affection are sustained. As children marry and bring in-laws into the family network, middle-aged parents, especially mothers, often become **kinkeepers.**

■ When family relationships are positive, grandparenthood is an important means of fulfilling personal and societal needs. In low-income families and in some subcultures, grandparents perform essential activities, including financial assistance and child care. Increasingly, grandparents have become primary caregivers when serious family problems exist.

■ Middle-aged adults reassess their relationships with aging parents, often becoming more appreciative. Yet the earlier quality of the parent–child relationship— positive or conflict ridden— usually persists.

■ Adults with ill or frail parents are often caught in a **middle-generation squeeze** as they face competing demands from their own children and their work life. The burden of caring for aging parents falls most heavily on adult daughters, but both men and women participate. As adults move from early to later middle age, the sex difference in parental caregiving declines.

Describe midlife sibling relationships and friendships, and discuss relationships across generations in the United States.

■ Bonds with siblings often strengthen in middle age, partly in response to major life events, such as launching and marriage of children and illness and death of parents. Sister–sister ties are typically closest in industrialized nations, where sibling relationships are voluntary. In nonindustrialized societies, where sibling relationships are basic to family functioning, other sibling attachments (such as brother–sister) may be unusually strong.

■ In midlife, friendships are fewer, more selective, and more deeply valued. Men continue to be less expressive with their friends than women, who have more close friendships and find them more satisfying. Viewing a spouse as a best friend can contribute greatly to marital happiness.

■ Recent research reveals strong supportive ties across generations in American families and communities. Most people do not resent government benefits for other age groups.

VOCATIONAL LIFE

Discuss job satisfaction and vocational development in middle adulthood, paying special attention to sex differences and experiences of ethnic minorities.

■ Vocational readjustments are common as middle-aged people seek to increase the personal meaning and self-direction of their work lives. Certain aspects of job performance improve. Overall job satisfaction increases at all occupational levels, though less so for women than for men.

■ **Burnout** is a serious occupational hazard more common today than in the past, especially in jobs involving work with people. Greater worker participation in planning and decision making, job teams, and work at home are likely to reduce job stress and improve satisfaction at all ages.

- Vocational development is vital during all phases of work life. It is less available to middle-aged and older workers because of negative stereotypes of aging, lack of encouragement from supervisors, and routine work assignments.

- Women and ethnic minorities face a **glass ceiling** as they advance in corporations, due to reduced access to formal and informal management training. Many further their careers by leaving the corporate world, often to start their own businesses.

Discuss career change and unemployment in middle adulthood.

- Most middle-aged people remain in the same vocation. Those who change careers usually leave one line of work for a related one. Radical career change often signals a personal crisis.

- Unemployment is especially difficult for middle-aged adults, who make up the majority of workers affected by corporate downsizing and layoffs. Social support is vital for reducing stress. Counseling can help midlife job seekers find alternative, gratifying work roles, but these rarely duplicate the status and pay of their previous positions.

Discuss the importance of planning for retirement, noting various issues that middle-aged adults should address.

- An increasing number of American workers are retiring from full-time work in midlife. Planning for retirement is important because the changes it brings—loss of income and status and increase in free time—are often stressful. Besides financial planning, other important issues include where to live, leisure activities, health insurance, and legal affairs.

IMPORTANT TERMS AND CONCEPTS

generativity versus stagnation (p. 518)
midlife crisis (p. 522)
possible selves (p. 525)
coping strategies (p. 526)

parental imperative theory (p. 527)
"big five" personality traits (p. 528)
femininization of poverty (p. 530)
kinkeeper (p. 531)

middle-generation squeeze (p. 535)
burnout (p. 540)
glass ceiling (p. 540)

FYI

FOR FURTHER INFORMATION AND HELP

MARRIAGE IN MIDLIFE

Association for Couples
in Marriage Enrichment
Winston–Salem, NC 27108
(800) 634-8325
Organization of married couples that aims to improve public understanding of marriage as a relationship capable of encouraging personal growth and mutual fulfillment. Conducts marriage retreats and growth groups.

National Marriage Encounter
4704 Jamerson Place
Orlando, FL 32807
(800) 828-3351
Web site: www.marriages.org
Offers weekend retreat programs aimed at encouraging communication between married partners and emphasizing personal and religious growth.

GRANDPARENTHOOD

Grandparents Raising
Grandchildren
P.O. Box 104
Colleyville, TX 76034
(817) 577-0435
Provides social support, legal assistance, and a unified voice for grandparents seeking to protect and care for their grandchildren. Works for the passage of legislation at state and national levels granting qualified grandparents legal custody rights without adoption.

CARING FOR AGING PARENTS

Children of Aging Parents
1609 Woodbourne Road,
Suite 302A
Levittown, PA 19057-1511
(800) 227-7294
A national information and referral service for caregivers of the elderly. Offers a speakers bureau for community organizations and conducts workshops, seminars, and conferences in the field of aging. Organizes and promotes caregivers' support groups.

VOCATIONAL LIFE

Career Planning and
Adult Development Network
4965 Sierra Road
San Jose, CA 95132
(408) 559-4946
Organization of professionals who focus on career planning and adult development issues. Keeps members informed of new knowledge in career decision making, dual-career families, midlife transitions, and preretirement counseling.

AGE	PHYSICAL	COGNITIVE	EMOTIONAL/SOCIAL

40–50 years

PHYSICAL

- Accommodative ability of the lens of the eye and color discrimination decline; sensitivity to glare increases.

- Sharp hearing loss at high frequency occurs.

- Hair continues to gray and thin.

- Lines on the face become more pronounced, and skin loses elasticity and begins to sag.

- Weight gain continues, accompanied by a rise in fatty deposits in the torso, whereas fat beneath the skin declines.

- Loss of lean body mass (muscle and bone) occurs.

- In women, production of estrogen drops, leading to shortening of and irregularity in the menstrual cycle.

- For men, quantity of semen and sperm declines.

- Intensity of sexual response declines, but sexual activity drops only slightly; stability is more typical than dramatic change.

- Proportion of women with no sexual partners increases dramatically.

- Rates of cancer and cardiovascular disease increase, at a higher rate for men than women.

COGNITIVE

- Consciousness of aging increases.

- A variety of crystallized and fluid abilities assessed by intelligence tests stabilize.

- The fluid ability of processing speed, assessed on reaction time tasks, declines; adults compensate by taking larger chunks of information into working memory and through practice and experience.

- On complex tasks, ability to divide and control attention declines; adults compensate through practice and experience.

- Amount of information that can be retained in working memory diminishes due to a decline in use of memory strategies.

- Retrieving information from working memory becomes more difficult.

- General factual knowledge, procedural knowledge, and knowledge related to one's occupation remain unchanged or increase.

- Gains in practical problem solving and expertise occur.

- Creativity shifts from a focus on generating unusual products to integrating experience and knowledge into unique ways of thinking and acting.

- If in a vocation offering challenge and autonomy, shows gains in cognitive flexibility.

EMOTIONAL/SOCIAL

- Desire to give to and guide the next generation intensifies.

- Re-evaluates life structure and tries to change components that are inadequate.

- Tries to reconcile four opposing tendencies within the self: young/old; destruction/creation; masculinity/femininity; engagement/separateness.

- Possible selves become fewer in number and more modest and concrete.

- Introspection increases as people contemplate the second half of life.

- Self-acceptance, autonomy, and environmental mastery increase.

- Coping strategies become more effective.

- Gender-role identity becomes more androgynous: "masculine" traits increase in women; "feminine" traits increase in men.

- May launch children.

- May enlarge the family network to include in-laws.

- May become a kinkeeper, especially if a mother.

- May care for a parent with a disability or chronic illness.

- Sibling bonds may strengthen.

- Number of friends generally declines.

- Job satisfaction increases.

AGE	PHYSICAL	COGNITIVE	EMOTIONAL/SOCIAL
50–60 years	■ Lens of the eye loses its accommodative ability entirely.	■ Changes in cognition described on the opposite page continue.	■ Emotional and social changes described on the opposite page continue.
	■ Hearing loss extends to all frequencies but remains greatest for highest tones.		■ Re-evaluates life structure and tries to change components that are inadequate.
	■ Skin continues to wrinkle and sag, and "age spots" appear.		■ Becomes concerned with "passing the torch"—that the positive aspects of the culture survive.
	■ Menopause occurs, usually between ages 50 and 55.		■ May become a grandparent.
	■ Continued loss of bone mass, accelerating especially for women after menopause and leading to high rates of osteoporosis.		■ Parent-to-child help-giving declines, and child-to-parent help-giving increases.
	■ Due to collapse of disks in the spinal column, height may drop by as much as 1 inch.		■ Sex difference in parental caregiving declines.
			■ May retire.

17

Physical and Cognitive Development in Late Adulthood

At age 64, Walt gave up his photography business and looked forward to leisure years ahead with 60-year-old Ruth, who retired from her position as a social worker at the same time. This culminating phase of Walt and Ruth's lives was filled with volunteer work, golfing three times a week, and joint summer vacations with Walt's older brother Dick and his wife Goldie. Walt also took up activities he had always loved but had little time to pursue—writing poems and short stories, attending theater performances, enrolling in a class on world politics, and cultivating a garden that became the envy of the neighborhood. Ruth read voraciously, served on the board of directors of an adoption agency, and had more time to visit her sister Ida in a nearby city.

Over the next 20 years, Walt and Ruth's energy and vitality was an inspiration to everyone who met them. Their warmth, concern for others, and generosity with their time led not just their own children and grandchildren, but former co-workers, nieces, nephews, and children of their friends to seek them out. On weekends, their home was alive with visitors.

Then, in their early eighties, the couple's lives changed profoundly. Walt had surgery to treat an enlarged, cancerous prostate gland and within 3 months was hospitalized again after a heart attack. He lingered for 6 weeks, with Ruth at his side, and then died. Ruth's grieving was interrupted by the need to care for Ida. Alert and spry at age 78, Ida deteriorated mentally in her seventy-ninth year, despite otherwise being in excellent physical health. In the meantime, Ruth's arthritis worsened and her vision and hearing weakened.

As Ruth turned 85, certain activities had become more difficult—"but not impossible. It just takes a little adjustment!" Ruth exclaimed in her usual upbeat manner. Reading was harder, so she checked out "talking books" from her local library. Her gait was slower and her eyesight less reliable, so she hesitated to go out alone. When her daughter and family took Ruth to dinner, the conversation moved so quickly in the noisy restaurant that Ruth felt overwhelmed and said little. But interacting with her on a one-to-one basis revealed that she was far from passive and withdrawn! In a calm environment, she showed the same intelligence, wit, and astute insights that she had displayed all her life.

Late adulthood stretches from age 60 to the end of the lifespan. Unfortunately, popular images of old age fail to capture the quality of these final decades. Instead, many myths prevail—that the elderly are feeble, senile, and sick; that they have little contact with their families, who set them aside in nursing homes; that they are no longer able to learn; and that they have entered a phase of deterioration and dependency. Young people who have little contact with older adults are often surprised that elders like Walt and Ruth even exist—active and involved in the world around them.

As we trace physical and cognitive development in old age, we will see that the balance of gains and declines does shift as death approaches. But the typical 60-year-old in industrialized nations can anticipate nearly two healthy, rewarding decades before this shift has meaning for everyday life. And as Ruth illustrates, even after members of the older generation become frail, they find ways of surmounting physical and cognitive challenges.

Late adulthood is best viewed as an extension of earlier periods, not a break with them. As long as social and cultural contexts grant the elderly support, respect, and purpose in life, these years are a time of continued potential.

PHYSICAL DEVELOPMENT IN LATE ADULTHOOD

If you were to guess the ages of older people on the basis of their appearance, chances are that you would frequently be wrong. Indeed, we often remark that an older person "looks young" or "looks old" for his or her age —a statement acknowledging that chronological age is an imperfect indicator of **functional age,** or actual competence and performance. Because people age biologically at different rates, experts distinguish between the **young-old** elderly, who appear physically young for their advanced years, and the **old-old** elderly, who show signs of decline.[1] According to this functional distinction, it is possible for an 80-year-old to be young-old and a 65-year-old to be old-old (Neugarten & Neugarten, 1987). Yet even these labels do not fully capture the wide variation in biological aging. Recall from Chapter 13 that within each person, change differs across parts of the body. For example, Ruth became infirm physically but remained active mentally, whereas Ida was physically fit for her age but showed signs of mental decline.

So much variation exists between and within individuals that as yet, researchers have not been able to identify any biological measure that predicts the overall rate at which an elderly person will age (Hayflick, 1994). But we do have estimates of how much longer older adults can expect to live, and our knowledge of factors affecting longevity in late adulthood has increased rapidly.

LONGEVITY

I wonder how many years I have left," Ruth asked herself each time a major life event, such as retirement and widowhood, occurred. In Chapter 13, we indicated that older adults are living longer than ever before. Recall that under the best of circumstances, the average human lives

---

[1] In the popular literature on aging, this distinction has been confused with chronological age: *young-old* referring to age 65 to 75, *old-old* to 75 to 85, and *oldest-old* to 85 to 99 (see, for example, Safire, 1997). In Western nations, persistence of the incorrect belief that age determines function is probably due to stereotypes of aging.

How old are these two elders? How old do they look and feel? Because people age biologically at different rates, the 88-year-old woman on the right does not appear much older than her 65-year-old companion on the left. Yet their birthdays are nearly a quarter century apart. (Bob Daemmrich/The Image Works)

around 85 or 90 years and the very longest-lived about 120, our maximum lifespan.

Largely due to medical advances and improved life conditions, the number of people age 65 and older in the United States has risen dramatically, from 3 million in 1900 to 31.6 million in 1990. From the beginning to the end of the twentieth century, elderly Americans increased from 4 percent to 12.5 percent of the population—a figure expected to reach 23 percent by the middle of the twenty-first century. The fastest-growing segment of senior citizens is the 85-and-older group, who currently make up 1.7 percent of the population. By the year 2020, the proportion of people age 85 and older will double; by 2050 it will quadruple, reaching 5.1 percent (U.S. Bureau of the Census, 1996).

Despite increasing numbers of senior citizens, the percentage of elderly in the United States is lower than in many other industrialized nations. The difference is largely due to high birthrates among America's poverty-stricken groups and less effective public policies ensuring health and well-being throughout the lifespan (see Chapter 2, page 65). In developing countries, high infant, childhood, and adult mortality rates mean that people age 65 and

older typically make up no more than 3 to 5 percent of the population. In these nations, the elderly population has grown at a slower rate (Central Intelligence Agency, 1996).

Americans reaching age 65 in 1990 can look forward, on the average, to 17 additional years of life—6 years more than a 65-year-old could expect in 1900, when living to age 65 was unusual. As at earlier ages, life expectancy continues to be greater for women (19 years) than for men (15 years). Today, the 65- to 69-year age group consists of 123 women for every 100 men; for people age 85 and older, the ratio climbs to 259 to 100. Discrepancies like these occur in all developed countries. They are not always present in the developing world because of high death rates of women in childbirth (Hayflick, 1994).

Although women outnumber men by a greater margin with advancing age, differences in average life expectancy between the sexes decline. In Chapter 13, we noted that a newborn girl can expect to live about 7 to 8 years longer than a newborn boy. At age 65, the difference narrows to about 3½ years; at age 85, to just over 1 year. Over age 100, the gender gap in life expectancy disappears. Similarly, differences in rates of chronic illness and in life expectancy between middle-class whites and low-income ethnic minorities (African-Americans, Hispanics, and Native Americans) decline with age (House et al., 1990a). Around age 85, a **life expectancy crossover** occurs, in that surviving members of ethnic minority groups live longer than members of the white majority (see Table 17.1). Researchers speculate that only the sturdiest males and economically disadvantaged ethnic minorities survive into very old age. Many are at the extreme of human health and vigor (Barer, 1994; Markides, 1989).

Of course, average life expectancy does not tell us how enjoyable living to a ripe old age is likely to be. Perhaps the real question Ruth meant to ask was "How many years of active, healthy life lie ahead?" We can get some idea by looking at the percentage of noninstitutionalized elderly who require assistance with everyday activities. As Figure 17.1 shows, most Americans age 65 and older are capable of living independent, productive lives, although with age (and especially after age 80) need for care increases.

TABLE 17.1

Average Life Expectancy for White and Black Elderly in the United States, Illustrating the Life Expectancy Crossover

AGE IN YEARS	WHITE MALE	AVERAGE LIFE EXPECTANCY IN YEARS BLACK MALE	WHITE FEMALE	BLACK FEMALE
65	15.2	13.6	19.0	17.0
70	12.1	11.0	15.3	13.9
75	9.4	8.8	11.9	11.0
80	7.1	6.9	8.9	6.5
85 and over	5.3	5.6	6.5	6.7 ◄── Life expectancy crossover

Note: Below age 85, life expectancy is greater for American whites than blacks. At age 85 and over, this trend reverses; average life expectancy is greater for blacks than whites. Notice, also, how the gender gap in life expectancy declines with age. Source: U. S. Bureau of the Census, 1996.

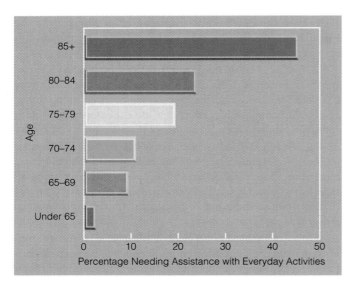

FIGURE 17.1

Percentage of Americans by age needing assistance with everyday activities, such as shopping and meal preparation. Although need for care increases with age, even at age 85 and beyond, over 50 percent of people lead fully independent lives. *(From U.S. Bureau of the Census, 1996.)*

Think about factors we have considered in earlier chapters that contribute to a long and healthy life. Much research shows that heredity and environment are interconnected in the aging process (Jazwinski, 1996). Identical twins typically die within 3 years of each other, whereas fraternal twins of the same sex differ by more than 6 years.

Furthermore, longevity runs in families. People with long-lived ancestors tend to survive longer and to be physically healthier in old age. And when both parents survive to age 70 or older, the chances that their children will live to 90 or 100 are double that of the general population (Hayflick, 1994; Sorensen et al., 1988). At the same time, evidence from studies of twins suggests that as we get older, the contribution of heredity to length of life decreases in favor of environmental factors (Vaillant, 1991).

Table 17.2 summarizes the wide array of variables that, in combination with genetic makeup, predict longevity. It also indicates where you can read about each in this book. The study of centenarians—people who cross the 100-year mark—offers special insights into how biological, psychological, and social influences work together to promote a long, satisfying life. Refer to the Lifespan Vista box on page 556 to find out more about the survival and successful adaptation of these longest-lived individuals.

PHYSICAL CHANGES

The programmed effects of specific genes and random cellular events believed to underlie aging (see Chapter 13) make physical declines more apparent in late adulthood. Compared to earlier phases, more organs and systems of the body are affected. Nevertheless, most structures are remarkably robust and can last into our eighties and beyond, if we take good care of them. For

TABLE 17.2

Factors Associated with Longevity

IF YOU ARE INTERESTED IN . . .	TURN TO . . .
Long-lived ancestors	Chapter 17, p. 554
Sex (female)	Chapter 13, p. 417; Chapter 17, p. 553
Social class (income, education, and occupational status)	Chapter 13, p. 425
Diet low in fat, high in essential nutrients	Chapter 13, p. 427; Chapter 17, p. 564
Body weight in normal range	Chapter 13, p. 426
Exercise	Chapter 13, p. 427; Chapter 15, p. 499; Chapter 17, p. 564
Abstinence from unhealthy alcohol and drug use	Chapter 13, p. 428
Self-rated health	Chapter 17, p. 563
Stable family life in childhood	Chapter 13, p. 418
Marriage	Chapter 13, p. 418
Conscientious, dependable personality	Chapter 13, p. 418
Hardiness (an optimistic outlook)	Chapter 15, p. 500
Low hostility	Chapter 15, p. 498
Low psychological stress	Chapter 13, p. 434; Chapter 15, p. 499
Social support	Chapter 13, p. 435; Chapter 15, p. 499

an overview of the physical changes we are about to discuss, return once again to Table 13.2 on page 422.

NERVOUS SYSTEM

On a routine office visit, 80-year-old Ruth responded to her doctor's query about how she was getting along by explaining, "During the last two days, I forgot the name of the family that just moved in next door, couldn't recall where I had put a pile of bills, and had trouble finding the right words to explain to a delivery service how to get to my house. Am I losing my mind?" Ruth asked anxiously.

"You're much too sharp for that," Dr. Wiley remarked. "Ruth, if you were losing your mind, you wouldn't be so concerned about forgetting." Ruth also wondered why extremes of hot and cold weather felt more uncomfortable than in earlier years. And she required more time to coordinate a series of movements and had become less sure of her balance.

Aging of the nervous system affects a wide range of complex thoughts and activities. Although brain weight declines throughout adulthood, the loss becomes greater after age 60 and may be as much as 5 to 10 percent, due to death of neurons and enlargement of ventricles (spaces) within the brain. Neuron loss occurs throughout the cerebral cortex but at different rates in different regions. In the visual, auditory, and motor areas, as many as 50 percent of neurons die. In contrast, parts of the cortex (such as the frontal lobe) responsible for integration of information, judgment, and reflective thought show far less change. Besides the cortex, the cerebellum (which controls balance and coordination) loses neurons—in all, about 25 percent. Glial cells, which myelinate neural fibers, decrease as well, contributing to diminished efficiency of the central nervous system (Arking, 1991; Whitbourne, 1996).

But the brain can overcome some of these declines. In several studies, growth of neural fibers in the brains of older adults unaffected by illness took place at the same rate as in middle-aged people. Aging neurons established new synapses after other neurons had degenerated (Cotman & Holets, 1985; Flood & Coleman, 1988). Furthermore, exercise can increase blood circulation to the brain, which helps preserve brain structures and behavioral capacities (Baylor & Spirduso, 1988).

The autonomic nervous system, involved in many life support functions, also performs less well in old age. For example, Ruth's lower tolerance for hot weather was due to decreased sweating. And her body found it more difficult to raise its core temperature during cold exposure. For these reasons, the elderly are at risk during heat waves and cold spells. The autonomic nervous system also releases higher levels of stress hormones into the bloodstream than it did at younger ages, perhaps in an effort to arouse body tissues that have become less responsive to these hormones over the years (Whitbourne, 1996). Later we will see that

this change may contribute to declines in immunity and to sleep problems among older adults.

VISION

In Chapter 15 (see page 489), we noted that structural changes in the eye make it harder to focus on nearby objects, see in dim light, and perceive color. In late adulthood, vision diminishes further. For example, the cornea (clear covering of the eye) becomes more translucent and scatters light, which blurs images and increases sensitivity to glare. The lens continues to yellow, leading to further impairment in color discrimination. From middle to old age, cloudy areas in the lens called **cataracts** increase, resulting in foggy vision and (without surgery) eventual blindness. The number of individuals with cataracts increases tenfold from middle to late adulthood; half of people in their eighties are affected (U.S. Bureau of the Census, 1996). Besides age, sun exposure and certain diseases, such as diabetes, increase the risk of cataracts. Fortunately, removal of the lens and replacement with an artificial lens implant or corrective eyewear are very successful in restoring vision (Larson, 1996).

Impaired eyesight in late adulthood is largely due to a reduction in light reaching the retina, caused by yellowing of the lens, shrinking of the pupil, and clouding of the vitreous (refer again to Chapter 15). Dark adaptation—moving from a brightly lit to a dim environment—is harder, making entering a movie theater after the show has started a challenge. In addition, depth perception is less reliable, since binocular vision (the brain's ability to combine images received from both eyes) declines (Garzia & Trick, 1992). And visual acuity (fineness of discrimination) worsens, with a sharp drop after age 70 (Whitbourne, 1996).

When blood flow to the macula, or central region of the retina, is restricted due to hardening of small blood vessels, older adults may develop **macular degeneration.** In the early stages of this condition, central vision blurs; gradually it is lost. Macular degeneration is the leading cause of blindness among older adults. If diagnosed early, it can sometimes be treated with laser therapy (Braus, 1995).

Visual difficulties have a profound impact on older people's self-confidence and everyday behavior. Ruth gave up driving in her mid-seventies, and she worried greatly when Walt had trouble making decisions behind the wheel. He found it hard to shift focus between the road and the dashboard and strained to make out pedestrians at dusk and night (Kline et al., 1992). On foot, older adults' problems with depth perception and dark adaptation increase their chances of stumbling.

When vision loss is extensive, it can affect leisure pursuits and be very isolating. Ruth could no longer enjoy museums, movies, playing bridge, and working crossword puzzles. Her poor vision also meant that she had to depend on others for help with housekeeping and shopping. Still,

A LIFESPAN VISTA

WHAT CAN WE LEARN ABOUT AGING FROM CENTENARIANS?

Born in 1893 in Croatia, Dr. Bogdan Stojic continued to practice medicine until he became a centenarian. After retiring at age 101, he spent his days reading, writing, going for walks, and enjoying his many friends. During his long life, he surmounted many difficult times. In his youth, he fought in three wars—the Balkan War against Turkey, the Serbo-Bulgarian War, and World War I. As a middle-aged soldier during World War II, he was taken prisoner by the Nazis, served as a medic in a prisoner-of-war camp, and narrowly escaped a death sentence. After the war, he entered private practice. At age 68, he and his wife packed their belongings and traveled thousands of miles, settling in Australia to live close to their daughter. There Dr. Stojic continued to treat patients for another 33 years.

About living a long and healthy life, Dr. Stojic reflected, "My parents were 90, but I never thought about it—how old I'd live. It just happened ....don't smoke; don't drink too much....My father and mother were very good. They spoke with us as if we were grown-up people when we were 10 years of age ... [and] my marriage has been very good." About his current trouble walking, he noted optimistically, "I'll be treated. Soon, I hope I will walk without the stick." He continued to have goals—among them, contributing to a cure for

Alzheimer's disease (Deveson, 1994, p. 218, 222–223).

Due to aging stereotypes and researchers' focus on the very old with the heaviest burden of illness and disability, popular images of the most senior members of the human species are ones of extreme frailty. Yet the past 40 years have seen a tenfold increase in centenarians in the industrialized world—a trend expected to accelerate

greatly in the coming century (see Figure 17.2). About one-third to one-half of centenarians lead active and independent lives; the rest have physical or mental impairments, ranging from mild to severe (Franceschi et al., 1995; McRae, 1995).

Robust centenarians are of special interest, since they represent the ultimate potential of the human species. What are they like? Case studies reveal

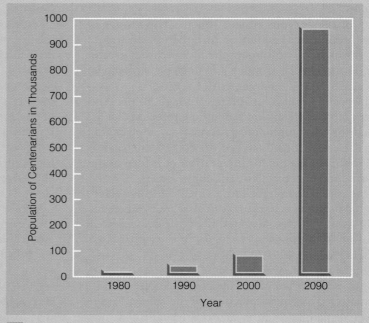

FIGURE **17.2**

Projected increase in centenarians in the United States from 1980 to 2090. *(From U.S. Bureau of the Census, 1996.)*

even among people age 85 and older, only 25 percent experience visual impairment severe enough to interfere with daily living (Crews, 1994). As in other aspects of aging, wide individual differences in visual capacity exist.

HEARING

At a Thanksgiving gathering with twelve family members present, 85-year-old Ruth had trouble hearing. "Mom, this is Leona, Joe's cousin. I'd like you to meet her," Ruth's daughter Sybil commented. In the clamor of boisterous

children, banging dishes, television sounds, and nearby verbal exchanges, Ruth didn't catch Leona's name or her relationship to Sybil's husband Joe.

"Tell me your name again," Ruth asked. "This is a mighty busy place right now, and it's hard for me to hear. Let's go into the next room so we can speak a bit."

Reduced blood supply and natural cell death in the inner ear, discussed in Chapter 15, in addition to stiffening of membranes (such as the ear drum) and degeneration of neural pathways lead hearing to decline in late adulthood. As Figure 17.3 on page 558 shows, hearing impairments

great diversity among them. Some have survived extreme illnesses in childhood and adulthood; others have rarely been sick throughout their lives. Growing up in an era when few were highly educated, most finished grade school. But some are illiterate, whereas others (like Dr. Stojik) hold advanced professional degrees. Although most describe themselves as middle class, a handful live in poverty or are millionaires (Beard, 1991).

At the same time, common threads characterize these centenarians' physical condition and life stories. Most have escaped age-related chronic illnesses, such as cardiovascular disease, cancer, diabetes, and dementia. Medical tests typically reveal a well-preserved and efficiently functioning immune system (Franceschi et al., 1995; Perls, 1995). As a group, they are of average or slender build and practice moderation in eating. Many have most or all of their own teeth—another sign of unusual physical health. Despite heavy tobacco use in their generation, the large majority report that they never smoked cigarettes. They also describe lifelong physical activity extending past age 100 (Beard, 1991; Kropf & Pugh, 1995).

In personality, these very senior citizens appear unusually forthright and genuine. They refuse to dwell on fears and tragedies. Instead, they focus on a better tomorrow. They are also assertive—willing to express themselves in bold and unusual ways. When asked about their longevity,

they often mention a long and happy marriage. And they have a history of community involvement, working for just causes that are central to their growth and happiness.

The activities of robust centenarians show that there is no age limit to stimulating work, leisure pursuits, and learning. Cooking; sewing; writing letters, poems, plays, and memoirs; making speeches; teaching music lessons and Sunday school; nursing the sick; chopping wood; selling merchandise, bonds, and insurance; painting; practicing medicine; and preaching sermons are among their varied involvements. In several cases, illiterate centenarians learned to read and write. One of the most impressive was a 105-year-old woman who enrolled in classes four nights a week. Within a short time, she could read road signs, newspaper headlines, and sections of the Bible. Although she found writing harder, she persisted, explaining, "I'm tired of having other people write letters for me" (Beard, 1991, p. 80).

Robust centenarians are often regarded as rare curiosities who do not represent the general population. As their numbers increase, they are likely to be viewed less as exceptions and more as people for whom normal development is at its best. These independent, mentally alert, fulfilled 100-year-olds illustrate how a healthy lifestyle, personal resourcefulness, and close ties to family and community can build on biological strengths, thereby pushing the limits of the active lifespan.

Dr. Bogdon Stojic, centenarian, defies stereotypes of the very old as extremely frail. He practiced medicine until age 101. (Sidney Fay)

are far more common than visual impairments, and they extend trends we described for middle adulthood in that more men than women are affected. Decrements are greatest at high frequencies, although detection of soft sounds diminishes throughout the frequency range (see page 490). In addition, responsiveness to startling noises lessens in old age (Ford et al., 1995).

Although hearing loss has less impact on self-care than does vision, it affects safety and enjoyment of life. In the din of traffic on city streets, 80-year-old Ruth didn't always interpret warnings correctly—both spoken ("Watch it,

don't step out yet") and nonspoken (the beep of a horn or a siren). And when she turned up the radio or television volume, she sometimes missed the ring of the telephone or a knock at the door (Gatehouse, 1990).

Of all hearing difficulties, the age-related decline in speech perception has the greatest impact on life satisfaction. Although Ruth used problem-centered coping to increase her chances of hearing conversation, she wasn't always successful. And sometimes people were inconsiderate. For example, on a dinner outing, Joe raised his voice impatiently when Ruth asked him to repeat himself. And

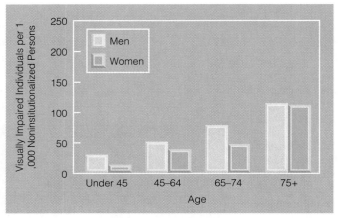

(a) Visual Impairments

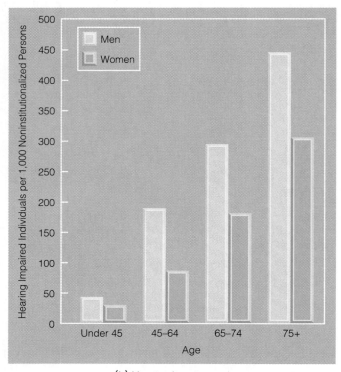

(b) Hearing Impairments

FIGURE 17.3

Number of men and women with (a) visual impairments and (b) hearing impairments per 1,000 noninstitutional- ized American adults of different ages. Hearing impairments are far more common than visual impairments in aging individuals, and more men than women are affected. Nevertheless, a visual impairment is more likely to interfere with self-care than is a hear- ing impairment. *(Adapted from U.S. Bureau of the Census, 1996.)*

he said to Sybil in Ruth's presence, "Be honest, Syb, Ruth's going deaf, isn't she?" This time, Ruth heard every word. At the family's Thanksgiving reunion, fewer relatives took time to talk with Ruth, and she felt some pangs of loneliness.

As with vision, most elders do not suffer from hearing loss great enough to disrupt their daily lives (Rudberg et al., 1993). Of those who do, compensating with a hearing aid and minimizing background noise are helpful. Fur- thermore, recall from Chapter 4 (page 139) that beginning at birth, our perception is intermodal (combines informa- tion from more than one sensory system). By attending to facial expressions, gestures, and lip movements, older adults can use vision to help interpret the spoken word. When interacting with the elderly, family members and others can seek a quiet environment. When they do, older people are far more likely to convey an image of alertness and competence than reduced sensitivity to the surround- ing world.

TASTE AND SMELL

Walt's brother Dick was a heavy smoker. In his sixties, he poured salt and pepper over his food, stirred extra spoonfuls of sugar into his coffee, and satisfied his fond- ness for spicy foods by asking for "extra hot" in Mexican and Indian restaurants.

Dick's reduced sensitivity to the four basic tastes—sweet, salty, sour, and bitter—is evident in many adults after age 60. In addition, cross-sectional research reveals that older adults have greater difficulty recognizing familiar foods by taste alone (Stevens, 1989; Stevens et al., 1991).

But the drop in taste sensitivity just described may be due to factors other than aging, since no change in the number or distribution of taste buds takes place late in life. Smoking, dentures, medications, and environmental pol- lutants can affect taste perception. When taste is harder to detect, food is less enjoyable, increasing the likelihood of deficiencies in the elderly person's diet. Flavor additives can help make food more attractive to older adults (Schiff- man & Warwick, 1989; Whitbourne, 1996).

Smell contributes to enjoyment of food and also has a self-protective function. An aging person who has diffi- culty detecting rancid food, gas fumes, or smoke may be in a life-threatening situation.

A decrease in the number of smell receptors after age 60 undoubtedly contributes to declines in odor sensitivity. Older adults are less accurate at linking odors with such descriptors as *floral, musky, fruity,* or *sweet.* Researchers believe that odor perception becomes distorted in late adulthood, a change that may promote complaints that "food no longer smells and tastes right" (Russell et al., 1993). But other factors may lead this decline to appear greater than it actually is. For example, some older adults have difficulty recalling odor labels, making odor recogni- tion tasks harder. Furthermore, practicing odor discrimi- nations affects their maintenance (Engen, 1982). Walt had been a wine enthusiast all his life. In his early eighties, he could still distinguish the aromas of fine wines.

TOUCH

Touch sensitivity is especially crucial for certain adults, such as the severely visually impaired who must read in Braille and people who make fine judgments about texture in their occupations or leisure pursuits (for example, in art and handicraft activities). To measure touch perception, researchers determine how close two stimuli on the skin must be before they are perceived as one. Findings indicate that aging brings a sharp decline on the hands, particularly the fingertips, and less of a drop on the arms. After age 70, the majority of elderly are affected (Stevens, 1992).

Why does deterioration in touch sensitivity occur? Since it varies with area stimulated, researchers believe that loss of touch receptors in certain regions of the skin is responsible. More sluggish blood circulation to the extremities may also contribute.

If we consider the spacing among elements of Braille letters, 45 percent of middle-aged and 100 percent of elderly people should have difficulty interpreting them (Stevens, 1992). Just as the sighted need new visual aids for reading as they age, so the visually impaired may need new tactile aids—an important consideration in responding to diversity in the aging population.

CARDIOVASCULAR AND RESPIRATORY SYSTEMS

Aging of the cardiovascular and respiratory systems proceeds gradually during early and middle adulthood, usually without notice. Signs of change are more likely to be apparent in late adulthood, and they prompt concern among aging individuals, who know these organ systems are vital for quality and length of life. In their sixties, Ruth and Walt noticed that they felt more physically stressed after running to catch a departing bus or to reach the far side of the street before the light changed.

As the years pass, the heart muscle becomes more rigid, and some of its cells die while others enlarge, leading the walls of the left ventricle (the largest heart chamber, from which blood is pumped to the body) to thicken. In addition, artery walls stiffen and accumulate some plaque (cholesterol and fats) due to normal aging (much more if the person has atherosclerosis). Finally, the heart muscle becomes less responsive to signals from pacemaker cells within the heart, which initiate each contraction (Arking, 1991; Whitbourne, 1996).

The combined result of these changes is that the heart pumps with less force, maximum heart rate decreases, and blood flow throughout the circulatory system slows. This means that sufficient oxygen may not be delivered to body tissues during high physical activity (Schulman & Gerstenblith, 1989). (Recall from Chapter 13 that a healthy heart supports typical levels of exertion well into old age.)

Changes in the respiratory system compound the reduced oxygenation just described. In Chapter 13, we noted that between ages 20 and 80, vital capacity (amount of air that can be forced in and out of the lungs) is reduced by half. The reason is that lung tissue loses its elasticity. As a result, the lungs fill and empty less efficiently, causing the blood to absorb less oxygen and give off less carbon dioxide. This explains why older people increase their breathing rate more and feel more out of breath when exercising.

Cardiovascular and respiratory deficiencies are more extreme in people who have smoked cigarettes throughout their lives, failed to reduce dietary fat, or had many years of exposure to environmental pollutants. In previous chapters, we noted that exercise is one of the most powerful ways of slowing the effects of aging on the cardiovascular system. Exercise also facilitates respiratory functioning, as we will see when we discuss health and fitness in a later section.

IMMUNE SYSTEM

As the immune system ages, T cells, which attack antigens (foreign substances) directly, become less effective (see Chapter 13, page 424). In addition, the immune system is more likely to malfunction by turning against normal body tissues in an **autoimmune response.** A less competent immune system can increase the elderly

Exercise is a powerful way of slowing the effects of aging on the cardiovascular system. By regularly doing Tai Chi, a Chinese system of exercise designed for meditation and self-defense, these elders protect their respiratory and immune systems, enhance flexibility and range of motion in their joints, and promote restful sleep. (A. Ramsey/Stock Boston)

person's risk for a variety of illnesses, including infectious diseases (such as the flu), cardiovascular disease, certain forms of cancer, and a variety of autoimmune disorders, such as rheumatoid arthritis and diabetes. But an age-related decline in immune functioning is not the cause of most illnesses among the elderly. It merely permits disease to progress when a stronger immune reaction would have rid the body of the disease agent (Ershler, 1993; Weksler, 1995).

We must keep in mind that immunity in old age varies tremendously. In longitudinal studies, a few elderly people continue to respond like young adults. But the responses of the majority range from partial to profound loss of function (Goodwin, 1995). The strength of the aging person's immune system seems to be a sign of overall physical vigor. Certain immune indicators, such as high T cell activity, predict survival over the next 2 years in very old people (Ferguson et al., 1995; Miller, 1996).

In Chapter 13, we emphasized that although impaired immune functioning in late life seems to be genetically determined, other physical changes of aging can intensify it. Recall that stress hormones undermine immunity. With age, the autonomic nervous system releases higher levels of these into the bloodstream (refer back to page 555). New findings suggest that a healthy diet and moderate exercise help protect the immune response in old age, whereas obesity aggravates the age-related decline (Lesourd, 1995; Moriguchi et al., 1995; Shephard & Shek, 1995).

SLEEP

When Walt climbed into bed at night, he usually lay awake for a half-hour to an hour before falling asleep, remaining in a drowsy state longer than when he was younger. During the night, he spent less time in the deepest phase of NREM sleep (see Chapter 4, page 106) and awoke multiple times. Again he sometimes lay awake for a half-hour or more before drifting back to sleep.

Older adults require about the same total sleep time as younger adults—around 7 hours per night. Yet as people age, they have more difficulty falling asleep, staying asleep, and sleeping deeply—a trend that begins after age 30 for men and after age 50 for women. The timing of sleep tends to change as well, toward earlier bedtime and morning awakenings (Frank et al., 1995). Changes in brain structures controlling sleep and higher levels of stress hormones in the bloodstream, which have an alerting effect on the central nervous system, are believed to be responsible (Prinz et al., 1990; Whitbourne, 1996).

Until age 70 or 80, men experience more sleep disturbances than do women for several reasons. First, enlargement of the prostate gland, which occurs in almost all aging men, constricts the ureters (tubes through which urine flows) and leads to a need to urinate more often, including during the night. (To see where the prostate gland is located, return to Chapter 2, page 47.) Second,

men are more prone to **sleep apnea,** a condition in which breathing ceases for 10 seconds or longer, resulting in many brief awakenings. The incidence of sleep apnea in elderly men is high; 30 to 50 percent have 20 or more episodes per night. Finally, periodic rapid movement of the legs sometimes accompanies sleep apnea but also occurs at other times of night. Called "restless legs," these movements may be due to muscle tension, reduced circulation, or age-related changes in motor areas of the brain. Although common among the elderly and not dangerous, they do disrupt sleep (Montplaisir & Godbout, 1989).

Older adults often express concern about sleep difficulties. About 45 percent report some degree of insomnia to their doctors (Bachman, 1992). Poor sleep can feed on itself, making matters worse. For example, Walt's nighttime wakings led to daytime fatigue and short naps, which made it harder to fall asleep the following evening. As Walt expected to have trouble sleeping, he worried about it, which also interfered with sleep.

Fortunately, there are many ways to foster restful sleep, such as establishing a consistent bedtime and waking time, exercising regularly, and using the bedroom only for sleep (not for eating, reading, or watching TV). Explaining that even very healthy older adults have trouble sleeping is vital. It lets people know that age-related change in the sleep–waking pattern is normal. The elderly receive 66 percent more prescription sedatives for sleep complaints than do 40- to 60-year-olds. Used briefly, these drugs can help relieve temporary insomnia. But long-term medication can make things worse by increasing the frequency and severity of sleep apnea and by inducing rebound insomnia after the drug is discontinued (Prinz et al., 1990). Finally, discomfort due to an enlarged prostate, including frequent urination at night, can be corrected with a new laser surgical procedure. It usually relieves all symptoms without complications (Larson, 1996).

PHYSICAL APPEARANCE AND MOBILITY

The inner physical declines we have considered are accompanied by many outward signs of growing older—involving the skin, hair, facial structure, and body build. In earlier chapters, we saw that changes leading to an aged appearance are underway as early as the decades of the twenties and thirties. Because these occur gradually, people may not take notice until a point is reached where the arrival of old age is obvious. From year to year, Walt and Ruth observed that Dick and Goldie's skin appeared more wrinkled. Their hair had turned from gray to white as all pigment was lost, and their bodies were rounder and arms and legs thinner. Each summer, Walt and Ruth returned from travels with Dick and Goldie more aware that they themselves were aging.

Creasing and sagging of the skin, described in Chapter 15, extends into old age. In addition, oil glands that lubricate the skin become less active, leading to dryness and roughness. "Age spots" increase; in some elderly individuals, the arms, backs of the hands, and face may be dotted with these pigmented marks. Moles and other small skin growths may also appear. Blood vessels can be seen beneath the more transparent skin, which has largely lost its layer of fatty support (Whitbourne, 1996). This loss further limits the older adult's ability to adapt to very hot and cold temperatures.

The face is especially likely to show these effects because it is frequently exposed to the sun, which accelerates aging. Other facial changes occur: The nose and ears broaden as new cells are deposited on the outer layer of the skeleton. Teeth may be yellowed, cracked, and chipped, and gums may recede; with better dental care, these outcomes are likely to be less pronounced in future generations. As hair follicles under the skin's surface die, hair on the head thins in both sexes, and the scalp may be visible. In men with hereditary pattern baldness, follicles do not die but, instead, begin to produce fine, downy hair (Whitbourne, 1996).

Body build changes as well. Height continues to decline, especially in women, as loss of bone mineral content leads to further collapse of the spinal column. Weight generally drops after age 60 due to additional loss in lean body mass (bone density and muscle), which is heavier than the fat deposits accumulating on the torso.

Several factors affect mobility. The first is muscle strength, which generally declines at a faster rate in late adulthood than it did in middle age. On the average, by 60 to 70 years of age, 10 to 20 percent of muscle power has been lost, a figure that climbs to 30 to 50 percent after ages 70 to 80 (Whitbourne, 1996). Second, bone strength deteriorates, as does flexibility and strength of the tendons and ligaments (which connect muscle to bone) and the joints. In her eighties, Ruth's reduced ability to flex her limbs and rotate her hips meant that walking at a steady, moderate pace, climbing stairs, and rising from a chair were difficult.

In Chapter 13, we noted that endurance athletes who continue training throughout adulthood retain their muscular physiques and much of their strength into their sixties and seventies. These especially active individuals lose fast-twitch muscle fibers like other aging individuals, but they compensate by strengthening remaining slow-twitch fibers so they work more efficiently.

Careful planning of an exercise program can also enhance joint flexibility and range of movement. When Ruth complained of joint stiffness and pain, her doctor pointed out that certain rhythmic and flexing exercises, including dance, could be helpful (Munn, 1981). "As a matter of fact, I'm taking a dance class myself," Dr. Wylie commented. "C'mon, Ruth, let me show you a few flamenco steps." As Dr. Wylie demonstrated, Ruth imitated. Outside the examining room, the nurses wondered what had sparked the sound of tapping shoes and peals of doctor–patient laughter.

ADAPTING TO PHYSICAL CHANGES OF LATE ADULTHOOD

Great diversity exists in older adults' adaptation to the physical changes of aging. Dick and Goldie took advantage of an enormous industry designed to stave off the appearance of old age, including cosmetics, wigs, and plastic surgery. In contrast, Ruth and Walt hardly gave a second thought to their thinning white hair and wrinkled skin. Their identities were not as bound up with their appearance as with the ability to remain actively engaged in their surroundings.

People vary in the aspects of physical aging that matter most to them. And because parts of the body age at different rates, older adults' sense of physical aging is multidimensional: they feel older in some domains than others. Compared to Dick and Goldie, Ruth and Walt approached aging with a more positive outlook and greater peace of mind. They did not try to hang on to a youthful identity but, instead, attempted to intervene in those aspects of aging that could be changed and accept those that could not.

Think back to our discussion of problem-centered and emotion-centered coping in Chapter 16. It applies here as well. As Walt and Ruth prevented and compensated for age-related declines through diet, exercise, environmental adjustments, and an active, stimulating lifestyle, they felt a sense of personal control over their fates. This prompted additional positive coping and improved physical functioning. In contrast, people who avoid confronting age-related change—who think it is inevitable and uncontrollable—tend to be passive when faced with it and to report more physical and psychological symptoms (Speake, 1987; Whitbourne & Primus, 1996).

In Western societies, where negative stereotypes of aging are widespread, adults of all ages regard signs of growing older negatively—a view that is strongest when men evaluate women. (See the section on the double standard of aging in Chapter 15, page 501.) Overall, the elderly respond less negatively than do younger people because they know that wrinkles, baldness, and a stockier body build describe themselves (Harris, 1994). The more older people endorse aging stereotypes, the less likely they are to cope adaptively with age-related change (Ryff, 1991). Also, the stereotype of a declining, incompetent elder encourages others to treat the elderly in condescending ways. This strengthens despondency in the face of aging.

In cultures where the elderly are treated with deference and respect, coping is much easier. An aging appearance may be a source of pride, signifying success in living to a ripe maturity, and elderly status is often cause for celebration. For example, a traditional Japanese ritual called the *kankrei* recognizes the older person's release from the responsibilities of middle age, new freedoms and competencies, and senior place in the family and society. In Japanese-American families, grandchildren often plan the kankrei as a surprise sixtieth birthday party, incorporating elements of both the traditional ritual (such as dress) and

CULTURAL INFLUENCES

CULTURAL VARIATIONS IN THE EXPERIENCE OF AGING

In a study of diverse communities around the world, a team of anthropologists found that older people fare best in contexts where the positive social meaning of aging transcends chronological age and physical decline (Keith et al., 1994). They do well when they have status and opportunities for social participation, even when they are frail. When elders are excluded from important social roles and infirmity brings separation from the community, aging leads to diminished well-being.

Consider the Herero, a pastoral people of Botswana, Africa. Elders who are strong and active spend their days just as younger adults do, tending the cattle and performing other chores. When older people decline physically, they retain positions of seniority and are treated with respect. A status hierarchy makes the oldest man and his wife leaders of the village. They are responsible for preserving the sacred flame of the ancestors, who remain significant family members after death. Children are sent to live in the homes of frail elders to provide care—an assignment that is a source of great pride and prestige.

Old age is also a gratifying time of life in Momence, Illinois, a small, working-class farming and manufacturing town. Since the population is highly stable, elders are granted positions of authority due to their length of residence and intimate knowledge of the community. Town, church, and club leaders tend to be older, and past leaders are included in decision making. Frail elders remain part of community life in a less direct way. Because they are embedded in family, neighborhood, and church networks that have persisted for many years, other citizens often inquire about them and monitor their situations.

In Swarthmore, a middle-class Philadelphia suburb with a less stable population, life for elders is less certain. Although 25 percent of residents are over age 60, most moved to the community after retirement, and their grown children live elsewhere. As a result, older people are granted no special status due to seniority, and they occupy few positions of importance. Unlike the Herero and the citizens of Momence, Swarthmore elders spend much time in age-segregated settings, such as bridge games and church groups for seniors. Townspeople tend to equate "being older" with chronological age—60 and beyond. In contrast, the Herero and the residents of Momence seldom refer to others in terms of their age. Rather, they mention knowledge and social position. When the elderly of Swarthmore become physically infirm, most have no network of support and must leave the community.

Among the Herero and in Momence, neither age nor physical decline threatens community ties. In Swarthmore, being old limits integration into community life, and frailty has profound, negative social consequences. It typically brings an end to community membership.

Status and opportunities for social participation are vital for elders' psychological well-being. These energetic seniors discuss issues and devise new city policies in Austin, Texas. (Bob Daemmrich/Stock Boston)

the Western birthday (such as a special cake) (Doi, 1991; Gelfand, 1994). Cultural valuing of aging prompts a welcoming approach to late adulthood, including some of its physical transitions. (See, also, the Cultural Influences box above.)

Although declines are inevitable, physical aging can be viewed with optimism or pessimism. As Walt commented, "You can think of your glass as half full or half empty." Today, a wealth of research supports the "half full" alternative. In the next section, we will see additional examples of the multiple ways that older people can control their own destiny.

BRIEF REVIEW

Chronological age is an imperfect indicator of older adults' physical capacities, which vary widely between and within individuals. The number of elderly, especially those 85 and older, is increasing rapidly in industrialized nations, less so in developing countries. As people age, the gender gap in life expectancy shrinks, and a life expectancy crossover occurs for ethnic minorities in relation to the white majority. Genetic makeup combines with a wide array of environmental variables to determine longevity.

In late adulthood, many organs and systems of the body function less effectively. Neuron loss occurs in many brain regions, and the autonomic nervous system performs less well. A reduction in light reaching the retina impairs a variety of aspects of vision. Hearing loss interferes with speech perception, making it more difficult to converse with others. Reduced taste and odor sensitivity can disrupt enjoyment of food. Touch sensitivity on the hands and fingertips decreases. Aging of the cardiovascular and respiratory systems leads older adults to feel more physically stressed after exercise, and the immune system is less effective in fighting disease. Changes in brain structures prompt sleep disturbances. Aging becomes apparent to others through wrinkled skin, thinning white hair, and a rounder body build.

Older adults can slow the rate of and compensate for many physical declines. Aging stereotypes in Western societies interfere with the ability to cope with physical aging adaptively.

ASK YOURSELF . . .

■ *Reread the story of Dr. Stojic, a robust centenarian, in the Lifespan Vista box on page 556. What aspects of his life history are consistent with research findings on factors that contribute to a long and healthy life?*

■ *When Joe insensitively commented in Ruth's presence that she must be "going deaf," he revealed a stereotype of aging. What is Joe's likely view of elderly individuals' capacities? Present a research-based argument that shows Joe is wrong.*

■ *Sixty-five-year-old Herman inspected his thinning hair and bare scalp in the mirror. "The best way to adjust to this is to learn to like it," he thought to himself. "I remember reading somewhere that bald older men are regarded as leaders." What type of coping is Herman using, and why is it an effective way to adapt to this aspect of physical aging?*

HEALTH, FITNESS, AND DISABILITY

At Walt and Ruth's fiftieth wedding anniversary, 77-year-old Walt thanked a roomful of well-wishers for joining in the celebration. Then he announced emotionally, "I'm so grateful Ruth and I are in good health and still able to give to our family, friends, and community."

As Walt's remarks affirm, health is central to psychological well-being in late life. When researchers ask the elderly about their possible selves (see Chapter 16, page 525), number of hoped-for physical selves declines and number of feared physical selves increases with age. Despite this realistic response to changes in physical capabilities, older adults are generally optimistic about their health. Because they judge themselves against same-age peers, they rate their health as favorably as do college students! And with respect to protecting their health, their sense of self-efficacy is as high as that of young adults and higher than that of middle-aged people (Hooker, 1992; Lachman, 1985).

This tendency to look on the bright side enables the elderly to keep striving in the face of physical difficulties instead of giving up and disengaging. The more optimistic elders are, the better they are at overcoming threats to health, which promotes further optimism and continued health-enhancing behaviors (Hooker & Kaus, 1992; Scheier & Carver, 1987). Furthermore, good health permits older adults to remain engaged in valued social roles, thereby preventing social isolation and fostering psychological well-being (Heidrich & Ryff, 1993). In this way, good physical and good mental health are intimately related in late life.

As we mentioned earlier, social-class and ethnic variations in health diminish in late adulthood. Nevertheless, before age 85, African-American and Hispanic elderly (over one-third of whom live in poverty) remain at greater risk for certain health problems (see Table 17.3). Native-American older adults are even worse off. More than 80 percent are poor, and physical aging is so advanced and chronic health conditions so widespread that the federal government grants Native Americans special health benefits. These begin as early as age 45, reflecting a much harder and shorter lifespan (Gelfand, 1994; Hopper, 1993).

TABLE 17.3

Poverty Rates and Health Problems among Elderly Ethnic Minorities

ETHNIC MINORITY	POVERTY RATE AGE 65 AND OVER[1]	HEALTH PROBLEMS GREATER THAN IN THE GENERAL POPULATION OF 65- TO 84-YEAR-OLDS
African–American	34%	Cardiovascular disease, stroke, a variety of cancers, diabetes
Hispanic	23%	Cardiovascular disease, diabetes
Native American	Over 80%	Diabetes, kidney disease, liver disease, tuberculosis, hearing and vision impairments

[1] The poverty rate for elderly Caucasian Americans is 10%; for all elderly people in the United States, it is 14%.
Sources: Gelfand, 1994; Hopper, 1993; U.S. Bureau of the Census, 1996.

Unfortunately, low-income, ethnic minority elders are less likely to seek medical treatment than their Caucasian counterparts. When they do, they often do not comply with the doctor's directions because they are less likely to believe they can control their health and that treatment will work. A high sense of self-efficacy does not apply to these older adults, further impairing their physical condition (Hopper, 1993).

The sex differences we noted in Chapter 15, with men prone to fatal diseases and women to non-life-threatening disabling conditions, extend into late adulthood. By very old age (85 and beyond), women are more impaired than men because only the sturdiest men have survived. In addition, with fewer physical limitations, older men are better able to remain independent and engage in exercise, hobbies, and involvement in the social world, all of which promote better health (Barer, 1994).

Widespread health-related optimism among the elderly means that substantial inroads into preventing disability can be made even in the last few decades of life. Yet the persistence of poverty and negative lifestyle factors means that as longevity increases, diseases of old age may rise as well, straining the nation's health care resources. Ideally, as life expectancy extends, we want the average period of diminished vigor before death to decrease. This public health goal is called the **compression of morbidity** (Fries, 1990). As we look closely at health, fitness, and disability in late adulthood, we will expand our discussion of health promotion in earlier chapters, taking up additional ways to reach this objective.

NUTRITION AND EXERCISE

The physical changes of late life lead to an increased need for certain nutrients—calcium and vitamin D to protect the bones; zinc and vitamins B_6, C, and E to protect the immune system; and vitamins C and E and beta-carotene to prevent free radicals (see Chapter 13, page 420). Yet declines in physical activity and in the senses of taste and smell reduce the quantity and quality of food eaten. In addition, the aging digestive system has greater difficulty absorbing certain nutrients, such as protein, calcium, and vitamin D. Together, these impairments increase the risk of dietary deficiencies (Blumberg, 1996).

Older adults who take vitamin-mineral supplements show enhanced health and physical functioning. In one study, a daily vitamin-mineral tablet resulted in 48 percent fewer days of infectious illness (Chandra, 1992). In another study, elderly men in a weight-bearing exercise class who took a daily protein-energy supplement showed a larger increase in muscle mass than did exercising classmates who did not take the supplement (Meredith et al., 1992). And longitudinal evidence reveals that older adults who reduce dietary fat and take vitamins C and E and beta-carotene lower their risk of cardiovascular disease. Vitamin supple-

ments also help prevent cataracts and macular degeneration (Rimm et al., 1993; Seddon et al., 1994a, 1994b; Stampfer et al., 1993).

Besides nutrition, exercise continues to be a powerful health intervention. Sedentary healthy older adults up to age 80 who begin endurance training (walking, exercycling, aerobic dance) show gains in vital capacity that compare favorably with those of much younger individuals (Spina et al., 1993; Stratton et al., 1994). And weight-bearing exercise begun in late adulthood—even as late as age 90—promotes muscle size and strength, blood flow to muscles, and ability of muscles to extract oxygen from blood. This translates into improved walking speed, balance, posture, and ability to carry out everyday activities independently, such as opening a stubborn jar lid, carrying an armload of groceries, or lifting a 30-pound grandchild. However, it takes frequent, strenuous exercise to increase the heart's pumping capacity in old age; low-intensity exercise is no longer sufficient (Fiatarone et al., 1990; Goldberg, Dengel, & Hagberg, 1996; Pyka et al., 1994).

Although good nutrition and physical activity are most beneficial when they are lifelong, it is never too late to change. Beginning in his sixties until his death at age 94, Walt's Uncle Louie played 1 to 2 hours of tennis a day and did ballroom dancing three nights a week. Exercise led Louie to sustain a high sense of physical self-esteem. As a dancer, he dressed nattily and moved gracefully. He often commented on how dance and other sports could transform an older person's appearance from dowdy to elegant, expressing the beauty of the inner self.

Before beginning an exercise regimen, the older person should have a medical checkup. Exercise can be dangerous

Although good nutrition and physical exercise are most beneficial when they are lifelong, it is never too late to change. This man enjoys planning and preparing tasty, nutritious meals, an interest he developed after retirement. (David Woo/Stock Boston)

if elders suffer from advanced cardiovascular disease, live in extreme climates, or are exposed to air pollution.

SEXUALITY

When Walt turned 60, he asked his 90-year-old Uncle Louie at what age sexual desire and activity cease, if they do. Walt's question stemmed from a widely held myth that sex drive disappears among the elderly (Deacon, Minichiello, & Plummer, 1995). As Louie's response indicates, this stereotype has not been successful in stamping out sexuality in late adulthood—a tribute to the toughness of elders.

"It's especially important to be reasonably rested and patient during sex," Louie explained to Walt. "I can't do it as often, and it's a quieter experience than it was in my youth, but my sexual interest has never gone away. Rachella and I have led a happy intimate life, and it's still that way."

Compared with other parts of the body, such as the eyes, ears, bones, and cardiovascular system, age-related changes in the reproductive organs, described in Chapter 15, are minimal. Although virtually all cross-sectional studies report a decline in sexual desire and frequency of sexual activity in older people, this trend may be exaggerated by cohort effects. A new generation of elders, accustomed to viewing sexuality positively, will probably be more sexually active. Furthermore, longitudinal evidence indicates that most healthy older married couples report continued, regu-

Most healthy older married couples report continued, regular sexual enjoyment. And even at the most advanced ages, there is more to sexuality than the sex act itself—feeling sensual, enjoying close companionship, and being loved and wanted. (Cleo/ PhotoEdit)

lar sexual enjoyment. The same generalization we discussed for midlife applies to late life: good sex in the past predicts good sex in the future. Among unmarried people over age 65, about 70 percent of men and 50 percent of women manage to have sex from time to time (Levy, 1994; Roughan, Kaiser, & Morley, 1993).

Too often, intercourse is used as the only measure of sexual activity—a circumstance that promotes a narrow view of pleasurable sex. Even at the most advanced ages, there is more to sexuality than the sex act itself—feeling sensual, enjoying close companionship, and being loved and wanted (Hodson & Skeen, 1994). Both older men and older women report that the male partner is usually the one who ceases to interact sexually. In a culture that emphasizes an erection as necessary for being sexual, a man may withdraw from all erotic activity when he finds that erections are harder to achieve and more time must elapse between them (Kaiser, 1991).

Disabilities that disrupt blood flow to the penis—most often, disorders of the autonomic nervous system, cardiovascular disease, and diabetes—are responsible for dampening sexuality in older men. Cigarette smoking, excessive alcohol intake, and a variety of prescription medications also lead to diminished sexual performance (Schiavi, Mandeli, & Schreiner-Engel, 1994). Among women, poor health and absence of a partner are major factors that reduce sexual activity. As the sex ratio increasingly favors females, aging heterosexual women have fewer and fewer opportunities for sexual encounters (Kaiser, 1991; U.S. Bureau of the Census, 1996).

Older people who know little about normal age-related changes in sexual functioning may react with anxiety when faced with sexual difficulties. Educational programs that dispel myths and foster a view of sex as extending throughout adulthood promote positive sexual attitudes in late life (Hillman & Stricker, 1994). In nursing homes, education for caregivers is vital for ensuring older residents' rights to privacy and other living conditions that permit sexual expression (Levy, 1994).

PHYSICAL DISABILITIES

Illness and disability climb as the end of the lifespan approaches. If you compare Table 17.4 with Table 15.1 on page 494, you will see that cardiovascular disease and cancer—illnesses we discussed in Chapter 15—increase dramatically from mid- to late life and remain the leading causes of death. Notice how, with increasing years since menopause, the sex difference in deaths due to cardiovascular disease declines. The gender gap for another vascular condition, *stroke*, also diminishes with age. It is caused by hemorrhage or blockage of blood flow in the brain and is a major cause of disability in late adulthood and death after age 75. *Emphysema*, due to extreme loss of elasticity in lung tissue, results in serious breathing difficulty. Although a few cases of emphysema are inherited, most are caused by long-term cigarette smoking (Larson, 1996).

Notice in Table 17.4 that death due to pneumonia increases nearly eighteenfold during late adulthood. As the longest-lived people escape chronic diseases or weaken physically because of them, the immune system eventually encounters an infection it cannot fight. Consequently, many of the very old succumb to one of the more than 50 lung inflammations classified as pneumonia. Doctors recommend that people age 65 and older be vaccinated against the most common type.

Other diseases are less frequent killers, but they limit older adults' ability to live fully and independently. Osteoporosis, discussed in Chapter 15 (see page 496), continues to rise in late adulthood; recall that it affects the majority of men and women after age 70. Yet another bone disorder—*arthritis*—adds to the physical limitations of many elders. And as Table 17.5 reveals, *adult-onset diabetes* and *unintentional injuries* multiply in late adulthood. In the following sections, we take up these latter three conditions.

Finally, an important point must be kept in mind as we discuss physical and mental disabilities of late adulthood: None are considered a normal part of the aging process, even though they increase with advancing age. In other words, just because certain physical and mental disabilities are *related to age* does not mean that they are *caused by aging*. To clarify this distinction, some experts distinguish between **primary aging** (another term for *biological aging*), or genetically influenced declines that affect all members of our species and take place even in the context of overall good health, and **secondary aging**, declines due to hereditary defects and negative environmental influences, such as poor diet, lack of exercise, disease, substance abuse, environmental pollution, and psychological stress.

Throughout this and earlier chapters, we have seen that it is often difficult to distinguish primary from secondary aging. But if we do not try to do so, we are in danger of magnifying decrements and promoting a false, stereotyped view of aging as illness and disease (Lemme, 1995). All the disabilities we are about to discuss are extreme conditions that fall within *secondary aging.*

■ ARTHRITIS. Beginning in her fifties, Ruth felt a slight morning stiffness in her neck, back, hips, and knees. In her sixties, she developed bony lumps on the end joints of her fingers. As the years passed, she experienced periodic joint swelling and some loss of flexibility—changes that affected her ability to move quickly and easily.

Arthritis, a condition of inflamed, painful, stiff, and sometimes swollen joints and muscles, becomes more common in late adulthood. It comes in several forms. Ruth has **osteoarthritis,** the most frequent type. Otherwise known as "wear-and-tear arthritis" or "degenerative joint disease," it is one of the few age-related disabilities where years of use makes a difference. Although a genetic proneness seems to exist, the disease usually does not appear until the forties and fifties. In frequently used joints, cartilage on the ends of the bones, which reduces friction during movement, gradually deteriorates. Almost all older adults show some degree of osteoarthritis on X-rays, although wide individual differences in severity exist (Doress-Worters & Siegal, 1994; Strange, 1996).

Unlike osteoarthritis, which is limited to certain joints, **rheumatoid arthritis** involves the whole body. An autoimmune response leads to inflammation of connective tissue, particularly the membranes that line the joints. The result is overall stiffness, inflammation, and aching. Tissue in the cartilage tends to grow, damaging surrounding ligaments, muscles, and bones. The result is deformed joints and often serious loss of mobility. Sometimes other organs, such as the heart and lungs, are affected.

Disability due to arthritis affects about 45 percent of American men over age 65. Among women, the incidence is higher; 50 percent between ages 65 and 74 and 60 percent over age 75 are affected. The reason for the sex difference is not clear. Although rheumatoid arthritis can strike at any age, it tends to increase after menopause. But unlike

TABLE 17.4

Death Rates[a] Among the Elderly by Leading Causes in the United States

CAUSE	65+YEARS TOTAL	65–74 YEARS MALES	65–74 YEARS FEMALES	75–84 YEARS MALES	75–84 YEARS FEMALES	85+ YEARS MALES	85+ YEARS FEMALES
Cardiovascular disease	1,845	1,179	579	2,745	1,776	7,158	6,264
Cancer	1,122	1,111	687	1,883	1,026	2,803	1,394
Stroke	389	156	119	510	443	1,501	1,591
Emphysema	242	200	121	479	233	831	318
Pneumonia	209	74	41	310	176	1,310	911
Adult-onset diabetes	116	79	74	151	138	269	248
Unintentional injury	83	61	31	131	75	344	220

[a]Number of deaths per 100,000 population in specified group.
Source: U.S. Bureau of the Census, 1996.

osteoporosis, it does not respond to estrogen replacement therapy (ERT), so something other than estrogen loss must be involved (Doress-Worters & Siegal, 1994). It may be due to a late-appearing genetic defect in the immune system (Gomolka et al., 1995).

Managing arthritis requires a balance of rest when the disease flares, pain relief (including, if necessary, certain drugs), and physical activity involving gentle stretching of all muscles to maintain mobility. Twice a week, 84-year-old Ruth attended a water-based exercise class. Within 2 months, her symptoms lessened, and she no longer needed a walker (Strange, 1996). Weight loss in obese people is helpful because it relieves the joints of added daily trauma from bearing an extra load (Eades, 1992).

Although osteoarthritis responds to treatment more easily than rheumatoid arthritis, the course of each varies greatly. With proper medication, joint protection, and lifestyle changes, many people with either form of the illness can lead long and productive lives. If hip or knee joints are badly damaged or deformed, they can be surgically rebuilt or replaced with plastic or metal devices.

■ ADULT-ONSET DIABETES. After we eat a meal, the body breaks down the food, releasing glucose (the primary energy source for cell activity) into the bloodstream. Insulin, produced by the pancreas, keeps the blood concentration of glucose within set limits by stimulating muscle and fat cells to absorb it. When this balance system fails, either because not enough insulin is produced or body cells become insensitive to it, *adult-onset diabetes* (otherwise known as *diabetes mellitus)* results. Over time, abnormally high blood glucose damages the blood vessels, increasing the risk of stroke, heart attack, circulatory problems in the legs, and injury to the eyes, kidneys, and nerves.

From middle to late adulthood, the incidence of adult-onset diabetes doubles, affecting 10 percent of the elderly in the United States and ranking as a leading cause of death (U.S. Bureau of the Census, 1996). Diabetes runs in families, suggesting that heredity is involved. But inactivity and abdominal fat deposits greatly increase the risk. Higher rates of adult-onset diabetes are found among African Americans, Mexican Americans, Puerto Ricans, and Native Americans for genetic as well as environmental reasons, including high-fat diets and obesity associated with poverty (Hopper, 1993; Markides, 1989).

Treating adult-onset diabetes requires lifestyle changes, including a carefully controlled diet, regular exercise, and weight loss. By promoting glucose absorption and reducing abdominal fat, physical activity lessens disease symptoms (Goldberg, Dengel, & Hagberg, 1996).

■ UNINTENTIONAL INJURIES. At age 65 and older, the death rate from unintentional injuries is at an all-time high—three times greater than in adolescence and young adulthood. Motor vehicle collisions and falls are largely responsible.

Motor Vehicle Accidents. Motor vehicle collisions account for only one-fourth of injury mortality in late life compared to one-half in middle adulthood. But a look at individual drivers tells a different story. Older adults have higher rates of traffic violations, accidents, and fatalities per mile driven of any age group, with the exception of drivers under age 25. The high rate of injury persists, even though many elders limit their driving after noticing that their ability to drive safely is slipping. Women are more likely to take these preventive steps. Deaths due to injuries—motor vehicle and otherwise—continue to be much higher for men than women in late adulthood (Stewart et al., 1993).

Recall that visual declines led Walt to have difficulty seeing the dashboard and identifying pedestrians at night. Most information used in driving is visual, and older drivers who rate their vision as poorer report more driving problems, such as surprise when other vehicles merge, trouble reading road signs, and discomfort at the speed of other vehicles (Kline et al., 1992). Compared to young drivers, the elderly are less likely to drive quickly and recklessly but more likely to fail to heed signs, give right-of-way, and turn appropriately. They often try to compensate for their difficulties by being more cautious. But slowed reaction time and indecisiveness can pose hazards as well (Garzia & Trick, 1992). Walt had several close calls when he took too long to change lanes, a maneuver that requires rapid, accurate judgment for safety.

The elderly are also at risk as pedestrians. Nearly 30 percent of deaths between ages 65 and 74 in the United States result from pedestrian accidents (U.S. Bureau of the Census, 1996). Confusing intersections and crossing signals that do not allow older people enough time to get to the other side of the street are often involved.

Falls. On one occasion, Ruth fell down the basement steps and lay there with a broken ankle until Walt arrived home an hour later. Ruth's tumble represents the leading type of accident among the elderly. About 30 percent of adults over age 65 and 40 percent over age 80 experienced a fall within the last year. Because of weakened bones and decline in ability to break a fall, about 10 percent of the time serious injury results. Among the most common is hip fracture. It increases twenty-fold from age 65 to 85 and is associated with a 12 to 20 percent increase in mortality. Of those who survive, half never regain the ability to walk without assistance (Garzia & Trick, 1992; Simoneau & Leibowitz, 1996).

Falling can also impair health indirectly—by promoting fear of falling. Almost half of older adults who have fallen admit that they purposefully avoid activities because they are afraid of falling again (Tinetti, Speechley, & Ginter, 1988). In this way, a fall can limit mobility and social contact, undermining both physical and psychological well-being. Although an active lifestyle may expose the elderly to more events and situations that can cause a fall, the health benefits of activity far outweigh the risk of serious injury due to falling (Tinetti, 1990).

CAREGIVING CONCERNS

Preventing Unintentional Injury in Late Adulthood

SUGGESTION	DESCRIPTION
FALLS	
Schedule regular medical checkups.	Eye exams to ensure that corrective lenses are up to date; physical exams to identify health risks that increase the chances of falling; review of medications for effects on attention and coordination.
Engage in regular exercise.	Strength and balance training to promote coordination and counteract fear of falling.
Use walking aids when necessary.	Canes and walkers to compensate for poor balance and unsteady gait.
Improve safety of the living environment.	Extra lighting in dim areas, such as entrances, hallways, and staircases; handrails in hallways and grab bars in bathrooms; loose rugs secured to floor or moved; furniture and other objects arranged so they are not obstacles.
Be alert and plan ahead in risky situations.	Watching for slippery pavement; carrying a flashlight at night; allowing extra time to cross streets; becoming familiar with new settings before moving about freely.
MOTOR VEHICLE COLLISIONS AND PEDESTRIAN ACCIDENTS	
Modify driving behavior in accord with visual and other limitations.	Driving fewer miles; reducing or eliminating driving during rush hour, at night, or in bad weather.
Modify pedestrian behavior in accord with visual and other limitations.	Wearing light-colored clothing at night; allowing extra time to cross streets; walking with a companion.
Attend training classes for older drivers; if not available, press for them in your community.	Practicing tracking vehicles and pedestrians in dim light, judging vehicle speed, and reading signs and dashboard displays; reviewing rules of the road.

Sources: Kline et al., 1992; Simoneau & Leibowitz, 1996; Stewart et al., 1993.

Preventing Unintentional Injuries. Many steps can be taken to reduce unintentional injury in late adulthood. Designing motor vehicles and street signs to accommodate visual needs of the elderly is a goal for the future. In the meantime, training that enhances visual and cognitive skills essential for safe driving and helps older adults avoid high-risk situations (such as busy intersections and rush hour) can save lives.

Similarly, efforts to prevent falls must address risks within the person and environment—through corrective eyewear, strength and balance training, and improved safety in homes and communities. The Caregiving Concerns table above summarizes ways to protect the elderly from injury.

MENTAL DISABILITIES

Normal age-related cell death in the brain, described earlier in this chapter, does not lead to loss of ability to engage in everyday activities. But when cell death and structural and chemical abnormalities are profound, serious deterioration of mental and motor functions occurs.

Dementia refers to a set of disorders occurring almost entirely in old age in which many aspects of thought and behavior are so impaired that everyday activities are disrupted (American Psychiatric Association, 1994). It rises sharply with age, striking adults of different ethnic and social-class groups about equally. Approximately 1 percent of people in their sixties are affected, a rate that doubles every 5 years until it stabilizes at 30 percent for people in their nineties (Ritchie, Kildea, & Robine, 1992).

About a dozen types of dementia have been identified. Some are reversible with proper treatment, but most are irreversible and incurable. A few forms, such as Parkinson's disease,[2] involve deterioration in subcortical brain regions (primitive structures below the cortex). But in the large majority of cases, progressive damage to the cortex—seat of human intelligence—occurs. *Cortical dementia* comes in two varieties: Alzheimer's disease and cerebrovascular dementia.

■ ALZHEIMER'S DISEASE. When Ruth took 79-year-old Ida to the ballet, an occasion that the two sisters looked forward to each year, Ida's behavior was different than before. Having forgotten the engagement, she reacted with anger at Ruth for arriving unannounced at her door. After Ida calmed down, she got lost driving in familiar parts of town on the way to the theater while insisting she knew her way perfectly well.

[2] In Parkinson's disease, neurons in the part of the brain that controls muscle movements deteriorate. Symptoms include tremors, shuffling gait, loss of facial expression, rigidity of limbs, difficulty maintaining balance, and stooped posture.

With no cure available for Alzheimer's disease, family caregivers like this adult son need assistance, encouragement, social support, and education about the disease to ensure the best possible adjustment for all concerned. (Griffin/The Image Works)

Settled in the balcony, Ida talked loudly and dug noisily in her purse as the lights dimmed and the music began.

"Shhhhhh," responded a dozen voices from the surrounding seats.

"It's just the music!" Ida snapped at full volume. "You can talk all you want until the dancing starts." Ruth was astonished and embarrassed at the behavior of her once socially sensitive sister.

Six months later, Ida was diagnosed with **Alzheimer's disease,** the most common form of dementia, accounting for 50 to 60 percent of all cases. Approximately 5 to 7 percent of people over age 65 have the disorder. Of those over age 80, about 15 percent are affected (Gatz, Kasl-Godley, & Karel, 1996; Helmchen et al., 1996). Each year in the United States, 100,000 people die of Alzheimer's, making it a leading cause of death in late adulthood (although on death certificates, more immediate causes, such as infection and respiratory failure, are usually listed).

Symptoms and Course of the Disease. Severe memory problems are often the earliest symptom—forgetting names, dates, appointments, familiar routes of travel, or the need to turn off the stove in the kitchen. At first, recent memory is most impaired, but as serious disorientation sets in, recall of distant events and such basic facts as time, date,

and place evaporates. Faulty judgment puts the person in danger. For example, Ida insisted on driving after she was no longer capable of doing so. Personality changes appear—loss of spontaneity and sparkle, anxiety and angry outbursts due to uncertainties created by mental problems, reduced initiative, and social withdrawal. Depression, which often appears in the early phase of Alzheimer's and other forms of dementia, seems to be part of the disease process (Reifler, 1994). However, depression may worsen as the older adult reacts to disturbing mental changes.

As the disease progresses, skilled and purposeful movements disintegrate. When Ruth took Ida into her home, she had to help her dress, bathe, eat, brush her teeth, and (eventually) walk and use the bathroom. Ruth also found that Ida's sleep was disrupted by delusions and imaginary fears. She often awoke in the night and banged on the wall, insisting that it was dinnertime. Sometimes she cried out that someone was choking her. Over time, Ida lost the ability to comprehend and produce speech. And when her brain ceased to process information, she could no longer recognize objects and familiar people. In the final months, Ida became increasingly vulnerable to infections, lapsed into a coma, and died.

The course of Alzheimer's varies greatly, from a few years to as many as 15 or more years. The average is about 7 years (Miller & Morris, 1993).

Brain Deterioration. A diagnosis of Alzheimer's disease is made through exclusion—after other causes of dementia have been ruled out by a physical examination and psychological testing. Doctors can only know for sure that a person has Alzheimer's by inspecting the brain after death for a set of abnormalities that either cause or result from the disease (Gatz, Kasl-Godley, & Karel, 1996). However, new developments in *functional brain-imaging techniques,* which yield three-dimensional pictures of brain activity, offer hope of accurately diagnosing Alzheimer's disease soon after symptoms appear (Burns, Howard, & Pettit, 1995).

Two major structural changes in the brain occur. The greater their distribution in the cortex, the more severe the Alzheimer's dementia tends to be. Inside neurons, **neurofibrillary tangles** appear—bundles of twisted threads that are the product of collapsed neural structures. Outside neurons, *plaques,* or deposits of material surrounded by clumps of dead nerve cells, develop. They contain *amyloid,* a protein associated with many other diseases that is deposited in tissue with reduced immunity and that may destroy surrounding cells. Although some neurofibrillatory tangles and **amyloid plaques** are present in the brains of normal middle-aged and older people and increase with age, they are far more abundant in Alzheimer's victims.

Accompanying structural changes in the brain are chemical changes—lowered levels of neurotransmitters necessary for communication between neurons. For example, a deficit of the neurotransmitter *acetylcholine* and destruction of its network of neurons, which transports messages between distant areas of the brain, disrupt perception, memory, reasoning, and judgment. A 50 to 75 percent drop in *serotonin*, a neurotransmitter that regulates arousal, may contribute to sleep disturbances and Alzheimer's-related depression (Cohen et al., 1993). As research continues, it is likely that other neurotransmitter imbalances will be discovered.

Risk and Protective Factors. A family history of Alzheimer's is the most consistent risk factor (Breitner et al., 1988). Researchers have identified genes on chromosomes 1, 14, 19, and 21 that increase the chances of developing the disease. Recall that chromosome 21 is involved in Down syndrome. Individuals with this chromosomal disorder who live past age 40 almost always have the brain abnormalities and symptoms of Alzheimer's disease. Nevertheless, 50 percent of Alzheimer's cases show no family history or currently known genetic marker (Mayeux, 1996; Plassman & Breitner, 1996).

Besides heredity, the roles of toxic substances, viruses, and defects in the blood–brain barrier (which protects the brain from harmful agents) are being explored. Alzheimer's disease may result from several factors working together, and different combinations may lead to somewhat different forms. Recently, researchers identified a blood protein called ApoE, which carries cholesterol, as a risk factor for Alzheimer's. Its impact on structural and chemical deterioration in the brain is currently the focus of much research. Although elevated levels of aluminum have been found in the brain tangles and plaques of Alzheimer's victims, the role of environmental exposure to aluminum remains uncertain (National Institute on Aging, 1996).

Finally, estrogen replacement therapy (ERT) and use of anti-inflammatory drugs (such as aspirin) may have protective effects. In several longitudinal studies, the incidence of Alzheimer's disease was substantially lower in women taking estrogen. In those with signs of the disease, beginning ERT led to improved cognitive functioning (Ohkura et al., 1995; Paganinihill & Henderson, 1996). Similar outcomes exist for anti-inflammatory medication. It may slow the course of the Alzheimer's by interrupting inflammation of brain tissue caused by amyloid and other abnormal proteins (McGeer, Schulzer, & McGeer, 1996). Yet another protective factor is education. Some researchers speculate that years of education lead to more synaptic connections. These act as a *cognitive reserve,* equipping the aging brain with greater tolerance for injury before it crosses the threshold into mental disability (Satz, 1993).

Interventions. As Ida's Alzheimer's worsened, the doctor prescribed a mild sedative and an antidepressant to help control her behavior. Drug treatments to correct neurotransmitter deficits have also been tried, but they are not effective, perhaps because Alzheimer's disease interferes

with the ability of neurons to respond to them (Burns, Howard, & Pettit, 1995).

With no cure available, family interventions ensure the best adjustment possible for the Alzheimer's victim, his or her spouse, and other relatives. Family caregivers need assistance and encouragement from extended family members, friends, and community agencies. Education about the disease is vital. In combination with social support, it helps caregivers respond with patience and compassion to the victim's frequent, repetitive questions and stories and socially inappropriate behavior. Return to page 536 to review strategies for relieving the stress of caring for an aging parent. These techniques are just as helpful to spouses and siblings, who serve as caregivers about as often as do adult children (Marks, 1996).

Communicating with Alzheimer's victims is especially challenging. The Caregiving Concerns table on the following page lists some helpful suggestions. Avoiding dramatic changes in living conditions, such as moving to a new location, rearranging furniture, or modifying daily routines, is important for helping the patient, whose cognitive world is gradually disintegrating, feel as secure as possible.

■ CEREBROVASCULAR DEMENTIA. In **cerebrovascular dementia,** a series of strokes leaves areas of dead brain cells, producing step-by-step degeneration of mental ability, with each step occurring abruptly after a stroke. About 5 to 10 percent of all cases of dementia are cerebrovascular, and about 10 percent are due to a combination of Alzheimer's and repeated strokes (Gatz, Kasl-Godley, & Karel, 1996; Ritchie, Kildea, & Robine, 1992).

A hereditary tendency toward cerebrovascular dementia appears to be indirect, through high blood pressure, cardiovascular disease, and diabetes, each of which increases the risk of stroke. But many environmental influences, including cigarette smoking, heavy alcohol use, high salt intake, very low dietary protein, obesity, and stress, also heighten stroke risk, so cerebrovascular dementia appears to be the combined result of genetic and environmental forces.

Because of their susceptibility to cardiovascular disease, more men than women have cerebrovascular dementia by their late sixties. Women are not at great risk until after age 75 (Miller & Morris, 1993). The disease also varies among countries. For example, deaths due to stroke are high in Japan. Although a low-fat diet reduces Japanese adults' risk of cardiovascular disease, high intake of alcohol and salt and a diet very low in animal protein increase the risk of stroke. As Japanese consumption of alcohol and salt declined and of meat rose in recent decades, the rate of cerebrovascular dementia and stroke-caused deaths dropped, although they remain higher than in other developed nations (Goldman & Takahashi, 1996; Myers, 1996).

Although Japan presents a unique, contradictory picture (a culture where cardiovascular disease is low and

CAREGIVING CONCERNS
Communicating Effectively with Alzheimer's Victims

PROBLEM BEHAVIOR	COMMUNICATION STRATEGY
Asking the same question over and over, telling ridiculous stories, or making socially inappropriate comments	Distract the person with other topics of discussion and activities; avoid arguing, responding with anger, or rebuking the person.
Failing to keep an appointment or complete a task	Use reminders and lists; make requests close to the time the task needs to be done; always ask for the desired behavior in the same setting (e.g., eating in the kitchen, dressing in the bedroom).
Denying memory problems, blaming others for mistakes	Do not force the person to face memory problems; respond with sympathy to the threat memory loss poses.
Reacting with anger to assistance from others	Avoid emphasizing the person's weaknesses; provide help in a kind manner; encourage and reinforce small successes and acceptance of help.
Restlessness, agitation, inability to sit still	Use a calm, reassuring approach; reduce noise and other distractions; involve the person in activities that relieve excess energy.
Failing to recognize familiar people, objects, and places	Agree that things look different, and calmly remind the person of people and objects in the situation.

Source: Gruetzner, 1992.

stroke is high), in most cases cerebrovascular dementia is caused by atherosclerosis. Prevention is the only effective way to stop the disease. Signs that a stroke might be coming are weakness, tingling, or numbness in an arm, a leg, or the face; sudden vision loss or double vision; speech difficulty; and severe dizziness and imbalance. Doctors may prescribe drugs to reduce the tendency of the blood to clot. Once strokes occur, paralysis and loss of speech, vision, coordination, memory, and other mental abilities are common.

■ MISDIAGNOSED AND REVERSIBLE DEMENTIA. Careful diagnosis of dementia is crucial, since other disorders can be mistaken for it. And some forms of dementia can be treated and a few reversed.

Depression is the disorder most often misdiagnosed as dementia. The depressed (but not demented) older adult is likely to exaggerate his or her mental difficulties, whereas the demented person minimizes them and is not fully aware of cognitive declines. Less than 1 percent of people over age 65 are severely depressed and another 2 percent are moderately depressed—rates that are lower than for young and middle-aged adults (Reiger et al., 1988). Most older people have probably learned how to adjust what they expect from life, resulting in fewer feelings of worthlessness. However, as we will see in Chapter 18, depression rises with age. It is often related to physical illness and pain and can lead to cognitive deterioration (Magni & Frisoni, 1996; Nussbaum et al., 1995). As at younger ages, the support of family members and friends; antidepressant medication; and individual, family, and group therapy can help relieve depression. How-

ever, the elderly in general and low-income, ethnic minorities in particular are unlikely to seek mental health services. This increases the chances that depression will deepen and be confused with dementia (Padgett et al., 1994).

The older we get, the more likely we are to be taking drugs that might have side effects resembling dementia. For example, some medications for coughs, diarrhea, and nausea inhibit the neurotransmitter acetylcholine, leading to Alzheimer's-like symptoms. Because tolerance for drugs decreases with age, these reactions are intensified in late adulthood. In addition, some diseases can cause temporary memory loss and mental symptoms, especially among the elderly, who often become confused and withdrawn when ill. Treatment of the underlying illness relieves the problem. Finally, environmental changes and social isolation can trigger mental declines. When supportive ties are restored, cognitive functioning usually bounces back (Greutzner, 1992).

HEALTH CARE

Health care professionals and lawmakers in industrialized nations worry about the economic consequences of rapid increase in the elderly population. Rising government-supported health care costs and demand for certain health care services, particularly long-term care, are of greatest concern.

■ COST OF HEALTH CARE FOR THE ELDERLY. Americans age 65 and older, who make up 12.5 percent of the population, account for 30 percent of federal health care expenditures. According to current estimates, the cost of Medicare (government-sponsored health insurance for the elderly)

SQUARING OF THE POPULATION PYRAMID AND INTERGENERATIONAL INEQUITY

Picture the population as a building, with the youngest at the bottom and the oldest at the top. Through most of the twentieth century, this structure resembled a pyramid, with fewer and fewer people at older ages. Aging of the "baby boom" generation, combined with declining birthrates and greater longevity, will cause a "senior boom" in the first half of the twenty-first century. By the year 2030, the percentage of young, middle-aged, and elderly people in the United States is expected to be about equal (see Figure 17.4). This **squaring of the population pyramid** has raised concerns about whether younger, employed citizens will be able to support the growing number of older, dependent

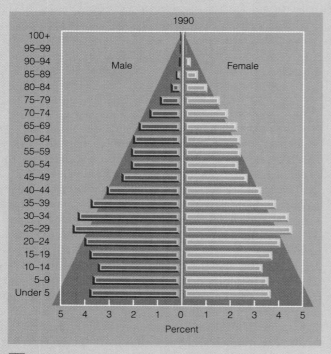

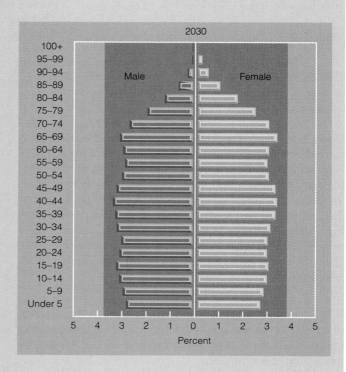

FIGURE 17.4

Population pyramids for 1990 and 2030. Notice the squaring of the population pyramid as the percentage of older adults rises and the percentage of young people shrinks. By 2030, the youngest and oldest generations are expected to be equally represented. *(From U.S. Bureau of the Census, 1992b.)*

is expected to double by the year 2020 and nearly triple by 2040 as the "baby boom" generation reaches late adulthood and average life expectancy extends further (Doty, 1992; Torrey, 1992).

Medicare expenses rise steeply with age. People age 80 and older receive, on the average, 77 percent more benefits than do younger senior citizens. Most of this increase is due to a need for long-term care—in hospitals and nursing homes—because of the age-related rise in disabling chronic diseases and acute illnesses.

Currently, the U.S. government pays about half of medical expenses for the aged. Some people argue that this degree of cost sharing must be drastically reduced. Otherwise, the elderly will become a burdensome responsibility, unfairly consuming national resources that ought to be given to other sectors of the population. Should society

seniors due to increasingly costly Social Security, Medicare, and Medicaid benefits.

In addition, some policymakers and citizens argue that population trends have led to **intergenerational inequity,** with the elderly receiving more than their fair share of public resources at the expense of the very young, who cannot vote and lobby for their needs (see Chapter 2). To create greater equity, a few people want to limit health care to aging individuals—a large part of the growing "public burden" of the elderly. They argue that after completion of a "natural life span," somewhere in the late seventies or early eighties, medical treatment should be provided only to relieve suffering, not to save lives (Callahan, 1987, 1990).

Even if we set aside serious ethical issues, age-related rationing of health services would be difficult to implement. Because people age at widely varying rates, it would be impossible to reach consensus on what age signifies the completion of a "natural life span." In addition, savings would largely come from a drop in very expensive, acute health care interventions. Yet the popular belief that the elderly more often use the most aggressive, costly medical technologies is wrong. These procedures are applied fairly evenly across age groups (Binstock, 1992).

Public policies that pit one age group against another promote intergenerational conflict and age discrimination. Recall from Chapter 16 that younger and older citizens currently recognize their interdependence and support benefits to one another. Indeed, throughout this book, we have seen how enhancing both younger and older citizens' well-being contributes to the welfare of all generations.

Other, more promising approaches to freeing up public resources so they are available for pressing social problems have been suggested:

- As the health and longevity of older Americans continues to increase, they are likely to remain in the work force for several additional years (U.S. Bureau of the Census, 1996). Responding to this trend, the age of initial eligibility for full Social Security benefits is scheduled to rise gradually from age 65 to 67 early in the next century. A similar suggestion for postponing the age of Medicare eligibility to age 67 or 70 has been made.

- For every dollar spent on medical care for Alzheimer's disease victims, less than two-tenths of one cent is spent on research aimed at finding a cure. Allocating additional resources to research on this and other major disabling diseases is the best way to prevent elderly health care costs from skyrocketing in the future (Schneider & Guralnik, 1990).

- Justice between the advantaged and disadvantaged at all ages is a far more promising principle of equity than justice between age groups, which trades off the value of one human life for another and diverts us from the larger goal of helping people of any age who are in need.

TRY THIS...

- Select one public policy for children and adolescents (such as government-subsidized day care or vocational training programs for non-college-bound youths), one public policy for young and middle-aged adults (such as employment leave to deal with family emergencies), and one public policy for the elderly (such as Social Security or Medicare). Explain how each benefits members of other generations.

limit health care payments to older people as a way of achieving a more equitable distribution of resources to different age groups? To consider this controversial issue, read the Social Issues box above.

■ LONG-TERM CARE. When Ida moved into Ruth's home, Ruth promised that she'd never place Ida in an institution. But as Ida's condition worsened and Ruth faced health problems of her own, she couldn't keep her word. Ida needed round-the-clock medical care. Reluctantly, Ruth placed her in a nursing home.

As Figure 17.5 reveals, advancing age is strongly associated with use of long-term care services, especially nursing homes. Among disorders of aging, hip fracture and dementia, especially Alzheimer's disease, most often lead to nursing home placement. Greater use of nursing homes is

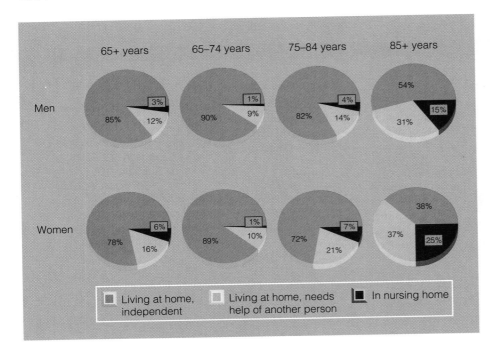

FIGURE 17.5

Percentage of elderly men and women by age who live at home independently; live at home but need the help of another person for tasks of daily living; and reside in a nursing home. Although the majority of older adults of all ages live at home, need for long-term assistance and nursing home care rises sharply with age. *(From National Center for Health Statistics, 1995; U.S. Bureau of the Census, 1996.)*

also prompted by loss of informal caregiving support through widowhood (which mostly affects women) and aging of adult children and other relatives. Of people age 85 and older in nursing homes, 69 percent have children who themselves are senior citizens (Doty, 1992).

Overall, only 5 percent of Americans age 65 and older are institutionalized, a rate about half that of other industrialized nations, such as Canada, the Netherlands, and Sweden. These countries provide more generous public financing of health services for the aged, including institutional care. In the United States, unless nursing home placement follows hospitalization for an acute illness, older adults must pay for it until their resources are exhausted. At that point, Medicaid (health insurance for the poor) takes over. Consequently, the largest users of nursing homes in the United States are people with very low and high incomes (Torrey, 1992). Middle-income elderly and their families are more likely to try to protect savings from being drained by high nursing home costs.

Nursing home use also varies across ethnic groups. For example, among people age 75 and older, Caucasian Americans are one-and-a-half times more likely to be institutionalized than are African-Americans. Large, closely knit extended families mean that over 70 percent of African-American elders do not live alone and over one-third reside with their adult children (Gibson & Jackson, 1992). Similarly, Asian Americans use nursing homes less often than do Caucasian Americans because of greater availability of family members to provide long-term care—a factor that also accounts for low rates of nursing home placement in Japan (Doty, 1992; McCormick et al., 1996). Neverthe-

less, at least 60 to 80 percent of all long-term care for older adults is provided by family members in Australia, Canada, the United States, and Western Europe. As we noted in Chapter 16, it is a myth that adult children, as well as other relatives, abandon the elderly.

To reduce institutionalized care of the elderly, some experts have advocated alternatives, such as publicly funded in-home help for family caregivers. However, evidence from other countries reveals that government financing of home care does not lower rates of nursing home placement much because the two types of care largely serve different populations. The elderly cared for at home are less impaired than those in nursing homes (Doty, 1992).

Still, public support for home care is important for relieving the financial and emotional burdens on family caregivers and enhancing the well-being of the elderly. In nursing homes, steps can be taken to improve the quality of services. For example, the Netherlands has established separate facilities for patients with mental and physical disabilities because each has distinctly different needs. And every elderly person, no matter how disabled, benefits from opportunities to maintain existing strengths and acquire new skills, which (as we will see in the next section) can compensate for declines. Among institutionalized elderly, health, sense of personal control, and gratifying social relationships strongly predict life satisfaction. These aspects of living are vital for all older people, whether they are institutionalized or not (Baltes, Wahl, & Reichert, 1992).

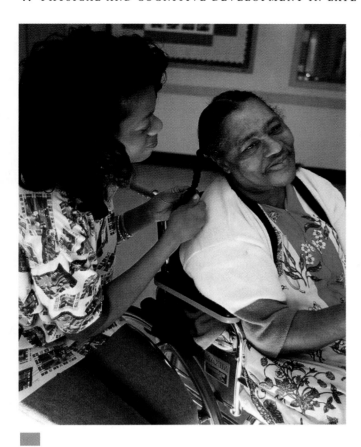

Health, a sense of personal control, and gratifying social relationships strongly predict life satisfaction for elderly people, even when they live in nursing homes. (Sondra Dawes/The Image Works)

BRIEF REVIEW

Older adults are generally optimistic about their health, a view that promotes health-enhancing behaviors. Nevertheless, poverty and negative lifestyle factors have interfered with the goal of compressing morbidity in late adulthood. Vitamin-mineral supplements, low dietary fat, and endurance and weight-bearing exercise promote many aspects of health in the elderly. Sexual enjoyment can continue into very old age.

Illness and disability increase in the final years of life. A variety of extreme conditions result from the combined impact of heredity defects and negative environmental influences. Among these are cardiovascular disease, cancer, stroke, emphysema, osteoarthritis and rheumatoid arthritis, adult-onset diabetes, and unintentional injuries. Prevention and treatment are possible for each.

Besides physical disabilities, dementia rises in old age. Alzheimer's disease, the most common form, is associated with structural and chemical alterations in the brain, often runs in families, and is linked to several genetic markers as well as environmental factors. In cerebrovascular dementia, a series of strokes causes abrupt, step-by-step mental impairment. In addition to heredity, many negative lifestyle factors heighten the risk of stroke. Non-dementia-related depression, medication side effects, and mental decline due to social isolation can easily be mistaken for dementia.

A rapid rise in the elderly population has led to widespread concern about public health care costs. Institutional care of the elderly is lower in the United States than in many other industrialized nations because of less generous public financing. Most older adults, especially ethnic minorities, receive long-term care from family members.

ASK YOURSELF . . .

■ *List as many ways as you can think of to achieve the public health goal of compression of morbidity in late life.*

■ *Marissa complained to a counselor that at age 68, her husband Wendell stopped initiating sex and no longer touched, stroked, or cuddled her. Why might Wendell have ceased to interact sexually? What interventions—both medical and educational—could be helpful to Marissa and Wendell?*

■ *Explain how depression can combine with physical illness and disability to promote cognitive deterioration in the elderly. Should cognitive declines due to physical limitations and depression be called dementia? Why or why not?*

COGNITIVE DEVELOPMENT IN LATE ADULTHOOD

When Ruth complained to her doctor about memory and verbal expression difficulties, she voiced common concerns about cognitive functioning in late life. Decline in speed of processing, underway throughout the adult years, is believed to affect many aspects of cognition in old age. In Chapter 15, we noted that reduced efficiency of thinking compromises attention, amount of information that can be held in working memory, use of memory strategies, and retrieval from long-term memory. These decrements continue in the final decades of life.

Recall that the more a mental ability depends on fluid intelligence (biologically based information-processing skills), the earlier it starts to decline. In contrast, mental abilities that rely on crystallized intelligence (culturally based knowledge) increase for a longer time. (Return to Figure 15.4 on page 504 to review these trends.) Gains in crystallized intelligence depend on continued opportunities to enhance cognitive skills. When these are available, crystallized abilities—general information and expertise in specific endeavors—can offset losses in fluid intelligence. Take a moment to list the various ways that adults compensate for intellectual declines, discussed in Chapter 15.

Look again at Figure 15.4 on page 504. In advanced old age, decrements in fluid intelligence eventually limit what people can accomplish with the help of cultural supports, including a rich background of experience, knowledge of how to remember and solve problems, and a stimulating daily life. Consequently, crystallized intelligence begins to decline as well (Baltes, 1997; Schaie, 1996b).

Overall, loss outweighs improvement and maintenance as people move closer to the end of life. But plasticity of development is still possible (Baltes & Carstensen, 1996). Research reveals large individual differences in cognitive functioning among the elderly—greater than at any other time of life (Hultsch & Dixon, 1990; Morse, 1993). Besides fuller expression of genetic and lifestyle influences, increased freedom to pursue self-chosen courses of action —some that enhance and others that undermine cognitive skills—may be responsible.

How can older adults make the most of their cognitive resources? According to one view, elders who sustain high levels of functioning engage in **selective optimization with compensation** (Baltes, 1997). That is, they narrow their goals, *selecting* personally valued activities as a way of *optimizing* (or maximizing) returns from their diminishing energy. They also come up with new ways of *compensating* for losses. One day, Ruth and Walt listened to a replay of an interview with 80-year-old concert pianist Arthur Rubenstein, who was asked how he managed to sustain such extraordinary piano playing, despite his advanced age. Rubenstein used each of the strategies just mentioned. He stated that he was *selective;* he played fewer pieces. This enabled him to *optimize* his energy; he could practice each piece more. Finally, he developed new, *compensatory* techniques for a decline in playing speed. For example, before a fast passage, he played extra slowly, so the fast section appeared to his audience to move more quickly.

As we review major changes in memory, language processing, and problem solving, we will consider ways that older adults can optimize and compensate in the face of

Concert pianist Arthur Rubenstein, performing before a sold-out audience in New York City at age 89, sustained his extraordinary piano playing through *selective optimization with compensation.* He was selective, playing fewer pieces so he could optimize his energy. He compensated for a decline in playing speed by playing extra slowly before a fast passage, so the fast section appeared to his audience to move more quickly. (UPI/Corbis-Bettmann)

declines. We will also see that certain abilities that depend on extensive life experience, not processing efficiency, are sustained or increase in old age. Last, we take up programs that recognize the elderly as lifelong learners—empowered by new knowledge, just as they were at earlier periods of development.

MEMORY

As older adults take in information more slowly and find it harder to apply strategies and retrieve relevant knowledge from long-term memory, the chances of memory failure increase (Salthouse, 1991b). A reduced capacity to hold material in working memory while operating on it means that memory problems are especially evident on complex tasks.

DELIBERATE VERSUS AUTOMATIC MEMORY

"Ruth, you know that movie we saw—with the little 5-year-old boy who did such a wonderful acting job. I'd like to suggest it to Dick and Goldie. Do you recall the name of it?" asked Walt.

"I can't think of it, Walt. We've seen several movies lately, and that one just doesn't ring a bell. Which theater was it at, and who'd we go with? Tell me more about the little boy, and maybe I'll think of it."

"It was at either University Cinema or TriValley Theater. We went by ourselves. Wait, maybe we saw it with John and Barbara," Walt replied.

Although all of us have had memory failures like this from time to time, difficulties with recall rise in old age. When Ruth and Walt watched the movie, their slower cognitive processing meant that they retained fewer details. And because their working memories could hold less at once, they attended poorly to context—who they went with and where they saw the movie (Craik & Jacoby, 1996; Spencer & Raz, 1995). When we try to remember, contextual cues serve as important retrieval cues. Because older adults take in less about a stimulus and its context, they sometimes cannot distinguish an experienced event from one they imagined (Hashtroudi, Johnson, & Chrosniak, 1990).

A few days later, Ruth saw a clip from the movie on TV, and she recognized it immediately. Compared to recall, recognition memory suffers far less in late adulthood because a multitude of environmental supports for remembering are present (Craik & Jennings, 1992). Greater decrement in recall than recognition suggests that providing more cues for remembering would enhance older adults' performance.

Age-related memory declines are largely limited to tasks that require deliberate processing. Because recognition is a fairly automatic type of memory that demands little mental effort, it does not change much in old age. Consider another automatic form of memory called **implicit memory,** or memory without conscious awareness. In a typical implicit memory task, you would be asked to fill in a word fragment (such as t _ _ k) after being shown a list of words. You would probably complete the sequence with a word you had just seen ("task") rather than other words ("took" or "teak"). Notice that you engaged in recall without trying to do so.

Age differences in implicit memory are much smaller than in explicit, or deliberate, memory. When memory depends on familiarity rather than conscious use of strategies, it is spared from impairment in old age (Kausler, 1994; Smith, 1996). The memory problems elders report—for names; places they put important objects; directions for getting from one place to another; and (as we will see later) appointments and medication schedules—all place substantial demands on their more limited working memories.

REMOTE MEMORY

Although older people often say that their **remote memory,** or very long-term recall, is clearer than their memory for recent events, research does not support this conclusion. In several studies, adults ranging in age from their twenties to their seventies were asked to recall names of grade school teachers and high school classmates and Spanish vocabulary from high school—information very well learned early in life. Memory declined rapidly for the first 3 to 6 years, then changed little for the next 20 years. After that,

additional modest forgetting occurred (Bahrick, 1984; Bahrick, Bahrick, & Wittlinger, 1975; Schonfeld, 1969).

How about *autobiographical memory,* or memory for personally meaningful events, such as what you did on your first date or how you celebrated your college graduation? To test for this type of memory, researchers typically give a series of words (such as *book, machine, sorry, surprised*) and ask adults to report a personal memory associated with each. Regardless of age, both remote and recent events are recalled more frequently than intermediate events, with recent events mentioned most often (Hyland & Ackerman, 1988; MacKinnon & Squire, 1989).

Why do older adults continue to forget the names of teachers and access recent autobiographical events more readily than remote ones? The likely answer is interference produced by years of additional experience (Kausler, 1994). As we accumulate more everyday memories, some resemble others. As a result, certain early memories become less clear than they once were.

PROSPECTIVE MEMORY

Elderly people often complain that they have become more absentminded about daily events. Because Ruth and Walt knew they were more likely to forget an appointment, they often asked about it repeatedly. "Sybil, what time is our dinner engagement?" Walt said several times during the 2 days before the event. His questioning was not a sign of dementia. He simply wanted to be sure to remember an important date.

So far, we have considered various aspects of *retrospective memory* (or remembrance of things past). **Prospective memory** refers to remembering to engage in planned actions at an appropriate time in the future. The amount of mental effort required determines whether older adults have trouble with prospective memory. Remembering the dinner date was challenging for Walt because he typically ate dinner with his daughter on Thursday evenings at 6 P.M., but this time, dinner was set for Tuesday at 7:15 P.M.

On laboratory tasks requiring prospective memory, such as remembering to ask for a red pen at a particular point in an experiment or to date a questionnaire after filling it out, adults over age 70 are more likely to forget the instructions (Dobbs & Rule, 1987). The difficulty does not always appear in real life, since older adults are better at setting up reminders for themselves, such as a buzzer ringing in the kitchen or a note tacked up in a prominent location. When trying to remember a future activity, younger adults rely more on strategies like rehearsal, older adults on external aids to memory (Loewen, Shaw, & Craik, 1990). In this way, the elderly compensate for their reduced-capacity working memories and the challenge of dividing attention between what they are doing now and what they must do in the future.

LANGUAGE PROCESSING

anguage and memory skills are closely related. In language comprehension (understanding the meaning of spoken or written prose), we recollect what we have heard or read without conscious awareness. Like implicit memory, language comprehension changes very little in late life as long as conversational partners do not speak very quickly (Wingfield & Stine, 1992). In contrast, two aspects of language production show age-related losses.

The first is retrieving words from long-term memory. When conversing with others, Ruth and Walt sometimes had difficulty coming up with just the right word to convey their thoughts. Consequently, their speech contained a greater number of pronouns and other unclear references than it did at younger ages. They also spoke more slowly and paused more often, partly because they needed time to search their memories for certain words (MacKay & Abrams, 1996).

Second, planning what to say and how to say it is harder in late adulthood. As a result, Walt and Ruth displayed more hesitations, false starts, word repetitions, and sentence fragments as they aged. Their statements were also less well organized than before (Kemper, 1992; Kemper, Kynette, & Norman, 1992).

What explains these changes? Once again, age-related limits on working memory seem to be responsible. Because less information can be held at once, the elderly have difficulty coordinating the multiple tasks required to produce speech. As a result, they sometimes have trouble remembering the nonverbal information they want to communicate, putting it into words, and conveying it in a coherent fashion.

As storyteller Willard Lape, Jr., plays the part of Salty Sam for attentive elementary school students at the Erie Canal Museum in Syracuse, he is likely to enrich his tales with inferences and moral lessons grounded in extensive life experience. (The Image Works)

As with memory, older adults develop compensatory techniques for their language production problems. For example, they simplify their grammatical structures so they can devote more effort to retrieving words and organizing their thoughts. In this way, they convey their message in more sentences, sacrificing efficiency for greater clarity (Kemper, Kynette, & Norman, 1992).

The elderly also compensate by representing information they want to communicate in terms of gist rather than details (Jepson & Labouvie-Vief, 1992). For example, when Walt told his granddaughter Marci fairy tales, he left out many concrete facts. Instead, he included personal inferences and a moral lesson—elements that appear less often in the storytelling of younger adults. Here is Walt's rendition of *Sleeping Beauty:*

> An evil fairy condemns Sleeping Beauty to death. But a kind fairy changes the curse from death to sleep. Then a handsome prince awakens the girl with a kiss. So you see, Marci, both good and bad exist in the world. The bad things instill in us the need to think of and care for others.

Older adults often make the best of their limited working memories by extracting the essence of a message. Then they enrich it with symbolic interpretations, drawing on their extensive life experience.

PROBLEM SOLVING

roblem solving is yet another cognitive skill that illustrates how aging brings not only deterioration, but important adaptive changes. In late adulthood, both traditional problem solving, which lacks a real-life context (as in playing Twenty Questions), and practical problem solving of the kind we discussed in Chapter 15 show declines. Older adults seem less willing to seek further information when a problem calls for it, and their memory limitations make it hard to keep all relevant facts in mind when dealing with a complex problem (Sinnott, 1989).

Yet the problematic situations the elderly encounter are often different from those experienced at earlier ages. Being retired, most do not have to deal with problems in the workplace. Even the social problems they confront at home may be reduced. Their children are typically grown and living on their own, and their marriages have endured long enough to have fewer difficulties (Heidrich & Denney, 1994). Instead, major concerns involve managing activities of daily living, such as preparing nutritious meals, handling finances, and attending to health concerns. Surveys in Germany and the United States reveal that older adults spend a third to half of a typical day on these issues (Willis, 1996).

How do the elderly solve problems of daily living? So far, most research has focused on health problem solving. Findings indicate that older adults make quicker decisions about whether they are ill and seek medical care sooner. In contrast, young and middle-aged adults are more likely to adopt a "wait and see" approach in favor of gathering more facts, even when the health problem is serious (Leventhal et al., 1993; Meyer, Russo, & Talbot, 1995).

This swift response of the elderly is interesting in view of their slower cognitive processing. Perhaps years of experience in coping with illnesses enables them to draw on extensive personal knowledge and move ahead with greater certainty (Labouvie-Vief & Hakim-Larson, 1989). Acting decisively when faced with health risks is particularly adaptive in old age.

WISDOM

We have seen that a wealth of life experience enhances the verbal messages and problem solving of the elderly. It also underlies another capacity believed to reach its height in old age: **wisdom.** When researchers ask people to describe wisdom, most mention breadth and depth of practical knowledge, ability to reflect on that knowledge in ways that make life more bearable and worthwhile, emotional maturity, and the integrative form of creativity we discussed in Chapter 15 (see page 508) (Simonton, 1990).

During her college years, Ruth and Walt's granddaughter Marci telephoned with a pressing personal dilemma. Ruth's advice reflected the elements of wisdom just mentioned. Unsure whether her love for her boyfriend Ken would endure, Marci began to date another student after Ken moved to another city to attend medical school. "I can't stand being pulled in two directions. I'm thinking of calling Ken and telling him about Steve," she exclaimed. "Do you think I should?"

"This is not a good time, Marci," Ruth advised. "You'll break Ken's heart before you've had a chance to size up your feelings for Steve. And you said Ken's taking some important exams in two weeks. If you tell him now and he's distraught, it could affect the rest of his life."

 Cultures around the world assume that age and wisdom go together. In village and tribal societies, the most important social positions, such as chieftain and shaman (religious leader), are reserved for the old. Similarly, in industrialized nations, people over age 65 are chief executive officers of large corporations, high-level religious leaders, members of legislatures, and supreme court justices. What explains this widespread trend? According to an ethological view, the genetic program of our species grants health, fitness, and strength to the young. Culture tames this youthful advantage in physical power with the insights of the old, ensuring balance and interdependence between generations (Assmann, 1994; Csikszentmihalyi & Rathunde, 1990).

Only a few efforts have been made to investigate the development of wisdom. In one series of studies, participants' responses to important but uncertain real-life situations revealed that age was no guarantee of wisdom. The wisest choices were made by a very small number of adults —some young, some middle aged, and some old. But type of life experience made a difference. People who had spent their careers in human services, where they received extensive training and practice in grappling with human problems, were well represented among the wise (Smith, Staudinger, & Baltes, 1994; Staudinger, Smith, & Baltes, 1992). Yet wisdom was not limited to people with formal training. Active elderly citizens who were highly regarded in their community and who came from many walks of life also excelled at wisdom tasks (Baltes et al., 1995).

FACTORS RELATED TO COGNITIVE CHANGE

As we saw in Chapter 15, a mentally active life— above-average education, stimulating leisure pursuits, community participation, and a flexible personality—predicts maintenance of mental abilities into advanced old age. In late adulthood, health status becomes an increasingly strong predictor of intellectual performance. A wide variety of chronic conditions, including vision and hearing impairments, cardiovascular disease, osteoporosis, and arthritis, are associated with cognitive declines (Baltes & Lindenberger, 1997; Schaie 1996a). But we must be cautious in interpreting this link between physical and cognitive deterioration. The relationship may be exaggerated by the fact that brighter adults are more likely to engage in health-protective behaviors, which postpone the onset of serious disease.

Retirement also affects cognitive change, both positively and negatively. When people leave routine jobs for stimulating leisure activities, outcomes are favorable. In contrast, retiring from a highly complex job without developing challenging substitutes accelerates intellectual declines (Schaie, 1996a).

Near the end of life, cognitive decrements are related to distance to death rather than chronological age. In the year before Walt died, those close to him noticed that he had become less active and more withdrawn. In the company of friends, he talked and moved less. At home, he spent more time looking out the window instead of immersing himself in creative writing and gardening.

Terminal decline refers to a steady, marked decrease in cognitive functioning prior to death. Researchers are not certain if it is limited to a few aspects of intelligence or if it affects all aspects, signifying general deterioration. Furthermore, studies vary greatly in estimated length of terminal decline. Some report that it lasts only 1 or 2 years; others, that it extends for as much as 10 years. The average is about 5 years (Berg, 1987; Siegler, McCarty, & Logue, 1982; White & Cunningham, 1988).

A mentally active life predicts maintenance of mental abilities into advanced old age. These men, intent on a game of dominoes, are enjoying a stimulating leisure pursuit and lively social participation—both key ingredients of favorable cognitive development. (Jeff Greenberg/The Picture Cube)

Perhaps the reason for these conflicting findings is that there are different kinds of terminal decline. One type might arise from disease processes. Another might be part of a general biological breakdown due to normal aging (Berg, 1996). At present, all we know for sure is that an extended, steep fall-off in cognitive functioning is a sign of loss of vitality and impending death.

COGNITIVE INTERVENTIONS

For most of late adulthood, cognitive declines are gradual. Although aging of the brain contributes to them, recall from our earlier discussion that the brain can compensate by growing new neural fibers. Furthermore, some cognitive decrements may be due to disuse of particular skills rather than biological aging. If plasticity of development is possible in old age, then interventions that train the elderly in cognitive strategies should at least partially reverse the age-related declines we have discussed.

The Adult Development and Enrichment Project (ADEPT) is the most extensive cognitive intervention program conducted to date. By using participants in the Seattle Longitudinal Study (see Chapter 13, page 439, and Chapter 15, page 504), researchers could do what no other investigation had yet done: assess the effects of cognitive training on long-term development (Schaie, 1996b).

Intervention began with adults over age 64, some of whom had maintained their scores on two mental abilities (inductive reasoning and spatial orientation) over the previous 14 years and others of whom had shown declines. After just five 1-hour training sessions in relevant cognitive skills, two-thirds of the participants improved their performance. The gains for decliners were dramatic. Forty percent returned to the level at which they had been functioning 14 years earlier!

A follow-up after 7 years revealed that although the scores of trained adults dropped somewhat, they were still doing better than untrained controls. Finally, "booster" training at this time led to further gains, although these were not as large as the earlier gains.

In other short-term studies, training resulted in improvements in memory and problem solving among the elderly (Kotler-Cope & Camp, 1990; Willis & Schaie, 1994). Clearly, a wide range of cognitive skills can be enhanced in old age. A vital goal is to transfer intervention from the laboratory to the community, weaving it into the daily experiences of elderly people. As we will see in the next section, a promising approach is to provide older adults with well-designed, highly interesting learning experiences in which cognitive training is an integral part.

LIFELONG LEARNING

Think about the competencies that older adults need to live in our complex, changing world. They are the same as those of younger people—communicating effectively through spoken and written systems; locating information, sorting through it, and selecting what is needed; using modern mathematical techniques, such as estimation; planning and organizing activities, including making good use of time and resources; mastering new technologies; and understanding past and current events and the relevance of each to their own lives. The elderly also need to develop new, problem-centered coping strategies—ways to sustain health and operate their households efficiently and safely; and self-employment skills for those who must continue their work lives.

Due to better health and earlier retirement, participation of the elderly in continuing education has increased substantially over the past few decades. Successful programs include a wide variety of offerings responsive to the diversity of senior citizens and teaching methods suited to their developmental needs (Knox, 1993).

TYPES OF PROGRAMS

One summer, Walt and Ruth attended an Elderhostel at a nearby university. After moving into a dormitory room, they joined thirty other senior citizens for 2 weeks of morning lectures on Shakespeare; afternoon visits to points of interest; and evening performances of plays at a nearby Shakespeare festival.

Elderhostel programs attract over a quarter million older Americans annually. Local educational institutions serve as hosts, combining stimulating 1- to 3-week courses taught by experts with recreational pursuits. Some programs make use of community resources through classes on local ecology or folk life. Others involve travel abroad. Still others focus on innovative topics and experiences—

Elderly participants in continuing education, like these women learning to explore the Internet at a local college, often come to see themselves differently. They abandon deeply ingrained stereotypes of aging as they realize that adults in their seventies and eighties can still engage in complex learning. (Wyman/ Monkmeyer Press)

writing one's own life story, discussing contemporary films with screenwriters, whitewater rafting, and Chinese painting and calligraphy.

Similar educational programs have sprung up around the world. Originating in France, the University of the Third Age[3] offers community-sponsored courses for elders that vary widely in content and style of presentation. These include open lectures, access to established university courses, workshops on special topics, excursions, and physical health programs. The model has spread to many countries. In Australia and Great Britain, elders often do the teaching, based on the idea that experts of all kinds retire (Swindell & Thompson, 1995).

Some programs foster intergenerational relations and community involvement. In Austria, for example, training in modern math, foreign languages, and other subjects is offered to grandparents to enable them to help grandchildren, in whose education they are encouraged to participate.

---

[3] The "third age" refers to the period after the "second age" of midlife, when older people are freed of work and parenting responsibilities and have more time to invest in lifelong learning.

Other routes to education among the elderly include returning to college. In the United States, nearly 80,000 adults age 65 and older take courses in institutions of higher education each fall (U.S. Department of Education, 1996). Additional learning opportunities are available through libraries, museums, religious institutions, community agencies, and private businesses.

Participants in programs mentioned so far tend to be active, well educated, and financially well off. Much less is available for elders with little education and limited income. Community senior centers with inexpensive offerings related to everyday living attract more low-income people than do programs such as Elderhostel (Knox, 1993). Regardless of course content and which seniors attend, using the techniques summarized in the Caregiving Concerns table on the page 582 increases the effectiveness of instruction for older adults.

■ BENEFITS OF CONTINUING EDUCATION

Elderly participants in continuing education report a rich array of benefits. These include learning new facts, understanding new ideas in many disciplines, making new friends, and developing a broader perspective on the world (Long & Zoller-Hodges, 1995). Furthermore, seniors come to see themselves differently. Many arrive with deeply ingrained stereotypes of aging, which they abandon when they realize that adults in their seventies and eighties, including themselves, can still engage in complex learning. In Elderhostel courses, participants with the least education report learning the most, an argument for recruiting less economically privileged people into these programs (Brady, 1984).

The educational needs of senior citizens are likely to be given greater attention in coming decades, as their numbers grow and they assert their right to lifelong learning. Once this happens, false stereotypes, such as "the elderly are too old to learn" and "education is for the young," are likely to weaken and, perhaps, disappear.

ASK YOURSELF . . .

■ *When Ruth couldn't recall which movie Walt was thinking about, she asked several questions: "Which theater was it at, and who'd we go with? Tell me more about the little boy [in the movie]." Which memory deficits of aging is Ruth trying to overcome?*

■ *Ruth complained to her doctor that she had trouble finding the right words to explain to a delivery service how to get to her house. What cognitive changes account for Ruth's difficulty?*

■ *Describe cognitive functions that are maintained and that improve in late adulthood. What aspects of aging contribute to them?*

Increasing the Effectiveness of Instruction for Older Adults

TECHNIQUE	DESCRIPTION
Providing a positive learning environment	Some elders have internalized negative stereotypes of their abilities and come to the learning environment with low self-esteem. A supportive group atmosphere, in which the instructor is viewed as a colleague, helps convince older adults that they can learn.
Allowing ample time to learn new information	Rate of learning varies widely among older adults, with some mastering new material at a fairly slow rate. Presenting information over several sessions or allowing for self-paced instruction aids mastery.
Presenting information in a well-organized fashion	Older adults do not organize information as effectively as younger adults. Material that is outlined, presented, and then summarized enhances memory and understanding. Introduction of irrelevant information and digressions make a presentation harder to comprehend.
Giving information related to elders' experiences	Relating new material to what has already been learned by drawing on elders' experiences and giving many vivid examples enhances recall.

Source: Thompson, 1992.

SUMMARY

PHYSICAL DEVELOPMENT IN LATE ADULTHOOD

LONGEVITY

Discuss aging and longevity among older adults.

- Vastly different rates of aging are apparent in late adulthood. In terms of **functional age,** older adults of any chronological age may be **young-old** or **old-old,** depending on their physical condition.

- In industrialized nations, the number of people aged 65 and over has risen dramatically over the twentieth century, a trend expected to continue. With advancing age, women outnumber men by a greater margin, but sex differences in average life expectancy decline. Differences between middle-class whites and low-income minorities also diminish until around age 80, when a **life expectancy crossover** occurs.

- Longevity runs in families, but environmental factors become increasingly important with age.

PHYSICAL CHANGES

Describe changes in the nervous system and the senses in late adulthood.

- Loss of neurons occurs throughout the cerebral cortex, with the visual, auditory, and motor areas most affected. However, the brain compensates by forming new synapses. The autonomic nervous system functions less well and releases more stress hormones.

- Older adults tend to suffer impaired eyesight, including **cataracts** and **macular degeneration.** Visual deficits affect elders' self-confidence and everyday behavior and can be very isolating.

- Hearing difficulties are more common than visual difficulties, especially in men. Impaired speech perception has the greatest impact on life satisfaction.

- Taste and odor sensitivity decline, making food less appealing. Touch sensitivity also deteriorates, particularly on the fingertips.

Describe the cardiovascular, respiratory, and immune system changes in late adulthood.

- Reduced capacity of the cardiovascular and respiratory systems becomes more apparent in late adulthood. As at earlier ages, not smoking, reducing dietary fat, avoiding environmental pollutants, and exercising can slow the effects of aging on these systems.

- The immune system functions less effectively in late life, permitting diseases to progress and making **autoimmune responses** more likely.

Discuss sleep difficulties in late adulthood.

- Older adults find it harder to fall asleep, stay asleep, and sleep deeply. Until age 70 or 80, men have more trouble sleeping than women, due to enlargement of the prostate leading to a need to urinate often, **sleep apnea,** and "restless legs."

Describe changes in physical appearance and mobility in late adulthood.

- Outward signs of aging, such as white hair, wrinkled and sagging skin, age spots, and decreased height and weight, become more noticeable. Mobility diminishes due to declining muscle and bone strength and joint flexibility.

■ Widespread negative stereotypes of aging in Western society make coping with physical aging more difficult. Older people fare best when they have status and opportunities for social participation, even when they are frail.

HEALTH, FITNESS, AND DISABILITY

Discuss health and fitness in late life, paying special attention to nutrition, exercise, and sexuality.

■ Most elders are optimistic about their health and with respect to protecting it, have a high sense of self-efficacy. Low-income ethnic minority elders remain at greater risk for certain health problems and are less likely to believe they can control their health.

■ As in early adulthood, in late life men are more prone to fatal diseases and women to disabling conditions. By very old age, women are more impaired than surviving men.

■ Poverty and negative lifestyle factors make the public health goal **compression of morbidity** hard to attain.

■ Because risk of dietary deficiencies increases, vitamin-mineral supplements are beneficial in late life. Exercise continues to be a powerful health intervention, even when begun in late adulthood.

■ Compared with other parts of the body, the reproductive organs undergo minimal change in late adulthood. Sexual desire and sexual activity decline but need not disappear.

Discuss common physical disabilities in late adulthood.

■ Illness and disability increase toward the end of life. Cardiovascular disease, cancer, stroke, and emphysema claim many lives. Because of declines in immune system functioning, vaccination against the most common type of pneumonia is advisable.

■ **Secondary aging,** rather than **primary aging,** is responsible for many disabilities in late adulthood. **Osteoarthritis** and **rheumatoid arthritis** strike at least half of older adults, especially women. Adult-onset diabetes increases.

■ Motor vehicle collisions per mile driven, pedestrian accidents, and falls increase in late adulthood. Changes in elders' lifestyles and environments can help prevent these causes of unintentional injury.

Discuss common mental disabilities in late adulthood.

■ **Alzheimer's disease** is the most common form of **dementia.** Often starting with severe memory problems, it brings personality changes, depression, loss of ability to comprehend and produce speech, disintegration of purposeful movements, and death. Underlying these changes are abundant **neurofibrillary tangles** and **amyloid plaques** and lowered neurotransmitter levels in the brain. The most consistent risk factor for Alzheimer's is a family history of the disease.

■ Both genetic and environmental factors contribute to **cerebrovascular dementia.** Because of their greater susceptibility to cardiovascular disease, men are at greater risk than women.

■ Treatable problems, such as depression, side effects of medication, and reactions to social isolation, can be mistaken for dementia. Therefore, careful diagnosis is essential.

Discuss health care issues that affect senior citizens.

■ Recent **squaring of the population pyramid** has prompted concern about **intergenerational inequity**—elders receiving more than their fair share of public resources. Fears that younger generations will not be able to support the rising cost of Social Security, Medicare, and Medicaid benefits have arisen.

■ Only 5 percent of American seniors are institutionalized, a rate about half that of other industrialized nations. Family members provide most long-term care, especially among ethnic minorities with closely knit extended families. Older adults cared for at home are generally less impaired than those in nursing homes.

COGNITIVE DEVELOPMENT IN LATE ADULTHOOD

Describe overall changes in cognitive functioning in late adulthood.

■ Individual differences in cognitive functioning are greater in late adulthood than at any other time of life. Although both fluid intelligence and crystallized intelligence decline in advanced old age, plasticity of development is still possible. Older adults can make the most of their cognitive resources through **selective optimization with compensation.**

MEMORY

How does memory change in late life?

■ Age-related limitations on working memory make memory difficulties more apparent on tasks that are complex and require deliberate processing. Automatic forms of memory, such as recognition and **implicit memory,** are largely spared.

■ Contrary to what older people sometimes report, **remote memory** is not clearer than recent memory. Older adults compensate for problems in **prospective memory** by using external memory aids.

LANGUAGE PROCESSING

Describe changes in language processing in late adulthood.

■ Although language comprehension changes little in late life, two aspects of language production—finding the right words and planning what to say and how to say it—show age-related losses. Older adults compensate by simplifying their grammatical structures and communicating gist rather than details.

PROBLEM SOLVING

How does problem solving change in late life?

■ Both traditional and practical problem solving decline in late adulthood. Yet older adults often respond more decisively than younger people in matters of health, perhaps because of greater experience in coping with illness.

WISDOM

Do age and wisdom go together, as cultures around the world assume?

■ Although older adults occupy important positions in societies around the world, age does not guarantee **wisdom.** Adults of all ages with extensive experience in dealing with human problems, including elders highly regarded in their community, are most likely to be represented among the wise.

FACTORS RELATED TO COGNITIVE CHANGE

List factors related to cognitive change in late adulthood.

■ Mentally active people are likely to maintain their cognitive abilities into advanced old age. A wide array of chronic conditions are associated with cognitive decline. Retirement can bring about either positive or negative changes. As death approaches, **terminal decline** becomes evident.

COGNITIVE INTERVENTIONS

Can cognitive interventions help older adults sustain their mental abilities?

■ The Adult Development and Enrichment Project (ADEPT) and various short-term studies show that training can enhance cognitive skills in older adults, including those who have suffered declines.

LIFELONG LEARNING

Discuss types of programs and benefits of continuing education in late life.

■ Better health and earlier retirement permit increasing numbers of older people to continue their education through college courses, community offerings, and programs such as Elderhostel. Elderly participants are enriched by new knowledge, new friends, a broader perspective on the world, and an image of themselves as more competent. Unfortunately, fewer continuing education opportunities are available to low-income, less educated seniors.

IMPORTANT TERMS AND CONCEPTS

functional age (p. 552)
young–old (p. 552)
old-old (p. 552)
life expectancy crossover (p. 553)
cataracts (p. 555)
macular degeneration (p. 555)
autoimmune response (p. 559)
sleep apnea (p. 560)
compression of morbidity (p. 564)
primary aging (p. 566)

secondary aging (p. 566)
osteoarthritis (p. 566)
rheumatoid arthritis (p. 566)
dementia (p. 568)
Alzheimer's disease (p. 569)
neurofibrillary tangles (p. 569)
amyloid plaques (p. 569)
cerebrovascular dementia (p. 571)
squaring of the population pyramid (p. 572)

intergenerational inequity (p. 573)
selective optimization with compensation (p. 576)
implicit memory (p. 577)
remote memory (p. 577)
prospective memory (p. 577)
wisdom (p. 579)
terminal decline (p. 579)

FOR FURTHER INFORMATION AND HELP

HEALTH AND FITNESS

National Health
Information Center
P.O. Box 1133
Washington, DC 20013-1133
(800) 336-4797
Web site: *www.nhic-nt.*
health.org
Publishes materials on a variety of health
topics, has about 100 hot lines on various
subjects, and offers a data base of over
1,100 health information resources. Oper-
ated by the U.S. Public Health Service.

United Seniors Health Cooperative
1331 H Street, N.W.
Washington, DC 20005
(202) 393-6222
Web site: *www.ushc.org*
Assists the elderly with independent liv-
ing and other health issues.

ARTHRITIS

Arthritis Foundation
1330 W. Peachtree Street
Atlanta, GA 30309
(800) 283-7800
Web site: *www.arthritis.org*
Publishes materials on arthritis and how
to live with it. Provides prescription dis-
count services for members. Local chap-
ters offer water-based exercise classes.

National Institute of Arthritis,
Musculoskeletal, and Skin Disease
Information Clearinghouse
1 AMS Circle
Bethesda, MD 20892-3675
(301) 496-8188
Web site: *www.nih.gov/niams*
Publishes materials on arthritis and pro-
vides information about federally funded
research and treatment centers.

DIABETES

American Diabetes Association
National Service Center
1660 Duke Street
Alexandria, VA 22314
(800) 232-3472
Web site: *www.diabetes.org*
Offers information on all aspects of dia-
betes, from research to self-care, including
a wide variety of publications.

ALZHEIMER'S DISEASE

Alzheimer's Association
919 N. Michigan Avenue, Suite 1000
Chicago, IL 60611
(800) 272-3900
Web site: *www.alz.org*
Organization of family members of
Alzheimer's disease victims. Promotes
research, provides educational programs
for the public, and works to develop fam-
ily support systems for relatives of
Alzheimer's sufferers. Has state and local
chapters that sponsor support groups and
provide lists of community resources.

LIFELONG LEARNING

Elderhostel
75 Federal Street
Boston, MA 02110
(617) 426-7788
Web site: *www.elderhostel.org*
Sponsors short-term, intensive campus-
based experiences for older adults in the
United States and many other countries.
Catalogues are available.

18

Emotional and Social Development in Late Adulthood

With Ruth at his side, Walt disclosed to guests on their fiftieth anniversary, "Even when there were hard times, it's the period of life I was in at the moment that I always liked the most. I adored playing baseball as a kid and learning the photography business when I was in my twenties. And I recall our wedding, the most memorable day of all," Walt continued, glancing affectionately at Ruth. "Then came the Depression, when professional picture taking was a luxury few people could afford. But we found ways to have fun without money, such as singing in the church choir and acting in community theater. A short time later, Sybil was born. It meant so much to me to be a father—and now a grandfather and a great-grandfather. Looking back at my parents and grandparents and forward at Sybil, Marci, and Marci's son Jamel, I feel a sense of unity with past and future generations."

Walt and Ruth greeted old age with calm acceptance, grateful for the gift of long life and loved ones. Yet not all older adults find peace of mind in these final decades. Walt's brother Dick was contentious and critical, often over petty issues, at other times over major disappointments. "Goldie, why did you serve cheesecake? No one eats cheesecake on birthdays" and "You know the reason we've got these financial worries? Uncle Louie wouldn't lend me the money to keep the bakery going, so I *had* to retire."

A mix of gains and losses characterizes these twilight years, extending the multidirectionality of development begun early in life. On the one hand, old age is an Indian summer—a time of pleasure and tranquility. Children are grown, life's work is nearly done, and responsibilities are lightened. On the other hand, it brings concerns about declining physical functions, unwelcome loneliness, and the growing specter of imminent death.

In this chapter, we consider how older adults reconcile these opposing forces. Although some are weary and discontented, most elders traverse this phase with poise and calm composure. They attach deeper significance to life and reap great benefits from family and friendship bonds, leisure activities, and community involvements. We will see how personal attributes and life history combine with home, neighborhood, community, and societal conditions to mold emotional and social development in late life.

Late adulthood is a time when individuals come to terms with their lives. Elders who arrive at a sense of integrity feel whole, complete, and satisfied with their achievements. Erik Erikson and his wife Joan provided ideal role models for Erikson's final stage: a couple who had grown old gracefully and who could often be seen walking hand in hand, deeply in love. (Palmer/Kane/Tony Stone Images)

ERIKSON'S THEORY: EGO INTEGRITY VERSUS DESPAIR

The final psychological conflict of Erikson's (1950) theory, **ego integrity versus despair,** involves coming to terms with one's life. Adults who arrive at a sense of integrity feel whole, complete, and satisfied with their achievements. They have adapted to the mix of triumphs and disappointments that are an inevitable part of love relationships, child rearing, work, friendships, and community involvement. They realize that the paths they followed, abandoned, and never selected were necessary for fashioning a meaningful life course.

The capacity to view one's life in the larger context of all humanity—as the chance combination of one person and one segment in history—contributes to the serenity and contentment that accompanies integrity. "These last few decades have been the happiest," Walt murmured as he clasped Ruth's hand, only weeks before the heart attack that would end his life. He was at peace with himself, his wife, and his children, having accepted his life course as something that had to be the way it was.

As he scanned the newspaper one day, Walt pondered, "I keep reading these percentages: 1 out of 5 people will get heart disease, 1 out of 3 will get cancer. But in truth, 1 out

Test Bank Items 18.1 through 18.3

of 1 will die. We are all mortal and must accept this fate." The year before, Walt had given his granddaughter Marci his collection of prized photos, which had absorbed him for over a half century. With the realization that the integrity of one's own life is part of an extended chain of human existence, death loses its sting (Vaillant, 1994; Wrightsman, 1988).

The negative outcome of this stage, despair, occurs when elders feel they have made many wrong decisions, yet time is too short to find an alternate route to integrity. Without another chance, the despairing person finds it hard to accept that death is near and is overwhelmed with bitterness, defeat, and hopelessness. According to Erikson, these attitudes are often expressed as anger and contempt for others, which disguise contempt for oneself. Dick's argumentative, faultfinding behavior, tendency to blame others for his personal failures, and regretful view of his own life reflect this deep sense of despair (Hamachek, 1990).

OTHER THEORIES OF PSYCHOSOCIAL DEVELOPMENT IN LATE ADULTHOOD

As with Erikson's stages of early and middle adulthood, other theorists have clarified and refined his vision of late adulthood, specifying the developmental tasks, thought processes, and activities that contribute to a sense of ego integrity. All agree that successful development in the later years involves greater integration and deepening of the personality.

PECK'S THEORY

According to Robert Peck (1968), Erikson's conflict of ego integrity versus despair comprises three distinct tasks. Each must be resolved for integrity to develop:

- *Ego differentiation versus work-role preoccupation.* This task results from retirement. It requires aging people who have invested heavily in their vocations to find other ways of affirming their self-worth. The person striving for integrity must *differentiate* a set of family, friendship, and community roles that are equally as satisfying as his or her vocation.

- *Body transcendence versus body preoccupation.* As we saw in Chapter 17, old age brings changes in appearance and declines in physical capacities and resistance to disease. For people whose psychological well-being is heavily bound up with the state of their bodies, late adulthood can be a difficult time. Older adults need to *transcend* physical limitations by emphasizing cognitive and social powers, which can compensate and

offer alternative rewards. Jamel captured his great-grandparents' success at this task when, as a third grader, he wrote about Ruth and Walt in a school assignment: "My grandparents love girls and boys and other people. They have grown old outside, but they are really young."

- *Ego transcendence versus ego preoccupation.* Whereas middle-aged people realize that life is finite, the elderly are reminded of the absolute certainty of death as siblings, friends, and peers in their community die. They must find a constructive way of facing this reality—through investing in a longer future than their own lifespan. Although the generative years of early and middle adulthood prepare people for a satisfying old age, attaining ego integrity requires a continuing effort to make life more secure, meaningful, and gratifying for those who will go on after one dies.

In Peck's theory, ego integrity requires that older adults move beyond their life's work, their bodies, and their separate identities. Drawing on their wisdom, they must reach out to others in ways that make the world better. Their success is apparent in their own inner state of contentment and their beneficial impact on others.

LABOUVIE-VIEF'S THEORY

In Chapter 13, we discussed Gisella Labouvie-Vief's theory of cognitive change, noting how in early adulthood abstract thought becomes *pragmatic*—a tool for solving real-world problems. Labouvie-Vief has also explored the development of adults' reasoning about emotion. She believes that older and more psychologically mature individuals are more in touch with their feelings. They describe emotional reactions in more complex and personalized ways, perhaps as a result of reflecting on life experiences and using coping strategies that permit fuller acknowledgment of emotion.

In one study, participants from adolescence to old age were asked to describe personal experiences in which they were happy, angry, fearful, and sad and to indicate how they knew they felt the emotion. Younger people explained feelings technically, as if they were observing them from the outside. For example, they were likely to say, "My adrenaline was high" or "My heart rate increased." They also emphasized how they *should* feel rather than how they *did* feel. In contrast, older adults gave vivid descriptions that viewed both mind and body as contributing to feeling states and that integrated subjective and objective aspects of emotion. Here are two examples:

Everything was accelerated, my adrenaline started pumping, my mind started thinking what to do next. I was flashing things through my mind, I'm sure my heartbeat sped up. I did feel fearful.

How will this grandfather put this moment into words? Elders describe emotional reactions in more complex and personalized ways. (Palmer/Kane/Tony Stone Images)

You have sunshine in your heart. During the wedding the candles were glowing. And that's just how I felt. I was glowing too. It was kind of dull outside. But that isn't how I felt. Everybody in the church felt like they were glowing. It was that kind of feeling. (Labouvie-Vief, DeVoe, & Bulka, 1989, p. 429)

With maturity, adults arrive at a fuller understanding of their emotions. Labouvie-Vief (1990) comments that they also acknowledge periods of intense rumination and are better able to interpret negative events in a positive light—factors that may contribute to the self-acceptance inherent in ego integrity.

REMINISCENCE AND LIFE REVIEW

When we think of the elderly, we often picture them engaged in **reminiscence**—telling stories about people and events from their past and reporting associated thoughts and feelings. Indeed, the image of a reminiscing elder is so widespread that it ranks among the negative stereotypes of aging. A common view is that older people live in the past to escape the realities of a shortened future and the nearness of death. Researchers do not yet have a full understanding of why older people reminisce more often than younger people do (Boden & Bielby, 1983; Lamme & Baars, 1993). But current theory and research indicate that reflecting on the past can be positive and adaptive.

Return to Walt's commentary about major events in his life at the beginning of this chapter. Walt engaged in a special form of reminiscence called **life review,** in which the person calls up, reflects on, and reconsiders past experiences, contemplating their meaning with the goal of achieving greater self-understanding. According to Robert Butler (1968), most older adults engage in life review as part of attaining ego integrity, preventing despair, and accepting the end of life. Butler's ideas have been so influential that many therapists encourage the elderly to engage in life-review reminiscence. In a recent study, homebound 61- to 99-year-olds who participated in a counselor-led life-review process showed gains in life satisfaction still apparent a year later. This did not occur for participants who simply enjoyed friendly visits or who received no treatment (Haight, 1992).

Although life review often prompts self-awareness and self-respect, many elders do not examine their pasts spontaneously. When they do, outcomes vary from positive to negative. As the Lifespan Vista box on pages 592–593 illustrates, recollections of distant events are colored by many factors, including personality, subsequent life experiences, and the timing and nature of the events themselves. In sum, large individual differences in reminiscence exist. Despairing elders tend to ruminate on negative content from the past, and some benefit from help in identifying worthwhile aspects of their lives (Sherman, 1993; Wallace, 1992).

Besides life review, reminiscence in old age serves other functions. Elders may talk about their past because others, from social service professionals to curious grandchildren and great-grandchildren, expect them to! As they discuss life experiences, older adults may satisfy social needs, such as developing or reestablishing friendships and seeking confirmation from others. Finally, reminiscence—for young and old alike—often occurs during times of life transition (Parker, 1995). The older adult who has recently retired, been widowed, or moved to a new residence may turn to the past

Reflecting on the past can be positive and adaptive. Many older adults engage in reminiscence and life review as part of attaining ego integrity. (John Eastcott/The Image Works)

to sustain a sense of personal continuity. Indeed, elders who feel their current lives are adrift reminisce especially often to recapture feelings of meaning and purpose (Fry, 1991).

CHANGES IN SELF-CONCEPT AND PERSONALITY TRAITS

Consider the ingredients of ego integrity: wholeness, contentment, and image of the self as part of a larger world order. These attributes are reflected in several late-life changes in self-concept and personality.

STRENGTHENING OF SELF-CONCEPT

Older adults have accumulated a lifetime of self-knowledge, leading to more secure and complex conceptions of themselves than at earlier ages (Perlmutter, 1988). Ruth, for example, knew with certainty that she was good at counseling others, growing a flower garden, giving dinner parties, budgeting money, and figuring out who could be trusted and who couldn't. This strengthening of her self-definition permitted her to compensate for lack of skill in domains she had never tried or did not master. Consequently, it allowed for self-acceptance—a key feature of integrity.

In Chapter 17, we noted that as the future shortens, hoped-for possible selves decline and feared selves increase in the realm of health. This shift also occurs in other domains. But most elders mention some hoped-for selves and are very active in pursuing them. These grant older adults goals in life and a sense of further development. They are vital for psychological well-being (Holahan, 1988; Markus & Herzog, 1992).

AGREEABLENESS, SOCIABILITY, AND ACCEPTANCE OF CHANGE

Longitudinal research reveals continuing stability of the "big five" personality traits from mid- to late life (see Chapter 16, page 528). During the years of late adulthood, shifts in three personality characteristics take place.

Rating open-ended interviews with elders in their sixties and, again, when they reached their eighties and nineties, researchers found that scores on adjectives that make up *agreeableness*—generous, acquiescent, and good-natured—were higher on the second occasion than the first for over one-third of the sample. These qualities seem to characterize people who have come to terms with life despite its imperfections.

In addition, participants showed a slight dip in *sociability* as they aged (Field & Millsap, 1991). Perhaps this reflects a narrowing of social contacts as people become more selective about relationships and as family members and friends die—trends we will take up in a later section.

A third, related development is greater *acceptance of change*—an attribute the elderly frequently mention as important to psychological well-being. That many older adults adjust well to change is evident in what they say when asked about dissatisfactions in their lives. They often respond that they are not unhappy about anything! This positive assessment contradicts the stereotype of old age as a time of low morale. To the contrary, despite physical declines and social losses, most elders are reasonably satisfied (Ryff, 1989). A capacity to accept life's twists and turns, many of which are beyond one's control, is vital for positive functioning in late adulthood.

SPIRITUALITY AND RELIGIOSITY

How are older adults able to accept change, feel whole and complete, and anticipate death with calm composure? One possibility is that they develop a more mature sense of spirituality. That is, they actively seek a higher meaning for life, knowing that it will end in the foreseeable future. Spirituality is not the same as religion. A transcendent sense of truth and beauty can be found in art, nature, and relationships with others. But religion provides many people with beliefs, symbols, and rituals that guide this quest for meaning.

Older adults attach great value to religious beliefs and behaviors. According to a recent national survey, 76 percent of Americans age 65 and older say that religion is very important in their lives, and 16 percent describe it as fairly important. Over half attend religious services weekly, nearly two-thirds watch religious TV programs, and about

Older adults attach great value to religious beliefs and behaviors, and faith may advance to a higher level—away from prescribed beliefs to a more reflective approach that emphasizes links to others and is at ease with mystery and uncertainty. (Michael Schwarz/The Image Works)

WORLD WAR II REFUGEE AND EVACUEE CHILDREN LOOK BACK FROM THE VANTAGE POINT OF OLD AGE

How do elders look back on a traumatic period occurring early in their lives? To find out, Glen Palmer (1997) interviewed people who had come to Australia just before or during World War II as Jewish refugees from Austria, Germany, and Poland or as evacuees from Great Britain. All had been under age 16 when their parents put them on boats, giving them up in anguish to protect them. Most parents of the Polish and German children were murdered by the Nazis. After kissing them good-bye, these children never saw their parents again. In contrast, the British evacuees were able to return to their homeland and families at the end of the war.

Looking back, these elders were largely supportive of their parents' decision to send them away. For the refugees, it meant that their lives had been spared (see Chapter 10, page 340). Beyond this, recollections of their experiences and assessments of its lifelong impact varied. Among factors that made a difference were the quality of care they received, their age at separation (child or adolescent), and their temperament.

Care situations included relatives, foster homes, group homes, and boarding schools. For children placed with families, being loved and wanted led to new attachments and pleasant memories, often at the expense of former bonds. As one interviewee noted, "If you send a child away in the formative years, you take the risk of losing the affection of that child." When relatives and foster families were unkind or outright rejecting, painful recollections of an "emotional limbo" emerged. Without new attachments, these children clung to fading images of their families, idealized their parents, and longed to return to them. Separation and inadequate care had a more profound, lasting impact on the memories of temperamentally withdrawn, anxious individuals than on those of outgoing individuals, who were better able to develop supportive social ties beyond the family.

Compared to interviewees who were children at the time of separation, those who were adolescents were less affected by the absence of affectional bonds with adults. "I knew in my mind that I had plenty of family support, even though they weren't around," one Polish refugee remarked. On the threshold of adult-hood, these young people had little desire for substitute parents. Instead, they fared best in group homes and boarding schools that provided appropriate oversight, balancing security with independence. Often their warmest memories were of bonds established with one another. In contrast, when children were placed in group homes and denied family attachments, their sadness was often still apparent in old age: "There was a desolation because there was no love. . . . You need to be able to say to a child, 'You're my special one.'"

By the end of the war, the majority of interviewees had become young adults. Most of the British evacuees were reunited with their families, but rarely with "happily ever after" endings. Children and parents had changed during the intervening years. As one interviewee put it, "It was easier to go than to come back." Still, reunion was vital, since its absence usually brings a lifetime of unresolved grief.

For most Polish and German refugees, loss of immediate family was total. More than 50 years later, their searching and sorrow continues. One interviewee commented, "The Holo-

---

one-fourth pray at least three times a day (Princeton Religion Research Center, 1994).

Although declining health and transportation difficulties reduce organized religious participation in advanced old age, informal religious activities remain prominent in the lives of today's elders (Ainlay, Singleton, & Swigert, 1992). Does involvement in religion increase with age? Longitudinal research suggests that it is fairly stable throughout adulthood. Even so, the need for religious affiliation may rise among the elderly. In one study, over one-third of older people said they would devote more time to religion if they could (DeGenova, 1992).

Furthermore, spirituality and faith may advance to a higher level in late adulthood—away from prescribed beliefs to a more reflective approach that emphasizes links to others and is at ease with mystery and uncertainty (Birren, 1990). In his theory of faith development, James Fowler (1981) posits five stages, summarized in Table 18.1 on page 594, which have been confirmed in several studies. Stages 4 and 5 characterize mature adults. Notice how individuals become aware of their own belief system as one of many possible world views, contemplate the deeper significance of religious symbols and rituals, and open themselves to other religious perspectives as sources of inspiration (McFadden, 1996). For example, as a complement to his Catholicism, Walt became intensely interested in Buddhism, especially its focus on attaining perfect peace and happiness by mastering thoughts and

caust pursues us daily. It is like having a flea in your head." This suggests that rather than healing with the passage of time, the pain of childhood loss and trauma can intensify in later life. Reunions with other refugees helped soothe persisting wounds as elders shared experiences with those best able to understand their continued suffering.

Although many interviewees remained deeply affected by their childhood experiences, very few were maladjusted. To the contrary, most found turning points through their families, careers, or other interests and led productive, fulfilling lives. Some seemed to draw on inner strengths developed during early years of separation when late-life adversities arose. As one Holocaust survivor commented after her husband died, "[W]hen you've been through what I've been through, you can survive anything."

Nevertheless, these victims of genocide—and other similarly traumatized individuals around the world—probably cannot make a complete recovery. None of the refugees returned to live in Europe, although some had visited their countries of birth in recent years. Rarely were these journeys comfortable. Almost always, they brought heartaches to the surface—yet another reminder of the pervasive effects of ethnic and political violence throughout the lifespan.

Below are the German Jewish children, who boarded the *Orama* for Australia in June of 1939, just before the outbreak of World War II. Seven-year-old Laurie, shown on the left with his parents 2 days before his departure, kissed them good-bye and never saw them again. When the war ended, Laurie learned that his mother had been deported to a concentration camp and murdered by the Nazis. A tracing agency located a grave for Laurie's father, although Laurie never found out exactly how he died. *(Courtesy of Glen Palmer, Griffith University, Queensland, Australia.)*

feelings, never harming others, and resisting attachment to worldly objects.

Involvement in both organized and informal religious activities is especially high among low-income ethnic minority elders, including African Americans, Hispanics, and Native Americans. In African-American communities, churches not only provide contexts for deriving meaning from life, but are centers for education, health, social welfare, and political activities aimed at improving life conditions. African-American elders look to religion as a powerful resource for social support beyond the family and for the inner strength to withstand daily stresses and physical impairments (Gelfand, 1994; Levin, Taylor, & Chatters, 1994). Asked about her philosophy of life, an

African-American 65-year-old reveals how faith enabled her to do more than survive. As she tells it, she has much to be thankful for:

We've had lots of misfortunes . . . but we always knowed that it could be worse. . . . I know somedays I get up I'd be stiff and my knees aching and my back is aching and my head is hurting and I can get up and go in the bathroom. I say, "Thank you Lord because I have the activities of my limbs. . . ." And then we get a meal on the table and we have . . . at least a portion of health and strength because when we get old we know we are going to have aches and pains, but we just have to bear with them and try to thank the Lord

TABLE 18.1

Fowler's Stages of Faith Development

STAGE OF FAITH	PERIOD OF DEVELOPMENT	DESCRIPTION
1. Intuitive-projective	3–7 years	Children's fantasy and imitation lead them to be powerfully influenced by stories, moods, and behaviors demonstrating the faith of adults. They become aware of right and wrong actions.
2. Mythic-literal	7–11 years	Children begin to internalize the stories, beliefs, and observances of their religious community, which they take literally. For example, they often hold concrete images of God living on top of the world and watching over everybody.
3. Synthetic-conventional	Adolescence	Adolescents have a coherent set of deeply felt beliefs and values, which provides a basis for identity. They have not yet examined this ideology systematically.
4. Individuative-reflective	Adulthood	Adults who reach this stage critically reflect on their beliefs and values, recognizing that their world view is only one of many possible world views. They actively shape a personal ideology, forming and reforming it over time. About religious rituals and symbols, they ask, "What does this really mean?"
5. Conjunctive	Late Adulthood	The few people who attain this stage form an enlarged vision of an all-inclusive human community. They act to bring it about by standing up against persecution and injustice and by promoting a common good that serves the needs of diverse groups. Great religious leaders, such as Mahatma Gandhi and Martin Luther King, Jr., illustrate conjunctive faith.

Source: Fowler, 1981.

for things being as well as they are. . . . I accept myself as God made me and he made me the person I am and that's the one I'm going to try and stay . . . until God calls me. (Nye, 1993, p. 109)

Sex differences in religious involvement are evident throughout adulthood. Women are more likely to be church or synagogue members and to engage in a variety of religious activities, from attending services to private prayer and Bible reading (Levin, Taylor, & Chatters, 1994). Women's greater experience of poverty and participation in caregiving, including for chronically ill family members, expose them to higher levels of stress. As with ethnic minorities, they may turn to religion for social support and for a larger vision of community that places life's challenges in perspective.

INDIVIDUAL DIFFERENCES IN PSYCHOLOGICAL WELL-BEING

In this and the previous chapter, we have seen that the overall direction of development is one in which adults adapt constructively to old age. Still, considerable variation exists in how people fare. Most elders regard themselves as active, capable, and valuable, despite physical, psychological, and social challenges. Others feel dependent, incompetent, and worthless. Identifying personal and environmental influences on late-life psychological well-being is vital for designing interventions that foster favorable adjustment.

CONTROL VERSUS DEPENDENCY

As Ruth's eyesight, hearing, and mobility declined in her eighties, Sybil visited daily to assist with self-care and household tasks. During the hours mother and daughter were together, Sybil interacted most often with Ruth when she asked for help with activities of daily living. When Ruth handled tasks on her own, Sybil usually withdrew.

Observations of older adults interacting with others in both private homes and institutions reveal a typical pattern in which dependency behaviors are attended to immediately, whereas independent behaviors are mostly ignored, regardless of the elder's health status. Notice how these sequences, so predictable that they are called the **dependency–support script** and the **independence–ignore script,** reinforce dependent behavior at the expense of independent behavior, regardless of the older person's competencies. Even a self-reliant elder like Ruth did not always resist Sybil's unnecessary help, since it brought about pleasurable social contact (Baltes, 1995, 1996).

Why do friends, family members, and caregivers respond in ways that promote excessive dependency in old age? A stereotype of the elderly as passive and incompetent appears to be responsible. Older adults seem well aware of others' low expectations for them. They are quick to take personal responsibility for their independent behavior, but

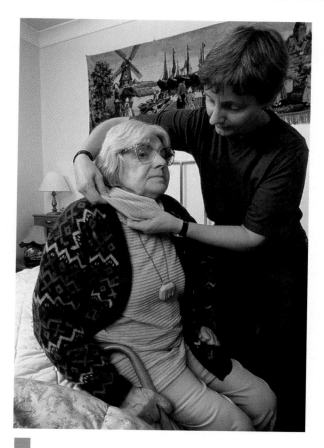

Is this daughter encouraging her mother to be dependent by helping her dress? The answer lies in whether the mother assumes personal control over her dependency. Dependency can be adaptive if it permits older people to conserve their strength and invest it in highly valued activities. (Penny Tweedie/Tony Stone Images)

they say their dependency is due either to overresponsive social partners or to illness (Wahl, 1991).

Many elders fear becoming dependent on others. In Western societies, which place a high value on independence, older adults often say that when they can no longer care for themselves, it is time to die. Does this mean we should encourage elders to be as independent as possible? According to Mary Baltes (1996), this alternative is as counterproductive as promoting passivity and incompetence. Aging brings diminished energy at a time when people confront many challenging developmental tasks. Dependency can be adaptive if it permits older people to conserve their strength by investing it in highly valued activities, using a set of strategies we considered in Chapter 17: *selective optimization with compensation.*

For dependency to foster well-being, elders need to assume personal control over it. For example, although she could handle the activities on her own, Ruth permitted Sybil to assist with dressing, financial matters, shopping, and food preparation so she would have more stamina for pleasurable reading. To *optimize* her energies, Ruth *selected*

certain domains in which to become dependent. This permitted her to *compensate* for her poor eyesight by taking extra time to use a magnifying glass while reading or listening to a book on tape.

Notice how Ruth chose dependency in some areas so she could maintain an important function in danger of decline. In many instances, dependency behaviors are not signs of helplessness. Instead, they grant elders autonomy—a means for managing their own aging.

But many times, the social world of the elderly fosters dependent behaviors at the cost of independent behaviors through an overresponsive dependency-support script (Baltes, 1996). When we intervene with older adults, we must ask ourselves, What kind of assistance are we providing? Help that is not wanted or needed or that exaggerates weaknesses can undermine mental health. It can also accelerate physical disability due to nonuse of existing skills. In contrast, help that frees up energy for pursuits that are personally satisfying and that lead to growth enhances elders' quality of life.

HEALTH

As we noted in Chapter 16, health is a powerful predictor of psychological well-being in late adulthood. Physical declines and chronic disease can be highly stressful, leading to a sense of loss of personal control—a major factor in adult mental health. Furthermore, physical illness is among the strongest risk factors for late-life depression (Gatz, Kasl-Godley, & Karel, 1996). Although fewer older than young and middle-aged adults are depressed (see Chapter 17), profound feelings of hopelessness rise with age as physical disability and social losses increase (George, 1989, 1994).

The relation between physical and mental health problems can become a vicious cycle, each intensifying the other. At times, the rapid decline of a sick elder is the work of despondency and "giving up" (Cohen, 1990). This downward spiral can be hastened by a move to a nursing home, requiring the older person to adjust to distance from family and friends and to a new self-definition as "a person who can survive only in an institution." In the month after admission, many residents deteriorate rapidly and become severely depressed. The stress of illness together with institutionalization is associated with heightened health problems and mortality (Tobin, 1989).

Depression in old age is often lethal. People age 65 and older have the highest suicide rate of any age group. As the Social Issues box on page 596 indicates, white males, particularly those who are chronically ill, show a greatly increased risk. What factors enable elders like Ruth to surmount the physical impairment–depression relationship, remaining optimistic and content? Personal characteristics we discussed in earlier chapters—effective coping and a sense of self-efficacy—are vitally important. But for frail elders to display these attributes, families and caregivers must grant them autonomy by avoiding the dependency-

ELDER SUICIDE

When 65-year-old Abe's wife died, he withdrew from life. Living far from his two daughters, he spent his nonworking days alone, watching television and reading mystery novels. As grandchildren were born, Abe visited his daughters' homes from time to time. When he did, he carried his despondent behavior with him. "Look at my new pajamas, Grandpa!" Abe's 6-year-old grandson Tony exclaimed on one occasion. Abe did not lift his head to acknowledge the little boy.

Abe retired after arthritis made walking difficult. With more empty days, his depression deepened. Gradually, Abe developed painful digestive difficulties, but he refused to go to the doctor. "Don't need to," he said abruptly when one of his daughters called and begged him to get medical attention. Answering her invitation to Tony's tenth birthday party, Abe wrote, "Maybe—if I'm still around next month. By the way, when I go, I want my body cremated." Two weeks later, Abe died from an intestinal blockage. His body was found in the living room chair where he habitually spent his days.

FACTORS RELATED TO ELDER SUICIDE. As Figure 18.1 shows, suicide peaks in late life. It climbs further during the elder years, reaching its highest rate among people age 75 and older. Although the incidence of suicide varies among nations, older adults are at increased risk around the world (World Health Organization, 1991).

Recall from Chapter 12 that the suicide rate is much higher among adolescent males than females. This sex difference persists throughout the lifespan, increasing in old age (refer again to Figure 18.1). In addition, compared to the white majority, most American ethnic minority elders (including African-Americans, Hispanics, and Native Americans) have very low suicide rates. What explains these trends? Despite the lifelong pattern of higher rates of depression and more suicide attempts among females, elderly women's closer ties to family and friends, greater willingness to seek social support, and religiosity prevent many from taking their own lives. High levels of social support through extended families and church affiliations may also prevent suicide among ethnic minorities.

As in earlier years, the method favored by elder males (firearms) offers less chance of revival than that favored by elder females (poisoning or drug overdose). Nevertheless, failed suicides are much rarer in old age than in adolescence. The ratio of attempts to completions for the young is as high as 300 to 1; for the elderly, it is 4 to 1 or lower. When elders decide to die, they seem especially determined to succeed (Bille-Brahe, 1993; Platt, 1992).

Underreporting of suicides probably occurs at all ages, but it is more

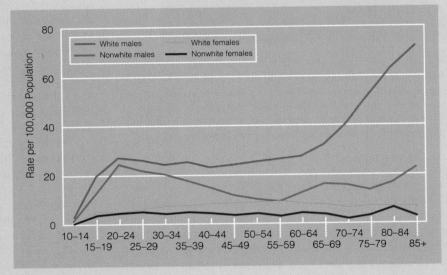

IGURE 18.1

Suicide rates over the lifespan by ethnicity and sex in the United States. Suicide peaks in late life, climbing further during the elder years and reaching its highest rate among people age 75 and older. Suicide is more frequent among males than females, a difference that increases in old age. Compared to the white majority, nonwhite ethnic minority elders have very low suicide rates. *(U.S. Bureau of the Census, 1996.)*

support script. When older adults remain in charge of personally important areas of their lives, they retain essential aspects of their identity in the face of change and report a more favorable outlook on their past and future (Brandtstädter & Rothermund, 1994).

NEGATIVE LIFE CHANGES

Ruth lost Walt to a heart attack, cared for Ida as her Alzheimer's symptoms worsened, and faced health problems of her own—all within a span of a few years. Elders are at risk for a variety of negative life changes—death of

As comfort and quality of life diminish, feelings of hopelessness and helplessness deepen. Very old people, especially men, are particularly likely to take their own lives under these conditions. (John Eastcott/The Image Works)

common in old age. Medical examiners are less likely to pursue suicide as a cause of death when a person is old. And many elders, like Abe, engage in indirect self-destructive acts rarely classified as suicide, such as deciding not to go to a doctor when they are ill and refusing to eat or take prescribed medications. Among institutionalized elders, these efforts to hasten death are widespread (Osgood, Brant, & Lipman, 1991). Consequently, elder

suicide is an even larger problem than official statistics indicate.

Two types of events prompt suicide in late life. Losses—retirement from a highly valued occupation, widowhood, and social isolation—place elders who have difficulty coping with change at risk. A second set of risks are chronic and terminal illnesses that cause severe physical pain. As comfort and quality of life diminish, feelings of hopelessness and helplessness deepen. Very old people, especially men, are particularly likely to take their own lives under these conditions. The chances are even greater when a sick elder is socially isolated—living alone or in a nursing home with high staff turnover and minimal caregiver support.

PREVENTION AND TREATMENT. Warning signs of suicide in late adulthood overlap with those at earlier ages. They include efforts to put personal affairs in order, statements about dying, despondency, and sleep and appetite changes. Family members, friends, and caregivers must also watch for indirect self-destructive acts, such as refusing food or medical treatment, which are unique to old age (Osgood, 1992).

When suicidal elders are depressed, the most effective treatment combines antidepressant medication with therapy. At-risk individuals need help in coping with role transitions, such as retirement, widowhood, and dependency brought about by illness. Distorted ways of thinking, such as "I'm old, and nothing can be done about my problems," must be countered and revised. Meeting with the family to find ways to reduce loneliness and desperation can

be helpful. And group approaches offer peer support and can lead to gratifying new relationships (Devons, 1996; McIntosh et al., 1994).

Although youth suicide has risen (see Chapter 12, page 406), elder suicide has declined during the past 50 years, due to increased economic security among older adults, improved medical care and social services, and more favorable cultural attitudes toward retirement. Communities are beginning to recognize the importance of additional preventive steps, such as telephone hot lines with trained volunteers who provide emotional support and agencies that arrange for regular home visitors or "buddy system" phone calls. In institutions, providing residents with privacy, autonomy, and space helps prevent self-destructive behavior (McIntosh et al., 1994; Osgood, 1992).

Finally, elder suicide raises a controversial ethical issue: Do people with incurable illnesses have the right to take their own lives? We will take up this topic in Chapter 19.

TRY THIS...

■ What warning signs of suicide did Abe display? Why was it hard for Abe's daughters to identify these signs? Contact agencies that serve the elderly in your community. Describe services they offer that might have prevented Abe's suicide.

spouse, siblings, and friends; illness and physical disabilities; declining income; and greater dependency.

Negative life changes are difficult for all people. But these events may actually evoke less stress and depression in older than in younger adults (Gatz, Kasl-Godley, &

Karel, 1996). In one study, a diagnosis of breast cancer was stressful for women of any age. But younger women had greater adjustment problems than older women did, perhaps because chronic disease is more expected in old age than earlier (Penman et al., 1987). Also, many elders have learned to cope with hard times and come to accept loss as part of human existence.

Still, when negative changes pile up, they test the coping skills of older adults. And some events, such as the death of a loved one and severe physical disability, are more likely to prompt depression and lasting mental health difficulties (Wykle & Musil, 1993). Under these circumstances, elders need extra attention and care from their families and communities.

SOCIAL SUPPORT AND SOCIAL INTERACTION

In late adulthood, social support continues to play a powerful role in reducing stress, thereby promoting physical health and psychological well-being. Most of the time, elders receive informal assistance from family members—first from their spouse, but if none exists, from children and then siblings. If these individuals are not available, other relatives and friends may step in.

Nevertheless, many older adults place such high value on independence that they do not want a great deal of support from people close to them unless they can reciprocate. Perhaps for this reason, adult children express a deeper sense of obligation toward their aging parents than their parents expect from them (see Chapter 16, page 538). Formal support—a paid helper or agency-provided services—as a complement to informal assistance not only helps relieve caregiving burdens, but prevents elders from feeling overly dependent in their close relationships (Chappell & Guse, 1989; Krause, 1990).

Ethnic minority elders, however, do not readily accept formal agency services. But when social and medical interventions use culturally sensitive practices and are connected to a neighborhood organization such as the church, they are more willing to participate. Although African-American seniors say they rely on their families more than the church for assistance, church affiliation contributes to their psychological well-being. Elderly blacks who report support and meaningful role involvement in both contexts score highest in mental health (Coke, 1992; Walls & Zarit, 1991).

Finally, having a sociable personality is related to high morale in old age (Adkins, Martin, & Poon, 1996). Extroverted elders are more likely to take advantage of opportunities to socialize. Interactions with others reduce loneliness and depression and foster self-esteem and life satisfaction. But as we will see in the next section, supportive communication in old age has little to do with quantity of contact. Instead, high-quality relationships, involving expressions of caring, encouragement, and respect, have the greatest impact on mental health in late life.

BRIEF REVIEW

Erikson's critical psychological conflict of late life is ego integrity versus despair. Older adults with a sense of integrity accept their life course with its imperfections, view it in the larger context of human existence, and do not fear death. In contrast, despairing elders are dissatisfied with their past yet realize time is too short to overcome their failures. Peck specifies three tasks essential to attaining integrity: differentiating new roles that are just as satisfying as one's vocation; transcending physical declines by emphasizing cognitive and social powers; and facing death constructively by taking steps to leave the world a better place. According to Labouvie-Vief, striving for integrity is supported by mature adults' fuller understanding of their emotions. Reminiscence is common in old age, sometimes serves the function of life review, and is affected by elders' personality and life circumstances.

In personality, older adults tend to become more agreeable, less extroverted, and more accepting of change. Although involvement in religion is fairly stable throughout adulthood, spirituality and faith may advance to a higher, more reflective level that looks to other religious perspectives for inspiration. Both organized and informal religious activities are vital sources of social support for low-income ethnic minority elderly. Older adults who are in control of their dependency needs, who cope effectively with social losses and physical declines, and who have access to informal and formal social supports report high psychological well-being.

ASK YOURSELF . . .

- *Compared to a few years earlier, 80-year-old Miriam took longer to get dressed in the morning. Joan, her paid home helper, said, "I'd like you to wait until I arrive before dressing. Then I can help you and it won't take so long." What impact is Joan's approach likely to have on Miriam's personality? Why?*

- *Suppose Miriam had said to Joan, "I'd like you to come 20 minutes earlier in the morning to help me get dressed. That way, I'll have more energy to make phone calls, pay the bills, and have lunch with some friends." Explain why, in this case, depending on Joan for dressing is likely to have a different impact.*

- *Many elders adapt effectively to negative life changes. List personal and environmental factors that facilitate this generally positive outcome.*

A CHANGING SOCIAL WORLD

Walt and Ruth's outgoingness led many family members and friends to seek them out, and they often reciprocated. In contrast, Dick's stubborn nature meant that his and Goldie's network of social ties was far more restricted, as it had been for many years, consisting of a just few relatives and long-time friends.

In late adulthood, extroverts (like Walt and Ruth) continue to interact with a wider range of people than do introverts and people (like Dick) with poor social skills. Nevertheless, both cross-sectional and longitudinal research reveal that size of social networks and therefore amount of social interaction decline for virtually everyone (Carstensen, 1992). This finding presents a curious paradox. If social interaction and social support are necessary for mental health, how is it possible for elders to interact less yet be generally satisfied with life and less depressed than younger adults?

SOCIAL THEORIES OF AGING

Social theories of aging offer explanations of the decline in social relations just described. Two prominent perspectives—disengagement and activity theory—interpret it in opposite ways. Each of these theories has been criticized. A recent approach—selectivity theory—accounts for a wider range of findings on social contacts in old age.

■ DISENGAGEMENT THEORY. According to **disengagement theory,** mutual withdrawal between elders and society takes place in anticipation of death (Cumming & Henry, 1961). Older people decrease their activity levels and interact less frequently, becoming more preoccupied with their inner lives. At the same time, society frees elders from employment and family responsibilities. The result is regarded as beneficial for both sides. Elders are granted a life of tranquility. Once they disengage, their deaths are less disruptive to society.

Think back to our discussion of wisdom in Chapter 17. Because of their long life experience, older adults in many cultures move into new roles of prestige and power. Clearly, not everyone disengages! Even after retirement, some people sustain certain aspects of their work, and others develop new, rewarding social roles in the community and in leisure. In tribal and village societies, most elders continue to hold important positions in their families. Their value to the larger community increases as they are called on more than ever for advice and decision making (Holmes & Holmes, 1995). Consequently, when old people disengage, it may not represent their personal preference. Instead, it may be due to a failure of the social world to

Young people seek advice and learn about the land and history of their people from this Navajo elder. Because of his long life experience, he is granted increased prestige and power in his cultural community. The greater social responsibility of many older adults presents a major challenge to disengagement theory of aging. (Suzi Moore/Woodfin Camp & Associates)

provide opportunities for engagement. (Return to the Cultural Influences box in Chapter 17, page 562, for some striking examples of community variations in meaningful roles available to elders.)

As we will see shortly, elders' retreat from interaction is considerably more complex than disengagement theory implies. Instead of disengaging from all social ties, they let go of unsatisfying contacts and maintain satisfying ones. And sometimes, they put up with less than satisfying relationships to remain engaged! For example, Ruth reluctantly agreed to travel with Dick and Goldie because she wanted to share the experience with Walt. But she often complained about Dick's insensitive behavior.

■ ACTIVITY THEORY. **Activity theory** tries to overcome the flaws of disengagement theory. It states that social barriers to engagement, not the desires of elders, cause declining rates of interaction. When elders lose certain roles (for example, through retirement or widowhood), they do their best to find others in an effort to stay active and busy. According to this view, arranging conditions that permit elders to remain engaged in roles and relationships is vital for life satisfaction (Havighurst & Albrecht, 1953; Maddox, 1963).

Although many people seek alternative sources of meaning and gratification in response to social losses, activity theory fails to acknowledge any psychological change in old age. Many studies show that merely offering opportunities for social contact does not lead to greater social activity on the part of elders. Indeed, the majority do not "take advantage" of such opportunities. In nursing homes, for example, social partners are abundant, but social interaction is very low even among the healthiest residents—

a circumstance that we will examine when we discuss housing arrangements for the elderly. Finally, especially troubling for activity theory is the repeated finding that when health status is controlled, elders who have larger social networks are not necessarily happier (Lee & Markides, 1990). Instead, recall that quality, not quantity, of relationships predicts psychological well-being in old age.

■ SELECTIVITY THEORY. A new approach asserts that our social networks become more selective as we age. In **selectivity theory,** social interaction does not decline suddenly in late adulthood. Rather, it extends lifelong selection processes. In middle adulthood, marital relationships deepen, siblings become closer, and number of friendships declines. In old age, contacts with family and long-term friends are sustained until the eighties, when they diminish in favor of a few very close relationships. In contrast, contacts with acquaintances and willingness to form new social ties fall off steadily from middle through late adulthood (Field & Minkler, 1988; Lang & Carstensen, 1994).

Why does this happen? According to selectivity theory, physical and psychological aspects of aging lead to changes in the functions of social interaction most important to us. Consider the reasons you interact with members of your social network. At times, you approach them to get information. At other times, you seek affirmation of your uniqueness and worth as a person. And you also choose social partners to regulate emotion, approaching those who evoke positive feelings and avoiding those who make you feel sad, angry, or uncomfortable.

In old age, the information-gathering and self-affirmation functions of social interaction become less significant. Because older adults have gathered a lifetime of information, there are fewer people with knowledge that they desire. And elders realize it is risky to approach people they do not know for self-affirmation. Stereotypes of aging increase the odds of receiving a condescending, hostile, or indifferent response.

Instead, elders emphasize the emotion-regulating function of interaction. Physical fragility makes it more important to avoid stress and maintain emotional equilibrium (Carstensen & Turk, 1994). In one study, younger and older adults were asked to categorize their social partners. Younger people more often used information seeking and future contact as the basis for sorting. In contrast, older people stressed anticipated feelings (Frederickson & Carstensen, 1990). They appeared highly motivated to select associates on the basis of emotion, approaching pleasant relationships and avoiding unpleasant ones.

Interacting mostly with relatives and friends makes it more likely that older adults' self-concepts and emotional equilibrium will be preserved. Elders state, sometimes quite explicitly, that their days are numbered and they make careful choices about how to allocate their time and energy. Indeed, when young people are asked whom they would choose to spend time with if they knew they would soon be leaving their community, they respond as older people do—with a preference for familiar partners (Carstensen, 1992; Frederickson & Carstensen, 1990).

In sum, selectivity theory views the reduction in social activity in old age as the result of changing life conditions, which lead older people to view social relationships differently than they did at earlier ages. As a result, they much prefer partners with whom they have developed pleasurable, rewarding relationships.

■ SOCIAL CONTEXTS OF AGING: COMMUNITIES, NEIGHBORHOODS, AND HOUSING

Elders live in contexts—both physical and social—that affect their social ties and, consequently, their development and adjustment. Communities, neighborhoods, and housing arrangements vary in the extent to which they enable aging residents to satisfy their social needs.

■ COMMUNITIES AND NEIGHBORHOODS. About half of American ethnic minority older adults live in inner cities, whereas slightly less than one-third of Caucasian Americans do. The majority of senior citizens reside in suburbs, where they moved earlier in their lives and usually remain after retirement. Suburban elders have higher incomes, report better health, and have easier access to social services than do inner-city elders. Inner-city elders, however, are better off in these ways than are the one-fourth of American senior citizens who live in small towns and rural areas. In addition, small-town and rural elderly are less likely to live near their children, who often leave the community in early adulthood (Coward, Lee, & Dwyer, 1993; U.S. Bureau of the Census, 1996).

These men gather regularly to play Chinese chess in Monterey Park, California. In both urban and rural areas, older adults report greater life satisfaction when many senior citizens reside in their neighborhood and are available as like-minded companions. (A. Ramsey/Woodfin Camp & Associates)

Yet rural elderly compensate for distance from family members and social services by interacting more with neighbors and friends. Positive aspects of smaller communities—stability of residents and shared values and lifestyles—foster gratifying social relationships.

Regardless of whether they live in urban or rural areas, older adults report greater life satisfaction when many senior citizens reside in their neighborhood and are available as like-minded companions (Lawton, Moss, & Moles, 1984). Presence of family is not as crucial, so long as neighbors and nearby friends provide social support. This does not mean that neighbors replace family relationships. But elders are content as long as their children and other relatives arrange for visits, in either the elder's or the family member's home (Hooyman & Kiyak, 1993).

Neighborhoods sometimes serve as ethnic enclaves, where minority older adults develop relationships based on common ethnic values and immigrants can interact in their native language. In ethnically diverse communities, minority elders must seek ways of affirming their ethnicity. They often do so through ethnically based churches and clubs (Gelfand, 1994).

Compared to older adults in urban areas, elders residing in quiet neighborhoods in small and mid-sized communities are more satisfied with life. The major reason is that smaller communities have lower crime rates (Scheidt & Windley, 1985). As we will see in the next section, fear of crime has profound, negative consequences for older adults' sense of security and comfort.

■ VICTIMIZATION AND FEAR OF CRIME. Walt and Ruth's single-dwelling house stood in an urban neighborhood with a large Greek population, five blocks from the business district where Walt's photography shop had been prior to his retirement. When leaving home for more than a few hours, Walt and Ruth telephoned their next-door neighbor and asked her to keep an eye on the property. As the neighborhood aged, some homes fell into disrepair, and the population became more transient. Although shops were open Thursday and Friday evenings, Walt and Ruth saved their errands for bright daylight hours. Even then, they were circumspect. Although they had never been victimized, crime was on their minds and affected their behavior.

Media attention has led to a widely held belief that crime against the elderly is common. In reality, older adults are less often targets of crime, especially violent crime, than are other age groups. However, in urban areas, purse snatching and pickpocketing are more often committed against elders (especially women) than younger people, probably because perpetrators feel that they can easily overpower older and female victims (U.S. Department of Justice, 1996). A single incident can strike intense fear in the hearts of elders, given its financial consequences for those with low incomes and its potential for physical injury.

For older adults living alone and in inner-city areas, fear of crime is sometimes greater than worries about income, health, and housing. It quickly undermines morale and neighborhood satisfaction (Bazargan, 1994; Watson, 1991). Neighborhood Watch and other programs that encourage residents to get to know and look out for one another reduce fear and feelings of isolation from the community. Some cities have established special police units to investigate and prevent crimes against the elderly. They improve life satisfaction of elderly victims (Zevitz & Gurnack, 1991).

■ HOUSING ARRANGEMENTS. Overwhelmingly, older adults want to stay in their own homes and neighborhoods where they spent their adult lives, and the large majority (88 percent) do so. In the United States, less than 5 percent relocate to other communities. These moves are usually motivated by a desire to live closer to children or, among more economically advantaged and healthy elders, a desire for a more temperate climate and a place to satisfy leisure interests. Most elder relocations occur within the same community, are prompted by declining health, widowhood, or disability, and increase with age (U.S. Senate Special Committee on Aging, 1991).

As we look at housing arrangements for the elderly, we will see that the more a setting deviates from home life, the harder it is for elders to adjust. Many frail older people resist moving, even though they know they cannot keep up their homes and need a safer environment. They realize that it takes stamina to adjust to a move, and they worry about how well the new situation will suit their need for autonomy (Parmelee & Lawton, 1990). Finally, the high cost of most senior citizen residential communities prevents many elders from relocating when functional declines warrant it.

More elders in Australia, Canada, the United States, and Western Europe live on their own today than ever before—a trend resulting from improved health and economic well-being. (Steve Hansen/Stock Boston)

Ordinary Homes. For the majority of elders, who are not physically impaired, staying in their own homes affords the greatest possible personal control—freedom to arrange space and schedule daily events as one chooses. More elders in Australia, Canada, the United States, and Western Europe live on their own today than ever before —a trend due to improved health and economic well-being (Glaser, 1997). But when health and mobility problems appear, independent living poses risks. Most homes are designed for younger people. They are seldom modified to suit the physical capacities of their elder residents (Parmelee & Lawton, 1990).

When Ruth reached her mid-eighties, Sybil begged her to move into her home. Like many adult children of Southern, Central, and Eastern European descent (Greek, Italian, Polish, and others), Sybil felt an especially strong obligation to care for her frail mother. Older adults of these cultural backgrounds, as well as African Americans, Asian Americans, Hispanics, and Native Americans, more often live in extended families (Angel et al., 1996; Clarke & Neidert, 1992; Kamo & Zhou, 1994; Worobey & Angel, 1990).

Yet compared to immigrants, American-born ethnic minority elders are less content living with their children. Many prefer to live on their own, but poverty prevents them from doing so. With sufficient income to keep her home, Ruth refused to move in with Sybil, despite her doctor's conviction that doing so would be in her best interests. Why do many elders react this way, even after health problems accumulate? Usually they are deeply attached to their homes as sites of memorable life events. They also value their independence, privacy, and network of nearby friends and neighbors (Fogel, 1992).

Over the past half century, the number of unmarried, divorced, and widowed elders living alone has risen dramatically. Currently, one-third of older Americans live on their own, a figure that rises to 47 percent for those age 85 and older. This trend is evident in all segments of the elderly population. However, it is not as pronounced for ethnic minorities (due to extended-family living) and for men, who are far more likely than women to be living with a spouse into advanced old age (U.S. Senate Special Committee on Aging, 1991).

Over 40 percent of American elders who live alone are poverty stricken—a rate five times greater than that of elderly couples. The large majority are widowed women. Although they draw on informal social supports as best they can, many have unmet needs. Some arrive at old age poor, whereas others become poor for the first time, often because they have outlived a spouse who suffered a lengthy, costly illness. With age, their financial status worsens as their assets shrink and their own health care costs rise. Under these conditions, isolation, loneliness, and depression can pile up.

As Figure 18.2 shows, elderly women living alone in other Western industrialized nations are spared these dire outcomes because of more generous government-sponsored income and health benefits (Hardy & Hazelrigg, 1993). In the United States, feminization of poverty deepens in old age.

Residential Communities. About 8 percent of American senior citizens live in residential communities, which come in great variety (Pynoos & Golant, 1996). Housing developments for the aged, either single-dwelling or apartment complexes, differ from ordinary homes only in that they have been modified to suit elders' capacities (for example,

FIGURE 18.2

Percentage of 65- to 74-year-old men and women living alone and earning less than 50 percent of the country's median income in six industrialized nations, based on the most recent available data. Although this criterion is actually above the U.S. poverty line, it marks a very low income and serves as an effective basis of comparison. Not only are the U.S. low-income rates much higher, but the U.S. gender gap is much larger. Because of less generous government-sponsored income and health benefits, elderly women living alone in the United States often suffer from poverty. *(Adapted from Hardy & Hazelrigg, 1993.)*

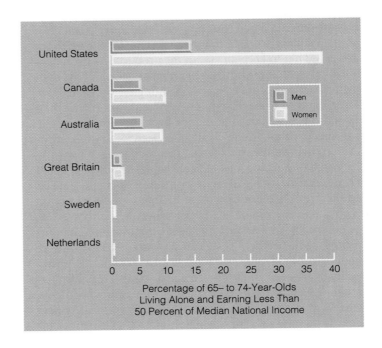

Percentage of 65- to 74-Year-Olds
Living Alone and Earning Less Than
50 Percent of Median National Income

single-level living space and grab bars in bathrooms). Some are federally subsidized units for the elderly poor, but most are privately developed retirement villages with adjoining recreational facilities. For elders who desire more assistance with daily living, **congregate housing** adds a variety of support services, including meals in a common dining room. Finally, **life care communities** offer a range of options, from independent or congregate housing to full nursing home care. For a large initial payment and additional monthly fees, life care guarantees that elders' needs will be met in one place as they age (Parmelee & Lawton, 1990).

In contrast to Ruth and Walt, who remained in their own home, Dick and Goldie entered congregate housing in their late sixties. For Dick, the move was a positive turn of events that permitted him to relate to peers on the basis of their current life together, setting aside past failures in the outside world. Dick found gratifying leisure pursuits—leading an exercise class, organizing a charity drive with Goldie, and using his skills as a baker to make cakes for birthday and anniversary celebrations.

Studies of diverse residential communities for the aged reveal that they can have positive effects on physical and mental health. A specially designed physical space helps elders overcome mobility limitations, thereby permitting greater social participation. And in societies where old age leads to reduced status, age-segregated living is gratifying to most elders who choose it. It may open up useful roles and leadership opportunities, which result in a more vigorous social life (Boyer, 1980; Holmes & Holmes, 1995). Furthermore, living in residential communities does not isolate elders from other relationships. The more friendships they have within, the more likely they are to maintain familiar ties outside the community (Legesse, 1979).

Yet a collection of elders does not guarantee the creation of a comfortable, content community. Shared values and goals among residents with similar backgrounds, a small enough facility to promote frequent communication, and availability of meaningful roles enhance life satisfaction. Finally, older adults in planned housing fare best when a good fit between their functional abilities and environmental supports exists (Moos & Lemke, 1985).

Special housing for the aged is rare in developing countries, where the elderly are almost always cared for by their families. Yet residential communities do appear in cultures changing from agricultural to industrial economies, such as Burma, China, Mexico, Saudi Arabia, and Sri Lanka. As nations modernize, they begin to take responsibility for care of their oldest citizens (Holmes & Holmes, 1995).

Nursing Homes. The 5 percent of Americans age 65 and older who live in nursing homes experience the most extreme restriction of autonomy. As we noted in Chapter 17, sense of personal control and social relationships are as vital for the mental health of nursing home residents as they are for community-dwelling elders.

Potential social partners are abundant in nursing homes, but interaction is low. To regulate emotion in social interaction (so important to elders), personal control over social experiences is vital. Yet nursing home residents have little opportunity to choose their social partners, and timing of contact is generally determined by the institution rather than by personal preference. Social withdrawal is an adaptive response to these overcrowded, hospital-like settings (Carstensen, 1992). Research reveals that interaction with people in the outside world predicts nursing home residents' life satisfaction; interaction within the institution does not (Baltes, Wahl, & Reichert, 1992).

Designing nursing homes to be less institutional and more homelike could do much to increase residents' sense of security and control over their social experiences (Schwarz, 1996). In most European facilities, residents live in private suites or small apartments furnished in part with their own belongings. When an elder's condition worsens, caregivers modify the existing space rather than move the individual to more medically oriented quarters. In this way, they preserve the person's identity, sense of place, and social relationships as best as possible. Because of high costs and stringent government regulations, efforts to modify American nursing homes in these ways have not met with much success.

BRIEF REVIEW

According to disengagement theory, social interaction declines in old age because elders and society withdraw from each other in anticipation of death. In contrast, activity theory identifies social barriers to participation as responsible for disengagement. Selectivity theory points out that elders emphasize the emotion-regulating function of interaction. As a result, they tend to limit social contacts to familiar people with whom they have formed pleasurable, rewarding ties.

Communities, neighborhoods, and housing arrangements vary in the extent to which they enable elders to satisfy their social needs. Because of lower crime rates, older adults are more content in quiet neighborhoods in small and mid-sized communities. The majority of senior citizens stay in the homes and neighborhoods where they spent their adult lives. An increasing number of elders are living alone. In the United States, many are poverty-stricken widowed women with unmet needs.

Elders who choose to live in age-segregated residential communities often benefit from the availability of like-minded companions and meaningful roles. Severe restriction of autonomy in American nursing homes leads residents to withdraw from social interaction.

Test Bank Items 18.59 through 18.62

ASK YOURSELF . . .

■ *Sam lives by himself in the same home he has occupied for over 30 years. His adult children can't understand why he won't move across town to a modern apartment, which would be easier to care for than his old, dilapidated house. Cite as many possible reasons as you can think of that Sam prefers to stay where he is.*

■ *Vera, a nursing home resident, speaks to her adult children and a close friend on the phone every day. In contrast, she seldom attends scheduled social events or interacts with the woman who shares her semi-private room. Using selectivity theory, explain Vera's behavior.*

RELATIONSHIPS IN LATE ADULTHOOD

The **social convoy** is an influential model of changes in our social networks as we move through life. Picture yourself in the midst of a cluster of ships traveling together, granting one another safety and support. Ships in the inner circle represent people closest to you, such as a spouse, best friend, parent, or child. Those less close but still important travel on the outside. With age, ships exchange places in the convoy, and some drift off while others join the procession (Antonucci, 1990). But as long as the convoy continues to exist, you adapt positively.

In the following sections, we examine the ways elders with diverse lifestyles sustain social networks of family members and friends. We will see that as ties are lost, elders draw others closer and even add replacements, although not at the rate they did at younger ages. Tragically, for some older adults the social convoy breaks down. We will explore the circumstances in which elders experience abuse and neglect at the hands of those close to them.

MARRIAGE

Even with the high divorce rate, 1 in every 5 first marriages in the United States is expected to survive for at least 50 years. Walt's comment to Ruth that "the last few decades have been the happiest" characterizes the attitudes and behaviors of many elderly couples who have spent their adult lives together. Marital satisfaction rises from middle to late adulthood, when it is at its peak (Brubaker, 1985; Condie, 1989; Levenson, Carstensen, & Gottman, 1993). Several changes in life circumstance and couples' communication underlie this trend.

First, perceptions of fairness in the relationship increase as men participate more in household tasks after retirement. In today's elders, who experienced little social pressure for gender equality in their youth, division of labor in the home still reflects traditional roles. Men take on more home maintenance projects, whereas women's duties continue as before. Nevertheless, men's involvement in caring for the home results in a greater sense of equity in marriage than before (Condie, 1989; Vinick & Ekerdt, 1991).

Second, with extra time together, the majority of couples engage in more joint leisure activities. Ruth and Walt walked, worked in the garden, played golf, and took frequent day trips. In interviews with a diverse sample of retired couples, women often stated that more time with their husbands enhanced marital closeness (Vinick & Ekerdt, 1991).

Finally, greater emotional understanding and emphasis on regulating emotion in relationships lead to more positive interactions among older couples. Observations of husband–wife communication reveal that couples married for at least 35 years resolve conflicts in ways that are less negative and more affectionate than do middle-aged couples. Even in unhappy marriages, elders are less likely to let disagreements escalate into expressions of anger and resentment (Carstensen, Gottman, & Levenson, 1995; Levenson, Carstensen, & Gottman, 1994). For example, when Dick complained about Goldie's cooking, Goldie often responded in a neutral way to diffuse Dick's anger: "All right, Dick, next birthday I'll give some thought to not having cheesecake." And when Goldie brought up Dick's bickering and criticism, Dick usually said, "I know, dear," and retreated to another room. As in other relationships, the elderly protect themselves from stress by molding marital ties to make them as pleasant as possible.

Although most older couples describe happy and supportive marriages, when marital dissatisfaction is present, it continues to take a greater toll on women than men. Recall from Chapter 14 that women tend to confront marital problems and try to solve them. In old age, the energy expended is especially taxing on their physical and mental health. Husbands, in contrast, often protect themselves by withdrawing, as they did when they were in their twenties and thirties (Levensen, Carstensen, & Gottman, 1993).

DIVORCE AND REMARRIAGE

When Walt's uncle Louie was 61, he divorced his wife Sandra, to whom he had been married for 17 years. Although she knew the marriage was far from perfect, Sandra had lived with Louie long enough that the divorce came as a shock. A year later, Louie married Rachella, a divorcee who shared his enthusiasm for sports and dance.

Couples who divorce in late adulthood constitute a very small proportion of all divorces in any given year—about 1 percent in the United States. But the divorce rate among people age 65 and older is increasing as new generations of elders become more accepting of marital breakup and as the divorce risk rises for second and subsequent marriages

Remarriage rates are low in late adulthood and decline with age, although they are considerably higher among divorced than widowed elders. Older men's opportunities to remarry are considerably greater than women's. *(Joel Gordon)*

(Uhlenberg, Cooney, & Boyd, 1990). When asked about the reasons for divorce, elderly men typically mention lack of shared interests and activities, whereas women frequently refer to their partner's refusal to communicate and emotional distance. "We never talked. I felt isolated," Sandra said (Weingarten, 1988).

Recall from Chapter 16 that the stress of divorce increases in midlife, a trend that continues into old age. The long-time married have given their adult lives to the relationship. Consequently, they find it harder to separate their identity from that of their former spouse, and they suffer more from a sense of personal failure. Relationships with family and friends shift at a time when close bonds are crucial for psychological well-being. Finally, women suffer most from late-life divorce because they are more likely than men to spend their remaining years living alone. The financial consequences are severe—greater than for widowhood because many accumulated assets are lost in property settlements (Wright & Maxwell, 1991).

In younger individuals, divorce often leads to greater awareness of and resolve to change negative patterns of behavior. In contrast, self-criticism in divorced elders heightens guilt and depression because their self-worth depends more on past than future accomplishments. Louie and Sandra blamed each other. "I was always miserable with Sandra," Louie claimed, even though the couple's earlier days had been reasonably happy. Blaming the partner and believing oneself to have acted in the only way possible may distort the marital history, but it is a frequent coping strategy that enables older adults to preserve integrity and self-esteem (Weingarten, 1989).

Remarriage rates are low in late adulthood and decline with age, although they are considerably higher among divorced than widowed elders. Older men's opportunities for remarriage are far greater than women's. Nevertheless, the gender gap in elder remarriage is much smaller after divorce than after widowhood (see Figure 18.3). Perhaps divorcees find it easier than widows to enter a new relationship because their previous one was disappointing. Also, compared to widows, divorced older women may be highly motivated to remarry because of their more extreme economic circumstances. Finally, some divorcees (like Louie and Rachella) leave their marriages only after a new bond is underway (Bowers & Bahr, 1989; Huyck, 1995).

Compared to their younger counterparts, remarried elders enter more stable relationships, as their divorce rate is much lower. In Louie and Rachella's case, the second marriage lasted for 32 years! Perhaps late-life marriages are more

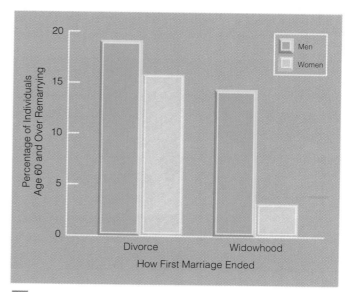

FIGURE 18.3

Rates of remarriage after divorce and widowhood among men and women age 65 and older in the United States. More divorcees than widows remarry. In addition, the gender gap in remarriage is much smaller after divorce than after widowhood. *(U.S. Bureau of the Census, 1994.)*

successful because they involve a better balance of romantic with practical concerns. Older couples who remarry are very satisfied with their new relationships, although men continue to be more content than women (Brubaker, 1985). With fewer potential mates, perhaps women who remarry in late life must settle for less desirable partners.

WIDOWHOOD

Walt died shortly after Ruth turned 80. Like over 70 percent of widowed elders, Ruth described the loss of her spouse as the most stressful event of her life (Lund, Caserta, & Dimond, 1993). She felt lonely, anxious, and depressed for several months after the funeral. For a time, she wondered whether she could continue to live independently because of her poor vision and arthritis.

Widows make up 33 percent of the elderly population in the United States. Because women live longer than men and are less likely to remarry, nearly 50 percent of women age 65 and older are widowed, whereas only 14 percent of men are. African Americans, Hispanics, and other ethnic minorities with higher rates of poverty and chronic disease are more likely to be widowed (U.S. Bureau of the Census, 1994).

Earlier we mentioned that most widows and widowers live alone rather than in extended families, a trend that is stronger for whites than ethnic minorities. Although they are less well off financially than married elders and are concerned about who will care for them in an emergency, most widowed older adults want to live on their own. Reasons they mention are retaining control over their time and living space and avoiding disagreements with their

adult children. When widowed elders relocate because they cannot make mortgage payments or keep up their homes, they usually move closer to family rather than into the same residence (Lopata, 1996; O'Bryant & Hansson, 1995).

Although loss of a spouse is felt very deeply, wide variations in adaptation exist. Age, social support, and personality make a difference. Elders have fewer lasting problems than do younger individuals, probably because death in later life is expected and viewed as less unfair (Stroebe & Stroebe, 1993). In-laws and former couple friends often withdraw from the widowed person's support system. After Walt died, Ruth saw less of Dick and Goldie. But adult children usually step in, as Sybil did for Ruth. Widowed elders with daughters tend to receive more social support than those with sons (Horowitz, 1985; O'Bryant, 1988a). Although loneliness is a problem, most widowed elders are not socially isolated. Many, especially those with outgoing, assertive personalities, compensate by increasing contacts with friends and neighbors. And some relish their solitude!

Widowed individuals must reorganize their lives, reconstructing an identity that is separate from the deceased spouse. The task is harder for wives whose roles depended on their husbands' than for those who developed rewarding roles of their own. But overall, men find it more difficult to adjust to widowhood. Most relied on their wives for social connectedness and essential household tasks. They mention these areas as major sources of strain and depression. Women point to financial worries as their greatest ongoing concern (Umberson, Wortman, & Kessler, 1992).

Although availability of potential partners overwhelmingly favors widowers, sex differences in the experience of widowhood contribute to men's higher remarriage rate. Women's kinkeeper role (see Chapter 16, page 531) and ability to form close friendships may lead them to feel less need to remarry. In addition, because many elderly women share the widowed state, they probably offer one another helpful advice and sympathy. In contrast, men often lack skills for maintaining family relationships, forming emotionally satisfying ties outside marriage, and handling the chores of their deceased wives.

Although widowhood affects lifestyle, in the long run most widowed elders in Western nations fare well. When followed up after several years, they do not differ from their married counterparts in psychological well-being or mortality (McCrae & Costa, 1988; Morycz, 1992). Once again, the elderly display a remarkable ability to cope with traumatic life events. Nevertheless, as the Cultural Influences box on the following page illustrates, adjustment to widowhood varies widely across cultures; not all families, communities, and societies are supportive. The Caregiving Concerns table on page 608 suggests a variety of ways to foster adaptation to loss of a spouse in old age.

These elderly women share the experience of widowhood and offer one another social support, helpful advice, and sympathy. In contrast, men often lack skills for maintaining family relationships and forming emotionally satisfying ties outside marriage. (Jeff Greenberg/The Picture Cube)

CULTURAL VARIATIONS IN THE EXPERIENCE OF WIDOWHOOD

A common assumption is that families in village and tribal societies always take care of their elders. Another is that industrialization inevitably leads to benefits for senior citizens. Yet the realities of widows' lives in traditional and rapidly changing societies reveal that neither of these beliefs is necessarily true. Instead, the welfare of women who outlive their husbands depends on favorable cultural attitudes toward widowhood and warm bonds with others—regardless of the culture's state of technological advancement.

Consider the Hindu religion in India, which historically regarded widowhood as punishment for a crime the widow committed in her previous life. In rural areas, women live in the husband's parents' household after marriage and become financially dependent on their in-laws. Belief that the husband's death is the wife's fault can justify very harsh treatment. Until the nineteenth century, widows were expected to engage in *suttee*, or self-sacrifice on the funeral pyre of the husband's body. Celebrated much like a wedding, the practice brought honor to the widow and her and her husband's families.

Suttee has long been forbidden by the Indian government, although it reappears occasionally in rural areas. Recent inheritance laws also ensure a widowed woman a portion of her deceased husband's property, denied under traditional Hindu customs. But in rural areas, these laws are not always followed, because they go against deeply ingrained cultural norms. In addition, many widowed women are illiterate, ignorant of the laws, and therefore unable to protect themselves (Lopata, 1996; Vlassoff, 1990).

In contrast to rural India, widows in Korean villages benefit from the Confucian value of *filial piety* (respect for parents). The oldest son takes his aging parents into his home and continues to care for his mother after his father's death. She enjoys a lively social life with familiar neighbors until she dies.

As Korea modernized, rural young adults moved to the cities. Unable to maintain the family farm and cut off from contact with grandchildren, many widows left their villages and moved in with their son's family (and sometimes their daughter's). There, they often earned their keep through household tasks and child care while both son and daughter-in-law were at work. And they experienced intense loneliness as village neighbors were replaced by crowded city streets where they knew few people outside their son's home (Koo, 1987). Without societal efforts to create adequate substitutes, rapid cultural change can lead to a breakdown in social supports for widows.

NEVER-MARRIED, CHILDLESS OLDER ADULTS

Shortly after Ruth and Walt's marriage in their twenties, Ruth's father died. Her sister Ida continued to live with and care for their mother, who was in ill health until she died 16 years later. When, at age 25, Ida received a marriage proposal, she responded, "I can't marry anybody while my mother lives. I'm expected to look after her." Ida's decision was not unusual for a daughter of her day. She never married or had children.

About 5 percent of older Americans remain unmarried and childless throughout their lives. Almost all are conscious of being different from the norm, but most have developed alternative meaningful relationships. Ida, for example, formed a strong bond with a neighbor's son. In his childhood, she provided emotional support and financial assistance, which helped him overcome a stressful home life. He included Ida in all significant family events and visited her regularly until she died. Other nonmarried elders also speak of the centrality of younger people—often nieces and nephews—in their social networks and of influencing them in enduring ways (Rubinstein et al., 1991). In addition, same-sex friendships are key in never-married elderly women's lives. These tend to be unusually close and often involve joint travel, periods of coresidence, and associations with one another's extended families.

Never-married elderly women report a level of life satisfaction equivalent to that of married elders and greater than that of divorcees and recently widowed elders. Only when they agree with the widespread belief that "life is empty without a partner" or cannot maintain personal contacts due to declining health do they report feeling lonely (Dykstra, 1995; Rubinstein, 1987). These single women often state that they avoided many problems associated with being a wife and mother, and they view their enhanced friendships as an advantage of not marrying. At the same time, they realize that friendships are not the same as blood ties when it comes to caregiving in old age, and they worry about how care will be provided if it is needed (Jerrome, 1990a).

Never-married elders are more likely than other elders to have siblings, other relatives, and nonrelatives living

CAREGIVING CONCERNS

Fostering Adaptation to Widowhood in Late Adulthood

SUGGESTION	DESCRIPTION
FAMILY AND FRIENDS	
Social support and interaction	Social support and interaction need to extend beyond the grieving period to ongoing assistance and caring relationships. Family members and friends can be of most help by making support available while encouraging the widowed elder to develop skills to cope independently.
COMMUNITY	
Senior centers	About 5,000 senior centers in communities around the United States offer communal meals and other social activities, enabling widowed and other elders to connect with people in similar circumstances and to access other community resources, such as listings of part-time employment and available housing.
Support groups	Special support groups can be found in senior citizens centers, churches, and other agencies. Besides new relationships, they offer an accepting atmosphere for coming to terms with loss, effective role modeling, and assistance with developing skills for daily living.
Religious activities	Involvement in church or synagogue can help relieve the loneliness associated with loss of a spouse and offer social support, new relationships, and meaningful roles.
Volunteer activities	One of the best ways widowed elders can find meaningful roles is through volunteer activities. Some are sponsored by formal service organizations, such as the American Red Cross or the Retired and Senior Volunteer Program. Other volunteer programs exist in hospitals, senior centers, schools, and charitable organizations.

Sources: Lopata, 1996; O'Bryant & Hansson, 1995.

in their households (Stull & Scarisbrick-Hauser, 1989). Although little is known about never-married men, reaching old age without having married is clearly not the same as being alone and forsaken!

SIBLINGS

Nearly 80 percent of Americans over age 60 have at least one living sibling. Most elder siblings live within 100 miles of each other, communicate regularly, and visit each other at least several times a year. In one study, 77 percent of a sample of Canadian elders considered at least one sibling to be a close friend (Connidis, 1989). Both men and women perceive bonds with sisters to be closer than bonds with brothers. Perhaps because of women's greater emotional expressiveness and nurturance, the closer the tie to a sister, the higher elders' psychological well-being (Cicirelli, 1989; O'Bryant, 1988b).

Elderly siblings in industrialized nations are more likely to socialize than provide each other with direct assistance, as most older adults turn to their spouse and children before they turn to their siblings. Nevertheless, siblings seem to be an important "insurance policy" in late adulthood. Most elders say they would turn to a sibling for help in a crisis, less often in other situations (Connidis, 1994). Widowed and never-married elders have more con-

tacts with siblings, perhaps due to stronger feelings of obligation, since they have fewer competing family relationships. They are also more likely to receive sibling support during illness (Connidis & Campbell, 1995). For example, when Ida's Alzheimer's symptoms worsened, Ruth came to her aid. Although Ida had many friends, Ruth was her only living biological relative.

Because siblings share a long and unique history, joint reminiscing about earlier times increases in late adulthood. Walt and Dick often talked about their boyhood days, evoking the warmth of early family life. These discussions helped them appreciate the lifelong significance of the sibling bond and contributed to a sense of family harmony—important aspects of ego integrity (Cicerelli, 1995).

FRIENDSHIPS

As family responsibilities and vocational pressures lessen in late adulthood, friendships take on increasing importance. Having friends is an especially strong predictor of mental health among the elderly (Blieszner & Adams, 1992; Nussbaum, 1994). In one study, retired adults were paged at random intervals and asked to write down what they were doing, whom they were with, and how they felt (much like the adolescents described on page 355 of Chapter 11). They reported more favorable experiences with friends than family members, a difference partly due to the many pleasurable leisure activities shared with friends. But unique qualities of friendship interaction that generate positive emotion—openness, spontaneity, mutual caring, and common interests—seemed especially influential (Larson, Mannell, & Zuzanek, 1986).

■ FUNCTIONS OF ELDER FRIENDSHIPS. The diverse functions of friendship in late adulthood clarify its profound significance:

■ *Intimacy and companionship are basic to meaningful elder friendships.* As Ida and her best friend Rosie took walks, went shopping, attended outdoor concerts in the summer, or just visited each other, the two women disclosed their deepest sources of happiness and worry. They also engaged in pleasurable conversation, laughed, and had fun (Crohan & Antonucci, 1989).

■ *Elderly women mention acceptance as a primary aspect of close friendship.* Late-life friends shield one another from others' negative judgments about their capabilities and worth as a person, which frequently stem from stereotypes of aging (Adams, 1985-1986). "Where's your cane, Rosie?" Ida asked when the two women were about to leave for a restaurant. "Come on, don't be self-conscious. When y'get one of those 'you're finished' looks from someone, just remember: In the Greek village where my mother grew up, there was no separation between generations, so the young ones got used to wrinkled skin and weak knees and recognized older women as the wise ones. Why, they were midwives, matchmakers, experts in herbal medicine; they knew about everything!" (Deveson, 1994).

■ *Friendships link elderly people to the larger community.* When elders cannot go out as often, interactions between friends can keep them abreast of events in the wider world. "Rosie," Ida reported, "did you know that the Thompson girl was named high school valedictorian... and the business community is putting its support behind Jesse for mayor." Elders with friends and older neighbors report a greater sense of integration into society than do those who mostly depend on family members for social contacts (Peterson, 1989). Friends can also open up new experiences that older adults might not take part in alone. Often a first trip to a senior citizens' center takes place within the context of friendship (Nussbaum, 1994).

■ *Friendships can help protect elders from the psychological consequences of loss.* Older adults in declining health who remain in contact with friends through phone calls and visits show improved psychological well-being. Similarly, when close relatives die, friends offer compensating social supports (Newsom & Schulz, 1996; Nussbaum, 1994).

Friendships link elderly people to the larger community. When elders cannot go out as often, interactions between friends can keep them abreast of events in the wider world. (Bob Daemmrich/The Image Works)

■ CHARACTERISTICS OF ELDER FRIENDSHIPS. Although older adults prefer familiar, established relationships over new ones, friendship formation continues throughout life. Ties to old and dear friends who live far away are maintained, but practical restrictions foster more frequent interaction with friends in the immediate environment. With age, elders report that the friends they feel closest to are fewer in number and live in the same community. As in earlier years, they tend to choose friends whose age, sex, race, ethnicity, and values are like their own. However, as age peers die, the very old report more intergenerational friendships (Johnson & Troll, 1994). In her eighties, Ruth spent time with Margaret, a 55-year-old widow she met while serving on the board of directors of an adoption agency. Two or three times a month, Margaret came to Ruth's home for tea and several hours of lively conversation.

Sex differences in friendship, discussed in previous chapters, extend into late adulthood. Women are more likely to have intimate friends; men depend on their wives and, to a lesser extent, their sisters for warm, open communication. Also, older women have more **secondary friends** —people who are not intimates but with whom they spend time occasionally, such as a group that meets for lunch, bridge, or museum tours. Through these associates, elders meet new people, remain socially involved, and gain in psychological well-being (Gupta & Korte, 1994). Perhaps women's interrupted careers and involvement in child rearing prepare them to develop more diverse social worlds in late adulthood, since they had to form new relationships with each life change.

In elder friendships, affection and emotional support are both given and received to maintain a healthy balance in the relationship. Although friends call on each other for help with tasks of daily living, they generally do so only in emergencies or for occasional, limited assistance. As elders avoid excessive dependency on friends, they register their own autonomy. Since friendships do not carry the obligations typically associated with family ties, tensions can arise when a friend tries to help too much. An inability to return the favor can undermine older adults' sense of independence and self-worth (Rawlins, 1992).

RELATIONSHIPS WITH ADULT CHILDREN

About 80 percent of ever-married older adults in the United States are parents of living children, most of whom are middle aged. In Chapter 16, we noted that contact between aging parents and adult children is frequent, and exchanges of help vary with the closeness of the parent–child bond and the parent and adult child's needs. Recall, also, that over time, parent-to-child help declines, whereas child-to-parent assistance increases.

Elders and their adult children are often in touch, even when they live far from each other. Nevertheless, as with other ties, quality rather than quantity of these interactions affects older adults' life satisfaction. As people grow older, children usually continue to provide rich rewards, including love, companionship, and stimulation. These warm bonds reduce the negative impact of physical impairments and other losses (such as death of a spouse) on psychological well-being. Alternatively, conflict or unhappiness with adult children contributes to poor physical and mental health (Peterson, 1989; Silverstein & Bengtson, 1991).

Although aging parents and adult children in the United States provide each other with various forms of help, the level of assistance is typically modest. In interviews with a nationally representative sample, elders reported that exchanges of advice were most common. Less than one-fifth said their children had assisted with household tasks and transportation within the past month. To avoid dependency, older parents expect more emotional support than practical assistance and usually do not seek help from children in the absence of pressing need (Eggebeen, 1992). But when the need arises, most children willingly assist (see Chapter 16).

 Sex differences in older parent–adult child interaction are evident. As kinkeepers, adult daughters are primary agents of their aging parents' family contacts, telephoning, writing letters, and arranging visits more often than do sons. Mother–daughter ties are particularly warm (Choi, 1995; Suitor et al., 1995). In a study of African-American elders, men (especially those living alone) experienced substantially fewer phone calls and visits from their adult children than did women. Perhaps the high rate of African-American father-absent homes weakens fathers' relationships with their adult children many years later (Spitze & Miner, 1992). As large numbers of divorced, noncustodial

fathers reach late adulthood in the next few decades, the number of elderly men with limited family contact may increase (Cooney & Uhlenberg, 1990).

As social networks shrink in size, relationships with adult children become more important sources of family involvement. Elders 85 years and older with children have substantially more contacts with relatives than do those without children (Johnson & Troll, 1992). Why is this so? Consider Ruth, whose daughter Sybil linked her to grandchildren, great-grandchildren, and relatives by marriage. When childless elders reach their eighties, siblings, other same-age relatives, and close friends may have become frail or died and no longer be available as companions. Aging seems to increase the role of children in elders' social lives.

RELATIONSHIPS WITH ADULT GRANDCHILDREN AND GREAT-GRANDCHILDREN

Older adults with adult grandchildren and great-grandchildren benefit from a wider potential network of support. Ruth and Walt saw their granddaughter Marci at family gatherings. At other times, Marci telephoned, visited, and sent greeting cards, expressing deep affection for her aging grandparents. In the few studies available on grandparent–adult grandchild relationships, the overwhelming majority of grandchildren felt obligated to assist grandparents in need. Grandparents expected affection (but not practical help) from grandchildren, and in most cases they received it. They regarded the adult grandchild tie as very gratifying and as a vital link between themselves and the future (Langer, 1990).

About 40 percent of Americans age 65 and older have great-grandchildren. Most describe their new role as limited and a sign of advancing age. Nevertheless, they welcome it with enthusiasm, commenting that it reaffirms the continuance of their families. Parents mediate great-grandchild contact, just as they did grandchild contact (see Chapter 16). Relationships are often closer on the maternal side, especially through daughters and granddaughters (Doka & Mertz, 1988).

Most great-grandparents believe they should be free of child-rearing responsibilities. But like other family roles, expectations vary with family context. In families with several generations of teenage childbearing, entry into parenthood, grandparenthood, and great-grandparenthood is accelerated. If mothers and grandmothers do not care for the baby because they are caught up in adolescent and early adulthood tasks, the responsibility may be pushed up the generation ladder. As one 53-year-old great-grandmother expressed it:

> My girl and grandgirl had these babies young. . . I takes care of babies, grown children, and the old peoples. I work too. I get so tired.

I don't know if I'll ever get to do somethin' for myself. (Burton, 1996, p. 206)

In some instances, off-time great-grandparenthood can lead to a return to parenting, disrupting the anticipated freedom of the late adulthood years.

ELDER MALTREATMENT

Although the majority of older adults enjoy positive relationships with family members, friends, and professional caregivers, some suffer maltreatment at the hands of these individuals. Recent media attention has led elder maltreatment to become a serious public concern in Western nations.

Estimates indicate that about 1.5 million American adults age 60 and older are mistreated by people closest to them each year (Baron & Welty, 1996). Reports from Australia, Canada, Finland, Great Britain, and the United States reveal surprisingly similar rates of maltreatment—about 3 to 7 percent of all elders (Kingston & Reay, 1996). Yet these figures underestimate the actual incidence, since most acts take place in private and victims are often unable or unwilling to complain.

Elder maltreatment takes the following forms:

1. *Physical abuse*—intentional infliction of pain, discomfort, or injury, through hitting, cutting, burning, physical force, restraint, sexual assault, and other acts.

2. *Physical neglect*—intentional or unintentional failure to fulfill caregiving obligations, which results in lack of food, medication, or health services or in the elderly person being left alone or isolated.

3. *Psychological abuse*—verbal assaults (such as name calling), humiliation (being treated as a child), and intimidation (threats of isolation or placement in a nursing home).

4. *Financial abuse*—illegal or improper exploitation of the elder's property or financial resources, through theft or use without the elder's consent.

Although these four types are widely recognized, a common definition of elder maltreatment does not exist. Like child maltreatment (see Chapter 8, page 268), an agreed-on definition is vital for finding out why elder maltreatment occurs and preventing it. A person with multiple bruises and head injuries has surely been abused, but maltreatment is harder to detect in other circumstances. For example, when a frail elder is left alone, how much time must elapse to constitute neglect? And when a relative demands that an elder give back a treasured gift received years earlier and the elder complies in fear of an angry outburst, has financial abuse occurred?

Reported cases suggest that financial abuse is the most common form, followed by psychological abuse and

neglect. Often several types occur in combination (Baron & Welty, 1996; Neale et al., 1996). The perpetrator is usually a person the older adult loves, trusts, and depends on for care and assistance. Most are spouses (usually men), followed by children of both sexes, then other relatives, friends, neighbors, and paid caregivers (Hornick, McDonald, & Robertson, 1992; Pillemer & Finkelhor, 1988).

RISK FACTORS. Characteristics of the victim, the abuser, their relationship, and its social context are related to the incidence and severity of elder maltreatment. The more of the following risk factors that are present, the greater the likelihood that abuse and neglect will occur.

Dependency of the Victim. Very old, frail, and mentally and physically impaired elders are more vulnerable to maltreatment. This does not mean that physical declines cause abuse, since most older adults with disabilities do not experience it. Rather, when other conditions are ripe for maltreatment, elders with severe disabilities are least able to protect themselves (Glendenning, 1993).

Dependency of the Perpetrator. Many abusers are dependent, emotionally or financially, on their victims. This dependency, experienced as powerlessness, can lead to aggressive, exploitative behavior. Often the perpetrator–victim relationship is one of mutual dependency. The abuser needs the older person for money or housing, and the older person needs the abuser for assistance with everyday tasks or to relieve loneliness (Kingston & Reay, 1996).

Psychological Disturbance and Stress of the Perpetrator. Abusers are more likely than other caregivers to have psychological problems and to have alcohol or other drug problems. Often they are socially isolated, have difficulties at work, or are unemployed, with resulting financial worries. These factors increase the likelihood that they will lash out when caregiving is highly demanding or the behavior of an elder with dementia is irritating or hard to manage.

History of Family Violence. Elder abuse is often part of a long history of family violence. In Chapter 8, we showed how aggressive cycles between family members can easily become self-sustaining, leading to the development of individuals who cope with anger through hostility toward others. When the husband is the perpetrator, elder abuse may simply be an extension of many years of spouse abuse (see Chapter 14, page 466).

Institutional Conditions. Elder maltreatment is more likely to occur in nursing homes that are run down and overcrowded and that have staff shortages, minimal staff supervision, high staff turnover, and few visitors. When highly stressful work conditions combine with minimal oversight of caregiving quality, the stage is set for abuse and neglect (Glendenning, 1993; Pillemer & Moore, 1989).

PREVENTING ELDER MALTREATMENT. Preventing elder maltreatment by family members is especially challenging. Once abuse or neglect is discovered, intervention must be negotiated with the perpetrator and the victim, neither of whom may desire help. Each must be convinced that assistance can improve their situation—treatment and social support aimed at reducing stress for the caregiver and protection and provision of unmet needs for the elder.

Prevention programs offer counseling and education for caregivers and respite (relief) services, such as elder day care and in-home help. Sometimes trained volunteer "buddies" make visits to the home to combat social isolation and assist elders with problem solving to avoid further harm. Support groups help seniors identify abusive acts, practice appropriate responses, and form new relationships. And agencies that provide informal financial services to older adults unable to manage on their own, such as writing and cashing checks and holding valuables in a safe, reduce financial abuse (Reis & Nahmiash, 1995; Wolf & Pillemer, 1994).

When elder abuse is extreme, legal action is the best way to shield elders from abusers, yet it seldom happens. Many victims are reluctant to initiate court procedures or, because of mental impairments, cannot do so (Griffiths, Roberts, & Williams, 1993). In these instances, social service professionals need to help caregivers rethink their role, even if it means the aging person might be institutionalized. In nursing homes, abuse and neglect can be prevented by improving staff selection, training, and working conditions.

Combating elder maltreatment also requires efforts at the level of the larger society. Public education to encourage reporting of suspected cases and improved understanding of the needs of older people is vital (Baron & Welty, 1996). Finally, countering negative stereotypes of aging reduces maltreatment, since recognition of elders' dignity, individuality, and autonomy is incompatible with acts of physical and psychological harm.

BRIEF REVIEW

Marital satisfaction peaks in late adulthood, due to increased perceptions of fairness in household tasks, more time for joint leisure activities, and more positive couple interaction. Although elder divorces are few, they generate more stress than earlier divorces as ties to family and friends shift. Remarriage rates are also low, but they are much higher after divorce than widowhood and for men than women. Remarriage in late adulthood yields more stable relationships than at earlier ages. The majority of older adults describe widowhood as highly stressful. Nevertheless, they adjust better than adults widowed at younger ages, especially if they have their children's social support and an outgoing, assertive personality. Never-married childless elders, who are mostly women, have formed alternative meaningful relationships, especially with same-sex friends.

Most siblings communicate regularly, and bonds with sisters are particularly close. Elders' closest friends typically live in the same community. They offer intimacy and companionship, acceptance, links to the wider world, and protection from the psychological consequences of loss. Positive ties with adult children limit the impact of age-related losses on psychological well-being. To maintain a sense of independence, most older adults expect more emotional support than practical assistance from their children. Mother–daughter bonds are especially warm. Relationships with adult grandchildren are usually gratifying and provide a vital link with the future. Although elders are less directly involved with great-grandchildren, these ties reaffirm their sense of family continuity.

About 3 to 7 percent of elders are maltreated by people close to them. Mutual dependency of perpetrator and victim, perpetrator psychological disturbance, a history of family violence, and nursing homes with poor working conditions increase the chances that elder abuse will occur.

ASK YOURSELF . . .

- *Seventy-year-old Sean says his 40-year marriage to Caitlin is the happiest it's ever been. Caitlin agrees that she's more content than before, but she isn't quite as positive as Sean. What might account for Sean and Caitlin's high marital satisfaction and Sean's especially favorable response?*

- *Lottie, a never-married elder, lives by herself. Curt, a college student who just moved in next door, is certain Lottie must be very lonely. Why is Curt's assumption probably wrong?*

- *Mae, who lost her job at age 51 and couldn't afford her own apartment, moved in with her 78-year-old widowed mother, Beryl, who was glad to have Mae's companionship. Mae grew depressed and spent her days watching TV and drinking heavily. Although Beryl tried to be patient, she complained about Mae's failure to look for work. Mae became belligerent, pushed her mother against the wall, and slapped her. Explain why this mother–daughter relationship led to elder abuse.*

RETIREMENT AND LEISURE

In Chapter 16, we noted that the period of retirement has lengthened due to increased life expectancy and a steady decline in average age of retirement—trends occurring in all Western industrialized nations. These changes have also led to a blurring of the distinction between work and retirement as older adults work part-time and, occasionally, return to full-time work both to support themselves and to introduce interest and challenge into the retirement years. Recent evidence indicates that almost one-third of American retirees re-enter the labor force, most within 1 year of initial retirement, although the likelihood of return declines with age (Hayward, Hardy, & Liu, 1994; Ruhm, 1989).

Because mandatory retirement no longer exists for most workers in Western countries, older adults have more choices about when to retire and how they spend their time. In the following sections, we examine factors that affect the decision to retire, happiness during the retirement years, and leisure pursuits. We will see that the process of retirement and retired life reflect an increasingly diverse retired population.

THE DECISION TO RETIRE

When Walt and Ruth retired, both had worked for enough years to be eligible for comfortable income-replacement benefits—Walt's through the government-sponsored Social Security program, Ruth's through a private pension plan. In addition, Walt and Ruth had planned for retirement (see Chapter 16, page 542) and decided when they would leave the work force. They wanted to retire early enough to pursue leisure activities while they were both in good health and could enjoy them together. In contrast, Walt's brother Dick was forced to retire as the operating costs of his bakery rose and his clientele dropped off. His wife Goldie kept her part-time job as a bookkeeper to help cover their living expenses.

Affordability of retirement is usually the first consideration in the decision to retire. Yet even in the face of eco-nomic concerns, many preretirees favor letting go of a steady work life for alternative, personally meaningful work or leisure activities. As one retired automobile worker said, "I was working since I was 10 years old. I thought of all the years I've been working, and I wanted a rest." Exceptions to this favorable outlook are people like Dick—forced into retirement or anticipating serious financial difficulties (Bossé, Spiro, & Kressin, 1996).

Figure 18.4 summarizes factors in addition to income that influence the decision to retire. People in good health, for whom vocational life is central to self-esteem, and whose work environments are pleasant and stimulating are likely to keep on working. For these reasons, individuals in professional occupations usually retire later than those in blue-collar or clerical jobs (Moen, 1996). Self-employed elders also stay with their jobs longer, probably because they can flexibly adapt their working hours to changing needs. In contrast, people in declining health; who are engaged in routine, boring work; and who have compelling leisure interests often opt for retirement.

Societal factors also affect retirement decisions. When many younger, less costly workers are available to replace older workers, industries are likely to offer added incentives for people to retire, such as increments to pension plans and earlier benefits. When concern increases about the burden on younger generations of a rising population of retirees, eligibility for retirement benefits may be postponed to a later age.

Retirement decisions vary with gender and ethnicity. On the average, women tend to retire earlier than men, largely because family events—husband's retirement or the need to care for an ill spouse or parent—play larger roles in their decisions (Hatch & Thompson, 1992; Quadagno & Hardy, 1996). However, women in or near poverty are an exception. Many find themselves without the financial

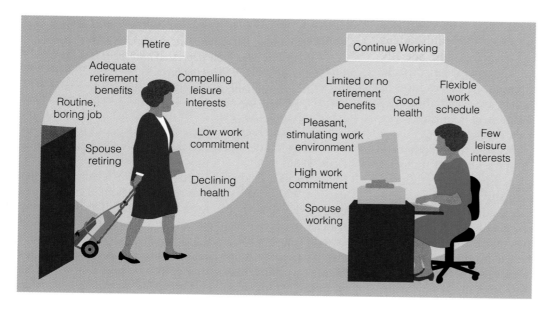

FIGURE 18.4

Factors that influence the decision to retire.

Retire

Adequate retirement benefits
Routine, boring job
Compelling leisure interests
Low work commitment
Spouse retiring
Declining health

Continue Working

Limited or no retirement benefits
Good health
Flexible work schedule
Pleasant, stimulating work environment
Few leisure interests
High work commitment
Spouse working

resources to retire and must continue working into old age. This trend is especially pronounced among African-American women, who are more likely to have minimal retirement benefits and to be caring for other family members (Choi, 1994; Perkins, 1993). In other Western nations, higher minimum pension benefits make retirement feasible for the economically disadvantaged. In addition, some countries have retirement policies sensitive to women's more interrupted work lives. In Canada, France, and Germany, for example, time devoted to child rearing is given some credit when figuring retirement benefits (O'Grady-LeShane & Williamson, 1992).

In sum, individual preferences shape retirement decisions. At the same time, older adults' opportunities and limitations greatly affect their choices.

ADJUSTMENT TO RETIREMENT

Because retirement involves giving up roles that are a vital part of identity and self-esteem, it is often thought of as a stressful process that contributes to declines in physical and mental health. Yet consider Dick, who reacted to the closing of his bakery with anxiety and depression. His adjustment difficulties were not very different from those of younger people experiencing job loss (see Chapter 16, page 542). Also, recall that Dick had a cranky, disagreeable personality. In this respect, his psychological well-being after retirement was similar to what it had been before!

We must be careful not to assume a cause-and-effect relationship each time retirement and an unfavorable reaction are paired. For example, a wealth of evidence confirms that health problems lead elders to retire, rather than the reverse. And for most people, mental health is fairly stable from the pre- to postretirement years, with little change prompted by retirement itself (Bossé, Spiro, & Kressin, 1996).

The widely held belief that retirement inevitably leads to adjustment problems is contradicted by countless research findings indicating that most older adults adapt well. They describe themselves as active and socially involved—major determinants of retirement satisfaction. But the opposite view—that retirement never produces ill effects—is also wrong. When elders are asked about the stresses of retirement, about 30 percent mention some adjustment difficulties (Bossé et al., 1990).

Factors associated with reluctance to retire—high work involvement, having to give up one's job, and financial worries—predict stress following retirement. In addition, work pressures make a difference. Moving out of a high-stress job is associated with positive adaptation to retirement among men, whereas leaving a pleasant, low-stress job is linked to greater difficulties (Wheaton, 1990). Furthermore, people who find it hard to give up the predictable schedule and social contacts of the work setting often react negatively to their less structured way of life (Fletcher & Hansson, 1991).

Social support reduces stress associated with retirement, just as it helps relieve the stress of other major life events. Although size of the social network typically shrinks as relationships with co-workers decline, quality of social support (number of people elders "can count on") remains fairly stable. In Dick's case, entering congregate housing eased a difficult postretirement period. It led to new friends and rewarding leisure pursuits, some of which he shared with Goldie. Besides friends, spouses are a vital source of support in fostering retirement adjustment. The number of leisure activities couples enjoy together predicts retirement satisfaction (Dorfman et al., 1988; Reeves & Darville, 1994).

Finally, earlier in this chapter we noted that marital happiness tends to rise after retirement (Atchley, 1992; Gilford & Bengtson, 1979). When a couple's relationship is generally positive, retirement can increase satisfaction by granting husband and wife more time for companionship. Consequently, a good marriage not only promotes adjustment to retirement, but benefits from the greater freedom of the retirement years. Return to Chapter 16, page 543, for ways adults can plan ahead to increase the chances of a favorable transition to retirement.

LEISURE ACTIVITIES

With retirement, most older adults have more time for leisure pursuits than ever before. After a "honeymoon period" of trying out new activities, many find that leisure interests and skills do not develop suddenly. Instead meaningful leisure pursuits are usually formed earlier and sustained or expanded during retirement (Cutler & Hendricks, 1990). For example, Walt's fondness for writing, theater, and gardening dated back to his youth. And Ruth's strong focus on her vocation of social work led her to become an avid community volunteer. The stability of leisure activities over the lifespan suggests that the best preparation for leisure in late life is to develop rewarding interests at a young age.

Involvement in leisure activities is positively related to elders' psychological well-being. But simply participating does not explain this relationship. Instead, elders select leisure pursuits because they permit self-expression, new achievements, the rewards of helping others, or pleasurable social interaction. These factors account for gains in well-being (Cutler & Hendricks, 1990; Kelly, Steinkamp, & Kelly, 1986).

With age, frequency and variety of leisure pursuits decline, with travel, outdoor recreation, and exercise especially likely to drop off. After age 75, mobility limitations lead leisure activities to become more sedentary and home based (Kelly & Ross, 1989). Elders in residential communities participate more than do those in ordinary homes because activities are conveniently available. But regardless of living arrangements, older adults do not spend much time in programs designed just for them. Rather, they choose activities on the basis of whether they are personally gratifying. Partly for this reason, organized activities in community senior centers attract only about 15 percent of elders who live nearby (Krout, Cutler, & Coward, 1990). Nevertheless, these structured opportunities are important for elders with limited incomes and for those who lack daily companionship.

Older adults make a vital contribution to society through volunteer work—in hospitals, senior centers, schools, charitable organizations, and other community settings (U.S. Senate Special Committee on Aging, 1992). Younger, better educated, and financially secure elders with social interests are more likely to volunteer, and women do so more often than men. Volunteering also varies with ethnicity. African Americans are more likely to assist others informally, often through their church. In Native American communities, providing food and lodging to older people is common. Elders reap great personal benefits from giving to others (Hooyman & Kiyak, 1993). However, volunteer work is seldom begun in late adulthood. Like other leisure pursuits, it originates earlier, usually during the individual's working years.

Finally, when Walt and Ruth got together with Dick and Goldie, the two couples could often be heard discussing politics. Older adults report greater awareness of and interest in public affairs and vote at a higher rate than any age group. And their political knowledge shows no sign of decline, even in late old age. Why is this so? After retiring, elders have more time to read and watch TV, through which they keep abreast of current events. They also have a major stake in politics as debate, sparked by intergenerational inequity, continues over policies central to their welfare (see Chapter 17, page 572). But elders' political concerns are far broader than those that serve their own age group, and their voting behavior is not driven by self-interest (Binstock & Day, 1996). Instead, their political involvement may stem from a deep desire for a safer, more secure world for future generations.

SUCCESSFUL AGING

Walt, Ruth, Dick, Goldie, and Ida, and the findings of research they illustrate, reveal great diversity in development during the final decades of life. Experts would undoubtedly apply the term **successful aging** to Walt and Ruth. They were reasonably healthy until an advanced age, were happily married, and had daily lives filled with rewarding activities and relationships. And Ruth, especially, displayed a remarkable capacity to adapt to negative life changes. Ida, too, was a successful ager until the onset of Alzheimer's symptoms overwhelmed her ability to manage life's challenges. As a single adult, she built a rich social network that sustained her into old age, despite the hardship of having spent many years caring for her ailing mother. In contrast, Dick and Goldie reacted with

despondency to physical aging and other losses (such as Dick's forced retirement). And Dick's angry outbursts restricted their social contacts, which intensified dissatisfaction with life.

Successful agers are people for whom growth, vitality, and striving limit and, at times, overcome physical, cognitive, and social declines. Researchers want to know more about their characteristics and development so they can help more seniors age successfully. Yet theorists disagree on the precise ingredients of successful aging. Some focus on easily measurable outcomes, such as excellent cardiovascular functioning, absence of disability, superior cognitive performance, and creative accomplishments. Yet this view has been heavily criticized (Baltes & Carstensen, 1996). Not everyone can become an outstanding athlete, a Nobel laureate, or an acclaimed artist. Each of us is limited by our genetic potential as it combines with environments we encounter and select for ourselves. Furthermore, outcomes valued in one culture may not be valued in others.

Recent definitions of successful aging have turned away from specific achievements toward processes people use to reach personally valued goals (Rowe & Kahn, 1987; Schultz & Heckhausen, 1996). This perspective avoids identifying one set of standards as "successful." Instead, it focuses on how people minimize losses while maximizing gains. Take a moment to list the many ways, considered in this and the previous chapter, that older adults realize their goals. Here are the most important ones, with page references so you can review:

- Effective coping with physical changes and negative life events (page 561)

- Optimism about improving health and physical functioning (page 563)

- Selective optimization with compensation to make the most of limited physical energies and cognitive resources (pages 576 and 595)

- Strengthening of self-concept, which promotes self-acceptance and pursuit of hoped-for possible selves (page 591)

- Acceptance of change, which fosters life satisfaction (page 591)

- A mature sense of spirituality and faith, permitting anticipation of death with calmness and composure (page 591)

- Personal control over domains of dependency and independence (page 595)

- High-quality relationships, which offer social support and pleasurable companionship (pages 598 and 600)

Together, these aspects of successful aging permit elders—even those who are suffering enormous hardships—to retain a sense of future, prepare for further challenges, and feel that life has purpose.

Successful aging is facilitated by societal contexts that permit elders to manage life changes effectively. As we have noted throughout our discussion, older adults need adequate health care, safe housing, and social services. In the United States, the federal government guarantees all citizens age 60 and older access to a wide variety of services tailored to their needs. (See the description of the Area Agencies on Aging in Chapter 2, page 66.) Nevertheless, inadequate funding and difficulties reaching rural communities mean that many older adults are not served. And isolated elders with little education often do not know how to seek available assistance. Furthermore, we have seen that America's system of sharing health care costs strains the financial resources of many senior citizens. And housing that adjusts to changes in elders' capacities, permitting them to age in place without disruptive and disorienting moves, is available only to the economically well-off in the United States.

Besides improving policies that meet older adults' basic needs, new future-oriented approaches must prepare for increased aging of the population. More emphasis on life-long learning for workers of all ages would help people maintain as well as increase skills and productivity as they grow older. Otherwise, society has little to gain from keep-ing more elders in the work force by raising the age of eligibility for full Social Security benefits over the next few decades. Finally, reforms that prepare for expected growth in the number of frail elders are vital, including affordable adapted housing, help for family caregivers, and sensitive nursing home care (Hicks, 1997).

All these changes involve recognizing, improving, and supporting the contributions senior citizens make to society—both the elders of today and those now aging to be the future elderly. A nation that takes care of its senior citizens and grants them a multitude of opportunities for personal growth maximizes the chances that each of us, when our time comes to be old, will age successfully.

ASK YOURSELF . . .

■ *Nate, happily married to Gladys, adjusted well to retirement. He also found that after he retired, his marriage became even happier. How can a good marriage ease the transition to retirement? How can retirement enhance marital satisfaction?*

 **UMMARY**

ERIKSON'S THEORY: EGO INTEGRITY VERSUS DESPAIR

According to Erikson, how does personality change in late adulthood?

■ The final psychological conflict of Erikson's theory, **ego integrity versus despair,** involves coming to terms with one's life. Adults who arrive at a sense of integrity feel whole and satisfied with their achievements. Despair occurs when elders feel they have made many wrong decisions, yet time is too short for change.

OTHER THEORIES OF PSYCHOSOCIAL DEVELOPMENT IN LATE ADULTHOOD

Describe Peck's and Labouvie-Vief's views of development in late adulthood, and discuss the functions of reminiscence and life review in older adults' lives.

■ According to Robert Peck, the conflict of ego integrity versus despair comprises three distinct tasks: (1) ego differentiation versus work-role preoccupation; (2) body transcendence versus body preoccupation; and (3) ego transcendence versus ego preoccupation.

■ Gisella Labouvie-Vief addresses the development of adults' reasoning about emotion, pointing out that older, more psychologically mature individuals are more in touch with their feelings.

■ Researchers do not yet fully understand why older people engage in **reminiscence** more often than do younger people. In a special form of reminiscence called **life review,** elders call up, reflect on, and reconsider past experiences with the goal of achieving greater self-understanding.

CHANGES IN SELF-CONCEPT AND PERSONALITY TRAITS

Describe changes in self-concept and personality traits in late adulthood.

■ Elders have accumulated a lifetime of self-knowledge, leading to more secure and complex self-concepts than at earlier ages.

■ During late adulthood, shifts in three personality traits take place. Agreeableness and acceptance of change tend to rise, whereas sociability dips slightly.

Discuss spirituality and religiosity in late adulthood, and trace the development of faith.

■ Although organized religious participation declines, informal religious activities remain common in late adulthood. Religious involvement is especially high among low-income ethnic minority elders and women.

- Faith and spirituality may advance to a higher level in late adulthood, away from prescribed beliefs to a more reflective approach that is at ease with uncertainty.

INDIVIDUAL DIFFERENCES IN PSYCHOLOGICAL WELL-BEING

Discuss individual differences in psychological well-being as older adults respond to increased dependency, declining health, and negative life changes.

- Friends, family members, and caregivers often promote excessive dependency in elders. In sequences of behavior called the **dependency-support script** and **independence-ignore script,** older adults' dependency behaviors are attended to immediately and their independent behaviors are ignored. For dependency to foster well-being, elders need to assume personal control over it.

- Health is a powerful predictor of psychological well-being in late adulthood. The relation between physical and mental health problems can become a vicious cycle, each intensifying the other.

- Although elders are at risk for a variety of negative life changes, these events may evoke less stress and depression than they do in younger adults. Many seniors have learned to cope with hard times.

Describe the role of social support and social interaction in promoting physical health and psychological well-being in late adulthood.

- In late adulthood, social support from family, friends, and paid helpers reduces stress, thereby promoting physical health and psychological well-being. At the same time, older adults do not want a great deal of support from people close to them unless they can reciprocate.

A CHANGING SOCIAL WORLD

Describe social theories of aging, including disengagement theory, activity theory, and selectivity theory.

- **Disengagement theory** holds that social interaction declines due to mutual withdrawal between elders and society in anticipation of death. However, not everyone disengages, and elders' retreat from interaction is more complex than this theory implies.

- According to **activity theory,** social barriers to engagement, not the desires of elders, cause declining rates of interaction. Yet offering older adults opportunities for social contact does not lead to greater social activity.

- **Selectivity theory** states that social networks become more selective as we age. As older adults emphasize the emotion-regulating function of interaction, they tend to limit their contacts to familiar partners with whom they have developed pleasurable, rewarding relationships.

How do communities, neighborhoods, and housing arrangements affect elders' social lives and adjustment?

- Elders residing in suburbs are better off in terms of income, health, and access to social services than those in inner cities. Those in small towns and rural areas are least well off in these ways. Older adults report greater life satisfaction when many senior citizens reside in their neighborhood.

- For elders living alone and in inner-city areas, fear of crime is sometimes greater than worries about income, health, and housing.

- The housing arrangement offering seniors the greatest personal control is their own home. Nevertheless, many older adults who live alone, especially widowed women, are poverty stricken and have unmet needs.

- Most residential communities for senior citizens are privately developed retirement villages with adjoining recreational facilities. **Congregate housing** offers a variety of support services, including meals in a common-dining room. **Life care communities** provide a range of options, from independent or congregate housing to full nursing home care. Only 5 percent of Americans age 65 and older live in nursing homes.

RELATIONSHIPS IN LATE ADULTHOOD

Describe changes in social relationships in late adulthood, including marriage, divorce, remarriage, and widowhood, and discuss never-married, childless older adults.

- As we move through life, a cluster of family members and friends, or **social convoy,** provides safety and support. With age, some bonds become closer, others more distant, while still others drift away.

- Marital satisfaction rises from middle to late adulthood as perceptions of fairness in the relationship increase, couples engage in joint leisure activities, and communication becomes more positive.

- When divorce occurs, stress is higher for older than younger adults. Although remarriage rates are low in late adulthood, those who do remarry enter into more stable relationships.

- Wide variations in adaptation to widowhood exist. Elders fare better than younger individuals. Women—especially those who developed rewarding roles outside the marital relationship—adjust more easily than men.

- Most older adults who remain unmarried and childless

throughout their lives develop alternative meaningful relationships. At the same time, they often worry about how care will be provided in old age if it is needed.

How do sibling relationships and friendship change in late life?

■ Because siblings share a long and unique history, joint reminiscing increases in late adulthood. Elder siblings continue to serve as vital sources of social support.

■ Among the functions of friendship in late adulthood are intimacy and companionship, acceptance, a link to the larger community, and protection from the psychological consequences of loss. Women are more likely than men to have both intimate friends and **secondary friends**—people with whom they spend time occasionally.

Describe older adults' relationships with adult children, adult grandchildren, and greatgrandchildren.

■ Elders and their adult children are often in touch, typically exchanging advice rather than direct assistance. Adult daughters are primary agents of elderly parents' family contacts.

■ Seniors with adult grandchildren and great-grandchildren benefit from a wider network of support. Most often, grandparents expect and receive affection rather than practical help from their grandchildren.

Discuss elder maltreatment, including risk factors and strategies for prevention.

■ Some elders suffer maltreatment at the hands of family members, friends, and professional caregivers. Risk factors include a mutually dependent perpetrator–victim relationship; perpetrator psychological disturbance and stress; a history of family violence; and overcrowded nursing homes with staff shortages and turnover.

■ Prevention of elder maltreatment requires both perpetrator and victim to recognize that help can improve their situation.

RETIREMENT AND LEISURE

Discuss the decision to retire, adjustment to retirement, and involvement in leisure activities.

■ The decision to retire depends on affordability, health status, opportunities to pursue meaningful activities, societal factors such as early retirement benefits, gender, and ethnicity.

■ Such factors as involvement in and enjoyment of work, social support, and marital happiness affect adjustment to retirement.

■ Engaging in meaningful and pleasurable leisure activities is positively related to elders' psychological well-being. The best preparation for leisure experiences in late life is to develop rewarding interests at a young age.

SUCCESSFUL AGING

Discuss the meaning of successful aging.

■ Elders who experience **successful aging** have developed many effective ways to minimize losses and maximize gains.

IMPORTANT TERMS AND CONCEPTS

ego integrity versus despair (p. 588)
reminiscence (p. 590)
life review (p. 590)
dependency-support script (p. 594)
independence-ignore script (p. 594)

disengagement theory (p. 599)
activity theory (p. 599)
selectivity theory (p. 600)
congregate housing (p. 603)
life care communities (p. 603)

social convoy (p. 604)
secondary friends (p. 609)
successful aging (p. 615)

FOR FURTHER INFORMATION AND HELP

NEGATIVE STEREOTYPES OF AGING

Gray Panthers
2025 Pennsylvania Avenue, N.W.
Suite 821
Washington, DC 20006
(800) 280-5362
A coalition of young, middle-aged, and older Americans dedicated to overcoming discrimination on the basis of age.

HOUSING ARRANGEMENTS

American Association of Homes
and Services for the Aged
901 E St., N.W., Suite 500
Washington, DC 20004-2037
(202) 783-2242
Web site: *www.aahsa.org*
Organization of not-for-profit nursing homes, housing, retirement communities, and health-related facilities and services for the elderly. Provides a coordinated means of protecting and advancing the interests of residents.

National Institute
of Senior Housing
c/o National Council on the Aging
409 3rd Street, S.W., Second Floor
Washington, DC 20024
(202) 479-6654
Web site: *www.ncoa.org*
Works to promote a national response to the growing need for affordable, decent housing and living arrangements for older adults.

WIDOWHOOD

Widowed Persons Service
c/o American Association
of Retired Persons
601 E Street, N.W.
Washington, DC 20049
(202) 434-2260
A national outreach group of volunteers, widowed for at least 18 months, who listen to and support people adjusting to widowhood.

ELDER MALTREATMENT

National Committee for the
Prevention of Elder Abuse
Medical Center of Central
Massachusetts
Institute on Aging
119 Belmont Street
Worcester, MA 01605
(508) 793-6611
Promotes understanding of elder maltreatment and development of services to protect older adults from abuse and neglect.

RETIREMENT

American Association
of Retired Persons
601 E Street, N.W.
Washington, DC 20049
(202) 434-2277
Web site: *www.aarp.org*
An active advocacy organization for older people, with 32 million members 50 years of age or older, working or retired. Aims to enhance quality of life, independence, and dignity of the elderly.

VOLUNTEERISM

National Senior Service Corps
1201 New York Avenue, N.W.
Washington, DC 20025
(202) 606-5000
Web site: *www.cns.gov/senior.html*
Brings retired people more fully into community life through volunteer services that vary according to preference and community needs. Projects are organized and operated at the local level in schools, courts, and health care, day care, youth, and other community centers.

AGE	PHYSICAL	COGNITIVE	EMOTIONAL/SOCIAL

60–80 years

PHYSICAL

- Neurons die at a faster rate, but the brain compensates through growth of new synapses.

- Autonomic nervous system performs less well, impairing adaptation to hot and cold weather.

- Declines in vision continue, in terms of increased sensitivity to glare and impaired color discrimination, dark adaptation, depth perception, and visual acuity.

- Declines in hearing continue throughout the frequency range.

- Taste and odor sensitivity may decline.

- Touch sensitivity declines on the hands, particularly the fingertips, less so on the arms.

- Declines in cardiovascular and respiratory functioning lead to greater physical stress during exercise.

- Aging of the immune system increases risk for a variety of illnesses.

- Sleep difficulties increase, especially for men.

COGNITIVE

- Processing speed continues to decline; crystallized abilities are largely sustained.

- Amount of information that can be retained in working memory diminishes further; memory problems are greatest on complex tasks requiring deliberate processing.

- Modest forgetting of remote memories occurs.

- Use of external aids for prospective memory increases.

- Retrieving words from long-term memory and planning what to say and how to say it become more difficult.

- Information is more likely to be remembered in terms of gist than details.

- Traditional and practical problem solving declines; health problem solving becomes more decisive.

- May hold one of the most important positions in society, such as chief executive officer, religious leader, or Supreme Court justice.

EMOTIONAL/SOCIAL

- Comes to terms with life, developing ego integrity.

- Describes emotional reactions in more complex and personalized ways.

- May engage in reminiscence and life review.

- Self-concept strengthens, becoming more secure and complex.

- Agreeableness and acceptance of change increase.

- Faith and spirituality may advance to a higher level.

- Size of social network and amount of social interaction decline.

- Selects social partners on the basis of emotion, approaching pleasant and avoiding unpleasant relationships.

- Marital satisfaction increases.

- May be widowed.

- Sibling bonds may strengthen further.

- Number of friends generally declines.

AGE	PHYSICAL	COGNITIVE	EMOTIONAL/SOCIAL
60–80 years (continued)	■ Graying and thinning of the hair continue; the skin wrinkles further and becomes more transparent as it loses its fatty layer of support. ■ Height and weight (due to loss of lean body mass) decline. ■ Loss of bone mass continues, leading to rising rates of osteoporosis. ■ Intensity of sexual response and sexual activity decline, although most healthy married couples report regular sexual enjoyment.	■ May excel at wisdom. ■ Can improve a wide range of cognitive skills through training.	■ May become a great-grandparent. ■ May retire. ■ More likely to vote and be knowledgeable about politics.

| 80 years and older | ■ Physical changes described above continue.

■ Mobility diminishes, due to loss of muscle and bone strength and joint flexibility. | ■ Cognitive changes described above continue.

■ Fluid abilities decline further; crystallized abilities begin to drop as well. | ■ Emotional and social changes described above continue.

■ As relatives and friends die, may develop friendships with younger individuals.

■ Relationships with adult children become more important.

■ Frequency and variety of leisure activities decline. |

Death, Dying, and Bereavement

As every life is unique, so each death is unique. The final forces of the human spirit separate themselves from the body in manifold ways. My mother Sofie's death was the culmination of a 5-year battle against cancer. In her last months, the disease invaded organs throughout her body, attacking the lungs in its final fury. She withered slowly, having been granted the mixed blessing of time to prepare against certain knowledge that death was just around the corner. My father, Philip, lived another 18 years. Outwardly healthy, active, and about to depart on a long-awaited vacation at age 80, a heart attack snuffed out his life suddenly, without time for last words or deathbed reconciliations.

As I set to work on this chapter, my 65-year-old neighbor Nicholas gambled for a higher quality of life. To be eligible for a kidney transplant, he elected bypass surgery to strengthen his heart. Doctors warned that his body might not withstand the operation. But Nicholas knew that without taking a chance, he would live only a few years in debilitated condition. Shortly after the surgery, infection set in, traveling throughout his system and so weakening him that only extreme measures—a respirator to sustain breathing and powerful drugs to elevate his fading blood pressure—could keep him alive.

"Come on, Dad, you can do it," encouraged Nicholas's daughter Sasha, standing by his bedside and stroking his hand. But Nicholas could not. After 2 months in intensive care, he experienced brain seizures and slipped into a coma. Three doctors met with his wife Giselle to tell her there was no hope. She asked them to disconnect the respirator, and within a half hour Nicholas drifted away.

Death is essential for the survival of our species. We die so that our own children and the children of others may live. When it comes to this fate, nature treats humankind, with all its unique capabilities, just as it treats every other living creature. We are granted the miracle of life because trillions of beings before us have lived and then died (Nuland, 1993). As hard as it is to accept the reality that we too will die, our greatest solace lies in the knowledge that death is part of ongoing life.

In this chapter, we address the culmination of lifespan development. Over the past century, technology has provided us with so much to keep death at bay that many people regard it as a forbidden topic. But pressing social and economic dilemmas that are an outgrowth of the dramatic increase in life expectancy are forcing us to attend to life's end—its quality, its timing, and ways to help people adjust to their own and others' final leavetaking.

Our discussion addresses the physical changes of dying, hopelessly ill patients' right to die, and the thoughts and feelings of people as they stand face to face with death. The experiences of Sofie, Philip, Nicholas, their families, and others illustrate how each person's life history joins with social and cultural contexts to shape death and dying, lending great diversity to this universal experience.

HOW WE DIE

Our vast literature on death is largely aimed at helping people cope with the emotional trauma of dying and its aftermath. Few people are aware of the physical aspects of death, since opportunities to witness them are less available than in previous generations. Today, the large majority of people in industrialized nations die in hospitals, where doctors and nurses, not loved ones, typically attend their last moments. Nevertheless, many people want to know how we die, either to anticipate their own end or grasp what is happening to a dying loved one.

PHYSICAL CHANGES

My father's fatal heart attack came suddenly during the night. On being told the news, I longed for reassurance that his death had been swift and without suffering.

When asked how they would like to die, most people say they want "death with dignity"—either a quick, agony-free end during sleep or a clear-minded final few moments in which they can say farewell and review their lives. In reality, death is the culmination of a straightforward biological process. In about 20 percent of people, it is gentle—especially when narcotic drugs ease pain and mask the destructive events taking place (Nuland, 1993). But most of the time it is not.

Recall that unintentional injuries are the leading cause of death in childhood and adolescence, cardiovascular disease and cancer in adulthood. Of the one-quarter of people in industrialized nations who die suddenly, within a few hours of symptoms, 80 to 90 percent are victims of heart attacks (Nuland, 1993). My yearning for a painless death for my father was probably not granted. Undoubtedly he felt the sharp, crushing sensation of a heart deprived of oxygen. As his heart twitched uncontrollably (called *fibrillation*) or stopped entirely, blood circulation slowed and ceased, and he was thrust into unconsciousness. A brain starved of oxygen for more than 2 to 4 minutes is irreversibly damaged—an outcome indicated by the pupils of the eye becoming unresponsive to light and widening into large, black circles. Other oxygen-deprived organs stop functioning as well.

Death is long and drawn out for three-fourths of people—many more than in times past, due to life-saving medical technology (Benoliel & Degner, 1995). They succumb in many different ways. Of those with heart disease, most have congestive heart failure, the cause of Nicholas's death. His scarred heart could no longer contract with the force needed to deliver enough oxygen to his tissues. As it tried harder, its muscle weakened further. Without sufficient blood pressure, fluid backed up in Nicholas's lungs. This hampered his breathing and created ideal conditions for inhaled bacteria to multiply, enter the bloodstream, and run rampant in his system, leading multiple organs to fail.

Cancer, as well, chooses diverse paths to wreak its damage. When it *metastasizes*, bits of tumor travel through the bloodstream and implant and grow in vital organs, disrupting their functioning—perhaps the liver, the kidneys, the intestinal tract, the lungs, or the brain. Medication made my mother's final days as comfortable as possible, granting a relatively easy death. But the preceding weeks involved physical suffering, including impaired breathing and digestion and turning and twisting to find a comfortable position in bed.

In general, dying takes place in three phases:

1. The **agonal phase:** The Greek word *agon* means struggle. Here *agonal* refers to gasps and muscle spasms during the first moments in which the body cannot sustain life anymore.

2. **Clinical death:** A short interval follows in which heartbeat, circulation, breathing, and brain functioning stop, but resuscitation is still possible.

3. **Mortality:** The individual passes into permanent death. Within a few hours, the newly lifeless being appears shrunken, not at all like the person he or she was when alive.

DEFINING DEATH

Think about what we have said so far, and you will see that death is not an event that happens at a single point in time. Rather, it is a process in which organs stop functioning in a sequence that varies from person to person. Because the dividing line between life and death is fuzzy, societies need a definition of death to help doctors decide when life-saving measures should be terminated, to signal survivors that they must begin to grieve their loss and reorganize their lives, and to establish when donated organs can be removed.

Several decades ago, loss of heartbeat and respiration signified death. But these criteria are no longer adequate, since resuscitation techniques frequently permit vital signs to be restored. Today, **brain death,** defined as irreversible cessation of all activity in the brain and the brain stem (which controls reflexes), is used in most industrialized nations.

It is important to note that not all countries accept this standard. In Japan, for example, no legal definition of death exists. Instead, doctors rely on traditional criteria (absence of heartbeat and respiration) that fit with Japanese laypeople's views. This approach stands in the way of a

Japanese elders visit a Buddhist shrine to pray for a happy death. Buddhist, Confucian, and Shinto beliefs about death, which stress ancestor worship, may be partly responsible for Japanese people's discomfort with organ donation. (Fujifotos/The Image Works)

national organ transplant program, since few organs can be salvaged from bodies without artificially maintaining vital signs. Buddhist, Confucian, and Shinto beliefs about death, which stress ancestor worship, may be partly responsible for Japan's discomfort with brain death and organ donation. Marring the body to harvest organs violates respect for the deceased. Although public attitudes are changing, there is less consensus on defining death in Japan than in other developed countries (Feldman, 1994).

Furthermore, in a great many cases the brain death standard does not solve the dilemma of when to halt treatment. Consider Nicholas, whose brain stem, but not cortex, registered electrical activity. Although not brain dead, he had entered a **persistent vegetative state.** Doctors were sure they could not restore consciousness or body movement. Medical intervention could do no more than postpone the natural process of dying. Because approximately 10,000 Americans are in a persistent vegetative state, with health care costs totaling $1.2 billion annually, many experts believe that absence of activity in the cortex should be sufficient to declare a person dead (Maier & Newman, 1995). In other instances, a fully conscious but very sick, suffering person refuses life-saving measures—an issue we will consider shortly when we take up the right to die.

DEATH WITH DIGNITY

Our brief look at the process of dying indicates that nature rarely delivers the idealized, easy end most people want, and medical science cannot guarantee it. Therefore, the greatest dignity in death can be found in the integrity of

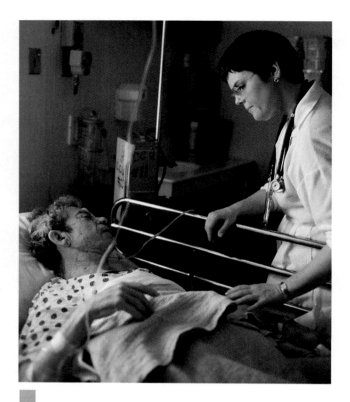

Doctors and nurses can help dying people learn enough about their condition to make reasoned choices about whether to fight on or say no to further treatment. (Goldberg/Monkmeyer Press)

the life that precedes it—an integrity we can foster by the way we communicate with and care for the dying person.

First, we can assure the majority of dying people, who succumb gradually, that we will support them through their physical and psychological distress. We can do everything possible to provide the utmost in humane and compassionate care.

Second, we can be candid about death's certainty. Unless people are aware that they are dying and understand (so far as possible) the likely circumstances of their death, they cannot share the sentiments that bring closure to relationships they hold most dear. Because Sofie knew how and when her death would probably take place, she chose a time when she, Philip, and her children could express what their lives had meant to one another. Among those precious bedside exchanges was Sofie's memorable last wish that Philip remarry after her death so he would not live out his final years alone. Openness about impending death granted Sofie a final generative act, helped her let go of the person closest to her, and offered comfort as she faced death.

Finally, doctors and nurses can help dying people learn enough about their condition to make reasoned choices about whether to fight on or say no to further treatment. An understanding of how the normal body works simplifies comprehension of how disease affects it—education that can begin as early as the childhood years.

In sum, although the conditions of illness often do not permit a graceful, serene death, we can ensure the most dignified exit possible by offering the dying person care, affection, and companionship; the truth about his or her diagnosis; and the maximum personal control over this final phase of life. These are essential ingredients of a "good death," and we will revisit them throughout this chapter.

THE RIGHT TO DIE

In 1976, the parents of Karen Ann Quinlan, a young woman who took drugs at a party and fell into an irreversible coma, sued to have her respirator turned off. The New Jersey Supreme Court, calling on Karen's right to privacy and the power of her parents as guardians, complied with this request. Although Karen was expected to die quickly, she breathed independently, continued to be fed intravenously, and lived another 10 years in a persistent vegetative state.

In 1983, 25-year-old Nancy Cruzan fell asleep at the wheel and was thrown from her car. After resuscitation, her heart and lungs worked but her brain was badly damaged. Like Karen, Nancy lay in a persistent vegetative state. Her parents wanted to end her meaningless existence by halting artificial feeding—a preference consistent with Nancy's statement to a friend a year earlier. But the Missouri courts refused, claiming that no written evidence of Nancy's wishes existed. The U.S. Supreme Court upheld this ruling, but it encouraged the state of Missouri to reconsider. Eventually, Nancy's doctor agreed that it was in her best interests to terminate treatment, and a county judge honored this request. She died a week later, nearly 8 years after losing consciousness.

Before the 1950s, medical science could do little to extend the lives of terminal patients, so the right to die was of little concern. Today, medical advances mean that the same procedures that preserve life can prolong inevitable death, diminishing both quality of life and personal dignity.

The Quinlan and Cruzan cases brought right-to-die issues to the forefront of public attention, sparking over 40 states to pass legislation honoring patients' desires concerning withdrawal of treatment in case of terminal illness and, in a few instances, a persistent vegetative state. But the Supreme Court accepts variation among the states in standards of proof of a patient's intentions. In the United States, no uniform right-to-die policy exists, and controversy persists over how to handle the diverse circumstances in which patients and families make requests (Glick, 1992).

Euthanasia refers to the practice of ending the life of a person suffering from an incurable condition. Its forms, depicted in Figure 19.1, vary along a continuum from least active to most active. As we will see when we discuss the

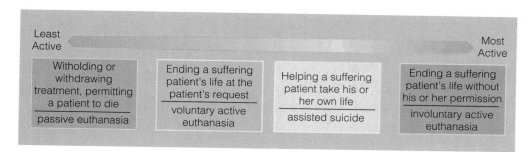

Forms of euthanasia, from least active and highest public acceptance to most active and lowest public acceptance.

most common types in the following sections, public acceptance declines as euthanasia moves from permitting a natural death to deliberately killing a suffering person.

P ASSIVE EUTHANASIA

In **passive euthanasia,** life-sustaining treatment is withheld or withdrawn, permitting a patient to die naturally. Should Nancy Cruzan have been allowed to die sooner? Was it right for Nicholas's doctors to have turned off his respirator at Giselle's request? Consider an Alzheimer's victim, whose disease has progressed to the point where he has lost all awareness and body functions. Should life support be withheld?

Recent polls reveal that the majority of people answer yes to these questions. When there is no hope of recovery, 85 percent of Americans support the patient's right to end treatment and 80 percent support family members' right to do so (Glick, 1992; Maier & Newman, 1995). In 1986, the American Medical Association endorsed withdrawing all forms of treatment from the terminally ill when death is imminent, and from those in a permanent vegetative state. Consequently, passive euthanasia is widely practiced as part of ordinary medical procedure, in which doctors exercise professional judgment.

Still, a minority of citizens do not endorse passive euthanasia. Religious denomination has surprisingly little effect on people's opinions. For example, most Catholics hold favorable views, despite slow official church acceptance due to fears that passive euthanasia might be a first step toward government-approved mercy killing. And because of controversial court cases, some doctors and health care institutions are unwilling to end treatment without legal protection.

In the absence of a national consensus on passive euthanasia, people can best ensure that their wishes will be followed by preparing an **advance medical directive**—a written statement of desired medical treatment should they become incurably ill. Two types of advance directives are recognized in most states.

The first is a **living will,** which specifies the treatments a person does or does not want in case of a terminal illness, coma, or other near-death situation (see Figure 19.2). For example, a person might state that without reasonable expectation of recovery, he or she should not be kept alive through medical intervention of any kind. In addition,

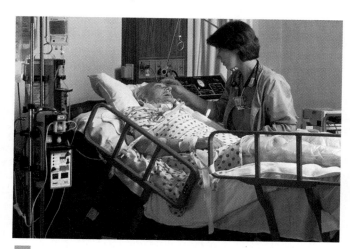

Today, medical advances mean that the same procedures that preserve life can prolong inevitable death, diminishing both quality of life and personal dignity. (Medichrome)

All medical facilities receiving federal funds must provide information at admission about state laws and institutional policies on patients' rights and advance medical directives. Most states recognize two types of advance directives: the living will and the durable power of attorney for health care. (Richard Pasley/ Stock Boston)

Living Will

THIS DECLARATION is made this _____ day of _____ , 19 _____ .

I, _____ , being of sound mind, willfully and voluntarily make known my desires that my moment of death shall not be artificially postponed. If at any time I should have an incurable and irreversible injury, disease, or illness judged to be a terminal condition by my attending physician who has personally examined me and has determined that my death is imminent except for death delaying procedures. I direct that such procedures which would only prolong the dying process be withheld or withdrawn, and that I be permitted to die naturally with only the administration of medication, sustenance, or the performance of any medical procedure deemed necessary by my attending physician to provide me with comfort care.

In the absence of my ability to give directions regarding the use of such death delaying procedures, it is my intention that this declaration shall be honored by my family and physician as the final expression of my legal right to refuse medical or surgical treatment and accept the consequences from such refusal.

Signed: _____

City, County and State of Residence: _____

_____

_____

The declarant is personally known to me and I believe him or her to be of sound mind. I saw the declarant sign the declaration in my presence (or the declarant acknowledged in my presence that he or she had signed the declaration) and I signed the declaration as a witness in the presence of the declarant. At the date of this instrument, I am not entitled to any portion of the estate of the declarant according to the laws of intestate succession or, to the best of my knowledge and belief, under any will of declarant or other instrument taking effect at declarant's death, or directly financially responsible for declarant's medical care.

Witness: _____

Witness: _____

FIGURE 19.2

Example of a living will. This document is legal in the State of Illinois. Each person completing a living will should use a form specific to the state in which he or she resides, since state laws vary widely. *(Courtesy of Office of the Attorney General, State of Illinois.)*

living wills sometimes specify that pain-relieving medication be given, even though this may shorten life. In Sofie's case, her doctor administered a powerful narcotic to relieve labored breathing and quiet her fear of suffocation. The narcotic suppressed respiration, leading death to occur hours or days earlier than if the medication had not been prescribed, but without distress. Such **comfort care** is accepted as appropriate and ethical medical practice.

Although living wills help ensure personal control, they do not guarantee it. Recognition of living wills is usually limited to patients who are terminally ill or otherwise expected to die shortly. Only a few states cover people in a persistent vegetative state due to injury or elders who linger with many chronic problems, including Alzheimer's disease—none of which are classified as terminal.

Because state laws vary widely and living wills cannot anticipate all future medical conditions and treatment

options, a second form of advance directive—the **durable power of attorney for health care**—has become more common. It authorizes appointment of another person (usually, although not always, a family member) to make health care decisions on one's behalf and requires only a short signed and witnessed statement like this:

> I hereby appoint [name] to act for me and in my name (in any way I could act in person) to make any and all decisions for me concerning my personal care, medical treatment, hospitalization, and health care and to require, withhold, or withdraw any type of medical treatment or procedure, even though my death may ensue. (Courtesy of *Office of Attorney General, State of Illinois.*)

The durable power of attorney for medical care is more flexible than the living will, since it permits a trusted spokesperson to confer with the doctor as medical circumstances arise. Because authority to speak for the patient is not limited to terminal illnesses, more latitude exists for dealing with unexpected situations. The majority of states have been receptive to this option, since it is a simple extension of existing durable powers of attorney, such as those that cover disposition of property when people cannot act for themselves (Maier & Newman, 1995).

Whether or not a person advocates passive euthanasia, it is important to have a living will, durable power of attorney, or both, since most deaths occur in hospitals. Surprisingly, less than 20 percent of Americans have executed such documents, perhaps because of widespread uneasiness about bringing up the topic of death, especially with relatives. To encourage people to make decisions about potential treatment while they are able, federal law now requires all medical facilities receiving federal funds to provide information at admission about state laws and institutional policies on patients' rights and advance directives.

Health care professionals, unclear about a patient's intent and fearing liability, will probably decide to continue treatment regardless of cost and a person's prior oral statements. Perhaps for this reason, a few states permit appointment of a *health care proxy,* or substitute decision maker, if a patient failed to provide an advance medical directive while competent (Suhr, 1991). Proxies are an important means for covering children and adolescents, who cannot legally execute advance medical directives.

VOLUNTARY ACTIVE EUTHANASIA

In recent years, the right-to-die debate has shifted from withdrawal of treatment for the hopelessly ill to more active alternatives. In **voluntary active euthanasia,** doctors or others act directly, at a patient's request, to end suffering before a natural end to life. The practice is a form of mercy killing and is a criminal offense in most countries. Nevertheless, support for it is growing. In recent polls, about 70 percent of American and Dutch citizens approved of it, although more disagreement exists about the conditions in which it is warranted (Caddell & Newton, 1995; Davis et al., 1993). When voluntary active euthanasia occurs in the United States, judges are usually lenient, granting suspended sentences or probation. These trends reflect rising public interest in self-determination in death as in life.

Nevertheless, attempts to legalize voluntary active euthanasia have prompted heated controversy. Supporters believe that it represents the most compassionate option for terminally ill people in severe pain. Opponents stress the moral difference between "letting die" and "killing" and point out that at times, even very sick patients recover. They also argue that involving doctors in taking the lives of suffering patients may impair people's trust in health professionals. Finally, a fear exists that legalizing this practice—even when strictly monitored to make sure it does not arise out of depression, loneliness, coercion, or a desire to diminish the burden of illness on others—could lead to a broadening of euthanasia. Initially limited to the terminally ill, it might be applied involuntarily to the frail, demented, or disabled—outcomes that most people find unacceptable and immoral (Kerridge & Mitchell, 1996).

Will legalizing voluntary active euthanasia lead us down a "slippery slope" to the killing of vulnerable people who did not ask to die? The Social Issues box on page 630 presents lessons from the Australian state of the Northern Territory, where voluntary active euthanasia was legalized in 1995, and from the Netherlands, where doctors can practice it without fear of prosecution.

ASSISTED SUICIDE

After checking Diane's blood count, Dr. Timothy Quill gently broke the news: leukemia. If she were to have any hope of survival, a strenuous course of treatment with only a 25 percent success rate would have to begin immediately. Convinced that she would suffer unspeakably from side effects and lack of control over her body, Diane chose not to undergo chemotherapy and a bone marrow transplant.

Dr. Quill made sure that Diane understood her options. As he adjusted to her decision, she opened another issue: Diane wanted no part of a lingering death. She calmly insisted that when the time came, she desired to take her own life in the least painful way possible—a choice she had discussed with her husband and son, who respected it. Realizing that Diane could get the most out of the time she had left only if her fears of prolonged pain were allayed, Dr. Quill granted her request for sleeping pills, making sure she knew the amounts needed for both sleep and suicide.

In a prestigious medical journal, Dr. Timothy Quill explained how and why he assisted a terminally ill patient in taking her own life. With over half the public in favor of it, assisted suicide may become a more widespread, unofficial practice in the United States. Yet grave dilemmas surround it. (Bob Mahoney)

SOCIAL ISSUES

VOLUNTARY ACTIVE EUTHANASIA: LESSONS FROM AUSTRALIA AND THE NETHERLANDS

In Australia's Northern Territory, one of the world's smallest legislatures passed a law allowing a terminally ill patient of sound mind, suffering from pain or other distress, to ask a doctor to end his or her life. Two other doctors must agree that the patient cannot be cured, and a psychiatrist must confirm that he or she is not suffering from treatable depression.

Since its passage, four deaths have occurred under the Northern Territory euthanasia statute, and it has been heavily criticized. The Aborigines, valuing harmony and balance with nature, regard it as culturally inappropriate. Their leaders claim the law will discourage Aboriginal elders, many of whom have experienced a lifetime of persecution at the hands of European settlers, from seeking medical care (Kerridge & Mitchell, 1996). Others consider the law to be a national issue, since patients traveled from other states to make use of it. In 1997, the Northern Territory legislation was overturned by the Australian Parliament, which claimed that assemblies do not have the right to legislate intentional killing.

In the Netherlands, voluntary active euthanasia remains a criminal offense. However, doctors who engage in it are exempt from punishment as long as they follow certain rules. They must make sure that the patient's physical or mental suffering is severe, with no prospect of relief; that no doubt exists about the patient's desire to die;

that the patient's decision is free, well informed, and stable over time; that all other options for care have been exhausted or refused; and that another doctor has been consulted.

Over 50 percent of Dutch doctors say they practice euthanasia, with cancer patients being the majority of their cases. Despite safeguards, a government report revealed that both voluntary and involuntary euthanasia have occurred. In over a thousand cases in a single year, doctors admitted actively causing death when a patient did not ask for it. In 30 percent of these, the reason given was the impossibility of treating pain. The remaining 70 percent ranged in justification, from a low quality of life to a terminal patient not dying after withdrawal of treatment (van der Maas, van Delders, & Pijnenborg, 1991). The elderly may be at special risk, since recent evidence suggests that a sharp decline in Dutch elder suicide has been compensated for by a rise in active euthanasia (Hendin, 1995).

The Northern Territory and Dutch examples reveal that legalizing voluntary active euthanasia can spark both the fear and the reality of death without consent. Nevertheless, moving pleas for such laws have been made by terminally ill individuals in severe pain. Recently, Canada and Great Britain established committees to investigate reforming laws on euthanasia. Each recommended against both active euthanasia and assisted suicide (Coatney, 1997).

Probably all would agree that when doctors feel compelled by respect for self-determination and relief of suffering to assist a patient in dying, they should be subject to the closest possible professional and legal oversight.

TRY THIS...

- Ask several people of different ages for their opinion on legalization of voluntary active euthanasia, including (if possible) a family member or health professional who has cared for a dying person. Note the diversity of opinion. Then formulate your own view.

- In a recent ruling, the U.S. Supreme Court affirmed that state governments have the right to outlaw doctor-assisted suicide. (However, it did not stop states from allowing the practice if they so choose.) According to the Court, a state may realistically fear that permitting assisted suicide could lead to voluntary and perhaps involuntary euthanasia. Do you agree with the Court's decision and reasoning? Why or why not?

Diane's next few months were busy and fulfilling. Her son took leave from college to be with her, and her husband worked at home as much as possible. Gradually, bone pain, fatigue, and fever set in. Saying good-bye to her family and friends, Diane asked to be alone for an hour, took a lethal dose of medication, and died at home (Quill, 1991).

Assisting a suicide is illegal in many, but not all, U.S. states. Only Oregon has passed a law that explicitly allows physicians to prescribe drugs so terminally ill patients can end their lives, but it has been put on hold due to a consti-

tutional challenge (Annas, 1994; Quill, 1995). Less public consensus exists for assisted suicide than for passive and voluntary active euthanasia. About 60 percent of Americans approve of it (Brody, 1992; Miller et al., 1994).

Interest in assisted suicide has been sparked by Dr. Jack Kevorkian's "suicide machines," which permit terminally ill patients, after brief counseling, to self-administer lethal drugs and carbon monoxide. By 1997, Dr. Kevorkian had participated in over fifty such deaths. Less publicity surrounded Dr. Quill's decision to assist Diane—a patient he knew well after serving as her doctor for many years—in taking her own life. After he told her story in a prestigious medical journal, letters to the editor were mixed: "Dr. Quill provided . . . deep concern for the patient's well-being and respect for her choices," wrote one respondent. "It is never the role of a physician to perform such an act . . . [but rather] to apply his or her knowledge, skill, and caring to save lives," stated another (Cardo, 1991; Freer, 1991, p. 658).

With over half of the public in favor of it, assisted suicide may become a more widespread, unofficial practice in the United States, as it was in Diane's case and is in the Netherlands. Yet grave dilemmas, like those we discussed for voluntary active euthanasia, surround it. Analyzing the practice, one group of medical experts concluded that it is warranted only when the following conditions are met:

■ The patient requests assisted suicide repeatedly and freely and is suffering intolerably, with no satisfactory options.

■ The doctor thoroughly explores alternatives for comfort care with the patient.

■ The practice is consistent with the doctor's fundamental values. (If not, the doctor should recommend transfer of care.)

■ Even when the doctor and patient agree that there is no other acceptable choice, independent monitoring is crucial for preventing abuse.

In trials for assisted suicide, no jury has ever returned a conviction when a doctor showed that he or she acted compassionately and competently (Stevens, 1997).

Earlier we noted that public opinion favors voluntary active euthanasia over assisted suicide. Yet in assisted suicide, the final act is solely the patient's, reducing the possibility of coercion. For this reason, some experts believe that legalizing assisted suicide is preferable to legalizing voluntary active euthanasia. However, in an atmosphere of intense pressure to contain health care costs and of high family caregiving burdens (see Chapter 17), legalizing either practice can be dangerous (Quill, Cassel, & Meier, 1992). Recently, the U.S. Supreme Court decided unanimously not to recognize the right of terminally ill patients to a doctor's help in ending their lives. At the same time, the majority of the justices indicated that they might change their minds in the future, after they have more evidence on the practical impact of such a right (Dworkin, 1997). Helping incurable, suffering patients who yearn for death is a profound moral and legal problem.

BRIEF REVIEW

Of the minority of people who die suddenly, most are victims of heart attacks. Usually, death is long and drawn out and occurs in many different ways. Most industrialized nations rely on brain death to signify that life has come to an end. However, when people are in a persistent vegetative state or ask to die because of unbearable suffering, the brain death standard does not resolve controversies over whether to halt life-saving treatment.

Although over forty states have passed legislation permitting passive euthanasia at the request of terminally ill patients or their families, no uniform right-to-die policy exists in the United States. People can best ensure that their wishes will be followed by executing either a living will or durable power of attorney for health care. For incompetent patients who did not engage in advance planning, a few states permit appointment of a health care proxy, or substitute decision maker.

Active forms of euthanasia remain the subject of heated controversy. The majority of Americans favor voluntary active euthanasia and assisted suicide for the terminally ill. Yet concern exists that legalizing these practices will lead us down a "slippery slope" to involuntary mercy killing.

ASK YOURSELF . . .

■ *Noreen knows that the majority of people die gradually. Thinking ahead to the day she dies, she imagines a peaceful scene in which she says good-bye to loved ones. What social and medical practices are likely to increase Noreen's chances of dying in the manner she desires?*

■ *If he should ever fall terminally ill, Ramon is certain that he wants doctors to halt life-saving treatment. To best ensure that his wish will be granted, what should Ramon do? Why is it impossible to guarantee that Ramon's desires will be honored in the United States at this time?*

UNDERSTANDING OF AND ATTITUDES TOWARD DEATH

A century ago, deaths most often occurred at home. People of all ages, including children, helped with care of the dying family member and were present at the moment of death. They saw their loved one buried on family property or in the local churchyard, where the grave could be visited regularly. Since infant and childhood mortality rates were high, virtually everyone was likely to know someone the same age as or younger than themselves who had died. And it was common for children to experience the death of a parent.

Compared to earlier generations, today's children and adolescents in developed nations are insulated from death. Residents of inner-city ghettos, where violence is part of everyday existence, are exceptions (see Chapter 10, page 339). But overall, more young people reach adulthood without having experienced the death of someone they know well. When they do, professionals in hospitals and funeral homes take care of most tasks that involve confronting death directly (Wass, 1995).

This distance from death undoubtedly contributes to a sense of uneasiness about it. Despite constant images of death on television and in movies, we are a death-denying society. Adults are often reluctant to talk about death with children and adolescents. And a variety of substitute expressions, such as such as "passing away," "going out," and "departing," permit us to avoid acknowledging it candidly. In the following sections, we examine the development of conceptions of and attitudes toward death, along with ways to foster increased understanding and acceptance.

CHILDHOOD

Five-year-old Miriam arrived at our university laboratory preschool the day after her dog Pepper died. Leslie, her teacher, noticed that instead of joining the other children as she usually did, Miriam stood by herself, looking anxious and unhappy. "What's wrong, Miriam?" Leslie asked.

"Daddy said Pepper had a sick tummy. He fell asleep and died." For a moment, Miriam looked hopeful. "When I get home, Pepper might be up."

Leslie answered directly, "No, Pepper won't get up again. He's not asleep. He's dead, and that means he can't sleep, eat, run, or play anymore."

Miriam wandered off. Later, she returned and confessed to Leslie, "I chased Pepper too hard," tears streaming from her eyes.

Leslie put her arm around Miriam. "Pepper didn't die because you chased him. He was very old and very sick," she explained.

Over the next few days, Miriam asked many questions: "When I go to sleep, will I die?" "Can a tummy ache make you die?" "Will Mommy and Daddy die?" "Can Pepper see me?" "Does Pepper feel better now?" (Goldman, 1996).

■ DEVELOPMENT OF THE DEATH CONCEPT. A realistic understanding of death is based on three ideas:

1. **Permanence:** Once a living thing dies, it cannot be brought back to life.

2. **Universality:** All living things eventually die.

3. **Nonfunctionality:** All living functions, including thought, feeling, movement, and body processes, cease at death.

Examining this dead mouse will help these children understand the *permanence* of death. Appreciation of two additional components of the death concept—*universality* and *nonfunctionality*—will come later. (A. Carey/The Image Works)

Without clear explanations, preschoolers rely on egocentric and magical thinking to make sense of death. They may believe, as Miriam did, that they are responsible for a relative or pet's death. And they can easily arrive at incorrect conclusions—in Miriam's case, that sleeping or a stomachache can cause someone to die.

Preschoolers grasp the three components of the death concept in the order just given, with most mastering them by age 7. *Permanence* is the first and most easily understood idea. When Leslie explained that Pepper would not get up again, Miriam accepted this fact quickly, perhaps because she had seen it in other less emotionally charged situations, such as the dead butterflies and beetles that she picked up and inspected while playing outside (Furman, 1990). Appreciation of *universality* comes slightly later. At first, children think that certain people do not die, especially those with whom they have close emotional ties or who are like themselves—other children. Finally, *nonfunctionality* is the most difficult component for children to grasp. Many preschoolers view dead things as retaining living capacities. When they first comprehend nonfunctionality, they do so in terms of its most visible aspects, such as heartbeat and breathing. Only later do they understand that thinking, feeling, and dreaming cease (Lazar & Torney-Purta, 1991; Speece & Brent, 1992).

Although children usually have an accurate conception of death by middle childhood, there are wide individual differences. Experiences with death make a difference (Speece & Brent, 1996). For example, terminally ill children under age 6 often have a well-developed concept of death. If parents and health professionals have not been forthright with them, children find out that they are deathly ill in other ways—through nonverbal communication, eavesdropping, and talking with other child patients (O'Halloran & Altmaier, 1996).

■ ENHANCING CHILDREN'S UNDERSTANDING. Parents often worry that discussing death candidly with children when the topic arises will fuel their fears, but this is not so. Instead, children with a good grasp of the facts of death have an easier time accepting it (Essa & Murray, 1994). Simple, direct explanations, like Leslie's, work best. When adults use clichés or make misleading statements, preschoolers often take these literally and react with confusion. For example, after a parent said to one 5-year-old, "Grandpa went on a long trip," the child wondered, "Why didn't he take me?" Responding to the statement "Grandpa's watching over you," the child thought, "I can't do anything without Grandpa seeing me."

Sometimes children ask very difficult questions, such as "Will I die? Will you die?" Parents can be truthful as well as comforting by taking advantage of children's sense of time. They can say something like "Not for many, many years. First I'm going to enjoy you as a grown-up and be a grandparent." As we will see later, open, honest discussions with children not only contribute to a realistic understanding of death, but facilitate grieving after a child has experienced a loss.

■ ADOLESCENCE

Adolescents can easily voice the permanence, universality, and nonfunctionality of death, but their understanding is largely limited to the realm of possibility. Recall that teenagers have difficulty integrating logical insights with the realities of everyday life (Corr, 1995). Consequently, their understanding of death is not yet fully mature.

■ THE GAP BETWEEN LOGIC AND REALITY. Teenagers' difficulty applying what they know about death is evident in both their reasoning and their behavior. Although they can explain the permanence and nonfunctionality of death, they are attracted to alternatives. For example, they often describe death as an enduring abstract state, such as "darkness," "eternal light," "transition," or "nothingness" (Pettle & Britten, 1995; Wenestam & Wass, 1987). They also formulate personal theories about life after death. Besides images of heaven and hell influenced by religious back-

This adolescent knows that death happens to everyone and can occur at any time, but his risk taking suggests otherwise. Teenagers' difficulty integrating logic with reality extends to their understanding of death. (D. & I. MacDonald/The Picture Cube)

ground, they speculate about reincarnation, transmigration of souls, and spiritual survival on earth or at another level (Wass, 1991).

Although mortality in adolescence is low compared to infancy and adulthood, teenage deaths are typically sudden and human-induced; unintentional injuries, homicide, and suicide are leading causes. Adolescents are clearly aware that death happens to everyone and can occur at any time. But this does not make them more safety conscious. Rather, their high-risk activities suggest that they do not take death personally. When asked to comment on death, teenagers often make statements like these:

Why dwell on it? There is nothing I can do about it.

I plan to live a long, fruitful life and if I get worried about dying at all, it won't be until I'm old. (Wass, 1991, p. 27)

What explains teenagers' difficulty integrating logic with reality in the domain of death? First, adolescence is a period of rapid growth in size and strength and onset of reproductive capacity—attainments that are the opposite of death! Second, recall the adolescent personal fable, which leads teenagers to be so wrapped up in their own uniqueness that they conclude they are beyond the reach of death while engaged in risky behavior. Finally, as teenagers construct a personal identity and experience their first freely chosen love relationships, they may be strongly attracted to romantic notions of death, which challenge logic (Noppe & Noppe, 1991, 1996). Not until early adulthood are they capable of the relativistic thinking needed to handle these conflicting ideas (see Chapter 13, page 436).

■ ENHANCING ADOLESCENTS' UNDERSTANDING. By encouraging adolescents to discuss concerns about death, adults can help them build a bridge between death as a logical concept and their personal experiences. In Chapter 12, we noted that teenagers with authoritative parents are more likely to turn to adults for guidance on important issues. In one study of 12- to 15-year-olds, most wanted to talk with parents rather than peers about the "meaning of life and death" and "what happens when you die." But 60 percent thought their parents would not be genuinely interested, and the majority of parents felt inadequately prepared for the task (McNeil, 1986).

Taking up adolescents' thoughts and feelings about death does not require waiting for a special occasion. It can be part of everyday conversation, sparked by a news report or the death of an acquaintance. Parents can capitalize on these moments to express their own views, listen closely, accept teenagers' feelings, and correct misconceptions. Such mutual sharing deepens bonds of love and provides the basis for further exploration when the need arises. The Caregiving Concerns table below suggests ways to discuss concerns about death with adolescents.

ADULTHOOD

In early adulthood, many people brush aside thoughts of death (Gresser, Wong, & Reker, 1987). This avoidance may be prompted by death anxiety, which we will consider in the next section. Alternatively, it may be due to relative disinterest in death-related issues, as young adults typically do not know many people who have died and (like adolescents) think of their own death as a long way off.

In Chapters 15 and 16, we described midlife as a time of stock taking in which people begin to view the lifespan in terms of time left to live, focusing on tasks to be completed. Middle-aged people no longer have a vague conception of their own death. They know that in the not-too-distant future it will be their turn to grow old and die.

In late adulthood, adults think and talk more about death, as it is much closer. Increasing evidence of mortality comes from physical changes, higher rates of illness and disability, and loss of relatives and friends (see Chapter 17). Compared to middle-aged people, older adults spend more time pondering the process and circumstances of dying rather than the state of death. Nearness to death seems to lead to a practical concern with how and when it might happen (de Vries, Bluck, & Birren, 1993; Kastenbaum, 1992).

Finally, although we have traced age-related changes, we must keep in mind that large individual differences exist. Some people focus on life and death issues early on, whereas others are less reflective, moving into advanced old age having given these matters little attention.

DEATH ANXIETY

As you read the following statements, do you find yourself agreeing, disagreeing, or reacting neutrally?

CAREGIVING CONCERNS

Discussing Concerns About Death with Adolescents

SUGGESTION	DESCRIPTION
Take the lead.	Be alert to the adolescent's nonverbal behaviors, bringing up the subject sympathetically, especially after a death-related situation has occurred.
Listen perceptively.	Grant full attention to the adolescent and the feelings underlying his or her words. When adults pretend to listen while thinking about other things, teenagers quickly pick up this sign of indifference and withdraw their confidence.
Acknowledge feelings.	Accept the adolescent's emotions as real and important; avoid being judgmental. For example, paraphrase sentiments you detect, such as "I see you're very puzzled about that. Let's talk more about it."
Engage in joint problem solving.	Convey your belief in the adolescent's worth by indicating that you do not want to impose an answer but rather to help him or her come to personally satisfying conclusions. To questions you cannot answer, say, "I don't know." Such honesty shows a willingness to generate and evaluate solutions jointly.

Source: McNeil, 1986.

"Never feeling anything again after I die upsets me."

"I hate the idea that I will be helpless after I die."

"The total isolation of death is frightening to me."

"The feeling that I will be missing out on so much after I die disturbs me." (Thorson & Powell, 1994, pp. 38–39)

Items like these appear on questionnaires used to measure **death anxiety**—fear and apprehension of death. Even people who are very accepting of the reality of death may be afraid of it (Firestone, 1994).

What predicts whether thoughts of our own demise trigger intense distress, relative calm, or something in between? To answer this question, researchers measure both general death anxiety and a variety of specific factors—fear of no longer existing, loss of control, a painful death, the body decaying, being separated from loved ones, the unknown, and others (Neimeyer, 1994). Findings reveal large individual and cultural variations in fear-arousing aspects of death. For example, in a study of devout Islamic Saudi Arabians, certain factors that appear repeatedly in the responses of Westerners, such as fear of the body decaying and the unknown, were entirely absent (Long, 1985).

Among Westerners, spirituality—a sense of life's meaning—seems to be more important than religious commitment in limiting death anxiety (Rasmussen & Johnson, 1994). Having a well-developed personal philosophy of death also reduces fear. In one study, both devout Christians and devout atheists viewed death as less threatening than did people with ambivalent religious views (Moore, 1992).

From what you have learned about adult psychosocial development, how do you think death anxiety might

change with age? If you predicted it would decline, reaching its lowest level in late adulthood, you are correct. Recall from Chapter 18 that the attainment of ego integrity and a more mature sense of spirituality reduce fear of death. This age-related drop has been found in many cultures and ethnic groups. And it is greatest among adults with deep faith in some form of higher being (Rasmussen & Johnson, 1994; Thorson & Powell, 1991).

Regardless of age, women appear more anxious about death than men are—a difference apparent in both Eastern and Western cultures (Neimeyer & Van Brunt, 1995). Researchers are not sure why this is so. Perhaps women are more likely to admit and men more likely to avoid troubled feelings about mortality—an explanation consistent with females' greater emotional expressiveness throughout the lifespan.

Experiencing some anxiety about death is normal. But like other fears, when it is very intense, it can undermine effective adjustment. Although physical health in adulthood is not related to death anxiety, mental health clearly is. People who are depressed or generally anxious are likely to have more severe death concerns (Neimeyer & Van Brunt, 1995). A large gap between their actual and ideal self-concepts leaves these individuals with a sense of incompleteness when they contemplate death. In contrast, self-confident individuals with a clear sense of purpose in life express very little fear of death.

Death anxiety is largely limited to adolescence and adulthood. Children rarely display it unless they live in war-torn areas where they are in constant danger (see Chapter 10, page 437). Terminally ill children are also at risk for high death anxiety. Compared to other same-age patients, children with cancer express more destructive thoughts and negative feelings about death (Malone, 1982). For those whose parents make the mistake of not telling them they are going to die, loneliness and death anxiety can be extreme (O'Halloran & Altmaier, 1996). With this in mind, let's take a closer look at the thinking and emotions of dying people.

What do people fear about death? Findings reveal large individual and cultural variations. For example, in a study of devout Islamic Saudi Arabians, fear of the decaying body and the unknown—common among Westerners—was entirely absent. (Topham/The Image Works)

BRIEF REVIEW

Understanding of death develops gradually in childhood, with permanence being mastered first, followed by universality and then nonfunctionality. Experiences with death foster earlier awareness. Although adolescents grasp the logic of death, they have difficulty integrating it with the realities of everyday life. Both children and adolescents benefit from opportunities to discuss their death-related concerns with parents in a candid, accepting atmosphere. As people age, they spend more time thinking and talking about death. Nevertheless, death anxiety declines over adulthood, although it is higher for women than men. Extreme death anxiety is linked to poor mental health.

ASK YOURSELF . . .

■ *When 4-year-old Chloe's aunt died, Chloe asked, "Where's Aunt Susie?" Her mother explained, "Aunt Susie is taking a long, peaceful sleep." For the next 2 weeks, Chloe refused to go to bed, and when finally coaxed into her room, she lay awake for hours. Explain the likely reason for Chloe's behavior, and suggest a better way of answering Chloe's question.*

■ *Explain why older adults think and talk more about death than do younger people but nevertheless feel less anxious about it.*

THINKING AND EMOTIONS OF DYING PEOPLE

In the year before her death, Sofie did everything possible to surmount her illness. In between treatments to control the cancer, she tested her strength. She continued to teach high school, traveled to visit her children, cultivated a garden, and took weekend excursions with Philip. Hope pervaded Sofie's approach to her deadly condition, and she spoke often about the disease—so much so that her friends wondered how she could confront it so directly.

As Sofie deteriorated physically, she moved in and out of a range of mental and emotional states. She was frustrated, and at times angry and depressed, about her inability to keep on fighting. I recall her lamenting anxiously on a day when she was in pain, "I'm sick, so very sick! I'm trying so hard, but I can't keep on." Once she asked when my husband and I, who were newly married, would have children. "If only I could live long enough to hold them in my arms!" she cried. In the last week, she appeared tired but free of her struggle. Occasionally, she spoke of her love for us and commented on the beauty of the hills outside her window. But mostly, she looked and listened rather than actively participated in conversation. One afternoon, she fell permanently unconscious.

DO STAGES OF DYING EXIST?

As dying people move closer to death, are their reactions predictable? Do they go through a series of changes that are the same for everyone, or are their thoughts and feelings unique?

■ KÜBLER-ROSS'S THEORY. Although her theory has been heavily criticized, Elisabeth Kübler-Ross (1969) is

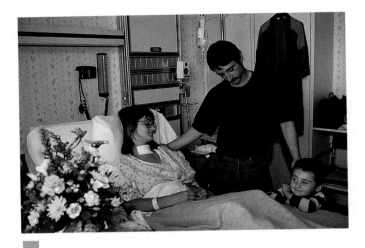

What is this woman—a young wife and mother—thinking and feeling during her final weeks? Dying people move in and out of a range of mental and emotional states affected by many personal and situational variables. (Jerry Koontz/The Picture Cube)

credited with awakening society's sensitivity to the psychological needs of dying patients. From interviews with over 200 terminally ill people, she devised a stage theory of typical responses to the prospect of death and the ordeal of dying. Progress can be arrested at any stage, and some people move back and forth between stages. But overall, five reactions characterize dying people. According to Kübler-Ross, when family members and health professionals understand these, they are in a better position to provide compassionate support.

1. **Denial.** On learning of the terminal illness, the person denies its seriousness to escape from the prospect of death. While the patient still feels reasonably well, denial is self-protective, allowing the individual to invest in rewarding activities and deal with the illness at his or her own pace. In reality, most people move in and out of denial, making great plans one day and acknowledging that death is near the next (Smith, 1993).

   Kübler-Ross recommends accepting the dying patient's denial as a coping strategy he or she may need initially. But family members and health professionals should not act in ways that prolong denial by distorting the truth about the person's condition. In doing so, they can hinder both medical treatment and drawing up an advance directive.

2. **Anger.** As the illness worsens, it is harder to maintain denial. The realization that time is short promotes anger at having to die without being given a chance to do all one wants to do. Family members and health professionals are often targets of the patient's rage, resentment, and envy of those who will go on living. Still, they must tolerate rather than lash out at the

patient's behavior, recognizing that the underlying cause is the unfairness of death.

3. **Bargaining.** Realizing the inevitability of death, the terminally ill person attempts to forestall it by bargaining for extra time—a deal he or she may try to strike with family members, friends, doctors, nurses, or God. Listening sympathetically is the best response to these efforts to sustain hope, as one doctor did to the pleas of a young AIDS-stricken father to live long enough to dance with his 8-year-old daughter at her wedding (Selwyn, 1996). At times, bargains are altruistic acts. Witness the following request from Tony, a 15-year-old leukemia patient, expressed to his mother:

> "I don't want to die yet. Gerry [youngest brother] is only 3 and is not old enough to understand. I've been able to talk to each of my older brothers to prepare them and they'll be okay. I can leave a letter for Gerry but it's not the same. If I could live just one more year, I could explain it to him myself and he will understand. Three is just too young." (Komp, 1996, pp. 69–70)

Although many dying patients' bargains are unrealistic and impossible to fulfill, Tony lived for exactly one year—a gift to those who survived him.

4. **Depression.** When denial, anger, and bargaining fail to postpone the course of illness, the person becomes depressed about the loss of his or her life. Kübler-Ross regards depression as necessary preparation for the last stage, acceptance. Unfortunately, many experiences associated with dying, including physical and mental deterioration, pain, lack of control, and being hooked to machines, promote despondency (Maier & Newman, 1995). Health care that responds humanely to the patient's wishes can limit hopelessness and despair.

5. **Acceptance.** Most people who reach acceptance, a state of peace and quiet about upcoming death, do so only in the last weeks or days. The weakened patient yields to death with its release from pain and anxiety. Disengagement from all but a few family members, friends, and caregivers occurs. Some dying people, in an attempt to pull away from all they have loved, withdraw into themselves for long periods of time. Often they seem totally caught up in their own thoughts. "I'm getting my mental and emotional house in order," one patient explained (Samarel, 1995, p. 101).

As in the other stages, people who have reached acceptance usually maintain some hope of living, if only a flicker. When hope is entirely gone, Kübler-Ross observed, death follows quickly.

■ **EVALUATION OF KÜBLER-ROSS'S THEORY.** Kübler-Ross cautioned that her five stages should not be viewed as a fixed sequence and that not all people display each of them—warnings that might have been better heeded had she not called them stages! Too often her theory has been interpreted simplistically, as a series of steps a "normal" dying person follows. Some health professionals, unaware of diversity in dying experiences, have insensitively tried to push patients through Kübler-Ross's sequence. And caregivers, through callousness or ignorance, can too easily dismiss a dying patient's legitimate complaints about treatment as "just what you would expect in Stage 2" (Corr, 1993; Kastenbaum, 1995).

Research confirms, in line with Kübler-Ross's observations, that dying people are more likely to display denial after learning of their condition and acceptance shortly before death (Kalish, 1985). But rather than stages, the five reactions Kübler-Ross observed are best viewed as coping strategies that anyone may call on in the face of threat. Furthermore, her list is much too limited. Dying people react in many additional ways—for example, through efforts to conquer the illness, as Sofie displayed; through an overwhelming need to control what happens to their bodies, as Dr. Quill's patient Diane showed; and through acts of generosity and caring, as Tony's concern for his 3-year-old brother, Gerry, illustrates.

Perhaps the most serious drawback to Kübler-Ross's theory is that dying patients' thoughts and feelings are removed from the contexts that grant them meaning. As we will see in the next section, adaptations to impending death can only be understood in relation to each individual's personality and life course, which patients bring with them to their final hours. Qualities of the immediate caregiving environment also make a difference (Doka, 1995).

INDIVIDUAL ADAPTATIONS TO IMPENDING DEATH

From the moment of her diagnosis, Sofie spent little time denying the deadliness of her disease. Instead, she met it head on, just as she had dealt with other challenges of life. Her impassioned plea to hold her grandchildren in her arms was less a bargain with fate than an expression of profound defeat that on the threshold of late adulthood, she would not live to enjoy its rewards. At the end, her quiet, withdrawn demeanor was probably resignation, not acceptance. She had been a person with a fighting spirit all her life, unwilling to give in to challenge.

According to recent theorists, a single strategy, such as acceptance, is not best for every dying patient. Instead, an **appropriate death** is one that makes sense in terms of the individual's pattern of living and values and, at the same time, preserves or restores significant relationships and is as free of suffering as possible (Pattison, 1978; Samarel, 1995; Weisman, 1984). New evidence reveals that personal and situational variables combine to affect the way people cope with their own dying—reactions that are as diverse as the life paths considered in previous chapters.

■ PERSONALITY AND COPING STYLE. Understanding the way individuals view stressful life events and have coped with them in the past helps us appreciate the way they manage the dying process. In one study, terminally ill patients were asked to describe their image of dying. Each regarded it differently—for example, as a responsibility, an insurmountable obstacle, a punishment, or an act of courage. These meanings helped explain their responses to their worsening illness (Paige, 1980, as reported by Samarel, 1995). Poorly adjusted individuals—those with conflict-ridden relationships and many disappointments in life—are usually more distressed (Kastenbaum, 1992).

■ NATURE OF THE DISEASE. The course of the illness and its symptoms affect the dying person's reactions. When cancer spread to Sofie's lungs and she could not catch her breath, she was agitated and fearful until oxygen and medication relieved her uncertainty about being able to breathe. In contrast, Nicholas's weakened heart and failing kidneys so depleted his strength that he responded only with passivity as death approached.

■ FAMILY MEMBERS' AND HEALTH PROFESSIONALS' BEHAVIOR. Earlier we noted that a candid approach, in which everyone close to and caring for the dying person acknowledges the terminal illness, is best. Yet this also introduces the burden of participating in the work of dying with the patient—bringing relationships to closure, reflecting on life, and dealing with fears and regrets.

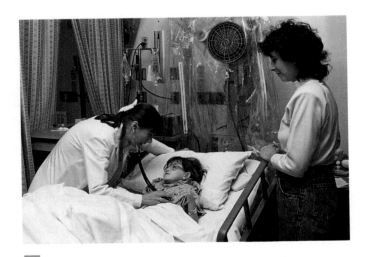

This young patient in a hospital unit for dying children benefits from a candid approach in which family members and health professionals acknowledge her terminal illness. (Frank Siteman/ The Picture Cube)

Because some people find it hard to engage in these tasks, they pretend that the disease is not as bad as it is. In patients inclined toward denial, a game of mutual pretense can be set in motion, in which all participants know the patient is dying but act as if it were not so. The game softens psychological pain for the moment but makes dying much more difficult (Samarel, 1995). At other times, the patient comes to suspect what he or she has not been told. In one instance, a terminally ill child flew into a rage because his doctor and a nurse spoke to him in ways that denied the fact that he would not grow up. Trying to get the child to cooperate with a medical procedure, the doctor said,

> "I thought you would understand, Sandy. You told me once you wanted to be a doctor."
>
> He screamed back, "I'm not going to be anything!" and threw an empty syringe at her.
>
> The nurse standing nearby asked, "What are you going to be?"
>
> "A ghost," said Sandy, and turned away from them. (Bluebond-Langner, 1977, p. 59)

The behavior of health professionals impeded Sandy's efforts to form a realistic time perspective and intensified his anger at the injustice of his premature death.

Care of the terminally ill is demanding and stressful. In one study, nurses trained to respond to the psychological needs of dying patients and their families consistently provided empathetic and supportive care. Keys to their success were staff meetings aimed at strengthening interpersonal skills; day-to-day mutual support among staff; and development of a personal philosophy of living and dying, which ensured that each nurse did not feel especially threatened by his or her own death (Samarel, 1991). Research indicates that highly death-anxious health professionals are more likely to delay informing terminal patients of their diagnoses and to prolong life-saving treatment when death is near (Eggerman & Dustin, 1985; Schulz & Aderman, 1979).

Effective communication with the dying person is open and honest, fostering a trusting relationship. At the same time, it is oriented toward maintaining hope. Dying patients often move through a hope trajectory—at first, hope for a cure; later, hope for prolonging life; and finally, hope for a peaceful death with as few burdens as possible for themselves and family members (Fanslow, 1981). Once a patient near death stops expressing hope, those close to them need to accept. Family members who find letting go very difficult may benefit from the sensitive guidance of health professionals and counselors. The Caregiving Concerns table on the following page offers suggestions for communicating with the dying.

■ SPIRITUALITY, RELIGION, AND CULTURE. Earlier we noted that a strong sense of spirituality reduces fear of

Communicating with Dying People

SUGGESTION	DESCRIPTION
Be truthful about the diagnosis and course of the disease.	Be honest about what the future is likely to hold, thereby permitting the dying person to bring closure to his or her life by expressing sentiments and wishes and participating in decisions about treatment.
Listen perceptively and acknowledge feelings.	Be truly present, focusing full attention on what the dying person has to say and accepting the patient's feelings. Patients who sense another's presence and concern are more likely to relax physically and emotionally and express themselves.
Maintain realistic hope.	Assist the dying person in maintaining hope by encouraging him or her to focus on a realistic goal that might yet be achieved—for example, resolution of a troubled relationship or special moments with a loved one. Knowing the dying person's hope, family members and health professionals can often help fulfill it.
Assist in the final transition.	Assure the dying person that he or she is not alone, offering a sympathetic touch, a caring thought, or just a calm presence. Some patients who struggle may benefit from being granted *permission to die*—the message that giving up and letting go is all right.

Sources: Benoliel & Degner, 1995; Samarel, 1995.

death. Informal reports from health professionals suggest that this is as true for dying patients as people in general. One experienced nurse commented,

> At the end, those [patients] with a faith—it doesn't really matter in what, but a faith in something—find it easier. Not always, but as a rule. I've seen people with faith panic and I've seen those without faith accept it [death]. But, as a rule, it's much easier for those with faith. (Samarel, 1991, pp. 64–65)

Vastly different cultural attitudes, guided by religious beliefs, also shape peoples' dying experiences. Buddhism, widely practiced in China, India, and Southeast Asia, fosters acceptance of death. By reading sutras (teachings of Buddha) to the dying person to calm the mind and emphasizing that dying leads to rebirth, Buddhists believe it is possible to reach Nirvana, a state beyond the world of suffering (Truitner & Truitner, 1993). In many Native American societies, death is met with stoic self-control, an approach taught at an early age through stories that emphasize a circular, rather than linear, relationship between life and death and the importance of making way for others (Lewis, 1990). Among African Americans, a dying loved one signals a crisis that unites family members in caregiving. The terminally ill person remains an active and vital force within the family until he or she no longer can carry out this role—an attitude of respect that undoubtedly eases the dying process (Brown, 1990).

In sum, dying prompts a multitude of thoughts and emotions. Which ones are selected and emphasized depends on a wide array of influences—personal, situational, and cultural. Does this remind you of vital aspects of the lifespan perspective—that development is multidimensional and multidirectional? These assumptions are just as relevant to this final phase as they are to each earlier period.

NEAR-DEATH EXPERIENCES

What are people's inner experiences like as they make the transition from life to death? Each of us has probably wondered about this question, as it addresses one of the most profound mysteries of human existence: what it is like to die. The closest researchers can come to answering it is to interview people who have been resuscitated after being pronounced clinically dead. Findings reveal that **near-death experiences** described by many individuals have common features: feelings of peace and well-being, leaving the body behind, traveling through a dark tunnel toward a distant white light, engaging in a life review, and encountering loved ones (Moody, 1975; Ring, 1980; Sabom, 1982).

Near-death experiences have been interpreted in diverse ways. Some take them as evidence of an afterlife. Others believe they are hallucinations induced by abnormal brain activity, since they resemble out-of-body experiences evoked by hallucinogenic drugs, such as LSD. Still others claim near-death experiences are due to release of hormones in the brain called *endorphins,* which block pain and induce a sense of remoteness from current experience. The response is believed to be adaptive, in that it enables us to stay calm when severely threatened and heightens chances of survival.

Near-death experiences are not universal. Many people have no memories of any kind linked to their close call. In addition, individuals who have not come close to death often have sensations similar to near-death experiences, and circumstances affect their dreamlike images. People in severe pain are more likely to report distance from their bodies, those under anesthetic a brilliant light (Kastenbaum, 1995). Finally, culture affects the events described. For example, the dark tunnel followed by life review appears to be specific to Western societies (Kellehear, 1993).

Diversity in near-death experiences means that many factors must be considered to explain them. And since we hear about them only from survivors, we cannot tell if they resemble what happens when a person really dies.

A PLACE TO DIE

Unlike times past, when most deaths occurred at home, today about 80 percent take place in hospitals (Nuland, 1996). In the large, impersonal hospital environment, meeting the human needs of dying patients and their families is secondary, not because professionals lack concern, but because the work to be done focuses on saving lives. A dying patient represents a failure.

In the 1960s, a death awareness movement arose as a reaction to the death-avoiding practices of hospitals— complicated machinery hooked to patients with no chance of survival and lack of communication with dying patients. It led to medical care suited to the needs of dying people and to hospice programs, which have spread throughout the world. Let's visit each of these settings for dying.

HOME

Had Sofie and Nicholas been asked where they wanted to die, undoubtedly each would have responded, "At home" —the preference of 9 out of 10 Americans (National Hospice Organization, 1992). The reason is clear; the home offers an atmosphere of intimacy and loving care in which the terminally ill person is unlikely to feel abandoned or humiliated by physical decline or dependence on others.

Nevertheless, it is important not to romanticize home death. Because of dramatic improvements in medicine, dying people are sicker or much older than they used to be. Consequently, their bodies may be extremely frail, making ordinary activities—eating, sleeping, taking a pill, toileting, and bathing—major ordeals (Sankar, 1993). Health problems of elderly spouses, work and other responsibilities of family members, and the physical, psychological, and financial strain of providing home care can make it difficult to honor a terminally ill person's wish to die at home.

For many people, the chance to be with the dying person until the very end is a rewarding tradeoff for the high stress of caregiving—a view reflected in the dramatic decline in AIDS deaths in hospitals over the past decade. Currently, nearly half of AIDS victims die at home (Kelly, Chu, & Buehler, 1993). But the advantages and disadvantages of home death should be carefully weighed before undertaking it. Adequate support for the caregiver is essential. A home health aide is often necessary—a service (as we will see shortly) that hospice programs have made more accessible. When family relationships are conflict ridden, a dying patient introduces additional strains, negating the benefits of home death. Finally, even with professional help, most homes are poorly equipped to handle the medical and comfort care needs of the dying.

HOSPITAL

Hospital dying takes many forms. Each pattern is affected by the physical state of the dying person, the hospital unit in which it takes place, and the goal and quality of care.

Sudden deaths, due to injury or critical illness, typically occur in emergency rooms. Doctors and nurses must evaluate the problem and take action quickly. Little time is available for contact with family members. When staff break the news of death in a sympathetic manner and provide explanations, family members are grateful. Otherwise, feelings of anger, frustration, and confusion can add to their grief (Benoliel & Degner, 1995). Crisis intervention services are needed to help survivors cope with sudden death, and some hospitals are doing a better job of providing them.

Nicholas died on an intensive care ward, organized to prevent death in patients whose condition can worsen quickly. Privacy and communicating with the family were secondary to monitoring his condition. To prevent disruption of nurses' activities, Giselle and Sasha could be at Nicholas's side only at scheduled times. Dying in intensive care is an experience unique to technologically sophisticated societies. It is especially depersonalizing for patients like Nicholas, who linger between life and death while hooked to machines for months.

Cancer patients, who account for most cases of prolonged dying, typically die in general or specialized cancer care hospital units. When hospitalized for a long time, they reach out for help with physical and emotional needs, usually with mixed success. In these hospital settings, as in intensive care, a conflict of values is apparent. The tasks associated with dying must be performed efficiently, so all patients can be served and health professionals are not drained emotionally by repeated attachments and separations. But we have seen that dying is a transition with profound personal, social, and cultural meanings. The hospice approach has done much to resolve this caregiving contradiction.

THE HOSPICE APPROACH

In medieval times, a **hospice** was a place where travelers could find rest and shelter. In the nineteenth and twentieth centuries, the word referred to homes for dying patients. Today, a hospice is not a place, but a comprehensive program of support services for terminally ill people and their families. It aims to provide a caring community sensitive to the dying person's needs so patients and family members can prepare for death in ways that are satisfying to them. Quality of life is central to the hospice approach. Here are its main features:

- The patient and family as a unit of care

- Emphasis on meeting the patient's physical, emotional, social, and spiritual needs, including controlling pain, retaining dignity and self-worth, and feeling cared for and loved

- Care provided by an interdisciplinary team, including the patient's doctor; a nurse or nurse's aid; a chaplain, counselor, or social worker; and a pharmacist

- The patient kept at home or in an inpatient setting with a homelike atmosphere where coordination of care is possible

- Focus on improving the quality of remaining life with comfort care, not life-prolonging measures

- In addition to regularly scheduled home care visits, on-call services available 24 hours a day, 7 days a week

- Follow-up bereavement services offered to families in the year after a death (National Hospice Organization, 1993b)

Because hospice care is a philosophy, not a facility, it can be applied in diverse ways. In Great Britain, care in a special inpatient unit, sometimes associated with a hospital, is typical. Home care has been emphasized in the United States. But hospice programs everywhere have expanded to include a continuum of care, from home to inpatient options (Latanzi-Licht & Connor, 1995). Central to the hospice approach is that the dying person and his or her family should be offered choices that guarantee an appropriate death. Some programs offer hospice day care, which permits caregivers to continue working or be relieved of the stresses of long-term care (Corr & Corr, 1992). Contact with others facing terminal illness is a supportive byproduct of many hospice arrangements.

Currently, the United States has approximately 2,000 hospices, serving 37 percent of people dying of cancer and 31 percent dying of AIDS. Programs also admit patients suffering from a variety of other terminal illnesses (National Hospice Organization, 1993a). Because the federal government recognizes that hospice care is a cost-effective alternative to expensive life-saving treatments, Medicare and Medicaid benefits are available, making it affordable for most dying patients and their families. Hospices also serve dying children—a tragedy so devastating that social support and bereavement intervention are vital (Armstrong-Dailey, 1991).

As a long-range goal, hospice organizations are striving for broader acceptance of their patient and family-centered approach. At present, the majority of Americans are unfamiliar with the philosophy, although three out of four say they are very interested when it is described to them (National Hospice Organization, 1992). The highly personalized attention hospices give to terminally ill patients and their families could be adapted for chronically ill patients and people faced with bereavement for any type of death—prolonged or sudden, expected or unexpected.

This 92-year-old woman, terminally ill with congestive heart failure, is spending her last weeks in a hospital wing with a homelike atmosphere, where she receives hospice care. A team of specialists sees that her physical, emotional, social, and spiritual needs are met. Here a volunteer pays a regular visit, bringing a bouquet of autumn flowers. (The Image Works)

BRIEF REVIEW

In Kübler-Ross's theory, terminally ill patients display five mental and emotional reactions as they die: denial, anger, bargaining, depression, and acceptance. However, this list of responses has been criticized as incomplete and removed from contexts that influence people's adaptations. Personality and coping style, nature of the disease, family members' and health professionals' behavior, spirituality, religion, and culture combine to yield highly individual reactions. Because near-death experiences are not universal, vary in the events described, and are reported by survivors, they tell us very little about what it is like to make the transition from life to death.

Although the majority of people want to die at home, caring for a dying patient is highly stressful. The advantages and disadvantages of home death should be carefully weighed. Dying in hospitals takes many forms, but too often the experience is depersonalizing and insensitive for both patients and family members. The hospice approach offers a comprehensive program of support services sensitive to the needs of dying people and their families. Whether dying at home or in an inpatient setting, the patient is granted comfort care and emotional support. After a death, families receive follow-up bereavement services.

ASK YOURSELF . . .

■ *Reread the description of Sofie's mental and emotional reactions to dying on page 636. Then review the story of Sofie's life on page 4. To what extent did Sofie follow Kübler-Ross's stages? How were Sofie's responses consistent with her personality and lifelong style of coping with adversity?*

■ *Return to Chapter 3, pages 96–98, and review the sections on natural childbirth and home delivery. What parallels do you see between these approaches to childbirth and the death awareness movement, including the hospice philosophy?*

■ *When 5-year-old Timmy's kidney failure was diagnosed as terminal, his parents could not accept the tragic news. Their hospital visits became shorter, and they evaded his anxious questions. Timmy tried to figure out what he had done to drive his parents away. Eventually, he blamed himself. He died with little physical pain but very much alone, and his parents suffered prolonged guilt after his death. Explain how hospice care could have helped Timmy and his family.*

BEREAVEMENT: COPING WITH THE DEATH OF A LOVED ONE

Loss is an inevitable part of existence throughout the lifespan. Even when change takes place for the better, we let go of some aspects of experience so we can embrace others. Consequently, our development prepares us for profound loss.

Bereavement is the experience of losing a loved one by death. The root of this word means to be robbed, suggesting unjust and injurious stealing of something valuable. Consistent with this image, we respond to loss with **grief**—intense physical and psychological distress. When we say someone is grief-stricken, we imply that his or her total way of being is affected.

Because grief can be overwhelming, cultures have devised ways of helping their members move beyond it to deal with the life changes demanded by death of a loved one. **Mourning** is the culturally specified expression of the

bereaved person's thoughts and feelings. Customs—such as gathering with family and friends, dressing in black, attending the funeral, and observing a prescribed mourning period with special rituals—vary greatly among societies and ethnic groups. But all have in common the goal of helping people work through their grief and learn to live in a world that does not include the deceased.

Clearly, grief and mourning are closely linked. In fact, we often use the two words interchangeably in everyday language. Let's look closely at how people respond to the death of a loved one.

PHASES OF GRIEVING

Many theorists studying a wide range of types of loss and bereaved individuals—both children and adults—have concluded that grieving usually takes place in three phases, each characterized by a different set of responses (Bowlby, 1980; Parkes & Weiss, 1983; Pollock, 1987; Rando, 1995):

1. **Avoidance.** On hearing the news, the survivor experiences shock followed by disbelief, which may last from hours to weeks. A numbed feeling serves as "emotional anesthesia" while the bereaved person begins to experience painful awareness of the loss. The loved one has been a part of daily life for so long that his or her death is too much to comprehend immediately.

2. **Confrontation.** In this phase, the mourner begins to confront the reality of the loss. Consequently, grief is experienced most intensely. The bereaved person may

Grief is experienced most intensely in the confrontation phase, as the mourner begins to face the reality of the loss. (Robert Brenner/PhotoEdit)

display a range of emotional reactions, including anxiety, sadness, protest, anger, helplessness, frustration, abandonment, and yearning for the loved one. In addition, the person may be unable to concentrate, absent-minded, and preoccupied with thoughts of the deceased. At times, self-destructive behaviors, such as taking drugs or driving too fast, occur. Loss of sleep and appetite are common.

Although the confrontation phase is the most difficult, each pang of grief that results from an unmet wish to be reunited with the deceased brings the mourner closer to the realization that the loved one is gone. After hundreds, perhaps thousands, of these painful moments, the bereaved person comprehends that a cherished relationship will have to be given up.

3. **Accommodation.** As grief subsides, attention starts to shift to the surrounding world. The survivor must transform the relationship with the deceased from physical presence to an inner representation. Although the loved one is not forgotten, emotional energies are freed for meeting everyday responsibilities, investing in new activities and goals, and seeking the companionship and love of others. On certain days, such as family celebrations or the anniversary of death, grief reactions may resurface and require attention, but they do not interfere with a healthy, positive approach to life.

How long does grieving last? No single answer can be given. Confrontation, the period of acute grief, may continue for several months to over a year. An occasional upsurge of grief, typical during the accommodation phase, may persist for a lifetime and is a normal response to losing a much-loved spouse, partner, child, or friend (Shuchter & Zisook, 1995).

PERSONAL AND SITUATIONAL VARIATIONS

Like dying, grieving is affected by many factors that make the duration of each phase and the reactions within it unique. Once again, personality, coping style, and religious and cultural background are influential. Sex differences are also evident, with women better prepared to deal with grief because they express feelings and seek social support more readily (Rando, 1995; Stroebe & Stroebe, 1987). Furthermore, the quality of the mourner's relationship with the deceased is important. For example, an end to a loving, fulfilling bond may lead to anguished grieving, but it is unlikely to leave the residue of guilt and regret that often follows a conflict-ridden, ambivalent tie.

Circumstances surrounding the death also shape mourners' responses. Avoidance is especially pronounced after a sudden death. In contrast, shock and disbelief are usually minimal after prolonged dying because the be-

reaved person has had time to engage in **anticipatory grieving**—acknowledging that the loss is inevitable and preparing emotionally for it. Research also reveals that adjusting to a death is made easier when the survivor understands the reasons for it. Without explanations that make sense, mourners often remain anxious and confused (Rando, 1995). This barrier to confronting loss is tragically apparent in cases of sudden infant death syndrome (SIDS), where doctors cannot tell parents exactly why their apparently healthy baby died (see Chapter 4, page 134).

■ BEREAVED PARENTS. The death of a child, whether unexpected or foreseen, is the most difficult loss an adult can face (Stillion, 1995). It brings special grieving problems, since children are extensions of parents' feelings about themselves—the focus of hopes and dreams, including the parents' sense of immortality. Also, since children depend on, admire, and appreciate their parents in a deeply gratifying way, they are an unmatched source of love. Finally, the unnaturalness of a child's death complicates it. Children are not supposed to die before their parents.

Research reveals that parental bereavement is especially intense and extended. Parents often report considerable distress many years later, along with frequent thoughts of the deceased (Rando, 1986). The guilt triggered by parents' outliving their child frequently becomes a tremendous burden, even when parents "know" better. For example, a psychologist who understood that his daughter's cancer was not hereditary nevertheless said to a therapist, "Her genes allowed her to develop cancer. I gave her her genes. Therefore, I killed my daughter" (Rando, 1991b, p. 239).

Although a child's death sometimes leads to marital breakup, this is likely to happen only when the relationship was unsatisfactory before. If parents can reorganize the family system and reestablish a sense of life's meaning by investing in other children and activities, then the result can be firmer family commitments and personal growth.

■ BEREAVED CHILDREN AND ADOLESCENTS. The loss of an attachment figure has longstanding consequences for children. When a parent dies, children's basic sense of security and being cared for is threatened. Death of a sibling not only deprives children of a close emotional tie but informs them, often for the first time, of their own vulnerability.

Children's grieving after a family loss is usually long-lasting. In several longitudinal studies, children described frequent crying, trouble concentrating in school, sleep difficulties, headaches, and other physical symptoms several months to years after a death (Davies, 1987; Silverman & Worden, 1992). At the same time, many children said they actively maintained mental contact with their dead parent or sibling, dreaming about and speaking to them regularly. In a follow-up 7 to 9 years after sibling loss, thinking about the deceased brother or sister at least once a day was

CULTURAL INFLUENCES

CULTURAL VARIATIONS IN MOURNING BEHAVIOR

The ceremonies that commemorated Sofie and Nicholas's deaths—the first Jewish, the second Quaker—were strikingly different. Yet they served common goals: providing a setting in which loss could be shared and mobilizing the community in support of the bereaved.

At the funeral home, Sofie's body was washed and shrouded, a Jewish ritual signifying return to a state of purity. Then it was placed in a plain wooden (not metal) coffin, so as not to impede the natural process of decomposition. To return the body quickly to the life-giving earth from which it sprang, the funeral was scheduled 3 days after death, as soon as relatives could gather. As the service began, Sofie's husband and children symbolized their anguish by cutting a black ribbon and pinning it to their clothing. The rabbi recited psalms of comfort, followed by a eulogy in which he related memories of Sofie that family members had

shared with him. The service continued at the gravesite. Once the coffin had been lowered into the ground, relatives and friends took turns shoveling earth on it, underscoring the finality of Sofie's death. The service concluded with the "homecoming" prayer called *Kaddish,* which affirms life while accepting death.

The family returned home to light a memorial candle, which burned throughout *shiva,* the 7-day period of mourning. A meal of consolation prepared by others followed, creating a warm feeling of community. Jewish custom prescribes that after 30 days, life should gradually return to normal. When a parent dies, the mourning period is extended to 12 months.

In the tradition of Quaker simplicity, Nicholas's death did not require elaborate preparation of the body, a casket, or a hearse to carry him to a cemetery. He was cremated promptly. During the next week, relatives and close friends gathered with Giselle and Sasha at their home.

Together, they planned a memorial service uniquely suited to celebrating Nicholas's life.

On the appointed day, people who had known Nicholas sat in chairs arranged in concentric circles. Standing at the center, a clerk of the Friends (Quaker) Meeting extended a welcome and explained to newcomers the custom of worshipping silently, with those who feel moved to speak rising and sharing thoughts and feelings at any time. Over the next hour, many people offered personal statements about Nicholas or read poems and selections from Scripture. Giselle and Sasha provided concluding comments. Then everyone joined hands to close the service, and a reception for the family followed.

Religious and cultural variations in mourning behavior are vast—both within and across societies. For most Jews and Christians, extensive ritual accompanies a funeral and burial. In contrast, the Quaker memorial service is among the least ritualized. In

Children's grieving after a family death is usually long-lasting. Honesty, affection, and reassurance help them tolerate painful feelings of loss. Funerals and other family bereavement rituals assist mourners of all ages in resolving grief. (John Ficara/Woodfin Camp & Associates)

common (Martinson, Davies, & McClowry, 1987: Silverman & Nickman, 1996). These images seem to facilitate coping with loss and are sometimes reported by bereaved adults as well.

Cognitive development contributes to the ability to grieve. For example, children with an immature understanding of death may believe the dead parent left voluntarily, perhaps in anger, and that the other parent may also disappear. For these reasons, young children need careful, repeated explanations assuring them that the parent did not want to die and was not angry at them (Furman, 1984). Keeping the truth from children isolates them and often leads to profound regrets. One 8-year-old who learned only a half-hour in advance that his sick brother was dying reflected, "If only I'd known, I could have said good-bye."

Regardless of children's level of understanding, honesty, affection, and reassurance help them tolerate painful feelings of loss. Grief-stricken school-age children are usually more willing to confide in parents than are adolescents. To appear normal, teenagers tend to keep their

some groups, grief is expressed freely. For example, venting deep emotion is often part of African-American funerals, particularly in the southern United States. Christians and Jews of European descent are usually restrained in their display of sorrow (Cytron, 1993; Perry, 1993). And in some societies, letting any feeling show is actively discouraged. The Balinese of Indonesia must remain calm in the face of death if the gods are to hear their prayers. With the help of supporters who joke, tease, and distract, mourners work hard to keep their composure, although the Balinese acknowledge the existence of underlying grief (Rosenblatt, 1993).

Religions also render accounts of the aftermath of death—teachings that console both dying and bereaved individuals. Jewish tradition affirms personal survival, although its form is not clearly specified. Greater emphasis is placed on the survival of a people—living on by granting life and care to others. Unlike other Christian faiths, Quakers focus almost exclusively on the here and now—"salvation by character" through working for peace, justice, and a loving community. Little attention is given to fear of hell or hope of heaven. The religions of tribal and village cultures typically include elaborate beliefs about ancestor spirits and the afterlife and customs designed to ease the journey of the deceased to the spiritual world.

By announcing that a death has occurred, ensuring social support, memorializing the deceased, and conveying a philosophy of life after death, funerals and memorial services are of great assistance to the bereaved. Yet some evidence indicates that these customs may be on the decline in the United States (Fulton, 1995). In large cities, more deaths are being followed by disposal of the body, without a public or religious observance of any kind. The trend is worrisome, since it deprives grief-stricken people of community assurance of love and concern. And in a death-denying society, it robs new generations of a crucial opportunity to learn about life's most basic fact: the inevitability of human mortality.

Whether at a state funeral or a private family burial, Jewish mourning customs are the same. Here, at a gathering of world leaders, the "homecoming" prayer called Kaddish is recited over the body of Israeli Prime Minister Yitzhak Rabin. The Kaddish does not mention death. Rather it reminds the bereaved that the departed loved one lives on in the hearts of others. (Reuters/David Silverman/Archive Photos)

---

grieving from both adults and peers. Consequently, they are more likely than children to become depressed or to escape from grief through acting-out behavior (Corr & Balk, 1996).

■ BEREAVEMENT OVERLOAD. When a person experiences several deaths in succession, *bereavement overload* can occur. Multiple losses deplete the coping resources of even well-adjusted people, leaving them emotionally overwhelmed and unable to resolve their grief. For many young adults, especially members of the gay community who have lost partners and friends, AIDS presents this special challenge. In a study of over 700 homosexual men, those experiencing two or more losses in close succession reported more distress and substance use than did those with only a single loss (Martin & Dean, 1993).

Since old age often brings the death of spouse, siblings, and friends in close succession, elders are also at risk for bereavement overload. But recall from Chapter 18 that compared to young people, older adults are often better equipped to handle these losses. They know that decline and death are expected in late adulthood, and they have had a lifetime of experience through which to develop effective coping strategies.

Funerals and other bereavement rituals, illustrated in the Cultural Influences box on pages 644–645, assist mourners of all ages in resolving grief with the help of family and friends. Bereaved individuals who cannot resume interest in everyday activities benefit from special interventions designed to help them adjust to loss.

BEREAVEMENT INTERVENTIONS

Social support is crucial during the aftermath of death. Yet it is often difficult to provide. Parents, caught up in their own grief, may not be available to their children. And relatives and friends trying to be helpful may not always succeed. When bereaved people were asked to describe contacts that were supportive, they mentioned freedom to express feelings in a receptive, nonjudgmental atmosphere.

Suggestions for Resolving Grief after a Loved One Dies

SUGGESTION	DESCRIPTION
Give yourself permission to feel the loss.	Clarify the reasons for the death and permit yourself to confront all thoughts and emotions associated with it. Make a conscious decision to overcome your grief, recognizing that this will take time.
Accept social support.	In the early part of grief, let others reach out to you by making meals, running errands, and keeping you company. Be assertive; ask for what you need so people who would like to help know what to do.
Be realistic about the course of grieving.	Expect to have some negative and intense reactions, such as feeling anguished, sad, and angry, that last from weeks to months and may occasionally resurface years after the death. Since there is no one way to grieve, find the best way for you.
Remember the deceased.	Review your relationship to and experiences with the deceased, permitting yourself to see that you can no longer be with him or her as before. Form a new relationship based on memories, keeping it alive through photographs, commemorative donations, prayers, and other symbols and actions.
At the appropriate time, invest in new activities and relationships.	Determine which roles you must give up and assume as a consequence of the death and take deliberate steps to incorporate these into your life. Set small goals at first, such as a night at the movies, a dinner date with a friend, or a week's vacation.

Source: Rando, 1991a.

They regarded giving advice and encouraging recovery negatively. Nongrieving individuals were aware of what they should and should not say to the bereaved. Nevertheless, mourners often reported they did not get as much help as they needed, perhaps because of others' inexperience or fear of doing the wrong thing (Lehman, Ellard, & Wortman, 1986).

Experts agree that sympathy and understanding are sufficient for most people to undertake the tasks necessary to recover from grief (see the Caregiving Concerns table above). But relatives and friends could clearly benefit from training in how to respond to the bereaved. Often mourners are too overcome to acknowledge well-meaning interactions, causing others to withdraw (Stylianos & Vachon, 1993). Listening patiently and "just being there" are among the best ways to help.

Bereavement interventions typically assist people in drawing on their existing social network and provide extra social support. Many self-help groups exist, bringing together mourners who have experienced the same type of loss for regular sessions over several months. Research supports their effectiveness in reducing stress and, in the case of a child's death, assisting couples in reworking a damaged relationship (Lieberman, 1993). Groups for bereaved children and adolescents provide valuable peer support that teenagers, especially, are unwilling to seek on their own (Bacon, 1996; Corr & Balk, 1996).

At times, mourning does not proceed smoothly. An unanticipated death, especially when violent or random; the loss of a child; a death that the mourner feels he or she could have prevented; or an ambivalent or dependent relationship with the deceased make it harder for bereaved people to overcome their loss (Raphael et al., 1993). In these instances, *grief therapy,* or individual counseling with a specially trained professional, may be necessary.

DEATH EDUCATION

Preparatory steps can be taken to help people of all ages cope with death more effectively. The death awareness movement that sparked increased sensitivity to the needs of dying patients has also led to the rise of death education. Courses in death, dying, and bereavement are now a familiar part of offerings in colleges and universities. Instruction has been integrated into the professional training of doctors, nurses, psychologists, social workers, and other helping professionals. It can also be found in adult education programs in many communities. And it has filtered down to a few elementary and secondary schools.

Death education at all levels has the following goals:

■ Increasing students' understanding of the physical and psychological changes that accompany dying

■ Helping students learn how to cope with the death of a loved one

■ Preparing students to be informed consumers of medical and funeral services

■ Promoting understanding of social and ethical issues involving death

Courses vary widely in format. Some focus on conveying information, whereas others are experiential and include many activities—role-playing, discussions with the terminally ill, visits to mortuaries and cemeteries, and personal awareness exercises. Research reveals that using a lecture style leads to gains in knowledge, but it often leaves students more uncomfortable about death than when they entered. In contrast, experiential programs that help people confront their own mortality are less likely to heighten death anxiety and sometimes reduce it (Durlak & Riesenberg, 1991; Maglio & Robinson, 1994). These findings suggest that to reach students cognitively and emotionally, death educators need to be more than well informed. They must be sensitive and responsive, skilled at communication, and able to help people deal with distress.

Whether acquired in the classroom or in our daily lives, our ideas and feelings about death are forged through interactions with others. Becoming more aware of how we die and our own mortality, we encounter our greatest loss, but we also gain. Dying people have at times confided in those close to them that awareness of the limits of their lifespan permitted them to dispense with superficial distractions and wasted energies and focus on what is truly important in their lives. As one AIDS patient summed up, "[It's] kind of like life, just speeded up"— an accelerated process in which over a period of weeks to months one grapples with issues that normally would have taken years to decades (Selwyn, 1996, p. 36). Applying this lesson to ourselves, we learn that by being in touch with death and dying, we can live ever more fully.

ASK YOURSELF . . .

■ *Compare the phases of grieving with terminally ill patients' thoughts and feelings as they move closer to death, described on pages 636–637. Can a dying person's reactions be viewed as a form of grieving? Explain.*

SUMMARY

HOW WE DIE

Describe the physical changes of dying, along with their implications for defining death and the meaning of death with dignity.

■ Of the one-quarter of people in industrialized nations who die suddenly, the overwhelming majority are victims of heart attacks. Death is long and drawn out for three-fourths of people—many more than in times past, due to life-saving medical technology.

■ In general, dying takes place in three phases: the **agonal phase,** during which the body first cannot sustain life any more; **clinical death,** a short interval in which resuscitation is still possible; and **mortality,** or permanent death.

■ In most industrialized nations, **brain death** is accepted as the definition of death. However, the thousands of patients who remain in a **persistent vegetative state** reveal that the brain death standard does not always solve the dilemma of when to halt treatment of the incurably ill.

■ Most people will not experience an easy death. Therefore, we can best ensure death with dignity by supporting dying patients through their physical and psychological distress, being candid about death's certainty, and helping them learn enough about their condition to make reasoned choices about treatment.

THE RIGHT TO DIE

Discuss controversies surrounding euthanasia and assisted suicide.

■ Today, the same medical procedures that preserve life can prolong inevitable death, diminishing quality of life and personal dignity. The various forms of **euthanasia** vary along a continuum from least active to most active.

■ **Passive euthanasia,** withholding or withdrawing life-sustaining treatment from a hopelessly ill patient, is widely accepted and practiced. Because U.S. state laws vary widely, people can best ensure that their wishes will be followed by preparing an **advance medical directive**— either a **living will** that includes directions for treatment, including **comfort care,** or a **durable power of attorney for health care.** When no advance medical directive exists, a few U.S. states permit appointment of a **health care proxy.**

■ Public support for **voluntary active euthanasia,** in which doctors or others comply with a suffering patient's request to die before a natural end to life, is growing. Nevertheless, the practice has sparked heated controversy, fueled by fears that it will undermine trust in health professionals and lead down a "slippery slope" to killing of vulnerable people who did not ask to die.

- Less public consensus exists for assisted suicide. Like voluntary active euthanasia, grave ethical dilemmas surround it.

UNDERSTANDING OF AND ATTITUDES TOWARD DEATH

Discuss age-related changes in conceptions of and attitudes toward death, and cite factors that influence death anxiety.

- We are a death-denying society. Compared to earlier generations, more young people reach adulthood having had little contact with death, and adults are often reluctant to bring up the topic.

- By age 7, children typically grasp the three components of the death concept—**permanence, universality,** and **nonfunctionality**—in the order just given. When parents talk about death candidly, children usually have a good grasp of the facts of death and an easier time accepting it.

- Adolescents often fail to apply their understanding of death to everyday life. Aware that death happens to everyone and can occur at any time, teenagers are nevertheless high risk takers and do not take death personally. By discussing concerns about death with adolescents, adults can help them build a bridge between death as a logical concept and their personal experiences.

- As people pass from early to middle adulthood, they become more conscious of the finiteness of their lives. Compared to middle-aged adults, elders focus more on practical matters of how and when death might happen.

- Large individual and cultural variations in **death anxiety** exist. Overall, fear of death declines with age, reaching its lowest level in late adulthood and in adults with deep faith in some form of higher being. People with mental health problems generally express more severe death concerns.

THINKING AND EMOTIONS OF DYING PEOPLE

Describe and evaluate Kübler-Ross's stage theory, citing factors that influence the responses of dying patients.

- According to Elizabeth Kübler-Ross's stage theory, five responses are typical as dying patients move closer to death: **denial, anger, bargaining, depression,** and **acceptance.** However, these reactions do not necessarily occur, and dying people often display other coping strategies.

- An **appropriate death** is one that makes sense in terms of the individual's pattern of living and values. A host of personal and situational variables—coping style, spirituality, religion, cultural background, nature of the disease, and family members' and health professionals' truthfulness and sensitivity—combine to affect the way people respond to their own dying.

Evaluate insights about the transition from life to death gained from near-death experiences .

- Not all people who come close to death report **near-death experiences,** and some who are not near death describe near-death-like sensations. Therefore, we cannot be sure these images resemble what actually happens when a person makes the transition from life to death.

A PLACE TO DIE

Evaluate the extent to which homes, hospitals, and the hospice approach meet the needs of dying people and their families.

- Although the overwhelming majority of people want to die at home, caring for a dying patient is highly stressful. Even with professional help, most homes are poorly equipped to handle the medical and comfort care needs of the dying.

- In hospitals, sudden deaths typically occur in emergency rooms. Sympathetic explanations from staff can reduce family members' anger, frustration, and confusion. Intensive care is especially depersonalizing for patients who are hooked to machines for months. Even in general or specialized cancer care units, emphasis on efficiency usually interferes with a dignified death.

- Whether a person dies at home or in a hospital, the **hospice** approach emphasizes quality of life over life-prolonging measures and strives to meet the dying person's physical, emotional, social, and spiritual needs. In the year after death, bereavement services are offered to families.

BEREAVEMENT: COPING WITH THE DEATH OF A LOVED ONE

Describe the phases of grieving, factors that underlie individual variations, and bereavement interventions.

- **Bereavement** refers to the experience of losing a loved one by death, **grief** to the intense physical and psychological distress that accompanies loss. **Mourning** customs are culturally prescribed expressions of thoughts and feelings designed to help people work through their grief.

- Grieving usually takes place in three phases: **avoidance,** involving shock and disbelief; **confrontation,** in which the mourner confronts the reality of the loss and experiences grief most intensely; and **accommodation,** or shifting attention to the surrounding world as grief subsides.

- Like dying, grieving is affected by many personal and situational factors that make the duration of each phase and reactions within it unique. Avoidance is typically minimal after prolonged dying because the bereaved person has had time to engage in **anticipatory grieving.** When a parent loses a child or a child loses a parent or sibling, grieving is especially intense.

■ Young adult members of the gay community and the elderly are at risk for bereavement overload. Sympathy and understanding are usually sufficient for people to recover from grief. Self-help groups can provide extra social support. When a bereaved individual finds it very hard to overcome loss, grief therapy may be necessary.

DEATH EDUCATION

Explain how death education can help people cope with death more effectively.

■ Today, instruction in death, dying, and bereavement can be found in colleges and universities; training programs for doctors, nurses, and helping professionals; adult education programs; and some elementary and secondary schools. When courses include an experiential component, they are more likely to reach students cognitively and emotionally.

IMPORTANT TERMS AND CONCEPTS

agonal phase (p. 625)

clinical death (p. 625)

mortality (p. 625)

brain death (p. 625)

persistent vegetative state (p. 625)

euthanasia (p. 626)

passive euthanasia (p. 627)

advance medical directive (p. 627)

living will (p. 627)

comfort care (p. 628)

durable power of attorney for health care (p. 628)

voluntary active euthanasia (p. 629)

permanence (p. 632)

universality (p. 632)

nonfunctionality (p. 632)

death anxiety (p. 635)

denial (p. 636)

anger (p. 636)

bargaining (p. 637)

depression (p. 637)

acceptance (p. 637)

appropriate death (p. 637)

near-death experiences (p. 639)

hospice (p. 640)

bereavement (p. 642)

grief (p. 642)

mourning (p. 642)

avoidance (p. 642)

confrontation (p. 642)

accommodation (p. 643)

anticipatory grieving (p. 643)

FYI FOR FURTHER INFORMATION AND HELP

EUTHANASIA

Choice in Dying—The National Council for the Right to Die
200 Varick Street
New York, NY 10014-2810
(212) 366-3540
Web site: *www.choices.org*
An organization that engages in legal research, produces educational publications and videos, and sponsors professional conferences on choices in dying. On request, distributes one free set of state-specific advance medical directives.

Hemlock Society
P.O. Box 101810
Denver, CO 80250-1810
(800) 247-7421
Web site: *www.hemlock.org/hemlock*
A nonprofit educational organization that advocates carefully safeguarded laws that would permit terminally ill, mentally competent adults to receive help in dying from a doctor.

HOSPICE CARE

Children's Hospice International
2202 Mt. Vernon Avenue, Suite 3C
Alexandria, VA 22301
(800) 242-4453
Web site: *www.chionline.org/twl.htm*
Organization of professionals, volunteers, and students who work with, or are interested in extending the hospice approach to, terminally ill children and their families.

National Hospice Organization
1901 N. Moore Street, Suite 901
Arlington, VA 22209
(800) 338-8619
Web site: *www.nho.org*
An organization of hospices and individuals interested in the promotion of the hospice philosophy and program of care.

BEREAVEMENT

The Compassionate Friends
P.O. Box 3696
Oak Brook, IL 60522-3696
(708) 990-0246
Web site: *www.allenwayne.com/washcares/comp.htm*
A self-help organization that supports parents and siblings in resolving grief on the death of their child or brother/sister.

They Help Each Other Spiritually (THEOS)
322 Boulevard of the Allies,
Suite 105
Pittsburgh, PA 15222-1919
(412) 471-7779
Assists in planning educational and practical programs for the widowed in Canada and the United States, with a special focus on overcoming grief.

DEATH EDUCATION

Association for Death Education and Counseling
638 Prospect Avenue
Hartford, CT 06105-4298
(860) 586-7503
Web site: *www.adec.org*
Association of individuals and institutions committed to effective death education and counseling. Aims to upgrade the quality of death education, patient care, and counseling in the areas of death, dying, and bereavement.

academic preschools Preschools in which teachers structure the program, training children in academic skills through repetition and drill. Distinguished from *child-centered preschools*.

acceptance Kübler-Ross's stage of dying in which the person, in the last weeks or days, enters a state of peace and quiet about upcoming death.

accommodation In Piaget's theory, that part of adaptation in which new schemes are created and old ones adjusted to produce a better fit with the environment. Distinguished from *assimilation*. Also, the third phase of grieving, in which grief subsides and emotional energies are freed for meeting everyday responsibilities, investing in new activities and goals, and seeking the companionship and love of others.

achieving stage Schaie's stage of early adulthood in which people adapt their cognitive skills to situations that have profound implications for achieving long-term goals. Consequently, they focus less on acquiring knowledge and more on applying it to everyday life.

acquisitive stage Schaie's stage of childhood and adolescence in which the goal of mental activity is knowledge acquisition.

active lifespan The number of years of vigorous, healthy life. Distinguished from *average life expectancy* and *maximum lifespan*.

activity theory A social theory of aging that states that the decline in social interaction in late adulthood is due to failure of the social environment to offer opportunities for social contact, not the desires of elders. Distinguished from *disengagement theory* and *selectivity theory*.

adaptation In Piaget's theory, the process of building schemes through direct interaction with the environment. Made up of two complementary processes: *assimilation* and *accommodation*.

advance medical directive A written statement of desired medical treatment should a person become incurably ill.

age-graded influences Influences on lifespan development that are strongly related to age and therefore fairly predictable in when they occur and how long they last.

age of viability The age at which the fetus can first survive if born early. Occurs sometime between 22 and 26 weeks.

agonal phase The phase of dying in which gasps and muscle spasms occur during the first moments in which the body cannot sustain life anymore. Distinguished from *clinical death* and *mortality*.

Alzheimer's disease The most common form of dementia, in which structural and chemical deterioration in the brain is associated with gradual loss of many aspects of thought and behavior, including memory, skilled and purposeful movements, and comprehension and production of speech.

amnion The inner membrane that forms a protective covering around the prenatal organism.

amniotic fluid The fluid that fills the amnion, helping to keep temperature constant and to provide a cushion against jolts caused by the mother's movement.

amyloid plaques A structural change in the brain associated with Alzheimer's disease in which deposits of the protein amyloid are surrounded by clumps of dead nerve cells, which amyloid appears to destroy.

anal stage Freud's second psychosexual stage, in which toddlers take pleasure in retaining and releasing urine and feces at will.

androgyny A type of gender-role identity in which the person scores high on both masculine and feminine personality characteristics.

anger Kübler-Ross's stage of dying in which the person expresses anger at having to die without being given a chance to do all he or she wants to do.

animistic thinking The belief that inanimate objects have lifelike qualities, such as thoughts, wishes, feelings, and intentions.

anorexia nervosa An eating disorder in which individuals (usually females) starve themselves because of a compulsive fear of getting fat.

anoxia Inadequate oxygen supply.

anticipatory grieving Before a prolonged, expected death, acknowledging that the loss is inevitable and preparing emotionally for it.

Apgar Scale A rating used to assess the newborn baby's physical condition immediately after birth.

applied behavior analysis A set of practical procedures that combine conditioning and modeling to eliminate undesirable behaviors and increase socially acceptable responses.

appropriate death A death that makes sense in terms of the individual's pattern of living and values and, at the same time, preserves or restores significant relationships and is as free of suffering as possible.

assimilation That part of adaptation in which the external world is interpreted in terms of current schemes. Distinguished from *accommodation*.

associative play A form of true social participation in which children are engaged in separate activities, but they interact by exchanging toys and comment on one another's behavior. Distinguished from *nonsocial activity, parallel play,* and *cooperative play.*

attachment The strong, affectional tie that humans feel toward special people in their lives.

attention-deficit hyperactivity disorder (ADHD) A childhood disorder involving inattentiveness, impulsivity, and excessive motor activity. Often leads to academic failure and social problems.

authoritarian style A child-rearing style that is demanding but low in responsiveness to children's rights and needs. Conformity and obedience are valued over open communication with the child. Distinguished from *authoritative* and *permissive styles.*

authoritative style A child-rearing style that is demanding and responsive. A rational, democratic approach in which parents' and children's rights are respected. Distinguished from *authoritarian* and *permissive styles.*

autoimmune response An abnormal response of the immune system in which it turns against normal body tissues.

autonomous morality Piaget's second stage of moral development, in which children view rules as flexible, socially agreed-on principles that can be revised when there is a need to do so. Begins around age 10.

autonomy At adolescence, a sense of oneself as a separate, self-governing individual. Involves relying more on oneself and less on parents for direction and guidance and engaging in careful, well-reasoned decision making.

autonomy versus shame and doubt In Erikson's theory, the psychological conflict of toddlerhood, which is resolved positively if parents provide young children with suitable guidance and appropriate choices.

autosomes The 22 matching chromosome pairs in each human cell.

average life expectancy The number of years an individual born in a particular year can expect to live. Distinguished from *maximum lifespan* and *active lifespan.*

avoidance The first phase of grieving, in which the survivor experiences shock followed by disbelief. Distinguished from *confrontation* and *accommodation.*

avoidant attachment The quality of insecure attachment characterizing infants who are usually not distressed by parental separation and who avoid the parent when she returns. Distinguished from *secure, resistant,* and *disorganized/disoriented attachment.*

babbling Repetition of consonant–vowel combinations in long strings, beginning around 6 months of age.

bargaining Kübler-Ross's stage of dying in which the person attempts to forestall death by bargaining for extra time—a deal he or she may try to strike with family members, friends, caregivers, or God.

basal metabolic rate (BMR) The amount of energy the body uses at complete rest.

basic emotions Emotions that can be directly inferred from facial expressions, such as happiness, interest, surprise, fear, anger, sadness, and disgust.

basic skills approach An approach to beginning reading instruction that emphasizes training in phonics—the basic rules for translating written symbols into sounds—and simplified reading materials. Distinguished from *whole-language approach.*

basic trust versus mistrust In Erikson's theory, the psychological conflict of infancy, which is resolved positively if caregiving, especially during feeding, is sympathetic and loving.

behaviorism An approach that views directly observable events—stimuli and responses—as the appropriate focus of study and the development of behavior as taking place through classical and operant conditioning.

bereavement The experience of losing a loved one by death.

"big five" personality traits Five basic factors, into which hundreds of personality traits have been organized: neuroticism, extroversion, openness to experience, agreeableness, and conscientiousness.

biological aging Genetically influenced, age-related declines in the functioning of organs and systems that are universal in all members of our species. Sometimes called *primary aging.*

blastocyst The zygote 4 days after fertilization, when the tiny mass of cells forms a hollow, fluid-filled ball.

body image Conception of and attitude toward one's physical appearance.

bonding Parents' feelings of affection and concern for the newborn baby.

brain death Irreversible cessation of all activity in the brain and the brain stem. The definition of death accepted in most industrialized nations.

brain plasticity The ability of other parts of the brain to take over functions of damaged regions.

breech position A position of the baby in the uterus that would cause the buttocks or feet to be delivered first.

bulimia An eating disorder in which individuals (mainly females) go on eating binges followed by deliberate vomiting, other purging techniques such as heavy doses of laxatives, and strict dieting.

burnout A condition in which long-term job stress leads to emotional exhaustion, a sense of loss of personal control, and feelings of reduced accomplishment.

canalization The tendency of heredity to restrict the development of some characteristics to just one or a few outcomes.

cardinality The mathematical principle that the last number in a counting sequence indicates the quantity of items counted.

carrier A heterozygous individual who can pass a recessive gene to his or her children.

cataracts Cloudy areas in the lens of the eye that increase from middle to old age, resulting in foggy vision and (without surgery) eventual blindness.

centration The tendency to focus on one aspect of a situation and neglect other important features. Distinguished from *decentration*.

cephalocaudal trend An organized pattern of physical growth and motor control that proceeds from head to tail. Distinguished from *proximodistal trend*.

cerebellum A brain structure that aids in balance and control of body movements.

cerebral cortex The largest structure of the human brain, which accounts for the highly developed intelligence of the human species.

cerebrovascular dementia A form of dementia in which a series of strokes leaves dead brain cells, producing step-by-step degeneration of mental ability, with each step occurring abruptly after a stroke.

cesarean delivery A surgical delivery in which the doctor makes an incision in the mother's abdomen and lifts the baby out of the uterus.

child-centered preschools Preschools in which teachers provide a wide variety of activities from which children select, and most of the day is devoted to free play. Distinguished from *academic preschools*.

child-directed speech A form of language used by adults to speak to infants and toddlers that consists of short sentences with exaggerated expression and very clear pronunciation. Also called *motherese*.

chorion The outer membrane that forms a protective covering around the prenatal organism. It sends out tiny fingerlike villi, from which the placenta begins to emerge.

chromosomes Rodlike structures in the cell nucleus that store and transmit genetic information.

circular reaction In Piaget's theory, a means of building schemes in which infants try to repeat a chance event caused by their own motor activity.

classical conditioning A form of learning that involves associating a neutral stimulus with a stimulus that leads to a reflexive response.

climacteric Midlife transition in which fertility declines. Brings an end to reproductive capacity in women and diminished fertility in men.

clinical death The phase of dying in which heartbeat, circulation, breathing, and brain functioning stop, but resuscitation is still possible. Distinguished from *agonal phase* and *mortality*.

clinical interview A method that uses a flexible, conversational style to probe for the subject's point of view.

clinical method A method that attempts to understand the unique individual child by combining interview data, observations, and sometimes test scores.

clique A small group of about five to seven members who are either close or good friends.

codominance A pattern of inheritance in which both genes, in a heterozygous combination, are expressed.

cognitive-developmental theory An approach introduced by Piaget that views the child as actively building psychological structures and cognitive development as taking place in stages.

cohabitation The lifestyle of unmarried couples who have an intimate, sexual relationship and share a residence.

cohort effects The effects of history-graded influences on the generalizability of findings: Children born in one period of time are influenced by particular historical and cultural conditions.

comfort care Care for terminally ill, suffering patients that provides sufficient and often potent medication to relieve pain, even though the treatment may shorten life.

companionate love Love based on warm, trusting affection and caregiving. Distinguished from *passionate love*.

compliance Voluntary obedience to requests and commands.

compression of morbidity The public health goal of reducing the average period of diminished vigor before death as life expectancy extends. So far, persistence of poverty and negative lifestyle factors have interfered with progress toward this goal.

concordance rate The percentage of instances in which both twins show a trait when it is present in one twin. Used to study the role of heredity in emotional and behavior disorders, which can be judged as either present or absent.

concrete operational stage Piaget's third stage, during which thought is logical, flexible, and organized in its application to concrete information. However, the capacity for abstract thinking is not yet present. Spans the years from 7 to 11.

conditioned response (CR) In classical conditioning, an originally reflexive response that is produced by a conditioned stimulus (CS).

conditioned stimulus (CS) In classical conditioning, a neutral stimulus that through pairing with an unconditioned stimulus (UCS) leads to a new response (CR).

confrontation The second phase of grieving, in which the mourner begins to confront the reality of the loss and grief is experienced most intensely. Distinguished from *avoidance* and *accommodation*.

congregate housing Housing for the elderly that adds a variety of support services, including meals in a common dining room.

conservation The understanding that certain physical characteristics of objects remain the same, even when their outward appearance changes.

continuous development A view that regards development as a cumulative process of adding on more of the same types of skills that were there to begin with. Distinguished from *discontinuous development*.

contrast sensitivity A general principle accounting for early pattern preferences, which states that if babies can detect a difference in contrast between two patterns, they will prefer the one with more contrast.

control processes, or **mental strategies** In information processing, procedures that operate on and transform information, increasing the efficiency of thinking as well as the chances that information will be retained.

controversial children Children who get a large number of positive and negative votes on sociometric measures of peer acceptance. Distinguished from *popular, neglected,* and *rejected* children.

conventional level Kohlberg's second level of moral development, in which moral understanding is based on conforming to social rules to ensure positive human relationships and societal order.

convergent thinking The generation of a single correct answer to a problem. The type of cognition emphasized on intelligence tests. Distinguished from *divergent thinking.*

cooing Pleasant vowel-like noises made by infants beginning around 2 months of age.

cooperative play A form of true social participation in which children's actions are directed toward a common goal. Distinguished from *nonsocial activity, parallel play,* and *associative play.*

coping strategies Thoughts and behaviors used to manage stress and the emotions associated with stress.

coregulation A transitional form of supervision in which parents exercise general oversight, while permitting children to be in charge of moment-by-moment decision making.

corpus callosum A large bundle of fibers that connects the two hemispheres of the brain.

correlation coefficient A number, ranging from +1.00 to –1.00, that describes the strength and direction of the relationship between two variables.

correlational design A research design that gathers information without altering participants' experiences and examines relationships between variables. Cannot determine cause and effect.

cross-linkage theory A theory of biological aging asserting that the formation of bonds, or links, between normally separate protein fibers causes the body's connective tissue to become less elastic over time, leading to many negative physical consequences.

cross-sectional design A research design in which groups of participants of different ages are studied at the same point in time. Distinguished from *longitudinal design.*

crowd A large, loosely organized group consisting of several cliques with similar normative characteristics.

crystallized intelligence Intellectual skills that depend on accumulated knowledge and experience, good judgment, and mastery of social conventions. Together, these capacities represent abilities acquired because they are valued by the individual's culture. Distinguished from *fluid intelligence.*

death anxiety Fear and apprehension of death.

decentration The ability to focus on several aspects of a problem at once and relate them. Distinguished from *centration.*

deferred imitation The ability to remember and copy the behavior of models who are not immediately present.

dementia A set of disorders occurring almost entirely in old age in which many aspects of thought and behavior are so impaired that everyday activities are disrupted.

denial Kübler-Ross's stage of dying in which the person denies the seriousness of his or her terminal condition in an effort to escape from the prospect of death.

deoxyribonucleic acid (DNA) Long, double-stranded molecules that make up chromosomes.

dependency-support script A typical pattern of interaction in which elders' dependency behaviors are attended to immediately, thereby reinforcing those behaviors. Distinguished from *independence-ignore script.*

dependent variable The variable the researcher expects to be influenced by the independent variable in an experiment.

depression Kübler-Ross's stage of dying in which the person becomes depressed, grieving the ultimate loss of his or her life.

deprivation dwarfism A growth disorder observed between 2 and 15 years of age. Characterized by very short stature, light weight in proportion to height, immature skeletal age, and decreased GH secretion. Caused by emotional deprivation.

developmental quotient, or DQ A score on an infant intelligence test, based primarily on perceptual and motor responses. Computed in the same manner as an IQ.

developmentally appropriate practice A set of standards devised by the National Association for the Education of Young Children that specify program characteristics that meet the developmental and individual needs of young children of varying ages, based on current research and the consensus of experts.

differentiation theory The view that perceptual development involves the detection of increasingly fine-grained invariant features in the environment.

difficult child A child whose temperament is such that he or she is irregular in daily routines, is slow to accept new experiences, and tends to react negatively and intensely. Distinguished from *easy child* and *slow-to-warm-up child.*

dilation and effacement of the cervix Widening and thinning of the cervix during the first stage of labor.

discontinuous development A view in which new and different ways of interpreting and responding to the world emerge at particular time periods. Distinguished from *continuous development.*

disengagement theory A social theory of aging that states that the decline in social interaction in late adulthood is due to mutual withdrawal between elders and society in anticipation of death. Distinguished from *activity theory* and *selectivity theory.*

dishabituation Increase in responsiveness after stimulation changes.

disorganized/disoriented attachment The quality of insecure attachment characterizing infants who respond in a confused, contradictory fashion when reunited with the parent. Distinguished from *secure, avoidant,* and *resistant attachment.*

distributive justice Beliefs about how to divide up material goods fairly.

divergent thinking The generation of multiple and unusual possibilities when faced with a task or problem. Associated with creativity. Distinguished from *convergent thinking*.

divorce mediation A series of meetings between divorcing adults and a trained professional, who tries to help them settle disputes. Aimed at reducing family conflict during the period surrounding divorce.

dominant cerebral hemisphere The hemisphere of the brain responsible for skilled motor action. The left hemisphere is dominant in right-handed individuals. In left-handed individuals, the right hemisphere may be dominant, or motor and language skills may be shared between the hemispheres.

dominant–recessive inheritance A pattern of inheritance in which, under heterozygous conditions, the influence of only one gene is apparent.

dual-earner marriage A family form in which both husband and wife are employed.

dualistic thinking In Perry's theory, the cognitive approach of younger college students, who search for absolute truth and therefore divide information, values, and authority into right and wrong, good and bad, we and they. Distinguished from *relativistic thinking*.

durable power of attorney for health care A written statement that authorizes appointment of another person (usually, although not always, a family member) to make health care decisions on one's behalf in case of incompetence.

dynamic systems of action In motor development, combinations of previously acquired abilities that lead to more advanced ways of exploring and controlling the environment. Each new skill is a joint product of maturation, movement possibilities of the body, environmental supports for the skill, and the task the child has in mind.

easy child A child whose temperament is such that he or she quickly establishes regular routines in infancy, is generally cheerful, and adapts easily to new experiences. Distinguished from *difficult child* and *slow-to-warm-up child*.

ecological systems theory Bronfenbrenner's approach, which views the person as developing within a complex system of relationships affected by multiple levels of the environment, from immediate settings of family and school to broad cultural values and programs.

educational self-fulfilling prophecy The idea that children may adopt teachers' positive or negative attitudes toward them and start to live up to these views.

egalitarian marriage A form of marriage in which husband and wife share power and authority. Both try to balance the time and energy they devote to the workplace, the children, and their relationship. Distinguished from *traditional marriage*.

ego In Freud's theory, the rational part of personality that reconciles the demands of the id, the external world, and conscience.

ego integrity versus despair In Erikson's theory, the psychological conflict of late adulthood, which is resolved positively when elders feel whole, complete, and satisfied with their achievements, having accepted their life course as something that had to be the way it was.

egocentrism The inability to distinguish the symbolic viewpoints of others from one's own.

elaboration The memory strategy of creating a relation between two or more items that are not members of the same category.

embryo The prenatal organism from 2 to 8 weeks after conception, during which time the foundations of all body structures and internal organs are laid down.

embryonic disk A small cluster of cells on the inside of the blastocyst, from which the embryo will develop.

emotional self-regulation Strategies for adjusting our emotional state to a comfortable level of intensity so we can accomplish our goals.

empathy The ability to understand and respond sympathetically to the feelings of others.

estrogen replacement therapy (ERT) Treatment in which a woman takes tablets containing estrogen or wears an estrogen skin patch during the climacteric and after menopause. Relieves menopausal symptoms (hot flashes and vaginal dryness) and prevents cardiovascular disease and loss of bone mass.

ethological theory of attachment A theory formulated by Bowlby, which views the infant's emotional tie to the mother as an evolved response that promotes survival.

ethology An approach concerned with the adaptive, or survival, value of behavior and its evolutionary history.

euthanasia The practice of ending the life of a person suffering from an incurable condition.

executive stage In Schaie's theory, a more advanced form of the responsibility stage that characterizes people at the helm of large organizations, whose responsibilities have become highly complex.

exosystem In ecological systems theory, settings that do not contain children but that affect their experiences in immediate settings.

expansions Adult responses that elaborate on children's speech, increasing its complexity.

experimental design A research design in which the investigator randomly assigns participants to treatment conditions. Permits inferences about cause and effect.

expertise Acquisition of extensive knowledge in a field or endeavor, supported by the specialization that begins with selecting a college major or an occupation in early adulthood.

expressive style A style of early language learning in which toddlers use language mainly to talk about the feelings and needs of themselves and other people. Distinguished from *referential style*.

extended family household A household in which three or more generations live together.

fantasy period Period of vocational development in which young children fantasize about career options through make-believe play. Distinguished from *tentative period* and *realistic period*.

fast-mapping Connecting a new word with an underlying concept after only a brief encounter.

feminization of poverty A trend in which women who support themselves or their families have become the majority of the adult poverty population, regardless of age and ethnic group.

fetal alcohol effects (FAE) The condition of children who display some but not all of the defects of fetal alcohol syndrome. Usually their mothers drank alcohol in smaller quantities during pregnancy than did mothers of children with fetal alcolhol syndrome (FAS).

fetal alcohol syndrome (FAS) A set of defects that results when women consume large amounts of alcohol during most or all of pregnancy. Includes mental retardation, slow physical growth, and facial abnormalities.

fetal monitors Electronic instruments that track the baby's heart rate during labor.

fetus The prenatal organism from the beginning of the third month to the end of pregnancy, during which time completion of body structures and dramatic growth in size takes place.

fluid intelligence Intellectual skills that largely depend on basic information-processing skills—the speed with which we can analyze information, the capacity of working memory, and the ability to detect relationships among stimuli. Distinguished from *crystallized intelligence*.

fontanels Six soft spots that separate the bones of the skull at birth.

formal operational stage Piaget's final stage, in which adolescents develop the capacity for abstract, scientific thinking. Begins around 11 years of age.

fraternal, or dizygotic, twins Twins resulting from the release and fertilization of two ova. They are genetically no more alike than ordinary siblings. Distinguished from *identical,* or *monozygotic, twins*.

free radicals Naturally occurring, highly reactive chemicals that form in the presence of oxygen and destroy cellular material, including DNA, proteins, and fats essential for cell functioning.

functional age Actual competence and performance of an older adult (as distinguished from chronological age).

functional play A type of play involving pleasurable motor activity with or without objects. Enables infants and toddlers to practice sensorimotor schemes.

gametes Human sperm and ova, which contain half as many chromosomes as a regular body cell.

gender constancy The understanding that sex remains the same even if clothing, hairstyle, and play activities change.

gender intensification Increased gender stereotyping of attitudes and behavior. Occurs in early adolescence.

gender-role identity An image of oneself as relatively masculine or feminine in characteristics.

gender schema theory An information-processing approach to gender typing that combines social learning and cognitive-developmental features to explain how environmental pressures and children's cognitions work together to shape gender-role development.

gender typing The process of developing gender roles, or gender-linked preferences and behaviors valued by the larger society.

gene A segment of a DNA molecule; contains hereditary instructions.

general growth curve A curve that represents overall changes in body size—rapid growth during infancy, slower gains in early and middle childhood, and rapid growth once more during adolescence.

generativity versus stagnation In Erikson's theory, the psychological conflict of midlife, which is resolved positively if the adult can integrate personal goals with the welfare of the larger social environment. The resulting strength is the capacity to give to and guide the next generation.

genetic counseling Counseling that helps couples assess the likelihood of giving birth to a baby with a hereditary disorder.

genetic–environmental correlation The idea that heredity influences the environments to which children are exposed.

genetic imprinting A pattern of inheritance in which genes are imprinted, or chemically marked, in such a way that one pair member is activated, regardless of its makeup.

genital stage Freud's psychosexual stage of adolescence, in which instinctual drives are reawakened and shift to the genital region, upsetting the delicate balance between id, ego, and superego established during middle childhood.

genotype The genetic makeup of the individual. Distinguished from *phenotype*.

giftedness Exceptional intellectual ability. Includes high IQ, high creativity, and specialized talent.

glass ceiling Invisible barrier, faced by women and ethnic minorities, to advancement up the corporate ladder.

glial cells Cells serving the function of myelinization.

goodness of fit An effective match between child-rearing practices and a child's temperament, leading to favorable adjustment.

grief Intense physical and psychological distress following a loss.

growth hormone (GH) A pituitary hormone that affects the development of almost all body tissues, except the central nervous system and (possibly) the genitals.

growth spurt Rapid gain in height and weight during adolescence.

habituation A gradual reduction in the strength of a response as the result of repetitive stimulation.

hardiness A set of three personal qualities—control, commitment, and challenge—that help people cope with stress adaptively, thereby reducing its impact on illness and mortality.

heritability estimate A statistic that measures the extent to which individual differences in complex traits, such as intelligence or personality, are due to genetic factors.

heteronomous morality Piaget's first stage of moral development, in which children view moral rules as permanent features of the external world that are handed down by authorities and cannot be changed. Extends from about 5 to 10 years of age.

heterozygous Having two different genes at the same place on a pair of chromosomes. Distinguished from *homozygous*.

hierarchical classification The organization of objects into classes and subclasses on the basis of similarities and differences between the groups.

history-graded influences Influences on lifespan development that are unique to a particular historical era and explain why people born around the same time (called a *cohort*) tend to be alike in ways that set them apart from people born at other times.

Home Observation for Measurement of the Environment (HOME) A checklist for gathering information about the quality of children's home lives through observation and parental interview.

homozygous Having two identical genes at the same place on a pair of chromosomes. Distinguished from *heterozygous*.

horizontal décalage Development within a Piagetian stage. Gradual mastery of logical concepts during the concrete operational stage is an example.

hospice A comprehensive program of support services that focuses on meeting terminally ill patients' physical, emotional, social, and spiritual needs and that offers follow-up bereavement services to families.

hostile aggression Aggression intended to harm another individual. Distinguished from *instrumental aggression*.

human development A field of study devoted to understanding human constancy and changes throughout the lifespan.

hypothetico-deductive reasoning A formal operational problem-solving strategy in which adolescents begin with a general theory of all possible factors that could affect an outcome in a problem and deduce specific hypotheses, which they test systematically.

id In Freud's theory, the part of personality that is the source of basic biological needs and desires.

identical, or monozygotic, twins Twins that result when a zygote, during the early stages of cell duplication, divides in two. They have the same genetic makeup. Distinguished from *fraternal, or dizygotic, twins*.

identification In Freud's theory, the process leading to formation of the superego in which children take the same-sex parent's characteristics into their personality.

identity A well-organized conception of the self, made up of values, beliefs, and goals to which the individual is solidly committed.

identity achievement The identity status of individuals who have explored and committed themselves to self-chosen values and occupational goals. Distinguished from *moratorium, identity foreclosure*, and *identity diffusion*.

identity diffusion The identity status of individuals who do not have firm commitments to values and goals and are not actively trying to reach them. Distinguished from *identity achievement, moratorium,* and *identity foreclosure*.

identity foreclosure The identity status of individuals who have accepted ready-made values and goals that authority figures have chosen for them. Distinguished from *identity achievement, moratorium,* and *identity diffusion*.

identity versus identity diffusion In Erikson's theory, the psychological conflict of adolescence, which is resolved positively when adolescents attain an identity after a period of exploration and inner soul-searching.

imaginary audience Adolescents' belief that they are the focus of everyone else's attention and concern.

imitation Learning by copying the behavior of another person. Also called *modeling* or *observational learning*.

implantation Attachment of the blastocyst to the uterine lining 7 to 9 days after fertilization.

implicit memory Memory without conscious awareness.

independence-ignore script A typical pattern of interaction in which elders' independent behaviors are mostly ignored, thereby leading them to occur less often. Distinguished from *dependency-support script*.

independent variable The variable manipulated by the researcher in an experiment.

induction A type of discipline in which the effects of the child's misbehavior on others are communicated to the child.

industry versus inferiority In Erikson's theory, the psychological conflict of middle childhood, which is resolved positively when experiences lead children to develop a sense of competence at useful skills and tasks.

infant mortality rate The number of deaths in the first year of life per 1,000 live births.

information-loss view A view that attributes age-related slowing of cognitive processing to greater loss of information as it moves through the system. As a result, the whole system must slow down to inspect and interpret the information. Distinguished from *neural network* view.

information processing An approach that views the human mind as a symbol-manipulating system through which information flows and that regards cognitive development as a continuous process.

initiative versus guilt In Erikson's theory, the psychological conflict of early childhood, which is resolved positively through play experiences that foster a healthy sense of initiative and through development of a superego, or conscience, that is not overly strict and guilt-ridden.

instrumental aggression Aggression aimed at obtaining an object, privilege, or space with no deliberate intent to harm another person. Distinguished from *hostile aggression*.

intelligence quotient, or IQ A score that permits an individual's performance on an intelligence test to be compared to the performances of other individuals of the same age.

intentional, or goal-directed, behavior A sequence of actions in which schemes are deliberately combined to solve a problem.

interactional synchrony A sensitively tuned "emotional dance," in which the caregiver responds to infant signals in a well-timed, appropriate fashion and both partners match emotional states, especially the positive ones.

intermodal perception Perception that combines information from more than one sensory system.

internal working model A set of expectations derived from early caregiving experiences concerning the availability of attachment figures and their likelihood of providing support during times of stress. Becomes a model, or guide, for all future close relationships.

intimacy versus isolation In Erikson's theory, the psychological conflict of young adulthood, which is resolved positively when young adults give up some of their newfound independence and make a permanent commitment to an intimate partner.

invariant features Features that remain stable in a constantly changing perceptual world.

irreversibility The inability to mentally go through a series of steps in a problem and then reverse direction, returning to the starting point. Distinguished from *reversibility*.

joint custody A child custody arrangement following divorce in which the court grants both parents say in important decisions about the child's upbringing.

kinkeeper Role assumed by members of the middle generation, especially mothers, who take responsibility for gathering the family for celebrations and making sure everyone stays in touch.

kinship studies Studies comparing the characteristics of family members to determine the importance of heredity in complex human characteristics.

kwashiorkor A disease usually appearing between 1 and 3 years of age that is caused by a diet low in protein. Symptoms include an enlarged belly, swollen feet, hair loss, skin rash, and irritable, listless behavior.

language acquisition device (LAD) In Chomsky's theory, a biologically based innate system that permits children, no matter which language they hear, to understand and speak in a rule-oriented fashion as soon as they have picked up enough words.

lanugo A white, downy hair that covers the entire body of the fetus, helping the vernix stick to the skin.

latency stage Freud's psychosexual stage of middle childhood, in which the sexual instincts lie dormant.

lateralization Specialization of functions of the two hemispheres of the cortex.

learned helplessness Attributions that credit success to luck and failure to low ability. Leads to anxious loss of control in the face of challenging tasks. Distinguished from *mastery-oriented attributions*.

learning disabilities Specific learning disorders that lead children to achieve poorly in school, despite an average or above-average IQ. Believed to be due to faulty brain functioning.

life care communities Housing for the elderly that offers a range of options, from independent or congregate housing to full nursing home care. For a large initial payment and monthly fees, guarantees that elders' needs will be met in one place as they age.

life expectancy crossover An age-related reversal in life expectancy of sectors of the population. For example, members of ethnic minorities who survive to age 85 live longer than members of the white majority.

life review The process of calling up, reflecting on, and reconsidering past experiences, contemplating their meaning with the goal of achieving greater self-understanding.

life structure In Levinson's theory, the underlying pattern or design of a person's life at a given time. Consists of relationships with significant others (the most important of which have to do with marriage/family and occupation) that are reorganized during each period of adult development.

lifespan perspective A balanced perspective that assumes development is lifelong, multidimensional and multidirectional, highly plastic, and embedded in multiple contexts.

living will A written statement that specifies the treatments a person does or does not want in case of a terminal illness, coma, or other near-death situation.

loneliness Feelings of unhappiness that result from a gap between actual and desired social relationships.

long-term memory In information processing, the part of the mental system that contains our permanent knowledge base.

longitudinal design A research design in which in which one group of participants is studied repeatedly at different ages. Distinguished from *cross-sectional design*.

longitudinal-sequential design A research design with both longitudinal and cross-sectional components in which groups of participants born in different years are followed over time.

macrosystem In ecological systems theory, the values, laws, and customs of a culture that influence experiences and interactions at inner levels of the environment.

macular degeneration Blurring and eventual loss of central vision due to hardening of small blood vessels and restriction of blood flow to the macula, or central region of the retina.

mainstreaming The integration of pupils with learning difficulties into regular classrooms for part or all of the school day.

make-believe play A type of play in which children pretend, acting out everyday and imaginary activities.

marasmus A disease usually appearing in the first year of life that is caused by a diet low in all essential nutrients. Leads to a wasted condition of the body.

mastery-oriented attributions Attributions that credit success to high ability and failure to insufficient effort. Leads to high self-esteem and a willingness to approach challenging tasks. Distinguished from *learned helplessness*.

maturation A genetically determined, naturally unfolding course of growth.

maximum lifespan The genetic limit to length of life for a person free of external risk factors. Distinguished from *average life expectancy* and *active lifespan*.

mechanistic theories Theories that regard the person as a passive reactor to environmental inputs. Distinguished from *organismic theories.*

meiosis The process of cell division through which gametes are formed and in which the number of chromosomes in each cell is halved.

memory strategies Deliberate mental activities that improve the likelihood of remembering.

menarche First menstruation.

menopause The end of menstruation and, therefore, reproductive capacity in women. Usually occurs between 50 and 55, although the age range is large—from 40 to 59.

mental representation An internal image of an absent object or a past event.

mesosystem In ecological systems theory, connections among microsystems.

metacognition Thinking about thought; awareness of mental activities.

microsystem In ecological systems theory, the activities and interaction patterns in the person's immediate surroundings.

middle-generation squeeze Pressure experienced by middle-aged adults who must care for ill or frail parents while they deal with competing demands of children (some of whom are under age 18 and still at home) and employment.

midlife crisis Inner turmoil, self-doubt, and major restructuring of the personality during the transition to middle adulthood. Characterizes the experiences of only a minority of adults.

mild mental retardation Substantially below-average intellectual functioning, resulting in an IQ between 55 and 70 and problems in adaptive behavior (social and self-help skills in everyday life). Affected individuals can be educated to the level of an average sixth grader and can live independently in adulthood.

mitosis The process of cell duplication, in which each new cell receives an exact copy of the original chromosomes.

moratorium The identity status of individuals who are exploring alternatives in an effort to find values and goals to guide their life. Distinguished from *identity achievement, identity foreclosure,* and *identity diffusion.*

mortality The phase of dying in which the individual passes into permanent death. Distinguished from *agonal phase* and *clinical death.*

motherese A form of language used by adults to speak to infants and toddlers that consists of short sentences with exaggerated expression and very clear pronunciation. Also called *child-directed speech.*

mourning The culturally specified expression of the bereaved person's thoughts and feelings through funerals and other rituals.

mutation A sudden but permanent change in a segment of DNA.

myelinization A process in which neural fibers are coated with an insulating fatty sheath called *myelin* that improves the efficiency of message transfer.

natural, or prepared, childbirth An approach designed to reduce pain and medical intervention and to make childbirth a rewarding experience for parents.

naturalistic observation A method in which the researcher goes into the natural environment to observe the behavior of interest. Distinguished from *structured observation.*

nature–nurture controversy Disagreement among theorists about whether genetic or environmental factors are the most important determinants of development and behavior.

near-death experiences Experiences reported by people who have been resuscitated after being pronounced clinically dead.

neglected children Children who are seldom chosen, either positively or negatively, on sociometric measures of peer acceptance. Distinguished from *popular, rejected,* and *controversial children.*

Neonatal Behavioral Assessment Scale (NBAS) A test developed to assess the behavior of the infant during the newborn period.

neural network view A view that attributes age-related slowing of cognitive processing to breaks in neural networks as neurons die. The brain forms bypasses—new synaptic connections that go around the breaks but are less efficient. Distinguished from *information-loss view.*

neural tube The primitive spinal cord that develops from the ectoderm, the top of which swells to form the brain.

neurofibrillary tangles A structural change in the brain associated with Alzheimer's disease in which abnormal bundles of threads run through the body of the neuron and into fibers establishing synaptic connections with other neurons.

neurons Nerve cells that store and transmit information.

niche-picking A type of genetic–environmental correlation in which individuals actively choose environments that complement their heredity.

noble savage Rousseau's view of the child as naturally endowed with an innate plan for orderly, healthy growth.

nocturnal enuresis Repeated bedwetting during the night.

nonfunctionality The component of the death concept specifying that all living functions, including thought, feeling, movement, and body processes, cease at death.

nonnormative influences Influences on lifespan development that are irregular, in that they happen to just one or a few individuals and do not follow a predictable timetable.

nonorganic failure to thrive A growth disorder usually present by 18 months of age that is caused by lack of affection and stimulation.

non-rapid-eye-movement (NREM) sleep A "regular" sleep state in which the body is quiet and heart rate, breathing, and brain wave activity are slow and regular. Distinguished from *rapid-eye-movement (REM) sleep.*

nonsocial activity Unoccupied, onlooker behavior and solitary play. Distinguished from *parallel, associative,* and *cooperative play.*

obesity A greater-than-20-percent increase over average body weight, based on the individual's age, sex, and physical build.

object permanence The understanding that objects continue to exist when they are out of sight.

old-old Showing signs of physical decline in old age. Distinguished from *young-old.*

open classroom An elementary school classroom based on the educational philosophy that children are active agents in their own development and learn at different rates. Teachers share decision making with pupils. Pupils are evaluated in relation to their own prior development. Distinguished from *traditional classroom*.

operant conditioning A form of learning in which a spontaneous behavior is followed by a stimulus that changes the probability that the behavior will occur again.

oral stage Freud's first psychosexual stage, during which infants obtain pleasure through the mouth.

organismic theories Theories that assume the existence of psychological structures within the person that underlie and control development. Distinguished from *mechanistic theories*.

organization In Piaget's theory, the internal rearrangement and linking together of schemes so that they form a strongly interconnected cognitive system. In information processing, the memory strategy of grouping together related items.

osteoarthritis A form of arthritis characterized by deteriorating cartilage on the ends of bones of frequently used joints. Leads to swelling, stiffness, and loss of flexibility. Distinguished from *rheumatoid arthritis*.

osteoporosis A severe version of age-related bone loss. Porous bones are easily fractured and when very extreme, lead to a slumped-over posture, a shuffling gait, and a "dowager's hump" in the upper back.

overextension An early vocabulary error in which a word is applied too broadly, to a wider collection of objects and events than is appropriate. Distinguished from *underextension*.

overregularization Application of regular grammatical rules to words that are exceptions.

parallel play A form of limited social participation in which the child plays near other children with similar materials but does not interact with them. Distinguished from *nonsocial, associative,* and *cooperative play*.

parental imperative theory A theory that claims traditional gender roles are maintained during the active parenting years to help ensure the survival of children. After children reach adulthood, parents are free to express the "other-gender" side of their personalities.

passionate love Love based on intense sexual attraction. Distinguished from *companionate love*.

passive euthanasia The practice of withholding or withdrawing life-sustaining treatment, permitting a patient to die naturally. Distinguished from *voluntary active euthanasia*.

peer group Peers who form a social unit by generating shared values and standards of behavior and a social structure of leaders and followers.

perception bound Being easily distracted by the concrete, perceptual appearance of objects.

permanence The component of the death concept specifying that once a living thing dies, it cannot be brought back to life.

permissive style A child-rearing style that is responsive but undemanding. An overly tolerant approach to child rearing. Distinguished from *authoritative* and *authoritarian styles*.

persistent vegetative state A state produced by absence of brain wave activity in the cortex, in which the person is unconscious, displays no voluntary movements, and has no hope of recovery.

personal fable Adolescents' belief that they are special and unique. Leads them to conclude that others cannot possibly understand their thoughts and feelings and may promote a sense of invulnerability to danger.

perspective taking The capacity to imagine what other people may be thinking and feeling.

phallic stage Freud's psychosexual stage of early childhood, in which sexual impulses transfer to the genital region of the body and the Oedipus and Electra conflicts are resolved.

phenotype The individual's physical and behavioral characteristics, which are determined by both genetic and environmental factors. Distinguished from *genotype*.

pituitary gland A gland located near the base of the brain that releases hormones affecting physical growth.

placenta The organ that separates the mother's bloodstream from the embryo or fetal bloodstream but permits exchange of nutrients and waste products.

polygenic inheritance A pattern of inheritance in which many genes determine a characteristic.

popular children Children who get many positive votes on sociometric measures of peer acceptance. Distinguished from *rejected, controversial,* and *neglected children*.

possible selves Future-oriented representations of what one hopes to become and is afraid of becoming. The temporal dimension of self-concept.

postconventional level Kohlberg's highest level of moral development, in which individuals define morality in terms of abstract principles and values that apply to all situations and societies.

postformal thought Cognitive development beyond Piaget's formal operational stage.

practical problem solving Problem solving that requires people to size up real-world situations and analyze how best to achieve goals that have a high degree of uncertainty.

pragmatic thought In Labouvie-Vief's theory, adult thought in which logic becomes a tool to solve real-world problems and inconsistencies and imperfections are accepted.

pragmatics The practical, social side of language that is concerned with how to engage in effective and appropriate communication with others.

preconventional level Kohlberg's first level of moral development, in which moral understanding is based on rewards, punishments, and the power of authority figures.

preformationism Medieval view of the child as a miniature adult.

premenstrual syndrome (PMS) An array of physical and psychological symptoms that usually appear 6 to 10 days prior to menstruation. The most common are abdominal cramps, fluid retention, diarrhea, tender breasts, backache, headache, fatigue, tension, irritability, and depression.

prenatal diagnostic methods Medical procedures that permit detection of developmental problems before birth.

preoperational stage Piaget's second stage, in which rapid growth in representation takes place. However, thought is not yet logical. Spans the years from 2 to 7.

presbycusis Age-related hearing impairments that involve a sharp loss at high frequencies around age 50, which gradually extends to all frequencies.

presbyopia Condition of aging in which, around age 60, the lens of the eye loses its capacity to accommodate entirely to nearby objects.

preterm Infants born several weeks or more before their due date.

primary aging Genetically influenced age-related declines in the functioning of organs and systems that affect all members of our species and take place even in the context of overall good health. Also called *biological aging*. Distinguished from *secondary aging*.

primary sexual characteristics Physical features that involve the reproductive organs directly (ovaries, uterus, and vagina in females; penis, scrotum, and testes in males). Distinguished from *secondary sexual characteristics*.

principle of mutual exclusivity The assumption by children in the early stages of vocabulary growth that words mark entirely separate (nonoverlapping) categories.

private speech Self-directed speech that children use to plan and guide their own behavior.

Project Head Start A federal program that provides low-income children with a year or two of preschool education before school entry and that encourages parent involvement in children's development.

propositional thought A type of formal operational reasoning in which adolescents evaluate the logic of verbal statements without referring to real-world circumstances.

prosocial, or altruistic, behavior Actions that benefit another person without any expected reward for the self.

prospective memory Recall that involves remembering to engage in planned actions at an appropriate time in the future.

proximodistal trend An organized pattern of physical growth and motor control that proceeds from the center of the body outward. Distinguished from *cephalocaudal trend*.

psychoanalytic perspective An approach to personality development introduced by Freud that assumes children move through a series of stages in which they confront conflicts between biological drives and social expectations. The way these conflicts are resolved determines psychological adjustment.

psychosexual theory Freud's theory, which emphasizes that how parents manage children's sexual and aggressive drives during the first few years is crucial for healthy personality development.

psychosocial theory Erikson's theory, which emphasizes that the demands of society at each Freudian stage not only promote the development of a unique personality, but also ensure that individuals acquire attitudes and skills that help them become active, contributing members of their society.

puberty Biological changes at adolescence that lead to an adult-sized body and sexual maturity.

public policies Laws and government programs designed to improve current conditions.

punishment In operant conditioning, removing a desirable stimulus or presenting an unpleasant one in order to decrease the occurrence of a response.

range of reaction Each person's unique, genetically determined response to a range of environmental conditions.

rape Intercourse by force, by threat of harm, or when the victim is incapable of giving consent (due to mental illness, mental retardation, or intoxication).

rapid-eye-movement (REM) sleep An "irregular" sleep state in which brain wave activity is similar to that of the waking state; eyes dart beneath the lids, heart rate, blood pressure, and breathing are uneven, and slight body movements occur. Distinguished from *non-rapid-eye-movement (NREM) sleep*.

realistic period Period of vocational development in which older adolescents and young adults focus on a general career category and, slightly later, settle on a single occupation. Distinguished from *fantasy period* and *tentative period*.

recall A type of memory that involves remembering a stimulus that is not present.

recasts Adult responses that restructure children's incorrect speech into a more mature form.

recognition A type of memory that involves noticing whether a stimulus is identical or similar to one previously experienced.

referential style A style of early language learning in which toddlers use language mainly to label objects. Distinguished from *expressive style*.

reflex An inborn, automatic response to a particular form of stimulation.

rehearsal The memory strategy of repeating information.

reinforcer In operant conditioning, a stimulus that increases the occurrence of a response.

reintegrative stage Schaie's stage of late adulthood, in which people reexamine and reintegrate their interests, attitudes, and values, using them as a guide for what knowledge to acquire and apply.

rejected children Children who are actively disliked and get many negative votes on sociometric measures of peer acceptance. Distinguished from *popular, controversial,* and *neglected children*.

rejected-aggressive children A subgroup of rejected children who engage in high rates of conflict, hostility, and hyperactive, inattentive, and impulsive behavior. Distinguished from *rejected-withdrawn children*.

rejected-withdrawn children A subgroup of rejected children who are passive and socially awkward. Distinguished from *rejected-aggressive children*.

relativistic thinking In Perry's theory, the cognitive approach of older college students, who favor multiple truths, each relative to its context of evaluation. Distinguished from *dualistic thinking*.

reminiscence The process of telling stories about people and events from the past and reporting associated thoughts and feelings.

remote memory Recall of events that happened long ago.

resistant attachment The quality of insecure attachment characterizing infants who remain close to the parent before departure and display angry, resistive behavior when she returns. Distinguished from *secure, avoidant,* and *disorganized/disoriented attachment*.

respiratory distress syndrome A disorder of preterm infants in which the lungs are so immature that the air sacs collapse, causing serious breathing difficulties.

responsibility stage Schaie's stage of middle adulthood in which people adapt their cognitive skills to the expansion of responsibilities to others that takes place on the job, in the community, and at home.

reticular formation A brain structure that maintains alertness and consciousness.

reversibility The ability to think through a series of steps in a problem and then mentally reverse direction, returning to the starting point. Distinguished from *irreversibility*.

Rh factor A protein that, when present in the fetus's blood but not in the mother's, can cause the mother to build up antibodies. If these return to the fetus's system, they destroy red blood cells, reducing the oxygen supply to organs and tissues.

rheumatoid arthritis A form of arthritis in which the immune system attacks the body, resulting in inflammation of connective tissue, particularly the membranes that line the joints. The result is stiffness, inflammation, aching, deformed joints, and serious loss of mobility. Distinguished from *osteoarthritis*.

scheme In Piaget's theory, a specific structure, or organized way of making sense of experience, that changes with age.

school phobia Severe apprehension about attending school, often accompanied by physical complaints that disappear once the child is allowed to remain home.

scripts General descriptions of what occurs and when it occurs in a particular situation. A basic means through which children organize and interpret their everyday experiences.

secondary aging Declines due to hereditary defects and environmental influences, such as poor diet, lack of exercise, substance abuse, environmental pollution, and psychological stress. Distinguished from *primary aging*.

secondary friends People who are not intimates but with whom the individual spends time occasionally, such as a group that meets for lunch, bridge, or museum tours.

secondary sexual characteristics Features visible on the outside of the body that serve as signs of sexual maturity but do not involve the reproductive organs (for example, breast development in females, appearance of underarm and pubic hair in both sexes). Distinguished from *primary sexual characteristics*.

secular trends in physical growth Changes in body size and rate of growth from one generation to the next. For example, in industrialized nations, age of menarche declined from 1860 to 1970, signifying faster physical maturation in modern young people.

secure attachment The quality of attachment characterizing infants who are distressed by parental separation and easily comforted by the parent when she returns. Distinguished from *avoidant, resistant,* and *disorganized/disoriented attachment*.

secure base The use of the familiar caregiver as a base from which the infant confidently explores the environment and returns for emotional support.

selective optimization with compensation A set of strategies that permits the elderly to sustain high levels of functioning. They *select* personally valued activities as a way of *optimizing* returns from their diminishing energies and come up with new ways of *compensating* for losses.

selectivity theory A social theory of aging that states that the decline in social interaction in late adulthood is due to physical and psychological changes, which lead elders to select associates largely on the basis of emotion. Consequently, they prefer familiar partners with whom they have developed pleasurable relationships. Distinguished from *disengagement theory* and *activity theory*.

self-care children Children who look after themselves while their parents are at work.

self-concept The sum total of attributes, abilities, attitudes, and values that an individual believes defines who he or she is.

self-conscious emotions Emotions that involve injury to or enhancement of the sense of self. Examples are shame, embarrassment, guilt, envy, and pride.

self-control The capacity to resist an impulse to engage in socially disapproved behavior.

self-esteem An aspect of self-concept that involves judgments about one's own worth and the feelings associated with those judgments.

self-regulation The process of continuously monitoring progress toward a goal, checking outcomes, and redirecting unsuccessful efforts.

sensitive period A time span during which the person is biologically prepared to acquire certain capacities but needs the support of an appropriately stimulating environment.

sensorimotor stage Piaget's first stage, during which infants and toddlers "think" with their eyes, ears, hands, and other sensorimotor equipment. Spans the first 2 years of life.

sensory register In information processing, that part of the mental system in which sights and sounds are held briefly before they decay or are transferred to working, or short-term, memory.

separation anxiety An infant's distressed reaction to the departure of the familiar caregiver.

separation–individuation In Mahler's theory, the process of separating from the mother and becoming aware of the self, which is triggered by crawling and walking.

seriation The ability to order items along a quantitative dimension, such as length or weight.

sex chromosomes The twenty-third pair of chromosomes, which determines the sex of the child. In females, called *XX*; in males, called *XY*.

sleep apnea A condition during sleep in which breathing ceases for 10 seconds or longer, resulting in many brief awakenings.

slow-to-warm-up child A child whose temperament is such that he or she is inactive, shows mild, low-key reactions to environmental stimuli, is negative in mood, and adjusts slowly when faced with new experiences. Distinguished from *easy child* and *difficult child.*

small-for date Infants whose birth weight is below normal when length of pregnancy is taken into account.

social clock Age-graded expectations for life events, such as beginning a first job, getting married, birth of the first child, buying a home, and retiring.

social convoy A model of age-related changes in social networks, which views the individual within a cluster of relationships moving through life. Close ties are in the inner circle, less close ties on the outside. With age, people change places in the convoy, new ties are added, and some are lost entirely.

social learning theory An approach that emphasizes the role of modeling, or observational learning, in the development of behavior.

social referencing Relying on a trusted person's emotional reaction to decide how to respond in an uncertain situation.

social smile The smile evoked by the stimulus of the human face. First appears between 6 and 10 weeks.

social systems perspective A view of the family as a complex system in which the behaviors of each family member affect those of others.

sociocultural theory Vygotsky's theory, in which children acquire the ways of thinking and behaving that make up a community's culture through cooperative dialogues with more knowledgeable members of society.

sociodramatic play The make-believe play with peers that first appears around age 2½ and increases rapidly until 4 to 5 years.

spermarche First ejaculation of seminal fluid.

stability versus change Disagreement among theorists about whether there are stable individual differences throughout the lifespan that are largely influenced by heredity and early experience, or whether change is possible and likely if new experiences support it.

stage A qualitative change in thinking, feeling, and behaving that characterizes a particular time period of development.

Stanford-Binet Intelligence Scale An individually administered intelligence test that is the modern descendent of Alfred Binet's first successful test for children. Measures general intelligence and four factors: verbal reasoning, quantitative reasoning, spatial reasoning, and short-term memory.

states of arousal Different degrees of sleep and wakefulness.

states versus transformations The tendency to treat the initial and final states in a problem as completely unrelated.

Strange Situation A procedure involving short separations from and reunions with the parent that assesses the quality of the attachment bond.

stranger anxiety The infant's expression of fear in response to unfamiliar adults. Appears in many babies after 6 months of age.

structured interview A method in which each participant is asked the same questions in the same way.

structured observation A method in which the investigator sets up a cue for the behavior of interest and observes it in a laboratory. Distinguished from *naturalistic observation.*

subculture A group of people with beliefs and customs that differ from those of the larger culture.

successful aging Aging in which gains are maximized and losses minimized.

sudden infant death syndrome (SIDS) Death of a seemingly healthy baby, who stops breathing, usually during the night, without apparent cause.

superego In Freud's theory, the part of personality that is the seat of conscience and is often in conflict with the id's desires.

symbiosis In Mahler's theory, the baby's intimate sense of oneness with the mother, encouraged by warm, physical closeness and gentle handling.

synapses The gaps between neurons, across which chemical messages are sent.

tabula rasa Locke's view of the child as a blank slate whose character is shaped by experience.

telegraphic speech Toddlers' two-word utterances that, like a telegram, leave out smaller and less important words.

temperament Stable individual differences in quality and intensity of emotional reaction.

tentative period Period of vocational development in which adolescents weigh vocational options against their interests, abilities, and values. Distinguished from *fantasy period* and *realistic period.*

teratogen Any environmental agent that causes damage during the prenatal period.

terminal decline A steady, marked decrease in cognitive functioning prior to death.

theory An orderly, integrated set of statements that describes, explains, and predicts behavior.

theory of multiple intelligences Gardner's theory, which identifies seven independent intelligences on the basis of distinct sets of processing operations that permit individuals to engage in a wide range of culturally valued activities.

thyroid-stimulating hormone (TSH) A pituitary hormone that stimulates the thyroid gland to release thyroxine, which is necessary for normal brain development and body growth.

time out A form of mild punishment in which children are removed from the immediate setting until they are ready to act appropriately.

traditional classroom An elementary school classroom based on the educational philosophy that children are passive learners who acquire information presented by teachers. Pupils are evaluated in terms of how well they keep up with a uniform set of standards for all children in their grade. Distinguished from *open classroom.*

traditional marriage A form of marriage involving clear division of husband's and wife's roles. The man is the head of household and economic provider. The woman devotes herself to caring for her husband and children and creating a nurturant, comfortable home. Distinguished from *egalitarian marriage.*

transductive reasoning Reasoning from one particular event to another particular event, instead of from general to particular or particular to general.

transition Climax of the first stage of labor, in which the frequency and strength of contractions are at their peak and the cervix opens completely.

transitive inference The ability to seriate—or order items along a quantitative dimension—mentally.

triarchic theory of intelligence Sternberg's theory, which states that information-processing skills, prior experience with tasks, and contextual (or cultural) factors interact to determine intelligent behavior.

trimesters Three equal time periods in the prenatal period, each of which lasts three months.

Type A behavior pattern A behavior pattern consisting of extreme competitiveness, ambition, impatience, hostility, angry outbursts, and a sense of time pressure.

umbilical cord The long cord connecting the prenatal organism to the placenta that delivers nutrients and removes waste products.

unconditioned response (UCR) In classical conditioning, a reflexive response that is produced by an unconditioned stimulus (UCS).

unconditioned stimulus (UCS) In classical conditioning, a stimulus that leads to a reflexive response.

underextension An early vocabulary error in which a word is applied too narrowly, to a smaller number of objects and events than is appropriate. Distinguished from *overextension.*

universality The component of the death concept specifying that all living things eventually die.

vernix A white, cheeselike substance covering the fetus and preventing the skin from chapping due to constant exposure to the amniotic fluid.

visual acuity Fineness of visual discrimination.

voluntary active euthanasia The practice of ending a patient's suffering, at the patient's request, before a natural end to life. A form of mercy killing. Distinguished from *passive euthanasia.*

Wechsler Intelligence Scale for Children–III (WISC-III) An individually administered intelligence test that includes both a measure of general intelligence and a variety of verbal and performance scores.

whole-language approach An approach to beginning reading instruction that parallels children's natural language learning and keeps reading materials whole and meaningful. Distinguished from *basic skills approach.*

wisdom A form of cognition that combines breadth and depth of practical knowledge, ability to reflect on that knowledge in ways that make life more bearable and worthwhile, emotional maturity, and creative integration of experience and knowledge into new ways of thinking and acting.

working, or **short-term, memory** In information processing, the conscious part of the mental system, where we actively "work" on a limited amount of information to ensure that it will be retained.

X-linked inheritance A pattern of inheritance in which a recessive gene is carried on the X chromosome. Males are more likely to be affected.

young-old Appearing physically young for one's advanced years. Distinguished from *old-old.*

zone of proximal development In Vygotsky's theory, a range of tasks that the child cannot yet handle alone but can do with the help of more skilled partners.

zygote The newly fertilized cell formed by the union of sperm and ovum at conception.

Aaron, R., & Powell, G. (1982). Feedback practices as a function of teacher and pupil race during reading groups instruction. *Journal of Negro Education, 51,* 50–59.

Abbott, S. (1992). Holding on and pushing away: Comparative perspectives on an eastern Kentucky child-rearing practice. *Ethos, 20,* 33–65.

Abra, J. (1989). Changes in creativity with age: Data, explanations, and further predictions. *International Journal of Aging and Human Development, 28,* 105–126.

Abrass, I. B. (1990). The biology and physiology of aging. *West Journal of Medicine, 153,* 641–645.

Achenbach, T. M., Phares, V., Howell, C. T., Rauh, V. A., & Nurcombe, B. (1990). Seven-year outcome of the Vermont program for low-birthweight infants. *Child Development, 61,* 1672–1681.

Ackerman, B. P. (1978). Children's understanding of speech acts in unconventional frames. *Child Development, 49,* 311–318.

Adams, R. G. (1985–1986). Emotional closeness and physical distance between friends: Implications for elderly women living in age-segregated and age-integrated settings. *International Journal of Aging and Human Development, 22,* 55–76.

Adams, R. J. (1987). An evaluation of color preference in early infancy. *Infant Behavior and Development, 10,* 143–150.

Adams, S. H. (1994). Role of hostility in women's health during midlife: A longitudinal study. *Health Psychology, 13,* 488–495.

Adelman, C. (1991). *Women at thirty-something: Paradoxes of attainment.* Washington, DC: U.S. Government Printing Office.

Adkins, G., Martin, P., & Poon, L. W. (1996). Personality traits and states as predictors of subjective well-being in centenarians, octogenarians, and sexagenarians. *Psychology and Aging, 11,* 408–416.

Adler, N. E., Boyce, T., Chesney, M. A., Cohen, S., Folkman, S., Kahn, R. L., & Syme, L. (1994). Socioeconomic status and health. *American Psychologist, 49,* 15–24.

Adolph, K. E., Eppler, M. A., & Gibson, E. J. (1993). Development of perception of affordances. In C. Rovee-Collier & L. P. Lipsitt (Eds.), *Advances in infancy research* (Vol. 8, pp. 51–98). Norwood, NJ: Ablex.

Ahlsten, G., Cnattingius, S., & Lindmark, G. (1993). Cessation of smoking during pregnancy improves fetal growth and reduces infant morbidity in the neonatal period: A population-based prospective study. *Acta Paediatrica, 82,* 177–181.

Ainlay, S. C., Singleton, R., & Swigert, V. L. (1992). Aging and religious participation: Reconsidering the effects of health. *Journal for the Scientific Study of Religion, 31,* 175–188.

Ainsworth, M. D. S., Blehar, M. C., Waters, E., & Wall, S. (1978). *Patterns of attachment.* Hillsdale, NJ: Erlbaum.

Albanes, D., Blair, A., & Taylor, P. R. (1989). Physical activity and risk of cancer in the NHANES I population. *American Journal of Public Health, 79,* 744–750.

Albert, R. S. (1994). The achievement of eminence: A longitudinal study of exceptionally gifted boys and their families. In R. F. Sobotnik & K. D. Arnold (Eds.), *Beyond Terman: Contemporary studies of giftedness and talent* (pp. 282–315). Norwood, NJ: Ablex.

Ales, K. L., Druzin, M. L., & Santini, D. L. (1990). Impact of advanced maternal age on the outcome of pregnancy. *Surgery, Gynecology & Obstetrics, 171,* 209–216.

Alessandri, S. M., & Wozniak, R. H. (1987). The child's awareness of parental beliefs concerning the child: A developmental study. *Child Development, 58,* 316–323.

Allen, J. P., Hauser, S. T., Bell, K. L., & O'Connor, T. G. (1994). Longitudinal assessment of autonomy and related-ness in adolescent–family interactions as predictors of adolescent ego development and self-esteem. *Child Development, 65,* 179–194.

Allen, K., Calsyn, D. A., Fehrenbach, P. A., & Benton, G. (1989). A study of the interpersonal behaviors of male batterers. *Journal of Interpersonal Violence, 4,* 79–89.

Allen, W. A., Piccone, N. L., & D'Amanda, C. (1993). *How drugs can affect your life.* Springfield, IL: Charles C Thomas.

Allison, D. B., & Pi-Sunyer, X. (1994, May–June). *Fleshing out obesity, 34*(3), 38–43.

Allison, J. A., & Wrightsman, L. S. (1993). *Rape: The misunderstood crime.* Newbury Park, CA: Sage.

Alpert-Gillis, L. J., & Connell, J. P. (1989). Gender and sex-role influences on children's self-esteem. *Journal of Personality, 57,* 97–114.

Alsaker, F. D. (1995). Timing of puberty and reactions to pubertal changes. In M. Rutter (Ed.), *Psychosocial disturbances in young people* (pp. 37–82). New York: Cambridge University Press.

Alter-Reid, K., Gibbs, M. S., Lachenmeyer, J. R., Sigal, J., & Massoth, N. A. (1986). Sexual abuse of children: A review of empirical findings. *Clinical Psychology Review, 6,* 249–266.

Altshuler, J. L., & Ruble, D. N. (1989). Developmental changes in children's awareness of strategies for coping with uncontrollable stress. *Child Development, 60,* 1337–1349.

Amaro, H. (1995). Love, sex and power: Considering women's realities in HIV prevention. *American Psychologist, 50,* 437–447.

Ambry, M. (1992). Childless chances. *American Demographics, 14,* 55.

American Academy of Pediatrics. (1993). *Caring for your baby and young child: Birth to age 5.* New York: Bantam.

American Cancer Society. (1996). *Cancer facts and figures—1996.* Atlanta, GA: Author.

American College of Sports Medicine. (1991). *Guidelines for exercise testing and prescription* (4th ed.). Philadelphia: Lea & Febiger.

American Psychiatric Association. (1994). *Diagnostic and statistical manual of mental disorders* (4th ed.). Washington, DC: Author.

American Psychological Association. (1992). Ethical principles of psychologists and code of conduct. *American Psychologist, 44,* 1597–1611.

Anand, K. J. S. (1990) The biology of pain perception in newborn infants. In D. Tyler & E. Krane (Eds.) *Advances in pain research therapy* (pp. 113–155). New York: Raven Press.

Anderson, E. S. (1984). The acquisition of sociolinguistic knowledge: Some evidence from children's verbal role play. *Western Journal of Speech Communication, 48,* 125–144.

Anderson, G. C. (1991). Current knowledge about skin-to-skin (kangaroo) care for preterm infants. *Journal of Perinatology, 11,* 216–226.

Anderson, J. J. B., Rondano, P., & Holmes, A. (1996). Roles of diet and physical activity in the prevention of osteoporosis. *Scandinavian Journal of Rheumatology, S103,* 65–74.

Anderson, P. J., & Graham, S. M. (1994). Issues in second-language phonological acquisition among children and adults. *Topics in Language Disorders, 14,* 84–100.

Anderson, S. A., Russell, C. S., & Schumm, W. R. (1983). Perceived marital quality and family life-cycle categories: A further analysis. *Journal of Marriage and the Family, 45,* 127–139.

Andersson, B-E. (1989). Effects of public day care—A longitudinal study. *Child Development, 60,* 857–866.

Andersson, B-E. (1992). Effects of day care on cognitive and socioemotional competence of thirteen-year-old Swedish schoolchildren. *Child Development, 63,* 20–36.

Angel, J. L., Angel, R. J., McClellan, J. L., & Markides, K. S. (1996). Nativity, declining health, and preferences in living arrangements among elderly Mexican Americans: Implications for long-term care. *Gerontologist, 36,* 464–473.

Angle, J., & Wissmann, D. A. (1980). The epidemiology of myopia. *American Journal of Epidemiology, 111,* 220–228.

Anglin, J. M. (1993). Vocabulary development: A morphological analysis. *Monographs of the Society for Research in Child Development, 58*(10, Serial No. 238).

Annas, G. J. (1994). Death by prescription: The Oregon initiative. *New England Journal of Medicine, 331,* 1240–1243.

Anthenelli, R. M., & Schuckit, M. A. (1992). Genetics. In J. H. Lowinson, P. Ruiz, & R. B. Millman (Eds.), *Substance abuse* (2nd ed., pp. 39–50). Baltimore: Williams & Wilkins.

Antonarakis, S. E. (1992). The meiotic stage of nondisjunction in trisomy 21: Determination by using DNA polymorphisms. *American Journal of Human Genetics, 50,* 544–550.

Antonucci, T. (1990). Social supports and social relationships. In R. Binstock & L. K. George (Eds.), *Handbook of aging and the social sciences* (3rd ed., pp. 205–227). New York: Academic Press.

Antonucci, T. C. (1994). A life-span view of women's social relations. In B. F. Turner & L. E. Troll (Eds.), *Women growing older* (pp. 239–269). Thousand Oaks, CA: Sage.

Antonucci, T. C., & Akiyama, H. (1995). Convoys of social relations: Family and friendships within a life span context. In R. Blieszner & V. H. Bedford (Eds.), *Handbook of aging and the family* (pp. 355–371). Westport, CT: Greenwood Press.

Apgar, V. (1953). A proposal for a new method of evaluation in the newborn infant. *Current Research in Anesthesia and Analgesia, 32,* 260–267.

Archer, S. L. (1982). The lower age boundaries of identity development. *Child Development, 53,* 1551–1556.

Archer, S. L. (1989). The status of identity: Reflections on the need for intervention. *Journal of Adolescence, 12,* 345–359.

Archer, S. L., & Waterman, A. S. (1990). Varieties of identity diffusions and foreclosures: An exploration of subcategories of the identity statuses. *Journal of Adolescent Research, 5,* 96–111.

Archer, S. L., & Waterman, A. S. (1994). Adolescent identity development: Contextual perspectives. In C. B. Fisher & R. M. Lerner (Eds.), *Applied developmental psychology* (pp. 76–100). New York: McGraw-Hill.

Arcus, D., & Kagan, J. (1995). Temperament and craniofacial variation in the first two years. *Child Development, 66,* 1529–1540.

Ariès, P. (1962). *Centuries of childhood.* New York: Random House.

Arking, R. (1991). *Biology of aging.* Englewood Cliffs, NJ: Prentice Hall.

Arlin, P. K. (1984). Adolescent and adult thought: A structural interpretation. In M. Commons, F. Richards, & C. Armom (Eds.), *Beyond formal operations: Late adolescent and adult cognitive development* (pp. 258–271). New York: Praeger.

Arlin, P. K. (1989). Problem solving and problem finding in young artists and young scientists. In M. L. Commons, J. D. Sinnott, F. A. Richards, & C. Armon (Eds.), *Adult development: Vol 1. Comparisons and applications of developmental models* (pp. 197–216). New York: Praeger.

Armstrong-Daily, A. (1991). Hospice care for children: Their families and health care providers. In D. Papadatou & C. Papadatos (Eds.), *Children and death* (pp. 225–229). New York: Hemisphere.

Arnold, K. (1994). The Illinois Valedictorian Project: Early adult careers of academically talented male and female high school students. In R. F. Subotnik & K. D. Arnold (Eds.), *Beyond Terman: Contemporary longitudinal studies of giftedness and talent* (pp. 24–51). Norwood, NJ: Ablex.

Arterberry, M. E., Craton, L. G., & Yonas, A. (1993). Infants' sensitivity to motion-carried information for depth and object properties. In C. E. Granrud (Ed.), *Visual perception and cognition in infancy* (pp. 215–234). Hillsdale, NJ: Erlbaum.

Artman, L., & Cahan, S. (1993). Schooling and the development of transitive inference. *Developmental Psychology, 29,* 753–759.

Ashmead, D. H., Davis, D. L., Whalen, T., & Odom, R. D. (1991). Sound localization and sensitivity to interaural time differences in human infants. *Child Development, 62,* 1211–1226.

Ashmead, D. H., McCarty, M. E., Lucas, L. S., & Belvedere, M. C. (1993). Visual guidance in infants' reaching toward suddenly displaced targets. *Child Development, 64,* 1111–1127.

Ashmead, D. H., & Perlmutter, M. (1980). Infant memory in everyday life. In M. Perlmutter (Ed.), *New directions for child development* (Vol. 10, pp. 1–16). San Francisco: Jossey-Bass.

Aslin, R. N. (1993). Perception of visual direction in human infants. In C. E. Granrud (Ed.), *Visual perception and cognition in infancy* (pp. 91–119). Hillsdale, NJ: Erlbaum.

Assmann, A. (1994). Wholesome knowledge: Concepts of wisdom in a historical and cross-cultural perspective. In D. L. Featherman, R. M. Lerner, & M. Perlmutter (Eds.), *Lifespan development and behavior* (pp. 187–224). Hillsdale, NJ: Erlbaum.

Asso, D., & Magos, A. (1992). Psychological and physiological changes in severe premenstrual syndrome. *Biological Psychology, 33,* 115–132.

Astley, S. J., Clarren, S. K., Little, R. E., Sampson, P. D., & Daling, J. R. (1992). Analysis of facial shape in children gestationally exposed to marijuana, alcohol, and/or cocaine. *Pediatrics, 89*, 67–77.

Aström, L. (1986). *I kvinnoled [In womenline]*. Lund: Liber Förlag.

Atchley, R. C. (1991). *Social forces and aging* (6th ed.). Belmont, CA: Wadsworth.

Atchley, R. C. (1992). Retirement and marital satisfaction. In M. E. Szinovacz, D. J. Ekerdt, & B. H. Vinick (Eds.), *Families in retirement* (pp. 145–158). Newbury Park, CA: Sage.

Atkin, C. (1978). Observation of parent–child interaction in supermarket decision making. *Journal of Marketing, 42*, 41–45.

Atkinson, R. C., & Shiffrin, R. M. (1968). Human memory: A proposed system and its control processes. In K. W. Spence & J. T. Spence (Eds.), *Advances in the psychology of learning and motivation* (Vol. 2, pp. 90–195). New York: Academic Press.

Au, T. K., Sidle, A. L., & Rollins, K. B. (1993). Developing an intuitive understanding of conservation and contamination: Invisible particles as a plausible mechanism. *Developmental Psychology, 29*, 286–299.

August, D., & Garcia, E. E. (1988). *Language minority education in the United States.* Springfield, IL: Thomas.

Axia, G., & Baroni, R. (1985). Linguistic politeness at different age levels. *Child Development, 56*, 918–927.

Azmitia, M. (1988). Peer interaction and problem solving: When are two heads better than one? *Child Development, 59*, 87–96.

Bachman, D. L. (1992). Sleep disorders with aging: Evaluation and treatment. *Geriatrics, 47*, 53–61.

Bacon, J. B. (1996). Support groups for bereaved children. In C. A. Corr & D. M. Corr (Eds.), *Handbook of childhood death and bereavement* (pp. 285–304). New York: Springer-Verlag.

Bahrick, H. P. (1984). Semantic memory content in permastore: Fifty years of memory for Spanish learned in school. *Journal of Experimental Psychology: General, 113*, 1–29.

Bahrick, H. P., Bahrick, P. O., & Wittlinger, R. P. (1975). Fifty years of memory for names and faces: A cross-sectional approach. *Journal of Experimental Psychology: General, 104*, 54–75.

Bahrick, L. E. (1988). Intermodal learning in infancy: Learning on the basis of two kinds of invariant relations in audible and visible events. *Child Development, 59*, 197–209.

Bahrick, L. E. (1992). Infants' perceptual differentiation of amodal and modal-ity-specific audio-visual relations. *Journal of Experimental Child Psychology, 53*, 180–199.

Bai, D. L., & Bertenthal, B. L. (1992). Locomotor status and the development of spatial search skills. *Child Development, 63*, 215–226.

Bailey, J. M., Bobrow, D., Wolfe, M., & Mikach, S. (1995). Sexual orientation of adult sons of gay fathers. *Developmental Psychology, 31*, 124–129.

Bailey, J. M., & Pillard, R. C. (1991). A genetic study of male sexual orientation. *Archives of General Psychology, 43*, 808–812.

Bailey, J. M., Pillard, R. C., Neale, M. C., & Agyei, Y. (1993). Heritable factors influence sexual orientation in women. *Archives of General Psychiatry, 50*, 217–223.

Bailey, T. (1993). Can youth apprenticeship thrive in the United States? *Educational Researcher, 22*(3), 4–10.

Baillargeon, R. (1987). Object permanence in 3½- and 4½-month-old infants. *Developmental Psychology, 23*, 655–664.

Baillargeon, R. (1994). Physical reasoning in infancy. In M. S. Gazzaniga (Ed.), *The cognitive neurosciences* (pp. 181–204). Cambridge, MA: MIT Press.

Baillargeon, R. (1995). A model of physical reasoning in infancy. In C. Rovee-Collier & L. P. Lipsitt (Eds.), *Advances in infancy research* (Vol. 9, pp. 305–371). Norwood, NJ: Ablex.

Baillargeon, R., & DeVos, J. (1991). Object permanence in young infants: Further evidence. *Child Development, 62*, 1227–1246.

Baillargeon, R., Graber, M., DeVos, J., & Black, J. (1990). Why do young infants fail to search for hidden objects? *Cognition, 36*, 255–284.

Baird, P. A., & Sadovnick, A. D. (1987). Life expectancy rates in Down Syndrome. *Journal of Pediatrics, 110*, 849–854.

Baker-Ward, L., Gordon, B. N., Ornstein, P. A., Larus, D. M., & Clubb, P. A. (1993). Young children's long-term retention of a pediatric examination. *Child Development, 64*, 1519–1533.

Ballard, B. D., Gipson, M. T. Guttenberg, W., & Ramsey, K. (1980). Palatability of food as a factor influencing obese and normal-weight children's eating habits. *Behavior Research and Therapy, 18*, 598–600.

Baltes, M. M. (1995, February). Dependency in old age: Gains and losses. *Psychological Science, 4*(1), 14–19.

Baltes, M. M. (1996). *The many faces of dependency in old age.* New York: Cambridge University Press.

Baltes, M. M., & Carstensen, L. L. (1991). Commentary. *Human Development, 34*, 256–260.

Baltes, M. M., & Carstensen, L. L. (1996). The process of successful ageing. *Ageing and Society, 16*, 397–422.

Baltes, M. M., Wahl, H.-W., & Reichert, M. (1992). Successful aging in long-term care institutions. In K. W. Schaie & M. P. Lawton (Eds.), *Annual review of gerontology and geriatrics* (pp. 311–337). New York: Springer.

Baltes, P. B. (1983). Life-span developmental psychology: Observations on history and theory revisited. In R. M. Lerner (Ed.), *Developmental psychology: Historical and philosophical perspectives* (pp. 79–111). Hillsdale, NJ: Erlbaum.

Baltes, P. B. (1987). Theoretical propositions of life-span developmental psychology: On the dynamics between growth and decline. *Developmental Psychology, 23*, 611–626.

Baltes, P. B. (1997). On the incomplete architecture of human ontogeny: Selection, optimization, and compensation as foundation of developmental theory. *American Psychologist, 52*, 366–380.

Baltes, P. B., Dittmann-Kohli, F., & Dixon, R. A. (1984). New perspectives on the development of intelligence in adulthood: Toward a dual-process conception and a model of selective optimization with compensation. In P. B. Baltes & O. G. Brim, Jr. (Eds.), *Life-span development and behavior* (Vol. 6, pp. 33–76). San Diego, CA: Academic Press.

Baltes, P. B., & Lindenberger, U. (1997). Emergence of a powerful connection between sensory and cognitive functions across the adult life span: A new window to the study of cognitive aging? *Psychology and Aging, 12*, 12–21.

Baltes, P. B., Lindenberger, U., & Staudinger, U. M. (1997). Life-span theory in developmental psychology. In R. M. Lerner (Ed.), *Handbook of child psychology: Vol. 1. Theoretical models of human development.* New York: Wiley.

Baltes, P. B., Staudinger, U. M., Maercker, A., & Smith, J. (1995). People nominated as wise: A comparative study of wisdom-related knowledge. *Psychology and Aging, 10*, 155–166.

Band, E. B., & Weisz, J. R. (1988). How to feel better when it feels bad: Children's perspectives on coping with everyday stress. *Developmental Psychology, 24*, 247–253.

Bandura, A. (1977). *Social learning theory.* Englewood Cliffs, NJ: Prentice Hall.

Bandura, A. (1986). *Social foundations of thought and action: A social cognitive theory.* Englewood Cliffs, NJ: Prentice Hall.

Bandura, A. (1989). Social cognitive theory. In R. Vasta (Ed.), *Annals of child development* (Vol. 6, pp. 1–60). Greenwich, CT: JAI Press.

Bandura, A. (1997). *Self-efficacy: The exercise of control.* New York: Freeman.

Banks, M. S. (1980). The development of visual accommodation during early infancy. *Child Development, 51,* 646–666.

Banks, M. S., & Bennett, P. J. (1988). Optical and photoreceptor immaturities limit the spacial and chromatic vision of human neonates. *Journal of the Optical Society of America, 5,* 2059–2097.

Banks, M. S., & Salapatek, P. (1981). Infant pattern vision: A new approach based on the contrast sensitivity function. *Journal of Experimental Child Psychology, 31,* 1–45.

Banks, M. S., & Salapatek, P. (1983). Infant visual perception. In M. M. Haith & J. J. Campos (Eds.), *Handbook of child psychology: Vol. 2. Infancy and developmental psychobiology* (4th ed., pp. 435–571). New York: Wiley.

Barer, B. M. (1994). Men and women aging differently. *International Journal of Aging & Human Development, 38,* 29–40.

Barinaga, M. (1991). How long is the human life span? *Science, 254,* 936–938.

Barker, D. J. P., Gluckman, P. D., Godfrey, K. M., Harding, J. E., Owens, J. A., & Robinson, J. S. (1993). Fetal nutrition and cardiovascular disease in adult life. *Lancet, 341,* 938–941.

Barker, R. G. (1955). *Midwest and its children.* Stanford, CA: Stanford University Press.

Barkley, R. A. (1994). Impaired delayed responding: A unified theory of attention-deficit hyperactivity disorder. In R. A. Barkley (Ed.), *Disruptive behavior disorders in childhood* (pp. 11–57). New York: Plenum.

Barkley, R. A. (1997). Behavioral inhibition, sustained attention, and executive functions: Constructing a unifying theory of ADHD. *Psychological Bulletin, 121,* 65–94.

Barkley, R. A., DuPaul, G. J., & Costello, A. J. (1993). Stimulant medications. In J. Werry & M. Aman (Eds.), *Handbook of pediatric psychopharmacology* (pp. 205–237). New York: Plenum.

Barnes, S., Gutfreund, M., Satterly, D., & Wells, D. (1983). Characteristics of adult speech which predict children's language development. *Journal of Child Language, 10,* 65–84.

Barnett, D., Manly, J., & Cicchetti, D. (1993). Defining child maltreatment: The interface between policy and research. In D. Cicchetti & S. Toth (Eds.), *Child abuse, child development, and social policy* (pp. 7–73). Norwood, NJ: Ablex.

Barnett, R. C., & Rivers, C. (1996). *She works/he works.* San Francisco: Harper.

Barnett, W. S. (1993). New wine in old bottles: Increasing the coherence of early childhood care and educational policy. *Early Childhood Research Quarterly, 8,* 519–558.

Baron, S., & Welty, A. (1996). Elder abuse. *Journal of Gerontological Social Work, 25,* 33–57.

Barr, H. M., Streissguth, A. P., Darby, B. L., & Sampson, P. D. (1990). Prenatal exposure to alcohol, caffeine, tobacco, and aspirin: Effects on fine and gross motor performance in 4-year-old children. *Developmental Psychology, 26,* 339–348.

Barr, S. (1996, September). Up against the glass. *Management Review, 85*(9), 12–17.

Barrera, M. E., & Maurer, D. (1981a). Discrimination of strangers by the three-month-old. *Child Development, 52,* 559–563.

Barrera, M. E., & Maurer, D. (1981b). Recognition of mother's photographed face by the three-month-old infant. *Child Development, 52,* 714–716.

Barrett, D. E., & Yarrow, M. R. (1977). Prosocial behavior, social inferential ability, and assertiveness in children. *Child Development, 48,* 475–481.

Barrett, K. C., & Campos, J. J. (1987). Perspectives on emotional development: II. A functionalist approach to emotion. In J. D. Osofsky (Ed.), *Handbook of infant development* (2nd ed., pp. 1101–1149). New York: Wiley.

Barringer, F. (1993, April 1). Viral sexual diseases are found in 1 of 5 in the U.S. *The New York Times,* pp. A1, B9.

Barth, R. P., Petro, J. V., & Leland, N. (1992). Preventing adolescent pregnancy with social and cognitive skills. *Journal of Adolescent Research, 7,* 208–222.

Bartholomew, K., & Horowitz, L. M. (1991). Attachment styles among young adults: A test of a four-category model. *Journal of Personality and Social Psychology, 61,* 226–244.

Bass, B. M., & Avolio, B. J. (1994). Shatter the glass ceiling: Women may make better managers. *Human Resource Management, 33,* 549–560.

Bastian, H. (1993). Personal beliefs and alternative childbirth choices: A survey of 552 women who planned to give birth at home. *Birth, 20,* 186–192.

Bates, E. (1979). *The emergence of symbols: Cognition and communication in infancy.* New York: Academic Press.

Bates, E., Bretherton, I., & Snyder, L. (1988). *From first words to grammar.* Cambridge, England: Cambridge University Press.

Bauer, P. J. (1996). What do infants recall of their lives? Memory for specific events by one- to two-year-olds. *American Psychologist, 51,* 29–41.

Bauer, P. J., & Mandler, J. M. (1989). Taxonomies and triads: Conceptual organization in one- to two-year-olds. *Cognitive Psychology, 21,* 156–184.

Baumeister, R. F. (1990). Identity crisis. In R. M. Lerner, A. C. Petersen, & J. Brooks-Gunn (Eds.), *The encyclopedia of adolescence* (Vol. 1, pp. 518–521). New York: Garland.

Baumrind, D. (1967). Child care practices anteceding three patterns of preschool behavior. *Genetic Psychology Monographs, 75,* 43–88.

Baumrind, D. (1971). Current patterns of parental authority. *Developmental Psychology Monograph, 4*(No. 1, Pt. 2).

Baumrind, D. (1983). Rejoinder to Lewis's reinterpretation of parental firm control effects: Are authoritative families really harmonious? *Psychological Bulletin, 94,* 132–142.

Baumrind, D. (1991). The influence of parenting style on adolescent competence and substance use. *Journal of Early Adolescence, 11,* 56–95.

Baumrind, D. (1995). Commentary on sexual orientation: Research and social policy implications. *Developmental Psychology, 31,* 130–136.

Baumrind, D., & Black, A. E. (1967). Socialization practices associated with dimension of competence in preschool boys and girls. *Child Development, 38,* 291–327.

Bayley, N. (1969). *Bayley Scales of Infant Development.* New York: Psychological Corporation.

Bayley, N. (1993). *Bayley Scales of Infant Development* (2nd ed.). San Antonio, TX: Psychological Corporation.

Baylor, A. M., & Spirduso, W. W. (1988). Systematic aerobic exercise and components of reaction time in older women. *Journal of Gerontology, 43,* P121–P126.

Bazargan, M. (1994). The effects of health, environmental, and socio-psychological variables on fear of crime and its consequences among urban black elderly individuals. *International Journal of Aging and Human Development, 38,* 99–115.

Beal, C. R. (1990). The development of text evaluation and revision skills. *Child Development, 61,* 247–258.

Beard, B. B. (1991). *Centenarians: The new generation.* New York: Greenwood Press.

Beatty, W. W. (1992). Gonadal hormones and sex differences in nonreproductive behaviors. In A. A. Gerall, H. Moltz, & I. L. Ward (Eds.), *Handbook of behavioral neurobiology: Vol. 11. Sexual differentiation* (pp. 85–128). New York: Plenum.

Beauchamp, G. K., Cowart, B. J., Mennella, J. A., & Marsh, R. R. (1994). Infant salt taste: Developmental, methodological, and contextual factors. *Developmental Psychobiology, 27,* 353–365.

Beautrais, A. L., Fergusson, D. M., & Shannon, F. T. (1982). Life events and childhood morbidity: A prospective study. *Pediatrics, 70,* 935–940.

Beck, M. (1992, March 16). Finding work after 50. *Newsweek,* pp. 58–60.

Beck, M. (1994, January 17). How far should we push mother nature? *Newsweek,* pp. 54–57.

Beebe, D. K. (1991). Emergency management of the adult female rape victim. *American Family Physician, 43,* 2041–2046.

Begley, S. (1995, February 13). Three is not enough. *Newsweek,* pp. 67–69.

Behrend, D. A. (1988). Overextensions in early language comprehension: Evidence from a signal detection approach. *Journal of Child Language, 15,* 63–75.

Behrend, D. A., Rosengren, K. S., & Perlmutter, M. (1992). The relation between private speech and parental interactive style. In R. M. Diaz & L. E. Berk (Eds.), *Private speech: From social interaction to self-regulation* (pp. 85–100). Hillsdale, NJ: Erlbaum.

Behrman, R. E., & Vaughan, V. C. (1987). *Nelson textbook of pediatrics* (13th ed.). Philadelphia: Saunders.

Beidel, D. (1991). Social phobia and overanxious disorder in school-age children. *Journal of the American Academy of Child and Adolescent Psychiatry, 30,* 545–552.

Beilin, H. (1992). Piaget's enduring contribution to developmental psychology. *Developmental Psychology, 28,* 191–204.

Belkin, L. (1992, July 28). Childless couples hang on to last hope, despite law. *The New York Times,* pp. B1–B2.

Bell, M. A., & Fox, N. A. (1992). The relations between frontal brain electrical activity and cognitive development during infancy. *Child Development, 63,* 1142–1163.

Bell, M. L. (1995). Attitudes toward menopause among Mexican American women. *Health Care for Women International, 16,* 425–435.

Bellinger, D., Leviton, A., Waternaux, C., Needleman, H., & Rabinowitz, M. (1987). Longitudinal analysis of prenatal and postnatal lead exposure and early cognitive development. *New England Journal of Medicine, 316,* 1037–1043.

Belsky, J. (1993). Etiology of child maltreatment: A developmental–ecological analysis. *Psychological Bulletin, 114,* 413–434.

Belsky, J., & Braungart, J. M. (1991). Are insecure-avoidant infants with extensive day-care experience less stressed by and more independent in the Strange Situation? *Child Development, 62,* 567–571.

Belsky, J., Rovine, M., & Taylor, D. G. (1984). The Pennsylvania Infant and Family Development Project: III. The origins of individual differences in infant–mother attachment: Maternal and infant contributions. *Child Development, 55,* 718–728.

Belsky, J., Youngblade, L., Rovine, M., & Volling, B. (1991). Patterns of marital change and parent–child interaction. *Journal of Marriage and the Family, 53,* 487–498.

Bem, S. L. (1974). The measurement of psychological androgyny. *Journal of Consulting and Clinical Psychology, 42,* 155–162.

Bem, S. L. (1984). Androgyny and gender schema theory: A conceptual and empirical integration. In R. A. Dienstbier & T. B. Sondregger (Eds.), *Nebraska Symposia on Motivation* (Vol. 34, pp. 179–226). Lincoln: University of Nebraska Press.

Bem, S. L. (1989). Genital knowledge and gender constancy in preschool children. *Child Development, 60,* 649–662.

Benacerraf, B. R., Green, M. F., Saltzman, D. H., Barss, V. A., Penso, C. A., Nadel, A. S., Heffner, L. J., Stryker, J. M., Sandstrom, M. M., & Frigoletto, F. D., Jr. (1988). Early amniocentesis for prenatal cytogenetic evaluation. *Radiology, 169,* 709–710.

Benbow, C. P. (1986). Physiological correlates of extreme intellectual precocity. *Neuropsychologia, 24,* 719–725.

Benbow, C. P., & Arjmand, O. (1990). Predictors of high academic achievement in mathematics and science by mathematically talented students: A longitudinal study. *Journal of Educational Psychology, 82,* 430–441.

Benbow, C. P., & Lubinski, D. (1993). Psychological profiles of the mathematically talented: Some sex differences and evidence supporting their biological basis. In G. R. Bock & K. Ackrill (Eds.), *The origins and development of high ability* (Ciba Foundation Symposium

178, pp. 44–59). Chichester: Wiley.

Benbow, C. P., & Stanley, J. C. (1980). Sex differences in mathematical ability: Fact or artifact? *Science, 210,* 1262–1264.

Benbow, C. P., & Stanley, J. C. (1983). Sex differences in mathematical reasoning: More facts. *Science, 222,* 1029–1031.

Bench, R. J., Collyer, Y., Mentz, L., & Wilson, I. (1976). Studies in infant behavioural audiometry: I. Neonates. *Audiology, 15,* 85–105.

Benedict, R. (1934). Anthropology and the abnormal. *Journal of Genetic Psychology, 10,* 59–82.

Benenson, J. F. (1993). Greater preference among females than males for dyadic interaction in early childhood. *Child Development, 64,* 544–555.

Bengtson, V. L., & Harootyan, R. A. (1994). *Intergenerational linkages: Hidden connections in American society.* New York: Springer.

Bengston, V. L., Reedy, M. N., & Gordon, C. (1985). Aging and self-conceptions: Personality processes and social contexts. In J. E. Birren & K. W. Schaie (Eds.), *Handbook of the psychology of aging* (pp. 544–593). New York: Van Nostrand Reinhold.

Bengtson, V. L., & Robertson, J. F. (Eds.). (1985). *Grandparenthood.* Beverly Hills, CA: Sage.

Bengtson, V. L., Rosenthal, C. L., & Burton, L. (1990). Families and aging: Diversity and heterogeneity. In R. H. Binstock & L. K. George (Eds.), *Handbook of aging and the social sciences* (3rd ed., pp. 263–287). San Diego: Academic Press.

Benoliel, J. Q., & Degner, L. F. (1995). Institutional dying: A convergence of cultural values, technology, and social organization. In H. Wass & R. A. Neimeyer (Eds.), *Dying: Facing the facts* (pp. 117–162). Washington, DC: Taylor and Francis.

Berardo, D., Shehan, C., & Leslie, G. (1987). A residue of tradition: Jobs, careers, and spouses' time in housework. *Journal of Marriage and the Family, 49,* 381–390.

Berchuck, A., Cirisano, F. Lancaster, J. M., Schildkraut, J. M., Wiseman, R. W., Futreal, A., & Marks, J. R. (1996). Role of BRCA1 mutation screening in the management of familial ovarian cancer. *American Journal of Obstetrics and Gynecology, 175,* 738–746.

Berg, S. (1987). Intelligence and terminal decline. In G. L. Maddox & E. W. Busse (Eds.), *Aging: The universal human experience* (pp. 411–416). New York: Springer-Verlag.

Berg, S. (1996). Aging, behavior, and terminal decline. In J. E. Birren & K. W. Schaie (Eds.), *Handbook of the psychology of aging* (4th ed., pp. 323–337). San Diego: Academic Press.

Berg, W. K., & Berg, K. M. (1987). Psychophysiological development in infancy: State, startle, and attention. In J. Osofsky (Ed.), *Handbook of infant development* (2nd ed., pp. 238–317). New York: Wiley.

Bergkvist, L., & Persson, I. (1996). Hormone replacement therapy and breast cancer—A review of current knowledge. *Drug Safety, 15,* 360–370.

Berk, L. E. (1985). Relationship of caregiver education to child-oriented attitudes, job satisfaction, and behaviors toward children. *Child Care Quarterly, 14,* 103–129.

Berk, L. E. (1992a). Children's private speech: An overview of theory and the status of research. In R. M. Diaz & L. E. Berk (Eds.), *Private speech: From social interaction to self-regulation* (pp. 17–53). Hillsdale, NJ: Erlbaum.

Berk, L. E. (1992b). The extracurriculum. In P. W. Jackson (Ed.), *Handbook of research on curriculum* (pp. 1002–1043). New York: Macmillan.

Berk, L. E. (1994a, November). Vygotsky's theory: The importance of make-believe play. *Young Children, 50*(1), 30–38.

Berk, L. E. (1994b, November). Why children talk to themselves. *Scientific American, 271*(5), 78–83.

Berk, L. E., & Landau, S. (1993). Private speech of learning disabled and normally achieving children in classroom academic and laboratory contexts. *Child Development, 64,* 556–571.

Berk, L. E., & Spuhl, S. (1995). Maternal interaction, private speech, and task performance in preschool children. *Early Childhood Research Quarterly, 10,* 145–169.

Berko Gleason, J. (1997). Language development: An overview and a preview. In J. Berko Gleason (Ed.), *The development of language* (4th ed., pp. 1–39). Boston: Allyn & Bacon.

Berkowitz, M. W., & Gibbs, J. C. (1983). Measuring the developmental features of moral discussion. *Merrill-Palmer Quarterly, 29,* 399–410.

Berman, P. W. (1980). Are women more responsive than men to the young? A review of developmental and situational variables. *Psychological Bulletin, 88,* 668–695.

Berman, P. W., & Pedersen, F. A. (Eds.). (1987). *Men's transition to parenthood: Longitudinal studies and early family experience.* Hillsdale, NJ: Erlbaum.

Bermejo, V. (1996). Cardinality development and counting. *Developmental Psychology, 32,* 263–268.

Berndt, T. J., Cheung, P. C., Lau, S., Hau, K-T., & Lew, W. J. F. (1993). Perceptions of parenting in mainland China, Taiwan, and Hong Kong: Sex differences and societal differences. *Developmental Psychology, 29,* 156–164.

Berndt, T. J., & Keefe, K. (1995). Friends' influence on adolescents' adjustment to school. *Child Development, 66,* 1312–1329.

Berndt, T. J., & Perry, T. B. (1990). Distinctive features and effects of early adolescent friendships. In R. Montemayor, G. R. Adams, & T. P. Gullotta (Eds.), *From childhood to adolescence: A transitional period?* (pp. 269–287). Newbury Park, CA: Sage.

Berne, P. H., & Savary, L. M. (1993). *Building self-esteem in children.* New York: Continuum.

Berney, B. (1993). Round and round it goes: The epidemiology of childhood lead poisoning, 1950–1990. *Milbank Quarterly, 71,* 3–39.

Bernier, J. C., & Siegel, D. H. (1994). Attention-deficit hyperactivity disorder: A family ecological systems perspective. *Families in Society, 75,* 142–150.

Berscheid, E. (1988). Some comments on love's anatomy: Or whatever happened to old-fashioned lust? In R. J. Sternberg & M. L. Barnes (Eds.), *The psychology of love* (pp. 359–374). New Haven, CT: Yale University Press.

Bertenthal, B. I. (1993). Infants' perception of biomechanical motions: Instrinsic image and knowledge-based constraints. In C. Granrud (Ed.), *Visual perception and cognition in infancy* (pp. 175–214). Hillsdale, NJ: Erlbaum.

Bertenthal, B. I., & Campos, J. J. (1987). New directions in the study of early experience. *Child Development, 58,* 560–567.

Bertenthal, B. I., Campos, J. J., & Barrett, K. (1984). Self-produced locomotion: An organizer of emotional, cognitive, and social development in infancy. In R. Emde & R. Harmon (Eds.), *Continuities and discontinuities in development* (pp. 174–210). New York: Plenum.

Bertenthal, B. I., Campos, J. J., & Haith, M. (1980). Development of visual organization: The perception of subjective contours. *Child Development, 51,* 1077–1080.

Bertenthal, B. I., Proffitt, D. R., Spetner, N. B., & Thomas, M. A. (1985). The development of infant sensitivity to biomechanical motions. *Child Development, 56,* 531–543.

Best, D. L., Williams, J. E., Cloud, J. M.,

Davis, S. W., Robertson, L. S., Edwards, J. R., Giles, H., & Fowles, J. (1977). Development of sex-trait stereotypes among young children in the United States, England, and Ireland. *Child Development, 48,* 1375–1384.

Betz, C. (1994, March). Beyond time-out: Tips from a teacher. *Young Children, 49*(3), 10–14.

Betz, N. E. (1993). Women's career development. In F. L. Denmark & M. A. Paludi (Eds.), *Psychology of women* (pp. 627–684). Westport, CT: Greenwood Press.

Betz, N. E., & Fitzgerald, L. F. (1987). *The career psychology of women.* New York: Academic Press.

Beyene, Y. (1992). Menopause: A biocultural event. In A. J. Dan & L. L. Lewis (Eds.), *Menstrual health in women's lives* (pp. 169–177). Urbana, IL: University of Illinois Press.

Bhatt, R. S., Rovee-Collier, C., & Weiner, S. (1994). Developmental changes in the interface between perception and memory retrieval. *Developmental Psychology, 30,* 151–162.

Bibace, R., & Walsh, M. E. (1980). Development of children's concepts of illness. *Pediatrics, 66,* 912–917.

Biemiller, A. (1994). Some observations on beginning reading instruction. *Educational Psychologist, 29,* 203–209.

Biernat, M. (1991). Gender stereotypes and the relationship between masculinity and femininity: A developmental analysis. *Journal of Personality and Social Psychology, 61,* 351–365.

Bigler, R. S., & Liben, L. S. (1990). The role of attitudes and interventions in gender-schematic processing. *Child Development, 61,* 1440–1452.

Bigler, R. S., & Liben, L. S. (1992). Cognitive mechanisms in children's gender stereotyping: Theoretical and educational implications of a cognitive-based intervention. *Child Development, 63,* 1351–1363.

Bigner, J. J., & Jacobsen, R. B. (1989a). Parenting behaviors of homosexual and heterosexual fathers. *Journal of Homosexuality, 18,* 173–186.

Bigner, J. J., & Jacobsen, R. B. (1989b). The value of children to gay and straight fathers. *Journal of Homosexuality, 18,* 163–172.

Bijeljac-Babic, R., Bertoncini, J., & Mehler, J. (1993). How do 4-day-old infants categorize multisyllable utterances? *Developmental Psychology, 29,* 711–721.

Bille-Brahe, U. (1993). The role of sex and age in suicidal behavior. *Acta Psychiatrica Scandinavica, 87*(Suppl. 371), 21–27.

Billson, J. M., & Terry, M. B. (1987). A student retention model for higher education. *College and University, 62,* 290–305.

Billy, J. O. G., Tanfer, K., Grady, W. R., & Klepinger, D. H. (1993). The sexual behavior of men in the United States. *Family Planning Perspectives, 25,* 52–60.

Bily, S., & Manoochehri, G. (1995). Breaking the glass ceiling. *American Business Review, 13*(2), 33–39.

Binstock, R. H. (1992). The oldest old and "intergenerational inequity." In R. M. Suzman, D. P. Willis, & K. G. Manton (Eds.), *The oldest old* (pp. 394–417). New York: Oxford University Press.

Binstock, R. H., & Day, C. L. (1996). Aging and politics. In R. H. Binstock & L. K. George (Eds.), *Handbook of aging and the social sciences* (pp. 362–387). San Diego: Academic Press.

Birch, E. E. (1993). Stereopsis in infants and its developmental relation to visual acuity. In K. Simons (Ed.), *Early visual development: Normal and abnormal* (pp. 224–236). New York: Oxford University Press.

Birch, L. L., & Fisher, J. A. (1995). Appetite and eating behavior in children. *Pediatric Clinics of North America, 42,* 931–953.

Birch, L. L., Johnson, S. L., & Fisher, J. A. (1995, January). Children's eating: The development of food acceptance patterns. *Young Children, 50*(2), 71–78.

Birdwood, G. (1996). *Understanding osteoporosis and its treatment.* New York: Pathenon.

Birenbaum-Carmeli, D. (1995). Maternal smoking during pregnancy: Social, medical, and legal perspectives on the conception of a human being. *Health Care for Women International, 16,* 57–73.

Birkel, R. C. (1987). Toward a social ecology of the home-care household. *Psychology and Aging, 2,* 294–301.

Birren, J. E. (1990). Spiritual maturity in psychological development. In J. J. Seeber (Ed.), *Spiritual maturity in later years* (pp. 41–53). New York: Haworth.

Bischof-Köhler, D. (1991). The development of empathy in infants. In M. E. Lamb & H. Keller (Eds.), *Infant development: Perspectives from German-speaking countries* (pp. 1–33). Hillsdale, NJ: Erlbaum.

Bishop, S. M., & Ingersoll, G. M. (1989). Effects of marital conflict and family structure on the self-concepts of pre- and early adolescents. *Journal of Youth and Adolescence, 18,* 25–38.

Bivens, J. A., & Berk, L. E. (1990). A longitudinal study of the development of elementary school children's private speech. *Merrill-Palmer Quarterly, 36,* 443–463.

Bjorklund, D. F., & Muir, J. E. (1988). Children's development of free recall memory: Remembering on their own. In R. Vasta (Ed.), *Annals of child development* (Vol. 5, pp. 79–123). Greenwich, CT: JAI Press.

Bjorklund, D. F., Schneider, W., Cassel, W. S., & Ashley, E. (1994). Training and extension of a memory strategy: Evidence for utilization deficiencies in the acquisition of an organizational strategy in high- and low-IQ children. *Child Development, 65,* 951–965.

Blair, S. N., Kohl, H. W., III, Paffenbarger, R. S., Jr., Clark, D. G., Cooper, K. H., & Gibbons, L. W. (1989). Physical fitness and all-cause mortality: A prospective study of healthy men and women. *Journal of the American Medical Association, 262,* 2395–2401.

Blake, J. (1989). *Family size and achievement.* Berkeley, CA: University of California Press.

Blasi, A. (1990). Kohlberg's theory and moral motivation. In D. Schrader (Ed.), *New directions for child development* (No. 47, pp. 51–57). San Francisco: Jossey-Bass.

Blass, E. M., Ganchrow, J. R., & Steiner, J. E. (1984). Classical conditioning in newborn humans 2–48 hours of age. *Infant Behavior and Development, 7,* 223–235.

Blieszner, R., & Adams, R. G. (1992). *Adult friendship.* Newbury Park, CA: Sage.

Block, J. (1971). *Lives through time.* Berkeley, CA: Bancroft.

Block, J. (1982). Assimilation, accommodation, and the dynamics of personality development. *Child Development, 53,* 281–295.

Block, J. (1995). A contrarian view of the five-factor approach to personality description. *Psychological Bulletin, 117,* 187–215.

Block, J., Block, J. H., & Gjerde, P. F. (1988). Parental functioning and home environment in families of divorce: Prospective and concurrent analyses. *Journal of the American Academy of Child and Adolescent Psychiatry, 27,* 207–213.

Block, J., & Robins, R. W. (1994). A longitudinal study of consistency and change in self-esteem from early adolescence to early adulthood. *Child Development, 64,* 909–923.

Bloom, B. S. (Ed.). (1985). *Developing talent in young people.* New York: Ballantine Books.

Bloom, M. V. (1987). Leaving home: A family transition. In J. Bloom-Feshbach & S. Bloom-Feshbach (Eds.), *The psychology of separation and loss* (pp. 232–266). San Francisco: Jossey-Bass.

Blotner, R., & Bearison, D. J. (1984). Developmental consistencies in sociomoral knowledge: Justice reasoning and altruistic behavior. *Merrill-Palmer Quarterly, 30,* 349–367.

Bluebond-Langner, M. (1977). Meanings of death to children. In H. Feifel (Ed.), *New meanings of death* (pp. 47–66). New York: McGraw-Hill.

Blumberg, J. B. (1996). Status and functional impact of nutrition in older adults. In E. L. Schneider & J. W. Rowe (Eds.), *Handbook of the biology of aging* (4th ed., pp. 393–414). San Diego: Academic Press.

Blumenstein, P., & Schwartz, P. (1983). *American couples.* New York: Morrow.

Boden, D., & Bielby, D. D. V. (1983). The past as resource: A conversational analysis of elderly talk. *Human Development, 26,* 308–319.

Boer, F., Goedhart, A. W., & Treffers, P. D. A. (1992). Siblings and their parents. In F. Boer & J. Dunn (Eds.), *Children's sibling relationships* (pp. 41–54). Hillsdale, NJ: Erlbaum.

Bogatz, G. A., & Ball, S. (1972). *The second year of Sesame Street: A continuing evaluation.* Princeton, NJ: Educational Testing Service.

Bogartz, R. S., Shinskey, J. L., & Speaker, C. J. (1997). Interpreting infant looking: The event set x event set design. *Developmental Psychology, 33,* 408–422.

Bohannon, J. N., III. (1993). Theoretical approaches to language acquisition. In J. Berko Gleason (Ed.), *The development of language* (3rd ed., pp. 239–297). New York: Macmillan.

Bohannon, J. N., III, & Stanowicz, L. (1988). The issue of negative evidence: Adult responses to children's language errors. *Developmental Psychology, 24,* 684–689.

Bohr, L., Pascarella, E., Nora, A., Zusman, B., Jacobs, M., Desler, M., & Bulakowski, C. (1994). Cognitive effects of two-year and four-year institutions: A preliminary study. *Community College Review, 22,* 4–11.

Bokovoy, J. L., & Blair, S. N. (1994). Aging and exercise. *Journal of Aging and Physical Activity, 2,* 243–260.

Boldizar, J. P. (1991). Assessing sex typing and androgyny in children: The children's sex role inventory. *Developmental Psychology, 27,* 505–515.

Borja-Alvarez, T., Zarbatany, L., & Pepper, S. (1991). Contributions of male and female guests and hosts to peer group entry. *Child Development, 62,* 1079–1090.

Borke, H. (1975). Piaget's mountains revisited: Changes in the egocentric landscape. *Developmental Psychology, 11,* 240–243.

Borkowski, J. G., Carr, M., Rellinger, E., & Pressley, M. (1990). Self-regulated cognition: Interdependence of metacognition, attributions, and self-esteem. In B. F. Jones & L. Idol (Eds.), *Dimensions of thinking and cognitive instruction* (pp. 53–92). Hillsdale, NJ: Erlbaum.

Bornholt, L. J., Goodnow, J. J., & Cooney, G. H. (1994). Influences of gender stereotypes on adolescents' perceptions of their own achievement. *American Educational Research Journal, 31,* 675–692.

Bornstein, M. H. (1989). Sensitive periods in development: Structural characteristics and causal interpretations. *Psychological Bulletin, 105,* 179–197.

Bornstein, M. H. (1992). Perception across the life cycle. In M. H. Bornstein & M. E. Lamb (Eds.), *Developmental psychology: An advanced textbook* (3rd ed., pp. 155–209). Hillsdale, NJ: Erlbaum.

Bornstein, M. H., Vibbert, M., Tal, J., & O'Donnell, K. (1992). Toddler language and play in the second year: Stability, covariation, and influences of parenting. *First Language, 12,* 323–338.

Borrine, M. L., Handal, P. J., Brown, N. Y., & Searight, H. R. (1991). Family conflict and adolescent adjustment in intact, divorced, and blended families. *Journal of Consulting and Clinical Psychology, 59,* 753–755.

Borstelmann, L. J. (1983). Children before psychology: Ideas about children from antiquity to the late 1800s. In W. Kessen (Ed.), *Handbook of child psychology: Vol. 1. History, theory, and methods* (pp. 1–40). New York: Wiley.

Bossé, R., Aldwin, C. M., Levenson, M. R., Workman-Daniels, K., & Ekerdt, D. J. (1990). Differences in social support among retirees and workers: Findings from the Normative Aging Study. *Psychology and Aging, 5,* 41–47.

Bossé, R., Spiro, A., III, & Kressin, N. R. (1996). The psychology of retirement. In R. T. Woods (Ed.), *Handbook of the clinical psychology of aging* (pp. 141–157). Chicester, England: Wiley.

Bostock, C. (1994). Does the expansion of grandparent visitation rights promote the best interests of the child?: A survey of grandparent visitation laws in fifty states. *Columbia Journal of Law and Social Problems, 27,* 319–373.

Boston Women's Health Book Collective. (1992). *The new Our Bodies, Ourselves: A book by and for women* (2nd ed.). New York: Simon & Schuster.

Bouchard, C. (1994). *The genetics of obesity.* Boca Raton, FL: CRC Press.

Bouchard, T. J., Jr. (1994). Genes, environment, and personality. *Science, 264,* 1700–1701.

Bouchard, T. J., Jr., Lykken, D. T., McGue, M., Segal, N. L., & Tellegen, A. (1990). Sources of human psychological differences: The Minnesota Study of Twins Reared Apart. *Science, 250,* 223–228.

Bouchard, T. J., Jr., & McGue, M. (1981). Familial studies of intelligence: A review. *Science, 212,* 1055–1058.

Boukydis, C. F. Z. (1985). Perception of infant crying as an interpersonal event. In B. M. Lester & C. F. Z. Boukydis (Eds.), *Infant crying* (pp. 187–215). New York: Plenum.

Boukydis, C. F. Z., & Burgess, R. L. (1982). Adult physiological response to infant cries: Effects of temperament of infant, parental status and gender. *Child Development, 53,* 1291–1298.

Bower, C., & Stanley, F. J. (1992). Periconceptional vitamin supplementation and neural tube defects: Evidence from a case-control study in Western Australia and a review of recent publications. *Journal of Epidemiological and Community Health, 46,* 157-162.

Bowers, I. H., & Bahr, S. J. (1989). Remarriage among the elderly. In S. J. Bahr & E. T. Peterson (Eds.), *Aging and the family* (pp. 83–95). Lexington, MA: Lexington Books.

Bowlby, J. (1969). *Attachment and loss: Vol. 1. Attachment.* New York: Basic Books.

Bowlby, J. (1979). *The making and breaking of affectional bonds.* London: Tavistock.

Bowlby, J. (1980). *Attachment and loss: Vol. 3. Loss: Sadness and depression.* New York: Basic Books.

Boyer, E. (1980). Health perception in the elderly: Its cultural and social aspects. In C. L. Fry (Ed.), *Aging in culture and society: Comparative viewpoints and strategies* (pp. 198–216). New York: Bergin.

Boyer, K., & Diamond, A. (1992). Development of memory for temporal order in infants and young children. In A. Diamond (Ed.), *Development and neural bases of higher cognitive function* (pp. 267–317). New York: New York Academy of Sciences.

Boyes, M. C., & Allen, S. G. (1993). Styles of parent–child interaction and moral reasoning in adolescence. *Merrill-Palmer Quarterly, 39,* 551–570.

Boyes, M. C., & Chandler, M. (1992). Cognitive development, epistemic doubt, and identity formation in adolescence. *Journal of Youth and Adolescence, 21,* 277–304.

Boysson-Bardies, B. de, & Vihman, M. M. (1991). Adaptation to language: Evidence from babbling and first words in four languages. *Language, 67,* 297–319.

Brackbill, Y., McManus, K., & Woodward, L. (1985). *Medication in maternity: Infant exposure and maternal information.* Ann Arbor: University of Michigan Press.

Bradburn, E. M., Moen, P., & Dempster-McClain, D. (1995). Women's return to school following the transition to motherhood. *Social Forces, 73,* 1517–1551.

Braddock, J. H., & McPartland, J. M. (1982). Assessing school desegregation effects: New directions in research. *Research in Sociology of Education and Socialization, 3,* 259–282.

Braddock, J. H., & McPartland, J. M. (1987). How minorities continue to be excluded from equal employment opportunities: Research on labor market and institutional barriers. *Journal of Social Issues, 43,* 5–39.

Bradley, R. H., & Caldwell, B. M. (1979). Home Observation for Measurement of the Environment: A revision of the preschool scale. *American Journal of Mental Deficiency, 84,* 235–244.

Bradley, R. H., & Caldwell, B. M. (1981). The HOME Inventory: A validation of the preschool scale for black children. *Child Development, 52,* 708–710.

Bradley, R. H., & Caldwell, B. M. (1982). The consistency of the home environment and its relation to child development. *International Journal of Behavioral Development, 5,* 445–465.

Bradley, R. H., Caldwell, B. M., & Rock, S. L. (1988). Home environment and school performance: A ten-year follow-up and examination of three models of environmental action. *Child Development, 59,* 852–867.

Bradley, R. H., Caldwell, B. M., Rock, S. L., Ramey, C. T., Barnard, D. E., Gray, C., Hammond, M. A., Mitchell, S., Gottfried, A., Siegel, L., & Johnson, D. L. (1989). Home environment and cognitive development in the first 3 years of life: A collaborative study involving six sites and three ethnic groups in North America. *Developmental Psychology, 25,* 217–235.

Bradley, R. H., Whiteside, L., Mundfrom, D. J., Casey, P. H., Kelleher, K. J., & Pope, S. K. (1994). Early indications of resilience and their relation to experiences in the home environments of low birthweight, premature children living in poverty. *Child Development, 65,* 346–360.

Brady, E. M. (1984). Demographic and educational correlates of self-reported learning among older students. *Educational Gerontology, 10*, 27–38.

Braine, L. G., Schauble, L., Kugelmass, S., & Winter, A. (1993). Representation of depth by children: Spatial strategies and lateral biases. *Developmental Psychology, 29*, 466–479.

Braine, M. D. S. (1976). Children's first word combinations. *Monographs of the Society for Research in Child Development, 41*(1, Serial No. 164).

Brainerd, C. J. (1978). *Piaget's theory of intelligence.* Englewood Cliffs, NJ: Prentice Hall.

Brand, E., Clingempeel, W. E., & Bowen-Woodward, K. (1988). Family relationships and children's psychological adjustment in stepmother and stepfather families: Findings and conclusions from the Philadelphia Stepfamily Research Project. In E. M. Hetherington & J. D. Arasteh (Eds.), *Impact of divorce, single-parenting, and stepparenting on children* (pp. 299–324). Hillsdale, NJ: Erlbaum.

Brandtstädter, J., & Rothermund, K. (1994). Self-percepts of control in middle and later adulthood: Buffering losses by rescaling goals. *Psychology and Aging, 9*, 265–273.

Braungart, J. M., Plomin, R. DeFries, J. C., & Fulker, D. W. (1992). Genetic influence on tester-rated infant temperament as assessed by Bayley's Infant Behavior Record: Nonadoptive and adoptive siblings and twins. *Developmental Psychology, 28*, 40–47.

Braus, P. (1995, June). Vision in an aging America. *American Demographics, 17*, 34–39.

Braverman, P. K., & Strasburger, V. C. (1993). Adolescent sexual activity. *Clinical Pediatrics, 32*, 658–668.

Braverman, P. K., & Strasburger, V. C. (1994). Sexually transmitted diseases. *Clinical Pediatrics, 33*, 26–37.

Bray, J. (1992). Family relationships and children's adjustment in clinical and nonclinical stepfather families. *Journal of Family Psychology, 6*, 60–68.

Brazelton, T. B. (1984). *Neonatal Behavioral Assessment Scale.* Philadelphia: Lippincott.

Brazelton, T. B., Koslowski, B., & Tronick, E. (1976). Neonatal behavior among urban Zambians and Americans. *Journal of the American Academy of Child Psychiatry, 15*, 97–107.

Brazelton, T. B., Nugent, J. K., & Lester, B. M. (1987). Neonatal Behavioral Assessment Scale. In J. D. Osofsky (Ed.), *Handbook of infant development* (2nd ed., pp. 780–817). New York: Wiley.

Bread for the World Institute. (1994). *Hunger 1994.* Silver Spring, MD: Author.

Brechtelsbauer, D. A. (1990). Adult hearing loss. *Disorders of the Ears, Nose, and Throat, 17*, 249–266.

Bredekamp, S., & Copple, C. (eds.). (1997). *Developmentally appropriate practice in early childhood programs* (rev. ed.). Washington, DC: National Association for the Education of Young Children.

Brehm, S. S. (1992). *Intimate relationships* (2nd ed.). New York: McGraw-Hill.

Breitner, J. C. S., Silverman, J. M., Mohs, R. C., & Davis, K. L. (1988). Familial aggregation in Alzheimer's disease: Comparison of risk among relatives of early- and late-onset cases, and among male and female relatives in successive generations. *Neurology, 38*, 207–212.

Brennan, K. A., & Shaver, P. R. (1995). Dimensions of adult attachment, affect regulation, and romantic relationship functioning. *Personality and Social Psychology Bulletin, 21*, 267–283.

Brennan, W. M., Ames, E. W., & Moore, R. W. (1966). Age differences in infants' attention to patterns of different complexities. *Science, 151*, 354–356.

Brenner, D., & Hinsdale, G. (1978). Body build stereotypes and self-identification in three age groups of females. *Adolescence, 13*, 551–562.

Breslau, N. (1993). Daily cigarette consumption in early adulthood: Age of smoking initiation and duration of smoking. *Drug and Alcohol Dependence, 33*, 287–291.

Bretherton, I. (1992). The origins of attachment theory: John Bowlby and Mary Ainsworth. *Developmental Psychology, 28*, 759–775.

Bretherton, I., Fritz, J., Zahn-Waxler, C., & Ridgeway, D. (1986). Learning to talk about emotions: A functionalist perspective. *Child Development, 57*, 529–548.

Briere, J. N. (1992). *Child abuse trauma.* Newbury Park, CA: Sage.

Brody, E. M., Kleban, M. H., Johnsen, P. T., Hoffman, C., & Schoonover, C. B. (1987). Patterns of parent-care when adult daughters work and when they do not. *Gerontologist, 26*, 372–381.

Brody, G. H., Stoneman, Z., & McCoy, J. K. (1992). Associations of maternal and paternal direct and differential behavior with sibling relationships: Contemporaneous and longitudinal analyses. *Child Development, 63*, 82–92.

Brody, G. H., Stoneman, Z., & McCoy, J. K. (1994). Forecasting sibling relationships in early adolescence from child temperaments and family processes in middle childhood. *Child Development, 65*, 771–784.

Brody, G. H., Stoneman, Z., McCoy, J. K., & Forehand, R. (1992). Contemporaneous and longitudinal associations of sibling conflict with family relationship assessments and family discussions about sibling problems. *Child Development, 63*, 391–400.

Brody, H. (1992). Assisted death–a compassionate response to a medical failure. *New England Journal of Medicine, 327*, 1384–1388.

Brody, J. E. (1992, November 11). PMS is a worldwide phenomenon. *The New York Times*, p. C14.

Brody, N. (1992). *Intelligence* (2nd ed.). San Diego: Academic Press.

Brody, N. (1994). Heritability of traits. *Psychological Inquiry, 5*, 117–119.

Broman, S. H. (1983). Obstetric medications. In C. C. Brown (Ed.), *Childhood learning disabilities and prenatal risk* (pp. 56–64). New York: Johnson & Johnson.

Bronfenbrenner, U. (1979). *The ecology of human development: Experiments by nature and design.* Cambridge, MA: Harvard University Press.

Bronfenbrenner, U. (1989). Ecological systems theory. In R. Vasta (Ed.), *Annals of child development* (Vol. 6, pp. 187–251). Greenwich, CT: JAI Press.

Bronfenbrenner, U. (1993). The ecology of cognitive development: Research models and fugitive findings. In R. H. Wozniak & K. W. Fischer (Eds.), *Development in context* (pp. 3–44). Hillsdale, NJ: Erlbaum.

Bronfenbrenner, U. (1995). The bioecological model from a life course perspective: Reflections of a participant observer. In P. Moen, G. H. Elder, Jr., & K. Lüscher (Eds.), *Examining lives in context* (pp. 599–618). Washington, DC: American Psychological Association.

Bronfenbrenner, U. (1997). The ecology of developmental processes. In R. M. Lerner (Ed.), *Handbook of child psychology: Vol. 1. Theoretical models of human development.* New York: Wiley.

Bronfenbrenner, U., & Ceci, S. J. (1994). Nature–nurture reconceptualized in developmental perspective: A bioecological model. *Psychological Review, 101*, 568–586.

Bronfenbrenner, U., & Crouter, A. C. (1983). The evolution of environmental models in developmental research. In W. Kessen (Ed.), *Handbook of child psychology: Vol. 1. History, theory and methods* (Vol. 1, pp. 357–476). New York: Wiley.

Bronson, G. W. (1991). Infant differences in rate of visual encoding. *Child Development, 62*, 44–54.

Brooks, P. H., & Roberts, M. C. (1990, Spring). Social science and the prevention of children's injuries. *Social Policy Report of the Society for Research in Child Development, 4*(1).

Brooks-Gunn, J. (1986). The relationship of maternal beliefs about sex typing to maternal and young children's behavior. *Sex Roles, 14,* 21–35.

Brooks-Gunn, J. (1988a). Antecedents and consequences of variations in girls' maturational timing. *Journal of Adolescent Health Care, 9,* 365–373.

Brooks-Gunn, J. (1988b). The impact of puberty and sexual activity upon the health and education of adolescent girls and boys. *Peabody Journal of Education, 64,* 88–113.

Brooks-Gunn, J., McCarton, C. M., Casey, P. H., McCormick, M. C., Bauer, C. R., Bernbaum, J. C., Tyson, J., Swanson, M., Bennett, F. C., Scott, D. T., Tonascia, J., & Meinert, C. L. (1994). Early intervention in low-birth-weight premature infants. *Journal of the American Medical Association, 272,* 1257–1262.

Brooks-Gunn, J., & Petersen, A. C. (1991). Studying the emergence of depression and depressive symptoms during adolescence. *Journal of Youth and Adolescence, 20,* 115–119.

Brooks-Gunn, J., & Ruble, D. N. (1980). Menarche: The interaction of physiology, cultural, and social factors. In A. J. Dan, E. A. Graham, & C. P. Beecher (Eds.), *The menstrual cycle: A synthesis of interdisciplinary research* (pp. 141– 159). New York: Springer-Verlag.

Brooks-Gunn, J., & Ruble, D. N. (1983). The experience of menarche from a developmental perspective. In J. Brooks-Gunn & A. C. Petersen (Eds.), *Girls at puberty* (pp. 155–177). New York: Plenum.

Brooks-Gunn, J., Warren, M. P., Samelson, M., & Fox, R. (1986). Physical similarity of and disclosure of menarcheal status to friends: Effects of grade and pubertal status. *Journal of Early Adolescence, 6,* 3–14.

Broomhall, H. S., & Winefield, A. H. (1990). A comparison of the affective well-being of young and middle-aged unemployed men matched for length of employment. *British Journal of Medical Psychology, 63,* 43–52.

Brown, A. L., Bransford, J. D., Ferrara, R. A., & Campione, J. C. (1983). Learning, remembering, and understanding. In J. H. Flavell & E. M. Markman (Eds.), *Handbook of child psychology: Vol. 3. Cognitive development* (4th ed., pp. 77–166). New York: Wiley.

Brown, A. M. (1990). Development of visual sensitivity to light and color

vision in human infants: A critical review. *Vision Research, 30,* 1159–1188.

Brown, B. B., Clasen, D., & Eicher, S. (1986). Perceptions of peer pressure, peer conformity dispositions, and self-reported behavior among adolescents. *Developmental Psychology, 22,* 521–530.

Brown, B. B. (1990). Peer groups. In S. Feldman & G. Elliott (Eds.), *At the threshold: The developing adolescent* (pp. 171–196). Cambridge, MA: Cambridge University Press.

Brown, B. B., Lohr, M. J., & McClenahan, E. L. (1986). Early adolescents' perceptions of peer pressure. *Journal of Early Adolescence, 6,* 139–154.

Brown, J. A. (1990). Social work practice with the terminally ill in the black community. In J. K. Parry (Ed.), *Social work practice with the terminally ill: A transcultural perspective* (pp. 67–82). Springfield, IL: Charles C Thomas.

Brown, R. W. (1973). *A first language: The early stages.* Cambridge, MA: Harvard University Press.

Browne, A. (1991). The victim's experience: Pathways to disclosure. *Psychotherapy, 28,* 150–156.

Browne, A. (1993). Violence against women by male partners: Prevalence, outcomes and policy implications. *American Psychologist, 48,* 1077–1087.

Brownell, C. A., & Carriger, M. S. (1990). Changes in cooperation and self-other differentiation during the second year. *Child Development, 61,* 1164–1174.

Brownell, K. D., & Wadden, T. A. (1992). Etiology and treatment of obesity: Understanding a serious, prevalent, and refractory disorder. *Journal of Consulting and Clinical Psychology, 60,* 505–517.

Brubaker, T. (1985). *Later life families.* Beverly Hills, CA: Sage.

Bruch, M. A., Gorsky, J. M., Collins, T. M., & Berger, P. A. (1989). Shyness and sociability examined: A multicomponent analysis. *Journal of Personality and Social Psychology, 57,* 904–915.

Bruck, M., Ceci, S. J., Francouer, E., & Renick, A. (1995). Anatomically detailed dolls do not facilitate preschoolers' reports of a pediatric examination involving genital touching. *Journal of Experimental Psychology: Applied, 1,* 95–109.

Bruner, J. (1990). *Acts of meaning.* Cambridge, MA: Harvard University Press.

Bryant, B. K. (1985). The neighborhood walk: Sources of support in middle childhood. *Monographs for the Society for Research in Child Development, 50* (3, Serial No. 210).

Bryant, D. M., & Ramey, C. T. (1987). An analysis of the effectiveness of early

intervention programs for environmentally at-risk children. In M. J. Guralnick & F. C. Bennett (Eds.), *The effectiveness of early intervention for at-risk handicapped children* (pp. 33–78). Orlando, FL: Academic Press.

Buchanan, C. M., Eccles, J. S., & Becker, J. B. (1992). Are adolescents the victims of raging hormones? Evidence for activational effects of hormones on moods and behavior at adolescence. *Psychological Bulletin, 111,* 62–107.

Buck, G. M., Cookfair, D. L., Michalek, A. M., Nasca, P. C., Standfast, S. J., Sever, L. E., & Kramer, A. A. (1989). Intrauterine growth retardation and risk of sudden infant death syndrome (SIDS). *American Journal of Epidemiology, 129,* 874–884.

Bugental, D. B., Blue, J., & Cruzcosa, M. (1989). Perceived control over caregiving outcomes: Implications for child abuse. *Developmental Psychology, 25,* 532–539.

Buhrmester, D., & Furman, W. (1987). The development of companionship and intimacy. *Child Development, 58,* 1101–1115.

Buhrmester, D., & Furman, W. (1990). Perceptions of sibling relationships during middle childhood and adolescence. *Child Development, 61,* 1387–1398.

Bulatao, R. A., & Arnold, F. (1977). *Relationships between the value and cost of children and fertility: Cross-cultural evidence.* Paper presented at the General Conference of the International Union for the Scientific Study of Population, Mexico City.

Bullock, M., & Lutkenhaus, P. (1990). Who am I? The development of self-understanding in toddlers. *Merrill-Palmer Quarterly, 36,* 217–238.

Burhans, K. K., & Dweck, C. S. (1995). Helplessness in early childhood: The role of contingent worth. *Child Development, 66,* 1719–1738.

Burkett, G., Yasin, S. Y., Palow, D., La Voie, L., & Martinez, M. (1994). Patterns of cocaine binging: Effect on pregnancy. *American Journal of Obstetrics and Gynecology, 171,* 372–379.

Burkhardt, S. A., & Rotatori, A. F. (1995). *Treatment and prevention of child hood sexual abuse.* Washington, DC: Taylor & Francis.

Burns, A., Howard, R., & Pettit, W. (1995). *Alzheimer's disease: A medical companion.* Oxford, England: Oxford University Press.

Burns, S. M., & Brainerd, C. J. (1979). Effects of constructive and dramatic play on perspective taking in very young children. *Developmental Psychology, 15,* 512–521.

Burton, B. K. (1992). Limb anomalies associated with chorionic villus sampling. *Obstetrics and Gynecology, 79* (Pt. 1), 726–730.

Burton, L. M. (1996). Age norms, the timing of family role transitions, and intergenerational caregiving among aging African American women. *Gerontologist, 36,* 199–208.

Bushnell, E. W. (1985). The decline of visually guided reaching during infancy. *Infant Behavior and Development, 8,* 139–155.

Bushnell, E. W., & Boudreau, J. P. (1993). Motor development and the mind: The potential role of motor abilities as a determinant of aspects of perceptual development. *Child Development, 64,* 1005–1021.

Buss, A. H., & Plomin, R. (1984). *Temperament: Early developing personality traits.* Hillsdale, NJ: Erlbaum.

Bussey, K., & Bandura, A. (1992). Self-regulatory mechanisms governing gender development. *Child Development, 63,* 1236–1250.

Butler, R. N. (1968). The life review: An interpretation of reminiscence in the aged. In B. Neugarten (Ed.), *Middle age and aging* (pp. 486–496). Chicago: University of Chicago Press.

Buunk, B. P., & Driel, B. van. (1989). *Variant lifestyles and relationships.* Newbury Park, CA: Sage.

Byrne, M. C., & Hayden, E. (1980). *Topic maintenance and topic establishment in mother–child dialogue.* Paper presented at the meeting of the American Speech and Hearing Association, Detroit, MI.

Byrnes, J. P. (1993). Analyzing perspectives on rationality and critical thinking: A commentary on the *Merrill-Palmer Quarterly* invitational issue. *Merrill-Palmer Quarterly, 39,* 159–171.

Byrnes, J. P., & Takahira, S. (1993). Explaining gender differences on SAT-math items. *Developmental Psychology, 29,* 805–810.

Byrnes, J. P., & Wasik, B. A. (1991). Role of conceptual knowledge in mathematical procedural learning. *Developmental Psychology, 27,* 777–786.

Caddell, D. P., & Newton, R. R. (1995). Euthanasia: American attitudes toward the physician's role. *Social Science and Medicine, 40,* 1671–1681.

Cadoff, J. (1995, March). Can we prevent SIDS? *Parents, 70*(3), 30–31, 35.

Cain, K. M., & Dweck, C. S. (1995). The relation between motivational patterns and achievement cognitions through the elementary school years. *Merrill-Palmer Quarterly, 41,* 25–52.

Caine, N. (1986). Behavior during puberty and adolescence. In G. Mitchell & J. Erwin (Eds.), *Comparative primate biology: Vol. 2A. Behavior, conservation, and ecology* (pp. 327–361). New York: Liss.

Cairns, E. (1996). *Children and political violence.* Cambridge: Blackwell.

Cairns, R. B., Leung, M-C., Buchanan, L., & Cairns, B. D. (1995). Friendships and social networks in childhood and adolescence: Fluidity, reliability, and interrelations. *Child Development, 66,* 1330–1345.

Caldas, S. J. (1993). Current theoretical perspectives on adolescent pregnancy and childbearing in the United States. *Journal of Adolescent Research, 8,* 4–20.

Calhoun, D. A. (1992). Hypertension in blacks: Socioeconomic stress and sympathetic nervous system activity. *American Journal of the Medical Sciences, 304,* 306–311.

Caliso, J., & Milner, J. (1992). Childhood history of abuse and child abuse screening. *Child Abuse and Neglect, 16,* 647–659.

Callahan, D. (1987). *Setting limits: Medical goals in an aging society.* New York: Simon & Schuster.

Callahan, D. (1990). *What kind of life: The limits of medical progress.* New York: Simon & Schuster.

Camara, K. A., & Resnick, G. (1988). Interparental conflict and cooperation: Factors moderating children's post-divorce adjustment. In E. M. Hetherington & J. D. Arasteh (Eds.), *Impact of divorce, single parenting, and step-parenting on children* (pp. 169–195). Hillsdale, NJ: Erlbaum.

Campbell, F. A., & Ramey, C. T. (1991). *The Carolina Abecedarian Project.* Paper presented at the biennial meeting of the Society for Research in Child Development, Seattle, WA.

Campbell, F. A., & Ramey, C. T. (1994). Effects of early intervention on intellectual and academic achievement: A follow-up study of children from low-income families. *Child Development, 65,* 684–698.

Campos, J., & Bertenthal, B. (1989). Locomotion and psychological development. In F. Morrison, K. Lord, & D. Keating (Eds.), *Applied developmental psychology* (Vol. 3, pp. 229–258). New York: Academic Press.

Campos, J. J., Caplovitz, K. B., Lamb, M. E., Goldsmith, H. H., & Stenberg, C. (1983). Socioemotional development. In M. M. Haith & J. J. Campos (Eds.), *Handbook of child psychology: Vol. 2. Infancy and developmental psychobiology* (4th ed., pp. 783–915). New York: Wiley.

Campos, R. G. (1989). Soothing pain-elicited distress in infants with swaddling and pacifiers. *Child Development, 60,* 781–792.

Camras, L. A., Oster, H., Campos, J. J., Miyake, K., & Bradshaw, D. (1992). Japanese and American infants' responses to arm restraint. *Developmental Psychology, 28,* 578–583.

Cangemi, J. P. (1983). Interpersonal relations on the college campus and declining enrollment: Any relationship? In C. J. Kowalski & J. P. Cangemi (Eds.), *Perspectives in higher education* (pp. 81–85). New York: Philosophical Library.

Canick, J. A., & Saller, D. N., Jr. (1993). Maternal serum screening for aneuploidy and open fetal defects. *Obstetrics and Gynecology Clinics of North America, 20,* 443–454.

Cannella, G. S. (1993). Learning through social interaction: Shared cognitive experience, negotiation strategies, and joint concept construction for young children. *Early Childhood Research Quarterly, 8,* 427–444.

Capaldi, D. M., & Patterson, G. R. (1991). Relation of parental transitions to boys' adjustment problems: I. A linear hypothesis. II. Mothers at risk for transitions and unskilled parenting. *Developmental Psychology, 27,* 489–504.

Capelli, C. A., Nakagawa, N., & Madden, C. M. (1990). How children understand sarcasm: The role of context and intonation. *Child Development, 61,* 1824–1841.

Caplan, M., Vespo, J., Pedersen, J., & Hay, D. F. (1991). Conflict and its resolution in small groups of one- and two-year-olds. *Child Development, 62,* 1513–1524.

Capuzzi, D. (1989). *Adolescent suicide prevention.* Ann Arbor, MI: ERIC Counseling and Personnel Services Clearinghouse.

Carden, A. D. (1994). Wife abuse and the wife abuser: Review and recommendations. *Counseling Psychologist, 22,* 539–582.

Cardo, L. M. (1991). Death and dignity: The case of Diane [Letter to the Editor]. *New England Journal of Medicine, 325,* 658.

Carey, S. (1985). *Conceptual change in childhood.* Cambridge, MA: MIT Press.

Carlson, V., Cicchetti, D., Barnett, D., & Braunwald, K. (1989). Disorganized/disoriented attachment relationship in maltreated infants. *Child Development, 25,* 525–531.

Carpenter, C. J. (1983). Activity structure and play: Implications for socialization. In M. Liss (Eds.), *Social and cognitive skills: Sex roles and children's play* (pp. 117–145). New York: Academic Press.

Carstensen, L. L. (1992). Selectivity theory: Social activity in life-span context. In K. W. Schaie & M. P. Lawton (Eds.),

Annual review of gerontology and geriatrics (pp. 195–217). New York: Springer.

Carstensen, L. L., Gottman, J. M., & Levenson, R. W. (1995). Emotional behavior in long-term marriages. *Psychology and Aging, 10,* 140–149.

Carstensen, L. L., & Turk, C. S. (1994). The salience of emotion across the adult life span. *Psychology and Aging, 9,* 315–324.

Carus, F. A. (1808). *Psychologie. Zweiter Teil: Specialpsychologie.* Leipzig: Barth & Kummer.

Carver, C. S., Scheier, M. F., & Weintraub, J. K. (1989). Assessing coping strategies: A theoretically based approach. *Journal of Personality and Social Psychology, 56,* 267–283.

Casaer, P. (1993). Old and new facts about perinatal brain development. *Journal of Child Psychology and Psychiatry, 34,* 101–109.

Case, R. (1985). *Intellectual development: A systematic reinterpretation.* New York: Academic Press.

Case, R. (1992). *The mind's staircase: Exploring the conceptual underpinnings of children's thought and knowledge.* Hillsdale, NJ: Erlbaum.

Casey, M. B. (1986). Individual differences in selective attention among prereaders: A key to mirror-image confusions. *Developmental Psychology, 22,* 824–831.

Casey, M. B., Nuttall, R., Pezaris, E., & Benbow, C. P. (1995). The influence of spatial ability on gender differences in mathematics college entrance test scores across diverse samples. *Developmental Psychology, 31,* 697–705.

Caspi, A., Elder, G. H., Jr., & Bem, D. J. (1987). Moving against the world: Life-course patterns of explosive children. *Developmental Psychology, 23,* 308–313.

Caspi, A., Elder, G. H., Jr., & Bem, D. J. (1988). Moving away from the world: Life-course patterns of shy children. *Developmental Psychology, 24,* 824–831.

Caspi, A., Elder, G. H., Jr., & Herbener, E. S. (1990). Childhood personality and the prediction of life-course patterns. In L. N. Robins & M. Rutter (Eds.), *Straight and devious pathways from childhood to adulthood* (pp. 13–35). Cambridge, England: Cambridge University Press.

Caspi, A., & Herbener, E. (1990). Continuity and change: Assortative marriage and the consistency of personality in adulthood. *Journal of Personality and Social Psychology, 58,* 250–258.

Caspi, A., Lynam, D., Moffitt, T. E., & Silva, P. A. (1993). Unraveling girls' delinquency: Biological, dispositional, and contextual contributions to adolescent misbehavior. *Developmental Psychology, 29,* 19–30.

Caspi, A., & Silva, P. A. (1995). Temperamental qualities at age three predict personality traits in young adulthood: Longitudinal evidence from a birth cohort. *Child Development, 66,* 486–498.

Cassel, C. K. (1987). Informed consent for research in geriatrics: History and concepts. *Journal of the American Geriatrics Society, 35,* 542–544.

Cassel, C. K. (1988). Ethical issues in the conduct of research in long-term care. *Gerontologist, 28*(Suppl.), 90–96.

Cassidy, J., & Berlin, L. J. (1994). The insecure/ambivalent pattern of attachment: Theory and research. *Child Development, 65,* 971–991.

Cassidy, S. B. (1995). Uniparental disomy and genomic imprinting as causes of human genetic disease. *Environmental and Molecular Mutagenesis, 25,* 13–20.

Caudill, W. (1973). Psychiatry and anthropology: The individual and his nexus. In L. Nader & T. W. Maretzki (Eds.), *Cultural illness and health: Essays in human adaptation* (Anthropological Studies 9, pp. 67–77). Washington, DC: American Anthropological Association.

Cazden, C. (1984). *Effective instructional practices in bilingual education.* Washington, DC: National Institute of Education.

Cazenave, N. A., & Straus, M. A. (1990). Race, class, network embeddedness, and family violence: A search for potent support systems. In M. A. Straus & R. J. Gelles (Eds.), *Physical violence in American families* (pp. 321–339). New Brunswick, NJ: Transaction.

Ceci, S. J. (1990). *On intelligence . . . More or less.* Englewood Cliffs, NJ: Prentice-Hall.

Ceci, S. J. (1991). How much does schooling influence general intelligence and its cognitive components? A reassessment of the evidence. *Developmental Psychology, 27,* 703–722.

Ceci, S. J., & Bruck, M. (1993). Suggestibility of the child witness: A historical review and synthesis. *Psychological Bulletin, 113,* 403–439.

Ceci, S. J., & Bruck, M. (1995). *Jeopardy in the courtroom: A scientific analysis of children's testimony.* Washington, DC: American Psychological Association.

Ceci, S. J., Leichtman, M. D., & Bruck, M. (1994). The suggestibility of children's eyewitness reports: Methodological issues. In F. Weinert & W. Schneider (Eds.), *Memory development: State of the art and future directions* (pp. 323–347). Englewood Cliffs, NJ: Erlbaum.

Celano, M. P., & Geller, R. J. (1993). Learning, school performance, and children with asthma: How much risk? *Journal of Learning Disabilities, 26,* 23–32.

Central Intelligence Agency (1996). *The world fact book.* Washington, DC: U.S. Government Printing Office.

Cerella, J. (1990). Aging and information processing rate. In J. E. Birren & K. W. Schaie (Eds.), *Handbook of the psychology of aging* (3rd ed.), pp. 201–221). San Diego: Academic Press.

Cernoch, J. M., & Porter, R. H. (1985). Recognition of maternal axillary odors by infants. *Child Development 56,* 1593–1598.

Chadwick, E. G., & Yogev, R. (1995). Pediatric AIDS. In G. E. Gaull (Ed.), *Pediatric Clinics of North America, 42,* 969–992.

Chalmers, J. B., & Townsend, M. A. R. (1990). The effects of training in social perspective taking on socially maladjusted girls. *Child Development, 61,* 178–190.

Chamberlain, M. C., Nichols, S. L., & Chase, C. H. (1991). Pediatric AIDS: Comparative cranial MRI and CT scans. *Pediatric Neurology, 7,* 357–362.

Chanarin, I. (1994). Adverse effects of increased dietary folate: Relation to measures to reduce the incidence of neural tube defects. *Clinical and Investigative Medicine, 17,* 244–252.

Chandler, M. J. (1973). Egocentrism and antisocial behavior: The assessment and training of social perspective-taking skills. *Developmental Psychology, 9,* 326–332.

Chandra, R. K. (1991). Interactions between early nutrition and the immune system. In *Ciba Foundation Symposium No. 156* (pp. 77–92). Chichester, England: Wiley.

Chandra, R. K. (1992). Effect of vitamin and trace-element supplementation on immune responses and infection in elderly subjects. *Lancet, 340,* 1124–1127.

Chao, R. K. (1994). Beyond parental control and authoritarian parenting style: Understanding Chinese parenting through the cultural notion of training. *Child Development, 65,* 1111–1119.

Chapman, M., & Lindenberger, U. (1988). Functions, operations, and décalage in the development of transitivity. *Developmental Psychology, 24,* 542–551.

Chapman, M., & Skinner, E. A. (1989). Children's agency beliefs, cognitive performance, and conceptions of effort and ability: Individual and developmental differences. *Child Development, 60,* 1229–1238.

Chappell, N. L., & Guse, L. W. (1989). Linkages between informal and formal support. In K. S. Markides & C. L. Cooper (Eds.), *Aging, stress and health* (pp. 219–237). Chichester, England: Wiley.

Charo, R. A. (1994). USA: New York surrogacy law. *Lancet, 440,* 361.

Chase-Lansdale, P. L., & Brooks-Gunn, J. (1994a). Correlates of adolescent pregnancy and parenthood. In C. B. Fisher & R. M. Lerner (Eds.), *Applied developmental psychology* (pp. 207–236). New York: McGraw-Hill.

Chase-Lansdale, P. L., & Brooks-Gunn, J. (Eds.). (1994b). *Escape from poverty: What makes a difference for children?* New York: Cambridge University Press.

Chase-Lansdale, P. L., Brooks-Gunn, J., & Zamsky, E. S. (1994). Young African-American multigenerational families in poverty: Quality of mothering and grandmothering. *Child Development, 65,* 373–393.

Chase-Lansdale, P. L., Cherlin, A. J., & Kiernan, K. E. (1995). The long-term effects of parental divorce on the mental health of young children. *Child Development, 66,* 1614–1634.

Chassin, L., Curran, P. J., Hussong, A. M., & Colder, C. R. (1996). The relation of parent alcoholism to adolescent substance use: A longitudinal follow-up study. *Journal of Abnormal Psychology, 105,* 70–80.

Chasteen, A. L. (1994). "The world around me": The environment and single women. *Sex Roles, 31,* 309–328.

Chatkupt, S., Mintz, M., Epstein, L. G., Bhansali, D., & Koenigsberger, M. R. (1989). Neuroimaging studies in children with human immunodeficiency virus type 1 infection. *Annals of Neurology, 26,* 453.

Chatterjee, S., Handcock, M. S., & Simonoff, J. S. (1995). *A casebook for a first course in statistics and data analysis.* New York: Wiley.

Cheek, J. M., & Busch, C. M. (1981). The influence of shyness on loneliness in a new situation. *Personality and Social Psychology Bulletin, 7,* 572–577.

Chen, X., Rubin, K. H., & Li, Z. (1995). Social functioning and adjustment in Chinese children: A longitudinal study. *Developmental Psychology, 31,* 531–539.

Cheng, M., & Hannah, M. (1993). Breech delivery at term: A critical review of the literature. *Obstetrics & Gynecology, 82,* 605–618.

Cherlin, A. J. (1992). *Marriage, divorce, remarriage* (rev. ed.). Cambridge, MA: Harvard University Press.

Cherlin, A. J., & Furstenberg, F. F., Jr. (1986). *The new American grandparent.* New York: Basic Books.

Cherlin, A. J., Furstenberg, F. F., Jr., Chase-Lansdale, P. L., Kiernan, K. E., Robins, P. K., Morrison, D. R., & Teitler, J. O. (1991). Longitudinal studies of effects of divorce on children in Great Britain and the United States. *Science, 252,* 1386–1389.

Cherlin, A. J., Kiernan, K. E., & Chase-Lansdale, P. L. (1995). Parental divorce in childhood and demographic outcomes in young adulthood. *Demography, 32,* 299–318.

Chess, S., & Thomas, A. (1984). *Origins and evolution of behavior disorders.* New York: Brunner/Mazel.

Chess, S., & Thomas, A. (1990). Continuities and discontinuities in temperament. In L. N. Robins & M. Rutter (Eds.), *Straight and devious pathways from childhood to adulthood* (pp. 205–220). Cambridge, England: Cambridge University Press.

Chi, M. T. H. (1978). Knowledge structures and memory development. In R. S. Siegler (Ed.), *Children's thinking: What develops?* (pp. 73–96). Hillsdale, NJ: Erlbaum.

Chi, M. T. H., Glaser, R., & Farr, M. J. (Eds.). (1988). *The nature of expertise.* Hillsdale, NJ: Erlbaum.

Children's Defense Fund. (1992). *The health of America's children.* Washington, DC: Author.

Children's Defense Fund. (1994). *The state of America's children: Yearbook 1994.* Washington, DC: Author.

Children's Defense Fund. (1997). *The state of America's children: Yearbook 1997.* Washington, DC: Author.

Childs, C. P., & Greenfield, P. M. (1982). Informal modes of learning and teaching: The case of Zinacanteco weaving. In N. Warren (Ed.), *Advances in cross-cultural psychology* (Vol. 2, pp. 269–316). London: Academic Press.

Chilmonczyk, B. A., Salmun, L. M., Megathlin, K. N., Neveus, L. M., Palomaki, G. E., Knight, G. J., Pulkkinen, A. J., & Haddow, J. E. (1993). Association between exposure to environmental tobacco smoke and exacerbations of asthma in children. *New England Journal of Medicine, 328,* 1665–1669.

Chisholm, J. S. (1989). Biology, culture, and the development of temperament: A Navajo example. In J. K. Nugent, B. M. Lester, & T. B. Brazelton (Eds.), *Biology, culture, and development* (Vol. 1, pp. 341–364). Norwood, NJ: Ablex.

Chodorow, N. (1978). *The reproduction of mothering.* Berkeley: University of California Press.

Choi, N. G. (1994). Racial differences in timing and factors associated with retirement. *Journal of Sociology & Social Welfare, 21,* 31–52.

Choi, N. G. (1995). Racial differences in the determinants of the coresidence of and contacts between elderly parents and their children. *Journal of Gerontological Social Work, 24,* 77–95.

Chollar, S. (1995, June). The psychological benefits of exercise. *American Health, 14*(5), 72–75.

Chomsky, C. (1969). *The acquisition of syntax in children from five to ten.* Cambridge, MA: MIT Press.

Chomsky, N. (1957). *Syntactic structures.* The Hague: Mouton.

Chu, G. C. (1985). The changing concept of self in contemporary China. In A. J. Marsella, G. DeVos, & F. L. K. Hsu (Eds.), *Culture and self: Asian and Western perspectives* (pp. 252–277). London: Tavistock.

Churchill, S. R. (1984). Disruption: A risk in adoption. In P. Sachdev (Ed.), *Adoption: Current issues and trends* (pp. 115–127). Toronto: Butterworth.

Cicchetti, D., & Aber, J. L. (1986). Early precursors of later depression: An organizational perspective. In L. P. Lipsitt & C. Rovee-Collier (Eds.), *Advances in infancy research* (Vol. 4, pp. 87–137). Norwood, NJ: Ablex.

Cicerelli, V. G. (1989). Feelings of attachment to siblings and well-being in later life. *Psychology and Aging, 4,* 211–216.

Cicerelli, V. G. (1995). *Sibling relationships across the life span.* New York: Plenum.

Clark, D. C., & Mokros, H. R. (1993). Depression and suicidal behavior. In P. H. Tolan & B. J. Cohler (Eds.), *Handbook of clinical research and practice with adolescents* (pp. 333–358). New York: Wiley.

Clark, E. V. (1983). Meanings and concepts. In J. H. Flavell & E. M. Markman (Eds.), *Handbook of child psychology: Vol. 3. Cognitive development* (pp. 787–840). New York: Wiley.

Clark, E. V. (1995). The lexicon and syntax. In J. L. Miller & P. D. Eimas (Eds.), *Speech, language, and communication* (pp. 303–337). San Diego: Academic Press.

Clarke, A. (Ed.). (1994). *Genetic counseling: Practice and principles.* London: Routledge.

Clarke, C. J., & Neidert, L. J. (1992). Living arrangements of the elderly: An examination of differences according to ancestry and generation. *Gerontologist, 32,* 796–804.

Clarke-Stewart, K. A. (1989). Infant day care: Maligned or malignant? *American Psychologist, 44,* 266–273.

Clausen, J. A. (1975). The social meaning of differential physical and sexual maturation. In S. E. Dragastin & G. H. Elder (Eds.), *Adolescence in the life cycle: Psychological change and the social context* (pp. 25–47). New York: Halsted.

Cleveland, J., & Shore, L. (1992). Self- and supervisory perspectives on age and work attitudes and performance. *Journal of Applied Psychology, 77,* 469–484.

Clifford, C., & Kramer, B. (1993). Diet as risk and therapy for cancer. *Medical Clinics of North America, 77,* 725–744.

Clifton, R. K., Muir, D. W., Ashmead, D. H., & Clarkson, M. G. (1993). Is visually guided reaching in early infancy a myth? *Child Development, 64,* 1099–1110.

Coakley, J. (1990). *Sport and society: Issues and controversies* (4th ed.). St. Louis: Mosby.

Coatney, C. (1997, January 8). Australia diver's world's first dose of legal euthanasia. *The Christian Science Monitor,* pp. 1, 18.

Cohen, D., Eisdorfer, C., Gorelick, P., Pavesa, G., Luchins, D. J., Freels, S., Ashford, J. W., Semla, T., Levy, P., & Hirschman, R. (1993). Psycho-pathology associated with Alzheimer's disease and related disorders. *Journal of Gerontology, 48,* M255–M260.

Cohen, F. L. (1984). *Clinical genetics in nursing practice.* Philadelphia: Lippincott.

Cohen, F. L. (1993a). Epidemiology of HIV infection and AIDS in children. In F. L. Cohen & J. D. Durham (Eds.), *Women, children, and HIV/AIDS* (pp. 137–155). New York: Springer.

Cohen, F. L. (1993b). HIV infection and AIDS: An overview. In F. L. Cohen & J. D. Durham (Eds.), *Women, children, and HIV/AIDS* (pp. 3–30). New York: Springer.

Cohen, G. D. (1990). Psychopathology and mental health in the mature and elderly adult. In J. E. Birren & K. W. Schaie (Eds.), *Handbook of the psychology of aging* (3rd ed., pp. 359–371). San Diego: Academic Press.

Cohen, S., & Williamson, G. M. (1991). Stress and infectious disease in humans. *Psychological Bulletin, 109,* 5–24.

Coie, J. D., Dodge, K. A., & Coppotelli, H. (1982). Dimensions and types of social status: A cross-age perspective. *Developmental Psychology, 18,* 557–570.

Coie, J. D., & Kreihbel, G. (1984). Effects of academic tutoring on the social status of low-achieving, socially rejected children. *Child Development, 55,* 1465–1478.

Coke, M. M. (1992). Correlates of life satisfaction among elderly African Americans. *Journal of Gerontology, 47,* P316–P320.

Colby, A., Kohlberg, L., Gibbs, J., & Lieberman, M. (1983). A longitudinal study of moral judgment. *Monographs of the Society for Research in Child Development, 48*(1–2, Serial No. 200).

Cole, E. R., & Stewart, A. J. (1996). Meanings of political participation among black and white women: Political identity and social responsibility. *Journal of Personality and Social Psychology, 71,* 130–140.

Cole, M. (1990). Cognitive development and formal schooling: The evidence from cross-cultural research. In L. C. Moll (Ed.), *Vygotsky and education* (pp. 89–110). New York: Cambridge University Press.

Collea, J. V., Chein, C., & Quilligan, E. J. (1980). The randomized management of term frank breech presentations: A study of 208 cases. *American Journal of Obstetrics and Gynecology, 137,* 235–244.

Collins, J. A. (1994). Reproductive technology—The price of progress. *New England Journal of Medicine, 331,* 270–271.

Collins, N. L., & Read, S. J. (1990). Adult attachment, working models, and relationship quality in dating couples. *Journal of Personality and Social Psychology, 58,* 644–663.

Collins, W. A., Wellman, H., Keniston, A. H., & Westby, S. D. (1978). Age-related aspects of comprehension and inference from a televised dramatic narrative. *Child Development, 49,* 389–399.

Colman, L. L., & Colman, A. D. (1991). *Pregnancy: The psychological experience.* New York: Noonday Press.

Colombo, J. (1995). On the neural mechanism underlying developmental and individual differences in visual fixation in infancy. *Developmental Review, 15,* 97–135.

Coltrane, S. (1990). Birth timing and the division of labor in dual-earner families. *Journal of Family Issues, 11,* 157–181.

Comstock, G. A. (1993). The medium and society: The role of television in American life. In G. L. Berry & J. K. Asamen (Eds.), *Children and television* (pp. 117–131). Newbury Park, CA: Sage.

Condie, S. J. (1989). Older married couples. In S. J. Bahr & E. T. Peterson (Eds.), *Aging and the family* (pp. 143–158). Lexington, MA: Lexington Books.

Condry, J., & Ross, D. F. (1985). Sex and aggression: The influence of gender label on the perceptions of aggression in children. *Child Development, 56,* 225–233.

Conel, J. L. (1959). *The postnatal development of the human cerebral cortex.* Cambridge, MA: Harvard University Press.

Conger, R. D., Conger, K. J., Elder, G. H., Jr., Lorenz, F. O., Simons, R. L., & Whitbeck, L. B. (1992). A family process model of economic hardship and adjustment of early adolescent boys. *Child Development, 63,* 527–541.

Connell, J. P., Spencer, M. B., & Aber, J. L. (1994). Educational risk and resilience in African-American youth: Context, self, action, and outcomes in school. *Child Development, 65,* 493–503.

Connidis, I. A. (1989). Siblings as friends in later life. *American Behavioral Scientist, 33,* 81–93.

Connidis, I. A. (1994). Sibling support in older age. *Journal of Gerontology, 49,* S309–S317.

Connidis, I. A., & Campbell, L. D. (1995). Closeness, confiding, and contact among siblings in middle and late adulthood. *Journal of Family Issues, 16,* 722–745.

Connidis, I. A., & McMullin, J. A. (1993). To have or have not: Parent status and the subjective well-being of older men and women. *The Gerontologist, 33,* 630–636.

Connolly, J. A., & Doyle, A. B. (1984). Relations of social fantasy play to social competence in preschoolers. *Developmental Psychology, 20,* 797–806.

Connolly, J. A., Doyle, A. B., & Reznick, E. (1988). Social pretend play and social interaction in preschoolers. *Journal of Applied Developmental Psychology, 9,* 301–313.

Constanzo, P. R., & Woody, E. Z. (1979). Externality as a function of obesity in children: Pervasive style or eating-specific attribute? *Journal of Personality and Social Psychology, 37,* 2286–2296.

Cooke, R. A. (1982). The ethics and regulation of research involving children. In B. B. Wolman (Ed.), *Handbook of developmental psychology* (pp. 149–172). Englewood Cliffs, NJ: Prentice-Hall.

Cooney, T. M., & Uhlenberg, P. (1990). The role of divorce in men's relations with their adult children after mid-life. *Journal of Marriage and the Family, 52,* 677–688.

Cooper, C. L., Cooper, R. D., & Eaker, L. (1988). *Living with stress.* New York: Penguin.

Cooper, R. P., & Aslin, R. N. (1990). Preference for infant-directed speech in the first month after birth. *Child Development, 61,* 1584–1595.

Cooper, R. P., & Aslin, R. N. (1994). Developmental differences in infant attention to the spectral properties of infant-directed speech. *Child Development, 65,* 1663–1677.

Cooperative Human Linkage Center (1996). A gene map of the human genome. *Science, 274,* 540–546.

Coplan, R. J., Rubin, K. H., Fox, N. A., Calkins, S. D., & Stewart, S. L. (1994). Being alone, playing alone, and acting alone: Distinguishing among reticence and passive and active solitude in young children. *Child Development, 65,* 129–137.

Copper, R. L., Goldenberg, R. L., Creasy, R. K., DuBard, M. B., Davis, R. O., Entman, S. S., Iams, J. D., & Cliver, S. P. (1993). A multicenter study of preterm birthweight and gestational age-specific neonatal mortality. *American Journal of Obstetrics and Gynecology, 168,* 78–84.

Coren, S., & Halpern, D. F. (1991). Left-handedness: A marker for decreased survival fitness. *Psychological Bulletin, 109,* 90–106.

Cornelius, S. W., & Caspi, A. (1987). Everyday problem solving in adulthood and old age. *Psychology and Aging, 2,* 144–153.

Cornell, E. H., & Gottfried, A. W. (1976). Intervention with premature human infants. *Child Development, 47,* 32–39.

Corpas, E., Harman, S. M., & Blackman, M. R. (1993). Human growth hormone and human aging. *Endocrine Reviews, 14,* 20–39.

Corr, C. A. (1993). Coping with dying: Lessons that we should and should not learn from the work of Elisabeth Kübler-Ross. *Death Studies, 17,* 69–83.

Corr, C. A. (1995). Entering into adolescent understandings of death. In E. A. Grollman (Eds.), *Bereaved children and teens* (pp. 21–35). Boston: Beacon Press.

Corr, C. A., & Balk, D. E. (1996). *Handbook of adolescent death and bereavement.* New York: Springer.

Corr, C. A., & Corr, D. M. (1992). Adult hospice day care. *Death Studies, 16,* 155–172.

Corrigan, R. (1987). A developmental sequence of actor–object pretend play in young children. *Merrill-Palmer Quarterly, 33,* 87–106.

Corwin, M. J., Lester, B. M., Sepkoski, C., Peucker, M., Kayne, H., & Golub, H. L. (1995). Newborn acoustic cry characteristics of infants subsequently dying of sudden infant death syndrome. *Pediatrics, 96,* 73–77.

Costa, P. T., Jr., & McCrae, R. R. (1994). Set like plaster? Evidence for the stability of adult personality. In T. F. Heatherton & J. L. Weinberger (Eds.), *Can personality change?* (pp. 21–40). Washington, DC: American Psychological Association.

Cotman, C. W., & Holets, V. K. (1985). Structural changes at synapses with age: Plasticity and regeneration. In C. E. Finch & E. L. Schneider (Eds.), *Handbook of the biology of aging* (2nd ed., pp. 617–644). New York: Van Nostrand Reinhold.

Cotton, P. (1990). Sudden infant death syndrome: Another hypothesis offered but doubts remain. *Journal of the American Medical Association, 263,* 2865, 2869.

Cotton, P. (1994). Smoking cigarettes may do developing fetus more harm than ingesting cocaine, some experts say. *Journal of the American Medical Association, 271,* 576–577.

Coulton, C. J., Korbin, J. E., Su, M., & Chow, J. (1995). Community level factors and child maltreatment rates. *Child Development, 66,* 1262–1276.

Courage, M. L., & Adams, R. J. (1990). Visual acuity assessment from birth to three years using the acuity card procedures: Cross-sectional and longitudinal samples. *Optometry and Vision Science, 67,* 713–718.

Cowan, C. P., & Cowan, P. A. (1988). Changes in marriage during the transition to parenthood: Must we blame the baby? In G. Y. Michaels & W. A. Goldberg (Eds.), *The transition to parenthood* (pp. 114–154). New York: Cambridge University Press.

Cowan, C. P., & Cowan, P. A. (1992). *When partners become parents.* New York: Basic Books.

Coward, R. T., Lee, G. R., & Dwyer, J. W. (1993). The family relations of rural elders. In C. N. Bull (Ed.), *Aging in rural America* (pp. 216–231). Newbury Park, CA: Sage.

Cox, D. R., Green, E. D., Lander, E. S., Cohen, D., & Myers, R. M. (1994), Assessing mapping progress in the Human Genome Project. *Science, 265,* 2031.

Cox, K., & Schwartz, J. D. (1990). *The well-informed patient's guide to caesarean births.* New York: Dell.

Cox, M. J., Owen, M. T., Henderson, V. K., & Margand, N. A. (1992). Prediction of infant–father and infant–mother attachment. *Developmental Psychology, 28,* 474–483.

Craik, F. I. M., & Jacoby, L. L. (1996). Aging and memory: Implications for skilled performance. In W. A. Rogers, A. D. Fisk, & N. Walker (Eds.), *Aging and skilled performance* (pp. 113–137). Mahwah, NJ: Erlbaum.

Craik, F. I. M., & Jennings, J. M. (1992). Human memory. In F. I. M. Craik & T. A. Salthouse (Eds.), *Handbook of aging and cognition* (pp. 51–110). Hillsdale, NJ: Erlbaum.

Craik, F. I. M., & Lockhart, R. S. (1972). Levels of processing: A framework for memory research. *Journal of Verbal Learning and Verbal Behavior, 11,* 671–684.

Crain-Thoreson, C., & Dale, P. S. (1992). Do early talkers become early readers? Linguistic precocity, preschool language, and emergent literacy. *Developmental Psychology, 28,* 421–429.

Cramond, B. (1994). The Torrance Tests of Creative Thinking: From design through establishment of predictive validity. In R. F. Subotnik & K. D. Arnold (Eds.), *Beyond Terman: Contemporary longitudinal studies of giftedness and talent* (pp. 229–254). Norwood, NJ: Ablex.

Cratty, B. J. (1986). *Perceptual and motor development in infants and children* (3rd ed.), Englewood Cliffs, NJ: Prentice-Hall.

Creatsas, G. K., Vekemans, M., Horejsi, J., Uzel, R., Lauritzen, C., & Osler, M. (1995). Adolescent sexuality in Europe a multicentric study. *Adolescent and Pediatric Gynecology, 8,* 59–63.

Crews, J. E. (1994). The demographic, social, and conceptual contexts of aging and vision loss. *Journal of the American Optometric Association, 65,* 63–68.

Crick, N. R., & Grotpeter, J. K. (1995). Relational aggression, gender, and social-psychological adjustment. *Child Development, 66,* 710–722.

Crick, N. R., & Ladd, G. W. (1993). Children's perceptions of their peer experiences: Attributions, loneliness, social anxiety, and social avoidance. *Developmental Psychology, 29,* 244–254.

Crockett, L. J. (1990). Sex role and sex-typing in adolescence. In R. M. Lerner, A. C. Petersen, & J. Brooks-Gunn (Eds.), *The encyclopedia of adolescence* (Vol. 2, pp. 1007–1017). New York: Garland.

Crohan, S. E., & Antonucci, T. C. (1989). Friends as a source of social support in old age. In R. G. Adams & R. Blieszner (Eds.), *Older adult friendship* (pp. 129–146). Newbury Park, CA: Sage.

Cross, S., & Markus, H. (1991). Possible selves across the life span. *Human Development, 34,* 230–255.

Crouter, A. C., Manke, B. A., & McHale, S. M. (1995). The family context of gender intensification in early adolescence. *Child Development, 66,* 317–329.

Crystal, D. S., Chen, C., Fuligni, A. J., Stevenson, H. W., Hsu, C-C., Ko, H-J., Kitamura, S., & Kimura, S. (1994). Psychological maladjustment and academic achievement: A cross-cultural study of Japanese, Chinese, and American high school students. *Child Development, 65,* 738–753.

Csikszentmihalyi, M., & Larson, R. (1984). *Being adolescent: Conflict and growth in the teenage years.* New York: Basic Books.

Csikszentmihalyi, M., & Rathunde, K. (1990). The psychology of wisdom: An evolutionary interpretation. In R. J. Sternberg (Ed.), *Wisdom: Its nature, origins, and development* (pp. 25–51). New York: Cambridge University Press.

Culbertson, F. M. (1997). Depression and gender. *American Psychologist, 52,* 25–31.

Cumming, E., & Henry, W. E. (1961). *Growing old: The process of disengagement.* New York: Basic Books.

Cummings, E. M., Iannotti, R. J., & Zahn-Waxler, C. (1985). Influence of conflict between adults on the emotions and aggression of young children. *Developmental Psychology, 21,* 495–507.

Cunningham, J. D., & Antill, J. K. (1994). Cohabitation and marriage: Retrospective and predictive comparisons. *Journal of Social and Personal Relationships, 11,* 77–93.

Cutler, S. J., & Hendricks, J. (1990). Leisure and time use across the life course. In R. Binstock & L. George (Eds.), *Aging and the social sciences* (3rd ed., pp. 169–185). New York: Academic Press.

Cytron, B. D. (1993). To honor the dead and comfort the mourners: Traditions in Judaism. In D. P. Irish, K. F. Lundquist, & V. J. Nelson (Eds.), *Ethnic variations in dying, death, and grief* (pp. 113–124). Washington, DC: Taylor & Francis.

Czeizel, A. E., & Dudas, I. (1992). Prevention of the first occurrence of neural tube defects by periconceptional vitamin supplementation. *New England Journal of Medicine, 327,* 1832-1835.

Dacey, J. S. (1989). Peak periods of creative growth across the life span. *Journal of Creative Behavior, 23,* 224–248.

Dahl, R.E., Scher, M. S., Williamson, D. E., Robles, N., & Day, N. (1995). A longitudinal study of prenatal marijuana use: Effects on sleep and arousal at age 3 years. *Archives of Pediatric and Adolescent Medicine, 149,* 145-150.

Dahlberg, G. (1994). The parent–child relationship and socialization in the context of modern childhood: The case of Sweden. In J. L. Roopnarine & D. B. Carter (Eds.), *Parent–child socialization in diverse cultures* (pp. 121–137). Norwood, NJ: Ablex.

Dalton, S. T. (1992). Lived experience of never-married women. *Issues in Mental Health Nursing, 13,* 69–80.

Damon, W. (1977). *The social world of the child.* San Francisco: Jossey-Bass.

Damon, W. (1988). *The moral child.* New York: Free Press.

Damon, W. (1990). Self-concept, adolescent. In R. M. Lerner, A. C. Petersen, & J. Brooks-Gunn (Eds.), *The encyclopedia of adolescence* (Vol. 2, pp. 87–91). New York: Garland.

Damon, W., & Hart, D. (1988). *Self-understanding in childhood and adolescence.* New York: Cambridge University Press.

Danforth, J. S., Barkley, R. A., & Stokes, T. F. (1990). Observations of parent–child interactions with hyperactive children: Research and clinical applications. *Clinical Psychology Review, 11,* 703–727.

Dannemiller, J. L., & Stephens, B. R. (1988). A critical test of infant pattern preference models. *Child Development, 59,* 210–216.

Danziger, S., & Danziger, S. (1993). Child poverty and public policy: Toward a comprehensive antipoverty agenda. *Daedalus, 122,* 57–84.

Darwin, C. (1936). *On the origin of species by means of natural selection.* New York: Modern Library. (Original work published 1859)

Daugirdas, J. T. (1992). *Sexually transmitted diseases.* Hinsdale, IL: Medtext.

DaVanzo, J., & Goldscheider, F. K. (1990). Coming home again: Returns to the parental home of young adults. *Population Studies, 44,* 241–255.

Davies, E. B. (1987). After a sibling dies. In M. A. Morgan (Ed.), *Bereavement: Helping the survivors. Proceedings of the 1987 King's College Conference.* London, Ontario: King's College.

Davis, A. J., Davidson, B., Hirschfield, M., Lauri, S., Lin, J. Y., Phillips, L., Pitman, E., Hui Shen, C., Laan, R. V., Zhang, H. L., & Ziv, L. (1993). An international perspective of active euthanasia: Attitudes of nurses in seven countries. *International Journal of Nursing Studies, 30,* 301–310.

Davis, B., & Aron, A. (1989). Perceived causes of divorce and postdivorce adjustment among recently divorced midlife women. *Journal of Divorce, 12,* 41–55.

Dawkins, M. P. (1983). Black students' occupational expectations: A national study of the impact of school desegregation. *Urban Education, 18,* 98–113.

Day, S. (1993, May). Why genes have a gender. *New Scientist, 138* (1874), 34–38.

Deacon, S., Minichiello, V., & Plummer, D. (1995). Sexuality and older people: Revisiting the assumptions. *Educational Gerontology, 21,* 497–513.

Deary, I. J. (1995). Auditory inspection time and intelligence: What is the direction of causation? *Developmental Psychology, 31,* 237–250.

DeCasper, A. J., & Spence, M. J. (1986). Prenatal maternal speech influences newborns' perception of speech sounds. *Infant Behavior and Development, 9,* 133–150.

Declercq, E. R. (1992). The transformation of American midwifery: 1975 to 1988. *American Journal of Public Health, 82,* 680–684.

Deffenbacher, J. L. (1994). Anger reduction: Issues, assessment, and intervention strategies. In A. W. Siegman & T. W. Smith (Eds.), *Anger, hostility, and the heart* (pp. 239–269). Hillsdale, NJ: Erlbaum.

DeGenova, M. K. (1992). If you had your life to live over again: What would you do differently? *International Journal of Aging and Human Development, 34,* 135–143.

Dekovic, M., & Gerris, J. R. M. (1994). Developmental analysis of social cognitive and behavioral differences between popular and rejected children. *Journal of Applied Developmental Psychology, 15,* 367–386.

De Lisi, R., & Gallagher, A. M. (1991). Understanding gender stability and constancy in Argentinean children. *Merrill-Palmer Quarterly, 37,* 483–502.

DeLoache, J. S., & Todd, C. M. (1988). Young children's use of spatial categorization as a mnemonic strategy. *Journal of Experimental Child Psychology, 46,* 1–20.

DeMarie-Dreblow, D. (1991). Relation between knowledge and memory: A reminder that correlation does not imply causality. *Child Development, 62,* 484–498.

Dembroski, T. M., MacDougall, J. M., Costa, P. T., & Grandits, G. A. (1989). Components of hostility as predictors of sudden death and myocardial infarction in the Multiple Risk Factor Intervention Trial. *Psychosomatic Medicine, 51,* 514–522.

Demetriou, A., Efklides, A., Papadaki, M., Papantoniou, G., & Economou, A. (1993). Structure and development of causal–experimental thought: From early adolescence to youth. *Developmental Psychology, 29,* 480–497.

Demetriou, A., Efklides, A., & Platsidou, M. (1993). The architecture and dynamics of developing mind. *Monographs of the Society for Research in Child Development, 58* (No. 5–6, Serial No. 234).

Denckla, M. B. (1991). Attention-deficit hyperactivity disorder—residual type. *Journal of Child Neurology, 6,* S44–S50.

Denham, S. A., Renwick, S. M., & Holt, R. W. (1991). Working and playing together: Prediction of preschool social-emotional competence from mother-child interaction. *Child Development, 62,* 242–249.

Denney, N. W. (1990). Adult age differences in traditional and practical problem solving. *Advances in Psychology, 72,* 329–349.

Denney, N. W., & Pearce, K A. (1989). A developmental study of practical problem solving in adults. *Psychology and Aging, 4,* 438–442.

Dennis, W. (1960). Causes of retardation among institutionalized children: Iran. *Journal of Genetic Psychology, 96,* 47–59.

Denollet, J., Sys, S. U., & Brutsaert, D. L. (1995). Personality and mortality after myocardial infarction. *Psychosomatic Medicine, 57,* 582–591.

Denollet, J., Sys, S. U., Stroobant, N., Rombouts, H., Gillebert, T. C., & Brutsaert, D. L. (1996). Personality as independent predictor of long-term mortality in patients with coronary heart disease. *Lancet, 347,* 417–421.

Derdeyn, A. P. (1985). Grandparent visitation rights: Rendering family dissension more pronounced? *American Journal of Orthopsychiatry, 55,* 277–287.

DeRosier, M. E., Kupersmidt, J. B., & Patterson, C. J. (1994). Children's academic and behavioral adjustment as a function of the chronicity and proximity of peer rejection. *Child Development, 65,* 1799–1813.

Deutsch, F. M., Ruble, D. N., Fleming, A., Brooks-Gunn, J., & Stangor, C. (1988). Information-seeking and maternal self-definition during the transition to motherhood. *Journal of Personality and Social Psychology, 55,* 420–431.

Deutsch, W., & Pechmann, T. (1982). Social interaction and the development of definite descriptions. *Cognition, 11,* 159–184.

Deveson, A. (1994). *Coming of age: Twenty-one interviews about growing older.* Newham, Australia: Scribe.

de Villiers, J. G., & de Villiers, P. A. (1973). A cross-sectional study of the acquisition of grammatical morphemes in child speech. *Journal of Psycholinguistic Research, 2,* 267–278.

de Villiers, P. A., & de Villiers, J. G. (1992). Language development. In M. H. Bornstein & M. E. Lamb (Eds.), *Developmental psychology: An advanced textbook* (3rd ed., pp. 337–418). Hillsdale, NJ: Erlbaum.

Devons, C. A. J. (1996). Suicide in the elderly: How to identify and treat patients at risk. *Geriatrics, 51*(3), 67–72.

de Vries, B., Bluck, S., & Birren, J. E. (1993). The understanding of death and dying in a life-span perspective. *Gerontologist, 33,* 366–372.

Diamond, A., Cruttenden, L., & Neiderman, D. (1994). AB with multiple wells: 1. Why are multiple wells sometimes easier than two wells? 2. Memory or memory + inhibition. *Developmental Psychology, 30,* 192–205.

Dickinson, D. K. (1984). First impressions: Children's knowledge of words gained from a single exposure. *Applied Psycholinguistics, 5,* 359–373.

Dick-Read, G. (1959). *Childbirth without fear.* New York: Harper & Row.

DiClemente, R. J. (1993). Preventing HIV/AIDS among adolescents. *Journal of the American Medical Association, 270,* 760–762.

Diekstra, R. F. W., Kienhorst, C. W. M., & de Wilde, E. J. (1995). Suicide and suicidal behaviour among adolescents. In M. Rutter & D. J. Smith (Eds.), *Psychosocial disorders in young people* (pp. 686–761). Chicester: Wiley.

Dietz, W. H., Jr., Bandini, L. G., & Gortmaker, S. (1990). Epidemiologic and metabolic risk factors for childhood obesity. *Klinische Pädiatrie, 202,* 69–72.

DiLalla, L. F., Kagan, J., & Reznick, J. S. (1994). Genetic etiology of behavioral inhibition among 2-year-old children. *Infant Behavior and Development, 17,* 405–412.

DiNitto, D. M. (1995). *Social welfare: Politics and public policy.* Boston: Allyn & Bacon.

Dion, K. L., & Dion, K. K. (1988). Romantic love: Individual and cultural perspectives. In R. J. Sternberg & M. L. Barnes (Eds.), *The psychology of love* (pp. 264–292). New Haven, CT: Yale University Press.

Dion, K. L., & Dion, K. K. (1993). Gender and ethnocultural comparisons in styles of love. *Journal of Social Issues, 49,* 53–69.

Dirks, J. (1982). The effect of a commercial game on children's Block Design scores on the WISC–R test. *Intelligence, 6,* 109–123.

Dittrichova, J., Brichacek, V., Paul, K., & Tautermannova, M. (1982). The structure of infant behavior: An analysis of sleep and waking in the first months of life. In W. W. Hartup (Ed.), *Review of child development research* (Vol. 6, pp. 73–100). Chicago: University of Chicago Press.

Dixon, J. A., & Moore, C. F. (1990). The development of perspective taking: Understanding differences in information and weighting. *Child Development, 61,* 1502–1513.

Dixon, R. A., & Lerner, R. M. (1992). A history of systems in developmental psychology. In M. H. Bornstein & M. E. Lamb (Eds.), *Developmental psychology: An advanced textbook* (3rd ed., pp. 3–58). Hillsdale, NJ: Erlbaum.

Dlugosz, L., & Bracken, M. B. (1992). Reproductive effects of caffeine: A review and theoretical analysis. *Epidemiological Review, 14,* 83–100.

Dobbs, A. R., & Rule, B. G. (1987). Prospective memory and self-reports of memory abilities in older adults. *Canadian Journal of Psychology, 41,* 209–222.

Dobkin, P. L., Tremblay, R. E., Másse, L. C., & Vitaro, F. (1995). Individual and peer characteristics in predicting boys' early onset of substance abuse: A seven-year longitudinal study. *Child Development, 66,* 1198–1214.

Dodge, K. A., Pettit, G. S., & Bates, J. E. (1994). Socialization mediators of the relation between socioeconomic status and child conduct problems. *Child Development, 65,* 649–655.

Dodge, K. A., & Somberg, D. R. (1987). Hostile attributional biases among aggressive boys are exacerbated under conditions of threats to the self. *Child Development, 58,* 213–224.

Dodwell, P. C., Humphrey, G. K., & Muir, D. W. (1987). Shape and pattern perception. In P. Salapatek & L. Cohen (Eds.), *Handbook of infant perception* (Vol. 2, pp. 1–77). Orlando, FL: Academic Press.

Doherty, W. J., & Needle, R. H. (1991). Psychological adjustment and substance use among adolescents before and after parental divorce. *Child Development, 62,* 328–337.

Doi, L. T. (1973). *The anatomy of dependence.* Tokyo: Kodansha International.

Doi, M. (1991). A transformation of ritual: The Nisei 60th birthday. *Journal of Cross-Cultural Gerontology, 6,* 153–161.

Doka, K. J. (1995). Coping with life-threatening illness: A task model. *Omega, 32,* 111–122.

Doka, K. J., & Mertz, M. E. (1988). The meaning and significance of great-grandparenthood. *Gerontologist, 28,* 192–197.

Dolgin, K. G., & Behrend, D. A. (1984). Children's knowledge about animates and inanimates. *Child Development, 55,* 1646–1650.

Donatelle, R. J., & Davis, L. G. (1997). *Health: The basics* (2nd ed.). Englewood Cliffs, NJ: Prentice Hall.

Donnerstein, E., Slaby, R. G., & Eron, L. D. (1994). The mass media and youth aggression. In L. D. Eron, J. H. Gentry, & P. Schlegel (Eds.), *Reason to hope: A psychosocial perspective on violence and youth* (pp. 219–250). Washington, DC: American Psychological Association.

Dontas, C., Maratsos, O., Fafoutis, M., & Karangelis, A. (1985). Early social development in institutionally reared Greek infants: Attachment and peer interaction. In I. Bretherton & E. Waters (Eds.), Growing points of attachment theory and research. *Monographs of the Society for Research in Child Development, 50* (1–2, Serial No. 209).

Doress-Worters, P. B., & Siegal, D. L. (1994). *The new ourselves, growing older.* New York: Touchstone.

Dorfman, L. T., Heckert, D. A., Hill, E. A., & Kohout, F. J. (1988). Retirement satisfaction in rural husbands and wives. *Rural Sociology, 53,* 25–39.

Dornbusch, S. Carlsmith, J., Gross, R., Martin, J., Jennings, D., Rosenberg, A., & Duke, P. (1981). Sexual development, age, and dating: A comparison of biological and social influences upon one set of behaviors. *Child Development, 52,* 179–185.

Dornbusch, S. M., Carlsmith, J. M., Bushwall, S. J., Ritter, P. L., Leiderman, H., Hastorf, A. H., & Gross, R. T. (1985). Single parents, extended households, and the control of adolescents. *Child Development, 56,* 326–341.

Dornbusch, S. M., Glasgow, K. L., & Lin, I-C. (1996). The social structure of schooling. *Annual Review of Psychology, 47,* 401–429.

Dornbusch, S. M., Ritter, P. L., Liederman, P. H., Roberts, D. F., & Fraleigh, M. J. (1987). The relation of parenting style to adolescent school performance. *Child Development, 58,* 1244–1257.

Dornbusch, S. M., Ritter, P. L., Mont-Reynaud, R., & Chen, Z. (1990). Family decision making and academic performance in a diverse high school population. *Journal of Adolescent Research, 5,* 143–160.

Dorris, M. (1989). *The broken cord.* New York: Harper & Row.

Doty, P. J. (1992). The oldest old and the use of institutional long-term care from an international perspective. In R. M. Suzman, D. P. Willis, & K. G. Manton (Eds.), *The oldest old* (pp. 251–267). New York: Oxford University Press.

Douvan, E., & Adelson, J. (1966). *The adolescent experience.* New York: Wiley.

Doyle, C. (1994). *Child sexual abuse.* London: Chapman & Hall.

Drabman, R. S., Cordua, G. D., Hammer, D., Jarvie, G. J., & Horton, W. (1979). Developmental trends in eating rates of normal and overweight preschool children. *Child Development, 50,* 211–216.

Droege, K. L., & Stipek, D. J. (1993). Children's use of dispositions to predict classmates' behavior. *Developmental Psychology, 29,* 646–654.

Drotar, D. (1992). Personality development, problem solving, and behavior problems among preschool children with early histories of nonorganic failure-to-thrive: A controlled study. *Developmental and Behavioral Pediatrics, 13,* 266–273.

Drotar, D., & Sturm, L. (1988). Prediction of intellectual development in young children with early histories of nonorganic failure-to-thrive. *Journal of Pediatric Psychology, 13,* 281–296.

Dubas, J. S., & Petersen, A. C. (1993). Female pubertal development. In M. Sugar (Ed.), *Female adolescent development* (2nd ed., pp. 3–26). New York: Brunner/Mazel.

DuBois, D. L., & Hirsch, B. J. (1990). School and neighborhood friendship patterns of blacks and whites in early adolescence. *Child Development, 61,* 524–536.

Duck, S. (1994). *Meaningful relationships.* Thousand Oaks, CA: Sage.

Duncan, G. J., Brooks-Gunn, J., & Klebanov, P. K. (1994). Economic deprivation and early childhood development. *Child Development, 65,* 296–318.

Dunham, P., & Dunham, F. (1992). Lexical development during middle infancy: A mutually driven infant–caregiver process. *Developmental Psychology, 28,* 414–420.

Dunham, P. J., Dunham, F., & Curwin, A. (1993). Joint-attentional states and lexical acquisition at 18 months. *Developmental Psychology, 29,* 827–831.

Dunn, J. (1989). Siblings and the development of social understanding in early childhood. In P. G. Zukow (Ed.), *Sibling interaction across cultures* (pp. 106–116). New York: Springer-Verlag.

Dunn, J. (1992). Sisters and brothers: Current issues in developmental research. In F. Boer & J. Dunn (Eds.), *Children's sibling relationships* (pp. 1–17). Hillsdale, NJ: Erlbaum.

Dunn, J. (1994). Temperament, siblings, and the development of relationships. In W. B. Carey & S. C. McDevitt (Eds.), *Prevention and early intervention* (pp. 50–58). New York: Brunner/Mazel.

Dunn, J., & Kendrick, C. (1982). *Siblings: Love, envy and understanding.* Cambridge, MA: Harvard University Press.

Dunn, J., Slomkowski, C., & Beardsall, L. (1994). Sibling relationships from the preschool period through middle childhood and early adolescence. *Developmental Psychology, 30,* 315–324.

Dunn, J. T. (1993). Iodine supplementation and the prevention of cretinism. *Annals of the New York Academy of Sciences, 678,* 158–168.

Durbin, D. L., Darling, N., Steinberg, L., & Brown, B. B. (1993). Parenting style and peer group membership among European-American adolescents. *Journal of Research on Adolescence, 3,* 87–100.

Durlak, J. A., & Riesenberg, L. A. (1991). The impact of death education. *Death Studies, 15,* 39–58.

Dusek, J. B. (1987). Sex roles and adjustment. In D. B. Carter (Ed.), *Current conceptions of sex roles and sex typing* (pp. 211–222). New York: Praeger.

Duvall, E. M. (1977). *Marriage and family development* (5th ed.). New York: Harper & Row.

Duxbury, L., & Higgins, C. (1994). Interference between work and family: A status report on dual-career and dual-earner mothers and fathers. *Employee Assistance Quarterly, 9,* 55–80.

Dweck, C. S., Davidson, W., Nelson, S., & Enna, B. (1978). Sex differences in learned helplessness: III. An experimental analysis. *Developmental Psychology, 14,* 268–276.

Dworkin, R. (1997, September 25). Assisted suicide: What the court really said. *New York Review of Books,* pp. 40–44.

Dye-White, E. (1986). Environmental hazards in the work setting: Their effect on women of child-bearing age. *American Association of Occupational Health and Nursing Journal, 34,* 76–78.

Dykstra, P. A. (1995). Loneliness among the never and formerly married: The importance of supportive friendships and a desire for independence. *Journal of Gerontology, 50B,* S321–S329.

Dyson, A. H. (1984). Emerging alphabetic literacy in school contexts: Toward defining the gap between school curriculum and child mind. *Written Communication, 1,* 5–55.

Eades, M. D. (1992). *Arthritis: Reducing your risk.* New York: Bantam.

Ebeling, K. S., & Gelman, S. A. (1994). Children's use of context in interpreting "big" and "little." *Child Development, 65,* 1178–1192.

Eberhart-Phillips, J. E., Frederick, P. D., & Baron, R. C. (1993). Measles in pregnancy: A descriptive study of 58 cases. *Obstetrics and Gynecology, 82,* 797–801.

Eccles, J. (1987). Adolescence: Gateway to gender-role transcendence. In D. B.

Carter (Ed.), *Current conceptions of sex roles and sex typing: Theory and research* (pp. 225–241). New York: Praeger.

Eccles, J. S. (1990). Academic achievement. In R. M. Lerner, A. C. Petersen, & J. Brooks-Gunn (Eds.), *The encyclopedia of adolescence* (pp. 1–5). New York: Garland.

Eccles, J. S., & Harold, R. D. (1991). Gender differences in sport involvement: Applying the Eccles' expectancy-value model. *Journal of Applied Sport Psychology, 3,* 7–35.

Eccles, J. S., Jacobs, J. E., & Harold, R. D. (1990). Gender-role stereotypes, expectancy effects, and parents' role in the socialization of gender differences in self-perceptions and skill acquisition. *Journal of Social Issues, 46,* 183–201.

Eccles, J. S., Midgley, C., Wigfield, A., Buchanan, C. M., Reuman, D., Flanagan, C., & Mac Iver, D. (1993a). Development during adolescence: The impact of stage–environment fit on young adolescents' experiences in schools and in families. *American Psychologist, 48,* 90–101.

Eccles, J. S., Wigfield, A., Midgley, C., Reuman, D., Mac Iver, D., & Feldlaufer, H. (1993b). Negative effects of traditional middle schools on students' motivation. *Elementary School Journal, 93,* 553–574.

Eckenrode, J., Laird, M., & Doris, J. (1993). School performance and disciplinary problems among abused and neglected children. *Developmental Psychology, 29,* 53–62.

Edelstein, B. L., & Douglass, C. W. (1995). Dispelling the myth that 50 percent of U.S. schoolchildren have never had a cavity. *Public Health Reports, 110,* 522–530.

Eder, R. A. (1989). The emergent personologist: The structure and content of 3½, 5½, and 7½-year-olds' concepts of themselves and other persons. *Child Development, 60,* 1218–1228.

Eder, R. A. (1990). Uncovering young children's psychological selves: Individual and developmental differences. *Child Development, 61,* 849–863.

Edwards, J. N. (1991). New conceptions: Biosocial innovations and the family. *Journal of Marriage and the Family, 53,* 349–360.

Edwards, J. N., & Booth, A. (1994). Sexuality, marriage, and well-being: The middle years. In A. S. Rossi (Ed.), *Sexuality across the life course* (pp. 233–259). Chicago: University of Chicago Press.

Edwards, M., Cangemi, J. P., & Kowalski,

C. J. (1990). The college dropout and institutional responsibility. *Education, 111,* 107–116.

Egeland, B., & Hiester, M. (1995). The long-term consequences of infant day-care and mother–infant attachment. *Child Development, 66,* 474–485.

Egeland, B., Jacobvitz, D., & Sroufe, L. A. (1988). Breaking the cycle of abuse. *Child Development, 59,* 1080–1088.

Egeland, B., & Sroufe, L. A. (1981). Developmental sequelae of maltreatment in infancy. In R. Rizley & D. Cicchetti (Eds.), *New directions for child development* (No. 11, pp. 77–92). San Francisco: Jossey-Bass.

Eggebeen, D. J. (1992). Parent–child support in aging American families. *Generations, 16*(3), 45–49.

Eggerman, S., & Dustin, D. (1985). Death orientation and communication with the terminally ill. *Omega, 16,* 255–265.

Eilers, R. E., & Oller, D. K. (1994). Infant vocalizations and the early diagnosis of severe hearing impairment. *Journal of Pediatrics, 124,* 199–203.

Eisenberg, N., Fabes, R. A., Carolo, G., Speer, A. L., Switzer, G., Karbon, M., & Troyer, D. (1993). The relations of empathy-related emotions and maternal practices to children's comforting behavior. *Journal of Experimental Child Psychology, 55,* 131–150.

Eisenberg, N., Fabes, R. A., Murphy, B., Maszk, P., Smith, M., & Karbon, M. (1995). The role of emotionality and regulation in children's social functioning: A longitudinal study. *Child Development, 66,* 1360–1384.

Eisenberg, N., & Miller, P. A. (1987). The relation of empathy to prosocial and related behaviors. *Psychological Bulletin, 101,* 91–119.

Eisenberg, N., Shell, R., Pasternack, J., Lennon, R., Beller, R., & Mathy, R. M. (1987). Prosocial development in middle childhood: A longitudinal study. *Developmental Psychology, 23,* 712–718.

Eisler, R. (1987). *The chalice and the blade.* New York: Harper & Row.

Elardo, R., & Bradley, R. H. (1981). The Home Observation for Measurement of the Environment (HOME) Scale: A review of research. *Developmental Review, 1,* 113–145.

Elder, G. H., Jr. (1974). *Children of the Great Depression.* Chicago: University of Chicago Press.

Elder, G. H., Jr., Caspi, A. (1988). Human development and social change: An emerging perspective on the life course. In N. Bolger, A. Caspi, G. Downey, & M. Moorehouse (Eds.), *Persons in context: Developmental processes* (pp.

77-113). Cambridge, MA: Cambridge University Press.

Elder, G. H., Jr., Caspi, A., & Van Nguyen, T. (1986). Resourceful and vulnerable children: Family influences in hard times. In R. K. Silbereisen, K. Eysferth, & G. Rodinger (Eds.), *Development as action in context: Problem behavior and normal youth development* (pp. 167–186). New York: Springer-Verlag.

Elder, G. H., Jr., Liker, J. K., & Cross, C. E. (1984). Parent–child behavior in the Great Depression: Life course and intergenerational influences. In P. B. Baltes & O. G. Brim (Eds.), *Life-span development and behavior* (Vol. 6, pp. 109–158). New York: Academic Press.

Elder, G. H., Jr., Van Nguyen, T., & Caspi, A. (1985). Linking family hardship to children's lives. *Child Development, 56,* 361–375.

Elicker, J., Englund, M., & Sroufe, L. A. (1992). Predicting peer competence and peer relationships in childhood from early parent–child relationships. In R. D. Parke & G. W. Ladd (Eds.), *Family–peer relationships: Modes of linkage* (pp. 77–106). Hillsdale, NJ: Erlbaum.

Elkind, D. (1994). *A sympathetic understanding of the child: Birth to sixteen* (3rd ed.). Boston: Allyn & Bacon.

Elkind, D., & Bowen, R. (1979). Imaginary audience behavior in children and adolescents. *Developmental Psychology, 15,* 33–44.

Ellicott, A. M. (1985). Psychosocial changes as a function of family-cycle phase. *Human Development, 28,* 270–274.

Elliott, E. S., & Dweck, C. S. (1988). Goals: An approach to motivation and achievement. *Journal of Personality and Social Psychology, 54,* 5–12.

Ellis, A. L., & Vasseur, R. B. (1993). Prior interpersonal contact with and attitudes toward gays and lesbians in an interviewing context. *Journal of Homosexuality, 25,* 31–45.

Else, L., Wonderlich, S. A., Beatty, W. W., Christie, D. W., & Staton, R. D. (1993). Personality characteristics of men who physically abuse women. *Hospital and Community Psychiatry, 44,* 54–58.

Emde, R. N. (1992). Individual meaning and increasing complexity: Contributions of Sigmund Freud and René Spitz to developmental psychology. *Developmental Psychology, 28,* 347–359.

Emde, R. N., Plomin, R., Robinson, J., Corley, R., DeFries, J., Fulker, D. W., Reznick, J. S., Campos, J., Kagan, J., & Zahn-Waxler, C. (1992). Temperament, emotion, and cognition at fourteen months: The MacArthur Longitudinal Twin Study. *Child Development, 63,* 1437–1455.

Emery, R. E., Mathews, S. G., & Kitzmann, K. M. (1994). Child custody mediation and litigation: Parents' satisfaction and functioning a year after settlement. *Journal of Consulting and Clinical Psychology, 62,* 124–129.

Emory, E. K., & Toomey, K. A. (1988). Environmental stimulation and human fetal responsibility in late pregnancy. In W. P. Smotherman & S. R. Robinson (Eds.), *Behavior of the fetus* (pp. 141–161). Caldwell, NJ: Telford.

Engen, T. (1982). *The perception of odors.* New York: Academic Press.

Enright, R. D., Gassin, E. A., & Wu, C. (1992). Forgiveness: A developmental view. *Journal of Moral Education, 21,* 99–114.

Enright, R. D., Lapsley, D. K., & Shukla, D. (1979). Adolescent egocentrism in early and late adolescence. *Adolescence, 14,* 687–695.

Epstein, C. J. (Ed.). (1993). The phenotypic mapping of Down syndrome and other aneuploid conditions. New York: Wiley-Liss.

Epstein, H. T. (1980). EEG developmental stages. *Developmental Psychobiology, 13,* 629–631.

Epstein, L. H., McCurley, J., Wing, R. R., & Valoski, A. (1990). Five-year follow-up of family-based treatments for childhood obesity. *Journal of Consulting and Clinical Psychology, 58,* 661–664.

Epstein, L. H., Wing, R. R., Koeske, R., & Valoski, A. (1987). Long-term effects of family-based treatment of childhood obesity. *Journal of Consulting and Clinical Psychology, 55,* 91–95.

Ericsson, K. A. (1990). Peak performance and age: An examination of peak performance in sports. In P. B. Baltes & M. M. Baltes (Eds.), *Successful aging* (pp. 164–196). Cambridge: Cambridge University Press.

Erikson, E. H. (1950). *Childhood and society.* New York: Norton.

Erikson, E. H. (1964). *Insight and responsibility.* New York: Norton.

Erikson, E. H. (1968). *Identity, youth, and crisis.* New York: Norton.

Ernst, N. D., & Harlan, W. R. (1991). Obesity and cardiovascular disease in minority populations: Executive summary. Conference highlights, conclusions, and recommendations. *American Journal of Clinical Nutrition, 53,* 1507S–1511S.

Ershler, W. B. (1993). The influence of an aging immune system on cancer incidence and progression. *Journal of Gerontology, 48,* B3–B7.

Eskenazi, B. (1993). Caffeine during pregnancy: Grounds for concern? *Journal of the American Medical Association, 270,* 2973–2974.

Espenschade, A., & Eckert, H. (1974). Motor development. In W. R. Johnson & E. R. Buskirk (Eds.), *Science and medicine of exercise and sport* (pp. 322–333). New York: Harper & Row.

Essa, E. L., & Murray, C. I. (1994). Young children's understanding and experience with death. *Young Children, 49*(4), 74–81.

Esterberg, K. G., Moen, P., & Dempster-McClain, D. (1994). Transition to divorce: A life-course approach to women's marital duration and dissolution. *Sociological Quarterly, 35,* 289–307.

Estes, D. (1994). Young children's understanding of the mind: Imagery, introspection, and some implications. *Journal of Applied Developmental Psychology, 15,* 529–548.

Etaugh, C., Williams, B., & Carlson, P. (1996). Changing attitudes toward day care and maternal employment as portrayed in women's magazines: 1977–1990. *Early Childhood Research Quarterly, 11,* 207–218.

Eveleth, P. B., & Tanner, J. M. (1976). *Worldwide variation in human growth.* Cambridge, England: Cambridge University Press.

Eyer, D. E. (1992). *Mother–infant bonding: A scientific fiction.* New Haven, CT: Yale University Press.

Fabes, R. A., Eisenberg, N., McCormick, S. E., & Wilson, M. S. (1988). Preschoolers' attributions of the situational determinants of others' naturally occurring emotions. *Developmental Psychology, 24,* 376–385.

Fabricius, W. V., & Wellman, H. M. (1993). Two roads diverged. Young children's ability to judge distance. *Child Development, 64,* 399–414.

Fabsitz, R. R., Sholinsky, P., & Carmelli, D. (1994). Genetic influences on adult weight gain and maximum body mass index in male twins. *American Journal of Epidemiology, 140,* 711–720.

Facchinetti, F., Battaglia, C., Benatti, R., Borella, P., & Genazzani, A. R. (1992). Oral magnesium supplementation improves fetal circulation. *Magnesium Research, 3,* 179–181.

Fagan, J., Slaughter, E., & Hartstone, E. (1987). Blind justice? The impact of race on the juvenile justice process. *Crime & Delinquency, 33,* 259–286.

Fagan, J. F., III. (1973). Infants' delayed recognition memory and forgetting. *Journal of Experimental Child Psychology, 16,* 424–450.

Fagan, J. F., III. (1977). Infant recognition memory: Studies in forgetting. *Child Development, 45,* 351–356.

Fagot, B. I. (1977). Consequences of moderate cross-gender behavior in preschool children. *Child Development, 48,* 902–907.

Fagot, B. I. (1978). The influence of sex of child on parental reactions to toddler children. *Child Development, 49,* 459–465.

Fagot, B. I., & Hagan, R. I. (1991). Observations of parent reactions to sex-stereotyped behaviors: Age and sex effects. *Child Development, 62,* 617–628.

Fagot, B. I., & Kavanaugh, K. (1990). The prediction of antisocial behavior from avoidant attachment classifications. *Child Development, 61,* 864–873.

Fagot, B. I., & Leinbach, M. D. (1989). The young child's gender schema: Environmental input, internal organization. *Child Development, 60,* 663–672.

Fagot, B. I., Leinbach, M. D., & O'Boyle, C. (1992). Gender labeling, gender stereotyping, and parenting behaviors. *Developmental Psychology, 28,* 225–230.

Fagot, B. I., & Patterson, G. R. (1969). An in vivo analysis of reinforcing contingencies for sex-role behaviors in the preschool child. *Developmental Psychology, 1,* 563–568.

Fahrmeier, E. D. (1978). The development of concrete operations among the Hausa. *Journal of Cross-Cultural Psychology, 9,* 23–44.

Fairburn, C. G., & Belgin, S. J. (1990). Studies of the epidemiology of bulimia nervosa. *American Journal of Psychiatry, 147,* 401–408.

Falbo, T. (1992). Social norms and the one-child family: Clinical and policy implications. In F. Boer & J. Dunn (Eds.), *Children's sibling relationships* (pp. 71–82). Hillsdale, NJ: Erlbaum.

Falbo, T., & Polit, D. (1986). A quantitative review of the only-child literature: Research evidence and theory development. *Psychological Bulletin, 100,* 176–189.

Falbo, T., & Poston, D. L., Jr. (1993). The academic, personality, and physical outcomes of only children in China. *Child Development, 64,* 18–35.

Faller, K. C. (1990). *Understanding child sexual maltreatment.* Newbury Park, CA: Sage.

Fanslow, C. A. (1981). Death: A natural facet of the life continuum. In D. Krieger (Ed.), *Foundations for holistic health nursing practices: The renaissance nurse* (pp. 249–272). Philadelphia: Lippincott.

Fantz, R. L. (1961, May). The origin of form perception. *Scientific American,* 204 (5), 66–72.

Farrington, D. P. (1987). Epidemiology. In H. C. Quay (Ed.), *Handbook of juvenile delinquency* (pp. 33–61). New York: Wiley.

Farver, J. M. (1993). Cultural differences in scaffolding pretend play: A comparison of American and Mexican mother–child and sibling–child pairs. In K. MacDonald (Ed.), *Parent–child play* (pp. 349–366). Albany, NY: SUNY Press.

Farver, J. M., & Wimbarti, S. (1995). Indonesian children's play with their mothers and older siblings. *Child Development, 66,* 1493–1503.

Featherman, D. (1980). Schooling and occupational careers: Constancy and change in worldly success. In O. Brim, Jr., & J. Kagan (Eds.), *Constancy and change in human development* (pp. 675–738). Cambridge, MA: Harvard University Press.

Feingold, A. (1992). Gender differences in mate selection preferences: A test of the parental investment model. *Psychological Bulletin, 112,* 125–139.

Feingold, A. (1993). Cognitive gender differences: A developmental perspective. *Sex Roles, 29,* 91–112.

Feingold, A. (1994). Gender differences in personality: A meta-analysis. *Psychological Bulletin, 116,* 429–456.

Feldman, D. H. (1991). *Nature's gambit.* New York: Teachers College Press.

Feldman, E. A. (1994). Culture, conflict, and cost: Perspectives on brain death in Japan. *International Journal of Technology Assessment in Health Care, 10,* 447–463.

Feldman, S. S., & Weinberger, D. A. (1994). Self-restraint as a mediator of family influences on boys' delinquent behavior: A longitudinal study. *Child Development, 65,* 195–211.

Felner, R. D., & Adan, A. M. (1988). The School Transitional Environment Project: An ecological intervention and evaluation. In R. H. Price, E. L. Cowan, R. P. Lorion, & J. Ramos-McKay (Eds.), *14 ounces of prevention: A casebook for practitioners* (pp. 111–122). Washington, DC: American Psychological Association.

Fendrich, J. M. (1993). *Ideal citizens: The legacy of the civil rights movement.* Albany: State University of New York Press.

Fenson, L., Dale, P. S., Reznick, J. S., Bates, E., Thal, D. J., & Pethick, S. J. (1994). Variability in early communicative development. *Monographs of the Soci-*

ety for Research in Child Development, 59(5, Serial No. 242).

Fergus, M. A. (1995, March 19). 99 years and . . . a joy to be around. *The Pantagraph,* pp. C1–C2.

Ferguson, F. G., Wikby, A., Maxson, P., Olsson, J., & Johansson, B. (1995). Immune parameters in a longitudinal study of a very old population of Swedish people: A comparison between survivors and nonsurvivors. *Journal of Gerontology, 50,* B378–B382.

Ferguson, L. R. (1978). The competence and freedom of children to make choices regarding participation in research: A statement. *Journal of Social Issues, 34,* 114–121.

Fergusson, D. M., Horwood, L. J., & Lynskey, M. T. (1993). Maternal smoking before and after pregnancy: Effects on behavioral outcomes in middle childhood. *Pediatrics, 92,* 815–822.

Fergusson, D. M., Horwood, L. J., & Shanon, F. T. (1987). Breast-feeding and subsequent social adjustment in six- to eight-year-old children. *Journal of Child Psychology and Psychiatry, 28,* 378–386.

Fernald, A. (1993). Approval and disapproval: Infant responsiveness to vocal affect in familiar and unfamiliar languages. *Child Development, 64,* 637–656.

Fernald, A., & Morikawa, H. (1993). Common themes and cultural variations in Japanese and American mothers' speech to infants. *Child Development, 64,* 637–656.

Fernald, A., Taeschner, T., Dunn, J., Papousek, M., Boyssen-Bardies, B., & Fukui, I. (1989). A cross-language study of prosodic modifications in mothers' and fathers' speech to preverbal infants. *Journal of Child Language, 16,* 477–502.

Feshbach, N. D., & Feshbach, S. (1982). Empathy training and the regulation of aggression: Potentialities and limitations. *Academic Psychology Bulletin, 4,* 399–413.

Fiatarone, M. A., Marks, E. C., Meredith, C. N., Lipsitz, L. A., & Evans, W. J. (1990). High intensity strength training in nonagenarians: Effects on skeletal muscle. *Journal of the American Medical Association, 263,* 3029–3034.

Field, D., & Millsap, R. E. (1991). Personality in advanced old age: Continuity or change? *Journal of Gerontology, 46,* 299–308.

Field, D., & Minkler, M. (1988). Continuity and change in social support between young–old, old–old, and very-old adults. *Journal of Gerontology, 43,* P100–P106.

Field, T. M. (1994). The effects of mother's physical and emotional unavailability on emotion regulation. In N. A. Fox (Ed.), The development of emotion regulation: Biological and behavioral considerations. *Monographs of the Society for Research in Child Development, 59*(2–3, Serial No. 240).

Field, T. M., Schanberg, S. M., Scafidi, F., Bauer, C. R., Vega-Lahr, N., Garcia, R. Nystrom, J., & Kuhn, C. M. (1986). Effects of tactile/kinesthetic stimulation on preterm neonates. *Pediatrics, 77,* 654–658.

Field, T. M., Woodson, R., Greenberg, R., & Cohen, D. (1982). Discrimination and imitation of facial expressions by neonates. *Science, 218,* 179–181.

Fiese, B. (1990). Playful relationships: A contextual analysis of mother-toddler interaction and symbolic play. *Child Development, 61,* 1648–1656.

Filipovic, Z. (1994). *Zlata's diary: A child's life in Sarajevo.* New York: Penguin.

Fine, G. A. (1980). The natural history of preadolescent male friendship groups. In H. C. Foot, A. J. Chapman, & J. R. Smith (Eds.), *Friendship and social relations in children* (pp. 293–320). Chichester, England: Wiley.

Finnigan, L. P. (1996). The NIH Women's Health Initiative: Its evolution and expected contributions to women's health. *American Journal of Preventive Medicine, 12,* 292–293.

Firestone, R. W. (1994). Psychological defenses against death anxiety. In R. A. Neimeyer (Ed.), *Death anxiety handbook* (pp. 217–241). Washington, DC: Taylor & Francis.

Fischer, C. S., & Phillips, S. L. (1982). Who is alone? Social characteristics of people with small networks. In L. A. Peplau & D. Perlman (Eds.), *Loneliness: A sourcebook of current theory, research, and therapy* (pp. 21–39). New York: Wiley.

Fischer, J. L., & Narus, L. R. (1981). Sex roles and intimacy in same sex and other sex relationships. *Psychology of Women Quarterly, 5,* 444–455.

Fischer, K. W. (1980). A theory of cognitive development: The control and construction of hierarchies of skills. *Psychological Review, 87,* 477–531.

Fischer, K. W., & Farrar, M. J. (1987). Generalizations about generalizations: How a theory of skill development explains both generality and specificity. *International Journal of Psychology, 22,* 643–677.

Fischer, K. W., & Pipp, S. L. (1984). Processes of cognitive development: Optimal level and skill acquisition. In R. J. Sternberg (Ed.), *Mechanisms of cognitive development* (pp. 45–80). New York: Freeman.

Fischer, K. W., & Rose, S. P. (1995, Fall). Concurrent cycles in the dynamic development of brain and behavior. *SRCD Newsletter*, pp. 3–4, 15–16.

Fischer, P. J., & Breakey, W. R. (1991). The epidemiology of alcohol, drug, and mental disorders among homeless persons. *American Psychologist, 46*, 1115–1128.

Fisher, C. B. (1993, Winter). Integrating science and ethics in research with high-risk children and youth. *Social Policy Report of the Society for Research in Child Development, 4*(4).

Fishler, K., & Koch, R. (1991). Mental development in Down syndrome mosaicism. American Journal on Mental Retardation, 96, 345–351.

Fivush, R. (1995). Language, narrative, and autobiography. *Consciousness and Cognition, 4*, 100–103.

Fivush, R., Haden, C., & Adam, S. (1995). Structure and coherence of pre-schoolers' personal narratives over time: Implications for childhood amnesia. *Journal of Experimental Child Psychology, 60*, 32–56.

Fivush, R., Kuebli, J., & Clubb, P. A. (1992). The structure of events and event representations: A developmental analysis. *Child Development, 63*, 188–201.

Fivush, R., & Reese, E. (1992). The social construction of autobiographical memory. In M. A. Conway, D. C. Rubin, H. Spinnler, & W. A. Wagenaar (Eds.), *Theoretical perspectives on autobiographical memory* (pp. 115–132). Ultrecht, Netherlands: Kluwer.

Flack, J. M., Amaro, H., Jenkins, W., Kunitz, S., Levy, J., Mixon, M., & Yu, E. (1995). Panel I: Epidemiology of minority health. *Health Psychology, 14*, 592–600.

Flavell, J. H. (1993). The development of children's understanding of false belief and the appearance–reality distinction. *International Journal of Psychology, 28*, 595–604.

Flavell, J. H., Green, F. L., & Flavell, E. R. (1987). Development of knowledge about the appearance–reality distinction. *Monographs of the Society for Research in Child Development, 51*(1, Serial No. 212).

Flavell, J. H., Green, F. L., & Flavell, E. R. (1993). Children's understanding of the stream of consciousness. *Child Development, 64*, 387–398.

Flavell, J. H., Green, F. L., & Flavell, E. R. (1995). Young children's knowledge about thinking. *Monographs of the Society for Research in Child Development, 60* (1, Serial No. 243).

Fletcher, A. C., Darling, N. E., Steinberg, L., & Dornbusch, S. M. (1995). The company they keep: Relation of adolescents' adjustment and behavior to their friends' perceptions of authoritative parenting in the social network. *Developmental Psychology, 31*, 300–310.

Fletcher, W. L., & Hansson, R. O. (1991). Assessing the social components of retirement anxiety. *Psychology and Aging, 6*, 76–85.

Flood, D. G., & Coleman, P. D. (1988). Cell type heterogeneity of changes in dendritic extent in the hippocampal region of the human brain in normal aging and in Alzheimer's disease. In T. L. Petit & G. O. Ivy (Ed.), *Neural plasticity: A lifespan approach* (pp. 265–281). New York: Alan R. Liss.

Florian, V., Mikulincer, M., & Taubman, O. (1995). Does hardiness contribute to mental health during a stressful real-life situation? The roles of appraisal and coping. *Journal of Personality and Social Psychology, 68*, 687–695.

Fogel, A. Toda, S., & Kawai, M. (1988). Mother–infant face-to-face interaction in Japan and the United States: A laboratory comparison using 3-month-old infants. *Developmental Psychology, 24*, 398–406.

Fogel, B. S. (1992). Psychological aspects of staying at home. *Generations, 16*, 15–19.

Food Research and Action Center. (1991). *Community Childhood Hunger Identification Project.* Washington, DC: Author.

Ford, J. M., Roth, W. T., Isaacks, B. G., White, P. M., Hood, S. H., & Pfefferbaum, A. (1995). *Biological Psychology, 39*, 57–80.

Ford, K., & Labbok, M. (1993). Breast-feeding and child health in the United States. *Journal of Biosocial Science, 25*, 187–194.

Fordham, S., & Ogbu, J. U. (1986). Black students' school success: Coping with the "burden of 'acting white.'" *Urban Review, 18*, 176–206.

Forehand, R., Wierson, M., Thomas, A. M., Fauber, R., Armistead, L., Kempton, T., & Long, N. (1991). A short-term longitudinal examination of young adolescent functioning following divorce: The role of family factors. *Journal of Abnormal Child Psychology, 19*, 97–111.

Forgatch, M. S., Patterson, G. R., & Ray, J. A. (1996). Divorce and boys' adjustment problems: Two paths with a single model. In E. M. Hetherington (Ed.), *Stress, coping, and resiliency in children and the family* (pp. 67–105). Hillsdale, NJ: Erlbaum.

Forman, E. A., Minick, N., & Stone, C. A. (1993). *Contexts for learning.* New York: Oxford University Press.

Fortier, I., Marcoux, S., & Beaulac-Baillargeon, L. (1993). Relation of caffeine intake during pregnancy to intrauterine growth retardation and preterm birth. *American Journal of Epidemiology, 137*, 931–940.

Fortier, I., Marcoux, S., & Brisson, J. (1994). Passive smoking during pregnancy and the risk of delivering a small-for-gestational-age infant. *American Journal of Epidemiology, 139*, 294–301.

Fowler, J. W. (1981). *Stages of faith.* San Francisco: Harper & Row.

Fox, C. H. (1994). Cocaine use in pregnancy. *Journal of the American Board of Family Practice, 1*, 225–228.

Fox, M., Gibbs, M., & Auerbach, D. (1985). Age and gender dimensions of friendship. *Psychology of Women Quarterly, 9*, 489–501.

Fox, N. A. (1991). If it's not left, it's right: Electroencephalograph asymmetry and the development of emotion. *American Psychologist, 46*, 863–872.

Fox, N. A. (1994). Dynamic cerebral processes underlying emotion regulation. In N. A. Fox (Ed.), The development of emotion regulation. *Monographs of the Society for Research in Child Development, 59*(2–3, Serial No. 240), pp. 152–166.

Fox, N. A. (1995). Of the way we were: Adult memories about attachment experiences and their role in determining infant–parent relationships: A commentary on van IJzendoorn. *Psychological Bulletin, 117*, 404–410.

Fox, N. A., Bell, M. A., & Jones, N. A. (1992). Individual differences in response to stress and cerebral asymmetry. *Developmental Neuropsychology, 8*, 161–184.

Fox, N. A., Calkins, S. D., & Bell, M. A. (1994). Neural plasticity and development in the first two years of life: Evidence from cognitive and socioemotional domains of research. *Development and Psychopathology, 6*, 677–696.

Fox, N. A., & Davidson, R. J. (1986). Taste-elicited changes in facial signs of emotion and the asymmetry of brain electrical activity in newborn infants. *Neuropsychologia, 24*, 417–422.

Fozard, J. L., Vercruyssen, M., Reynolds, S. L., Hancocke, P. A., & Quilter, R. E. (1994). Age differences and changes in reaction time: The Baltimore Longitudinal Study of Aging. *Journal of Gerontology, 49*, P179–P189.

Fracasso, M. P., & Busch-Rossnagel, N. A. (1992). Parents and children of Hispanic origin. In M. E. Procidano & C. B. Fisher (Eds.), *Contemporary families* (pp. 83–98). New York: Teachers College Press.

Framo, J. L. (1994). The family life cycle: Impressions. *Contemporary Family Therapy, 16,* 87–117.

Franceschi, C., Monti, D., Sansoni, P., & Cossarizza, A. (1995). The immunology of exceptional individuals: The lesson of centenarians. *Immunology Today, 16*(1), 12–16.

Francis, P. L., & McCroy, G. (1983). *Bimodal recognition of human stimulus configurations.* Paper presented at the biennial meeting of the Society for Research in Child Development, Detroit.

Frank, S. A., Roland, D. C., Sturis, J., Byrne, M. M. Refetoff, S., Polonsky, K. S., & van Cauter, E. (1995). Effects of aging on glucose regulation during wakefulness and sleep. *American Journal of Physiology, 269,* E1006–E1016.

Frankel, K. A., & Bates, J. E. (1990). Mother-toddler problem solving: Antecedents in attachment, home behavior, and temperament. *Child Development, 61,* 810–819.

Franklin, M. (1995). The effects of differential college environments on academic learning and student perceptions of cognitive development. *Research in Higher Education, 36,* 127–153.

Frazier, M. M. (1994). Issues, problems and programs in nurturing the disadvantaged and culturally different talented. In K. A. Heller, F. J., Jonks, & H. A. Passow (Eds.), *International handbook of research and development of giftedness and talent* (pp. 685–692). Oxford: Pergamon Press.

Frederickson, B. L., & Carstensen, L. L. (1990). Relationship classification using grade of membership analysis: A typology of sibling relationships in later life. *Journal of Gerontology, 45,* S43–S51.

Freedman, D. G., & Freedman, N. (1969). Behavioral differences between Chinese-American and European-American newborns. *Nature, 224,* 1227.

Freeman, D. (1983). *Margaret Mead and Samoa: The making and unmaking of an anthropological myth.* Cambridge, MA: Harvard University Press.

Freer, J. P. (1991). Death and dignity: The case of Diane [Letter to the Editor]. *New England Journal of Medicine, 325,* 658.

Freud, S. (1973). *An outline of psychoanalysis.* London: Hogarth. (Original work published 1938)

Freud, S. (1974). *The ego and the id.* London: Hogarth. (Original work published 1923)

Fried, P. A. (1993). Prenatal exposure to tobacco and marijuana: Effects during pregnancy, infancy, and early childhood. *Clinical Obstetrics and Gynecology, 36,* 319–337.

Fried, P. A., & Makin, J. E. (1987). Neonatal behavioral correlates of prenatal exposure to marijuana, cigarettes, and alcohol in a low risk population. *Neurobehavioral Toxicology and Teratology, 9,* 1–7.

Fried, P. A., & Watkinson, B. (1990). 36- and 48-month neurobehavioral follow-up of children prenatally exposed to marijuana, cigarettes, and alcohol. *Journal of Developmental and Behavioral Pediatrics, 11,* 49–58.

Friedman, A. G., Greene, P. G., & Stokes, T. (1991). Improving dietary habits of children: Effects of nutrition education and correspondence training. *Behavior Therapy and Experimental Psychiatry, 21,* 263–268.

Friedman, H. S., Tucker, J. S., Schwartz, J. E., Martin, L. R., Tomlinson-Keasey, C., Wingard, D. L., & Criqui, M. H. (1995a). Childhood conscientiousness and longevity: Health behaviors and cause of death. *Journal of Personality and Social Psychology, 68,* 696–703.

Friedman, H. S., Tucker, J. S., Schwartz, J. E., Tomlinson-Keasey, C., Martin, L. R., Wingard, D. L., & Criqui, M. H. (1995b). Psychosocial and behavioral predictors of longevity: The aging and death of the "Termites." *American Psychologist, 50,* 69–78.

Friedman, J. A., & Weinberger, H. L. (1990). Six children with lead poisoning. *American Journal of Diseases of Children, 144,* 1039–1044.

Fries, J. F. (1990). Medical perspectives on successful aging. In P. B. Baltes & M. M. Baltes (Eds.), *Successful aging* (pp. 35–49). Cambridge: Cambridge University Press.

Frodi, A. (1985). When empathy fails: Aversive infant crying and child abuse. In B. M. Lester & C. F. Z. Boukydis (Eds.), *Infant crying: Theoretical and research perspectives* (pp. 263–277). New York: Plenum.

Froggatt, P., Beckwith, J. B., Schwartz, P. J., Valdes-Dapena, M., & Southall, D. P. (1988). Cardiac and respiratory mechanisms that might be responsible for sudden infant death syndrome. In P. J. Schwartz, D. P. Southall, & M. Valdes-Dapena (Eds.), *The sudden infant death syndrome* (Annals of the New York Academy of Sciences, Vol. 533, pp. 421–426). New York: The New York Academy of Sciences.

Fry, C. L. (1985). Culture, behavior, and aging in the comparative perspective. In J. E. Birren & K. W. Schaie (Eds.), *Handbook of the psychology of aging* (2nd ed., pp. 216–244). New York: Van Nostrand Reinhold.

Fry, P. S. (1991). Individual differences in reminiscence among older adults: Predictors of frequency and pleasantness ratings of reminiscence activity. *International Journal of Aging and Human Development, 33,* 311–326.

Fuchs, I., Eisenberg, N., Hertz-Lazarowitz, R., & Sharabany, R. (1986). Kibbutz, Israeli city, and American children's moral reasoning about prosocial moral conflicts. *Merrill-Palmer Quarterly, 32,* 37–50.

Fuligni, A. J., & Eccles, J. S. (1993). Perceived parent–child relationships and early adolescents' orientation toward peers. *Developmental Psychology, 29,* 622–632.

Fuligni, A. J., & Stevenson, H. W. (1995). Time use and mathematics achievement among American, Chinese, and Japanese high school students. *Child Development, 66,* 830–842.

Fulmer, R. (1989). Lower income and professional families: A comparison of structure and life cycle process. In B. Carter & M. McGoldrick (Eds.), *The changing family life cycle* (2nd ed., pp. 545–578). Boston: Allyn & Bacon.

Fulton, R. (1995). The contemporary funeral: Functional or dysfunctional? In H. Wass & R. A. Neimeyer (Eds.), *Dying: Facing the facts* (3rd ed., pp. 185–209). Washington, DC: Taylor and Francis.

Furman, E. (1984). Children's patterns in mourning the death of a loved one. In H. Wass & C. A. Corr (Eds.), *Childhood and death* (pp. 185–203). Washington, DC: Hemisphere.

Furman, E. (1990, November). Plant a potato—learn about life (and death). *Young Children, 46*(1), 15–20.

Furman, W., & Buhrmester, D. (1992). Age and sex differences in perceptions of networks of personal relationships. *Child Development, 63,* 103–115.

Furstenberg, F. F., Jr., Brooks-Gunn, J., & Chase-Lansdale, L. (1989). Teenaged pregnancy and childbearing. *American Psychologist, 44,* 313–320.

Furstenberg, F. F., Jr., Brooks-Gunn, J., & Morgan, S. P. (1987). *Adolescent mothers and their children in later life.* Cambridge, England: Cambridge University Press.

Furstenberg, F. F., Jr. & Cherlin, A. J. (1991). *Divided families.* Cambridge, MA: Harvard University Press.

Furstenberg, F. F., Jr., & Crawford, D. B. (1978). Family support: Helping teenagers to cope. *Family Planning Perspectives, 10,* 322–333.

Furstenberg, F. F., Jr., Levine, J. A., & Brooks-Gunn, J. (1990). The children of teenage mothers: Patterns of early childbearing in two generations. *Family Planning Perspectives, 22,* 54–61.

Furstenberg, F. F., Jr., & Nord, C. W. (1985). Parenting apart: Patterns of childrearing after marital disruption. *Journal of Marriage and the Family, 47,* 893–904.

Furuno, S., O'Reilly, K., Inatsuka, T., Hosaka, C., Allman, T., & Zeisloft-Falbey, B. (1987). *Hawaii Early Learning Profile.* Palo Alto, CA: VORT Corporation.

Fuson, K. C. (1988). *Children's counting and concepts of number.* New York: Springer-Verlag.

Fuson, K. C., & Kwon, Y. (1992). Korean children's understanding of multidigit addition and subtraction. *Child Development, 63,* 491–506.

Gable, R. K., Thompson, D. L., & Iwanicki, E. F. (1983). The effects of voluntary desegregation on occupational outcomes. *Vocational Guidance Quarterly, 31,* 230–239.

Gaddis, A., & Brooks-Gunn, J. (1985). The male experience of pubertal change. *Journal of Youth and Adolescence, 14,* 61–69.

Galambos, N. L., Almeida, D. M., & Petersen, A. C. (1990). Masculinity, femininity, and sex role attitudes in early adolescence: Exploring gender intensification. *Child Development, 61,* 1905–1914.

Galambos, S. J., & Goldin-Meadow, S. (1990). The effects of learning two languages on levels of metalinguistic awareness. *Cognition, 34,* 1–56.

Galin, D., Johnstone, J., Nakell, L., & Herron, J. (1979). Development of the capacity for tactile information transfer between hemispheres in normal children. *Science, 204,* 1330–1332.

Galinsky, E., Howes, C., Kontos, S., & Shinn, M. (1994). *The study of children in family child care and relative care: Highlights of findings, selected initiatives to improve the quality of family child care.* New York: Families and Work Institute.

Galler, J. R., Ramsey, C. F., Morley, D. S., Archer, E., & Salt, P. (1990). The long-term effects of early kwashiorkor compared with marasmus. IV. Performance on the National High School Entrance Examination. *Pediatric Research, 28,* 235–239.

Galler, J. R., Ramsey, F., & Solimano, G. (1985a). A follow-up study of the effects of early malnutrition on subsequent development: I. Physical growth

and sexual maturation during adolescence. *Pediatric Research, 19,* 518–523.

Galler, J. R., Ramsey, F., & Solimano, G. (1985b). A follow-up study of the effects of early malnutrition on subsequent development: II. Fine motor skills in adolescence. *Pediatric Research, 19,* 524–527.

Galler, J. R., Ramsey, F., Solimano, G., Kucharski, L. T., & Harrison, R. (1984). The influence of early malnutrition on subsequent behavioral development: IV. Soft neurological signs. *Pediatric Research, 18,* 826–832.

Galotti, K. M., Kozberg, S. F., & Farmer, M. C. (1991). Gender and developmental differences in adolescents' conceptions of moral reasoning. *Journal of Youth and Adolescence, 20,* 13–30.

Gannon, L., & Ekstrom, B. (1993). Attitudes toward menopause: The influence of sociocultural paradigms. *Psychology of Women Quarterly, 17,* 275–288.

Gannon, S., & Korn, S. J. (1983). Temperament, cultural variation, and behavior disorder in preschool children. *Child Psychiatry and Human Development, 13,* 203–212.

Ganong, L. H., & Coleman, M. (1992). Gender differences in expectations of self and future partner. *Journal of Family Issues, 13,* 55–64.

Ganong, L. H., & Coleman, M. (1994). *Remarried family relationships.* Thousand Oaks, CA: Sage.

Ganong, L. H., Coleman, M., & Fine, M. (1995). Remarriage and stepfamilies. In W. Burr & R. Day (Eds.), *Advanced family science.* Provo, UT: Brigham Young University Press.

Garbarino, J., & Kostelny, K. (1993). Neighborhood and community influences on parenting. In T. Luster & L. Okagaki (Eds.), *Parenting: An ecological perspective* (pp. 203–226). Hillsdale, NJ: Erlbaum.

Gardner, H. (1980). *Artful scribbles: The significance of children's drawings.* New York: Basic Books.

Gardner, H. (1983). *Frames of mind: The theory of multiple intelligences.* New York: Basic Books.

Gardner, M. J., Snee, M. P., Hall, A. J., Powell, C. A., Downes, S., & Terrell, J. D. (1990). Leukemia cases linked to fathers' radiation dose. *Nature, 343,* 423–429.

Garland, A. F., & Zigler, E. (1993). Adolescent suicide prevention: Current research and social policy implications. *American Psychologist, 48,* 169–182.

Garmezy, N. (1993). Children in poverty: Resilience despite risk. *Psychiatry, 56,* 127–136.

Garmon, L. C., Basinger, K. S., Gregg, V. R., & Gibbs, J. C. (1996). Gender differences in stage and expression of moral judgment. *Merrill-Palmer Quarterly, 42,* 418–437.

Garner, D. M. (1993). Pathogenesis of anorexia nervosa. *Lancet, 341,* 1631–1635.

Garnier, H. E., Stein, J. A., & Jacobs, J. K. (1997). The process of dropping out of high school: A 19-year perspective. *American Educational Research Journal, 34,* 395–410.

Garrett, P., Ng'andu, N., & Ferron, J. (1994). Poverty experiences of young children and the quality of their home environments. *Child Development, 65,* 331–345.

Garrison, C., Schluchter, M., Schoenbach, V., & Kaplan, B. (1989). Epidemiology of depressive symptoms in young adolescents. *Journal of the American Academy of Child and Adolescent Psychiatry, 28,* 343–351.

Garvey, C. (1975). Requests and responses in children's speech. *Journal of Child Language, 2,* 41–63.

Garvey, C. (1990). *Play.* Cambridge, MA: Harvard University Press.

Garwood, S. G., Phillips, D., Hartman, A., & Zigler, E. F. (1989). As the pendulum swings: Federal agency programs for children. *American Psychologist, 44,* 434–440.

Garzia, R. P., & Trick, L. R. (1992). Vision in the 90's: The aging eye. *Journal of Optometric Vision Development, 23,* 4–41.

Gatehouse, S. (1990). Determinants of self-reported disability in older subjects. *Ear and Hearing, 11*(Suppl.), 57S–65S.

Gathercole, S. E., Adams, A-M., & Hitch, G. (1994). Do young children rehearse? An individual-differences analysis. *Memory & Cognition, 22,* 201–207.

Gatz, M., Bengtson, V. L., & Blum, M. J. (1990). Caregiving families. In J. E. Birren & K. W. Schaie (Eds.), *Handbook of the psychology of aging* (3rd ed., pp. 404–426). San Diego, CA: Academic Press.

Gatz, M., Kasl-Godley, J. E., & Karel, M. J. (1996). Aging and mental disorders. In J. E. Birren & K. W. Schaie (Eds.), *Handbook of the psychology of aging* (pp. 365–382). Sand Diego: Academic Press.

Gauvain, M., & Rogoff, B. (1989). Ways of speaking about space: The development of children's skill in communicating spatial knowledge. *Cognitive Development, 4,* 295–307.

Ge, X., Best, K. M., Conger, R. D., & Simons, R. L. (1996). Parenting behaviors and the occurrence and co-occurrence of adolescent depressive symptoms and conduct problems. *Developmental Psychology, 32,* 717–731.

Ge, X., Conger, R. D., & Elder, G. H., Jr. (1996). Coming of age too early: Pubertal influences on girls' vulnerability to psychological distress. *Child Development, 67,* 3386–3400.

Geary, D. C. (1995). *Children's mathematical development: Research and practical applications.* Washington, DC: American Psychological Association.

Geary, D. C., Bow-Thomas, C. C., Fan, L., & Siegler, R. S. (1993). Even before formal instruction, Chinese children outperform American children in mental addition. *Cognitive Development, 8,* 517–529.

Geary, D. C., Bow-Thomas, C. C., Liu, F., & Siegler, R. S. (1996). Development of Arithmetical competencies in Chinese and American children: Influence of age, language, and schooling. *Child Development, 67,* 2022–2044.

Geary, D. C., & Burlingham-Dubree, M. (1989). External validation of the strategy choice model for addition. *Journal of Experimental Child Psychology, 47,* 175–192.

Gelfand, D. E. (1994). *Aging and ethnicity.* New York: Springer.

Gellatly, A. R. H. (1987). Acquisition of a concept of logical necessity. *Human Development, 30,* 32–47.

Gelman, R. (1972). Logical capacity of very young children: Number invariance rules. *Child Development, 43,* 75–90.

Gelman, R., & Shatz, M. (1978). Appropriate speech adjustments: The operation of conversational constraints on talk to two-year-olds. In M. Lewis & L. A. Rosenblum (Eds.), *Interaction, conversation, and the development of language* (pp. 27–61). New York: Wiley.

Gentner, D. (1982). Why nouns are learned before verbs: Linguistic relativity versus natural partitioning. In S. A. Kuczaj, II (Ed.), *Language development: Vol. 2. Language, thought, and culture* (pp. 301–322). Hillsdale, NJ: Erlbaum.

Gentry, J. R. (1981, January). Learning to spell developmentally. *The Reading Teacher, 35*(2), 378–381.

George, C., Kaplan, N., & Main, M. (1985). *The Adult Attachment Interview.* Unpublished manuscript, University of California at Berkeley.

George, L. K. (1989). Stress, social support, and depression over the life-course. In K. S. Markides & C. L. Cooper (Eds.), *Aging, stress and health* (pp. 241–268). Chichester, England: Wiley.

George, L. K. (1994). Social factors and depression in late life. In L. S. Schneider, C. F. Reynolds, III, B. D. Lebowitz, & A. J. Friedhoff (Eds.), *Diagnosis and treatment of depression in late life* (pp. 131–153). Washington, DC: American Psychiatric Press.

Gershon, E. S., Targum, S. D., Kessler, L. R., Mazure, C. M., & Bunney, W. E., Jr. (1977). Genetics studies and biologic strategies in affective disorders. *Progress in Medical Genetics, 2,* 103–164.

Gesell, A. (1933). Maturation and patterning of behavior. In C. Murchison (Ed.), *A handbook of child psychology.* Worcester, MA: Clark University Press.

Gesell, A., & Ilg, F. L. (1949a). The child from five to ten. In A. Gesell & F. Ilg (Eds.), *Child development* (pp. 394–454). New York: Harper & Row. (Original work published 1946)

Gesell, A., & Ilg, F. L. (1949b). The infant and child in the culture of today. In A. Gesell & F. Ilg (Eds.), *Child development* (pp. 1–393). New York: Harper & Row. (Original work published 1943)

Gibbs, J. C. (1991). Toward an integration of Kohlberg's and Hoffman's theories of morality. In W. M. Kurtines & J. L. Gewirtz (Eds.), *Handbook of moral behavior and development* (Vol. 1, pp. 183–222). Hillsdale, NJ: Erlbaum.

Gibbs, J. C. (1995). The cognitive developmental perspective. In W. M. Kurtines & J. L. Gewirtz (Eds.), *Moral development: An introduction* (pp. 27–48). Boston: Allyn & Bacon.

Gibson, E. J. (1970). The development of perception as an adaptive process. *American Scientist, 58,* 98–107.

Gibson, E. J. (1988). Exploratory behavior in the development of perceiving, acting, and the acquiring of knowledge. *Annual Review of Psychology, 39,* 1–41.

Gibson, E. J., & Walk, R. D. (1960). The "visual cliff." *Scientific American, 202,* 64–71.

Gibson, J. J. (1979). *The ecological approach to visual perception.* Boston: Houghton Mifflin.

Gibson, R. C., & Burns, C. J. (1991, Winter). The health, labor force, and retirement experiences of aging minorities. *Generations, 15*(4), 31–35.

Gibson, R. C., & Jackson, J. S. (1992). The black oldest old: Health, functioning, and informal support. In R. M. Suzman, D. P. Willis, & K. G. Manton (Eds.), *The oldest old* (pp. 321–340). New York: Oxford University Press.

Gilfillan, M. C., Curtis, L., Liston, W. A., Pullen, I., Whyte, D. A., & Brock, J. J. H. (1992). Prenatal screening for cystic fibrosis. *Lancet, 340,* 214–216.

Gilford, R., & Bengtson, V. L. (1979). Measuring marital satisfaction in three generations: Positive and negative dimensions. *Journal of Marriage and the Family 41,* 387–398.

Gilligan, C. F. (1982). *In a different voice.* Cambridge, MA: Harvard University Press.

Gillmore, M. R., Hawkins, J. D., Day, L. E., & Catalano, R. F. (1997). Friendship and deviance: New evidence on an old controversy. *Journal of Early Adolescence, 16.*

Ginsburg, H. P., & Opper, S. (1988). *Piaget's theory of intellectual development* (3rd ed.). Englewood Cliffs, NJ: Prentice Hall.

Ginzberg, E. (1972). Toward a theory of occupational choice: A restatement. *Vocational Guidance Quarterly, 20,* 169–176.

Ginzberg, E. (1988). Toward a theory of occupational choice. *Career Development Quarterly, 36,* 358–363.

Glaser, K. (1997). The living arrangements of elderly people. *Reviews in Clinical Gerontology, 7,* 63–72.

Gleitman, L. R. (1990). The structural sources of verb meanings. *Language Acquisition, 1,* 3–55.

Glendenning, F. (1993). What is elder abuse and neglect? In P. Decalmer & F. Glendenning (Eds.), *The mistreatment of elderly people* (pp. 1–34). London: Sage.

Glick, H. R. (1992). *The right to die.* New York: Columbia University Press.

Glick, P. C. (1990). American families: As they are and were. *Sociology and Social Research, 74,* 139–145.

Glidden, L. M., & Pursley, J. T. (1989). Longitudinal comparisons of families who have adopted children with mental retardation. *American Journal on Mental Retardation, 94,* 272–277.

Gnepp, J. (1983). Children's social sensitivity: Inferring emotions from conflicting cues. *Developmental Psychology, 19,* 805–814.

Goffin, S. G. (1988, March). Putting our advocacy efforts into a new context. *Young Children, 43* (3), 52–56.

Gold, D. T. (1996). Continuities and discontinuities in sibling relationships across the life span. In V. L. Bengtson (Ed.), *Adulthood and aging: Research on continuities and discontinuities* (pp. 228–243). New York: Springer.

Goldberg, A. P., Dengel, D. R., & Hagberg, J. M. (1996). Exercise physiology and aging. In E. L. Schneider & J. W. Rowe (Eds.), *Handbook of the biology of aging* (pp. 331–354). San Diego: Academic Press.

Goldberg, G. S., & Kremen, E. (1990). *The feminization of poverty: Only in America?* New York: Praeger.

Goldin-Meadow, S., & Morford, M. (1985). Gesture in early language: Studies of deaf and hearing children. *Merrill-Palmer Quarterly, 31,* 145–176.

Goldman, L. E. (1996, September). We can help children grieve: A child-oriented model for memorializing. *Young Children, 51*(6), 69–73.

Goldman, N., & Takahashi, S. (1996). Old-age mortality in Japan: Demographic and epidemiological perspectives. In G. Caselli & A. D. Lopez (Eds.), *Health and mortality among elderly populations* (pp. 157–181). New York: Oxford University Press.

Goldscheider, F. K., & Goldscheider, C. (1993). *Leaving home before marriage: Ethnicity, familism, and generational relationships.* Madison: University of Wisconsin Press.

Goldsmith, H. H. (1987). Roundtable: What is temperament? Four approaches. *Child Development, 58,* 505–529.

Goldsmith, H. H., & Rothbart, M. (1990). *The Laboratory Temperament Assessment Battery.* Eugene: University of Oregon.

Goldstein, E. (1979). Effect of same-sex and cross-sex role models on the subsequent academic productivity of scholars. *American Psychologist, 34,* 407–410.

Golomb, C., & Galasso, L. (1995). Make believe and reality: Explorations of the imaginary realm. *Developmental Psychology, 31,* 800–810.

Golombok, S., & Tasker, F. L. (1996). Do parents influence the sexual orientation of their children? Findings from a longitudinal study of lesbian families. *Developmental Psychology, 32,* 3–11.

Gomez-Schwartz, B., Horowitz, J. M., & Cardarelli, A. P. (1990). *Child sexual abuse: Initial effects.* Newbury Park, CA: Sage.

Gomolka, M., Menninger, H., Saal, J. E., Lemmel, E.-M., Albert, E. D., Niwa, O., Epplen, J. T., & Epplen, C. (1995). Immunoprinting: Various genes are associated with increased risk to develop rheumatoid arthritis in different groups of adult patients. *Journal of Molecular Medicine, 73,* 19–29.

Göncü, A. (1993). Development of intersubjectivity in the dyadic play of preschoolers. *Early Childhood Research Quarterly, 8,* 99–116.

Gonzalez, N. M., & Campbell, M. (1994). Cocaine babies: Does prenatal exposure to cocaine affect development? *Journal of the American Academy of Child and Adolescent Psychiatry, 33,* 16–19.

Good, T. L., & Brophy, J. E. (1994). *Looking in classrooms.* New York: HarperCollins.

Goodman, G. S., Hirschman, J. E., Hepps, D., & Rudy, L. (1991). Children's memory for stressful events. *Merrill-Palmer Quarterly, 37,* 109–158.

Goodman, G. S., & Tobey, A. E. (1994). Memory development within the context of child sexual abuse investigations. In C. B. Fisher & R. M. Lerner (Eds.), *Applied developmental psychology* (pp. 46–75). New York: McGraw-Hill.

Goodman, K. S. (1986). *What's whole in whole language?* Portsmouth, NH: Heinemann.

Goodman, L. A., Koss, M. P., & Russo, N. F. (1993). Violence against women: Physical and mental health effects. *Applied and Preventive Psychology, 2,* 79–89.

Goodman, R. A., & Whitaker, H. A. (1985). Hemispherectomy: A review (1928–1981) with special reference to the linguistic abilities and disabilities of the residual right hemisphere. In R. A. Goodman & H. A. Whitaker (Eds.), *Hemispheric functions and collaboration in the child* (pp. 121–155). New York: Academic Press.

Goodman, S. H., Gravitt, G. W., Jr., & Kaslow, N. J. (1995). Social problem solving: A moderator of the relation between negative life stress and depressive symptoms in children. *Journal of Abnormal Child Psychology, 23,* 473–485.

Goodwin, D. W. (1992). Alcohol: Clinical aspects. In J. H. Lowinson, P. Ruiz, & R. B. Millman (Eds.), *Substance abuse* (2nd ed., pp. 144–151). Baltimore: Williams & Wilkins.

Goodwin, J. S. (1995). Decreased immunity and increased morbidity in the elderly. *Nutrition Reviews, 53,* S41–S46.

Gopnik, A., & Meltzoff, A. N. (1986). Relations between semantic and cognitive development in the one-word stage: The specificity hypothesis. *Child Development, 57,* 1040–1053.

Gopnik, A., & Meltzoff, A. N. (1987). The development of categorization in the second year and its relation to other cognitive and linguistic developments. *Child Development, 58,* 1523–1531.

Gopnik, A., & Meltzoff, A. N. (1992). Categorization and naming: Basic-level sorting in eighteen-month-olds and its relation to language. *Child Development, 63,* 1091–1103.

Gorman, J., Leifer, M., & Grossman, G. (1993). Nonorganic failure to thrive: Maternal history and current maternal functioning. *Journal of Clinical Child Psychology, 22,* 327–336.

Gorn, G. J., & Goldberg, M. E. (1982). Behavioral evidence of the effects of televised food messages on children. *Journal of Consumer Research, 9,* 200–205.

Gortmaker, S. L., Dietz, W. H., Jr., & Cheung, L. W. Y. (1990). Inactivity, diet, and the fattening of America. *Journal of the American Dietetic Association, 90,* 1247–1252.

Gortmaker, S. L., Must, A., Perrin, J. M., Sobol, A. M., & Dietz, W. H. (1993). Social and economic consequences of overweight in adolescence and young adulthood. *New England Journal of Medicine, 329,* 1008–1012.

Gottesman, I. I. (1963). Genetic aspects of intelligent behavior. In N. Ellis (Ed.), *Handbook of mental deficiency* (pp. 253–296). New York: McGraw-Hill.

Gottfried, A. E. (1991). Maternal employment in the family setting: Developmental and environmental issues. In J. V. Lerner & N. L. Galambos (Eds.), *Employed mothers and their children* (pp. 63–84). New York: Garland.

Gottlieb, G. (1991). Experiential canalization of behavioral development: Theory. *Developmental Psychology, 27,* 4–13.

Gottman, J. M., Gonso, J., & Rasmussen, B. (1975). Social interaction, social competence, and friendship in children. *Child Development, 46,* 709–718.

Graber, J. A., Brooks-Gunn, J., Paikoff, R. L., & Warren, M. P. (1994). Prediction of eating problems: An 8-year study of adolescent girls. *Developmental Psychology, 30,* 823–834.

Graham, S., Doubleday, C., & Guarino, P. A. (1984). The development of relations between perceived controllability and the emotions of pity, anger, and guilt. *Child Development, 55,* 561–565.

Gralinski, J. H., & Kopp, C. B. (1993). Everyday rules for behavior: Mothers' requests to young children. *Developmental Psychology, 29,* 573–584.

Grant, J. P. (1995). *The state of the world's children 1995.* New York: Oxford University Press (in cooperation with UNICEF).

Grantham-McGregor, S., Powell, C., Walker, S., Chang, S., & Fletcher, P. (1994). The long-term follow-up of severely malnourished children who participated in an intervention program. *Child Development, 65,* 428–439.

Grantham-McGregor, S., Schofield, W., & Powell, C. (1987). Development of severely malnourished children who received psychosocial stimulation: Six-year follow-up. *Pediatrics, 79,* 247–254.

Graves, S. B. (1993). Television, the portrayal of African Americans, and the development of children's attitudes. In G. L. Berry & J. K. Asamen (Eds.), *Children and television* (pp. 179–190). Newbury Park, CA: Sage.

Green, G. D., & Bozett, F. W. (1991). Lesbian mothers and gay fathers. In J. C. Gonsiorek & J. D. Weinrich (Eds.), *Homosexuality: Research implications for public policy* (pp. 197–214). Newbury Park, CA: Sage.

Green, J. A., Jones, L. E., & Gustafson, G. E. (1987). Perception of cries by parents and nonparents: Relation to cry acoustics. *Developmental Psychology, 23,* 370–382.

Greenberg, P. (1990). Why not academic preschool? *Young Children, 45*(2), 70–80.

Greenberger, E., & Chen, C. (1996). Perceived family relationships and depressed mood in early and late adolescence: A comparison of European and Asian Americans. *Developmental Psychology, 32,* 707–716.

Greenberger, E., & Goldberg, W. A. (1989). Work, parenting, and the socialization of children. *Developmental Psychology, 25,* 22–35.

Greenberger, E., O'Neil, R., & Nagel, S.K. (1994). Linking workplace and homeplace: Relations between the nature of adults' work and their parenting behavior. *Developmental Psychology, 30,* 990–1002.

Greenberger, E., & Steinberg, L. (1986). *When teenagers work.* New York: Basic Books.

Greenfield, P. (1992, June). *Notes and references for developmental psychology.* Conference on Making Basic Texts in Psychology More Culture-Inclusive and Culture-Sensitive, Western Washington University, Bellingham, WA.

Greenfield, P. M. (1994). Independence and interdependence as developmental scripts: Implications for theory, research, and practice. In P. M. Greenfield & R. R. Cocking (Eds.), *Cross-cultural roots of minority child development* (pp. 1–37). Hillsdale, NJ: Erlbaum.

Greeno, J. G. (1989). A perspective on thinking. *American Psychologist, 44,* 134–141.

Greenough, W. T., & Black, J. E. (1992). Induction of brain structure by experience: Substrates for cognitive development. In M. R. Gunnar & C. A. Nelson (Eds.), *Minnesota Symposia on Child Psychology* (pp. 155–200). Hillsdale, NJ: Erlbaum.

Greenough, W. T., Wallace, C. S., Alcantara, A. A., Anderson, B. J., Hawrylak, N., Sirevaag, A. M., Weiler, I. J., & Withers, G. S. (1993). Experience affects the structure of neurons, glia, and blood vessels. In N. J. Anastasiow & S. Harel (Eds.), *At-risk infants: Interventions, family, and research* (pp. 175–185). Baltimore: Paul H. Brookes.

Greenstein, D., Carlson, J., & Howell, C. W. (1993). Counseling with interfaith couples. *Individual Psychology, 49,* 428–437.

Gresser, G., Wong, P., & Reker, G. (1987). Death attitudes across the life-span: The development and validation of the Death Attitude Profile (DAP). *Omega, 18,* 113–128.

Griffiths, A., Roberts, G., & Williams, J. (1993). Elder abuse and the law. In P. Decalmer & F. Glendenning (Eds.), *The mistreatment of elderly people* (pp. 62–75). London: Sage.

Grolnick, W. S., & Slowiaczek, M. L. (1994). Parents' involvement in children's schooling: A multidimensional conceptualization and motivational model. *Child Development, 65,* 237–252.

Grossmann, K., Grossmann, K. E., Spangler, G., Suess, G., & Unzner, L. (1985). Maternal sensitivity and newborns' orientation responses as related to quality of attachment in Northern Germany. In I. Bretherton & E. Waters (Eds.), Growing points of attachment theory and research. *Monographs of the Society for Research in Child Development, 50* (1–2, Serial No. 209).

Grotevant, H. D., & Cooper, C. R. (1988). The role of family experience in career exploration during adolescence. In P. Baltes, D. Featherman, & R. Lerner (Eds.), *Life-span development and behavior* (Vol. 8, pp. 231–258). Hillsdale, NJ: Erlbaum.

Grotevant, H. D., & Cooper, C. R. (1985). Patterns of interaction in family relationships and the development of identity exploration in adolescence. *Child Development, 56,* 415–428.

Gruetzner, H. (1992). *Alzheimer's.* New York: Wiley.

Grusec, J. E. (1988). *Social development: History, theory, and research.* New York: Springer.

Guerra, N. G., Tolan, P. H., & Hammond, W. R. (1994). Prevention and treatment of adolescent violence. In L. D. Eron, J. H. Gentry, & P. Schlegel (Eds.), *Reason to hope: A psychosocial perspective on violence and youth* (pp. 383–403). Washington, DC: American Psychological Association.

Guidubaldi, J., & Cleminshaw, H. K. (1985). Divorce, family health and child adjustment. *Family Relations, 34,* 35–41.

Guilford, J. P. (1985). The structure-of-intellect model. In B. B. Wolman (Ed.), *Handbook of intelligence* (pp. 225–266). New York: Wiley.

Gullette, M. M. (1988). *Safe at last in the middle years.* Berkeley: University of California Press.

Gunnar, M. R., & Nelson, C. A. (1994). Event-related potentials in year-old infants: Relations with emotionality and cortisol. *Child Development, 65,* 80–94.

Gupta, V., & Korte, C. (1994). The effects of a confidant and a peer group on the well-being of single elders. *International Journal of Aging and Human Development, 39,* 293–302.

Gustafson, G. E., & Harris, K. L. (1990). Women's responses to young infants' cries. *Developmental Psychology, 26,* 144–152.

Gutmann, D. (1977). The cross-cultural perspective: Notes toward a comparative psychology of aging. In J. E. Birren & K. W. Schaie (Eds.), *Handbook of the psychology of aging* (pp. 302–326). New York: Van Nostrand Reinhold.

Gutmann, D. L. (1987). *Reclaimed powers: Toward a new psychology of men and women in later life.* New York: Basic Books.

Gutmann, D. L., & Huyck, M. H. (1994). Development and pathology in postparental men: A community study. In E. Thompson, Jr. (Ed.), *Older men's lives* (pp. 65–84). Thousand Oaks, CA: Sage.

Guyer, B., Strolino, D. M., Ventura, S. J., & Singh, G. K. (1995). Annual summary of vital statistics—1995. *Pediatrics, 96,* 1029–1039.

Haan, N. (1972). Personality development from adolescence to adulthood in the Oakland Growth and Guidance Studies. *Seminars in Psychiatry, 4,* 399–414.

Haan, N., Aerts, E., & Cooper, B. (1985). *On moral grounds: The search for practical morality.* New York: New York University Press.

Hack, M. B., Taylor, H. G., Klein, N., Eiben, R., Schatschneider, C., & Mercuri-Minich, N. (1994). School-age outcomes in children with birth weights under 750 g. *New England Journal of Medicine, 331,* 753–759.

Hackel, L. S., & Ruble, D. N. (1992). Changes in the marital relationship after the first baby is born: Predicting the impact of expectancy disconfirmation. *Journal of Personality and Social Psychology, 62,* 944–957.

Haden, C. A., Haine, R. A., & Fivush, R. (1997). Developing narrative structure in parent–child reminiscing across the preschool years. *Developmental Psychology, 33,* 295–307.

Hadjistavropoulos, H. D., Craig, K. D., Grunau, R. V. E., & Johnston, C. C. (1994). Judging pain in newborns: Facial and cry determinants. *Journal of Pediatric Psychology, 19,* 485–491.

Hagberg, J. M. (1987). Effects of training on the decline of VO$_{2max}$ with aging. *Federation Proceedings, 46,* 1830–1833.

Hagberg, J. M., Allen, W. K., Seals, D. R., Hurley, B. F., Ehsani, A. A., & Holloszy, J. O. (1985). A hemodynamic comparison of young and older endurance athletes during exercise. *Journal of Applied Physiology, 58,* 2041–2046.

Hager, M. (1996, December 23). The cancer killer. *Newsweek,* pp. 42–47.

Hahn, W. K. (1987). Cerebral lateralization of function: From infancy through childhood. *Psychological Bulletin, 101,* 376–392.

Haight, B. K. (1992). Long-term effects of a structured life review process. *Journal of Gerontology, 47,* P312–P315.

Haight, W. L., & Miller, P. J. (1993). *Pretending at home: Early development in a sociocultural context.* Albany, NY: State University of New York Press.

Hainline, L. (1993). Conjugate eye movements of infants. In K. Simons (Ed.), *Early visual development: Normal and abnormal* (pp. 47–55). New York: Oxford University Press.

Hakuta, K., Ferdman, B. M., & Diaz, R. M. (1987). Bilingualism and cognitive development: Three perspectives. In S. Rosenberg (Ed.), *Advances in applied psycholinguistics: Vol. 2. Reading, writing, and language learning* (pp. 284–319). New York: Cambridge University Press.

Halford, G. S. (1992). *Children's understanding: The development of mental models.* Hillsdale, NJ: Erlbaum.

Hall, D. G. (1996). Preschoolers' default assumptions about word meaning: Proper names designate unique individuals. *Developmental Psychology, 32,* 177–186.

Hall, G. S. (1904). *Adolescence.* New York: Appleton.

Hall, G. S. (1922). *Senescence: The last half of life.* New York: Appleton.

Hall, J. G., Sybert, V. P., Williamson, R. A., Fisher, N. L., & Reed, S. D. (1982). Turner's syndrome. *West Journal of Medicine, 137,* 32–44.

Hall, W. S. (1989). Reading comprehension. *American Psychologist, 44,* 157–161.

Halliday, J. L., Watson, L. F., Lumley, J., Danks, D. M., & Sheffield, L. S. (1995). New estimates of Down syndrome risks of chorionic villus sampling, amniocentesis, and live birth in women of advanced maternal age from a uniquely defined population. *Prenatal Diagnosis, 15,* 455–465.

Halloran, C. M., & Altmaier, E. M. (1996). Awareness of death among children: Does a life-threatening illness alter the process of discovery? *Journal of Counseling & Development, 74,* 259–262.

Halmi, K. A. (1987). Anorexia nervosa and bulimia. In V. B. Van Hasselt & M. Hersen (Eds.), *Handbook of adolescent psychology* (pp. 265–287). New York: Pergamon.

Halpern, D. F. (1992). *Sex differences in cognitive abilities* (2nd ed.). Hillsdale, NJ: Erlbaum.

Halverson, H. M. (1931). An experimental study of prehension in infants by means of systematic cinema records. *Genetic Psychology Monographs, 10,* 107–286.

Hamachek, D. (1990). Evaluating self-concept and ego status in Erikson's last three psychosocial stages. *Journal of Counseling & Development, 68,* 677–683.

Hamberger, L. K., Saunders, D. G., & Hovey, M. (1992). The prevalence of domestic violence in community practice and rate of physician inquiry. *Family Medicine, 24,* 283–287.

Hamelin, K., & Ramachandran, C. (1993, June). Kangaroo care. *Canadian Nurse, 89*(6), 15–17.

Hamer, D. H., Hu, S., Magnuson, V. L., Hu, N., & Pattatucci, A. M. L. (1993). A linkage between DNA markers on the X chromosome and male sexual orientation. *Science, 261,* 321–327.

Hamilton, S. F. (1990). *Apprenticeship for adulthood: Preparing youth for the future.* New York: Free Press.

Hamilton, S. F. (1993). Prospects for an American-style youth apprenticeship system. *Educational Researcher, 22*(3), 11–16.

Hamilton, S. F., & Hurrelmann, K. (1994). The school-to-career transition in Germany and the United States. *Teachers College Record, 96,* 329–344.

Hammill, D. D. (1990). On defining learning disabilities: An emerging consensus. *Journal of Learning Disabilities, 23,* 74–84.

Hanna, E., & Meltzoff, A. N. (1993). Peer imitation by toddlers in laboratory, home, and day-care contexts: Implications for social learning and memory. *Developmental Psychology, 29,* 701–710.

Hardy, M. A., & Hazelrigg, L. E. (1993). The gender of poverty in an aging population. *Research on Aging, 15,* 243–278.

Hare, J. (1994). Concerns and issues faced by families headed by a lesbian couple. *Families in Society, 43,* 27–35.

Hare, J., & Richards, L. (1993). Children raised by lesbian couples: Does context of birth affect father and partner involvement? *Family Relations, 42,* 249–255.

Harlow, H. F., & Zimmerman, R. (1959). Affectional responses in the infant monkey. *Science, 130,* 421–432.

Harootyan, R. A., & Vorek, R. E. (1994). Volunteering, helping and gift giving in families and communities. In V. L. Bengtson & R. A. Harootyan (Eds.), *Intergenerational linkages: Hidden connections in American society* (pp. 77–111). New York: Springer.

Harpaz, D., Benderly, M., Goldbourt, U., Kishan, Y., & Behar, S. (1996). Effect of aspirin on mortality in women with symptomatic or silent myocardial ischemia. *American Journal of Cardiology, 78,* 1215–1219.

Harris, M. B. (1994). Growing old gracefully: Age concealment and gender. *Journal of Gerontology, 49,* P149–P158.

Harris, M. J., & Rosenthal, R. (1985). Mediation of interpersonal expectancy effects: 31 meta-analyses. *Psychological Bulletin, 97,* 363–386.

Harris, R. L., Ellicott, A. M., & Holmes, D. S. (1986). The timing of psychosocial transitions and changes in women's lives: An examination of women aged 45 to 60. *Journal of Personality and Social Psychology, 51,* 409–416.

Harris, R. T. (1991, March–April). Anorexia nervosa and bulimia nervosa in female adolescents. *Nutrition Today, 26*(2), 30–34.

Harris, S., Mussen, P. H., & Rutherford, E. (1976). Some cognitive, behavioral, and personality correlates of maturity of moral judgment. *Journal of Genetic Psychology, 128,* 123–135.

Harrison, A. O., Wilson, M. N., Pine, C. J., Can, S. Q., & Buriel, R. (1994). Family ecologies of ethnic minority children. In G. Handel & G. G. Whitchurch (Eds.), *The psychosocial interior of the family* (pp. 187–210). New York: Aldine de Gruyter.

Harrison, M. R. (1993). Fetal surgery. *Western Journal of Medicine, 159,* 341–349.

Hart, B., & Risley, T. R. (1995). *Meaningful differences in the everyday experience of young American children.* Baltimore: Paul H. Brookes.

Harter, S. (1982). The perceived competence scale for children. *Child Development, 53,* 87–97.

Harter, S. (1990). Issues in the assessment of the self-concept of children and adolescents. In A. LaGreca (Ed.), *Through the eyes of a child* (pp. 292–325). Boston: Allyn & Bacon.

Harter, S., & Buddin, B. J. (1987). Children's understanding of the simultaneity of two emotions: A five-stage developmental acquisition sequence. *Developmental Psychology, 23,* 388–399.

Harter, S., & Whitesell, N. (1989). Developmental changes in children's understanding of simple, multiple, and blended emotion concepts. In C. Saarni & P. Harris (Eds.), *Children's understanding of emotion* (pp. 81–116). Cambridge, England: Cambridge University Press.

Harter, S., Wright, K., & Bresnick, S. (1987). *A developmental sequence of the emergence of self affects.* Paper presented at the biennial meeting of the Society for Research in Child Development, Baltimore.

Hartley, A. A., & McKenzie, C. R. M. (1991). Attentional and perceptual contributions to the identification of extrafoveal stimuli: Adult age comparisons. *Journal of Gerontology, 46,* P202–P206.

Hartley, R. (1995). *Families and cultural diversity in Australia.* Sydney: Allen & Unwin.

Hartman, M., & Hasher, L. (1991). Aging and suppression: Memory for previously relevant information. *Psychology and Aging, 6,* 587–594.

Hartup, W. W. (1996). The company they keep: Friendships and their developmental significance. *Child Development, 67,* 1–13.

Hartup, W. W., French, D. C., Laursen, B., Johnston, M. K., & Ogawa, J. R. (1993). Conflict and friendship relations in middle childhood: Behavior in a closed-field situation. *Child Development, 64,* 445–454.

Harway, M., & Hansen, M. (1994). *Spouse abuse.* Sarasota, FL: Professional Resource Press.

Hashima, P. Y., & Amato, P. R. (1994). Poverty, social support, and parental behavior. *Child Development, 65,* 394–403.

Hashimoto, K., Noguchi, M., & Nakatsuji, N. (1992). Mouse offspring derived from fetal ovaries or reaggregates which were cultured and transplanted into adult females. *Development: Growth & Differentiation, 34,* 233–238.

Hashtroudi, S., Johnson, M. K., & Chrosniak, L. D. (1990). Aging and qualitative characteristics of memories for perceived and imagined complex events. *Psychology and Aging, 5,* 119–126.

Hatch, L. R., & Thompson, A. (1992). Family responsibilities and women's retirement. In M. E. Szinovacz, D. J. Ekerdt, & B. H. Vinick (Eds.), *Families and retirement* (pp. 99–113). Newbury Park, CA: Sage.

Hatch, M. C., Shu, X-O., McLean, D. E., Levin, B., Begg, M., Reuss, L., & Susser, M. (1993). Maternal exercise during pregnancy, physical fitness, and fetal growth. *American Journal of Epidemiology, 137,* 1105–1114.

Hatcher, L., & Crook, J. C. (1988). First-job surprises for college graduates: An exploratory investigation. *Journal of College Student Personnel, 29,* 441–448.

Hatfield, E. (1988). Passionate and companionate love. In R. J. Sternberg & M. L. Barnes (Eds.), *The psychology of love* (pp. 191–217). New Haven, CT: Yale University Press.

Hatfield, E. (1993). *Love, sex, and intimacy: Their psychology, biology, and history.* New York: HarperCollins.

Hauser, S. T. (1995, March). *Exceptional outcomes: Negotiating a perilous adolescence.* Paper presented at the biennial meeting of the Society for Research in Child Development, Indianapolis.

Havighurst, R., & Albrecht, R. (1953). *Older people.* New York: Longmans, Green.

Haviland, J., & Lelwica, M. (1987). The induced affect response: 10-week-old infants' responses to three emotion expressions. *Developmental Psychology, 23,* 97–104.

Hawkins, A. J., Christiansen, S. L., Sargent, K. P., & Hill, E. J. (1993). Rethinking fathers' involvement in child care: A developmental perspective. *Journal of Family Issues, 14,* 531–549.

Hawkins, J. D., & Lam, T. (1987). Teacher practices, social development, and delinquency. In J. D. Burchard & S. N. Burchard (Eds.), *Prevention of delinquent behavior* (pp. 241–274). Newbury Park, CA: Sage.

Hawkins, J. D., Catalano, R. F., & Miller, J. Y. (1992). Risk and protective factors for alcohol and other drug problems in adolescence and early adulthood: Implications for substance abuse prevention. *Psychological Bulletin, 112,* 64–105.

Hayflick, L. (1994). *How and why we age.* New York: Ballantine.

Hayne, H., Rovee-Collier, C., & Perris, E. E. (1987). Categorization and memory retrieval by three-month-olds. *Child Development, 58,* 750–767.

Hayslip, B., Jr. (1994). Stability of intelligence. In R. J. Sternberg (Ed.), *Encyclopedia of human intelligence* (Vol. 2, pp. 1019–1026). New York: Macmillan.

Hayward, M. D., Hardy, M. A., & Liu, M. (1994). Work after retirement: The experiences of older men in the U.S. *Social Science Research, 23,* 82–107.

Hazan, C., & Shaver, P. R. (1987). Romantic love conceptualized as an attachment process. *Journal of Personality and Social Psychology, 52,* 511–524.

Heath, S. B. (1982). Questioning at home and at school: A comparative study. In G. Spindler (Ed.), *Doing the ethnography of schooling: Educational anthropology in action* (pp. 102–127). New York: Holt.

Heath, S. B. (1989). Oral and literate traditions among black Americans living in poverty. *American Psychologist, 44,* 367–373.

Heath, S. B. (1990). The children of Trackton's children: Spoken and written in social change. In J. Stigler, G. Herdt, & R. A. Shweder (Eds.), *Cultural psychology: Essays on comparative human development* (pp. 496–519). New York: Cambridge University Press.

Hecht, M., Marston, P. J., & Larkey, L. K. (1994). Love ways and relationship quality. *Journal of Social and Personal Relationships, 11,* 25–43.

Heidrich, S. M., & Denney, N. W. (1994). Does social problem solving differ from other types of problem solving during the adult years? *Experimental Aging Research, 20,* 105–126.

Heidrich, S. M., & Ryff, C. D. (1993). Physical and mental health in later life: The self-system as mediator. *Psychology and Aging, 8,* 327–338.

Heinl, T. (1983). *The baby massage book.* London: Coventure.

Heinonen, O. P., Slone, D., & Shapiro, S. (1977). *Birth defects and drugs in pregnancy.* Littleton, MA: PSG Publishing Company.

Helburn, S. W. (Ed.). (1995). *Cost, quality and child outcomes in child care centers.* Denver: University of Colorado.

Helmchen, H., Baltes, M. M., Geiselmann, B., Kanowski, S., Linden, M., Reischies, F. M., Wagner, M., & Wilms, H.-U. (1996). Psychische Erkrankungen im Alter [Psychiatric illnesses in old age]. In K. U. Mayer & P. B. Baltes (Eds.), *Die Berliner Altersstudie* (pp. 185–220). Berlin, Germany: Akademie Verlag.

Helmers, K. F., Posluszny, D. M., & Krantz, D. S. (1994). Associations of hostility and coronary artery disease: A review of studies. In A. W. Siegman & T. W. Smith (Eds.), *Anger, hostility, and the heart* (pp. 67–96). Hillsdale, NJ: Erlbaum.

Helmrich, S. P., Ragland, D. R., & Paffenbarger, R. S., Jr. (1994). Prevention of non-insulin-dependent diabetes mellitus with physical activity. *Medicine and Science in Sports and Exercise, 26,* 824–830.

Helson, R., Mitchell, V., & Moane, G. (1984). Personality and patterns of adherence and nonadherence to the social clock. *Journal of Personality and Social Psychology, 46,* 1079–1096.

Helson, R., & Moane, G. (1987). Personality change in women from college to midlife. *Journal of Personality and Social Psychology, 53,* 176–186.

Helson, R., & Picano, J. (1990). Is the traditional role bad for women? *Journal of Personality and Social Psychology, 59,* 311–320.

Helson, R., & Roberts, B. W. (1994). Ego development and personality change in adulthood. *Journal of Personality and Social Psychology, 66,* 911–920.

Helson, R., & Stewart, A. (1994). Personality change in adulthood. In T. F. Heatherton & J. L. Weinberger (Eds.), *Can personality change?* (pp. 201–225). Washington, DC: American Psychological Association.

Helson, R., & Wink, P. (1992). Personality change in women from the early 40s to the early 50s. *Psychology and Aging, 7,* 46–55.

Hendin, H. (1995). Assisted Suicide, euthanasia, and suicide prevention: The implications of the Dutch experience. *Suicide and Life-Threatening Behavior, 25,* 193–204.

Hendrick, C., & Hendrick, S. (1986). A theory and method of love. *Journal of Personality and Social Psychology, 50,* 392–402.

Hendrick, S. S., & Hendrick, C. (1992). *Romantic love.* Newbury Park, CA: Sage.

Herek, G. M., & Glunt, E. K. (1993). Interpersonal contact and heterosexuals' attitudes toward gay men: Results from a national survey. *Journal of Sex Research, 30,* 239–244.

Hergenrather, J. R., & Rabinowitz, M. (1991). Age-related differences in the organization of children's knowledge of illness. *Developmental Psychology, 27,* 952–959.

Herman, J. L. (1992). *Trauma and recovery.* New York: Basic Books.

Herrnstein, R. J., & Murray, C. (1994). *The bell curve.* New York: Free Press.

Hershberger, S. L., & D'Augelli, A. R. (1995). The impact of victimization on the mental health and suicidality of lesbian, gay, and bisexual youths. *Developmental Psychology, 31,* 65–74.

Hetherington, E. M. (1989). Coping with family transitions: Winners, losers, and survivors. *Child Development, 60,* 1–14.

Hetherington, E. M. (1991). The role of individual differences and family relationships in children's coping with divorce and remarriage. In P. A. Cowan & E. M. Hetherington (Eds.), *Family transitions* (pp. 165–194). Hillsdale, NJ: Erlbaum.

Hetherington, E. M. (1993). An overview of the Virginia Longitudinal Study of Divorce and Remarriage: A focus on early adolescence. *Journal of Family Psychology, 7,* 39–56.

Hetherington, E. M. (1995). *The changing American family and the well-being of children.* Master lecture presented at the biennial meeting of the Society for Research in Child Development, Indianapolis, March.

Hetherington, E. M., & Clingempeel, W. G. (1992). Coping with marital transitions: A family systems perspective. *Monographs of the Society for Research in Child Development, 57* (2–3, Serial No. 227).

Hetherington, E. M., Cox, M., & Cox, R. (1985). Long-term effects of divorce and remarriage on the adjustment of children. *Journal of the American Academy of Child Psychiatry, 24,* 518–530.

Hetherington, E. M., & Jodl, K. M. (1994). Stepfamilies as settings for child development. In A. Booth & J. Dunn (Eds.), *Stepfamilies: Who benefits? Who does not?* (pp. 55–79). Hillsdale, NJ: Erlbaum.

Hetherington, E. M., Law, T. C., & O'Connor, T. G. (1994). Divorce: Challenges, changes, and new chances. In F. Walsh (Ed.), *Normal family processes* (2nd ed., pp. 208–234). New York: Guilford.

Hetherington, S. E. (1990). A controlled study of the effect of prepared childbirth classes on obstetric outcomes. *Birth, 17,* 86–90.

Hewlett, B. S. (1992). Husband–wife reciprocity and the father–infant relationship among Aka pygmies. In B. S. Hewlett (Ed.), *Father–child relations: Cultural and biosocial contexts* (pp. 153–176). New York: Aldine de Gruyter.

Heyman, G. D., & Dweck, C. S. (1992). Achievement goals and intrinsic motivation: Their relation and their role in adaptive motivation. *Motivation and Emotion, 16,* 231–247.

Heyman, G. D., Dweck, C. S., & Cain, K. M. (1992). Young children's vulnerability to self-blame and helplessness: Relationship to beliefs about goodness. *Child Development, 63,* 401–415.

Hicks, P. (1997, January). The impact of aging on public policy. *OECD Observer, No. 203,* pp. 19–21.

Higgins, C. A., Duxbury, L. E., & Irving, R. H. (1992). Work–family conflict in the dual-career family. *Organizational Behavior and Human Decision Processes, 51,* 51–75.

Hill, E. M., Young, J. P., & Nord, J. L. (1994). Childhood adversity, attachment security, and adult relationships: A preliminary study. *Ethology and Sociobiology, 15,* 323–338.

Hill, J. P., & Holmbeck, G. (1986). Attachment and autonomy during adolescence. In G. Whitehurst (Ed.), *Annals of child development* (Vol. 3, pp. 145–189). Greenwich, CT: JAI Press.

Hill, J. P., & Holmbeck, G. N. (1987). Family adaptation to biological change during adolescence. In R. M. Lerner & T. T. Foch (Eds.), *Biological-psychosocial interactions in early adolescence* (pp. 207–224). Hillsdale, NJ: Erlbaum.

Hillier, L., Hewitt, K. L., & Morrongiello, B. A. (1992). Infants' perception of illusions in sound localization: Reaching to sounds in the dark. *Journal of Experimental Child Psychology, 53,* 159–179.

Hillman, J. L., & Stricker, G. (1994). A linkage of knowledge and attitudes toward elderly sexuality: Not necessarily a uniform relationship. *Gerontologist, 34,* 256–260.

Hills-Banczyk, S. G., Avery, M. D., Savik, K., Potter, S., & Duckett, L. J. (1993). Women's experiences with combining breast-feeding and employment. *Journal of Nurse-Midwifery, 38,* 257–266.

Hinde, R. A. (1989). Ethological and relationships approaches. In R. Vasta (Ed.), *Annals of child development* (Vol. 6, pp. 251–285). Greenwich, CT: JAI Press.

Hines, M., & Green, R. (1991). Human hormonal and neural correlates of sex-typed behaviors. *Review of Psychiatry, 10,* 536–555.

Hirsch, C. (1996). Understanding the influence of gender role identity on the assumption of family caregiving roles by men. *International Journal of Aging and Human Development, 42,* 103–121.

Hirsh-Pasek, K., Kemler Nelson, D. G., Jusczyk, P. W., Cassidy, K. W., Druss, B., & Kennedy, L. (1987). Clauses are perceptual units for young infants. *Cognition, 26,* 269–286.

Hiscock, M., & Kinsbourne, M. (1987). Specialization of the cerebral hemispheres: Implications for learning. *Journal of Learning Disabilities, 20,* 130–143.

Hobart, C., & Brown, D. (1988). Effects of prior marriage children on adjustment in remarriages: A Canadian study.

Journal of Comparative Family Studies, 19, 381–396.

Hodges, J., & Tizard, B. (1989). Social and family relationships of ex-institutional adolescents. *Journal of Child Psychology and Psychiatry, 30*, 77–97.

Hodges, R. M., & French, L. A. (1988). The effect of class and collection labels on cardinality, class-inclusion, and number conservation tasks. *Child Development, 59*, 1387–1396.

Hodson, D. S., & Skeen, P. (1994). Sexuality and aging: The hammerlock of myths. *Journal of Applied Gerontology, 13*, 219–235.

Hoff-Ginsburg, E. (1986). Function and structure in maternal speech: Their relation to the child's development of syntax. *Developmental Psychology, 22*, 155–163.

Hoffman, L. W. (1989). Effects of maternal employment in the two-parent family. *American Psychologist, 44*, 283–292.

Hoffman, L. W., Thornton, A., & Manis, J. D. (1978). The value of children to parents in the United States. *Journal of Population, 1*, 91–131.

Hoffman, M. L. (1988). Moral development. In M. H. Bornstein & M. E. Lamb (Eds.), *Developmental psychology: An advanced textbook* (2nd ed., pp. 497–548). Hillsdale, NJ: Erlbaum.

Hofstede, G. (1980). *Culture's consequences: International differences in work-related values.* London: Sage.

Holahan, C. K. (1984). Marital attitudes over 40 years: A longitudinal cohort analysis. *Journal of Gerontology, 39*, 49–57.

Holahan, C. K. (1988). Relation of life goals at age 70 to activity participation and health and psychological well-being among Terman's gifted men and women. *Psychology and Aging, 3*, 286–291.

Holden, G. W. (1983). Avoiding conflict: Mothers as tacticians in the supermarket. *Child Development, 54*, 233–240.

Holden, G. W., & West, M. J. (1989). Proximate regulation by mothers: A demonstration of how differing styles affect young children's behavior. *Child Development, 60*, 64–69.

Holland, J. L. (1966). *The psychology of vocational choice.* Waltham, MA: Blaisdell.

Holland, J. L. (1985). *Making vocational choices: A theory of vocational personalities and work environments.* Englewood Cliffs, NJ: Prentice Hall.

Holmbeck, G. N., & Hill, J. P. (1991). Conflictive engagement, positive affect, and menarche in families with seventh-grade girls. *Child Development, 62*, 1030–1048.

Holmbeck, G. N., & Wandrei, M. L. (1993). Individual and relational predictors of adjustment in first-year college students. *Journal of Counseling Psychology, 40*, 73–78.

Holmbeck, G. N., Waters, K. A., & Brookman, R. R. (1990). Psychosocial correlates of sexually transmitted diseases and sexual activity in black adolescent females. *Journal of Adolescent Research, 5*, 431–448.

Holmes, E. R., & Holmes, L. D. (1995). *Other cultures, elder years* (2nd ed.). Thousand Oaks, CA: Sage.

Holmes, L. B. (1993). Report on the National Institute of Child Health and Human Development workshop on chorionic villus sampling and limb and other defects. *Teratology, 48*, 7–13.

Holzman, C., & Paneth, N. (1994). Maternal cocaine use during pregnancy and perinatal outcomes. *Epidemiologic Reviews, 16*, 315–334.

Honeycutt, J. M. (1991). The endorsement of myths about marriage as a function of gender, age, religious denomination, and educational level. *Communication Research Reports, 8*, 101–111.

Hook, E. B. (1988). Evaluation and projection of rates of chromosome abnormalities in chorionic villus studies (c.v.s.). *American Journal of Human Genetics Supplement, 43*, A108.

Hooker, K. (1992). Possible selves and perceived health in older adults and college students. *Journal of Gerontology, 47*, P85–P89.

Hooker, K., & Kaus, C. R. (1992). Possible selves and health behaviors in later life. *Journal of Aging & Health, 4*, 390–411.

Hooker, K., & Kaus, C. R. (1994). Health-related possible selves in young and middle adulthood. *Psychology and Aging, 9*, 126–133.

Hooyman, N. R., & Kiyak, H. A. (1993). *Social gerontology: A multidisciplinary perspective* (3rd ed.). Boston: Allyn & Bacon.

Hopkins, B., & Westra, T. (1988). Maternal handling and motor development: An intracultural study. *Genetic, Social and General Psychology Monographs, 14*, 377–420.

Hopper, S. V. (1993). The influence of ethnicity on the health of older women. *Clinics in Geriatric Medicine, 9*, 231–259.

Horgan, D. (1978). The development of the full passive. *Journal of Child Language, 5*, 65–80.

Horn, J. L. (1982). The aging of human abilities. In B. B. Wolman (Ed.), *Handbook of developmental psychology* (pp. 847–870). Englewood Cliffs, NJ: Prentice-Hall.

Horn, J. L., & Donaldson, G. (1980). Cognitive development in adulthood. In O. G. Brim, Jr., & J. Kagan (Eds.), *Constancy and change in human development* (pp. 445–529). Cambridge, MA: Harvard University Press.

Horn, J. L., Donaldson, G., & Engstrom, R. (1981). Apprehension, memory, and fluid intelligence decline through the "vital years" of adulthood. *Research on Aging, 3*, 33–84.

Horn, J. M. (1983). The Texas Adoption Project: Adopted children and their intellectual resemblance to biological and adoptive parents. *Child Development, 54*, 268–275.

Horn, T. S. (1987). The influence of teacher–coach behavior on the psychological development of children. In D. Gould & M. R. Weiss (Eds.), *Advances in pediatric sport sciences* (Vol. 2, pp. 121–142). Champaign, IL: Human Kinetics.

Horn, W. F., O'Donnell, J. P., & Vitulano, L. A. (1983). Long-term follow-up studies of learning disabled persons. *Journal of Learning Disabilities, 16*, 542–555.

Horner, T. M. (1980). Two methods of studying stranger reactivity in infants: A review. *Journal of Child Psychology and Psychiatry, 21*, 203–219.

Hornick, J. P., McDonald, L., & Robertson, G. B. (1992). Elder abuse in Canada and the United States: Prevalence, legal, and service issues. In R. D. Peters, R. J. McMahon, & V. L. Quinsey (Eds.), *Aggression and violence throughout the life span* (pp. 301–335). Newbury Park, CA: Sage.

Horowitz, A. (1985). Sons and daughters as caregivers to older parents: Differences in role performance and consequences. *Gerontologist, 25*, 612–617.

Horowitz, F. D. (1987). *Exploring developmental theories: Toward a structural/behavioral model of child development.* Hillsdale, NJ: Erlbaum.

Horowitz, F. D., & O'Brien, M. (1986). Gifted and talented children: State of knowledge and directions for research. *American Psychologist, 41*, 1147–1152.

Horowitz, F. D. (1992). John B. Watson's legacy: Learning and environment. *Developmental Psychology, 28*, 360–367.

Horton, R. (1995, July 13). Is homosexuality inherited? *New York Review of Books*, pp. 36–41.

Hotaling, G. T., & Sugarman, D. B. (1986). An analysis of risk markers in husband to wife violence: The current state of knowledge. *Violence and Victims, 1*, 101–124.

House, J. S., Kessler, R. C., Herzog, A. R., Mero, R. P., Kinney, A. M., & Breslow, M. J. (1990a). Age, socioeconomic status, and health. *Milbank Quarterly, 68,* 383–411.

House, J. S., Kessler, R. C., Herzog, A. R., Mero, R. P., Kinney, A. M., & Breslow, M. J. (1990b). *The social stratification of aging and health.* Ann Arbor, MI: Institute for Social Research.

House, J. S., Kessler, R. C., Herzog, A. R., Mero, R., Kinney, A., & Breslow, M. (1991). Social stratification, age, and health. In K. W. Schaie, D. Blazer, & J. S. House (Eds.), *Aging, health behaviors, and health outcomes* (pp. 1–32). Hillsdale, NJ: Erlbaum.

Houseknecht, S. K. (1987). Voluntary childlessness. In M. B. Sussman & S. K. Steinmetz (Eds.), *Handbook of marriage and the family* (pp. 369–392). New York: Plenum.

Houts, A. C. (1991). Nocturnal enuresis as a biobehavioral problem. *Behavior Therapy, 22,* 133–151.

Howard, A., & Bray, D. W. (1988). *Managerial lives in transition: Advancing age and changing times.* New York: Guilford Press.

Howard, M., & McCabe, J. B. (1990). Helping teenagers postpone sexual involvement. *Family Planning Perspectives, 22,* 21–26.

Howe, M. L., & Courage, M. L. (1993). On resolving the enigma of infantile amnesia. *Psychological Bulletin, 113,* 305–326.

Howe, N., & Ross, H. S. (1990). Socialization, perspective-taking, and the sibling relationship. *Developmental Psychology, 26,* 160–165.

Howes, C. (1988). Relations between early child care and schooling. *Developmental Psychology, 24,* 53–57.

Howes, C. (1990). Can the age of entry into child care and the quality of child care predict adjustment in kindergarten? *Developmental Psychology, 26,* 292–303.

Howes, C. (1992). *The collaborative construction of pretend.* Albany, NY: SUNY Press.

Howes, C., Hamilton, C. E., & Matheson, C. C. (1994). Children's relationships with peers: Differential associations with aspects of the teacher–child relationship. *Child Development, 65,* 253–263.

Howes, C., & Matheson, C. C. (1992). Sequences in the development of competent play with peers: Social and social pretend play. *Developmental Psychology, 28,* 961–974.

Howes, C., Phillips, D. A., & Whitebook, M. (1992). Thresholds of quality: Implications for the social development of children in center-based child care. *Child Development, 63,* 449–460.

Howes, C., Rodning, C., Galluzzo, D. C., & Myers, L. (1988). Attachment and child care: Relationships with mother and caregiver. *Early Childhood Research Quarterly, 3,* 403–416.

Howes, P., & Markman, H. J. (1989). Marital quality and child functioning: A longitudinal investigation. *Child Development, 60,* 1044–1051.

Hoyseth, K. S., & Jones, P. J. H. (1989). Ethanol induced teratogenesis: Characterization, mechanisms, and diagnostic approaches. *Life Sciences, 44,* 643–649.

Hsu, F. L. K. (1981). *Americans and Chinese: Passage to difference* (3rd ed.). Honolulu: University of Hawaii Press.

Hudson, J. A. (1988). Children's memory for atypical actions in script-based stories: Evidence for a disruption effect. *Journal of Experimental Child Psychology, 46,* 159–173.

Hudson, J. A., Fivush, R., & Kuebli, J. (1992). Scripts and episodes: The development of event memory. *Applied Cognitive Psychology, 6,* 483–505.

Hudspeth, W. J., & Pribram, K . H. (1992). Psychophysiological indices of cerebral maturation. *International Journal of Psychophysiology, 12,* 19–29.

Huesmann, L. R. (1986). Psychological processes promoting the relation between exposure to media violence and aggressive behavior by the viewer. *Journal of Social Issues, 42,* 125–139.

Hugick, L. (1992, June). Public opinion divided on gay rights. *Gallup Poll Monthly, 321,* 2–6.

Hultsch, D. F. (1975). Adult age differences in retrieval: Trace-dependent and cue-dependent forgetting. *Developmental Psychology, 11,* 197–201.

Hultsch, D. F., & Dixon, R. A. (1990). Learning and memory in aging. In J. E. Birren & K. W. Schaie (Eds.), *Handbook of the psychology of aging* (3rd ed., pp. 258–274). San Diego: Academic Press.

Humphrey, T. (1978). Function of the nervous system during prenatal life. In U. Stave (Ed.), *Perinatal physiology* (pp. 651–683). New York: Plenum.

Hurme, H. (1991). Dimensions of the grandparent role in Finland. In P. K. Smith (Ed.), *The psychology of grandparenthood: An international perspective* (pp. 19–31). London: Routledge.

Huston, A. C. (1983). Sex-typing. In E. M. Hetherington (Ed.), *Handbook of child psychology: Vol. 4. Socialization, personality, and social development* (4th ed., pp. 387–467). New York: Wiley.

Huston, A. C. (1994, Summer). *Children in poverty: Child development and public policy.* Cambridge, England: Cambridge University Press.

Huston, A. C., & Alvarez, M. M. (1990). The socialization context of gender role development in early adolescence. In R. Montemayor, G. R. Adams, & T. P. Gullotta (Eds.), *From childhood to adolescence: A transitional period?* (pp. 156–179). Newbury Park, CA: Sage.

Huston, A. C., Donnerstein, E., Fairchild, H., Feshbach, N. D., Katz, P. A., Murray, J. P., Rubinstein, E. A., Wilcox, B. L., & Zuckerman, D. (1992). *Big world, small screen: The role of television in American society.* Lincoln: University of Nebraska Press.

Huston, P., McHale, S., & Crouter, A. (1986). When the honeymoon's over: Changes in the marriage relationship over the first year. In R. Gilmour & S. Duck (Eds.), *The emerging field of personal relationships* (pp. 109–132). Hillsdale, NJ: Erlbaum.

Huttenlocher, P. R. (1994). Synaptogenesis in the human cerebral cortex. In G. Dawson & K. W. Fischer (Eds.), *Human behavior and the developing brain* (pp. 137–152). New York: Guilford.

Huyck, M. H. (1990). Gender differences in aging. In J. E. Birren & K. W. Schaie (Eds.), *Handbook of the psychology of aging* (3rd ed., pp. 124–134). New York: Academic Press.

Huyck, M. H. (1995). Marriage and close relationships of the marital kind. In R. Blieszner & V. H. Bedford (Eds.), *Handbook of aging and the family* (pp. 181–200). Westport, CT: Greenwood Press.

Huyck, M. H. (1996). Continuities and discontinuities in gender identity. In V. L. Bengtson (Ed.), *Adulthood and aging* (pp. 98–121). New York: Springer-Verlag.

Hyde, J. S., Fenema, E., & Lamon, S. J. (1990). Gender differences in mathematics performance: A meta-analysis. *Psychological Bulletin, 107,* 139–155.

Hyde, J. S., & Linn, M. C. (1988). Gender differences in verbal ability: A meta-analysis. *Psychological Bulletin, 104,* 53–69.

Hyland, D. T., & Ackerman, A. M. (1988). Reminiscence and autobiographical memory in the study of the personal past. *Journal of Gerontology, 41,* P35–P39.

Hynd, G. W., Horn, K. L., Voeller, K. K., & Marshall, R. M. (1991). Neurobiological basis of attention-deficit hyperactivity disorder (ADHD). *School Psychology Review, 20,* 174–186.

Inhelder, B., & Piaget, J. (1958). *The growth of logical thinking from childhood to adolescence: An essay on the construction of formal operational structures.* New York: Basic Books. (Original work published 1955)

International Education Association. (1988). *Science achievement in seventeen countries: A preliminary report.* Oxford, England: Pergamon Press.

Intons-Peterson, M. J. (1988). *Gender concepts of Swedish and American youth.* Hillsdale, NJ: Erlbaum.

Irgens, L. M., Markestad, T. Baste, V., Schreuder, P., Skjaerven, R., & Oyen, N. (1995). Sleeping position and sudden infant death syndrome in Norway 1967–1991. *Archives of Disease in Childhood, 72,* 478–482.

Irvine, J. J. (1986). Teacher–student interactions: Effects of student race, sex, and grade level. *Journal of Educational Psychology, 78,* 14–21.

Isabella, R. (1993). Origins of attachment: Maternal interactive behavior across the first year. *Child Development, 64,* 605–621.

Isabella, R., & Belsky, J. (1991). Interactional synchrony and the origins of infant–mother attachment: A replication study. *Child Development, 62,* 373–384.

Ishii-Kuntz, M. (1990). Social interaction and psychological well-being: Comparison across stages of adulthood. *International Journal of Aging and Human Development, 30,* 15–36.

Izard, C. E. (1991). *The psychology of emotions.* New York: Plenum.

Izard, C. E., Fantauzzo, C. A., Castle, J. M., Haynes, O. M., Rayias, M. F., & Putnam, P. H. (1995). The ontogeny and significance of infants' facial expressions in the first 9 months of life. *Developmental Psychology, 31,* 997–1013.

Jacklin, C. N., & Maccoby, E. E. (1983). Issues of gender differentiation in normal development. In M. D. Levine, W. B. Carey, A. C. Crocker, & R. T. Gross (Eds.), *Developmental-behavioral pediatrics* (pp. 175–184). Philadelphia: Saunders.

Jacobson, C. K., & Heaton, T. B. (1989). Voluntary childlessness among American men and women in the late 1980s. *Social Biology, 38,* 79–93.

Jacobson, J. L., Jacobson, S. W., Fein, G., Schwartz, P. M., & Dowler, J. (1984). Prenatal exposure to an environmental toxin: A test of the multiple effects model. *Developmental Psychology, 20,* 523–532.

Jacobson, J. L., Jacobson, S. W., & Humphrey, H. E. B. (1990). Effects of in utero exposure to polychlorinated biphenyls on cognitive functioning in young children. *Journal of Pediatrics, 116,* 38–45.

Jacobson, J. L., Jacobson, S. W., Padgett, R. J., Brumitt, G. A., & Billings, R. L. (1992). Effects of prenatal PCB exposure on cognitive processing efficiency and sustained attention. *Developmental Psychology, 28,* 297–306.

Jacobson, S. W., Fein, G. G., Jacobson, J. L., Schwartz, P. M., & Dowler, J. (1985). The effect of intrauterine PCB exposure on visual recognition memory. *Child Development, 56,* 853–860.

Jacobson, S. W., Jacobson, J. L., Sokol, R. J., Martier, S. S., & Ager, J. W. (1993). Prenatal alcohol exposure and infant information processing ability. *Child Development, 64,* 1706–1721.

Jadack, R. A., Hyde, J. S., Moore, C. F., & Keller, M. L. (1995). Moral reasoning about sexually transmitted diseases. *Child Development, 66,* 167–177.

Jakobi, P., Weissman, A., Peretz, B. A., & Hocherman, I. (1993). Evaluation of prognostic factors for vaginal delivery after cesarean section. *Journal of Reproductive Medicine, 38,* 729–733.

James, J. B., Lewkowicz, C., Libhaber, J., & Lachman, M. (1995). Rethinking the gender identity crossover hypothesis: A test of a new model. *Sex Roles, 32,* 185–207.

Jameson, S. (1993). Zinc status in pregnancy: The effect of zinc therapy on perinatal mortality, prematurity, and placental ablation. *Annals of the New York Academy of Sciences, 678,* 178–192.

Jarvik, M. E., & Schneider, N. G. (1992). Caffeine. In J. H. Lowinson, P. Ruiz, & R. B. Millman (Eds.), *Substance abuse* (2nd ed., pp. 334–356). Baltimore: Williams & Wilkins.

Jaskiewicz, J. A., & McAnarney, E. R. (1994). Pregnancy during adolescence. *Pediatrics in Review, 15,* 32–38.

Jazwinski, S. M. (1996). Longevity, genes, and aging. *Science, 273,* 54–59.

Jenkins, S. R., & Maslach, C. (1994). Psychological health and involvement in interpersonally demanding occupations: A longitudinal perspective. *Journal of Organizational Behavior, 15,* 101–127.

Jensen, A. R. (1969). How much can we boost IQ and scholastic achievement? *Harvard Educational Review, 39,* 1–123.

Jensen, A. R. (1980). *Bias in mental testing.* New York: Free Press.

Jensen, A. R. (1985a). Methodological and statistical techniques for the chronometric study of mental abilities. In C. R. Reynolds & V. L. Willson (Eds.), *Methodological and statistical advances in the study of individual difference* (pp. 51–116). New York: Plenum.

Jensen, A. R. (1985b). The nature of the black–white difference on various psychometric tests: Spearman's hypothesis. *Behavioral and Brain Sciences, 8,* 193–219.

Jensen, A. R., & Figueroa, R. A. (1975). Forward and backward digit-span interaction with race and IQ: Predictions from Jensen's theory. *Journal of Educational Psychology, 67,* 882–893.

Jensen, A. R., & Whang, P. A. (1994). Speed of accessing arithmetic facts in long-term memory: A comparison of Chinese-American and Anglo-American children. *Contemporary Educational Psychology, 19,* 1–12.

Jepson, K. L., & Labouvie-Vief, G. (1992). Symbolic processing of youth and elders. In R. L. West and J. D. Sinnott (Eds.), *Everyday memory and aging* (pp. 124–137). New York: Springer.

Jerrome, D. (1990a). Frailty and friendship. *Journal of Cross-Cultural Gerontology, 5,* 51–64.

Jerrome, D. (1990b). Intimate relationships. In J. Bond & P. Coleman (Eds.), *Aging in society* (pp. 181–208). London: Sage.

Johnson, A. M., Wadsworth, J., Wellings, K., Bradshaw, S., & Field, J. (1992). Sexual lifestyles and HIV risk. *Nature, 360,* 420–426.

Johnson, C. L. (1985). *Growing up and growing old in Italian-American families.* New Brunswick, NJ: Rutgers University Press.

Johnson, C. L., & Troll, L. (1992). Family functioning in late life. *Journal of Gerontology, 47,* S66–S72.

Johnson, C. L., & Troll, L. E. (1994). Constraints and facilitators to friendships in late life. *Gerontologist, 34,* 79–87.

Johnson, D. W., Johnson, R. T., & Maruyama, G. (1984). Goal interdependence and interpersonal attraction in heterogeneous classrooms: A meta-analysis. In N. Miller & M. B. Brewer (Eds.), *Groups in contact: The psychology of desegregation* (pp. 187–212). New York: Academic Press.

Johnson, J. E., & Hooper, F. E. (1982). Piagetian structuralism and learning: Two decades of educational application. *Contemporary Educational Psychology, 7,* 217–237.

Johnson, J. S., & Newport, E. L. (1989). Critical period effects in second language learning: The influence of maturational state on the acquisition of English as a second language. *Cognitive Psychology, 21,* 60–99.

Johnston, J. R., Kline, M., & Tschann, J. M. (1989). Ongoing post-divorce conflict. *American Journal of Orthopsychiatry, 57,* 587–600.

Jones, E. F., Forrest, J. D., Goldman, N., Henshaw, S. K., Lincoln, R., Rosoff, J. I., Westoff, C. F., & Wulf, D. (1988). Teenage pregnancy in developed countries: Determinants and policy implications. *Family Planning Perspectives, 17,* 53–63.

Jones, G. P., & Dembo, M. H. (1989). Age and sex role differences in intimate friendships during childhood and adolescence. *Merrill-Palmer Quarterly, 35,* 445–462.

Jones, M. C. (1965). Psychological correlates of somatic development. *Child Development, 36,* 899–911.

Jones, M. C., & Bayley, N. (1950). Physical maturing among boys as related to behavior. *Journal of Educational Psychology, 41,* 129–148.

Jones, M. C., & Mussen, P. H. (1958). Self-conceptions, motivations, and interpersonal attitudes of early- and late-maturing girls. *Child Development, 29,* 491–501.

Jones, S. S., & Raag, T. (1989). Smile production in older infants: The importance of a social recipient for the facial signal. *Child Development, 60,* 811–818.

Jones, W. H. (1990). Loneliness and social exclusion. *Journal of Social and Clinical Psychology, 9,* 214–220.

Jonsen, A., & Stryker, J. (Eds.). (1992). *The social impact of AIDS in the United States.* Washington, DC: National Academy Press.

Jordan, B. (1993). *Birth in four cultures.* Prospect Heights, IL: Waveland.

Jordan, P. (1990). Laboring for relevance: The male experience of expectant and new parenthood. *Nursing Research, 39,* 15–19.

Jorgensen, M., & Keiding, K. (1991). Estimation of spermarche from longitudinal spermaturia data. *Biometrics, 47,* 177–193.

Josselson, R. (1994). The theory of identity development and the question of intervention. In S. L. Archer (Ed.), *Interventions for adolescent identity development* (pp. 12–25). Thousand Oaks, CA: Sage.

Jouriles, E. N., & Le Compte, S. H. (1991). Husbands' aggression toward wives and mothers' and fathers' aggression toward children: Moderating effects of child gender. *Journal of Consulting and Clinical Psychology, 59,* 190–192.

Jung, C. (1933). *Modern man in search of a soul.* New York: Harcourt.

Jusczyk, P. (1995). Language acquisition: Speech sounds and phonological development. In J. L. Miller & P. D. Eimas (Eds.), *Handbook of perception and cognition: Vol. 2. Speech, language, and communication* (pp. 263–301). Orlando, FL: Academic Press.

Jusczyk, P. W., Cutler, A., & Redanz, N. J. (1993). Infants' preference for the predominant stress patterns of English words. *Child Development, 64,* 675–687.

Kagan, J. (1989). *Unstable ideas: Temperament, cognition, and self.* Cambridge, MA: Harvard University Press.

Kagan, J. (1994). *Galen's prophecy.* New York: Basic Books.

Kagan, J., Arcus, D., Snidman, N., Feng, W. Y. Hendler, J., & Greene, S. (1994). Reactivity in infants: A cross-national comparison. *Developmental Psychology, 30,* 342–345.

Kagan, J., Kearsley, R. B., & Zelazo, P. R. (1978). *Infancy: Its place in human development.* Cambridge, MA: Harvard University Press.

Kagan, J., & Snidman, N. (1991). Temperamental factors in human development. *American Psychologist, 46,* 856–862.

Kahana, E., Biegel, D., & Wykle, M. (1994). *Family caregiving across the lifespan.* Thousand Oaks, CA: Sage.

Kahill, S. (1988). Symptoms of professional burnout: A review of the empirical evidence. *Canadian Psychology, 29,* 284–297.

Kahn, P. H., Jr. (1992). Children's obligatory and discretionary moral judgments. *Child Development, 63,* 416–430.

Kaiser, F. E. (1991). Sexuality and impotence in the aging man. *Clinics in Geriatric Medicine, 7,* 63–72.

Kaitz, M., Good, A., Rokem, A. M., & Eidelman, A. I. (1987). Mothers' recognition of their newborns by olfactory cues. *Developmental Psychobiology, 20,* 587–591.

Kaitz, M., Good, A., Rokem, A. M., & Eidelman, A. I. (1988). Mothers' and fathers' recognition of their newborns' photographs during the postpartum period. *Journal of Developmental and Behavioral Pediatrics, 9,* 223–226.

Kaitz, M., Meirov, H., Landman, I., & Eidelman, A. I. (1993). Infant recognition by tactile cues. *Infant Behavior and Development, 16,* 333–341.

Kalb, C. (1997, May 5). How old is too old? *Newsweek,* p. 64.

Kaler, S. B., & Kopp, C. B. (1990). Compliance and comprehension in very young toddlers. *Child Development, 61,* 1997–2003.

Kalish, R. A. (1985). The social context of death and dying. In R. H. Binstock & E. Shanas (Eds.), *Handbook of aging and the social sciences* (2nd ed., pp. 149–170). New York: Van Nostrand Reinhold.

Kalleberg, A. L., & Loscocco, K. A. (1983). Aging, values, and rewards: Explaining age differences in job satisfaction. *American Sociological Review, 48,* 78–90.

Kamerman, S. B. (1993). International perspectives on child care policies and programs. *Pediatrics, 91,* 248–252.

Kamo, Y., & Zhou, M. (1994). Living arrangements of elderly Chinese and Japanese in the United States. *Journal of Marriage and the Family, 56,* 544–558.

Kandall, S. R., & Gaines, J. (1991). Maternal substance use and subsequent sudden infant death syndrome (SIDS) in offspring. *Neurotoxicology and Teratology, 13,* 235–240.

Kandel, D. B., Raveis, V. H., & Davies, M. (1991). Suicidal ideation in adolescence: Depression, substance use, and other risk factors. *Journal of Youth and Adolescence, 20,* 289–309.

Kanner, A. D., Feldman, S. S., Weinberger, D. A., & Ford, M. E. (1987). Uplifts, hassles, and adaptational outcomes in early adolescents. *Journal of Early Adolescence, 7,* 371–394.

Kantor, D., & Lehr, W. (1975). *Inside the family.* San Francisco: Jossey-Bass.

Kanugo, M. S. (1994). *Genes and aging.* New York: Cambridge University Press.

Kaplan, R. M. (1985). The controversy related to the use of psychological tests. In B. B. Wolman (Ed.), *Handbook of intelligence* (pp. 465–504). New York: Wiley.

Karadsheh, R. (1991). *This room is a junkyard!: Children's comprehension of metaphorical language.* Paper presented at the biennial meeting of the Society for Research in Child Development, Seattle, WA.

Kaslow, F. W., Hansson, K., & Lundblad, A. (1994). Long-term marriages in Sweden: And some comparisons with similar couples in the United States. *Contemporary Family Therapy, 16,* 521–537.

Kassebaum, N. L. (1994). Head Start: Only the best for America's children. *American Psychologist, 49,* 123–126.

Kastenbaum, R. (1992). *The psychology of death* (2nd ed.). New York: Springer.

Kastenbaum, R. J. (1995). *Death, society, and human experience* (5th ed.). Boston: Allyn & Bacon.

Katchadourian, H. (1977). *The biology of adolescence.* San Francisco: Freeman.

Katchadourian, H. (1990). Sexuality. In S. S. Feldman & G. R. Elliott (Eds.), *At the threshold: The developing adolescent* (pp. 330–351). Cambridge, MA: Harvard University Press.

Kaufman, J., & Zigler, E. (1989). The intergenerational transmission of child abuse. In D. Cicchetti & V. Carlson (Eds.), *Child maltreatment: Theory and research on the causes and consequences of child abuse and neglect* (pp. 129–150). Cambridge: Cambridge University Press.

Kausler, D. H. (1991). *Experimental psychology, cognition, and human aging.* New York: Springer.

Kausler, D. H. (1994). *Learning and memory in normal aging.* San Diego: Academic Press.

Kavanaugh, R. D., & Harris, P. L. (1994). Imagining the outcome of pretend transformations: Assessing the competence of normal children and children with autism. *Developmental Psychology, 30,* 847–854.

Kaye, K., & Marcus, J. (1981). Infant imitation: The sensory-motor agenda. *Developmental Psychology, 17,* 258–265.

Kaye, K., Elkind, L., Goldberg, D., & Tytun, A. (1989). Birth outcomes for infants of drug abusing mothers. *New York State Journal of Medicine, 89,* 256–261.

Kayman, S., Bruvold, W., & Stern, J. S. (1990). Maintenance and relapse after weight loss in women: Behavioral aspects. *American Journal of Clinical Nutrition, 52,* 800–807.

Kearins, J. M. (1981). Visual spatial memory in Australian aboriginal children of desert regions. *Cognitive Psychology, 13,* 434–460.

Keasey, C. B. (1971). Social participation as a factor in the moral development of preadolescents. *Developmental Psychology, 5,* 216–220.

Keating, D. (1979). Adolescent thinking. In J. Adelson (Ed.), *Handbook of adolescent psychology* (pp. 211–246). New York: Wiley.

Keen, C. L., & Zidenberg-Cherr, S. (1994). Should vitamin-mineral supplements be recommended for all women with childbearing potential? *American Journal of Clinical Nutrition, 59,* 532S–539S.

Keil, F. C. (1986). Conceptual domains and the acquisition of metaphor. *Cognitive Development, 1,* 73–96.

Keith, J., Fry, C. L., Glascock, A. P., Ikels, C., Dickerson-Putman, J., Harpending, H. C., & Draper, P. (1994). *The aging experience: Diversity and commonality across cultures.* Thousand Oaks, CA: Sage.

Keith, P. M., & Schafer, R. B. (1991). *Relationships and well-being over the life stages.* New York: Praeger.

Kellehear, A. (1993). Culture, biology, and the near-death experience: A reappraisal. *Journal of Nervous and Mental Disease, 181,* 148–156.

Keller, A., Ford, L. H., & Meacham, J. A. (1978). Dimensions of self-concept in preschool children. *Developmental Psychology, 14,* 483–489.

Kelley, M. L., Power, T. G., & Wimbush, D. D. (1992). Determinants of disciplinary practices in low-income black mothers. *Child Development, 63,* 573–582.

Kelly, H. (1981). Viewing children through television. In H. Kelly & H. Gardner (Eds.), *New directions for child development* (No. 13, pp. 59–71). San Francisco: Jossey-Bass.

Kelly, J. J., Chu, S. Y., & Buehler, J. W. (1993). AIDS deaths shift from hospital to home. *American Journal of Public Health, 83,* 1433–1437.

Kelly, J. R., & Ross, J-E. (1989). Later-life leisure: Beginning a new agenda. *Leisure Sciences, 11,* 47–59.

Kelly, J. R., Steinkamp, M. W., & Kelly, J. (1986). Later life satisfaction: Does leisure contribute? *Leisure Sciences, 9,* 189–200.

Kempe, C. H., Silverman, B. F., Steele, P. W., Droegemueller, P. W., & Silver, H. K. (1962). The battered-child syndrome. *Journal of the American Medical Association, 181,* 17–24.

Kemper, S. (1992). Adults' sentence fragments: Who, what, when, where, and why. *Communication Research, 19,* 444–458.

Kemper, S., Kynette, D., & Norman, S. (1992). Age differences in spoken language. In R. L. West & J. D. Sinnott (Eds.), *Everyday memory and aging* (pp. 138–152). New York: Springer-Verlag.

Kendall-Tackett, K. A., Williams, L. M., & Finkelhor, D. (1993). Impact of sexual abuse on children: A review and synthesis of recent empirical studies. *Psychological Bulletin, 113,* 164–180.

Kendler, K. S., & Robinette, C. D. (1983). Schizophrenia in the National Academy of Sciences–National Research Council twin registry: A 16-year update. *American Journal of Psychiatry, 140,* 1551–1563.

Kendrick, A. S., Kaufmann, R., & Messenger, K. P. (1991). *Healthy young children: A manual for programs.* Washington, DC: National Association for the Education of Young Children.

Kennedy, G. E., & Kennedy, C. E. (1993). Grandparents: A special resource for children in stepfamilies. *Journal of Divorce & Remarriage, 19,* 45–68.

Kennedy, R. E. (1993). Depression as a disorder of social relationships: Implications for school policy and preven-

tion. In R. M. Lerner (Ed.), *Early adolescence: Perspectives on research, policy, and intervention* (pp. 383–398). Hillsdale, NJ: Erlbaum.

Kennell, J., Klaus, M., McGrath, S., Robertson, S., & Hinkley, C. (1991). Continuous emotional support during labor in a U.S. hospital. *Journal of the American Medical Association, 265,* 2197–2201.

Kenrick, D. T., & Keefe, R. C. (1992). Age preferences in mates reflect sex differences in human reproductive strategies. *Behavioral and Brain Sciences, 15,* 75–133.

Keogh, B. K. (1988). Improving services for problem learners. *Journal of Learning Disabilities, 21,* 6–11.

Kerckhoff, A. C., & Macrae, J. (1992). Leaving the parental home in Great Britain: A comparative perspective. *Sociological Quarterly, 33,* 281–301.

Kermoian, R., & Campos, J. J. (1988). Locomotor experience: A facilitator of spatial cognitive development. *Child Development, 59,* 908–917.

Kerr, B. A. (1983). Raising the career aspirations of gifted girls. *Vocational Guidance Quarterly, 32,* 37–43.

Kerr, M., Lambert, W. W., Stattin, H., & Klackenberg-Larsson, I. (1994). Stability of inhibition in a Swedish longitudinal sample. *Child Development, 65,* 138–146.

Kerridge, I. H., & Mitchell, K. R. (1996). The legislation of active voluntary euthanasia in Australia: Will the slippery slope prove fatal? *Journal of Medical Ethics, 22,* 273–278.

Ketterlinus, R. D., Henderson, S. H., & Lamb, M. E. (1990). Maternal age, sociodemographics, prenatal health and behavior: Influences on neonatal risk status. *Journal of Adolescent Health Care, 11,* 423–431.

Kimmel, D. C., & Moody, H. R. (1990). Ethical issues in gerontological research and services. In J. E. Birren & K. W. Schaie (Eds.), *Handbook of the psychology of aging* (3rd ed., pp. 489–501). San Diego, CA: Academic Press.

Kingston, P., & Reay, A. (1996). Elder abuse and neglect. In R. T. Woods (Ed.), *Handbook of the clinical psychology of ageing* (pp. 423–438). Chichester, England: Wiley.

Kinzie, J. D., Sack, W., Angell, R., Clarke, G., & Ben, R. (1989). A three-year follow-up of Cambodian young people traumatized as children. *Journal of the American Academy of Child and Adolescent Psychiatry, 28,* 501–504.

Kirby, D. (1992). School-based programs to reduce sexual risk taking. *Journal of School Health, 62,* 280–287.

Kirby, D., Short, L., Collins, J., Rugg, D., Kolbe, L., Howard, M., Miller, B., Sonenstein, F., & Zabin, L. S. (1994). School-based programs to reduce sexual behaviors: A review of effectiveness. *Public Health Reports, 109*(3), 339–360.

Kirk, W. G. (1993). *Adolescent suicide.* Champaign, IL: Research Press.

Kirkpatrick, L. A., & Davis, K. E. (1994). Attachment style, gender, and relationship stability: A longitudinal analysis. *Journal of Personality and Social Psychology, 66,* 502–512.

Kitchener, K. S., Lynch, C. L., Fischer, K. W., & Wood, P. K. (1993). Developmental range of reflective judgment: The effect of contextual support and practice on developmental stage. *Developmental Psychology, 29,* 893–906.

Kivnick, H. Q. (1983). Dimensions of grandparenthood meaning: Deductive conceptualization and empirical derivation. *Journal of Personality and Social Psychology, 44,* 1056–1068.

Klahr, D. (1992). Information-processing approaches in cognitive development. In M. H. Bornstein & M. E. Lamb (Eds.), *Developmental psychology: An advanced textbook* (3rd ed., pp. 273–335). Hillsdale, NJ: Erlbaum.

Klaus, M. H., & Kennell, J. H. (1982). *Parent–infant bonding.* St. Louis: Mosby.

Klimes-Dougan, B., & Kistner, J. (1990). Physically abused preschoolers' responses to peers' distress. *Developmental Psychology, 26,* 599–602.

Kline, D. W., Kline, T. J. B., Fozard, J. L., Kosnik, W., Schieber, F., & Sekuler, R. (1992). Vision, aging, and driving: The problems of older drivers. *Journal of Gerontology, 47,* P27–P34.

Klinnert, M. D., Gavin, L. A., Wamboldt, F. S., & Mrazek, D. A. (1992). Marriages with children at medical risk: The transition to parenthood. *Journal of the American Academy of Child and Adolescent Psychiatry, 31,* 334–342.

Klonoffcohen, H. S., Edelstein, S. L., Lefkowitz, E. S., Srinivasan, I. P., Kaegi, D., Chang, J. C., & Wiley, K. J. (1995). The effect of passive smoking and tobacco exposure through breast milk on sudden infant death syndrome. *Journal of the American Medical Association, 273,* 795–798.

Klungness, L. (1990). Diagnosis and behavioral treatment of children who refuse to attend school. In P. A. Keller & S. R. Heyman (Eds.), *Innovations in clinical practice: A source book* (Vol. 9, pp. 107–118). Sarasota, FL: Professional Resource Exchange, Inc.

Knapp, M. L., & Taylor, E. H. (1994). Commitment and its communication in romantic relationships. In A. L. Weber & J. H. Harvey (Eds.), *Perspectives on close relationships* (pp. 153–175). Boston: Allyn & Bacon.

Knobloch, H., & Pasamanick, B. (Eds.), (1974). *Gesell and Amatruda's Developmental Diagnosis.* Hagerstown, MD: Harper & Row.

Knox, A. B. (1993). *Strengthening adult and continuing education.* San Francisco: Jossey-Bass.

Kobak, R. R., & Sceery, A. (1988). The transition to college: Working models of attachment, affect regulation, and perceptions of self and others. *Child Development, 88,* 135–146.

Kobasa, S. C. (1979). Stressful life events, personality, and health: An inquiry into hardiness. *Journal of Personality and Social Psychology, 37,* 1–11.

Kochanek, K. D., Maurer, J. D., & Rosenberg, H. M. (1994). Why did black life expectancy decline from 1984 through 1989 in the United States? *American Journal of Public Health, 84,* 938–944.

Kochanska, G. (1993). Toward a synthesis of parental socialization and child temperament in early development of conscience. *Child Development, 64,* 325–347.

Kochanska, G. (1995). Children's temperament, mothers' discipline, and security of attachment: Multiple pathways to emerging internalization. *Child Development, 66,* 597–615.

Kochanska, G. (1997). Multiple pathways to conscience for children with different temperaments: From toddlerhood to age 5. *Developmental Psychology, 33,* 228–240.

Kochanska, G., Aksan, N., & Koenig, A. L. (1995). A longitudinal study of the roots of preschoolers' conscience: Committed compliance and emerging internalization. *Child Development, 66,* 1752–1769.

Kochanska, G., Casey, R. J., & Fukumoto, A. (1995). Toddlers' sensitivity to standard violations. *Child Development, 66,* 643–656.

Kochanska, G., & Radke-Yarrow, M. (1992). Inhibition in toddlerhood and the dynamics of the child's interaction with an unfamiliar peer at age five. *Child Development, 63,* 325–335.

Koff, T. H., & Park, R. W. (1993). *Aging public policy: Bonding the generations.* Amityville, NY: Baywood.

Kogan, N. (1979). A study of categorization. *Journal of Gerontology, 34,* 358–367.

Kogan, N., & Mills, M. (1992). Gender influences on age cognitions and preferences: Sociocultural or sociobiological? *Psychology and Aging, 7,* 98–106.

Kohlberg, L. (1966). A cognitive-developmental analysis of children's sex-role concepts and attitudes. In E. E. Maccoby (Ed.), *The development of sex differences* (pp. 82–173). Stanford, CA: Stanford University Press.

Kohlberg, L. (1969). Stage and sequence: The cognitive-developmental approach to socialization. In D. A. Goslin (Ed.), *Handbook of socialization theory and research* (pp. 347–480). Chicago: Rand McNally.

Kohlberg, L., Levine, C., & Hewer, A. (1983). *Moral stages: A current formulation and a response to critics.* Basel, Switzerland: Karger.

Kohn, M. L., Naoi, A., Schoenbach, C., Schooler, C., & Slomczynski, K. M. (1990). Position in the class structure and psychological functioning in the United States, Japan, and Poland. *American Journal of Sociology, 95,* 964–1008.

Kohn, M. L., & Schooler, C. (1978). The reciprocal effects of the substantive complexity of work and intellectual flexibility: A longitudinal assessment. *American Journal of Sociology, 84,* 24–52.

Kohn, M. L., & Slomczynski, D. M. (1990). *Social structure and self-direction: A comparative analysis of the United States and Poland.* Cambridge: Blackwell.

Kojima, H. (1986). Childrearing concepts as a belief–value system of the society and the individual. In H. Stevenson, H. Azuma, & K. Hakuta (Eds.), *Child development and education in Japan* (pp. 39–54). New York: Freeman.

Kolata, G. (1992, April 26). A parents' guide to kids' sports. *New York Times Magazine,* pp. 12–15, 40, 44, 46.

Kolberg, R. (1993). Human embryo cloning reported. *Science, 262,* 652–653.

Komp, D. M. (1996). The changing face of death in children. In H. M. Spiro, M. G. M. Curnen, & L. P. Wandel (Eds.), *Facing death: Where culture, religion, and medicine meet* (pp. 66–76). New Haven: Yale University Press.

Koo, J. (1987). Widows in Seoul, Korea. In H. Z. Lopata (Ed.), *Widows: The Middle East, Asia and the Pacific* (pp. 56–78). Durham, NC: Duke University Press.

Kopp, C. B. (1987). The growth of self-regulation: Caregivers and children. In N. Eisenberg (Ed.), *Contemporary topics in developmental psychology* (pp. 34–55). New York: Wiley.

Kopp, C. B. (1994). Infant assessment. In C. B. Fisher & R. M. Lerner (Eds.), *Applied developmental psychology* (pp. 265–293). New York: McGraw-Hill.

Kornhaber, A., & Woodward, K. L. (1981).

Grandparent/grandchildren: The vital connection. Garden City, NJ: Anchor.

Koss, M. P. (1993). Rape: Scope, impact, interventions, and public policy responses. *American Psychologist, 48,* 1062–1069.

Koss, M. P., Gidycz, C. A., & Wisniewski, N. (1987). The scope of rape: Incidence and prevalence of sexual aggression and victimization in a national sample of higher education students. *Journal of Consulting and Clinical Psychology, 55,* 162–170.

Koss, M. P., & Harvey, M. (1991). *The rape victim: Clinical and community interventions.* Newbury Park, CA: Sage.

Koss, M. P., Koss, P., & Woodruff, W. J. (1991). Deleterious effects of criminal victimization on women's health and medical utilization. *Archives of Internal Medicine, 151,* 342–357.

Kotler-Cope, S., & Camp, C. J. (1990). Memory interventions in aging populations. In E. A. Lovelace (Ed.), *Aging and cognition* (pp. 231–261). Amsterdam: North-Holland.

Kotre, J. (1984). *Outliving the self: Generativity and the interpretation of lives.* Baltimore: Johns Hopkins University Press.

Kramer, M. S. (1993). Effects of energy and protein intakes on pregnancy outcome: An overview of the research evidence from controlled clinical trials. *American Journal of Clinical Nutrition, 58,* 627–635.

Krause, N. (1990). Perceived health problems, formal/informal support, and life satisfaction among older adults. *Journal of Gerontology, 45,* S193–S205.

Kreutzer, M. A., Leonard, C., & Flavell, J. H. (1975). An interview study of children's knowledge about memory. *Monographs of the Society for Research in Child Development, 40*(1, Serial No. 159).

Krevans, J., & Gibbs, J. C. (1996). Parents' use of inductive discipline: Relations to children's empathy and prosocial behavior. *Child Development, 67,* 3263–3277.

Kriska, A. M., Blair, S. N., & Pereira, M. A. (1994). The potential role of physical activity in the prevention of non-insulin-dependent diabetes mellitus: The epidemiological evidence. *Exercise and Sport Sciences Reviews, 22,* 121–143.

Kroger, J. (1995). The differentiation of "firm" and "developmental" foreclosure identity statuses: A longitudinal study. *Journal of Adolescent Research, 10,* 317–337.

Kronenfeld, J. J., & Glik, D. C. (1995). Unintentional injury: A major health problem for young children and youth. *Journal of Family & Economic Issues, 16,* 365–393.

Kropf, N. P., & Pugh, K. L. (1995). Beyond life expectancy: Social work with centenarians. *Journal of Gerontological Social Work, 23,* 121–137.

Krout, J., Cutler, S. J., & Coward, R. T. (1990). Correlates of senior center participation: A national analysis. *Gerontologist, 30,* 72–79.

Ku, L., Sonenstein, F. L., & Pleck, J. H. (1993). Factors influencing first intercourse for teenage men. *Public Health Reports, 108,* 680–694.

Kübler-Ross, E. (1969). *On death and dying.* New York: Macmillan.

Kuczynski, L. (1984). Socialization goals and mother–child interaction: Strategies for long-term and short-term compliance. *Developmental Psychology, 20,* 1061–1073.

Kuczynski, L., Kochanska, G., Radke-Yarrow, M., & Girnius-Brown, O. (1987). A developmental interpretation of young children's noncompliance. *Developmental Psychology, 23,* 799–806.

Kuebli, J., & Fivush, R. (1992). Gender differences in parent–child conversations about past emotions. *Sex Roles, 27,* 683–698.

Kuhl, P. K., Williams, K. A., Lacerda, F., Stevens, K. N., & Lindblom, B. (1992). Linguistic experience alters phonetic perception in infants by 6 months of age. *Science, 255,* 606–608.

Kuhn, D. (1992). Cognitive development. In M. H. Bornstein & M. E. Lamb (Eds.), *Developmental psychology: An advanced textbook* (3rd ed., pp. 211–272). Hillsdale, NJ: Erlbaum.

Kuhn, D., Amsel, E., & O'Loughlin, M. (1988). *The development of scientific thinking skills.* Orlando, FL: Academic Press.

Kuhn, D., Garcia-Mila, M., Zohar, A., & Andersen, C. (1995). Strategies of knowledge acquisition. *Monographs of the Society for Research in Child Development, 60*(245, Serial No. 4).

Kunkel, D. (1993). Policy and the future of children's television. In G. L. Berry & J. K. Asamen (Eds.), *Children & television* (pp. 273–290). Newbury Park, CA: Sage.

Kunzinger, E. L., III. (1985). A short-term longitudinal study of memorial development during early grade school. *Developmental Psychology, 21,* 642–646.

Kurdek, L. A., & Fine, M. A. (1994). Family acceptance and family control as predictors of adjustment in young adolescents: Linear, curvilinear, or interactive effects? *Child Development, 65,* 1137–1146.

Kurth, A. (1993). Reproductive issues, pregnancy, and childbearing in HIV-infected women. In F. L. Cohen & J. D. Durham (Eds.), *Women, children, and HIV/AIDS* (pp. 137–155). New York: Springer.

Kutner, L. (1993, June). Getting physical. *Parents,* pp. 96–98.

Labouvie-Vief, G. (1980). Beyond formal operations: Uses and limits of pure logic in life-span development. *Human Development, 23,* 141–160.

Labouvie-Vief, G. (1985). Logic and self-regulation from youth to maturity: A model. In M. Commons, F. Richards, & C. Armon (Eds.), *Beyond formal operations: Late adolescent and adult cognitive development* (pp. 158–180). New York: Praeger.

Labouvie-Vief, G. (1990). Modes of knowledge and the organization of development. In M. L. Commons, C. Armon, L. Kohlberg, F. A. Richards, T. A. Grotzer, & J. D. Sinnott (Eds.), *Adult development* (Vol. 2, pp. 43–62). New York: Praeger.

Labouvie-Vief, G., DeVoe, M., & Bulka, D. (1989). Speaking about feelings: Conceptions of emotion across the life span. *Psychology and Aging, 4,* 425–437.

Labouvie-Vief, G., & Hakim-Larson, J. (1989). Developmental shifts in adult thought. In S. Hunter & M. Sundel (Eds.), *Midlife myths* (pp. 69–96). Newbury Park, CA: Sage.

Lachman, M. E. (1985). Personal efficacy in middle and old age: Differential and normative patterns of change. In G. H. Elder, Jr. (Ed.), *Life course dynamics* (pp. 188–213). Ithaca, NY: Cornell University Press.

Ladd, G. W., & Cairns, E. (1996). Children: Ethnic and political violence. *Child Development, 67,* 14–18.

Ladd, G. W., & Mize, J. (1983). A cognitive-social learning model of social skill training. *Psychological Review, 90,* 127–157.

Ladd, G. W., & Price, J. M. (1987). Predicting children's social and school adjustment following the transition from preschool to kindergarten. *Child Development, 58,* 1168–1189.

Lagercrantz, H., & Slotkin, T. A. (1986). The "stress" of being born. *Scientific American, 254,* 100–107.

Lamaze, F. (1958). *Painless childbirth.* London: Burke.

Lamb, M. (1994). Infant care practices and the application of knowledge. In C. B. Fisher & R. M. Lerner (Eds.), *Applied developmental psychology* (pp. 23–45). New York: McGraw-Hill.

Lamb, M. E. (1976). Interaction between eight-month-old children and their fathers and mothers. In M. E. Lamb (Ed.), *The role of the father in child development* (pp. 307–327). New York: Wiley.

Lamb, M. E. (1987). *The father's role: Cross-cultural perspectives.* Hillsdale, NJ: Erlbaum.

Lamb, M. E., & Oppenheim, D. (1989). Fatherhood and father–child relationships: Five years of research. In S. H. Cath, A. Gurwitt, & L. Gunsberg (Eds.), *Fathers and their families* (pp. 11–26). Hillsdale, NJ: Erlbaum.

Lamb, M. E., Sternberg, K. J., & Prodromidis, M. (1992). Nonmaternal care and the security of infant–mother attachment: A reanalysis of the data. *Infant Behavior and Development, 15,* 71–83.

Lamb, M. E., Thompson, R. A., Gardner, W., Charnov, E. L., & Connell, J. P. (1985). Infant–mother attachment: The origins and developmental significance of individual differences in the Strange Situation: Its study and biological interpretation. *Behavioral and Brain Sciences, 7,* 127–147.

Lamb, S. (1991). First moral sense: Aspects of and contributors to a beginning morality in the second year of life. In W. M. Kurtines & J. L. Gewirtz (Eds.), *Handbook of moral behavior and development* (Vol. 2, pp. 171–189). Hillsdale, NJ: Erlbaum.

Lamborn, S. D., Mounts, N. S., Steinberg, L., & Dornbusch, S. M. (1991). Patterns of competence and adjustment among adolescents from authoritative, authoritarian, indulgent, and neglectful families. *Child Development, 62,* 1049–1065.

Lamme, S., & Baars, J. (1993). Including social factors in the analysis of reminiscence in elderly individuals. *International Journal of Aging and Human Development 37,* 297–311.

Lampl, M. (1993). Evidence of saltatory growth in infancy. *American Journal of Human Biology, 5,* 641–652.

Lampl, M., Veldhuis, J. D., & Johnson, M. L. (1992). Saltation and stasis: A model of human growth. *Science, 258,* 801–803.

Landesman, S., & Ramey, C. (1989). Developmental psychology and mental retardation: Integrating scientific principles with treatment practices. *American Psychologist, 44,* 409–415.

Lang, F. R., & Carstensen, L. L. (1994). Close emotional relationships in late life: Further support for proactive aging in the social domain, *Psychology and Aging, 9,* 315–324.

Langer, N. (1990). Grandparents and adult grandchildren: What do they do for one another? *International Journal of Aging and Human Development, 31,* 101–110.

Langlois, J. H., & Stephan, C. W. (1981). Beauty and the beast: The role of physical attractiveness in peer relationships and social behavior. In S. S. Brehm, S. M. Kassin, & S. X. Gibbons (Eds.), *Developmental social psychology: Theory and research* (pp. 152–168). New York: Oxford University Press.

Langreth, R. (1993, November). Can we live to 150? *Popular Science,* pp. 77–82.

Laosa, L. M. (1981). Maternal behavior: Sociocultural diversity in modes of family interaction. In R. W. Henderson (Ed.), *Parent–child interaction: Theory, research, and prospects* (pp. 125–167). New York: Academic Press.

Lapointe, A. E., Askew, J. M., & Mead, N. A. (1992). *Learning mathematics.* Princeton, NJ: Educational Testing Service.

Lapointe, A. E., Mead, N. A., & Askew, J. M. (1992). *Learning science.* Princeton NJ: Educational Testing Service.

Lapsley, D. K. (1993). Toward an integrated theory of adolescent ego development: The "new look" at adolescent egocentrism. *American Journal of Orthopsychiatry, 63,* 562–571.

Lapsley, D. K., Jackson, S., Rice, K., & Shadid, G. (1988). Self-monitoring and the "new look" at the imaginary audience and personal fable: An ego-developmental analysis. *Journal of Adolescent Research, 3,* 17–31.

Lapsley, D. K., Milstead, M., Quintana, S., Flannery, D., & Buss, R. (1986). Adolescent egocentrism and formal operations: Tests of a theoretical assumption. *Developmental Psychology, 22,* 800–807.

Lapsley, D. K., Rice, K. G., & FitzGerald, D. P. (1990). Adolescent attachment, identity, and adjustment to college: Implications for the continuity of adaptation hypothesis. *Journal of Counseling and Development, 68,* 561–565.

Larson, D. E. (1996). *Mayo Clinic family health book.* New York: Morrow.

Larson, J. H. (1988). The marriage quiz: College students' beliefs in selected myths about marriage. *Family Relations, 37,* 3–11.

Larson, R., & Ham, M. (1993). Stress and "storm and stress" in early adolescence: The relationship of negative events with dysphoric affect. *Developmental Psychology, 29,* 130–140.

Larson, R., & Lampman-Petraitis, C. (1989). Daily emotional states as reported by children and adolescents. *Child Development, 60,* 1250–1260.

Larson, R., Mannell, R., & Zuzanek, J. (1986). Daily well-being of older adults with friends and family. *Psychology and Aging, 1,* 117–126.

Larson, R. W., Richards, M. H., Moneta, G., Holmbeck, G., & Duckett, E. (1996). Changes in adolescents' daily interactions with their families from ages 10 to 18: Disengagement and transformation. *Developmental Psychology, 32,* 744–754.

Latanzi-Licht, M., & Connor, S. (1995). Care of the dying: The hospice approach. In H. Waas & R. A. Neimeyer (Eds.), *Dying: Facing the facts* (pp. 143–162). Washington, DC: Taylor & Francis.

Laumann, E. O., Gagnon, J. H., Michael, R. T., & Michaels, S. (1994). *The social organization of sexuality.* Chicago: University of Chicago Press.

Lawrence, B. S. (1980, Summer). The myth of the midlife crisis. *Sloan Management Review, 21*(4), 35–49.

Lawton, L., Silverstein, M., & Bengtson, V. L. (1994). Solidarity between generations in families. In V. L. Bengtson & R. A. Harootyan (Eds.), *Intergenerational linkages: Hidden connections in American society* (pp. 19–42). New York: Springer-Verlag.

Lawton, M. P. (1980). Environment and aging. Monterey, CA: Brooks/Cole.

Lawton, M. P., Moss, M., & Moles, E. (1984). The supra-personal neighborhood context of older people: Age heterogeneity and well-being. *Environment and Behavior, 16,* 89–109.

Lazar, A., & Torney-Purta, J. (1991). The development of the subconcepts of death in young children: A short-term longitudinal study. *Child Development, 62,* 1321–1333.

Lazar, I., & Darlington, R. (1982). Lasting effects of early education: A report from the Consortium for Longitudinal Studies. *Monographs of the Society for Research in Child Development, 47*(2–3, Serial No. 195).

Lazarus, R. (1985). The trivialization of distress. In J. Rosen & L. Solomon (Eds.), *Preventing health risk behaviors and promoting coping with illness* (Vol. 8, pp. 279–298). Hanover, NH: University Press of New England.

Lazarus, R. S. (1991). *Emotion and adaptation.* New York: Oxford University Press.

Lazarus, R. S., & Lazarus, B. N. (1994). *Passion and reason.* New York: Oxford University Press.

Leaper, C. (1991). Influence and involvement in children's discourse. *Child Development, 62,* 797–811.

Lee, A. M. (1980). Child-rearing practices and motor performance of Black and White children. *Research Quarterly for Exercise and Sport, 51,* 494–500.

Lee, C. L., & Bates, J. E. (1985). Mother–child interaction at age two years and perceived difficult temperament. *Child Development, 56,* 1314–1325.

Lee, D. J., & Markides, K. S. (1990). Activity and morality among aged persons over an eight-year period. *Journal of Gerontology, 45,* S39–S42.

Lee, R. T., & Ashforth, B. E. (1990). On the meaning of Maslach's three dimensions of burnout. *Journal of Applied Psychology, 75,* 743–747.

Lee, S. H., Ewert, D. P., Frederick, P. D., & Mascola, L. (1992). Resurgence of congenital rubella syndrome in the 1990s. *Journal of the American Medical Association, 267,* 2616–2620.

Lee, V. E., Brooks-Gunn, J., & Schnur, E. (1988). Does Head Start work? A 1-year follow-up of disadvantaged children attending Head Start, no preschool. *Developmental Psychology, 24,* 210–222.

Lee, V. E., Brooks-Gunn, J., Schnur, E., & Liaw, F. (1990). Are Head Start effects sustained? A longitudinal follow-up comparison of disadvantaged children attending Head Start, no preschool, and other preschool programs. *Child Development, 61,* 495–507.

Lee, V. E., & Loeb, S. (1994). *Where do Head Start attendees end up? One reason why preschool effects fade out.* Urbana, IL; *ERIC* (Document No. ED3685100).

Legesse, A. (1979). Age sets and retirement communities. In J. Keith (Ed.), The ethnography of old age. *Anthropological Quarterly, 52,* 61–69.

Lehman, D., Ellard, J., & Wortman, C. (1986). Social support for the bereaved: Recipients' and providers' perspectives on what is helpful. *Journal of Consulting and Clinical Psychology, 54,* 438–446.

Lehman, D. R., & Nisbett, R. E. (1990). A longitudinal study of the effects of undergraduate training on reasoning. *Developmental Psychology, 26,* 952–960.

Lehnert, K. L., Overholser, J. C., & Spirito, A. (1994). Internalized and externalized anger in adolescent suicide attempters. *Journal of Adolescent Research, 9,* 105–119.

Leichtman, M. D., & Ceci, S. J. (1995). The effect of stereotypes and suggestions on preschoolers' reports. *Developmental Psychology, 31,* 568–578.

Lemme, B. H. (1995). *Development in adulthood.* Boston: Allyn and Bacon.

Lempert, H. (1989). Animacy constraints on preschoolers' acquisition of syntax. *Child Development, 60,* 237–245.

Leonard, M. F., Rhymes, J. P., & Solnit, A. J. (1986). Failure to thrive in infants: A family problem. *American Journal of Diseases of Children, 111,* 600–612.

Lerner, J. V., & Abrams, A. (1994). Developmental correlates of maternal employment influences on children. In C. B. Fisher & R. M. Lerner (Eds.), *Applied developmental psychology* (pp. 174–206). New York: McGraw-Hill.

Lerner, J. W. (1989). Educational interventions in learning disabilities. *Journal of the American Academy of Child and Adolescent Psychiatry, 28,* 326–331.

Lerner, R. M., & Schroeder, C. (1971). Physique identification, preference, and aversion in kindergarten children. *Developmental Psychology, 5,* 538.

Lesourd, B. (1995). Protein undernutrition as the major cause of decreased immune function in the elderly: Clinical and functional implications. *Nutrition Reviews, 53,* S86–S94.

Lester, B. M. (1985). Introduction: There's more to crying than meets the ear. In B. M. Lester & C. F. Z. Boukydis (Eds.), *Infant crying* (pp. 1–27). New York: Plenum.

Lester, B. M. (1987). Developmental outcome prediction from acoustic cry analysis in term and preterm infants. *Pediatrics, 80,* 529–534.

Levenson, R. W., Carstensen, L. L., & Gottman, J. M. (1993). Long-term marriage: Age, gender, and satisfaction. *Psychology and Aging, 8,* 301–313.

Levenson, R. W., Carstensen, L. L., & Gottman, J. M. (1994). The influence of age and gender on affect, physiology, and their interrelations: A study of long-term marriages. *Journal of Personality and Social Psychology, 67,* 56–68.

Leventhal, E. A., Leventhal, H., Schaefer, P. M., & Easterling, D. (1993). Conservation of energy, uncertainty reduction, and swift utilization of medical care among the elderly. *Journal of Gerontology, 48,* P78–P86.

Levin, J. S., Taylor, R. J., & Chatters, L. M. (1994). Race and gender differences in religiosity among older adults: Findings from four national surveys. *Journal of Gerontology, 49,* S137–S145.

Levine, L. E. (1983). Mine: Self-definition in 2-year-old boys. *Developmental Psychology, 19,* 544–549.

Levine, L. J. (1995). Young children's understanding of the causes of anger and sadness. *Child Development, 66,* 697–709.

LeVine, R. A., Dixon, S., LeVine, S., Richman, A., Leiderman, P. H., Keefer, C. H., & Brazelton, T. B. (1994). *Child care and culture: Lessons from Africa.* New York: Cambridge University Press.

Levine, R. L., & Stadtman, E. R. (1992). Oxidation of proteins during aging. *Generations, 16*(4), 39–42.

Levinson, D. J. (1978). *The seasons of a man's life.* New York: Knopf.

Levinson, D. J. (1986). A conception of adult development. *American Psychologist, 41,* 3–13.

Levinson, D. J. (1996). *The seasons of a woman's life.* New York: Knopf.

Levy, G. D., Taylor, M. G., & Gelman, S. A. (1995). Traditional and evaluative aspects of flexibility in gender roles, social conventions, moral rules, and physical laws. *Child Development, 66,* 515–531.

Levy, J. A. (1994). Sex and sexuality in later life stages. In A. S. Rossi (Ed.), *Sexuality across the life course* (pp. 287–309). Chicago: University of Chicago Press.

Levy-Shiff, R. (1994). Individual and contextual correlates of marital change across the transition to parenthood. *Developmental Psychology, 30,* 591–601.

Levy-Shiff, R., & Israelashvili, R. (1988). Antecedents of fathering: Some further exploration. *Developmental Psychology, 24,* 434–440.

Lewin, E. (1993). *Lesbian mothers.* Ithaca, NY: Cornell University Press.

Lewinsohn, P. M., Hops, H., Roberts, R. E., Seeley, J. R., & Andrews, J. A. (1993). Adolescent psychopathology: I. Prevalence and incidence of depression and other DSM–III–R disorders in high school students. *Journal of Abnormal Psychology, 102,* 133–144.

Lewis, C. C. (1981). The effects of parental firm control: A reinterpretation of findings. *Psychological Bulletin, 90,* 547–563.

Lewis, L. L. (1992). PMS and progesterone: The ongoing controversy. In A. J. Dan & L. L. Lewis (Eds.), *Menstrual health in women's lives* (pp. 61–74). Urbana: University of Illinois Press.

Lewis, M. (1991). Ways of knowing: Objective self-awareness or consciousness. *Developmental Review, 11,* 231–243.

Lewis, M. (1992a). The self in self-conscious emotions (commentary on self-evaluation in young children). *Monographs of the Society for Research in Child Development, 57*(Serial No. 226, No. 1).

Lewis, M. (1992b). *Shame: The exposed self.* New York: Free Press.

Lewis, M., & Brooks-Gunn, J. (1979). *Social cognition and the acquisition of self.* New York: Plenum.

Lewis, M., Ramsay, D. S., & Kawakami, K. (1993). Differences between Japanese infants and Caucasian American infants in behavioral and cortisol response to inoculation. *Child Development, 64,* 1722–1731.

Lewis, M., Sullivan, M. W., Stanger, C., & Weiss, M. (1989). Self development and self-conscious emotions. *Child Development, 60,* 146–156.

Lewis, R. (1990). Death and dying among the American Indians. In J. K. Parry (Ed.), *Social work practice with the terminally ill: A transcultural perspective* (pp. 23–32). Springfield, IL: Charles C Thomas.

Li, C. Q., Windsor, R. A., Perkins, L. (1993). The impact on infant birth weight and gestational age of cotinine-validated smoking reduction during pregnancy. *Journal of the American Medical Association, 269,* 1519–1524.

Liaw, F., & Brooks-Gunn, J. (1993). Patterns of low-birth-weight children's cognitive development. *Developmental Psychology, 29,* 1024–1035.

Liben, L. S., & Signorella, M. L. (1993). Gender-schematic processing in children: The role of initial interpretations of stimuli. *Developmental Psychology, 29,* 141–149.

Licastro, F., Savorani, G., Sarti, G., & Salsi, A. (1990). Zinc and thymic hormone-dependent immunity in normal aging and in patients with senile dementia of the Alzheimer type. *Journal of Neuroimmunology, 27,* 201–208.

Lidz, C. S. (1991). *Practitioner's guide to dynamic assessment.* New York: Guilford.

Lie, S. O. (1990). Children in the Norwegian health care system. *Pediatrics, 86*(6, Pt. 2), 1048–1052.

Lieberman, M. A. (1993). Bereavement self-help groups: A review of conceptual and methodological issues. In M. S. Stroebe, W. Stroebe, & R. O. Hansson (Eds.), *Handbook of bereavement* (pp. 427–453). New York: Cambridge University Press.

Liefbroer, A. C., & de Jong-Gierveld, J. (1990). Age differences in loneliness among young adults with and without a partner relationship. In P. J. D. Drenth, J. A. Sergeant, & R. J. Takens (Eds.), *European perspectives in psychology* (Vol. 3, pp. 265–278). Chicester: Wiley.

Lifschitz, M., Berman, D., Galili, A., & Gilad, D. (1977). Bereaved children: The effects of mother's perception and social system organization on their short range adjustment. *Journal of Child Psychiatry, 16,* 272–284.

Light, P., & Perrett-Clermont, A-N. (1989). Social context effects in learning and testing. In A. Gellatly, D. Rogers, & J. Sloboda (Eds.), *Cognition and social worlds* (pp. 99–112). Oxford: Clarendon Press.

Lin, C. C., & Fu, V. R. (1990). A comparison of child-rearing practices among Chinese, immigrant Chinese, and Caucasian-American parents. *Child Development, 61,* 429–433.

Linde, E. V., Morrongiello, B. A., & Rovee-Collier, C. (1985). Determinants of retention in 8-week-old infants. *Developmental Psychology, 21,* 601–613.

Lindell, S. G. (1988). Education for childbirth: A time for change. *Journal of Obstetrics, Gynecology, and Neonatal Nursing, 17,* 108–112.

Linn, S., Lieberman, E., Schoenbaum, S. C., Monson, R. R., Stubblefield, P. G., & Ryan, K. J. (1988). Adverse outcomes of pregnancy in women exposed to diethylstilbestrol in utero. *Journal of Reproductive Medicine, 33,* 3–7.

Lipsitt, L. P. (1990). Learning and memory in infants. *Merrill-Palmer Quarterly, 36,* 53–66.

Little, J., & Thompson, B. (1988) Descriptive epidemiology. In I. MacGillvray, D. M. Campbell, & B. Thompson (Eds.), *Twinning and twins* (pp. 37–66). New York: Wiley.

Livesley, W. J., & Bromley, D. B. (1973). *Person perception in childhood and adolescence.* London: Wiley.

Lochman, J. E., Coie, J. D., Underwood, M. K., & Terry, R. (1993). Effectiveness of a social relations intervention program for aggressive and nonaggressive, rejected children. *Journal of Consulting and Clinical Psychology, 61,* 1053–1058.

Locke, J. (1892). Some thoughts concerning education. In R. H. Quick (Ed.), *Locke on education* (pp. 1–236). Cambridge, England: Cambridge University Press. (Original work published 1690)

Loewen, E. R., Shaw, R. J., & Craik, F. I. M. (1990). Age differences in components of metamemory. *Experimental Aging Research, 16,* 43–48.

Lombardo, J. P., & Lavine, L. O. (1981). Sex-role stereotyping and patterns of self-disclosure. *Sex Roles, 7,* 403–411.

Long, D. D. (1985). A cross-cultural examination of fears of death among Saudi Arabians. *Omega, 16,* 43–50.

Long, H. B., & Zoller-Hodges, D. (1995). Outcomes of Elderhostel participation. *Educational Gerontology, 21,* 113–127.

Looney, M. A., & Plowman, S. A. (1990). Passing rates of American children and youth on the FITNESSGRAM criterion-referenced physical fitness standards. *Research Quarterly of Exercise and Sport, 61,* 215–223.

Lopata, H. Z. (1996). *Current widowhood: Myths and realities.* Thousand Oaks, CA: Sage.

Lopez, F. G. (1989). Current family dynamics, trait anxiety, and academic adjustment: Test of a family-based model of vocational identity. *Journal of Vocational Behavior, 35,* 76–87.

Lord, S. E., Eccles, J. S., & McCarthy, K. A. (1994). Surviving the junior high transition: Family processes and self-perceptions as protective and risk factors. *Journal of Early Adolescence, 14,* 162–199.

Lorenz, K. (1952). *King Solomon's ring.* New York: Crowell.

Lorenz, K. Z. (1943). Die angeborenen Formen möglicher Erfanhrung. *Zeitschrift für Tierpsychologie, 5,* 235–409.

Losey, K. M. (1995). Mexican-American students and classroom interaction: An overview and critique. *Review of Educational Research, 65,* 283–318.

Lowinson, J. H., Ruiz, P. & Millman, R. B. (Eds.) (1992). *Substance abuse.* Baltimore: Williams & Wilkins.

Lozoff, B., Jordan, B., & Malone, S. (1988). Childbirth in cross-cultural perspective. *Marriage and Family Review, 13,* 35–60.

Lozoff, B., Wolf, A., Latz, S., & Paludetto, R. (1995, March). *Cosleeping in Japan, Italy, and the U.S.: Autonomy versus interpersonal relatedness.* Paper presented at the biennial meeting of the Society for Research in Child Development, Indianapolis.

Ludemann, P. M. (1991). Generalized discrimination of positive facial expressions by seven- and ten-month-old infants. *Child Development, 62,* 55–67.

Lummis, M., & Stevenson, H. W. (1990). Gender differences in beliefs about achievement: A cross-cultural study. *Developmental Psychology, 26,* 254–263.

Lund, D. A., Caserta, M. S., & Dimond, M. F. (1993). The course of spousal bereavement in later life. In M. S. Stroebe, W. Stroebe, & R. O. Hansson (Eds.), *Handbook of bereavement* (pp. 240–245). New York: Cambridge University Press.

Luster, T., Rhoades, K., & Haas, B. (1989). The relation between parental values and parenting behavior. *Journal of Marriage and the Family, 51,* 139–147.

Lutz, S. E., & Ruble, D. N. (1995). Children and gender prejudice: Context, motivation, and the development of gender constancy. In R. Vasta (Ed.), *Annals of child development* (Vol. 10, pp. 131–166). London: Jessica Kingsley.

Lyon, T. D., & Flavell, J. H. (1993). Young children's understanding of forgetting. *Child Development, 64,* 789–900.

Lyon, T. D., & Flavell, J. H. (1994). Young children's understanding of "remember" and "forget." *Child Development, 65,* 1357–1371.

Lyons-Ruth, K., Alpern, L., & Repacholi, B. (1993). Disorganized infant attachment classification and maternal psychosocial problems as predictors of hostile–aggressive behavior in the preschool classroom. *Child Development, 64,* 572–585.

Lyons-Ruth, K., Connell, D. B., Grunebaum, H. U., & Botein, S. (1990). Infants at social risk: Maternal depression and family support services as mediators of infant development and security of attachment. *Child Development, 61,* 85–98.

Lytton, H., & Romney, D. M. (1991). Parents' sex-related differential socialization of boys and girls: A meta-analysis. *Psychological Bulletin, 109,* 267–296.

Maccoby, E. E. (1980). Sex differences and sex typing. In E. Maccoby (Ed.), *Social development: Psychological growth and the parent–child relationship* (pp. 203–250). San Diego: Harcourt Brace Jovanovich.

Maccoby, E. E. (1984a). Middle childhood in the context of the family. In W. A. Collins (Ed.), *Development during middle childhood* (pp. 184–239). Washington, DC: National Academy Press.

Maccoby, E. E. (1984b). Socialization and developmental change. *Child Development, 55,* 317–328.

Maccoby, E. E. (1990). Gender and relationships. *American Psychologist, 45,* 513–520.

Maccoby, E. E., & Jacklin, C. N. (1987). Gender segregation in childhood. In E. H. Reese (Ed.), *Advances in child development and behavior* (Vol. 20, pp. 239–287). New York: Academic Press.

Maccoby, E. E., & Martin, J. A. (1983). Socialization in the context of the family: Parent–child interaction. In E. M. Hetherington (Ed.), *Handbook of child psychology: Vol. 4. Socialization, personality, and social development* (4th ed., pp. 1–101). New York: Wiley.

MacEwen, K. E., Barling, J., Kelloway, E. K., & Higginbottom, S. F. (1995). Predicting retirement anxiety: The roles of parental socialization and personal planning. *Journal of Social Psychology, 135,* 203–213.

Macfarlane, J. (1971). From infancy to adulthood. In M. C. Jones, N. Bayley, J. W. Macfarlane, & M. P. Honzik (Eds.), *The course of human development* (pp. 406–410). Waltham, MA: Xerox College Publishing.

MacFarlane, J., Smith, D. M., & Garrow, D. H. (1978). The relationship between mother and neonate. In S. Kitzinger (Ed.), *The place of birth* (pp. 185–200). New York: Oxford University Press.

MacKay, D. G., & Abrams, L. (1996). Language, memory, and aging: Distributed deficits and the structure of new-versus-old connections. In J. E. Birren & K. W. Schaie (Eds.), *Handbook of the psychology of aging* (pp. 251–265). San Diego: Academic Press.

MacKinnon, C. E. (1989). An observational investigation of sibling interactions in married and divorced families. *Developmental Psychology, 25,* 36–44.

MacKinnon, D. F., & Squire, L. R. (1989). Autobiographical memory and amnesia. *Psychobiology, 17,* 247–256.

Mackinnon, L. T. (1992). *Exercise and immunology.* Champaign, IL: Human Kinetics.

MacMillan, D. L., Keogh, B. K., & Jones, R. L. (1986). Special educational research on mildly handicapped learners. In M. C. Wittrock (Ed.), *Handbook of research on teaching* (3rd ed., pp. 686–724). New York: Macmillan.

Madden, N., & Slavin, R. (1983). Mainstreaming students with mild handicaps: Academic and social outcomes. *Review of Educational Research, 53,* 519–659.

Maddi, S. R., & Kobasa, S. C. (1984). *The hardy executive: Health under stress.* Homewood, IL: Dow Jones–Irwin.

Maddox, G. L. (1963). Activity and morale: A longitudinal study of selected elderly subjects. *Social Forces, 42,* 195–204.

Maglio, C. J., & Robinson, S. E. (1994). The effects of death education on death anxiety: A meta-analysis. *Omega, 29,* 319–335.

Magni, E., & Frisoni, G. B. (1996). Depression and somatic symptoms in the elderly: The role of cognitive function. *International Journal of Geriatric Psychiatry, 11,* 517–522.

Mahler, M. S., Pine, F., & Bergman, A. (1975). *The psychological birth of the human infant.* New York: Basic Books.

Mahoney, J. L., & Cairns, R. B. (1997). Do extracurricular activities protect against early school dropout? *Developmental Psychology, 33,* 241–253.

Maier, D. M., & Newman, M. J. (1995). Legal and psychological considerations in the development of a euthanasia statute for adults in the United States.

Behavioral Sciences and the Law, 13, 3–25.

Maier, S. F., Watkins, L. R., & Fleshner, M. (1994). Psychoneuroimmunology: The interface between behavior, brain, and immunity. *American Psychologist, 49,* 1004–1017.

Main, M., & Solomon, J. (1990). Procedures for identifying infants as disorganized/disoriented during the Ainsworth Strange Situation. In M. Greenberg, D. Cicchetti, & M. Cummings (Eds.), *Attachment in the preschool years: Theory, research, and intervention* (pp. 121–160). Chicago: University of Chicago Press.

Makin, J., Fried, P. A., & Watkinson, B. (1991). A comparison of active and passive smoking during pregnancy: Long-term effects. *Neurotoxicology and Teratology, 13,* 5–12.

Makin, J. W., & Porter, R. H. (1989). Attractiveness of lactating females' breast odors to neonates. *Child Development, 60,* 803–810.

Malatesta, C. Z., Grigoryev, P., Lamb, C., Albin, M., & Culver, C. (1986). Emotion socialization and expressive development in preterm and full-term infants. *Child Development, 57,* 316–330.

Malatesta, C. Z., & Haviland, J. M. (1982). Learning display rules: The socialization of emotion expression in infancy. *Child Development, 53,* 991–1003.

Malatesta-Magai, C. Z., Izard, C. E., & Camras, L. (1991). Conceptualizing early infant affect: Emotions as fact, fiction or artifact? In K. Strongman (Ed.), *International review of studies on emotion* (pp. 1–36). New York: Wiley.

Malina, R. M. (1975). *Growth and development: The first twenty years in man.* Minneapolis: Burgess.

Malina, R. M. (1990). Physical growth and performance during the transitional years (9–16). In R. Montemayor, G. R. Adams, & T. P. Gullotta (Eds.), *From childhood to adolescence: A transitional period?* (pp. 41–62). Newbury Park, CA: Sage.

Malina, R. M., & Bouchard, C. (1991). *Growth, maturation, and physical activity.* Champaign, IL: Human Kinetics.

Mallinckrodt, B., & Fretz, B. R. (1988). Social support and the impact of job loss on older professionals. *Journal of Counseling Psychology, 35,* 281–286.

Malloy, M. H., & Hoffman, H. J. (1995). Prematurity, sudden infant death syndrome, and age of death. *Pediatrics, 96,* 464–471.

Malone, M. M. (1982). Consciousness of dying and projective fantasy of young children with malignant disease.

Developmental and Behavioral Pediatrics, 3, 55–60.

Maloney, M., & Kranz, R. (1991). *Straight talk about eating disorders.* New York: Facts on File.

Mandler, J. M. (1992). The foundations of conceptual thought in infancy. *Cognitive Development, 7,* 273–285.

Mandler, J. M., Bauer, P. J., & McDonough, L. (1991). Separating the sheep from the goats: Differentiating global categories. *Cognitive Psychology, 23,* 263–298.

Mandler, J. M., & McDonough, L. (1993). Concept formation in infancy. *Cognitive Development, 8,* 291–318.

Mangelsdorf, S., Gunnar, M., Kestenbaum, R., Lang, S., & Andreas, D. (1990). Infant proneness-to-distress temperament, maternal personality, and mother–infant attachment: Associations and goodness of fit. *Child Development, 61,* 820–831.

Manson, J. E., Hosteson, H., Satterfield, S., Hebert, P., O'Connor, G. T., Buring, J. F., & Hennekens, C. H. (1992). The primary prevention of myocardial infarction. *New England Journal of Medicine, 326,* 1406–1416.

Manson, J. E., Willett, W. C., Stampfer, M. J., Colditz, G. A., Hunter, D. J., Hankinson, S. E., Hennekens, C. H., & Speizer, F. E. (1995). Body weight and mortality among women. *New England Journal of Medicine, 333,* 678–685.

Maratsos, M. P., & Chalkley, M. A. (1980). The internal language of children's syntax: The ontogenesis and representation of syntactic categories. In K. Nelson (Ed.), *Children's language* (Vol. 2, pp. 127–214). New York: Gardner Press.

Marcia, J. E. (1980). Identity in adolescence. In J. Adelson (Ed.), *Handbook of adolescent psychology* (pp. 159–187). New York: Wiley.

Marcia, J. E., Waterman, A. S., Matteson, D. R., Archer, S. L., & Orlofsky, J. L. (1993). *Ego identity: A handbook for psychosocial research.* New York: Springer-Verlag.

Marcus, G. F. (1993). Negative evidence in language acquisition. *Cognition, 46,* 53–85.

Marcus, G. F. (1995). Children's over-regularization of English plurals: A quantitative analysis. *Journal of Child Language, 22,* 447–459.

Marcus, G. F., Pinker, S., Ullman, M., Hollander, M., Rosen, T. J., & Xu, F. (1992). Overregularization in language acquisition. *Monographs of the Society for Research in Child Development, 57*(4, Serial No. 228).

Margolin, B. H., Morrison, H. I., & Hulka, B. S. (1994). Cigarette smoking and

sperm density: A meta-analysis. *Fertility and Sterility, 61,* 35–43.

Markides, K. S. (1989). Aging, gender, race/ethnicity, class, and health: A conceptual overview. In K. S. Markides (Ed.), *Aging and health* (pp. 9–21). Newbury Park, CA: Sage.

Markides, K. S. (1994). Gender and ethnic diversity in aging. In R. J. Manheimer (Ed.), *Older Americans almanac* (pp. 49–69). Detroit: Gale Research, Inc.

Markides, K. S., & Cooper, C. L. (1989). Aging, stress, social support and health: An overview. In K. S. Markides & C. L. Cooper (Eds.), *Aging, stress and health* (pp. 1–10). Chicester: Wiley.

Markman, E. M. (1989). *Categorization and naming in children.* Cambridge, MA: MIT Press.

Markman, E. M. (1992). Constraints on word learning: Speculations about their nature, origins, and domain specificity. In M. R. Gunnar & M. P. Maratsos (Eds.), *Minnesota Symposia on Child Psychology* (Vol. 25, pp. 59–101). Hillsdale, NJ: Erlbaum.

Markovits, H., & Vachon, R. (1989). Reasoning with contrary-to-fact propositions. *Journal of Experimental Child Psychology, 47,* 398–412.

Markovits, H., & Vachon, R. (1990). Conditional reasoning, representation, and level of abstraction. *Developmental Psychology, 26,* 942–951.

Marks, N. (1995). Midlife marital status differences in social support relationships with adult children and psychological well-being. *Journal of Family Issues, 16,* 5–28.

Marks, N. F. (1996). Caregiving across the lifespan: National prevalence and predictors. *Family Relations, 45,* 27–36.

Markstrom-Adams, C., & Adams, G. R. (1995). Gender, ethnic group, and grade differences in psychosocial functioning during middle adolescence? *Journal of Youth and Adolescence, 24,* 397–417.

Markus, H. R., & Herzog, A. R. (1992). The role of self-concept in aging. In K. W. Schaie & M. P. Lawton (Eds.), *Annual review of gerontology and geriatrics* (pp. 110–143). New York: Springer.

Marsh, D. T., Serafica, F. C., & Barenboim, C. (1981). Interrelationships among perspective taking, interpersonal problem solving, and interpersonal functioning. *Journal of Genetic Psychology, 138,* 37–48.

Marsh, H. W. (1989). Sex differences in the development of verbal and mathematics constructs: The high school and beyond study. *American Educational Research Journal, 26,* 191–225.

Marsh, H. W. (1990). The structure of academic self-concept: The Marsh/Shavelson model. *Journal of Educational Psychology, 82,* 623–636.

Marsh, H. W., Smith, I. D., & Barnes, J. (1985). Multidimensional self-concepts: Relations with sex and academic achievement. *Journal of Educational Psychology, 77,* 581–596.

Martin, C. L. (1989). Children's use of gender-related information in making social judgments. *Developmental Psychology, 25,* 80–88.

Martin, C. L., & Halverson, C. F. (1981). A schematic processing model of sex typing and stereotyping in children. *Child Development, 52,* 1119–1134.

Martin, C. L., & Halverson, C. F. (1987). The role of cognition in sex role acquisition. In D. B. Carter (Ed.), *Current conceptions of sex roles and sex typing: Theory and research* (pp. 123–137). New York: Praeger.

Martin, C. L., & Little, J. K. (1990). The relation of gender understanding to children's sex-typed preferences and gender stereotypes. *Child Development, 61,* 1427–1439.

Martin, J. A. (1981). A longitudinal study of the consequences of early mother–infant interaction: A micro-analytic approach. *Monographs of the Society for Research in Child Development, 46*(3, Serial No. 190).

Martin, J. B. (1987). Molecular genetics: Applications to the clinical neurosciences. *Science, 298,* 765–772.

Martin, J. E., & Dean, L. (1993). Bereavement following death from AIDS: Unique problems, reactions, and special needs. In M. S. Stroebe, W. Stroebe, & R. O. Hansson (Eds.), *Handbook of bereavement* (pp. 317–330). Cambridge, England: Cambridge University Press.

Martin, J. E., Dubbert, P. M., Katell, A. D., Thompson, J. K., Raczynski, J. R., Lake, M., Smith, P. O., Webster, J. S., Sikora, T., & Cohen, R. E. (1984). Behavioral control of exercise in sedentary adults: Studies 1 through 6. *Journal of Consulting and Clinical Psychology, 52,* 795–811.

Martin, R. J., White, B. D., & Hulsey, M. G. (1991). The regulation of body weight. *American Scientist, 79,* 528–541.

Martin, R. M. (1975). Effects of familiar and complex stimuli on infant attention. *Developmental Psychology, 11,* 178–185.

Martin, S. L., Ramey, C. T., & Ramey, S. (1990). The prevention of intellectual impairment in children of impoverished families: Findings of a randomized trial of educational day care.

American Journal of Public Health, 80, 844–847.

Martinson, I. M., Davies, E., & McClowry, S. G. (1987). The long-term effect of sibling death on self-concept. *Journal of Pediatric Nursing, 2,* 227–235.

Mason, C. A., Cauce, A. M., Gonzales, N., & Hiraga, Y. (1996). Neither too sweet nor too sour: Problem peers, maternal control, and problem behavior in African American adolescents. *Child Development, 67,* 2115–2130.

Mason, M. G., & Gibbs, J. C. (1993). Social perspective taking and moral judgment among college students. *Journal of Adolescent Research, 8,* 109–123.

Massad, C. M. (1981). Sex role identity and adjustment during adolescence. *Child Development, 52,* 1290–1298.

Masur, E. F., McIntyre, C. W., & Flavell, J. H. (1973). Developmental changes in apportionment of study time among items in a multi-trial free recall task. *Journal of Experimental Child Psychology, 15,* 237–246.

Matas, L., Arend, R., & Sroufe, L. A. (1978). Continuity of adaptation in the second year: The relationship between quality of attachment and later competence. *Child Development, 49,* 547–556.

Matheny, A. P., Jr. (1991). Children's unintentional injuries and gender: Differentiation and psychosocial aspects. *Children's Environment Quarterly, 8,* 51–61.

Matute-Bianchi, M. E. (1986). Ethnic identities and patterns of school success and failure among Mexican-descent and Japanese-American students in a California high school: An ethnographic analysis. *American Journal of Education, 95,* 233–255.

Maurer, T., & Tarulli, B. (1994). Perceived environment, perceived outcome, and person variables in relationship to voluntary development activity by employees. *Journal of Applied Psychology, 79,* 3–14.

Mayberry, R. I. (1993). First-language acquisition after childhood differs from second-language acquisition: The case of American Sign Language. *Journal of Speech and Hearing Research, 36,* 1258–1270.

Mayes, L. C., & Zigler, E. (1992). An observational study of the affective concomitants of mastery in infants. *Journal of Child Psychology and Psychiatry, 33,* 659–667.

Mayeux, R. (1996). Understanding Alzheimer's disease: Expect more genes and other things. *Annals of Neurology, 30,* 689–690.

Maylor, E., & Valentine, T. (1992). Linear and nonlinear effects of aging on categorizing and naming faces. *Psychology and Aging, 7,* 317–323.

Mazin, A. L. (1993). The genome loses all 5-methylcytosine during the lifespan: How is this related to accumulation of mutations with aging? *Molecular Biology, 27,* 96–104.

Mazur, E. (1993). Developmental differences in children's understanding of marriage, divorce, and remarriage. *Journal of Applied Developmental Psychology, 14,* 191–212.

Mazzocco, M. M. M., Nord, A. M., van Doorninck, W., Green, C. L., Kovar, C. G., & Pennington, B. F. (1994). Cognitive development among children with early-treated phenylketonuria. *Developmental Neuropsychology, 10,* 133–151.

McAdams, D. P. (1985). *Power, intimacy, and the life story: Personological inquiries into identity.* New York: Guilford Press.

McAdams, D. P. (1988). *Power, intimacy, and the life story.* New York: Guilford.

McAdams, D. P. (1993). *The stories we live by: Personal myths and the making of the self.* New York: Morrow.

McAdams, D. P. (1994). Can personality change? Levels of stability and growth in personality across the life span. In T. F. Heatherton & J. L. Weinberger (Eds.), *Can personality change?* (pp. 299–313). Washington, DC: American Psychological Association.

McAdams, D. P., & de St. Aubin, E. (1992). A theory of generativity and its assessment through self-report, behavioral acts, and narrative themes in autobiography. *Journal of Personality and Social Psychology, 62,* 1003–1015.

McAdams, D. P., de St. Aubin, E., & Logan, R. L. (1993). Generativity among young, midlife, and older adults. *Psychology and Aging, 8,* 221–230.

McAdams, R. P. (1993). *Lessons from abroad: How other countries educate their children.* Lancaster, PA: Technomic.

McAnarney, E. R., Kreipe, R. E., Orr, D. P., & Comerci, G. D. (1992). *Textbook of adolescent development.* Philadelphia: Saunders.

McAuley, E., Bane, S. M., & Mihalko, S. L. (1995). Exercise in middle-aged adults: Self-efficacy and self-presentational outcomes. *Preventive Medicine, 24,* 319–328.

McAuley, E., & Jacobson, L. (1991). Self-efficacy and exercise participation in sedentary adult females. *American Journal of Health Promotion, 5,* 185–191, 207.

McCabe, A. E., & Peterson, C. (1988). A comparison of adults' versus children's spontaneous use of *because* and *so. Journal of Genetic Psychology, 149,* 257–268.

McCall, R. B., Appelbaum, M. I., & Hogarty, P. S. (1973). Developmental changes in mental performance. *Monographs of the Society for Research in Child Development, 42* (3, Serial No. 171).

McCall, R. B., & Carriger, M. S. (1993). A meta-analysis of infant habituation and recognition memory performance as predictors of later IQ. *Child Development, 64,* 57–79.

McCartney, K. (1984). The effect of quality of day care environment upon children's language development. *Developmental Psychology, 20,* 244–260.

McCartney, K., Scarr, S., Phillips, D., & Grajek, S. (1985). Day care as intervention: Comparisons of varying quality programs. *Journal of Applied Developmental Psychology, 6,* 247–260.

McConaghy, M. J. (1979). Gender permanence and the genital basis of gender: Stages in the development of constancy of gender identity. *Child Development, 50,* 1223–1226.

McCormick, W. C., Uomoto, J., Young, H., Graves, A. B., Vitaliano, P., Mortimer, J. A., Edland, S. D., & Larson, E. B. (1996). Attitudes toward use of nursing homes and home care in older Japanese-Americans. *Journal of the American Gerontological Society, 44,* 769–777.

McCrae, R. R., & Costa, P. T., Jr. (1988). Psychological resilience among widowed men and women: A 10-year follow-up of a national sample. *Journal of Social Issues, 44,* 129–142.

McCrae, R. R., & Costa, P. T., Jr. (1990). *Personality in adulthood.* New York: Guilford.

McCullough, P., & Rutenberg, S. (1989). Launching children and moving on. In B. Carter & M. McGoldrick (Eds.), *The changing family life cycle* (pp. 285–309). Boston: Allyn and Bacon.

McCune, L. (1993). The development of play as the development of consciousness. In M. H. Bornstein & A. O'Reilly (Eds.), *New directions for child development* (No. 59, pp. 67–79). San Francisco: Jossey-Bass.

McFadden, S. H. (1996). Religion, spirituality, and aging. In J. E. Birren & K. W. Schaie (Eds.), *Handbook of the psychology of aging* (pp. 162–177). San Diego: Academic Press.

McFalls, J. A., Jr. (1990). The risks of reproductive impairment in the later years of childbearing. *Annual Review of Sociology, 16,* 491–519.

McGee, L. M., & Richgels, D. J. (1989, December). "K is Kristen's": Learning the alphabet from a child's perspective. *The Reading Teacher, 43*(3), 216–225.

McGee, L. M., & Richgels, D. J. (1990). *Literacy's beginnings: Supporting young readers and writers.* Boston: Allyn and Bacon.

McGee, L. M., & Richgels, D. J. (1996). *Literacy's beginnings: Supporting young readers and writers (2nd ed.)* Boston: Allyn and Bacon.

McGeer, P. L., Schulzer, M., & McGeer, E. G. (1996). Arthritis and anti-inflammatory agents as possible protective factors for Alzheimer's disease: A review of 17 epidemiologic studies. *Neurology, 47,* 425–432.

McGillicuddy-De Lisi, A. V., Watkins, C., & Vinchur, A. J. (1994). The effect of relationship on children's distributive justice reasoning. *Child Development, 65,* 1694–1700.

McGinnis, J. M., & Lee, P. R. (1995). Healthy people 2000 at mid decade. *Journal of the American Medical Association, 273,* 1123–1129.

McGinty, M. J., & Zafran, E. I. (1988). *Surrogacy: Constitutional and legal issues.* Cleveland, OH: The Ohio Academy of Trial Lawyers.

McGoldrick, M. (1989). Women and the family life cycle. In B. Carter & M. McGoldrick (Eds.), *The changing family life cycle* (2nd ed., pp. 31–69). Boston: Allyn and Bacon.

McGoldrick, M., Heiman, M., & Carter, B. (1993). The changing family life cycle: A perspective on normalcy. In F. Walsh (Ed.), *Normal family processes* (pp. 405–443). New York: Guilford.

McGroarty, M. (1992, March). The societal context of bilingual education. *Educational Researcher, 21*(2), 7–9.

McGue, M., Bouchard, T. J., Jr., Iacono, W. G., & Lykken, D. T. (1993). Behavioral genetics of cognitive ability: A life-span perspective. In R. Plomin & G. E. McClearn (Eds.), *Nature, nurture, and psychology* (pp. 59–76). Washington, DC: American Psychological Association.

McGuinness, D., & Pribram, K. H. (1980). The neuropsychology of attention: Emotional and motivational controls. In M. C. Wittcock (Ed.), *The brain and psychology* (pp. 95–139). New York: Academic Press.

McGuire, E. J., & Savashino, J. A. (1984). Urodynamic studies in enuresis and the non-neurogenic-neurogenic bladder. *Journal of Neurology, 132,* 299–302.

McHale, S. M., Bartko, W. T., Crouter, A. C., & Perry-Jenkins, M. (1990). Children's housework and psycho-social functioning: The mediating effects of parents' sex-role behaviors and attitudes. *Child Development, 61,* 1413–1426.

McIntosh, J. L., Santos, J. F., Hubbard, R. W., & Overholser, J. C. (1994). *Elder suicide: Research, theory, and treatment.* Washington, DC: American Psychological Association.

McKusick, V. A. (1995). *Mendelian inheritance in man: Catalogs of autosomal dominant, autosomal recessive, and X-linked phenotypes* (10th ed.). Baltimore: Johns Hopkins University Press.

McLean, D. F., Timajchy, K. H., Wingo, P. A., & Floyd, R. L. (1993). Psychosocial measurement: Implications of the study of preterm delivery in black women. *American Journal of Preventive Medicine, 9,* 39–81.

McLoyd, V. (1989). Socialization and development in a changing economy: The effects of paternal job and income loss on children. *American Psychologist, 44,* 293–302.

McLoyd, V.C. (1990). The impact of economic hardship on black families and chldren: Psychological distress, parenting, and socioemotional development. *Child Development, 61,* 311–346.

McLoyd, V. C., Jayaratne, T. E., Ceballo, R., & Borquez, J. (1994). Unemployment and work interruption among African-American single mothers: Effects on parenting and adolescent socioemotional functioning. *Child Development, 65,* 562–589.

McManus, I. C., Sik, G., Cole, D. R., Mellon, A. F., Wong, J., & Kloss, J. (1988). The development of handedness in children. *British Journal of Developmental Psychology, 6,* 257–273.

McNamee, S., & Peterson, J. (1986). Young children's distributive justice reasoning, behavior, and role taking: Their consistency and relationship. *Journal of Genetic Psychology, 146,* 399–404.

McNeil, J. N. (1986). Talking about death: Adolescents, parents, and peers. In C. A. Corr & J. N. McNeil (Eds.), *Adolescence and death* (pp. 185–201). New York: Springer.

MCR Vitamin Study Research Group. (1991). Prevention of neural tube defects: Results of the Medical Research Council Vitamin Study. *Lancet, 338,* 131–137.

McRae, M. J. (1993). Litigation, electronic fetal monitoring, and the obstetric nurse. *Journal of Obstetric, Gynecologic, and Neonatal Nursing, 22,* 410–419.

McRae, T. D. (1995). "But doctor, my left knee is also 100." In M. M. Seltzer (Ed.), *The impact of increased life expectancy* (pp. 148–164). New York: Springer.

McWilliams, M. (1986). *The parents' nutrition book.* New York: Wiley.

Mead, G. H. (1934). *Mind, self, and society.* Chicago: University of Chicago Press.

Mead, M. (1928). *Coming of age in Samoa.* Ann Arbor, MI: Morrow.

Mead, M., & Newton, N. (1967). Cultural patterning of perinatal behavior. In S. Richardson & A. Guttmacher (Eds.), *Childbearing: Its social and psychological aspects* (pp. 142–244). Baltimore: Williams & Wilkins.

Mediascope, Inc. (1996). *National Television Violence Study: Executive summary 1994–1995.* Studio City, CA: Author.

Medrich, E. A., Rosen, J., Rubin, V., & Buckley, S. (1982). *The serious business of growing up.* Berkeley: University of California Press.

Meehan, A. M. (1984). A meta-analysis of sex differences in formal operational thought. *Child Development, 55,* 1110–1124.

Meilman, P. W. (1979). Cross-sectional age changes in ego identity status during adolescence. *Developmental Psychology, 15,* 230–231.

Melnikow, J., & Alemagno, S. (1993). Adequacy of prenatal care among inner-city women. *Journal of Family Practice, 37,* 575–582.

Meltzoff, A. N. (1988). Infant imitation and memory: Nine-month-olds in immediate and deferred tests. *Child Development, 59,* 217–255.

Meltzoff, A. N. (1990). Towards a developmental cognitive science. *Annals of the New York Academy of Sciences, 608,* 1–37.

Meltzoff, A. N. (1995). Understanding the intentions of others: Re-enactment of intended acts by 18-month-old children. *Developmental Psychology, 31,* 838–850.

Meltzoff, A. N., & Borton, R. W. (1979). Intermodal matching by human neonates. *Nature, 282,* 403–404.

Meltzoff, A.N., & Kuhl, P. K. (1994). Faces and speech: Intermodal processing of biologically relevant signals in infants and adults. In D. J. Lewkowicz & R. Lickliter (Eds.), *The development of intersensory perception* (pp. 335–369). Hillsdale, NJ: Erlbaum.

Meltzoff, A. N., & Moore, M. K. (1977). Imitation of facial and manual gestures by human neonates. *Science, 198,* 75–78.

Meltzoff, A. N., & Moore, M. K. (1992). Early imitation within a functional framework: The importance of person identity, movement, and development. *Infant Behavior and Development, 15,* 479–505.

Meltzoff, A. N., & Moore, M. K. (1994). Imitation, memory, and the representation of persons. *Infant Behavior and Development, 17,* 83–99.

Mercer, R. T., Nichols, E. G., & Doyle, G. C. (1989). *Transitions in a woman's life: Major life events in developmental context.* New York: Springer-Verlag.

Meredith, C. N., Frontera, W. R., O'Reilly, K. P., & Evans, W. J. (1992). Body composition in elderly men: Effect of dietary modification during strength training. *Journal of the American Geriatrics Society, 40,* 155–162.

Mergenhagen, P. (1996). Her own boss. *American Demographics, 18,* 36–41.

Mervis, C. (1985). On the existence of prelinguistic categories: A case study. *Infant Behavior and Development, 8,* 293–300.

Mervis, C. B. (1987). Child-basic object categories and early lexical development. In U. Neisser (Ed.), *Concepts and conceptual development: Ecological and intellectual factors in categorization* (pp. 201–233). Cambridge, England: Cambridge University Press.

Meyer, B. J. F., Russo, C., & Talbot, A. (1995). Discourse comprehension and problem solving: Decisions about the treatment of breast cancer by women across the life-span. *Psychology and Aging, 10,* 84–103.

Meyer, D. R., & Garasky, S. (1993). Custodial fathers: Myths, realities, and child support policy. *Journal of Marriage and the Family, 55,* 73–79.

Meyer, M. H., & Bellas, M. L. (1995). U.S. old-age policy and the family. In R. Blieszner & V. H. Bedford (Eds.), *Handbook of aging and the family* (pp. 263–283). Westport, CT: Greenwood Press.

Michael, R. T., Gagnon, J. H., Laumann, E. O., & Kolata, G. (1994). *Sex in America.* Boston: Little, Brown.

Michaels, G. Y. (1988). Motivational factors in the decision and timing of pregnancy. In G. Y. Michaels & W. A. Goldberg (Eds.), *The transition to parenthood: Current theory and research* (pp. 23–61). New York: Cambridge University Press.

Michel, C. (1989). Radiation embryology. *Experientia, 45,* 69–77.

Mikulincer, M., & Erev, I. (1991). Attachment style and the structure of romantic love. *British Journal of Social Psychology, 30,* 273–291.

Milburn, N., & D'Ercole, A. (1991). Homeless women, children, and families. *American Psychologist, 46,* 1159–1160.

Miller, E., & Morris, R. (1993). *The psychology of dementia.* New York: Wiley.

Miller, F. G., Quill, T. E., Brody, H., Fletcher, J. C., Gostin, L. O., & Meier, D. E. (1994). Regulating physician-assisted death. *New England Journal of Medicine, 118,* 119–123.

Miller, G. A. (1991). *The science of words.* New York: Scientific American Library.

Miller, J., Slomczynski, K. M., & Kohn, M. L. (1985). Continuity of learning-generalization: The effect of job on men's intellective process in the United States and Poland. *American Journal of Sociology, 91,* 593–615.

Miller, J. M., Boudreaux, M. C., & Regan, F. A. (1995). A case-control study of cocaine use in pregnancy. *American Journal of Obstetrics and Gynecology, 172,* 180–185.

Miller, K. A., & Kohn, M. L. (1983). The reciprocal effects of job conditions and the intellectuality of leisure-time activities. In M. L. Kohn & C. Schooler (Eds.), *Work and personality: An inquiry into the impact of social stratification* (pp. 217–241). Norwood, NJ: Ablex.

Miller, K. F., & Baillargeon, R. (1990). Length and distance: Do preschoolers think that occlusion brings things together? *Developmental Psychology, 26,* 103–114.

Miller, N. (1989). *In search of gay America: Men and women in a time of change.* New York: Atlantic Monthly.

Miller, N., & Maruyama, G. (1976). Ordinal position and peer popularity. *Journal of Personality and Social Psychology, 33,* 123–131.

Miller, N. B., Cowan, P. A., Cowan, C. P., Hetherington, E. M., & Clingempeel, W. G. (1993). Externalizing in pre-schoolers and early adolescents: A cross-study replication of a family model. *Developmental Psychology, 29,* 3–16.

Miller, P. H. (1993). *Theories of developmental psychology* (3rd ed.). New York: Freeman.

Miller, P. H., & Bigi, L. (1979). The development of children's understanding of attention. *Merrill-Palmer Quarterly, 25,* 235–250.

Miller, P. H., & Seier, W. L. (1994). Strategy utilization deficiencies in children: When, where, and why. In H. W. Reese (ed.), *Advances in child development and behavior* (Vol. 24, pp. 107–156). New York: Academic Press.

Miller, P. H., & Zalenski, R. (1982). Preschoolers' knowledge about attention. *Developmental Psychology, 18,* 871–875.

Miller, R. A. (1996). The aging immune system: Primer and prospectus. *Science, 273,* 70–74.

Miller, S. S., & Cavanaugh, J. C. (1990). The meaning of grandparenthood and its relationship to demographic, relationship, and social participation variables. *Journal of Gerontology, 45,* P244–P246.

Miller-Jones, D. (1989). Culture and testing. *American Psychologist, 44,* 360–366.

Mills, R., & Grusec, J. (1989). Cognitive, affective, and behavioral consequences of praising altruism. *Merrill-Palmer Quarterly, 35,* 299–326.

Millstein, S. G., & Irwin, C. E. (1988). Accident-related behaviors in adolescents: A biosocial view. *Alcohol, Drugs, and Driving, 4,* 21–29.

Millstein, S. G., & Litt, I. F. (1990). Adolescent health. In S. S. Feldman & G. R. Elliott (Eds.), *At the threshold: The developing adolescent* (pp. 431–456). Cambridge, MA: Harvard University Press.

Minuchin, P. P. (1988). Relationships within the family: A systems perspective on development. In R. A. Hinde & J. Stevenson-Hinde (Eds.), *Relationships within families: Mutual influences* (pp. 7–26). New York: Oxford University Press.

Mischel, W., & Liebert, R. M. (1966). Effects of discrepancies between observed and imposed reward criteria on their acquisition and transmission. *Journal of Personality and Social Psychology, 3,* 45–53.

Mitchell, V., & Helson, R. (1990). Women's prime of life. *Psychology of Women Quarterly, 14,* 451–470.

Miyake, K., Chen, S., & Campos, J. J. (1985). Infant temperament, mother's mode of interaction, and attachment in Japan: An interim report. In I. Bretherton & E. Waters (Eds.), Growing points of attachment theory and research. *Monographs of the Society for Research in Child Development, 50* (1–2, Serial No. 209).

Mize, J., & Ladd, G. W. (1990). A cognitive–social learning approach to social skill training with low-status preschool children. *Developmental Psychology, 26,* 388–397.

Moen, P. (1996). Gender, age, and the life course. In R. H. Binstock & L. K. George (Eds.), *Handbook of aging and the social sciences* (pp. 171–187). San Diego: Academic Press.

Moilanen, I. (1989). The growth, development, and education of Finnish twins: A longitudinal follow-up study in a birth cohort from pregnancy to nineteen years of age. *Growth, Development and Aging, 18,* 302–306.

Monagle, L., Dan, A., Krogh, V., Jossa, F., Farinaro, E., & Trevisan, M. (1993). Premenstrual symptom prevalence rates: An Italian-American comparison. *American Journal of Epidemiology, 138,* 1070–1081.

Monroe, S., Goldman, P., & Smith, V. E. (1988). *Brothers: Black and poor—A true story of courage and survival.* New York: Morrow.

Montemayor, R., & Eisen, M. (1977). The development of self-conceptions from childhood to adolescence. *Developmental Psychology, 13,* 314–319.

Montepare, J., & Lachman, M. (1989). "You're only as old as you feel": Self-perceptions of age, fears of aging, and life satisfaction from adolescence to old age. *Psychology and Aging, 4,* 73–78.

Montplaisir, J., & Godbout, R. (1989). Restless legs syndrome and periodic movements during sleep. In M. H. Kryger, T. Roth, & W. C. Dement (Eds.), *Principles and practice of sleep medicine* (pp. 402–409). Philadelphia: Saunders.

Moody, R. A., Jr. (1975). *Life after life.* New York: Bantam.

Moody, R. A., Jr. (1988). *The light beyond.* New York: Bantam.

Moon, S. M., & Feldhusen, J. F. (1994). The Program for Academic and Creative Enrichment (PACE): A follow-up study ten years later. In R. F. Subotnik & K. D. Arnold (Eds.), *Beyond Terman: Contemporary longitudinal studies of giftedness and talent* (pp. 375–400). Norwood, NJ: Ablex.

Moore, E. G. J. (1986). Family socialization and the IQ test performance of traditionally and transracially adopted black children. *Developmental Psychology, 22,* 317–326.

Moore, K. L., & Persaud, T. V. N. (1993). *Before we are born* (4th ed.). Philadelphia: Saunders.

Moore, M. K. (1992). An empirical investigation of the relationship between religiosity and death concern. *Dissertation Abstracts International, 53*(2–A), 527.

Moorehouse, M. J. (1991). Linking maternal employment patterns to mother–child activities and children's school competence. *Developmental Psychology, 27,* 295–303.

Moos, R. H., & Lemke, S. (1985). Specialized living environments for older people. In J. E. Birren & K. W. Schaie (Eds.), *Handbook of the psychology of aging* (2nd ed., pp. 865–889). New York: Van Nostrand Reinhold.

Moran, G. F., & Vinovskis, M. A. (1986). The great care of godly parents: Early childhood in Puritan New England. *Monographs of the Society for Research in Child Development, 50*(4–5, Serial No. 211).

Morell, C. M. (1994). *Unwomanly conduct: The challenges of intentional childlessness.* New York: Routledge.

Morelli, G., Rogoff, B., Oppenheim, D., & Goldsmith, D. (1992). Cultural variation in infants' sleeping arrangements: Questions of independence. *Developmental Psychology, 28,* 604–613.

Morgan, J. L., & Saffran, J. R. (1995). Emerging integration of sequential and suprasegmental information in preverbal speech segmentation. *Child Development, 66,* 911–936.

Morgan, L. A. (1991). *After marriage ends: Economic consequences for midlife women.* Newbury Park, CA: Sage.

Morgane, P. J., Austin-LaFrance, R., Bronzino, J., Tonkiss, J., Diaz-Cintra, S., Cintra, L., Kemper, T., & Galler, J. R. (1993). Prenatal malnutrition and development of the brain. *Neuroscience and Biobehavioral Reviews, 17,* 91–128.

Moriguchi, S., Oonishi, K., Kato, M., & Kishino, Y. (1995). Obesity is a risk factor for deteriorating cellular immune functions decreased with aging. *Nutrition Research, 15,* 151–160.

Moroney, J. T., & Allen, M. H. (1994). Cocaine and alcohol use in pregnancy. In O. Devinsky, F. Feldmann, & B. Hainline (Eds.), *Neurological complications of pregnancy* (pp. 231–242).

Morrison, A. M. (1992). Women and minorities in management. *American Psychologist, 45,* 200–208.

Morrongiello, B. A. (1986). Infants' perception of multiple-group auditory patterns. *Infant Behavior and Development, 9,* 307–319.

Morse, C. K. (1993). Does variability increase with age? An archival study of cognitive measures. *Psychology and Aging, 8,* 156–164.

Mortimer, J. T., & Borman, K. M. (Eds.). (1988). *Work experience and psychological development through the lifespan.* Boulder, CO: Westview.

Morycz, R. K. (1992). Widowhood and bereavement in late life. In V. B. Van Hasselt & M. Hersen (Eds.), *Handbook of social development* (pp. 545–582). New York: Plenum.

Moshman, D., & Franks, B. A. (1986). Development of the concept of inferential validity. *Child Development, 57,* 153–165.

Mott, S. R., James, S. R., & Sperhac, A. M. (1990). *Nursing care of children and families.* Redwood City, CA: Addison-Wesley.

Moyer, M. S. (1992). Sibling relationships among older adults. *Generations, 16*(3), 55–58.

Muecke, L., Simons-Morton, B., Huang, I. W., & Parcel, G. (1992). Is childhood obesity associated with high-fat foods and low physical activity? *Journal of School Health, 62,* 19–23.

Mullen, M. K. (1994). Earliest recollections of childhood: A demographic analysis. *Cognition, 52,* 55–79.

Mullis, I. V. S., Dossey, J. A., Campbell, J. R., Gentile, C. A., O'Sullivan, C., & Latham, A. S. (1994). *NAEP 1992 trends in academic progress.* Washington, DC: U.S. Government Printing Office.

Munn, K. (1981). Effects of exercise on the range of motion in elderly subjects. In E. Smith & R. Serfass (Eds.), *Exercise and aging* (pp. 167–186). Hillside, NJ: Enslow.

Munro, G., & Adams, G. R. (1977). Ego identity formation in college students and working youth. *Developmental Psychology, 13,* 523–524.

Murray, A. D. (1985). Aversiveness is in the mind of the beholder. In B. M. Lester & C. F. Z. Boukydis (Eds.), *Infant crying* (pp. 217–239). New York: Plenum.

Murray, A. D., Johnson, J., & Peters, J. (1990). Fine-tuning of utterance length to preverbal infants: Effects on later language development. *Journal of Child Language, 17,* 511–525.

Murray, M. J., & Meacham, R. B. (1993). The effect of age on male reproductive function. *World Journal of Urology, 11,* 137–140.

Murrett-Wagstaff, S., & Moore, S. G. (1989). The Hmong in America: Infant behavior and rearing practices. In J. K. Nugent, B. M. Lester, & T. B. Brazelton (Eds.), *Biology, culture, and development* (Vol. 1, pp. 319–339). Norwood, NJ: Ablex.

Murry, V. M. (1992). Incidence of first pregnancy among black adolescent females over three decades. *Youth & Society, 23,* 478–506.

Mussen, P., & Eisenberg-Berg, N. (1977). *Roots of caring, sharing, and helping.* San Francisco: Freeman.

Myers, G. C. (1996). Comparative mortality trends among older persons in developed countries. In G. Caselli & A. D. Lopez (Eds.), *Health and mortality among elderly populations* (pp. 87–111). Oxford, England: Clarendon Press.

Myerson, J., Hale, S., Wagstaff, D., Poon, L. W., & Smith, G. A. (1990). The information-loss model: A mathematical theory of age-related cognitive slowing. *Psychological Review, 97,* 475–487.

Nachtigall, R. D. (1993). Secrecy: An unresolved issue in the practice of donor insemination. *American Journal of Obstetrics and Gynecology, 168,* 1846–1851.

Naigles, L. G., & Gelman, S. A. (1995). Overextensions in comprehension and production revisited: Preferential-looking in a study of dog, cat, and cow. *Journal of Child Language, 22,* 19–46.

Nánez, J., Sr., & Yonas, A. (1994). Effects of luminance and texture motion on infant defensive reactions to optical collision. *Infant Behavior and Development, 17,* 165–174.

National Academy on Aging (1992, November/December). Poverty and income security among older persons. *Issues in Aging,* pp. 13–19.

National Association for the Education of Young Children. (1991). *Accreditation criteria and procedures of the National Academy of Early Childhood Programs* (rev. ed.) Washington, DC: Author.

National Center for Health Statistics. (1995). *Advance data, An overview of nursing homes and their current residents: Data from the 1995 National Nursing Home Survey.* Washington, DC: U.S. Government Printing Office.

National Hospice Organization. (1992). Poll says hospice leads. *Hospice, 3*(3), 4.

National Hospice Organization. (1993a). *1992 provider census data.* Arlington, VA: Author.

National Hospice Organization. (1993b). *Standards of a hospice program of care.* Arlington, VA: Author.

National Institute on Aging, National Institutes of Health. (1996). *Alzheimer's disease: Unraveling the mystery.* Washington, DC: U.S. Government Printing Office.

National Victims Center. (1992, April). *Rape in America: A report to the nation.* Arlington, VA: Author.

Neale, A. V., Hwalek, M. A., Goodrich, C. S., & Quinn, K. M. (1996). The Illinois Elder Abuse System: Program description and administrative findings. *Gerontologist, 36,* 502–511.

Needleman, H. L., Schell, A., Bellinger, D., Leviton, A., & Allred, E. N. (1990). The long-term effects of exposure to low doses of lead in childhood. *New England Journal of Medicine, 322,* 83–88.

Neimeyer, R. A. (Ed.). (1994). *Death anxiety handbook.* Washington, DC: Taylor & Francis.

Neimeyer, R. A., & Van Brunt, D. (1995). Death anxiety. In H. Waas & R. A. Neimeyer (Eds.), *Dying: Facing the facts* (3rd ed., pp. 49–88). Washington, DC: Taylor & Francis.

Nelson, C. A. (1995). The ontogeny of human memory: A cognitive neuro-science perspective. *Developmental Psychology, 31,* 723–738.

Nelson, K. (1973). Structure and strategy in learning to talk. *Monographs of the Society for Research in Child Development, 38*(1–2, Serial No. 149).

Nelson, K. (1993). The psychological and social origins of autobiographical memory. *Psychological Science, 1,* 1–8.

Nelson, K. E., Dinninger, M., Bonvillian, J., Kaplan, B., & Baker, N. (1984). Maternal adjustments and non-adjustments as related to children's linguistic advances and language acquisition theories. In A. Pelligrini & T. Yawkey (Eds.), *The development of oral and written languages: Readings in developmental and applied linguistics* (pp. 31–56). Norwood, NJ: Ablex.

Netley, C. T. (1986). Summary overview of behavioral development in individuals with neonatally identified X and Y aneuploidy. *Birth Defects, 22,* 293–306.

Neugarten, B. L. (1968). Adult personality: Toward a psychology of the life cycle. In B. Neugarten (Ed.), *Middle age and aging* (pp. 137–147). Chicago: University of Chicago Press.

Neugarten, B., & Neugarten, D. (1987, May). The changing meanings of age. *Psychology Today, 21*(5), 29–33.

Neugarten, B. L. (1968). The awareness of middle aging. In B. L. Neugarten (Ed.), *Middle age and aging* (pp. 93–98). Chicago: University of Chicago Press.

Neugarten, B. L. (1979). Time, age, and the life cycle. *American Journal of Psychiatry, 136,* 887–894.

Neugarten, B. L., & Datan, N. (1973). Sociological perspectives on the life course. In P. B. Baltes & K. W. Schaie (Eds.), *Life-span developmental psychology: Personality and socialization* (pp. 53–69). San Diego, CA: Academic Press.

Newacheck, P. W., & Starfield, B. (1988). Morbidity and use of ambulatory care services among poor and nonpoor children. *American Journal of Public Health, 78,* 927–933.

Newborg, J., Stock, J. R., & Wnek, L. (1984). *Batelle Developmental Inventory.* Allen, TX: LINC Associates.

Newcomb, A. F., Bukowski, W. M., & Pattee, L. (1993). Children's peer relations: A meta-analytic review of popular, rejected, neglected, controversial, and average sociometric status. *Psychological Bulletin, 113,* 99–128.

Newcomb, M. D., & Bentler, P. M. (1988). Consequences of adolescent substance use on young adult health status and utilization of health services: A structural equation model over four years. *Social Science and Medicine, 24,* 71–82.

Newcomb, M. D., & Bentler, P. M. (1989). Substance use and abuse among

children and teenagers. *American Psychologist, 44,* 242–248.

Newcombe, N., & Fox, N. A. (1994). Infantile amnesia: Through a glass darkly. *Child Development, 65,* 31–40.

Newcombe, N., & Huttenlocher, J. (1992). Children's early ability to solve perspective-taking problems. *Developmental Psychology, 28,* 635–643.

Newsom, J. T., & Schulz, R. (1996). Social support as a mediator in the relation between functional status and quality of life in older adults. *Psychology and Aging, 11,* 34–44.

Nichols, M. R. (1993). Paternal perspectives of the childbirth experience. *Maternal–Child Nursing Journal, 21,* 99–108.

Nicolopoulou, A. (1993). Play, cognitive development, and the social world: Piaget, Vygotsky, and beyond. *Human Development, 36,* 1–23.

Nidorf, J. F. (1985). Mental health and refugee youths: A model for diagnostic training. In T. C. Owen (Ed.), *Southeast Asian mental health: Treatment, prevention, services, training, and research* (pp. 391–427). Washington, DC: National Institute of Mental Health.

Nieman, D. (1994). Exercise: Immunity from respiratory infections. *Swimming Technique, 31*(2), 38–43.

Nilsson, L., & Hamberger, L. (1990). *A child is born.* New York: Delacorte.

Noe, R., & Wilk, S. (1993). Investigation of factors that influence employees' participation in development activities. *Journal of Applied Psychology, 78,* 291–302.

Noppe, L. D., & Noppe, I. C. (1991). Dialectical themes in adolescent conceptions of death. *Journal of Adolescent Research, 6,* 28–42.

Noppe, L. D., & Noppe, I. C. (1996). Ambiguity in adolescent understanding of death. In C. A. Corr & D. E. Balk (Eds.), *Handbook of adolescent death and bereavement* (pp. 25–41). New York: Springer.

Norbeck, J. S., & Tilden, V. P. (1983). Wife stress, social support, and emotional disequilibrium in complications of pregnancy: A prospective, multivariate study. *Journal of Health and Social Behavior, 24,* 30–46.

Nottelmann, E. D. (1987). Competence and self-esteem during transition from childhood to adolescence. *Developmental Psychology, 23,* 441–450.

Nottelmann, E. D., & Jensen, P. S. (1995). Comorbidity of disorders in children and adolescents. In T. H. Ollendick & R. J. Prinz (Eds.), *Advances in clinical child psychology* (pp. 109–155). New York: Plenum.

Notzon, F. C. (1990). International differences in the use of obstetric interventions. *Journal of the American Medical Association, 263,* 3286–3291.

Novak, M., & Thacker, C. (1991). Satisfaction and strain among middle-aged women who return to school: Replication and extension of findings in a Canadian context. *Educational Gerontology, 17,* 323–342.

Nowakowski, R. S. (1987). Basic concepts of CNS development. *Child Development, 58,* 568–595.

Nucci, L., & Turiel, E. (1993). God's word, religious rules, and their relation to Christian and Jewish children's concepts of morality. *Child Development, 64,* 1475–1491.

Nuckolls, K. B., Cassel, J., & Kaplan, B. H. (1972). Psychosocial assets, life crisis, and the prognosis of pregnancy. *American Journal of Epidemiology, 95,* 431–441.

Nuland, S. B. (1993). *How we die.* New York: Random House.

Nuland, S. B. (1996). The doctor's role in death. In H. M. Spiro, M. G. M. Curnen, & L. P. Wandel (Eds.), *Facing death: Where culture, religion, and medicine meet* (pp. 38–43). New Haven, CT: Yale University Press.

Nussbaum, J. F. (1994). Friendship in older adulthood. In M. L. Hummer, J. M. Wiemann, & J. F. Nussbaum (Eds.), *Interpersonal communication in older adulthood* (pp. 209–225). Thousand Oaks, CA: Sage.

Nussbaum, P. D., Kaszniak, A. W., Allender, J., & Rapcsak, S. (1995). Depression and cognitive decline in the elderly: A follow-up study. *Clinical Neurologist, 9,* 101–111.

Nye, W. P. (1993). Amazing grace: Religion and identity among elderly black individuals. *International Journal of Aging and Human Development, 36,* 103–114.

O'Bryant, S. L. (1988a). Sex-differentiated assistance in older widows' support networks. *Sex Roles, 19,* 91–106.

O'Bryant, S. L. (1988b). Sibling support and older widows' well-being. *Journal of Marriage and the Family, 50,* 173–183.

O'Bryant, S. L., & Hansson, R. O. (1995). Widowhood. In R. Blieszner & V. H. Bedford (Eds.), *Handbook of aging and the family* (pp. 440–458). Westport, CT: Greenwood Press.

O'Connor, B. P. (1995). Identity development and perceived parental behavior as sources of adolescent egocentrism. *Journal of Youth and Adolescence, 24,* 205–227.

O'Connor, P. (1992). *Friendships between women.* New York: Guilford.

O'Grady-LeShane, R., & Williamson, J. B. (1992). Family provisions in old-age pensions. In M. E. Szinovacz, D. J. Ekerdt, & B. H. Vinick (Eds.), *Families and retirement* (pp. 64–77). Newbury Park, CA: Sage.

O'Halloran, C. M., & Altmaier, E. M. (1996). Awareness of death among children: Does a life-threatening illness alter the process of discovery? *Journal of Counseling & Development, 74,* 259–262.

O'Malley, P. M., Johnston, L. D., & Bachman, J. G. (1995). Adolescent substance use: Epidemiology and implications for public policy. *Pediatric Clinics of North America, 42,* 241–260.

O'Reilly, A. W. (1995). Using representations: Comprehension and production of actions with imagined objects. *Child Development, 66,* 999–1010.

O'Reilly, A. W., & Bornstein, M. H. (1993). Caregiver–child interaction in play. In M. H. Bornstein & A. W. O'Reilly (Eds.), *New directions for child development* (No. 59, pp. 55–66). San Francisco: Jossey-Bass.

Oakes, J., Gamoran, A., & Page, R. N. (1992). Curriculum differentiation: Opportunities, outcomes, and meanings. In P. W. Jackson (Ed.), *Handbook of research on curriculum* (pp. 570–608). New York: Macmillan.

Oakes, L. M., Madole, K. L., & Cohen, L. B. (1991). Infants' object examining: Habituation and categorization. *Cognitive Development, 6,* 377–392.

Oates, R. K. (1984). Similarities and differences between nonorganic failure to thrive and deprivation dwarfism. *Child Abuse and Neglect, 8,* 438–445.

Oates, R. K., Peacock, A., & Forrest, D. (1985). Long-term effects of nonorganic failure to thrive. *Pediatrics, 75,* 36–40.

Obler, L. K. (1993). Language beyond childhood. In J. Berko Gleason (Ed.), *The development of language* (3rd ed., pp. 421–449). Columbus, OH: Merrill.

Ochse, R., & Plug, C. (1986). Cross-cultural investigation of the validity of Erikson's theory of personality development. *Journal of Personality and Social Psychology, 50,* 1240–1252.

Oden, M. H., & Terman, L. M. (1968). The fulfillment of promise—40-year follow-up of the Terman gifted group. *Genetic Psychology Monographs, 77,* 3–93.

Offer, D. (1988). *The teenage world: Adolescents' self-image in ten countries.* New York: Plenum.

Office of Educational Research and Improvement. (1993). *Youth indicators 1993: Trends in the well-being of American youth.* Washington, DC: U.S. Government Printing Office.

Office of Technology Assessment, U.S. Congress. (1994). *Public information about osteoporosis: What's available, what's needed?* Washington, DC: U.S. Government Printing Office.

Ogbu, J. (1988). Black education: A cultural-ecological perspective. In H. P. McAdoo (Ed.), *Black families* (pp. 169–186). Beverly Hills, CA: Sage.

Ogbu, J. U. (1985). A cultural ecology of competence among inner-city blacks. In M. B. Spencer, G. K. Brookins, & W. R. Allen (Eds.), *Beginnings: The social and affective development of black children* (pp. 45–66). Hillsdale, NJ: Erlbaum.

Ohkura, T., Isse, K., Akazawa, K., Hamamoto, M., Yaoi, Y., & Hagino, N. (1995). Long-term estrogen replacement therapy in female patients with dementia of the Alzheimer type: 7 case reports. *Dementia, 6,* 99–107.

Okagaki, L., & Sternberg, R. J. (1993). Parental beliefs and children's school performance. *Child Development, 64,* 36–56.

Oliker, S. J. (1989). *Best friends and marriage: Exchange among women.* Berkeley: University of California Press.

Oliver, M. B., & Hyde, J. S. (1993). Gender differences in sexuality: A meta-analysis. *Psychological Bulletin, 114,* 29–51.

Olshansky, S. J., Carnes, B. A., & Cassel, C. (1990). In search of Methuselah: Estimating the upper limits to human longevity. *Science, 250,* 634–640.

Omer, H., & Everly, G. S. (1988). Psychological factors in preterm labor: Critical review and theoretical synthesis. *American Journal of Psychiatry, 145,* 1507–1513.

Ornstein, P. A., Naus, M. J., & Liberty, C. (1975). Rehearsal and organizational processes in children's memory. *Child Development, 46,* 818–830.

Ornstein, S., & Isabella, L. (1990). Age vs. stage models of career attitudes of women: A partial replication and extension. *Journal of Vocational Behavior, 36,* 1–19.

Osgood, N. J. (1992). *Suicide in later life: Recognizing the warning signs.* New York: Lexington Books.

Osgood, N. J., Brant, B. A., & Lipman, A. (1991). *Suicide among the elderly in long-term care facilities.* New York: Greenwood Press.

Osherson, D. N., & Markman, E. M. (1975). Language and the ability to evaluate contradictions and tautologies. *Cognition, 2,* 213–226.

Ostling, I., & Kelloway, E. K. (1992, June). *Predictors of life satisfaction in retire-*

ment: A mediational model. Paper presented at the annual conference of the Canadian Psychological Association, Quebec City.

Owen, M. T., & Cox, M. (1988). Maternal employment and the transition to parenthood. In A. E. Gottfried & A. W. Gottfried (Eds.), Maternal employment and children's development: Longitudinal research (pp. 85–119). New York: Plenum.

Owens, T. (1982). Experience-based career education: Summary and implications of research and evaluation findings. Child and Youth Services Journal, 4, 77–91.

Padgett, D. K., Patrick, C., Bruns, B. J., & Schlesinger, H. J. (1994). Women and outpatient mental health services: Use by black, Hispanic, and white women in a national insured population. Journal of Mental Health Administration, 2, 347–360.

Padgham, J. J., & Blyth, D. A. (1990). Dating during adolescence. In R. M. Lerner, A. C. Petersen, & J. Brooks-Gunn (Eds.), The encyclopedia of adolescence (Vol. 1, pp. 196–198). New York: Garland.

Padilla, M. L., & Landreth, G. L. (1989). Latchkey children: A review of the literature. Child Welfare, 68, 445–454.

Paffenbarger, R. S., Jr., Hyde, R. T., Wing, A. L., Lee, I-M., Jung, D. L., & Kampert, J. B. (1993). The association of changes in physical-activity level and other lifestyle characteristics with mortality among men. New England Journal of Medicine, 329, 538–545.

Paganinihill, A., & Henderson, V. W. (1996). Estrogen replacement therapy and risk of Alzheimer's disease. Archives of Internal Medicine, 156, 2213–2217.

Paikoff, R. L., & Brooks-Gunn, J. (1991). Do parent–child relationships change during puberty? Psychological Bulletin, 110, 47–66.

Palkovitz, R., & Copes, M. (1988). Changes in attitudes, beliefs and expectations associated with the transition to parenthood. Marriage and Family Review, 6, 183–199.

Palmer, G. (1997). Reluctant refugee. Sydney: Kangaroo Press.

Pan, H. W. (1994). Children's play in Taiwan. In J. L. Roopnarine, J. E. Johnson, & F. H. Hooper (Eds.), Children's play in diverse cultures (pp. 31–50). Albany, NY: SUNY Press.

Papernow, P. (1993). Becoming a stepfamily: Patterns of development in remarried families. New York: Gardner.

Papini, D. R. (1994). Family interventions. In S. L. Archer (Ed.), Interventions for adolescent identity development (pp. 47–61). Thousand Oaks, CA: Sage.

Paris, S. G., & Newman, R. S. (1990). Developmental aspects of self-regulated learning. Educational Psychologist, 25, 87–102.

Parke, R. D., & Tinsley, B. R. (1981). The father's role in infancy: Determinants of involvement in caregiving and play. In M. E. Lamb (Ed.), The role of the father in child development (pp. 429–458). New York: Wiley.

Parker, J. G., & Asher, S. R. (1987). Peer relations and later personal adjustment: Are low-accepted children at risk? Psychological Bulletin, 102, 357–389.

Parker, J. G., & Asher, S. R. (1993). Friendship and friendship quality in middle childhood: Links with peer group acceptance and feelings of loneliness and social dissatisfaction. Developmental Psychology, 29, 611–621.

Parker, R. G. (1995). Reminiscence: A continuity theory framework. Gerontologist, 35, 515–525.

Parkes, C. M., & Weiss, R. S. (1983). Recovery from bereavement. New York: Basic Books.

Parmelee, P. A., & Lawton, M. P. (1990). The design of special environments for the aged. In J. E. Birren & K. W. Schaie (Eds.), Handbook of the psychology of aging (3rd ed., pp. 464–488). San Diego, CA: Academic Press.

Parsons, J. E., Adler, T. F., & Kaczala, C. M. (1982). Socialization of achievement attitudes and beliefs: Parental influences. Child Development, 53, 310–321.

Parten, M. (1932). Social participation among preschool children. Journal of Abnormal and Social Psychology, 27, 243–269.

Pascarella, E., Bohr, L., Nora, A., Zusman, B., Inman, P., & Desler, M. (1993). Cognitive impacts of living on campus versus commuting to college. Journal of College Student Development, 34, 216–220.

Pascarella, E. T., & Terenzini, P. T. (1991). How college affects students. San Francisco: Jossey-Bass.

Patterson, C. J. (1995). Sexual orientation and human development: An overview. Developmental Psychology, 31, 3–11.

Patterson, G. R. (1982). Coercive family processes. Eugene, OR: Castilia Press.

Patterson, G. R., DeBaryshe, B. D., & Ramsey, E. (1989). A developmental perspective on antisocial behavior. American Psychologist, 44, 329–335.

Patterson, G. R., Reid, J. B., & Dishion, T. J. (1992). Antisocial boys. Eugene, OR: Castalia.

Patterson, M. M., & Lynch, A. Q. (1988). Menopause: Salient issues for counselors. Journal of Counseling and Development, 67, 185–188.

Patteson, D. M., & Barnard, K. E. (1990). Parenting of low birth weight infants: A review of issues and interventions. Infant Mental Health Journal, 11, 37–56.

Pattison, E. M. (1978). The living–dying process. In C. A. Garfield (Ed.), Psychological care of the dying patient (pp. 133–168). New York: McGraw-Hill.

Pearson, J. D., Morell, C. H., Gordon-Salant, S., Brant, L. J., Metter, E. J., Klein, L. L., & Fozard, J. L. (1995). Gender differences in a longitudinal study of age-associated hearing loss. Journal of the Acoustical Society of America, 97, 1196–1205.

Pearson, J. L., Hunter, A. G. Ensminger, M. E., & Kellam, S. G. (1990). Black grandmothers in multigenerational households: Diversity in family structure and parenting involvement in the Woodlawn community. Child Development, 61, 434–442.

Peck, R. C. (1968). Psychological developments in the second half of life. In B. L. Neugarten (Ed.), Middle age and aging (pp. 88–92). Chicago: University of Chicago Press.

Peckham, C. S., & Logan, S. (1993). Screening for toxoplasmosis during pregnancy. Archives of Disease in Childhood, 68, 3–5.

Pederson, D. R., & Moran, G. (1995). A categorical description of infant–mother relationships in the home and its relation to Q-sort measures of infant–mother interaction. In Waters, E., Vaughn, B. E., Posada, G., & Kondo-Ikemura K. (Eds.). (1995). Caregiving, cultural, and cognitive perspectives on secure-base behavior and working models: New growing points of attachment theory and research. Monographs of the Society for Research in Child Development, 60(2–3, Serial No. 244).

Pedlow, R., Sanson, A., Prior, M., & Oberklaid, F. (1993). Stability of maternally reported temperament from infancy to 8 years. Developmental Psychology, 29, 998–1007.

Penman, D. T., Bloom, J. R., Fotopoulos, S., Cook, M. R., Holland, J. C., Gates, C., Flamer, D., Murawski, B., Ross, R., Brandt, U., Muenz, L. R., & Pee, D. (1987). The impact of mastectomy on self-concept and social function: A combined cross-sectional and longitudinal study with comparison groups.

In S. D. Stellman (Ed.), *Women and cancer* (pp. 101–129). New York: Haworth Press.

Pennington, B. F., Bender, B., Puck, M., Salbenblatt, J., & Robinson, A. (1982). Learning disabilities in children with sex chromosome abnormalities. *Child Development, 53,* 1182–1192.

Pennington, B. F., & Smith, S. D. (1988). Genetic influences on learning disabilities: An update. *Journal of Consulting and Clinical Psychology, 56,* 817–823.

Peplau, L. A. (1991). Lesbian and gay relationships. In J. C. Gonsiorek & J. D. Weinrich (Eds.), *Homosexuality* (pp. 177–196). Newbury Park, CA: Sage.

Perelle, I. B., & Ehrman, L. (1994). An international study of human handedness: The data. *Behavior Genetics, 24,* 217–227.

Perfetti, C. A. (1988). Verbal efficiency in reading ability. In M. Daneman, G. E. MacKinnon, & T. G. Waller (Eds.), *Reading research: Advances in theory and practice* (Vol. 6, pp. 109–143). San Diego, CA: Academic Press.

Perkins, K. (1993). Working-class women and retirement. *Journal of Gerontological Social Work, 20,* 129–146.

Perleth, C., & Heller, K. A. (1994). The Munich Longitudinal Study of Giftedness. In R. F. Subotnik & K. D. Arnold (Eds.), *Beyond Terman: Contemporary studies of giftedness and talent* (pp. 77–114). Norwood, NJ: Ablex.

Perlmutter, M. (1984). Continuities and discontinuities in early human memory: Paradigms, processes, and performances. In R. V. Kail, Jr., & N. R. Spear (Eds.), *Comparative perspectives on the development of memory* (pp. 253–287). Hillsdale, NJ: Erlbaum.

Perlmutter, M. (1988). Cognitive potential throughout life. In J. E. Birren & V. L. Bengtson (Eds.), *Emergent theories of aging* (pp. 247–268). New York: Springer.

Perlmutter, M., Kaplan, M., & Nyquist, L. (1990). Development of adaptive competence in adulthood. *Human Development, 33,* 185–197.

Perls, T. T. (1995, January). The oldest old. *Scientific American, 272*(1), 70–75.

Perry, D. G., Perry, L. C., & Weiss, R. J. (1989). Sex differences in the consequences that children anticipate for aggression. *Developmental Psychology, 25,* 312–319.

Perry, H. L. (1993). Dying, death, and grief among selected ethnic communities. In D. P. Irish, K. F. Lundquist, & V. J. Nelson (Eds.), *Ethnic variations in dying, death, and grief* (pp. 51–65). Washington, DC: Taylor & Francis.

Perry, W. G., Jr. (1970). *Forms of intellectual and ethical development in the college years.* New York: Holt, Rinehart & Winston.

Perry, W. G., Jr. (1981). Cognitive and ethical growth. In A. Chickering (Ed.), *The modern American college* (pp. 76–116). San Francisco: Jossey-Bass.

Pery, F. (1995, November–December). Careers: Retirement planning essentials. *Healthcare Executive, 10*(6), 42–43.

Peterson, A. C., Sarigiani, P. A., & Kennedy, R. E. (1991). Adolescent depression: Why more girls? *Journal of Youth and Adolescence, 20,* 247–271.

Peterson, B. E., & Klohnen, E. C. (1995). Realization of generativity in two samples of women at midlife. *Psychology and Aging, 10,* 20–29.

Peterson, E. T. (1989). Elderly parents and their offspring. In S. J. Bahr & E. T. Peterson (Eds.), *Aging and the family* (pp. 175–191). Lexington, MA: Lexington Books.

Peterson, J. W. (1990). Age of wisdom: Elderly black women in family and church. In J. Sokolovsky (Ed.), *The cultural context of aging* (pp. 213–227). New York: Bergin & Garvey.

Peterson, L. (1989). Latchkey children's preparation for self-care: Overestimated, underrehearsed, and unsafe. *Journal of Clinical Child Psychology, 18,* 36–43.

Peterson, L., & Brown, D. (1994). Integrating child injury and abuse–neglect research: Common histories, etiologies, and solutions. *Psychological Bulletin, 116,* 293–315.

Pettle, S. A., & Britten, C. M. (1995). Talking with children about death and dying. *Child: Care, Health, and Development, 21,* 395–404.

Phelps, K. E., & Woolley, J. D. (1994). The form and function of young children's magical beliefs. *Developmental Psychology, 30,* 385–394.

Phillips, D. A. (1987). Socialization of perceived academic competence among highly competent children. *Child Development, 58,* 1308–1320.

Phillips, D. A., McCartney, K., & Scarr, S. (1987). Child-care quality and children's social development. *Developmental Psychology, 23,* 537–543.

Phillips, D. A., Voran, M., Kisker, E., Howes, C., & Whitebook, M. (1994). Child care for children in poverty: Opportunity or inequity? *Child Development, 65,* 472–492.

Phillips, D. A., & Zimmerman, M. (1990). The developmental course of perceived competence and incompetence among competent children. In R. Sternberg

& J. Kolligian (Eds.), *Competence considered* (pp. 41–66). New Haven, CT: Yale University Press.

Phillips, O. P., & Elias, S. (1993). Prenatal genetic counseling issues in women of advanced reproductive age. *Journal of Women's Health, 2,* 1–5.

Phinney, J. S. (1989). Stages of ethnic identity development in minority group adolescents. *Journal of Early Adolescence, 9,* 34–49.

Phinney, J. S. (1993). Multiple group identities: Differentiation, conflict, and integration. In J. Kroger (Ed.), *Discussions on ego identity* (pp. 47–73). Hillsdale, NJ: Erlbaum.

Phinney, J., & Alipuria, L. (1990). Ethnic identity in college students from four ethnic groups. *Journal of Adolescence, 13,* 171–183.

Piaget, J. (1926). *The language and thought of the child.* New York: Harcourt, Brace & World. (Original work published 1923)

Piaget, J. (1930). The child's conception of the world. New York: Harcourt, Brace, & World. (Original work published 1926)

Piaget, J. (1950). *The psychology of intelligence.* New York: International Universities Press.

Piaget, J. (1951). *Play, dreams, and imitation in childhood.* New York: Norton. (Original work published 1945)

Piaget, J. (1952). *The origins of intelligence in children.* New York: International Universities Press. (Original work published 1936)

Piaget, J. (1965). *The moral judgment of the child.* New York: Free Press. (Original work published 1932)

Piaget, J. (1967). *Six psychological studies.* New York: Vintage.

Piaget, J. (1971). *Biology and knowledge.* Chicago: University of Chicago Press.

Piaget, J. (1985). *The equilibration of cognitive structures: The central problem of intellectual development.* Chicago: University of Chicago Press.

Piaget, J., & Inhelder, B. (1956). *The child's conception of space.* London: Routledge & Kegan Paul. (Original work published 1948)

Pianta, R., Egeland, B., & Erickson, M. F. (1989). The antecedents of maltreatment: Results of the Mother–Child Interaction Research Project. In D. Cicchetti & V. Carlson (Eds.), *Child maltreatment* (pp. 203–253). New York: Cambridge University Press.

Picariello, M. L., Greenberg, D. N., & Pillemer, D. B. (1990). Children's sex-related stereotyping of colors. *Child Development, 61,* 1453–1460.

Pick, H. L., Jr. (1989). Motor development: The control of action. *Developmental Psychology, 25,* 867–870.

Pickens, J., Field, T., Nawrocki, T., Martinez, A., Soutullo, D., & Gonzalez, J. (1994). Full-term and preterm infants' perception of face–voice synchrony. *Infant Behavior and Development, 17,* 447–455.

Pierce, W. D., & Epling, W. F. (1995). *Behavior analysis and learning.* Englewood Cliffs, NJ: Prentice Hall.

Pilkington, C. L., & Piersel, W. C. (1991). School phobia: A critical analysis of the separation anxiety theory and an alternative conceptualization. *Psychology in the Schools, 28,* 290–303.

Pillemer, K. A., & Finkelhor, D. (1988). The prevalence of elder abuse: A random sample survey. *Gerontologist, 28,* 51–57.

Pillemer, K. A., & Moore, D. W. (1989). Abuse of patients in nursing homes: Findings from a survey of staff. *Gerontologist, 29,* 314–320.

Pillow, B. H. (1988). The development of children's beliefs about the mental world. *Merrill-Palmer Quarterly, 34,* 1–32.

Pinel, J. P. J. (1997). *Biopsychology* (3rd ed.) Boston: Allyn and Bacon.

Pines, A., & Aronson, E. (1988). *Career burnout.* New York: Free Press.

Pinker, S., Lebeaux, D. S., & Frost, L. A. (1987). Productivity and constraints in the acquisition of the passive. *Cognition, 26,* 195–267.

Pipes, P. L. (1989). *Nutrition in infancy and childhood* (4th ed.). St. Louis: Mosby.

Pipp, S., Easterbrooks, M. A., & Brown, S. R. (1993). Attachment status and complexity of infants' self- and other-knowledge when tested with mother and father. *Social Development, 2,* 1–14.

Plassman, B. L., & Breitner, C. S. (1996). Recent advances in the genetics of Alzheimer's disease and vascular dementia with an emphasis on gene–environment interactions. *Journal of the American Genetics Society, 44,* 1242–1250.

Platt, S. (1992). Epidemiology of suicide and parasuicide. *Journal of Psychopharmacology, 6*(2, Suppl.), 291–299.

Pleck, J. (1985). *Working wives/working husbands.* Beverly Hills, CA: Sage.

Plomin, R. (1994a). The Emanuel Miller Memorial Lecture 1993: Genetic research and identification of environmental influences. *Journal of Child Psychology and Psychiatry, 35,* 817–834.

Plomin, R. (1994b). *Genetics and experience: The interplay between nature and nurture in development.* Newbury Park, CA: Sage.

Plomin, R. (1994c). Nature, nurture, and social development. *Social Development, 3,* 37–53.

Plomin, R., Reiss, D., Hetherington, E. M., & Howe, G. W. (1994). Nature and nurture: Genetic contributions to measures of the family environment. *Developmental Psychology, 30,* 32–43.

Plude, D. J., & Doussard-Roosevelt, J. A. (1989). Aging, selective attention, and feature integration. *Psychology and Aging, 4,* 98–105.

Plude, D. J., & Hoyer, W. J. (1985). Attention and performance: Identifying and localizing age deficits. In N. Charness (Ed.), *Aging and human performance* (pp. 47–99). Chichester, England: Wiley.

Plumert, J. M., Pick, H. L., Jr., Marks, R. A., Kintsch, A. S., & Wegesin, D. (1994). Locating objects and communicating about locations: Organizational differences in children's searching and direction-giving. *Developmental Psychology, 30,* 443–453.

Podrouzek, W., & Furrow, D. (1988). Preschoolers' use of eye contact while speaking: The influence of sex, age, and conversational partner. *Journal of Psycholinguistic Research, 17,* 89–93.

Pohl, J. M., Given, C. W., Collins, C. E., & Given, B. A. (1994). Social vulnerability and reactions to caregiving in daughters and daughters-in-law caring for disabled aging parents. *Health Care for Women International, 15,* 385–395.

Poindron, P., & Le Neindre, P. (1980). Endocrine and sensory regulation of maternal behavior in the ewe. In J. S. Rosenblatt, R. A. Hinde, C. Beer, & M. Busnel (Eds.), *Advances in the study of behavior* (pp. 76–119). New York: Academic Press.

Polansky, N. A., Gaudin, J. M., Ammons, P. W., & Davis, K. B. (1985). The psychological ecology of the neglectful mother. *Child Abuse & Neglect, 9,* 265–275.

Polka, L., & Werker, J. F. (1994). Developmental changes in perception of non-native vowel contrasts. *Journal of Experimental Psychology: Human Perception and Performance, 20,* 421–435.

Pollitt, E. (1996). A reconceptualization of the effects of undernutrition on children's biological, psychosocial, and behavioral development. *Social Policy Report of the Society for Research in Child Development, 10*(5).

Pollitt, E., Gorman, K. S., Engle, P. L., Martorell, R., & Rivera, J. (1993). Early supplementary feeding and cognition. *Monographs of the Society for Research in Child Development, 58*(7, Serial No. 235).

Pollock, G. H. (1987). The mourning–liberation process in health and disease. *Psychiatric Clinics of North America, 10,* 345–354.

Pollock, L. (1987). *A lasting relationship: Parents and children over three centuries.* Hanover, NH: University Press of New England.

Porter, R. H., Makin, J. W., Davis, L. B., & Christensen, K. M. (1992). An assessment of the salient olfactory environment of formula-fed infants. *Physiology & Behavior, 50,* 907–911.

Posner, J. K., & Vandell, D. L. (1994). Low-income children's after-school care: Are there beneficial effects of after-school programs? *Child Development, 58,* 568–595.

Poulin-Dubois, D., Serbin, L. A., Kenyon, B., & Derbyshire, A. (1994). Infants' intermodal knowledge about gender. *Developmental Psychology, 30,* 436–442.

Powell, B., & Steelman, L. C. (1993). The educational benefits of being spaced out: Sibship density and educational progress. *American Sociological Review, 58,* 367–381.

Powell, D. R. (1986, March). Parent education and support programs. *Young Children, 41*(3), 47–53.

Powers, S. I., Hauser, S. T., & Kilner, L. A. (1989). Adolescent mental health. *American Psychologist, 44,* 200–208.

Prager, K. J., & Bailey, J. M. (1985). Androgyny, ego development, and psychological crisis resolution. *Sex Roles, 13,* 525–535.

Prechtl, H. F. R., & Beintema, D. (1965). *The neurological examination of the full-term newborn infant.* London: Heinemann Medical Books.

Prentice, A., & Lind, T. (1987). Fetal heart rate monitoring during labor—Too frequent intervention, too little benefit? *Lancet, 2,* 1375–1377.

Pressley, M. (1994). State-of-the-science primary-grades reading instruction or whole language? *Educational Psychologist, 29,* 211–215.

Pressley, M. (1995). More about the development of self-regulation: Complex, long-term, and thoroughly social. *Educational Psychologist, 30,* 207–212.

Previc, F. H. (1991). A general theory concerning the prenatal origins of cerebral lateralization. *Psychological Review, 98,* 299–334.

Princeton Religion Research Center. (1994). *Religion in America.* Princeton, NJ: Gallup Poll.

Prinz, P. N., Vitiello, M. V., Raskind, M. A., & Thorpy, M. J. (1990). Geriatrics:

Sleep disorders and aging. *New England Journal of Medicine, 323,* 520–526.

Prochaska, J. O., DiClemente, C. C., & Norcross, J. C. (1992). In search of how people change: Applications to addictive behaviors. *American Psychologist, 47,* 1102–1114.

Pryor, J. B., & Reeder, G. D. (1993). Collective and individual representations of HIV/AIDS stigma. In J. B. Pryor & G. D. Reeder (Eds.), *The social psychology of HIV infection* (pp. 263–286). Hillsdale, NJ: Erlbaum.

Pyka, G., Lindenberger, E., Charette, S., & Marcus, R. (1994). Muscle strength and fiber adaptations to a year-long resistance training program in elderly men and women. *Journal of Gerontology, 49,* M22–M27.

Pynoos, J., & Golant, S. (1996). Housing and living arrangements for the elderly. In R. H. Binstock & L. K. George (Eds.), *Handbook of aging and the social sciences* (pp. 303–324). San Diego: Academic Press.

Qazi, Q. H., Sheikh, T. M., Fikrig, S., & Menikoff, H. (1988). Lack of evidence for craniofacial dysmorphism in perinatal human immunodeficiency virus infection. *Journal of Pediatrics, 11,* 7–11.

Quadagno, J., & Hardy, M. (1996). Work and retirement. In R. H. Binstock & L. K. George (Eds.), *Handbook of aging and the social sciences* (pp. 325–345). San Diego: Academic Press.

Quay, H. C. (1987). Institutional treatment. In H. C. Quay (Ed.), *Handbook of juvenile delinquency* (pp. 244–265). New York: Wiley.

Quill, T. E. (1991). Death and dignity: A case of individualized decision making. *New England Journal of Medicine, 324,* 691–694.

Quill, T. E. (1995). The Oregon Death with Dignity Act [Letter to the Editor]. *New England Journal of Medicine, 332,* 1174–1175.

Quill, T. E., Cassel, C. K., & Meier, D. E. (1992). Care of the hopelessly ill: Proposed clinical criteria for physician-assisted suicide. *New England Journal of Medicine, 327,* 1380–1383.

Quint, J. (1991). Project Redirection: Making and measuring a difference. *Evaluation and Program Planning, 14,* 75–86.

Quintero, R. A., Puder, K. S., & Cotton, D. B. (1993). Embryoscopy and fetoscopy. *Obstetrics and Gynecology Clinics of North America, 20,* 563–581.

Rabiner, D., & Coie, J. (1989). Effect of expectancy inductions on rejected children's acceptance by unfamiliar peers. *Developmental Psychology, 25,* 450–457.

Rabiner, D. L., Keane, S. P., & MacKinnon-Lewis, C. (1993). Children's beliefs about familiar and unfamiliar peers in relation to their sociometric status. *Developmental Psychology, 29,* 236–243.

Radke-Yarrow, M., Cummings, E. M., Kuczynski, I., & Chapman, M. (1985). Patterns of attachment in two- and three-year-olds in normal families with parental depression. *Child Development, 56,* 884–893.

Radke-Yarrow, M., & Zahn-Waxler, C. (1984). Roots, motives, and patterns in children's prosocial behavior. In J. Reykowski, J. Karylowski, D. Bar-Tel, & E. Staub (Eds.), *The development and maintenance of prosocial behaviors: International perspectives on positive morality* (pp. 81–99). New York: Plenum.

Radziszewska, B., & Rogoff, B. (1988). Influence of adult and peer collaboration on the development of children's planning skills. *Developmental Psychology, 24,* 840–848.

Rafferty, Y., & Shinn, M. (1991). The impact of homelessness on children. *American Psychologist, 46,* 1170–1179.

Räihä, N. C. R., & Axelsson, I. E. (1995). Protein nutrition in infancy. *Pediatric Clinics of North America, 42,* 745–763.

Ramey, C. T., & Campbell, F. A. (1984). Preventive education for high-risk children: Cognitive consequences of the Carolina Abecedarian Project. *American Journal of Mental Deficiency, 88,* 515–523.

Ramsay, S. (1995). IVF successes. *Lancet, 345,* 246.

Ramsøy, N. R. (1994). Non-marital cohabitation and change in norms: The case of Norway. *Acta Sociologica, 37,* 23–37.

Rando, T. A. (1986). *Parental loss of a child.* Champaign, IL: Research Press.

Rando, T. A. (1991a). *How to go on living when someone you love dies.* New York: Bantam.

Rando, T. A. (1991b). Parental adjustment to the loss of a child. In D. Papadatou & C. Papadatos (Eds.), *Children and death* (pp. 233–253). New York: Hemisphere.

Rando, T. A. (1995). Grief and mourning: Accommodating to loss. In H. Wass & R. A. Neimeyer (Eds.), *Dying: Facing the facts* (3rd ed., pp. 211–241). Washington, DC: Taylor & Francis.

Raphael, B., Middleton, W., Martinek, N., & Misso, V. (1993). Counseling and therapy of the bereaved. In M. S. Stroebe, W. Stroebe, & R. O. Hansson (Eds.), *Handbook of bereavement* (pp. 427–453). New York: Cambridge University Press.

Rappaport, L. (1993). The treatment of nocturnal enuresis—Where are we now? *Pediatrics, 92,* 465–466.

Rasmussen, C. H., & Johnson, M. E. (1994). Spirituality and religiosity: Relative relationships to death anxiety. *Omega, 29,* 313–318.

Rast, M., & Meltzoff, A. N. (1995). Memory and representation in young children with Down syndrome: Exploring deferred imitation and object permanence. *Development and Psychopathology, 7,* 393–407.

Ratner, N., & Bruner, J. S. (1978). Social exchange and the acquisition of language. *Journal of Child Language, 5,* 391–402.

Raup, J. L., & Myers, J. E. (1989). The empty nest syndrome: Myth or reality? *Journal of Counseling & Human Development, 68,* 180–183.

Rawlins, W. K. (1992). *Friendship matters.* New York: Aldine De Gruyter.

Rayner, K., & Pollatsek, A. (1989). *The psychology of reading.* Englewood Cliffs, NJ: Prentice Hall.

Read, C. R. (1991). Achievement and career choices: Comparisons of males and females. *Roeper Review, 13,* 188–193.

Receputo, G., Di Stefano, S., Fornaro, D., Malaguarnera, M., & Motta, L. (1994). Comparison of tactile sensitivity in a group of elderly and young adults and children using a new instrument called a "Tangoceptometer." *Archives of Gerontology and Geriatrics, 18,* 207–214.

Redding, N. P., & Dowling, W. D. (1992). Rites of passage among women reentering higher education. *Adult Education Quarterly, 42,* 221–236.

Redl, F. (1966). *When we deal with children.* New York: The Free Press.

Rees, M. (1993). Menarche when and why? *Lancet, 342,* 1375–1376.

Reese, E., Haden, C. A., & Fivush, R. (1996). Mothers, fathers, daughters, sons: Gender differences in autobiographical reminiscing. *Research on Language and Social Interaction, 29,* 27–56.

Reeves, J. B., & Darville, R. L. (1994). Social contact patterns and satisfaction with retirement of women in dual-career/earner families. *International Journal of Aging and Human Development, 39,* 163–175.

Reich, P. A. (1986). *Language development.* Englewood Cliffs, NJ: Prentice Hall.

Reid, H. M., & Fine, A. (1992). Self-disclosure in men's friendships: Variations associated with intimate relations. In P. M. Nardi (Ed.), *Men's* friendships (pp. 153–171). Newbury Park, CA: Sage.

Reid, I. R.(1996). Therapy of osteoporosis —calcium, vitamin D and exercise. *American Journal of the Medical Sciences, 312,* 278–286.

Reifler, B. V. (1994). Depression: Diagnosis and comorbidity. In L. S. Schneider, C. F. Reynolds, III, B. D. Lebowitz, & A. J. Friedhoff (Eds.), *Diagnosis and treatment of depression in late life* (pp. 55–59). Washington, DC: American Psychiatric Press.

Reiger, D. A., Boyd, J. H., Burke, J. D., Rae, D. S., Myers, J. K., Dramer, M., Robins, L. N., George, L. K., Karno, M., & Locke, B. Z. (1988). One-month prevalence of mental disorders in the United States. *Archives of General Psychiatry, 45,* 977–986.

Reik, W. (1992). Imprinting in leukemia. *Nature, 359,* 362–363.

Reinke, B. J., Holmes, D. S., & Harris, R. L. (1985). The timing of psychosocial changes in women's lives: The years 25 to 45. *Journal of Personality and Social Psychology, 48,* 1353-1364.

Reis, M., & Nahmiash, D. (1995). When seniors are abused: An intervention model. *Gerontologist, 35,* 666–671.

Reiser, J., Yonas, A., & Wikner, K. (1976). Radial localization of odors by human neonates. *Child Development, 47,* 856–859.

Reisman, J. E. (1987). Touch, motion, and proprioception. In P. Salapatek & L. Cohen (Eds.), Handbook of infant perception: Vol. 1. *From sensation to perception* (pp. 265–303). Orlando, FL: Academic Press.

Repke, J. T. (1992). Drug supplementation in pregnancy. *Current Opinion in Obstetrics and Gynecology, 4,* 802–806.

Resnick, H. S., & Newton, T. (1992). Assessment and treatment of post-traumatic stress disorder in adult survivors of sexual assault. In D. Fox (Ed.), *Treating PTSD: Procedures for combat veterans, battered women, adult and child sexual assaults* (pp. 99–126). New York: Guilford.

Ressler, E. M. (1993). *Children in war.* New York: United Nations Children's Fund.

Rest, J. R. (1979). *Development in judging moral issues.* Minneapolis: University of Minnesota Press.

Rest, J. R., & Narvaez, D. (1991). The college experience and moral development. In W. M. Kurtines & J. L. Gewirtz (Eds.), *Handbook of moral behavior and development* (Vol. 2, pp. 229–245). Hillsdale, NJ: Erlbaum.

Reznick, J. S., & Goldfield, B. A. (1992). Rapid change in lexical development in comprehension and production.

Developmental Psychology, 28, 406–413.

Rhein, J. von (1997, January 19). Ardis Krainik, Lyric Opera's life force, dies. *Chicago Tribune,* pp. 1, 16.

Ricciardelli, L. A. (1992). Bilingualism and cognitive development: Relation to threshold theory. *Journal of Psycholinguistic Research, 21,* 301–316.

Rice, F. P. (1993). *The adolescent: Development, relationships, and culture* (7th ed.). Boston: Allyn and Bacon.

Rice, J. K. (1994). Reconsidering research on divorce, family life cycle, and the meaning of family. *Psychology of Women Quarterly, 18,* 559–584.

Rice, M. L., Huston, A. C., Truglio, R., & Wright, J. (1990). Words from "Sesame Street": Learning vocabulary while viewing. *Developmental Psychology, 26,* 421–428.

Richards, D. D., & Siegler, R. S. (1986). Children's understandings of the attributes of life. *Journal of Experimental Child Psychology, 42,* 1–22.

Richards, M. H., & Duckett, E. (1994). The relationship of maternal employment to early adolescent daily experience with and without parents. *Child Development, 65,* 225–236.

Richardson, S. A. Koller, H., & Katz, M. (1986). Factors leading to differences in the school performance of boys and girls. *Developmental and Behavioral Pediatrics, 7,* 49–55.

Richedwards, J. W., & Hennekens, C. H. (1996). Postmenopausal hormones and coronary heart disease. *Current Opinion in Cardiology, 11,* 440–446.

Richgels, D. J., McGee, L. M, & Slaton, E. A. (1989). Teaching expository text structure in reading and writing. In K. D. Muth (Ed.), *Children's comprehension of text* (pp. 167–184). Newark, DE: International Reading Association.

Richman, A. L., Miller, P. M., & LeVine, R. A. (1992). Cultural and educational variations in maternal responsiveness. *Developmental Psychology, 28,* 614–621.

Riese, M. L. (1987). Temperament stability between the neonatal period and 24 months. *Developmental Psychology, 23,* 216–222.

Rimm, E. B., Stampfer, M. J., Ascherio, A., Giovannucci, E., Colditz, G. A., & Willett, W. C. (1993). Vitamin E consumption and the risk of coronary heart disease in men. *New England Journal of Medicine, 328,* 1450–1456.

Ring, K. (1980). *Life at death.* New York: Coward, McCann & Geoghegan.

Ritchie, K., Kildea, D., & Robine, J.-M. (1992). The relationship between age and the prevalence of senile dementia:

A meta-analysis of recent data. *International Journal of Epidemiology, 21,* 763–769.

Roberto, K. A., & Kimboko, P. J. (1989). Friendships in later life: Definitions and maintenance patterns. *International Journal of Aging and Human Development, 28,* 9–19.

Roberton, M. A. (1984). Changing motor patterns during childhood. In J. R. Thomas (Ed.), *Motor development during childhood and adolescence* (pp. 48–90). Minneapolis: Burgess.

Roberts, M. C., Alexander, K., & Knapp, L. G. (1990). Motivating children to use safety belts: A program combining rewards and "flash for life." *Journal of Community Psychology, 18,* 110–119.

Roberts, M. C., Elkins, P. D., & Royal, G. P. (1984). Psychological applications to the prevention of accidents and injuries. In M. C. Roberts & L. Peterson (Eds.), *Prevention of problems in childhood: Psychological research and applications* (pp. 173–199). New York: Wiley.

Roberts, P., & Newton, P. M. (1987). Levinsonian studies of women's adult development. *Psychology and Aging, 2,* 154–163.

Roberts, R. J., Jr., & Aman, C. J. (1993). Developmental differences in giving directions: Spatial frames of reference and mental rotation. *Child Development, 64,* 1258–1270.

Robertson, D. L. (1991). Gender differences in the academic progress of adult undergraduates: Patterns and policy implications. *Journal of College Student Development, 32,* 490–496.

Robertson, J. F. (1995). Grandparenting in an era of rapid change. In R. Blieszner & V. H. Bedford (Eds.), *Handbook of aging and the family* (pp. 243–260). Westport, CT: Greenwood Press.

Robin, D. J., Berthier, N. E., & Clifton, R. K. (1996). Infants' predictive reaching for moving objects in the dark. *Developmental Psychology, 32,* 824–835.

Robinson, E. J., & Mitchell, P. (1994). Young children's false-belief reasoning: Interpretation of messages is not easier than the classic task. *Developmental Psychology, 30,* 67–72.

Robinson, J. L., Kagan, J., Reznick, J. S., & Corley, R. (1992). The heritability of inhibited and uninhibited behavior: A twin study. *Developmental Psychology, 28,* 1030–1037.

Robison, J. I., Hoerr, S. L., Strandmark, J., & Mavis, B. (1993). Obesity, weight loss, and health. *Journal of the American Dietetic Association, 93,* 445–449.

Rochat, P., & Goubet, N. (1995). Development of sitting and reaching in 5- to 6-

month-old infants. *Infant Behavior and Development, 18,* 53–68.

Roche, A. F. (1979). Secular trends in stature, weight, and maturation. In A. F. Roche (Ed.), Secular trends in human growth, maturation, and development. *Monographs of the Society for Research in Child Development, 44*(3–4, Serial No. 179).

Roche, A. F. (1981). The adipocyte-number hypothesis. *Child Development, 52,* 31–43.

Rockwell, R. C., Elder, G. H., & Ross, D. J. (1979). Psychological patterns in marital timing and divorce. *Social Psychology Quarterly, 42,* 399–404.

Roffwarg, H. P., Muzio, J. N., & Dement, W. C. (1966). Ontogenetic development of the human sleep–dream cycle. *Science, 152,* 604–619.

Roggman, L. A., Langlois, J. H., Hubbs-Tait, L., & Rieser-Danner, L. A. (1994). Infant day-care, attachment, and the "file drawer problem." *Child Development, 65,* 1429–1443.

Rogoff, B. (1986). The development of strategic use of context in spatial memory. In M. Perlmutter (Ed.), *Perspectives on intellectual development* (pp. 107–123). Hillsdale, NJ: Erlbaum.

Rogoff, B. (1990). *Apprenticeship in thinking.* New York: Oxford University Press.

Rogoff, B., & Chavajay, P. (1995). What's become of research on the cultural basis of cognitive development? *American Psychologist, 50,* 859–877.

Rogoff, B., Malkin, C., & Gilbride, K. (1984). Interaction with babies as guidance in development. In B. Rogoff & J. V. Wertsch (Eds.), *New directions for child development* (No. 23, pp. 31–44). San Francisco: Jossey-Bass.

Rogoff, B., Mosier, C., Mistry, J., & Göncü, A. (1993). Toddlers' guided participation with their caregivers in cultural activity. In E. A. Forman, N. Minick, & C. A. Stone (Eds.), *Contexts for learning* (pp. 230–253). New York: Oxford University Press.

Romaine, S. (1984).*The language of children and adolescents: The acquisition of communicative competence.* Oxford, England: Blackwell.

Rook, K. S., Catalano, R., & Dooley, D. (1989). The timing of major life events: Effects of departing from the social clock. *American Journal of Community Psychology, 17,* 233–258.

Roopnarine, J. L., Talukder, E., Jain, D., Joshi, P., & Srivastav, P. (1990). Characteristics of holding, patterns of play, and social behaviors between parents and infants in New Delhi, India. *Developmental Psychology, 26,* 667–673.

Roper Starch Worldwide. (1994). *A national survey of 252 males and 251 females.* New York: Author.

Rosa, R. W. (1993). Retinoid embropathy in humans. In G. Koren (Ed.), *Retinoids in clinical practice* (pp. 77–109). New York: Marcel Dekker.

Roscoe, B., Diana, M. S., & Brooks, R. H. (1987). Early, middle, and late adolescents' views on dating and factors influencing partner selection. *Adolescence, 22,* 59–68.

Rose, R. J. (1995) Genes and human behavior. *Annual Review of Psychology, 46,* 625–654.

Rose, S. A. (1980). Enhancing visual recognition memory in preterm infants. *Developmental Psychology, 16,* 85–92.

Rose, S. A., & Feldman, J. F. (1995). Prediction of IQ and specific cognitive abilities at 11 years from infancy measures. *Developmental Psychology, 31,* 685–696.

Rosen, A. B., & Rozin, P. (1993). Now you see it, now you don't: The preschool child's conception of invisible particles in the context of dissolving. *Developmental Psychology, 29,* 300–311.

Rosen, M. G., & Dickinson, J. C. (1993). Management of post-term pregnancy. *New England Journal of Medicine, 326,* 1628–1629.

Rosen, W. D., Adamson, L. B., & Bakeman, R. (1992). An experimental investigation of infant social referencing: Mothers' messages and gender differences. *Developmental Psychology, 28,* 1172–1178.

Rosenberg, M. (1979). *Conceiving the self.* New York: Basic Books.

Rosenberg, M., Schooler, C., & Schoenbach, C. (1989). Self-esteem and adolescent problems: Modeling reciprocal effects. *American Sociological Review, 54,* 1004–1018.

Rosenblatt, J. S., & Lehrman, D. (1963). Maternal behavior of the laboratory rat. In H. R. Rheingold (Ed.), *Maternal behavior in mammals* (pp. 8–57). New York: Wiley.

Rosenblatt, P. C. (1993). Cross-cultural variation in the experience, expression, and understanding of grief. In D. P. Irish, K. F. Lundquist, & V. J. Nelsen (Eds.), *Ethnic variations in dying, death, and grief* (pp. 13–19). Washington, DC: Taylor & Francis.

Rosenman, R. H., Brand, R. J., Jenkins, C. D., Friedman, M., Strauss, R., & Wurm, M. (1975). Coronary heart disease in the Western Collaborative Group Study: Final follow-up experience of 8½ years. *Journal of the American Medical Association, 223,* 872–877.

Roskos, K., & Neuman, S. B. (1993). Descriptive observations of adults' facilitation of literacy in young children's play. *Early Childhood Research Quarterly, 8,* 77–98.

Ross, G. S. (1980). Categorization in 1- to 2-year-olds. *Developmental Psychology, 16,* 391–396.

Rossi, A. S. (1980). Life-span theories and women's lives. *Signs: Journal of Women in Culture and Society, 6,* 4–32.

Rossi, A. S., & Rossi, P. H. (1990). *Of human bonding: Parent–child relations across the life course.* New York: Aldine De Gruyter.

Rossi, A. S., & Rossi, P. H. (1991). Normative obligations and parent–child help exchange across the life course. In K. Pillemer & K. McCartney (Eds.), *Parent–child relations throughout life* (pp. 201–223). Hillsdale, NJ: Erlbaum.

Rothbaum, F., Rosen, K. S., Pott, M., & Beatty, M. (1995). Early parent–child relationships and later problem behavior: A longitudinal study. *Merrill-Palmer Quarterly, 41,* 133–151.

Rotheram-Borus, M. J. (1993). Biculturalism among adolescents. In M. Bernal & G. Knight (Eds.), *Ethnic identity* (pp. 81–102). Albany, NY: State University of New York Press.

Rothman, K. J., Moore, L. L., Singer, M. R., Nguyen, U. S., Manneno, S., & Milunsky, A. (1995). Teratogenicity of high vitamin A intake. *New England Journal of Medicine, 333,* 1369–1373.

Roughan, P. A., Kaiser, F. E., & Morley, J. E. (1993). Sexuality and the older woman. *Clinics in Geriatric Medicine, 9,* 87–106.

Rourke, B. P. (1988). Socioemotional disturbances of learning disabled children. *Journal of Consulting and Clinical Psychology, 56,* 801–810.

Rousseau, J. J. (1955). *Emile.* New York: Dutton. (Original work published 1762)

Rovee-Collier, C. (1991). The "memory system" of prelinguistic infants. In A. Diamond (Ed.), *Annals of the New York Academy of Sciences* (Vol. 608, pp. 517–536). New York: New York Academy of Sciences.

Rovee-Collier, C., Patterson, J., & Hayne, H. (1985). Specificity in the reactivation of infant memory. *Developmental Psychobiology, 18,* 559–574.

Rovee-Collier, C. K. (1987). Learning and memory. In J. D. Osofsky (Ed.), *Handbook of infant development* (2nd ed., pp. 98–148). New York: Wiley.

Rowe, J. W., & Kahn, R. L. (1987). Human aging: Usual and successful. *Science, 237,* 143–149.

Rozin, P. (1990). Development in the food domain. *Developmental Psychology, 26,* 555–562.

Rubenstein, C. M., & Shaver, P. (1982). *In search of intimacy.* New York: Delacorte Press.

Rubin, J. Z., Provenzano, F. J., & Luria, Z. (1974). The eye of the beholder: Parents' views on sex of newborns. *American Journal of Orthopsychiatry, 44,* 512–519.

Rubin, K. H. (1982). Nonsocial play in preschoolers: Necessarily evil? *Child Development, 53,* 651–657.

Rubin, K. H., Fein, G. G., & Vandenberg, B. (1983). Play. In E. M. Hetherington (Ed.), *Handbook of child psychology: Vol. 4. Socialization, personality, and social development* (4th ed., pp. 693–744). New York: Wiley.

Rubin, K. H., Maioni, T. L., & Hornung, M. (1976). Free play behaviors in middle- and lower-class preschoolers: Parten and Piaget revisited. *Child Development, 47,* 414–419.

Rubin, K. H., Watson, K. S., & Jambor, T. W. (1978). Free-play behaviors in preschool and kindergarten children. *Child Development, 49,* 539–536.

Rubinstein, R. L. (1987). Never married elderly as a social type: Re-evaluating some images. *Gerontologist, 27,* 108–113.

Rubinstein, R. L., Alexander, B. B., Goodman, M., & Luborsky, M. (1991). Key relationships of never married, childless older women: A cultural analysis. *Journal of Gerontology, 46,* S270–S277.

Rudberg, M. A., Furner, S. E., Dunn, J. E., & Cassel, C. K. (1993). The relationship of visual and hearing impairments to disability: An analysis using the longitudinal study of aging. *Journal of Gerontology, 48,* M261–M265.

Ruff, H. A., & Lawson, K. R. (1990). Development of sustained, focused attention in young children during free play. *Developmental Psychology, 26,* 85–93.

Ruff, H. A., Lawson, K. R., Parrinello, R., & Weissberg, R. (1990). Long-term stability of individual differences in sustained attention in the early years. *Child Development, 61,* 60–75.

Ruff, H. A., & Rothbart, M. K. (1996). *Attention in early development: Themes and variations.* New York: Oxford University Press.

Ruffman, T., Perner, J., Olson, D. R., & Doherty, M. (1993). Reflecting on scientific thinking: Children's understanding of the hypothesis–evidence relation. *Child Development, 64,* 1617–1636.

Ruhm, C. (1989). Why older Americans stop working. *Gerontologist, 29,* 294–300.

Ruhm, C. J. (1996). Gender differences in employment behavior during late middle age. *Journal of Gerontology, 51B,* S11–S17.

Rumberger, R. W. (1990). Second chance for high school dropouts: Dropout recovery programs in the United States. In D. Inbar (Ed.), *Second chance in education: An interdisciplinary and international perspective* (pp. 227–250). Philadelphia: Falmer.

Rumberger, R. W., Ghatak, R., Poulos, G., Ritter, P. L., & Dornbusch, S. M. (1990). Family influences on dropout behavior in one California high school. *Sociology of Education, 63,* 283–299.

Runco, M. A. (1992). Children's divergent thinking and creative ideation. *Developmental Review, 12,* 233–264.

Rushton, H. G. (1989). Nocturnal enuresis: Epidemiology, evaluation, and currently available treatment options. *Journal of Pediatrics, 114,* 691–696.

Russell, J. A. (1990). The preschooler's understanding of the causes and consequences of emotion. *Child Development, 61,* 1872–1881.

Russell, M. J., Cummings, B. J., Proffitt, B. F., Wysocki, C. J., Gilbert, A. N., & Cotman, C. W. (1993). Life span changes in the verbal categorization of odors. *Journal of Gerontology, 48,* P49–P53.

Russell, R. J. H., & Wells, P. A. (1994). Predictors of happiness in married couples. *Personality and Individual Differences, 17,* 313–321.

Rutter, M. (1979). Protective factors in children's responses to stress and disadvantage. In M. W. Kent & J. Rolf (Eds.), *Primary prevention of psychopathology: Vol. III. Social competence in children* (pp. 49–74). Hanover, NH: University Press of New England.

Rutter, M. (1987). Psychosocial resilience and protective mechanisms. *American Journal of Orthopsychiatry, 57,* 316–331.

Ryan, K. J. (1989). Ethical issues in reproductive endocrinology and infertility. *American Journal of Obstetrics and Gynecology, 160,* 1415–1417.

Ryff, C. D. (1989). In the eye of the beholder: Views of psychological well-being among middle-aged and older adults. *Psychology and Aging, 4,* 195–210.

Ryff, C. D. (1991). Possible selves in adulthood and old age: A tale of shifting horizons. *Psychology and Aging, 6,* 286–295.

Ryff, C. D., & Heincke, S. G. (1983). Subjective organization of personality in adulthood and aging. *Journal of Personality and Social Psychology, 44,* 807–816.

Ryff, C. D., Lee, Y. H., Essex, M. J., & Schmutte, P. S. (1994). My children and me: Midlife evaluations of grown children and of self. *Psychology and Aging, 9,* 195–205.

Ryff, C. D., & Migdal, S. (1984). Intimacy and generativity: Self-perceived transitions. *Signs: Journal of Women in Culture and Society, 9,* 470–481.

Ryynänen, M., Kirkinen, P., Mannermaa, A., & Saarikoski, S. (1995). Carrier diagnosis of the fragile X syndrome— A challenge in antenatal clinics. *American Journal of Obstetrics and Gynecology, 172,* 1236–1239.

Saarni, C. (1989). Children's understanding of strategic control of emotional expression in social transactions. In C. Saarni & P. L. Harris (Eds.), *Children's understanding of emotion* (pp. 181–208). Cambridge, England: Cambridge University Press.

Saarni, C., Mumme, D., & Campos, J. J. (1997). Emotional development. In W. Damon (Ed.), *Handbook of child psychology: Vol. 3. Social, emotional, and personality development* (5th ed.). New York: Wiley.

Sabatier, J. P., Guaydiersouquieres, G., Laroche, D., Benmalek, A., Fournier, L., Guillonmetz, F., Delavenne, J., & Denis, A. Y. (1996). Bone mineral acquisition during adolescence and early adulthood—a study in 574 healthy females 10–24 years of age. *Osteoporosis International, 6,* 141–148.

Sabom, M. B. (1982). *Recollections of death.* New York: Simon & Schuster.

Sadler, T. W. (1995). *Langman's medical embryology* (7th ed.). Baltimore: Williams & Wilkins.

Safire, W. (1997, March 9). The young old. *New York Times Magazine,* p. 14.

Safman, P. C. (1988). Women from special populations: The challenge of reentry. In L. H. Lewis (Ed.), *Addressing the needs of returning women* (pp. 79–94). San Francisco: Jossey-Bass.

Salapatek, P. (1975). Pattern perception in early infancy. In L. B. Cohen & P. Salapatek (Eds.), *Infant perception: From sensation to cognition* (pp. 133–248). New York: Academic Press.

Salapatek, P., & Cohen, L. (Eds.). (1987). *Handbook of infant perception: Vol. 2.*

From perception to cognition. Orlando, FL: Academic Press.

Salthouse, T. A. (1984). Effects of age and skill in typing. *Journal of Experimental Psychology: General, 113,* 345–371.

Salthouse, T. A. (1985). Speed of behavior and its implications for cognition. In J. E. Birren & K. W. Schaie (Eds.), *Handbook of the psychology of aging* (2nd ed., pp. 400–426). New York: Van Nostrand Reinhold.

Salthouse, T. A. (1991a). Cognitive facets of aging well. *Generations, 51*(1), 35–38.

Salthouse, T. A. (1991b). *Theoretical perspectives in cognitive aging.* Hillsdale, NJ: Erlbaum.

Salthouse, T. A., & Babcock, R. L. (1991). Decomposing adult age differences in working memory. *Developmental Psychology, 27,* 763–776.

Salthouse, T. A., & Maurer, T. J. (1996). Aging, job performance, and career development. In J. E. Birren & K. W. Schaie (Eds.), *Handbook of the psychology of aging* (pp. 353–364). San Diego, CA: Academic Press.

Salthouse, T. A., & Meinz, E. J. (1995). Aging, inhibition, working memory, and speed. *Journal of Gerontology, 50,* P297–P306.

Salthouse, T. A., Rogan, J. D., & Prill, K. A. (1984). Division of attention: Age differences on a visually presented memory task. *Memory & Cognition, 15,* 507–516.

Salthouse, T. A., & Skovronek, E. (1992). Within-context assessment of working memory. *Journal of Gerontology, 47,* P110–P129.

Samarel, N. (1991). *Caring for life and death.* Washington, DC: Hemisphere.

Samarel, N. (1995). The dying process. In H. Wass & R. A. Neimeyer (Eds.), *Dying: Facing the facts* (3rd ed., pp. 89–116). Washington, DC: Taylor & Francis.

Sameroff, A. J., Seifer, R., Baldwin, A., & Baldwin, C. (1993). Stability of intelligence from preschool to adolescence: The influence of social and family risk factors. *Child Development, 64,* 80–97.

Samson, L. F. (1988). Perinatal viral infections and neonates. *Journal of Perinatal Neonatal Nursing, 1,* 56–65.

Samson, R. J., & Laub, J. H. (1993). *Crime in the making: Pathways and turning points through life.* Cambridge, MA: Harvard University Press.

Sandberg, D. E., Ehrhardt, A. A., Ince, S. E., & Meyer-Bahlberg, H. F. L. (1991). Gender differences in children's and adolescents' career aspirations. *Journal of Adolescent Research, 6,* 371–386.

Sanderson, J. A., & Siegal, M. (1988). Conceptions of moral and social rules in rejected and nonrejected pre-schoolers. *Journal of Clinical Child Psychology, 17,* 66–72.

Sandqvist, K. (1992). Sweden's sex-role scheme and commitment to gender equality. In S. Lewis, D. N. Izraeli, & H. Hottsmans (Eds.), *Dual-earner families: International perspective.* London: Sage.

Sanford, J. P. (1985). *Comprehension-level tasks in secondary classrooms.* Austin: Research and Development Center for Teacher Education, University of Texas at Austin.

Sankar, A. (1993). Images of home death and the elderly patient: Romantic versus real. *Generations, 27*(2), 59–63.

Santrock, J. W., & Warshak, R. A. (1986). Development of father custody relationships and legal/clinical considerations in father-custody families. M. E. Lamb (Ed.), *The father's role: Applied perspectives* (pp. 135–166). New York: Wiley.

Sarason, S. B. (1977). *Work, aging, and social change.* New York: Free Press.

Sasser-Coen, J. A. (1993). Qualitative changes in creativity in the second half of life: A life-span developmental perspective. *Journal of Creative Behavior, 27,* 18–27.

Satz, P. (1993). Brain reserve capacity on symptom onset after brain injury: A formulation and review of evidence for threshold theory. *Neuropsychology, 7,* 273–295.

Saxe, G. B. (1988, August–September). Candy selling and math learning. *Educational Researcher, 17*(6), 14–21.

Saywitz, K. J., & Nathanson, R. (1993). Children's testimony and their perceptions of stress in and out of the courtroom. *Child Abuse & Neglect, 17,* 613–622.

Scarr, S. (1985). Constructing psychology: Making facts and fables for our times. *American Psychologist, 40,* 499–512.

Scarr, S., & Kidd, K. K. (1983). Developmental behavior genetics. In M. M. Haith & J. J. Campos (Eds.), *Handbook of child psychology: Vol. 2. Infancy and developmental psychobiology* (pp. 345–433). New York: Wiley.

Scarr, S., & McCartney, K. (1983). How people make their own environments: A theory of genotype → environment effects. *Child Development, 54,* 424–435.

Scarr, S., Phillips, D. A., & McCartney, K. (1990). Facts, fantasies, and the future of child care in America. *Psychological Science, 1,* 26–35.

Scarr, S., Phillips, D., McCartney, K., & Abbott-Shim, M. (1993). Quality of child care as an aspect of family and child care policy in the United States. *Pediatrics, 91,* 182–188.

Scarr, S., & Weinberg, R. A. (1983). The Minnesota Adoption Studies: Genetic differences and malleability. *Child Development, 54,* 260–267.

Schachter, F. F., & Stone, R. K. (1985). Difficult sibling–easy sibling: Temperament and the within-family environment. *Child Development, 56,* 1335–1344.

Schaefer, M., Hatcher, R. P., & Bargelow, P. D. (1980). Prematurity and infant stimulation. *Child Psychiatry and Human Development, 10,* 199–212.

Schaie, K. W. (1977/1978). Toward a stage theory of adult cognitive development. *Aging and Human Development, 8,* 129–138.

Schaie, K. W. (1988a). Ageism in psychological research. *American Psychologist, 43,* 179–183.

Schaie, K. W. (1988b). Variability in cognitive functioning in the elderly. In M. A. Bender, R. C. Leonard, & A. D. Woodhead (Eds.), *Phenotypic variation in populations* (p. 191–211). New York: Plenum.

Schaie, K. W. (1989). Individual differences in rate of cognitive change in adulthood. In V. L. Bengtson & K. W. Schaie (Eds.). In A. Woodhead, M. Bender, & R. Leonard (Eds.), *Phenotypic variation in populations* (pp. 191–212). New York: Plenum.

Schaie, K. W. (1994). The course of adult intellectual development. *American Psychologist, 49,* 304–313.

Schaie, K. W. (1996a). Intellectual development in adulthood. In J. E. Birren & K. W. Schaie (Eds.), *Handbook of the psychology of aging* (4th ed, pp. 266–286). San Diego: Academic Press.

Schaie, K. W. (1996b). *Intellectual development in adulthood: The Seattle Longitudinal Study.* New York: Cambridge University Press.

Schaie, K. W., & Willis, S. L. (1992). *Adult development and aging* (3rd ed.). New York: HarperCollins.

Schanberg, S., & Field, T. M. (1987). Sensory deprivation stress and supplemental stimulation in the rat pup and preterm human neonate. *Child Development, 58,* 1431–1447.

Schauble, L. (1990). Belief revision in children: The role of prior knowledge and strategies for generating evidence. *Journal of Experimental Child Psychology, 49,* 31–57.

Schauble, L. (1996). The development of scientific reasoning in knowledge-rich contexts. *Developmental Psychology, 32,* 102–119.

Scheidt, R. J., & Windley, P. G. (1985). The ecology of aging. In J. E. Birren & K. W. Schaie (Eds.), *Handbook of the psychology of aging* (pp. 245–258). New York: Van Nostrand Reinhold.

Scheier, M. F., & Carver, C. S. (1987). Dispositional optimism and physical well-being: The influence of generalized outcome expectancies on health. *Journal of Personality, 55,* 169–210.

Schiavi, R. C., Mandeli, J., & Schreiner-Engel, P. (1994). Sexual satisfaction in healthy aging men. *Journal of Sex and Marital Therapy, 20,* 3–13.

Schiffman, S. S., & Warwick, Z. S. (1989). Use of flavor-amplified foods to improve nutritional status in elderly patients. *Annals of the New York Academy of Sciences, 561,* 267–276.

Schinke, S. P., Blythe, B. J., & Gilchrist, D. (1981). Cognitive-behavioral prevention of adolescent pregnancy. *Journal of Counseling Psychology, 28,* 451–454.

Schlegel, A., & Barry, H., III (1991). *Adolescence: An anthropological inquiry.* New York: Free Press.

Schlesinger, M., & Kronebusch, K. (1994). The sources of intergenerational burdens and tensions. In V. L. Bengtson & R. A. Harootyan (Eds.), *Intergenerational linkages: Hidden connections in American society* (pp. 185–209). New York: Springer.

Schneider, E. L. (1992). Biological theories of aging. *Generations, 16*(4), 7–10.

Schneider, E. L., & Guralnik, J. M. (1990). The aging of America: Impact on health care costs. *Journal of the American Medical Association, 263,* 2335–2340.

Schneider, W., & Pressley, M. (1989). *Memory development between 2 and 20.* New York: Springer-Verlag.

Schneirla, T. C., Rosenblatt, J. S., & Tobach, E. (1963). Maternal behavior in the cat. In H. R. Rheingold (Ed.), *Maternal behavior in mammals* (pp. 122–168). New York: Wiley.

Scholl, T. O., Heidiger, M. L., & Belsky, D. H. (1996). Prenatal care and maternal health during adolescent pregnancy: A review and meta-analysis. *Journal of Adolescent Health, 15,* 444–456.

Schonfeld, A. E. D. (1969, July). *In search of early memories.* Paper presented at the International Congress of Gerontology, Washington, DC.

Schor, E. L. (1987). Unintentional injuries. *American Journal of Diseases of Children, 141,* 1280–1284.

Schramm, W., Barnes, D., & Bakewell, J. (1987). Neonatal mortality in Missouri home births. *American Journal of Public Health, 77,* 930–935.

Schroeder, K. A., Blood, L. L., & Maluso, D. (1993). Gender differences and similarities between male and female undergraduate students regarding expectations for career and family roles. *College Student Journal, 27,* 237–249.

Schroots, J., & Birren, J. (1990). Concept of time and aging in science. In J. E. Birren & K. W. Schaie (Eds.), *Handbook of the psychology of aging* (3rd ed., pp. 45–64). San Diego: Academic Press.

Schulman, S. P., & Gerstenblith, G. (1989). Cardiovascular changes with aging: The response to exercise. *Journal of Cardiopulmonary Rehabilitation, 9,* 12–16.

Schultz, R., & Heckhausen, J. (1996). A life span model of successful aging. *American Psychologist, 51,* 702–714.

Schulz, R., & Aderman, D. (1979). Physicians' death anxiety and patient outcomes. *Omega, 9,* 327–332.

Schulz, R., & Curnow, C. (1988). Peak performance and age among super-athletes: Track and field, swimming, baseball, tennis, and golf. *Journal of Gerontology, 43,* P113–P120.

Schulz, R., Visintainer, P., & Williamson, G. M. (1990). Psychiatric and physical morbidity effects of caregiving. *Journal of Gerontology, 45,* 181–191.

Schunk, D. H. (1990). Goal setting and self-efficacy during self-regulated learning. *Educational Psychologist, 25,* 71–86.

Schwartz, J. E., Friedman, H. S., Tucker, J. S., Tomlinson-Keasey, C., Wingard, D. L., & Criqui, M. H. (1995). Childhood sociodemographic and psychosocial factors as predictors of longevity across the life-span. *American Journal of Public Health, 85,* 1237–1245.

Schwarz, B. (1996). *Nursing home design: Consequences of employing the medical model.* New York: Garland.

Schwebel, M., Maher, C. A., & Fagley, N. S. (1990). Introduction: The social role in promoting cognitive growth over the life span. In M. Schwebel, C. A. Maher, & N. S. Fagley (Eds.), *Promoting cognitive growth over the life span* (pp. 1–20). Hillsdale, NJ: Erlbaum.

Schweinhart, L. J., Barnes, H. V., & Weikart, D. P. (1993). Significant benefits: The High/Scope Perry Preschool Study through age 27. *Monographs of the High/Scope Educational Research Foundation* (No. 10). Ypsilanti, MI: High/Scope Press.

Scruggs, T. E., & Mastropieri, M. A. (1994). Successful mainstreaming in elementary science classes: A qualitative study of three reputational cases. *American Educational Research Journal, 31,* 785–811.

Scully, D. (1990). *Understanding sexual violence: A study of convicted rapists.* Boston: Unwin Hyman.

Sears, P. S., & Barbie, A. H. (1977). Career and life satisfaction among Terman's gifted women. In J. C. Stanley, W. George, & C. Solano (Eds.), *The gifted and creative: Fifty year perspective* (pp. 154–172). Baltimore: Johns Hopkins University Press.

Sears, R. R., Maccoby, E. E., & Levin, H. (1957). *Patterns of child rearing.* New York: Harper & Row.

Sebald, H. (1986). Adolescents' shifting orientation toward parents and peers: A curvilinear trend over recent decades. *Journal of Marriage and the Family, 48,* 5–13.

Seddon, J. M., Ajani, U. A., Sperduto, R. D., Hiller, R., Blair, H. N., Burton, T. C., Farber, D. T., Yannuzzi, L. A., & Willett, W. (1994a). Dietary carotenoids, vitamins A, C, and E, and advanced age-related macular degeneration. *Journal of the American Medical Association, 272,* 1413–1420.

Seddon, J. M., Christen, W. G., Manson, J. E., LaMotte, F. S., Glynn, R. J., Buring, J. E., & Hennekens, C. H. (1994b). The use of vitamin supplements and the risk of cataract among U.S. male physicians. *American Journal of Public Health, 84,* 788–792.

Seidman, E., Allen, L., Aber, J. L., Mitchell, C., & Feinman, J. (1994). The impact of school transitions in early adolescence on the self-system and perceived social context of poor urban youth. *Child Development, 65,* 507–522.

Seifer, R., Sameroff, A. J., Barrett, L. C., & Krafchuk, E. (1994). Infant temperament measured by multiple observations and mother report. *Child Development, 65,* 1478–1490.

Seifer, R., & Schiller, M. (1995). The role of parenting sensitivity, infant temperament, and dyadic interaction in attachment theory and assessment. In E. Waters, B. E. Vaughn, G. Posada, & K. Kondo-Ikemura (Eds.), *Caregiving, cultural, and cognitive perspectives on secure-base behavior and working models: New growing points of attachment theory and research. Monographs of the Society for Research in Child Development, 60*(2–3, Serial No. 244).

Seifer, R., Schiller, M., Sameroff, A. J., Resnick, S., & Riordan, K. (1996). Attachment, maternal sensitivity, and infant temperament during the first year of life. *Developmental Psychology, 32,* 12–25.

Seitz, V., & Apfel, N. H. (1993). Adolescent mothers and repeated childbearing: Effects of a school-based intervention program. *American Journal of Orthopsychiatry, 63,* 572–581.

Seitz, V., & Apfel, N. H. (1994). Effects of a school for pregnant students on the incidence of low-birthweight deliveries. *Child Development, 65,* 666–676.

Seligman, L. (1994). *Developmental career counseling and assessment* (2nd ed.). Thousand Oaks, CA: Sage.

Seligman, M. E. P. (1975). *Helplessness: On depression, development, and death.* San Francisco: Freeman.

Seligmann, J. (1994, May 2). The pressure to lose. *Newsweek,* pp. 60–61.

Selman, R. L. (1976). Social-cognitive understanding: A guide to educational and clinical practice. In T. Lickona (Ed.), *Moral development and behavior: Theory, research, and social issues* (pp. 299–316). New York: Holt, Rinehart and Winston.

Selman, R. L. (1980). *The growth of interpersonal understanding.* New York: Academic Press.

Selman, R. L., & Byrne, D. F. (1974). A structural-developmental analysis of levels of role taking in middle childhood. *Child Development, 45,* 803–806.

Seltzer, M. M., & Ryff, C. D. (1994). Parenting across the life span: The normative and nonnormative cases. In D. L. Featherman, R. M. Lerner, & M. Perlmutter (Eds.), *Life-span development and behavior* (pp. 1–40). Hillsdale, NJ: Erlbaum.

Seltzer, R. (1993). AIDS, homosexuality, public opinion, and changing correlates over time. *Journal of Homosexuality, 26,* 85-87.

Seltzer, V., & Benjamin, F. (1990). Breastfeeding and the potential for human immunodeficiency virus transmission. *Obstetrics and Gynecology, 75,* 713–715.

Selwyn, P. A. (1996). Before their time: A clinician's reflections on death and AIDS. In H. M. Spiro, M. G. M. Curnen, & L. P. Wandel (Eds.), *Facing death: Where culture, religion, and medicine meet* (pp. 33–37). New Haven, CT: Yale University Press.

Serbin, L. A., Powlishta, K. K., & Gulko, J. (1993). The development of sex typing in middle childhood. *Monographs of the Society for Research in Child Development, 58*(2, Serial No. 232).

Sever, J. L. (1983). Maternal infections. In C. C. Brown (Ed.), *Childhood learning disabilities and prenatal risk* (pp. 31–38). New York: Johnson & Johnson.

Shagle, S. C., & Barber, B. K. (1993). Effects of family, marital, and parent–child conflict on adolescent self-derogation and suicidal ideation. *Journal of Marriage and the Family, 55,* 964–974.

Shainess, N. (1961). A re-evaluation of some aspects of femininity through a study of menstruation: A preliminary report. *Comparative Psychiatry, 2,* 20–26.

Shantz, C. U. (1987). Conflicts between children. *Child Development, 58,* 283–305.

Shapiro, L. R., Crawford, P. B., Clark, M. J., Pearson, D. L., Raz, J., & Huenemann, R. (1984). Obesity prognosis: A longitudinal study of children from the age of 6 months to 9 years. *American Journal of Public Health, 74,* 968–972.

Shaver, P., Furman, W., & Buhrmester, D. (1985). Transition to college: Network changes, social skills, and loneliness. In S. Duck & D. Perlman (Eds.), *Understanding personal relationships: An interdisciplinary approach* (pp. 193–219). London: Sage.

Shaver, P. R., & Brennan, K. A. (1992). Attachment styles and the "big five" personality traits: Their connections with each other and with romantic relationship outcomes. *Personality and Social Psychology Bulletin, 18,* 536–545.

Shaw, G. M., Schaffer, D., Velie, E. M., Morland, K., & Harris, J. A. (1995). Periconceptional vitamin use, dietary folate, and the occurrence of neural tube defects. *Epidemiology, 6,* 219–226.

Shedler, J., & Block, J. (1990). Adolescent drug use and psychological health: A longitudinal inquiry. *American Psychologist, 45,* 612–630.

Sheiman, D. L., & Slomin, M. (1988). *Resources for middle childhood.* New York: Garland.

Shephard, R. J., & Shek, P. N. (1995). Exercise, aging and immune function. *International Journal of Sports Medicine, 16,* 1–6.

Sherman, E. (1993). Mental health and successful adaptation in later life. *Generations, 17*(1), 43–46.

Shiffrin, R. M., & Atkinson, R. C. (1969). Storage and retrieval processes in long-term memory. *Psychological Review, 76,* 179–193.

Shimamura, A. P., Berry, J. M., Mangels, J. A., Rusting, C. L., & Jurica, P. J. (1995). Memory and cognitive abilities in university professors: Evidence for sucessful aging. *Psychological Science, 6,* 271–277.

Shock, N. W. (1977). Biological theories of aging. In J. E. Birren & K. W. Schaie (Eds.), *Handbook of the psychology of aging* (pp. 103–115). New York: Van Nostrand Reinhold.

Shostak, A. B. (1987). Singlehood. In M. B. Sussmann & S. K. Steinmetz (Eds.), *Handbook of marriage and the family* (pp. 355–367). New York: Plenum.

Shuchter, S. R., & Zisook, S. (1995). The course of normal grief. In M. S. Stroebe, W. Stroebe, & R. O. Hansson (Eds.), *Handbook of bereavement* (pp. 44–61). Cambridge: Cambridge University Press.

Shulman, S., Elicker, J., & Sroufe, A. (1994). Stages of friendship growth in preadolescence as related to attachment history. *Journal of Social and Personal Relationships, 11,* 341–361.

Shurtleff, D. B., & Lemire, R. J. (1995). Epidemiology, etiologic factors, and prenatal diagnosis of openspinal dysraphism. *Neurosurgery Clinics of North America, 6,* 183–193.

Shweder, R. A., Mahapatra, M., & Miller, J. G. (1990). Culture and moral development. In J. Stigler, R. A. Shweder, & G. Herdt (Eds.), *Cultural psychology: Essays on comparative human development* (pp. 130–204). New York: Cambridge University Press.

Siebert, J. M., Garcia, A., Kaplan, M., & Septimus, A. (1989). Three model pediatric AIDS programs: Meeting the needs of children, families, and communities. In J. M. Siebert & R. A. Olson (Eds.), *Children, adolescents, and AIDS* (pp. 25–60). Lincoln, NB: University of Nebraska Press.

Siegler, I. C., McCarty, S. M., & Logue, P. E. (1982). Wechsler memory scale scores, selective attrition, and distance from death. *Journal of Gerontology, 37,* 176–181.

Siegler, R. S. (1978). The origins of scientific reasoning. In R. S. Siegler (Ed.), *Children's thinking: What develops?* (pp. 109–149). Hillsdale, NJ: Erlbaum.

Siegler, R. S. (1983). Information processing approaches to development. In W. Kessen (Ed.), *Handbook of child psychology: Vol. 1. History, theory, and methods* (pp. 129–212). New York: Wiley.

Siegler, R. S. (1998). *Children's thinking* (3rd ed.). Englewood Cliffs, NJ: Prentice Hall.

Siegman, A. W. (1994). Cardiovascular consequences of expressing and repressing anger. In A. W. Siegman & T. W. Smith (Eds.), *Anger, hostility, and the heart* (pp. 173–198). Hillsdale, NJ: Erlbaum.

Sigelman, C., Maddock, A., Epstein, J., & Carpenter, W. (1993). Age differences in understandings of disease causality: AIDS, colds, and cancer. *Child Development, 64,* 272–284.

Signorella, M., & Liben, L. S. (1984). Recall and reconstruction of gender-related pictures: Effects of attitude, task difficulty, and age. *Child Development, 55*, 393–405.

Signorielli, N. (1993). Television, the portrayal of women, and children's attitudes. In G. L. Berry & J. K. Asamen (Eds.), *Children and television: Images in a changing sociocultural world* (pp. 229–242). Newbury Park, CA: Sage.

Sill, A. M., Stick, M. J., Prenger, V. L., Phillips, S. L., Boughman, J. A., & Arnos, K. S. (1994). Genetic epidemiologic study of hearing loss in an adult population. *American Journal of Medical Genetics, 54*, 149–153.

Silver, L. B. (1989). Psychological and family problems associated with learning disabilities: Assessment and intervention. *Journal of the American Academy of Child and Adolescent Psychiatry, 28*, 319–325.

Silverberg, S. B., & Steinberg, L. (1990). Psychological well-being of parents with early adolescent children. *Developmental Psychology, 26*, 658–666.

Silverman, P. R., & Nickman, S. L. (1996). Children's construction of their dead parents. In D. Klass, P. R. Silverman, & S. L. Nickman (Ed.), *Continuing bonds: New understandings of grief* (pp. 73–86). Washington, DC: Taylor & Francis.

Silverman, P. R., & Worden, J. M. (1992). Children's reactions in the early months after the death of a parent. *American Journal of Orthopsychiatry, 62*, 93–104.

Silverman, W. K., La Greca, A. M., & Wasserstein, S. (1995). What do children worry about? Worries and their relation to anxiety. *Child Development, 66*, 671–686.

Silverstein, M., & Bengtson, V. L. (1991). Do close parent–child relations reduce the mortality risk of older parents? *Journal of Health and Social Behavior, 32*, 382–395.

Simmons, R. G., Black, A., & Zhou, Y. (1991). African-American versus white children and the transition to junior high school. *American Journal of Education, 99*, 481–520.

Simmons, R. G., & Blyth, D. A. (1987). *Moving into adolescence.* New York: Aldine De Gruyter.

Simoneau, G. G., & Leibowitz, H. W. (1996). Posture, gait, and falls. In J. Birren & K. W. Schaie (Eds.), *Handbook of the psychology of aging* (4th ed., pp. 204–217). San Diego: Academic Press.

Simons, R. L., Conger, R. D., & Whitbeck, L. B. (1988). A multistage social learning model of the influences of

family and peers upon adolescent substance use. *Journal of Drug Issues, 18*, 293–316.

Simons, R. L., Lorenz, F. O., Conger, R. D., & Wu, C-I. (1992). Support from spouse as a mediator and moderator of the disruptive influence of economic strain on parenting. *Child Development, 63*, 1282–1301.

Simons, R. L., Lorenz, F. O., Wu, C-I., & Conger, R. D. (1993). Social network and marital support as mediators and moderators of the impact of stress and depression on parental behavior. *Developmental Psychology, 29*, 368–381.

Simons, R. L., Whitbeck, L. B., Conger, R. D., & Wu, C-I. (1991). Intergenerational transmission of harsh parenting. *Developmental Psychology, 27*, 159–171.

Simons, R. L., Whitbeck, L. B., Conger, R. D., & Melby, J. N. (1990). Husband and wife differences in determinants of parenting. *Journal of Marriage and the Family, 52*, 375–392.

Simonton, D. K. (1988). Age and creative productivity: What do we know after a century of research? *Psychological Bulletin, 104*, 251–267.

Simonton, D. K. (1990). Creativity and wisdom in aging. In J. E. Birren & K. W. Schaie (Eds.), *Handbook of the psychology of aging* (3rd ed., pp. 320–329). San Diego: Academic Press.

Simonton, D. K. (1991). Creative productivity through the adult years. *Generations, 15*(2), 13–16.

Simpson, J. A., & Harris, B. A. (1994). Interpersonal attraction. In A. L. Weber & J. H. Harvey (Eds.), *Perspectives on close relationships* (pp. 45–66). Boston: Allyn and Bacon.

Simpson, J. A., Rholes, W. S., & Nelligan, J. S. (1992). Support-seeking and support-giving within couple members in an anxiety-provoking situation: The role of attachment styles. *Journal of Personality and Social Psychology, 62*, 434–446.

Simpson, S. A., & Harding. A. E. (1993). Predictive testing for Huntington's disease after the gene. *Journal of Medical Genetics, 30*, 1036–1038.

Singer, D. G., & Singer, J. L. (1990). *The house of make-believe.* Cambridge, MA: Harvard University Press.

Sinnott, J. D. (1989). A model for solution of ill-structured problems: Implications for everyday and abstract problem solving. In J. D. Sinnott (Ed.), *Everyday problem solving: Theory and applications* (pp. 72–99). New York: Praeger.

Sivard, R. L. (1993). *World military and

social expenditures* (16th ed.). Leesburg, VA: WMSE Publications.

Skinner, B. F. (1957). *Verbal behavior.* New York: Appleton-Century-Crofts.

Skinner, B. F. (1983). Intellectual self-management in old age. *American Psychologist, 38*, 239–244.

Skinner, E. A., & Belmont, M. J. (1993). Motivation in the classroom: Reciprocal effects of teacher behavior and student engagement across the school year. *Journal of Educational Psychology, 85*, 571–581.

Skolnick, A. (1981). Married lives: Longitudinal perspectives on marriage. In D. Eichorn, J. Clausen, N. Haan, M. Honzig, & P. Mussen (Eds.), *Present and past in middle age* (pp. 270–300). New York: Academic Press.

Slaby, R. G., & Frey, K. S. (1975). Development of gender constancy and selective attention to same-sex models. *Child Development, 46*, 849–856.

Slaby, R. G., Roedell, W. C., Arezzo, D., & Hendrix, K. (1995). *Early violence prevention.* Washington, DC: National Association for the Education of Young Children.

Slade, A. (1987). A longitudinal study of maternal involvement and symbolic play during the toddler period. *Child Development, 58*, 367–375.

Smetana, J. (1988). Concepts of self and social convention: Adolescents' and parents' reasoning about hypothetical and actual family conflicts. In M. Gunnar & W. A. Collins (Eds.), *Minnesota Symposia on Child Psychology* (Vol. 21, pp. 79–122). Hillsdale, NJ: Erlbaum.

Smetana, J. G. (1989). Toddlers' social interactions in the context of moral and conventional transgressions in the home. *Developmental Psychology, 25*, 499–508.

Smetana, J. G., & Asquith, P. (1994). Adolescents' and parents' conceptions of parental authority and personal autonomy. *Child Development, 65*, 1147–1162.

Smetana, J. G., & Braeges, J. L. (1990). The development of toddlers' moral and conventional judgments. *Merrill-Palmer Quarterly, 36*, 329–346.

Smilansky, S. (1968). *The effects of sociodramatic play on disadvantaged children: Preschool children.* New York: Wiley.

Smiley, P. A., & Dweck, C. S. (1994). Individual differences in achievement goals among young children. *Child Development, 65*, 1723–1743.

Smith, A. D. (1996). Memory. In J. E. Birren & K. W. Schaie (Eds.), *Handbook *of the psychology of aging* (4th ed., pp. 236–250). San Diego: Academic Press.

Smith, B. A., & Blass, E. M. (1996). Taste-mediated calming in premature, preterm, and full-term human infants. *Developmental Psychology, 32,* 1084–1089.

Smith, C., & Lloyd, B. (1978). Maternal behavior and perceived sex of infant: Revisited. *Child Development, 49,* 1263–1266.

Smith, D. C. (1993). The terminally ill patient's right to be in denial. *Omega, 27,* 115–121.

Smith, H. (1992). The detrimental health effects of ionizing radiation. *Nuclear Medicine Communications, 13,* 4–10.

Smith, H. O., Kammererdoak, D. N., Barbo, D. M., & Sarto, G. E. (1996). Hormone replacement therapy in the menopause—a pro opinion. *Cancer Journal for Clinicians, 46,* 343–364.

Smith, J., & Baltes, P. B. (1992). A life-span perspective on thinking and problem-solving. In M. Schwebel, C. A. Maher, & N. S. Fagley (Eds.), *Promoting cognitive growth over the life span* (pp. 47–69). Hillsdale, NJ: Erlbaum.

Smith, J., & Prior, M. (1995). Temperament and stress resilience in school-age children: A within-families study. *Journal of the American Academy of Child and Adolescent Psychiatry, 34,* 168–179.

Smith, J., Staudinger, U. M., & Baltes, P. B. (1994). Occupational settings facilitating wisdom-related knowledge: The sample case of clinical psychologists. *Journal of Consulting and Clinical Psychology, 66,* 989–999.

Smith, P. (1991). Introduction: The study of grandparenthood. In P. K. Smith (Ed.), *The psychology of grandparenthood: An international perspective* (pp. 1–16). London: Routledge.

Smith, P. K. (1978). A longitudinal study of social participation in preschool children: Solitary and parallel play reexamined. *Developmental Psychology, 14,* 517–523.

Smith, S. (Ed.). (1995). *Advances in applied developmental psychology: Vol. 6. Two-generation programs for families in poverty.* Norwood, NJ: Ablex.

Snarey, J. (1995). In a communitarian voice: The sociological expansion of Kohlbergian theory, research, and practice. In W. M. Kurtines & J. L. Gewirtz (Eds.), *Moral development: An introduction* (pp. 109–134). Boston: Allyn and Bacon.

Snarey, J., Son, L., Kuehne, V. S., Hauser, S., & Vaillant, G. (1987). The role of parenting in men's psychosocial development: A longitudinal study of early adulthood infertility and midlife generativity. *Developmental Psychology, 23,* 593–603.

Snarey, J. R., Reimer, J., & Kohlberg, L. (1985). The development of social–moral reasoning among kibbutz adolescents: A longitudinal cross-cultural study. *Developmental Psychology, 21,* 3–17.

Snow, C. E. (1993). Families as social contexts for literacy development. In C. Daiute (Ed.), *New directions for child development* (No. 61, pp. 11–24). San Francisco: Jossey-Bass.

Society for Research in Child Development (1993). Ethical standards for research with children. In *Directory of Members* (pp. 337–339). Ann Arbor, MI: Author.

Sodian, B., Taylor, C., Harris, P. L., & Perner, J. (1991). Early deception and the child's theory of mind: False trails and genuine markers. *Child Development, 62,* 468–483.

Sodian, B., & Wimmer, H. (1987). Children's understanding of inference as a source of knowledge. *Child Development, 58,* 424–433.

Soken, H. H., & Pick, A. D. (1992). Intermodal perception of happy and angry expressive behaviors by seven-month-old infants. *Child Development, 63,* 787–795.

Sommer, K., Whitman, T. L., Borkowski, J. G., Schellenbach, C., Maxwell, S., & Keogh, D. (1993). Cognitive readiness and adolescent parenting. *Developmental Psychology, 29,* 389–398.

Sommerville, J. (1982). *The rise and fall of childhood.* Beverly Hills, CA: Sage.

Sonenstein, F. L., Pleck, J. H., & Ku, L. C. (1991). Levels of sexual activity among adolescent males in the United States. *Family Planning Perspectives, 23,* 162–167.

Song, M., & Ginsburg, H. P. (1987). The development of informal and formal mathematical thinking in Korean and U.S. children. *Child Development, 58,* 1286–1296.

Sonnenschein, S. (1986). Development of referential communication skills: How familiarity with a listener affects a speaker's production of redundant messages. *Developmental Psychology, 22,* 549–552.

Sontag, S. (1979). The double standard of aging. In J. Williams (Ed.), *Psychology of women* (pp. 462–478). San Diego, CA: Academic Press.

Sorce, J., Emde, R., Campos, J., & Klinnert, M. (1985). Maternal emotional signaling: Its effect on the visual cliff behavior of 1-year-olds. *Developmental Psychology, 21,* 195–200.

Sorensen, T. I. A., Nielsen, G. G., Andersen, P. K., & Teasdale, T. W. (1988). Genetic and environmental influences on premature death in adult adoptees. *New England Journal of Medicine, 318,* 727–732.

Sorenson, E. S. (1993). *Children's stress and coping.* New York: Guilford.

Sosa, R., Kennell, J., Klaus, M., Robertson, S., & Urrutia, J. (1980). The effect of a supportive companion on perinatal problems, length of labor, and mother-infant interaction. *New England Journal of Medicine, 303,* 597–600.

Southard, B. (1985). Interlimb movement control and coordination in children. In J. E. Clark & J. E. Humphrey (Eds.), *Motor development* (Vol. 1, pp. 55–66). Princeton, NJ: Princeton Books.

Southern, W. T., Jones, E. D., & Stanley, J. C. (1994). Acceleration and enrichment: The context and development of program options. In K. A. Heller, F. J. Jonks, & H. A. Passow (Eds.), *International handbook of research and development of giftedness and talent* (pp. 387–409). Oxford: Pergamon Press.

Speake, D. L. (1987). Health promotion activities and the well elderly. *Health Values, 11,* 25–30.

Speece, M. W., & Brent, S. B. (1992). The acquisition of a mature understanding of three components of the concept of death. *Death Studies, 16,* 211–229.

Speece, M. W., & Brent, S. B. (1996). The development of children's understanding of death. In C. A. Corr & D. M. Corr (Eds.), *Handbook of childhood death and bereavement* (pp. 29–50). New York: Springer.

Speicher, B. (1994). Family patterns of moral judgment during adolescence and early adulthood. *Developmental Psychology, 30,* 624–632.

Spelke, E. S. (1987). The development of intermodal perception. In P. Salapatek & L. Cohen (Eds.), *Handbook of infant perception: Vol. 2. From perception to cognition* (pp. 233–273). Orlando, FL: Academic Press.

Spelke, E. S. (1991). Physical knowledge in infancy: Reflections on Piaget's theory. In S. Carey & R. Gelman (Eds.), *The epigenesis of mind: Essays on biology and cognition* (pp. 133–169). Hillsdale, NJ: Erlbaum.

Spelke, E. S. (1994). Initial knowledge: Six suggestions. *Cognition, 50,* 431–445.

Spellacy, W. N., Miller, S. J., & Winegar, A. (1986). Pregnancy after 40 years of age. *Obstetrics and Gynecology, 68,* 452–454.

Spence, M. J., & DeCasper, A. J. (1987). Prenatal experience with low-frequency maternal voice sounds influences neonatal perception of maternal voice samples. *Infant Behavior and Development, 10,* 133–142.

Spencer, M. B., & Dornbusch, S. M. (1990). Challenges in studying minority youth. In S. Feldman & G. R. Elliott (Eds.), *At the threshold: The developing adolescent* (pp. 123–146). Cambridge, MA: Harvard University Press.

Spencer, W. D., & Raz, N. (1995). Differential effects of aging on memory for context and context: A meta-analysis. *Psychology and Aging, 10,* 527–539.

Spina, R. J., Ogawa, T., Miller, T. R., Hohrt, W. M., & Ehsani, A. A. (1993). Effect of exercise training on left ventricular performance in older women free of cardiopulmonary disease. *American Journal of Cardiology, 71,* 99–104.

Spinetta, J., & Rigler, D. (1972). The child-abusing parent: A psychological review. *Psychological Bulletin, 77,* 296–304.

Spira, A. (1992). *Les comportements sexuels en France.* Paris: La documentation Française.

Spitz, R. A. (1945). Hospitalism: An inquiry into the genesis of psychiatric conditions in early childhood. *Psychoanalytic Study of the Child, 1,* 113–117.

Spitz, R. A. (1946). Anaclitic depression. *Psychoanalytic Study of the Child, 2,* 313–342.

Spitze, G., & Miner, S. (1992). Gender differences in adult–child contact among black elderly parents. *Gerontologist, 32,* 213–218.

Spivack, G., & Shure, M. B. (1974). *Social adjustment of young children: A cognitive approach to solving real life problems.* San Francisco: Jossey-Bass.

Spock, B., & Rothenberg, M. B. (1992). *Dr. Spock's baby and child care.* New York: Pocket Books.

Sroufe, L. A. (1979). Socioemotional development. In J. D. Osofsky (Ed.), *Handbook of infant development* (pp. 462–516). New York: Wiley.

Sroufe, L. A. (1988). A developmental perspective on day care. *Early Childhood Research Quarterly, 3,* 283–292.

Sroufe, L. A., Egeland, B., & Kreutzer, T. (1990). The fate of early experience following developmental change: Longitudinal approaches to individual adaptation. *Child Development, 61,* 1363–1373.

Sroufe, L. A., & Waters, E. (1976). The ontogenesis of smiling and laughter: A perspective on the organization of development in infancy. *Psychological Review, 83,* 173–189.

Sroufe, L. A., & Wunsch, J. P. (1972). The development of laughter in the first year of life. *Child Development, 43,* 1324–1344.

Stagner, R. (1985). Aging in industry. In J. E. Birren & K W. Schaie (Eds.), *Handbook of the psychology of aging* (pp. 789–817). New York: Van Nostrand Reinhold.

Stahl, S. A. (1992). Saying the "P" word: Nine guidelines for effective phonics instruction. *The Reading Teacher, 45,* 618–625.

Stahl, S. A., McKenna, M. C., & Pagnucco, J. R. (1994). The effects of whole-language instruction: An update and a reappraisal. *Educational Psychologist, 29,* 175–185.

Stampfer, M. J., Hennekens, C. H., Manson, J. E., Colditz, G. A., Rosner, B., & Willett, W. C. (1993). Vitamin E consumption and the risk of coronary disease in women. *New England Journal of Medicine, 328,* 1444–1449.

Standing, T. S., & Glazer, G. (1992). -Attitudes of low-income clinic patients toward menopause. *Health Care for Women International, 13,* 271–280.

Stanley, B., & Seiber, J. E. (Eds.). (1992). *Social research on children and adolescents: Ethical issues.* Newbury Park, CA: Sage.

Starrels, M. E. (1994). Husbands' involvement in female gender-typed household chores. *Sex Roles, 31,* 473–491.

Stattin, H., & Magnusson, D. (1990). *Pubertal maturation in female development.* Hillsdale, NJ: Erlbaum.

Staub, E. (1996). Cultural–societal roots of violence. *American Psychologist, 51,* 117–132.

Staudinger, U. M., Smith, J., & Baltes, P. B. (1992). Wisdom-related knowledge in a life-review task: Age differences and the role of professional specialization. *Psychology and Aging, 7,* 271–281.

Steinberg, L. (1984). The varieties and effects of work during adolescence. In M. Lamb, A. Brown, & B. Rogoff (Eds.), *Advances in developmental psychology* (pp. 1–37). Hillsdale, NJ: Erlbaum.

Steinberg, L. (1986). Latchkey children and susceptibility to peer pressure: An ecological analysis. *Developmental Psychology, 22,* 433–439.

Steinberg, L. (1988). Simple solutions to a complex problem: A response to Rodman, Pratto, & Nelson. *Developmental Psychology, 24,* 295–296.

Steinberg, L., & Dornbusch, S. M. (1991). Negative correlates of part-time employment during adolescence: Replication and elaboration. *Developmental Psychology, 27,* 304–313.

Steinberg, L., Fletcher, A., & Darling, N. (1994). Parental monitoring and peer influences on adolescent substance use. *Pediatrics, 93,* 1060–1064.

Steinberg, L., Lamborn, S. D., Darling, M., Mounts, N. S., & Dornbusch, S. M. (1994). Over-time changes in adjustment and competence among adolescents from authoritative, authoritarian, indulgent, and neglectful families. *Child Development, 65,* 754–770.

Steinberg, L., Lamborn, S. D., Dornbusch, S. M., & Darling, N. (1992). Impact of parenting practices on adolescent achievement: Authoritative parenting, school involvement, and encouragement to succeed. *Child Development, 63,* 1266–1281.

Steinberg, L., & Silverberg, S. (1986). The vicissitudes of autonomy in early adolescence. *Child Development, 57,* 841–851.

Steinberg, L. D. (1987). The impact of puberty on family relations: Effects of pubertal status and pubertal timing. *Developmental Psychology, 23,* 451–460.

Steinberg, L. D. (1990). Interdependence in the family: Autonomy, conflict, and harmony in the parent–adolescent relationship. In S. S. Feldman & G. R. Elliott (Eds.), *At the threshold: The developing adolescent* (pp. 255–276). Cambridge, MA: Harvard University Press.

Steiner, J. E. (1979). Human facial expression in response to taste and smell stimulation. In H. W. Reese & L. P. Lipsitt (Eds.), *Advances in child development and behavior* (Vol. 13, pp. 257–295). New York: Academic Press.

Steinhardt, M. A. (1992). Physical education. In P. W. Jackson (Ed.), *Handbook of research on curriculum* (pp. 964–1001). New York: Macmillan.

Stenberg, C., & Campos, J. (1990). The development of anger expressions in infancy. In N. Stein, B. Leventhal, & T. Trabasso (Eds.), *Psychological and biological approaches to emotion* (pp. 247–282). Hillsdale, NJ: Erlbaum.

Stephen, E. H., Freedman, V. A., & Hess, J. (1993). Near and far: Contact of children with their non-residential fathers. *Journal of Divorce & Remarriage, 20,* 171–191.

Stern, D. N. (1985). *The interpersonal world of the infant: A view from psychoanalysis and developmental psychology.* New York: Basic Books.

Stern, M., & Karraker, K. H. (1989). Sex stereotyping of infants: A review of gender labeling studies. *Sex Roles, 20,* 501–522.

Sternberg, K. J., Lamb, M. E., Greenbaum, C., Cicchetti, D., Dawaud, S., Cortes, R. M., Krispin, O., & Lorey, F. (1993). Effects of domestic violence on children's behavior problems and depression. *Developmental Psychology, 29,* 44–52.

Sternberg, R., & Lubart, T. I. (1995). *Defying the crowd.* New York: Basic Books.

Sternberg, R. J. (1985). *Beyond IQ: A triarchic theory of human intelligence.* New York: Cambridge University Press.

Sternberg, R. J. (1987). Liking versus loving: A comparative evaluation of theories. *Psychological Bulletin, 102,* 331–345.

Sternberg, R. J. (1988a). Triangulating love. In R. J. Sternberg & M. L. Barnes (Eds.), *The psychology of love* (pp. 119–138). New Haven, CT: Yale University Press.

Sternberg, R. J. (1988b). A triarchic view of intelligence in cross-cultural perspective. In S. H. Irvine & J. W. Berry (Eds.), *Human abilities in cultural context* (pp. 60–85). New York: Cambridge University Press.

Sternberg, R. J., & Lubart, T. I. (1991). Creating creative minds. *Phi Delta Kappan, 72*(8), 608–614.

Sternberg, R. J., & Lubart, T. I. (1995). *Defying the crowd.* New York: Basic Books.

Sternberg, R. J., & Odagaki, L. (1989). Continuity and discontinuity in intellectual development are not a matter of "either–or." *Human Development, 32,* 159–166.

Stevens, J. C. (1989). Food quality reports from noninstitutionalized aged. *Annals of the New York Academy of Sciences, 561,* 87–93.

Stevens, J. C. (1992). Aging and spatial acuity of touch. *Journal of Gerontology, 27,* P35–P40.

Stevens, J. C., Cain, W. S., Demarque, A., & Ruthruff, A. M. (1991). On the discrimination of missing ingredients: Aging and salt flavor. *Appetite, 16,* 129–140.

Stevens, M. L. T. (1997, March–April). What Quinlan can tell Kevorkian about the right to die. *The Humanist, 57*(2), 10–14.

Stevenson, D. L., & Baker, D. P. (1987). The family–school relation and the child's school performance. *Child Development, 58,* 1348–1357.

Stevenson, H. W. (1992, December). Learning from Asian schools. *Scientific American, 267*(6), 32–38.

Stevenson, H. W. (1994). Extracurricular programs in East Asian schools. *Teachers College Record, 95,* 389–407.

Stevenson, H. W., & Baker, D. P. (1987). The family–school relation and the child's school performance. *Child Development, 58,* 1348–1357.

Stevenson, H. W., Chen, C., & Lee, S-Y. (1993). Mathematics achievement of Chinese, Japanese, and American children: Ten years later. *Science, 259,* 53–58.

Stevenson, H. W., & Lee, S-Y. (1990). Contexts of achievement: A study of American, Chinese, and Japanese children. *Monographs of the Society for Research in Child Development, 55*(1–2, Serial No. 221).

Stevenson, H. W., Stigler, J. W., Lee, S–Y., Lucker, G. W., Litamura, S., & Hsu, C. (1985). Cognitive performance and academic achievement of Japanese, Chinese, and American children. *Child Development, 56,* 718–734.

Stevenson, R., & Pollitt, C. (1987). The acquisition of temporal terms. *Journal of Child Language, 14,* 533–545.

Stewart, A. J., & Gold-Steinberg, S. (1990). Case studies of psychosocial development and political commitment. *Psychology of Women Quarterly, 14,* 543–566.

Stewart, A. J., & Healy, J. M. (1989). Linking individual development and social changes. *American Psychologist, 44,* 30–42.

Stewart, A. J., & Vandewater, E. A. (1992, August). *Combining tough and tender methods to study women's lives.* Paper presented at the annual meeting of the American Psychological Association, Washington, DC.

Stewart, D. A. (1982). *Children with sex chromosome aneuploidy: Follow-up studies.* New York: Liss.

Stewart, R. B. (1983). Sibling attachment relationships: Child–infant interactions in the Strange Situation. *Developmental Psychology, 19,* 192–199.

Stewart, R. B., Moore, M. T., Marks, R. G., May, F. E., & Hale, W. E. (1993). *Driving cessation and accidents in the elderly: An analysis of symptoms, diseases, cognitive dysfunction and medications.* Washington, DC: AAA Foundation for Traffic Safety.

Stewart, S. L., & Rubin, K. H. (1995). The social problem-solving skills of anxious-withdrawn children. *Development and Psychopathology, 7,* 323–336.

Stice, E., & Barrera, M., Jr. (1995). A longitudinal examination of the reciprocal relations between perceived parenting and adolescents' substance use and externalizing behaviors. *Developmental Psychology, 31,* 322–334.

Stillion, J. M. (1995). Death in the lives of adults: Responding to the tolling of the bell. In H. Wass & R. A. Neimeyer (Eds.), *Dying: Facing the facts* (pp. 303–322). New York: Taylor & Francis.

Stillman, R. J. (1982). In utero exposure to diethylstilbestrol: Adverse effects on the reproductive tract and reproductive performance in male and female offspring. *American Journal of Obstetrics and Gynecology, 142,* 905–921.

Stipek, D., & Mac Iver, D. (1989). Developmental change in children's assessment of intellectual competence. *Child Development, 60,* 531–538.

Stipek, D., Recchia, S., & McClintic, S. (1992). Self-evaluation in young children. *Monographs of the Society for Research in Child Development, 57* (1, Serial No. 226).

Stipek, D. J., Gralinski, J. H., & Kopp, C. B. (1990). Self-concept development in the toddler years. *Developmental Psychology, 26,* 972–977.

Stipek, D. J., & Kowalski, P. S. (1989). Learned helplessness in task-orienting versus performance-orienting testing conditions. *Journal of Educational Psychology, 81,* 384–391.

Stith, S. M., & Farley, S. C. (1993). A predictive model of male spousal violence. *Journal of Family Violence, 8,* 183–201.

Stoch, M. B., Smythe, P. M., Moodie, A. D., & Bradshaw, D. (1982). Psychosocial outcome and CT findings after growth undernourishment during infancy: A 20-year developmental study. *Developmental Medicine and Child Neurology, 24,* 419–436.

Stocker, C., & Dunn, J. (1994). Sibling relationships in childhood and adolescence. In J. C. DeFries, R. Plomin, & D. W. Fulker (Eds.), *Nature and nurture in middle childhood* (pp. 214–232). Cambridge, MA: Blackwell.

Stodolsky, S. S. (1974). How children find something to do in preschools. *Genetic Psychology Monographs, 90,* 245–303.

Stodolsky, S. S. (1988). *The subject matters.* Chicago: University of Chicago Press.

Stoel-Gammon, C., & Otomo, K. (1986). Babbling development of hearing-impaired and normal hearing subjects. *Journal of Speech and Hearing Disorders, 51,* 33–41.

Stone, R., Cafferata, G. L., & Sangl, J. (1987). Caregivers of the frail elderly: A national profile. *Gerontologist, 27,* 616–626.

Stoneman, Z., Brody, G. H., & MacKinnon, C. E. (1986). Same-sex and cross-sex siblings: Activity choices, roles, behavior, and gender stereotypes. *Sex Roles, 15,* 495–511.

Stoney, C. M., & Engebretson, T. O. (1994). Anger and hostility: Potential mediators of the gender difference in coronary artery disease. In A. W. Siegman & T. W. Smith (Eds.), *Anger, hostility, and the heart* (pp. 215–238). Hillsdale, NJ: Erlbaum.

Story, M., French, S. A., Resnick, M. D., & Blum, R. W. (1995). Ethnic/racial and socioeconomic differences in dieting behaviors and body image perceptions in adolescents. *International Journal of Eating Disorders, 18,* 173–179.

Strange, C. J. (1996). *Coping with arthritis in its many forms.* Washington, DC: U.S. Government Printing Office. (Reprint from *FDA Consumer Magazine.*)

Strasburger, V. C. (1989). Adolescent sexuality and the media. *Adolescent Gynecology, 36,* 747–773.

Strassberg, Z., Dodge, K., Pettit, G. S., & Bates, J. E. (1994). Spanking in the home and children's subsequent aggression toward kindergarten peers. *Development and Psychopathology, 6,* 445–461.

Stratton, J. R., Levy, W. C., Cereueira, M. D., Schwartz, R. S., & Abrass, I. B. (1994). Cardiovascular responses to exercise: Effects of aging and exercise training in healthy men. *Circulation, 89,* 1648–1655.

Strayer, J. (1993). Children's concordant emotions and cognitions in response to observed emotions. *Child Development, 64,* 188–201.

Streissguth, A. P., Barr, H. M., Sampson, P. D., & Bookstein, F. L. (1994). Prenatal alcohol and offspring development: The first fourteen years. *Drug & Alcohol Dependence, 36,* 89–99.

Streissguth, A. P., Barr, H. M., Sampson, P. D., Darby, B. L., & Martin, D. C. (1989). IQ at age 4 in relation to maternal alcohol use and smoking during pregnancy. *Developmental Psychology, 25,* 3–11.

Streissguth, A. P., Treder, R., Barr, H. M., Shepard, T., Bleyer, W. A., Sampson, P. D., & Martin, D. G. (1987). Aspirin and acetaminophen use by pregnant women and subsequent child IQ and attention decrements. *Teratology, 35,* 211–219.

Streitmatter, J. (1993). Gender differences in identity development: An examination of longitudinal data. *Adolescence, 28,* 55–66.

Streitmatter, J. L., & Pate, G. S. (1989). Identity status development and cognitive prejudice in early adolescents. *Journal of Early Adolescence, 9,* 142–152.

Strober, M., McCracken, J., & Hanna, G. (1990). Affective disorders. In R. M. Lerner, A. C. Petersen, & J. Brooks-Gunn (Eds.), *The encyclopedia of adolescence* (Vol. 1, pp. 18–25). New York: Garland.

Stroebe, W., & Stroebe, M. (1987). *Bereavement and health.* New York: Cambridge University Press.

Stroebe, W., & Stroebe, M. S. (1993). Determinants of adjustment to bereavement in younger widows and widowers. In M. S. Stroebe, W. Stroebe, & R. O. Hansson (Eds.), *Handbook of bereavement* (pp. 208–226). New York: Cambridge University Press.

Stroh, L. K., Brett, J. M., & Reilly, A. H. (1996). Family structure, glass ceiling, and traditional explanations for the differential rate of turnover of female and male managers. *Journal of Vocational Behavior, 49,* 99–118.

Stryker, J., Coates, T. J., DeCarlo, P., Haynes-Sanstad, K., Shriver, M., & Makadon, H. J. (1995). Prevention of HIV infection: Looking back, looking ahead. *Journal of the American Medical Association, 273,* 1143–1148.

Stull, D., & Scarisbrick-Hauser, A. (1989). Never-married elderly. *Research on Aging, 11,* 124–139.

Stunkard, A. J., & Sørensen, T. I. A. (1993). Obesity and socioeconomic status—a complex relation. *New England Journal of Medicine, 329,* 1036–1037.

Stunkard, A. J., Sørenson, T. I. A., Hanis, C., Teasdale, T. W., Chakraborty, R., Schull, W. J., & Schulsinger, F. (1986). An adoption study of human obesity. *New England Journal of Medicine, 314,* 193–198.

Stylianos, S. K., & Vachon, M. L. S. (1993). The role of social support in bereavement. In M. S. Stroebe, W. Stroebe, & R. O. Hansson (Eds.), *Handbook of bereavement* (pp. 397–410). New York: Cambridge University Press.

Subbotsky, E. V. (1994). Early rationality and magical thinking in preschoolers: Space and time. *British Journal of Developmental Psychology, 12,* 97–108.

Suhr, J. N. (1991). *Cruzan v. Director, Missouri Department of Health:* A clear and convincing call for comprehensive legislation to protect incompetent patients' rights. *American University Law Review, 40,* 1477–1519.

Suitor, J. J., Pillemer, K., Keeton, S., & Robison, J. (1995). Aging parents and aging children: Determinants of relationships quality. In R. Blieszner & V. H. Bedford (Eds.), *Handbook of aging and the family* (pp. 223–242). Westport, CT: Greenwood Press.

Sullivan, L. W. (1987). The risks of the sickle-cell trait: Caution and common sense. *New England Journal of Medicine, 317,* 830–831.

Sullivan, M. L. (1993). Culture and class as determinants of out-of-wedlock childbearing and poverty during late adolescence. *Journal of Research on Adolescence, 3,* 295–316.

Sullivan, S. A., & Birch, L. L. (1990). Pass the sugar, pass the salt: Experience dictates preference. *Developmental Psychology, 26,* 546–551.

Sullivan, S. A., & Birch, L. L. (1994). Infant dietary experience and acceptance of solid foods. *Pediatrics, 93,* 271–277.

Suls, J., & Mullen, B. (1982). From the cradle to the grave: Comparison and self-evaluation across the life span. In J. Suls (Ed.), *Psychological perspectives on the self* (Vol. 1, pp. 97–128). Hillsdale, NJ: Erlbaum.

Sulzby, E. (1985). Children's emergent reading of favorite books: A developmental study. *Reading Research Quarterly, 20,* 458–481.

Suominen, H., Heikkinen, E., Parkatti, T., Forsberg, S., & Kiiskinen, A. (1980). Effect of lifelong physical training on functional aging in men. *Scandinavian Journal of the Society of Medicine, 14*(Suppl.), 225–240.

Super, C. M. (1981). Behavioral development in infancy. In R. H. Monroe, R. L. Monroe, & B. B. Whiting (Eds.), *Handbook of cross-cultural human development* (pp. 181–270). New York: Garland.

Super, C. M., & Harkness, S. (1982). The infant's niche in rural Kenya and metropolitan America. In L. L. Adler (Ed.), *Cross-cultural research at issue* (pp. 247–255). New York: Academic Press.

Super, D. (1980). A life-span, life-space approach to career development. *Journal of Vocational Behavior, 16,* 282–298.

Super, D. (1984). Career and life development. In D. Brown & L. Brooks (Eds.), *Career choice and development* (pp. 192–234). San Francisco: Jossey-Bass.

Swain, S. O. (1992). Men's friendships with women: Intimacy, sexual boundaries, and the informant role. In P. M. Nardi (Eds.), *Men's friendships* (pp. 153–171). Newbury Park, CA: Sage.

Swanson, H. S. W. (1993). Donor anonymity in artificial insemination: Is it still necessary? *Columbia Journal of Law and Social Problems, 27,* 151–190.

Swindell, R., & Thompson, J. (1995). An international perspective on the University of the Third Age. *Educational Gerontology, 21,* 429–447.

Symons, D. (1987). An evolutionary approach: Can Darwin's view of life shed light on human sexuality? In J. H. Geer & W. T. O'Donohue (Eds.), *Theories of human sexuality* (pp. 91–126). New York: Plenum.

Szepkouski, G. M., Gauvain, M., & Carberry, M. (1994). The development of planning skills in children with and without mental retardation. *Journal of Applied Developmental Psychology, 15,* 187–206.

Tager-Flusberg, H. (1993). Putting words together: Morphology and syntax in the preschool years. In J. Berko Gleason (Ed.), *The development of language* (pp. 151–193). New York: Macmillan.

Takahashi, K. (1990). Are the key assumptions of the "Strange Situation" procedure universal? A view from Japanese research. *Human Development, 33,* 23–30.

Tamir, L. M. (1980). Men at middle age. In D. G. McGuigan (Ed.), *Women's lives: New theory, research, and policy.* Ann Arbor: University of Michigan Center for Continuing Education of Women.

Tamis-LeMonda, C. S., & Bornstein, M. H. (1989). Habituation and maternal encouragement of attention in infancy as predictors of toddler language, play, and representational competence. *Child Development, 60,* 738–751.

Tamis-Le Monda, C. S., & Bornstein, M. H. (1994). Specificity in mother–toddler language–play relations across the second year. *Developmental Psychology, 30,* 283-292.

Tanner, J. M. (1990). *Foetus into man* (2nd ed.). Cambridge, MA: Harvard University Press.

Taylor, A. R., Asher, S. R., & Williams, G. A. (1987). The social adaptation of mainstreamed mildly retarded children. *Child Development, 58,* 1321–1334.

Taylor, B. J. (1991). A review of epidemiological studies of sudden infant death syndrome in southern New Zealand. *Journal of Paediatric Child Health, 27,* 344–348.

Taylor, M. C., & Hall, J. A. (1982). Psychological androgyny: Theories, methods, and conclusions. *Psychological Bulletin, 92,* 347–366.

Taylor, R. D., & Roberts, D. (1995). Kinship support and maternal and adolescent well-being in economically disadvantaged African-American families. *Child Development, 66,* 1585–1597.

Taylor, R. J. (1985). The extended family as a source of support to elderly blacks. *Gerontologist, 25,* 488–495.

Teberg, A. J., Walther, F. J., & Pena, I. C. (1988). Mortality, morbidity, and outcome of the small-for-gestational-age infant. *Seminar in Perinatology, 12,* 84–94.

Tedder, J. L. (1991). Using the Brazelton neonatal assessment scale to facilitate the parent–infant relationship in a primary care setting. *Nurse Practitioner, 16,* 27–36.

Teikari, J. M., O'Donnell, J. O., Kaprio, J., & Koskenvuo, M. (1991). Impact of heredity in myopia. *Human Heredity, 41,* 151–156.

Tertinger, D. A., Greene, B. F., & Lutzker, J. R. (1984). Home safety: Development and validation of one component of an ecobehavioral treatment program for abused and neglected children. *Journal of Applied Behavior Analysis, 17,* 159–174.

Tetens, J. N. (1777). *Philosophische Versuche über die menschliche Natur und ihre Entwicklung.* Leipzig: Weidmanns Erben & Reich.

Teti, D. M., Gelfand, D. M., Messinger, D. S., & Isabella, R. (1995). Maternal depression and the quality of early attachment: An examination of infants, preschoolers, and their mothers. *Developmental Psychology, 31,* 364–376.

Teyber, E. (1992). *Helping children cope with divorce.* New York: Lexington Books.

Thackwray, D. E., Smith, M. C., Bodfish, J. W., & Meyers, A. W. (1993). A comparison of behavioral and cognitive-behavioral interventions for bulimia nervosa. *Journal of Consulting and Clinical Psychology, 61,* 639–645.

Thapar, A., Gottlesman, I. I., Owen, M. J., O'Donovan, M. C., & McGuffin, P. (1994). The genetics of mental retardation. *British Journal of Psychiatry, 164,* 747–758.

Tharp, R. G. (1993). Institutional and social context of educational practice and reform. In E. A. Forman, N. Minick, & C. A. Stone (Eds.), *Contexts for learning* (pp. 269–282). New York: Oxford University Press.

Tharp, R. G., & Gallimore, R. (1988). *Rousing minds to life: Teaching, learning, and schooling in social context.* Cambridge, England: Cambridge University Press.

Thatcher, R. W. (1991). Maturation of human frontal lobes: Physiological evidence for staging. *Developmental Neuropsychology, 7,* 397–419.

Thatcher, R. W. (1994). Cyclic cortical reorganization: Origins of human cognitive development. G. Dawson & K. W. Fischer (Eds.), *Human behavior and the developing brain* (pp. 232–266). New York: Guilford.

Thatcher, R. W., Walker, R. A., & Giudice, S. (1987). Human cerebral hemispheres develop at different rates and ages. *Science, 236,* 1110–1113.

Theisen, S. C., Mansfield, P. K., Seery, B. L., & Voda, A. (1995). Predictors of midlife women's attitudes toward menopause. *Health Values, 19,* 22–31.

Thelen, E. (1989). The (re)discovery of motor development: Learning new things from an old field. *Developmental Psychology, 25,* 946–949.

Thelen, E. (1995). Motor development: A new synthesis. *American Psychologist, 50,* 79–95.

Thelen, E., & Adolph, K. E. (1992). Arnold Gesell: The paradox of nature and nurture. *Developmental Psychology, 28,* 368–380.

Thelen, E., Fisher, D. M., & Ridley-Johnson, R. (1984). The relationship between physical growth and a newborn reflex. *Infant Behavior and Development, 7,* 479–493.

Thelen, E., & Smith, L. B. (1997). Dynamic systems theories. In R. M. Lerner (Ed.), *Handbook of child psychology: Vol. 1 Theoretical models of human development* (5th ed.). New York: Wiley.

Theorell, K., Prechtl, H., & Vos, J. (1974). A polygraphic study of normal and abnormal newborn infants. *Neuropaediatrie, 5,* 279–317.

Thomas, A., & Chess, S. (1977). *Temperament and development.* New York: Brunner/Mazel.

Thomas, A., Chess, S., & Birch, H. G. (1970, August). The origins of personality. *Scientific American, 223*(2), 102–109.

Thomas, J. L. (1986a). Age and sex differences in perceptions of grandparenthood. *Journal of Gerontology, 41,* 417–423.

Thomas, J. L. (1986b). Gender differences in satisfaction with grandparenting. *Psychology and Aging, 1,* 215–219.

Thomas, J. L. (1989). Gender and perceptions of grandparenthood. *International Journal of Aging and Human Development, 29,* 269–282.

Thomas, J. R. (1984). Children's motor skill development. In J. R. Thomas (Ed.), *Motor development during childhood and adolescence* (pp. 91–104). Minneapolis, MN: Burgess.

Thompson, D. N. (1992). Applications of psychological research for the instruction of elderly adults. In R. L. West & J. D. Sinnott (Eds.), *Everyday memory and aging* (pp. 173–181). New York: Springer-Verlag.

Thompson, R. A. (1990a). On emotion and self-regulation. In R. A. Thompson (Ed.), *Nebraska Symposia on Motivation* (Vol. 36, pp. 383–483). Lincoln: University of Nebraska Press.

Thompson, R. A. (1990b). Vulnerability in research: A developmental perspective on research risk. *Child Development, 61*, 1–16.

Thompson, R. A. (1994). Emotion regulation: A theme in search of a definition. In N. A. Fox (Ed.), The development of emotion regulation: Biological and behavioral considerations. *Monographs of the Society for Research in Child Development, 59*(2–3, Serial No. 240), pp. 25–52.

Thompson, R. A., Lamb, M., & Estes, D. (1982). Stability of infant–mother attachment and its relationship to changing life circumstances in an unselected middle-class sample. *Child Development, 53*, 144–148.

Thompson, R. A., & Limber, S. (1991). "Social anxiety" in infancy: Stranger wariness and separation distress. In H. Leitenberg (Ed.), *Handbook of social and evaluation anxiety* (pp. 85–137). New York: Plenum.

Thompson, R. A., Tinsley, B. R., Scalora, M. J., & Parke, R. D. (1989). Grandparents' visitation rights: Legalizing the ties that bind. *American Psychologist, 44*, 1217–1222.

Thornbury, J. M., & Mistretta, CD. M. (1981). Tactile sensitivity as a function of age. *Journal of Gerontology, 36*, 34–39.

Thorndike, R. L., Hagen, E. P., & Sattler, J. M. (1986). *The Stanford-Binet Intelligence Scale: Guide for administering and scoring* (4th ed.). Chicago: Riverside Publishing.

Thornton, A. (1991). Influence of the marital history of parents on the marital and cohabitational experiences of children. *American Journal of Sociology, 96*, 868–894.

Thornton, A., Axinn, W. G., & Hill, D. H. (1992). Reciprocal effects of religiosity, cohabitation, and marriage. *American Journal of Sociology, 98*, 628–651.

Thorson, J. A., & Powell, F. C. (1991). Constructions of death among those high in intrinsic religious motivation: A factor analytic study. *Death Studies, 15*, 131–138.

Thorson, J. A., & Powell, F. C. (1994). A Revised Death Anxiety Scale. In R. A. Neimeyer (Ed.), *Death anxiety handbook* (pp. 31–43). Washington, DC: Taylor & Francis.

Tinetti, M. E. (1990). Falls. In C. K. Cassel, D. E. Risenberg, L. B. Sorensen, & J. R. Walsh (Eds.), *Geriatric medicine* (2nd ed., pp. 528–534). New York: Springer-Verlag.

Tinetti, M. E., Speechley, M., & Ginter, S. F. (1988). Risk factors for falls among elderly persons living in the community. *New England Journal of Medicine, 319*, 1701–1707.

Tishman, S., Perkins, D. N., & Jay, E. (1995). *The thinking classroom.* Boston: Allyn and Bacon.

Tizard, B., & Hodges, J. (1978). The effect of early institutional rearing on the development of eight-year-old children. *Journal of Child Psychology and Psychiatry, 19*, 99–118.

Tizard, B., & Rees, J. (1975). The effect of early institutional rearing on the behaviour problems and affectional relationships of four-year-old children. *Journal of Child Psychology and Psychiatry, 16*, 61–73.

Tobin, J. J., Wu, D. Y. H., & Davidson, D. H. (1989). *Preschool in three cultures.* New Haven, CT: Yale University Press.

Tobin, S. S. (1989). The effects of institutionalization. In K. S. Markides & C. L. Cooper (Eds.), *Aging, stress and health* (pp. 139–164). Chichester, England: Wiley.

Toda, S., & Fogel, A. (1993). Infant response to the still-face situation at 3 and 6 months. *Developmental Psychology, 29*, 532–538.

Toder, F. A. (1994). *Your kids are grown: Moving on with and without them.* New York: Plenum.

Todres, R., & Bunston, T. (1993). Parent education program evaluation: A review of the literature. *Canadian Journal of Community Mental Health, 12*, 225–257.

Tolman, R. M., & Bennett, L. W. (1990). A review of quantitative research on men who batter. *Journal of Interpersonal Violence, 5*, 87–118.

Tolson, T. F. J., & Wilson, M. N. (1990). The impact of two- and three-generational black family structure on perceived family climate. *Child Development, 61*, 416–428.

Tomasello, M. (1995). Language is not an instinct. *Cognitive Development, 10*, 131–156.

Tomasello, M., & Akhtar, N. (1995). Two-year-olds use pragmatic cues to differentiate reference to objects and actions. *Cognitive Development, 10*, 201–224.

Tomasello, M., & Barton, M. (1994). Learning words in nonostensive contexts. *Developmental Psychology, 30*, 639–650.

Torfs, C. P., Berg, B. van den, Oechsli, F. W., & Cummins, S. (1990). Prenatal and perinatal factors in the etiology of cerebral palsy. *Journal of Pediatrics, 116*, 615–619.

Torrance, E. P. (1980). *The Torrance Tests of Creative Thinking.* New York: Scholastic Testing Service.

Torrance, E. P. (1988). The nature of creativity as manifest in its testing. In R. J. Sternberg (Ed.), *The nature of creativity: Contemporary psychological perspectives* (pp. 43–75). New York: Cambridge University Press.

Torrey, B. B. (1992). Sharing increasing costs on declining income: The visible dilemma of the invisible aged. In R. M. Suzman, D. P. Willis, & K. G. Manton (Eds.), *The oldest old* (pp. 381–393). New York: Oxford University Press.

Touris, M., Kromelow, S., & Harding, C. (1995). Mother-firstborn attachment and the birth of a sibling. *American Journal of Orthopsychiatry, 65*, 293–297.

Touwen, B. C. L. (1984). Primitive reflexes—Conceptual or semantic problem? In H. F. R. Prechtl (Ed.), *Continuity of neural functions from prenatal to postnatal life* (Clinics in Developmental Medicine No. 94, pp. 115–125). Philadelphia: Lippincott.

Tower, R. B., Singer, D. G., Singer, J. L., & Biggs, A. (1979). Differential effects of television programming on preschoolers' cognition, imagination, and social play. *American Journal of Orthopsychiatry, 49*, 265–281.

Trent, W. (1991). *Desegregation analysis report.* New York: Legal Defense and Educational Fund.

Trevethan, S. D., & Walker, L. J. (1989). Hypothetical versus real-life moral reasoning among psychopathic and delinquent youth. *Development and Psychopathology, 1*, 91–103.

Triandis, H. C. (1989). The self and social behavior in differing cultural contexts. *Psychological Review, 96*, 506–520.

Triandis, H. C., Bontempo, R., Villareal, M. J., Asai, M., & Lucca, N. (1988). Individualism and collectivism: Cross-cultural perspectives on self–ingroup relationships. *Journal of Personality and Social Psychology, 54*, 323–338.

Trickett, P. K., Aber, J. L., Carlson, V., & Cicchetti, D. (1991). Relationship of socioeconomic status to the etiology and developmental sequelae of physical child abuse. *Developmental Psychology, 27*, 148–158.

Trinkoff, A., & Parks, P. L. (1993). Prevention strategies for infant walker-related injuries. *Public Health Reports, 108,* 784–788.

Tronick, E. Z. (1989). Emotions and emotional communication in infants. *American Psychologist, 44,* 112–119.

Truitner, K., & Truitner, N. (1993). Death and dying in Buddhism. In D. P. Irish, K. F. Lundquist, & V. J. Nelsen (Eds.), *Ethnic variations in dying, death, and grief* (pp. 125–136). Washington, DC: Taylor & Francis.

Tseng, W., Kuotai, T., Hsu, J., Jinghua, C., Lian, Y., & Kameoka, V. (1988). Family planning and child mental health in China: The Nanjing Survey. *American Journal of Psychiatry, 145,* 1396–1403.

Tudge, J. (1990). Vygotsky, the zone of proximal development, and peer collaboration: Implications for classroom practice. In L. C. Moll (Ed.), *Vygotsky and education* (pp. 155–172). New York: Cambridge University Press.

Tur, E., Yosipovitch, F., & Oren-Vulfs, S. (1992). Chronic and acute effects of cigarette smoking on skin blood flow. *Angiology, 43,* 328–335.

Turiel, E. (1983). *The development of social knowledge: Morality and convention.* New York: Cambridge University Press.

Turiel, E., Smetana, J. G., & Killen, M. (1991). Social contexts in social cognitive development. In W. M. Kurtines & J. L. Gewirtz (Eds.), *Handbook of moral behavior and development* (Vol. 2, pp. 307–332). Hillsdale, NJ: Erlbaum.

Turkheimer, E., & Gottesman, I. I. (1991). Individual differences and the canalization of human behavior. *Developmental Psychology, 27,* 18–22.

Turner, B. F. (1982). Sex-related differences in aging. In B. B. Wolman (Ed.), *Handbook of developmental psychology* (pp. 912–936). Englewood Cliffs, NJ: Prentice Hall.

Turner, M. J., Bailey, W. C., & Scott, J. P. (1994). Factors influencing attitude toward retirement and retirement planning among midlife university employees. *Journal of Applied Gerontology, 13,* 143–156.

Turner, P. J., & Gervai, J. (1995). A multidimensional study of gender typing in preschool children and their parents: Personality, attitudes, preferences, behavior, and cultural differences. *British Journal of Developmental Psychology, 11,* 323–342.

U:S. Bureau of the Census. (1960). *Statistical abstract of the United States* (81st ed.). Washington, DC: U.S. Government Printing Office.

U.S. Bureau of the Census. (1992a). *Marriage, divorce, and remarriage in the 1990s.* Washington, DC: U.S. Government Printing Office.

U.S. Bureau of the Census. (1992b). *Population projections of the United States, by age, sex, race, and Hispanic origin: 1992 to 2050.* Washington, DC: U.S. Government Printing Office.

U.S. Bureau of the Census. (1994). *Marital status and living arrangements: March 1994. Current population reports, P20-484.* Washington, DC: U.S. Government Printing Office.

U.S. Bureau of the Census. (1996). *Statistical abstract of the United States* (118th ed.). Washington, DC: U.S. Government Printing Office.

U.S. Centers for Disease Control. (1992, January 3). Sexual behavior among high school students—United States, 1990. *Morbidity and Mortality Weekly Report, 40,* 885–888.

U.S. Centers for Disease Control. (1996, January). *HIV/AIDS surveillance.* Atlanta, GA: Author.

U.S. Department of Education, National Center for Education Statistics. (1996). *Digest of education statistics 1994.* Washington, DC: U.S. Government Printing Office.

U.S. Department of Health and Human Services. (1994a). *Health United States 1993.* Washington, DC: U.S. Government Printing Office.

U.S. Department of Health and Human Services. (1994b). *Vital statistics of the United States: Vol. 2. Mortality (Part B).* Washington, DC: U.S. Government Printing Office.

U.S. Department of Health and Human Services. National Institute on Drug Abuse (1995). *National survey results on drug use from Monitoring the Future study: Vol. 1. Secondary school students.* Washington, DC: U.S. Government Printing Office.

U.S. Department of Health and Human Services. (1996a). *Advance data from vital and health statistics of the National Center for Health Statistics.* Washington, DC: U.S. Government Printing Office.

U.S. Department of Health and Human Services. (1996b). *Healthy people 2000.* Washington, DC: U.S. Government Printing Office.

U.S. Department of Health and Human Services. (1996c). *Vital Statistics of the United States, 1993.* Washington, DC: U.S. Government Printing Office.

U.S. Department of Justice. (1996). *Crime in the United States.* Washington, DC: U.S. Government Printing Office.

U.S. Department of Labor, Bureau of Labor Statistics. (1996, April). Consumer Price Index. *Monthly Labor Review, 118*(4), 116.

U.S. Public Health Service. (1996). *Vital statistics of the United States: 1996.*

Washington, DC: U.S. Government Printing Office.

U.S. Senate Special Committee on Aging. (1991). *Aging America: Trends and projections.* Washington, DC: U.S. Government Printing Office.

Udry, J. R. (1990). Hormonal and social determinants of adolescent sexual initiation. In J. Bancroft & J. M. Reinisch (Eds.), *Adolescence and puberty* (pp. 70–87). New York: Oxford University Press.

Uhlenberg, P., Cooney, T., & Boyd, R. (1990). Divorce for women after midlife. *Journal of Gerontology, 45,* S3–311.

Ulrich, B. D., & Ulrich, D. A. (1985). The role of balancing in performance of fundamental motor skills in 3-, 4-, and 5-year-old children. In J. E. Clark & J. H. Humphrey (Eds.), *Motor development* (Vol. 1, pp. 87–98). Princeton, NJ: Princeton Books.

Umberson, D., Wortman, C. B., & Kessler, R. C. (1992). Widowhood and depression: Explaining long-term gender differences in vulnerability. *Journal of Health and Social Behavior, 33,* 10–24.

United Nations. (1991a). *Women: Challenges to the year 2000.* New York: Author.

United Nations. (1991b). *World population trends and policies: 1991 monitoring report.* New York: Author.

United Nations. (1991c). *The world's women: 1970–1990.* New York: Author.

United Nations. (1995). *Women in a changing global economy.* New York: Author.

United Nations. (1996). *World social situation in the 1990s.* New York: Author.

Urberg, K. A., Degirmencioglue, S. M., Tolson, J. M., & Halliday-Scher, K. (1995). The structure of adolescent peer networks. *Developmental Psychology, 31,* 540–547.

Vaillant, G. E. (1976). Natural history of male psychological health: V. Relation of choice of ego mechanisms of defense to adult adjustment. *Archives of General Psychiatry, 33,* 535–545.

Vaillant, G. E. (1977). *Adaptation to life.* Boston: Little, Brown.

Vaillant, G. E. (1991). The association of ancestral longevity with successful aging. *Journal of Gerontology, 46,* P292–P298.

Vaillant, G. E. (1994). "Successful aging" and psychosocial well-being. In E. H. Thompson, Jr. (Ed.), *Older men's lives* (pp. 22–41). Thousand Oaks, CA: Sage.

Vaillant, G. E., & Koury, S. H. (1994). Late midlife development. In G. H. Pollock & S. I. Greenspan (Eds.), *The course of life* (pp. 1–22). Madison, CT: International Universities Press.

Vaillant, G. E., & Vaillant, C. O. (1990). Determinants and consequences of creativity in a cohort of gifted women. *Psychology of Women Quarterly, 14,* 607–616.

Valian, V. V. (1993). *Parental replies: Linguistic status and didactic role.* Cambridge, MA: MIT Press.

Vandell, D., & Powers, C. (1983). Day care quality and children's free play activities. *American Journal of Orthopsychiatry, 53,* 293–300.

Vandenberg, J., Johannsson, O., Hakansson, S., Olsson, H., & Borg, A. (1996). Allelic loss at chromosome 13Q12-Q13 is associated with poor prognosis in familial and sporadic breast cancer. *British Journal of Cancer, 74,* 1615–1619.

van den Boom, D. C. (1995). Do first-year intervention effects endure? Follow-up during toddlerhood of a sample of Dutch irritable infants. *Child Development, 66,* 1798–1816.

van den Boom, D. C., & Hoeksma, J. B. (1994). The effect of infant irritability on mother–infant interaction: A growth-curve analysis. *Developmental Psychology, 30,* 581–590.

Van de Perre, P., Simonon, A., Hitimana, D., Davis, F., Msellati, P., Mukamabano, J., Van Goethem, C., Karita, E., & Lepage, P. (1993). Infective and anti-infective properties of breastmilk from HIV-1-infected women. *Lancet, 341,* 914–918.

van der Maas, P. J., van Delders, J. J. M., & Pijnenborg, L. (1991). Euthanasia and other medical decisions concerning the end of life. *Health Policy Monograph.* New York: Elsevier.

Vanfossen, B., Jones, J., & Spade, J. (1987). Curriculum tracking and status maintenance. *Sociology of Education, 60,* 104–122.

van IJzendoorn, M. H. (1995a). Adult attachment representations, parental responsiveness, and infant attachment: A meta-analysis on the predictive validity of the Adult Attachment Interview. *Psychological Bulletin, 117,* 387–403.

van IJzendoorn, M. H. (1995b). Of the way we are: On temperament, attachment, and the transmission gap: A rejoinder to Fox (1995). *Psychological Bulletin, 117,* 411–415.

van IJzendoorn, M. H., & Bakermans-Kranenburg, M. J. (1995). Adult Attachment Interview classifications in mothers, fathers, adolescents, and clinical groups: A meta-analytic search for normative data. *Journal of*

Consulting and Clinical Psychology, 64, 8–21.

van IJzendoorn, M. H., & Kroonenberg, P. M. (1988). Cross-cultural patterns of attachment: A meta-analysis of the Strange Situation. *Child Development, 59,* 147–156.

Vaselle-Augenstein, R., & Ehrlich, A. (1992). The male batterer. In E. C. Viano (Eds.), *Intimate violence: Interdisciplinary perspectives* (pp. 139–154). Washington, DC: Hemisphere.

Vasudev, J., & Hummel, R. C. (1987). Moral stage sequence and principled reasoning in an Indian sample. *Human Development, 30,* 103–118.

Vaughn, B. E., Kopp, C. B., & Krakow, J. B. (1984). The emergence and consolidation of self-control from eighteen to thirty months of age: Normative trends and individual differences. *Child Development, 55,* 990–1004.

Vaughn, B. E., Stevenson-Hinde, J., Waters, E., Kotsaftis, A., Lefever, G. B., Shouldice, A., Trudel, M., & Belsky, J. (1992). Attachment security and temperament in infancy and early childhood: Some conceptual clarifications. *Developmental Psychology, 28,* 463–473.

Vaughn, B. E., & Waters, E. (1990). Attachment behavior at home and in the lab: Q-sort observations and Strange Situation classifications of one-year-olds. *Child Development, 61,* 1965–1973.

Ventura, S. J. (1989). Trends and variations in first births to older women in the United States, 1970–1986. *Vital and Health Statistics* (Series 21). Hyattsville, MD: U.S. Department of Health and Human Services.

Verbrugge, H. P. (1990a). The national immunization program of the Netherlands. *Pediatrics, 86*(6, Pt. 2), 1060–1063.

Verbrugge, H. P. (1990b). Youth health care in the Netherlands: A bird's eye view. *Pediatrics, 86*(6, Pt. 2), 1044–1047.

Verhulst, F. C., & Versluis-Den Bieman, H. J. M. (1995). Developmental course of problem behaviors in adolescent adoptees. *Journal of the American Academy of Child and Adolescent Psychiatry, 34,* 151–159.

Vernon, P. A. (1993). Intelligence and neural efficiency. In D. K. Detterman (Ed.), *Current topics in human intelligence* (Vol. 3, pp. 171–187). Norwood, NJ: Ablex.

Vijg, J., & Gossen, J. A. (1993). Somatic mutations and cellular aging. *Comparative Biochemistry and Physiology, 104,* 429–437.

Vinick, B. H., & Ekerdt, D. J. (1991). Retirement: What happens to husband–wife relationships? *Journal of Geriatric Psychiatry, 24,* 23–40.

Vlassoff, C. (1990). The value of sons in an Indian village: How widows see it. *Population Studies, 44,* 5–20.

Vohr, B. R., & Garcia-Coll, C. T. (1988). Follow-up studies of high-risk low-birth-weight infants: Changing trends. In H. E. Fitzgerald, B. M. Lester, & M. W. Yogman (Eds.), *Theory and research in behavioral pediatrics* (pp. 1–65). New York: Plenum.

Volling, B. L., & Belsky, J. (1992). Contribution of mother–child and father–child relationships to the quality of sibling interaction: A longitudinal study. *Child Development, 63,* 1209–1222.

Vorhees, C. V. (1986). Principles of behavioral teratology. In E. P. Riley & C. V. Vorhees (Eds.), *Handbook of behavioral teratology* (pp. 23–48). New York: Plenum.

Vorhees, C. V., & Mollnow, E. (1987). Behavioral teratogenesis: Long-term influences on behavior from early exposure to environmental agents. In J. D. Osofsky (Ed.), *Handbook of infant development* (2nd ed., pp. 913–971). New York: Wiley.

Vormbrock, J. K. (1993). Attachment theory as applied to wartime and job-related marital separation. *Psychological Bulletin, 114,* 122–144.

Vuchinich, S., Hetherington, E. M., Vuchinich, R. A., & Clingempeel, W. G. (1991). Parent–child interaction and gender differences in early adolescents' adaptation to stepfamilies. *Developmental Psychology, 27,* 618–626.

Vygotsky, L. S. (1978). *Mind in society: The development of higher psychological processes.* Cambridge, MA: Harvard University Press. (Original works published 1930, 1933, and 1935)

Vygotsky, L. S. (1987). Thinking and speech. In R. W. Rieber, & A. S. Carton (Eds.), & N. Minick (Trans.), *The collected works of L. S. Vygotsky: Vol. 1. Problems of general psychology* (pp. 37–285). New York: Plenum. (Original work published 1934.)

Wachs, T. D. (1975). Relation of infants' performance on Piagetian scales between twelve and twenty-four months and their Stanford-Binet performance at thirty-one months. *Child Development, 46,* 929–935.

Wachs, T. D. (1995). Relation of mild-to-moderate malnutrition to human development: Correlational studies. *Journal of Nutrition Supplement, 125,* 22455–22545.

Waddington, C. H. (1957). *The strategy of the genes.* London: Allen & Unwin.

Wagner, R. K., & Sternberg, R. J. (1985). Practical intelligence in real-world pursuits: The role of tacit knowledge. *Journal of Personality and Social Psychology, 49,* 436–458.

Wahl, H.-W. (1991). Dependence in the elderly from an interactional point of view: Verbal and observational data. *Psychology and Aging, 6,* 238–246.

Wakat, D. K. (1978). Physiological factors of race and sex in sport. In L. K. Bunker & R. J. Rotella (Eds.), *Sport psychology: From theory to practice* (pp. 194–209). Charlotte, VA: University of Virginia. (Proceedings of the 1978 Sport Psychology Institute)

Walden, T. A., & Ogan, T. A. (1988). The development of social referencing. *Child Development, 59,* 1230–1240.

Walk, R. D., & Gibson, E. J. (1961). A comparative and analytic study of visual depth perception. *Psychological Monographs, 75*(15, Whole No. 519).

Walker, D., Greenwood, C., Hart, B., & Carta, J. (1994). Prediction of school outcomes based on early language production and socioeconomic factors. *Child Development, 65,* 606–621.

Walker, L. (1995). Sexism in Kohlberg's moral psychology? In W. M. Kurtines & J. L. Gewirtz (Eds.), *Moral development: An introduction* (pp. 83–107). Boston: Allyn and Bacon.

Walker, L. E. (1979). *The battered woman.* New York: Harper & Row.

Walker, L. J. (1989). A longitudinal study of moral reasoning. *Child Development, 60,* 157–166.

Walker, L. J., & Taylor, J. H. (1991a). Family interactions and the development of moral reasoning. *Child Development, 62,* 264–283.

Walker, L. J., & Taylor, J. H. (1991b). Stage transitions in moral reasoning: A longitudinal study of developmental processes. *Developmental Psychology, 27,* 330–337.

Wallace, J. B. (1992). Reconsidering the life review: The social construction of talk about the past. *Gerontologist, 32,* 120–125.

Waller, M. B. (1993, January). Helping crack-affected children succeed. *Educational Leadership, 50*(4), 57–60.

Wallerstein, J. S., & Blakeslee, S. (1995). *The good marriage.* Boston: Houghton Mifflin.

Wallerstein, J., & Corbin, S. B. (1989). Daughters of divorce: Report from a ten-year follow-up. *American Journal of Orthopsychiatry, 59,* 593–604.

Wallerstein, J. S., Corbin, S. G., & Lewis, J. M. (1988). Children of divorce: A ten-year study. In E. M. Hetherington & J.

Arasteh (Eds.), *Impact of divorce, single parenting, and stepparenting on children* (pp. 198–214). Hillsdale, NJ: Erlbaum.

Wallerstein, J. S., & Kelly, J. B. (1980). *Surviving the break-up: How children and parents cope with divorce.* New York: Basic Books.

Walls, C. T., & Zarit, S. H. (1991). Informal support from black churches and the well-being of elderly blacks. *Gerontologist, 31,* 490–495.

Wannamethee, G., Shaper, A. G., & Macfarlane, P. W. (1993). Heart rate, physical activity, and mortality from cancer and other noncardiovascular diseases. *American Journal of Epidemiology, 137,* 735–748.

Ward, R. A., LaGory, M., Sherman, S., & Traynor, D. (1981). *Neighborhood age structure and support networks.* Paper presented at the annual meeting of the Gerontological Society of America, Toronto.

Warr, P. B. (1987). *Work, unemployment and mental health.* Oxford, England: Clarendon Press.

Warr, P. B. (1992). Age and occupational well-being. *Psychology and Aging, 7,* 37–45.

Warr, P. B. (1994). Age and employment. In M. D. Dunnette, L. Hough, & H. Triandis (Eds.), *Handbook of industrial and organizational psychology* (pp. 485–550). Palo Alto, CA: Consulting Psychologists Press.

Warren, A. R., & Tate, C. S. (1992). Egocentrism in children's telephone conversations. In R. M. Diaz & L. E. Berk (Eds.), *Private speech: From social interaction to self-regulation* (pp. 245–264). Hillsdale, NJ: Erlbaum.

Wass, H. (1991). Helping children cope with death. In D. Papadatou & C. Papadatos (Eds.), *Children and death* (pp. 11–32). New York: Hemisphere.

Wass, H. (1995). Death in the lives of children and adolescents. In H. Wass & R. A. Neimeyer (Eds.), *Dying: Facing the facts* (3rd ed., pp. 269–301). Washington, DC: Taylor & Francis.

Wasserman, G., Graziano, J. H., Factor-Litvac, P., Popovac, D., Morina, N., & Musabegovic, A. (1992). Independent effects of lead exposure and iron deficiency anemia on developmental outcome at age 2 years. *Journal of Pediatrics, 121,* 695–703.

Waterman, A. S. (1985). Identity in context of adolescent psychology. In A. S. Waterman (Ed.), *New directions for child development* (No. 30, pp. 5–24). San Francisco: Jossey-Bass.

Waterman, A. S. (1989). Curricula interventions for identity change: Substantive and ethical considera-

tions. *Journal of Adolescence, 12,* 389–400.

Waterman, A. S., & Whitbourne, S. K. (1982). Androgyny and psychosocial development among college students and adults. *Journal of Personality, 50,* 121–133.

Watson, D. J. (1989). Defining and describing whole language. *Elementary School Journal, 90,* 129–141.

Watson, J. B., & Raynor, R. (1920). Conditioned emotional reactions. *Journal of Experimental Psychology, 3,* 1–14.

Watson, W. H. (1991). Ethnicity, crime, and aging: Risk factors and adaptation. *Generations, 15,* 53–57.

Waxman, S. R., & Hall, D. G. (1993). The development of a linkage between count nouns and object categories: Evidence from fifteen- to twenty-one-month-old infants. *Child Development, 64,* 1224–1241.

Waxman, S. R., & Hatch, T. (1992). Beyond the basics: Preschool children label objects flexibly at multiple hierarchical levels. *Journal of Child Language, 19,* 153–166.

Waxman, S. R., & Senghas, A. (1992). Relations among word meanings in early lexical development. *Developmental Psychology, 28,* 862–873.

Weale, R. A. (1992). *The senescence of human vision.* New York: Oxford University Press.

Wechsler, D. (1989). *Manual for the Wechsler Preschool and Primary Scale of Intelligence–Revised.* New York: Psychological Corporation.

Wechsler, D. (1991). *Manual for the Wechsler Intelligence Test for Children–III.* New York: Psychological Corporation.

Wegman, M. E. (1994). Annual summary of vital statistics—1993. *Pediatrics, 94,* 792–803.

Wehren, A., DeLisi, R., & Arnold, M. (1981). The development of noun definition. *Journal of Child Language, 8,* 165–175.

Weiland, S. (1993). Erik Erikson: Ages, stages, and stories. *Generations, 17*(2), 17–22.

Weinberg, M. K., & Tronick, E. Z. (1994). Beyond the face: An empirical study of infant affective configurations of facial, vocal, gestural, and regulatory behaviors. *Child Development, 65,* 1503–1515.

Weinberg, R. A., Scarr, S., & Waldman, I. D. (1992). The Minnesota transracial adoption study: A follow-up of IQ test performance at adolescence. *Intelligence, 16,* 117–135.

Weingarten, H. R. (1988). Late life divorce and the life review. *Journal of Gerontological Social Work, 12*(3–4), 83–97.

Weingarten, H. R. (1989). The impact of late life divorce: A conceptual and empirical study. *Journal of Divorce, 12,* 21–38.

Weinstein, R. S., Marshall, H. H., Sharp, L., & Botkin, M. (1987). Pygmalion and the student: Age and classroom differences in children's awareness of teacher expectations. *Child Development, 58,* 1079–1093.

Weisman, A. D. (1984). *The coping capacity: On the nature of being mortal.* New York: Sciences Press.

Weisner, T. S. (1993). Ethnographic and ecocultural perspectives on sibling relationships. In Z. Stoneman & P. W. Berman (Eds.), *The effects of mental retardation, disability, and illness on sibling relationships* (pp. 51–83). Baltimore: Paul H. Brookes.

Weisner, T. S., & Wilson-Mitchell, J. E. (1990). Nonconventional family lifestyles and sex typing in six-year-olds. *Child Development, 61,* 1915–1933.

Weiss, G., & Hechtman, L. T. (1993). *Hyperactive children grown up* (2nd ed.). New York: Guilford.

Weitzman, M., Gortmaker, S., & Sobol, A. (1990). Racial, social, and environmental risks for childhood asthma. *American Journal of Diseases of Children, 144,* 1189–1194.

Weksler, M. E. (1995). Immune senescence: Deficiency or dysregulation. *Nutrition Reviews, 53,* S3–S7.

Wellings, K., Field, J., Johnson, A., & Wadsworth, J. (1994). *Sexual behavior in Britain: The National Survey of Sexual Attitudes and Lifestyles.* New York: Penguin.

Wellman, B. (1992). Men in networks: Private communities, domestic friendships. In P. Nardi (Ed.), *Men's friendships* (pp. 74–114). Newbury Park, CA: Sage.

Wellman, H. M. (1988). The early development of memory strategies. In F. F. Weinert & M. Perlmutter (Eds.), *Memory development: Universal changes and individual differences* (pp. 3–29). Hillsdale, NJ: Erlbaum.

Wellman, H. M. (1990). *The child's theory of mind.* Cambridge, MA: MIT Press.

Wellman, H. M., & Hickling, A. K. (1994). The mind's "I": Children's conception of the mind as an active agent. *Child Development, 65,* 1564–1580.

Wellman, H. M., Somerville, S. C., & Haake, R. J. (1979). Development of search procedures in real-life spatial environments. *Developmental Psychology, 15,* 530–542.

Wells, A. S., & Crain, R. L. (1994). Perpetuation theory and the long-term effects of school desegregation. *Review of Educational Research, 64,* 531–555.

Wells, A. S., Crain, R. L., & Uchetelle, S. (1995). *Stepping over the color line: Black inner-city students in suburban schools.* New Haven: Yale University Press.

Wenestam, C., & Wass, H. (1987). Swedish and U.S. children's thinking about death: A qualitative study and cross-cultural comparison. *Death Studies, 11,* 99–121.

Wentzel, K. R., & Asher, S. R. (1995). The academic lives of neglected, rejected, popular, and controversial children. *Child Development, 66,* 754–763.

Werner, E. E. (1989, April). Children of the garden island. *Scientific American, 260* (4), 106–111.

Werner, E. E. (1991). Grandparent–grandchild relationships amongst U.S. ethnic groups. In P. K. Smith (Ed.), *The psychology of grandparenthood: An international perspective* (pp. 68–82). London: Routledge.

Werner, E. E., & Smith, R. S. (1992). *Overcoming the odds: High risk children from birth to adulthood.* Ithaca, NY: Cornell University Press.

Werner, S. J., & Siqueland, E. R. (1978). Visual recognition memory in the preterm infant. *Infant Behavior and Development, 1,* 79–94.

Wertsch, J. V., & Tulviste, P. (1992). L. S. Vygotsky and contemporary developmental psychology. *Developmental Psychology, 28,* 548–557.

Weyerer, S., & Kupfer, B. (1994). Physical exercise and psychological health. *Sports Medicine, 17,* 108–116.

Whalen, C. K., & Henker, B. (1991). Therapies for hyperactive children: Comparisons, combinations, and compromises. *Journal of Consulting and Clinical Psychology, 59,* 126–137.

Whalen, C. K., Henker, B., Burgess, S., & O'Neil, R. (1995). Young people talk about AIDS: "When you get sick, you stay sick." *Journal of Clinical Child Psychology, 24,* 338–345.

Wheaton, B. (1990). Life transitions, role, histories, and mental health. *American Sociological Review, 55,* 209–223.

Wheeler, M. D. (1991). Physical changes of puberty. *Endocrinology and Metabolism Clinics of North America, 20,* 1–14.

Whitbeck, L., Hoyt, D. R., & Huck, S. M. (1994). Early family relationships, intergenerational solidarity, and support provided to parents by their adult children. *Journal of Gerontology, 49,* 585–594.

Whitbourne, S. K. (1985). *The aging body.* New York: Springer.

Whitbourne, S. K. (1996). *The aging individual: Physical and psychological perspectives.* New York: Springer.

Whitbourne, S. K., & Primus, L. (1996). Identity, physical. In J. E. Birren (Ed.), *Encyclopedia of aging* (pp. 733–742). San Diego: Academic Press.

Whitbourne, S. K., & Weinstock, C. S. (1979). *Adult development: The differentiation of experience.* New York: Holt, Rinehart & Winston.

Whitbourne, S. K., Zuschlag, M. K., Elliot, L. B., & Waterman, A. S. (1992). Psychosocial development in adulthood: A 22-year sequential study. *Journal of Personality and Social Psychology, 63,* 260–271.

White, B., & Held, R. (1966). Plasticity of sensorimotor development in the human infant. In J. F. Rosenblith & W. Allinsmith (Eds.), *The causes of behavior* (pp. 60–70). Boston: Allyn and Bacon.

White, K. R., Taylor, M. J., & Moss, V. D. (1992). Does research support claims about the benefits of involving parents in early intervention programs? *Review of Educational Research, 62,* 91–125.

White, L. K. (1994). Coresidence and leaving home: Young adults and their parents. *Annual Review of Sociology, 20,* 81–102.

White, L. K., & Edwards, J. N. (1990). Emptying the nest and parental well-being: An analysis of national panel data. *American Sociological Review, 55,* 235–242.

White, L. K., & Kim, H. (1987). The family-building process: Childbearing choices by parity. *Journal of Marriage and the Family, 49,* 271–279.

White, L. K., & Riedmann, A. (1992). Ties among adult siblings. *Social Forces, 71,* 85–102.

White, N., & Cunningham, W. R. (1988). Is terminal drop pervasive or specific? *Journal of Gerontology, 43,* P141–P144.

White, S. H. (1992). G. Stanley Hall: From philosophy to developmental psychology. *Developmental Psychology, 28,* 25–34.

Whitehurst, G., & Vasta, R. (1975). Is language acquired through imitation? *Journal of Psycholinguistic Research, 4,* 37–59.

Whitehurst, G. J. (1982). Language development. In B. B. Wolman (Ed.), *Handbook of developmental psychology* (pp. 367–386). New York: Wiley.

Whitehurst, G. J., Arnold, D. S., Epstein, J. N., Angell, A. L., Smith, M., & Fischel, J. E. (1994). A picture book reading intervention in day care and home for children from low-income families. *Developmental Psychology, 30,* 679–689.

Whiting, B., & Edwards, C. P. (1988a). *Children in different worlds.* Cambridge, MA: Harvard University Press.

Whiting, B., & Edwards, C. P. (1988b). A cross-cultural analysis of sex differences in the behavior of children aged 3 through 11. In G. Handel (Ed.), *Childhood socialization* (pp. 281–297). New York: Aldine de Gruyter.

Whitley, B. E., Jr., & Kite, M. E. (1995). Sex differences in attitudes toward homosexuality: A comment on Oliver and Hyde (1993). *Psychological Bulletin, 117*, 146–154.

Whitney, M. P., & Thoman, E. B. (1994). Sleep in premature and full-term infants from 24-hour home recordings. *Infant Behavior and Development, 17*, 223–234.

Wichmann, S., & Martin, D. R. (1992). Heart disease; Not for men only. *The Physician and Sports Medicine, 20*, 138–148.

Wiebe, D. J., & Williams, P. G. (1992). Hardiness and health: A social psychophysiological perspective on stress and adaptation. *Journal of Social and Clinical Psychology, 11*, 238–262.

Wigfield, R. E., Fleming, P. J., Berry, P. J., Rudd, P. T., & Golding, J. (1992). Can the fall in Avon's sudden infant death rate be explained by changes in sleeping position? *British Medical Journal, 304*, 282–283.

Wilcox, A. J., Weinberg, C. R., & Baird, D. D. (1995). Timing of sexual intercourse in relation to ovulation: Effects on the probability of conception, survival of the pregnancy, and sex of the baby. *New England Journal of Medicine, 333*, 1517–1519.

Wilcox, A. J., & Skjoerven, R. (1992). Birthweight and perinatal mortality: The effect of gestational age. *American Journal of Public Health, 83*, 378–382.

Wilensky, H. L. (1983). Evaluating research and politics: Political legitimacy and consensus as missing variables in the assessment of social policy. In E. Spiro & E. Yuchtman-Yaar (Eds.), *Evaluating the welfare state: Social and political perspectives* (pp. 51–74). New York: Academic Press.

Wilke, C. J., & Thompson, C. A. (1993). First-year reentry women's perceptions of their classroom experiences. *Journal of the Freshman Year Experience, 5*, 69–90.

Wille, D. E. (1991). Relation of preterm birth with quality of infant–mother attachment at one year. *Infant Behavior and Development, 14*, 227–240.

Willer, B., Hofferth, S. L., Kisker, E. E., Divine-Hawkins, P., Farquhar, E., & Glantz, F. B. (1991). *The demand and supply of child care in 1990: Joint findings from the National Child Care Survey 1990 and A Profile of Child Care Settings.* Washington, DC: National

Association for the Education of Young Children.

Williams, B. C., & Kotch, J. B. (1990). Excess injury mortality among children in the United States: Comparison of recent international statistics. *Pediatrics, 86*(6, Pt. 2), 1067–1073.

Williams, C. S., Buss, K. A., & Eskenazi, B. (1992). Infant resuscitation is associated with an increased risk of left-handedness. *American Journal of Epidemiology, 136*, 277–286.

Williams, E., & Radin, N. (1993). Paternal involvement, maternal employment, and adolescents' academic achievement: An 11-year follow-up. *American Journal of Orthopsychiatry, 63*, 306–312.

Williams, E., Radin, N., & Allegro, T. (1992). Sex-role attitudes of adolescents reared primarily by their fathers: An 11-year follow-up. *Merrill-Palmer Quarterly, 38*, 457–476.

Williams, F. T. (1992). Aging versus disease. *Generations, 16*(4), 21–25.

Williams, P. G., Wiebe, D. J., & Smith, T. W. (1992). Coping processes as mediators of the relationship between hardiness and health. *Journal of Behavioral Medicine, 20*, 138–148.

Williamson, D. F., Kahn, H. S., Remington, P. L., & Anda, R. F. (1990). The 10-year incidence of overweight and major weight gain in U.S. adults. *Archives of Internal Medicine, 150*, 665–672.

Willinger, M. (1995). Sleep position and sudden infant death syndrome. *Journal of the American Medical Association, 273*, 818–819.

Willis, S. (1996). Everyday problem solving. In J. E. Birren & K. W. Schaie (Eds.), *Handbook of the psychology of aging* (4th ed., pp. 287–307). San Diego: Academic Press.

Willis, S. L., & Schaie, K. W. (1994). Cognitive training in the normal elderly. In F. Forette, Y. Christen, & F. Boller (Eds.), *Plasticité cérébrale et stimulation cognitive* [Cerebral plasticity and cognitive stimulation] (pp. 91–113). Paris: Fondation Nationale de Gérontologie.

Wilson, M. N. (1986). The black extended family: An analytical consideration. *Developmental Psychology, 22*, 246–258.

Wilson, R., & Cairns, E. (1988). Sex-role attributes, perceived competence and the development of depression in adolescence. *Journal of Child Psychology and Psychiatry, 29*, 635–650.

Wilson, W. J. (1991). Studying inner-city social dislocations: The challenge of public agenda research. *American Sociological Review, 56*, 1–14.

Windle, M. A. (1994). A study of friendship characteristics and problem behaviors among middle adolescents. *Child Development, 65*, 1764–1777.

Wingfield, A., & Stine, E. A. L. (1992). Age differences in perceptual processing and memory for spoken language. In R. L. West & J. D. Sinnott (Eds.), *Everyday memory and aging* (pp. 101–123). New York: Springer-Verlag.

Wink, P., & Helson, R. (1993). Personality change in women and their partners. *Journal of Personality and Social Psychology, 65*, 597–605.

Winn, S., Roker, D., & Coleman, J. (1995). Knowledge about puberty and sexual development in 11–16 year olds: Implications for health and sex education in schools. *Educational Studies, 21*, 187–201.

Winner, E. (1986, August). Where pelicans kiss seals. *Psychology Today, 20*(8), 25–35.

Winner, E. (1988). *The point of words: Children's understanding of metaphor and irony.* Cambridge, MA: Harvard University Press.

Wintre, M. G., & Vallance, D. D. (1994). A developmental sequence in the comprehension of emotions: Intensity, multiple emotions, and valence. *Developmental Psychology, 30*, 509–514.

Witelson, S. F., & Kigar, D. L. (1988). Anatomical development of the corpus callosum in humans: A review with reference to sex and cognition. In D. L. Molfese & S. J. Segalowitz (Eds.), *Brain lateralization in children* (pp. 35–57). New York: Guilford Press.

Wolf, A. W., Jimenez, E., & Lozoff, B. (1994). No evidence of developmental III effects of low-level lead exposure in a developing country. *Developmental and Behavioral Pediatrics, 15*, 224–231.

Wolf, R., & Pillemer, K. (1994). What's new in elder abuse programming: Four bright ideas. *Gerontologist, 34*(1), 126–129.

Wolf, R. M. (1993, November). The National Assessment of Educational Progress: The Nation's Report Card. *NASSP Bulletin, 77*(556), 36–45.

Wolff, P. H. (1966). The causes, controls and organization of behavior in the neonate. *Psychological Issues, 5*(1, Serial No. 17).

Wolfner, G., Faust, D., & Dawes, R. (1993). The use of anatomical dolls in sexual abuse evaluations: The state of the science. *Applied and Preventive Psychology, 2*, 1–11.

Wood, D. J. (1989). Social interaction as tutoring. In M. H. Bornstein & J. S. Bruner (Eds.), *Interaction in human development* (pp. 59–80). Hillsdale, NJ: Erlbaum.

Woolley, J. D., & Wellman, H. M. (1990). Young children's understanding of realities, nonrealities, and appearances. *Child Development, 61,* 946–961.

World Health Organization. (1991). *World health statistics annual 1990.* Geneva: Author.

Worobey, J. L., & Angel, R. J. (1990). Functional capacity and living arrangements of unmarried elderly persons. *Journal of Gerontology, 45,* S95–S101.

Wright, C. L., & Maxwell, J. W. (1991). Social support during adjustment to later-life divorce: How adult children help parents. *Journal of Divorce & Remarriage, 15,* 21–48.

Wright, J. C., Huston, A. C., Reitz, A. L., & Piemyat, S. (1994). Young children's perceptions of television reality: Determinants and developmental differences. *Developmental Psychology, 30,* 229–239.

Wright, P. H. (1982). Men's friendships, women's friendships and the alleged inferiority of the latter. *Sex Roles, 8,* 1–20.

Wrightsman, L. S. (1988). *Personality development in adulthood.* Newbury Park, CA: Sage.

Wrightsman, L. S. (1994). *Adult personality development: Vol. 1. Theories and concepts.* Thousand Oaks, CA: Sage.

Wykle, M. L., & Musil, C. M. (1993). Mental health of older persons: Social and cultural factors. *Generations, 17*(1), 7–12.

Yang, B., Ollendick, T. H., Dong, Q., Xia, Y., & Lin, L. (1995). Only children and children with siblings in the People's Republic of China: Levels of fear, anxiety, and depression. *Child Development, 66,* 1301–1311.

Yarrow, M. R., Scott, P. M., & Waxler, C. Z. (1973). Learning concern for others. *Developmental Psychology, 8,* 240–260.

Yazigi, R. A., Odem, R. R., & Polakoski, K. L. (1991). Demonstration of specific binding of cocaine to human spermatozoa. *Journal of the American Medical Association, 266,* 1956–1959.

Yip, R., Scanlon, K., & Trowbridge, F. (1993). Trends and patterns in height and weight status of low-income U.S. children. *Critical Reviews in Food Science and Nutrition, 33,* 409–421.

Yogman, M. W. (1981). Development of the father–infant relationship. In H. Fitzgerald, B. Lester, & M. W. Yogman (Eds.), *Theory and research in behavioral pediatrics* (Vol. 1, pp. 221–279). New York: Plenum.

Yonas, A., Granrud, E. C., Arterberry, M. E., & Hanson, B. L. (1986). Infants' distance perception from linear per-spective and texture gradients. *Infant Behavior and Development, 9,* 247–256.

Yonas, A., & Hartman, B. (1993). Perceiving the affordance of contact in four- and five-month-old infants. *Child Development, 64,* 298–308.

Young, K. T. (1990). American conceptions of infant development from 1955 to 1984: What the experts are telling parents. *Child Development, 61,* 17–28.

Young, P. (1991). Families with adolescents. In F. H. Brown (Ed.), *Reweaving the family tapestry* (pp. 131–168). New York: Norton.

Younger, B. A. (1985). The segregation of items into categories by ten-month-old infants. *Child Development, 56,* 1574–1583.

Younger, B. A. (1993). Understanding category members as "the same sort of thing": Explicit categorization in ten-month infants. *Child Development, 64,* 309–320.

Youniss, J. (1980). *Parents and peers in social development: A Piagetian-Sullivan perspective.* Chicago: University of Chicago Press.

Yuill, N., & Perner, J. (1988). Intentionality and knowledge in children's judgments of actor's responsibility and recipient's emotional reaction. *Developmental Psychology, 24,* 358–365.

Zabin, L. S., & Hayward, S. C. (1993). *Adolescent sexual behavior and childbearing.* Newbury Park, CA: Sage.

Zahn-Waxler, C., Kochanska, G., Krupnick, J., & McKnew, D. (1990). Patterns of guilt in children of depressed and well mothers. *Developmental Psychology, 26,* 51–59.

Zahn-Waxler, C., Radke-Yarrow, M., & King, R. M. (1979). Child-rearing and children's prosocial initiations toward victims of distress. *Child Development, 50,* 319–330.

Zahn-Waxler, C., Radke-Yarrow, M., Wagner, E., & Chapman, M. (1992). Development of concern for others. *Developmental Psychology, 28,* 126–136.

Zahn-Waxler, C., & Robinson, J. (1995). Empathy and guilt: Early origins of feelings of responsibility. In J. P. Tangney & K. W. Fischer (Eds.), *Self-conscious emotions* (pp. 143–173). New York: Guilford.

Zajonc, R. B., & Mullally, P. R. (1997). Birth order: Reconciling conflicting effects. *American Psychologist, 52,* 685–699.

Zametkin, A. J. (1995). Attention-deficit disorder: Born to be hyperactive? *Journal of the American Medical Association, 273,* 1871–1874.

Zani, B. (1993). Dating and interpersonal relationships in adolescence. In S. Jackson & H. Rodriguez-Tomé (Eds.), *Adolescence and its social worlds* (pp. 95–119). Hillsdale, NJ: Erlbaum.

Zedeck, S., Maslach, C., Mosier, K., & Skitka, L. (1988). Affective response to work and quality of family life: Employee and spouse perspectives. *Journal of Social Behavior and Personality, 3,* 135–157.

Zelazo, N. A., Zelazo, P. R., Cohen, K. M., & Zelazo, P. D. (1993). Specificity of practice effects on elementary neuromotor patterns. *Developmental Psychology, 29,* 686–691.

Zelazo, P. R. (1983). The development of walking: New findings on old assumptions. *Journal of Motor Behavior, 2,* 99–137.

Zenger, T., & Lawrence, B. (1989). Organizational demography: The differential effects of age and tenure distributions on technical communication. *Academy of Management Journal, 32,* 353–376.

Zevitz, R. G., & Gurnack, A. M. (1991). Factors related to elderly crime victims' satisfaction with police service: The impact of Milwaukee's Gray Squad. *Gerontologist, 31,* 92–101.

Zhang, J., Cai, W., & Lee, D. J. (1992). Occupational hazards and pregnancy outcomes. *American Journal of Industrial Medicine, 21,* 397–408.

Ziegler, C. B., Dusek, J. B., & Carter, D. B. (1984). Self-concept and sex-role orientation: An investigation of multidimensional aspects of personality development in adolescence. *Journal of Early Adolescence, 4,* 25–39.

Zigler, E., & Hall, N. W. (1989). Physical child abuse in America: Past, present, and future. In D. Cicchetti & V. Carlson (Eds.), *Child maltreatment* (pp. 203–253). New York: Cambridge University Press.

Zigler, E., & Styfco, S. J. (1994a). Head Start: Criticisms in a constructive context. *American Psychologist, 49,* 127–132.

Zigler, E., & Styfco, S. J. (1994b). Is the Perry Preschool better than Head Start? Yes and no. *Early Childhood Research Quarterly, 9,* 269–287.

Zillman, D., Bryant, J., & Huston, A. C. (1994). *Media, family, and children.* Hillsdale, NJ: Erlbaum.

Zimmerman, B. J. (1990). Self-regulation learning and academic achievement: An overview. *Educational Psychologist, 25,* 3–18.

Zorilla, E. P., DeRubeis, R. J., & Redei, E. (1995). High self-esteem, hardiness and affective stability are associated with higher basal pituitary–adrenal hormone levels. *Psychoneuroendocrinology, 20,* 591–601.

Important Terms and Concepts are indicated in **boldface type**.
Boldface numbers indicate the pages where they are defined.

Nyansongo of Kenya, gender typing among, 329-330

O

Oakland Growth Study, 34-35
Obesity, 124-125
  arthritis and, 567
  causes of, 280-281, 426
  consequences of, 281, 426
  defined, 280
  diabetes and, 567
  early adulthood, 426-427
  immune response and, 560
  incidence of, 280
  information resources, 448
  middle childhood, 280-281
  strokes and, 571
  treatment of, 281-282, 426
Object labeling, in vocabulary development, 238
Object permanence, 149
  intelligence predictions based on tasks of, 160
  language development and, 167
  research on, 151-152
  sensorimotor stage of cognitive development, 149-152
Object relations theory, symbiosis and separation-individuation, 177-178
Object sorting, 155
  preoperational stage of cognitive development, 225
Object words, in language development, 238
Object-hiding tasks, in sensorimotor stage of cognitive development, 149
Objects
  attachment objects, 187
  categorization. See Categorization abilities
  conservation principle. See Conservation
  self-concept and, 247-248
  seriation, 290
Observation research methods, 28-29
Observational learning. See Modeling
Occupations. See Employment; Job entries; Vocational entries
Odor. See Smell
Oedipus conflict, 16, 246-247, 254
Older adults. See Late adulthood
Old-old elderly, **552**
One-child families, 331-333
Only children, 331-333
On-the-job training, 540. See also Work-study programs
Open classrooms, 306
Openness to experience, as personality trait, 528
Operant conditioning, 18, **132**-133
  language development and, 164
  memory studies, 154-155
  moral development and, 255
Opposite-sex friendships, 460-461
Optimism. See also Hardiness
  health perceptions in late adulthood, 563
  successful aging and, 615
Oral stage of development (Freud), 16, **177**
Organ donation, 625
Organismic theories, 6, 8
  Rousseau's "noble savage," 12
  stance of major developmental theories, 27

Organization
  cognitive-developmental theory, **147**
  memory strategy, 293, 506
Organized games with rules, 287
Organs. See also specific entries (e.g., Cardiovascular system; Respiratory system)
  aging of, 420
  prenatal development of, 80
Osteoarthritis, 566-567
Osteoporosis, 493, **496**-498, 566
  cognitive deterioration and, 579
  information resources, 515
Other-sex friendships, 460-461
Out-of-wedlock births, 475
  illiteracy and, 523
  teenage pregnancies, 361-362
Ovaries, 46. See also Reproductive system
Overextension, in language development, **167**
Overregularization, in grammar development, **238**
Over-the-counter drugs. See Medications
Overview of lifespan, 10-11. See also specific periods (e.g., Prenatal development; Early childhood; Middle adulthood)
Overweight. See Obesity; Weight
Ovulation, 46. See also Menarche; Menstrual cycle
Ovum, 45-46
  aging process, 52-53
  donor banks, 56-57
Oxygenation, aging process and, 559

P

Pacific Islanders. See also Hawaiians
  childbirth practices, 96
  child-rearing practices, 267
  sibling relationships among, 537
Pain, newborns' response to, 105
Pain-relievers. See Analgesics
Palmar grasp reflex, 103-104, 130
Parallel play, 251-252
Parent education, 470
  information resources, 483
Parent-adolescent relationship, 400-401, 470, 531. See also Child-rearing practices; Families; Home environment
  abstract thinking, handling consequences of, 372-373
  anorexia nervosa and, 359
  conflict in, 355-356
  extended family households and, 65
  identity development and, 391, 393
Parent-adult-child relationship
  aging parents. See Aging parents
  early adulthood, 462-463
  homosexuality and, 476
  leaving home, 462-463, 531
  middle adulthood, 531, 538
  promoting positive ties with adult children, 532
  return home, of young adults, 463
Parental imperative theory, 527
Parental leave, 337
Parent-child relationship, 469-470. See also Child-rearing practices; Families; Father-child relationship; Home environment; Mother-child

relationship
  cultural influences on, 264-265
  divorce and, 332, 334
  extended family households and, 65
  make-believe play and, 159
  middle childhood, 330-331
  only children, 332-333
  schools, involvement with, 311-312
  stepparents, 474-475
Parenthood, 467-470. See also Child-rearing practices; Families; Father entries; Mother entries
  additional births, 469
  adjustment to, 468-469
  advantages and disadvantages, 468, 470
  bereaved parents, 643-644
  costs of child rearing, 468
  decision to have children, 467-468
  gender-role identity and, 527
  generativity and, 518, 520
  homosexual parents, 475-476
  information resources, 483
  middle adulthood, 529
  never-married parents, 475
  single parenting. See Single-parent families
  teenage parenting. See Teenage pregnancy and parenthood
  variant styles of, 474-476
Parent-infant relationship. See also Child-rearing practices; Families; Father-infant relationship; Home environment; Mother-infant relationship
  attachment, 191-194. See also Attachment
  cosleeping arrangements, 122
Parenting practices. See Child-rearing practices
Parent-in-law relationships, 531
Parents Anonymous, 270, 273
Parents magazine, 14
Parent-teacher interactions, 63
Parkinson's disease, 568
Part-time work
  adolescents, 378-380
  maternal employment, 337
Passionate love, 457-459
Passive correlation, 72-73
Passive euthanasia, 627-629
Paternal age
  chromosomal abnormalities and, 53
  delay in parenthood, 469
Paternal relationship. See Father entries; Parent entries
Paternity leave, 337
Pattern perception, 138-140
Pavlov's theory of classical conditioning, 18, 131-132
PCBs, prenatal development and, 86
PCP. See also Drug use and abuse
  teenage use, 365
Peck's theory of psychosocial development, 589
Pedestrian accidents, in late adulthood, 567
Pedigrees, in genetic counseling, 54
Peer acceptance, 327-328
  delinquent behavior and, 407
Peer culture, 326
Peer groups, 336
  adolescence, 402
  delinquent behavior and, 407
  middle childhood, 326-327

Peer pressure, 403
Peer relations. See also Friendships; Peer groups; Play
  academic achievement in adolescence and, 377-378
  ADHD and, 294
  adolescence, 401-403
  adult learners, 511
  aggression and, 258
  child maltreatment and, 269-270
  cross-gender behavior and, 329
  early adulthood, 457-462
  early childhood, 246, 251-253
  ethnic identity and, 392
  exercise groups, 500
  gender typing and, 261-262
  intergenerational relationships, 537-538, 581
  late adulthood, 615
  mainstreaming of students with learning difficulties and, 309
  maternal employment and, 336
  maturational timing and, 356-357
  middle childhood, 326-328
  moral development and, 256-257, 325, 397
  self-concept and, 319
  self-esteem and, 319-320
  sexual abuse victims, 341
  substance abuse and, 366
Pendulum problem, and abstract thinking, 369-371
Penis. See Genitals; Reproductive system
Pension plans, 542
People's Republic of China. See Chinese
Perception-bound thinking, **222**
Perceptual development. See also specific senses (e.g., Hearing; Touch; Vision)
  canalization, 72
  differentiation theory of, 140-141
  infancy and toddlerhood, 135-141
  intermodal perception, 139-140
  newborns, 104-106
Permanence
  component of death concept, **632**-633
  object permanence. See Object permanence
Permissive style parenting, **266**-267
  academic achievement in adolescence and, 377
Perry's theory of postformal thought, 436
Persistence
  child-rearing practices and, 267
  early childhood, 248
  learned helplessness and, 321
Persistent vegetative state, 625-626
Personal fables, and abstract thinking, **372**-373
  adolescent concept of death and, 634
Personality development. See Emotional and social development; Self entries
Personality traits. See also Temperament; specific traits (e.g., Aggression; Shyness)
  Alzheimer's disease and, 569
  "big five" personality traits, 528, 591
  centenarians, 557
  cognitive development and, 504, 579